Molecular Biology

Molecular Biology

Second Edition

Robert F. Weaver
University of Kansas

Boston   Burr Ridge, IL   Dubuque, IA   Madison, WI   New York   San Francisco   St. Louis
Bangkok   Bogotá   Caracas   Kuala Lumpur   Lisbon   London   Madrid   Mexico City
Milan   Montreal   New Delhi   Santiago   Seoul   Singapore   Sydney   Taipei   Toronto

McGraw-Hill Higher Education
A Division of The McGraw-Hill Companies

MOLECULAR BIOLOGY, SECOND EDITION

1 2 3 4 5 6 7 8 9 0 QPV/QPV 0 9 8 7 6 5 4 3 2 1

ISBN 0–07–234517–9
ISBN 0–07–112287–7 (ISE)

Publisher: *James M. Smith*
Senior developmental editor: *Jean Sims Fornango*
Associate marketing manager: *Tami Petsche*
Project manager: *Jane E. Matthews*
Senior production supervisor: *Sandy Ludovissy*
Design manager: *Stuart D. Paterson*
Freelance cover/interior designer: *Maureen McCutcheon*
Cover image: *Poul Nissen, Ph.D., Yale University, with permission of The American Association for the Advancement of Science.*
Senior photo research coordinator: *Lori Hancock*
Photo research: *Chris Hammond*
Supplement producer: *Jodi K. Banowetz*
Media technology producer: *Lori A. Welsh*
Compositor: *Shepherd, Inc.*
Typeface: *10/12 Sabon*
Printer: *Quebecor World Versailles Inc.*

Cover image
Active site cleft of the large ribosomal subunit of *Haloarcula*. Synthesis of peptide bonds is catalyzed by ribosomal RNA in the active site, which contains the aminoacyl ends of the transfer RNAs (in contrasting colors).

Library of Congress Cataloging-in-Publication Data

Weaver, Robert Franklin, 1942–
    Molecular biology / Robert F. Weaver.—2nd ed.
        p. ; cm.
    Includes bibliographical references and index.
    ISBN 0–07–234517–9 (alk. paper)—ISBN 0–07–112287–7 (ISE)
    1. Molecular biology. I. Title.
    [DNLM: 1. Molecular Biology. QH 506 W363m 2002]

QH506 .W43 2002
572.8—dc21

                            2001030688
                            CIP

www.mhhe.com

To my parents

Rob Weaver was born in Topeka, Kansas, and grew up in Arlington, Virginia. He received his bachelor's degree in chemistry from The College of Wooster in Wooster, Ohio, in 1964. He earned his Ph.D. in boichemistry at Duke University in 1969, then spent two years doing postdoctoral research at the University of California, San Francisco, where he studied the structure of eukayotic RNA polymerases with William J. Rutter.

He joined the faculty of the University of Kansas as an assistant professor of biochemistry in 1971, was promoted to associate professor, and then to full professor in 1981. In 1984, he became chair of the Department of Biochemistry, and served in that capacity until he was named Associate Dean of the College of Liberal Arts and Sciences in 1995.

Prof. Weaver is the divisional dean for the science and mathematic departments within the college, which includes supervising 14 different departments and programs. As a professor of molecular biosciences, he teaches courses in introductory molecular biology and the molecular biology of cancer. He directs a research laboratory in which undergraduates and graduate students participate in research on the molecular biology of a baculovirus that infects caterpillars.

Prof. Weaver is the author of many scientific papers resulting from research funded by the National Institutes of Health, the National Science Foundation, and the American Cancer Society. He has also coauthored two genetics textbooks and has written two articles on molecular biology in the *National Geographic Magazine*. He has spent two years performing research in European Laboratories as an American Cancer Society Research Scholar, one year in Zurich, Switzerland, and one year in Oxford, England.

BRIEF CONTENTS

C O N T E N T S

PART I
Introduction

PART III

Transcription in Prokaryotes

CHAPTER 6

The Transcription Apparatus of Prokaryotes 133

CHAPTER 7

Operons: Fine Control of Prokaryotic Transcription 175

CHAPTER 8

Major Shifts in Prokaryotic Transcription 205

This textbook is designed for an introductory course in molecular biology. But what is molecular biology? The definition of this elusive term depends on who is doing the defining. In this book, I consider molecular biology to be the study of genes and their activities at the molecular level.

When I was a student in college and graduate school I found that I became most excited about science, and learned best, when the instructor emphasized the experimental strategy and the data that led to the conclusions, rather than just the conclusions themselves. Thus, when I began teaching an introductory molecular biology course in 1972, I adopted that teaching strategy and have used it ever since. I have found that my students react as positively as I did.

One problem with this approach, however, was that no textbook placed as great an emphasis on experimental data as I would have liked. So I tried assigning reading from the literature in lieu of a textbook. Although this method was entirely appropriate for an advanced course, it was a relatively inefficient process and not practical for a first course in molecular biology. To streamline the process, I augmented the literature readings with hand-drawn cartoons of the data I wanted to present. Later, when technology became available, I made transparencies of figures from the journal articles. But I really wanted a textbook that presented the concepts of molecular biology, along with experiments that led to those concepts. I finally decided that the best way to get such a book would be to write it myself. I had already coauthored a successful introductory genetics text in which I took an experimental approach—as much as possible with a book at that level. That gave me the courage to try writing an entire book by myself and to treat the subject as an adventure in discovery.

Organization

The book begins with a four-chapter sequence that should be a review for most students. Chapter 1 is a brief history of genetics. Chapter 2 discusses the structure and chemical properties of DNA. Chapter 3 is an overview of gene expression, and Chapter 4 deals with the nuts and bolts of gene cloning. All these are topics that the great majority of molecular biology students have already learned in an introductory genetics course. Still, students of molecular biology need to have a grasp of these concepts and may

need to refresh their understanding of them. I do not deal specifically with these chapters in class; instead, I suggest students consult them if they need more work on these topics. These chapters are written at a more basic level than the rest of the book.

Chapter 5 describes a number of common techniques used by molecular biologists. It would not have been possible to include all the techniques described in this book in one chapter, so I tried to include the most common or, in a few cases, valuable techniques that are not mentioned elsewhere in the book. When I teach this course, I do not lecture on Chapter 5 as such. Instead, I refer students to it when we first encounter a technique in a later chapter. I do it that way to avoid boring my students with technique after technique. I also realize that the concepts behind some of these techniques are rather sophisticated, and the students' appreciation of them is much deeper after they've acquired more experience in molecular biology.

Chapters 6–9 describe transcription in prokaryotes. Chapter 6 introduces the basic transcription apparatus, including promoters, terminators, and RNA polymerase, and shows how transcripts are initiated, elongated, and terminated. Chapter 7 describes the control of transcription in four different operons, then Chapter 8 shows how bacteria and their phages control transcription of many genes at a time, often by providing alternative sigma factors. Chapter 9 discusses the interaction between prokaryotic DNA-binding proteins, mostly helix-turn-helix proteins, and their DNA targets.

Chapters 10–13 present control of transcription in eukaryotes. Chapter 10 deals with the three eukaryotic RNA polymerases and the promoters they recognize. Chapter 11 introduces the general transcription factors that collaborate with the three RNA polymerases and points out the unifying theme of the TATA-box-binding protein, which participates in transcription by all three polymerases. Chapter 12 explains the functions of gene-specific transcription factors, or activators. This chapter also illustrates the structures of several representative activators and shows how they interact with their DNA targets. Chapter 13 describes the structure of eukaryotic chromatin and shows how activators can interact with histones to activate or repress transcription.

Chapters 14–16 introduce some of the posttranscriptional events that occur in eukaryotes. Chapter 14 deals with RNA splicing. Chapter 15 describes capping and polyadenylation, and Chapter 16 introduces a collection of fascinating "other posttranscriptional events," including rRNA and tRNA processing, *trans*-splicing, and RNA editing. This chapter also discusses two kinds of posttranscriptional control of gene expression: (1) RNA interference; and (2) modulating mRNA stability (using the transferrin receptor gene as the prime example).

Chapters 17–19 describe the translation process in both prokaryotes and eukaryotes. Chapter 17 deals with initiation of translation, including the control of translation at the initiation step. Chapter 18 shows how polypeptides are elongated, with the emphasis on elongation in prokaryotes. Chapter 19 provides details on the structure and function of two of the key players in translation: ribosomes and tRNA.

Chapters 20–23 describe the mechanisms of DNA replication, recombination, and translocation. Chapter 20 introduces the basic mechanism of DNA replication, and some of the proteins (including the DNA polymerases) involved in replication. Chapter 21 provides details of the initiation, elongation, and termination steps in DNA replication in prokaryotes and eukaryotes. Chapters 22 and 23 describe DNA rearrangements that occur naturally in cells. Chapter 22 discusses homologous recombination and Chapter 23 deals with site-specific recombination and translocation.

New to the Second Edition

After the Introduction, all the chapters of this second edition have been extensively updated and include new information. Three major expansions of the first edition are evident:

- First, Chapter 24 on genomics is new. Since the first edition appeared, the total sequences of many complex genomes have been obtained, including the genomes of a worm, a fly, and a mustard plant. We even have a rough draft of the human genome, and a polished human genome sequence is on the horizon. These historic developments have already begun to revolutionize molecular biology. Now we can examine the activities of all the genes in an organism at once and see how they respond to various perturbations, including development and disease. The sequence of any genetic loci, including those involved in disease states, is instantly available from the sequence database. Genomes of related organisms can be compared to detect the effects of evolution and to pinpoint conserved DNA sequences that are likely to be keys to the function of gene products.

- Second, Chapter 22 of the first edition has been split in two (Chapters 22 and 23). Trying to force all of recombination and translocation into one chapter did not allow for enough treatment of either topic. Now Chapter 22 is devoted exclusively to homologous recombination, so the mechanism of that vital process can be analyzed in detail. A new discussion of meiotic recombination in yeast has also been added. Chapter 23 covers site-specific recombination—with λ integration and excision as the major example—and translocation. The translocation section has been considerably expanded and now includes new discussions of retroviruses, non-LTR retrotransposons such as LINES, nonautonomous retrotransposons such as Alu elements, and transposable group II introns.

- Third, Chapter 20 includes a new section on DNA damage and repair. These are important elements of molecular biology, and they also play a major role in human disease, especially cancer.

Supplements

- Visual Resource Library
  The presentation CD-ROM contains digital files for all of the line art, tables, and most of the photographs in the text in an easy-to-use format. This format is compatible with either PC or Macintosh.

- Text-Specific Website
  The following website, specific to this text, provides access to digital image files, updates, and web links for both students and instructors. Separate message boards for both instructor and student discussion are also available:

  www.mhhe.com/weaver2

In writing this book, I have been aided immeasurably by the advice of many editors and reviewers. They have contributed greatly to the accuracy and readability of the book, but they cannot be held accountable for any remaining errors or ambiguities. For those, I take full responsibility. I would like to thank the following people for their help.

Second Edition Reviewers

Anne Britt
University of California, Davis

Mark Bolyard
Southern Illinois University

M. Suzanne Bradshaw
University of Cincinnati

Robert Brunner
University of California, Berkeley

Stephen J. D'Surney
University of Mississippi

Caroline J. Decker
Washington State University

Jeffery DeJong
University of Texas, Dallas

John S. Graham
Bowling Green State University

Ann Grens
Indiana University

Ulla M. Hansen
Boston University

Laszlo Hanzely
Northern Illinois University

Robert B. Helling
University of Michigan

Martinez J. Hewlett
University of Arizona

David C. Hinkle
University of Rochester

Barbara C. Hoopes
Colgate University

Richard B. Imberski
University of Maryland

Cheryl Ingram-Smith
Pennsylvania State University

Alan Kelly
University of Oregon

Robert N. Leamnson
University of Massachusetts, Dartmouth

Karen A. Malatesta
Princeton University

Robert P. Metzger
San Diego State University

David A. Mullin
Tulane University

Brian K. Murray
Brigham Young University

Michael A. Palladino
Monmouth University

James G. Patton
Vanderbilt University

Martha Peterson
University of Kentucky

Marie Pizzorno
Bucknell University

Florence Schmieg
University of Delaware

Zhaomin Yang
Auburn University

First Edition Reviewers

Rodney P. Anderson
Ohio Northern University

Kevin L. Anderson
Mississippi State University

Prakash H. Bhuta
Eastern Washington University

Dennis Bogyo
Valdosta State University

Richard Crawford
Trinity College

Christopher A. Cullis
Case Western Reserve University

Beth De Stasio
Lawrence University

R. Paul Evans
Brigham Young University

Edward R. Fliss
Missouri Baptist College

Michael A. Goldman
San Francisco State University

Robert Gregerson
Lyon College

Eileen Gregory
Rollins College

Barbara A. Hamkalo
University of California, Irvine

Mark L. Hammond
Campbell University

Terry L. Helser
State University of New York, Oneonta

Carolyn Herman
Southwestern College

Andrew S. Hopkins
Alverno College

Carolyn Jones
Vincennes University

Teh-Hui Kao
Pennsylvania State University

Mary Evelyn B. Kelley
Wayne State University

Harry van Keulen
Cleveland State University

Leo Kretzner
University of South Dakota

Charles J. Kunert
Concordia University

Robert N. Leamnson
University of Massachusetts, Dartmouth

James D. Liberatos
Louisiana Tech University

Cran Lucas
Louisiana State University

James J. McGivern
Gannon University

James E. Miller
Delaware Valley College

Robert V. Miller
Oklahoma State University

George S. Mourad
Indiana University-Purdue University

David A. Mullin
Tulane University

James R. Pierce
Texas A&M University, Kingsville

Joel B. Piperberg
Millersville University

John E. Rebers
Northern Michigan University

Florence Schmieg
University of Delaware

Brian R. Shmaefsky
Kingwood College

Paul Keith Small
Eureka College

David J. Stanton
Saginaw Valley State University

Francis X. Steiner
Hillsdale College

Amy Cheng Vollmer
Swarthmore College

Dan Weeks
University of Iowa

David B. Wing
New Mexico Institute of Mining & Technology

GUIDE TO EXPERIMENTAL TECHNIQUES IN MOLECULAR BIOLOGY

A Brief History

The garden pea plant. This was the experimental subject of Gregor Mendel's genetic investigation. *Holt Studios International (Nigel Cattlin)/Photo Researchers, Inc.*

Whathat is molecular biology? The term has more than one definition. Some define it very broadly as the attempt to understand biological phenomena in molecular terms. But this definition makes molecular biology difficult to distinguish from another well-known discipline, biochemistry. Another definition is more restrictive and therefore more useful: the study of gene structure and function at the molecular level. This attempt to explain genes and their activities in molecular terms is the subject matter of this book.

Molecular biology grew out of the disciplines of genetics and biochemistry. In this chapter we will review the major early developments in the history of this hybrid discipline, beginning with the earliest genetic experiments performed by Gregor Mendel in the mid-19th century. In Chapters 2 and 3 we will add more substance to this brief outline. By definition, the early work on genes cannot be considered molecular biology, or even molecular genetics, because early geneticists did not know the molecular nature of genes. Instead, we call it transmission genetics because it deals with the transmission of traits from parental organisms to their offspring. In fact, the chemical composition of genes was not known until 1944. At that point, it became possible to study genes as molecules, and the discipline of molecular biology was born. ■

1.1   Transmission Genetics

In 1865, Gregor Mendel (Figure 1.1) published his findings on the inheritance of seven different traits in the garden pea. Before Mendel's research, scientists thought inheritance occurred through a blending of each trait of the parents in the offspring. Mendel concluded instead that inheritance is particulate. That is, each parent contributes particles, or genetic units, to the offspring. We now call these particles **genes.** Furthermore, by carefully counting the number of progeny plants having a given **phenotype,** or observable characteristic (e.g., yellow seeds, white flowers), Mendel was able to make some important generalizations. The word *phenotype*, by the way, comes from the same Greek root as *phenomenon,* meaning *appearance.* Thus, a tall pea plant exhibits the tall phenotype, or appearance. *Phenotype* can also refer to the whole set of observable characteristics of an organism.

Mendel's Laws of Inheritance

Mendel saw that a gene can exist in different forms called **alleles.** For example, the pea can have either yellow or green seeds. One allele of the gene for seed color gives rise

to yellow seeds, the other to green. Moreover, one allele can be **dominant** over the other, **recessive,** allele. Mendel demonstrated that the allele for yellow seeds was dominant when he mated a green-seeded pea with a yellow-seeded pea. All of the progeny in the first filial generation (F_1) had yellow seeds. However, when these F_1 yellow peas were allowed to self-fertilize, some green-seeded peas reappeared. The ratio of yellow to green seeds in the second filial generation (F_2) was very close to 3:1.

The term *filial* comes from the Latin: *filius,* meaning son; *filia,* meaning daughter. Therefore, the first filial generation (F_1) contains the offspring (sons and daughters) of the original parents. The second filial generation (F_2) is the offspring of the F_1 individuals.

Mendel concluded that the allele for green seeds must have been preserved in the F_1 generation, even though it did not affect the seed color of those peas. His explanation was that each parent plant carried two copies of the gene; that is, the parents were **diploid,** at least for the characteristics he was studying. According to this concept, **homozygotes** have two copies of the same allele, either two alleles for yellow seeds or two alleles for green seeds. **Heterozygotes** have one copy of each allele. The two parents in the first mating were homozygotes; the resulting F_1 peas were all heterozygotes. Further, Mendel reasoned that sex cells contain only one copy of the gene; that is, they are **haploid.** Homozygotes can therefore produce sex cells, or **gametes,** that have only one allele, but heterozygotes can produce gametes having either allele.

This is what happened in the matings of yellow with green peas: The yellow parent contributed a gamete with a gene for yellow seeds; the green parent, a gamete with a gene for green seeds. Therefore, all the F_1 peas got one allele for yellow seeds and one allele for green seeds. They had not lost the allele for green seeds at all, but because yellow is dominant, all the seeds were yellow. However, when these heterozygous peas were self-fertilized, they produced gametes containing alleles for yellow and green color in equal numbers, and this allowed the green phenotype to reappear.

Here is how that happened. Assume that we have two sacks, each containing equal numbers of green and yellow marbles. If we take one marble at a time out of one sack and pair it with a marble from the other sack, we will wind up with the following results: ¼ of the pairs will be yellow/yellow; ¼ will be green/green; and the remaining ½ will be yellow/green. The alleles for yellow and green peas work the same way. Recalling that yellow is dominant, you can see that only ¼ of the progeny (the green/green ones) will be green. The other ¾ will be *yellow* because they have at least one allele for yellow seeds. Hence, the ratio of yellow to green peas in the second (F_2) generation is 3:1.

Mendel also found that the genes for the seven different characteristics he chose to study operate indepen-

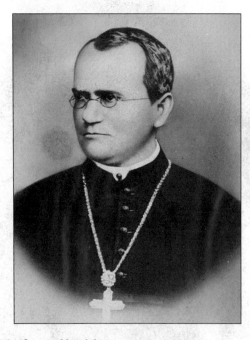

Figure 1.1   Gregor Mendel. (*Source:* Courtesy of Dept. of Library Services/American Museum of Natural History, Neg. no. 219467.)

dently of one another. Therefore, combinations of alleles of two different genes (e.g. yellow or green peas with round or wrinkled seeds, where yellow and round are dominant and green and wrinkled are recessive) gave ratios of 9:3:3:1 for yellow/round, yellow/wrinkled, green/round, and green/wrinkled, respectively. Inheritance that follows the simple laws that Mendel discovered can be called **Mendelian inheritance.**

SUMMARY Genes can exist in several different forms, or alleles. One allele can be dominant over another, so heterozygotes having two different alleles of one gene will generally exhibit the characteristic dictated by the dominant allele. The recessive allele is not lost; it can still exert its influence when paired with another recessive allele in a homozygote.

The Chromosome Theory of Inheritance

Other scientists either did not know about or uniformly ignored the implications of Mendel's work until 1900 when three botanists, who had arrived at similar conclusions independently, rediscovered it. After 1900, most geneticists accepted the particulate nature of genes, and the field of genetics began to blossom. One factor that made it easier for geneticists to accept Mendel's ideas was a growing understanding of the nature of chromosomes, which had begun in the latter half of the 19th century. Mendel had predicted that gametes would contain only one allele of each gene instead of two. If chromosomes carry the genes, their numbers should also be reduced by half in the gametes—and they are. Chromosomes therefore appeared to be the discrete physical entities that carry the genes.

This notion that chromosomes carry genes is the **chromosome theory of inheritance.** It was a crucial new step in genetic thinking. No longer were genes disembodied factors; now they were observable objects in the cell nucleus. (See Figure B1.1 to review key structures in the cell.) Some geneticists, particularly Thomas Hunt Morgan (Figure 1.2), remained skeptical of this idea. Ironically, in 1910 Morgan himself provided the first definitive evidence for the chromosome theory.

Morgan worked with the fruit fly *(Drosophila melanogaster)*, which was in many respects a much more convenient organism than the garden pea for genetic studies because of its small size, short generation time, and large number of offspring. When he mated red-eyed flies (dominant) with white-eyed flies (recessive), most, but not all, of the F_1 progeny were red-eyed. Furthermore, when Morgan mated the red-eyed males of the F_1 generation with their red-eyed sisters, they produced about ¼ white-eyed males, but no white-eyed females. In other words, the eye color

phenotype was **sex-linked.** It was transmitted along with sex in these experiments. How could this be?

We now realize that sex and eye color are transmitted together because the genes governing these characteristics are located on the same chromosome—the X chromosome. (Most chromosomes, called **autosomes,** occur in pairs in a given individual, but the X chromosome is an example of a **sex chromosome,** of which the female fly has two copies and the male has one.) However, Morgan was reluctant to draw this conclusion until he observed the same sex linkage with two more phenotypes, miniature wing and yellow body, also in 1910. That was enough to convince him of the validity of the chromosome theory of inheritance. (See Figure B1.2 to review the cell cycle and the process of mitosis, which explains how traits are transmitted from parent to daughter cells.)

Before we leave this topic, let us make two crucial points. First, every gene has its place, or **locus,** on a chromosome. Figure 1.3 depicts a hypothetical chromosome and the positions of three of its genes, called *A, B,* and *C.* Second, diploid organisms such as human beings normally have two copies of all chromosomes (except sex chromosomes). That means that they have two copies of most genes, and that these copies can be the same alleles, in which case the organism is **homozygous,** or different alleles, in which case it is **heterozygous.** For example, Figure 1.3b shows a diploid pair of chromosomes with different alleles at one locus *(Aa)* and the same alleles at the other two loci *(BB and cc).* The **genotype,** or allelic constitution, of this organism with respect to these three genes, is *AaBBcc.* Because this organism has two different alleles (*A* and *a*) in its two chromosomes at the *A* locus, it

Figure 1.2 Thomas Hunt Morgan. *(Source:* National Library of Medicine.)

BOX 1.1

Cell Structure

Before we proceed with our discussion of transmission genetics and molecular genetics, it is important to review cellular structures that are of particular interest in the study of molecular biology (Figure B1.1).

The **nucleus** is a discrete structure within the cell that contains the genetic material, DNA. The DNA is usually present in the cell as a network of fibers called **chromatin.** During mitosis, the DNA molecules form the condensed chromosome structure by coiling and supercoiling around specialized proteins.

Other structures of importance to molecular biologists are the centrosome, the endoplasmic reticulum, the Golgi apparatus, ribosomes, and the mitochondrion. The **centrosome** is a structure composed largely of microtubules. Paired centrosomes organize the formation of spindle fibers during mitosis. The **endoplasmic reticulum (ER)** is a network of membranes throughout the cell; cells have both smooth and rough ER. The rough ER is studded with ribosomes, giving it a rough appearance. The bound ribosomes of the ER synthesize proteins destined for secretion from the cell or that are to be incorporated into the membrane or into specific vacuoles. The **Golgi apparatus** is a continuation of the membrane network of the endoplasmic reticulum. **Ribosomes** that are free in the cytosol synthesize proteins that remain in the cell and are not transported through the endoplasmic reticulum and Golgi apparatus. The **mitochondrion** is the site of energy production in the cell. The mitochondrion consists of a double membrane system. The outer membrane is smooth and the inner membrane has convoluted folds. The mitochondrion can divide independently of the cell and contains its own circular, double-stranded DNA.

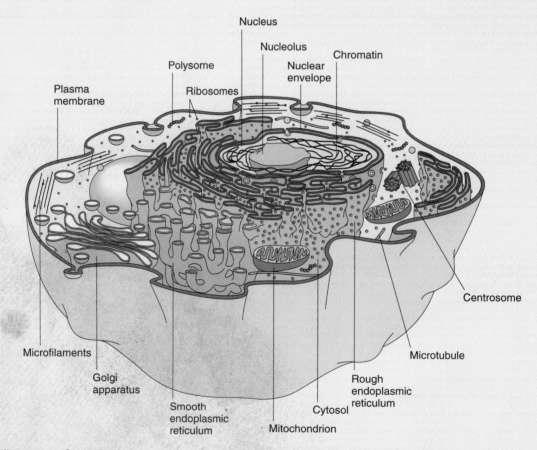

Figure B1.1   Cell structure. Generalized representation of an animal (eukaryotic) cell showing internal structures.

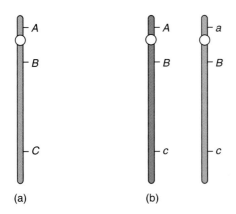

Figure 1.4   Recombination in *Drosophila*. The two X chromosomes of the female are shown schematically. One of them (red) carries two wild-type genes: (m^+), which results in normal wings, and (w^+), which gives red eyes. The other (blue) carries two mutant genes: *miniature* (*m*) and *white* (*w*). During egg formation, a recombination, or crossing over, indicated by the crossed lines, occurs between these two genes on the two chromosomes. The result is two recombinant chromosomes with mixtures of the two parental genes. One is m^+ *w*, the other is *m* w^+.

Figure 1.3   Location of genes on chromosomes. (a) A schematic diagram of a chromosome, indicating the positions of three genes: *A, B,* and *C*. **(b)** A schematic diagram of a diploid pair of chromosomes, indicating the positions of the three genes—*A, B,* and *C*—on each, and the genotype (*A* or *a*; *B* or *b*; and *C* or *c*) at each locus.

is heterozygous at that locus (Greek: *hetero,* meaning different). Since it has the same, dominant *B* allele in both chromosomes at the *B* locus, it is homozygous dominant at that locus (Greek: *homo,* meaning same). And because it has the same, recessive *c* allele in both chromosomes at the *C* locus, it is homozygous recessive there. Finally, because the *A* allele is dominant over the *a* allele, the phenotype of this organism would be the dominant phenotype at the *A* and *B* loci and the recessive phenotype at the *C* locus.

This discussion of varying phenotypes in *Drosophila* gives us an opportunity to introduce another important genetic concept: **wild-type** versus **mutant.** The wild-type phenotype is the most common, or at least the generally accepted standard, phenotype of an organism. To avoid the mistaken impression that a wild organism is automatically a wild-type, some geneticists prefer the term **standard type.** In *Drosophila,* red eyes and full-size wings are wild-type. Mutations in the *white* and *miniature* genes result in mutant flies with white eyes and miniature wings, respectively. Mutant alleles are usually recessive, as in these two examples, but not always.

Genetic Recombination and Mapping

It is easy to understand that genes on separate chromosomes behave independently in genetic experiments, and that genes on the same chromosome—like the genes for miniature wing (*miniature*) and white eye (*white*)— behave as if they are linked. However, genes on the same chromosome usually do not show perfect **genetic linkage.** In fact, Morgan discovered this phenomenon when he examined the behavior of the sex-linked genes he had found. For example, although *white* and *miniature* are both on the X chromosome, they remain linked in offspring only 65.5% of the time. The other offspring have a new com-

bination of alleles not seen in the parents and are therefore called **recombinants.**

How are these recombinants produced? The answer was already apparent by 1910, because microscopic examination of chromosomes during meiosis (gamete formation) had shown crossing over between **homologous** chromosomes (chromosomes carrying the same genes, or alleles of the same genes). This resulted in the exchange of genes between the two homologous chromosomes. In the previous example, during formation of eggs in the female, an X chromosome bearing the *white* and *miniature* alleles experienced crossing over with a chromosome bearing the red eye and normal wing alleles (Figure 1.4). Because the crossing-over event occurred between these two genes, it brought together the *white* and normal wing alleles on one chromosome and the red (normal eye) and *miniature* alleles on the other. Because it produced a new combination of alleles, we call this process **recombination.** (See Figure B1.3 to review the process of meiosis and the occurrence of recombination.)

Morgan assumed that genes are arranged in a linear fashion on chromosomes, like beads on a string. This, together with his awareness of recombination, led him to propose that the farther apart two genes are on a chromosome, the more likely they are to recombine. This makes sense because there is simply more room between widely spaced genes for crossing over to occur. A. H. Sturtevant extended this hypothesis to predict that a mathematical relationship exists between the distance separating two genes on a chromosome and the frequency of recombination between these two genes. Sturtevant collected data on recombination in the fruit fly that supported his hypothesis. This established the rationale for **genetic mapping** techniques still in use today. Simply stated, if two loci recombine with a frequency of 1%, we say that they are separated by a map distance of one **centimorgan** (named for Morgan himself). By the 1930s, other investigators found that the same rules applied to other **eukaryotes** (nucleus-containing organisms), including the mold *Neurospora,* the garden pea, maize (corn), and even human beings. These rules also apply to **prokaryotes,** organisms in which the genetic material is not confined to a nuclear compartment.

BOX 1.2

Cell Cycle and Mitosis

Cell Cycle

Cellular reproduction is a cyclic process in which daughter cells are produced through nuclear division (mitosis) and cellular division (cytokinesis). Mitosis and cytokinesis are part of the growth–division cycle called the **cell cycle** (Figure B1.2a). Mitosis, which is divided into four stages (Figure B1.2b), is a relatively small part of the total cell cycle; the majority of the time, a cell is in a growth stage called **interphase**. Interphase is divided into three parts: G_1, S, and G_2. The first gap phase, G_1, follows mitosis and is a period of growth and metabolic activity. The S phase follows G_1 and is a period of DNA synthesis, in which the DNA is replicated. Another gap phase, G_2, follows DNA synthesis and precedes the next mitotic division. Certain mature cell types do not continue to divide but remain in interphase (in G_0).

Mitosis

Mitosis is the period in the cell cycle during which the nucleus divides and gives rise to two daughter cells that have chromosome numbers identical to that of the original cell (see Figure B1.2b). Mitosis can be divided into four distinct stages (**prophase, metaphase, anaphase,** and **telophase**) characterized by the appearance and orientation of the homologous pairs of chromosomes.

Prophase

At the end of the G_2 interphase, the chromosomes have replicated, and each has two sister chromatids attached to each other at the centromere. Prophase indicates the beginning of mitosis, when the replicated chromosomes coil and condense. The centrosomes divide and the spindle apparatus also appears.

Metaphase

During metaphase, spindle fibers from the spindle apparatus attach to the centromeres of the chromosomes. The chromosomes gradually migrate to the midline of the cell, oriented between the two centrosomes.

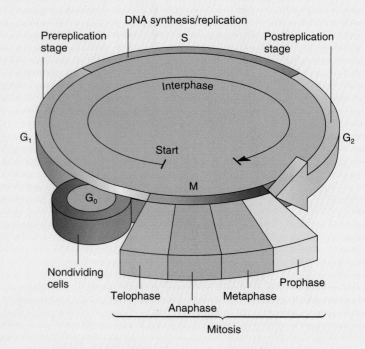

Figure B1.2a Cell cycle. Phases in the cell cycle of a typical eukaryotic cell. The events that occur from the start of one cell division (mitosis), when two daughter cells are produced, to the next cell division are collectively termed the **cell cycle.**

Anaphase

Anaphase begins with the separation of the centromeres and the contraction of the spindle fibers pulling the centromeres toward the centrosomes. As a result, the sister chromatids, now called daughter chromosomes, separate, and the chromosomes move in opposite directions toward the centrosomes.

Telophase

The two sets of chromosomes reach opposite poles of the cell, where they begin to uncoil. Telophase ends with cytokinesis (cellular division) and the formation of two daughter cells.

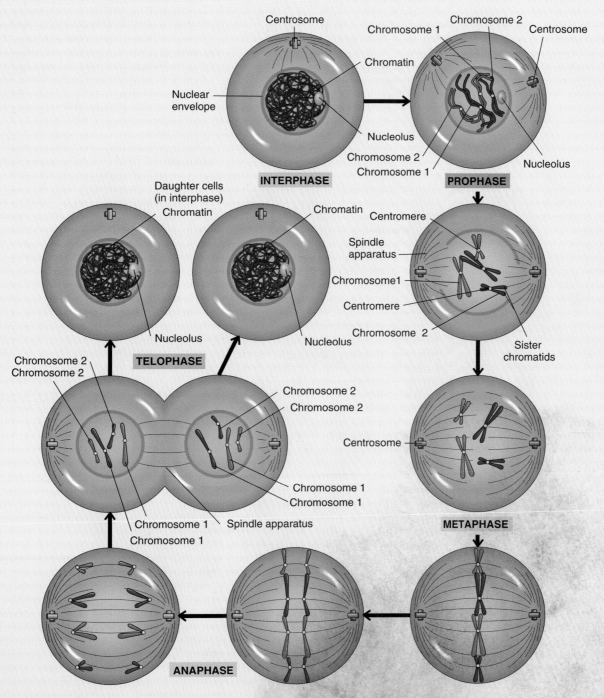

Figure B1.2b **Mitosis.** Mitosis in an animal cell with two homologous pairs of chromosomes.

BOX 1.3

Meiosis

The process of meiosis involves two cell divisions that result in daughter cells (gametes) with half the parental number of chromosomes. If the number of chromosomes were not reduced prior to fertilization, the chromosome number would double in each generation.

Meiosis also allows the exchange of genetic material on homologous chromosomes through recombination, which occurs at the end of prophase I. Recombination results in new combinations of alleles of different genes on a chromosome.

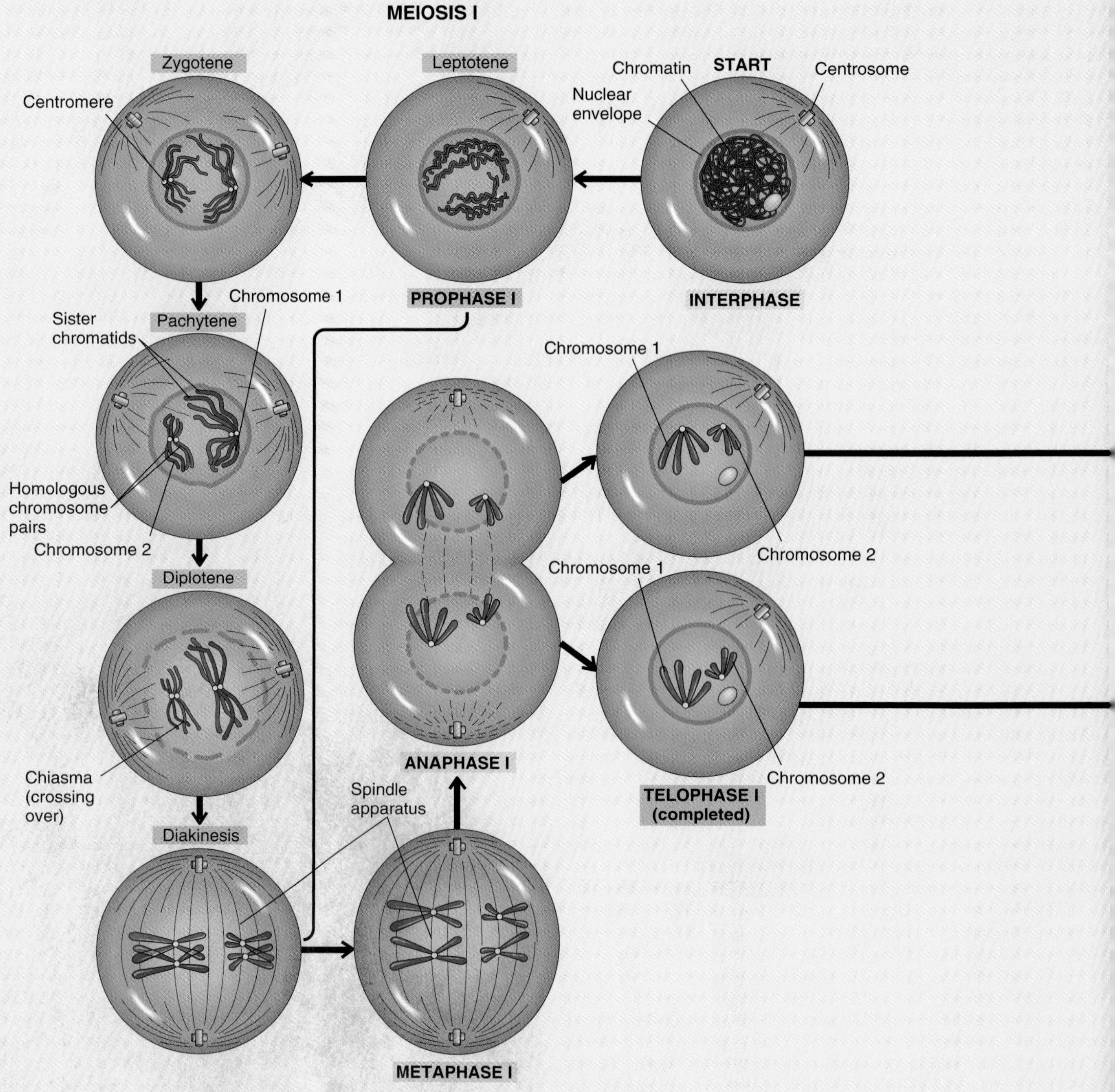

MEIOSIS I

Figure B1.3   **Meiosis.** A diagrammatic representation of meiosis in a cell with two chromosomes.

Meiosis, like mitosis, can be divided into distinct phases (Figure B1.3). **Meiosis I** includes a multipart prophase I (during which crossing over occurs), metaphase I, anaphase I (unlike in mitosis, homologous pairs of chromosomes, not sister chromatids, separate to the opposite poles of the cell), and telophase I. **Meiosis II** begins with prophase II, continues on to metaphase II (the centromeres attach to the spindle apparatus), anaphase II (as in mitosis, the centromeres separate and the chromosomes begin to migrate to opposite poles), and telophase II, which completes meiosis, resulting in four daughter cells with half the number of chromosomes as the parental cell. As shown in Figure B1.3, the parental cell contains four chromosomes (two homologous pairs), and the resulting four daughter cells have two chromosomes each.

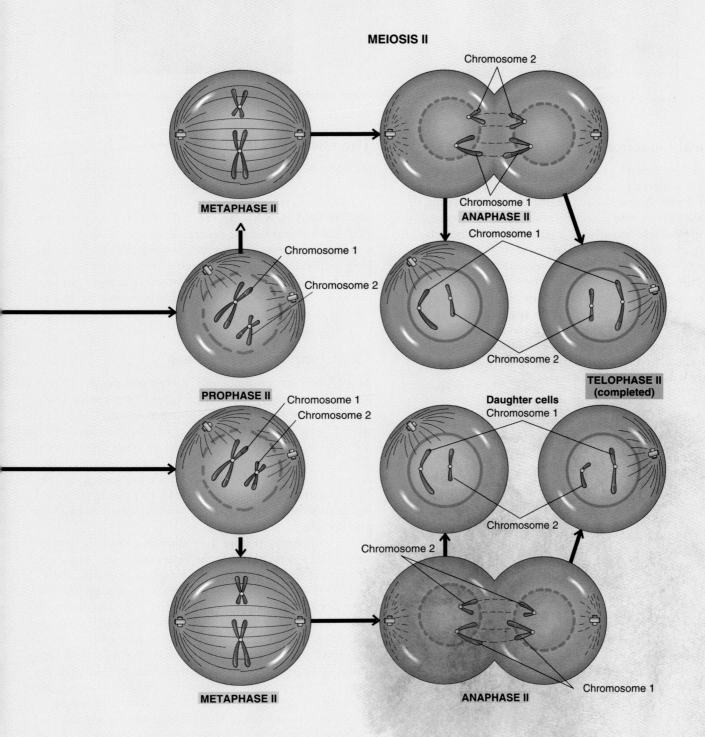

MEIOSIS II

METAPHASE II

PROPHASE II

METAPHASE II

ANAPHASE II

ANAPHASE II

TELOPHASE II (completed)

Daughter cells

Chromosome 1
Chromosome 2

Figure 1.5   Barbara McClintock. (*Source:* Bettmann Archive.)

Figure 1.6   Friedrich Miescher. (*Source:* National Library of Medicine.)

Physical Evidence for Recombination

Barbara McClintock (Figure 1.5) and Harriet Creighton provided a direct physical demonstration of recombination in 1931. By examining maize chromosomes microscopically, they could detect recombinations between two easily identifiable features of a particular chromosome (a knob at one end and a long extension at the other). Furthermore, whenever this physical recombination occurred, they could also detect recombination genetically. Thus, they established a direct relationship between a region of a chromosome and a gene. Shortly after McClintock and Creighton performed this work on maize, Curt Stern observed the same phenomenon in *Drosophila*. So recombination could be detected both physically and genetically in animals as well as plants.

SUMMARY The chromosome theory of inheritance holds that genes are arranged in linear fashion on chromosomes. The reason that certain traits tend to be inherited together is that the genes governing these traits are on the same chromosome. However, recombination between two chromosomes during meiosis can scramble the parental alleles to give nonparental combinations. The farther apart two genes are on a chromosome the more likely such recombination between them will be.

1.2   Molecular Genetics

The studies just discussed tell us important things about the transmission of genes and even about how to map genes on chromosomes, but they do not tell us what genes are made of or how they work. This has been the province of molecular genetics, which also happens to have its roots in Mendel's era.

The Discovery of DNA

In 1869, Friedrich Miescher (Figure 1.6) discovered in the cell nucleus a mixture of compounds that he called nuclein. The major component of nuclein is **deoxyribonucleic acid (DNA)**. By the end of the 19th century, chemists had learned the general structure of DNA and of a related compound, **ribonucleic acid (RNA)**. Both are long polymers—chains of small compounds called nucleotides. Each nucleotide is composed of a sugar, a phosphate group, and a base. The chain is formed by linking the sugars to one another through their phosphate groups.

The Composition of Genes

By the time the chromosome theory of inheritance was generally accepted, geneticists agreed that the chromosome must be composed of a polymer of some kind. This would agree with its role as a string of genes. But which polymer is it? Essentially, the choices were three: DNA, RNA, and **protein**. Protein was the other major component of Miescher's nuclein; its chain is composed of links called **amino acids**. The amino acids in protein are joined by **peptide bonds,** so a single protein chain is called a **polypeptide.**

Oswald Avery (Figure 1.7) and his colleagues demonstrated in 1944 that DNA is the right choice (see Chapter 2). These investigators built on an experiment performed earlier by Frederick Griffith in which he transferred a genetic trait from one strain of bacteria to another. The trait was virulence, the ability to cause a lethal infection, and it could be transferred simply by mixing dead virulent cells with live avirulent (nonlethal) cells. It was very likely that the substance that caused the transformation from avirulence to virulence in the recipient cells was the gene for virulence, because the recipient cells passed this trait on to their progeny.

What remained was to learn the chemical nature of the transforming agent in the dead virulent cells. Avery

Figure 1.7 Oswald Avery. (*Source:* National Academy of Sciences.)

(a)   (b)

Figure 1.8   (a) George Beadle; (b) E.L. Tatum. (*Source:* (*a, b*) AP/Wide World Photos.)

and his coworkers did this by applying a number of chemical and biochemical tests to the transforming agent, showing that it behaved just as DNA should, not as RNA or protein should.

The Relationship between Genes and Proteins

The other major question in molecular genetics is this: How do genes work? To lay the groundwork for the answer to this question, we have to backtrack again, this time to 1902. That was the year Archibald Garrod noticed that the human disease alcaptonuria seemed to behave as a Mendelian recessive trait. It was likely, therefore, that the disease was caused by a defective, or mutant, gene. Moreover, the main symptom of the disease was the accumulation of a black pigment in the patient's urine, which Garrod believed derived from the abnormal buildup of an intermediate compound in a biochemical pathway.

By this time, biochemists had shown that all living things carry out countless chemical reactions and that these reactions are accelerated, or catalyzed, by proteins called **enzymes.** Many of these reactions take place in sequence, so that one chemical product becomes the starting material, or substrate, for the next reaction. Such sequences of reactions are called **pathways,** and the products or substrates within a pathway are called **intermediates.** Garrod postulated that an intermediate accumulated to abnormally high levels in alcaptonuria because the enzyme that would normally convert this intermediate to the next was defective. Putting this idea together with the finding that alcaptonuria behaved genetically as a Mendelian recessive trait, Garrod suggested that a defective gene gives rise to a defective enzyme. To put it another way: A gene is responsible for the production of an enzyme.

Garrod's conclusion was based in part on conjecture; he did not really know that a defective enzyme was involved in

alcaptonuria. It was left for George Beadle and E. L. Tatum (Figure 1.8) to prove the relationship between genes and enzymes. They did this using the mold *Neurospora* as their experimental system. *Neurospora* has an enormous advantage over the human being as the subject of genetic experiments. By using *Neurospora,* scientists are not limited to the mutations that nature provides, but can use **mutagens** to introduce mutations into genes and then observe the effects of these mutations on biochemical pathways. Beadle and Tatum found many instances where they could create *Neurospora* mutants and then pin the defect down to a single step in a biochemical pathway, and therefore to a single enzyme (see Chapter 3). They did this by adding the intermediate that would normally be made by the defective enzyme and showing that it restored normal growth. By circumventing the blockade, they discovered where it was. In these same cases, their genetic experiments showed that a single gene was involved. Therefore, a defective gene gives a defective (or absent) enzyme. In other words, a gene seemed to be responsible for making one enzyme. This was the one-gene/one-enzyme hypothesis. This hypothesis was actually not quite right for at least three reasons: (1) An enzyme can be composed of more than one polypeptide chain, whereas a gene has the information for making only one polypeptide chain. (2) Many genes contain the information for making polypeptides that are not enzymes. (3) As we will see, the end products of some genes are not polypeptides, but RNAs. A modern restatement of the one-gene/one-enzyme hypothesis would be: Most genes contain the information for making one polypeptide.

Activities of Genes

Let us now return to the question at hand: How do genes work? This is really more than one question because genes do more than one thing. First, they are replicated

Figure 1.9   James Watson (left) and Francis Crick. (*Source:* Corbis/Bettmann Archive.)

(a)   (b)

Figure 1.10   (a) Rosalind Franklin; (b) Maurice Wilkins. (*Source:* (a) From *The Double Helix* by James D. Watson, 1968, Atheneum Press, NY. © Cold Spring Harbor Laboratory Archives. (b) Courtesy Professor M. H. F. Wilkins, Biophysics Dept., King's College, London.)

faithfully; second, they direct the production of RNAs and proteins; third, they accumulate mutations and so allow evolution. Let us look briefly at each of these activities.

How Genes Are Replicated   First of all, how is DNA replicated faithfully? To answer that question, we need to know the overall structure of the DNA molecule as it is found in the chromosome. James Watson and Francis Crick (Figure 1.9) provided the answer in 1953 by building models based on chemical and physical data that had been gathered in other laboratories, primarily x-ray diffraction data collected by Rosalind Franklin and Maurice Wilkins (Figure 1.10).

(a)   (b)

Figure 1.11   (a) Matthew Meselson; (b) Franklin Stahl. (*Source:* (a) Courtesy Dr. Matthew Meselson. (b) Cold Spring Harbor Laboratory Archives.)

Watson and Crick proposed that DNA is a **double helix**—two DNA strands wound around each other. More important, the bases of each strand are on the inside of the helix, and a base on one strand pairs with one on the other in a very specific way. DNA has only four different bases: adenine, guanine, cytosine, and thymine, which we abbreviate A, G, C, and T. Wherever we find an A in one strand, we always find a T in the other; wherever we find a G in one strand, we always find a C in the other. In a sense, then, the two strands are complementary. If we know the base sequence of one, we automatically know the sequence of the other. This complementarity is what allows DNA to be replicated faithfully. The two strands come apart, and enzymes build new partners for them using the old strands as templates and following the Watson–Crick base-pairing rules (A with T, G with C). This is called **semiconservative replication** because one strand of the parental double helix is conserved in each of the daughter double helices. In 1958, Matthew Meselson and Franklin Stahl (Figure 1.11) proved that DNA replication in bacteria follows the semiconservative pathway (see Chapter 20).

How Genes Direct the Production of Polypeptides   **Gene expression** is the process by which a gene product (an RNA or a polypeptide) is made. Two steps, called **transcription** and **translation,** are required to make a polypeptide from the instructions in a DNA gene. In the transcription step, an enzyme called RNA polymerase makes a copy of one of the DNA strands; this copy is not DNA, but its close cousin RNA. In the translation step, this RNA (**messenger RNA,** or **mRNA**) carries the genetic instructions to the cell's protein factories, called **ribosomes.** The ribosomes "read" the **genetic code** in the mRNA and put together a protein according to its instructions.

Actually, the ribosomes already contain molecules of RNA, called **ribosomal RNA (rRNA).** Francis Crick originally thought that this RNA residing in the ribosomes

(a)   (b)

Figure 1.12   (a) François Jacob; (b) Sydney Brenner. (*Source: (a, b)* Cold Spring Harbor Laboratory Archives.)

Figure 1.13   Gobind Khorana (left) and Marshall Nirenberg. (*Source:* Corbis/Bettmann Archive.)

carried the message from the gene. According to this theory, each ribosome would be capable of making only one kind of protein—the one encoded in its rRNA. François Jacob and Sydney Brenner (Figure 1.12) had another idea: The ribosomes are nonspecific translation machines that can make an unlimited number of different proteins, according to the instructions in the mRNAs that visit the ribosomes. Experiment has shown that this idea is correct (see Chapter 3).

What is the nature of this genetic code? Marshall Nirenberg and Gobind Khorana (Figure 1.13), working independently with different approaches, cracked the code in the early 1960s (see Chapter 18). They found that 3 bases constitute a code word, called a **codon,** that stands for one amino acid. Out of the 64 possible 3-base codons, 61 specify amino acids; the other three are stop signals.

The ribosomes scan a messenger RNA 3 bases at a time and bring in the corresponding amino acids to link to the growing protein chain. When they reach a stop signal, they release the completed protein.

How Genes Accumulate Mutations   Genes change in a number of ways. The simplest is a change of one base to another. For example, if a certain codon in a gene is GAG (for the amino acid called glutamate), a change to GTG converts it to a codon for another amino acid, valine. The protein that results from this mutated gene will have a valine where it ought to have a glutamate. This may be one change out of hundreds of amino acids, but it can have profound effects. In fact, this specific change has occurred in the gene for one of the human blood proteins and is responsible for the genetic disorder we call sickle cell disease.

Genes can suffer more profound changes, such as deletions or insertions of large pieces of DNA. Segments of DNA can even move from one locus to another. The more drastic the change, the more likely that the gene or genes involved will be totally inactivated.

Gene Cloning   In recent years, geneticists have learned to isolate genes, place them in new organisms, and reproduce them by a set of techniques collectively known as **gene cloning.** Cloned genes not only give molecular biologists plenty of raw materials for their studies, they also can be induced to yield their protein products. Some of these, such as human insulin or blood clotting factors, can be very useful. Cloned genes can also be transplanted to plants and animals, including humans. These transplanted genes can alter the characteristics of the recipient organisms, so they may provide powerful tools for agriculture and for intervening in human genetic diseases. We will examine gene cloning in detail in Chapter 4.

SUMMARY Most genes are made of DNA arranged in a double helix. This structure explains how genes engage in their three main activities: replication, carrying information, and collecting mutations. The complementary nature of the two DNA strands in a gene allows them to be replicated faithfully by separating and serving as templates for the assembly of two new complementary strands. The sequence of nucleotides in a typical gene is a genetic code that carries the information for making an RNA. Most of these are messenger RNAs that carry the information to protein-synthesizing ribosomes. The end result is a new protein chain made according to the gene's instructions. A change in the sequence of bases constitutes a mutation, which can change the sequence of amino acids in the gene's protein product. Genes can be cloned, allowing molecular biologists to harvest abundant supplies of their products.

Table 1.1   Molecular Biology Time Line

1859	Charles Darwin	Published *On the Origin of Species*
1865	Gregor Mendel	Advanced the principles of segregation and independent assortment
1869	Friedrich Miescher	Discovered DNA
1900	Hugo de Vries, Carl Correns, Erich von Tschermak	Rediscovered Mendel's principles
1902	Archibald Garrod	First suggested a genetic cause for a human disease
1902	Walter Sutton, Theodor Boveri	Proposed the chromosome theory
1908	G.H. Hardy, Wilhelm Weinberg	Formulated the Hardy–Weinberg principle
1910, 1916	Thomas Hunt Morgan, Calvin Bridges	Demonstrated that genes are on chromosomes
1913	A.H. Sturtevant	Constructed a genetic map
1927	H.J. Muller	Induced mutation by x-rays
1931	Harriet Creighton, Barbara McClintock	Obtained physical evidence for recombination
1941	George Beadle, E.L. Tatum	Proposed the one-gene/one-enzyme hypothesis
1944	Oswald Avery, Colin McLeod, Maclyn McCarty	Identified DNA as the material genes are made of
1953	James Watson, Francis Crick, Rosalind Franklin, Maurice Wilkins	Determined the structure of DNA
1958	Matthew Meselson, Franklin Stahl	Demonstrated the semiconservative replication of DNA
1961	Sydney Brenner, François Jacob, Matthew Meselson	Discovered messenger RNA
1966	Marshall Nirenberg, Gobind Khorana	Finished unraveling the genetic code
1970	Hamilton Smith	Discovered restriction enzymes that cut DNA at specific sites, which made cutting and pasting DNA easy, thus facilitating DNA cloning
1972	Paul Berg	Made the first recombinant DNA in vitro
1973	Herb Boyer, Stanley Cohen	First used a plasmid to clone DNA
1977	Walter Gilbert and Frederick Sanger	Worked out methods to determine the sequence of bases in DNA
1977	Frederick Sanger	Determined the base sequence of an entire viral genome (ϕX174)
1977	Phillip Sharp, Richard Roberts, and others	Discovered interruptions (introns) in genes
1990	Lap-Chee Tsui, Francis Collins, John Riordan	Found the gene that is responsible for cystic fibrosis
1990	James Watson and many others	Launched the Human Genome Project to map the entire human genome and, ultimately, to determine its base sequence
1993	The Huntington's Disease Group	Identified the Huntington's disease gene
1995	Craig Venter, Hamilton Smith	Determined the base sequences of the genomes of two bacteria: *Hemophilus influenzae* and *Mycoplasma genitalium,* the first genomes of free-living organisms to be sequenced
1996	Many investigators	Determined the base sequence of the genome of brewer's yeast, *Saccharomyces cerevisiae,* the first eukaryotic genome to be sequenced
1997	Fred Blattner, Takashi Honuchi and many colleagues	Determined the base sequence of the genome of *Escherichia coli*
1997	Ian Wilmut and colleagues	Cloned a sheep (Dolly) from an adult sheep udder cell
1998	Many investigators	Determined the base sequence of the genome of the roundworm *Caenorhabditis elegans,* the first animal genome to be sequenced
1999	Many investigators	Determined the base sequence of human chromosome 22
2000	Many investigators	Determined the base sequence of the genome of the fruit fly, *Drosophila melanogaster,* a mainstay of genetic research
2000	Many investigators	Determined the base sequence of the genome of the mustard, *Arabidopsis thaliana,* the first plant genome to be sequenced
2000	Alain Fischer and colleagues	First clearly successful gene therapy trial—on two patients with severe combined immunodeficiency ("bubble boy" syndrome)
2001	Craig Venter, Francis Collins, and many others	Reported working draft of human genome sequence

This concludes our brief chronology of molecular biology. Table 1.1 reviews some of the milestones. Although it is a very young discipline, it has an exceptionally rich history, and molecular biologists are now adding new knowledge at an explosive rate. Indeed, the pace of discovery in molecular biology, and the power of its techniques, has led many commentators to call it a revolution. Because some of the most important changes in medicine and agriculture over the next few decades are likely to depend on the manipulation of genes by molecular biologists, this revolution will touch everyone's life in one way or another. Thus, you are embarking on a study of a subject that is not only fascinating and elegant, but one that has practical importance as well. F.H. Westheimer, professor emeritus of chemistry at Harvard University put it well: "The greatest intellectual revolution of the last 40 years may have taken place in biology. Can anyone be considered educated today who does not understand a little about molecular biology?" Happily, after this course you should understand more than a little.

SUMMARY

Genes can exist in several different forms called alleles. A recessive allele can be masked by a dominant one in a heterozygote, but it does not disappear. It can be expressed again in a homozygote bearing two recessive alleles.

Genes exist in a linear array on chromosomes. Therefore, traits governed by genes that lie on the same chromosome can be inherited together. However, recombinations between homologous chromosomes occur during meiosis, so that gametes bearing nonparental combinations of alleles can be produced. The farther apart two genes lie on a chromosome, the more likely such recombination between them will be.

Most genes are made of double-stranded DNA arranged in a double helix. One strand is the complement of the other, which means that faithful gene replication requires that the two strands separate and acquire complementary partners. The linear sequence of bases in a typical gene carries the information for making a protein.

The process of making a gene product is called gene expression. It occurs in two steps: transcription and translation. In the transcription step, RNA polymerase makes a messenger RNA, which is a copy of the information in the gene. In the translation step, ribosomes "read" the mRNA and make a protein according to its instructions. Thus, a change (mutation) in a gene's sequence may cause a corresponding change in the protein product.

SUGGESTED READINGS

Creighton, H. B., and B. McClintock. 1931. A correlation of cytological and genetical crossing-over in *Zea mays*. *Proceedings of the National Academy of Sciences* 17:492–97.

Mirsky, A. E. 1968. The discovery of DNA. *Scientific American* 218 (June):78–88.

Morgan, T. H. 1910. Sex-limited inheritance in *Drosophila*. *Science* 32:120–22.

Sturtevant, A. H. 1913. The linear arrangement of six sex-linked factors in *Drosophila*, as shown by their mode of association. *Journal of Experimental Zoology* 14:43–59.

The Molecular Nature of Genes

Before we begin to study in detail the structure and activities of genes, and the experimental evidence underlying those concepts, we need a fuller outline of the adventure that lies before us. Thus, in the next two chapters, we will flesh out the brief history of molecular biology presented in Chapter 1. In this chapter we will begin this task by considering the behavior of genes as molecules.

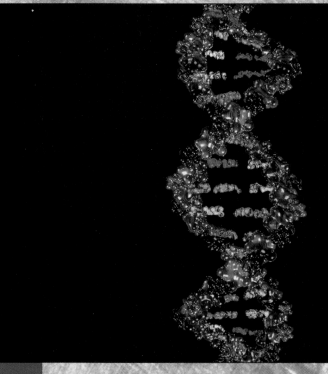

Computer-generated image of DNA structure. © Douglas Struthers/Tony Stone Images.

2.1   The Nature of Genetic Material

The studies that eventually revealed the chemistry of genes began in Tübingen, Germany, in 1869. There, Friedrich Miescher isolated nuclei from pus cells (white blood cells) in waste surgical bandages. He found that these nuclei contained a novel phosphorus-bearing substance that he named *nuclein.* Nuclein is mostly **chromatin,** which is a complex of **deoxyribonucleic acid (DNA)** and chromosomal proteins.

By the end of the 19th century, both DNA and **ribonucleic acid (RNA)** had been separated from the protein that clings to them in the cell. This allowed more detailed chemical analysis of these **nucleic acids.** (Notice that the term *nucleic acid* and its derivatives, *DNA* and *RNA,* come directly from Miescher's term *nuclein.*) By the beginning of the 1930s, P. Levene, W. Jacobs, and others had demonstrated that RNA is composed of a sugar (ribose) plus four nitrogen-containing bases, and that DNA contains a different sugar (deoxyribose) plus four bases. They discovered that each base is coupled with a sugar–phosphate to form a nucleotide. We will return to the chemical structures of DNA and RNA later in this chapter. First, let us examine the evidence that genes are made of DNA.

Transformation in Bacteria

Frederick Griffith laid the foundation for the identification of DNA as the genetic material in 1928 with his experiments on **transformation** in the bacterium pneumococcus, now known as *Streptococcus pneumoniae.* The wild-type organism is a spherical cell surrounded by a mucous coat called a capsule. The cells form large, glistening colonies, characterized as smooth (S) (Figure 2.1a). These cells are **virulent,** that is, capable of causing lethal infections upon injection into mice. A certain mutant strain of *S. pneumoniae* has lost the ability to form a capsule. As a result, it grows as small, rough (R) colonies (Figure 2.1b). More importantly, it is **avirulent;** because it has no protective coat, it is engulfed by the host's white blood cells before it can proliferate enough to do any damage.

The key finding of Griffith's work was that heat-killed virulent colonies of *S. pneumoniae* could **transform** avirulent cells to virulent ones. Neither the heat-killed virulent bacteria nor the live avirulent ones by themselves could cause a lethal infection. Together, however, they were deadly. Somehow the virulent trait passed from the dead cells to the live, avirulent ones. This transformation phenomenon is illustrated in Figure 2.2. Transformation was not transient; the ability to make a capsule and therefore to kill host animals, once conferred on the avirulent bacteria, was passed to their descendants as a heritable trait. In other words, the avirulent cells somehow gained the gene for vir-

(a)

(b)

Figure 2.1   Variants of *Streptococcus pneumoniae.* (a) The large, glossy colonies contain smooth (S) virulent bacteria; **(b)** the small, mottled colonies are composed of rough (R) avirulent bacteria. (*Source:* (*a, b*) Harriet Ephrussi-Taylor.)

ulence during transformation. This meant that the transforming substance in the heat-killed bacteria was probably the gene for virulence itself. The missing piece of the puzzle was the chemical nature of the transforming substance.

DNA: The Transforming Material   Oswald Avery, Colin MacLeod, and Maclyn McCarty supplied the missing piece in 1944. They used a transformation test similar to the one that Griffith had introduced, and they took pains to define the chemical nature of the transforming substance from virulent cells. First, they removed the protein from the extract with organic solvents and found that the extract still transformed. Next, they subjected it to digestion with various enzymes. Trypsin and chymotrypsin, which destroy protein, had no effect on transformation. Neither did ribonuclease, which degrades RNA. These experiments ruled out protein or RNA as the transforming material. On the other hand, Avery and his coworkers found that the enzyme deoxyribonuclease (DNase), which breaks down DNA, destroyed the transforming ability of the virulent cell extract. These results suggested that the transforming substance was DNA.

Direct physical-chemical analysis supported the hypothesis that the purified transforming substance was DNA. The analytical tools Avery and his colleagues used were the following:

1. *Ultracentrifugation*   They spun the transforming substance in an ultracentrifuge (a very high-speed

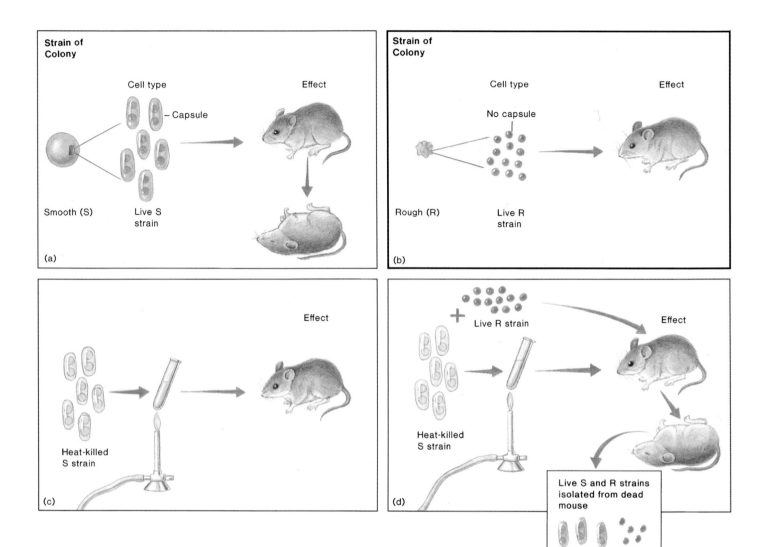

Figure 2.2   Griffith's transformation experiments. (a) Virulent strain S *S. pneumoniae* bacteria kill their host; **(b)** avirulent strain R bacteria cannot infect successfully, so the mouse survives; **(c)** strain S bacteria that are heat-killed can no longer infect; **(d)** a mixture of strain R and heat-killed strain S bacteria kills the mouse. The killed virulent (S) bacteria have transformed the avirulent (R) bacteria to virulent (S).

centrifuge) to estimate its size. The material with transforming activity sedimented rapidly (moved rapidly toward the bottom of the centrifuge tube), suggesting a very high molecular weight, characteristic of DNA.

2. *Electrophoresis*   They placed the transforming substance in an electric field to see how rapidly it moved. The transforming activity had a relatively high mobility, also characteristic of DNA because of its high charge-to-mass ratio.

3. *Ultraviolet Absorption Spectrophotometry*   They placed a solution of the transforming substance in a spectrophotometer to see what kind of ultraviolet light it absorbed most strongly. Its absorption spectrum matched that of DNA. That is, the light it absorbed

most strongly had a wavelength of about 260 nanometers (nm), in contrast to protein, which absorbs maximally at 280 nm.

4. *Elementary Chemical Analysis*   This yielded an average nitrogen-to-phosphorus ratio of 1.67, about what one would expect for DNA, which is rich in both elements, but vastly lower than the value expected for protein, which is rich in nitrogen but poor in phosphorus. Even a slight protein contamination would have raised the nitrogen-to-phosphorus ratio.

Further Confirmation   These findings should have settled the issue of the nature of the gene, but they had little immediate impact. The mistaken notion, from early chemical analyses, that DNA was a monotonous repeat of a

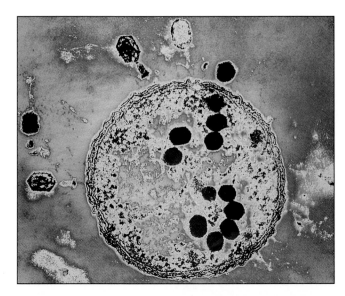

Figure 2.3   A false color transmission electron micrograph of T2 phages infecting an *E. coli* cell. Phage particles at left and top appear ready to inject their DNA into the host cell. Another T2 phage has already infected the cell, however, and progeny phage particles are being assembled. The progeny phage heads are readily discernible as dark polygons inside the host cell. (*Source:* © Lee Simon/Photo Researchers, Inc.)

four-nucleotide sequence, such as ACTG-ACTG-ACTG, and so on, persuaded many geneticists that it could not be the genetic material. Furthermore, controversy persisted about possible protein contamination in the transforming material, whether transformation could be accomplished with other genes besides those governing R and S, and even whether bacterial genes were like the genes of higher organisms.

Yet, by 1953, when James Watson and Francis Crick published the double-helical model of DNA structure, most geneticists agreed that genes were made of DNA. What had changed? For one thing, Erwin Chargaff had shown in 1950 that the bases were not really found in equal proportions in DNA, as previous evidence had suggested, and that the base composition of DNA varied from one species to another. In fact, this is exactly what one would expect for genes, which also vary from one species to another. Furthermore, Rollin Hotchkiss had refined and extended Avery's findings. He purified the transforming substance to the point where it contained only 0.02% protein and showed that it could still change the genetic characteristics of bacterial cells. He went on to show that such highly purified DNA could transfer genetic traits other than R and S.

Finally, in 1952, A.D. Hershey and Martha Chase performed another experiment that added to the weight of evidence that genes were made of DNA. This experiment involved a **bacteriophage** (bacterial virus) called T2 that infects the bacterium *Escherichia coli* (Figure 2.3). (The term *bacteriophage* is usually shortened to *phage.*) During

infection, the phage genes enter the host cell and direct the synthesis of new phage particles. The phage is composed of protein and DNA only. The question is this: Do the genes reside in the protein or in the DNA? The Hershey–Chase experiment answered this question by showing that, on infection, most of the DNA entered the bacterium, along with only a little protein. The bulk of the protein stayed on the outside (Figure 2.4). Because DNA was the major component that got into the host cells, it likely contained the genes. Of course, this conclusion was not unequivocal; the small amount of protein that entered along with the DNA could conceivably have carried the genes. But taken together with the work that had gone before, this study helped convince geneticists that DNA, and not protein, is the genetic material.

The Hershey–Chase experiment depended on radioactive labels on the DNA and protein—a different label for each. The labels used were phosphorus-32 (^{32}P) for DNA and sulfur-35 (^{35}S) for protein. These choices make sense, considering that DNA is rich in phosphorus but phage protein has none, and that protein contains sulfur but DNA does not.

Hershey and Chase allowed the labeled phages to attach by their tails to bacteria and inject their genes into their hosts. Then they removed the empty phage coats by mixing vigorously in a blender. Because they knew that the genes must go into the cell, their question was: What went in, the ^{32}P-labeled DNA or the ^{35}S-labeled protein? As we have seen, it was the DNA. In general, then, genes are made of DNA. On the other hand, as we will see later in this chapter, other experiments showed that some viral genes consist of RNA.

SUMMARY Genes are made of nucleic acid, usually DNA. Some simple genetic systems such as viruses have RNA genes.

The Chemical Nature of Polynucleotides

By the mid-1940s, biochemists knew the fundamental chemical structures of DNA and RNA. When they broke DNA into its component parts, they found these constituents to be nitrogenous **bases, phosphoric acid,** and the sugar **deoxyribose** (hence the name deoxyribonucleic acid). Similarly, RNA yielded bases and phosphoric acid, plus a different sugar, **ribose.** The four bases found in DNA are **adenine** (A), **cytosine** (C), **guanine** (G), and **thymine** (T). RNA contains the same bases, except that **uracil** (U) replaces thymine. The structures of these bases, shown in Figure 2.5, reveal that adenine and guanine are related to the parent molecule, purine. Therefore, we refer to these compounds as **purines.** The other bases resemble pyrimidine, so they are called **pyrimidines.** These structures constitute the alphabet of genetics.

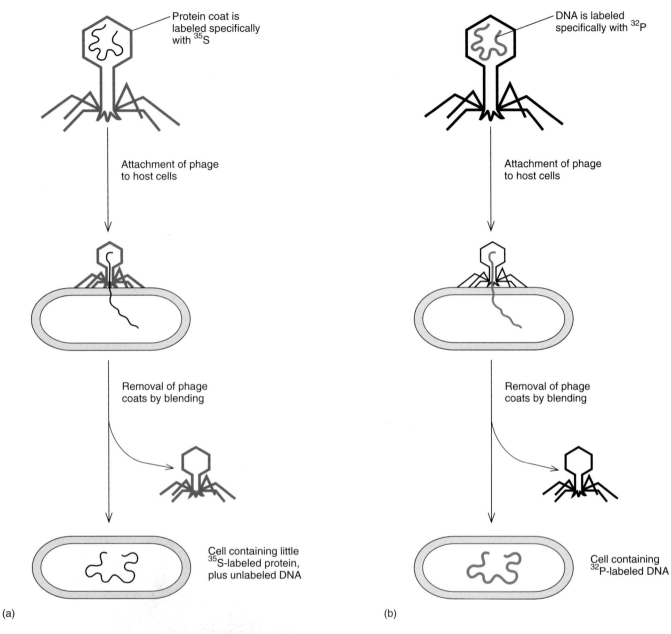

(a)

(b)

Figure 2.4 The Hershey—Chase experiment. Phage T2 contains genes that allow it to replicate in *E. coli.* Because the phage is composed of DNA and protein only, its genes must be made of one of these substances. To discover which, Hershey and Chase performed a two-part experiment. In the first part (**a**), they labeled the phage protein with ^{35}S (red), leaving the DNA unlabeled (black). In the second part (**b**), they labeled the phage DNA with ^{32}P (red), leaving the protein unlabeled (black). Since the phage genes must enter the cell, the experimenters reasoned that the type of label found in the infected cells would indicate the nature of the genes. Most of the labeled protein remained on the outside and was stripped off the cells by use of a blender (**a**), whereas most of the labeled DNA entered the infected cells (**b**). The conclusion was that the genes of this phage are made of DNA.

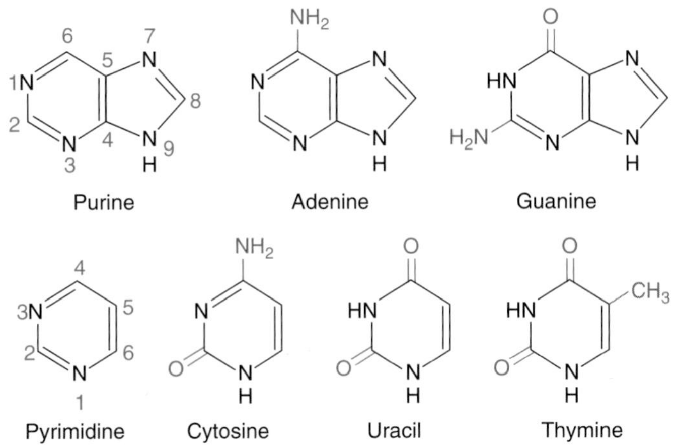

Figure 2.5   The bases of DNA and RNA. The parent bases, purine and pyrimidine, on the left, are not found in DNA and RNA. They are shown for comparison with the other five bases.

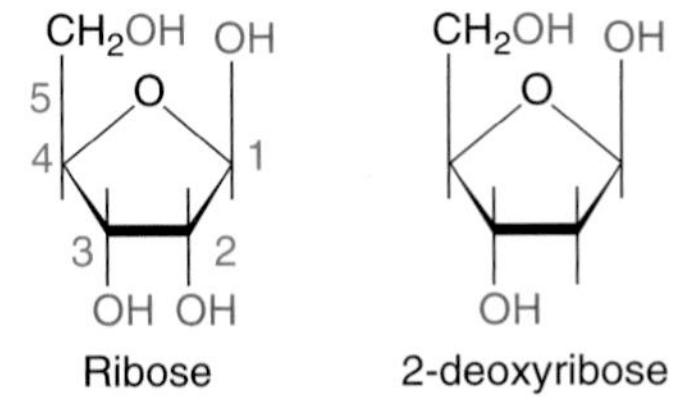

Figure 2.6   The sugars of nucleic acids. Note the OH in the 2-position of ribose and its absence in deoxyribose.

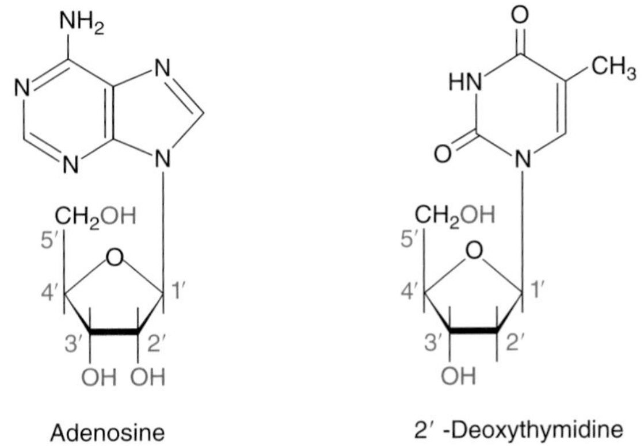

Figure 2.7   Two examples of nucleosides.

Figure 2.6 depicts the structures of the sugars found in nucleic acids. Notice that they differ in only one place. Where ribose contains a hydroxyl (OH) group in the 2-position, deoxyribose lacks the oxygen and simply has a hydrogen (H), represented by the vertical line. Hence the name *deoxyribose*. The bases and sugars in RNA and DNA are joined together into units called **nucleosides** (Figure 2.7). The names of the nucleosides derive from the corresponding bases:

Base	Nucleoside (RNA)	Deoxynucleoside (DNA)
Adenine	Adenosine	Deoxyadenosine
Guanine	Guanosine	Deoxyguanosine
Cytosine	Cytidine	Deoxycytidine
Uracil	Uridine	Not usually found
Thymine	Not usually found	(Deoxy)thymidine

Because thymine is not usually found in RNA, the "deoxy" designation for its nucleoside is frequently assumed, and the deoxynucleoside is simply called **thymidine**. The numbering of the carbon atoms in the sugars of the nucleosides (see Figure 2.7) is important. Note that the ordinary numbers are used in the bases, so the carbons in the sugars are called by primed numbers. Thus, for example, the base is linked to the 1′-position of the sugar, the 2′-position is deoxy in deoxynucleosides, and

the sugars are linked together in DNA and RNA through their 3′ and 5′-positions.

The structures in Figure 2.5 were drawn using an organic chemistry shorthand that leaves out certain atoms for simplicity's sake. Figures 2.6 and 2.7 use a slightly different convention, in which a straight line with a free end denotes a C-H bond with a hydrogen atom at the end. Figure 2.8 shows the structures of adenine and deoxyribose, first in shorthand, then with every atom included.

The subunits of DNA and RNA are **nucleotides,** which are nucleosides with a phosphate group attached through a phosphoester bond (Figure 2.9). An ester is an organic compound formed from an alcohol (bearing a hydroxyl group) and an acid. In the case of a nucleotide, the alcohol group is the 5′-hydroxyl of the sugar, and the acid is phosphoric acid, which is why we call the ester a *phosphoester*. Figure 2.9 also shows the structure of one of the four DNA precursors, deoxyadenosine-5′-triphosphate (dATP). When synthesis of DNA takes place, two phosphate groups are removed from dATP, leaving deoxyadenosine-5′-monophosphate (dAMP). The other three nucleotides in

DNA (dCMP, dGMP, and dTMP) have analogous structures and names.

We will discuss the synthesis of DNA in detail in Chapters 20 and 21. For now, notice the structure of the bonds that join nucleotides together in DNA and RNA (Figure 2.10). These are called **phosphodiester bonds** because they involve phosphoric acid linked to *two* sugars: one through a sugar 5′-group the other through a sugar 3′-group. You will notice that the bases have been rotated in this picture, relative to their positions in previous figures. This more closely resembles their geometry in DNA or RNA. Note also that this **trinucleotide**, or string of three nucleotides, has polarity: The top of the molecule bears a free 5′-phosphate group, so it is called the **5′-end.** The bottom, with a free 3′-hydroxyl group, is called the **3′-end.**

Figure 2.11 introduces a shorthand way of representing a nucleotide or a DNA chain. This notation presents

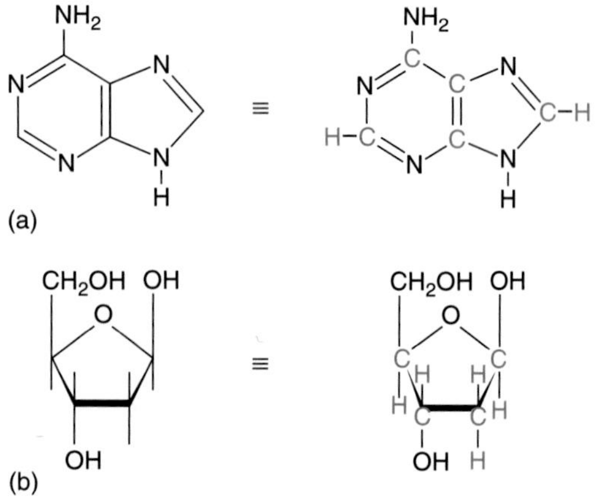

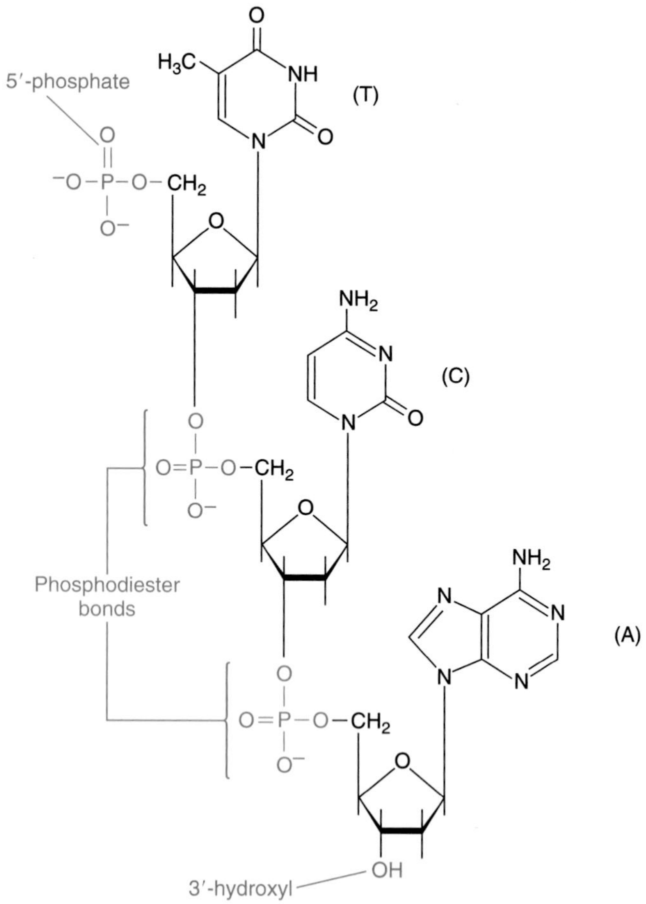

Figure 2.10   A trinucleotide. This little piece of DNA contains only three nucleotides linked together by phosphodiester bonds (red) between the 5′- and 3′-hydroxyl groups of the sugars. The 5′-end of this DNA is at the top, where a free 5′-phosphate group (blue) is located; the 3′-end is at the bottom, where a free 3′-hydroxyl group (also blue) appears. The sequence of this DNA could be read as 5′pdTpdCpdA3′. This would usually be simplified to TCA.

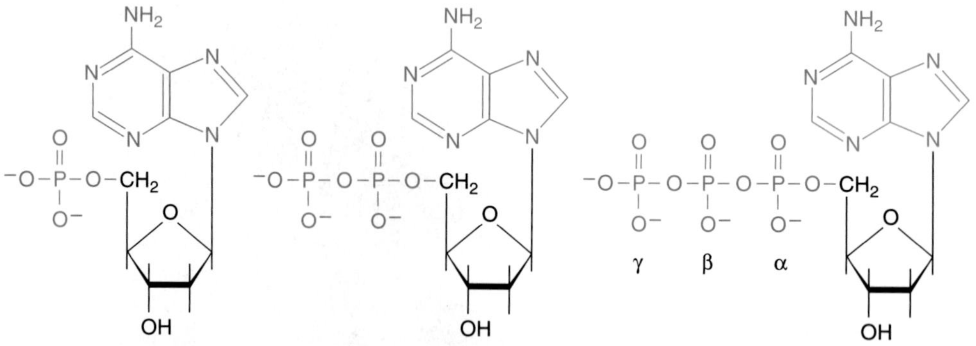

Figure 2.8   The structures of adenine and deoxyribose. Note that the structures on the left do not designate most or all of the carbons and some of the hydrogens. These designations are included in the structures on the right, in red and blue, respectively.

Deoxyadenosine-5′-monophosphate (dAMP)

Deoxyadenosine-5′-diphosphate (dADP)

Deoxyadenosine-5′-triphosphate (dATP)

Figure 2.9   Three nucleotides. The 5′-nucleotides of deoxyadenosine are formed by phosphorylating the 5′-hydroxyl group. The addition of one phosphate results in deoxyadenosine-5′-monophosphate (dAMP). One more phosphate yields deoxyadenosine-5′-diphosphate (dADP). Three phosphates (designated α, β, γ) give deoxyadenosine-5′-triphosphate (dATP).

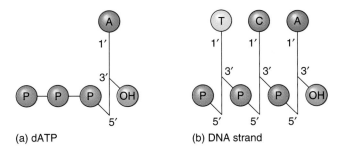

(a) dATP   (b) DNA strand

Figure 2.11 Shorthand DNA notation. (a) The nucleotide dATP. This illustration highlights four features of this DNA building block: (1) The deoxyribose sugar is represented by the vertical black line. (2) At the top, attached to the 1′-position of the sugar is the base, adenine (green). (3) In the middle, at the 3′-position of the sugar is a hydroxyl group (OH, orange). (4) At the bottom, attached to the 5′-position of the sugar is a triphosphate group (purple). (b) A short DNA strand. The same trinucleotide (TCA) illustrated in figure 2.10 is shown here in shorthand. Note the 5′-phosphate and the phosphodiester bonds (purple), and the 3′-hydroxyl group (orange). According to convention, this little piece of DNA is written 5′ to 3′ left to right.

the deoxyribose sugar as a vertical line, with the base joined to the 1′-position at the top and the phosphodiester links to neighboring nucleotides through the 3′-(middle) and 5′-(bottom) positions.

> **SUMMARY** DNA and RNA are chain-like molecules composed of subunits called nucleotides. The nucleotides contain a base linked to the 1′-position of a sugar (ribose in RNA or deoxyribose in DNA) and a phosphate group. The phosphate joins the sugars in a DNA or RNA chain through their 5′- and 3′-hydroxyl groups by phosphodiester bonds.

2.2   DNA Structure

All the facts about DNA and RNA just mentioned were known by the end of the 1940s. By that time it was also becoming clear that DNA was the genetic material and that it therefore stood at the very center of the study of life. Yet the three-dimensional structure of DNA was unknown. For these reasons, several researchers dedicated themselves to finding this structure.

Experimental Background

One of the scientists interested in DNA structure was Linus Pauling, a theoretical chemist at the California Institute of Technology. He was already famous for his studies on chemical bonding and for his elucidation of the α-helix; an important feature of protein structure. Indeed, the α-helix, held together by hydrogen bonds, laid the intellectual groundwork for the double-helix model of DNA

proposed by Watson and Crick. Another group trying to find the structure of DNA included Maurice Wilkins, Rosalind Franklin, and their colleagues at King's College in London. They were using x-ray diffraction to analyze the three-dimensional structure of DNA. Finally, James Watson and Francis Crick entered the race. Watson, still in his early 20s, but already holding a Ph.D. degree from Indiana University, had come to the Cavendish Laboratories in Cambridge, England, to learn about DNA. There he met Crick, a physicist who at age 35 was retraining as a molecular biologist. Watson and Crick performed no experiments themselves. Their tactic was to use other groups' data to build a DNA model.

Erwin Chargaff was another very important contributor. We have already seen how his 1950 paper helped identify DNA as the genetic material, but the paper contained another piece of information that was even more significant. Chargaff's studies of the base compositions of DNAs from various sources revealed that the content of purines was always roughly equal to the content of pyrimidines. Furthermore, the amounts of adenine and thymine were always roughly equal, as were the amounts of guanine and cytosine. These findings, known as Chargaff's rules, provided a valuable foundation for Watson and Crick's model. Table 2.1 presents Chargaff's data. You will notice some deviation from the rules due to incomplete recovery of some of the bases, but the overall pattern is clear.

Perhaps the most crucial piece of the puzzle came from an x-ray diffraction picture of DNA taken by Franklin in 1952—a picture that Wilkins shared with James Watson in London on January 30, 1953. The x-ray technique worked as follows: The experimenter made a very concentrated, viscous solution of DNA, then reached in with a needle and pulled out a fiber. This was not a single molecule, but a whole batch of DNA molecules, forced into side-by-side alignment by the pulling action. Given the right relative humidity in the surrounding air, this fiber was enough like a crystal that it diffracted x-rays in an interpretable way. In fact, the x-ray diffraction pattern in Franklin's picture (Figure 2.12) was so simple—a series of spots arranged in an X shape—that it indicated that the DNA structure itself must be very simple. By contrast, a complex, irregular molecule like a protein gives a complex x-ray diffraction pattern with many spots, rather like a surface peppered by a shotgun blast. Because DNA is very large, it can be simple only if it has a regular, repeating structure. And the simplest repeating shape that a long, thin molecule can assume is a corkscrew, or helix.

The Double Helix

Franklin's x-ray work strongly suggested that DNA was a helix. Not only that, it gave some important information about the size and shape of the helix. In particular, the spacing between adjacent spots in an arm of the X is

Table 2.1   Composition of DNA in moles of base per mole of phosphate

	Human Sperm		Human Thymus	Human Liver Carcinoma	Yeast		Avian Tubercle Bacilli	Bovine Thymus			Bovine Spleen	
	#1	#2			#1	#2		#1	#2	#3	#1	#2
A:	0.29	0.27	0.28	0.27	0.24	0.30	0.12	0.26	0.28	0.30	0.25	0.26
T:	0.31	0.30	0.28	0.27	0.25	0.29	0.11	0.25	0.24	0.25	0.24	0.24
G:	0.18	0.17	0.19	0.18	0.14	0.18	0.28	0.21	0.24	0.22	0.20	0.21
C:	0.18	0.18	0.16	0.15	0.13	0.15	0.26	0.16	0.18	0.17	0.15	0.17
Recovery:	0.96	0.92	0.91	0.87	0.76	0.92	0.77	0.88	0.94	0.94	0.84	0.88

Source: E. Chargaff "Chemical Specificity of Nucleic Acids and Mechanism of Their Enzymatic Degradation," *Experientia* 6:206, 1950.

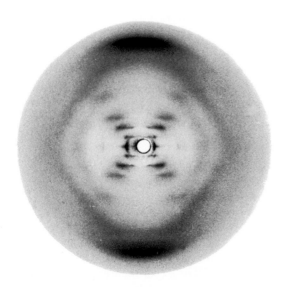

Figure 2.12   Franklin's x-ray picture of DNA. The regularity of this pattern indicated that DNA is a helix. The spacing between the strong bands at the top and bottom gave the spacing between elements of the helix (base pairs) as 3.4 Å. The spacing between neighboring lines in the pattern gave the overall repeat of the helix (the length of one helical turn) as 34 Å. (*Source:* Courtesy Professor M. H. F. Wilkins, Biophysics Dept., King's College, London.)

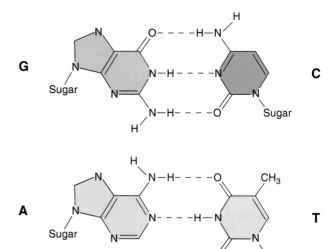

Figure 2.13   The base pairs of DNA. A guanine–cytosine pair (G—C), held together by three hydrogen bonds (dashed lines), has almost exactly the same shape as an adenine-thymine pair (A—T), held together by two hydrogen bonds.

inversely related to the overall repeat distance in the helix, 34 angstroms (34 Å), and the spacing from the top of the X to the bottom is inversely related to the spacing (3.4 Å) between the repeated elements (**base pairs**) in the helix. However, even though the Franklin picture told much about DNA, it presented a paradox: DNA was a helix with a regular, repeating structure, but for DNA to serve its genetic function, it must have an *irregular* sequence of bases.

Watson and Crick saw a way to resolve this contradiction and satisfy Chargaff's rules at the same time: DNA must be a **double helix** with its sugar–phosphate backbones on the outside and its bases on the inside. Moreover, the bases must be paired, with a purine in one strand always across from a pyrimidine in the other. This way the helix would be uniform, it would not have bulges where two large purines were paired or constrictions where two small pyrimidines were paired. Watson has

joked about the reason he seized on a double helix: "I had decided to build two-chain models. Francis would have to agree. Even though he was a physicist, he knew that important biological objects come in pairs."

But Chargaff's rules went further than this. They decreed that the amounts of adenine and thymine were equal and so were the amounts of guanine and cytosine. This fit very neatly with Watson and Crick's observation that an adenine–thymine base pair held together by hydrogen bonds has almost exactly the same shape as a guanine–cytosine base pair (Figure 2.13). So Watson and Crick postulated that adenine must always pair with thymine, and guanine with cytosine. This way, the double-stranded DNA will be uniform, composed of very similarly shaped base pairs (bp), regardless of the unpredictable sequence of either DNA strand by itself. This was their crucial insight, the key to the structure of DNA.

The double helix, often likened to a twisted ladder, is presented in three ways in Figure 2.14. The curving sides of

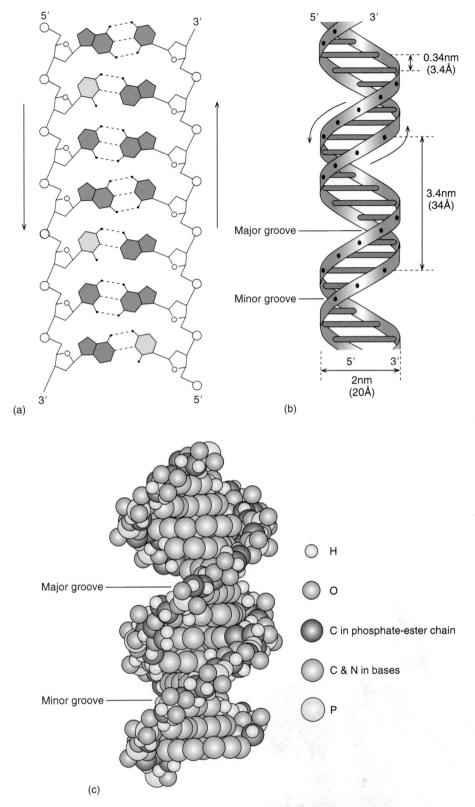

Figure 2.14 Three models of DNA structure. (a) The helix is straightened out to show the base pairing in the middle. Each type of base is represented by a different color, with the sugar–phosphate backbones in black. Note the three hydrogen bonds in the G—C pairs and the two in the A—T pairs. The vertical arrows beside each strand point in the 5′ → 3′ direction and indicate the antiparallel nature of the two DNA strands. The left strand runs 5′ → 3′, top to bottom; the right strand runs 5′ → 3′, bottom to top. The deoxyribose rings (white pentagons with O representing oxygen) also show that the two strands have opposite orientations: The rings in the right strand are inverted relative to those in the left strand. **(b)** The DNA double helix is presented as a twisted ladder whose sides represent the sugar–phosphate backbones of the two strands and whose rungs represent base pairs. The curved arrows beside the two strands indicate the 5′ → 3′ orientation of each strand, further illustrating that the two strands are antiparallel. **(c)** A space-filling model. The sugar–phosphate backbones appear as strings of black, red, white, and yellow spheres, whereas the base pairs are rendered as horizontal flat plates composed of blue spheres. Note the major and minor grooves in the helices depicted in parts **(b)** and **(c)**.

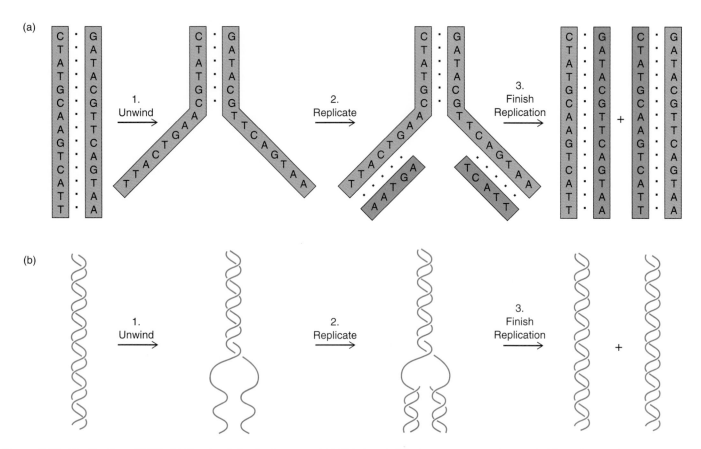

Figure 2.15 Replication of DNA. (a) For simplicity, the two parental DNA strands (blue) are represented as parallel lines. Step 1: During replication these parental strands separate, or unwind. Step 2: New strands (pink) are built with bases complementary to those of the separated parental strands. Step 3: Replication is finished, with the parental strands totally separated and the new strands completed. The end result is two double-stranded molecules of DNA identical to the original. Therefore, each daughter duplex gets one parental strand (blue) and one new strand (pink). Because only one parental strand is conserved in each of the daughter duplexes, this mechanism of replication is called semiconservative. **(b)** A more realistic portrayal of the same process. Here the strands are shown in a double helix instead of as parallel lines. Notice again that two daughter duplexes are generated, each with one parental strand (blue) and one new strand (pink).

the ladder represent the sugar–phosphate backbones of the two DNA strands; the rungs are the base pairs. The spacing between base pairs is 3.4 Å, and the overall helix repeat distance is about 34 Å, meaning that there are about 10 bp per turn of the helix. (One angstrom [Å] is one ten-billionth of a meter or one-tenth of a nanometer [nm].) The arrows indicate that the two strands are **antiparallel.** If one has $5' \rightarrow 3'$ polarity from top to bottom, the other must have $3' \rightarrow 5'$ polarity from top to bottom. In solution, DNA has a structure very similar to the one just described, but the helix contains about 10.5 bp per turn.

Watson and Crick published the outline of their model in the journal *Nature,* back-to-back with papers by Wilkins and Franklin and their coworkers showing the x-ray data. The Watson–Crick paper is a classic of simplicity—only 900 words, barely over a page long. It was published very rapidly, less than a month after it was submitted. Actually, Crick wanted to spell out the biological implications of the

model, but Watson was uncomfortable doing that. They compromised on a sentence that is one of the great understatements in scientific literature: "It has not escaped our notice that the specific base pairing we have proposed immediately suggests a possible copying mechanism for the genetic material."

As this provocative sentence indicates, Watson and Crick's model does indeed suggest a copying mechanism for DNA. Because one strand is the **complement** of the other, the two strands can be separated, and each can then serve as the template for building a new partner. Figure 2.15 shows schematically how this is accomplished. Notice how this mechanism, known as semiconservative replication, ensures that the two daughter DNA duplexes will be exactly the same as the parent, preserving the integrity of the genes as cells divide. In 1958, Matthew Meselson and Franklin Stahl demonstrated that this really is how DNA replicates (Chapter 20).

SUMMARY The DNA molecule is a double helix, with sugar–phosphate backbones on the outside and base pairs on the inside. The bases pair in a specific way: adenine (A) with thymine (T), and guanine (G) with cytosine (C).

Solved Problems

Problem 1

The following drawings are the outlines of two DNA base pairs, with the bases identified as *a, b, c,* and *d.* What are the real identities of these bases?

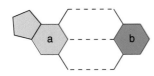

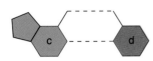

Solution

Base *a* is a purine (composed of a six-membered ring and a five-member ring), so we know it must be either adenine (A) or guanine (G). Furthermore, it is joined to its partner by three H bonds (dashed lines). G–C pairs are joined by three H bonds, whereas A–T pairs are joined by only two. Base *a* must therefore be guanine (G). Its partner, *b,* must therefore be cytosine (C). Base *c* is a purine that is joined to its partner by only two H bonds. Because A–T pairs are joined by two H bonds, base *c* must be adenine (A). Its partner, *d,* must therefore be thymine (T). ■

Problem 2

Here is the sequence of one strand of a DNA fragment:

5′ATGCGTGACTAATTCG3′

Write the sequence of this DNA in double-stranded form.

Solution

Following the Watson–Crick base-pairing rules, and remembering that the two DNA strands are antiparallel, the double-stranded DNA would look like this:

5′ATGCGTGACTAATTCG3′
3′TACGCACTGATTAAGC5′   ■

2.3   Genes Made of RNA

The genetic system explored by Hershey and Chase was a phage, a bacterial virus. A virus particle by itself is essentially just a package of genes. It has no life of its own, no metabolic activity; it is inert. But when the virus infects a host cell, it seems to come to life. Suddenly the host cell begins making viral proteins. Then the viral genes are replicated and the newly made genes, together with viral coat proteins, assemble into progeny virus particles. Because of their behavior as inert particles outside, but life-like agents inside their hosts, viruses resist classification. Some scientists refer to them as "living things" or even "organisms." Others prefer a label that, although more cumbersome, is also more descriptive of a virus's less-than-living status: "genetic system." This term applies to any entity—be it a true organism or a virus—that contains genes capable of replicating.

Most genetic systems studied to date contain genes made of DNA. But some viruses, including several phages, plant and animal viruses (e.g., HIV, the AIDS virus), have RNA genes. Sometimes viral RNA genes are double-stranded, but usually they are single-stranded.

We have already encountered one famous example of the use of viruses in molecular biology research. We will see many more in subsequent chapters. In fact, without viruses, the field of molecular biology would be immeasurably poorer.

SUMMARY Certain viruses contain genes made of RNA instead of DNA.

2.4   Physical Chemistry of Nucleic Acids

DNA and RNA molecules can assume several different structures. Let us examine these and the behavior of DNA under conditions that encourage the two strands to separate and then come together again.

A Variety of DNA Structures

The structure for DNA proposed by Watson and Crick (see Figure 2.14) represents the sodium salt of DNA in a fiber produced at very high relative humidity (92%). This is called the **B form** of DNA. Although it is probably close to the conformation of most DNA in the cell, it is not the only conformation available to double-stranded nucleic acids. If we reduce the relative humidity surrounding the DNA fiber to 75%, the sodium salt of DNA assumes the

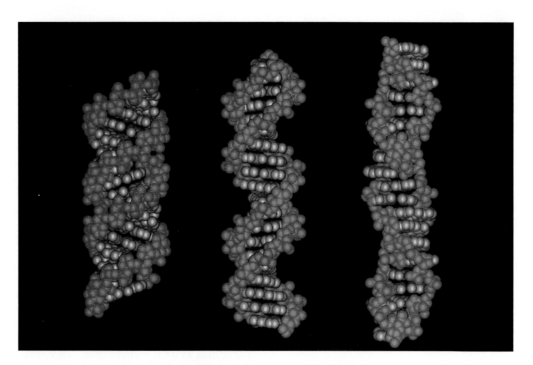

Figure 2.16   Computer graphic models of A-, B-, and Z-DNA. (left) A-DNA. Note the base pairs (blue), whose tilt up from right to left is especially apparent in the wide grooves at the top and near the bottom. Note also the right-handed helix traced by the sugar–phosphate backbone (red). **(middle)** B-DNA. Note the familiar right-handed helix, with roughly horizontal base pairs. **(right)** Z-DNA. Note the left-handed helix. All these DNAs are depicted with the same number of base pairs, emphasizing the differences in compactness of the three DNA forms. (*Source:* Courtesy Fusao Takusagawa.)

A form (Figure 2.16, left). This differs from the B form (Figure 2.16, middle) in several respects. Most obviously, the plane of a base pair is no longer roughly perpendicular to the helical axis, but tilts 20° away from horizontal. Also, the A helix packs in 11 bp per helical turn instead of the 10 found in the B form crystal structure, and each turn occurs in only 31 instead of 34 Å. This means that the pitch, or distance required for one complete turn of the helix, is only 28 instead of 34 Å, as in B-DNA. A hybrid polynucleotide containing one DNA and one RNA strand assumes the A form in solution, as does a double-stranded RNA. Table 2.2 presents these helical parameters for A and B form DNA, and for a left-handed Z-form of DNA, discussed in the next paragraph.

Both the A and B form DNA structures are right-handed: The helix turns clockwise away from you whether you look at it from the top or the bottom. Alexander Rich and his colleagues discovered in 1979 that DNA does not always have to be right-handed. They showed that double-stranded DNA containing strands of alternating purines and pyrimidines (e.g., poly[dG-dC] · poly[dG-dC]):

—GCGCGCGC—
—CGCGCGCG—

can exist in an extended left-handed helical form. Because of the zigzag look of this DNA's backbone when viewed from the side, it is often called **Z-DNA.** Figure 2.16, right, presents a picture of Z-DNA. The helical parameters of

Table 2.2   Forms of DNA

Form	Pitch Å	Residues per Turn	Inclination of Base Pair from Horizontal
A	24.6	~11	+19°
B	33.2	~10	−1.2°
Z	45.6	12	−9°

Source: Watson, et al., The Molecular Biology of the Gene, 4th edition, Benjamin/Cummings Publishing.

this structure are given in Table 2.2. Although Rich discovered Z-DNA in studies of model compounds like poly[dG-dC] · poly[dG-dC], this structure seems to be more than just a laboratory curiosity. Evidence suggests that living cells contain a small proportion of Z-DNA, although we still do not know its function.

SUMMARY In the cell, DNA may exist in the common B form, with base pairs horizontal. A small fraction of the DNA may assume an extended left-handed helical form called Z-DNA (at least in eukaryotes). An RNA–DNA hybrid assumes a third helical shape, called the A form, with base pairs tilted away from the horizontal.

Table 2.3   Relative G + C Contents of Various DNAs

Sources of DNA	Percent (G + C)
Dictyostelium (slime mold)	22
Streptococcus pyogenes	34
Vaccinia virus	36
Bacillus cereus	37
B. megaterium	38
Hemophilus influenzae	39
Saccharomyces cerevisiae	39
Calf thymus	40
Rat liver	40
Bull sperm	41
Streptococcus pneumoniae	42
Wheat germ	43
Chicken liver	43
Mouse spleen	44
Salmon sperm	44
B. subtilis	44
T1 bacteriophage	46
Escherichia coli	51
T7 bacteriophage	51
T3 bacteriophage	53
Neurospora crassa	54
Pseudomonas aeruginosa	68
Sarcina lutea	72
Micrococcus lysodeikticus	72
Herpes simplex virus	72
Mycobacterium phlei	73

Source: From Davidson, *The Biochemistry of the Nucleic Acids,* 8th edition revised by Adams et al. Lippencott.

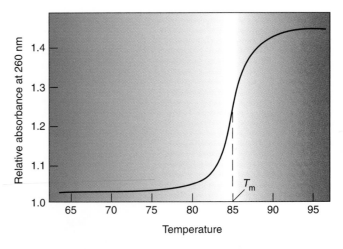

Figure 2.17   Melting curve of *Streptococcus pneumoniae* DNA. The DNA was heated, and its melting was measured by the increase in absorbance at 260 nm. The point at which the melting is half complete is the melting temperature, or T_m. The T_m for this DNA under these conditions is about 85° C.

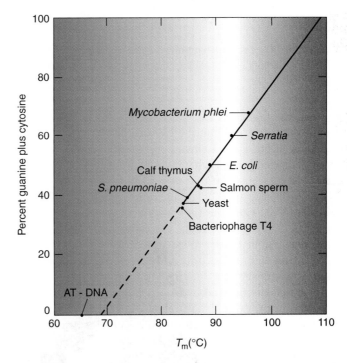

Figure 2.18   Relationship between DNA melting temperature and GC content. AT-DNA refers to synthetic DNAs composed exclusively of A and T (GC content = 0).

Separating the Two Strands of a DNA Double Helix   Although the ratios of G to C and A to T in an organism's DNA are fixed, the GC content (percentage of G + C) can vary considerably from one DNA to another. Table 2.3 lists the GC contents of DNAs from several organisms and viruses. The values range from 22 to 73%, and these differences are reflected in differences in the physical properties of DNA.

When a DNA solution is heated enough, the noncovalent forces that hold the two strands together weaken and finally break. When this happens, the two strands come apart in a process known as **DNA denaturation,** or **DNA melting.** The temperature at which the DNA strands are half denatured is called the melting temperature, or T_m. Figure 2.17 contains a melting curve for DNA from *Streptococcus pneumoniae.* The amount of strand separation, or melting, is measured by the absorbance of the DNA solution at 260 nm. Nucleic acids absorb light at this wavelength because of the electronic structure in their bases,

but when two strands of DNA come together, the close proximity of the bases in the two strands quenches some of this absorbance. When the two strands separate, this quenching disappears and the absorbance rises 30%–40%. This is called the **hyperchromic shift.** The precipitous rise in the curve shows that the strands hold fast until the temperature approaches the T_m and then rapidly let go.

The GC content of a DNA has a significant effect on its T_m. In fact, as Figure 2.18 shows, the higher a DNA's GC

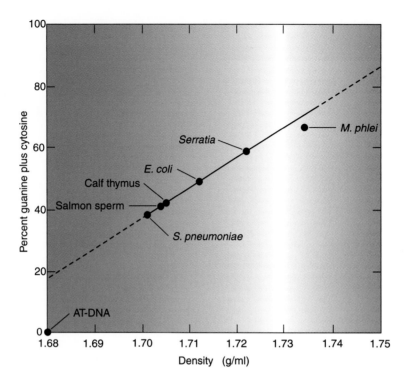

Figure 2.19 The relationship between the GC contents and densities of DNAs from various sources. AT-DNA is a synthetic DNA that is pure A + T; its GC content is therefore zero. (*Source:* P. Doty, *The Harvey Lectures* 55:121, 1961. Copyright 1961 Academic Press, Orlando, FL. Reprinted with permission.)

content, the higher its T_m. Why should this be? Recall that one of the forces holding the two strands of DNA together is hydrogen bonding. Remember also that G–C pairs form three hydrogen bonds, whereas A–T pairs have only two. It stands to reason, then, that two strands of DNA rich in G and C will hold to each other more tightly than those of AT-rich DNA. Consider two pairs of embracing centipedes. One pair has 200 legs each, the other 300. Naturally the latter pair will be harder to separate.

Heating is not the only way to denature DNA. Organic solvents such as dimethyl sulfoxide and formamide, or high pH, disrupt the hydrogen bonding between DNA strands and promote denaturation. Lowering the salt concentration of the DNA solution also aids denaturation by removing the ions that shield the negative charges on the two strands from each other. At low ionic strength, the mutually repulsive forces of these negative charges are strong enough to denature the DNA at a relatively low temperature.

The GC content of a DNA also affects its density. Figure 2.19 shows a direct, linear relationship between GC content and density, as measured by density gradient centrifugation in a CsCl solution (see Chapter 20). Part of the reason for this dependence of density on base composition seems to be real: the larger molar volume of an A–T base pair, compared with a G–C base pair. But part may be an artifact of the method of measuring density using CsCl: A G–C base pair seems to have a greater tendency to bind to

CsCl than does an A–T base pair. This makes its density seem even higher than it actually is.

> **SUMMARY** The GC content of a natural DNA can vary from less than 25% to almost 75%. This can have a strong effect on the physical properties of the DNA, in particular on its melting temperature and density, each of which increases linearly with GC content. The melting temperature (T_m) of a DNA is the temperature at which the two strands are half-dissociated, or denatured. Low ionic strength, high pH, and organic solvents also promote DNA denaturation.

Reuniting the Separated DNA Strands Once the two strands of DNA separate, they can, under the proper conditions, come back together again. This is called **annealing** or **renaturation**. Several factors contribute to renaturation efficiency. Here are three of the most important:

1. *Temperature* The best temperature for renaturation of a DNA is about 25° C below its T_m. This temperature is low enough that it does not promote denaturation, but high enough to allow rapid diffusion of DNA molecules and to weaken the transient bonding between mismatched sequences and short intrastrand base-paired regions. This suggests that

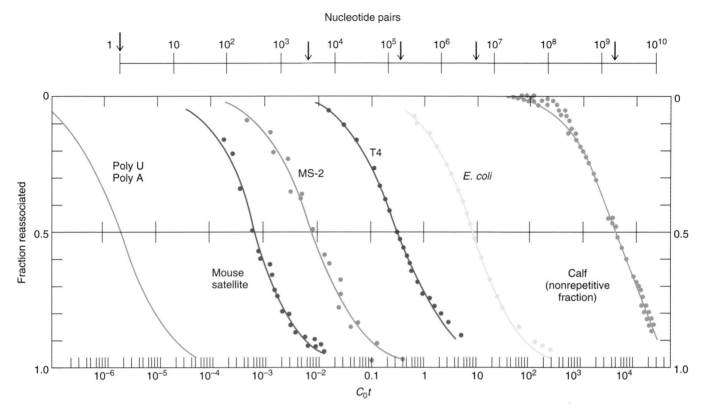

Figure 2.20 C_0t curves of various polynucleotides. The fraction of strands reassociated is plotted versus C_0t. Note that the descending curves actually represent reassociation. The complexity of the polynucleotides, measured in nucleotide pairs, is plotted on the upper *x*-axis. (*Source:* R. J. Britten and D. E. Kohne, *Science* 161:530, 1968. Copyright 1968 American Association for the Advancement of Science, Washington, D.C. Reprinted by permission.)

rapid cooling following denaturation would prevent renaturation. Indeed, a common procedure to ensure that denatured DNA stays denatured is to plunge the hot DNA solution into ice. This is called *quenching*.

2. *DNA Concentration*   The concentration of DNA in the solution is also important. Within reasonable limits, the higher the concentration, the more likely it is that two complementary strands will encounter each other within a given time. In other words, the higher the concentration, the faster the annealing.

3. *Renaturation Time*   Obviously, the longer the time allowed for annealing, the more will occur.

R.J. Britten and D. E. Kohne invented a term, C_0t, to encompass the latter two factors, DNA concentration and time. C_0t (pronounced "cot") is the product of the initial DNA concentration (C_0) in moles of nucleotides per liter and time (t) in seconds. All other factors being equal, the extent of renaturation of complementary strands in a DNA solution will depend on C_0t, as shown by the C_0t curves in Figure 2.20. Note that the fraction reassociated *increases* in the *downward* direction on the *y*-axis, so that the falling curves represent renaturing DNAs.

This set of C_0t curves also points to a final factor in DNA annealing: the complexity of DNA. Look at the curve farthest to the left. It represents the annealing of

two synthetic RNA strands: a strand of poly(U) with a strand of poly(A). Because these RNA strands are complementary no matter where they happen to hit each other—perfectly aligned or off-center—their effective complexity, or length, is one base. This means that they will anneal very readily, or at a low C_0t. The C_0t at which poly(U) and poly(A) are half-annealed (the **half-C_0t**, or $C_0t_{1/2}$) is only about 2×10^{-6}.

Near the middle, we see the C_0t curve for bacteriophage T4 DNA. This DNA is much more complex than poly(U)·poly(A); it has about 2×10^5 bp (see the top *x*-axis), and no repetitive regions, so the two strands have to contact each other exactly right to renature. This results in a $C_0t_{1/2}$ about 2×10^5 higher than that of poly(U)·poly(A), or about 0.4.

Finally, consider calf DNA (the nonrepetitive fraction). This DNA, which constitutes about 70% of the total calf genome, is very complex indeed. Every haploid set of calf-chromosomes has about 2×10^9 bp of such unique DNA—about 10,000 times as much as in a T4 phage. This means that complementary regions of two strands of calf DNA will encounter many more unrelated sequences before they finally hit each other. In fact, they will take 10,000 times as long as T4 DNA at the same concentration. Put another way, in order to "find" each other in the same time it takes T4 DNA, their concentration would

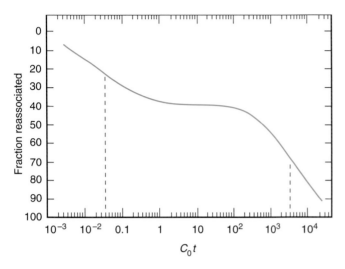

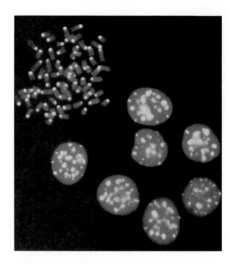

Figure 2.21 C_0t curve of calf thymus DNA. Note that this curve has two phases, corresponding to two main abundance classes of DNA. The first anneals with a $C_0t_{1/2}$ of about 0.04 (dotted line on left). This is midrepetitive DNA; it accounts for about 30% of the total DNA, according to the drop on the y-axis in this segment of the curve (from about 10% to about 40% reassociated). The second class of DNA is the unique, or single-copy, fraction. This anneals with a $C_0t_{1/2}$ of about 4000 (dotted line on right) and composes about 60% of the total DNA. A third class, highly repetitive DNA, reassociated at very low C_0t even before the experiment began. This composes about 10% of the total DNA, which is implied by the fact that about 10% of the DNA is already reassociated at the lowest C_0t in the experiment. (*Source:* R. J. Britten and D. E. Kohne, *Science* 161:530, 1968. Copyright 1968 American Association for the Advancement of Science, Washington, D.C. Reprinted by permission.)

Figure 2.22   Satellite DNA is located in centromeres. A mouse satellite DNA probe was coupled to a molecule called biotin, then hybridized to either mouse metaphase chromosomes (upper left) or interphase cells (lower right). Then a biotin-binding protein called avidin, tagged with a yellow-green fluorescent dye, was added. Wherever the mouse satellite probe had hybridized, the fluorescent avidin would bind and could then be detected by fluorescence microscopy. This micrograph shows the fluorescence, and therefore the satellite DNA, in the centromeres of the metaphase chromosomes. The bulk DNA in the chromosomes was counterstained with a red fluorescent dye for contrast. In the whole cells, we can still see the fluorescent centromeres, even though the individual chromosomes **cannot be distinguished.** (*Source:* Weier et al., *BioTechniques* 10, no. 4 (April 1991) cover. Used with permission of Eaton Publishing.)

have to be 10,000 times higher than that of the T4 DNA. This is just another way of saying that the $C_0t_{1/2}$ of calf DNA should be 10,000 times higher than that of T4 DNA. And it is—approximately 4000, compared with 0.4.

Thus we see a linear relationship between DNA complexity and $C_0t_{1/2}$: The more complex a DNA, the higher its $C_0t_{1/2}$. Of course, not all eukaryotic DNA is nonrepetitive. Forty percent of a mouse's DNA is repetitive, present in 100 to 1 million copies per haploid set of chromosomes. At the upper end of this range is the **highly repetitive DNA,** which constitutes about 10% of the total. This consists of a short sequence of DNA repeated over and over, with some variations, 1 million times per mouse genome. The C_0t curve of this DNA alone is shown as "mouse satellite DNA," the second curve from the left in Figure 2.20. Notice the vast difference in its $C_0t_{1/2}$ and that of **single-copy** (nonrepetitive) mammalian DNA, shown on the far right on the same graph.

The balance of the repetitive DNA is termed **moderately repetitive** or **mid-repetitive.** It constitutes approximately 30% of the mouse genome. The combination of repetitive and single-copy DNA in the mammalian genome yields a complex C_0t curve shown in Figure 2.21. At very low C_0t, the highly repetitive DNA anneals, followed by the moderately repetitive DNA, and finally by the single-copy DNA. Eukaryotic genes that contain the information for making proteins are generally composed of single-copy

DNA; genes for ribosomal RNA (Chapter 10) are moderately repetitive; and the centromere of the chromosome contains highly repetitive DNA (Figure 2.22).

> **SUMMARY** Separated DNA strands can be induced to renature, or anneal. Several factors influence annealing; among them are (1) temperature, (2) DNA concentration, (3) time, and (4) DNA complexity. The product of initial DNA concentration and time is C_0t which, other factors being equal, determines the extent of renaturation. DNAs with high complexity anneal at a higher C_0t than do simpler DNAs.

Solved Problem

Problem 3

Figure 2.23 shows the C_0t curves of two different DNAs. The blue curve (left) represents the renaturation of phage λ-DNA; the red curve (right) depicts the renaturation of the DNA from a nuclear polyhedrosis virus (NPV), a baculovirus that infects a caterpillar. (*a*) What are the $C_0t_{1/2}$'s of the two DNAs? (*b*) Given that the size of the λ-DNA is 50 kb, what is the size of the NPV DNA, assuming no repetitive sequences in either DNA, and an equivalent GC content in both?

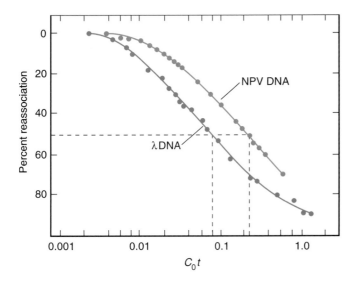

Figure 2.23 C_0t **curves of a nuclear polyhedrosis virus (NPV) DNA (red) and λ phage DNA (blue).** In our laboratory, we followed the renaturation of the two DNAs by observing the drop in absorbance at 260 nm.

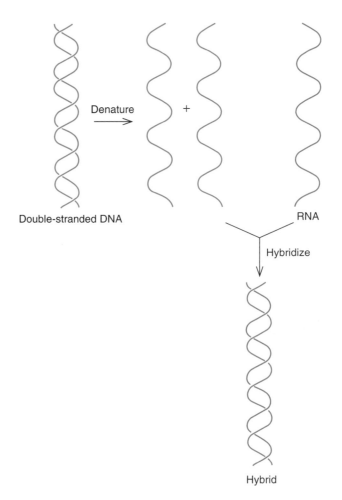

Figure 2.24 Hybridizing DNA and RNA. First, the DNA at upper left is denatured to separate the two DNA strands (blue). Then the DNA strands are mixed with a strand of RNA (red) that is complementary to one of the DNA strands. This hybridization reaction is carried out at a relatively high temperature, which favors RNA–DNA hybridization over DNA–DNA duplex formation. This hybrid has one DNA strand (blue) and one RNA strand (red).

Solution

a. The $C_0t_{1/2}$ of a DNA is the C_0t at which it is one-half renatured. To find this value from the C_0t curve, we can draw a dashed line through the 50% point of the y-axis, as in Figure 2.23. Then, at the point each C_0t curve crosses this line, we read the C_0t on the x-axis. The $C_0t_{1/2}$'s for these two DNAs are therefore 0.08 (λ-DNA) and 0.22 (NPV DNA). Notice that the x-axis is a logarithmic scale, so the vertical tick marks between 0.01 and 0.1 are 0.02, 0.04, 0.06, and 0.08, respectively. Because the $C_0t_{1/2}$ of the NPV DNA does not fall directly on a tick mark, we can approximate its $C_0t_{1/2}$ by interpolation.

b. If two DNAs have no repetitive sequences and do not differ in their GC contents, then their sizes are simply proportional to their $C_0t_{1/2}$'s. Because we know the size of one DNA (λ), we can calculate the size of the other by multiplying by the ratio of their $C_0t_{1/2}$'s. Thus, size of NPV DNA

$$= \frac{0.22}{0.08} \times 50 \text{ kb} \left(\text{size of } \lambda \text{ DNA}\right)$$

$$\approx 140 \text{ kb}$$

Hybridization of Two Different Polynucleotide Chains
So far, we have dealt only with two separated DNA strands simply getting back together again, but other possibilities exist. Consider, for example, a strand of DNA and a strand of RNA getting together to form a double helix. This could happen if we separated the two strands

of a gene, isolated one strand, and placed it together with a complementary RNA strand (Figure 2.24). We would not refer to this as annealing; instead, we would call it **hybridization,** because we are putting together a **hybrid** of two different nucleic acids. The two chains do not have to be as different as DNA and RNA. If we put together two different strands of DNA having complementary, or nearly complementary, sequences we could still call it hybridization—as long as the strands are of different origin. The difference between the two complementary strands may be very subtle; for example, one may be radioactive and the other not. As we will see later in this book, hybridization is an extremely valuable technique. In fact, it would be difficult to overestimate the importance of hybridization to molecular biology.

Table 2.4   Sizes of various DNAs

Source	Molecular weight	Base pairs	Length
Subcellular Genetic Systems:			
SV40 (mammalian tumor virus)	3.5×10^6	5226	1.7 μm
Bacteriophage φX174 (double-stranded form)	3.2×10^6	5386	1.8 μm
Bacteriophage λ	3.3×10^6	5×10^4	13 μm
Bacteriophage T2 or T4	1.3×10^8	2×10^5	50 μm
Human mitochondria	9.5×10^6	16,596	5 μm
Prokaryotes:			
Hemophilus influenzae	1.2×10^9	1.83×10^6	620 μm
Escherichia coli	3.1×10^9	4.65×10^6	1.6 mm
Salmonella typhimurium	8×10^9	1.1×10^7	3.8 mm
Eukaryotes (content per haploid nucleus):			
Saccharomyces cerevisiae (yeast)	7.9×10^9	1.2×10^7	4.1 mm
Neurospora crassa (pink bread mold)	~1.9×10^{10}	~2.7×10^7	~9.2 mm
Drosophila melanogaster (fruit fly)	~1.2×10^{11}	~1.8×10^8	~6.0 cm
Rana pipiens (frog)	~1.4×10^{13}	~2.3×10^{10}	~7.7 m
Mus musculus (mouse)	~1.5×10^{12}	~2.2×10^9	~75 cm
Homo sapiens (human)	~2.3×10^{12}	~3.5×10^9	~120 cm
Zea mays (corn, or maize)	~4.4×10^{12}	~6.6×10^9	~2.2 m
Lilium longiflorum (lily)	~2×10^{14}	~3×10^{11}	~100 m

DNAs of Various Sizes and Shapes

Table 2.4 shows the sizes of the haploid genomes of several organisms and viruses. The sizes are expressed three ways: molecular weight, number of base pairs, and length. These are all related, of course. We already know how to convert number of base pairs to length, because about 10.5 bp occur per helical turn, which is 34 Å long. To convert base pairs to molecular weight, we simply need to multiply by 660, which is the approximate molecular weight of one average nucleotide pair.

How do we measure these sizes? For small DNAs, this is fairly easy. For example, consider phage PM2 DNA, which contains a double-stranded, circular DNA. How do we know it is circular? The most straightforward way to find out is simply by looking at it. We can do this using an electron microscope, but first we have to treat the DNA so that it stops electrons and will show up in a micrograph just as bones stop x-rays and therefore show up in an x-ray picture. The most common way of doing this is by *shadowing* the DNA with a heavy metal such as platinum. We place the DNA on an electron microscope grid and bombard it with minute droplets of metal from a shallow angle. This makes the metal pile up beside the DNA like snow behind a fence. We rotate the DNA on the grid so it becomes shadowed all around. Now the metal will stop the electrons in the electron microscope

and make the DNA appear as light strings against a darker background. Printing reverses this image to give a picture such as Figure 2.25, which is an electron micrograph of PM2 DNA in two forms: an open circle (lower left) and a **supercoil** (upper right), in which the DNA coils around itself rather like a twisted rubber band. We can also use pictures like these to measure the length of the DNA. This is more accurate if we include a standard DNA of known length in the same picture.

The size of a DNA can also be estimated by gel electrophoresis, a topic we will discuss in Chapter 5.

SUMMARY Natural DNAs come in sizes ranging from several kilobases to hundreds of megabases. The size of a small DNA can be estimated by electron microscopy. This technique can also reveal whether a DNA is circular or linear, and whether it is supercoiled.

Solved Problem

Problem 4

Phage P1 has a double-stranded DNA with 91,500 bp (91.5 kb). (*a*) How many full double-helical turns does this

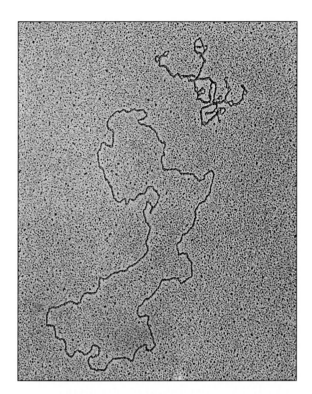

Figure 2.25   Electron micrograph of phage PM2 DNA. The open circular form is shown on the lower left and the supercoiled form is shown at the upper right. (*Source:* © Jack Griffith.)

DNA contain? (*b*) How long is the DNA in microns (1 micron = 10^4 Å)? (*c*) What is the molecular mass of this DNA? (*d*) How many phosphorus atoms does the DNA contain?

Solution

a. One double-helical turn encompasses 10.5 bp. Thus, all we have to do is divide the total number of base pairs by 10.5:

91,500 bp × 1 helical turn/10.5 bp ≈ 8710 helical turns.

b. The spacing between base pairs is about 3.4 Å, or 3.4×10^{-4} μm along the helical axis, so we just need to multiply the total number of base pairs by 3.4×10^{-4}:

91,500 bp × 3.4×10^{-4} μm/bp ≈ 31 μm.

c. One base pair has a molecular mass of about 660 daltons (D), so we need to multiply the total number of base pairs by 660 D:

91,500 bp × 660 D/bp ≈ 6.0×10^7 D (6.0×10^4 kD).

d. Each base pair has two phosphorus atoms, one on each strand. Therefore, we simply multiply the total number of base pairs by 2:

91,500 bp × 2 phosphorus atoms/bp
= 183,000 phosphorus atoms.   ■

The Relationship Between DNA Size and Genetic Capacity
How many genes are in a given DNA? It is impossible to tell just from the size of the DNA, because we do not know how much of a given DNA is devoted to genes and how much is space between genes, or even intervening sequences within genes. We can, however, estimate an upper limit on the number of genes a DNA can hold. We start with the assumption that an average protein has a molecular mass of about 40,000 D. How many amino acids does this represent? The molecular masses of amino acids vary, but they average a bit over 100 D. To simplify our calculation, let us assume that the average is 100. That means our average protein contains 40,000/100, or 400 amino acids. Because each amino acid requires 3 bp of DNA to code for it, a protein containing 400 amino acids needs a gene of about 1200 bp.

Consider a few of the DNAs listed in Table 2.4. The *E. coli* chromosome contains 4.6×10^6 bp, so it could encode about 3800 average proteins. Phage λ, which infects *E. coli,* has only 5×10^4 bp, so it can code for only about 40 proteins. The smallest double-stranded DNA on the list, belonging to the mammalian tumor virus SV40, has a mere 5226 bp. In principle, that is only enough to code for about four proteins, but the virus squeezes in some extra information by overlapping its genes.

Solved Problem

Problem 5

Table 2.4 shows that the DNAs of bacteriophages T2 and T4 contain 2×10^5 bp. How many genes of average size (encoding proteins of about 40,000 molecular weight) can these phages contain?

Solution

Because the average molecular mass of an amino acid is about 100 D, a protein with a molecular mass of 40,000 D contains about 400 amino acids. It takes 3 bp to encode one amino acid, so each protein will be encoded in about 1200 bp ($3 \times 400 = 1200$). Thus, the number of genes of average size in a DNA having 2×10^5 bp will be $2 \times 10^5/1.2 \times 10^3 = 1.67 \times 10^2$, or 167.   ■

DNA Content and the C-Value Paradox   You would probably predict that complex organisms such as vertebrates need more genes than simple organisms like yeast. Therefore, they should have higher **C-Values,** or DNA content per haploid cell. In general, your prediction would be right; mouse and human haploid cells contain more than 100 times more DNA than yeast haploid cells. Furthermore, yeast cells have about five times more DNA than *E. coli* cells, which are even simpler. However, this correspondence between an organism's physical complexity and the DNA content of its cells is not perfect.

Consider, for example, the frog. Intuitively, you would not suspect that an amphibian would have a higher C-value than a human, yet the frog has seven times more DNA per cell. Even more dramatic is the fact that the lily has 100 times more DNA per cell than a human.

This perplexing situation is called the **C-value paradox.** It becomes even more difficult to explain when we look at organisms within a group. For example, some amphibian species have C-values 100 times higher than those of others, and the C-values of flowering plants vary even more widely. Does this mean that one kind of higher plant has 100 times more genes than another? That is simply unbelievable. It would raise questions about what all those extra genes are good for and why we do not notice tremendous differences in physical complexity among these organisms. The more plausible explanation of the C-value paradox is that organisms with extraordinarily high C-values simply have a great deal of extra, noncoding DNA. The function, if any, of this extra DNA is still mysterious.

In fact, even mammals have much more DNA than they need for genes. Applying our simple rule (dividing the number of base pairs by 1200) to the human genome yields an estimate of about 2 million for the maximum number of genes, which is far too high.

If we cannot use the C-value as a reliable guide to the number of genes in higher organisms, how can we make a reasonable estimate? One approach has been to count the number of different mRNAs in a given human cell type, which gives an estimate of about 10,000. Of course, humans have over 100 different cell types, but they all produce many common mRNAs. Therefore, we cannot simply multiply 10,000 times 100. Indeed, the overlap in genes expressed in different human cells is so extensive that we probably need to multiply 10,000 by a number less than 10. In fact, the working draft of the human genome suggests that there are only 25–40,000 genes. This means that human cells contain over 50 times more DNA than they need to code for even as many as 40,000 proteins. Much of this extra DNA is found in intervening sequences within eukaryotic genes (Chapter 14). The rest is in noncoding regions outside of genes, including most of the moderately repetitive DNA and all of the highly repetitive DNA detected by C_0t analysis.

SUMMARY There is a rough correlation between the DNA content and the number of genes in a cell or virus. However, this correlation breaks down in several cases of closely related organisms where the DNA content per haploid cell (C-value) varies widely. This C-value paradox is probably explained, not by extra genes, but by extra noncoding DNA in some organisms.

SUMMARY

Genes of all true organisms are made of DNA; certain viruses have genes made of RNA. DNA and RNA are chain-like molecules composed of subunits called nucleotides. DNA has a double-helical structure with sugar–phosphate backbones on the outside and base pairs on the inside. The bases pair in a specific way: adenine (A) with thymine (T) and guanine (G) with cytosine (C). When DNA replicates, the parental strands separate; each then serves as the template for making a new, complementary strand.

The G + C content of a natural DNA can vary from 22 to 73%, and this can have a strong effect on the physical properties of DNA, particularly its melting temperature. The melting temperature (T_m) of a DNA is the temperature at which the two strands are half-dissociated, or denatured. Separated DNA strands can be induced to renature, or anneal. The product of initial DNA concentration and time is the C_0t, which, other things being equal, determines the extent of renaturation. DNAs with high complexity anneal at a higher C_0t than do simpler DNAs. Complementary strands of polynucleotides (either RNA or DNA) from different sources can form a double helix in a process called hybridization. Natural DNAs vary widely in length. The size of a small DNA can be estimated by electron microscopy.

A rough correlation occurs between the DNA content and the number of genes in a cell or virus. However, this correlation does not hold in several cases of closely related organisms in which the DNA content per haploid cell (C-value) varies widely. This C-value paradox is probably explained by extra noncoding DNA in some organisms. This noncoding DNA includes most of the repetitive DNA detected by C_0t analysis.

REVIEW QUESTIONS

1. Compare and contrast the experimental approaches used by Avery and colleagues, and by Hershey and Chase, to demonstrate that DNA is the genetic material.

2. Draw the general structure of a deoxynucleoside monophosphate. Show the sugar structure in detail and indicate the positions of attachment of the base and the phosphate. Also indicate the deoxy position.

3. Draw the structure of a phosphodiester bond linking two nucleotides. Show enough of the two sugars that the sugar positions involved in the phosphodiester bond are clear.

4. Which DNA purine forms three H bonds with its partner in the other DNA strand? Which forms two H-bonds? Which

DNA pyrimidine forms three H bonds with its partner? Which forms two H bonds?

5. Draw a typical DNA melting curve. Label the axes and point out the melting temperature.

6. Use a graph to illustrate the relationship between the GC content of a DNA and its melting temperature. What is the explanation for this relationship?

7. Use a graph to illustrate the relationship between the GC content of a DNA and its density. What is the explanation for this relationship?

8. On a single graph, show the C_0t curves of T4 phage DNA, *E. coli* DNA, and mammalian nonrepetitive DNA. What accounts for the different half-C_0ts of these three DNAs?

9. On a single graph, show the C_0t curves of mammalian (e.g., mouse) satellite DNA and nonrepetitive DNA. What is the ratio of the two half-C_0ts? What does this ratio tell us about the extent of repetition of the satellite DNA?

10. Use a drawing to illustrate the principle of nucleic acid hybridization.

11. Describe and illustrate the results of an experiment to show that mouse satellite DNA is located in the centromeres of chromosomes.

12. How many proteins of average size could be encoded in a virus with a DNA genome having 12,000 bp, assuming no overlap of genes?

SUGGESTED READINGS

Adams, R.L.P., R.H. Burdon, A.M. Campbell, and R.M.S. Smellie, eds. 1976. *Davidson's The Biochemistry of the Nucleic Acids,* 8th ed. The structure of DNA, chapter 5. New York: Academic Press.

Avery, O.T., C. M. McLeod, and M. McCarty. 1944. Studies on the chemical nature of the substance-inducing transformation of pneumococcal types. *Journal of Experimental Medicine* 79:137–58.

Chargaff, E. 1950. Chemical specificity of the nucleic acids and their enzymatic degradation. *Experientia* 6:201–9.

Dickerson, R. E. 1983. The DNA helix and how it reads. *Scientific American* 249 (December): 94–111.

Hershey, A. D., and M. Chase. 1952. Independent functions of viral protein and nucleic acid in growth of bacteriophage. *Journal of General Physiology* 36:39–56.

Watson, J. D., and F. H. C. Crick. 1953. Molecular structure of the nucleic acids: A structure for deoxyribose nucleic acid. *Nature* 171:737–38.

———. 1953. Genetical implications of the structure of deoxyribonucleic acid. *Nature* 171:964–67.

An Introduction to Gene Function

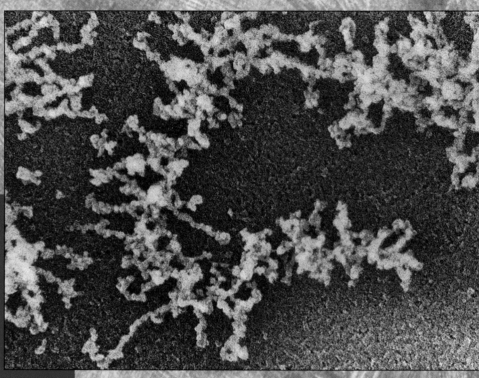

Color-enhanced scanning electron micrograph of DNA strands showing transcription detail (×61,500). © *Biophoto Associates/Photo Researchers, Inc.*

As we learned in Chapter 1, a gene participates in three major activities:

1. A gene is a repository of information. That is, it holds the information for making one of the key molecules of life, an RNA. The sequence of bases in the RNA depends directly on the sequence of bases in the gene. Most of these RNAs, in turn, serve as templates for making other critical cellular molecules, proteins. The production of a protein from a DNA blueprint is called gene expression. Chapters 6–19 deal with various aspects of gene expression.

2. A gene can be replicated. This duplication is very faithful, so the genetic information can be passed essentially unchanged from generation to generation. We will discuss gene replication in Chapters 20–21.

3. A gene can accept occasional changes, or mutations. This allows organisms to evolve. Sometimes, these changes involve recombination, exchange of DNA between chromosomes or sites within a chromosome. A subset of recombination

events involve pieces of DNA (transposable elements) that move from one place to another in the genome. We will deal with recombination and transposable elements in Chapters 22 and 23.

This chapter outlines the three activities of genes and provides some background information that will be useful in our deeper explorations in subsequent chapters. ■

3.1 Storing Information

Let us begin by examining the gene expression process, starting with a brief overview, followed by an introduction to protein structure and an outline of the two steps in gene expression.

Overview of Gene Expression

As we have seen, producing a protein from information in a DNA gene is a two-step process. The first step is synthesis of an RNA that is complementary to one of the strands of DNA. This is called **transcription.** In the second step, called **translation,** the information in the RNA is used to make a polypeptide. Such an informational RNA is called a **messenger RNA** (**mRNA**) to denote the fact that it carries information—like a message—from a gene to the cell's protein factories.

Like DNA and RNA, proteins are polymers—long, chain-like molecules. The monomers, or links, in the protein chain are called amino acids. DNA and protein have this informational relationship: Three nucleotides in the DNA gene stand for one amino acid in a protein.

Figure 3.1 summarizes the gene expression process and introduces the nomenclature we apply to the strands of DNA. Notice that the mRNA has the same sequence (except that U's substitute for T's) as the top strand (blue) of the DNA. An mRNA holds the information for making a polypeptide, so we say it "codes for" a polypeptide, or "encodes" a polypeptide. (Note: It is redundant to say "encodes for" a polypeptide.) In this case, the mRNA codes for the following string of amino acids: methionine-serine-asparagine-alanine, which is abbreviated Met-Ser-Asn-Ala. We can see that the codeword (or **codon**) for methionine in this mRNA is the triplet AUG; similarly, the codons for serine, asparagine, and alanine are AGU, AAC, and GCG, respectively.

Because the bottom DNA strand is complementary to the mRNA, we know that it served as the template for making the mRNA. Thus, we call the bottom strand the **template strand,** or the transcribed strand. For the same reason, the top strand is the **nontemplate strand,** or the nontranscribed strand. Because the top strand in our example has essentially the same coding properties as the corresponding mRNA, many geneticists call it the *coding strand.* The opposite

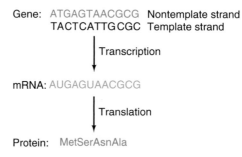

Gene: ATGAGTAACGCG Nontemplate strand
TACTCATTGCGC Template strand

↓ Transcription

mRNA: AUGAGUAACGCG

↓ Translation

Protein: MetSerAsnAla

Figure 3.1 Outline of gene expression. In the first step, transcription, the template strand (black) is transcribed into mRNA. Note that the nontemplate strand (blue) of the DNA has the same sequence (except for the T—U change) as the mRNA (red). In the second step, the mRNA is translated into protein (green). This little "gene" is only 12 bp long and codes for only four amino acids (a tetrapeptide). Real genes are much larger.

strand would therefore be the *anticoding strand.* Also, since the top strand has the same sense as the mRNA, this same system of nomenclature refers to this top strand as the *sense strand,* and to the bottom strand as the *antisense strand.* However, many other geneticists use the "coding strand" and "sense strand" conventions in exactly the opposite way. From now on, to avoid confusion, we will use the unambiguous terms *template strand* and *nontemplate strand.*

Protein Structure

Since we are seeking to understand gene expression, and because proteins are the final products of most genes, let us take a brief look at the nature of proteins. Proteins, like nucleic acids, are chain-like polymers of small subunits. In the case of DNA and RNA, the links in the chain are nucleotides. The chain links of proteins are **amino acids.** Whereas DNA contains only four different nucleotides, proteins contain 20 different amino acids. The structures of these compounds are shown in Figure 3.2. Each amino acid has an amino group (NH_3^+), a carboxyl group (COO^-), a hydrogen atom (H), and a side chain. The only difference between any two amino acids is in their different side chains. Thus, it is the arrangement of amino acids, with their distinct side chains, that gives each protein its unique character. The amino acids join together in proteins via **peptide bonds,** as shown in Figure 3.3. This gives rise to the name **polypeptide** for a chain of amino acids. A protein can be composed of one or more polypeptides.

A polypeptide chain has polarity, just as the DNA chain does. The *dipeptide* (two amino acids linked together) shown on the right in Figure 3.3 has a free amino group at its left end. This is the **amino terminus,** or **N-terminus.** It also has a free carboxyl group at its right end, which is the **carboxyl terminus,** or **C-terminus.**

The linear order of amino acids constitutes a protein's **primary structure.** The way these amino acids interact

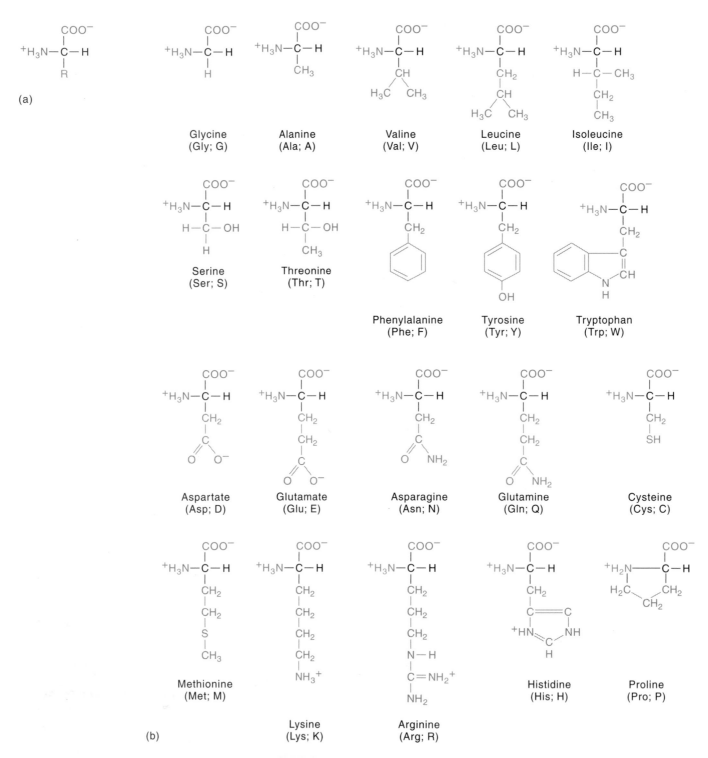

Figure 3.2 Amino acid structure. (a) The general structure of an amino acid. It has both an amino group (NH_3^+; red) and an acid group (COO^-; blue); hence the name. Its other two positions are occupied by a proton (H) and a side chain (R, gray). (b) Each of the 20 different amino acids has a different side chain. All of them are illustrated here. Three-letter and one-letter abbreviations are in parentheses.

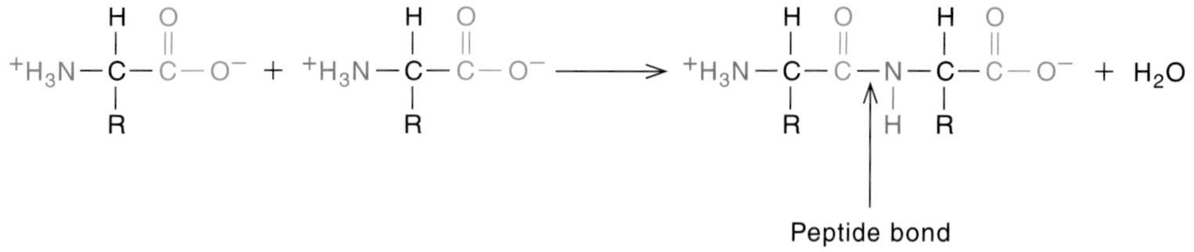

$$^+H_3N-\underset{\underset{R}{|}}{\overset{\overset{H}{|}}{C}}-\underset{}{\overset{\overset{O}{\|}}{C}}-O^- \;+\; {}^+H_3N-\underset{\underset{R}{|}}{\overset{\overset{H}{|}}{C}}-\underset{}{\overset{\overset{O}{\|}}{C}}-O^- \;\longrightarrow\; {}^+H_3N-\underset{\underset{R}{|}}{\overset{\overset{H}{|}}{C}}-\underset{}{\overset{\overset{O}{\|}}{C}}-\underset{\underset{H}{|}}{N}-\underset{\underset{R}{|}}{\overset{\overset{H}{|}}{C}}-\underset{}{\overset{\overset{O}{\|}}{C}}-O^- \;+\; H_2O$$

Peptide bond

Figure 3.3   Formation of a peptide bond. Two amino acids with side chains R and R′ combine through the acid group of the first and the amino group of the second to form a dipeptide, two amino acids linked by a peptide bond. One molecule of water also forms as a by-product.

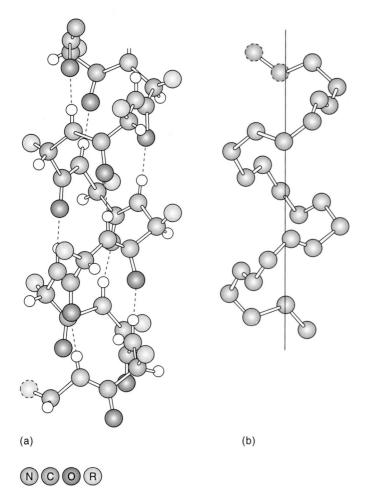

(a)   (b)

Figure 3.4   An example of protein secondary structure: The α-helix. **(a)** The positions of the amino acids in the helix are shown, with the helical backbone in black and blue. The dashed lines represent hydrogen bonds between hydrogen and oxygen atoms on nearby amino acids. The small white circles represent hydrogen atoms. **(b)** A simplified rendition of the α-helix, showing only the atoms in the helical backbone.

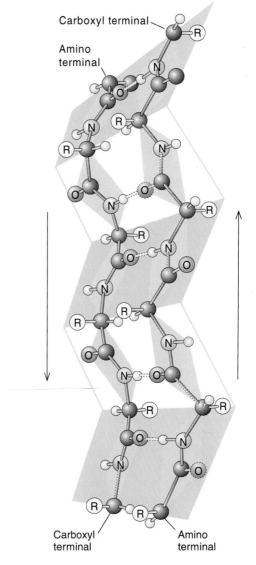

Figure 3.5   An antiparallel β-sheet. Two polypeptide chains are arranged side by side, with hydrogen bonds (dashed lines) between them. The green and white planes show that the β-sheet is pleated. The chains are antiparallel in that the amino terminus of one and the carboxyl terminus of the other are at the top. The arrows indicate that the two β-strands run from amino to carboxyl terminal in opposite directions. Parallel β-sheets, in which the β-strands run in the same direction, also exist.

with their neighbors gives a protein its **secondary structure**. The *α-helix* is a common form of secondary structure. It results from hydrogen bonding among near-neighbor amino acids, as shown in Figure 3.4. Another common secondary structure found in proteins is the β-pleated sheet (Figure 3.5). This involves extended protein chains, packed side by side, that interact by hydrogen

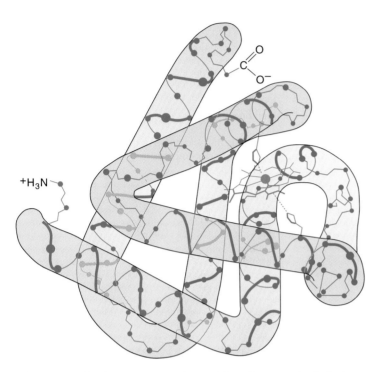

Figure 3.6 Tertiary structure of myoglobin. The several α-helical regions of this protein are represented by turquoise corkscrews. The overall molecule seems to resemble a sausage, twisted into a roughly spherical or globular shape. The heme group is shown in red, bound to two histidines (turquoise polygons) in the protein.

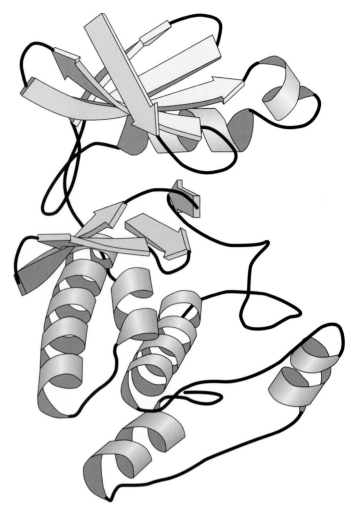

Figure 3.7 Tertiary structure of the core region common to all protein kinases. Secondary structure elements, including α-helices (coiled ribbon), β-pleated sheets (flat arrows), and turns (between β-sheet elements) are apparent. The structure is divided into two globular domains (beige and blue), with the catalytic cleft in between. (*Source:* S.S. Taylor, *Science,* 253/Cover illustration, 1991. Copyright 1991 American Association for the Advancement of Science. Reprinted by permission.)

bonding. The packing of the chains next to each other creates the sheet appearance. Silk is a protein very rich in β-pleated sheets. A third example of secondary structure is simply a turn. Such turns connect the α-helices and β-pleated sheet elements in a protein.

The total three-dimensional shape of a polypeptide is its **tertiary structure.** Figure 3.6 illustrates how the protein myoglobin folds up into its tertiary structure. Elements of secondary structure are apparent, especially the several α-helices of the molecule. Note the overall roughly spherical shape of myoglobin. Most polypeptides take this form, which we call *globular.*

Figure 3.7 is a different representation of protein structure called a ribbon model. This model depicts the tertiary structure of the catalytic core common to all members of a group of enzymes known as protein kinases. Here we can clearly see three types of secondary structure: α-helices, represented by helical ribbons, and β-pleated sheets, represented by flat arrows laid side by side, and turns between elements of the β-pleated sheets.

Figure 3.7 also shows that a protein can contain more than one compact structural region. Each of these regions (beige and blue in Figure 3.7) is called a **domain.** Antibodies (the proteins that white blood cells make to repel invaders) provide another good example of domains. Each of the four polypeptides in the IgG-type antibody contains globular do-

mains, as shown in Figure 3.8. When we study protein–DNA binding in Chapter 9, we will see that domains can contain common structural–functional **motifs.** For example, a finger-shaped motif called a zinc finger is involved in DNA binding. Figure 3.8 also illustrates the highest level of protein structure—**quaternary structure**—which is the way two or more individual polypeptides fit together in a complex protein. It has long been assumed that a protein's amino acid sequence determines all of its higher levels of structure, much as the linear sequence of letters in this book determines word, sentence, and paragraph structure. However, this analogy is an oversimplification. Most proteins cannot fold properly by themselves outside their normal cellular environment. Some cellular factors besides the protein itself seem to be required in these cases, and folding often must occur during synthesis of a polypeptide.

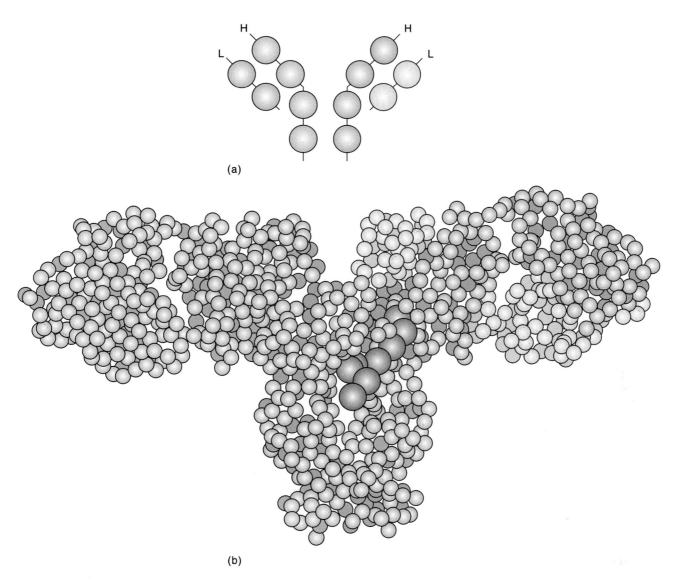

Figure 3.8   The globular domains of an immunoglobulin. (a) Schematic diagram, showing the four polypeptides that constitute the immunoglobulin: two light chains (L) and two heavy chains (H). The light chains each contain two globular regions, and the heavy chains have four globular domains apiece. **(b)** Space-filling model of an immunoglobulin. The colors correspond to those in part **(a)**. Thus, the two H chains are in red and blue; the L chains are in green and yellow. A complex sugar attached to the protein is shown in gray. Note the globular domains in each of the polypeptides. Also note how the four polypeptides fit together to form the quaternary structure of the protein.

What forces hold a protein in its proper shape? Some of these are covalent bonds, but most are noncovalent. The principal covalent bonds within and between polypeptides are disulfide (S–S) bonds between cysteines. The noncovalent bonds are primarily hydrophobic and hydrogen bonds. Predictably, hydrophobic amino acids cluster together in the interior of a polypeptide, or at the interface between polypeptides, so they can avoid contact with water (*hydrophobic,* meaning water-fearing). Hydrophobic interactions play a major role in tertiary and quaternary structures of proteins.

SUMMARY Proteins are polymers of amino acids linked through peptide bonds. The sequence of amino acids in a polypeptide (primary structure) gives rise to that molecule's local shape (secondary structure), overall shape (tertiary structure), and interaction with other polypeptides (quaternary structure).

Protein Function

Why are proteins so important? Some proteins provide the structure that helps give cells integrity and shape. Others serve as *hormones* to carry signals from one cell to another. For example, the pancreas secretes the hormone insulin that signals liver and muscle cells to take up the sugar glucose from the blood. Proteins can also bind and carry substances. The protein hemoglobin carries oxygen from the lungs to remote areas of the body; myoglobin stores oxygen in muscle tissue until it is used. Proteins also control the activities of genes, as we will see many times in this book. And proteins serve as enzymes that catalyze the hundreds of chemical reactions necessary for life.

The Relationship between Genes and Proteins Our knowledge of the gene–protein link dates back as far as 1902, when a physician named Archibald Garrod noticed that a human disease, *alcaptonuria*, behaved as if it were caused by a single recessive gene. Fortunately, Mendel's work had been rediscovered 2 years earlier and provided the theoretical background for Garrod's observation. Patients with alcaptonuria excrete copious amounts of *homogentisic acid*, which has the startling effect of coloring their urine black, Garrod reasoned that the abnormal buildup of this compound resulted from a defective metabolic pathway. Somehow, a blockage somewhere in the pathway was causing the intermediate, homogentisic acid, to accumulate to abnormally high levels, much as a dam causes water to accumulate behind it. Several years later, Garrod postulated that the problem came from a defect in the pathway that degrades the amino acid phenylalanine (Figure 3.9).

By that time, metabolic pathways had been studied for years and were known to be controlled by enzymes—one enzyme catalyzing each step. Thus, it seemed that alcaptonuria patients carried a defective enzyme. And because the disease was inherited in a simple Mendelian fashion, Garrod concluded that a gene must control the enzyme's production. When that gene is defective, it gives rise to a defective enzyme. This suggested the crucial conceptual link between genes and proteins.

George Beadle and E. L. Tatum carried this argument a step further with their studies of a common bread mold, *Neurospora*, in the 1940s. They performed their experiments as follows: First, they bombarded the peritheca (spore-forming parts) of *Neurospora* with x-rays or ultraviolet radiation to cause mutations. Then, they collected the spores from the irradiated mold and germinated them separately to give pure strains of mold. They screened many thousands of strains to find a few mutants. The mutants revealed themselves by their inability to grow on minimal medium composed only of sugar, salts, inorganic nitrogen, and the vitamin biotin. Wild-type *Neurospora* grows readily on such a medium; the mutants had to be fed something extra—a vitamin, for example—to survive.

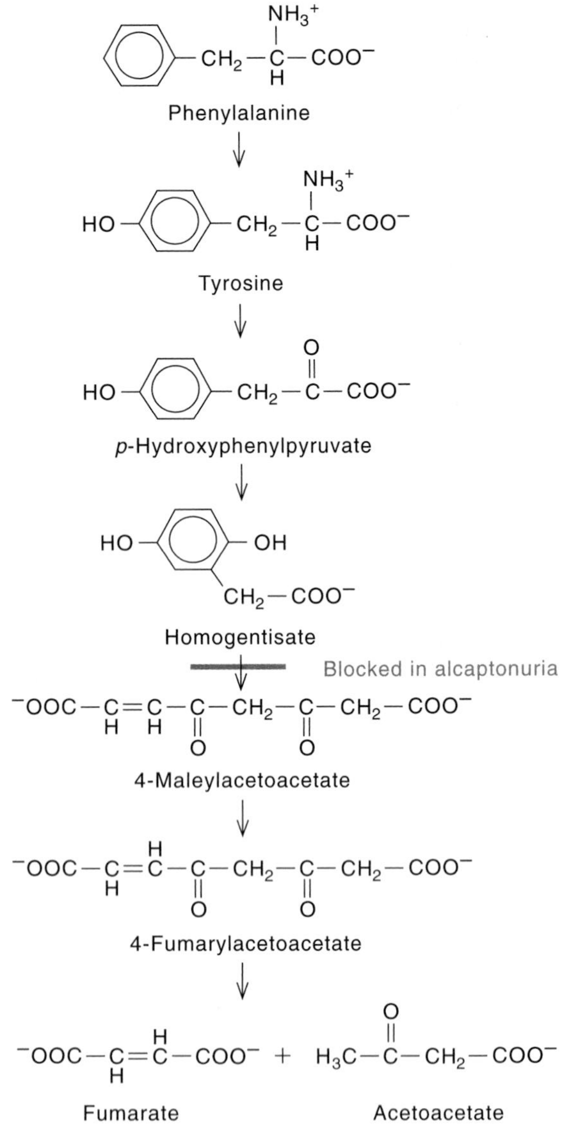

Figure 3.9 Pathway of phenylalanine breakdown. Alcaptonuria patients are defective in the enzyme that converts homogentisate to 4-maleylacetoacetate.

Next, Beadle and Tatum performed biochemical and genetic analyses on their mutants. By carefully adding substances, one at a time, to the mutant cultures, they pinpointed the biochemical defect. For example, the last step in the synthesis of the vitamin pantothenate involves putting together the two halves of the molecule: pantoate and β-alanine (Figure 3.10). One "pantothenateless" mutant would grow on pantothenate, but not on the two halves of the vitamin. This demonstrated that the last step (step 3) in the biochemical pathway leading to pantothenate was blocked, so the enzyme that carries out that step must have been defective.

The genetic analysis was just as straightforward. *Neurospora* is an *ascomycete,* in which nuclei of two different

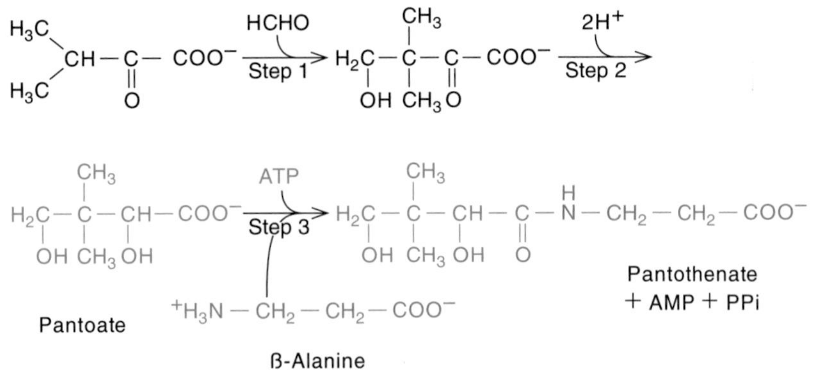

Figure 3.10   Pathway of pantothenate synthesis. The last step (step 3), formation of pantothenate from the two half-molecules, pantoate (blue) and β-alanine (red), was blocked in one of Beadle and Tatum's mutants. The enzyme that carries out this step must have been defective.

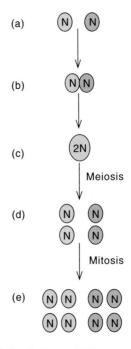

Figure 3.11   Sporulation in the mold *Neurospora crassa*. (a) Two haploid nuclei, one wild-type (yellow) and one mutant (blue), have come together in the immature fruiting body of the mold. **(b)** The two nuclei begin to fuse. **(c)** Fusion is complete, and a diploid nucleus (green) has formed. One haploid set of chromosomes is from the wild-type nucleus, and one set is from the mutant nucleus. **(d)** Meiosis occurs, producing four haploid nuclei. If the mutant phenotype is controlled by one gene, two of these nuclei (blue) should have the mutant allele and two (yellow) should have the wild-type allele. **(e)** Finally, mitosis occurs, producing eight haploid nuclei, each of which will go to one ascospore. Four of these nuclei (blue) should have the mutant allele and four (yellow) should have the wild-type allele. If the mutant phenotype is controlled by more than one gene, the results will be more complex.

mating types fuse and undergo meiosis to give eight haploid *ascospores,* borne in a fruiting body called an *ascus.* Therefore, a mutant can be crossed with a wild-type strain of the opposite mating type to give eight spores (Figure 3.11). If the mutant phenotype results from a mu-

tation in a single gene, then four of the eight spores should be mutant and four should be wild-type. Beadle and Tatum collected the spores, germinated them separately, and checked the phenotypes of the resulting molds. Sure enough, they found that four of the eight spores gave rise to mutant molds, demonstrating that the mutant phenotype was controlled by a single gene. This happened over and over again, leading these investigators to the conclusion that each enzyme in a biochemical pathway is controlled by one gene. Subsequent work has shown that many enzymes contain more than one polypeptide chain and that each polypeptide is usually encoded in one gene. This is the **one-gene/one-polypeptide hypothesis.**

SUMMARY Most genes contain the information for making one polypeptide.

Discovery of Messenger RNA

The concept of a messenger RNA carrying information from gene to ribosome developed in stages during the years following the publication of Watson and Crick's DNA model. In 1958, Crick himself proposed that RNA serves as an intermediate carrier of genetic information. He based his hypothesis in part on the fact that the DNA resides in the nucleus of eukaryotic cells, whereas proteins are made in the cytoplasm. This means that something must carry the information from one place to the other. Crick noted that ribosomes contain RNA and suggested that this ribosomal RNA (rRNA) is the information bearer. But rRNA is an integral part of ribosomes; it cannot escape. Therefore, Crick's hypothesis implied that each ribosome, with its own rRNA, would produce the same kind of protein over and over.

François Jacob and colleagues proposed an alternative hypothesis calling for nonspecialized ribosomes that translate unstable RNAs called *messengers.* The messengers are independent RNAs that bring genetic information from the genes to the ribosomes. In 1961, Jacob, along with Sydney Brenner and Matthew Meselson, published their proof of the messenger hypothesis. This study used the same bacteriophage (T2) that Hershey and Chase had employed almost a decade earlier to show that the genes were made of DNA (Chapter 2). The premise of the experiments was this: When phage T2 infects *E. coli,* it subverts its host from making bacterial proteins to making phage proteins. If Crick's hypothesis were correct, this switch to phage protein synthesis should be accompanied by the production of new ribosomes equipped with phage-specific RNAs.

To distinguish new ribosomes from old, these investigators labeled the ribosomes in uninfected cells with heavy isotopes of nitrogen (^{15}N) and carbon (^{13}C). This made "old" ribosomes heavy. Then they infected these cells with phage T2 and simultaneously transferred them to medium containing light nitrogen (^{14}N) and carbon (^{12}C). Any "new" ribosomes made after phage infection would therefore be light and would separate from the old, heavy ribosomes during density gradient centrifugation. Brenner and colleagues also labeled the infected cells with ^{32}P to tag any phage RNA as it was made. Then they asked this question: Was the radioactively labeled phage RNA associated with new or old ribosomes? Figure 3.12 shows that the phage RNA was found on old ribosomes whose rRNA was made before infection even began. Clearly, this old rRNA could not carry phage genetic information; by extension, it was very unlikely that it could carry host genetic information either. Thus, the ribosomes are constant. The nature of the polypeptides they make depends on the mRNA that associates with them. This relationship resembles that of a CD player and CD. The nature of the music (polypeptide) depends on the CD (mRNA), not the player (ribosome).

Other workers had already identified a better candidate for the messenger: a class of unstable RNAs that associate transiently with ribosomes. Interestingly enough, in phage T2-infected cells, this RNA had a base composition very similar to that of phage DNA—and quite different from that of bacterial DNA and RNA. This is exactly what we would expect of phage messenger RNA, and that is exactly what it is. On the other hand, host mRNA, unlike host rRNA, has a base composition similar to that of host DNA. This lends further weight to the hypothesis that mRNA, not rRNA, is the informational molecule.

SUMMARY Messenger RNAs carry the genetic information from the genes to the ribosomes, which synthesize polypeptides.

Transcription

As you might expect, transcription follows the same base-pairing rules as DNA replication: T, G, C, and A in the DNA pair with A, C, G, and U, respectively in the RNA product. (Notice that uracil appears in RNA in place of thymine in DNA.) This base-pairing pattern ensures that an RNA transcript is a faithful copy of the gene (Figure 3.13).

Of course, highly directed chemical reactions such as transcription do not happen at significant rates by themselves—they are enzyme-catalyzed. The enzyme that directs transcription is called **RNA polymerase.** Figure 3.14 presents a schematic diagram of *E. coli* RNA polymerase at work. Transcription has three phases: initiation, elongation, and termination. The following is an outline of these three steps in bacteria:

1. *Initiation*   First, the enzyme recognizes a region called a **promoter,** which lies just "upstream" of the gene. The polymerase binds tightly to the promoter and causes localized melting, or separation, of the two DNA strands within the promoter. At least 12 bp are melted. Next, the polymerase starts building the RNA chain. The substrates, or building blocks, it uses for this job are the four **ribonucleoside triphosphates:** ATP, GTP, CTP, and UTP. The first, or initiating, substrate is usually a purine nucleotide. After the first nucleotide is in place, the polymerase joins a second nucleotide to the first, forming the initial phosphodiester bond in the RNA chain. Several nucleotides may be joined before the polymerase leaves the promoter and elongation begins.

2. *Elongation*   During the elongation phase of transcription, RNA polymerase directs the sequential binding of ribonucleotides to the growing RNA chain in the 5′ → 3′ direction (from the 5′-end toward the 3′-end of the RNA). As it does so, it moves along the DNA **template,** and the "bubble" of melted DNA moves with it. This melted region exposes the bases of the template DNA one by one so that they can pair with the bases of the incoming ribonucleotides. As soon as the transcription machinery passes, the two DNA strands wind around each other again, reforming the double helix. This points to two fundamental differences between transcription and DNA replication: (a) RNA polymerase makes only one RNA strand during transcription, which means that it copies only one DNA strand in a given gene. (However, the opposite strand may be transcribed in another gene.) Transcription is therefore said to be **asymmetrical.** This contrasts with semiconservative DNA replication, in which both DNA strands are copied. (b) In transcription, DNA melting is limited and transient. Only enough strand separation occurs

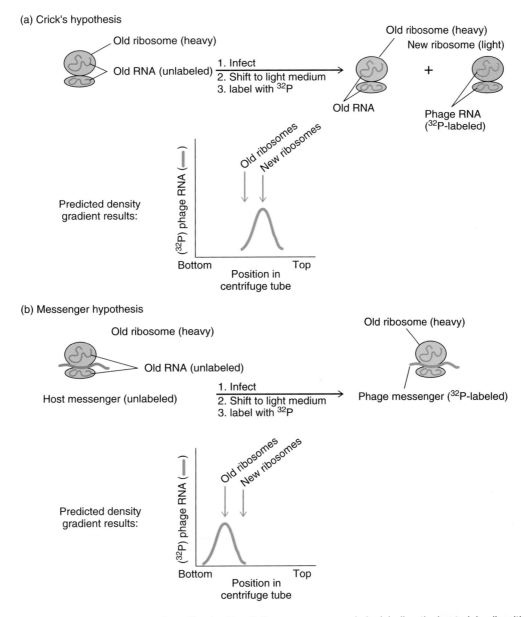

Figure 3.12 Experimental test of the messenger hypothesis. *E. coli* ribosomes were made by labeling the bacterial cells with heavy isotopes of carbon and nitrogen. The bacteria were then infected with T2 phage and simultaneously shifted to "light" medium containing the normal isotopes of carbon and nitrogen, plus some ^{32}P to make the phage RNA radioactive. **(a)** Crick had proposed that ribosomal RNA carried the message for making proteins. If this were so, then whole new ribosomes with phage-specific ribosomal RNA would have been made after phage infection. In that case, the new ^{32}P-labeled RNA (green) should have moved together with the new, light ribosomes (red). **(b)** Jacob and colleagues had proposed that a messenger RNA carried genetic information to the ribosomes. According to this hypothesis, phage infection would cause the synthesis of new, phage-specific messenger RNAs that would be ^{32}P-labeled (green). These would associate with old, heavy ribosomes (blue). The radioactive label would therefore move together with the old, heavy ribosomes in the density gradient. This was indeed what happened.

to allow the polymerase to "read" the DNA template strand. However, during replication, the two parental DNA strands separate permanently.

3. *Termination* Just as promoters serve as initiation signals for transcription, other regions at the ends of genes, called **terminators,** signal termination. These work in conjunction with RNA polymerase to loosen the association between RNA product and DNA template. The result is that the RNA dissociates from the RNA polymerase and DNA, thereby stopping transcription.

A final, important note about conventions: RNA sequences are usually written 5′ to 3′, left to right. This feels

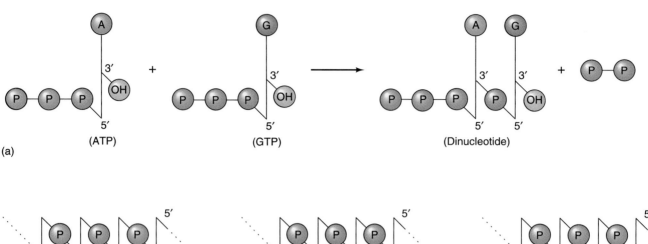

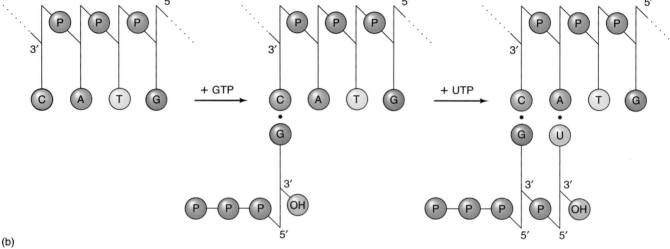

Figure 3.13   Making RNA. (a) Phosphodiester bond formation in RNA synthesis. ATP and GTP are joined together to form a dinucleotide. Note that the phosphorus atom closest to the guanosine is retained in the phosphodiester bond. The other two phosphates are removed as a by-product called pyrophosphate. (b) Synthesis of RNA on a DNA template. The DNA template at top contains the sequence 3′-dC-dA-dT-dG-5′ and extends in both directions, as indicated by the dashed lines. To start the RNA synthesis, GTP forms a base-pair with the dC nucleotide in the DNA template. Next, UTP provides a uridine nucleotide, which forms a base-pair with the dA nucleotide in the DNA template and forms a phosphodiester bond with the GTP. This produces the dinucleotide GU. In the same way, a new nucleotide joins the growing RNA chain at each step until transcription is complete. The pyrophosphate by-product is not shown.

natural to a molecular biologist because RNA is made in a 5′-to-3′ direction, and, as we will see, mRNA is also translated 5′ to 3′. Thus, because ribosomes read the message 5′ to 3′, it is appropriate to write it 5′ to 3′ so that we can read it like a sentence.

Genes are also usually written so that their transcription proceeds in a left-to-right direction. This "flow" of transcription from one end to the other gives rise to the term *upstream,* which refers to the DNA close to the start of transcription (near the left end when the gene is written conventionally). Thus, we can describe most promoters as lying just upstream of their respective genes. By the same convention, we say that genes generally lie *downstream* of their promoters. Genes are also conventionally written with their nontemplate strands on top.

SUMMARY Transcription takes place in three stages: initiation, elongation, and termination. Initiation involves binding RNA polymerase to the promoter, local melting, and forming the first few phosphodiester bonds; during elongation, the RNA polymerase links together ribonucleotides in the 5′ → 3′ direction to make the rest of the RNA; finally, in termination, the polymerase and RNA product dissociate from the DNA template.

(1) Initiation:

(a) RNA polymerase binds
to promoter.

(b) First few phosphodiester
bonds form.

(2) Elongation.

ppp

(3) Termination.

ppp

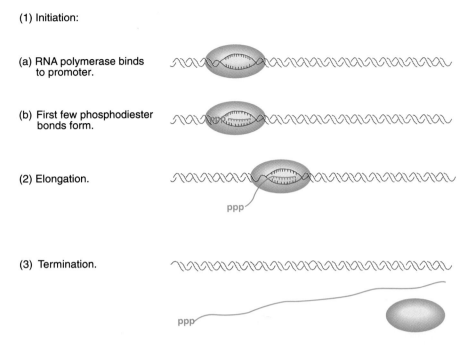

Figure 3.14   Transcription. (1a) In the first stage of initiation, RNA polymerase (red) binds tightly to the promoter and "melts" a short stretch of DNA. **(1b)** In the second stage of initiation, the polymerase joins the first few nucleotides of the nascent RNA (blue) through phosphodiester bonds. The first nucleotide retains its triphosphate group (ppp). **(2)** During elongation, the melted bubble of DNA moves with the polymerase, allowing the enzyme to "read" the bases of the DNA template strand and make complementary RNA. **(3)** Termination occurs when the polymerase reaches a termination signal, causing the RNA and the polymerase to fall off the DNA template.

Solved Problem

Problem

The following is the sequence of the template strand of a DNA fragment:

5′ATGCGTGACTAATTCG3′

a. Write the sequence of this DNA in conventional double-stranded form.
b. Assuming that transcription of this DNA begins with the first nucleotide and ends with the last, write the sequence of the transcript of this DNA in conventional form (5′ to 3′ left to right).

Solution

a. The template strand has been given in the 5′ → 3′ orientation. But the template strand is conventionally written in the 3′ → 5′ orientation, so let's begin by inverting it to read as follows:

3′-GCTTAATCAGTGCGTA-5′

Now we need to add the nontemplate strand, following the Watson–Crick base-pairing rules, and remembering that the nontemplate strand is conventionally presented on top:

5′-CGAATTAGTCACGCAT-3′
3′-GCTTAATCAGTGCGTA-5′

b. Because we know that the transcript of a gene has the same sequence and orientation as the nontemplate strand (with U's in place of T's), we simply need to write the sequence of the nontemplate strand, replacing all T's with U's as follows:

5′CGAAUUAGUCACGCAU3′

Translation

The mechanism of translation is also complex and fascinating. The details of translation will concern us in later chapters; for now, let us look briefly at two substances that play key roles in translation: ribosomes and transfer RNA.

Ribosomes: Protein-Synthesizing Machines   Figure 3.15 shows the approximate shapes of the *E. coli* **ribosome** and its two subunits: the **50S** and **30S** particles. The numbers 50S and 30S refer to the **sedimentation coefficients** of the two subunits. These coefficients are a measure of the speed with which the particles sediment through a solution when spun in an ultracentrifuge. The 50S particle, with a larger sedimentation coefficient, migrates more rapidly to the bottom of the centrifuge tube under the influence of a centrifugal force. The coefficients are functions of the mass and shape of the particles. Heavy particles sediment more rapidly than light ones; spherical particles migrate faster than extended or flattened ones—just

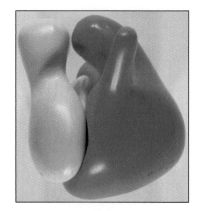

(a)

(b)

Figure 3.15 *E. coli* **ribosome structure.** (a) The 70S ribosome is shown from the "side" with the 30S particle (yellow) and the 50S particle (red) fitting together. (b) The 70S ribosome is shown rotated 90 degrees relative to the view in part (a). The 30S particle (yellow) is in front, with the 50S particle (red) behind. (*Source:* (a) Lake, J. Ribosome structure determined by electron microscopy of *Escherichia coli* small subunits, large subunits and monomeric ribosomes. *J. Mol. Biol.* 105 (1976) 131–59, by permission of Academic Press. (b) Lake, J. Ribosome structure determined by electron microcopy of *Escherichia coli* small subunits, large subunits and monomeric ribosomes. *J. Mol. Biol.* 105 (1976) p. 155, fig. 14, by permission of Academic Press.)

as a skydiver falls more rapidly in a tuck position than with arms and legs extended. The 50S particle is actually about twice as massive as the 30S. Together, the 50S and 30S subunits compose a **70S** ribosome. Notice that the numbers do not add up. This is because the sedimentation coefficients are not proportional to the particle weight; in fact, they are roughly proportional to the two-thirds power of the particle weight.

Each ribosomal subunit contains RNA and protein. The 30S particle includes one molecule of **ribosomal RNA (rRNA)** with a sedimentation coefficient of 16S, plus 21 ribosomal proteins. The 50S particle is composed of 2 rRNAs (23S + 5S) and 34 proteins (Figure 3.16). All these ribosomal proteins are of course gene products themselves. Thus, a ribosome is produced by dozens of different genes. Eukaryotic ribosomes are even more complex, with one more rRNA and many more proteins.

Note that rRNAs participate in protein synthesis but do not code for proteins. Transcription is the only step in expression of the genes for rRNAs, aside from some trimming of the transcripts. No translation of these RNAs occurs.

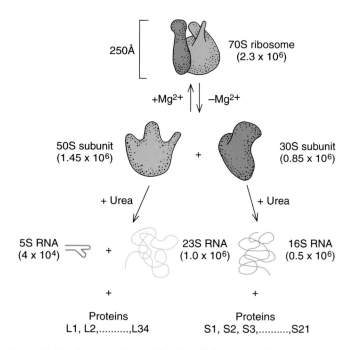

Figure 3.16 **Composition of the *E. coli* ribosome.** The arrows at the top denote the dissociation of the 70S ribosome into its two subunits when magnesium ions are withdrawn. The lower arrows show the dissociation of each subunit into RNA and protein components in response to the protein denaturant, urea. The masses (M_r, in daltons) of the ribosome and its components are given in parentheses.

SUMMARY Ribosomes are the cell's protein factories. Bacteria contain 70S ribosomes with two subunits, called 50S and 30S. Each of these contains ribosomal RNA and many proteins.

Transfer RNA: The Adapter Molecule The transcription mechanism was easy for molecular biologists to predict. RNA resembles DNA so closely that it follows the same base-pairing rules. By following these rules, RNA polymerase produces replicas of the genes it transcribes. But what rules govern the ribosome's translation of mRNA to protein? This is a true translation problem. A nucleic acid language must be translated to a protein language. Francis Crick suggested the answer to this problem in a 1958 paper before much experimental evidence was available to back it up. What is needed, Crick reasoned, is some kind of adapter molecule that can recognize the nucleotides in the RNA language as well as the amino acids in the protein language. He was right. He even noted that a type of small RNA of unknown function might play the adapter role. Again, he guessed right. Of course he made some bad guesses in this paper as well, but even they were important. By their very creativity, Crick's ideas stimulated the research (some from Crick's own laboratory) that led to solutions to the puzzle of translation.

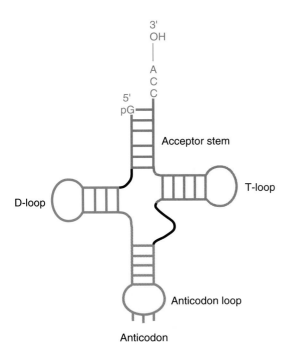

Figure 3.17   Cloverleaf structure of yeast tRNA^Phe. At top is the acceptor stem (red), where the amino acid binds to the 3′-terminal adenosine. At left is the dihydro U loop (D-loop, blue), which contains at least one dihydrouracil base. At bottom is the anticodon loop (green), containing the anticodon. The T-loop (right, gray) contains the virtually invariant sequence TψC. Each loop is defined by a base-paired stem of the same color.

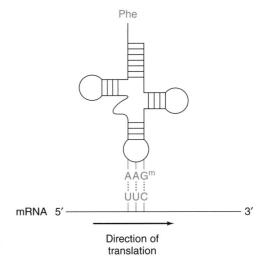

Figure 3.18   Codon–anticodon recognition. The recognition between a codon in an mRNA and a corresponding anticodon in a tRNA obeys essentially the same Watson–Crick rules as apply to other polynucelotides. Here, a 3′AAG^m5′ anticodon (blue) on a tRNA^Phe is recognizing a 5′UUC3′ codon (red) for phenylalanine in an mRNA. The G^m denotes a methylated G, which base-pairs like an ordinary G. Notice that the tRNA is pictured backwards (3′ → 5′) relative to normal convention, which is 5′ → 3′, left to right. That was done to put its anticodon in the proper orientation (3′ → 5′, left to right) to base-pair with the codon, shown conventionally reading 5′ → 3′, left to right. Remember that the two strands of DNA are antiparallel; this applies to any double-stranded polynucleotide, including one as small as the 3-bp codon–anticodon pair.

The adapter molecule in translation is indeed a small RNA that recognizes both RNA and amino acids; it is called **transfer RNA (tRNA).** Figure 3.17 shows a schematic diagram of a tRNA that recognizes the amino acid phenylalanine (Phe). In Chapter 19 we will discuss the structure and function of tRNA in detail. For the present, the *cloverleaf* model, though it bears scant resemblance to the real shape of tRNA, will serve to point out the fact that the molecule has two "business ends." One end (the top of the model) attaches to an amino acid. Because this is a tRNA specific for phenylalanine (tRNA^Phe), only phenylalanine will attach. An enzyme called *phenylalanine-tRNA synthetase* catalyzes this reaction. The generic name for such enzymes is **aminoacyl-tRNA synthetase.**

The other end (the bottom of the model) contains a 3-bp sequence that pairs with a complementary 3-bp sequence in an mRNA. Such a triplet in mRNA is called a **codon;** naturally enough, its complement in a tRNA is called an **anticodon.** The codon in question here has attracted the anticodon of a tRNA bearing a phenylalanine. That means that this codon tells the ribosome to insert one phenylalanine into the growing polypeptide. The recognition between codon and anticodon, mediated by the ribosome, obeys the same Watson–Crick rules as any other double-stranded polynucleotide, at least in the case of the first two base pairs. The third pair is allowed somewhat more freedom, as we will see in Chapter 18.

It is apparent from Figure 3.18 that UUC is a codon for phenylalanine. This implies that the **genetic code** contains three-letter words, as indeed it does. We can predict the number of possible 3-bp codons as follows: The number of permutations of 4 different bases taken 3 at a time is 4^3, which is 64. But only 20 amino acids exist. Are some codons not used? Actually, three of the possible codons (UAG, UAA, and UGA) code for termination; that is, they tell the ribosome to stop. All of the other codons specify amino acids. This means that most amino acids have more than one codon; the genetic code is therefore said to be **degenerate.** Chapter 18 presents a fuller description of the code and how it was broken.

SUMMARY Two important sites on tRNAs allow them to recognize both amino acids and nucleic acids. One site binds covalently to an amino acid. The other site contains an anticodon that base-pairs with a 3-bp codon in mRNA. The tRNAs are therefore capable of serving the adapter role postulated by Crick and are the key to the mechanism of translation.

Initiation of Protein Synthesis   We have just seen that three codons terminate translation. A codon (AUG) also usually initiates translation. The mechanisms of these two processes are markedly different. As we will see in Chapter 18, the three termination codons interact with protein factors, whereas the initiation codon interacts with a special aminoacyl-tRNA. In eukaryotes this is methionyl-tRNA (a tRNA with methionine attached); in prokaryotes it is a derivative called N-formylmethionyl-tRNA. This is just methionyl tRNA with a formyl group attached to the amino group of methionine.

We find AUG codons not only at the beginning of mRNAs, but also in the middle of messages. When they are at the beginning, AUGs serve as initiation codons, but when they are in the middle, they simply code for methionine. The difference is context. Prokaryotic messages have a special sequence, called a **Shine–Dalgarno sequence,** named for its discoverers, just upstream of the initiating AUG. The Shine–Dalgarno sequence attracts ribosomes to the nearby AUG so translation can begin. Eukaryotes, by contrast, do not have Shine–Dalgarno sequences. Instead, their mRNAs have a special methylated nucleotide called

a **cap** at their 5′ ends. A protein called the **cap-binding protein (CBP)** binds to the cap and then helps attract ribosomes. We will discuss these phenomena in greater detail in Chapter 17.

> **SUMMARY** AUG is usually the initiating codon. It is distinguished from internal AUGs by a Shine–Dalgarno ribosome-binding sequence near the beginning of prokaryotic mRNAs, and by a cap structure at the 5′ end of eukaryotic mRNAs.

Translation Elongation   At the end of the initiation phase of translation, the initiating aminoacyl-tRNA is bound to a site on the ribosome called the **P site.** For elongation to occur, the ribosome needs to add amino acids one at a time to the initiating amino acid. We will examine this process in detail in Chapter 18. For the moment, let us consider a simple overview of the elongation process in *E. coli* (Figure 3.19). Elongation begins with the binding of the second aminoacyl-tRNA to another site

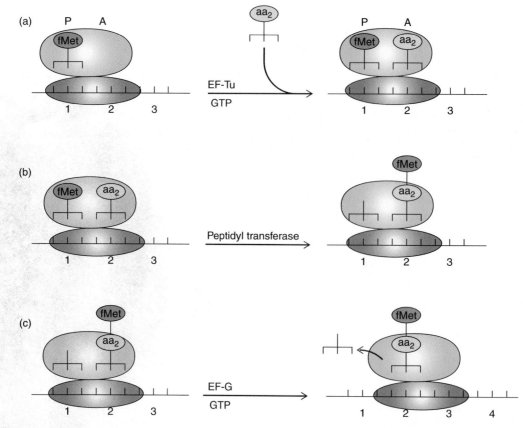

Figure 3.19   Summary of translation elongation. (a) EF-Tu, with help from GTP, transfers the second aminoacyl-tRNA to the A site. (The P and A sites are conventionally represented on the left and right halves of the ribosome, as indicated at the top.) (b) Peptidyl transferase, an integral part of the large rRNA in the 50S subunit, forms a peptide bond between fMet and the second aminoacyl-tRNA. This creates a dipeptidyl-tRNA in the A site. (c) EF-G, with help from GTP, translocates the mRNA one codon's length through the ribosome. This brings codon 2, along with the peptidyl-tRNA to the P site, and codon 3 to the A site. It also moves the deacylated tRNA out of the P site into the E site (not shown), from which it is ejected. The A site is now ready to accept another aminoacyl-tRNA to begin another round of elongation.

on the ribosome called the **A site.** This process requires an **elongation factor** called **EF-Tu,** where EF stands for "elongation factor," and energy provided by GTP.

Next, a peptide bond must form between the two amino acids. The large ribosomal subunit contains an enzyme known as **peptidyl transferase,** which forms a peptide bond between the amino acid or peptide in the P site (formylmethionine [fMet] in this case) and the amino acid part of the aminoacyl tRNA in the A site. The result is a dipeptidyl-tRNA in the A site. The dipeptide is composed of fMet plus the second amino acid, which is still bound to its tRNA. The large ribosomal RNA contains the peptidyl transferase active center.

The third step in elongation, **translocation,** involves the movement of the mRNA one codon's length through the ribosome. This maneuver transfers the dipeptidyl-tRNA from the A site to the P site and moves the deacetylated tRNA from the P site to another site, the **E site,** which provides an exit from the ribosome. Translocation requires another elongation factor called **EF-G** and GTP.

> **SUMMARY** Translation elongation involves three steps: (1) transfer of the second aminoacy-tRNA to the A site; (2) formation of a peptide bond between the first amino acid in the P site and the second aminoacyl-tRNA in the A site; and (3) translocation of the mRNA one codon's length through the ribosome, bringing the newly formed peptidyl-tRNA to the P site.

Termination of Translation and mRNA Structure   Three different codons (UAG, UAA, and UGA) cause termination of translation. Protein factors called **release factors** recognize these **termination codons** (or **stop codons**) and cause translation to stop, with release of the polypeptide chain. The initiation codon at one end, and the termination codon at the other end of a coding region of a gene identify an **open reading frame (ORF).** It is called "open" because it contains no internal termination codons to interrupt the translation of the corresponding mRNA. The "reading frame" part of the name refers to the way the ribosome can read the mRNA in three different ways, or "frames," depending on where it starts.

Figure 3.20 illustrates the reading frame concept. This minigene (shorter than any gene you would expect to find) contains a start codon (ATG) and a stop codon (TAG). (Remember that these DNA codons will be transcribed to mRNA with the corresponding codons AUG and UAG.) In between (and including these codons) we have a short open reading frame that can be translated to yield a pentapeptide (a peptide containing five amino acids): fMet-Gly-Tyr-Arg-Pro. In principle, translation could also begin four bases upstream at another AUG, but notice that translation would be in another **reading frame,** so the codons would be different: AUG, CAU, GGG, AUA, UAG. Translation in this second reading frame would therefore produce a tetrapeptide: fMet-His-Gly-Ile. The third reading frame has no initiation codon. A natural mRNA may also

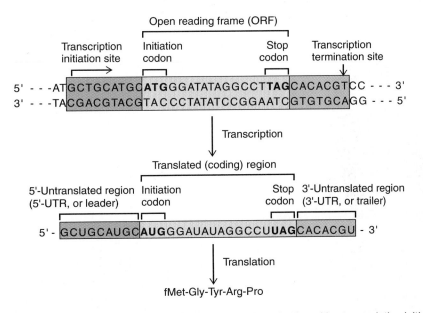

Figure 3.20   Simplified gene and mRNA structure. At top is a simplified gene that begins with a transcription initiation site and ends with a transcription termination site. In between are the translation initiation codon and the stop codon, which define an open reading frame that can be translated to yield a polypeptide (a very short polypeptide with only five amino acids, in this case). The gene is transcribed to give an mRNA with a coding region that begins with the initiation codon and ends with the termination codon. This is the RNA equivalent of the open reading frame in the gene. The material upstream of the initiation codon in the mRNA is the leader, or 5′-untranslated region. The material downstream of the termination codon in the mRNA is the trailer, or 3′-untranslated region. Note that this gene has another open reading frame that begins four bases farther upstream, but it codes for an even shorter polypeptide with only four amino acids. Notice also that this alternative reading frame is shifted 1 bp to the left relative to the other.

have more than one open reading frame, but the largest is usually the one that is used.

Figure 3.20 also shows that transcription and translation in this gene do not start and stop at the same places. Transcription begins with the first G and translation begins 9 bp downstream at the start codon (AUG). Thus, the mRNA produced from this gene has a 9-bp **leader,** which is also called the **5′-untranslated region,** or **5′-UTR.** Similarly, a **trailer** is present at the end of the mRNA between the stop codon and the translation termination site. The trailer is also called the **3′-untranslated region,** or **3′-UTR.** In a eukaryotic gene, the transcription termination site would probably be farther downstream, but the mRNA would be cleaved downstream of the translation stop codon and a string of A's [poly(A)] would be added to the 3′-end of the mRNA. In that case, the trailer would be the stretch of RNA between the stop codon and the poly(A).

> **SUMMARY** Translation terminates at a stop codon (UAG, UAA, or UGA). The genetic material including a translation initiation codon, a coding region, and a termination codon, is called an open reading frame. The piece of an mRNA between its 5′-end and the initiation codon is called a leader or 5′-UTR. The part between the 3′-end [or the poly(A)] and the termination codon is called a trailer or 3′-UTR.

3.2   Replication

A second characteristic of genes is that they replicate faithfully. The Watson–Crick model for DNA replication (introduced in Chapter 2) assumes that as new strands of DNA are made, they follow the usual base-pairing rules of A with T and G with C. This is essential because the DNA-replicating machinery must be capable of discerning a good pair from a bad one, and the Watson–Crick base pairs give the best fit. The model also presupposes that the two parental strands separate and that each then serves as a template for a new progeny strand. This is called **semiconservative** replication because each daughter double helix has one parental strand and one new strand (Figure 13.21a). In other words, one of the parental strands is "conserved" in each daughter double helix. This is not the only possibility. Another potential mechanism (Figure 13.21b) is **conservative** replication, in which the two parental strands stay together and somehow produce another daughter helix with two completely new strands. Yet another possibility is **dispersive** replication, in which the DNA becomes fragmented so that new and old DNA coexist in the same strand after replication (Figure 13.21c). As mentioned in Chapter 1, Matthew Mesel-

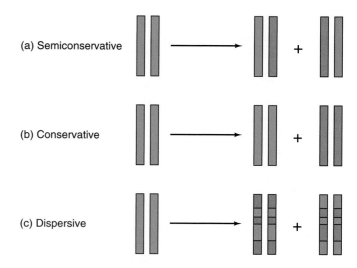

Figure 3.21   Three hypotheses for DNA replication.
(a) Semiconservative replication (see also Figure 2.15) gives two daughter duplex DNAs, each of which contains one old strand (blue) and one new strand (red). **(b)** Conservative replication yields two daughter duplexes, one of which has two old strands (blue) and one of which has two new strands (red). **(c)** Dispersive replication gives two daughter duplexes, each of which contains strands that are a mixture of old and new DNA.

son and Franklin Stahl proved that DNA really does replicate by a semiconservative mechanism. Chapter 20 will present this experimental evidence.

3.3   Mutations

A third characteristic of genes is that they accumulate changes, or mutations. By this process, life itself can change, because mutation is essential for evolution. Now that we know most genes are strings of nucleotides that code for polypeptides, which in turn are strings of amino acids, it is easy to see the consequences of changes in DNA. If a nucleotide in a gene changes, it is likely that a corresponding change will occur in an amino acid in that gene's protein product. Sometimes, because of the degeneracy of the genetic code, a nucleotide change will not affect the protein. For example, changing the codon AAA to AAG is a mutation, but it would probably not be detected because both AAA and AAG code for the same amino acid: lysine. Such innocuous alterations are called **silent mutations.** More often, a changed nucleotide in a gene results in an altered amino acid in the protein. This may be harmless if the amino acid change is **conservative** (e.g., a leucine changed to an isoleucine). But if the new amino acid is much different from the old one, the change frequently impairs or destroys the function of the protein.

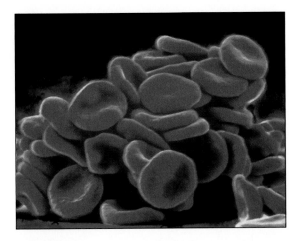

(a)

(b)

Figure 3.22   Normal red blood cells and sickle cells. These scanning electron micrographs contrast (**a**) the regular, biconcave shape of normal cells with (**b**) the distorted shape of red blood cells from a person with sickle cell disease. Magnifications: (**a**) ×3333 (**b**) ×5555. (*Source:* (a) © Jeroboam/Omikron/Photo Researchers, Inc. (b) © Omikron/Photo Researchers, Inc.)

Sickle Cell Disease

An excellent example of a disease caused by a defective gene is *sickle cell disease,* a true genetic disorder. People who are homozygous for this condition have normal-looking red blood cells when their blood is rich in oxygen. The shape of normal cells is a *biconcave disc;* that is, the disc is concave viewed from both the top and bottom. However, when these people exercise, or otherwise deplete the oxygen in their blood, their red blood cells change dramatically to a sickle, or crescent, shape (Figure 3.22). This has dire consequences. The sickle cells cannot fit through tiny capillaries, so they clog and rupture them, starving parts of the body for blood and causing internal bleeding and pain. Furthermore, the sickle cells are so fragile that they burst, leaving the patient anemic. Without medical attention, patients undergoing a sickling crisis are in mortal danger.

What causes this sickling of red blood cells? The problem is in *hemoglobin,* the red, oxygen-carrying protein in the red blood cells. Normal hemoglobin remains soluble under ordinary physiological conditions, but the hemoglobin in sickle cells precipitates when the blood oxygen level falls, forming long, fibrous aggregates that distort the blood cells into the sickle shape.

What is the difference between normal hemoglobin (HbA) and sickle cell hemoglobin (HbS)? Vernon Ingram answered this question in 1957 by determining the amino acid sequences of parts of the two proteins using a process that was invented by Frederick Sanger and is known as **protein sequencing.** Ingram focused on the β-globins of the two proteins. β-globin is one of the two different polypeptide chains found in the tetrameric (four-chain) hemoglobin protein. First, Ingram cut the two polypeptides into pieces with an enzyme that breaks selected peptide bonds. These pieces, called *peptides,* can be separated by a two-dimensional method called **fingerprinting** (Figure 3.23). The peptides are separated in the first dimension by paper electrophoresis. Then the paper is turned 90 degrees and the peptides are subjected to paper chromatography to separate them still farther in the second dimension. The peptides usually appear as spots on the paper; different proteins, because of their different amino acid compositions, give different patterns of spots. These patterns are aptly named **fingerprints.**

When Ingram compared the fingerprints of HbA and HbS, he found that all the spots matched except for one (Figure 3.24). This spot had a different mobility in the HbS fingerprint than in the normal HbA fingerprint, which indicated that it had an altered amino acid composition. Ingram checked the amino acid sequences of the two peptides in these spots. He found that they were the amino-terminal peptides located at the very beginning of both proteins. And he found that they differed in only one amino acid. The glutamic acid in the sixth position of HbA becomes a valine in HbS (Figure 3.25). This is the only difference in the two proteins, yet it is enough to cause a profound distortion of the protein's behavior.

Knowing the genetic code (Chapter 18), we can ask: What change in the β-globin gene caused the change Ingram detected in its protein product? The two codons for glutamic acid (Glu) are GAA and GAG; two of the four codons for valine (Val) are GUA and GUG. If the glutamic acid codon in the HbA gene is GAG, a single base change to GTG would alter the mRNA to GUG, and the amino acid inserted into HbS would be valine instead of glutamic acid. A similar argument can be made for a GAA → GTA change. Notice that, by convention, we are presenting the DNA strand that has the same sense as the mRNA (the nontemplate strand). Actually, the opposite strand (the template strand), reading CAC, is transcribed to give a GUG sequence in the mRNA. Figure 3.26 presents a summary of the mutation and its consequences.

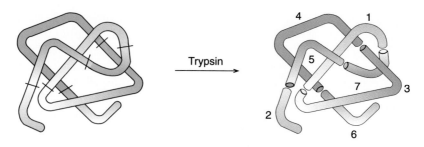

(a) Cutting protein to peptides

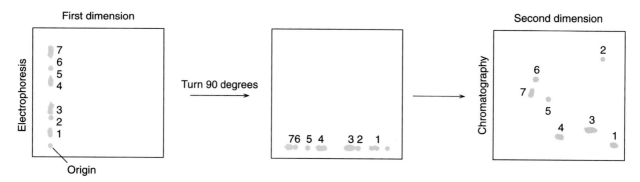

(b) Two-dimensional separation of peptides

Figure 3.23   Fingerprinting a protein. (a) A hypothetical protein, with six trypsin-sensitive sites indicated by slashes. After digestion with trypsin, seven peptides are released. **(b)** These tryptic peptides separate partially during electrophoresis in the first dimension, then fully after the paper is turned 90 degrees and chromatographed in the second dimension with another solvent.

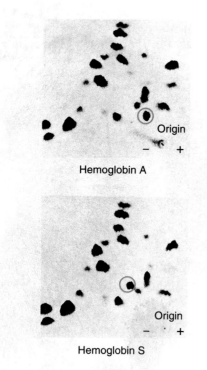

Figure 3.24   Fingerprints of hemoglobin A and hemoglobin S. The fingerprints are identical except for one peptide (circled), which shifts up and to the left in hemoglobin S. (*Source: (a, b)* Dr. Corrado Baglioni.)

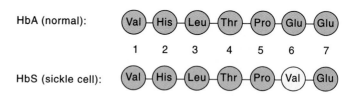

Figure 3.25   Sequences of amino-terminal peptides from normal and sickle cell β-globin. The numbers indicate the positions of the corresponding amino acids in the mature protein. The only difference is in position 6, where a valine (Val) in HbS replaces a glutamate (Glu) in HbA.

We can see how changing the blueprint does indeed change the product.

Sickle cell disease is a very common problem among people of central African descent. The question naturally arises: Why has this deleterious mutation spread so successfully through the population? The answer seems to be that although the homozygous condition can be lethal, heterozygotes have little if any difficulty because their normal allele makes enough product to keep their blood cells from sickling. Moreover, heterozygotes are at an advantage in central Africa, where malaria is rampant, because HbS helps protect against replication of the malarial parasite when it tries to infect their blood cells.

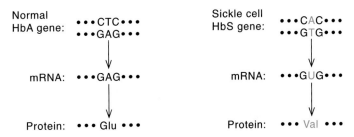

Figure 3.26 The sickle cell mutation and its consequences. The GAG in the sixth codon of the nontemplate strand of the normal gene changes to GTG. This leads to a change from GAG to GUG in the sixth codon of the β-globin mRNA of the sickle cells. This, in turn, results in insertion of a valine in the sixth amino acid position of sickle cell β-globin, where a glutamate ought to go.

SUMMARY Sickle cell disease is a human genetic disorder. It results from a single base change in the gene for β-globin. The altered base causes insertion of the wrong amino acid into one position of the β-globin protein. This altered protein results in distortion of red blood cells under low-oxygen conditions. This disease illustrates a fundamental genetic concept: A change in a gene can cause a corresponding change in the protein product of that gene.

SUMMARY

The three main activities of genes are information storing, replication, and accumulating mutations. Proteins, or polypeptides, are polymers of amino acids linked through peptide bonds. Most genes contain the information for making one polypeptide and are expressed in a two-step process: transcription or synthesis of an mRNA copy of the gene, followed by translation of this message to protein. Translation takes place on complex structures called ribosomes, the cell's protein factories. Translation also requires adapter molecules that can recognize both the genetic code in mRNA and the amino acids the mRNA encodes. Transfer RNAs (tRNAs), with their two business ends, fill this role.

Translation elongation involves three steps: (1) transfer of the second aminoacy-tRNA to the A site; (2) formation of a peptide bond between the first amino acid in the P site and the second aminoacyl-tRNA in the A site; and (3) translocation of the mRNA one codon's length through the ribosome, bringing the newly formed peptidyl-tRNA to the P site. Translation terminates at a stop codon (UAG, UAA, or UGA). A region of RNA or DNA including a translation initiation codon, a coding region, and a termination codon, is called an open reading frame. The piece of an mRNA between its 5′-end and the initiation codon is called a leader or 5′-UTR. The part between the 3′-end [or the poly(A)] and the termination codon is called a trailer or 3′-UTR.

DNA replicates in a semiconservative manner: When the parental strands separate, each serves as the template for making a new, complementary strand. A change, or mutation, in a gene frequently causes a change at a corresponding position in the polypeptide product. Sickle cell disease is an example of the deleterious effect of such mutations.

REVIEW QUESTIONS

1. Draw the general structure of an amino acid.

2. Draw the structure of a peptide bond.

3. Use a rough diagram to compare the structures of a protein α-helix and an antiparallel β-sheet. For simplicity, show only the backbone atoms of the protein.

4. What do we mean by primary, secondary, tertiary, and quaternary structures of proteins?

5. What was Garrod's insight into the relationship between genes and proteins, based on the disease alcaptonuria?

6. Describe Beadle and Tatum's experimental approach to demonstrating the relationship between genes and proteins.

7. What are the two main steps in gene expression?

8. Describe and give the results of the experiment of Jacob and colleagues that demonstrated the existence of mRNA.

9. What are the three steps in transcription? With a diagram, illustrate each one.

10. What ribosomal RNAs are present in *E. coli* ribosomes? To which ribosomal subunit does each rRNA belong?

11. Draw a diagram of the cloverleaf structure of a tRNA. Point out the site to which the amino acid attaches and the site of the anticodon.

12. How does a tRNA serve as an adaptor between the 3-bp codons in mRNA and the amino acids in protein?

13. Explain how a single base change in a gene could lead to premature termination of translation of the mRNA from that gene.

14. Explain how a single base deletion in the middle of a gene would change the reading frame of that gene.

15. Explain how a single base change in a gene can lead to a single amino acid change in that gene's polypeptide product. Illustrate with an example.

SUGGESTED READINGS

Beadle, G.W., and E.L. Tatum. 1941. Genetic control of biochemical reactions in *Neurospora*. *Proceedings of the National Academy of Sciences* 27:499–506.

Brenner, S., F. Jacob, and M. Meselson. 1961. An unstable intermediate carrying information from genes to ribosomes for protein synthesis. *Nature* 190:576–81.

Corwin, H.O., and J.B. Jenkins, eds. 1976. *Conceptual Foundations of Genetics: Selected Readings*. Boston: Houghton Mifflin.

Crick, F.H.C. 1958. On protein synthesis. *Symposium of the Society for Experimental Biology* 12:138–63.

Meselson, M., and F.W. Stahl. 1958. The replication of DNA in *Escherichia coli*. *Proceedings of the National Academy of Sciences* 44:671–82.

CHAPTER 4

Molecular Cloning Methods

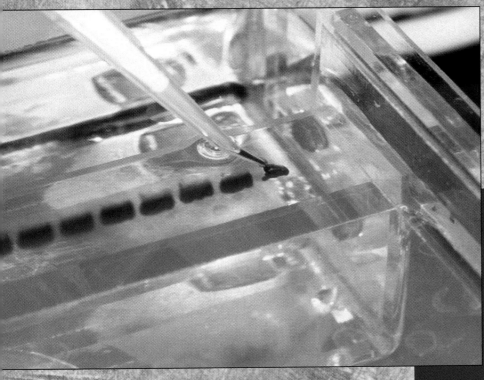

DNA electrophoresis. Samples of DNA are loaded into an agarose gel for separation. The yellow pipet dispenses a blue DNA sample into a well of the gel. © Klaus Guldbrandsen/SPL/Photo Researchers, Inc.

Now that we have reviewed the fundamentals of gene structure and function, we are ready to start a more detailed study of molecular biology. The main focus of our investigation will be the experiments that molecular biologists have performed to elucidate the structure and function of genes. For this reason, we need to pause at this point to discuss some of the major experimental techniques of molecular biology. Because it would be impractical to talk about them all at this early stage, we will deal with the common ones in the next two chapters and introduce the others as needed throughout the book. We will begin in this chapter with the technique that has revolutionized the discipline, gene cloning.

Imagine that you are a geneticist in the year 1972. You want to investigate the function of eukaryotic genes at the molecular level. In particular, you are curious about the molecular structure and function of the human growth hormone (hGH) gene. What is the base sequence of this gene? What does its promoter look like? How does RNA polymerase interact with this gene? What changes occur in this gene to cause conditions like hypopituitary dwarfism?

These questions cannot be answered unless you can purify enough of the gene to study—probably about a milligram's worth. A milligram does not sound like much, but it is an overwhelming amount when you imagine purifying it from whole human DNA. Consider that the DNA involved in one hGH gene is much less than one part per million in the human genome. And even if you could collect that much material somehow, you would not know how to separate the one gene you are interested in from all the rest of the DNA. In short, you would be stuck.

Gene cloning neatly solves these problems. By linking eukaryotic genes to small bacterial or phage DNAs and inserting these recombinant molecules into bacterial hosts, one can produce large quantities of these genes in pure form. In this chapter we will see how to clone genes in bacteria and in eukaryotes. ■

4.1   Gene Cloning

The purpose of any cloning experiment is to produce a **clone,** a group of identical cells or organisms. We know that some plants can be cloned simply by taking cuttings (Greek: *klon,* meaning twig), and that others can be cloned by growing whole plants from single cells collected from one plant. Even vertebrates can be cloned. John Gurdon produced clones of identical frogs by transplanting nuclei from a single frog embryo to many enucleate eggs, and a sheep named Dolly was cloned in Scotland in 1997 using an enucleate egg and a nucleus from an adult sheep mammary gland. Identical twins constitute a natural clone.

The usual procedure in a gene cloning experiment is to place a foreign gene into bacterial cells, isolate individual cells, and grow colonies from each of them. All the cells in each colony are identical and will contain the foreign gene. Thus, as long as we ensure that the foreign gene can replicate, we can clone the gene by cloning its bacterial host. Stanley Cohen, Herbert Boyer, and their colleagues performed the first cloning experiment in 1973.

The Role of Restriction Endonucleases

Cohen and Boyer's elegant plan depended on invaluable enzymes called **restriction endonucleases.** Stewart Linn and Werner Arber discovered restriction endonucleases in

E. coli in the late 1960s. These enzymes get their name from the fact that they prevent invasion by foreign DNA, such as viral DNA, by cutting it up. Thus, they "restrict" the host range of the virus. Furthermore, they cut at sites within the foreign DNA, rather than chewing it away at the ends, so we call them endonucleases (Greek: *endo,* meaning within) rather than exonucleases (Greek: *exo,* meaning outside). Linn and Arber hoped that their enzymes would cut DNA at specific sites, giving them finely honed molecular knives with which to slice DNA. Unfortunately, these particular enzymes did not fulfill that hope.

However, an enzyme from *Haemophilus influenzae* strain R_d, discovered by Hamilton Smith, did show specificity in cutting DNA. This enzyme is called *Hin*dII (pronounced Hin-dee-two). Restriction enzymes derive the first three letters of their names from the Latin name of the microorganism that produces them. The first letter is the first letter of the genus and the next two letters are the first two letters of the species (hence: *Haemophilus influenzae* yields *Hin*). In addition, the strain designation is sometimes included; in this case, the "d" from R_d is used. Finally, if the strain of microorganism produces just one restriction enzyme, the name ends with the Roman numeral I. If more than one enzyme is produced, they are numbered II, III, and so on.

*Hin*dII recognizes this sequence:

$$\downarrow$$
$$\text{GTPyPuAC}$$
$$\text{CAPuPyTG}$$
$$\uparrow$$

and cuts both DNA strands at the points shown by the arrows. Py stands for either of the pryimidines (T or C), and Pu stands for either purine (A or G). Wherever this sequence occurs, and *only* when this sequence occurs, *Hin*dII will make a cut. Happily for molecular biologists, *Hin*dII turned out to be only one of hundreds of restriction enzymes, each with its own specific recognition sequence. Table 4.1 lists the sources and recognition sequences for several popular restriction enzymes. Note that some of these enzymes recognize 4-bp sequences instead of the more common 6-bp sequences. As a result, they cut much more frequently. This is because a given sequence of 4 bp will occur about once in every $4^4 = 256$ bp, whereas a sequence of 6 bp will occur only about once in every $4^6 = 4096$ bp. Thus, a 6-bp cutter will yield DNA fragments of average length about 4000 bp, or 4 **kilobases** (**4 kb**). Some restriction enzymes, such as *Not*I, recognize 8-bp sequences, so they cut much less frequently (once in $4^8 \approx 65,000$ bp); they are therefore called **rare cutters.** In fact, *Not*I cuts even less frequently than you would expect in mammalian DNA, because its recognition sequence includes two copies of the rare dinucleotide CG. Notice also that the recognition sequences for *Sma*I and *Xma*I are

Table 4.1 Recognition Sequences and Cutting Sites of Selected Restriction Endonucleases

Enzyme	Recognition Sequence*
*Alu*I	A G ↓ C T
*Bam*HI	G ↓ G A T C C
*Bgl*II	A ↓ G A T C T
*Cla*I	A T ↓ C G A T
*Eco*RI	G ↓ A A T T C
*Hae*III	G G ↓ C C
*Hind*II	G T Py ↓ Pu A C
*Hind*III	A ↓ A G C T T
*Hpa*II	C ↓ C G G
*Kpn*I	G G T A C ↓ C
*Mbo*I	↓ G A T C
*Pst*I	C T G C A ↓ G
*Pvu*I	C G A T ↓ C G
*Sal*I	G ↓ T C G A C
*Sma*I	C C C ↓ G G G
*Xma*I	C ↓ C C G G G
*Not*I	G C ↓ G G C C G C

*Only one DNA strand, written 5′ → 3′ left to right is presented, but restriction endonucleases actually cut double-stranded DNA as illustrated in the text for *Eco*RI. The cutting site for each enzyme is represented by an arrow.

identical, although the cutting sites within these sequences are different. We call such enzymes that recognize identical sequences **isoschizomers** (Greek: *iso,* meaning equal; *schizo,* meaning split).

The main advantage of restriction enzymes is their ability to cut DNA strands reproducibly in the same places. This property is the basis of many techniques used to analyze genes and their expression. But this is not the only advantage. Many restriction enzymes make staggered cuts in the two DNA strands (see Table 4.1), leaving single-stranded overhangs, or **sticky ends,** that can base-pair together briefly. This makes it easier to stitch two different DNA molecules together, as we will see. Note, for example, the complementarity between the ends created by *Eco*RI (pronounced Eeko R-1 or Echo R-1):

$$
\begin{array}{ccc}
\downarrow \\
\text{5′---GAATC---3′} & & \text{---G3′} \text{5′AATTC---} \\
\text{3′---CTTAAG---5′} & \rightarrow & \text{---CTTAA5′} + \text{3′G---} \\
\uparrow
\end{array}
$$

Note also that *Eco*RI produces 4-base overhangs that protrude from the 5′-ends of the fragments. *Pst*I cuts at the 3′-ends of its recognition sequence, so it leaves 3′-overhangs. *Sma*I cuts in the middle of its sequence, so it produces blunt ends with no overhangs.

Restriction enzymes can make staggered cuts because the sequences they recognize usually display twofold symmetry. That is, they are identical after rotating them 180 degrees. For example, imagine inverting the *Eco*RI recognition sequence just described:

$$
\begin{array}{c}
\downarrow \\
\text{5′---GAATTC---3′} \\
\text{3′---CTTAAG---5′} \\
\uparrow
\end{array}
$$

You can see it will still look the same after the inversion. In a way, these sequences read the same forward and backward. Thus, *Eco*RI cuts between the G and the A in the top strand (on the left), and between the G and the A in the bottom strand (on the right), as shown by the vertical arrows.

Sequences with twofold symmetry are also called **palindromes.** In ordinary language, palindromes are sentences that read the same forward and backward. Examples are Napoleon's lament: "Able was I ere I saw Elba," or a wart remedy: "Straw? No, too stupid a fad; I put soot on warts," or a statement of preference in Italian food: "Go hang a salami! I'm a lasagna hog." DNA palindromes also read the same forward and backward, but you have to be careful to read the same sense (5′ → 3′) in both directions. This means that you read the top strand left to right and the bottom strand right to left.

One final question about restriction enzymes: If they can cut up invading viral DNA, why do they not destroy the host cell's own DNA? The answer is this: Almost all restriction endonucleases are paired with methylases that recognize and methylate the same DNA sites. The two enzymes—the restriction endonuclease and the methylase—are collectively called a **restriction–modification system,** or an **R-M system.** After methylation, DNA sites are protected against most restriction endonucleases so the methylated DNA can persist unharmed in the host cell. But what about DNA replication? Doesn't that create newly replicated DNA strands that are unmethylated, and therefore vulnerable to cleavage? Figure 4.1 explains how DNA continues to be protected during replication. Every time the cellular DNA replicates, one strand of the daughter duplex will be a newly made strand and will be unmethylated. But the other will be a parental strand and therefore be methylated. This half-methylation (hemimethylation) is enough to protect the DNA duplex against cleavage by the great majority of restriction endonucleases, so the methylase has time to find the site and methylate the other strand yielding fully methylated DNA.

Cohen and Boyer took advantage of the sticky ends created by a restriction enzyme in their cloning experiment (Figure 4.2). They cut two different DNAs with the same restriction enzyme. *Eco*RI. Both DNAs were **plasmids,** small, circular DNAs that are independent of the host chromosome. The first, called pSC101, carried a

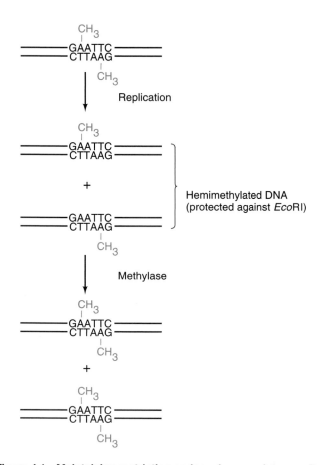

Figure 4.1 Maintaining restriction endonuclease resistance after DNA replication. We begin with an *Eco*RI site that is methylated (red) on both strands. After replication, the parental strand of each daughter DNA duplex remains methylated, but the newly made strand of each duplex has not been methylated yet. The one methylated strand in these hemimethylated DNAs is enough to protect both strands against cleavage by *Eco*RI. Soon, the methylase recognizes the unmethylated strand in each *Eco*RI site and methylates it, regenerating the fully methylated DNA.

gene that conferred resistance to the antibiotic tetracycline; the other, RSF1010, conferred resistance to both streptomycin and sulfonamide. Both plasmids had just one *Eco*RI **restriction site,** or cutting site for *Eco*RI. Therefore, when *Eco*RI cut these circular DNAs, it converted them to linear molecules and left them with the same sticky ends. These sticky ends then base-paired with each other, at least briefly. Of course, some of this base-pairing involved sticky ends on the same DNA, which simply closed up the circle again. But some base-pairing of sticky ends brought the two different DNAs together. Finally, **DNA ligase** completed the task of joining the two DNAs covalently. DNA ligase is an enzyme that forms covalent bonds between the ends of DNA strands.

The desired result was a **recombinant DNA,** two previously separate pieces of DNA linked together. This new, recombinant plasmid was probably outnumbered by the two parental plasmids that had been cut and then reli-

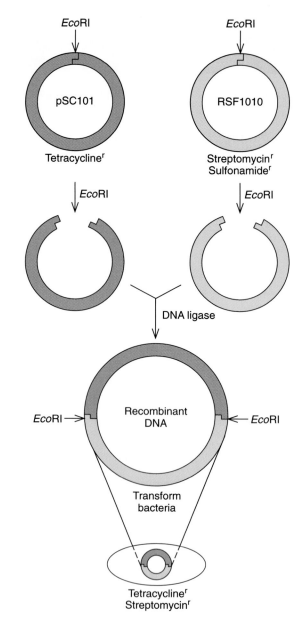

Figure 4.2 The first cloning experiment involving a recombinant DNA assembled in vitro. Boyer and Cohen cut two plasmids, pSC101 and RSF1010, with the same restriction endonuclease, *Eco*RI. This gave the two linear DNAs the same sticky ends, which were then linked in vitro using DNA ligase. The investigators reintroduced the recombinant DNA into *E. coli* cells by transformation and screened for clones that were resistant to both tetracycline and streptomycin. These clones were therefore harboring the recombinant plasmid.

gated, but it was easy to detect. When introduced into bacterial cells, it conferred resistance to both tetracycline, a property of pSC101, and to streptomycin, a property of RSF1010. Recombinant DNAs abound in nature, but this one differs from most of the others in that it was not created naturally in a cell. Instead, molecular biologists put it together in a test tube.

SUMMARY Restriction endonucleases recognize specific sequences in DNA molecules and make cuts in both strands. This allows very specific cutting of DNAs. Also, because the cuts in the two strands are frequently staggered, restriction enzymes can create sticky ends that help link together two DNAs to form a recombinant DNA in vitro.

Vectors

Both plasmids in the Cohen and Boyer experiment are capable of replicating in *E. coli*. Thus, both can serve as carriers to allow replication of recombinant DNAs. All gene cloning experiments require such carriers, which we call **vectors,** but a typical experiment involves only one vector, plus a piece of foreign DNA that depends on the vector for its replication. The foreign DNA has no **origin of replication,** the site where DNA replication begins, so it cannot replicate unless it is placed in a vector that does have an origin of replication. Since the mid-1970s, many vectors have been developed; these fall into two major classes: plasmids and phages.

Plasmids as Vectors   In the early years of the cloning era, Boyer and his colleagues developed a set of very popular vectors known as the pBR plasmid series. These early vectors are scarcely used anymore, but we describe them here because they provide a simple illustration of cloning methods. One of these plasmids, **pBR322** (Figure 4.3), contains genes that confer resistance to two antibiotics: ampicillin and tetracycline. Between these two genes lies the origin of replication. The plasmid was engineered to contain only one cutting site for each of several common restriction enzymes, including *Eco*RI, *Bam*HI, *Pst*I, *Hin*dIII, and *Sal*I. These unique sites are convenient because they allow one to use each enzyme to create a site for inserting foreign DNA without losing any of the plasmid DNA.

For example, let us consider cloning a foreign DNA fragment into the *Pst*I site of pBR322 (Figure 4.4). First, the investigators cut the vector with *Pst*I to generate the sticky ends characteristic of that enzyme. In this example, they also cut the foreign DNA with *Pst*I, thus giving it *Pst*I sticky ends. Next, they combine the cut vector with the foreign DNA and incubate them with DNA ligase. As the sticky ends on the vector and on the foreign DNA base-pair momentarily, DNA ligase seals the nicks, attaching the two DNAs together covalently. Once this is done, the DNAs cannot come apart again unless they are recut with *Pst*I. A nick is just a broken phosphodiester bond; a restriction enzyme like *Pst*I breaks the bond, and DNA ligase re-forms it, as shown here:

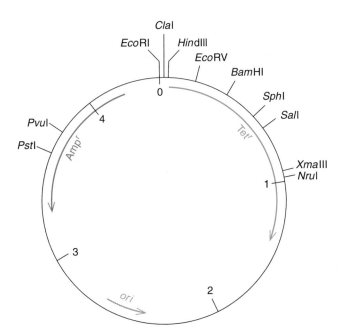

Figure 4.3   The plasmid pBR322, showing the locations of 11 unique restriction sites that can be used to insert foreign DNA. The locations of the two antibiotic resistance genes (Amp^r = ampicillin resistance; Tet^r = tetracycline resistance) and the origin of replication (*ori*) are also shown. Numbers refer to kilobase pairs (kb) from the *Eco*RI site.

In the next step, the investigators transform *E. coli* with the DNA mixture. The traditional way to do this is to incubate the cells in concentrated calcium salt solution to make their membranes leaky, then mix these permeable cells with the DNA to allow the DNA entrance to the leaky cells. Alternatively, one can use high voltage to drive the DNA into cells—a process called **electroporation.**

It would be simple if all the cut DNA had been ligated to plasmids to form recombinant DNAs, but that never happens. Instead, one gets a mixture of re-ligated plasmids and re-ligated inserts, along with the recombinants. How can these be sorted out? This is where the antibiotic resistance genes of the vector come into play. First one grows cells in the presence of tetracycline, which selects for cells that have taken up either the vector alone or the

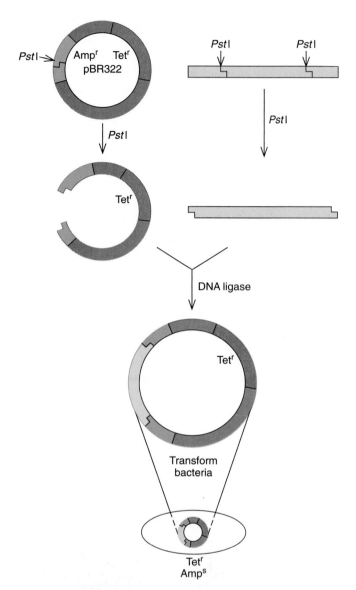

Figure 4.4   Cloning foreign DNA using the *Pst*I site of pBR322. Cut both the plasmid and the insert (yellow) with *Pst*I, then join them through these sticky ends with DNA ligase. Next, transform bacteria with the recombinant DNA and screen for tetracycline-resistant, ampicillin-sensitive cells. The recombinant plasmid no longer confers ampicillin resistance because the foreign DNA interrupts that resistance gene (blue).

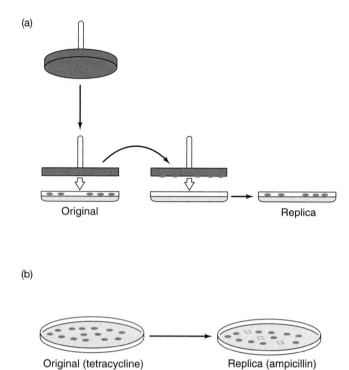

Figure 4.5   Screening bacteria by replica plating. (a) The replica plating process. Touch a velvet-covered circular tool to the surface of the first dish containing colonies of bacteria. Cells from each of these colonies stick to the velvet and can be transferred to the replica plate in the same positions relative to each other. **(b)** Screening for inserts in the pBR322 ampicillin resistance gene by replica plating. The original plate contains tetracycline, so all colonies containing pBR322 will grow. The replica plate contains ampicillin, so colonies bearing pBR322 with inserts in the ampicillin resistance gene will not grow (these colonies are depicted by dotted circles). The corresponding colonies from the original plate can then be picked.

vector with inserted DNA. Cells that received no DNA, or that received insert DNA only, will not be tetracycline-resistant and will fail to grow.

Next, our investigators would want to find the clones that have received recombinant DNAs. Figure 4.4 shows that the *Pst*I site, where DNA is being inserted in this experiment, lies within the ampicillin resistance gene. Therefore, inserting foreign DNA into the *Pst*I site inactivates this gene and leaves the host cell vulnerable to ampicillin. Thus, screening for clones that are both tetracycline-resistant and ampicillin-sensitive will find the ones with recombinant DNAs.

How does one do the screening? One way is **replica plating.** In this case, one would transfer copies of the clones from the original tetracycline plate to an ampicillin plate. This can be accomplished with a sterile velvet transfer tool as illustrated in Figure 4.5. Touching the tool lightly to the surface of the tetracycline plate to pick up cells from each clone, then touching the tool to a fresh ampicillin plate deposits cells from each original clone in the same relative positions as on the original plate, leaving behind part of each original clone. Investigators look for colonies that do *not* grow on the ampicillin plate and then find the corresponding growing clone on the tetracycline plate. Using that most sophisticated of scientific tools—a sterile toothpick—they transfer cells from this positive clone to fresh medium for storage or immediate use. Notice that we did *not* call this procedure a **selection,** because it does not remove unwanted clones automatically. Instead, each clone had to be examined individually. We call this more laborious process a **screen.**

Nowadays, one can choose from many plasmid cloning vectors besides the pBR plasmids. One useful, though somewhat dated, class of plasmids is the **pUC** series (Figure 4.6). These plasmids are based on pBR322, from which about 40% of the DNA, including the tetracycline

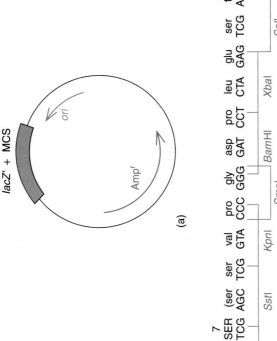

(a)

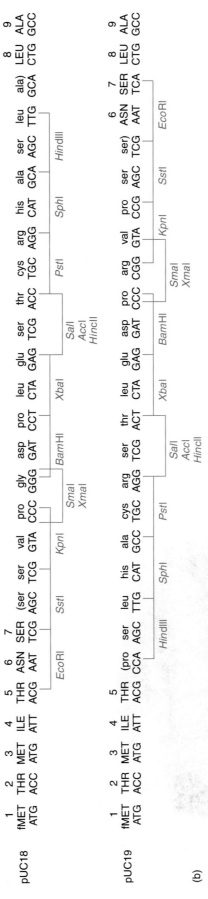

(b)

Figure 4.6 Architecture of a pUC plasmid. (a) The pUC plasmids retain the ampicillin resistance gene and the origin of replication of pBR322. In addition, they include a multiple cloning site (MCS) inserted into a gene encoding the amino terminal part of β-galactosidase (*lacZ'*). **(b)** The MCSs of pUC18 and pUC19. In pUC18, the MCS, containing 13 restriction sites, is inserted after the seventh codon of the *lacZ'* gene. In pUC19, it comes after the fifth codon and is reversed. The amino acids in lowercase letters enclosed by parentheses are encoded by the MCSs. Note that, in addition to containing the restriction sites, the MCSs preserve the reading frame of the *lacZ'* gene, so plasmids with no inserts still support active expression of this gene. Therefore, clones harboring the vector alone turn blue in the presence of the synthetic β-galactosidase substrate X-gal, whereas clones harboring the vector plus an insert remain white.

65

resistance gene, has been deleted. Furthermore, the pUC vectors have their cloning sites clustered into one small area called a **multiple cloning site (MCS).** The pUC vectors contain the pBR322 ampicillin resistance gene to allow selection for bacteria that have received a copy of the vector. Moreover, to compensate for the loss of the other antibiotic resistance gene, they have genetic elements that provide a more convenient way of screening for clones with recombinant DNAs.

Figure 4.6b shows the multiple cloning sites of pUC18 and pUC19. Notice that they lie within a DNA sequence coding for the amino terminal portion (the α-peptide) of the enzyme β-galactosidase (denoted *lacZ'*). The host bacteria used with the pUC vectors carry a gene fragment that encodes the carboxyl portion of β-galactosidase. By themselves, the β-galactosidase fragments made by these partial genes have no activity. But they can complement each other in vivo by so-called α-**complementation.** In other words, the two partial gene products can cooperate to form an active enzyme. Thus, when pUC18 by itself transforms a bacterial cell carrying the partial β-galactosidase gene, active β-galactosidase is produced. If these clones are plated on medium containing a β-galactosidase indicator, colonies with the pUC plasmid will turn color. The indicator X-gal, for instance, is a synthetic, colorless galactoside; when β-galactosidase cleaves X-gal, it releases galactose plus an indigo dye that stains the bacterial colony blue.

On the other hand, interrupting the plasmid's partial β-galactosidase gene by placing an insert into the multiple cloning site usually inactivates the gene. It can no longer make a product that complements the host cell's β-galactosidase fragment, so the X-gal remains colorless. Thus, it is a simple matter to pick the clones with inserts. They are the white ones; all the rest are blue. Notice that, in contrast to screening with pBR322, this is a one-step process. One looks simultaneously for a clone that (1) grows on ampicillin and (2) is white in the presence of X-gal. The multiple cloning sites have been carefully constructed to preserve the reading frame of β-galactosidase. Thus, even though the gene is interrupted by 18 codons a functional protein still results. But further interruption by large inserts is usually enough to destroy the gene's function.

Even with the color screen, cloning into pUC can give false-positives, that is, white colonies without inserts. This can happen if the vector's ends are "nibbled" slightly by nucleases before ligation to the insert. Then, if these slightly degraded vectors simply close up during the ligation step, chances are that the *lacZ'* gene has been changed enough that white colonies will result. This underscores the importance of using clean DNA and enzymes that are free of nuclease activity.

This phenomenon of a vector religating with itself can be a greater problem when we use vectors that do not have a color screen, because then it is more difficult to distinguish colonies with inserts from those without. Even with pUC and related vectors, we would like to minimize vector religation. A good way to do this is to treat the vector with alkaline phosphatase, which removes the 5-phosphates necessary for ligation. Without these phosphates, the vector cannot ligate to itself, but can still ligate to the insert that retains its 5'-phosphates. Figure 4.7b illustrates this process. Notice that, because only the insert has phosphates, two nicks (unformed phosphodiester bonds) remain in the ligated product. These are not a problem; they will be sealed by DNA ligase in vivo once the ligated DNA has made its way into a bacterial cell.

The multiple cloning site also allows one to cut it with two different restriction enzymes (say, *Eco*RI and *Bam*HI) and then to clone a piece of DNA with one *Eco*RI end and one *Bam*HI end. This is called **directional cloning,** because the insert DNA is placed into the vector in only one orientation. (The *Eco*RI and *Bam*HI ends of the insert have to match their counterparts in the vector.) Knowing the orientation of an insert has certain benefits, which we will explore later in this chapter. Directional cloning also has the advantage of preventing the vector from simply religating by itself because its two restriction sites are incompatible.

Notice that the multiple cloning site of pUC18 is just the reverse of that of pUC19. That is, the restriction sites are in opposite order with respect to the *lacZ'* gene. This means one can clone a fragment in either orientation simply by shifting from one pUC plasmid to the other. Even more convenient vectors than these are now available. We will discuss some of them later in this chapter.

SUMMARY The first two generations of plasmid cloning vectors were pBR322 and the pUC plasmids. The former has two antibiotic resistance genes and a variety of unique restriction sites into which one can introduce foreign DNA. Most of these sites interrupt one of the antibiotic resistance genes, making screening straightforward. Screening is even easier with the pUC plasmids. These have an ampicillin resistance gene and a multiple cloning site that interrupts a partial β-galactosidase gene. One screens for ampicillin-resistant clones that do not make active β-galactosidase and therefore do not turn the indicator, X-gal, blue. The multiple cloning site also makes it convenient to carry out directional cloning into two different restriction sites.

Phages as Vectors   Bacteriophages are natural vectors that transduce bacterial DNA from one cell to another. It was only natural, then, to engineer phages to do the same

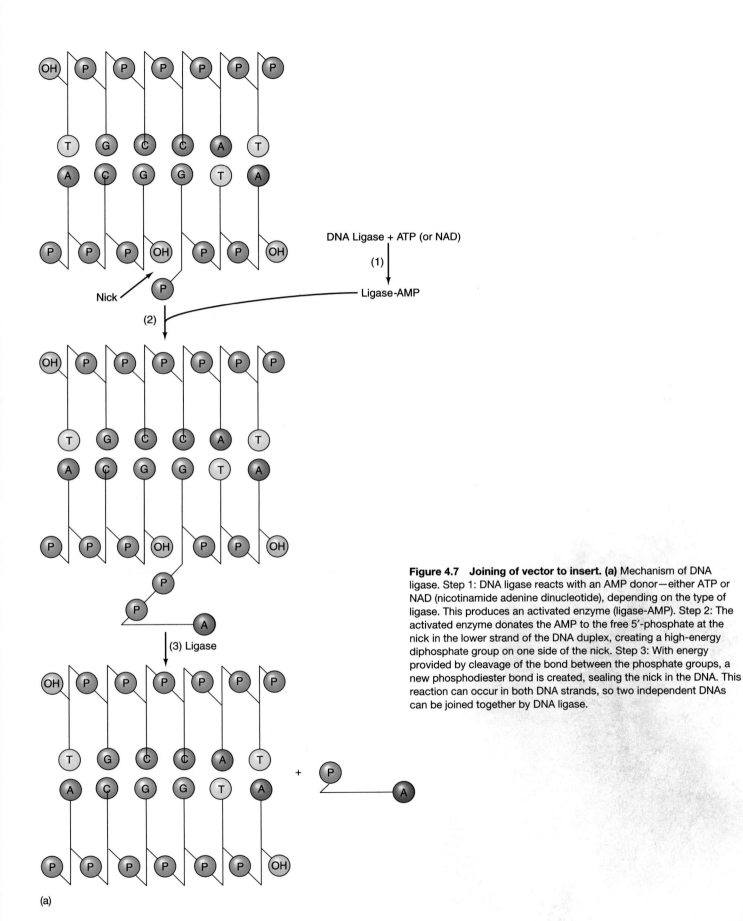

Figure 4.7 Joining of vector to insert. (a) Mechanism of DNA ligase. Step 1: DNA ligase reacts with an AMP donor—either ATP or NAD (nicotinamide adenine dinucleotide), depending on the type of ligase. This produces an activated enzyme (ligase-AMP). Step 2: The activated enzyme donates the AMP to the free 5′-phosphate at the nick in the lower strand of the DNA duplex, creating a high-energy diphosphate group on one side of the nick. Step 3: With energy provided by cleavage of the bond between the phosphate groups, a new phosphodiester bond is created, sealing the nick in the DNA. This reaction can occur in both DNA strands, so two independent DNAs can be joined together by DNA ligase.

DNA Ligase + ATP (or NAD)

(1)

Ligase-AMP

(2)

(3) Ligase

Nick

(a)

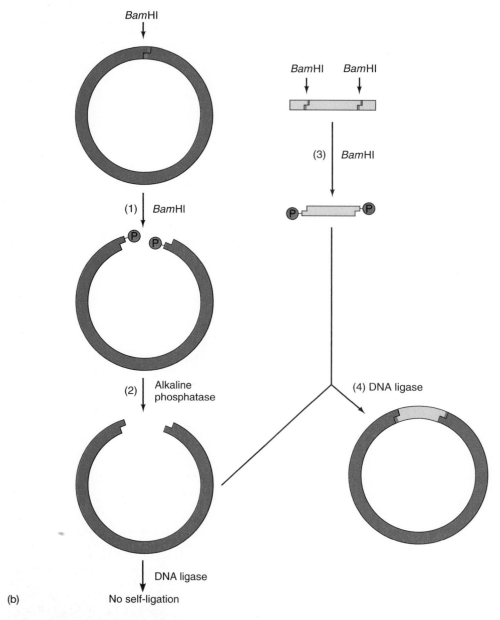

Figure 4.7   Joining of vector to insert—*continued*. (b) Alkaline phosphatase prevents vector religation. Step 1: Cut the vector (blue, top left) with *Bam*HI. This produces sticky ends with 5′-phosphates (red). Step 2: Remove the phosphates with alkaline phosphatase, making it impossible for the vector to religate with itself. Step 3: Also cut the insert (yellow, upper right) with *Bam*HI, producing sticky ends with phosphates that are not removed. Step 4: Finally, ligate the vector and insert together. The phosphates on the insert allow two phosphodiester bonds to form (red), but leave two unformed bonds, or nicks. These are completed once the DNA is in the transformed bacterial cell.

thing for *all* kinds of DNA. Phage vectors have a natural advantage over plasmids: They infect cells much more efficiently than plasmids transform cells, so the yield of clones with phage vectors is usually higher. With phage vectors, clones are not colonies of cells, but **plaques** formed when a phage clears out a hole in a lawn of bacteria. Each plaque derives from a single phage that infects a cell, producing progeny phages that burst out of the cell, killing it and infecting surrounding cells. This process continues until a visible patch, or plaque, of dead cells appears. Because all the phages in the plaque derive from one original phage, they are all genetically identical—a clone.

λ *Phage Vectors*   Fred Blattner and his colleagues con-structed the first phage vectors by modifying the well-known λ phage (Chapter 8). They took out the region in the middle of the phage DNA, but retained the genes needed for phage replication. The missing phage genes could then be replaced with foreign DNA. Blattner named these vectors **Charon phages after Charon, the boatman on the river Styx in classical mythology. Just as Charon carried souls to the underworld, the Charon phages carry foreign DNA into bacterial cells. Charon the boatman is pronounced "Karen," but Charon the phage is often pro-nounced "Sharon." A more general term for λ vectors

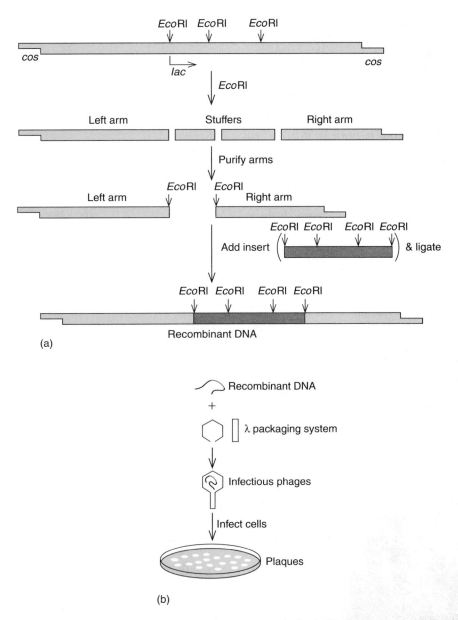

Figure 4.8   Cloning in Charon 4. (a) Forming the recombinant DNA. Cut the vector (yellow) with *Eco*RI to remove the stuffer fragment and save the arms. Next, ligate partially digested insert DNA (red) to the arms. The extensions of the ends are 12-base cohesive ends (*cos* sites), whose size is exaggerated here. **(b)** Packaging and cloning the recombinant DNA. Mix the recombinant DNA from part **(a)** with an in vitro packaging extract that contains λ phage head and tail components and all other factors needed to package the recombinant DNA into functional phage particles. Finally, plate these particles on *E. coli* and collect the plaques that form.

such as Charon 4 is **replacement vectors,** since λ DNA is removed and replaced with foreign DNA.

One clear advantage of the λ phages over plasmid vectors is that they can accommodate much more foreign DNA. For example, Charon 4 can accept up to about 20 kb of DNA, a limit imposed by the capacity of the λ phage head. By contrast, most plasmid vectors with inserts that large replicate poorly. When would one need such high capacity? A common use for λ replacement vectors is in constructing **genomic libraries.** Suppose we wanted to clone the entire human genome. This would obviously require a great many clones, but the larger the

insert in each clone, the fewer total clones would be needed. In fact, such genomic libraries have been constructed for the human genome and for genomes of a variety of other organisms, and λ replacement vectors have been popular vectors for this purpose.

Aside from their high capacity, some of the λ vectors have the advantage of a minimum size requirement for their inserts. Figure 4.8 illustrates the reason for this requirement: To get the Charon 4 vector ready to accept an insert, it can be cut with *Eco*RI. This cuts at three sites near the middle of the phage DNA, yielding two "arms" and two "stuffer" fragments. Next, the arms are purified by gel electrophoresis or ultracentrifugation and the stuffers are discarded. The final step is to ligate the arms to the insert, which then takes the place of the discarded stuffers.

At first glance, it may appear that the two arms could simply ligate together without accepting an insert. Indeed, this happens, but it does not produce a clone, because the two arms constitute too little DNA and will not be packaged into a phage. The packaging is done in vitro when the recombinant DNA is mixed with all the components needed to put together a phage particle. Nowadays one can buy the purified λ arms, as well as the packaging extract in cloning kits. The extract has rather stringent requirements as to the size of DNA it will package. It must have at least 12 kb of DNA in addition to λ arms, but no more than 20 kb, or it will be too large to fit into the phage head.

Because each clone has at least 12 kb of foreign DNA, the library does not waste space on clones that contain insignificant amounts of DNA. This is an important consideration because, even at 12–20 kb per clone, the library needs about half a million clones to ensure that each human gene is represented at least once. It would be much more difficult to make a human genomic library in pBR322 or a pUC vector because bacteria selectively take up and reproduce small plasmids. Therefore, most of the clones would contain inserts of a few thousand, or even just a few hundred base pairs. Such a library would have to contain many millions of clones to be complete.

Because *Eco*RI produces fragments with an average size of about 4 kb, but the vector will not accept any inserts smaller than 12 kb, the DNA cannot be completely cut with *Eco*RI, or most of the fragments will be too small to clone. Furthermore, *Eco*RI, and most other restriction enzymes, cut in the middle of most eukaryotic genes one or more times, so a complete digest would contain only fragments of most genes. One can avoid these problems by performing an incomplete digestion with *Eco*RI (using a low concentration of enzyme or a short reaction time, or both). If the enzyme cuts only about every fourth or fifth site, the average length of the resulting fragments will be about 16–20 kb, just the size the vector will accept and big enough to include the entirety of most eukaryotic genes. If we want a more random set of fragments, we can also use mechanical means such as ultrasound instead of a restriction endonuclease to shear the DNA to an appropriate size for cloning.

A genomic library is very handy. Once it is established, one can search for any gene of interest. The only problem is that no catalog exists for such a library to help find particular clones, so some kind of probe is needed to show which clone contains the gene of interest. An ideal probe would be a labeled nucleic acid whose sequence matches that of the gene of interest. One would then carry out a **plaque hybridization** procedure in which the DNA from each of the thousands of λ phages from the library is

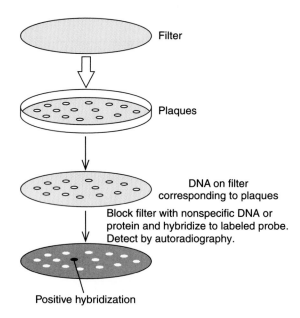

Figure 4.9 Selection of positive genomic clones by plaque hybridization. First, touch a nitrocellulose or similar filter to the surface of the dish containing the Charon 4 plaques from Figure 4.8. Phage DNA released naturally from each plaque sticks to the filter. Next, denature the DNA with alkali and hybridize the filter to a labeled probe for the gene under study, then use x-ray film to reveal the position of the label. Cloned DNA from one plaque near the center of the filter has hybridized, as shown by the dark spot on the film.

hybridized to the labeled probe. The plaque with the DNA that forms a labeled hybrid is the right one.

We have encountered hybridization before in Chapter 2, and we will discuss it again in Chapter 5. Figure 4.9 shows how plaque hybridization works. Thousands of plaques are grown on each of several Petri dishes (only a few plaques are shown here for simplicity). Next, a filter made of a DNA-binding material such as **nitrocellulose** or coated nylon is touched to the surface of the Petri dish. This transfers some of the phage DNA from each plaque to the filter. The DNA is then denatured with alkali and hybridized to the labeled probe. Before the probe is added, the filter is saturated with a nonspecific DNA or protein to prevent nonspecific binding of the probe. When the probe encounters complementary DNA, which should be only the DNA from the clone of interest, it will hybridize, labeling that DNA spot. This labeled spot is then detected with x-ray film. The black spot on the film shows where to look on the original Petri dish for the plaque containing the gene of interest. In practice, the original plate may be so crowded with plaques that it is impossible to pick out the right one, so several plaques can be picked from that area, replated at a much lower phage density, and the hybridization process can be repeated to find the positive clone.

We have introduced λ phage vectors as agents for genomic cloning. But other types of λ vectors are very useful for making another kind of library—a cDNA library—as we will learn later in this chapter.

Cosmids   Another vector designed especially for cloning large DNA fragments is called a **cosmid**. Cosmids behave both as plasmids and as phages. They contain the *cos* sites, or cohesive ends, of λ phage DNA, which allow the DNA to be packaged into λ phage heads (hence the "cos" part of the name "cosmid"). They also contain a plasmid origin of replication, so they can replicate as plasmids in bacteria (hence the "mid" part of the name).

Because almost the entire λ genome, except for the *cos* sites, has been removed from the cosmids, they have room for large inserts (40–50 kb). Once these inserts are in place, the recombinant cosmids are packaged into phage particles. These particles cannot replicate as phages because they have almost no phage DNA, but they are infectious, so they carry their recombinant DNA into bacterial cells. Once inside, the DNA can replicate as a plasmid because it has a plasmid origin of replication.

M13 phage vectors   Another phage used as a cloning vector is the filamentous (long, thin, filament-like) phage M13. Joachim Messing and his coworkers endowed the phage DNA with the same β-galactosidase gene fragment and multiple cloning sites found in the pUC family of vectors. In fact, the M13 vectors were engineered first; then the useful cloning sites were simply transferred to the pUC plasmids.

What is the advantage of the M13 vectors? The main factor is that the genome of this phage is a single-stranded DNA, so DNA fragments cloned into this vector can be recovered in single-stranded form. As we will see later in this chapter, single-stranded DNA can be an aid to site-directed mutagenesis, by which we can introduce specific, premeditated alterations into a gene. It also makes it easier to determine the sequence of a piece of DNA.

Figure 4.10 illustrates how to clone a double-stranded piece of DNA into M13 and harvest a single-stranded DNA product. The DNA in the phage particle itself is single-stranded, but after infecting an *E. coli* cell, the DNA is converted to a double-stranded replicative form (RF). This double-stranded replicative form of the phage DNA is used for cloning. After it is cut with one or two restriction enzymes at its multiple cloning site, foreign DNA with compatible ends can be inserted. This recombinant DNA is then used to transform host cells, giving rise to progeny phages that bear single-stranded recombinant DNA. The phage particles, containing phage DNA, are secreted from the transformed cells and can be collected from the growth medium.

Phagemids   Another class of vectors that produce single-stranded DNA has also been developed. These are like the cosmids in that they have characteristics of both phages

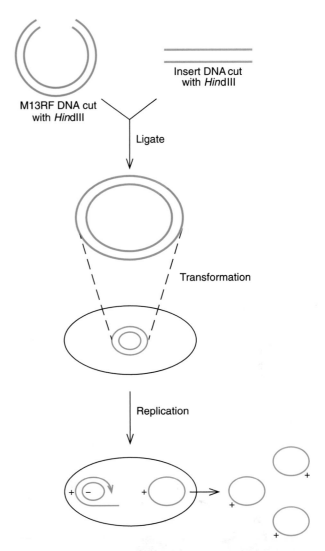

Figure 4.10   Obtaining single-stranded DNA by cloning in M13 phage. Foreign DNA (red), cut with *Hind*III, is inserted into the *Hind*III site of the double-stranded phage DNA. The resulting recombinant DNA is used to transform *E. coli* cells, whereupon the DNA replicates, producing many single-stranded product DNAs. The product DNAs are called positive (+) strands, by convention. The template DNA is therefore the negative (−) strand.

and plasmids; thus, they are called **phagemids**. One popular variety (Figure 4.11) goes by the trade name pBluescript (pBS). Like the pUC vectors, pBluescript has a multiple cloning site inserted into the *lacZ'* gene, so clones with inserts can be distinguished by white versus blue staining with X-gal. This vector also has the origin of replication of the single-stranded phage f1, which is related to M13. This means that a cell harboring a recombinant phagemid, if infected by f1 helper phage that supplies the single-stranded phage DNA replication machinery, will produce and package single-stranded phagemid DNA. A final useful feature of this class of vectors is that the multiple cloning site is flanked by two

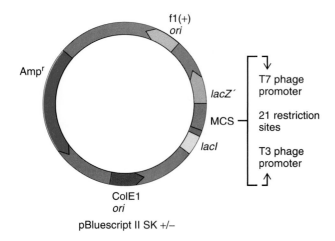

pBluescript II SK +/−

Figure 4.11 The pBluescript vector. This plasmid is based on pBR322 and has that vector's ampicillin resistance gene (green) and origin of replication (purple). In addition, it has the phage f1 origin of replication (orange). Thus, if the cell is infected by an f1 helper phage to provide the replication machinery, single-stranded copies of the vector can be packaged into progeny phage particles. The multiple cloning site (MCS, red) contains 21 unique restriction sites situated between two phage RNA polymerase promoters (T7 and T3). Thus, any DNA insert can be transcribed in vitro to yield an RNA copy of either strand, depending on which phage RNA polymerase is provided. The MCS is embedded in an *E. coli lacZ'* gene (blue), so the uncut plasmid will produce the β-galactosidase N-terminal fragment when an inducer such as isopropylthiogalactoside (IPTG) is added to counteract the repressor made by the *lacI* gene (yellow). Thus, clones bearing the uncut vector will turn blue when the indicator X-gal is added. By contrast, clones bearing recombinant plasmids with inserts in the MCS will have an interrupted *lacZ'* gene, so no functional β-galactosidase is made. Thus, these clones remain white.

different phage RNA polymerase promoters. For example, pBS has a T3 promoter on one side and a T7 promoter on the other. This allows one to isolate the double-stranded phagemid DNA and transcribe it in vitro with either of the phage polymerases to produce pure RNA transcripts corresponding to either strand.

SUMMARY Two kinds of phages have been especially popular as cloning vectors. The first of these is λ, from which certain nonessential genes have been removed to make room for inserts. Some of these engineered phages can accommodate inserts up to 20 kb, which makes them useful for building genomic libraries, in which it is important to have large pieces of genomic DNA in each clone. Cosmids can accept even larger inserts—up to 50 kb—making them a favorite choice for genomic libraries. The second major class of phage vectors consists of the M13 phages. These vectors have the convenience of a multiple cloning site and the further advantage of producing single-stranded recombinant DNA, which can be used for DNA se-

quencing and for site-directed mutagenesis. Plasmids called phagemids have also been engineered to produce single-stranded DNA in the presence of helper phages.

Eukaryotic Vectors Several very useful vectors have been designed for cloning genes into eukaryotic cells. Later in this chapter, we will consider some vectors that are designed to yield the protein products of genes in eukaryotes. We will also introduce vectors based on the Ti plasmid of *Agrobacterium tumefaciens* that can carry genes into plant cells. In Chapter 24 we will discuss vectors known as yeast artificial chromosomes (YACs) and bacterial artificial chromosomes (BACs) designed for cloning huge pieces of DNA (hundreds of thousands of base pairs).

Identifying a Specific Clone with a Specific Probe

We have already mentioned the need for a probe to identify a desired clone among the thousands of irrelevant ones. What sort of probe could be employed? Two different kinds are widely used: polynucleotides (or oligonucleotides) and antibodies. Both are molecules able to bind very specifically to other molecules. We will discuss polynucleotide probes here and antibody probes later in this chapter.

Polynucleotide Probes To probe for the gene we want, we might use the homologous gene from another organism if someone has already managed to clone it. For example, if a clone of the human insulin gene is desired and another research group had already cloned the rat insulin gene, we could ask them for their clone to use as a probe. We would hope the two genes have enough similarity in sequence that the rat probe could hybridize to the human gene. This hope is usually fulfilled. However, we generally have to lower the **stringency** of the hybridization conditions so that the hybridization reaction can tolerate some mismatches in base sequence between the probe and the cloned gene.

Researchers use several means to control stringency. High temperature, high organic solvent concentration, and low salt concentration all tend to promote the separation of the two strands in a DNA double helix. We can therefore adjust these conditions until only perfectly matched DNA strands will form a duplex; this is high stringency. By relaxing these conditions (lowering the temperature, for example), we lower the stringency until DNA strands with a few mismatches can hybridize.

Without homologous DNA from another organism, what could we use? There is still a way out if we know at least part of the sequence of the protein product of the gene. We faced a problem just like this in our lab when we cloned the gene for a plant toxin known as ricin. Fortunately, the entire amino acid sequences of both polypeptides of ricin were known. That meant we could examine the amino acid sequence and, using the genetic code, deduce a set of nucleotide sequences that would code for these amino acids. Then we could construct these nucleotide sequences chemically and use these synthetic probes to find the ricin gene by hybridization. The probes in this kind of procedure are strings of several nucleotides, so they are called **oligonucleotides.** Why did we have to use more than one oligonucleotide to probe for the ricin gene? The genetic code is degenerate, which means that most amino acids are encoded by more than one triplet codon. Thus, we had to consider several different nucleotide sequences for most amino acids.

Fortunately, we were spared much inconvenience because one of the polypeptides of ricin includes this amino acid sequence: Trp-Met-Phe-Lys-Asn-Glu. The first two amino acids in this sequence have only one codon each, and the next three only two each. The sixth gives us two extra bases because the degeneracy occurs only in the third base. Thus, we had to make only eight 17-base oligonucleotides (*17-mers*) to be sure of getting the exact coding sequence for this string of amino acids. This degenerate sequence can be expressed as follows:

$$
\begin{array}{ccccccc}
 & & U & G & U & & \\
UGG & AUG & UUC & AAA & AAC & GA \\
Trp & Met & Phe & Lys & Asn & Glu
\end{array}
$$

Using this mixture of eight 17-mers (UGGAUGUU-CAAAAACGA, UGGAUGUU<u>U</u>AAAAACGA, etc.), we quickly identified several ricin-specific clones. The following worked-out problem shows in detail how degenerate probes work.

Solved Problem

Problem

Here is the amino acid sequence of part of a hypothetical protein whose gene you want to clone:

Arg-Leu-Met-Glu-Trp-Ile-Cys-Pro-Met-Leu

a. What sequence of five amino acids would give a 17-mer probe (including two bases from the next codon) with the least degeneracy?

b. How many different 17-mers would you have to synthesize to be sure your probe matches the corresponding sequence in your cloned gene perfectly?

c. If you started your probe two codons to the right of the optimal one (the one you chose in part *a*), how many different 17-mers would you have to make?

Solution

a. Begin by consulting the genetic code (Chapter 18) to determine the coding degeneracy of each amino acid in the sequence. This yields

$$
\begin{array}{cccccccccc}
6 & 6 & 1 & 2 & 1 & 3 & 2 & 4 & 1 & 6 \\
\end{array}
$$
Arg-Leu-Met-Glu-Trp-Ile-Cys-Pro-Met-Leu

where the numbers above the amino acids represent the coding degeneracy for each. In other words, arginine has six codons, leucine six, methionine one, and so on. Now the task is to find the contiguous set of five codons with the lowest degeneracy. A quick inspection shows that Met-Glu-Trp-Ile-Cys works best.

b. To find how many different 17-mers we would have to prepare, multiply the degeneracies at all positions within the region covered by our probe. For the five amino acids we have chosen, this is $1 \times 2 \times 1 \times 3 \times 2 = 12$. Note that we can use the first two bases (CC) in the proline (Pro) codons without encountering any degeneracy because the fourfold degeneracy in coding for proline all occurs in the third base in the codon (CCU, CCA, CCC, CCG). Thus, our probe can be 17 bases long, instead of the 15 bases we get from the codons for the five amino acids we have selected.

c. If we had started two amino acids farther to the right, starting with Trp, the degeneracy would have been $1 \times 3 \times 2 \times 4 \times 1 = 24$, so we would have had to prepare 24 different probes instead of just 12. ◼

SUMMARY Specific clones can be identified using polynucleotide probes that bind to the gene itself. Knowing the amino acid sequence of a gene product, one can design a set of oligonucleotides that encode part of this amino acid sequence. This can be one of the quickest and most accurate means of identifying a particular clone.

4.2 The Polymerase Chain Reaction

We have now seen how to clone fragments of DNA generated by cleavage with restriction endonucleases, or by physical shearing of DNA. But a relatively new technique, called **polymerase chain reaction (PCR),** can also yield a DNA fragment for cloning and is especially useful for cloning cDNAs, which we will examine in the next section. First, let us look at the PCR method.

PCR was invented by Kary Mullis and his colleagues in the 1980s. As Figure 4.12 explains, this technique uses

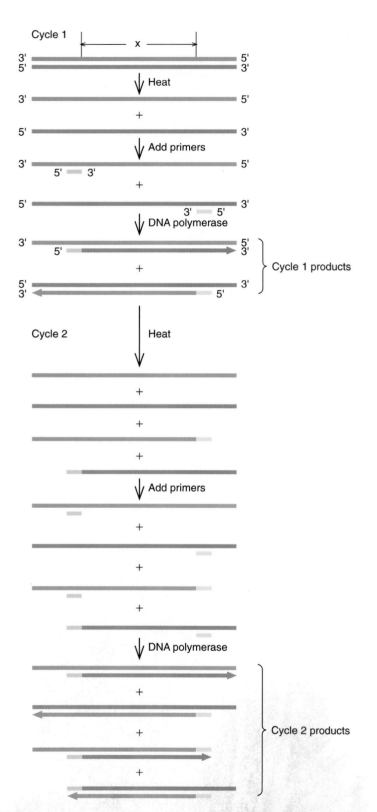

Cycle 1

x

Figure 4.12   Amplifying DNA by the polymerase chain reaction. Cycle 1: Start with a DNA duplex (top) and heat it to separate its two strands (red and blue). Then add short, single-stranded DNA primers (purple and yellow) complementary to sequences on either side of the region (X) to be amplified. The primers hybridize to the appropriate sites on the separated DNA strands; now a special heat-stable DNA polymerase uses these primers to start synthesis of complementary DNA strands. The arrows represent newly made DNA in which replication has stopped at the tip of the arrowhead. At the end of cycle 1, two DNA duplexes are present, including the region to be amplified, whereas we started with only one. The 5′ → 3′ polarities of all DNA strands and primers are indicated throughout cycle 1. The same principles apply in cycle 2. Cycle 2: Repeat the process, heating to separate DNA strands, cooling to allow annealing with primers, letting the heat-stable DNA polymerase make more DNA. Now each of the four DNA strands, including the two newly made ones, can serve as templates for complementary DNA synthesis. The result is four DNA duplexes that have the region to be amplified. Notice that each cycle doubles the number of molecules of DNA because the products of each cycle join the parental molecules in serving as templates for the next cycle. This exponential increase yields 8 molecules in the next cycle and 16 in the cycle after that. This process obviously leads to very high numbers in only a short time.

the enzyme DNA polymerase to make a copy of a selected region of DNA. Mullis and colleagues chose the part (X) of the DNA they wanted to amplify by putting in short pieces of DNA (primers) that hybridized to DNA sequences on each side of X and caused initiation (priming) of DNA synthesis through X. The copies of both strands of X, as well as the original DNA strands, then serve as templates for the next round of synthesis. In this way, the amount of the selected DNA region doubles over and over with each cycle—up to millions of times the starting amount—until enough is present to be seen by gel electrophoresis. Originally, workers had to add fresh DNA polymerase at every round because standard enzymes do not stand up to the high temperatures (over 90° C) needed to separate the strands of DNA before each round of replication. However, special heat-stable polymerases that can take the heat are now available. One of these, **Taq polymerase,** comes from *Thermus aquaticus,* a bacterium that lives in hot springs and therefore has heat-stable enzymes. All one has to do is mix the Taq polymerase with the primers and template DNA in a test tube, seal the tube, then place it in a **thermal cycler.** The thermal cycler is programmed to cycle over and over again among three different temperatures: first a high temperature (about 95° C) to separate the DNA strands; then a relatively low temperature (about 50° C) to allow the primers to anneal to the template DNA strands; then a medium temperature (about 72° C) to allow DNA synthesis. Each cycle takes as little as a few minutes, and it usually takes fewer than 20 cycles to produce as much amplified DNA as one needs. PCR is such a powerful amplifying device that it has even helped spawn science fiction stories such as *Jurassic Park* (see Box 4.1).

SUMMARY PCR amplifies a region of DNA between two predetermined sites. Oligonucleotides complementary to these sites serve as primers for synthesis of copies of the DNA between the sites. Each cycle of PCR doubles the number of copies of the amplified DNA until a large quantity has been made.

cDNA Cloning

A **cDNA** (short for **complementary DNA** or **copy DNA**) is a DNA copy of an RNA, usually an mRNA. Sometimes we want to make a **cDNA library,** a set of clones representing as many as possible of the mRNAs in a given cell type at a given time. Such libraries can contain tens of thousands of different clones. Other times, we want to make one particular cDNA—a clone containing a DNA copy of just one mRNA. The technique we use depends in part on which of these goals we wish to achieve.

Figure 4.13 illustrates one simple, yet effective method for making a cDNA library. The central part of any cDNA cloning procedure is synthesis of the cDNA from an mRNA template using **reverse transcriptase** (RNA-dependent DNA polymerase). Reverse transcriptase is like any other DNA-synthesizing enzyme in that it cannot initiate DNA synthesis without a primer. To get around this problem, we take advantage of the poly(A) tail at the 3′-end of most eukaryotic mRNAs and use oligo(dT) as the primer. The oligo(dT) is complementary to poly(A), so it binds to the poly(A) at the 3′-end of the mRNA and primes DNA synthesis, using the mRNA as the template.

After the mRNA has been copied, yielding a single-stranded DNA (the "first strand"), the mRNA is partially degraded with **ribonuclease H (RNase H).** This enzyme degrades the RNA strand of an RNA–DNA hybrid—just what we need to begin to digest the RNA base-paired to the first-strand cDNA. The remaining RNA fragments serve as primers for making the "second strand," using the first as the template. This phase of the process depends on a phenomenon called **nick translation,** which is illustrated in Figure 4.14. The net result is a double-stranded cDNA with a small fragment of RNA at the 5′-end of the second strand.

The essence of nick translation is the simultaneous removal of DNA ahead of a **nick** (a single-stranded DNA break) and synthesis of DNA behind the nick, rather like a road paving machine that tears up old pavement at its front end and lays down new pavement at its back end. The net result is to move, or "translate," the nick in the $5′ \rightarrow 3′$ direction. The enzyme usually used for nick translation is *E. coli* DNA polymerase I, which has a $5′ \rightarrow 3′$ exonuclease activity that allows the enzyme to degrade DNA ahead of the nick as it moves along.

The next task is to ligate the cDNA to a vector. This was easy with pieces of genomic DNA cleaved with restriction enzymes, but cDNAs have no sticky ends. It is true that blunt ends can be ligated together, even though the process is relatively inefficient. However, to get the efficient ligation afforded by sticky ends, one can create sticky ends (oligo[dC] in this case) on the cDNA, using an enzyme called **terminal deoxynucleotidyl transferase (TdT)** or simply **terminal transferase** and one of the deoxyribonucleoside triphosphates. In this case, dCTP was used. The enzyme adds dCMPs, one at a time, to the 3′-ends of the cDNA. In the same way, oligo(dG) ends can be added to a vector. Annealing the oligo(dC) ends of the cDNA to the oligo(dG) ends of the vector brings the vector and cDNA together in a recombinant DNA that can be used directly for transformation. The base pairing between the oligonucleotide tails is strong enough that no ligation is required before transformation. The DNA ligase inside the transformed cells finally performs the ligation, and DNA polymerase I removes any remaining RNA and replaces it with DNA.

BOX 4.1

Jurassic Park: More than a Fantasy?

In Michael Crichton's book *Jurassic Park*, and in the movie of the same name, a scientist and an entrepreneur collaborate in a fantastic endeavor: to generate living dinosaurs. Their strategy is to isolate dinosaur DNA, but not directly from dinosaur remains, from which DNA would be impossible to get. Instead they find Jurassic-period blood-sucking insects that had feasted on dinosaur blood and had then become mired in tree sap, which had turned to amber, entombing and preserving the insects. They reason that, because blood contains white blood cells that have DNA, the insect gut contents should contain this dinosaur DNA. The next step is to use PCR to amplify the dinosaur DNA, piece the fragments together, place them in an egg, and *voila!* A dinosaur is hatched.

This scenario sounds preposterous, and indeed certain practical problems keep it totally in the realm of science fiction. But it is striking that some parts of the story are already in the scientific literature. In June 1993, the same month that *Jurassic Park* opened in movie theaters, a paper appeared in the journal *Nature* describing the apparent PCR amplification and sequencing of part of a gene from an extinct weevil trapped in Lebanese amber for 120–135 million years. That takes us back to the Cretaceous period, not quite as ancient as the Jurassic, but a time when plenty of dinosaurs were still around. If this work is valid, it may indeed be possible to find preserved, blood-sucking insects with dinosaur DNA in their guts. Furthermore, it may be possible that this DNA is still intact enough that it can serve as a template for PCR amplification. After all, the PCR technique is powerful enough to start with a single molecule of DNA and amplify it to any degree we wish.

So what stands in the way of making dinosaurs? Leaving aside the uncharted territory of creating a vertebrate animal from naked DNA, we have to consider the simple limitations of the PCR process itself. The first of these is the present limit to the size of a DNA fragment that we can amplify by PCR: up to 40 kb. That is probably on the order of one-hundred thousandth the size of the whole dinosaur genome, which means that we would ultimately have to piece together at least a hundred thousand PCR fragments to reconstitute the whole genome. And that assumes that we know enough about the sequence of the dinosaur DNA, at the start, to make PCR primers for all of those fragments.

But what if we worked out a way to make PCR go much farther than 40,000 bp? What if PCR became so powerful that we could amplify whole chromosomes, up to hundreds of millions of base pairs at a time? Then we would run up against the fact that DNA is an inherently unstable molecule, and no full-length chromosomes would be expected to survive for millions of years, even in an insect embalmed in amber. PCR can amplify relatively short stretches because the primers need to find only one molecule that is unbroken over that short stretch. But finding a whole unbroken chromosome, or even an unbroken megabase-size stretch of DNA, would be extremely unlikely.

These considerations have generated considerable uncertainty about the few published examples of amplifying ancient DNA by PCR. Many scientists argue that it is simply not credible that a molecule as fragile as DNA can last for millions of years. They believe that dinosaur DNA would long ago have decomposed into nucleotides and be utterly useless as a template for PCR amplification. Indeed, this appears to be true for all ancient DNA, with the possible exception of that preserved in amber.

On the other hand, the PCR machine has amplified some kind of DNA from the ancient insect samples. If it is not ancient insect DNA, what is it? This brings us to the second limitation of the PCR method, which is also its great advantage: its exquisite sensitivity. As we have seen, PCR can amplify a single molecule of DNA, which is fine if that is the DNA we want to amplify. It can, however, also seize on tiny quantities—even single molecules—of contaminating DNAs in our sample and amplify them instead of the DNA we want. For this reason, the workers who examined the Cretaceous weevil DNA did all their PCR amplification and sequencing on that DNA before they even began work on modern insect DNA, to which they compared the weevil sequences. That way, they minimized the worry that they were amplifying trace contaminants of modern insect DNA left over from previous experiments, when they thought they were amplifying DNA from the extinct weevil. But DNA is everywhere, especially in a molecular biology lab, and eliminating every last molecule is agonizingly difficult. Furthermore, dinosaur DNA in the gut of an insect would be heavily contaminated with insect DNA, not to mention DNA from intestinal bacteria. And who is to say the insect fed on only one type of dinosaur before it died in the tree sap? If it fed on two, the PCR procedure would probably amplify both their DNAs together, and there would be no way to separate them.

In other words, some of the tools to create a real Jurassic Park are already in hand, but, as exciting as it is to imagine seeing a living dinosaur, the practical problems are still so overwhelming that they make it seem impossible.

On a far more realistic level, the PCR technique is already allowing us to compare the sequences of genes from extinct organisms with those of their present-day relatives. And this is spawning an exciting new field, which botanist Michael Clegg calls "molecular paleontology."

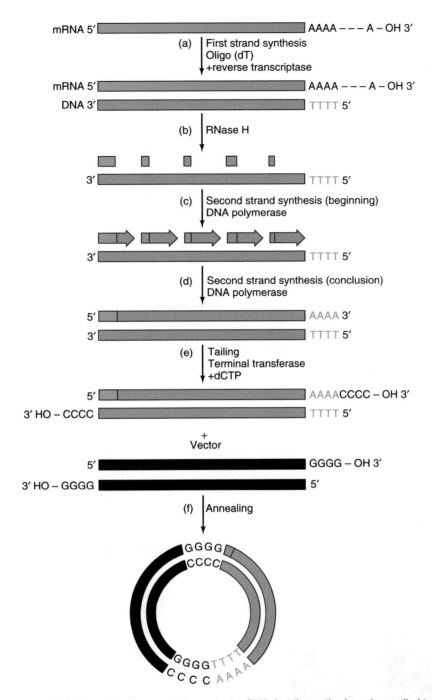

Figure 4.13 Making a cDNA library. This figure focuses on cloning a single cDNA, but the method can be applied to a mixture of mRNAs and produce a library of corresponding cDNAs. **(a)** Use oligo(dT) as a primer and reverse transcriptase to copy the mRNA (blue), producing a cDNA (red) that is hybridized to the mRNA template. **(b)** Use RNase H to partially digest the mRNA, yielding a set of RNA primers base-paired to the first-strand cDNA. **(c)** Use *E. coli* DNA polymerase I under nick translation conditions to build second-strand cDNAs on the RNA primers. **(d)** The second-strand cDNA growing from the leftmost primer (blue) has been extended all the way to the 3'-end of the oligo(dA) corresponding to the oligo(dT) primer on the first-strand cDNA. **(e)** To give the double-stranded cDNA sticky ends, add oligo(dC) with terminal transferase. **(f)** Anneal the oligo(dC) ends of the cDNA to complementary oligo(dG) ends of a suitable vector (black). The recombinant DNA can then be used to transform bacterial cells. Enzymes in these cells remove remaining nicks and replace any remaining RNA with DNA.

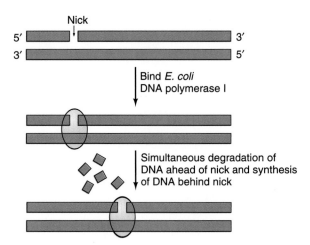

Figure 4.14   Nick translation. This illustration is a generic example with double-stranded DNA, but the same principles apply to an RNA–DNA hybrid. Beginning with a double-stranded DNA with a nick in the top strand, *E. coli* DNA polymerase I binds to this nick and begins elongating the DNA fragment on the top left in the 5′ → 3′ direction (left to right). At the same time, the 5′ → 3′ exonuclease activity degrades the DNA fragment to its right to make room for the growing fragment behind it. The small red rectangles represent nucleotides released by exonuclease digestion of the DNA.

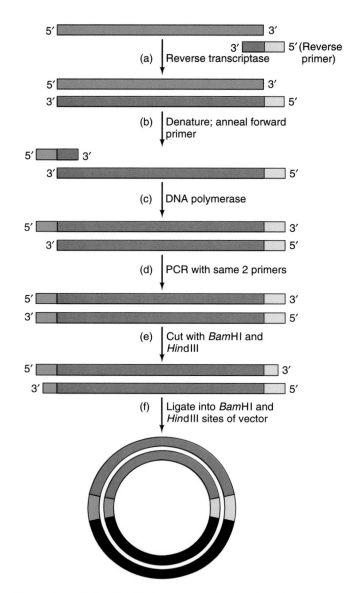

Figure 4.15   Using RT-PCR to clone a single cDNA. (a) Use a reverse primer (red) with a *Hind*III site (yellow) at its 5′-end to start first-strand cDNA synthesis, with reverse transcriptase to catalyze the reaction. **(b)** Denature the mRNA–cDNA hybrid and anneal a forward primer (red) with a *Bam*HI site (green) at its 5′-end. **(c)** This forward primer initiates second-strand cDNA synthesis, with DNA polymerase catalyzing the reaction. **(d)** Continue PCR with the same two primers to amplify the double-stranded cDNA. **(e)** Cut the cDNA with *Bam*HI and *Hind*III to generate sticky ends. **(f)** Ligate the cDNA to the *Bam*HI and *Hind*III sites of a suitable vector. Finally, transform cells with the recombinant cDNA to produce a clone.

Using RT-PCR in cDNA Cloning   If one wants to clone a cDNA from just one mRNA whose sequence is known, one can use a type of PCR called **reverse transcriptase PCR (RT-PCR)** as illustrated in Figure 4.15. The main difference between this procedure and the PCR method described earlier in this chapter is that this one starts with an mRNA instead of a double-stranded DNA. Thus, one begins by converting the mRNA to DNA. As usual, this RNA → DNA step can be done with reverse transcriptase: One reverse transcribes the mRNA to make a single-stranded DNA, then uses a primer to convert the single-stranded DNA to double-stranded. Then one can use standard PCR to amplify the cDNA until enough is available for cloning. One can even add restriction sites to the ends of the cDNA by using primers that contain these sites. In this example, a *Bam*HI site is present on one primer and a *Hind*III site is present on the other (placed a few nucleotides from the ends to allow the restriction enzymes to cut efficiently). Thus, the PCR product is a cDNA with these two restriction sites at its two ends. Cutting the PCR product with these two restriction enzymes creates sticky ends that can be ligated into the vector of choice. Having two *different* sticky ends allows directional cloning, so the cDNA will have only one of two possible orientations in the vector. This is very useful when a cDNA is cloned into an expression vector, because the cDNA must be in the same orientation as the promoter that drives transcription of the cDNA. A caveat is necessary, however: One must make sure that the cDNA itself has neither of the restriction sites that have been added to its ends. If it does, the restriction enzymes

will cut within the cDNA, as well as at the ends, and the products will be useless.

What kind of vector should be used to ligate to a cDNA or cDNAs? Several choices are available, depending on the method used to detect positive clones (those that bear the desired cDNA). A plasmid or phagemid vector such as pUC or pBS can be used; if so, positive clones are usually identified by **colony hybridization** with a radioactive DNA probe. This procedure is analogous to the plaque hybridization described previously. Or one can use a λ phage, such as λgt11, as a vector. This vector places the cloned cDNA under the control of a *lac* promoter, so that transcription and translation of the cloned gene can occur. One can then use an antibody to screen directly for the protein product of the correct gene. We will describe this procedure in more detail later in this chapter. Alternatively, a polynucleotide probe can be used to hybridize to the recombinant phage DNA.

Rapid Amplification of cDNA Ends Very frequently, our cDNA is not full-length, possibly because the reverse transcriptase, for whatever reason, did not make it all the way to the end of the mRNA. This does not mean we have to be satisfied with an incomplete cDNA, however. Fortunately, we can fill in the missing pieces of a cDNA, using a procedure called **rapid amplification of cDNA ends** (RACE). Figure 4.16 illustrates the technique (5′-RACE) for filling in the 5′-end of a cDNA (the usual problem), but an analogous technique (3′-RACE) can be used to fill in a missing 3′-end of a cDNA.

A 5′-RACE procedure begins with an RNA preparation containing the mRNA of interest and a partial cDNA whose 5′-end is missing. An incomplete strand of the cDNA can be annealed to the mRNA and then reverse transcriptase can be used to copy the rest of the mRNA. Then the completed cDNA can be tailed with oligo(dC) (for example), using terminal transferase and dCTP. Next, oligo(dG) is used to prime second-strand synthesis. This step produces a double-stranded cDNA that can be amplified by PCR, using oligo(dG) and a 3′-specific oligonucleotide as primers.

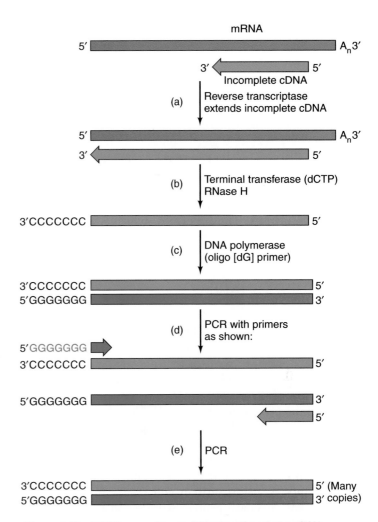

Figure 4.16   RACE procedure to fill in the 5′-end of a cDNA.
(a) Hybridize an incomplete cDNA (red), or an oligonucleotide segment of a cDNA to mRNA (green), and use reverse transcriptase to extend the cDNA to the 5′-end of the mRNA. **(b)** Use terminal transferase and dCTP to add C residues to the 3′-end of the extended cDNA; also, use RNase H to degrade the mRNA. **(c)** Use an oligo(dG) primer and DNA polymerase to synthesize a second strand of cDNA (blue). **(d)** and **(e)** Perform PCR with oligo(dG) as the forward primer and an oligonucleotide that hybridizes to the 3′-end of the cDNA as the reverse primer. The product is a cDNA that has been extended to the 5′-end of the mRNA. A similar procedure (3′-RACE) can be used to extend the cDNA in the 3′-direction. In that case, there is no need to tail the 3′-end of the cDNA with terminal transferase because the mRNA already contains poly(A); thus, the reverse primer would be oligo(dT).

SUMMARY To make a cDNA library, one can synthesize cDNAs one strand at a time, using mRNAs from a cell as templates for the first strands and these first strands as templates for the second strands. Reverse transcriptase generates the first strands and *E. coli* DNA polymerase I generates the second strands. One can endow the double-stranded cDNAs with oligonucleotide tails that base-pair with complementary tails on a cloning vector, then use these recombinant DNAs to transform bacteria. RT-PCR can be used to generate a cDNA from a single type of mRNA, but the sequence of the mRNA must be known so the primers for the PCR step can be designed. Restriction site sequences can be placed on the PCR primers, so these sites appear at the ends of the cDNA. This makes it easy to cleave them and then to ligate the cDNA into a vector. Particular clones can be detected by colony hybridization with radioactive DNA probes, or with antibodies if an expression vector such as λgt11 is used. Incomplete cDNA can be filled in by 5′- or 3′-RACE.

4.3   Methods of Expressing Cloned Genes

Why would we want to clone a gene? An obvious reason, suggested at the beginning of this chapter, is that cloning allows us to produce large quantities of particular DNA sequences so we can study them in detail. Thus, the gene itself can be a valuable product of gene cloning. Another goal of gene cloning is to make a large quantity of the gene's product, either for investigative purposes or for profit.

If the goal is to use bacteria to produce the protein product of a cloned eukaryotic gene—especially a higher eukaryotic gene—a cDNA will probably work better than a gene cut directly out of the genome. That is because most higher eukaryotic genes contain interruptions called introns (Chapter 14) that bacteria cannot deal with. Eukaryotic cells usually transcribe these interruptions, forming a pre-mRNA, and then cut them out and stitch the remaining parts (exons) of the pre-mRNA together to form the mature mRNA. Thus, a cDNA, which is a copy of an mRNA, already has its introns removed and can be expressed correctly in a bacterial cell.

Expression Vectors

The vectors we have examined so far are meant to be used primarily in the first stage of cloning—when we first put a foreign DNA into a bacterium and get it to replicate. By and large, they work well for that purpose, growing readily in *E. coli* and producing high yields of recombinant DNA. Some of them even work as **expression vectors** that can yield the protein products of the cloned genes. For example, the pUC and pBS vectors place inserted DNA under the control of the *lac* promoter, which lies upstream of the multiple cloning site. If an inserted DNA happens to be in the same reading frame as the *lacZ'* gene it interrupts, a **fusion protein** will result. It will have a partial β-galactosidase protein sequence at its amino end and another protein sequence, encoded in the inserted DNA, at its carboxyl end (Figure 4.17).

However, if we are interested in high expression of our cloned gene, specialized expression vectors usually work better. Bacterial expression vectors typically have two elements that are required for active gene expression: a strong promoter and a ribosome binding site near an initiating ATG codon.

Expression Vectors with Strong Promoters
The main function of an expression vector is to yield the product of a gene—usually, the more product the better. Therefore, expression vectors are ordinarily equipped with very strong promoters; the rationale is that the more mRNA that is produced, the more protein product will be made.

One such strong promoter is the *trp* (tryptophan operon) promoter. It forms the basis for several expression vectors, including *ptrpL1*. Figure 4.18 shows how this vector works. It has the *trp* promoter/operator region, followed by a ribosome binding site, and can be used directly as an expression vector by inserting a foreign gene into the *Cla*I site. Alternatively, the *trp* control region can be made "portable" by cutting it out with *Cla*I and *Hin*dIII and inserting it in front of a gene to be expressed in another vector.

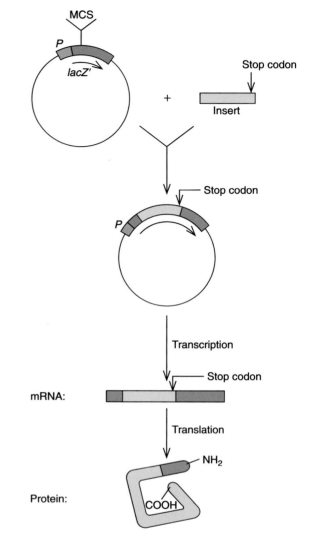

Figure 4.17   Producing a fusion protein by cloning in a pUC plasmid. Insert foreign DNA (yellow) into the multiple cloning site (MCS); transcription from the *lac* promoter (purple) gives a hybrid mRNA beginning with a few *lacZ'* codons, changing to insert sequence, then back to *lacZ'* (red). This mRNA is translated to a fusion protein containing a few β-galactosidase amino acids at the beginning (amino end), followed by the insert amino acids for the remainder of the protein. Because the insert contains a translation stop codon, the remaining *lacZ'* codons are not translated.

Inducible Expression Vectors
It is usually advantageous to keep a cloned gene repressed until it is time to express it. One reason is that eukaryotic proteins produced in large quantities in bacteria can be toxic. Even if these proteins are not actually toxic, they can build up to such great levels that they interfere with bacterial growth. In either case, if the cloned gene were allowed to remain turned on constantly, the bacteria bearing the gene would never grow to a great enough concentration to produce meaningful quantities of protein product. The solution is to keep the cloned gene turned off by placing it downstream of an inducible promoter that can be turned off.

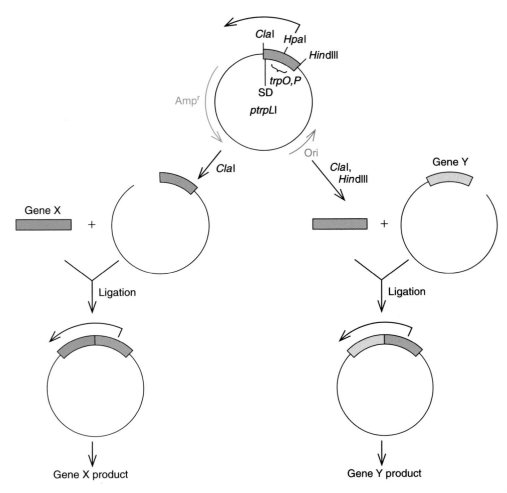

Figure 4.18   Two uses of the *ptrpL* 1 expression vector. The vector contains a *Cla*I cloning site, preceded by a Shine–Dalgarno ribosome binding site (SD) and the *trp* operator–promoter region (*trpO,P*). Transcription occurs in a counterclockwise direction as shown by the arrow (top). The vector can be used as a traditional expression vector (left) simply by inserting a foreign coding region (X, green) into the unique *Cla*I site. Alternatively (right), the *trp* control region (purple) can be cut out with *Cla*I and *Hind*III and inserted into another plasmid bearing the coding region (Y, yellow) to be expressed.

The *lac* promoter is inducible to a certain extent, presumably remaining off until stimulated by the synthetic inducer isopropylthiogalactoside (IPTG). However, the repression caused by the *lac* repressor is incomplete (leaky), and some expression of the cloned gene will be observed even in the absence of inducer. One way around this problem is to express a gene in a plasmid or phagemid that carries its own *lacI* (repressor) gene, as pBS does (see Figure 4.11). The excess repressor produced by such a vector keeps the cloned gene turned off until it is time to induce it with IPTG. (For a review of the *lac* operon, see Chapter 7.)

Another strategy is to use a tightly controlled promoter such as the λ phage promoter P_L. Expression vectors with this promoter–operator system are cloned into host cells bearing a temperature-sensitive λ repressor gene (*c1857*). As long as the temperature of these cells is kept relatively low (32° C), the repressor functions, and no ex-

pression takes place. However, when the temperature is raised to the nonpermissive level (42° C), the temperature-sensitive repressor can no longer function and the cloned gene is derepressed.

A very popular method of ensuring tight control, as well as high-level induced expression, is to place the gene to be expressed in a plasmid under control of a T7 phage promoter. Then this plasmid is placed in a cell that contains a tightly regulated gene for T7 RNA polymerase. For example, the T7 RNA polymerase gene may be under control of a modified *lac* promoter in a cell that also carries the gene for the *lac* repressor. Thus, the T7 polymerase gene is strongly repressed unless the *lac* inducer is present. As long as no T7 polymerase is present, transcription of the gene of interest cannot take place because the T7 promoter has an absolute requirement for its own polymerase. But as soon as a *lac* inducer is added, the cell begins to make T7 polymerase, which transcribes the gene of interest. And because many molecules of T7 polymerase are made, the gene is turned on to a very high level and abundant amounts of protein product are made.

> **SUMMARY** Expression vectors are designed to yield the protein product of a cloned gene, usually in the greatest amount possible. To optimize expression, these vectors include strong bacterial promoters and bacterial ribosome binding sites that would be missing on cloned eukaryotic genes. Most cloning vectors are inducible, which avoids premature overproduction of a foreign product that could poison the bacterial host cells.

Expression Vectors That Produce Fusion Proteins Most expression vectors produce fusion proteins. This might at first seem a disadvantage because the natural product of the inserted gene is not made. However, the extra amino acids on the fusion protein can be a great help in purifying the protein product.

Consider the oligohistidine expression vectors, one of which has the trade name pTrcHis (Figure 4.19). These have a short sequence just upstream of the multiple cloning site that encodes a stretch of six histidines. Thus, a protein expressed in such a vector will be a fusion protein with six histidines at its amino end. Why would one want to attach six histidines to a protein? Oligohistidine regions like this have a high affinity for metals like nickel, so proteins that have such regions can be purified using nickel **affinity chromatography.** The beauty of this method is its simplicity and speed. After the bacteria have made the fusion protein, one simply lyses them, adds the crude bacterial extract to a nickel affinity column, washes out all unbound proteins, then releases the fusion protein with histidine or a histidine analog called imidazole. This procedure allows one to harvest essentially pure fusion protein in only one step. This is possible because very few if any natural proteins have oligohistidine regions, so the fusion protein is essentially the only one that binds to the column.

What if the oligohistidine tag interferes with the protein's activity? The designers of these vectors have thoughtfully provided a way to remove it. Just before the multiple cloning site is a coding region for a stretch of amino acids recognized by the enzyme enterokinase (a protease, not really a kinase at all). So enterokinase can be used to cleave the fusion protein into two parts: the oligohistidine tag and the protein of interest. The site recognized by enterokinase is very rare, and the chance that it exists in any given protein is insignificant. Thus, the rest of the protein should not be chopped up as its oligohistidine tag is removed. The enterokinase-cleaved protein can be run through the nickel column once more to separate the oligohistidine fragments from the protein of interest.

λ phages have also served as the basis for expression vectors; one designed specifically for this purpose is λ**gt11.** This phage (Figure 4.20) contains the *lac* control region followed by the *lacZ* gene. The cloning sites are located within the *lacZ* gene, so products of a gene inserted correctly into this vector will be fusion proteins with a leader of β-galactosidase.

The expression vector λgt11 has become a popular vehicle for making and screening cDNA libraries. In the examples of screening presented earlier, the proper DNA sequence was detected by probing with a labeled oligonucleotide or polynucleotide. By contrast, λgt11 allows one to screen a group of clones directly for the expression of the right protein. The main ingredients required for this procedure are a cDNA library in λgt11 and an antiserum directed against the protein of interest.

Figure 4.21 shows how this works. Lambda phages with various cDNA inserts are plated, and the proteins released by each clone are blotted onto a support such as nitrocellulose. Once the proteins from each plaque have been transferred to nitrocellulose, they can be probed with antiserum. Next, antibody bound to protein from a particular plaque can be detected, using labeled protein A from *Staphylococcus aureus.* This protein binds tightly to antibody and labels the corresponding spot on the nitrocellulose. This label can be detected by autoradiography or by phosphorimaging (Chapter 5), then the corresponding plaque can be picked from the master plate. Note that a fusion protein is detected, not the protein of interest by itself. Furthermore, it does not matter if a whole cDNA has been cloned or not. The antiserum is a mixture of antibodies that will react with several different parts of the protein, so even a partial gene will do, as long as its coding region is cloned in the same orientation and reading frame as the β-galactosidase coding region.

Even partial cDNAs are valuable because they can be completed by RACE, as we saw earlier in this chapter. The β-galactosidase tag on the fusion proteins helps to stabilize them in the bacterial cell, and can even make them easy to purify by affinity chromatography on a column containing an anti-β-galactosidase antibody.

> **SUMMARY** Expression vectors frequently produce fusion proteins, with one part of the protein coming from coding sequences in the vector and the other part from sequences in the cloned gene itself. Many fusion proteins have the great advantage of being simple to isolate by affinity chromatography. The λgt11 vector produces fusion proteins that can be detected in plaques with a specific antiserum.

Eukaryotic Expression Systems Eukaryotic genes are not really "at home" in prokaryotic cells, even when they are

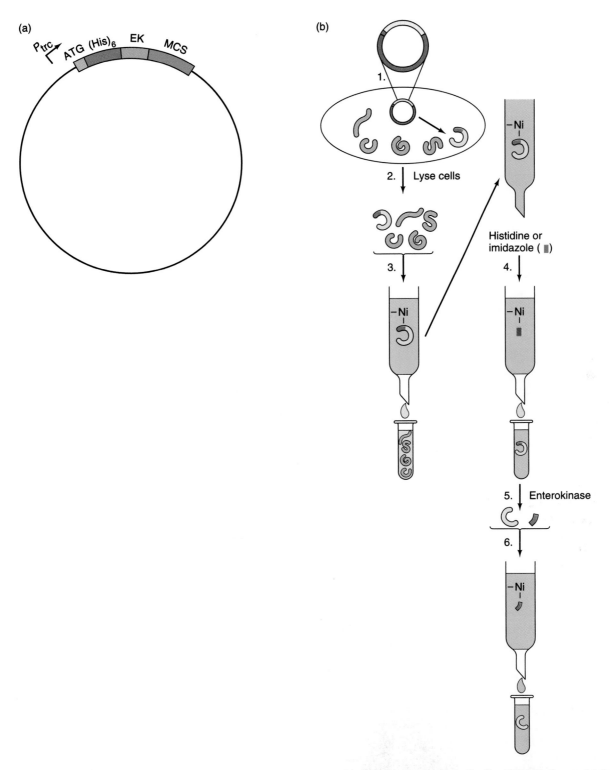

Figure 4.19 Using an oligohistidine expression vector. (a) Map of a generic oligohistidine vector. Just after the ATG initiation codon (green) lies a coding region (red) encoding six histidines in a row [(His)$_6$]. This is followed by a region (orange) encoding a recognition site for the proteolytic enzyme enterokinase (EK). Finally, the vector has a multiple cloning site (MCS, blue). Usually, the vector comes in three forms with the MCS sites in each of the three reading frames. One can select the vector that puts the gene in the right reading frame relative to the oligohistidine. **(b)** Using the vector. 1. Insert the gene of interest (yellow) into the vector in frame with the oligohistidine coding region (red) and transform bacterial cells with the recombinant vector. The cells produce the fusion protein (red and yellow), along with other, bacterial proteins (green). 2. Lyse the cells, releasing the mixture of proteins. 3. Pour the cell lysate through a nickel affinity chromatography column, which binds the fusion protein but not the other proteins. 4. Release the fusion protein from the column with histidine or with imidazole, a histidine analogue, which competes with the oligohistidine for binding to the nickel. 5. Cleave the fusion protein with enterokinase. 6. Pass the cleaved protein through the nickel column once more to separate the oligohistidine from the desired protein.

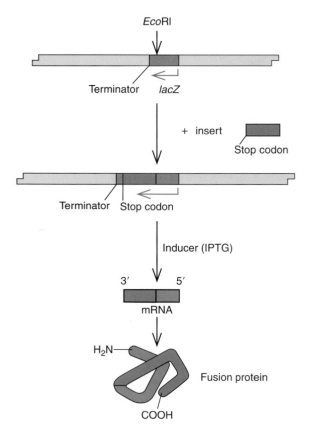

Figure 4.20   Synthesizing a fusion protein in λgt11. The gene to be expressed (green) is inserted into the *Eco*RI site near the end of the *lacZ* coding region (red) just upstream of the transcription terminator. Thus, on induction of the *lacZ* gene by IPTG, a fused mRNA results, containing the inserted coding region just downstream of the bulk of the coding region of β-galactosidase. This mRNA is translated by the host cell to yield a fusion protein.

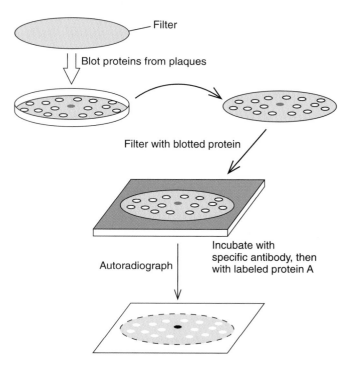

Figure 4.21   Detecting positive λgt11 clones by antibody screening. A filter is used to blot proteins from phage plaques on a Petri dish. One of the clones (red) has produced a plaque containing a fusion protein including β-galactosidase and a part of the protein of interest. The filter with its blotted proteins is incubated with an antibody directed against the protein of interest, then with labeled *Staphylococcus* protein A, which binds to most antibodies. It will therefore bind only to the antibody–antigen complexes at the spot corresponding to the positive clone. A dark spot on the film placed in contact with the filter reveals the location of the positive clone.

expressed under the control of their prokaryotic vectors. One reason is that *E. coli* cells sometimes recognize the protein products of cloned eukaryotic genes as outsiders and destroy them. Another is that prokaryotes do not carry out the same kinds of posttranslational modifications as eukaryotes do. For example, a protein that would ordinarily be coupled to sugars in a eukaryotic cell will be expressed as a bare protein when cloned in bacteria. This can affect a protein's activity or stability, or at least its response to antibodies. A more serious problem is that the interior of a bacterial cell is not as conducive to proper folding of eukaryotic proteins as the interior of a eukaryotic cell. Frequently, the result is improperly folded, inactive products of cloned genes. This means that one can frequently express a cloned gene at a stupendously high level in bacteria, but the product forms highly insoluble, inactive granules called **inclusion bodies**. These are of no use unless one can get the protein to refold and regain its activity. Fortunately, it is frequently possible to renature the proteins from inclusion bodies. In that case, the inclusion

bodies are an advantage because they can be separated from almost all other proteins by simple centrifugation.

To avoid the potential incompatibility between a cloned gene and its host, the gene can be expressed in a eukaryotic cell. In such cases, the initial cloning is usually done in *E. coli,* using a **shuttle vector** that can replicate in both bacterial and eukaryotic cells. The recombinant DNA is then transferred to the eukaryote of choice. One eukaryote suited for this purpose is yeast. It shares the advantages of rapid growth and ease of culture with bacteria, yet it is a eukaryote and thus it carries out some of the protein folding and glycosylation (adding sugars) characteristic of a eukaryote. In addition, by splicing a cloned gene to the coding region for a yeast export signal peptide, one can usually ensure that the gene product will be secreted to the growth medium. This is a great advantage in purifying the protein. The yeast cells are simply removed in a centrifuge, leaving relatively pure secreted gene product behind in the medium.

The yeast expression vectors are based on a plasmid, called the *2-micron plasmid*, that normally inhabits yeast cells. It provides the origin of replication needed by any vector that must replicate in yeast. Yeast–bacterial shuttle vectors also contain the pBR322 origin of replication, so they can also replicate in *E. coli*. In addition, of course, a yeast expression vector must contain a strong yeast promoter.

Another eukaryotic vector that has been remarkably successful is derived from the **baculovirus** that infects the caterpillar known as the alfalfa looper. Viruses in this class have a rather large circular DNA genome, approximately 130 kb in length. The major viral structural protein, polyhedrin, is made in copious quantities in infected cells. In fact, it has been estimated that when a caterpillar dies of a baculovirus infection, up to 10% of the dry mass of the dead insect is this one protein. This huge mass of protein indicates that the polyhedrin gene must be very active, and indeed it is—apparently due to its powerful promoter. Max Summers and his colleages, and Lois Miller and her colleagues first developed successful vectors using the polyhedrin promoter in 1983 and 1984, respectively. Since then, many other baculovirus vectors have been constructed using this and other viral promoters.

At their best, baculovirus vectors can produce up to half a gram per liter of protein from a cloned gene—a large amount indeed. Figure 4.22 shows how a typical baculovirus expression system works. First, the gene of interest is cloned in one of the vectors. In this example, let us consider a vector with the polyhedrin promoter. (The polyhedrin coding region has been deleted from the vector. This does not inhibit virus replication because polyhedrin is not required for transmission of the virus from cell to cell in culture.) Most such vectors have a unique *Bam*HI site directly downstream of the promoter, so they can be cut with *Bam*HI and a fragment with *Bam*HI-compatible ends can be inserted into the vector, placing the cloned gene under the control of the polyhedrin promoter. Next the recombinant plasmid (vector plus insert) is mixed with wild-type viral DNA that has been cleaved so as to remove a gene essential for viral replication, along with the polyhedrin gene. Cultured insect cells are then transfected with this mixture.

Because the vector has extensive homology with the regions flanking the polyhedrin gene, recombination can occur within the transfected cells. This transfers the cloned gene into the viral DNA, still under the control of the polyhedrin promoter. Now this recombinant virus can be used to infect cells and the protein of interest can be harvested after these cells enter the very late phase of infection, during which the polyhedrin promoter is most active. What about the nonrecombinant viral DNA that enters the transfected cells along with the recombinant vector? It cannot give rise to infectious virus because it lacks an essential gene that can only be supplied by the vector.

Notice the use of the term *transfected* with eukaryotic cells instead of *transformed,* which we use with bacteria. We make this distinction because *transformation* has another meaning in eukaryotes: the conversion of a normal cell to a cancer-like cell. To avoid confusion with this phenomenon, we use **transfection** to denote introducing new DNA into a eukaryotic cell.

Transfection in animal cells is conveniently carried out in at least two ways: (1) Cells can be mixed with DNA in a phosphate buffer, then a solution of a calcium salt can be added to form a precipitate of $Ca_3(PO_4)_2$. The cells take up the calcium phosphate crystals, which also include some DNA. (2) The DNA can be mixed with lipid, which forms **liposomes,** small vesicles that include some DNA solution inside. These DNA-bearing liposomes then fuse with the cell membranes, delivering their DNA into the cells. Plant cells are commonly transfected by a **biolistic** method in which small metal pellets are coated with DNA and literally shot into cells.

SUMMARY Foreign genes can be expressed in eukaryotic cells, and these eukaryotic systems have some advantages over their prokaryotic counterparts for producing eukaryotic proteins. Two of the most important advantages are (1) Eukaryotic proteins made in eukaryotic cells tend to be folded properly, so they are soluble, rather than aggregated into insoluble inclusion bodies. (2) Eukaryotic proteins made in eukaryotic cells are modified (phosphorylated, glycosylated, etc.) in a eukaryotic manner.

Other Eukaryotic Vectors

Some well-known eukaryotic vectors serve purposes other than expressing foreign genes. For example, yeast artificial chromosomes (YACs), bacterial artificial chromosomes (BACs), and PI phage artificial chromosomes (PACs) are capable of accepting huge chunks of foreign DNA and therefore find use in large sequencing programs such as the human genome project, where big pieces of cloned DNA are especially valuable. We will discuss the artificial chromosomes in chapter 24 in the context of genomics. Another important eukaryotic vector is the Ti plasmid, which can transport foreign genes into plant cells and ensure their replication there.

Using the Ti Plasmid to Transfer Genes to Plants

Genes can also be introduced into plants, using vectors that can replicate in plant cells. The common bacterial vectors do not serve this purpose because plant cells cannot recognize their prokaryotic promoters and replication origins. Instead, a plasmid containing so-called **T-DNA** can be used. This is a piece of DNA from a plasmid known as **Ti** (tumor-inducing).

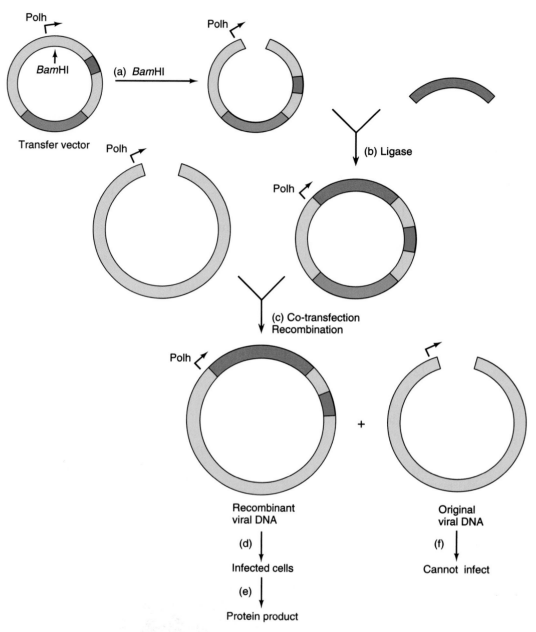

Figure 4.22   Expressing a gene in a baculovirus. First, insert the gene to be expressed (red) into a baculovirus transfer vector. In this case, the vector contains the powerful polyhedrin promoter (Polh), flanked by the DNA sequences (yellow) that normally surround the polyhedrin gene, including a gene (green) that is essential for virus replication; the polyhedrin coding region itself is missing from this transfer vector. Just downstream of the promoter is a *Bam*HI restriction site, which can be used to open up the vector (step **a**) so it can accept the foreign gene (red) by ligation (step **b**). In step **c**, mix the recombinant transfer vector with linear viral DNA that has been cut so as to remove the essential gene. Transfect insect cells with the two DNAs together. This process is known as co-transfection. The two DNAs are not drawn to scale; the viral DNA is actually almost 15 times the size of the vector. Inside the cell, the two DNAs recombine by a double crossover that inserts the gene to be expressed, along with the essential gene, into the viral DNA. The result is a recombinant virus DNA that has the gene of interest under the control of the polyhedrin promoter. Next, infect cells with the recombinant virus. Finally, in steps **d** and **e,** infect cells with the recombinant virus and collect the protein product these cells make. Notice that the original viral DNA is linear and it is missing the essential gene, so it cannot infect cells. **(f)** This lack of infectivity selects automatically for recombinant viruses; they are the only ones that can infect cells.

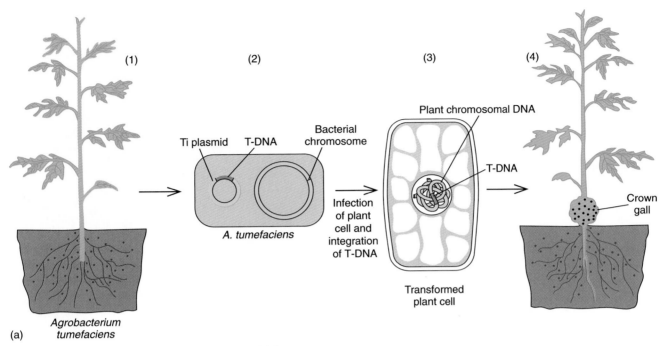

Figure 4.23   Crown gall tumors. (a) Formation of a crown gall 1. *Agrobacterium* cells enter a wound in the plant, usually at the crown, or the junction of root and stem. 2. The *Agrobacterium* contains a Ti plasmid in addition to the much larger bacterial chromosome. The Ti plasmid has a segment (the T-DNA, red) that promotes tumor formation in infected plants. 3. The bacterium contributes its Ti plasmid to the plant cell, and the T-DNA from the Ti plasmid integrates into the plant's chromosomal DNA. 4. The genes in the T-DNA direct the formation of a crown gall, which nourishes the invading bacteria. **(b)** Photograph of a crown gall tumor generated by cutting off the top of a tobacco plant and inoculating with *Agrobacterium.* This crown gall tumor is a teratoma, which generates normal as well as tumorous tissues springing from the tumor.
(*Source:* (*b*) Dr. Robert Turgeon and Dr. B. Gillian Turgeon, Cornell University.)

The Ti plasmid inhabits the bacterium *Agrobacterium tumefaciens,* which causes tumors called **crown galls** (Figure 4.23) in dicotyledonous plants. When this bacterium infects a plant, it transfers its Ti plasmid to the host cells, whereupon the T-DNA integrates into the plant DNA, causing the abnormal proliferation of plant cells that gives rise to a crown gall. This is advantageous for the invading

bacterium, because the T-DNA has genes directing the synthesis of unusual organic acids called **opines.** These opines are worthless to the plant, but the bacterium has enzymes that can break down opines so they can serve as an exclusive energy source for the bacterium.

The T-DNA genes coding for the enzymes that make opines (e.g., mannopine synthetase) have strong promoters.

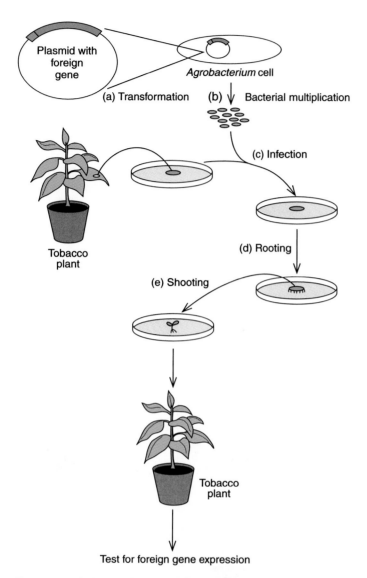

Figure 4.24   Using a T-DNA plasmid to introduce a gene into tobacco plants. (a) A plasmid is constructed with a foreign gene (red) under the control of the mannopine synthetase promoter (blue). This plasmid is used to transform *Agrobacterium* cells. (b) The transformed bacterial cells divide repeatedly. (c) A disk of tobacco leaf tissue is removed and incubated in nutrient medium, along with the transformed *Agrobacterium* cells. These cells infect the tobacco tissue, transferring the plasmid bearing the cloned foreign gene, which integrates into the plant genome. (d) The disk of tobacco tissue sends out roots into the surrounding medium. (e) One of these roots is transplanted to another kind of medium, where it forms a shoot. This plantlet grows into a transgenic tobacco plant that can be tested for expression of the transplanted gene. (*Source:* (a) M. Dell-Chilton, "A Vector for Introducing New Genes into Plants." Copyright 1983 Scientific American, Inc. Reprinted by permission.)

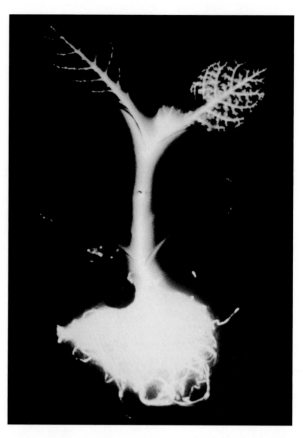

Figure 4.25   A tobacco plant that has been transformed with the firefly luciferase gene. The yellow light emitted by the plant indicates the activity of luciferase. Note the concentration of the enzyme in the roots and stem. (*Source:* Ow et al. Transient and stable expression of the firefly luciferase gene in plant cells and transgenic plants. *Science* 234 (14 Nov 1986) p. 858, f. 5. © American Association for the Advancement of Science (AAAS).)

Plant molecular biologists take advantage of them by putting T-DNA into small plasmids, then placing foreign genes under the control of one of these promoters. Figure 4.24 outlines the process used to transfer a foreign gene to a tobacco plant, producing a **transgenic plant.** One punches out a small disk (7 mm or so in diameter) from a tobacco leaf and places it in a dish with nutrient medium. Under these conditions, tobacco tissue will grow around the edge of the disk. Next, one adds *Agrobacterium* cells containing the foreign gene cloned into a Ti plasmid; these bacteria infect the growing tobacco cells and introduce the cloned gene.

When the tobacco tissue grows roots around the edge, those roots are transplanted to medium that encourages shoots to form. These plantlets give rise to full-sized tobacco plants whose cells contain the foreign gene. This gene can confer new properties on the plant, such as pesticide resistance, drought resistance, or disease resistance.

Perhaps the most celebrated success so far in plant genetic engineering has been the development of the "Flavr Savr" tomato. Calgene geneticists provided this plant with an antisense copy of a gene that contributes to fruit softening during ripening. The RNA product of this antisense gene is complementary to the normal mRNA, so it hybridizes to the mRNA and blocks expression of the gene. This allows tomatoes to ripen without softening as much, so they can ripen naturally on the vine instead of being picked green and ripened artificially.

Other plant molecular biologists have made additional strides, including the following: (1) conferring pesticide resistance on petunias and tobacco plants; (2) conferring virus resistance on tobacco plants by inserting a gene for the viral coat protein; and (3) inserting the gene for firefly luciferase into tobacco plants—this experiment has no practical value, but it does have the arresting effect of making the plant glow in the dark (Figure 4.25).

> **SUMMARY** Molecular biologists can transfer cloned genes to plants, creating transgenic organisms with altered characteristics, using a plant vector such as the Ti plasmid.

SUMMARY

To clone a gene, we must insert it into a vector that can carry the gene into a host cell and ensure that it will replicate there. The insertion is usually carried out by cutting the vector and the DNA to be inserted with the same restriction endonucleases to endow them with the same "sticky ends." Vectors for cloning in bacteria come in two major types: plasmids and phages.

Among the plasmid cloning vectors are pBR322 and the pUC plasmids. pBR322 has two antibiotic resistance genes and a variety of unique restriction sites into which one can introduce foreign DNA. Most of these sites interrupt one of the antibiotic resistance genes, making screening straightforward. Screening is even easier with the pUC plasmids and pBS phagemids. These vectors have an ampicillin resistance gene and a multiple cloning site that interrupts a partial β-galactosidase gene whose product is easily detected with a color test. The desired clones are ampicillin-resistant and do not make active β-galactosidase.

Two kinds of phages have been especially popular as cloning vectors. The first is λ, which has had certain nonessential genes removed to make room for inserts. In some of these engineered phages, inserts up to 20 kb in length can be accommodated. Cosmids, a cross between phage and plasmid vectors, can accept inserts as large as 50 kb. This makes these vectors very useful for building genomic libraries. The second major class of phage vectors is the M13 phages. These vectors have the convenience of a multiple cloning region and the further advantage of producing single-stranded recombinant DNA, which can be used for DNA sequencing and for site-directed mutagenesis. Plasmids called phagemids have an origin of replication for a single-stranded DNA phage, so they can produce single-stranded copies of themselves.

Expression vectors are designed to yield the protein product of a cloned gene, usually in the greatest amount possible. To optimize expression, bacterial expression vectors provide strong bacterial promoters and bacterial ribosome binding sites that would be missing from cloned eukaryotic genes. Most cloning vectors are inducible, to avoid premature overproduction of a foreign product that could poison the bacterial host cells. Expression vectors frequently produce fusion proteins, which can often be isolated quickly and easily. Eukaryotic expression systems have the advantages that the protein products are usually soluble, and these products are modified in a eukaryotic manner.

Cloned genes can also be transferred to plants, using a plant vector such as the Ti plasmid. This procedure can alter the plants' characteristics.

REVIEW QUESTIONS

1. Consulting Table 4.1, determine the length and the nature (5′ or 3′) of the overhang (if any) created by the following restriction endonucleases:
   a. *Alu*I
   b. *Bgl*II
   c. *Cla*I
   d. *Kpn*I
   e. *Mbo*I
   f. *Pvu*I
   g. *Not*I

2. Why do we need to attach DNAs to vectors to clone them?

3. Describe the process of cloning a DNA fragment into the *Pst*I site of the vector pBR322. How would you screen for clones that contain an insert?

4. Describe the process of cloning a DNA fragment into the *Bam*HI and *Pst*I sites of the vector pUC18. How would you screen for clones that contain an insert?

5. Describe the process of cloning a DNA fragment into the *Eco*RI site of the Charon 4 vector.

6. You want to clone a 1-kb cDNA. Which vectors discussed in this chapter would be appropriate to use? Which would be inappropriate? Why?

7. You want to make a genomic library with DNA fragments averaging about 45 kb in length. Which vector discussed in this chapter would be most appropriate to use? Why?

8. You want to make a library with DNA fragments averaging over 100 kb in length. Which vector discussed in this chapter would be most appropriate to use? Why?

9. You have constructed a cDNA library in a λ phagemid vector. Describe how you would screen the library for a particular gene of interest. Describe methods using oligonucleotide and antibody probes.

10. How would you obtain single-stranded cloned DNAs from an M13 phage vector? From a phagemid vector?

11. Diagram a method for creating a cDNA library.

12. Diagram the process of nick translation.

13. Outline the polymerase chain reaction (PCR) method for amplifying a given stretch of DNA.

14. What is the difference between reverse transcriptase PCR (RT-PCR) and standard PCR? For what purpose would you use RT-PCR?

15. Describe the use of a vector that produces fusion proteins with oligohistidine at one end. Show the protein purification scheme to illustrate the advantage of the oligohistidine tag.

16. What is the difference between a λ insertion vector and a λ replacement vector? What is the advantage of each?

17. Describe the use of a baculovirus system for expressing a cloned gene. What advantages over a prokaryotic expression system does the baculovirus system offer?

18. What kind of vector would you use to insert a transgene into a plant such as tobacco? Diagram the process you would use.

SUGGESTED READINGS

Capecchi, N.R. 1994. Targeted gene replacement. *Scientific American* 270 (March):52–59.

Chilton, M.-D. 1983. A vector for introducing new genes into plants. *Scientific American* 248 (June):50–59.

Cohen, S. 1975. The manipulation of genes. *Scientific American* 233 (July):24–33.

Cohen, S., A. Chang, H. Boyer, and R. Helling. 1973. Construction of biologically functional bacterial plasmids in vitro. *Proceedings of the National Academy of Sciences* 70:3240–44.

Gasser, C.S., and R.T. Fraley. 1992. Transgenic crops. *Scientific American* 266 (June):62–69.

Gilbert, W., and L. Villa-Komaroff. 1980. Useful proteins from recombinant bacteria. *Scientific American* 242 (April):74–94.

Nathans, D., and H.O. Smith. 1975. Restriction endonucleases in the analysis and restructuring of DNA molecules. *Annual Review of Biochemistry* 44:273–93.

Sambrook, J., et al. 2001. *Molecular cloning: A laboratory manual,* 3rd ed. Plainview, NY: Cold Spring Harbor Laboratory Press.

Watson, J.D., J. Tooze, and D.T. Kurtz. 1983. *Recombinant DNA: A Short Course.* New York: W.H. Freeman.

CHAPTER 5

Molecular Tools for Studying Genes and Gene Activity

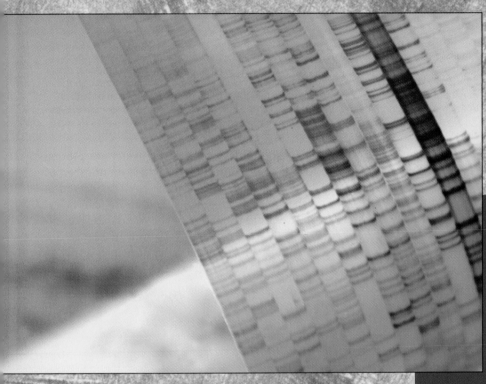

Autoradiograph of DNA sequencing gel. © *Peter Dazeley/Tony Stone Images.*

In this chapter we will describe the most popular techniques that molecular biologists use to investigate the structure and function of genes. Most of these start with cloned genes. Many use gel electrophoresis. Many also use labeled tracers, and many rely on nucleic acid hybridization. We have already examined gene cloning techniques. Let us continue by briefly considering three other mainstays of molecular biology research: molecular separations including gel electrophoresis, labeled tracers, and hybridization.

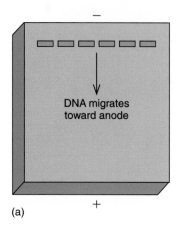

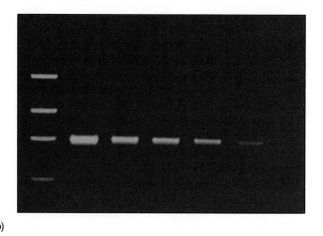

Figure 5.1   DNA gel electrophoresis. (a) Scheme of the method: This is a horizontal gel made of agarose (a substance derived from seaweed, and the main component of agar). The agarose melts at high temperature, then gels as it cools. A "comb" is inserted into the molten agarose; after the gel cools, the comb is removed, leaving slots, or wells (orange). The DNA is then placed in the wells, and an electric current is run through the gel. Because the DNA is an acid, it is negatively charged at neutral pH and electrophoreses, or migrates, toward the positive pole, or anode. **(b)** A photograph of a gel after electrophoresis showing the DNA fragments as bright bands. DNA binds to a dye that fluoresces orange under ultraviolet light, but the bands appear pink in this photograph.

5.1   Molecular Separations

Gel Electrophoresis

It is very often necessary in molecular biology research to separate proteins or nucleic acids from each other. For example, we may need to purify a particular enzyme from a crude cellular extract in order to use it or to study its properties. Or we may want to purify a particular RNA or DNA molecule that has been produced or modified in an enzymatic reaction, or we may simply want to separate a series of RNAs or DNA fragments from each other. We will describe here some of the most common techniques used in such molecular separations, including gel electrophoresis of both nucleic acids and proteins, ion exchange chromatography, and gel filtration chromatography.

Gel electrophoresis can be used to separate different nucleic acid or protein species. We will begin by considering DNA gel electrophoresis. In this technique one makes a gel with slots in it, as shown in Figure 5.1. One puts a little DNA in a slot and runs an electric current through the gel. The DNA is negatively charged because of the phosphates in its backbone, so it migrates toward the positive pole (the *anode*) at the end of the gel. The secret of the gel's ability to separate DNAs of different sizes lies in friction. Small DNA molecules experience little frictional drag from solvent and gel molecules, so they migrate rapidly. Large DNAs, by contrast, encounter correspondingly more friction, so their mobility is lower. The result is that the electric current will distribute the DNA fragments according to their sizes: the largest near the top, the smallest near the bottom. Finally, the DNA is stained with a fluorescent dye and the gel is examined under ultraviolet illumination. Figure 5.2 depicts the results of such analysis on fragments of phage DNA of

known size. The mobilities of these fragments are plotted versus the log of their molecular weights (or number of base pairs). Any unknown DNA can be electrophoresed in parallel with the standard fragments, and its size can be estimated if it falls within the range of the standards. For example, a DNA with a mobility of 20 mm in Figure 5.2 would contain about 910 bp. The same principles apply to electrophoresing RNAs of various sizes.

Solved Problem

Problem 1

Following is a graph showing the results of a gel electrophoresis experiment on double-stranded DNA fragments having sizes between 0.2 and 1.2 kb.

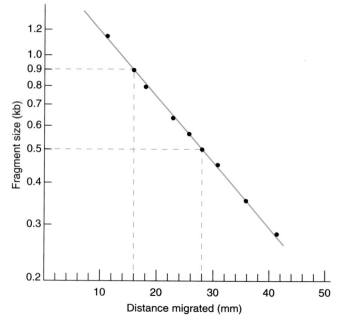

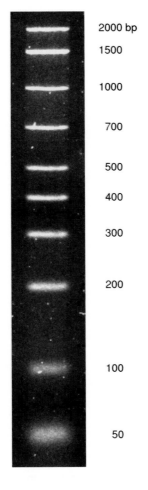

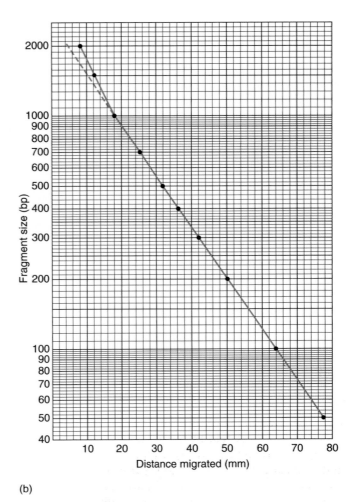

(a)

(b)

Figure 5.2   Analysis of DNA fragment size by gel electrophoresis. (a) Photograph of a stained gel of commercially prepared fragments after electrophoresis. The bands that would be orange in a color photo show up white in a black-and-white photo taken with an orange filter. The sizes of the fragments (in bp) are given at right. **(b)** Graph of the migration of the DNA fragments versus their sizes in base pairs. The vertical axis is logarithmic rather than linear, because the electrophoretic mobility (migration rate) of a DNA fragment is proportional to the log of its size. However, notice the departure from this proportionality at large fragment sizes, represented by the difference between the solid line (actual results) and the dashed line (theoretical behavior). This suggests the limitations of conventional electrophoresis for measuring the sizes of very large DNAs.

On the basis of this graph, answer the following questions:

a. What is the size of a fragment that migrated 16 mm in this experiment?

b. How far would a 0.5-kb fragment migrate in this experiment?

Solution

a. Draw a vertical dashed line from the 16-mm point on the *x* axis up to the experimental line. From the point where that vertical line intersects the experimental line, draw a horizontal dashed line to the *y* axis. This line intersects the *y* axis at the 0.9-kb point. This shows that fragments that migrate 16 mm in this experiment are 0.9 kb (or 900 bp) long.

b. Draw a horizontal dashed line from the 0.5-kb point on the *y* axis across to the experimental line. From the point where that horizontal line intersects the

experimental line, draw a vertical dashed line down to the *x* axis. This line intersects the *x* axis at the 28-mm point. This shows that 0.5-kb fragments migrate 28 mm in this experiment. ■

Determining the size of a large DNA by gel electrophoresis requires special techniques. One reason is that the relationship between the log of a DNA's size and its electrophoretic mobility deviates strongly from linearity if the DNA is very large. A hint of this deviation is apparent at the top left of Figure 5.2*b*. Another reason is that double-stranded DNA is a relatively rigid rod—very long and thin. The longer it is, the more fragile it is. In fact, large DNAs break very easily; even seemingly mild manipulations, like swirling in a beaker or pipetting, create shearing forces sufficient to fracture them. To visualize this, think of DNA as a piece of uncooked spaghetti. If it is short—say a centimeter or two—you can treat it roughly without harming it, but if it is long, breakage becomes almost inevitable.

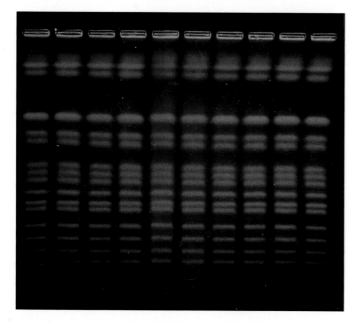

Figure 5.3   Pulsed-field gel electrophoresis of yeast chromosomes. Identical samples of yeast chromosomes were electrophoresed in 10 parallel lanes and stained with ethidium bromide. The bands represent chromosomes having sizes ranging from 0.2 Mb (at bottom) to 2.2 Mb (at top). Original gel is about 13 cm wide by 12.5 cm long.

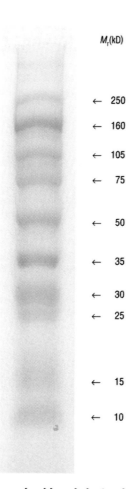

Figure 5.4   SDS-polyacrylamide gel electrophoresis. Polypeptides of the molecular masses shown at right were coupled to dyes and subjected to SDS-PAGE. The dyes allow us to see each polypeptide during and after electrophoresis. (*Source:* Courtesy of Amersham Pharmacia Biotech.)

In spite of these difficulties, molecular biologists have developed a kind of gel electrophoresis that can separate DNA molecules up to several million base pairs (**megabases, Mb**) long and maintain a relatively linear relationship between the log of their sizes and their mobilities. Instead of a constant current through the gel, this method uses pulses of current, with relatively long pulses in the forward direction and shorter pulses in the opposite, or even sideways, direction. This **pulsed-field gel electrophoresis (PFGE)** is valuable for measuring the sizes of DNAs even as large as some of the chromosomes found in yeast. Figure 5.3 presents the results of pulsed-field gel electrophoresis on yeast chromosomes. The 16 visible bands represent chromosomes containing 0.2–2.2 Mb.

Electrophoresis is also often applied to proteins, in which case the gel is usually made of polyacrylamide. We therefore call it **polyacrylamide gel electrophoresis, or PAGE.** To determine the polypeptide makeup of an enzyme, the experimenter must treat the enzyme so that the polypeptides, or enzyme subunits, will electrophorese independently. This is usually done by treating the enzyme with a detergent (**sodium dodecyl sulfate, or SDS**) to **denature** the subunits so they no longer bind to one another. The SDS has two added advantages: (1) It coats all the polypeptides with negative charges, so they all electrophorese toward the anode. (2) It masks the natural charges of the subunits, so they

all electrophorese according to their molecular masses and not by their native charges. Small polypeptides fit easily through the pores in the gel, so they migrate rapidly. Larger polypeptides migrate more slowly. Researchers also usually employ a reducing agent to break covalent bonds between subunits.

Figure 5.4 shows the results of **SDS-PAGE** on a series of polypeptides, each of which is attached to a dye so they can be seen during electrophoresis ordinarily, the polypeptides would all be stained after electrophoresis with a dye such as Coomassie Blue.

SUMMARY DNAs, RNAs, and proteins of various masses can be separated by gel electrophoresis. The most common gel used in nucleic acid electrophoresis is agarose, but polyacrylamide is usually used in protein electrophoresis. SDS-PAGE is used to separate polypeptides according to their masses.

Two-Dimensional Gel Electrophoresis

SDS-PAGE gives very good resolution of polypeptides, but sometimes a mixture of polypeptides is so complex that we need an even better method to resolve them all. For example, we may want to separate all of the thousands of polypeptides present at a given time in a given cell type. This is very commonly done now as part of a subfield of molecular biology known as proteomics, which we will discuss in Chapter 24.

To improve on the resolving power of a one-dimensional SDS-PAGE procedure, molecular biologists have developed two-dimensional methods. In one simple method, described in Chapter 19, one can simply run nondenaturing gel electrophoresis (no SDS) in one dimension at one pH and one polyacrylamide gel concentration, then in a second dimension at a second pH and a second polyacrylamide concentration. Proteins will electrophorese at different rates at different pH values because their net charges change with pH. They will also behave differently at different polyacrylamide concentrations according to their sizes. But individual polypeptides cannot be analyzed by this method because the lack of detergent makes it impossible to separate the polypeptides that make up a complex protein.

An even more powerful method is commonly known as **two-dimensional gel electrophoresis,** even though it involves a bit more than the name implies. In the first step, the mixture of proteins is electrophoresed through a narrow tube gel containing molecules called ampholytes that set up a pH gradient from one end of the tube to the other. A negatively charged molecule will electrophorese toward the anode until it reaches its **isoelectric point,** the pH at which it has no net charge. Without net charge, it is no longer drawn toward the anode, or the cathode, for that matter, so it stops. This step is called **isoelectric focusing** because it focuses proteins at their isoelectric points in the gel.

In the second step, the gel is removed from the tube and placed at the top of a slab gel for ordinary SDS-PAGE. Now the proteins that have been partially resolved by isoelectric focusing are further resolved according to their sizes by SDS-PAGE. Figure 5.5 presents two-dimensional gel electrophoresis separations of *E. coli* proteins grown in the presence and absence of benzoic acid. Proteins from the cells grown without benzoic acid were stained with the red fluorescent dye Cy3, and proteins from the cells grown with benzoic acid were stained with the blue fluorescent dye Cy5. Two-dimensional gel electrophoresis of these two sets of proteins, separately and together allows us to see which proteins are prevalent in the presence or absence of benzoic acid, and which are prevalent under both conditions.

SUMMARY High-resolution separation of polypeptides can be achieved by two-dimensional gel electrophoresis, which uses isoelectric focusing in the first dimension and SDS-PAGE in the second.

Ion-Exchange Chromatography

Chromatography is an old term that originally referred to the pattern one sees after separating colored substances on paper. Nowadays, many different types of chromatography exist for separating biological substances. **Ion-exchange chromatography** uses a resin to separate substances according to their charges. For example, DEAE-Sephadex chromatography uses an ion-exchange resin that contains positively charged diethylaminoethyl (DEAE) groups. These positive charges attract negatively charged substances, including proteins. The greater the negative charge, the tighter the binding.

We will see an example of DEAE-Sephadex chromatography in Chapter 10 in which the experimenters separated three forms of an enzyme called RNA polymerase. They made a slurry of DEAE-Sephadex and poured it into a column. After the resin had packed down, they loaded the sample, a crude cellular extract containing the RNA polymerases. Finally, they **eluted,** or removed, the substances that had bound to the resin in the column by passing a solution of gradually increasing ionic strength through the column. The purpose of this salt gradient was to use the negative ions in the salt solution to compete with the proteins for binding sites on the resin, thus removing the proteins one by one. This is why we call it ion-exchange chromatography. As the ionic strength of the elution buffer increases, fractions are collected and then assayed (tested) to determine how much of the substance of interest they contain. If the substance is an enzyme, the fractions are assayed for that particular enzyme activity. It is also useful to measure the ionic strength of each fraction to determine what salt concentration is necessary to elute each of the enzymes of interest.

One can also use a negatively charged resin to separate positively charged substances, including proteins. For example, phosphocellulose is commonly used to separate proteins by cation–exchange chromatography. Note that it is not essential for a protein to have a net positive charge to bind to a cation–exchange resin like phosphocellulose. Most proteins have a net negative charge, yet they can still bind to a cation exchange resin if they have a significant center of positive charge. Figure 5.6 depicts the results of a hypothetical ion-exchange chromatography experiment in which two forms of an enzyme are separated.

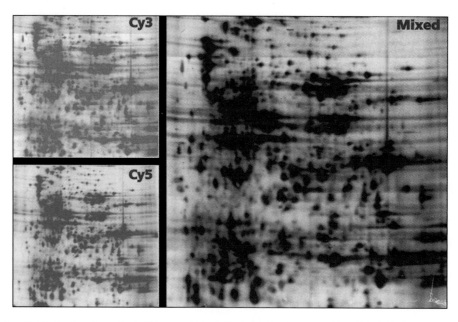

Figure 5.5   Two-dimensional gel electrophoresis. In this experiment, the investigators grew *E. coli* cells in the presence or absence of benzoic acid. Then they stained a lysate of the cells grown in the absence of benzoic acid with the red fluorescent dye Cy3, so the proteins from that lysate would fluoresce red. They stained a lysate of the cells grown in the presence of benzoic acid with the blue fluorescent dye Cy5, so those proteins would fluoresce blue. Finally, they performed two-dimensional gel electrophoresis on **(a)** the proteins from cells grown in the absence of benzoic acid, **(b)** on the proteins grown in the presence of benzoic acid, and **(c)** on a mixture of the two sets of proteins. In panel **(c)**, the proteins that accumulate only in the absence of benzoic acid fluoresce red, those that accumulate only in the presence of benzoic acid fluoresce blue, and those that accumulate under both conditions fluoresce both red and blue, and so appear purple. (*Source:* Courtesy of Amersham Pharmacia Biotech.)

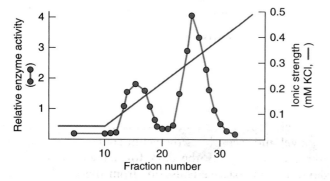

Figure 5.6   Ion-exchange chromatography. Begin by loading a cell extract containing two different forms of an enzyme onto an ion-exchange column. Then pass a buffer of increasing ionic strength through the column and collect fractions (32 fractions in this case). Assay each fraction for enzyme activity (red) and ionic strength (blue), and plot the data as shown. The two forms of the enzyme are clearly separated by this procedure.

> **SUMMARY** Ion-exchange chromatography can be used to separate substances, including proteins, according to their charges. Positively charged resins like DEAE-Sephadex are used for anion-exchange chromatography, and negatively charged resins like phosphocellulose are used for cation-exchange chromatography.

Gel Filtration Chromatography

Standard biochemical separations of proteins usually require more than one step and, because valuable protein is lost at each step, it is important to minimize the number of these steps. One way to do this is to design a strategy that enables each step to take advantage of a different property of the protein of interest. Thus, if anion-exchange chromatography is the first step and cation-exchange chromatography is the second, a third step that separates proteins on some other basis besides charge is needed. Protein size is an obvious next choice.

Gel filtration chromatography is one method that separates molecules based on their physical dimensions. Gel filtration resins are porous beads that can be likened to "whiffle balls," hollow plastic balls with holes in them. Imagine a column filled with tiny whiffle balls. When we pass a solution containing different size molecules through this column, the small molecules will easily enter the holes in the whiffle balls (the pores in the beads) and therefore flow through the column slowly. On the other hand, large molecules will not be able to enter any of the beads and will flow more quickly through the column. They emerge with the so-called **void volume**—the volume of buffer surrounding the beads, but not included in the beads. Intermediate-size molecules will enter some beads and not others and so will have an intermediate mobility.

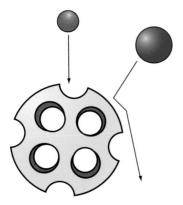

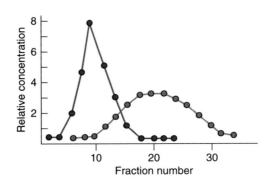

(b)

(a)

Figure 5.7 Gel filtration chromatography. (a) Principle of the method. A resin bead is schematically represented as a "Whiffle ball" (yellow). Large molecules (blue) cannot fit into the beads, so they are confined to the relatively small buffer volume outside the beads. Thus, they emerge quickly from the column. Small molecules (red), by contrast, can fit into the beads and so have a large buffer volume available to them. Accordingly, they take a longer time to emerge from the column. **(b)** Experimental results. Add a mixture of large and small molecules from panel **(a)** to the column, and elute them by passing buffer through the column. Collect fractions and assay each for concentration of the large (blue) and small (red) molecules. As expected, the large molecules emerge earlier than the small ones.

Thus, large molecules will emerge first from the column, and small molecules will emerge last. Many different resins with different size pores are available for separating different size molecules. Figure 5.7 illustrates this method.

> **SUMMARY** Gel filtration chromatography uses columns filled with porous resins that let in smaller substances, but exclude larger ones. Thus, the smaller substances are slowed in their journey through the column, but larger substances travel relatively rapidly through the column.

5.2 Labeled Tracers

Until recently, "labeled" has been virtually synonymous with "radioactive" because radioactive tracers have been available for decades, and they are easy to detect. Radioactive tracers allow vanishingly small quantities of substances to be detected. This is important in molecular biology because the substances we are trying to detect in a typical experiment are present in very tiny amounts. Let us assume, for example, that we are attempting to measure the appearance of an RNA product in a transcription reaction. We may have to detect RNA quantities of less than a picogram (pg; only one trillionth of a gram, or 10^{-12} g). Direct measurement of such tiny quantities by ultraviolet light absorption or by staining with dyes is not possible because of the limited sensitivities of these methods. On the other hand, if the RNA is radioactive we can

measure small amounts of it easily because of the great sensitivity of the equipment used to detect radioactivity. Let us now consider the favorite techniques molecular biologists use to detect radioactive tracers: autoradiography, phosphorimaging, and liquid scintillation counting.

Autoradiography

Autoradiography is a means of detecting radioactive compounds with a photographic emulsion. The form of emulsion favored by molecular biologists is a piece of x-ray film. Figure 5.8 presents an example in which the investigator electrophoreses some radioactive DNA fragments on a gel and then places the gel in contact with the x-ray film and leaves it in the dark for a few hours, or even days. The radioactive emissions from the bands of DNA expose the film, just as visible light would. Thus, when the film is developed, dark bands appear, corresponding to the DNA bands on the gel. In effect, the DNA bands take a picture of themselves, which is why we call this technique *auto*radiography.

To enhance the sensitivity of autoradiography, one can use an **intensifying screen.** This is a screen coated with a compound that fluoresces when it is excited by β-rays at low temperature. (β-rays are the radioactive emissions from the common radioisotopes used in molecular biology: ^{3}H, ^{14}C, ^{35}S, and ^{32}P. They are high-energy electrons.) Thus, one can put a radioactive gel (or other medium) on one side of a photographic film and the intensifying screen on the other. Some β-rays expose the film directly, but others pass right through the film and would be lost without the screen. When these high-energy

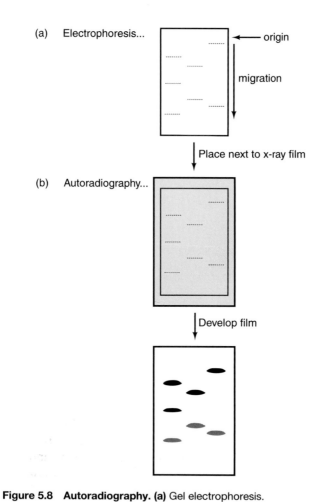

(a)   Electrophoresis...

migration

origin

Place next to x-ray film

(b)   Autoradiography...

Develop film

Figure 5.8   Autoradiography. (a) Gel electrophoresis. Electrophorese radioactive DNA fragments in three parallel lanes on a gel, either agarose or polyacrylamide, depending on the sizes of the fragments. At this point the DNA bands are invisible, but their positions are indicated here with dotted lines. **(b)** Autoradiography. Place a piece of x-ray film in contact with the gel and leave it for several hours, or even days if the DNA fragments are only weakly radioactive. Finally, develop the film to see where the radioactivity has exposed the film. This shows where the DNA bands are on the gel. In this case, the large, slowly migrating bands are the most radioactive, so the bands on the autoradiograph that correspond to them are the darkest.

electrons strike the screen, they cause fluorescence, which is detected by the film.

What if the goal is to measure the exact amount of radioactivity in a fragment of DNA? One can get a rough estimate by looking at the intensity of a band on an autoradiograph, and an even better estimate by scanning the autoradiograph with a **densitometer**. This instrument passes a beam of light through a sample—an autoradiograph in this case—and measures the absorbance of that light by the sample. If the band is very dark, it will absorb most of the light, and the densitometer records a large peak of absorbance (Figure 5.9). If the band is faint, most

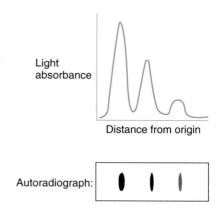

Light absorbance

Distance from origin

Autoradiograph:

Figure 5.9   Densitometry. An autoradiograph is pictured beneath a densitometer scan of the same film. Notice that the areas under the three peaks of the scan are proportional to the darkness of the corresponding bands on the autoradiograph.

of the light passes through, and the densitometer records only a minor peak of absorbance. By measuring the area under each peak one can get an estimate of the radioactivity in each band. This is still an indirect measure of radioactivity, however. To get a really accurate reading of the radioactivity in each band, one can scan the gel with a **phosphorimager**, or subject the DNA to liquid scintillation counting.

Phosphorimaging

The technique of **phosphorimaging** has several advantages over standard autoradiography, but the most important is that it is much more accurate in quantifying the amount of radioactivity in a substance. This is because its response to radioactivity is far more linear than that of an x-ray film. With standard autoradiography, a band with 50,000 radioactive **disintegrations per minute (dpm)** may look no darker than one with 10,000 dpm because the emulsion in the film is already saturated at 10,000 dpm. But the phosphorimager detects radioactive emissions and analyzes them electronically, so the difference between 10,000 dpm and 50,000 dpm would be obvious. Here is how this technique works: One starts with a radioactive sample—a blot with DNA bands that have hybridized with a labeled probe, for example. This sample is placed in contact with a phosphorimager plate, which absorbs β-rays. These rays excite molecules on the plate, and these molecules remain in an excited state until the phosphorimager scans the plate with a laser. At that point, the β-ray energy trapped by the plate is released and monitored by a computerized detector. The computer converts the energy it detects to an image such as the one in Figure 5.10. This is a false color image, in which

Figure 5.10　False color phosphorimager scan of an RNA blot.
After hybridizing a radioactive probe to an RNA blot and washing away unhybridized probe, the blot was exposed to a phosphorimager plate. The plate collected energy from β-rays from the radioactive probe bound to the RNA bands, then gave up this energy when scanned with a laser. A computer converted this energy into an image in which the colors correspond to radiation intensity according to the following color scale: yellow (lowest) < purple < magenta < light blue < green < dark blue < black (highest). (*Source:* © Jay Freis/Image Bank.)

the different colors represent different degrees of radioactivity, from the lowest (yellow) to the highest (black).

Liquid Scintillation Counting

Liquid scintillation counting uses the radioactive emissions from a sample to create photons of visible light that a photomultiplier tube can detect. To do this, one places the radioactive sample (a band cut out of a gel, for example), into a vial with **scintillation fluid.** This liquid contains a **fluor,** a compound that fluoresces when it is bombarded with radioactivity. In effect, it converts the invisible radioactivity into visible light. A liquid scintillation counter is an instrument that lowers the vial into a dark chamber with a photomultiplier tube. There, the tube detects the light resulting from the radioactive emissions exciting the fluor. The instrument counts these bursts of light, or **scintillations,** and records them as **counts per minute (cpm).** This is not the same as disintegrations per minute because the scintillation counter is not 100% efficient. One common radioisotope used by molecular biologists is ^{32}P. The β-rays emitted by this isotope are so energetic that they create photons even without a fluor, so a liquid scintillation counter can count them directly, though at a lower efficiency than with scintillation fluid.

SUMMARY Detection of the tiny quantities of substances used in molecular biology experiments generally requires the use of labeled tracers. If the tracer is radioactive one can detect it by autoradiography, using x-ray film or a phosphorimager, or by liquid scintillation counting.

Nonradioactive Tracers

As we pointed out earlier in this section, the enormous advantage of radioactive tracers is their sensitivity, but now nonradioactive tracers rival the sensitivity of their radioactive forebears. This can be a significant advantage because radioactive substances pose a potential health hazard and must be handled very carefully. Furthermore, radioactive tracers create radioactive waste, and disposal of such waste is increasingly difficult and expensive. How can a nonradioactive tracer compete with the sensitivity of a radioactive one? The answer is, by using the multiplier effect of an enzyme. An enzyme is coupled to a probe that detects the molecule of interest, so the enzyme will produce many molecules of product, thus amplifying the signal. This works especially well if the product of the enzyme is chemiluminescent (light-emitting, like the tail of a firefly), because each molecule emits many photons, amplifying the signal again. Figure 5.11 shows the principle behind one such tracer method. The light can be detected by autoradiography with x-ray film, or by a phosphorimager.

To avoid the expense of a phosphorimager or x-ray film, one can use enzyme substrates that change color instead of becoming chemiluminescent. These **chromogenic** substrates produce colored bands corresponding to the location of the enzyme and, therefore, to the location of the molecule of interest. The intensity of the color is directly related to the amount of that molecule, so this is also a quantitative method.

SUMMARY Some very sensitive nonradioactive labeled tracers are now available. Those that employ chemiluminescence can be detected by autoradiography or by phosphorimaging, just as if they were radioactive. Those that produce colored products can be detected directly, by observing the appearance of colored spots.

5.3　Using Nucleic Acid Hybridization

The phenomenon of hybridization—the ability of one single-stranded nucleic acid to form a double helix with

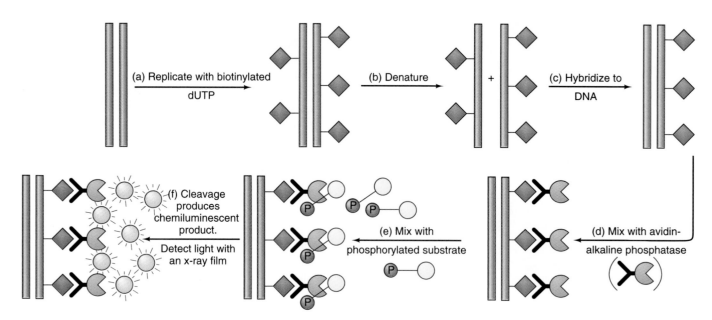

Figure 5.11   Detecting nucleic acids with a nonradioactive probe. This sort of technique is usually indirect; detecting a nucleic acid of interest by hybridization to a labeled probe that can in turn be detected by virtue of its ability to produce a colored or light-emitting substance. In this example, the following steps are executed. **(a)** Replicate the probe DNA in the presence of dUTP that is tagged with the vitamin biotin. This generates biotinylated probe DNA. **(b)** Denature this probe and **(c)** hybridize it to the DNA to be detected (pink). **(d)** Mix the hybrids with a bifunctional reagent containing both avidin and the enzyme alkaline phosphatase (green). The avidin binds tightly and specifically to the biotin in the probe DNA. **(e)** Add a phosphorylated compound that will become chemiluminescent as soon as its phosphate group is removed. The alkaline phosphatase enzymes attached to the probe cleave the phosphates from these substrate molecules, rendering them chemiluminescent (light-emitting). **(f)** Detect the light emitted from the chemiluminescent substrate with an x-ray film.

another single strand of complementary base sequence—is one of the backbones of modern molecular biology. We have already encountered plaque and colony hybridization in Chapter 4. Here we will illustrate several further examples of hybridization techniques.

Southern Blots: Identifying Specific DNA Fragments

Many eukaryotic genes are parts of families of closely related genes. How would one determine the number of family members in a particular gene family? If a member of that gene family—even a partial cDNA—has been cloned, one can estimate this number.

One begins by using a restriction enzyme to cut genomic DNA isolated from the organism. It is best to use a restriction enzyme such as *Eco*RI or *Hin*dIII that recognizes a 6-bp cutting site. These enzymes will produce thousands of fragments of genomic DNA, with an average size of about 4000 bp. Next, these fragments are electrophoresed on an agarose gel (Figure 5.12). The result, if the bands are visualized by staining, will be a blurred streak of thousands of bands, none distinguishable from the others (although Figure 5.12, for simplicity's sake shows just a few bands). Eventually, a labeled probe will be hybridized to these bands to see how many of them contain coding sequences for the gene of interest. First,

however, the bands are transferred to a medium on which hybridization is convenient.

Edward Southern was the pioneer of this technique; he transferred, or blotted, DNA fragments from an agarose gel to nitrocellulose by diffusion, as depicted in Figure 5.12. This process has been called **Southern blotting** ever since. Nowadays, blotting is frequently done by electrophoresing the DNA bands out of the gel and onto the blot. Figure 5.12 also illustrates this process. Before blotting, the DNA fragments are denatured with alkali so that the resulting single-stranded DNA can bind to the nitrocellulose, forming the Southern blot. Media superior to nitrocellulose are now available; some, such as Gene Screen, use nylon supports that are far more flexible than nitrocellulose. Next, the cloned DNA is labeled by adding DNA polymerase to it in the presence of labeled DNA precursors. Then this labeled probe is denatured and hybridized to the Southern blot. Wherever the probe encounters a complementary DNA sequence, it hybridizes, forming a labeled band corresponding to the fragment of DNA containing the gene of interest. Finally, these bands are visualized by autoradiography with x-ray film or by phosphorimaging.

If only one band is seen, the interpretation is relatively easy; probably only one gene has a sequence matching the cDNA probe. Alternatively, a gene (e.g., a histone or ribo-

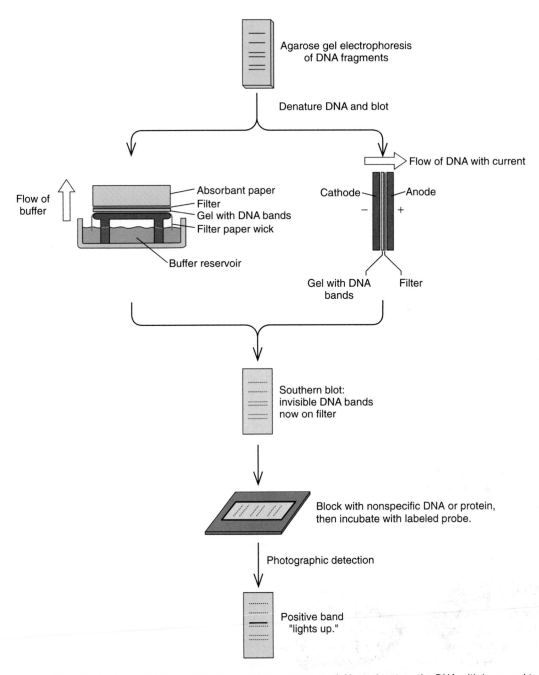

Figure 5.12   Southern blotting. First, electrophorese DNA fragments in an agarose gel. Next, denature the DNA with base and transfer the single-stranded DNA fragments from the gel (yellow) to a sheet of nitrocellulose (red) or similar material. One can do this in two ways: by diffusion, in which buffer passes through the gel, carrying the DNA with it (left), or by electrophoresis (right). Next, hybridize the blot to a labeled probe and detect the labeled bands by autoradiography or phosphorimaging.

somal RNA gene) could be repeated over and over again in tandem, with a single restriction site in each copy of the gene. This would yield a single very dark band. If multiple bands are seen, multiple genes are probably present, but it is difficult to tell exactly how many. One gene can give more than one band if it contains one or more cutting sites for the restriction enzyme used. One can minimize this problem by using a short probe, such as a 100–

200-bp restriction fragment of the cDNA, for example. Chances are, a restriction enzyme that cuts on average only every 4000 bp will not cut within the 100–200-bp region of the genes that hybridize to such a probe. If multiple bands are still obtained with a short probe, they probably represent a gene family whose members' sequences are similar or identical in the region that hybridizes to the probe.

SUMMARY Labeled DNA (or RNA) probes can be used to hybridize to DNAs of the same, or very similar, sequence on a Southern blot. The number of bands that hybridize to a short probe gives an estimate of the number of closely related genes in an organism.

DNA Fingerprinting and DNA Typing

Southern blots are not just a research tool. They are widely used in forensic laboratories to identify individuals who have left blood or other DNA-containing tissues at the scenes of crimes. Such **DNA typing** has its roots in a discovery by Alec Jeffreys and his colleagues in 1985. These workers were investigating a DNA fragment from the gene for a human blood protein, α-globin, when they discovered that this fragment contained a sequence of bases repeated several times. This kind of repeated DNA is called a **minisatellite**. More interestingly, they found similar minisatellite sequences in other places in the human genome, again repeated several times. This simple finding turned out to have far-reaching consequences, because individuals differ in the pattern of repeats of the basic sequence. In fact, they differ enough that two individuals have only a remote chance of having exactly the same pattern. That means that these patterns are like fingerprints; indeed, they are called **DNA fingerprints.**

A DNA fingerprint is really just a Southern blot. To make one, investigators first cut the DNA under study with a restriction enzyme such as *Hae*III. Jeffreys chose this enzyme because the repeated sequence he had found did not contain a *Hae*III recognition site. That means that *Hae*III will cut on either side of the minisatellite regions, but not inside, as shown in Figure 5.13a. In this case, the DNA has three sets of repeated regions, containing four, three, and two repeats, respectively. Thus, three different-size fragments bearing these repeated regions will be produced.

Next, the fragments are electrophoresed, denatured, and blotted. The blot is then probed with a labeled minisatellite DNA, and the labeled bands are detected with x-ray film, or by phosphorimaging. In this case, three labeled bands occur, so three dark bands will appear on the film (Figure 5.13d).

Real animals have a much more complex genome than the simple piece of DNA in this example, so they will have many more than three fragments that contain a minisatellite sequence that will react with the probe. Figure 5.14 shows an example of the DNA fingerprints of several unrelated people and a set of monozygotic twins. As we have already mentioned, this is such a complex pattern of fragments that the patterns for two individuals are extremely unlikely to be identical, unless they come from monozygotic twins. This complexity makes DNA fingerprinting a very powerful identification technique.

Forensic Uses of DNA Fingerprinting and DNA Typing

A valuable feature of DNA fingerprinting is the fact that, although almost all individuals have different patterns, parts of the pattern (sets of bands) are inherited in a Mendelian fashion. Thus, fingerprints can be used to establish parentage. An immigration case in England illustrates the power of this technique. A Ghanaian boy born in England had moved to Ghana to live with his father. When he wanted to return to England to be with his mother, British authorities questioned whether he was a son or a nephew of the woman. Information from blood group genes was equivocal, but DNA fingerprinting of the boy demonstrated that he was indeed her son.

In addition to testing parentage, DNA fingerprinting has the potential to identify criminals. This is because a person's DNA fingerprint is, in principle, unique, just like a traditional fingerprint. Thus, if a criminal leaves some of his cells (blood, semen, or hair, for example) at the scene of a crime, the DNA from these cells can identify him. As Figure 5.14 shows, however, DNA fingerprints are very complex. They contain dozens of bands, some of which smear together, which can make them hard to interpret.

To solve this problem, forensic scientists have developed probes that hybridize to a single DNA locus that varies from one individual to another, rather than to a whole set of DNA loci as in a classical DNA fingerprint. Each probe now gives much simpler patterns, containing only one or a few bands. This is an example of a restriction fragment length polymorphism (RFLP) disussed in detail in chapter 24. RFLPs occur because the pattern of restriction fragment sizes at a given locus varies from one person to another. Of course, each probe by itself is not as powerful an identification tool as a whole DNA fingerprint with its multitude of bands, but a panel of four or five probes can give enough different bands to be definitive. We sometimes still call such analysis DNA fingerprinting, but a better, more inclusive term is *DNA typing.*

One early, dramatic case of DNA typing involved a man who murdered a man and woman as they slept in a pickup truck, then about forty minutes later went back and raped the woman. This act not only compounded the crime, it also provided forensic scientists with the means to convict the perpetrator. They obtained DNA from the sperm cells he had left behind, typed it, and showed that the pattern matched that of the suspect's DNA. This evidence helped convince the jury to convict the defendant and recommend severe punishment. Figure 5.15 presents an example of DNA typing that was used to identify another rape suspect. The pattern from the suspect clearly matches that from the sperm DNA. This is the result from only one probe. The others also gave patterns that matched the sperm DNA.

One advantage of DNA typing is its extreme sensitivity. Only a few drops of blood or semen are sufficient to

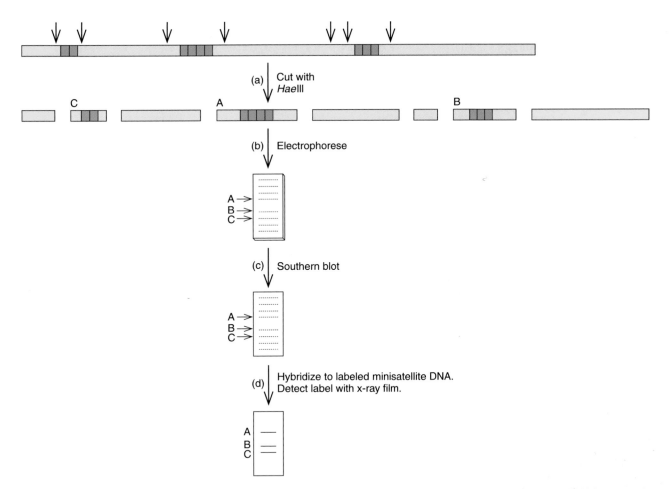

Figure 5.13   DNA fingerprinting. (a) First, cut the DNA with a restriction enzyme. In this case, the enzyme *Hae*III cuts the DNA in seven places (short arrows), generating eight fragments. Only three of these fragments (labeled A, B, and C according to size) contain the minisatellites, represented by blue boxes. The other fragments (yellow) contain unrelated DNA sequences. **(b)** Electrophorese the fragments from part **(a)**, which separates them according to their sizes. All eight fragments are present in the electrophoresis gel, but they remain invisible. The positions of all the fragments, including the three (A, B, and C) with minisatellites are indicated by dotted lines. **(c)** Denature the DNA fragments and Southern blot them. **(d)** Hybridize the DNA fragments on the Southern blot to a radioactive DNA with several copies of the minisatellite. This probe will bind to the three fragments containing the minisatellites, but with no others. Finally, use x-ray film to detect the three labeled bands.

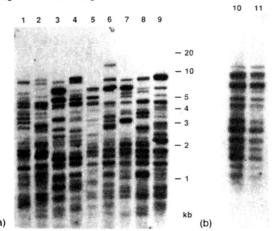

Figure 5.14   DNA fingerprint. (a) The nine lanes contain DNA from nine unrelated white subjects. Note that no two patterns are identical, especially at the upper end. **(b)** The two lanes contain DNA from monozygotic twins, so the patterns are identical (although there is more DNA in lane 10 than in lane 11). (*Source:* G. Vassart et al. A sequence in M13 phage detects hypervariable minisatellites in human and animal DNA. *Science* 235 (6 Feb 1987) p. 683, f. 1. © AAAS.)

perform a test. However, sometimes forensic scientists have even less to go on—a hair pulled out by the victim, for example. Although the hair by itself may not be enough for DNA typing, it can be useful if it is accompanied by hair follicle cells. Selected segments of DNA from these cells can be amplified by PCR and typed.

In spite of its potential accuracy, DNA typing has sometimes been effectively challenged in court, most famously in the O.J. Simpson trial in Los Angeles in 1995. Defense lawyers have focused on two problems with DNA typing: First, it is tricky and must be performed very carefully to give meaningful results. Second, there has been controversy about the statistics used in analyzing the data. This second question revolves around the use of the product rule in deciding whether the DNA typing result uniquely identifies a suspect. Let us say that a given probe detects a given allele (a set of bands in this case) in one in a hundred people in the general population. Thus, the chance of a match with a given person with this probe is one in a hundred, or 10^{-2}. If

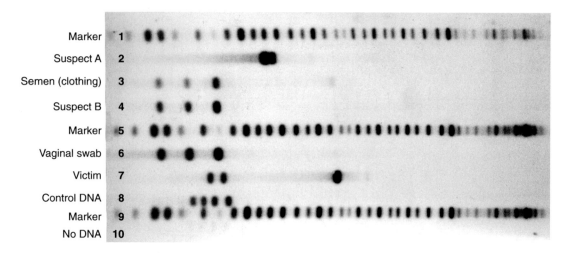

Figure 5.15   Use of DNA typing to help identify a rapist. Two suspects have been accused of attacking and raping a young woman, and DNA analyses have been performed on various samples from the suspects and the woman. Lanes 1, 5, and 9 contain marker DNAs. Lane 2 contains DNA from the blood cells of suspect A. Lane 3 contains DNA from a semen sample found on the woman's clothing. Lane 4 contains DNA from the blood cells of suspect B. Lane 6 contains DNA obtained by swabbing the woman's vaginal canal (Too little of the victim's own DNA was present to detect.). Lane 7 contains DNA from the woman's blood cells. Lane 8 contains a control DNA. Lane 10 is a control containing no DNA. Partly on the basis of this evidence, suspect B was found guilty of the crime. Note how his DNA fragments in lane 4 match the DNA fragments from the semen in lane 3 and the vaginal swab in lane 6. (*Source:* Courtesy Lifecodes Corporation, Stamford, CT.)

we use five probes, and all five alleles match the suspect, we might conclude that the chances of such a match are the product of the chances of a match with each individual probe, or $(10^{-2})^5$ or 10^{-10}. Because fewer than 10^{10} (10 billion) people are now on earth, this would mean this DNA typing would statistically eliminate everyone but the suspect. Prosecutors have used a more conservative estimate that takes into account the fact that members of some ethnic groups have higher probabilities of matches with certain probes. Still, probabilities greater than one in a million are frequently achieved, and they can be quite persuasive in court. Of course, DNA typing can do more than identify criminals. It can just as surely demonstrate a suspect's innocence (see suspect A in Figure 5.15).

SUMMARY Modern DNA typing uses a battery of DNA probes to detect variable sites in individual animals, including humans. As a forensic tool, DNA typing can be used to test parentage, to identify criminals, or to remove innocent people from suspicion.

Northern Blots: Measuring Gene Activity   Suppose you have cloned a cDNA (a DNA copy of an RNA) and want to know how actively the corresponding gene (gene X) is transcribed in a number of different tissues of organism Y. You could answer that question in several ways, but the method we describe here will also tell the size of the mRNA the gene produces.

You would begin by collecting RNA from several tissues of the organism in question. Then you electrophorese these

RNAs in an agarose gel and blot them to a suitable support. Because a similar blot of DNA is called a Southern blot, it was natural to name a blot of RNA a **Northern blot.** (By analogy, a blot of protein is called a **Western blot.**)

Next, you hybridize the Northern blot to a labeled cDNA probe. Wherever an mRNA complementary to the probe exists on the blot, hybridization will occur, resulting in a labeled band that you can detect with x-ray film. If you run marker RNAs of known size next to the unknown RNAs, you can tell the sizes of the RNA bands that "light up" when hybridized to the probe.

Furthermore, the Northern blot tells you how abundant the gene X transcript is. The more RNA the band contains, the more probe it will bind and the darker the band will be on the film. You can quantify this darkness by measuring the amount of light it absorbs in a densitometer. Figure 5.16 shows a Northern blot of RNA from eight different rat tissues, hybridized to a probe for a gene encoding G3PDH (glyceraldehyde-3-phosphate dehydrogenase), which is involved in sugar metabolism. Clearly, this gene is most actively expressed in the heart and skeletal muscle, and least actively expressed in the lung.

SUMMARY A Northern blot is similar to a Southern blot, but it contains electrophoretically separated RNAs instead of DNAs. The RNAs on the blot can be detected by hybridizing them to a labeled probe. The intensities of the bands reveal the relative amounts of specific RNA in each.

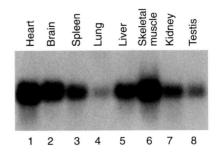

Figure 5.16   A Northern blot. Poly(A)⁺ RNA was isolated from the rat tissues indicated at the top, then equal amounts of RNA from each tissue were electrophoresed and Northern blotted. The RNAs on the blot were hybridized to a labeled probe for the rat glyceraldehyde-3-phosphate dehydrogenase (G3PDH) gene, and the blot was then exposed to x-ray film. The bands represent the G3PDH mRNA, and their intensities are indicative of the amounts of this mRNA in each tissue. (*Source:* Courtesy Clontech.)

In Situ Hybridization: Locating Genes in Chromosomes

This chapter has illustrated the use of probes to identify the band on a Southern blot that contains a gene of interest. Labeled probes can also be used to hybridize to chromosomes and thereby reveal which chromosome has the gene of interest. The strategy of such **in situ hybridization** is to spread the chromosomes from a cell and partially denature the DNA to create single-stranded regions that can hybridize to a labeled probe. One can use x-ray film to detect the label in the spread after it is stained and probed. The stain allows one to visualize and identify the chromosomes, and the darkening of the photographic emulsion locates the labeled probe, and therefore the gene to which it hybridized.

Other means of labeling the probe are also available. Figure 5.17 shows the localization of the muscle glycogen phosphorylase gene to human chromosome 11 using a DNA probe labeled with dinitrophenol, which can be detected with a fluorescent antibody. The chromosomes are counterstained with propidium iodide, so they will fluoresce red; against this background, the yellow fluorescence of the antibody probe on chromosome 11 is easy to see. This technique is known as **fluorescence in situ hybridization (FISH).**

> **SUMMARY** One can hybridize labeled probes to whole chromosomes to locate genes or other specific DNA sequences. This type of procedure is called in situ hybridization; if the probe is fluorescently labeled, the technique is called fluorescence in situ hybridization (FISH).

DNA Sequencing

In 1975, Frederick Sanger and his colleagues, and Alan Maxam and Walter Gilbert developed two different meth-

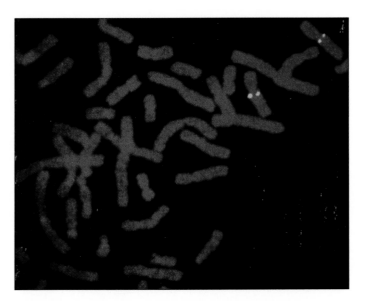

Figure 5.17   Using a fluorescent probe to find a gene in a chromosome by in situ hybridization. A DNA probe specific for the human muscle glycogen phosphorylase gene was coupled to dinitrophenol. A human chromosome spread was then partially denatured to expose single-stranded regions that can hybridize to the probe. The sites where the DNP-labeled probe hybridized were detected indirectly as follows: A rabbit anti-DNP antibody was bound to the DNP on the probe; then a goat antirabbit antibody, coupled with fluorescein isothiocyanate (FITC), which emits yellow fluorescent light, was bound to the rabbit antibody. The chromosomal sites where the probe hybridized show up as bright yellow fluorescent spots against a red background that arises from staining the chromosomes with the fluorescent dye propidium iodide. This analysis identifies chromosome 11 as the site of the glycogen phosphorylase gene. (*Source:* Courtesy Dr. David Ward, *Science* 247 (5 Jan 1990) cover.)

ods for determining the exact base sequence of a cloned piece of DNA. These spectacular breakthroughs revolutionized molecular biology. They have allowed molecular biologists to determine the sequences of thousands of genes, and they have even made it possible to sequence all 3 billion base pairs of the human genome.

The Sanger Chain-Termination Sequencing Method   The original method of sequencing a piece of DNA by the Sanger method (Figure 5.18) is presented here to explain the principles. In practice, it is rarely done manually this way anymore. In the next section we will see how the technique has been automated. The original method began with cloning the DNA into a vector, such as M13 phage or a phagemid, that would give the cloned DNA in single-stranded form. These days, one can start with double-stranded DNA and simply heat it to create single-stranded DNAs for sequencing. To the single-stranded DNA one hybridizes an oligonucleotide primer about 20 bases long. This synthetic primer is designed to hybridize to a sequence adjacent to the multiple cloning region of the vector and is oriented with its 3′-end pointing toward the insert in the multiple cloning region.

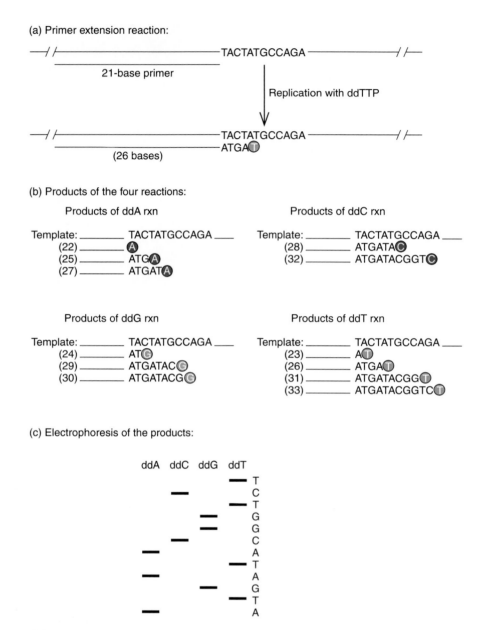

Figure 5.18   The Sanger dideoxy method of DNA sequencing. (a) The primer extension (replication) reaction. A primer, 21 bp long in this case, is hybridized to the single-stranded DNA to be sequenced, then mixed with the Klenow fragment of DNA polymerase and dNTPs to allow replication. One dideoxy NTP is included to terminate replication after certain bases; in this case, ddTTP is used, and it has caused termination at the second position where dTTP was called for. **(b)** Products of the four reactions (rxns). In each case, the template strand is shown at the top, with the various products underneath. Each product begins with the 21-base primer and has one or more nucleotides added to the 3′-end. The last nucleotide is always a dideoxy nucleotide (color) that terminated the chain. The total length of each product is given in parentheses at the left end of the fragment. Thus, fragments ranging from 22 to 33 nucleotides long are produced. **(c)** Electrophoresis of the products. The products of the four reactions are loaded into parallel lanes of a high-resolution electrophoresis gel and electrophoresed to separate them according to size. By starting at the bottom and finding the shortest fragment (22 bases in the A lane), then the next shortest (23 bases in the T lane), and so forth, one can read the sequence of the product DNA. Of course, this is the complement of the template strand.

Extending the primer using the Klenow fragment of DNA polymerase produces DNA complementary to the insert. The trick to Sanger's method is to carry out such DNA synthesis reactions in four separate tubes and to include in each tube a different chain terminator. The chain terminator is a **dideoxy nucleotide** such as **dideoxy ATP** (**ddATP**). Not only is this terminator 2′-deoxy, like a normal DNA precursor, it is 3′-deoxy as well. Thus, it cannot form a phosphodiester bond because it lacks the necessary 3′-hydroxyl group. That is why we call it a chain terminator; whenever a dideoxy nucleotide is incorporated into a growing DNA chain, DNA synthesis stops.

Dideoxy nucleotides by themselves do not permit any DNA synthesis at all, so an excess of normal deoxy nucleotides must be used, with just enough dideoxy nucleotide to stop DNA strand extension once in a while at random. This random arrest of DNA growth means that some strands will terminate early, others later. Each tube contains a different dideoxy nucleotide: ddATP in tube 1, so chain termination will occur with A's; ddCTP in tube 2, so chain termination will occur with C's; and so forth. Radioactive dATP is also included in all the tubes so the DNA products will be radioactive.

The result is a series of fragments of different lengths in each tube. In tube 1, all the fragments end in A; in tube 2, all end in C; in tube 3, all end in G; and in tube 4, all end in T. Next, all four reaction mixtures are electrophoresed in parallel lanes in a high-resolution polyacrylamide gel under denaturing conditions, so all DNAs are single-stranded. Finally, autoradiography is performed to visualize the DNA fragments, which appear as horizontal bands on an x-ray film.

Figure 5.18c shows a schematic of the sequencing film. To begin reading the sequence, start at the bottom and find the first band. In this case, it is in the A lane, so you know that this short fragment ends in A. Now move to the next longer fragment, one step up on the film; the gel electrophoresis has such good resolution that it can separate fragments differing by only one base in length, at least until the fragments become much longer than this. And the next fragment, one base longer than the first, is found in the T lane, so it must end in T. Thus, so far you have found the sequence AT. Simply continue reading the sequence in this way as you work up the film. The sequence is shown, reading bottom to top, at the right of the drawing. At first you will be reading just the sequence of part of the multiple cloning region of the vector. However, before very long, the DNA chains will extend into the insert—and unknown territory. An experienced sequencer can continue to read sequence from one film for hundreds of bases.

Figure 5.19 shows a typical sequencing film. The shortest band (at the very bottom) is in the C lane. After that, a series of six bands occurs in the A lane. So the sequence begins CAAAAAA. It is easy to read many more bases on this film.

The "manual" sequencing technique just described is powerful, but it is still relatively slow. If we are to sequence a really large amount of DNA, such as the 3 billion base pairs found in the human genome, then rapid, automated sequencing methods are required. Indeed, automated DNA sequencing has been in use for many years. Figure 5.20a describes one such technique, again based on Sanger's chain-termination method. This procedure uses dideoxynucleotides, just as in the manual method, with one important exception. The primers used in each of the four reactions are tagged with a different

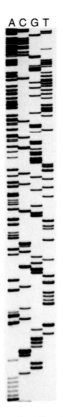

Figure 5.19   A typical sequencing film. The sequence begins CAAAAAACGG. You can probably read the rest of the sequence to the top of the film. (*Source:* Courtesy Life Technologies, Inc., Gaithersburg, MD.)

fluorescent molecule, so the products from each tube will emit a different color fluorescence when excited by light.

After the extension reactions and chain termination are complete, all four reactions are mixed and electrophoresed together in the same lane on a gel (Figure 5.20b). Near the bottom of the gel is an analyzer that excites the fluorescent oligonucleotides with a laser beam as they pass by. Then the color of the fluorescent light emitted from each oligonucleotide is detected electronically. This information then passes to a computer, which has been programmed to convert the color information to a base sequence. If it "sees" blue, for example, this might mean that this oligonucleotide came from the dideoxy C reaction, and therefore ends in C (actually a ddC). Green may indicate A; orange, G; and red, T. The computer gives a printout of the profile of each passing fluorescent band, color-coded for each base (Figure 5.20c), and stores the sequence of these bases in its memory for later use.

Nowadays, automated sequencers may simply print out the sequence or send it directly to a computer for analysis. Large genome projects use many automated sequencers running simultaneously to obtain millions or even billions of bases of sequence (Chapter 24).

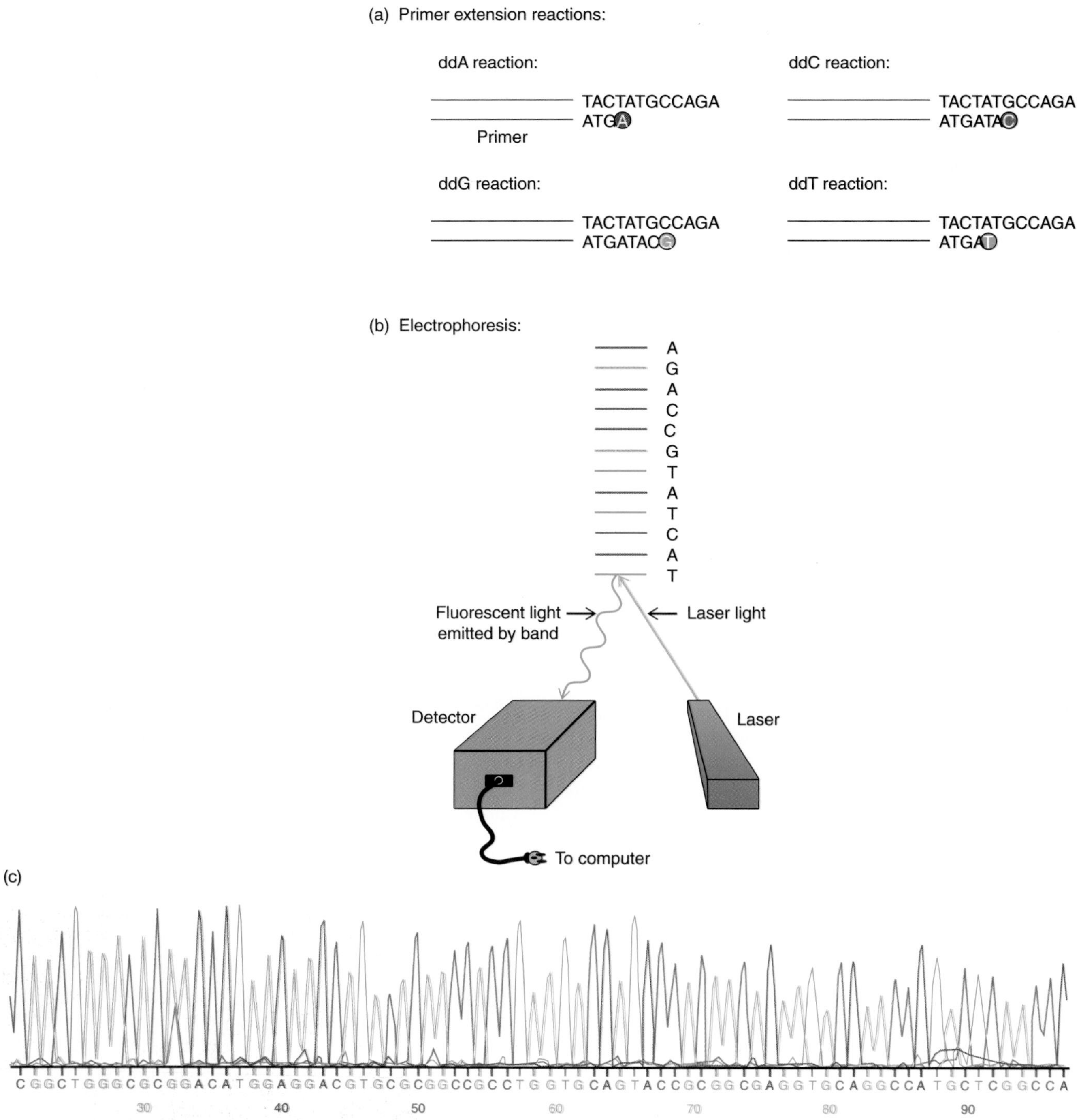

Figure 5.20 **Automated DNA sequencing. (a)** The primer extension reactions are run in the same way as in the manual method, except that the primers in each reaction are labeled with a different fluorescent molecule that emits light of a distinct color. Only one product is shown for each reaction, but all possible products are actually produced, just as in the manual sequencing. **(b)** Electrophoresis and detection of bands. The various primer extension reaction products separate according to size on gel electrophoresis. The bands are color-coded according to the termination reaction that produced them (e.g., green for oligonucleotides ending in ddA, blue for those ending in ddC, and so forth). A laser scanner excites the fluorescent tag on each band as it passes by, and a detector analyzes the color of the resulting emitted light. This information is converted to a sequence of bases and stored by a computer. **(c)** Sample printout of an automated DNA sequencing experiment. Each colored peak is a plot of the fluorescence intensity of a band as it passes through the laser beam. The colors of these peaks, and those of the bands in part **(b)** and the tags in part **(a)**, were chosen for convenience. They may not correspond to the actual colors of the fluorescent light. (*Source:* (c) Courtesy Amersham Pharmacia Biotech, Inc.)

SUMMARY The Sanger DNA sequencing method uses dideoxy nucleotides to terminate DNA synthesis, yielding a series of DNA fragments whose sizes can be measured by electrophoresis. The last base in each of these fragments is known, because we know which dideoxy nucleotide was used to terminate each reaction. Therefore, ordering these fragments by size—each fragment one (known) base longer than the next—tells us the base sequence of the DNA.

Maxam–Gilbert Sequencing The Maxam–Gilbert sequencing method follows a very different strategy. Instead of synthesizing DNA *in vitro* and stopping the synthesis reactions with chain terminators, this method starts with full-length, end-labeled DNA and cleaves it with base-specific reagents. Figure 5.21 illustrates how this method works with guanine bases, but the same principle applies to all four bases. First, we end-label a DNA fragment we want to sequence. This can be 5′- or 3′-end labeling as explained later in this chapter. Next, we modify one kind of base. Here, we use dimethyl sulfate (DMS) to methylate guanines. (Actually, this reagent also methylates adenines, but not in a way that leads to DNA strand cleavage.) As in the chain termination method, we do not want to affect every guanine, or we will produce only tiny fragments that will not allow us to determine the DNA's sequence. Therefore, we do the methylation under mild conditions that lead to an average of only one methylated guanine per DNA strand. Next, we use a reagent (piperidine) that does two things: It causes loss of the methylated base, then it breaks the DNA backbone at the site of the lost base (the **apurinic site**). In this case, the G in the middle of the sequence was methylated, so strand breakage occurred there, producing a labeled trimer. In another DNA molecule, the first G could be methylated, giving rise to a labeled phosphate (the base and sugar would be lost in the chemical cleavages). Finally, we electrophorese the products and detect them by autoradiography, just as in the chain-termination method.

Of course, we need to run three other reactions that cleave at the other three bases. There are several ways of doing this. For example, we can weaken the glycosidic bonds to both adenine and guanine with acid; then piperidine will cause depurination and strand breakage after both A's and G's. If we electrophorese this A + G reaction beside the G only reaction, we can obtain the A's by comparison. Similarly, hydrazine opens both thymine and cytosine rings, and piperidine can then remove these bases and break the DNA strand at the resulting **apyrimidinic sites**. In the presence of 1 M NaCl, hydrazine is specific for cytosine only, so we can run this reaction next to the C + T reaction and obtain the T's by comparison.

Maxam–Gilbert sequencing is used only in special cases these days, but the methods of modifying and then

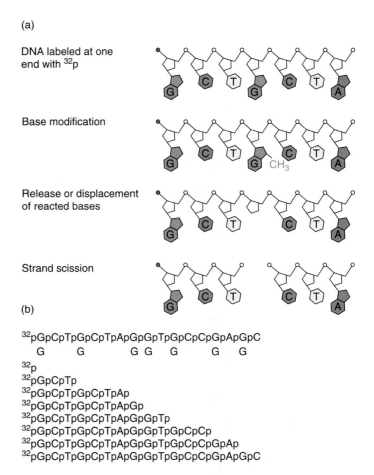

(a)

DNA labeled at one end with ^{32}p

Base modification

Release or displacement of reacted bases

Strand scission

(b)

^{32}pGpCpTpGpCpTpApGpGpGpTpGpCpCpGpApGpC
   G   G   G G   G   G   G
^{32}p
^{32}pGpCpTp
^{32}pGpCpTpGpCpTpAp
^{32}pGpCpTpGpCpTpApGp
^{32}pGpCpTpGpCpTpApGpGpTp
^{32}pGpCpTpGpCpTpApGpGpGpTpGpCpCp
^{32}pGpCpTpGpCpTpApGpGpGpTpGpCpCpGpAp
^{32}pGpCpTpGpCpTpApGpGpGpTpGpCpCpGpApGpC

Figure 5.21 Maxam–Gilbert sequencing. (a) The chain cleavage reaction. This reaction, specific for guanines, begins with an end-labeled DNA fragment (5′-end label denoted by a purple dot in this example). Guanines (red) are methylated with a mild dimethyl sulfate (DMS) treatment that methylates on average one guanine per DNA strand. Then the methylated DNA is treated with piperidine, which first removes the methylated base and then breaks the DNA strand at the apurinic site. This leaves a 3′-phosphate on the nucleotide that preceded the guanine nucleotide. **(b)** The fragments created by chain cleavage at guanines. At top is a 5′-end-labeled DNA fragment, with the positions of the Gs indicated just below. At bottom are all the possible fragments produced by cleavage at guanines in this DNA. Note that they do not end in guanine because the guanines that led to chain cleavage were lost. For this reason, the first fragment is just phosphate. (*Source:* From Maxim and Gilbert, *Methods in Enzymology* 65:5000, 1980. Copyright 1980 Academic Press, Orlando, FL. Reprinted with permission.)

breaking DNA used by Maxam and Gilbert are still useful for other purposes (see Chapter 6, for example).

SUMMARY The Maxam–Gilbert method allows us to break an end-labeled DNA strand at specific bases using base-specific reagents. We can then electrophorese the labeled fragments in a high-resolution polyacrylamide gel and detect them by autoradiography, just as in the chain-termination method. This procedure orders the fragments by size and therefore tells us the sequence of the DNA.

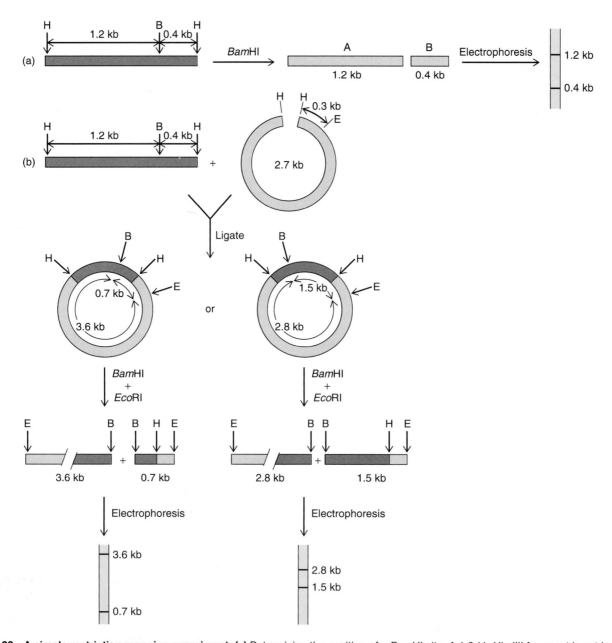

Figure 5.22   A simple restriction mapping experiment. (a) Determining the position of a *Bam*HI site. A 1.6-kb *Hind*III fragment is cut by *Bam*HI to yield two subfragments. The sizes of these fragments are determined by electrophoresis to be 1.2 kb and 0.4 kb, demonstrating that *Bam*HI cuts once, 1.2 kb from one end of the *Hind*III fragment and 0.4 kb from the other end. **(b)** Determining the orientation of the *Hind*III fragment in a cloning vector. The 1.6-kb *Hind*III fragment can be inserted into the *Hind*III site of a cloning vector, in either of two ways: (1) with the *Bam*HI site near an *Eco*RI site in the vector or (2) with the *Bam*HI site remote from an *Eco*RI site in the vector. To determine which, cleave the DNA with both *Bam*HI and *Eco*RI and electrophorese the products to measure their sizes. A short fragment (0.7 kb) shows that the two sites are close together (left). On the other hand, a long fragment (1.5 kb) shows that the two sites are far apart (right).

Restriction Mapping

Before sequencing a large stretch of DNA, some preliminary mapping is usually done to locate landmarks on the DNA molecule. These are not genes, but small regions of the DNA—cutting sites for restriction enzymes, for example. A map based on such physical characteristics is called, naturally enough, a **physical map.** (If restriction sites are the only markers involved, we can also call it a **restriction map.**)

To introduce the idea of restriction mapping, let us consider the simple example illustrated in Figure 5.22. We start with a *Hind*III fragment 1.6 kb (1600 bp) long

(Figure 5.22*a*). When this fragment is cut with another restriction enzyme (*Bam*HI), two fragments are generated, 1.2 and 0.4 kb long. The sizes of these fragments can be measured by electrophoresis, as pictured in Figure 5.22*a*. The sizes reveal that *Bam*HI cuts 0.4 kb from one end of the 1.6-kb *Hin*dIII fragment, and 1.2 kb from the other.

Now suppose the 1.6-kb *Hin*dIII fragment is cloned into the *Hin*dIII site of a hypothetical plasmid vector, as illustrated in Figure 5.22*b*. Because this is not directional cloning, the fragment will insert into the vector in either of the two possible orientations: with the *Bam*HI site on the right (left side of the figure), or with the *Bam*HI site on the left (right side of the figure). How can you determine which orientation exists in a given clone? To answer this question, locate a restriction site asymmetrically situated in the vector, relative to the *Hin*dIII cloning site. In this case, an *Eco*RI site is only 0.3 kb from the *Hin*dIII site. This means that if you cut the cloned DNA pictured on the left with *Bam*HI and *Eco*RI, you will generate two fragments: 3.6 and 0.7 kb long. On the other hand, if you cut the DNA pictured on the right with the same two enzymes, you will generate two fragments: 2.8 and 1.5 kb in size. You can distinguish between these two possibilities easily by electrophoresing the fragments to measure their sizes, as shown at the bottom of Figure 5.22*c*. Usually, DNA is prepared from several different clones, each of them is cut with the two enzymes, and the fragments are electrophoresed side by side with one lane reserved for marker fragments of known sizes. On average, half of the clones will have one orientation, and the other half will have the opposite orientation. You can use similar reasoning to solve the following mapping problem.

Solved Problem

Problem 2

You are mapping a double-stranded DNA circle 17 kb in size. As shown in Figure 5.23*a*, *Hin*dIII cuts this DNA twice, yielding two fragments (*Hin*dIII-A and *Hin*dIII-B) with sizes of 11 and 6 kb, respectively. *Pst*I also cuts twice, yielding two fragments (*Pst*I-A and *Pst*I-B) with sizes of 5 and 12 kb, respectively. Figure 5.23*b* shows the cleavage patterns of each of these four fragments cut with the other restriction enzyme. For example, the *Hin*dIII-A fragment can be cleaved with *Pst*I to yield two fragments: 7 and 4 kb in size.

Show the locations of all four restriction sites on the circular DNA. To simplify matters, start by placing the *Hin*dIII-B fragment in the upper right quadrant of the circle, with one of the *Hin*dIII sites at the top of the

circle (at 12 o'clock), and the other *Hin*dIII site at about 4 o'clock. Also, place the *Pst*I site in this fragment closer to the 4 o'clock end.

Solution

Two *Hin*dIII fragments can be arranged in a circle in only two possible ways. We just need to decide which of the two is correct. We have already decided arbitrarily to place the smaller of the two *Hin*dIII fragments (*Hin*dIII-B) in the top right quadrant of the circle, oriented with its *Pst*I site nearer the 4 o'clock end. The only question remaining is the orientation of the *Hin*dIII-A fragment. Should its *Pst*I site be closer to the 12 o'clock or 4 o'clock end of the *Hin*dIII-B fragment? To answer this question, we just need to consider what size *Pst*I fragments will result. Suppose we place *Hin*dIII-A with its *Pst*I site closer to 12 o'clock, as shown in Figure 5.24*a*. Because this site is 4 kb from the *Hin*dIII site, and the *Pst*I site of *Hin*dIII-B is another 5 kb from the *Hin*dIII site, the *Pst*I cleavage of the plasmid would generate a 9-kb fragment. Furthermore, because the whole plasmid is 17 kb, the other *Pst*I fragment would be $17 - 9 = 8$ kb. But this is not the result we observe. Therefore, the other possibility must be correct. Let us demonstrate that it is. If the *Hin*dIII-A fragment has its *Pst*I site closer to the 4 o'clock *Hin*dIII site, as in Figure 5.24*b*, then cleavage of the plasmid with *Pst*I will generate fragments of 5 $(4 + 1)$ and 12 $(7 + 5)$ kb. This is exactly what we observe, so this must be the correct arrangement. ∎

These examples are relatively simple, but we use the same kind of logic to solve much more complex mapping problems. Sometimes it helps to label (radioactively or nonradioactively) one restriction fragment and hybridize it to a Southern blot of fragments made with another restriction enzyme to help sort out the relationships among fragments. For example, consider the linear DNA in Figure 5.25. We might be able to figure out the order of restriction sites without the use of hybridization, but it is not simple. Consider the information we get from just a few hybridizations. If we Southern blot the *Eco*RI fragments and hybridize them to the labeled *Bam*HI-A fragment, for example, the *Eco*RI-A and *Eco*RI-C fragments will become labeled. This demonstrates that *Bam*HI-A overlaps these two *Eco*RI fragments. If we hybridize the blot to the *Bam*HI-B fragment, the *Eco*RI-A and *Eco*RI-D fragments become labeled. Thus, *Bam*HI-B overlaps *Eco*RI-A and *Eco*RI-D. Ultimately, we will discover that no other *Bam*HI fragments besides A and B hybridize to *Eco*RI-A, so *Bam*HI-A and *Bam*HI-B must be adjacent in *Eco*RI-A. Using this kind of approach, we can piece together the physical map of the whole 30-kb fragment.

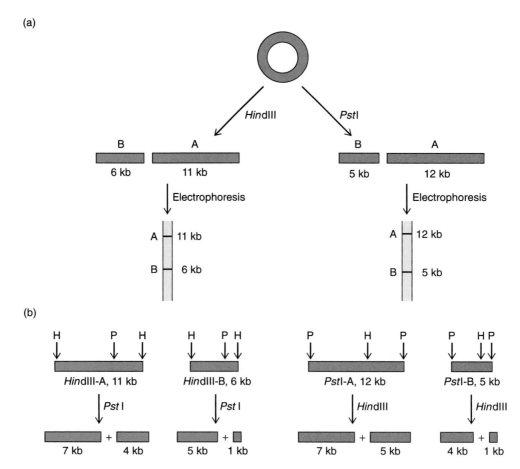

Figure 5.23 Restriction mapping of an unknown DNA. *Abbreviations:* H = *Hind*III site; P = *Pst*I site.

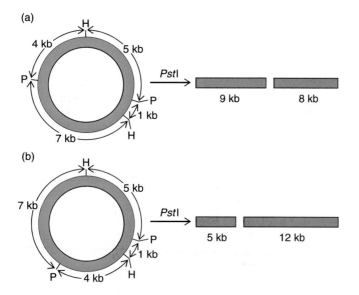

Figure 5.24 Two potential maps of the unknown DNA.
Abbreviations: H = *Hind*III site; P = *Pst*I site.

SUMMARY A physical map tells us about the spatial arrangement of physical "landmarks," such as restriction sites, on a DNA molecule. One important strategy in restriction mapping (mapping of restriction sites) is to cut the DNA in question with two or more restriction enzymes in separate reactions, measure the sizes of the resulting fragments, then cut each with another restriction enzyme and measure the sizes of the subfragments by gel electrophoresis. These sizes allow us to locate at least some of the recognition sites relative to the others. We can improve this process considerably by Southern blotting some of the fragments and then hybridizing these fragments to labeled fragments generated by another restriction enzyme. This strategy reveals overlaps between individual restriction fragments.

Protein Engineering with Cloned Genes: Site-Directed Mutagenesis

Traditionally, protein biochemists relied on chemical methods to alter certain amino acids in the proteins they

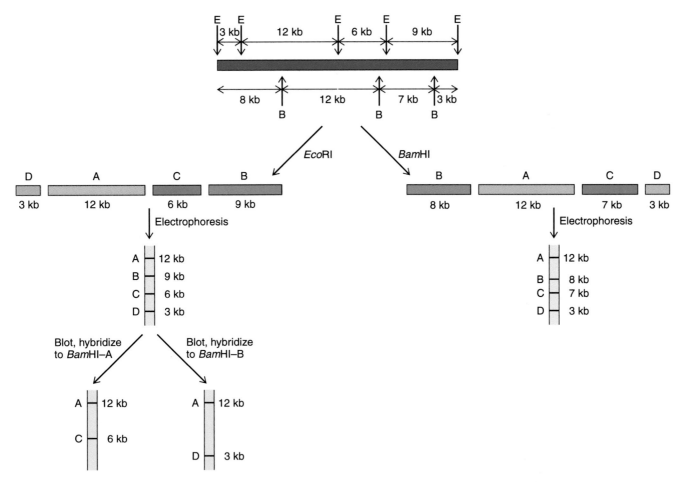

Figure 5.25   Using Southern blots in restriction mapping. A 30-kb fragment is being mapped. It is cut three times each by *Eco*RI (E) and *Bam*HI (B). To aid in the mapping, first cut with *Eco*RI, electrophorese the four resulting fragments (*Eco*RI-A, -B, -C, and -D), next, Southern blot the fragments and hybridize them to labeled, cloned *Bam*HI-A and -B fragments. The results, shown at lower left, demonstrate that the *Bam*HI-A fragment overlaps *Eco*RI-A and -C, and the *Bam*HI-B fragment overlaps *Eco*RI-A and -D. This kind of information, coupled with digestion of *Eco*RI fragments by *Bam*HI (and vice versa), allows the whole restriction map to be pieced together.

studied, so they could observe the effects of these changes on protein activities. But chemicals are rather crude tools for manipulating proteins; it is difficult to be sure that only one amino acid, or even one kind of amino acid, has been altered. Cloned genes make this sort of investigation much more precise, allowing us to perform microsurgery on a protein. By changing specific bases in a gene, we also change amino acids at corresponding sites in the protein product. Then we can observe the effects of those changes on the protein's function.

Let us suppose that we have a cloned gene in which we want to change a single codon. In particular, the gene codes for a sequence of amino acids that includes a tyrosine. The amino acid tyrosine contains a phenolic group:

—OH

To investigate the importance of this phenolic group, we can change the tyrosine codon to a phenylalanine codon. Phenylalanine is just like tyrosine except that it lacks the phenolic group; instead, it has a simple phenyl group:

If the tyrosine phenolic group is important to a protein's activity, replacing it with phenylalanine's phenyl group should diminish that activity.

In this example, let us assume that we want to change the DNA codon TAC (Tyr) to TTC (Phe). How do we perform such **site-directed mutagenesis?** An increasingly popular technique, depicted in Figure 5.26 relies on PCR (Chapter 4). We begin with a cloned gene containing a tyrosine codon (TAC) that we want to change to a phenylalanine codon (TTC). The CH_3 symbols indicate that this DNA, like DNAs isolated from most strains of *E. coli*, is methylated on 5′-GATC-3′-sequences. This methylated sequence happens to be the recognition site

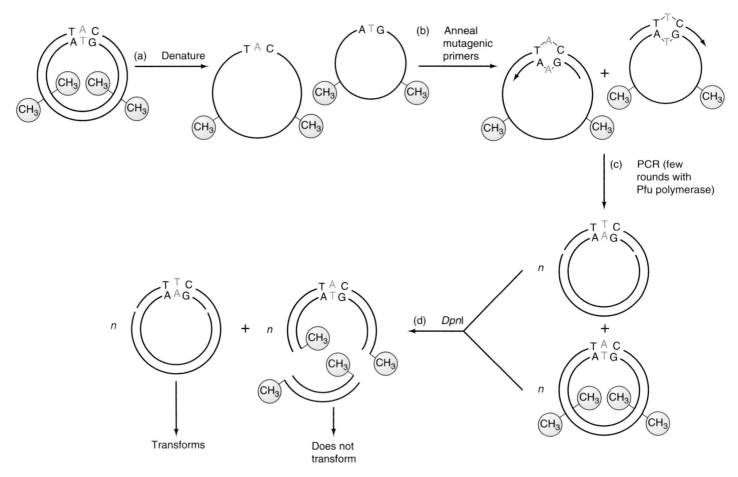

Figure 5.26 PCR-based site-directed mutagenesis. Begin with a plasmid containing a gene with a TAC tyrosine codon that is to be altered to a TTC phenylalanine codon. Thus, the A–T pair (blue) in the original must be changed to a T–A pair. This plasmid was isolated from a normal strain of *E. coli* that methylates the A's of GATC sequences. The methyl groups are indicated in yellow. **(a)** Heat the plasmid to separate its strands. The strands of the original plasmid are intertwined, so they don't completely separate. They are shown here separating completely for simplicity's sake. **(b)** Anneal mutagenic primers that contain the TTC codon, or its reverse complement, GAA. The altered base in each primer is indicated in red. **(c)** Perform a few rounds of PCR (about eight) with the mutagenic primers to amplify the plasmid with its altered codon. Use a faithful, heat-stable DNA polymerase, such as Pfu polymerase, to minimize mistakes in copying the plasmid. **(d)** Treat the DNA in the PCR reaction with *Dpn*I to digest the methylated wild-type DNA. Because the PCR product was made in vitro, it is not methylated and is not cut. Finally, transform *E. coli* cells with the treated DNA. In principle, only the mutated DNA survives to transform. Check this by sequencing the plasmid DNA from several clones.

for the restriction enzyme *Dpn*I, which will come into play later in this procedure. Two methylated *Dpn*I sites are shown, even though many more are usually present because GATC occurs about once every 250 bp in a random sequence of DNA.

The first step is to denature the DNA by heating. The second step is to anneal mutagenic primers to the DNA. One of these primers is a 25-base oligonucleotide (a 25-mer) with the following sequence:

3′-CGAGTCTGCCAAAGCATGTATAGTA-5′

This has the same sequence as a piece of the gene's non-template strand, except that the central triplet has been changed from ATG to AAG, with the altered base underlined. The other primer is the complementary 25-mer. Both primers incorporate the altered base to change the

codon we are targeting. The third step is to use a few rounds of PCR with these primers to amplify the DNA, and incorporate the change we want to make. We deliberately use just a few rounds of PCR to minimize other mutations that might creep in by accident during DNA replication. For the same reason, we use a very faithful DNA polymerase called Pfu polymerase. This enzyme is purified from bacteria called *Pyrococcus furiosus* (Latin: *furious fireball*), which live in the intensely hot water surrounding undersea vents. It has the ability to "proofread" the DNA it synthesizes, so it makes relatively few mistakes. A similar enzyme from another thermophilic (heat-loving) bacterium is called vent polymerase.

Once the mutated DNA is made, we must either separate it from the remaining wild-type DNA or destroy the latter. This is where the methylation of the wild-type

DNA comes in handy. *Dpn*I will cut only GATC sites that are methylated. Because the wild-type DNA is methylated, but the mutated DNA, which was made in vitro, is not, only the wild-type DNA will be cut. Once cut, it is no longer capable of transforming *E. coli* cells, so the mutated DNA is the only species that yields clones. We can check the sequence of DNA from several clones to make sure it is the mutated sequence and not the original, wild-type sequence. Usually, it is.

SUMMARY Using cloned genes, we can introduce changes at will, thus altering the amino acid sequences of the protein products. The mutagenized DNA can be made with double-stranded DNA, two complementary mutagenic primers, and PCR. Simply digesting the PCR product with *Dpn*I removes almost all of the wild-type DNA, so cells can be transformed primarily with mutagenized DNA.

5.4 Mapping and Quantifying Transcripts

One recurring theme in molecular biology has been mapping transcripts (locating their starting and stopping points) and quantifying them (measuring how much of a transcript exists at a certain time). Molecular biologists use a variety of techniques to map and quantify transcripts, and, in addition to the Northern blotting method discussed earlier, we will encounter several in this book.

You might think that the simplest way of finding out how much transcript is made at a given time would be to label the transcript by allowing it to incorporate labeled nucleotides in vivo or in vitro, then to electrophorese it and detect the transcript as a band on the electrophoretic gel by autoradiography. In fact, this has been done for certain transcripts, both in vivo and in vitro. However, it works in vivo only if the transcript in question is quite abundant and easy to separate from other RNAs by electrophoresis. Transfer RNA and 5S ribosomal RNA satisfy both these conditions and their synthesis has been traced in vivo by simple electrophoresis (Chapter 10). This direct method succeeds in vitro only if the transcript has a clearcut terminator, so a discrete species of RNA is made, rather than a continuum of species with different 3'-ends that would produce an unintelligible smear, rather than a sharp band. Again, in some instances this is true, most notably in the case of prokaryotic transcripts, but eukaryotic examples are rare. Thus, we frequently need to turn to other, less direct, but more specific methods. Several popular techniques are available for mapping the 5'-ends of transcripts, and one of these also locates the 3'-end. Two

of them can also tell how much of a given transcript is in a cell at a given time. Like Northern blotting, these two methods rely on the power of nucleic acid hybridization to detect just one kind of RNA among thousands.

S1 Mapping

S1 Mapping is used to locate the 5'- or 3'-ends of RNAs and to quantify the amount of a given RNA in cells at a given time. The principle behind this method is to label a single-stranded DNA probe that can hybridize only to the transcript of interest, rather than labeling the transcript itself. The probe must span the sequence where the transcript starts or ends. After hybridizing the probe to the transcript, we apply **S1 nuclease,** which degrades only single-stranded DNA and RNA. Thus, the transcript protects part of the probe from degradation. The size of this part can be measured by gel electrophoresis, and the extent of protection tells where the transcript starts or ends. Figure 5.27 shows in detail how S1 mapping can be used to find the transcription start site. First, the DNA probe is labeled at its 5'-end with ^{32}P-phosphate. The 5'-end of a DNA strand usually already contains a nonradioactive phosphate, so this phosphate is removed with an enzyme called alkaline phosphatase before the labeled phosphate is added. Then the enzyme polynucleotide kinase is used to transfer the ^{32}P-phosphate group from $[\gamma\text{-}^{32}\text{P}]$ATP to the 5'-hydroxyl group at the beginning of the DNA strand.

In this example, a *Bam*HI fragment has been labeled on both ends, which would yield two labeled single-stranded probes. However, this would needlessly confuse the analysis, so the label on the left end must be removed. That task is accomplished here by recutting the DNA with another restriction enzyme, *Sal*I, then using gel electrophoresis to separate the short, left-hand fragment from the long fragment that will produce the probe. Now the double-stranded DNA is labeled on only one end, and it can be denatured to yield a labeled single-stranded probe. Next, the probe DNA is hybridized to a mixture of cellular RNAs that contains the transcript of interest. Hybridization between the probe and the complementary transcript will leave a tail of single-stranded DNA on the left, and single-stranded RNA on the right. Next, S1 nuclease is used. This enzyme specifically degrades single-stranded DNA or RNA, but leaves double-stranded polynucleotides, including RNA–DNA hybrids, intact. Thus, the part of the DNA probe, including the terminal label, that is hybridized to the transcript will be protected. Finally, the length of the protected part of the probe is determined by high-resolution gel electrophoresis alongside marker DNA fragments of known length. Because the location of the right-hand end of the probe (the labeled *Bam*HI site) is known exactly, the length of the protected probe automatically tells the location of the left-hand end,

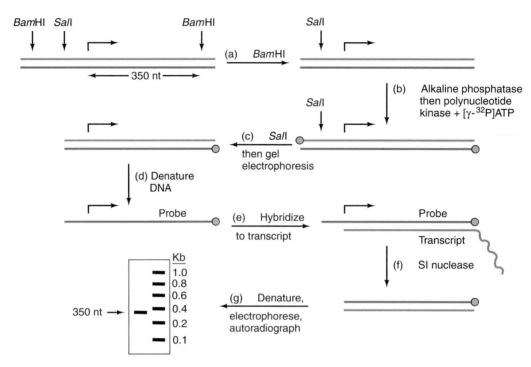

Figure 5.27   S1 mapping the 5′-end of a transcript. Begin with a cloned piece of double-stranded DNA with several known restriction sites. In this case, the exact position of the transcription start site is not known, even though it is marked here (⌐) based on what will be learned from the S1 mapping. It is known that the transcription start site is flanked by two *Bam*HI sites, and a single *Sal*I site occurs just upstream of the start site. In step **(a)** cut with *Bam*HI to produce the *Bam*HI fragment shown at upper right. In step **(b)** remove the unlabeled phosphates on this fragment's 5′-hydroxyl groups, then label these 5′-ends with polynucleotide kinase and [γ³²P] ATP. The orange circles denote the labeled ends. In step **(c)** cut with *Sal*I and separate the two resulting fragments by electrophoresis. This removes the label from the left end of the double-stranded DNA. In step **(d)** denature the DNA to generate a single-stranded probe that can hybridize with the transcript (red) in step **(e)**. In step **(f)** treat the hybrid with S1 nuclease. This digests the single-stranded DNA on the left and the single-stranded RNA on the right of the hybrid from step **(e)**. In step **(g)** denature the remaining hybrid and electrophorese the protected piece of the probe to see how long it is. DNA fragments of known length are included as markers in a separate lane. The length of the protected probe indicates the position of the transcription start site. In this case, it is 350 bp upstream of the labeled *Bam*HI site in the probe.

which is the transcription start site. In this case, the protected probe is 350 nt long, so the transcription start site is 350 bp upstream of the labeled *Bam*HI site.

One can also use S1 mapping to locate the 3′-end of a transcript. It is necessary to prepare a 3′-end-labeled probe and hybridize it to the transcript, as shown in Figure 5.28. All other aspects of the assay are the same as for 5′-end mapping. 3′-End-labeling is different from 5′-labeling because polynucleotide kinase will not phosphorylate 3′-hydroxyl groups on nucleic acids. One way to label 3′-ends is to perform **end-filling,** as shown in Figure 5.29. When a DNA is cut with a restriction enzyme that leaves a recessed 3′-end, that recessed end can be extended in vitro until it is flush with the 5′-end. If labeled nucleotides are included in this end-filling reaction, the 3′-ends of the DNA will become labeled.

S1 mapping can be used not only to map the ends of a transcript, but to tell the transcript concentration. Assuming that the probe is in excess, the intensity of the band on the autoradiograph is proportional to the concentration of the

transcript that protected the probe. The more transcript, the more protection of the labeled probe, so the more intense the band on the autoradiograph. Thus, once it is known which band corresponds to the transcript of interest its intensity can be used to measure the transcript concentration.

One important variation on the S1 mapping theme is **RNase mapping (RNase protection assay).** This procedure is analogous to S1 mapping and can yield the same information about the 5′- and 3′-ends and concentration of a specific transcript. The probe in this method, however, is made of RNA and can therefore be degraded with RNase instead of S1 nuclease. This technique is gaining in popularity, partly because of the relative ease of preparing RNA probes (**riboprobes**) by transcribing recombinant plasmids or phagemids in vitro with purified phage RNA polymerases. Another advantage of using riboprobes is that they can be labeled to very high specific activity by including a labeled nucleotide in the in vitro transcription reaction. The higher the specific activity of the probe, the more sensitive it is in detecting tiny quantities of transcripts.

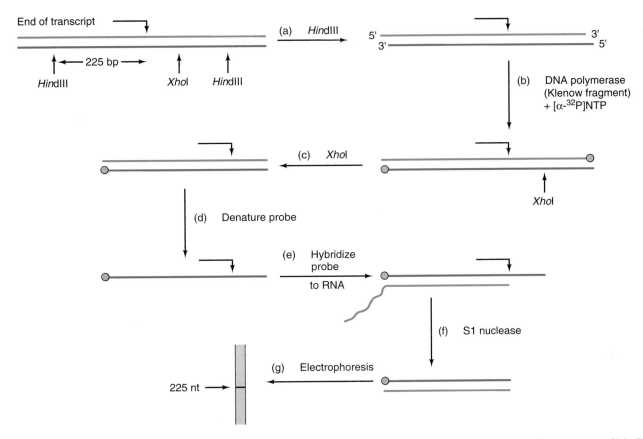

Figure 5.28   S1 mapping the 3′-end of a transcript. The principle is the same as in 5′-end mapping except that a different means of labeling the probe—at its 3′-end instead of its 5′-end, is used (detailed in Figure 5.29). In step **(a)** cut with *Hind*III, then in step **(b)** label the 3′-ends of the resulting fragment. The orange circles denote these labeled ends. In step **(c)** cut with *Xho*I and purify the left-hand labeled fragment by gel electrophoresis. In step **(d)** denature the probe and hybridize it to RNA (red) in step **(e)**. In step **(f)** remove the unprotected region of the probe (and of the RNA) with S1 nuclease. Finally, in step **(g)** electrophorese the labeled protected probe to determine its size. In this case it is 225 nt long, which indicates that the 3′-end of the transcript lies 225 bp downstream of the labeled *Hind*III site on the left-hand end of the probe.

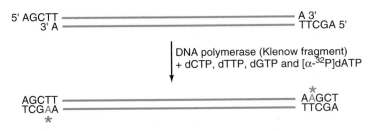

Figure 5.29   3′-end-labeling a DNA by end-filling. The DNA fragment at the top has been created by cutting with *Hind*III, which leaves 5′-overhangs at each end, as shown. These can be filled in with a fragment of DNA polymerase called the Klenow fragment. This enzyme fragment has an advantage over the whole DNA polymerase in that it lacks the normal 5′ → 3′ exonuclease activity, which could degrade the 5′-overhangs before they could be filled in. The end-filling reaction is run with all four nucleotides, one of which (dATP) is labeled, so the DNA end becomes labeled. If more labeling is desired, more than one labeled nucleotide can be used.

SUMMARY In S1 mapping a labeled DNA probe is used to detect the 5′- or 3′-end of a transcript. Hybridization of the probe to the transcript protects a portion of the probe from digestion by S1 nuclease, which specifically degrades single-stranded polynucleotides. The length of the section of probe protected by the transcript locates the end of the transcript, relative to the known location of an end of the probe. Because the amount of probe protected by the transcript is proportional to the concentration of transcript, S1 mapping can also be used as a quantitative method. RNase mapping is a variation on S1 mapping that uses an RNA probe and RNase instead of a DNA probe and S1 nuclease.

Primer Extension

S1 mapping has been used in some classic experiments we will introduce in later chapters, and it is the best method for mapping the 3′-end of a transcript, but it has some

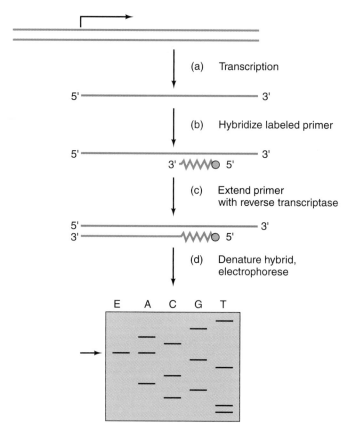

Figure 5.30   Primer extension. (a) Transcription occurs naturally within the cell, so begin by harvesting cellular RNA. **(b)** Knowing the sequence of at least part of the transcript, synthesize and label a DNA oligonucleotide that is complementary to a region not too far from the suspected 5'-end, then hybridize this oligonucleotide to the transcript. It should hybridize specifically to this transcript and to no others. **(c)** Use reverse transcriptase to extend the primer by synthesizing DNA complementary to the transcript, up to its 5'-end. If the primer itself is not labeled, or if it is desirable to introduce extra label into the extended primer, labeled nucleotides can be included in this step. **(d)** Denature the hybrid and electrophorese the labeled, extended primer. In separate lanes run sequencing reactions, performed with the same primer and a DNA from the transcribed region, as markers. In principle, this can indicate the transcription start site to the exact base. In this case, the extended primer (arrow) coelectrophoreses with a DNA fragment in the sequencing A lane. Because the same primer was used in the primer extension reaction and in all the sequencing reactions, this shows that the 5'-end of this transcript corresponds to the middle A (underlined) in the sequence TTCGACTGACAGT.

drawbacks. One is that the S1 nuclease tends to "nibble" a bit on the ends of the RNA–DNA hybrid, or even within the hybrid where A—T-rich regions can melt transiently. On the other hand, sometimes the S1 nuclease will not digest the single-stranded regions completely, so the transcript appears to be a little longer than it really is. These can be serious problems if we need to map the end of a transcript with one-nucleotide accuracy. But another method, called **primer extension** can tell the 5'-end (but not the 3'-end) to the very nucleotide.

Figure 5.30 shows how primer extension works. The first step, transcription, generally occurs naturally in vivo.

We simply harvest cellular RNA containing the transcript whose 5'-end we want to map, and whose sequence is known. Next, we hybridize a labeled oligonucleotide (the primer) of at least 12 nt to the cellular RNA. (We usually use primers at least 18 nt long.) Notice that the specificity of this method derives from the complementarity between the primer and the transcript, just as the specificity of S1 mapping comes from the complementarity between the probe and the transcript. In principle, this primer (or an S1 probe) will be able to pick out the transcript we want to map from a sea of other, unrelated RNAs.

Next, we use reverse transcriptase to extend the oligonucleotide primer to the 5'-end of the transcript. As presented in Chapter 4, reverse transcriptase is an enzyme that performs the reverse of the transcription reaction; that is, it makes a DNA copy of an RNA template. Hence, it is perfectly suited for the job we are asking it to do: making a DNA copy of the RNA we are mapping. Once this primer extension reaction is complete, we can denature the RNA–DNA hybrid and electrophorese the labeled DNA along with markers on a high-resolution gel such as the ones used in DNA sequencing. In fact, it is convenient to use the same primer used during primer extension to do a set of sequencing reactions with dideoxynucleotides. We can then use the products of these sequencing reactions as markers. In the example illustrated here, the product comigrates with a band in the A lane, indicating that the 5'-end of the transcript corresponds to the second A (underlined) in the sequence TTCGACTGACAGT. This is a very accurate determination of the transcription start site.

Just as with S1 mapping, primer extension can also give an estimate of the concentration of a given transcript. The higher the concentration of transcript, the more molecules of labeled primer will hybridize and therefore the more labeled reverse transcripts will be made. The more labeled reverse transcripts, the darker the band on the autoradiograph of the electrophoretic gel.

SUMMARY Using primer extension we can locate the 5'-end of a transcript by hybridizing an oligonucleotide primer to the RNA of interest, extending the primer with reverse transcriptase to the 5'-end of the transcript, and electrophoresing the reverse transcript to determine its size. The intensity of the signal obtained by this method is a measure of the concentration of the transcript.

Run-off Transcription and G-Less Cassette Transcription

Suppose you want to mutate a gene's promoter and observe the effects of the mutations on accuracy and efficiency of transcription. You would need a convenient

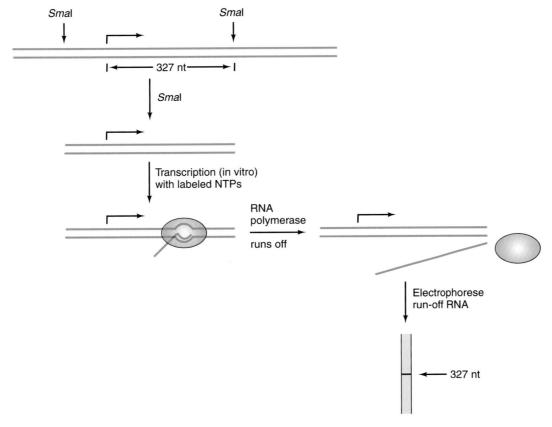

Figure 5.31 Run-off transcription. Begin by cutting the cloned gene, whose transcription is to be measured, with a restriction enzyme. Then transcribe this truncated gene in vitro. When the RNA polymerase (orange) reaches the end of the shortened gene, it falls off and releases the run-off transcript (red). The size of the run-off transcript (327 in this case) can be measured by gel electrophoresis and corresponds to the distance between the start of transcription and the known restriction site at the 3′-end of the shortened gene (a *Sma*I site in this case). The more actively this gene is transcribed, the stronger the 327-signal.

assay that would tell you two things: (1) whether transcription is accurate (i.e., it initiates in the right place, which you have already mapped in previous primer extension or other experiments); and (2) how much of this accurate transcription occurred. You could use S1 mapping or primer extension, but they are relatively complicated. A simpler method, called **run-off transcription,** will give answers more rapidly.

Figure 5.31 illustrates the principle of run-off transcription. You start with a DNA fragment containing the gene you want to map, then cut it with a restriction enzyme in the middle of the transcribed region. Next, you transcribe this truncated gene fragment in vitro with labeled nucleotides so the transcript becomes labeled. Because you have cut the gene in the middle, the polymerase reaches the end of the fragment and simply "runs off." Hence the name of this method. Now you can measure the length of the run-off transcript. Because you know precisely the location of the restriction site at the 3′-end of the truncated gene (a *Sma*I site in this case), the length of the run-off transcript (327 nt in this case) tells you that the start of transcription is 327 bp upstream of the *Sma*I site.

Notice that S1 mapping and primer extension are well suited to mapping transcripts made in vivo; by contrast, run-off transcription relies on transcription in vitro. Thus, it will work only with genes that are accurately transcribed in vitro and cannot give information about cellular transcript concentrations. However, it is a good method for measuring the rate of in vitro transcription. The more transcript is made, the more intense will be the run-off transcription signal. Indeed, run-off transcription is most useful as a quantitative method. Only after you have identified the physiological transcription start site by S1 mapping or primer extension do you use run-off transcription in vitro.

A variation on the run-off theme of quantifying accurate transcription in vitro is the **G-less cassette** assay (Figure 5.32). Here, instead of cutting the gene, a G-less cassette, or stretch of nucleotides lacking guanines in the nontemplate strand is inserted just downstream of the promoter. This template is transcribed in vitro with CTP, ATP, and UTP, one of which is labeled, but no GTP. Transcription will stop at the end of the cassette where the first G is required, yielding an aborted transcript of

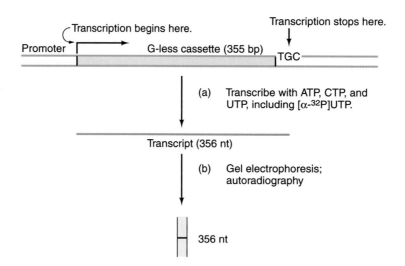

Figure 5.32   G-less cassette assay. (a) Transcribe a template with a G-less cassette (pink) inserted downstream of the promoter in vitro in the absence of GTP. This cassette is 355 bp long, contains no Gs in the nontemplate strand, and is followed by the sequence TGC, so transcription stops just before the G, producing a transcript 356 nt long. **(b)** Electrophorese the labeled transcript and autoradiograph the gel and measure the intensity of the signal, which indicates how actively the cassette was transcribed.

predictable size (based on the size of the G-less cassette, which is usually a few hundred base pairs long). These transcripts are electrophoresed, and the gel is autoradiographed to measure the transcription activity. The stronger the promoter, the more of these aborted transcripts will be produced, and the stronger the corresponding band on the autoradiograph will be.

SUMMARY Run-off transcription is a means of checking the efficiency and accuracy of in vitro transcription. A gene is truncated in the middle and transcribed in vitro in the presence of labeled nucleotides. The RNA polymerase runs off the end and releases an incomplete transcript. The size of this run-off transcript locates the transcription start site, and the amount of this transcript reflects the efficiency of transcription. In G-less cassette transcription, a promoter is fused to a double-stranded DNA cassette lacking G's in the nontemplate strand, then the construct is transcribed in vitro in the absence of GTP. Transcription aborts at the end of the cassette, yielding a predictable size band on gel electrophoresis.

5.5   Measuring Transcription Rates in Vivo

Primer extension, S1 mapping, and Northern blotting are useful for determining the concentrations of specific transcripts in a cell at a given time, but they do not necessarily tell us the rates of synthesis of the transcripts. That is because the transcript concentration depends not only on its rate of synthesis, but also on its rate of degradation. To measure transcription rates, we can employ other methods, including nuclear run-on transcription and reporter gene expression.

Nuclear Run-on Transcription

The idea of this assay (Figure 5.33a) is to isolate nuclei from cells, then allow them to extend in vitro the transcripts they had already started in vivo. This continuing transcription in isolated nuclei is called **run-on transcription** because the RNA polymerase that has already initiated transcription in vivo simply "runs on" or continues to elongate the same RNA chains. The run-on reaction is usually done in the presence of labeled nucleotides so the products will be labeled. Because initiation of new RNA chains in isolated nuclei does not generally occur, one can be fairly confident that any transcription observed in the isolated nuclei is simply a continuation of transcription that was already occurring in vivo. Therefore, the transcripts obtained in a run-on reaction should reveal not only transcription rates but also give an idea about which genes are transcribed in vivo. To eliminate the possibility of initiation of new RNA chains in vitro, one can add heparin, an anionic polysaccharide that binds to any free RNA polymerase and prevents reinitiation.

Once labeled run-on transcripts have been produced, they must be identified. Because few if any of them are complete transcripts, their sizes will not be meaningful. The easiest way to perform the identification is by dot blotting (see Figure 5.33b). Samples of known, denatured

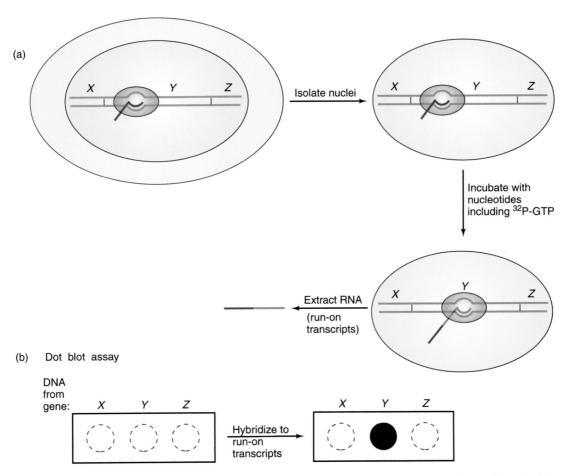

Figure 5.33 **Nuclear run-on transcription. (a)** The run-on reaction. Start with cells that are in the process of transcribing the *Y* gene, but not the *X* or *Z* genes. The RNA polymerase (orange) is making a transcript (blue) of the *Y* gene. Isolate nuclei from these cells and incubate them with nucleotides so transcription can continue (run-on). Also include a labeled nucleotide in the run-on reaction so the transcripts become labeled (red). Finally, extract the labeled run-on transcripts. **(b)** Dot blot assay. Spot single-stranded DNA from genes *X*, *Y*, and *Z* on nitrocellulose or another suitable medium, and hybridize the blot to the labeled run-on transcripts. Because gene *Y* was transcribed in the run-on reaction, its transcript will be labeled, and the gene *Y* spot becomes labeled. The more active the transcription of gene *Y*, the more intense the labeling will be. On the other hand, because genes *X* and *Z* were not active, no labeled *X* and *Z* transcripts were made, so the *X* and *Z* spots remain unlabeled.

DNAs are spotted on a filter and this **dot blot** is hybridized to the labeled run-on RNA. The RNA is identified by the DNA to which it hybridizes. Furthermore, the relative activity of a given gene is proportional to the degree of hybridization to the DNA from that gene. The conditions in the run-on reaction can also be manipulated, and the effects on the products can be measured. For example, inhibitors of certain RNA polymerases can be included to see if they inhibit transcription of a certain gene. If so, the RNA polymerase responsible for transcribing that gene can be identified.

SUMMARY Nuclear run-on transcription is a way of ascertaining which genes are active in a given cell by allowing transcription of these genes to continue in isolated nuclei. Specific transcripts can be identified by their hybridization to known DNAs on dot blots. The run-on assay can also be used to determine the effects of assay conditions on nuclear transcription.

Reporter Gene Transcription

Another way to measure transcription in vivo is to place a surrogate **reporter gene** under control of a specific promoter, and then measure the accumulation of the product of this reporter gene. For example, imagine that you want to examine the structure of a eukaryotic promoter. One way to do this is to make mutations in the DNA region that contains the promoter, then introduce the mutated DNA into cells and measure the effects of the mutations on promoter activity. You can use S1 mapping or primer extension analysis to do this measurement, but you can also substitute a reporter gene for the natural gene, and then assay the activity of the reporter gene product.

Why do it this way? The main reason is that reporter genes have been carefully chosen to have products that are very convenient to assay—more convenient than S1 mapping or primer extension. One of the most popular reporter genes is *lacZ*, whose product, β-galactosidase, can be measured using the substrate X-gal, which turns blue on cleavage. Another widely used reporter gene is the bacterial gene (*cat*) encoding the enzyme **chloramphenicol acetyl transferase** (**CAT**). The growth of most bacteria is inhibited by the antibiotic chloramphenicol (CAM), which blocks a key step in protein synthesis (Chapter 18). Some bacteria have developed a means of evading this antibiotic by acetylating it and therefore blocking its activity. The enzyme that carries out this acetylation is CAT. But eukaryotes are not susceptible to this antibiotic, so they have no need for CAT. Thus, the background level of CAT activity in eukaryotic cells is zero. That means that one can introduce a *cat* gene into these cells, under control of a eukaryotic promoter, and any CAT activity observed is due to the introduced gene.

How could you measure CAT activity in cells that have been transfected with the *cat* gene? In one of the most popular methods, an extract of the transfected cells is mixed with radioactive chloramphenicol and an acetyl donor (acetyl-CoA). Then thin-layer chromatography is used to separate the chloramphenicol from its acetylated products. The greater the concentrations of these products, the higher the CAT activity in the cell extract, and therefore the higher the promoter activity. Figure 5.34 outlines this procedure.

Another standard reporter gene is the **luciferase** gene from firefly lanterns. The enzyme luciferase, mixed with ATP and luciferin, converts the luciferin to a chemiluminescent compound that emits light. That is the secret of the firefly's lantern, and it also makes a convenient reporter because the light can be detected easily with x-ray film, or even in a scintillation counter.

In the experiments described here, we are assuming that the amount of reporter gene product is a reasonable measure of transcription rate (the number of RNA chain initiations per unit of time) and therefore of promoter activity. But the gene products come from a two-step process that includes translation as well as transcription. Ordinarily, we are justified in assuming that the translation rates do not vary from one DNA construct to another, as long as we are manipulating only the promoter. That is because the promoter lies outside the coding region. For this reason changes in the promoter cannot affect the structure of the mRNA itself and therefore should not affect translation. However, we can deliberately make changes in the region of a gene that will be transcribed into mRNA and then use our reporter to measure the ef-

fects of these changes on translation. Thus, depending on where the changes to a gene are made, a reporter gene can detect alterations in either transcription or translation rates.

SUMMARY To measure the activity of a promoter, one can link it to a reporter gene, such as the genes encoding β-galactosidase, CAT, or luciferase, and let the easily assayed reporter gene products indicate the activity of the promoter. One can also use reporter genes to detect changes in translational efficiency after altering regions of a gene that affect translation.

5.6   Assaying DNA–Protein Interactions

Another of the recurring themes of molecular biology is the study of DNA–protein interactions. We have already discussed RNA polymerase–promoter interactions, and we will encounter many more examples. Therefore, we need methods to quantify these interactions and to determine exactly what part of the DNA interacts with a given protein. We will consider here two methods for detecting protein–DNA binding and three examples of methods for showing which DNA bases interact with a protein.

Filter Binding

Nitrocellulose membrane filters have been used for decades to filter-sterilize solutions. Part of the folklore of molecular biology is that someone discovered by accident that DNA can bind to such nitrocellulose filters because they lost their DNA preparation that way. Whether this story is true or not is unimportant. What is extremely important is that nitrocellulose filters can indeed bind DNA, but only under certain conditions. Single-stranded DNA binds readily to nitrocellulose, but double-stranded DNA by itself does not. On the other hand, protein does bind, and if a protein is bound to double-stranded DNA the protein–DNA complex will bind. This is the basis of the assay portrayed in Figure 5.35.

In Figure 5.35a, labeled double-stranded DNA is poured through a nitrocellulose filter. The amount of label in the filtrate (the material that passes through the filter) and in the filter-bound material is measured, which shows that all the labeled material has passed through the filter into the filtrate. This confirms that double-stranded DNA does not bind to nitrocellulose. In Figure 5.35b, a solution of a labeled protein is filtered, showing that all the protein is bound to the filter. This demonstrates that proteins bind by themselves to the filter. In Figure 5.35c,

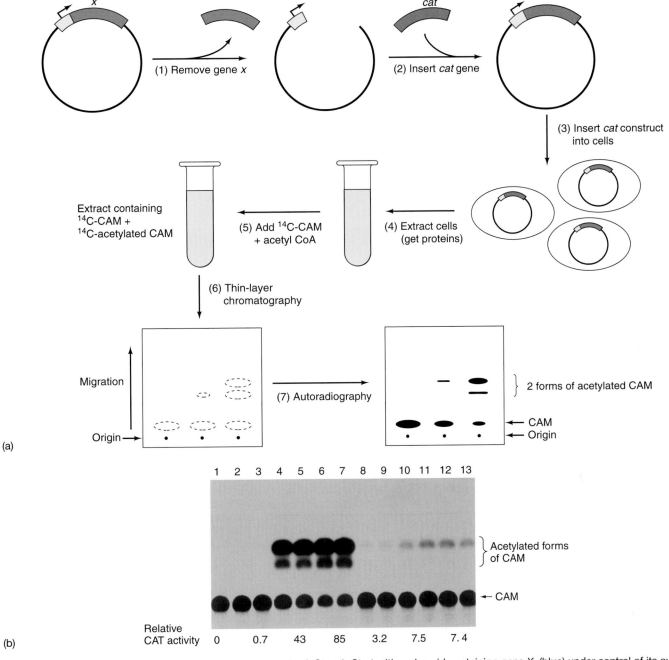

Figure 5.34 Using a reporter gene. (a) Outline of the method. Step 1: Start with a plasmid containing gene *X*, (blue) under control of its own promoter (yellow) and use restriction enzymes to remove the coding region of gene *X*. Step 2: Insert the bacterial *cat* gene under control of the *X* gene's promoter. Step 3: Place this construct into eukaryotic cells. Step 4: After a period of time, make an extract of the cells that contains soluble cellular proteins. Step 5: To begin the CAT assay, add ^{14}C-CAM and the acetyl donor acetyl-CoA. Step 6: Perform thin-layer chromatography to separate acetylated and unacetylated species of CAM. Step 7: Finally, subject the thin layer to autoradiography to visualize CAM and its acetylated derivatives. Here CAM is seen near the bottom of the autoradiograph and two acetylated forms of CAM, with higher mobility, are seen near the top. **(b)** Actual experimental results. Again, the parent CAM is near the bottom, and two acetylated forms of CAM are nearer the top. The experimenters scraped these radioactive species off of the thin-layer plate and subjected them to liquid scintillation counting, yielding the CAT activity values reported at the bottom (averages of duplicate lanes). Lane 1 is a negative control with no cell extract. *Abbreviations:* CAM = chloramphenicol; CAT = chloramphenicol acetyl transferase. (*Source:* (b) Qin, Liu, and Weaver. Studies on the control region of the p10 gene of the *Autographa californica* nuclear polyhedrosis virus. *J. General Virology* 70 (1989) f. 2, p. 1276. (Society for General Microbiology, Reading, England).)

(a) Double-stranded DNA   (b) Protein   (c) Protein-DNA complex

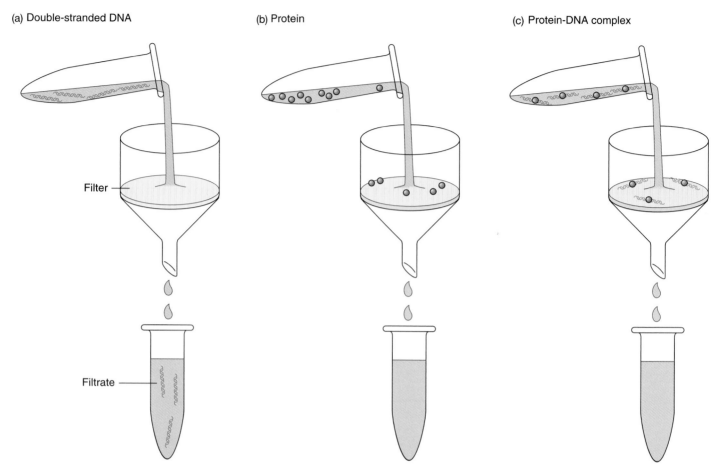

Figure 5.35   Nitrocellulose filter binding assay. (a) Double-stranded DNA. End-label double-stranded DNA (red), and pass it through a nitrocellulose filter. Then monitor the radioactivity on the filter and in the filtrate by liquid scintillation counting. None of the radioactivity sticks to the filter, indicating that double-stranded DNA does not bind to nitrocellulose. Single-stranded DNA, on the other hand, binds tightly. **(b)** Protein. Label a protein (green), and filter it through nitrocellulose. The protein binds to the nitrocellulose. **(c)** Double-stranded DNA-protein complex. Mix an end-labeled double-stranded DNA (red) with an unlabeled protein (green) to which it binds to form a DNA–protein complex. Then filter the complex through nitrocellulose. The labeled DNA now binds to the filter because of its association with the protein. Thus, double-stranded DNA–protein complexes bind to nitrocellulose, providing a convenient assay for association between DNA and protein.

double-stranded DNA is again labeled, but this time it is mixed with a protein to which it binds. Because the protein binds to the filter, the protein–DNA complex will also bind, and the radioactivity is found bound to the filter, rather than in the filtrate. Thus, filter binding is a direct measure of DNA–protein interaction.

> **SUMMARY** Filter binding as a means of measuring DNA–protein interaction is based on the fact that double-stranded DNA will not bind by itself to a nitrocellulose filter or similar medium, but a protein–DNA complex will. Thus, one can label a double-stranded DNA, mix it with a protein, and assay protein–DNA binding by measuring the amount of label retained by the filter.

Gel Mobility Shift

Another method for detecting DNA–protein interaction relies on the fact that a small DNA has a much higher mobility in gel electrophoresis than the same DNA does when it is bound to a protein. Thus, one can label a short, double-stranded DNA fragment, then mix it with a protein, and electrophorese the complex. Then one subjects the gel to autoradiography to detect the labeled species. Figure 5.36 shows the electrophoretic mobilities of three different species. Lane 1 contains naked DNA, which has a very high mobility because of its small size. Recall from earlier in this chapter that DNA electropherograms are conventionally depicted with their origins at the top, so high-mobility DNAs are found near the bottom, as shown here. Lane 2 contains the same DNA bound to a protein, and its mobility is greatly reduced. This is the origin of

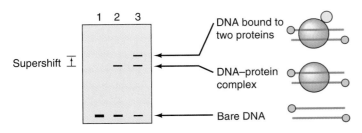

Figure 5.36 Gel mobility shift assay. Subject pure, labeled DNA or DNA–protein complexes to gel electrophoresis, then autoradiograph the gel to detect the DNAs and complexes. Lane 1 shows the high mobility of bare DNA. Lane 2 shows the mobility shift that occurs on binding a protein (red) to the DNA. Lane 3 shows the supershift caused by binding a second protein (yellow) to the DNA–protein complex. The orange dots at the ends of the DNAs represent terminal labels.

but they will not be detected because they are unlabeled. A control with DNA alone (no protein) is always included, and more than one protein concentration is usually used so the gradual disappearance of the bands in the footprint region reveals that protection of the DNA depends on the concentration of added protein. The footprint represents the region of DNA protected by the protein, and therefore tells where the protein binds.

DMS Footprinting and Other Footprinting Methods

DNase footprinting gives a good idea of the location of the binding site for the protein, but DNase is a macromolecule and is therefore a rather blunt instrument for probing the fine details of the binding site. That is, gaps may occur in the interaction between protein and DNA that DNase would not fit into and therefore would not detect. Moreover, DNA binding proteins frequently perturb the DNA within the binding region, distorting the double helix. These perturbations are interesting, but are not generally detected by DNase footprinting because the protein keeps the DNase away. More detailed footprinting, requires a smaller molecule that can fit into the nooks and crannies of the DNA–protein complex and reveal more of the subtleties of the interaction. A favorite tool for this job is the methylating agent **dimethylsulfate (DMS).**

Figure 5.38 illustrates DMS footprinting, which starts in the same way as DNase footprinting, with end-labeling the DNA and binding the protein. Then the DNA–protein complex is methylated with DMS, using a mild treatment so that on average only one methylation event occurs per DNA molecule. Next, the protein is dislodged, and the DNA is treated with piperidine, which removes methylated purines, creating apurinic sites (deoxyriboses without bases), then breaks the DNA at these apurinic sites. These reactions are the same as the Maxam–Gilbert DNA sequencing reactions described earlier in this chapter. Finally, the DNA fragments are electrophoresed, and the gel is autoradiographed to detect the labeled DNA bands. Each band ends next to a nucleotide that was methylated and thus unprotected by the protein. In this example, three bands progressively disappear as more and more protein is added. But one band actually becomes more prominent at high protein concentration. This suggests that binding the protein distorts the DNA double helix such that it makes the base corresponding to this band more vulnerable to methylation.

In addition to DNase and DMS, other reagents are commonly used to footprint protein–DNA complexes by breaking DNA except where it is protected by bound proteins. For example, organometallic complexes containing copper or iron act by generating **hydroxyl radicals** that attack and break DNA strands.

the name for this technique: **gel mobility shift assay.** Sometimes it is also called a "band shift assay." Lane 3 depicts the behavior of the same DNA bound to two proteins. The mobility is reduced still further because of the greater mass of protein clinging to the DNA. This is called a **supershift.** The protein could be another DNA-binding protein, or a second protein that binds to the first. It can even be an antibody that specifically binds to the first protein.

> **SUMMARY** A gel mobility shift assay detects interaction between a protein and DNA by the reduction of the electrophoretic mobility of a small DNA that occurs on binding to a protein.

DNase Footprinting

Footprinting is a method for detecting protein–DNA interactions that can tell where the target site lies on the DNA and even which bases are involved in protein binding. Several methods are available, but three are very popular: DNase, dimethylsulfate (DMS), and hydroxyl radical footprinting. **DNase footprinting** (Figure 5.37) relies on the fact that a protein, by binding to DNA, covers the binding site and so protects it from attack by DNase. In this sense, it leaves its "footprint" on the DNA. The first step in a footprinting experiment is to end-label the DNA. Either strand can be labeled, but only one strand per experiment. Next, the protein (yellow in the figure) is bound to the DNA. Then the DNA–protein complex is treated with DNase I under mild conditions (very little DNase), so that an average of only one cut occurs per DNA molecule. Next, the protein is removed from the DNA, the DNA strands are separated, and the resulting fragments are electrophoresed on a high-resolution polyacrylamide gel alongside size markers (not shown). Of course, fragments will arise from the other end of the DNA as well,

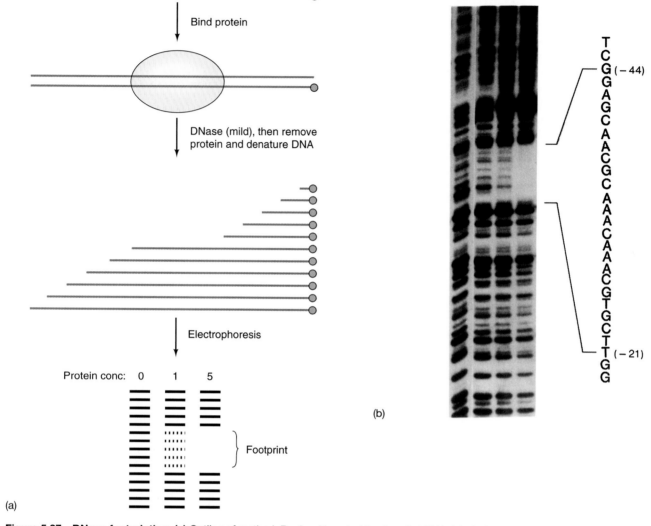

(a)

(b)

Figure 5.37   DNase footprinting. (a) Outline of method. Begin with a double-stranded DNA, labeled at one end (orange). Next, bind a protein to the DNA. Next, digest the DNA–protein complex under mild conditions with DNase I, so as to introduce approximately one break per DNA molecule. Next, remove the protein and denature the DNA, yielding the end-labeled fragments shown at center. Notice that the DNase cut the DNA at regular intervals except where the protein bound and protected the DNA. Finally, electrophorese the labeled fragments, and perform autoradiography to detect them. The three lanes represent DNA that was bound to 0, 1, and 5 units of protein. The lane with no protein shows a regular ladder of fragments. The lane with one unit of protein shows some protection, and the lane with five units of proteins shows complete protection in the middle. This protected area is called the footprint; it shows where the protein binds to the DNA. Sequencing reactions performed on the same DNA in parallel lanes are usually included. These serve as size markers, that show exactly where the protein bound. **(b)** Actual experimental results. Lanes 1–4 contained DNA bound to 0, 10, 18, and 90 pmol of protein, respectively (1 pmol = 10^{-12} mol). The DNA sequence was obtained previously by standard dideoxy sequencing. (*Source: (b)* Ho et al. Bacteriophage lambda protein *c*II binds promoters on the opposite face of the DNA helix from RNA polymerase. *Nature* 304 (25 Aug 1983) p. 705, f. 3, © Macmillan Magazines Ltd.)

SUMMARY Footprinting is a means of finding the target DNA sequence, or binding site, of a DNA-binding protein. DNase footprinting is performed by binding the protein to its end-labeled DNA target, then attacking the DNA–protein complex with DNase. When the resulting DNA fragments are electrophoresed, the protein binding site shows up as a gap, or "footprint" in the pattern where the protein protected the DNA from degradation. DMS footprinting follows a similar principle, ex-cept that the DNA methylating agent DMS, instead of DNase, is used to attack the DNA–protein complex. The DNA is then broken at the methylated sites. Unmethylated (or hypermethylated) sites show up on electrophoresis of the labeled DNA fragments and demonstrate where the protein bound to the DNA. Hydroxyl radical footprinting uses copper- or iron-containing organometallic complexes to generate hydroxyl radicals that break DNA strands.

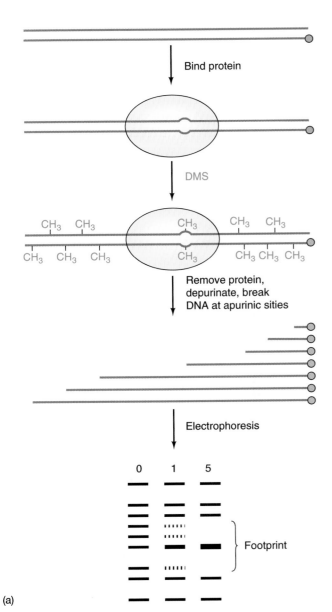

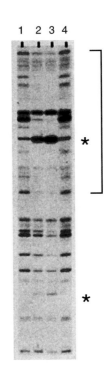

(b)

(a)

Figure 5.38   DMS footprinting. (a) Outline of the method. As in DNase footprinting, start with an end-labeled DNA, then bind a protein (yellow) to it. In this case, the protein causes some tendency of the DNA duplex to melt in one region, represented by the small "bubble." Next, methylate the DNA with DMS. This adds methyl groups (CH_3 red) to certain bases in the DNA. Do this under mild conditions so that, on average, only one methylated base occurs per DNA molecule (even though all seven methylations are shown together on one strand for convenience here). Next, use the Maxam–Gilbert sequencing reagents to remove methylated purines from the DNA, then to break the DNA at these apurinic sites. This yields the labeled DNA fragments depicted at center. Electrophorese these fragments and autoradiograph the gel to give the results shown at bottom. Notice that three sites are protected against methylation by the protein, but one site is actually made more sensitive to methylation (darker band). This is because of the opening up of the double helix that occurs in this position when the protein binds. **(b)** Actual experimental results. Lanes 1 and 4 have no added protein, whereas lanes 2 and 3 have increasing concentrations of a protein that binds to this region of the DNA. The bracket indicates a pronounced footprint region. The asterisks denote bases made *more* susceptible to methylation by protein binding. (*Source:* (b) Learned et al. Human rRNA transcription is modulated by the coordinate binding of two factors to an upstream control element. *Cell* 45 (20 June 1986) p. 849, f. 2a. Reprinted by permission of Elsevier Science.)

5.7   Knockouts

Most of the techniques we have discussed in this chapter are designed to probe the structures and activities of genes. But these frequently leave a big question about the role of the gene being studied: What purpose does the gene play in the life of the organism? We can often answer this question best by seeing what happens when we create deliberate mutations in a particular gene in a living organism. We now have techniques for targeted disruption of genes in several organisms. For example, we can disrupt genes in mice, and when we do, we call the products **knockout mice.**

Figure 5.39 explains one way to begin the process of creating a knockout mouse. We start with cloned DNA

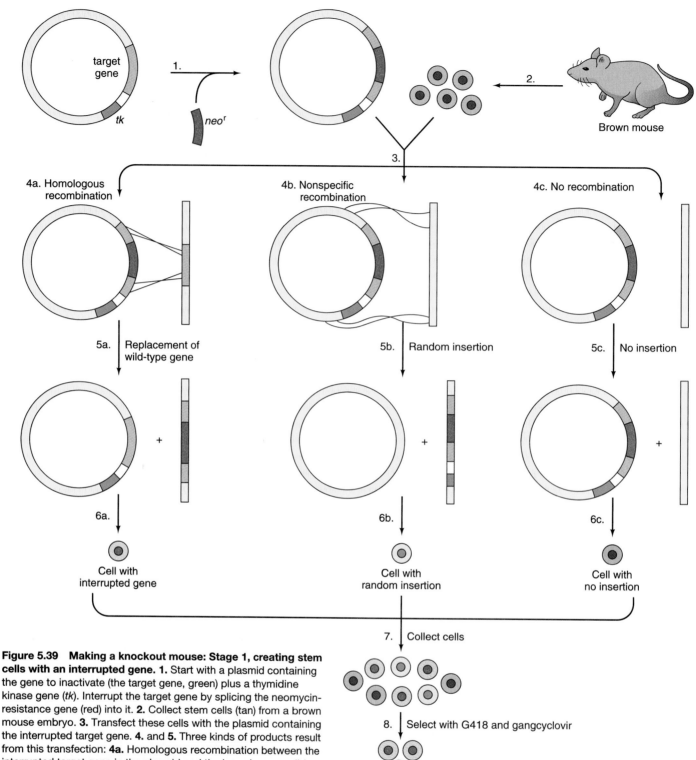

Figure 5.39 Making a knockout mouse: Stage 1, creating stem cells with an interrupted gene. 1. Start with a plasmid containing the gene to inactivate (the target gene, green) plus a thymidine kinase gene (*tk*). Interrupt the target gene by splicing the neomycin-resistance gene (red) into it. **2.** Collect stem cells (tan) from a brown mouse embryo. **3.** Transfect these cells with the plasmid containing the interrupted target gene. **4.** and **5.** Three kinds of products result from this transfection: **4a.** Homologous recombination between the interrupted target gene in the plasmid and the homologous, wild-type gene causes replacement of the wild-type gene in the cellular genome by the interrupted gene **(5a). 4b.** Nonspecific recombination with a nonhomologous sequence in the cellular genome results in random insertion of the interrupted target gene plus the *tk* gene into the cellular genome **(5b). 4c.** When no recombination occurs, the interrupted target gene is not integrated into the cellular genome at all **(5c). 6.** The cells resulting from these three events are color-coded as indicated: Homologous recombination yields a cell (red) with an interrupted target gene **(6a);** nonspecific recombination yields a cell (blue) with the interrupted target gene and the *tk* gene inserted at random **(6b);** no recombination yields a cell (tan) with no integration of the interrupted gene. **7.** Collect the transfected cells, containing all three types (red, blue, and tan). **8.** Grow the cells in medium containing the neomycin analog G418 and the drug gangcyclovir. The G418 kills all cells without a neomycin-resistance gene, namely those cells (tan) that did not experience a recombination event. The gangcyclovir kills all cells that have a *tk* gene, namely those cells (blue) that experienced a nonspecific recombination. This leaves only the cells (red) that experienced homologous recombination and therefore have an interrupted target gene.

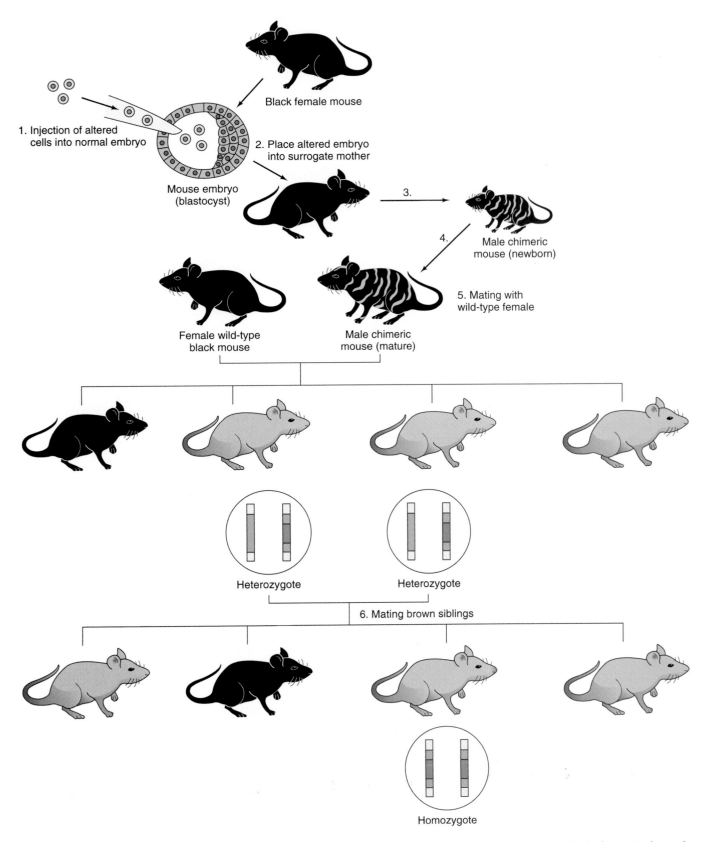

Figure 5.40 Making a knockout mouse: Stage 2, placing the interrupted gene in the animal. (1) Inject the cells with the interrupted gene (see Figure 5.39) into a blastocyst-stage embryo from black parent mice. **(2)** Transplant this mixed embryo to the uterus of a surrogate mother. **(3)** The surrogate mother gives birth to a chimeric mouse, which one can identify by its black and brown coat. (Recall that the altered cells came from an agouti [brown] mouse, and they were placed into an embryo from a black mouse.) **(4)** Allow the chimeric mouse (a male) to mature. **(5)** Mate it with a wild-type black female. Discard any black offspring, which must have derived from the wild-type blastocyst; only brown mice could have derived from the transplanted cells. **(6)** Select a brown brother and sister pair, both of which show evidence of an interrupted target gene (by Southern blot analysis), and mate them. Again, examine the DNA of the brown progeny by Southern blotting. This time, one animal that is homozygous for the interrupted target gene is found. This is the knockout mouse. Now observe this animal to determine the effects of knocking out the target gene.

containing the mouse gene we want to knock out. We interrupt this gene with another gene that confers resistance to the antibiotic neomycin. Elsewhere in the cloned DNA, outside the target gene, we introduce a thymidine kinase (*tk*) gene. Later, these extra genes will enable us to weed out those clones in which targeted disruption did not occur.

Next, we mix the engineered mouse DNA with stem cells from an embryonic brown mouse. In a small percentage of these cells, the interrupted gene will find its way into the nucleus and homologous recombination will occur between the altered gene and the resident, intact gene. This recombination places the altered gene into the mouse genome and removes the *tk* gene. Unfortunately, such recombination events are relatively rare, so many stem cells will experience no recombination and therefore will suffer no interruption of their resident gene. Still other cells will experience nonspecific recombination, in which the interrupted gene will insert randomly into the genome without replacing the intact gene.

The problem now is to eliminate the cells in which homologous recombination did not occur. This is where the extra genes we introduced earlier come in handy. Cells in which no recombination took place will have no neomycin resistance gene. Thus, we can eliminate them by growing the cells in medium containing the neomycin derivative G418. Cells that experienced nonspecific recombination will have incorporated the *tk* gene, along with the interrupted gene, into their genome. We can kill these cells with gangcyclovir, a drug that is lethal to *tk*+ cells. (The stem cells we used are *tk*−.) Treatment with these two drugs leaves us with engineered cells that have undergone homologous recombination and are therefore heterozygous for an interruption in the target gene.

Our next task is to introduce this interrupted gene into a whole mouse (Figure 5.40). We do this by injecting our engineered cells into a mouse blastocyst that is destined to develop into a black mouse. Then we place this altered embryo into a surrogate mother, who eventually gives birth to a chimeric mouse. We can recognize this mouse as a chimera by its patchy coat; the black zones come from the original black embryo, and the brown zones result from the transplanted engineered cells.

To get a mouse that is a true heterozygote instead of a chimera, we allow the chimera to mature, then mate it with a black mouse. Because brown (agouti) is dominant, some of the progeny should be brown. In fact, all of the offspring resulting from gametes derived from the engineered stem cells should be brown. But only half of these brown mice will carry the interrupted gene because the engineered stem cells were heterozygous for the knockout. Southern blots showed that two of the brown mice in our example carry the interrupted gene. We mate these and look for progeny that are homozygous for the knockout by examination of their DNA. In our example, one of the

mice from this mating is a knockout, and now our job is to look for its phenotype. Frequently, the phenotype is not obvious. But obvious or not, it can be very instructive.

In other cases, the knockout is lethal and the affected mouse fetuses die before birth. Still other knockouts have intermediate effects. For example, consider the tumor suppressor gene called *p53*. Humans with defects in this gene are highly susceptible to certain cancers. Mice with their *p53* gene knocked out develop normally but are afflicted with tumors at an early age.

SUMMARY To probe the role of a gene, molecular biologists can perform targeted disruption of the corresponding gene in a mouse, and then look for the effects of that mutation on the "knockout mouse."

SUMMARY

Methods of purifying proteins and nucleic acids are crucial in molecular biology. DNAs, RNAs, and proteins of various sizes can be separated by gel electrophoresis. The most common gel used in nucleic acid electrophoresis is agarose, and polyacrylamide is usually used in protein electrophoresis. Sodium dodecyl sulfate polyacrylamide gel electrophoresis (SDS-PAGE) is used to separate polypeptides according to their sizes. High-resolution separation of polypeptides can be achieved by two-dimensional gel electrophoresis, which uses isoelectric focusing in the first dimension and SDS-PAGE in the second. Ion-exchange chromatography can be used to separate substances, including proteins, according to their charges. Positively charged resins like DEAE-Sephadex are used for anion-exchange chromatography, and negatively charged resins like phosphocellulose are used for cation-exchange chromatography. Gel filtration chromatography uses columns filled with porous resins that let smaller substances in, but exclude larger substances. Thus, the smaller substances are slowed as they move, but larger substances travel relatively rapidly through the column.

Detection of the tiny quantities of substances in molecular biology experiments generally requires labeled tracers. If the tracer is radioactive it can be detected by autoradiography, using x-ray film or a phosphorimager, or by liquid scintillation counting. Some very sensitive nonradioactive labeled tracers are also now available. These produce light (chemiluminescence) or colored spots.

Labeled DNA (or RNA) probes can be hybridized to DNAs of the same, or very similar, sequence on a Southern blot. Modern DNA typing uses Southern blots and a battery of DNA probes to detect variable sites in individual animals, including humans. As a forensic tool, DNA typing can be used to test parentage, to identify criminals, or to remove innocent people from suspicion.

A Northern blot is similar to a Southern blot, but it contains electrophoretically separated RNAs instead of DNAs. The RNAs on the blot can be detected by hybridizing them to a labeled probe. The intensities of the bands reveal the relative amounts of specific RNA in each, and the positions of the bands indicate the lengths of the respective RNAs.

Labeled probes can be hybridized to whole chromosomes to locate genes or other specific DNA sequences. This type of procedure is called in situ hybridization; if the probe is fluorescently labeled, the technique is called fluorescence in situ hybridization (FISH).

The sequence of a DNA can be determined in two ways. The Sanger DNA sequencing method uses dideoxy nucleotides to terminate DNA synthesis, yielding a series of DNA fragments whose sizes can be measured by electrophoresis. The last base in each of these fragments is known because we know which dideoxy nucleotide was used to terminate each reaction. Therefore, ordering these fragments by size—each fragment one (known) base longer than the next—tells us the base sequence of the DNA. The Maxam–Gilbert method allows us to break an end-labeled DNA strand at specific bases using base-specific reagents. We can then electrophorese the labeled fragments in a high-resolution polyacrylamide gel and detect them by autoradiography, just as in the chain-termination method. This procedure orders the fragments by size and therefore tells the sequence of the DNA.

A physical map depicts the spatial arrangement of physical "landmarks," such as restriction sites, on a DNA molecule. We can improve a restriction map considerably by Southern blotting some of the fragments and then hybridizing these fragments to labeled fragments generated by another restriction enzyme. This reveals overlaps between individual restriction fragments.

Using cloned genes, we can introduce changes conveniently by site-directed mutagenesis, thus altering the amino acid sequences of the protein products. The mutagenized DNA can be made with double-stranded DNA, two complementary mutagenic primers, and PCR. Wild-type DNA can be eliminated by cleavage with *Dpn*I, so clones are transformed primarily with mutagenized DNA, not with wild-type.

In S1 mapping, a labeled DNA probe is used to detect the 5′- or 3′-end of a transcript. Hybridization of the probe to the transcript protects a portion of the probe from digestion by S1 nuclease, which specifically degrades single-stranded polynucleotides. The length of the section of probe protected by the transcript locates the end of the transcript, relative to the known location of an end of the probe. Because the amount of probe protected by the

transcript is proportional to the concentration of transcript, S1 mapping can also be used as a quantitative method. RNase mapping is a variation on S1 mapping that uses an RNA probe and RNase instead of a DNA probe and S1 nuclease.

Using primer extension one can locate the 5′-end of a transcript by hybridizing an oligonucleotide primer to the RNA of interest, extending the primer with reverse transcriptase to the 5′-end of the transcript, and electrophoresing the reverse transcript to determine its size. The intensity of the signal obtained by this method is a measure of the concentration of the transcript.

Run-off transcription is a means of checking the efficiency and accuracy of in vitro transcription. We truncate a gene in the middle and transcribe it in vitro in the presence of labeled nucleotides. The RNA polymerase runs off the end and releases an incomplete transcript. The size of this run-off transcript locates the transcription start site, and the amount of this transcript reflects the efficiency of transcription. G-less cassette transcription also produces a shortened transcript of predictable size, but does so by placing a G-less cassette just downstream of a promoter and transcribing this construct in the absence of GTP.

Nuclear run-on transcription is a way of ascertaining which genes are active in a given cell by allowing transcription of these genes to continue in isolated nuclei. Specific transcripts can be identified by their hybridization to known DNAs on dot blots. We can also use the run-on assay to determine the effects of assay conditions on nuclear transcription.

To measure the activity of a promoter, one can link it to a reporter gene, such as the genes encoding β-galactosidase, CAT, or luciferase, and let the easily assayed reporter gene products tell us indirectly the activity of the promoter. One can also use reporter genes to detect changes in translational efficiency after altering regions of a gene that effect translation.

Filter binding as a means of measuring DNA–protein interaction is based on the fact that double-stranded DNA will not bind by itself to a nitrocellulose filter, or similar medium, but a protein–DNA complex will. Thus, one can label a double-stranded DNA, mix it with a protein, and assay protein–DNA binding by measuring the amount of label retained by the filter. A gel mobility shift assay detects interaction between a protein and DNA by the reduction of the electrophoretic mobility of a small DNA that occurs when the DNA binds to a protein.

Footprinting is a means of finding the target DNA sequence, or binding site, of a DNA-binding protein. We perform DNase footprinting by binding the protein to its DNA target, then digesting the DNA–protein complex with DNase. When we electrophorese the resulting DNA fragments, the protein binding site shows up as a gap, or "footprint," in the pattern where the protein protected the DNA from degradation. DMS footprinting follows a similar principle, except that we use the DNA methylating agent DMS, instead of DNase, to attack the DNA–protein complex. Unmethylated (or hypermethylated) sites show up on electrophoresis and demonstrate where the protein is bound to the DNA. Hydroxyl radical footprinting uses organometallic complexes to generate hydroxyl radicals that break DNA strands.

To probe the role of a gene, we can perform targeted disruption of the corresponding gene in a mouse, and then look for the effects of that mutation on the "knockout mouse."

REVIEW QUESTIONS

1. Use a drawing to illustrate the principle of DNA gel electrophoresis. Indicate roughly the comparative electrophoretic mobilities of DNAs with 150, 600, and 1200 bp.

2. A certain double-stranded DNA fragment has a mobility of 30 mm in the electrophoresis experiment depicted in Figure 5.2b. What is the approximate size in base pairs?

3. Compare and contrast SDS-PAGE and modern two-dimensional gel electrophoresis of proteins.

4. Describe the principle of ion-exchange chromatography. Use a graph to illustrate the separation of three different proteins by this method.

5. Describe the principle of gel filtration chromatography. Use a graph to illustrate the separation of three different proteins by this method. Indicate on the graph the largest and smallest of these proteins.

6. Compare and contrast the principles of autoradiography and phosphorimaging. Which method provides more quantitative information?

7. Describe a nonradioactive method for detecting a particular nucleic acid fragment in an electrophoretic gel.

8. Diagram the process of Southern blotting and probing to detect a DNA of interest. Compare and contrast this procedure with Northern blotting.

9. Describe a DNA fingerprinting method using a minisatellite probe. Compare this method with a modern forensic DNA typing method using probes to detect single variable DNA loci.

10. What kinds of information can we obtain from a Northern blot?

11. Describe fluorescence in situ hybridization (FISH). When would you use this method, rather than Southern blotting?

12. Compare and contrast the principles of the Maxam–Gilbert method and Sanger chain-termination method of DNA sequencing.

13. Draw a diagram of an imaginary Sanger sequencing autoradiograph, and provide the corresponding DNA sequence.

14. Show how a manual DNA sequencing method can be automated.

15. Show how to use restriction mapping to determine the orientation of a restriction fragment ligated into a restriction site in a vector. Use fragment sizes different from those in the text.

16. Explain the principle of site-directed mutagenesis, then describe a method to carry out this process.

17. Compare and contrast the S1 mapping and primer extension methods for mapping the 5'-end of an mRNA. Which of these methods can be used to map the 3'-end of an mRNA. Why would the other method not work?

18. Describe the run-off transcription method. Why does this method not work with in vivo transcripts, as S1 mapping and primer extension do?

19. How would you label the 5'-ends of a double-stranded DNA? The 3'-ends?

20. Describe a nuclear run-on assay, and show how it differs from a run-off assay.

21. How does a dot blot differ from a Southern blot?

22. Describe the use of a reporter gene to measure the strength of a promoter.

23. Describe a filter-binding assay to measure binding between a DNA and a protein.

24. Compare and contrast the gel mobility shift and DNase footprinting methods of assaying specific DNA–protein interactions. What information does DNase footprinting provide that gel mobility shift does not?

25. Compare and contrast DMS and DNase footprinting. Why is the former method more precise than the latter?

26. Describe a method for creating a knockout mouse. Explain the importance of the thymidine kinase and neomycin resistance genes in this procedure. What information can a knockout mouse provide?

27. Design a Southern blot experiment to check a chimeric mouse's DNA for insertion of the neomycin resistance gene. You may assume any array of restriction sites you wish in the target gene and in the *neo*[r] gene. Show sample results for a successful and an unsuccessful insertion.

SUGGESTED READINGS

Galas, D. J., and A. Schmitz. 1978. DNAse footprinting: A simple method for the detection of protein-DNA binding specificity. *Nucleic Acids Research.* 5:3157–70.

Lichter, P. 1990. High resolution mapping of human chromosome 11 by in situ hybridization with cosmid clones. *Science* 247:64–69.

Maxam, A. M., and W. Gilbert. 1980. Sequencing end-labeled DNA with base-specific chemical cleavage. *Methods in Enzymology,* vol. 65, p. 499. Academic Press, Inc.

Sambrook, J., et al. 2001. *Molecular Cloning: A Laboratory Manual,* 3rd ed. Plainview, NY: Cold Spring Harbor Laboratory Press.

CHAPTER 6

The Transcription Apparatus of Prokaryotes

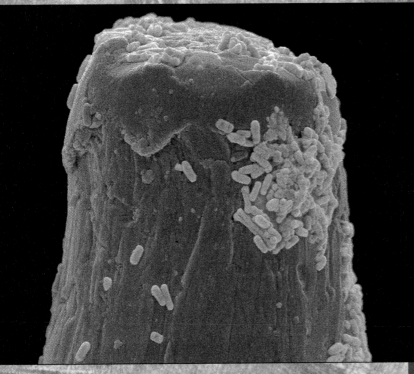

False color scanning electron micrograph of *E. coli* bacteria on the tip of a pin (×1700). © *Andrew Syred/Tony Stone Images.*

In Chapter 3 we learned that transcription is the first step in gene expression. Indeed, transcription is a vital control point in the expression of many genes. The next four chapters will examine in detail the mechanism of transcription and its control in prokaryotes. In this chapter we will focus on RNA polymerase, the enzyme that catalyzes transcription. We will also look at the interaction between RNA polymerase and DNA. This interaction begins when an RNA polymerase docks at a promoter (a specific polymerase binding site next to a gene), continues as the polymerase elongates the RNA chain, and ends when the polymerase reaches a terminator, or stopping point, and releases the finished transcript.

6.1   RNA Polymerase Structure

As early as 1960–1961, RNA polymerases were discovered in animals, plants, and bacteria. And, as you might anticipate, the bacterial enzyme was the first to be studied in great detail. By 1969, the polypeptides that make up the *E. coli* RNA polymerase had been identified by SDS polyacrylamide gel electrophoresis (SDS-PAGE) as described in Chapter 5.

Figure 6.1 is a cartoon representing the results of an SDS-PAGE experiment on the *E. coli* RNA polymerase. This enzyme is remarkable in that it contains two very large subunits: **beta (β)** and **beta-prime (β′)**, with molecular masses of 150 and 160 kD, respectively. The other RNA polymerase subunits are called **sigma (σ)** and **alpha (α)**, with molecular masses of 70 and 40kD, respectively. The subunit content of an RNA polymerase **holoenzyme** is β′, β, σ, α_2; in other words, two molecules of α and one of all the others are present.

When Richard Burgess and Andrew Travers subjected the RNA polymerase **holoenzyme** (the whole enzyme) to ion-exchange chromatography on a negatively charged resin called phosphocellulose, they discovered that they had separated the σ-subunit from the remainder of the enzyme, called the **core enzyme** (see Figure 6.1). In the process, they had caused a profound change in the enzyme's activity (Table 6.1). Whereas the holoenzyme could transcribe intact phage T4 DNA in vitro quite actively, the core enzyme had little ability to do this. On the other hand, core polymerase retained its basic RNA polymerizing function because it could still transcribe highly nicked templates very well. (As will be explained, transcription of nicked DNA is a laboratory artifact and has no biological significance.)

σ as a Specificity Factor

Adding σ back to the core reconstituted the enzyme's ability to transcribe unnicked T4 DNA. Even more significantly, Ekkehard Bautz and colleagues showed that the holoenzyme transcribed only a certain class of T4 genes (called immediate early genes), but the core showed no such specificity.

What do we mean by immediate early genes? Consider the replication cycle of the phage T4. Transcription of most viruses can be divided into three phases. The first phase, usually called **immediate early,** uses only the host transcription machinery, so it can occur immediately after infection begins. In T4 phage-infected cells, immediate early transcription takes place during the first 2 min of infection, although some immediate early transcription can continue beyond that time. Moreover, immediate early transcription does not rely on any viral proteins, and, indeed, no phage proteins enter the cell during infec-

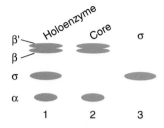

Figure 6.1   Cartoon illustrating separation of the subunits of *E. coli* **RNA polymerase by SDS-PAGE.** Lane 1, holoenzyme; lane 2, core enzyme after σ is removed by phosphocellulose chromatography; lane 3, σ.

Table 6.1 Ability of Core and Holoenzyme to Transcribe DNAs		
	Relative Transcription Activity	
DNA	**Core**	**Holoenzyme**
T4 (native, intact)	0.5	33.0
Calf thymus (native, nicked)	14.2	32.8

tion. Thus, immediate early transcription is not blocked by protein synthesis inhibitors added at the beginning of infection.

The second phase of viral transcription is called **delayed early.** This phase lasts from about 2–10 min after infection by T4 phage. By definition, delayed early transcription depends on at least one viral protein and can therefore be blocked by protein synthesis inhibitors added at the time infection begins. The last viral transcription phase is called **late.** It begins with the onset of phage DNA replication (about 10 min after infection by T4 phage) and continues until the end of the infection cycle (about 25 min after infection). Because late transcription cannot occur before phage DNA replication begins, it can be blocked by DNA synthesis inhibitors.

The timing of these three phases of transcription varies from one virus to another, and researchers working on diverse viruses have given them different names, but the three phases are a constant feature of the replication cycles of a wide variety of viruses. They are summarized in Table 6.2.

How did Bautz and coworkers show that σ restored specificity to the nonspecific core? Their experiment relied on a technique called **hybridization–competition,** which is described in Figure 6.2. First, a DNA containing two regions, X and Y, is transcribed *in vitro* in the presence of labeled nucleotides (*NTPs) to label the RNA product. The label used in such experiments is traditionally a radioactive atom, usually ^{3}H or [32]P. In this case, only region X of the DNA is transcribed, so only X RNA is produced.

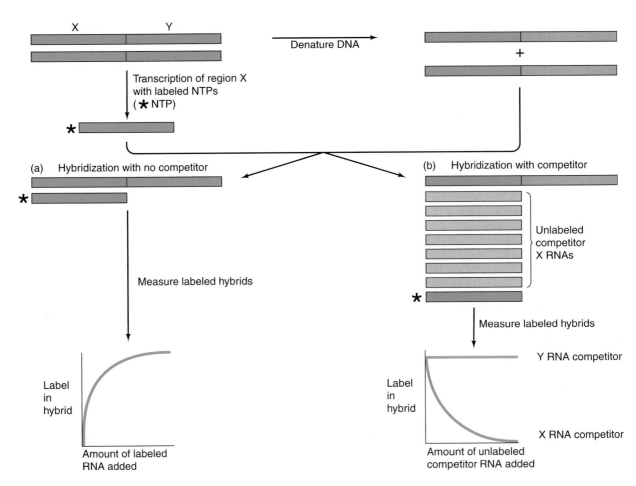

Figure 6.2 Hybridization–competition. At top left, transcribe a DNA containing regions X (blue) and Y (red) in vitro in the presence of radioactively labeled nucleotides (NTPs) to produce a labeled RNA (green). Hybridize this RNA to the DNA template strand, after denaturing the DNA duplex and immobilizing the separated strands on nitrocellulose filters. **(a)** Hybridize increasing amounts of the labeled RNA to the template DNA in the absence of competitor RNA. Next, treat the filters with RNase to remove unhybridized RNA and determine the radioactivity in the hybrids on the filters. The hybridization curve at the lower left shows the results. **(b)** This time hybridize a constant amount of labeled RNA to the denatured template DNA in the presence of increasing amounts of unlabeled X competitor RNA. Because a great excess of unlabeled competitor is added, the hybridization of labeled RNA to the DNA is almost completely blocked, as shown in the curve at the lower right. On the other hand, if only Y RNA competitor is used, no competition is seen because it hybridizes to another region of the DNA.

Table 6.2	**Viral Transcription Phases**		

Phase	Requirements	Blocked by	Timing in T4 (min)
Immediate early	Host proteins	—	0–2
Delayed early	At least one viral protein	Protein synthesis inhibitors	2–10
Late	Viral DNA replication and viral proteins	DNA or protein synthesis inhibitors	10–25

Next, this RNA is hybridized back to the template DNA. That is, the two strands of the DNA are separated (by heating, for example); then, as shown in Figure 6.2a, a transcript of one of the DNA strands base-pairs with that template strand. After the hybrids are treated with ri-bonuclease (**RNase**) to remove any unhybridized RNA, the extent of formation of labeled hybrids is measured. One can do all this by immobilizing the single-stranded DNA on a support such as a nitrocellulose filter, then hybridizing the radioactive X RNA to it, treating with

RNase, and measuring the radioactivity bound to the filter. The result is a hybridization curve such as the one at the lower left in Figure 6.2. Notice that the label in the hybrid reaches a saturation level. This occurs when all of the available X sites on the DNA have hybridized to X RNA and can form no more hybrid no matter how much more labeled X RNA is added.

So far, we know that we can hybridize a labeled RNA to its template DNA, but we do not know whether the labeled RNA was transcribed from region X, region Y, or both. To find out, we can do a competition, as shown in Figure 6.2b. Here, we start with a saturating amount of labeled RNA and add increasing amounts of unlabeled competitor RNA whose identity we know. If the competitor is X RNA, it will compete with the labeled RNA for X sites on the DNA. If we add a big enough excess of this competitor, we will see no labeled hybrids at all. On the other hand, if the competitor is Y RNA, we will see no competition because the labeled RNA and competitor hybridize to two separate regions of the DNA. This tells us that the labeled RNA is of the X type, and therefore that region X, not region Y, was transcribed.

Bautz and colleagues applied this method to transcription of T4 DNA in vitro by *E. coli* RNA polymerase holoenzyme and core. As Figure 6.3 shows, the transcripts made by the holoenzyme can be competed to a great extent by T4 immediate early RNA. That is, T4 immediate early RNA could prevent hybridization of these transcripts to T4 DNA. Thus, the holoenzyme is highly specific for the immediate early genes. By contrast, the product of the core polymerase can be competed to about the same extent by immediate early, delayed early, and late T4 RNAs. Thus, the core enzyme has no specificity; it transcribes the T4 DNA promiscuously.

Not only is the core enzyme indiscriminate about the T4 genes it transcribes, it also transcribes both DNA strands. Bautz and associates demonstrated this by hybridizing the labeled product of the holoenzyme or the core enzyme to authentic T4 phage RNA and then checking for RNase resistance (Figure 6.4). Because authentic T4 RNA is made **asymmetrically** (only one DNA strand in any given region is copied), it should not hybridize to T4 RNA made properly in vitro because this RNA is also made asymmetrically and is therefore identical, not complementary, to the authentic RNA. Bautz and associates did indeed observe this behavior with RNA made in vitro by the holoenzyme. However, if the RNA is made symmetrically in vitro, up to half of it will be complementary to the in vivo RNA and will be able to hybridize to it and thereby become resistant to RNase. In fact, Bautz and associates found that about 30% of the labeled RNA made by the core polymerase in vitro became RNase-resistant after hybridization to authentic T4 RNA. Thus, the core enzyme acts in an unnatural way by transcribing both DNA strands.

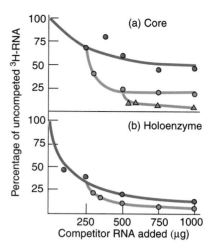

Figure 6.3   The σ factor confers specificity for the T4 immediate early genes. Bautz and associates transcribed T4 DNA in vitro with (**a**) *E. coli* core RNA polymerase or (**b**) holoenzyme, in the presence of ^{3}H-labeled ATP and GTP to label the T4 transcripts. They performed hybridization–competition by a technique very similar to that described in Figure 6.2 with immediate early, delayed early, and late T4 competitor RNAs isolated from *E. coli* cells 1 min (green), 5 min (red), and 17 min (blue) after infection with T4 phage, respectively. The only real difference between the technique used here and that described in Figure 6.2 was that the hybrids in this experiment were formed in solution and then collected by filtration through nitrocellulose. The hybrids stuck to the filter because the single-stranded DNA regions were much longer than the transcripts that hybridized to them. Thus, the long, unhybridized single-stranded DNA tails bound to the filter. The branching curves arise from the use of more than one competitor in a given reaction. For example, the first point on the delayed early curve (red) in panel (**a**) came from a competition reaction using 250 μg of immediate early RNA plus 50 μg of delayed early RNA; the first point on the late curve (blue) in panel (**a**) came from a competition reaction using 250 μg of immediate early RNA, 250 μg of delayed early RNA, and 50 μg of late RNA. A comparison of panel (**a**) and (**b**) shows that all three classes of T4 RNA compete well with the transcripts made by the core polymerase (**a**), but immediate early T4 RNA competes with the large majority of the transcripts made by the holoenzyme. Thus, σ confers specificity for the immediate early genes. (*Source:* Bautz, E.K.F., R.A. Bautz, and J.J. Dunn, "*E. coli* σ factor: A positive control element in phage T4 Development." Reprinted with permission from *Nature* 223:1023, 1969. Copyright Macmillan Magazines Limited.)

Clearly, depriving the holoenzyme of its σ-subunit leaves a core enzyme with basic RNA synthesizing capability, but lacking specificity. Adding σ back restores specificity. In fact, σ was named only after this characteristic came to light, and the σ, or Greek letter s, was chosen to stand for "specificity."

SUMMARY The key player in the transcription process is RNA polymerase. The *E. coli* enzyme is composed of a core, which contains the basic transcription machinery, and a σ-factor, which directs the core to transcribe specific genes.

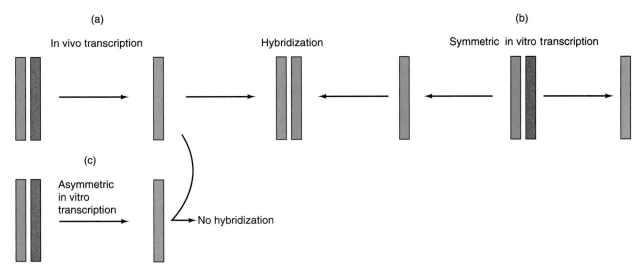

Figure 6.4 **Testing for asymmetric or symmetric transcription in vitro.** **(a)** In vivo transcription is asymmetric. Only one DNA strand (red) in the region pictured is transcribed, producing RNA (orange). **(b)** If transcription in vitro is symmetric, both DNA strands (red and blue) are transcribed, producing RNAs (orange and green, respectively). One of these products (green) is complementary to the in vivo RNA (orange) and therefore hybridizes. We can detect this hybrid by its resistance to RNase. **(c)** If transcription in vitro works the same is it does in vivo, it is asymmetric, and only the correct strand (red) transcribed, producing RNA (orange). This is identical to the in vivo RNA and therefore does not hybridize to it. One can demonstrate this by showing that the product remains RNase sensitive.

6.2   Promoters

In the T4 DNA transcription experiments presented in Table 6.1, why was core polymerase still capable of transcribing nicked DNA, but not intact DNA? Nicks and gaps in DNA provide ideal initiation sites for RNA polymerase, even core polymerase, but this kind of initiation is necessarily nonspecific. Few nicks or gaps occurred on the intact T4 DNA, so the core polymerase encountered only a few such artificial initiation sites and transcribed this DNA only weakly. On the other hand, when σ was present, the holoenzyme could recognize the authentic RNA polymerase binding sites on the T4 DNA and begin transcription there. These polymerase binding sites are called **promoters.** Transcription that begins at promoters in vitro is specific and mimics the initiation that would occur in vivo. Thus, σ operates by directing the polymerase to initiate at specific promoters. In this section, we will examine the interaction of bacterial polymerase with promoters, and the structures of these prokaryotic promoters.

Binding of RNA Polymerase to Promoters

How does σ change the way the core polymerase behaves toward promoters? David Hinkle and Michael Chamberlin used nitrocellulose filter binding studies (Chapter 5) to help answer this question. To measure how tightly holoenzyme and core enzyme bind to DNA, they isolated these enzymes from *E. coli* and bound them to ^{3}H-labeled T7 phage DNA, whose early promoters are recognized by the *E. coli* poly-

merase. Then they added a great excess of unlabeled T7 DNA, so that any polymerase that dissociated from a labeled DNA had a much higher chance of re-binding to an unlabeled DNA than to a labeled one. After varying lengths of time, they passed the mixture through nitrocellulose filters. The labeled DNA would bind to the filter only if it was still bound to polymerase. Thus, this assay measured the dissociation rate of the polymerase–DNA complex. As the last (and presumably tightest-bound) polymerase dissociated from the labeled DNA, that DNA would no longer bind to the filter, so the filter would become less radioactive.

Figure 6.5 shows the results of this experiment. Obviously, the polymerase holoenzyme binds much more tightly to the T7 DNA than does the core enzyme. In fact, the holoenzyme dissociates with a half time ($t_{1/2}$) of 30–60 h. This means that after 30–60 h, only half of the complex had dissociated, which indicates very tight binding indeed. By contrast, the core polymerase dissociated with a $t_{1/2}$ of less than a minute, so it bound much less tightly than the holoenzyme did. Thus, the σ-factor can promote tight binding, at least to certain DNA sites.

In a separate experiment, Hinkle and Chamberlin switched the procedure around, first binding polymerase to unlabeled DNA, then adding excess labeled DNA, and finally filtering the mixture at various times through nitrocellulose. This procedure measured the dissociation of the first (and loosest bound) polymerase, because a newly dissociated polymerase would be available to bind to the free labeled DNA and thereby cause it to bind to the filter. This assay revealed that the holoenzyme, as well as the core, had loose binding sites on the DNA.

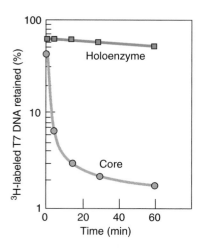

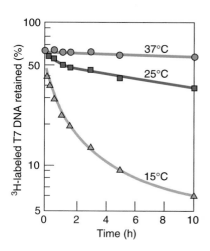

Figure 6.5 Sigma stimulates tight binding between RNA polymerase and promoter. Hinkle and Chamberlin allowed ³H-labeled T7 DNA to bind to *E. coli* core polymerase (blue) or holoenzyme (red). Next, they added an excess of unlabeled T7 DNA, so that any polymerase that dissociated from the labeled DNA would be likely to rebind to unlabeled DNA. They filtered the mixtures through nitrocellulose at various times to monitor the dissociation of the labeled T7 DNA–polymerase complexes. (As the last polymerase dissociates from the labeled DNA, the DNA will no longer bind to the filter, which loses radioactivity.) The much slower dissociation of the holoenzyme (red) relative to the core polymerase (blue) rate shows much tighter binding between T7 DNA and holoenzyme. (*Source:* Hinckle, D.C. and Chamberlin, M.J., "Studies of the Binding of *Escherichia coli* RNA Polymerase to DNA," *Journal of Molecular Biology,* Vol. 70, 157–185, 1972. By permission of Academic Press Limited, London.)

Figure 6.6 The effect of temperature on the dissociation of the polymerase-T7 DNA complex. Hinkle and Chamberlin formed complexes between *E. coli* RNA polymerase holoenzyme and ³H-labeled T7 DNA at three different temperatures: 37° C (red), 25° C (green), and 15° C (blue). Then they added unlabeled T7 DNA to compete with any polymerase that dissociated; they removed samples at various times and passed them through a nitrocellulose filter to monitor dissociation of the complex. The complex formed at 37° C was more stable than that formed at 25° C, which was much more stable than that formed at 15° C. Thus, higher temperature favors tighter binding between RNA polymerase holoenzyme and T7 DNA. (*Source:* Hinckle, D.C. and Chamberlin, M.J., "Studies of the Binding of *Escherichia coli* RNA Polymerase to DNA," *Journal of Molecular Biology,* Vol. 70, 157–815, 1972. By permission of Academic Press L:imited, London.)

Chamberlin interpreted these studies to mean that the holoenzyme finds two kinds of binding sites on T7 DNA: tight binding sites and loose ones. On the other hand, the core polymerase is only capable of binding loosely to the DNA. Because Bautz and coworkers had already shown that the holoenzyme, but not the core, can recognize promoters, it follows that the tight binding sites are probably promoters, and the loose binding sites represent the rest of the DNA. Chamberlin and colleagues also showed that the tight complexes between holoenzyme and T7 DNA could initiate transcription immediately on addition of nucleotides, which reinforces the conclusion that the tight binding sites are indeed promoters. If the polymerase had been tightly bound to sites remote from the promoters, a lag would have occurred while the polymerases searched for initiation sites. Furthermore, Chamberlin and coworkers titrated the tight binding sites on each molecule of T7 DNA and found only eight. This is not far from the number of early promoters on this DNA. By contrast, the number of loose binding sites for both holoenzyme and core enzyme is about 1300, which suggests that these loose sites are found virtually everywhere on the DNA and are therefore nonspecific. The inability of the core polymerase to bind to the tight (promoter) binding sites accounts for its inability to transcribe DNA specifically, which requires binding at promoters.

Hinkle and Chamberlin also tested the effect of temperature on binding of holoenzyme to T7 DNA and found

a striking enhancement of tight binding at elevated temperature. Figure 6.6 shows a significantly higher dissociation rate at 25° than at 37° C, and a much higher dissociation rate at 15° C. Because high temperature promotes DNA melting (strand separation, Chapter 2) this finding is consistent with the notion that tight binding involves local melting of the DNA. We will see direct evidence for this hypothesis later in this chapter.

Hinkle and Chamberlin summarized these and other findings with the following hypothesis for polymerase-DNA interaction (Figure 6.7): RNA polymerase holoenzyme binds loosely to DNA at first. It either binds initially at a promoter or scans along the DNA until it finds one. The complex with holoenzyme loosely bound at the promoter is called a **closed promoter complex** because the DNA remains in closed double-stranded form. Then the holoenzyme can melt a short region of the DNA at the promoter to form an **open promoter complex** in which the polymerase is bound tightly to the DNA. This is called an open complex because the DNA has to open up to form it.

It is this conversion of a loosely bound polymerase in a closed promoter complex to the tightly bound polymerase in the open promoter complex that requires σ, and this is also what allows transcription to begin. We can now appreciate how σ fulfills its role in determining specificity of transcription: It selects the promoters to which RNA polymerase will bind tightly. The genes adjacent to these promoters will then be transcribed.

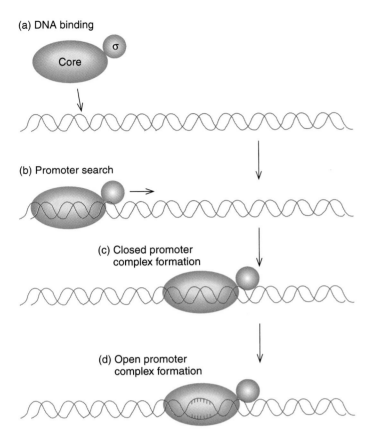

(a) DNA binding

(b) Promoter search

(c) Closed promoter complex formation

(d) Open promoter complex formation

Figure 6.7   RNA polymerase/promoter binding. (a) The polymerase holoenzyme binds nonspecifically to the DNA. **(b)** The holoenzyme slides along the DNA, searching for a promoter. **(c)** The holoenzyme has found a promoter and has bound loosely, forming a closed promoter complex. **(d)** The holoenzyme has bound tightly, melting a local region of DNA and forming an open promoter complex.

Electron microscopy studies on two-dimensional crystals of *E. coli* core polymerase and holoenzyme have given us an idea of the overall shape of the enzyme with and without σ. Figure 6.8 depicts these shapes, which are so dramatic that they help explain how the polymerase binds to DNA. The core polymerase Figure 6.8a resembles a right hand with the tips of the thumb and fingers just touching at upper left to form a circle. This produces a channel, labeled C, right through the enzyme. The diameter of this channel is about 25Å, just enough to accommodate a double-helical DNA. In the holoenzyme, by contrast, the "fingers" and "thumb" are not touching, so the channel is open to the left, as illustrated in Figure 6.8b.

Now we can make a plausible explanation of how the holoenzyme can bind tightly to DNA, whereas the core cannot: The "hand" of the core polymerase is closed, so it cannot get around the DNA to grasp it. Instead, the outside of the hand simply binds weakly and nonspecifically to the DNA. By contrast, the hand of the holoenzyme is open, so it can grasp the DNA. Figure 6.9 presents a model to explain the binding of holoenzyme to DNA. Once it has bound the DNA, the hand can close, and local DNA melting can occur

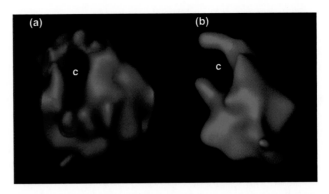

Figure 6.8   Overall shapes of *E. coli* RNA polymerase core (a) and holoenzyme (b) deduced from electron microscopy studies on two-dimensional crystals of the enzymes. The channel through the enzyme (denoted c) is closed in the core and open in the holoenzyme. (*Source:* Polyakov et al. Three-dimensional structure of E. coli core RNA polymerase: Promoter binding and elongation conformation of the enzyme. *Cell* 83 (3 Nov 1995) p. 368, f. 5b-c. Reprinted by permission of Elsevier Science.)

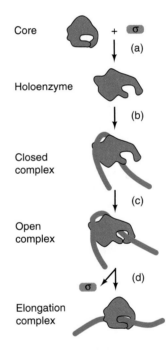

Core

Holoenzyme

Closed complex

Open complex

Elongation complex

Figure 6.9   Model of the interaction between *E. coli* RNA polymerase holoenzyme and a promoter. (a) Core (red) with its closed channel binds to σ (blue) to form holoenzyme with an open channel. **(b)** Holoenzyme binds loosely to DNA (green) to form a closed complex. **(c)** Holoenzyme grasps the DNA tightly, forming an open promoter complex. **(d)** Finally, σ dissociates, leaving core firmly clamped around the DNA, so it can transcribe the DNA processively. (*Source:* Polyakov et al., "Three-dimensional structure of *E. coli* core RNA polymerase: Promoter binding and elongation conformation of the enzyme," *Cell* 83: 368, 1995. Copyright 1995 Cell Press, Cambridge, MA. Reprinted by permission.)

to form the open promoter complex. Now that the σ-factor has done its job, it can dissociate, leaving the core clamped around the DNA in such a way that it cannot dissociate. So the closed hand that initially prevented the core from grasping the DNA now prevents it from falling off. This is important because RNA polymerase must *not* fall off during transcription. If it does, transcription terminates and the product will usually be a useless RNA fragment. Thus we say that RNA polymerase is highly **processive**. That is, once it begins its transcription job, it sticks to it to the end. Do not be confused by the fact that the hand is open in the closed promoter complex and closed in the open promoter complex. It is the closed and open DNAs, not the polymerases, that give the closed and open complexes their names.

One would expect that comparing the three-dimensional shapes of core and holoenzyme would allow us to see where σ lies with respect to holoenzyme because σ is the difference between the two forms of the enzyme. But the great shift in conformation between core and holoenzyme frustrated this expectation. The shape changes so much that it is difficult to relate structures in the core to those in the holoenzyme.

SUMMARY The σ-factor allows initiation of transcription by causing the RNA polymerase holoenzyme to bind tightly to a promoter. This tight binding depends on local melting of the DNA to form an open-promoter complex and is possible only in the presence of σ. The σ-factor can therefore select which genes will be transcribed. After σ has participated in initiation, it dissociates from the core polymerase, leaving the core to carry on with the elongation process. The structure of the *E. coli* core and holoenzyme help rationalize this mechanism: The holoenzyme has an open channel that can grasp the DNA. Once the DNA is in the channel, the "fingers" of the holoenzyme close around the DNA in helping to form the open-promoter complex. When the σ-factor dissociates, it leaves the core clamped around the DNA so it cannot dissociate during elongation.

Promoter Structure

What is the special nature of a prokaryotic promoter that attracts RNA polymerase? David Pribnow compared several *E. coli* and phage promoters and discerned two regions they held in common. The first was a sequence of 6 or 7 bp centered approximately 10 bp upstream of the start of transcription. This was originally dubbed the "Pribnow box," but is now usually called the **–10 box**. The second was another short sequence centered approximately 35 bp upstream of the transcription start site; it is known as the **–35 box**. More than 100 promoters have now been examined and a "typical" sequence for each of these boxes has emerged (Figure 6.10).

These so-called **consensus sequences** represent probabilities. The capital letters in Figure 6.10 denote bases that have a high probability of being found in the given position. The lowercase letters correspond to bases that are usually found in the given position, but at a lower frequency than those

denoted by capital letters. The probabilities are such that one rarely finds –10 or –35 boxes that match the consensus sequences perfectly. However, when such perfect matches are found, they tend to occur in very strong promoters that initiate transcription unusually actively. In fact, mutations that destroy matches with the consensus sequences tend to be **down mutations**. That is, they make the promoter weaker, resulting in less transcription. Mutations that make the promoter sequences more like the consensus sequences—usually make the promoters stronger; these are called **up mutations**. The spacing between promoter elements is also important, and deletions or insertions that move the –10 and –35 boxes unnaturally close together or far apart are deleterious. In Chapter 10 we will see that eukaryotic promoters have their own consensus sequences, one of which resembles the –10 box quite closely.

In addition to the –10 and –35 boxes, which we can call **core promoter elements,** some very strong promoters have an additional element further upstream called an **UP element**. *E. coli* cells have seven genes (*rrn* **genes**) that encode rRNAs. Under rapid growth conditions, when rRNAs are required in abundance, these seven genes by themselves account for the majority of the transcription occurring in the cell. Obviously, the promoters driving these genes are extraordinarily powerful, and their UP elements are part of the explanation. Figure 6.11 shows the structure of one of these promoters, the *rrnB* P1 promoter. Upstream of the core promoter (blue), it has an UP element (red) between positions –40 and –60. We know that the UP element is a true promoter element because it stimulates transcription of the *rrnB* P1 gene by a factor of 30 in the presence of RNA polymerase alone. Because it is recognized by the polymerase itself, we conclude that it is a promoter element.

This promoter is also associated with three so-called **Fis sites** (yellow) between positions –60 and –150, which are binding sites for the transcription-activator protein Fis. The Fis sites, because they do not bind to RNA polymerase itself, are not classical promoter elements, but instead are members of another class of transcription-activating DNA elements called enhancers. We will discuss prokaryotic enhancers in greater detail in Chapter 9.

SUMMARY Prokaryotic promoters contain two regions centered at –10 and –35 bp upstream of the transcription start site. In *E. coli,* these bear a greater or lesser resemblance to two consensus sequences: TATAAT and TTGACA, respectively. In general, the more closely regions within a promoter resemble these consensus sequences, the stronger that promoter will be. Some extraordinarily strong promoters contain an extra element (an UP element) upstream of the core promoter. This makes these promoters even more attractive to RNA polymerase.

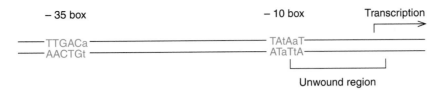

Figure 6.10   A prokaryotic promoter. The positions of –10 and –35 boxes and the unwound region are shown relative to the start of transcription for a typical *E. coli* promoter. Capital letters denote bases found in those positions in more than 50% of promoters examined; small letters denote bases found in those positions in 50% or fewer of promoters examined.

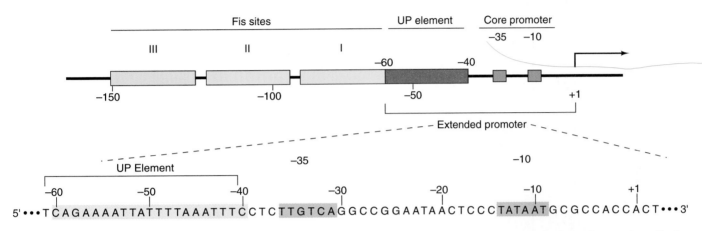

Figure 6.11   The *rrnB* P1 promoter. The core promoter elements (–10 and –35 boxes, blue) and the UP element (red) are shown schematically above, and with their complete base sequences (nontemplate strand) below, with the same color coding. The three Fis sites, which are not true promoter elements, are also shown schematically above (yellow). (*Source:* Ross et al., "A third recognition element in bacterial promoters: DNA binding by the alpha subunit of RNA polymerase." *Science* 262:1407, 1993. Copyright 1993 American Association for the Advancement of Science. Reprinted by permission.)

6.3   Transcription Initiation

Until 1980, it was a common assumption that transcription initiation ended when RNA polymerase formed the first phosphodiester bond, joining the first two nucleotides in the growing RNA chain. Then, Agamemnon Carpousis and Jay Gralla reported that initiation is more complex than that. They incubated *E. coli* RNA polymerase with DNA bearing an *E. coli* promoter known as the *lac* UV5 promoter. Along with the polymerase and DNA, they included heparin, a negatively charged polysaccharide that competes with DNA in binding tightly to free RNA polymerase. The heparin prevented any reassociation between DNA and polymerase released at the end of a cycle of transcription. These workers also included labeled ATP in their assay to label the RNA products. Then they subjected the products to gel electrophoresis to measure their sizes. They found several very small oligonucleotides, ranging in size from dimers to hexamers (2–6 nt long), as shown in Figure 6.12. The sequences of these oligonucleotides matched the sequence of the beginning of the expected transcript from the *lacZ* promoter. Moreover, when Carpousis and Gralla measured the amounts of these oligonucleotides and compared them to the number of RNA polymerases, they found many oligonucleotides per polymerase. Because the heparin in

the assay prevented free polymerase from reassociating with the DNA, this result implied that the polymerase was making many small, abortive transcripts without ever leaving the promoter. Other investigators have since verified this result and have found abortive transcripts up to 9 or 10 nt in size.

Thus, we see that transcription initiation is more complex than first supposed. It is now commonly represented in four steps, as depicted in Figure 6.13: (1) formation of a closed promoter complex; (2) conversion of the closed promoter complex to an open promoter complex; (3) polymerizing the first few nucleotides (up to 10) while the polymerase remains at the promoter; and (4) **promoter clearance,** in which the transcript becomes long enough to form a stable hybrid with the template strand. This helps to stabilize the transcription complex, and the polymerase changes to its elongation conformation, loses its σ-factor, and moves away from the promoter. In this section, we will examine the initiation process in more detail.

The Functions of σ

We have seen that σ selects genes for transcription by causing tight binding between RNA polymerase and promoters, and that tight binding depends on local melting of the DNA, which allows an open promoter complex to

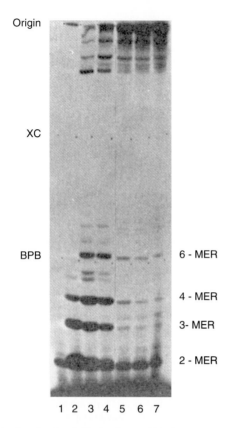

Figure 6.12   Synthesis of short oligonucleotides by RNA polymerase bound to a promoter. Carpousis and Gralla allowed *E. coli* RNA polymerase to synthesize ^{32}P-labeled RNA in vitro using a DNA containing the *lac* UV5 promoter, heparin to bind any free RNA polymerase, [^{32}P]ATP, and various concentrations of the other three nucleotides (CTP, GTP, and UTP). They electrophoresed the products on a polyacrylamide gel and visualized the oligonucleotides by autoradiography. Lane 1 is a control with no DNA; lane 2, ATP only; lanes 3–7; ATP with concentrations of CTP, GTP, and UTP increasing by twofold in each lane, from 25 μM in lane 3 to 400 μM in lane 7. The positions of two-mers through six-mers are indicated at right. The positions of two marker dyes (bromophenol blue [BPB] and xylene cyanol [XC]) are indicated at left. The apparent dimer in lane 1, with no DNA, is an artifact caused by a contaminant in the labeled ATP.

(*Source:* Carpousis and Gralla, *Biochemistry* 19 (8 Jul 1980) p. 3249, f. 2 © American Chemical Society.)

form. We have also seen that σ can dissociate from core after sponsoring polymerase–promoter binding. This implies that σ can stimulate initiation of transcription and then can be reused again and again. Let us now examine the evidence for each of these conclusions.

σ Stimulates Transcription Initiation   Because σ directs tight binding of RNA polymerase to promoters, it places the enzyme in a position to initiate transcription—just at the beginning of a gene. Therefore, we would expect σ to stimulate initiation of transcription. To test this, Travers and Burgess took advantage of the fact that the first nucleotide incorporated into an RNA retains all three of its phosphates (α, β, and γ), whereas all other nucleotides retain only their α-phosphate (Chapter 3). These investiga-

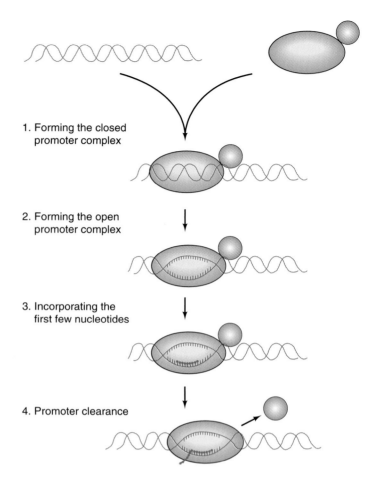

Figure 6.13   Stages of transcription initiation. 1. RNA polymerase binds to DNA in a closed promoter complex. 2. The σ-factor allows the polymerase to convert the closed promoter complex to an open promoter complex. 3. The polymerase incorporates the first nine or ten nt into the nascent RNA. Some abortive transcripts are pictured at left. 4. The polymerase clears the promoter, loses its σ factor, and begins the elongation phase.

tors incubated polymerase core in the presence of increasing amounts of σ in two separate sets of reactions. In the first reactions, the labeled nucleotide was [^{14}C]ATP, which is incorporated throughout the RNA and therefore measures elongation of RNA chains. In the other reactions, the labeled nucleotide was [γ-^{32}P]ATP or [γ-^{32}P]GTP, whose label should be incorporated only into the first position of the RNA, and therefore is a measure of transcription initiation. (They used ATP and GTP because transcription usually starts with a purine nucleotide—more often ATP than GTP.) The results in Figure 6.14 show that σ stimulated the incorporation of both ^{14}C- and γ-^{32}P-labeled nucleotides, which suggests that σ enhanced both initiation and elongation. However, initiation is the rate-limiting step in transcription (it takes longer to get a new RNA chain started than to extend one). Thus, σ could appear to stimulate elongation by stimulating initiation and thereby providing more initiated chains for core polymerase to elongate.

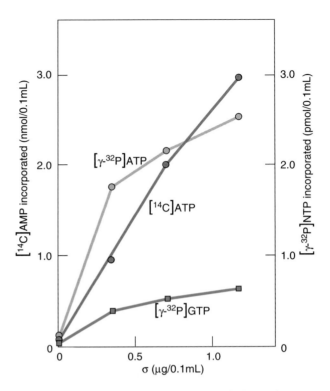

Figure 6.14   Sigma seems to stimulate both initiation and elongation. Travers and Burgess transcribed T4 DNA in vitro with *E. coli* RNA polymerase core plus increasing amounts of σ. In separate reactions, they included [^{14}C]ATP (red), [γ-^{32}P]ATP (blue), or [γ-^{32}P]GTP (green) in the reaction mix. The incorporation of the [^{14}C]ATP measured bulk RNA synthesis, or elongation; the incorporation of the γ-^{32}P-labeled nucleotides measured initiation. Because all three curves rise with increasing σ concentration, this experiment makes it appear that σ stimulates both elongation and initiation. (*Source:* Travers, A.A. and R.R. Burgess, "Cyclic re-use of the RNA polymerase sigma factor." *Nature* 222:537–40, 1969. Reprinted with permission. Copyright 1969 Macmillan Magazines Limited.)

Travers and Burgess proved that is the case by demonstrating that σ really does not accelerate the rate of RNA chain growth. To do this, they held the number of RNA chains constant and showed that under those conditions σ did not affect the length of the RNA chains. They held the number of RNA chains constant by allowing a certain amount of initiation to occur, then blocking any further chain initiation with the antibiotic **rifampicin,** which blocks prokaryotic transcription initiation, but not elongation. Then they used ultracentrifugation to measure the length of RNAs made in the presence or absence of σ. They found that σ made no difference in the lengths of the RNAs. If it really had stimulated the rate of elongation, it would have made the RNAs longer. Therefore, σ does not stimulate elongation, and the apparent stimulation in the previous experiment was simply an indirect effect of enhanced initiation.

SUMMARY Sigma stimulates initiation, but not elongation, of transcription.

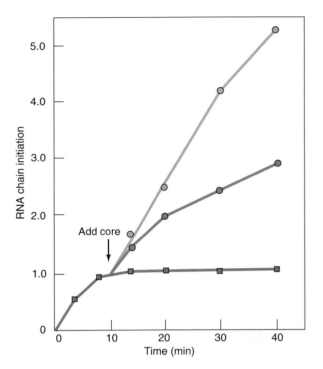

Figure 6.15   Sigma can be reused. Travers and Burgess allowed RNA polymerase holoenzyme to initiate and elongate RNA chains on a T4 DNA template at low ionic strength, so the polymerases could not dissociate from the template to start new RNA chains. The red curve shows the initiation of RNA chains, measured by [γ-^{32}P]ATP and [γ-^{32}P]GTP incorporation, under these conditions. After 10 min (arrow), when most chain initiation had ceased, the investigators added new, rifampicin-resistant core polymerase in the presence (green) or absence (blue) of rifampicin. The immediate rise of both curves showed that addition of core polymerase can restart RNA synthesis, which implied that the new core associated with σ that had been associated with the original core. In other words, the σ was recycled. The fact that transcription occurred even in the presence of rifampicin showed that the new core, which was from rifampicin-resistant cells, together with the old σ, which was from rifampicin-sensitive cells, could carry out rifampicin-resistant transcription. Thus, the core, not the σ, determines rifampicin resistance or sensitivity. (*Source:* Travers, A.A. and R.R. Burgess, "Cycle re-use of the RNA polymerase sigma factor." *Nature* 222–537–40, 1969. Reprinted with permission. Copyright 1969 Macmillan Magazines Limited.)

Reuse of σ   In the same 1969 paper, Travers and Burgess demonstrated that σ can be recycled. The key to this experiment was to run the transcription reaction at low ionic strength, which prevents RNA polymerase core from dissociating from the DNA template at the end of a gene. This caused transcription initiation (as measured by the incorporation of ^{32}P-labeled purine nucleotides into RNA) to slow to a stop, as depicted in Figure 6.15 (red line). Then, when they added new core polymerase, these investigators showed that transcription began anew (blue line). This meant that the new core was associating with σ that had been released from the original holoenzyme. In a separate experiment, they demonstrated that the new transcription could occur on a different kind of DNA added along with the new core polymerase. This supported the conclusion that σ had been released from the original core

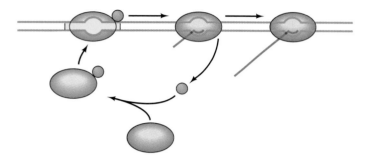

Figure 6.16   The σ cycle. RNA polymerase binds to the promoter (yellow) at left, causing local melting of the DNA. As the polymerase moves to the right, elongating the RNA, the σ-factor dissociates and joins with a new core polymerase (lower left) to initiate another RNA chain.

and was associating with a new core on a new DNA template. Accordingly, Travers and Burgess proposed that σ cycles from one core to another, as shown in Figure 6.16. They dubbed this the "σ cycle."

Figure 6.15 contains still another piece of valuable information. When Travers and Burgess added core polymerase from a rifampicin-resistant mutant, along with rifampicin, transcription still occurred (green line). Because the σ was from the original, rifampicin-sensitive polymerase, the rifampicin resistance in the renewed transcription must have been conferred by the newly added core. The fact that less initiation occurred in the presence of rifampicin probably means that the rifampicin-resistant core is still somewhat sensitive. We might have expected the σ-factor, not the core, to determine rifampicin sensitivity or resistance because rifampicin blocks initiation, and σ is the acknowledged initiation factor. Nevertheless, the core is the key to rifampicin sensitivity, and experiments to be presented later in this chapter will provide some clarification of why this is so.

SUMMARY Sigma can be reused by different core polymerases, and the core, not σ, governs rifampicin sensitivity or resistance.

Local DNA Melting at the Promoter   Chamberlin's studies on RNA polymerase–promoter interactions showed that such complexes were much more stable at elevated temperature. This suggested that local melting of DNA occurs on tight binding to polymerase, because high temperature would tend to stabilize melted DNA.

Tao-shih Hsieh and James Wang provided more direct evidence for local DNA melting in 1978. They bound *E. coli* RNA polymerase to a restriction fragment containing three phage T7 early promoters and measured the hyperchromic shift caused by such binding. This increase in the DNA's absorbance of 260-nm light is not only indica-

tive of DNA strand separation, its magnitude is directly related to the number of base pairs that are opened. Knowing the number of RNA polymerase holoenzymes bound to their DNA, Hsieh and Wang calculated that each polymerase caused a separation of about 10 bp.

In 1979, Ulrich Siebenlist, working in Walter Gilbert's lab, identified the base pairs that RNA polymerase melted in a T7 phage early promoter. Figure 6.17 shows the strategy of his experiment. First he end-labeled the promoter DNA, then added RNA polymerase to form an open promoter complex. As we have seen, this involves local DNA melting, and when the strands separate, the N_1 of adenine—normally involved in hydrogen bonding to a T in the opposite strand—becomes susceptible to attack by certain chemical agents. In this case, Siebenlist methylated the exposed adenines with dimethyl sulfate (DMS). Then, when he removed the RNA polymerase and the melted region closed up again, the methyl groups prevented proper base-pairing between these N_1-methyl-adenines and the thymines in the opposite strand and thus preserved at least some of the single-stranded character of the formerly melted region. Next, he treated the DNA with S1 nuclease, which specifically cuts single-stranded DNA. This enzyme should therefore cut wherever an adenine had been in a melted region of the promoter and had become methylated. In principle, this should produce a series of end-labeled fragments, each one terminating at an adenine in the melted region. Finally, Siebenlist electrophoresed the labeled DNA fragments to determine their precise lengths. Then, knowing these lengths and the exact position of the labeled end, he could calculate accurately the position of the melted region.

Figure 6.18 shows the results. Instead of the expected neat set of fragments, we see a blur of several fragments extending from position +3 to −9. The reason for the blur seems to be that the multiple methylations in the melted region so weakened base pairing that *no* strong base pairs could re-form; the whole melted region retained at least partially single-stranded character and therefore remained open to cutting by S1 nuclease. The length of the melted region detected by this experiment is 12 bp, roughly in agreement with Hsieh and Wang's estimate, although this may be an underestimate because the next base pairs on either side are G-C pairs whose participation in the melted region would not have been detected. This is because neither guanines nor cytosines are readily methylated under the conditions used in this experiment. It is also satisfying that the melted region is just at the place where RNA polymerase begins transcribing.

The experiments of both Hsieh and Wang, and of Siebenlist, as well as other early experiments, measured the DNA melting in a simple binary complex between polymerase and DNA. None of these experiments examined the size of a DNA bubble in complexes in which

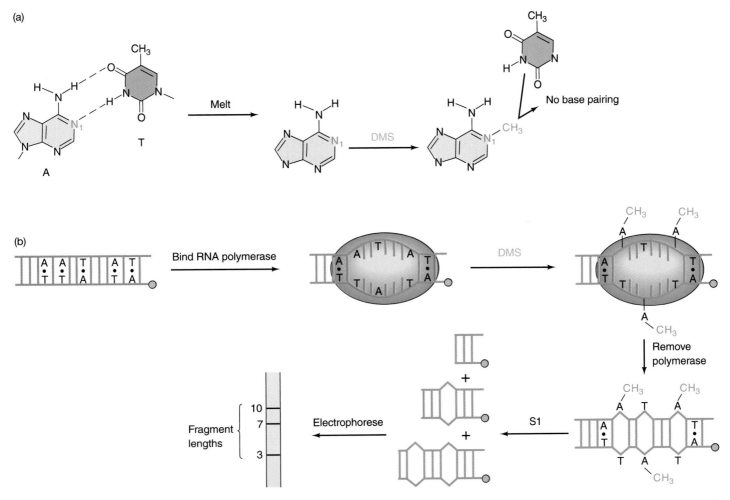

Figure 6.17   Locating the region of a T7 phage early promoter melted by RNA polymerase. (a) When adenine is base-paired with thymine (left) the N₁ nitrogen of adenine is hidden in the middle of the double helix and is therefore protected from methylation. On melting (right), the adenine and thymine separate; this opens the adenine up to attack by dimethyl sulfate (DMS, blue), and the N₁ nitrogen is methylated. Once this occurs, the methyl-adenine can no longer base-pair with its thymine partner. (b) A hypothetical promoter region containing five A-T base pairs is end-labeled (orange), then RNA polymerase (red) is bound, which causes local melting of the promoter DNA. The three newly exposed adenines are methylated with dimethyl sulfate (DMS). Then, when the polymerase is removed, the A-T base pairs cannot reform because of the interfering methyl groups (CH₃, blue). Now S1 nuclease can cut the DNA at each of the unformed base pairs because these are local single-stranded regions. Very mild cutting conditions are used so that only about one cut per molecule occurs. Otherwise, only the shortest product would be seen. The resulting fragments are denatured and electrophoresed to determine their sizes. These sizes tell how far the melted DNA region was from the labeled DNA end.

initiation or elongation of RNA chains was actually taking place. Thus, in 1982, Howard Gamper and John Hearst set out to estimate the number of base pairs melted by polymerases, not only in binary complexes, but also in actively transcribing complexes that also contained RNA (ternary complexes). They used SV40 DNA, which happens to have one promoter site recognized by the *E. coli* RNA polymerase. They bound RNA polymerase to the SV40 DNA at either 5° C or 37° C in the absence of nucleotides to form binary complexes, or in the presence of nucleotides to form ternary complexes. Under the conditions of the experiment, each polymerase initiated only once, and no polymerase terminated transcription, so all polymerases remained complexed to the DNA. This al-

lowed an accurate assessment of the number of polymerases bound to the DNA.

After binding a known number of *E. coli* RNA polymerases to the DNA, Gamper and Hearst relaxed any supercoils that had formed with a crude extract from human cells, then removed the polymerases from the relaxed DNA (Figure 6.19a). The removal of the protein left melted regions of DNA, which meant that the whole DNA was underwound. Because the DNA was still a covalently closed circle, this underwinding introduced strain into the circle that was relieved by forming supercoils. The higher the superhelical content, the greater the double helix unwinding that has been caused by the polymerase. The superhelical content of a DNA can be measured by

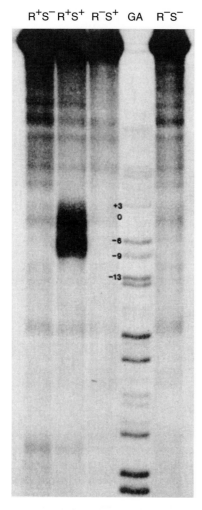

R⁺S⁻ R⁺S⁺ R⁻S⁺ GA R⁻S⁻

+3
0
−6
−9
−13

Figure 6.18   RNA polymerase melts the DNA in the −9 to +3 region of the T7 A3 promoter. Siebenlist performed a methylation-S1 assay as described in Figure 6.17. Lane R⁺S⁺ shows the results when both RNA polymerase (R) and S1 nuclease (S) were used. The other lanes were controls in which Siebenlist left out either RNA polymerase, or S1 nuclease, or both. The GA lane is the result of Maxam–Gilbert sequencing of the promoter fragment, showing Gs, and to a lesser extent As. This sequence allowed Siebenlist to locate the melted region approximately between positions −9 and +3. (*Source:* Siebenlist. RNA polymerase unwinds an 11-base pair segment of a phage T7 promoter. *Nature* 279 (14 June 1979) p. 652, f. 2, © Macmillan Magazines Ltd.)

gel electrophoresis because the more superhelical turns a DNA contains, the faster it will migrate in an electrophoretic gel.

Figure 6.19b is a plot of the change in the superhelicity as a function of the number of active polymerases per genome at 37° C. A linear relationship existed between these two variables, and one polymerase caused about 1.6 superhelical turns, which means that each polymerase unwound 1.6 turns of the DNA double helix. If a double helical turn contains 10.6 bp, then each polymerase melted 17 bp ($1.6 \times 10.6 = 17$). A similar calculation of the data from the 5° C experiment yielded a value of 18 bp melted by one polymerase. From these data, Gamper and Hearst

concluded that a polymerase binds at the promoter, melts 17 ± 1 bp of DNA to form a **transcription bubble,** and a bubble of this size moves with the polymerase as it transcribes the DNA. Of course, this melting is essential because it exposes the template strand so it can be transcribed.

> **SUMMARY** On binding to a promoter, RNA polymerase causes the melting of at least 10 bp, but probably about 17 bp in the vicinity of the transcription start site. This transcription bubble moves with the polymerase, exposing the template strand so it can be transcribed.

Extent of Polymerase Binding to Promoters

We have seen that at least 10 bp around the transcription start site and the −10 box in promoter are melted on RNA polymerase binding. But there are surely other regions—the −35 box, for example—that are important in polymerase–promoter interactions. A number of studies have shed light on this issue; let us consider one of them.

One way to define the limits of the search would be to bind RNA polymerase tightly to a promoter, use DNase to nibble away the DNA that is not protected by the polymerase, then look to see what DNA is left. This would tell us the extent of the DNA in intimate contact with the tightly bound polymerase. Figure 6.20 outlines how Walter Gilbert and his colleagues applied this procedure to the *E. coli lac* promoter. First, they labeled a piece of DNA containing the promoter at one of its 5′-ends. Next, they bound RNA polymerase to the promoter and then treated the complex with exonuclease III, which degrades DNA base by base from the 3′-end. This left a labeled DNA strand whose upstream end had been chewed back to the point where the polymerase blocked any further DNA degradation. Finally, they released the protected DNA, denatured, and electrophoresed it. They observed a major band whose length represented a fragment extending upstream to position −44. Thus, RNA polymerase extends from about position −44 at least to position +2 (the limit of melting induced by the polymerase). Other studies, in which the polymerase has been chemically crosslinked to specified DNA bases, have shown that the downstream reach of the polymerase extends at least to position +3. That long span of at least 46 bp is consistent with the large size of the enzyme.

> **SUMMARY** The *E. coli* RNA polymerase extends at least from position −44 to +3 in the open promoter complex.

(a)

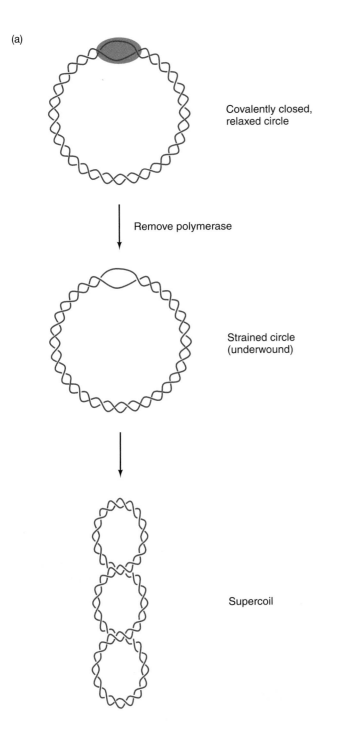

Covalently closed,
relaxed circle

Remove polymerase

Strained circle
(underwound)

Supercoil

(b)

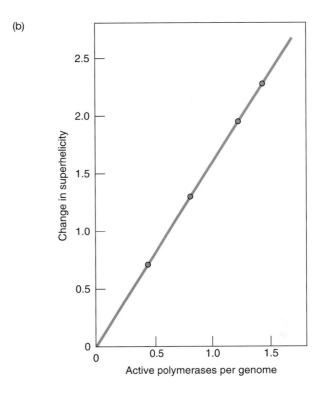

Active polymerases per genome

Figure 6.19   Measuring the melting of DNA by polymerase binding. (a) Principle of the experiment. Gamper and Hearst added *E. coli* RNA polymerase (red) to SV40 DNA, then relaxed any supercoils with a nicking-closing extract to produce the complexes shown at top. Then they removed the polymerase, leaving the DNAs strained (middle) because of the region that had been melted by the polymerase. This strain was quickly relieved by forming supercoils (bottom). The greater the superhelicity, the greater the unwinding caused by the polymerase. **(b)** Experimental results. Gamper and Hearst plotted the change in superhelicity of DNA as a function of the number of polymerases added. The plot was a straight line with a slope of 1.6 (1.6 superhelical turns introduced per polymerase).

Structure of σ

By the late 1980s, the genes encoding a variety of σ-factors from various bacteria had been cloned and sequenced. As we will see in Chapter 8, each bacterium has a primary σ-factor that transcribes its vegetative genes—those required for everyday growth. For example, the primary σ in *E. coli* is called σ^{70}, and the primary σ in *B. subtilis* is σ^{43}. These proteins are named for their molecular masses, 70 and 43 kD, respectively. In addition, bacteria have alternative σ-factors that transcribe specialized genes (heat shock genes, sporulation genes, and so forth). In 1988, Helmann and Chamberlin reviewed the literature on all these factors and analyzed the striking similarities in amino acid sequence among them, which are clustered in four regions (regions 1–4, see Figure 6.21). The conservation of sequence in these regions suggests that they are important in the function of σ, and Helmann and Chamberlin proposed the following functions for each region.

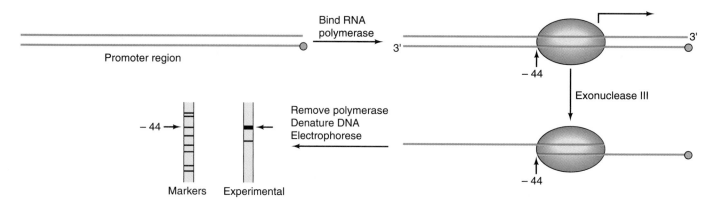

Figure 6.20 Defining the upstream extent of RNA polymerase bound to the *lac* promoter. Begin with a promoter-containing piece of DNA labeled at its downstream 5′-end. Bind RNA polymerase to the labeled DNA, then digest any unprotected 3′-ends with exonuclease III. Finally, release the protected DNA, separate the two strands, and electrophorese them to determine the length of the labeled protected fragment. As standard DNA length markers use a set of fragments generated by sequencing the same labeled DNA. This shows that the polymerase protected the bottom strand up to position –44.

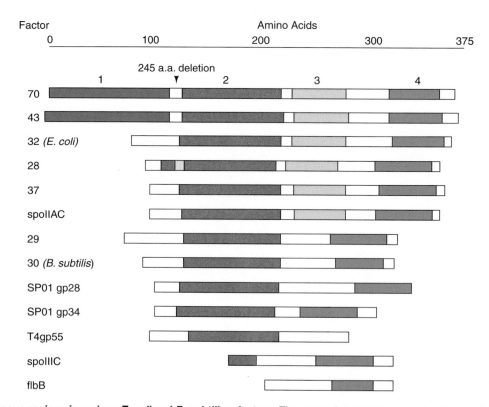

Figure 6.21 Homologous regions in various *E. coli* and *B. subtilis* σ factors. The σ proteins are represented as horizontal bars, with homologous regions aligned vertically. Only the top two, the primary σ factors of *E. coli* and *B. subtilis,* respectively, contain the first homologous region. Also, σ⁷⁰ contains a sequence of 245 amino acids between regions 1 and 2 that is missing in σ⁴³. This is marked above the σ⁷⁰ bar. Lighter shading denotes an area that is conserved only in some of the proteins.

Region 1 This region is found only in the primary σ's (σ^{70} and σ^{43}). Its role appears to be to prevent σ from binding by itself to DNA. We will see later in this chapter that a fragment of σ is capable of DNA binding, but region 1 prevents the whole polypeptide from doing that. This is important because σ binding to promoters could inhibit holoenzyme binding and thereby inhibit transcription.

The big difference in size between σ^{70} and σ^{43} is accounted for by an extra 245 amino acids between regions 1 and 2 of σ^{70}, not found in σ^{43}. This suggests that these extra amino acids may confer on σ^{70} the ability to do

Figure 6.22   Summary of regions of primary structure in *E. coli* σ⁷⁰. The four conserved regions are indicated, with subregions delineated in regions 1, 2, and 4. The cb region (purple) is thought to be involved in binding to the core polymerase. (*Source:* Dombroski, A. J., et al., "Polypeptides containing highly conserved regions of the transcription initiation factor σ⁷⁰ exhibit specificity of binding to promoter DNA." *Cell* 70:501–12, 1992. Copyright 1992 Cell Press, Cambridge, MA. Reprinted by permission.)

that σ⁴³ cannot. Later in this chapter we will see that σ⁷⁰ can loosen the binding between polymerase and nonpromoter regions, thereby enhancing the specificity of promoter binding. On the other hand, the *B. subtilis* σ⁴³ cannot do this. Instead, it relies on another polypeptide called δ to perform this task. Thus it is tempting to suggest that the extra amino acids in region 1 of σ⁷⁰ give it this δ-like activity.

Region 2   This region is found in all σ-factors and is the most highly conserved σ region, so it is probably responsible for at least one fundamental activity common to all the factors. Region 2 can be subdivided into four parts, 2.1–2.4 (Figure 6.22).

Region 2.1 seems to be involved in binding to the polymerase core. Deletion analysis between residues 361–390 of the *E. coli* σ-factor by Richard Burgess and colleagues have shown this region to be necessary and sufficient for core binding. Moreover, Seth Darst and coworkers have obtained the crystal structure of a large fragment of the *E. coli* σ⁷⁰ and have found a hydrophobic patch of amino acids in region 2.1. This hydrophobic region is exposed to the solvent in the isolated σ-factor and is a candidate for at least part of the core-binding domain of the σ-factor.

We have good evidence that region 2.4 is responsible for a crucial σ activity, recognition of the promoter's –10 box. First of all, if σ region 2.4 does recognize the –10 box, then σ's with similar specificities should have similar regions 2.4. This is demonstrable; σ⁴³ of *B. subtilis* and σ⁷⁰ of *E. coli* recognize identical promoter sequences, including –10 boxes. Indeed, these two σ's are interchangeable. And the regions 2.4 of these two σ's are 95% identical.

Richard Losick and colleagues performed genetic experiments that also link region 2.4 with –10 box binding. Region 2.4 of the σ-factor contains an amino acid sequence that suggests it can form an α-helix. We will learn in Chapter 9 that an α-helix is a favorite DNA-binding motif, which is consistent with a role for this part of the σ in promoter binding. Losick and colleagues reasoned as follows: If this potential α-helix is really a –10 box-recognition element, then the following experiment should be possible. First, they could make a single base change in a promoter's –10 box, which destroys its ability to bind to RNA polymerase. Then, they could make a compensating mutation in one of the amino acids in region 2.4 of the σ-factor. If the σ-factor mutation can sup-

press the promoter mutation, it provides strong evidence that there really is a relationship between the –10 box and region 2.4 of the σ. So Losick and colleagues caused a G → A transition in the –10 box of the *B. subtilis spo*VG promoter, which prevented binding between the promoter and RNA polymerase. Then they caused a Thr → Ile mutation at amino acid 100 in region 2.4 of σ^H, which normally recognizes the *spo*VG promoter. This σ mutation restored the ability of the polymerase to recognize the mutant promoter.

Region 3   Little is known about this σ region except that its sequence suggests that it can form a helix-turn-helix DNA-binding domain (Chapter 9). The significance of this is unclear.

Region 4   Like region 2, region 4 can be subdivided into subregions. Also like region 2, region 4 seems to play a key role in promoter recognition. Subregion 4.2 contains another helix-turn-helix DNA-binding domain, which suggests that it plays a role in polymerase–DNA binding. In fact, subregion 4.2 appears to govern binding to the –35 box of the promoter. As with the σ region 2.4 and the –10 box, genetic and other evidence supports the relationship between the σ region 4.2 and the –35 box. Again, we see that σ's that recognize promoters with similar –35 boxes have similar regions 4.2. And again, we can show suppression of mutations in the promoter (this time in the –35 box) by compensating mutations in region 4.2 of the σ-factor. For instance, Miriam Susskind and her colleagues showed that an Arg→ His mutation in position 588 of the *E. coli* σ⁷⁰ suppresses G → A or G → C mutations in the –35 box of the *lac* promoter. Figure 6.23 summarizes this and other interactions between regions 2.4 and 4.2 of σ and the –10 and –35 boxes, respectively, of prokaryotic promoters.

These results all suggest the importance of σ regions 2.4 and 4.2 in binding to the –10 and –35 boxes, respectively, of the promoter. The σ-factor even has putative DNA-binding domains in strategic places. But we are left with the perplexing fact that σ by itself does not bind to promoters, or to any other region of DNA. Only when it is bound to the core can σ bind to promoters. How do we resolve this apparent paradox?

Carol Gross and her colleagues suggested that regions 2.4 and 4.2 of σ are capable of binding to promoter regions on their own, but other domains in σ interfere with this binding. They further suggested that when σ associates with core it changes conformation, unmasking its DNA-binding domains, so it can bind to promoters. To test this hypothesis, these workers made fusion proteins (Chapter 4) containing glutathione-*S*-transferase (GST) and fragments of the *E. coli* σ-factor (region 2.4, or 4.2, or both). These fusion proteins are easy to purify because of the affinity of GST for glutathione. Then they showed that a fusion protein containing region 2.4 could bind to a

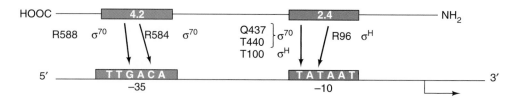

Figure 6.23   Specific interactions between σ regions and promoter regions. Arrows denote interactions revealed by mutation suppressor experiments. The designations beside the arrows show the amino acid mutated (e.g., R588 σ^{70} is amino acid number 588 of *E. coli* σ^{70}, which is an arginine [R]). This amino acid in σ region 4.2 appears to recognize the G in the –35 box of the promoter. Notice that the linear structure of the σ-factor (top) is written with the C-terminus at left, to match the promoter written conventionally, 5′ → 3′ left to right (bottom). (*Source:* Dombroski, A. J., et al., "Polypeptides containing highly conserved regions of transcription initiation factor σ^{70} exhibit specificity of binding to promoter DNA." *Cell* 70:501–12, 1992. Copyright 1992 Cell Press, Cambridge, MA. Reprinted by permission.)

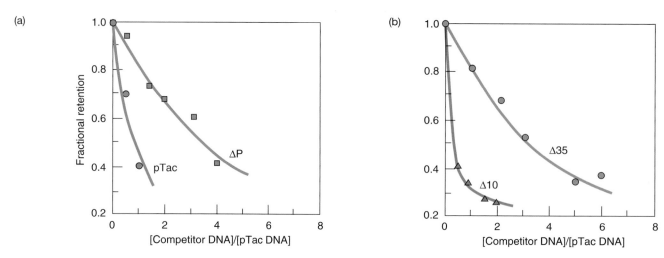

Figure 6.24   Analysis of binding between σ region 4.2 and the promoter –35 box. (a) Recognition of the promoter. Gross and colleagues measured binding between a σ fragment-GST fusion protein and a labeled DNA fragment (pTac) containing the *tac* promoter. The σ fragment in this experiment contained only the 108 amino acids at the C-terminus of the *E. coli* σ, which includes region 4, but not region 2. Gross and coworkers measured binding of the labeled DNA–protein complex to nitrocellulose filters in the presence of competitor DNA containing the *tac* promoter (p Tac), or lacking the *tac* promoter (ΔP). Because pTac DNA competes much better than ΔP DNA, they concluded that the fusion protein with region 4 can bind to the *tac* promoter. (b) Recognition of the –35 region. Gross and colleagues repeated the experiment but used two different competitor DNAs: One (Δ10) had a *tac* promoter with a 6-bp deletion in the –10 box; the other (Δ35) had a *tac* promoter with a 6-bp deletion in the –35 box. Because deleting the –35 box makes the competitor no better than a DNA with no *tac* promoter at all and removing the –10 box had no effect, it appears that the σ fragment with region 4 binds to the –35 box, but not to the –10 box. (*Source:* Dombroski, A. J., et al., "Polypeptides containing highly conserved regions of transcription initiation factor s70 exhibit specificity of binding to promoter DNA." *Cell* 70:501–12, 1992. Copyright 1992 Cell Press, Cambridge, MA. Reprinted by permission.)

DNA fragment containing a –10 box, but not a –35 box. Furthermore, a fusion protein containing region 4.2 could bind to a DNA fragment containing a –35 box, but not a –10 box.

To measure the binding between fusion proteins and promoter elements, Gross and coworkers used a nitrocellulose filter-binding assay. They labeled the target DNA containing one or both promoter elements from the composite *tac* promoter. The *tac* promoter has the –10 box of the *lac* promoter and the –35 box of the *trp* promoter. Then they added a fusion protein to the labeled target DNA in the presence of excess unlabeled competitor DNA and measured the formation of a labeled DNA–protein complex by nitrocellulose binding.

Figure 6.24a shows the results of an experiment in which Gross and colleagues bound a labeled *tac* promoter to a GST–σ-region 4 fusion protein. Because σ-region 4

contains a putative –35 box-binding domain, we expect this fusion protein to bind to DNA containing the *tac* promoter more strongly than to DNA lacking the *tac* promoter. Figure 6.24a demonstrates this is just what happened. Unlabeled DNA containing the *tac* promoter was an excellent competitor, whereas unlabeled DNA missing the *tac* promoter competed relatively weakly. Thus, the GST–σ region 4 protein binds weakly to nonspecific DNA, but strongly to *tac* promoter-containing DNA, as we expect.

Figure 6.24b shows that the binding between the GST–σ region 4 proteins and the promoter involves the –35 box, but not the –10 box. As we can see, a competitor from which the –35 box was deleted competed no better than nonspecific DNA, but a competitor from which the –10 box was deleted competed very well because it still contained the –35 box. Thus, σ region 4 can bind specifically to the −35 box, but not to the −10 box. Similar experiments with a GST–σ region 2 fusion protein showed that this protein can bind specifically to the −10 box, but not the −35 box.

> **SUMMARY** Comparison of the sequences of different σ genes reveals four regions of similarity among a wide variety of σ-factors. The best evidence for the functions of these regions shows that subregions 2.4 and 4.2 are involved in promoter −10 box and −35 box recognition, respectively.

Destabilizing Nonspecific Polymerase–DNA Interactions

As we mentioned in our discussion of σ region 1, the σ-factor not only stabilizes specific interactions between polymerase and promoters, it also destabilizes nonspecific interactions. The net result is a strong tendency of the holoenzyme to form specific rather than nonspecific interactions.

Hinkle and Chamberlin demonstrated the tendency of σ to loosen nonspecific interactions with a filter-binding assay described earlier in this chapter and in Chapter 5. First, they bound core polymerase to labeled T7 DNA, bound the complex to a nitrocellulose filter, then added excess unlabeled DNA. As the complex dissociated, the newly released core polymerase rebound to the excess unlabeled DNA, releasing the labeled DNA from the filter. Figure 6.25 shows this dissociation curve. In a parallel experiment, Hinkle and Chamberlin added σ one minute after the unlabeled DNA. We can see that the σ-factor significantly accelerated the dissociation of the nonspecifically bound core from the DNA.

By contrast, the *B. subtilis* σ^{43} does not have this ability to loosen nonspecific interactions between polymerase and DNA; another factor, δ, performs that function. Richard Losick and colleagues provided the first suggestion of this activity in 1977 when they showed that δ inhibited transcription of a template with no promoters (poly [dA-dT]), but had little effect on transcription of a specific template (φe phage DNA).

Matthew Hilton and H.R. Whitely extended this finding in 1985 when they showed directly by a cross-linking procedure that δ loosens nonspecific complexes between polymerase and DNA. These investigators prepared a short (92-bp) piece of ^{32}P-labeled DNA that contained a promoter recognized by the *B. subtilis* polymerase. The DNA was also substituted with BrdU so it would cross-link to bound polymerase. Next, they bound the DNA to either core or holoenzyme in the presence or absence of δ, then cross-linked the DNA to the protein. Finally, they separated the subunits of the

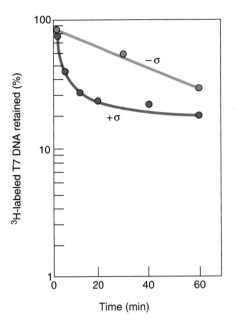

Figure 6.25   *E. coli* σ helps dissociate nonspecific binding between core polymerase and DNA. Hinkle and Chamberlin used a filter-binding assay, as described in Figure 5.35, to measure the dissociation rate of an *E. coli* core polymerase–T7 DNA complex in the presence (red) and absence (blue) of σ. Note the much faster dissociation that occurred when σ was present. The curve in the presence of σ leveled off at a point where 20% of polymerase–DNA complexes remained undissociated. This was probably an artifact that came from an excess of core polymerase that associated with σ, forming holoenzyme that could bind to labeled DNA and cause it to stick tightly to the filter. The initial rate of dissociation of the two curves is probably the best measure of the ability of σ to loosen the nonspecific interaction between core and DNA. (*Source:* Hinckle, D.C. and Chamberlin, M.J., "Studies of the Binding of Escherichia coli RNA Polymerase to DNA," *Journal of Molecular Biology,* Vol. 70, 157–185, 1972. By permission of Academic Press Limited, London.)

polymerase by SDS-PAGE and autoradiographed the gel to determine the degree of cross-linking between the polymerase subunits and the labeled DNA. As Figure 6.26 illustrates, the core by itself binds readily, though nonspecifically, to the DNA. By contrast, the core plus δ hardly binds the DNA at all. On the other hand, σ has relatively little effect on holoenzyme–DNA binding. Thus, δ serves to enhance the specificity of binding between RNA polymerase and DNA by loosening any nonspecific associations that happen to occur.

> **SUMMARY** The *E. coli* σ-factor, in addition to its promoter-binding capability, can loosen nonspecific interactions between polymerase and DNA and thus enhance the specificity of the polymerase–promoter interactions. The *B. subtilis* σ lacks this loosening activity. Instead, another protein factor called δ provides it.

Figure 6.26 The δ polypeptide enhances the specificity of binding between polymerase and DNA. Hilton and Whitely cross-linked labeled DNA to *B. subtilis* core polymerase, either alone or in the presence of other polypeptides. Then they dissociated the polymerase subunits and electrophoresed them by SDS-PAGE. They measured the association between polymerase and labeled DNA by autoradiography to detect the labeled DNA attached to each polymerase subunit. In the electrophoresis lanes shown here, core was bound to labeled DNA along with: lane 1, core alone; lane 2, δ; lane 3, σ; lane 4, σ plus δ. The δ-subunit greatly reduced the nonspecific binding of core (compare lanes 1 and 2), but not the specific binding of core plus σ (compare lanes 3 and 4). (*Source:* Hilton and Whitely, UV cross-linking of the *Bacillus subtilis* RNA polymerase DNA in promoter and nonpromoter complexes. *J. Biol. Chem.* 260, no. 13 (5 Jul 1985) p. 8123, f. 2 (sect #4-7). American Society for Biochemistry and Molecular Biology.)

The Role of the α-Subunit in UP Element Recognition

As we learned earlier in this chapter, RNA polymerase itself can recognize an upstream promoter element called an UP element. We know that the σ-factor recognizes the core promoter elements, but which polymerase subunit is responsible for recognizing the UP element? Based on the following evidence, it appears to be the α-subunit of the core polymerase.

Richard Gourse and colleagues made *E. coli* strains with mutations in the α-subunit and found that some of these were incapable of responding to the UP element—

they gave no more transcription from promoters with UP elements than from those without UP elements. To measure transcription, they placed a wild-type form of the very strong *rrnB* P1 promoter, or a mutant form that was missing its UP element, about 170 bp upstream of an *rrnB* P1 transcription terminator in a cloning vector. They transcribed these constructs with three different RNA polymerases, all of which had been reconstituted from purified subunits: (1) wild-type polymerase with a normal α-subunit; (2) α-235, a polymerase whose α-subunit was missing 94 amino acids from its C-terminus; and (3) R265C, a polymerase whose α-subunit contained a cysteine (C) in place of the normal arginine (R) at position 265. They included a labeled nucleotide to label the RNA, then subjected this RNA to gel electrophoresis, and finally performed autoradiography to visualize the RNA products. Figure 6.27a depicts the results with wild-type polymerase. The wild-type promoter (lanes 1 and 2) allowed a great deal more transcription than the same promoter with vector DNA substituted for its UP element (lanes 3 and 4), or having its UP element deleted (lanes 5 and 6). Figure 6.27b shows the same experiment with the polymerase with 94 C-terminal amino acids missing from its α-subunit. We see that this polymerase is just as active as the wild-type polymerase in transcribing a gene with a core promoter (compare panels a and b, lanes 3–6). However, in contrast to the wild-type enzyme, this mutant polymerase did not distinguish between promoters with and without an UP element (compare lanes 1 and 2 with lanes 3–6). The UP element provided no benefit at all. Thus, it appears that the C-terminal portion of the α-subunit enables the polymerase to respond to an UP element.

Figure 6.27c demonstrates that the polymerase with a cysteine in place of an arginine at position 265 of the α-subunit (R265C) does not respond to the UP element (lanes 7–10 all show modest transcription). Thus, this single amino acid change appears to destroy the ability of the α-subunit to recognize the UP element. This phenomenon was not an artifact caused by an inhibitor in the R265C polymerase preparation because a mixture of R265C and the wild-type polymerase still responded to the UP element (lanes 1–4 all show strong transcription).

To test the hypothesis that the α-subunit actually contacts the UP element, Gourse and coworkers performed DNase footprinting experiments with DNA containing the *rrnB* P1 promoter and either wild-type or mutant RNA polymerase. They found that the wild-type polymerase made a footprint in the core promoter and the UP element, but that the mutant polymerase lacking the C-terminal domain of the α-subunit made a footprint in the core promoter only (Figure 6.28). This indicates that the α-subunit C-terminal domain is required for interaction between polymerase and UP elements. Further evidence for this hypothesis came from an experiment in which Gourse and coworkers used purified α-subunit

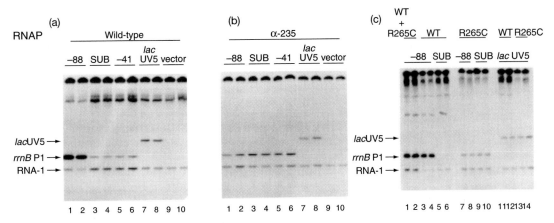

Figure 6.27 Importance of the α subunit of RNA polymerase in UP element recognition. Gourse and colleagues performed in vitro transcription on plasmids containing the promoters indicated at top. They placed the promoters between 100 and 200 nt upstream of a transcription terminator to produce a transcript of defined size. After the reaction, they subjected the labeled transcripts to gel electrophoresis and detected them by autoradiography. The promoters were as follows: −88 contained wild-type sequence throughout the region between positions −88 and +1; SUB contained an irrelevant sequence instead of the UP element between positions −59 and −41; −41 lacked the UP element upstream of position −41 and had vector sequence instead; lacUV5 is a *lac* promoter without an UP element; vector indicates a plasmid with no promoter inserted. The positions of transcripts from the *rrnB* P1 and lacUV5 promoters, as well as an RNA (RNA-1) transcribed from the plasmid's origin of replication, are indicated at left. RNAP at top indicates the RNA polymerase used, as follows: **(a)** Wild-type polymerase used throughout. **(b)** α-235 polymerase (missing 94 C-terminal amino acids of the α-subunit) used throughout. **(c)** Wild-type (WT) polymerase or R265C polymerase (with cysteine substituted for arginine 265) used, as indicated. (*Source:* Ross et al., A third recognition element in bacterial promoters: DNA binding by the alpha subunit of RNA polymerase. *Science* 262 (26 Nov 1993) f. 2, p. 1408. © AAAS.)

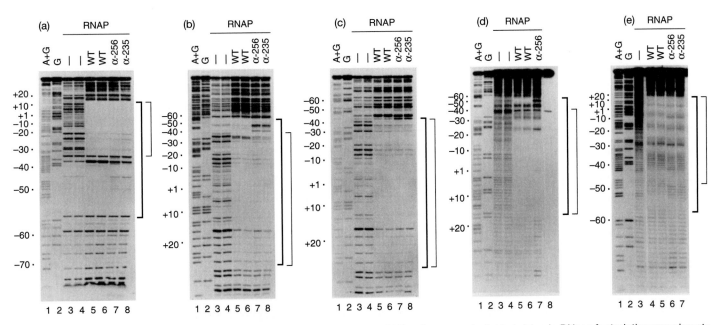

Figure 6.28 Footprinting the UP element. Gourse and colleagues used the RNA polymerases indicated at top in DNase footprinting experiments **(a–c)** or hydroxyl radical footprinting experiments **(d–e)**. The polymerases were wild-type (WT), or missing 73 or 94 C-terminal amino acids of the α subunit (α-256 and α-235, respectively). The lanes marked A+G and A are Maxam–Gilbert sequencing lanes in which the DNA was broken at A's and G's, or only at A's, respectively. This yielded a partial sequence the investigators could use to identify the protected regions. The lanes marked (−) contained no polymerase. The bold brackets at right indicate the region protected by the wild-type polymerase; the thin brackets indicate the region protected by the mutant polymerase. The DNAs labeled for footprinting were the following: **(a)** the nontemplate strand, **(b)** the template strand, **(c)** the template strand, **(d)** the template strand, and **(e)** the nontemplate strand. (*Source:* Ross et al., *Science* 262 (26 Nov 1993) f. 4, p. 1409. © AAAS.)

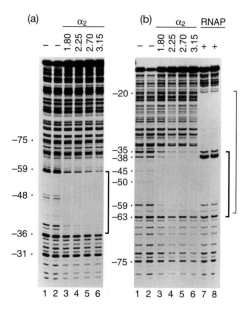

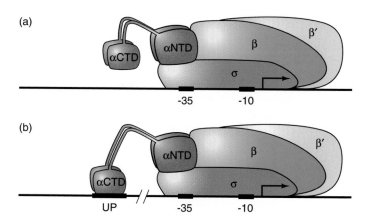

Figure 6.30 Model for the function of the C-terminal domain (CTD) of the polymerase α subunit. (a) In a core promoter, the α CTDs are not used, but (b) in a promoter with an UP element, the α CTDs contact the UP element. Notice that two α-subunits are depicted: one behind the other.

Figure 6.29 Footprinting the UP element with pure α subunit. Gourse and colleagues performed DNase footprinting with (a) end-labeled template strand (a) or (b) nontemplate strand from the *rrnB* P1 promoter. They used the amounts (in micrograms) of purified α-dimers, or 10 nM RNA polymerase holoenzyme (RNAP). The bold brackets indicate the footprints in the UP element caused by the α-subunit, and the thin bracket indicates the footprint caused by the holoenzyme. (*Source:* Ross et al., *Science* 262 (26 Nov 1993) f. 5, p. 1408. © AAAS.)

dimers to footprint the UP element of the *rrnB* P1 promoter. Figure 6.29 shows the results—a clear footprint in the UP element caused by the α-subunit dimer all by itself.

Richard Gourse, Richard Ebright, and their colleagues used **limited proteolysis** analysis to show that the α-subunit N-terminal and C-terminal domains (the α-**NTD** and α-**CTD,** respectively) fold independently to form two domains that are tethered together by a flexible linker. A protein **domain** is a part of a protein that folds independently to form a defined structure. Because of their folding, domains tend to resist proteolysis, so limited digestion with a proteolytic enzyme will attack unstructured elements between domains and leave the domains themselves alone. When Gourse and Ebright and collaborators performed limited proteolysis on the *E. coli* RNA polymerase α-subunit, they released a polypeptide of about 28 kD, and three polypeptides of about 8 kD. The sequences of the ends of these products showed that the 28-kD polypeptide contained amino acids 8–241, whereas the three small polypeptides contained amino acids 242–329, 245–329, and 249–329. This suggested that the α-subunit folds into two domains: a large N-terminal domain encompassing (approximately) amino acids 8–241, and a small C-terminal domain including (approximately) amino acids 249–329.

Furthermore, these two domains appear to be joined by an unstructured linker that can be cleaved in at least three

places by the protease used in this experiment (Glu-C). This linker seems at first glance to include amino acids 242–248. Because Glu-C requires three unstructured amino acids on either side of the bond that is cleaved, however, the linker is longer than it appears at first. In fact, it must be at least 13 amino acids long (residues 239–251).

These experiments suggest a model such as the one presented in Figure 6.30. RNA polymerase binds to a core promoter via its σ-factor, with no help from the C-terminal domains of its α-subunits, but it binds to a promoter with an UP element using σ plus the α-subunit C-terminal domains. This allows very strong interaction between polymerase and promoter and therefore produces a high level of transcription.

SUMMARY The RNA polymerase α-subunit has an independently folded C-terminal domain that can recognize and bind to a promoter's UP element. This allows very tight binding between polymerase and promoter.

6.4 Elongation

After initiation of transcription is accomplished and σ has left the scene, the core continues to elongate the RNA, adding one nucleotide after another to the growing RNA chain. In this section we will explore this elongation process.

Core Polymerase Functions in Elongation

So far we have been focusing on the role of σ because of the importance of this factor in determining the specificity of initiation. However, the core polymerase contains the RNA synthesizing machinery, so the core is the central player in elongation. In this section we will learn that the β-subunit is involved in phosphodiester bond formation, that the β- and β′-subunits participate in DNA binding, and that the α-subunit has several activities, including assembly of the core polymerase.

The Role of β in Phosphodiester Bond Formation Walter Zillig was the first to investigate the individual core subunits, in 1970. He began by separating the *E. coli* core polymerase into its three component polypeptides and then combining them again to reconstitute an active enzyme. The separation procedure worked as follows: Alfred Heil and Zillig electrophoresed the core enzyme on cellulose acetate in the presence of urea. Like SDS, urea is

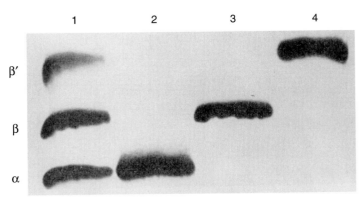

a denaturing agent that can separate the individual polypeptides in a complex protein. Unlike SDS, however, urea is a mild denaturant that is relatively easy to remove. Thus, it is easier to renature a urea-denatured polypeptide than an SDS-denatured one. After electrophoresis was complete, Heil and Zillig cut out the strips of cellulose acetate containing the polymerase subunits and spun them in a centrifuge to drive the protein out of the cellulose acetate. This gave them all three separated polypeptides, which they electrophoresed individually to demonstrate their purity (Figure 6.31).

Once they had separated the subunits, they recombined them to form active enzyme, a process that worked best in the presence of σ. Using this separation–reconstitution system, Heil and Zillig could mix and match the components from different sources to answer questions about their functions. For example, recall that the core polymerase determines sensitivity or resistance to the antibiotic rifampicin. Separation and reconstitution of the core allowed Heil and Zillig to ask which core subunit confers this antibiotic sensitivity or resistance. When they recombined the α-, β′-, and σ-subunits from a rifampicin-sensitive bacterium with the β-subunit from a rifampicin-resistant bacterium, the resulting polymerase was antibiotic-resistant (Figure 6.32). Conversely, when the β-subunit came from an antibiotic-sensitive bacterium, the reconstituted enzyme was antibiotic-sensitive, regardless of the origin of the other subunits. Thus, the β-subunit is obviously the determinant of rifampicin sensitivity or resistance.

Another antibiotic, known as *streptolydigin*, blocks RNA chain elongation. By the same separation and reconstitution strategy used for rifampicin, Heil and Zillig showed that the β-subunit also governed streptolydigin resistance or sensitivity. At first this seems paradoxical. How can the same core subunit be involved in both initiation and elongation? The probable answer comes from a

Figure 6.31 Purification of the individual subunits of *E. coli* RNA polymerase. Heil and Zillig subjected the *E. coli* core polymerase to urea gel electrophoresis on cellulose acetate, then collected the separated polypeptides. Lane 1, core polymerase after electrophoresis; lane 2, purified α; lane 3, purified β; lane 4, purified β′. (*Source:* Heil, A. and Zillig, W. Reconstitution of bacterial DNA-dependent RNA-polymerase from isolated subunits as a tool for the elucidation of the role of the subunits in transcription. *FEBS Letters* 11 (Dec 1970) p. 166, f. 1.)

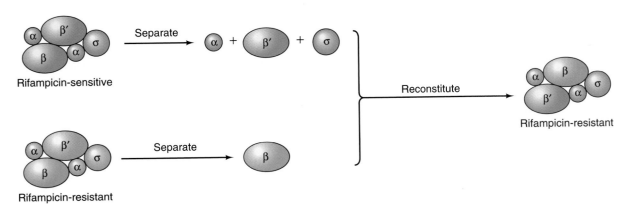

Figure 6.32 Separation and reconstitution of RNA polymerase to locate the determinant of antibiotic resistance. Start with RNA polymerases from rifampicin-sensitive and -resistant *E. coli* cells, separate them into their component polypeptides, and recombine them in various combinations to reconstitute the active enzyme. In this case, the α-, β′-, and σ-subunits came from the rifampicin-sensitive polymerase (blue), and the β-subunit came from the antibiotic-resistant enzyme (red). The reconstituted polymerase is rifampicin-resistant, which shows that the β-subunit determines sensitivity or resistance to this antibiotic.

consideration of the one thing that initiation and elongation have in common: phosphodiester bond formation. Initiation is not complete until several phosphodiester bonds have formed, so rifampicin can block initiation by preventing the formation of those bonds. And elongation obviously entails the formation of one phosphodiester bond after another, so streptolydigin could inhibit elongation by hindering that process. Thus, one explanation for the role of β in the response to these two antibiotics is that it is responsible for some aspect of forming phosphodiester bonds between nucleotides.

In 1987, M.A. Grachev and colleagues provided direct evidence for this postulated role of β, using a technique called **affinity labeling**. The idea behind this technique is to label an enzyme with a derivative of a normal substrate that can be cross-linked to protein. In this way, one can use the affinity reagent to seek out and then tag the active site of the enzyme. Finally, one can dissociate the enzyme to see which subunit the tag is attached to. Grachev and coworkers used 14 different affinity reagents, all ATP or GTP analogs. One of these, which was the first in the series, and therefore called I, has the structure shown in Figure 6.33a. When they added it to RNA polymerase, it went to the active site, as an ATP that is initiating transcription would normally do, and then formed a covalent bond with an amino group at the active site according to the reaction in Figure 6.33b.

In principle, these investigators could have labeled the affinity reagent itself and proceeded from there. However, they recognized a pitfall in that simple strategy: The affinity reagent might bind to other amino groups on the enzyme surface in addition to the one(s) in the active site. To circumvent this problem, they used an unlabeled affinity reagent, followed by a radioactive nucleotide ([α-^{32}P]UTP or CTP) that would form a phosphodiester bond with the affinity reagent in the active site and therefore label that site and no others on the enzyme. Finally, they dissociated the labeled enzyme and subjected the subunits to SDS-PAGE. The results are presented in Figure 6.34. Obviously, the β-subunit is the only core subunit labeled by any of the affinity reagents, suggesting that this subunit is responsible for nucleotide binding and phosphodiester bond formation. In some cases, we also see some labeling of σ, suggesting that it too may play a role in forming at least the first phosphodiester bond.

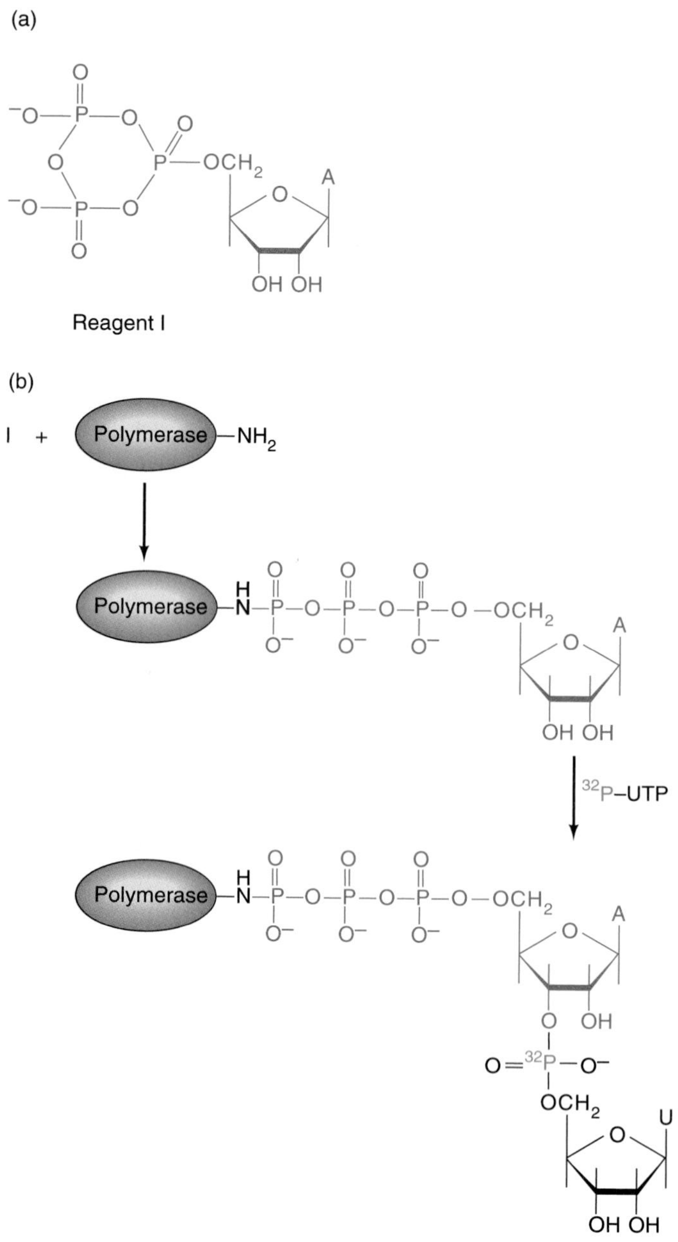

> **SUMMARY** The core subunit β binds nucleotides at the active site of the RNA polymerase where phosphodiester bonds are formed. The σ-factor is also near the nucleotide-binding site, at least during the initiation phase.

The Roles of β' and β in DNA Binding In 1996, Evgeny Nudler and colleagues used an elegant series of experi-

Figure 6.33 Affinity labeling RNA polymerase at its active site. (a) Structure of one of the affinity reagents (I), an ATP analog. (b) The affinity-labeling reactions. First, add reagent I to RNA polymerase. The reagent binds covalently to amino groups at the active site (and perhaps elsewhere). Next, add radioactive UTP, which forms a phosphodiester bond (blue) with the enzyme-bound reagent I. This reaction should occur only at the active site, so only that site becomes radioactively labeled.

ments to show that both the β- and β'-subunits are involved in DNA binding, and that two DNA binding sites are present: (1) an upstream, relatively weak site where DNA melting occurs and electrostatic forces predominate; and (2) a downstream, strong binding site where hydrophobic forces bind DNA and protein together.

Figure 6.35 illustrates the general approach Nudler and colleagues took. Using gene-cloning techniques

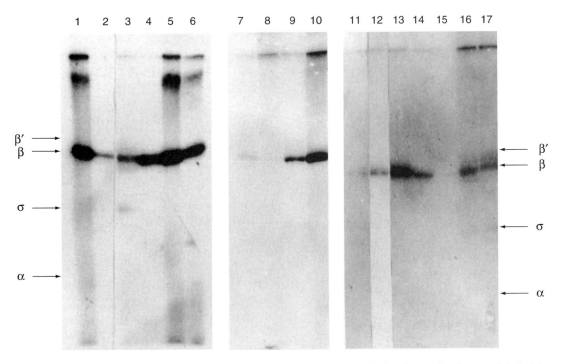

Figure 6.34 **The β-subunit is at the active site where phosphodiester bonds are formed.** Grachev and colleagues labeled the active site of *E. coli* RNA polymerase as described in Figure 6.33, then separated the polymerase subunits by electrophoresis to identify the subunits that compose the active site. Each lane represents labeling with a different nucleotide-affinity reagent plus radioactive UTP, except lanes 5 and 6, which resulted from using the same affinity reagent, but either radioactive UTP (lane 5) or CTP (lane 6). The autoradiograph of the separated subunits demonstrates labeling of the β-subunit with most of the reagents. In a few cases, σ was also faintly labeled. Thus, the β-subunit appears to participate in the phosphodiester bond-forming active site. (*Source:* Grachev et al., Studies on the functional topography of *Escherichia coli* RNA polymerase. *Eur. J. Biochem.* 163 (16 Dec 1987) p. 117, f. 2.)

described in Chapter 4, they engineered an RNA polymerase with six extra histidines at the C-terminus of the β′-subunit. This allowed them to tether the polymerase to a nickel resin so they could change templates and substrates rapidly by washing the resin and adding fresh reagents. Figure 6.35a shows how this allowed them to "walk" the polymerase to a particular position on the template by adding some of the nucleoside triphosphates (e.g., ATP, CTP, and GTP, but not UTP) and allowing the polymerase to synthesize a short RNA and then stop at the point where it needed the missing nucleotide. Washing away the first set of nucleotides and adding a second set walked the polymerase to a site further downstream, and so forth. Finally, they could add all four nucleotides and "chase" the polymerase off the template altogether, producing a run-off transcript.

Figure 6.35b shows a surprising phenomenon: When Nudler and colleagues chased the polymerase in the presence of a secondary template and ATP, GTP, and CTP, the polymerase jumped from the primary template (blue) to the secondary template (magenta) and continued elongating the transcript until it reached an A in the template strand, when it had to stop for lack of UTP. They knew it jumped from one template to another because they could chase it off the secondary template and get the expected

size run-off transcript, which was longer than the run-off transcript from the primary template would have been. Even more convincing, they could walk the polymerase along the secondary template and get arrested transcripts of the expected size.

Next, they varied the characteristics of the secondary template to see what was required for the polymerase to jump to it and resume transcription. They found the following: (1) A single-stranded secondary template would work, but only at low-salt concentrations. At high-salt concentrations, the complex fell apart, showing that the polymerase could not bind tightly to single-stranded DNA. (2) A stretch of only 9 bp of double-stranded DNA was required for salt-resistant polymerase binding, and this double-stranded DNA had to be between positions +2 and +11 relative to the growing 3′-end of the transcript. The fact that binding to this region was salt-resistant means that the binding there is hydrophobic, rather than electrostatic. In other words, it involves interactions between hydrophobic amino acids on the polymerase interacting with bases in one of the DNA grooves. (3) Another requirement for salt-stable binding was a stretch of at least 6 nt of single-stranded DNA upstream of the 3′-end of the RNA.

Next, these workers analyzed the salt-sensitive binding site by using primary templates in which the polymerase

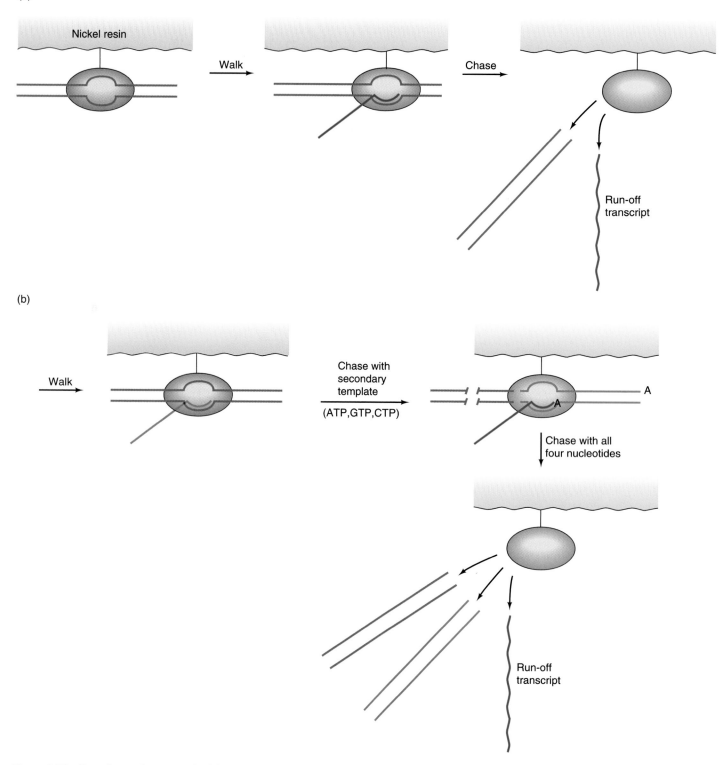

Figure 6.35 Experimental strategy for identifying template requirements for binding to RNA polymerase. (a) Walking and chasing the RNA polymerase. Nudler and colleagues bound RNA polymerase (red) to a nickel resin (yellow) through an oligo-His tag on the β′-subunit of the polymerase. Then they "walked" the polymerase along the template by adding sets of three or fewer nucleotides that would allow the polymerase to advance a few base pairs and then stop for lack of a nucleotide. Finally, they could add all four nucleotides and "chase" the polymerase off the end of the template, releasing the run-off transcript in the process. (b) Template transfer. Again, Nudler and colleagues walked the polymerase to a particular site on the primary template (blue), then chased the polymerase in the presence of a secondary template (magenta) and (in this case) ATP, GTP, and CTP. This allowed the polymerase to jump to the secondary template and walk to the point where the first UTP is needed (the first A in the template strand). Notice that jumping to the opposite strand would be blocked because UTP, the missing nucleotide, would be needed at the very first position if the top strand were the template strand. Finally, the polymerase can be chased off of the second template, releasing both templates and the lengthened run-off transcript.

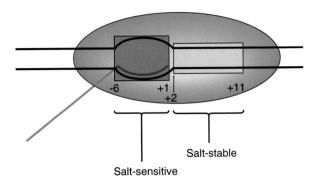

Figure 6.36   DNA-binding sites in _E. coli_ RNA polymerase holoenzyme. Based on template transfer experiments as described in Figure 6.35, but with many altered primary and secondary templates, Nudler and colleagues proposed that two sites on the template DNA interact with RNA polymerase. The downstream site (yellow), encompassing base pairs +2 through +11, makes salt-stable contacts with polymerase, suggesting that these interactions are hydrophobic in nature. The upstream site (blue), located between bp −6 and +1, makes salt-sensitive contacts with polymerase, indicating that these are electrostatic interactions.

had been walked to a position close to the downstream end. These complexes lacked the downstream double-stranded region required for salt-resistant binding, so the salt-sensitive binding site was all that was left. What was required for this salt-sensitive binding? The minimum requirement was a 6-nt stretch of single-stranded template strand in the region from position −6 to +1 relative to the 3′-end of the RNA.

Figure 6.36 summarizes these findings. It shows the downstream salt-stable binding site (yellow) between positions +2 and +11, and the upstream salt-sensitive binding site (blue) between positions −6 and +1. This salt-sensitive site can consist of single-stranded DNA, as long as it is the template strand, which reflects the fact that this region is melted in a transcription-competent complex with RNA polymerase. This model suggests that the interaction between the RNA transcript and the DNA is not the most important contributor to the stability of the ternary DNA–RNA–protein complex because it is involved in the salt-sensitive (weak) binding site. The most important contribution to the stability of the complex comes from the strong, salt-resistant binding site downstream. Indeed, this strong binding site is a vital part of a **sliding clamp** (see Figure 6.9d) that holds the polymerase tightly on the DNA.

These studies also allowed Nudler and coworkers to identify the polymerase subunits involved in both binding sites. Their strategy was to introduce 5-iododeoxyuridine, a thymidine analog, into a short, radioactive secondary template at defined positions. This nucleoside can be cross-linked to neighboring protein on activation by ultraviolet light. The sites chosen for study were at positions −3, +1, and +6, which should lie in the salt-sensitive zone, in the active site, and in the salt-stable zone, respectively.

Position +1 is at the active site by definition because that is opposite the 3′-end of the RNA, where the last nucleotide was just added.

After cross-linking, these workers subjected the complexes to SDS-PAGE and autoradiography to determine which polymerase subunits were bound to the radiolabeled DNA. Figure 6.37 demonstrates that the β-subunit was cross-linked to DNA at the −3 and +1 positions, as we would expect, because the β-subunit contains the active site of the enzyme. On the other hand, both β and β′ were cross-linked to DNA at the +6 position. Because this is in the strong binding site, this result is consistent with an earlier finding that the β′-subunit binds strongly to DNA on its own. It also showed, however, that β plays a role in tight binding, as well as in loose binding at the active site. Furthermore, when the +6 position of a single-stranded DNA was cross-linked to polymerase, less binding to β′ occurred. This reinforced the notion that tight binding via β′ requires double-stranded DNA.

Finally, Nudler and colleagues pinned down the sites within β and β′ that bound to DNA. They began by purifying from the SDS-PAGE gel the β- and β′-subunits cross-linked to labeled DNA. Then they treated these polypeptides with CNBr, which cleaves after methionine (Met) residues. This produced a series of peptides, some of which were cross-linked to the labeled template. Gel electrophoresis followed by autoradiography showed that the crosslink to the β-subunit occurred near the N-terminus, between Met 30 and Met 102, and that the cross-link to the β′-subunit occurred near the C-terminus, between Met 1230 and Met 1273.

SUMMARY Template transfer experiments have delineated two DNA sites that interact with polymerase. One is a weak binding site involving the melted DNA zone and the catalytic site on the β-subunit of the polymerase. Protein–DNA interactions at this site are primarily electrostatic and are salt-sensitive. The other is a strong binding site involving DNA downstream of the active site, and the β′- and β-subunits of the enzyme.

The Role of α in Polymerase Assembly

We have seen earlier in this chapter that the C-terminal domain of the RNA polymerase α-subunit is responsible for recognizing UP elements, and we will learn in Chapter 7 that it also interacts with activator proteins that bind to sites outside the promoter. But the N-terminal domain also plays a key role in transcription by helping to organize the RNA polymerase. The order of subunit assembly of the _E. coli_ RNA polymerase is $\alpha \rightarrow \alpha_2 \rightarrow \alpha_2\beta \rightarrow \alpha_2\beta\beta'$, which

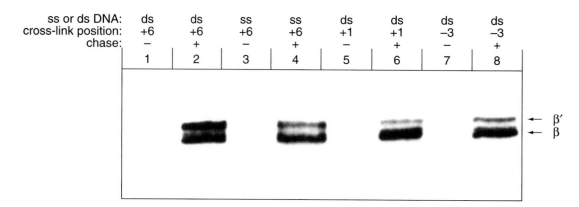

ss or ds DNA:	ds	ds	ss	ss	ds	ds	ds	ds
cross-link position:	+6	+6	+6	+6	+1	+1	−3	−3
chase:	−	+	−	+	−	+	−	+
	1	2	3	4	5	6	7	8

Figure 6.37   Cross-linking labeled DNA to RNA polymerase subunits. Nudler and colleagues placed a reactive nucleoside (5-iododeoxyuridine) at the cross-link position indicated at top, in a single-stranded (ss) or double-stranded (ds) labeled secondary template, also as indicated at top. They either chased (+) or did not chase (−) the polymerase from the primary template to this secondary template, as indicated at top. After the chase (or lack of it), they irradiated the polymerase–DNA complex with UV light to cross-link the reactive nucleoside to whichever polymerase subunit was nearby. Then they subjected the products to SDS-PAGE and autoradiography to visualize the labeled polymerase subunits. Only polymerases that were chased to the secondary template formed cross-links. The β-subunit was involved in all cross-links, and the β′-subunit was cross-linked to position +6 in double-stranded DNA, but less so in single-stranded DNA. (*Source:* Nudler et al., Transcription processivity: Protein-DNA interactions holding together the elongation complex. *Science* 273 (12 July 1996) f. 5c, p. 214. © AAAS.)

suggests that α plays a central role. But what part of α is involved in assembly?

Richard Hayward, working with Kazuhiko Igarashi and Akira Ishihama answered this question by examining mutant bacterial strains with deletions at the C-terminus of α. The mutants produced α-subunits terminating at amino acids 150, 176, 235, 256, and 296, whereas the wild-type protein terminates at amino acid 329. These workers tested RNA polymerases from all these strains for the ability to assemble in vivo into complexes that resembled core polymerase and holoenzyme. They found that the polymerase with an α-subunit ending at amino acid 235, 256, or 296 could assemble properly, but the polymerase with an α-subunit ending at amino acid 176 could not. Thus, up to 94 amino acids (but not 153 amino acids) could be removed from the C-terminus without hindering assembly of the polymerase. This finding suggested that the assembly function lies in the N-terminal domain of the α-subunit (the α-NTD).

What part of the α-NTD interacts with the β- and β′-subunits? To find out, Richard Ebright and colleagues identified the amino acids of the α-NTD that are protected against hydroxyl radical cleavage by the β- and β′-subunits. All these amino acids clustered into regions of the α-NTD known as motifs 1 and 2. Furthermore, Ishihama and colleagues demonstrated that all of the mutations that interfere with binding to the β- and β′-subunits also lie in motifs 1 and 2 of the α-NTD dimer. Gongyi Zhang and Seth Darst then performed x-ray crystallography on the α-NTD and showed that all of these sites in motifs 1 and 2 that are important in binding to β and β′ lie on one side of the α-NTD. Thus, the β- and β′-subunits appear to interact with one side of the α-NTD dimer,

which happens to be the side that faces away from the C-terminal domains of the α-subunit dimer (the α-CTD).

SUMMARY The α-subunit serves to assemble the RNA polymerase holoenzyme, and this assembly function resides in the N-terminal domain of α (the α-NTD). The points of contact between the α-NTD and the β- and β′-subunits lie on the opposite side of the α-NTD dimer from the dimer's C-terminal domains.

Structure of the Elongation Complex

In this section, we will see how well the predictions about the roles of the core polymerase subunits have been borne out by structural studies. We will also consider the topology of elongation: How does the polymerase deal with the problems of unwinding and rewinding its template, and of moving along its twisted (helical) template without twisting its RNA product around the template?

The RNA–DNA Hybrid Up to this point we have been assuming that the RNA product forms an RNA–DNA hybrid with the DNA template strand for a few bases before peeling off and exiting from the polymerase. But the length of this hybrid has been controversial, with estimates ranging from 3–12 bp, and some investigators even doubted whether it existed. But Nudler and Goldfarb and their colleagues applied their transcript walking technique, together with RNA–DNA cross-linking, to prove that an RNA–DNA hybrid really does occur within the elongation complex, and that this hybrid is 8–9 bp long.

(a)

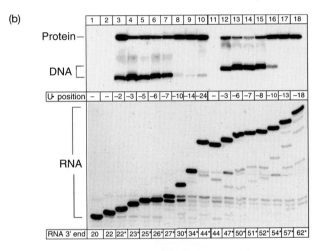

(b)

Protein—

DNA

U• position

RNA

RNA 3′ end

Figure 6.38   RNA–DNA and RNA–protein cross-linking in elongation complexes. (a) Structure of the cross-linking reagent U• base-paired with an A in the DNA template strand. The reagent is in position to form a covalent bond with the DNA as shown by the arrow. **(b)** Results of cross-linking. Nudler Goldfarb and colleagues incorporated U• at position 21 or 45 of a [^{32}P]nascent RNA in an elongation complex. Then they walked the U• to various positions between −2 and −24 with respect to the 3′-end (position −1) of the nascent RNA. Then they cross-linked the RNA to the DNA template (or the protein in the RNA polymerase). They then electrophoresed the DNA and protein in one gel (top) and the free RNA transcripts in another (bottom) and autoradiographed the gels. Lanes 1, 2, and 11 are negative controls in which the RNA contained no U•. Lanes 3–10 contained products from reactions in which the U• was in position 21; lanes 12–18 contained products from reactions in which the U• was in position 45 of the nascent RNA. Asterisks at bottom denote the presence of U• in the RNA. Cross-linking to DNA was prevalent only when U• was between positions −2 and −8. (*Source:* Nudler, E. et al. The RNA-DNA Hybrid Maintains the Register of Transcription by Preventing Backtracking of RNA Polymerase. *Cell* 89 (1997) f. 1, p. 34. Reprinted by permission of Elsevier Science.)

These workers incorporated a UMP derivative (U•) at either position 21 or 45 with respect to the 5′-end of a ^{32}P-labeled nascent RNA. U• is normally unreactive, but in the presence of NaBH$_4$ it becomes capable of cross-linking to a base-paired base, as shown in Figure 6.38a. Actually, U• can reach to a purine adjacent to the base-paired A in the DNA strand, but this experiment was designed to prevent that from happening. So cross-linking could occur only to an A in the DNA template strand that

was base-paired to the U• base in the RNA product. If no base-pairing occurred, no cross-linking would be possible.

Nudler, Goldfarb, and their colleagues walked the U• base in the transcript to various positions with respect to the 3′-end of the RNA, beginning with position −2 (the nucleotide next to the 3′-end, which is numbered −1) and extending to position −44. Then they tried to cross-link the RNA to the DNA template strand. Finally, they electrophoresed both the DNA and protein in one gel, and just the RNA in another. Note that the RNA will always be labeled, but the DNA or protein will be labeled only if the RNA has been cross-linked to them.

Figure 6.38b shows the results. The DNA was strongly labeled if the U• base was in position −2 through position −8, but only weakly labeled when the U• base was in position −10 and beyond. Thus, the U• base was base-paired to its A partner in the DNA template strand only when it was in position −2 through −8, but base-pairing was much decreased when the reactive base was in position −10. So the RNA–DNA hybrid extends from position −1 to position −8, or perhaps −9, but no farther. (The nucleotide at the very 3′-end of the RNA, at position −1, must be base-paired to the template to be incorporated correctly.) This conclusion was reinforced by the protein labeling results. Protein in the RNA polymerase became more strongly labeled when the U• was not within the hybrid region (positions −1 through −8). This presumably reflects the fact that the reactive group was more accessible to the protein when it was not base-paired to the DNA template.

Not only does the RNA–DNA hybrid exist in the elongation complex, Mikhail Kashlev and colleagues have shown that a hybrid at least 9 nt long appears to be essential for processivity of transcription. They built a synthetic elongation complex by combining *E. coli* RNA polymerase core with a single-stranded DNA that served as the "template strand" and annealing RNA primers of various sizes that were complementary to the single-stranded DNA. These served as the "transcripts." Then they added a second DNA strand, complementary to the first, which served as the "nontemplate strand." These synthetic complexes were capable of elongating the RNA primer processively, but only if the hybrid between the RNA and template strand was at least 9 nt long.

SUMMARY The RNA–DNA hybrid within the elongation complex extends from position −1 to position −8 or −9 with respect to the 3′-end of the nascent RNA. In vitro studies have suggested that processivity of transcription depends on an RNA–DNA hybrid at least 9 bp long.

Structural Studies   To get the clearest picture of the structure of the elongation complex, we need to know the

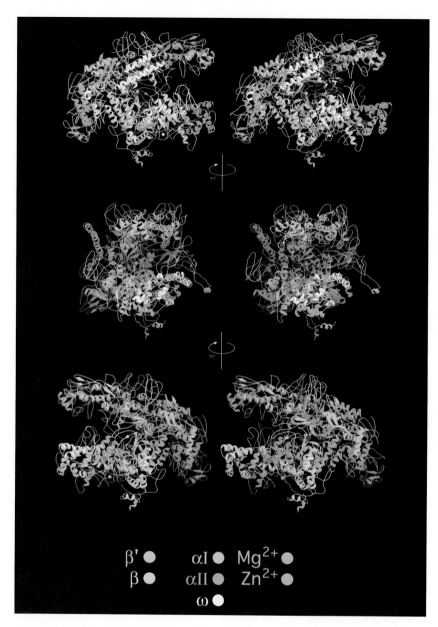

Figure 6.39   Crystal structure of the _Thermus aquaticus_ RNA polymerase core enzyme. Three different stereo views are shown, differing by 90° rotations. The subunits and metal ions in the enzyme are color-coded as indicated at the bottom. The metal ions are depicted as small colored spheres. The larger red and green dots denote unstructured regions of the β- and β′-subunits that are missing from these diagrams. (_Source:_ Zhang, G. et al. Crystal Structure of _Thermus aquaticus_ Core RNA Polymerase at 3.3 Å Resolution. _Cell_ 98 (1999) 811–824. Reprinted by permission of Elsevier Science.)

structure of the core polymerase. Figure 6.9 presented the best representation we have to date of the _E. coli_ core polymerase. However, the resolution of this structure, is limited by the technique: electron microscopy of two-dimensional crystals. X-ray crystallography would give much better resolution, but it requires three-dimensional crystals and, so far, no one has succeeded in preparing three-dimensional crystals of the _E. coli_ polymerase. However, Darst and colleagues, have crystallized the core polymerase from another bacterium, _Thermus aquaticus,_ and have obtained a crystal structure to a resolution of 3.3 Å. This structure is very similar in overall shape to the

lower resolution structure of the _E. coli_ core polymerase, so the detailed structures are probably also similar. In other words, the crystal structure of the _T. aquaticus_ core polymerase is our best window right now on the structure of a bacterial core polymerase.

Figure 6.39 depicts the overall shape of the enzyme in three different orientations. We notice first of all that it resembles an open crab claw, just as we observed for the _E. coli_ core in Figure 6.9. The four subunits (β, β′, and two α) are shown in different colors so we can distinguish them. This coloring reveals that half of the claw is composed primarily of the β-subunit, and the other half is

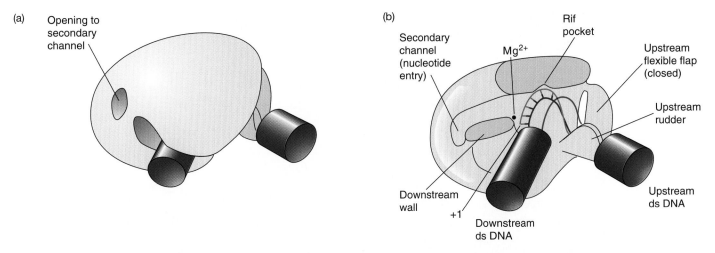

Figure 6.40   Schematic diagram of the core enzyme-DNA–RNA ternary complex. (a) The whole complex, with DNA lying in the main channel. **(b)** Cutaway drawing showing the predicted positions of DNA and RNA in the enzyme. The double-stranded DNA regions are depicted as blue cylinders. The DNA template strand is in blue and the nontemplate strand in cyan. The nascent RNA is in red, with the 3′-end labeled +1. The catalytic center is marked Mg^{2+} and the rifampicin-binding site is labeled Rif pocket. The downstream wall separates the downstream main channel from the secondary channel. The upstream rudder may help separate the two DNA strands. (*Source:* Zhang, E., E.A. Campbell, L. Minakhin, C. Richter, K. Severinov, and S. A. Darst, "Crystal Structure of *Thermus aquaticus* Core RNA Polymerase at 3.3 Å Resolution." *Cell* 98:811–24. Copyright 1999 Cell Press, Cambridge, MA. Reprinted by permission.)

composed primarily of the β′-subunit. The two α-subunits lie at the "hinge" of the claw, and a small ω-subunit is at the bottom, wrapped around the C-terminus of β′. The ω-subunit is not essential for enzyme activity and we have therefore ignored it until now, but it may help the β′-subunit fold properly.

Figure 6.40 shows a schematic diagram of the core polymerase, first intact, and then with part of the enzyme cut away so we can see the proposed positions of the DNA template and the RNA product inside. We see that the enzyme contains a channel, about 27 Å wide, between the two parts of the claw, and the template DNA lies in this channel. The **catalytic center** of the enzyme is marked Mg^{2+}. Three pieces of evidence place the Mg^{2+} at the catalytic center. First, an invariant string of amino acids (NADFDGD) occurs in the β′-subunit from all bacteria examined so far, which contains three aspartate residues (D) suspected of chelating Mg^{2+} ion. Second, mutations in any of these Asp residues are lethal. They create an enzyme that can form an open-promoter complex at a promoter, but is devoid of catalytic activity. Thus, these Asp residues are essential for catalytic activity, but not for tight binding to DNA. Finally, as Figure 6.41 demonstrates, the crystal structure of the *T. aquaticus* core polymerase shows that the side chains of the three Asp residues are indeed coordinated to a Mg^{2+} ion. Thus, the three Asp residues and a Mg^{2+} ion are at the catalytic center of the enzyme.

Figures 6.40 and 6.41 also identify a rifampicin-binding site in the part of the β-subunit that forms the ceiling of the channel through the enzyme. The amino acids whose alterations cause rifampicin resistance are tagged with pink dots in Figure 6.41. Clearly, these amino acids are tightly clustered in the three-dimensional structure, presumably at the site of rifampicin binding. We also know that rifampicin allows RNA synthesis to begin, but blocks elongation of the RNA chain beyond just a few nucleotides. On the other hand, the antibiotic has no effect on elongation once promoter clearance has occurred. How can we interpret the location of the rifampicin-binding site in terms of the antibiotic's activity? One explanation is that rifampicin bound in the channel blocks the exit through which the growing RNA should pass, and thus prevents growth of a short RNA. Once an RNA reaches a certain length, it might block access to the rifampicin binding site, or at least prevent effective binding of the antibiotic.

Figure 6.40 illustrates several other interesting features of the channel through the enzyme. First, the "downstream wall" separates the channel into two parts: the main channel, which holds the downstream double-stranded part of the DNA template; and a secondary channel that does not have the right geometry to accommodate even a single strand of the template DNA. This secondary channel could serve as the entry point for the nucleotides that will be polymerized into RNA. Second, the main channel makes a turn of greater than 90 degrees, which agrees with atomic force microscopy studies on ternary complexes involving core, DNA, and RNA. Figure 6.40 shows two regions of double-stranded DNA bound to the polymerase. The downstream region presumably corresponds to the salt-stable region identified by Nudler and colleagues (Figure 6.36). The model in Figure 6.40 is partly speculative, in that it proposes positions for DNA and RNA, even though those components

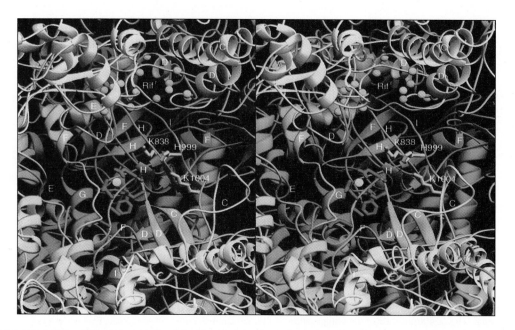

Figure 6.41   The catalytic center of the core polymerase. The Mg^{2+} ion is shown as a pink sphere, coordinated by three aspartate side chains (red) in this stereo image. The amino acids involved in rifampicin resistance are denoted by magenta spheres at the top of the channel, surrounding the presumed rifampicin-binding site, or Rif pocket, labeled Rif^r. The colors of the polymerase subunits are as in Figure 16.39 (β', rose; β, cyan; α yellow). (*Source:* Zhang G. et al. "Crystal Structure of *Thermus aquaticus* Core RNA Polymerase at 3.3 Ä Resolution." *Cell* 98 (1999) 811–824. Reprinted by permission of Elsevier Science.)

were not included in the crystal structure. To know for sure how the nucleic acids fit into the enzyme, we will need crystals of binary and ternary complexes involving enzyme, DNA, and RNA.

> **SUMMARY** X-ray crystallography on the *Thermus aquaticus* RNA polymerase core has revealed an enzyme shaped like a crab claw designed to grasp DNA. A channel through the enzyme includes the catalytic center (a Mg^{2+} ion coordinated by three Asp residues), and the rifampicin binding site.

Topology of Elongation   Does the core, moving along the DNA template, maintain the local melted region created during initiation? Common sense tells us that it does because this would help the RNA polymerase "read" the bases of the template strand and therefore insert the correct bases into the transcript. Experimental evidence also suggests that this is so. Jean-Marie Saucier and James Wang added nucleotides to an open-promoter complex, allowing the polymerase to move down the DNA as it began elongating an RNA chain, and found that the same degree of melting persisted. This leads to the summary of transcription elongation presented in Figure 6.42.

The static nature of Figure 6.42 is somewhat misleading. If we could see transcription as a dynamic process, we would observe the DNA double helix opening up in front of the moving "bubble" of melted DNA and closing up

again behind. In theory, RNA polymerase could accomplish this process in two ways, and Figure 6.43 presents both of them. One way would be for the polymerase and the growing RNA to rotate around and around the DNA template, following the natural twist of the double-helical DNA, as transcription progressed (Figure 6.43a). This would not twist the DNA at all, but it would require considerable energy to make the polymerase gyrate that much, and it would leave the transcript hopelessly twisted around the DNA template, with no known enzyme to untwist it.

The other possibility is that the polymerase moves in a straight line, with the template DNA rotating in one direction ahead of it to unwind, and rotating in the opposite direction behind it to wind up again (Figure 6.43b). But this kind of rotating of the DNA introduces strain. To visualize this, think of unwinding a coiled telephone cord, or actually try it if you have one available. You can feel (or imagine) the resistance you encounter as the cord becomes more and more untwisted, and you can appreciate that you would also encounter resistance if you tried to wind the cord more tightly than its natural state. It is true that the rewinding of DNA at one end of the melted region creates an opposite and compensating twist for the unwinding at the other. But the polymerase in between keeps this compensation from reaching across the melted region, and the long span of DNA around the circular chromosome insulates the two ends of the melted region from each other the long way around.

So if this second mechanism of elongation is valid, we have to explain how the strain of unwinding the DNA is

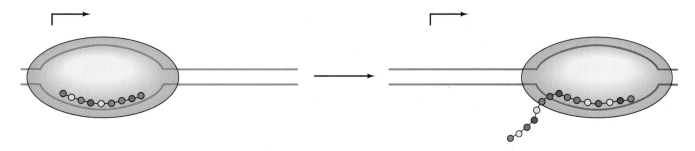

Figure 6.42 Summary of the elongation reaction. At left, RNA polymerase (pink) has bound to a promoter, formed an open-promoter complex, forged the first phosphodiester bonds (up to 9 or 10), and cleared the promoter. This is the end of the initiation phase of transcription. At right, the RNA polymerase has moved to the right, and the RNA chain has grown accordingly. Note that the "transcription bubble" of melted DNA has moved along with the polymerase and so has the region of RNA–DNA hybrid (about 9 bp).

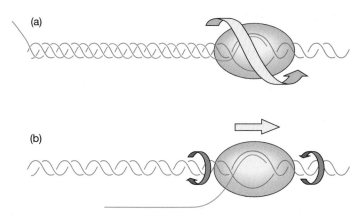

Figure 6.43 Two hypotheses of the topology of transcription of double-stranded DNA. (a) The RNA polymerase (pink) moves around and around the double helix, as indicated by the yellow arrow. This avoids straining the DNA, but it wraps the RNA product (red) around the DNA template. (b) The polymerase moves in a straight line, as indicated by the yellow arrow. This avoids twisting the RNA product (red) around the DNA, but it forces the DNA ahead of the moving polymerase to untwist and the DNA behind the polymerase to twist back up again. These two twists, represented by the green arrows, introduce strain into the DNA template that must be relieved by topoisomerases.

relaxed. As we will see in Chapter 20 when we discuss DNA replication, a class of enzymes called **topoisomerases** can introduce transient breaks into DNA strands and so relax this kind of strain. We will see that strain due to twisting a double-helical DNA causes the helix to tangle up like a twisted rubber band. This process is called **supercoiling,** and the supercoiled DNA is called a **supercoil** or **superhelix.** Unwinding due to the advancing polymerase causes a compensating overwinding ahead of the unwound region. (Compensating overwinding is what makes it difficult to unwind a coiled telephone cord.) The supercoiling due to overwinding is by convention called positive. Thus, positive supercoils build up in front of the advancing polymerase. Conversely, negative supercoils form behind the polymerase. One line of evidence that di-

rectly supports this model of transcription comes from studies with topoisomerase mutants that cannot relax supercoils. If the mutant cannot relax positive supercoils, these build up in DNA that is being transcribed. On the other hand, negative supercoils accumulate during transcription in topoisomerase mutants that cannot relax that kind of superhelix.

> **SUMMARY** Elongation of transcription involves the polymerization of nucleotides as the RNA polymerase core travels along the template DNA. As it moves, the polymerase maintains a short melted region of template DNA. This requires that the DNA unwind ahead of the advancing polymerase and close up again behind it. This process introduces strain into the template DNA that is relaxed by topoisomerases.

6.5 Termination of Transcription

When the polymerase reaches a **terminator** at the end of a gene it falls off the template, releasing the RNA. *E. coli* cells contain about equal numbers of two kinds of terminators. The first kind, known as **intrinsic terminators,** function with the RNA polymerase by itself without help from other proteins. The second kind depend on an auxiliary factor called **rho (ρ).** Naturally, these are called **rho-dependent terminators.** Let us consider the mechanisms of termination employed by these two systems, beginning with the simpler, intrinsic terminators.

Rho-Independent Termination

Rho-independent, or intrinsic, termination depends on terminators consisting of two elements: an inverted repeat followed immediately by a T-rich region in the nontemplate strand of the gene. The model of termination we will present later in this section depends on a "hairpin" structure in the RNA transcript of the inverted repeat. Before we get to the model, we should understand how an inverted repeat predisposes a transcript to form a hairpin.

Inverted Repeats and Hairpins   Consider this inverted repeat:

5′-TACGAAGTTCGTA-3′

•

3′-ATGCTTCAAGCAT-5′

Such a sequence is symmetrical around its center, indicated by the dot; it would read the same if rotated 180 degrees in the plane of the paper, and we always read the strand that runs 5′ → 3′ left to right. Now observe that a transcript of this sequence

UACGAA<u>G</u>UUCGUA

is self-complementary around its center (the underlined G). That means that the self-complementary bases can pair to form a hairpin as follows:

U • A
A • U
C • G
G • C
A • U
A U
G

The A and the U at the apex of the hairpin cannot form a base pair because of the physical constraints of the turn in the RNA.

The Structure of an Intrinsic Terminator   The *E. coli trp* operon (Chapter 7) contains a DNA sequence called an attenuator that causes premature termination of transcription. The *trp* attenuator contains the two elements (an inverted repeat and a string of T's in the nontemplate DNA strand) suspected to be vital parts of an intrinsic terminator, so Peggy Farnham and Terry Platt used attenuation as an experimental model for normal termination.

The inverted repeat in the *trp* attenuator is not perfect, but 8 bp are still possible, and 7 of these are strong G–C pairs, held together by three hydrogen bonds. The hairpin looks like this:

A • U
G • C
C • G
C • G
C • G
G • C
C • G ⟩ A
C • G
U   U
A   A

Notice that a small loop occurs at the end of this hairpin because of the U–U and A–A combinations that cannot base-pair. Furthermore, one A on the right side of the stem has to be "looped out" to allow 8 bp instead of just 7. Still, the hairpin should form and be relatively stable.

Farnham and Platt reasoned as follows: As the T-rich region of the attenuator is transcribed, eight A–U base pairs would form between the A's in the DNA template strand and the U's in the RNA product. They also knew that rU–dA base pairs are exceptionally weak; they have a melting temperature 20° C lower than even rU–rA or dT–rA pairs. This led the investigators to propose that the polymerase paused at the terminator, and then the weakness of the rU–dA base pairs allowed the RNA to dissociate from the template, terminating transcription.

What data support this model? If the hairpin and string of rU–dA base pairs in the *trp* attenuator are really important, we would predict that any alteration in the base sequence that would disrupt either one would be deleterious to attenuation. Farnham and Platt devised the following in vitro assay for attenuation (Figure 6.44): They started with a *Hpa*II restriction fragment containing the *trp* attenuator and transcribed it in vitro. If attenuation works, and transcription terminates at the attenuator, a short (140-nt) transcript should be the result. On the other hand, if transcription fails to terminate at the attenuator, it will continue to the end of the fragment, yielding a run-off transcript 260 nt in length. These two transcripts are easily distinguished by electrophoresis.

When these investigators altered the string of eight T's in the nontemplate strand of the terminator to the sequence TTTTGCAA, creating the mutant they called *trp a*1419, attenuation was weakened. This is consistent with the hypothesis that the weak rU–dA pairs are important in termination, because half of them would be replaced by stronger base pairs in this mutant.

Moreover, this mutation could be overridden by substituting the nucleotide iodo-CTP (I-CTP) for normal CTP in the in vitro reaction. The most likely explanation is that base-pairing between G and iodo-C is stronger than between G and ordinary C. Thus, the GC-rich hairpin should be stabilized by I-CMP, and this effect counteracts the loss of weak base pairs in the region following the hairpin. On the other hand, IMP (inosine monophosphate, a GMP analog) should weaken base-pairing in the hairpin because I–C pairs, with only two hydrogen bonds holding them together, are weaker than G–C pairs with three. Sure enough, substituting ITP for GTP in the transcription reaction weakened termination at the attenuator. Thus, all of these effects are consistent with the hypothesis.

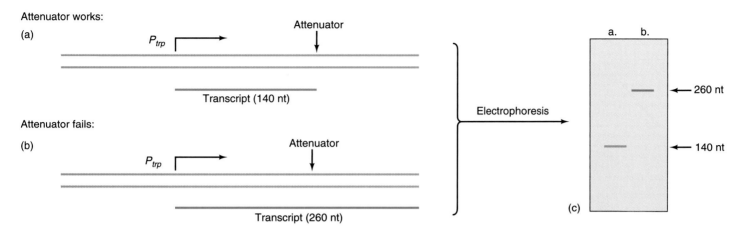

Figure 6.44   An assay for attenuation. (a) When the DNA fragment containing the *trp* promoter and attenuator is transcribed under conditions in which the attenuator works, transcription stops in the attenuator, and a 140-nt transcript (red) results. **(b)** When the same DNA fragment is transcribed under conditions that cause the attenuator to fail, a run-off transcript of 260 nt (green) is the result. **(c)** The transcripts from the two different reactions can be distinguished easily by electrophoresis. Using this assay, we can tell whether the attenuator works under a variety of conditions.

> **SUMMARY** Using the *trp* attenuator as a model terminator, Farnham and Platt showed that rho-independent terminators have two important features: (1) an inverted repeat that allows a hairpin to form at the end of the transcript; (2) a string of T's in the nontemplate strand that results in a string of weak rU–dA base pairs holding the transcript to the template strand.

A Model for Termination   Several hypotheses have been proposed for the roles of the hairpin and string of rU–dA base pairs in the mechanism of termination. Two important clues help narrow the field of hypotheses. First, hairpins are found to destabilize elongation complexes that are stalled artificially (not at strings of rU–dA pairs). Second, terminators in which half of the inverted repeat is missing still stall at the strings of rU–dA pairs, even though no hairpin can form. This leads to the following general hypothesis: The rU–dA pairs cause the polymerase to pause, allowing the hairpin to form and destabilize the already weak rU–dA pairs that are holding the DNA template and RNA product together. This destabilization results in dissociation of the RNA from its template, terminating transcription.

W. S. Yarnell and Jeffrey Roberts proposed a variation on this hypothesis in 1999, as illustrated in Figure 6.45. This model calls for the withdrawal of the RNA from the active site of the polymerase that has stalled at a terminator—either because the newly formed hairpin helps pull it out or because the polymerase moves downstream without elongating the RNA, thus leaving the RNA behind. To test their hypothesis, Yarnell and

Roberts used a DNA template that contained two mutant terminators (ΔtR2 and Δt82) downstream of a strong promoter. These terminators had a T-rich region in the nontemplate strand, but only half of an inverted repeat, so hairpins could not form. To compensate for the hairpin, these workers added an oligonucleotide that was complementary to the remaining half of the inverted repeat. They reasoned that the oligonucleotide would base-pair to the transcript and restore the function of the hairpin.

To test this concept, they attached magnetic beads to the template, so it could be easily removed from the mixture by centrifugation. Then they used *E. coli* RNA polymerase to synthesize labeled RNAs in vitro in the presence and absence of the appropriate oligonucleotides. Finally, they removed the template magnetically to form a pellet and electrophoresed the pellet and the supernatant and detected the RNA species by autoradiography.

Figure 6.46 shows the results. In lanes 1–6, no oligonucleotides were used, so little incomplete RNA was released into the supernatant (see faint bands at ΔtR2 and Δt82 markers in lanes 1, 3, and 5). However, pausing definitely did occur at both terminators, especially at short times (see stronger bands in lanes 2, 4, and 6). This was a clear indication that the hairpin is not required for pausing, though it is required for efficient release of the transcript. In lanes 7–9, Yarnell and Roberts included an oligonucleotide (t19) complementary to the remaining, downstream half of the inverted repeat in the ΔtR2 terminator. Clearly, this oligonucleotide stimulates termination at the mutant terminator, as the autoradiograph shows a dark band corresponding to a labeled RNA released into the supernatant. This labeled RNA is exactly the same size as an RNA released by the wild-type terminator would be. Similar results,

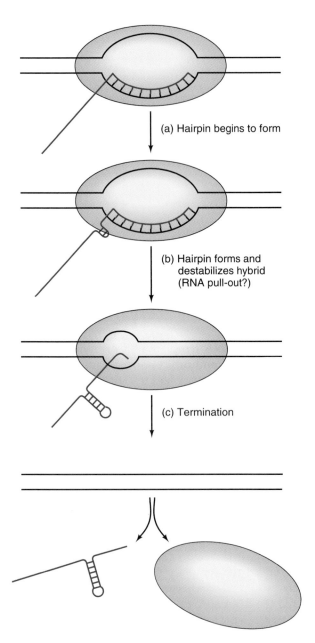

(a) Hairpin begins to form

(b) Hairpin forms and destabilizes hybrid (RNA pull-out?)

(c) Termination

Figure 6.45 A model for intrinsic termination. (a) The polymerase has paused at a string of weak rU–dA base pairs and a hairpin has started to form just downstream of these base pairs. **(b)** As the hairpin forms, it further destabilizes the RNA–DNA hybrid. This destabilization could take several forms: The formation of the hairpin could physically pull the RNA out of the polymerase; or it could simply cause the transcription bubble to collapse, expelling the RNA from the hybrid. **(c)** The RNA product and polymerase dissociate completely from the DNA template, terminating transcription.

though less dramatic, were obtained with an oligonucleotide (t18) that is complementary to the downstream half of the inverted repeat in the Δt82 terminator.

To test further the importance of base-pairing between the oligonucleotide and the half-inverted repeat, these workers mutated one base in the t19 oligonucleotide to

yield an oligonucleotide called t19H1. Lane 13 showed that this change caused a dramatic reduction in termination at ΔtR2. Then they made a compensating mutation in ΔtR2 and tested t19H1 again. Lane 14 shows that this restored strong termination at ΔtR2. This template also contained the wild-type Δt82 terminator, so abundant termination also occurred there. Lanes 15 and 16 are negative controls in which no t19H1 oligonucleotide was present, and, as expected, very little termination occurred at the ΔtR2 terminator.

Together, these results show that the hairpin itself is not required for termination. All that is needed is something to base-pair with the downstream half of the inverted repeat. Furthermore, the T-rich region is not required if transcription can be slowed to a crawl artificially. Yarnell and Roberts advanced the polymerase to a site that had neither an inverted repeat nor a T-rich region and made sure it paused there by washing away the nucleotides. Then they added an oligonucleotide that hybridized upstream of the artificial pause site. Under these conditions, they observed release of the nascent RNA.

SUMMARY The essence of a bacterial terminator is twofold: (1) base-pairing of something to the transcript; and (2) something that causes transcription to pause. A normal intrinsic terminator satisfies the first condition by causing a hairpin to form in the transcript, and the second by causing a string of U's to be incorporated just downstream of the hairpin.

Rho-Dependent Termination

Jeffrey Roberts discovered rho as a protein that caused an apparent depression of the ability of RNA polymerase to transcribe certain phage DNAs in vitro. This depression is simply the result of termination. Whenever rho causes a termination event, the polymerase has to reinitiate to begin transcribing again. And, because initiation is a time-consuming event, less net transcription can occur. To establish that rho is really a termination factor, Roberts performed the following experiments.

Rho Affects Chain Elongation, but not Initiation Just as Travers and Burgess used $[\gamma\text{-}^{32}\text{P}]\text{ATP}$ and $[^{14}\text{C}]\text{ATP}$ to measure transcription initiation and elongation, respectively, Roberts used $[\gamma\text{-}^{32}\text{P}]\text{GTP}$ and $[^{3}\text{H}]\text{UTP}$ for the same purposes. He carried out in vitro transcription reactions with these two labeled nucleotides in the presence of increasing concentrations of rho. Figure 6.47 shows the results. We see that rho had little effect on initiation; if anything, the rate of initiation went up. But rho caused a significant decrease in chain elongation. This is consistent

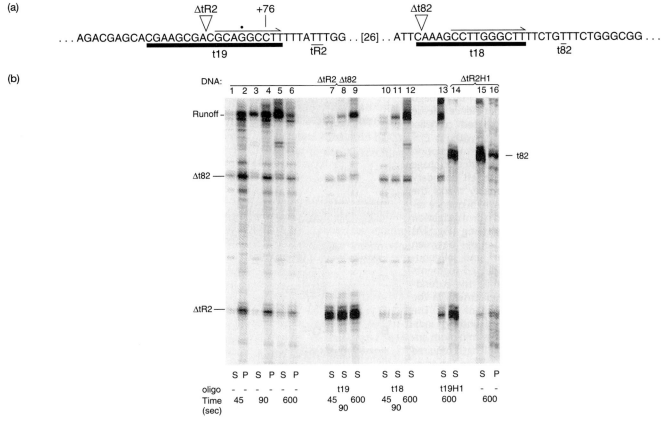

Figure 6.46   Release of transcripts from elongation complexes by oligonucleotides complementary to mutant terminators. (a) Scheme of the template used in these experiments. The template contained two mutant terminators, ΔtR2, and Δt82, situated as shown, downstream of a strong promoter. The normal termination sites for these two terminators are labeled with thin underlines. The black bars denote regions complementary to the oligonucleotides used (t19 and t18). The rightward arrows denote the half inverted repeats remaining in the mutant terminators. The dot indicates the site of a base altered in the t19HI oligonucleotide and of a compensating mutation in the DNA template in certain of the experiments. The template was attached to a magnetic bead so it could be removed from solution easily by centrifugation.
(b) Experimental results. Yarnell and Roberts synthesized labeled RNA in the presence of the template in panel **(a)** and; no oligonucleotide (lanes 1–6 and 15–16), the t19 oligonucleotide (lanes 7–9), the t18 oligonucleotide (lanes 10–12); and the t19HI oligonucleotide (lanes 13–14). They allowed transcription for the times given at bottom, then removed the template and any RNA attached to it by centrifugation. They electrophoresed the labeled RNA in the pellet (P) or supernatant (S), as indicated at bottom, and autoradiographed the gel. The positions of runoff transcripts, and of transcripts that terminated at the ΔtR2 and Δt82 terminators, are indicated at left. (*Source: (a–b)* Yarnell, W. S. and Roberts, J. W. Mechanism of Intrinsic Transcription Termination and Antitermination. *Science* 284 (23 April 1999) 611–612. © AAAS.)

with the notion that rho terminates transcription, thus forcing time-consuming reinitiation.

Rho Causes Production of Shorter Transcripts   It is relatively easy to measure the size of RNA transcripts by gel electrophoresis or, in 1969, when Roberts performed his experiments, by ultracentrifugation. But just finding short transcripts would not have been enough to conclude that rho was causing termination. It could just as easily have been an RNase that chopped up longer transcripts into small pieces.

To exclude the possibility that rho was simply acting as a nuclease, Roberts first made ³H-labeled λ RNA in the absence of rho, then added these relatively large pieces of

RNA to new reactions carried out in the presence of rho, in which ¹⁴C-UTP was the labeled RNA precursor. Finally, he measured the sizes of the ¹⁴C- and ³H-labeled λ RNAs by ultracentrifugation. Figure 6.48 presents the results. The solid curves show no difference in the size of the preformed ³H-labeled RNA even when it had been incubated with rho in the second reaction. Rho therefore shows no RNase activity. However, the ¹⁴C-labeled RNA made in the presence of rho (red line in Figure 6.48b) is obviously much smaller than the preformed RNA made without rho. Thus, rho is causing the synthesis of much smaller RNAs. Again, this is consistent with the role of rho in terminating transcription. Without rho, the transcripts grew to abnormally large size.

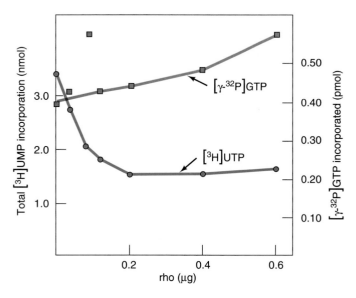

Figure 6.47 Rho decreases the net rate of RNA synthesis.
Roberts allowed *E. coli* RNA polymerase to transcribe λ phage DNA in the presence of increasing concentrations of rho. He used [γ-^{32}P]GTP to measure initiation (red) and [^{3}H]UTP to measure elongation (green). Rho depressed the elongation rate, but not initiation. (*Source:* Roberts, J.W. "Termination factor for RNA synthesis," *Nature* 224:1168–74, 1969. Copyright Macmillan Magazines Limited. Reprinted with permission.)

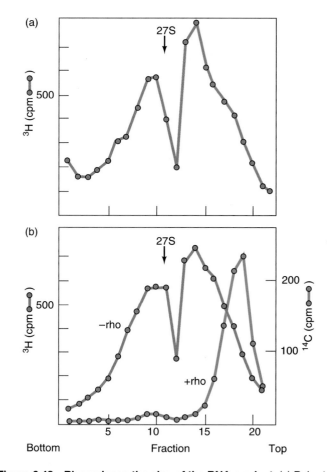

Figure 6.48 Rho reduces the size of the RNA product. (a) Roberts allowed *E. coli* RNA polymerase to transcribe λ DNA in the absence of rho. He included [^{3}H]UTP in the reaction to label the RNA. Finally, he used ultracentrifugation to separate the transcripts by size. He collected fractions from the bottom of the centrifuge tube, so low-numbered fractions, at left, contained the largest RNAs. **(b)** Roberts used *E. coli* RNA polymerase to transcribe λ DNA in the presence of rho. He also included [^{14}C]ATP to label the transcripts, plus the ^{3}H-labeled RNA from panel (**a**). Again, he ultracentrifuged the transcripts to separate them by size. The ^{14}C-labeled transcripts (red) made in the presence of rho were found near the top of the gradient (at right), indicating that they were relatively small. On the other hand, the ^{3}H-labeled transcripts (blue) from the reaction lacking rho were relatively large and the same size as they were originally. Thus, rho has no effect on the size of previously made transcripts, but it reduces the size of the transcripts made in its presence. (*Source:* Roberts, J.W. "Termination factor for RNA synthesis," *Nature* 224:1168–74, 1969. Copyright Macmillan Magazines Limited. Reprinted with permission.)

Rho Releases Transcripts from the DNA Template Finally, Roberts used ultracentrifugation to analyze the sedimentation properties of the RNA products made in the presence and absence of rho. The transcripts made without rho (Figure 6.49a) cosedimented with the DNA template, indicating that they had not been released from their association with the DNA. By contrast, the transcripts made in the presence of rho sedimented at a much lower rate, independent of the DNA. Thus, rho seems to release RNA transcripts from the DNA template. In fact, rho (the Greek letter ρ) was chosen to stand for "release."

The Mechanism of Rho How does rho do its job? Because we have seen that σ operates through RNA polymerase by binding tightly to the core enzyme, we might envision a similar role for rho. But rho seems not to have affinity for RNA polymerase. Whereas σ is a critical part of the polymerase holoenzyme, rho is not.

Instead (Figure 6.50), rho binds to the transcript upstream of the termination site at a **rho loading site,** a sequence of 60–100 nt that is relatively free of secondary structure and is rich in cytosine residues, Rho consists of a hexamer of identical subunits, each of which has ATPase activity. Binding of rho to the RNA activates the ATPase, which supplies the energy to propel the rho hexamer along the RNA in the 5′ → 3′ direction, following the RNA polymerase. This chase continues until the polymerase stalls in the terminator region just after making

the RNA hairpin. Then rho can catch up and release the transcript. In support of this hypothesis, Terry Platt and colleagues showed in 1987 that rho has RNA–DNA helicase activity that can unwind an RNA–DNA hybrid. Thus, when rho encounters the polymerase stalled at the terminator, it can unwind the RNA–DNA hyrbid within the transcription bubble, releasing the RNA and terminating transcription.

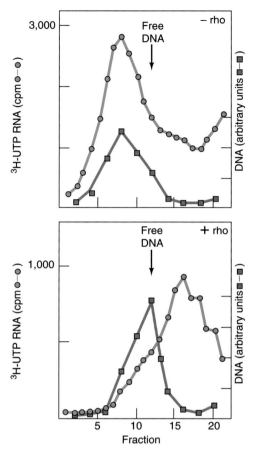

Figure 6.49   Rho releases the RNA product from the DNA template. Roberts transcribed λ DNA under the same conditions as in Figure 6.48, in the **(a)** absence or **(b)** presence of rho. Then he subjected the ^{3}H-labeled product (red) to ultracentrifugation to see whether the product was associated with the DNA template (blue). **(a)** The RNA made in the absence of rho sedimented together with the template in a complex that was larger than free DNA. **(b)** The RNA made in the presence of rho sedimented independently of DNA at a position corresponding to relatively small molecules. Thus, transcription with rho releases transcripts from the DNA template. (*Source:* Roberts, J.W. "Termination factor for RNA synthesis," *Nature* 224:1168–74, 1969. Copyright Macmillan Magazines Limited. Reprinted with permission.)

(a) Rho binds to transcript at rho loading site and pursues polymerase.

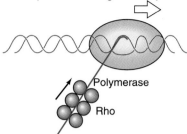

(b) Hairpin forms; polymerase pauses; rho catches up.

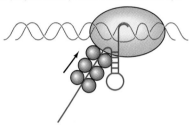

(c) Rho helicase releases transcript and causes termination.

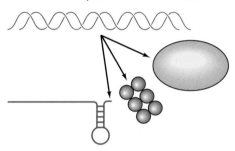

Figure 6.50   A model of rho-dependent termination. (a) As polymerase (red) makes RNA, rho (blue) binds to the transcript (green) at a rho loading site and pursues the polymerase. **(b)** When the hairpin forms in the transcript, the polymerase pauses, giving rho a chance to catch up. **(c)** rho helicase unwinds the RNA–DNA hybrid and releases the transcript.

SUMMARY Rho-dependent terminators consist of an inverted repeat, which can cause a hairpin to form in the transcript, but no string of T's. Rho binds to a growing transcript, follows the RNA polymerase, catches the polymerase when it stalls at the hairpin, and releases the transcript from the DNA–polymerase complex by unwinding the RNA–DNA hybrid. This terminates transcription.

SUMMARY

The catalytic agent in the transcription process is RNA polymerase. The *E. coli* enzyme is composed of a core, which contains the basic transcription machinery, and a σ-factor, which directs the core to transcribe specific genes. The σ-factor allows initiation of transcription by causing the RNA polymerase holoenzyme to bind tightly to a promoter. This σ-dependent tight binding requires local melting of 10–17 bp of the DNA in the vicinity of the transcription start site to form an open promoter complex. Thus, by binding only to certain promoters, a σ-factor can select which genes will be transcribed. The initiation process continues until 9 or 10 nt have been incorporated into the RNA, then σ dissociates from the core polymerase, the core changes to an elongation-specific conformation, leaves the promoter, and carries on with the elongation process. The σ-factor can be reused by different core polymerases. The core, not σ, governs rifampicin sensitivity or resistance.

Prokaryotic promoters contain two regions centered at –10 and –35 bp upstream of the transcription start site. In *E. coli,* these have the consensus sequences TATAAT and TTGACA, respectively. In general, the more closely regions within a promoter resemble these consensus sequences, the stronger that promoter will be. Some extraordinarily strong promoters contain an extra element (an UP element) upstream of the core promoter. This makes these promoters even more attractive to RNA polymerase. The *E. coli* RNA polymerase extends at least from position –44 to +3 in the open promoter complex at the *lac* promoter.

Four regions are similar among σ-factors, and subregions 2.4 and 4.2 are involved in promoter –10 box and –35 box recognition, respectively. The *E. coli* σ-factor, in addition to its promoter-binding capability, can loosen nonspecific interactions between polymerase and DNA and thus enhance the specificity of the polymerase–promoter interactions. In *B. subtilis*, another factor, δ, not σ, provides this loosening activity.

The core subunit β binds nucleotides at the active site of the RNA polymerase where phosphodiester bonds are formed. The σ-factor is also nearby during the initiation phase. Template transfer experiments have delineated two DNA sites that interact with polymerase. One is a weak binding site involving the melted DNA zone and the catalytic site on the β-subunit of the polymerase. Protein–DNA interactions at this site are primarily electrostatic and are salt-sensitive. The other is a strong binding site involving DNA downstream of the active site, and the β′- and β-subunits of the enzyme. The RNA polymerase α-subunit serves to assemble the RNA polymerase holoenzyme, and this function resides in the N-terminal domain of α (the α-NTD). The α-subunit also has an independently folded C-terminal domain that can recognize and bind to a promoter's UP element. This allows very tight binding between polymerase and promoter.

Elongation of transcription involves the polymerization of nucleotides as the RNA polymerase core travels along the template DNA. As it moves, the polymerase maintains a short melted region of template DNA. This transcription bubble is about 17 bases long and contains an RNA–DNA hybrid about 9 bp long. The movement of the transcription bubble requires that the DNA unwind ahead of the advancing polymerase and close up again behind it. This process introduces strain into the template DNA that is relaxed by topoisomerases.

The RNA polymerase core is shaped like a crab claw. The catalytic center, containing a Mg^{2+} ion coordinated by three Asp residues, lies in a channel that conducts DNA through the enzyme.

Intrinsic terminators have two important elements: (1) an inverted repeat that allows a hairpin to form at the end of the transcript; (2) a string of T's in the nontemplate strand that results in a string of weak rU–dA base pairs holding the transcript to the template. Together, these elements cause the polymerase to pause and the transcript to be released. Rho-dependent terminators consist of an inverted repeat, which can cause a hairpin to form in the transcript, but no string of T's. Rho binds to a growing transcript, catches up to the RNA polymerase when it stalls at the hairpin, and the RNA–DNA helicase activity of rho releases the transcript from the DNA–polymerase complex. This terminates transcription.

REVIEW QUESTIONS

1. Explain the following findings: (1) Core RNA polymerase transcribes intact T4 phage DNA only weakly, while holoenzyme transcribes this template very well; but (2) core polymerase can transcribe calf thymus DNA about as well as the holoenzyme can.

2. How did Bautz and colleagues demonstrate that the *E. coli* RNA polymerase holoenzyme is specific for the phage T4 immediate early genes, whereas the core is nonspecific?

3. How did Bautz and colleagues show that the holoenzyme transcribes phage T4 DNA asymmetrically, but the core transcribes this DNA symmetrically?

4. In general, how can you block viral delayed early and late transcription? How can you block late transcription only? Explain how these procedures work.

5. Describe an experiment to measure the dissociation rate of the tightest complex between a protein and a DNA. Show sample results of weak and tight binding. How do these results relate to the binding of core polymerase and holoenzyme to DNA that contains promoters.

6. What effect does temperature have on the dissociation rate of polymerase–promoter complexes? What does this suggest about the nature of the complex?

7. Diagram the difference between a closed and an open promoter complex.

8. Diagram a typical prokaryotic promoter, and a promoter with an UP element. Exact sequences are not necessary.

9. Describe and give the results of an experiment that demonstrates the formation of abortive transcripts by *E. coli* RNA polymerase.

10. Diagram the four-step transcription initiation process in *E. coli.*

11. Describe and show the results of an experiment that measures the effects of σ on transcription initiation and elongation rates.

12. How can you show that σ does not really accelerate the rate of transcription elongation?

13. What final conclusion can you draw from the experiments in the previous two questions?

14. Describe and show the results of an experiment that demonstrates the re-use of σ. On the same graph, show the results of an experiment that shows that the core polymerase determines resistance to rifampicin.

15. Draw a diagram of the "σ-cycle."

16. Describe and show the results of an experiment that shows which base pairs are melted when RNA polymerase binds to a promoter. Explain how this procedure works.

17. Describe and show the results of an experiment that gives the best estimate of the number of base pairs melted during transcription by *E. coli* RNA polymerase.

18. Describe and show the results of an experiment that shows the 5′-end of the polymerase–promoter interaction zone. How would you modify this experiment to detect the 3′-end of the interaction zone?

19. What regions of the σ-factor are thought to be involved in recognizing (1) the –10 box of the promoter and (2) the –35 box of the promoter? Without naming specific residues, describe the genetic evidence for these conclusions.

20. Describe a binding assay that provides biochemical evidence for interaction between σ-region 4.2 and the –35 box of the promoter.

21. The *B. subtilis* σ^{43} is considerably smaller than the *E. coli* σ^{70}. It also lacks one of the functions of the *E. coli* factor. What is this function, and how can you show that it is supplied by the *E. coli* σ- and by the *B. subtilis* δ-factor?

22. Cite evidence to support the hypothesis that the α-subunit of *E. coli* RNA polymerase is involved in recognizing a promoter UP element.

23. Describe how limited proteolysis can be used to define the domains of a protein such as the α-subunit of *E. coli* RNA polymerase.

24. Describe an experiment to determine which polymerase subunit is responsible for rifampicin and streptolydigin resistance or sensitivity.

25. Describe and give the results of an experiment that shows that the β-subunit of *E. coli* RNA polymerase contains the active site that forms phosphodiester bonds.

26. Describe and give the results of a template transfer experiment that shows that the β-subunit of *E. coli* RNA polymerase is involved in weak binding to DNA, whereas the β′- and β-subunits are involved in strong binding to DNA.

27. Describe an RNA–DNA cross-linking experiment that demonstrates the existence of an RNA–DNA hybrid at least 8 bp long within the transcription elongation complex.

28. Draw a rough sketch of the structure of a bacterial RNA polymerase core based on x-ray crystallography. Point out the positions of the subunits of the enzyme, the catalytic center, and the rifampicin binding site. Based on this structure, propose a mechanism for inhibition of transcription by rifampicin.

29. Present two models for the way the RNA polymerase can maintain the bubble of melted DNA as it moves along the DNA template. Which of these models is favored by the evidence? Cite the evidence in a sentence or two.

30. What are the two important elements of an intrinsic transcription terminator? How do we know they are important? (Cite evidence.)

31. Present evidence that a hairpin is not required for pausing at an intrinsic terminator.

32. Present evidence that base-pairing (of something) with the RNA upstream of a pause site is required for intrinsic termination.

33. What does a rho-dependent terminator look like? What role is rho thought to play in such a terminator?

34. Draw the structure of an RNA hairpin with an 8-bp stem and a 5-bp loop. Make up a sequence that will form such a structure. Show the sequence in the linear as well as the hairpin form.

35. How can you show that rho causes a decrease in net RNA synthesis, but no decrease in chain initiation? Describe and show the results of an experiment.

36. Describe and show the results of an experiment that demonstrates the production of shorter transcripts in the presence of rho. This experiment should also show that rho does not simply act as a nuclease.

37. Describe and show the results of an experiment that demonstrates that rho releases transcripts from the DNA template.

SUGGESTED READINGS

General References and Reviews

Busby, S., and R.H. Ebright. 1994. Promoter structure, promoter recognition, and transcription activation in prokaryotes. *Cell* 79:743–46.

Geiduschek, E.P. 1997. Paths to activation of transcription. *Science* 275:1614–16.

Helmann, J.D., and M.J. Chamberlin. 1988. Structure and function of bacterial sigma factors. *Annual Review of Biochemistry* 57:839–72.

Landick, R. 1999. Shifting RNA polymerase into overdrive. *Science* 284:598–99.

Landick, R. and J.W. Roberts. 1996. The shrewd grasp of RNA polymerase. *Science* 273:202–3.

Losick, R., and M.J. Chamberlin. 1976. *RNA Polymerase.* Plainview, NY: Cold Spring Harbor Laboratory Press.

Richardson, J.P. 1996. Structural organization of transcription termination factor rho. *Journal of Biological Chemistry* 271:1251–54.

Research Articles

Bautz, E.K.F., F.A. Bautz, and J.J. Dunn. 1969. *E. coli* σ factor: A positive control element in phage T4 development. *Nature* 223:1022–24.

Blatter, E.E., W. Ross, H. Tang, R.L. Gourse, and R.H. Ebright. 1994. Domain organization of RNA polymerase α subunit: C-terminal 85 amino acids constitute a domain capable of dimerization and DNA binding *Cell* 78:889–96.

Brennan, C.A., A.J. Dombroski, and T. Platt. 1987. Transcription termination factor rho is an RNA–DNA helicase *Cell* 48:945–52.

Burgess, R.R., A.A. Travers, J.J. Dunn, and E.K.F. Bautz. 1969. Factor stimulating transcription by RNA polymerase. *Nature* 221:43–46.

Carpousis, A.J., and J.D. Gralla. 1980. Cycling of ribonucleic acid polymerase to produce oligonucleotides during initiation in vitro at the *lac* UV5 promoter. *Biochemistry* 19:3245–53.

Darst, S.A., E.W. Kubalek, and R.D. Kornberg. 1989. Three-dimensional structure of *Escherichia coli* RNA polymerase holoenzyme determined by electron crystallography. *Nature* 340:730–32.

Dombroski, A.J., W.A. Walter, M.T. Record, Jr., D.A. Siegele, and C.A. Gross. (1992). Polypeptides containing highly conserved regions of transcription initiation factor σ70 exhibit specificity of binding to promoter DNA. *Cell* 70:501–12.

Farnham, P.J., and T. Platt. 1980. A model for transcription termination suggested by studies on the *trp* attenuator in vitro using base analogs. *Cell* 20:739–48.

Fukuda, R. and A. Ishihama. 1974. Subunits of RNA polymerase in function and structure: V. Maturation in vitro of core enzyme from *Escherichia coli.* Journal of Molecular Biology 87:523–40.

Gamper, H. and Hearst, J. 1982. A topological model for transcription based on unwinding angle analysis of *E. coli* RNA polymerase binary, initiation, and ternary complexes. *Cell* 29:81–90.

Grachev, M.A., T.I. Kolocheva, E.A. Lukhtanov, and A.A. Mustaev. 1987. Studies on the functional topography of *Escherichia coli* RNA polymerase: Highly selective affinity labelling of initiating substrates. *European Journal of Biochemistry* 163:113–21.

Hayward, R.S., K. Igarashi, and A. Ishihama. 1991. Functional specialization within the α-subunit of *Escherichia coli* RNA polymerase. *Journal of Molecular Biology* 221:23–29.

Heil, A. and W. Zillig. 1970. Reconstitution of bacterial DNA-dependent RNA polymerase from isolated subunits as a tool for the elucidation of the role of the subunits in transcription. *FEBS Letters* 11:165–71.

Hilton, M.D. and H.R. Whitely. 1985. UV cross-linking of the *Bacillus subtilis* RNA polymerase to DNA in promoter and nonpromoter complexes. *Journal of Biological Chemistry* 260:8121–27.

Hinkle, D.C. and M.J. Chamberlin. 1972. Studies on the binding of *Escherichia coli* RNA polymerase to DNA: I. The role of sigma subunit in site selection. *Journal of Molecular Biology* 70:157–85.

Hsieh, T.-s. and J.C. Wang. 1978. Physicochemical studies on interactions between DNA and RNA polymerase: Ultraviolet absorbance measurements. *Nucleic Acids Research* 5:3337–45.

Nudler, E., E. Avetissova, V. Makovtsov, and A. Goldfarb. 1996. Transcription processivity: Protein–DNA interactions holding together the elongation complex. *Science* 273:211–17.

Nudler, E., A. Mustaev, E. Lukhtanov, and A. Goldfarb. 1997. The RNA–DNA hybrid maintains the register of transcription by preventing backtracking of RNA polymerase. *Cell* 89:33–41.

Malhotra, A., E. Severinova, and S.A. Darst. 1996. Crystal structure of a σ70 subunit fragment from *E. coli* RNA polymerase. *Cell* 87:127–36.

Polyakov, A., E. Severinova, and S.A. Darst 1995. Three-dimensional structure of *E. coli* core RNA polymerase: Promoter binding and elongation conformations of the enzyme. *Cell* 83:365–73.

Roberts, J.W. 1969. Termination factor for RNA synthesis. *Nature* 224:1168–74.

Ross, W., K.K. Gosink, J. Salomon, K. Igarashi, C. Zou, A. Ishihama, K. Severinov, and R.L. Gourse. 1993. A third recognition element in bacterial promoters: DNA binding by the α subunit of RNA polymerase. *Science* 262:1407–13.

Saucier, J.-M. and J.C. Wang. 1972. Angular alteration of the DNA helix by *E. coli* RNA polymerase. *Nature New Biology* 239:167–70.

Sidorenkov, I., N. Komissarova, and M. Kashlev. 1998. Crucial role of the RNA:DNA hybrid in the processivity of transcription. *Molecular Cell* 2:55–64.

Siebenlist, U. 1979. RNA polymerase unwinds an 11-base pair segment of a phage T7 promoter. *Nature* 279:651–52.

Tjian, R., R. Losick, J. Pero, and A. Hinnebush. 1977. Purification and comparative properties of the delta and sigma subunits of RNA polymerase from *Bacillus subtilis.* *European Journal of Biochemistry* 74:149–54.

Travers, A.A. and R.R. Burgess. 1969. Cyclic re-use of the RNA polymerase sigma factor. *Nature* 222:537–40.

Yarnell, W.S. and J.W. Roberts. 1999. Mechanism of intrinsic transcription termination and antitermination. *Science* 284:611–15.

Zhang, G. and S.A. Darst. 1998. Structure of the *Escherichia coli* RNA polymerase α subunit amino terminal domain. *Science* 281:262–66.

Zhang, G., E.A. Campbell, L. Minakhin, C. Richter, K. Severinov, and S.A. Darst. 1999. Crystal structure of *Thermus aquaticus* core RNA polymerase at 3.3 Å resolution *Cell* 98:811–24.

CHAPTER 7

Operons: Fine Control of Prokaryotic Transcription

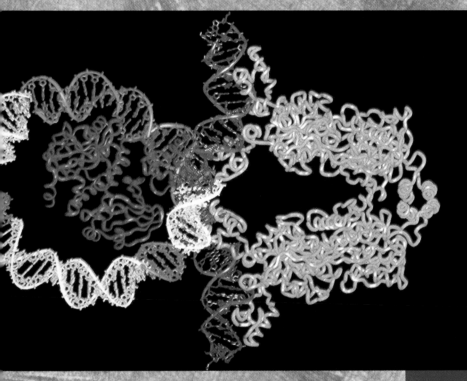

Computer graphic of *lac* repressor (pink) and CAP–cAMP dimer (blue) binding to DNA. © *Mitchell Lewis/SPL/Photo Researchers, Inc.*

The *E. coli* genome contains over 3000 genes. Some of these are active all the time because their products are in constant demand. But some of them are turned off most of the time because their products are rarely needed. For example, the enzymes required for the metabolism of the sugar arabinose would be useful only when arabinose is present and when *E. coli*'s favorite energy source, glucose, is absent. Such conditions are not common, so the genes encoding these enzymes are usually turned off. Why doesn't the cell just leave all its genes on all the time, so the right enzymes are always there to take care of any eventuality? The reason is that gene expression is an expensive process. It takes a lot of energy to produce RNA and protein. In fact, if all of an *E. coli* cell's genes were turned on all the time, production of RNAs and proteins would drain the cell of so much energy that it could not compete with more efficient organisms. Thus, control of gene expression is essential to life. In this chapter we will explore one strategy bacteria employ to control the expression of their genes: by grouping functionally related genes together so they can be regulated together easily. Such a group of contiguous, coordinately controlled genes is called an operon. ■

7.1   The *lac* Operon

The first operon to be discovered has become the prime example of the operon concept. It contains three genes that code for the enzymes that allow *E. coli* cells to use the sugar **lactose,** hence the name *lac* **operon.** Consider a flask of *E. coli* cells growing on a medium containing the sugars **glucose** and **lactose** (Figure 7.1). The cells exhaust the glucose and stop growing. Can they adjust to the new nutrient source? For a short time it appears that they cannot; but then, after a lag period of about an hour, growth resumes. During the lag, the cells have been turning on the *lac* operon and beginning to accumulate the enzymes they need to metabolize lactose. The growth curve in Figure 7.1 is called "diauxic" from the Latin *auxilium,* meaning help, because the two sugars help the bacteria grow.

What are these enzymes? First, *E. coli* needs an enzyme to transport the lactose into the cells. The name of this enzyme is **galactoside permease.** Next, the cells need an enzyme to break the lactose down into its two compo-

nent sugars: galactose and glucose. Figure 7.2 shows this reaction. Because lactose is composed of two simple sugars, we call it a *disaccharide.* These six-carbon sugars, galactose and glucose, are joined together by a linkage called a β-**galactosidic** bond. Lactose is therefore called a β-**galactoside,** and the enzyme that cuts it in half is called β-**galactosidase.** The genes for these two enzymes, galactoside permease and β-galactosidase, are found side by side in the *lac* operon, along with another structural gene—for **galactoside transacetylase**—whose function in lactose metabolism is still not clear.

The three genes coding for enzymes that carry out lactose metabolism are grouped together in the following order: β-galactosidase (*lacZ*), galactoside permease (*lacY*), galactoside transacetylase (*lacA*). They are all transcribed together to produce one messenger RNA, called a **polycistronic message,** starting from a single promoter. Thus, they can all be controlled together simply by controlling that promoter. The term *polycistronic* comes from *cistron,* which is a synonym for *gene.* Therefore, a polycistronic message is simply a message with information from more than one gene. Each cistron in the mRNA has its own ribosome binding site, so each cistron can be translated by separate ribosomes that bind independently of each other.

> **SUMMARY** Lactose metabolism in *E. coli* is carried out by two enzymes, with possible involvement by a third. The genes for all three enzymes are clustered together and transcribed together from one promoter, yielding a polycistronic message. These three genes, linked in function, are therefore also linked in expression. They are turned off and on together.

Negative Control of the *lac* Operon

Figure 7.3 illustrates one aspect of *lac* operon regulation: the classical version of negative control. We will see later in this chapter and in Chapter 9 that this classical view is oversimplified, but it is a useful way to begin consideration of the operon concept. The term **negative control** implies that the operon is turned on unless something intervenes to stop it. The "something" that can turn off the *lac* operon is the *lac* **repressor.** This repressor, the product of a **regulatory gene** called the *lacI* **gene** shown at the extreme left in Figure 7.3, is a tetramer of four identical polypeptides; it binds to the **operator** just to the right of the promoter. When the repressor is bound to the operator, the operon is **repressed.** That is because the operator and promoter are contiguous, and when the repressor occupies the operator, it appears to prevent RNA polymerase from binding to the promoter and transcribing the

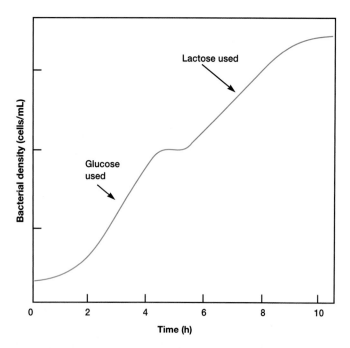

Figure 7.1   Diauxic growth. *E. coli* cells are grown on a medium containing both glucose and lactose, and the bacterial density (number of cells/mL) is plotted versus time in hours. The cells grow rapidly on glucose until that sugar is exhausted, then growth levels off while the cells induce the enzymes needed to metabolize lactose. As those enzymes appear, growth resumes.

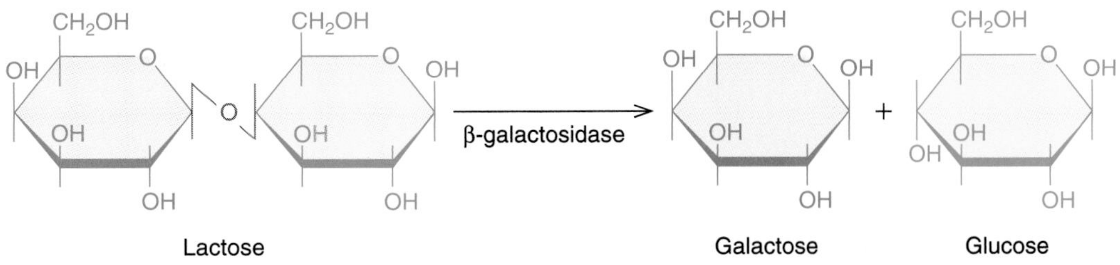

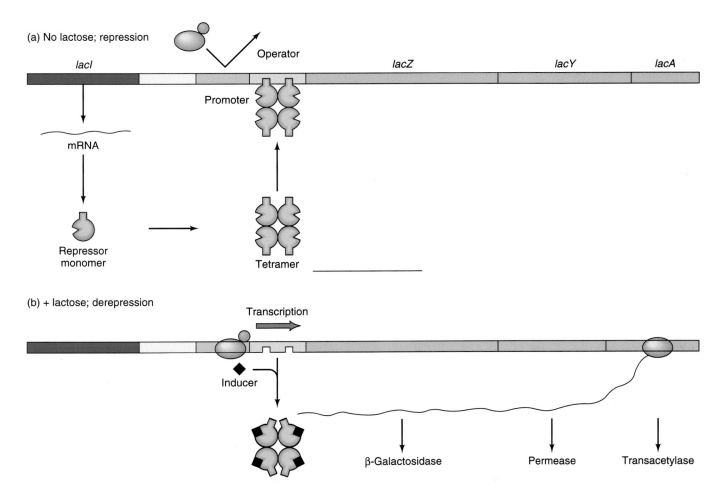

Figure 7.2   The β-galactosidase reaction. The enzyme breaks the β-galactosidic bond (gray) between the two sugars, galactose (red) and glucose (blue), that compose lactose.

Figure 7.3   Negative control of the *lac* operon. (a) No lactose; repression. The *lacI* gene produces repressor (green), which binds to the operator and blocks RNA polymerase from transcribing the *lac* genes. **(b)** Presence of lactose, derepression. The inducer (black) binds to repressor, changing it to a form (bottom) that no longer binds well to the operator. This removes the repressor from the operator, allowing RNA polymerase to transcribe the structural genes. This produces a multicistronic mRNA that is translated to yield β-galactosidase, permease, and transacetylase.

operon. Because its genes are not transcribed, the operon is off, or repressed.

The *lac* operon is repressed as long as no lactose is available. That is economical; it would be wasteful for the cell to produce enzymes needed to use an absent sugar. On the other hand, when all the glucose is gone and lactose is present, a mechanism should exist for removing the repressor so the operon can be derepressed to take advan-

tage of the new nutrient. How does this mechanism work? The repressor is a so-called **allosteric protein:** one in which the binding of one molecule to the protein changes the shape of a remote site on the protein and alters its interaction with a second molecule (Greek: *allos,* meaning other + *stereos,* meaning shape). The first molecule in this case is called the **inducer** of the *lac* operon because it binds to the repressor, causing the protein to

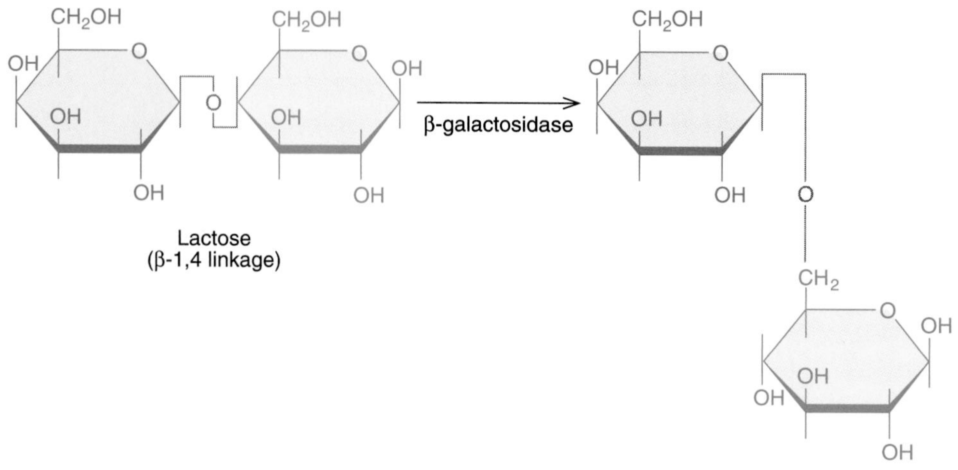

Figure 7.4   Conversion of lactose to allolactose. A side reaction carried out by β-galactosidase rearranges lactose to the inducer, allolactose. Note the change in the galactosidic bond from β-1,4 to β-1,6.

change to a conformation that favors dissociation from the operator (the second molecule), thus inducing the operon (Figure 7.3b).

What is the nature of this inducer? It is actually an alternative form of lactose called **allolactose** (again, Greek: *allos,* meaning other). When β-galactosidase cleaves lactose to galactose plus glucose, it rearranges a small fraction of the lactose to allolactose. Figure 7.4 shows that allolactose is just galactose linked to glucose in a different way than in lactose. (In lactose, the linkage is through a β-1,4 bond; in allolactose, the linkage is β-1,6.)

You may be asking yourself: How can lactose be metabolized to allolactose if no permease is present to get it into the cell and no β-galactosidase exists to perform the metabolizing because the *lac* operon is repressed? The answer is that repression is somewhat leaky, and a low basal level of the *lac* operon products is always present. This is enough to get the ball rolling by producing a little inducer. It does not take much inducer to do the job, because only about 10 tetramers of repressor are present per cell. Furthermore, the derepression of the operon will snowball as more and more operon products are available to produce more and more inducer.

Discovery of the Operon

The development of the operon concept by François Jacob and Jacques Monod and their colleagues was one of the classic triumphs of the combination of genetic and biochemical analysis. The story begins in 1940, when Monod began studying the inducibility of lactose metabolism in *E. coli.* Monod learned that an important feature of lactose metabolism was β-galactosidase, and that this enzyme was inducible by lactose and by other galactosides. Furthermore, he and Melvin Cohn had used an anti-

β-galactosidase antibody to detect β-galactosidase protein, and they showed that the amount of this protein increased on induction. Because more gene product appeared in response to lactose, the β-galactosidase gene itself was apparently being induced.

To complicate matters, certain mutants (originally called "cryptic mutants") were found that could make β-galactosidase but still could not grow on lactose. What was missing in these mutants? To answer this question, Monod and his coworkers added a radioactive galactoside to wild-type and mutant bacteria. They found that uninduced wild-type cells did not take up the galactoside, and neither did the mutants, even if they were induced. Induced wild-type cells did accumulate the galactoside. This revealed two things: First, a substance (galactoside permease) is induced along with β-galactosidase in wild-type cells and is responsible for transporting galactosides into the cells; second, the mutants seem to have a defective gene (Y^-) for this substance (Table 7.1).

Monod named this substance galactoside permease, and then endured criticism from his colleagues for naming a protein before it had been isolated. He later remarked, "This attitude reminded me of that of two traditional English gentlemen who, even if they know each other well by name and by reputation, will not speak to each other before having been formally introduced." In their efforts to purify galactoside permease, Monod and his colleagues identified another protein, galactoside transacetylase, which is induced along with β-galactosidase and galactoside permease.

Thus, by the late 1950s, Monod knew that three enzyme activities (and therefore presumably three genes) were induced together by galactosides. He had also found some mutants, called **constitutive mutants,** that needed no

Table 7.1 Effect of Cryptic Mutant (*lacY*⁻) on Accumulation of Galactoside

Genotype	Inducer	Accumulation of Galactoside
Z^+Y^+	–	–
Z^+Y^+	+	+
Z^+Y^- (cryptic)	–	–
Z^+Y^- (cryptic)	+	–

induction. They produced the three gene products all the time. Monod realized that further progress would be greatly accelerated by genetic analysis, so he teamed up with François Jacob, who was working just down the hall at the Pasteur Institute.

In collaboration with Arthur Pardee, Jacob and Monod created **merodiploids** (partial diploid bacteria) carrying both the wild-type (inducible) and constitutive alleles. The inducible allele proved to be dominant, demonstrating that wild-type cells produce some substance that keeps the *lac* genes turned off unless they are induced. Because this substance turned off the genes from the constitutive as well as the inducible parent, it made the merodiploids inducible. Of course, this substance is the *lac* repressor. The constitutive mutants had a defect in the gene (*lacI*) for this repressor. These mutants are therefore *lacI*⁻ (Figure 7.5a).

The existence of a repressor required that some specific DNA sequence exists to which the repressor would bind. Jacob and Monod called this the operator. The specificity of this interaction suggested that it should be subject to genetic mutation; that is, some mutations in the operator should abolish its interaction with the repressor. These would also be constitutive mutations, so how can they be distinguished from constitutive mutations in the repressor gene?

Jacob and Monod realized that they could make this distinction by determining whether the mutation was dominant or recessive. Jacob explained this with an analogy. He likened the two operators in a merodiploid bacterium to two radio receivers, each of which controls a separate door to a house. The repressor genes in a merodiploid are like radio transmitters, each sending out an identical signal to keep the doors closed. A mutation in one of the repressor genes would be like a breakdown in one of the transmitters; the other transmitter would still function and the doors would both remain closed. In other words, both *lac* operons in the merodiploid would still be repressible. Thus, such a mutation should be recessive (Figure 7.5a), and we have already observed that it is.

On the other hand, a mutation in one of the operators in a merodiploid would be like a breakdown in one of the receivers. Because it could no longer receive the radio signal, its door would swing open; the other, with a functioning receiver, would remain closed. In other words, the mutant *lac* operon in the merodiploid would be constantly derepressed, whereas the wild-type one would remain repressible. Therefore, this mutation would be dominant, but only with respect to the *lac* operon under the control of the mutant operator (Figure 7.5b). We call such a mutation **cis-dominant** because it is dominant only with respect to genes on the same DNA (*in cis;* Latin: *cis,* meaning here), not on the other DNA in the merodiploid. Jacob and Monod did indeed find such *cis*-dominant mutations, and they proved the existence of the operator. These mutations are called O^c, for **operator constitutive.**

What about mutations in the repressor gene that render the repressor unable to respond to inducer? Such mutations should make the *lac* operon uninducible and should be dominant both *in cis* and **in trans** (on different DNA molecules in a merodiploid; Latin: *trans,* meaning across) because the mutant repressor will remain bound to both operators even in the presence of inducer or of wild-type repressor (Figure 7.5c). We might liken this to a radio signal that cannot be turned off. Monod and his colleagues found two such mutants, and Suzanne Bourgeois later found many others. These are named I^s to distinguish them from constitutive repressor mutants (I⁻), which make a repressor that cannot recognize the operator.

Both of the common kinds of constitutive mutants (I⁻ and O^c) affected all three of the *lac* genes (Z, Y, and A) in the same way. The genes had already been mapped and were found to be adjacent on the chromosome. These findings strongly suggested that the operator lay near the three structural genes.

We now recognize yet another class of repressor mutants, those that are constitutive and dominant (I⁻ᵈ). This kind of mutant gene (Figure 7.5d) makes a defective product that can still form tetramers with wild-type repressor monomers. However, the defective monomers spoil the activity of the whole tetramer so it cannot bind to the operator. Hence the dominant nature of this mutation. These mutations are not just *cis*-dominant because the "spoiled" repressors cannot bind to either operator in a merodiploid. This kind of "spoiler" mutation is widespread in nature, and it is called by the generic name **dominant-negative.**

Thus, Jacob and Monod, by skillful genetic analysis, were able to develop the operon concept. They predicted the existence of two key control elements: the repressor gene and the operator. Deletion mutations revealed a third element (the promoter) that was necessary for expression of all three *lac* genes. Furthermore, they could conclude that all three *lac* genes (*lacZ, Y,* and *A*) were clustered into a single control unit: the *lac* operon. Subsequent biochemical studies have amply confirmed Jacob and Monod's beautiful hypothesis.

Merodiploid with one wild-type gene and one:

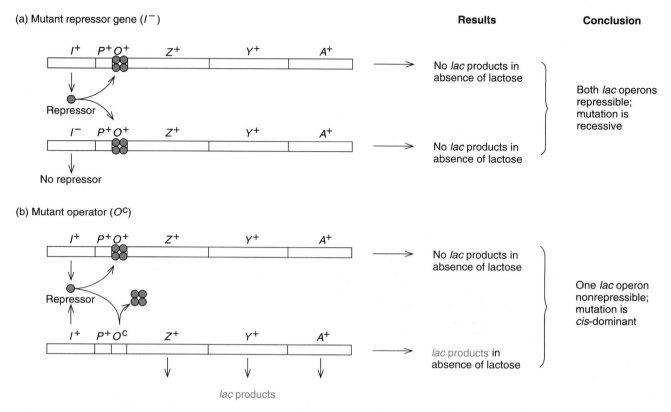

Figure 7.5   Effects of regulatory mutations in the *lac* operon in merodiploids. Jacob and Monod and others created merodiploid *E. coli* strains as described in panels **(a)–(d)** and tested them for *lac* products in the presence and absence of lactose. **(a)** This merodiploid has one wild-type operon (top) and one operon (bottom) with a mutation in the repressor gene (I^-). The wild-type repressor gene (I^+) makes enough normal repressor (green) to repress both operons, so the I^- mutation is recessive. **(b)** This merodiploid has one wild-type operon (top) and one operon (bottom) with a mutation in the operator (O^c) that makes it defective in binding repressor (green). The wild-type operon remains repressible, but the mutant operon is not; it makes *lac* products even in the absence of lactose. Because only the operon connected to the mutant operator is affected, this mutation is *cis*-dominant.

SUMMARY Negative control of the *lac* operon occurs as follows: The operon is turned off as long as the repressor binds to the operator, because the repressor keeps RNA polymerase from transcribing the three *lac* genes. When the supply of glucose is exhausted and lactose is available, the few molecules of *lac* operon enzymes produce a few molecules of allolactose from the lactose. The allolactose acts as an inducer by binding to the repressor and causing a conformational shift that encourages dissociation from the operator. With the repressor removed, RNA polymerase is free to transcribe the three *lac* genes. A combination of genetic and biochemical experiments revealed the two key elements of negative control of the *lac* operon: the operator and the repressor.

Repressor–Operator Interactions

After the pioneering work of Jacob and Monod, Walter Gilbert and Benno Müller-Hill succeeded in partially purifying the *lac* repressor. This work is all the more impressive, considering that it was done in the 1960s, before the advent of modern gene cloning. Gilbert and Müller-Hill's challenge was to purify a protein (the *lac* repressor) that is present in very tiny quantities in the cell, without an easy assay to identify the protein. The most sensitive assay available to them was binding a labeled synthetic inducer (isopropylthiogalactoside, or IPTG) to the repressor. But, with a crude extract of wild-type cells, the repressor was in such low concentration that this assay could not detect it. To get around this problem, Gilbert and Müller-Hill used a mutant *E. coli* strain with a repressor mutation ($lacI^t$) that causes the repressor to bind IPTG more tightly than normal. This tight binding allowed the mutant repressor to bind enough inducer that the protein could be detected even

Merodiploid with one wild-type gene and one:

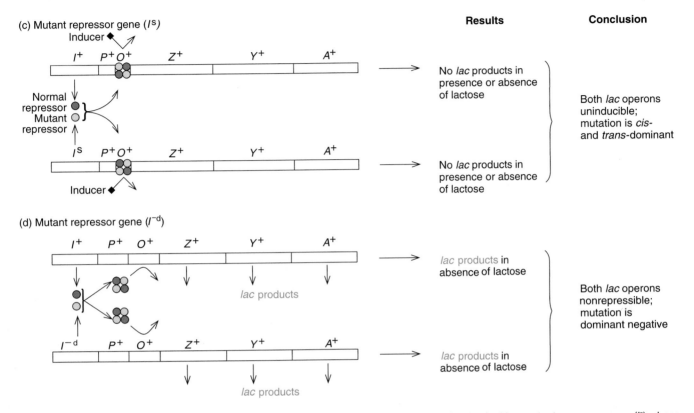

Figure 7.5 (*continued*) (**c**) This merodiploid has one wild-type operon (top) and one operon (bottom) with a mutant repressor gene (I^s) whose product (yellow) cannot bind inducer. The mutant repressor therefore binds irreversibly to both operators and renders both operons uninducible. This mutation is therefore dominant. Notice that these repressor tetramers containing some mutant and some wild-type subunits behave as mutant proteins. That is, they remain bound to the operator even in the presence of inducer. (**d**) This merodiploid has one wild-type operon (top) and one operon (bottom) with a mutant repressor gene (I^{-d}) whose product (yellow) cannot bind to the *lac* operator. Moreover, mixtures (heterotetramers) composed of both wild-type and mutant repressor monomers still cannot bind to the operator. Thus, because the operon remains derepressed even in the absence of lactose, this mutation is dominant. Furthermore, because the mutant protein poisons the activity of the wild-type protein, we call the mutation dominant-negative.

in very impure extracts. Because they could detect the protein, Gilbert and Müller-Hill could purify it.

Melvin Cohn and his colleagues used repressor purified by this technique in operator-binding studies. To assay repressor–operator binding, Cohn and colleagues used the nitrocellulose filter binding assay we discussed in Chapters 5 and 6. If repressor–operator interaction worked normally, we would expect it to be blocked by inducer. Indeed, Figure 7.6 shows a typical saturation curve for repressor–operator binding in the absence of inducer, but no binding in the presence of the synthetic inducer, IPTG. In another binding experiment (Figure 7.7), Cohn and coworkers showed that DNA containing the constitutive mutant operator (*lacO*^{c}) required a higher concentration of repressor to achieve full binding than did the wild-type operator. This was an important demonstration: What Jacob and Monod had defined genetically as the operator really was the binding site for repressor. If it were not, then mutating it should not have affected repressor binding.

SUMMARY Cohn and colleagues demonstrated with a filter-binding assay that *lac* repressor binds to *lac* operator. Furthermore, this experiment showed that a genetically defined constitutive *lac* operator has lower than normal affinity for the *lac* repressor, demonstrating that the sites defined genetically and biochemically as the operator are one and the same.

The Mechanism of Repression

For years it was assumed that the *lac* repressor acted by denying RNA polymerase access to the promoter, in spite of the fact that Ira Pastan and his colleagues had shown as early as 1971 that RNA polymerase could bind tightly to the *lac* promoter, even in the presence of repressor. Pastan's experimental plan was to incubate polymerase

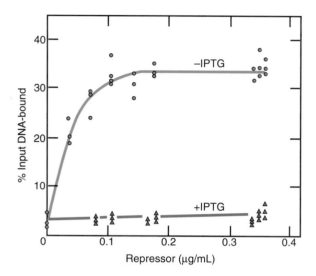

Figure 7.6   Assaying the binding between *lac* operator and *lac* repressor. Cohn and colleagues labeled *lacO*-containing DNA with ^{32}P and added increasing amounts of *lac* repressor. They assayed binding between repressor and operator by measuring the radioactivity attached to nitrocellulose. Only labeled DNA bound to repressor would attach to nitrocellulose. Red: repressor bound in the absence of the inducer IPTG. Green: repressor bound in the presence of 1 mM IPTG, which prevents repressor–operator binding. (*Source:* Riggs, A., et al., "On the Assay, Isolation, and Characterization of the *lac* Repressor," *Journal of Molecular Biology,* Vol. 34: 366. Copyright 1968 by permission of Academic Press Limited, London.)

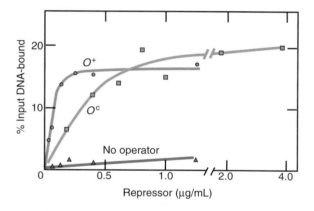

Figure 7.7   The O^c *lac* operator binds repressor with lower affinity than does the wild-type operator. Cohn and colleagues performed a *lac* operator–repressor binding assay as described in Figure 7.6, using three different DNAs as follows: red, DNA containing a wild-type operator (O^+); blue, DNA containing an operator-constitutive mutation (O^c) that binds repressor with a lower affinity; green, control, λφ80 DNA, which does not have a *lac* operator.

with DNA containing the *lac* operator in the presence of repressor, then to add inducer (IPTG) and rifampicin together. As we will see later in this chapter, rifampicin will inhibit transcription unless an open promoter complex has already formed. (Recall from Chapter 6 that an open promoter complex is one in which the RNA polymerase has caused local DNA melting at the promoter and is tightly bound there.) In this case, transcription did occur, showing that the *lac* repressor had not prevented the for-

mation of an open promoter complex. Thus, these results suggested that the repressor does not block access by RNA polymerase to the *lac* promoter. Susan Straney and Donald Crothers reinforced this view in 1987 by showing that polymerase and repressor can bind together to the *lac* promoter.

If we accept that RNA polymerase can bind tightly to the promoter, even with repressor occupying the operator, how do we explain repression? Straney and Crothers suggested that repressor blocks the formation of an open promoter complex, but that would be hard to reconcile with the rifampicin resistance of the complex observed by Pastan. Barbara Krummel and Michael Chamberlin proposed an alternative explanation: Repressor blocks the transition from the initial transcribing complex state (Chapter 6) to the elongation state. In other words, repressor traps the polymerase in a nonproductive state in which it spins its wheels making short oligonucleotides without ever moving away from the promoter.

Jookyung Lee and Alex Goldfarb provided some evidence for this idea. First, they used a run-off transcription assay (Chapter 5) to show that RNA polymerase is already engaged on the DNA template, even in the presence of repressor. The experimental plan was as follows: First, they incubated repressor with a 123-bp DNA fragment containing the *lac* control region plus the beginning of the *lacZ* gene. After allowing 10 min for the repressor to bind to the operator, they added polymerase. Then they added heparin—a polyanion that binds to any RNA polymerase that is free or loosely bound to DNA and keeps it from binding to DNA. They also added all the remaining components of the RNA polymerase reaction except CTP. Finally, they added labeled CTP with or without the inducer IPTG. The question is this: Will a run-off transcript be made? If so, the RNA polymerase has formed a heparin-resistant (open) complex with the promoter even in the presence of the repressor. In fact, as Figure 7.8 shows, the run-off transcript did appear, just as if repressor had not been present. Thus, under these conditions in vitro repressor does not seem to inhibit tight binding between polymerase and the *lac* promoter.

If it does not inhibit transcription of the *lac* operon by blocking access to the promoter, how would the *lac* repressor function? Lee and Goldfarb noted the appearance of shortened abortive transcripts (Chapter 6), only about 6 nt long, in the presence of repressor. Without repressor, the abortive transcripts reached a length of 9 nt. The fact that *any* transcripts—even short ones—were made in the presence of repressor reinforced the conclusion that, at least under these conditions, RNA polymerase really can bind to the *lac* promoter in the presence of repressor. This experiment also suggested that repressor may limit *lac* operon transcription by locking the polymerase into a nonproductive state in which it can make only abortive transcripts. Thus, extended transcription cannot get started.

1   2   3

} Run off

LacR: −   +   +
IPTG: −   −   +

Figure 7.8 RNA polymerase forms an open promoter complex with the *lac* promoter even in the presence of *lac* repressor in vitro. Lee and Goldfarb incubated a DNA fragment containing the *lac* UV5 promoter with (lanes 2 and 3) or without (lane 1) *lac* repressor (LacR). After repressor–operator binding had occurred, they added RNA polymerase. After allowing 20 min for open promoter complexes to form, they added heparin to block any further complex formation, along with all the other reaction components except CTP. Finally, after 5 more minutes, they added [α-³²P]CTP alone or with the inducer IPTG. They allowed 10 more minutes for RNA synthesis and then electrophoresed the transcripts. Lane 3 shows that transcription occurred even when repressor bound to the DNA before polymerase could. Thus, repressor did not prevent polymerase from binding and forming an open promoter complex. (*Source:* Lee and Goldfarb, *lac* repressor acts by modifying the initial transcribing complex so that it cannot leave the promoter. *Cell* 66 (23 Aug 1991) f. 1, p. 794. Reprinted by permission of Elsevier Science.)

One problem with the studies of Lee and Goldfarb and the others just cited is that they were performed in vitro under rather nonphysiological conditions. For example, the concentrations of the proteins (RNA polymerase and repressor) were much higher than they would be in vivo. To deal with such problems, Thomas Record and colleagues performed kinetic studies in vitro under conditions likely to prevail in vivo. They formed RNA polymerase/*lac* promoter complexes, then measured the rate of abortive transcript synthesis by these complexes alone, or after addition of either heparin or *lac* repressor. They measured transcription by using a UTP analog with a fluorescent tag on the γ-phosphate (*pppU). When UMP was incorporated into RNA, tagged pyrophosphate (*pp) was released, and the fluorescence intensity increased. Fig-

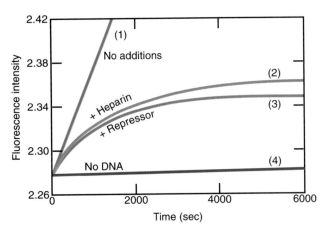

Figure 7.9 Effect of *lac* repressor on dissociation of RNA polymerase from the *lac* promoter. Record and colleagues made complexes between RNA polymerase and DNA containing the *lac* promoter–operator region. Then they allowed the complexes to synthesize abortive transcripts in the presence of a UTP analog fluorescently labeled in the γ-phosphate. As the polymerase incorporates UMP from this analog into transcripts, the labeled pyrophosphate released increases in fluorescence intensity. The experiments were run with no addition (curve 1, green), with heparin to block reinitiation by RNA polymerase that dissociates from the DNA (curve 2, blue), and with a low concentration of *lac* repressor (curve 3, red). A control experiment was run with no DNA (curve 4, purple). The repressor inhibited reinitiation of abortive transcription as well as heparin, suggesting that it blocks dissociated RNA polymerase from reassociating with the promoter. (*Source:* Schlax, P.J., Capp, M.W. and M.T. Record, Jr. "Inhibition of transcription initiation by lac repressor," *Journal of Molecular Biology* 245: 331–50. Copyright 1995 by permission of Academic Press Limited, London.)

ure 7.9 demonstrates that the rate of abortive transcript synthesis continued at a high level in the absence of competitor, but rapidly leveled off in the presence of either heparin or repressor.

Record and colleagues explained these results as follows: The polymerase–promoter complex is in equilibrium with free polymerase and promoter. Moreover, in the absence of competitor (curve 1), the polymerases that dissociate go right back to the promoter and continue making abortive transcripts. However, both heparin (curve 2) and repressor (curve 3) prevent such reassociation. Heparin does so by binding to the polymerase and preventing its association with DNA. But the repressor presumably does so by binding to the operator adjacent to the promoter and blocking access to the promoter by RNA polymerase. Thus, these data support the old hypothesis of a competition between polymerase and repressor.

We have seen that the story of the *lac* repressor mechanism has had many twists and turns. Have we seen the last twist? The latest results suggest that the original, competition hypothesis is correct, but we may not have heard the end of the story yet.

Another factor in repression of the *lac* operon is the presence of not one, but three operators: one **major operator** near the transcription start site and two **auxiliary operators** (one upstream and one downstream). Figure 7.10

(a)

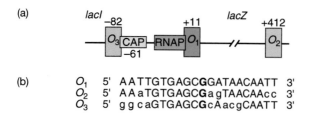

(b)

		5'		3'
O_1	5'	AATTGTGAGC**G**GATAACAATT	3'
O_2	5'	AAaTGTGAGC**G**agTAACAAcc	3'
O_3	5'	ggcaGTGAGC**G**cAacgCAATT	3'

Figure 7.10 The three *lac* operators. (a) Map of the *lac* control region. The major operator (O_1) is shown in red; the two auxiliary operators are shown in pink. The CAP and RNA polymerase binding sites are in yellow and blue, respectively. CAP is a positive regulator of the *lac* operon discussed in the next section of this chapter. **(b)** Sequences of the three operators. The sequences are aligned, with the central G of each in boldface. Sites at which the auxiliary operator sequences differ from the major operator are lower case in the O_2 and O_3 sequences.

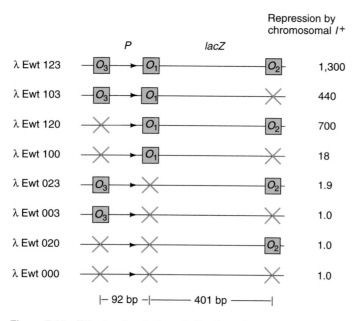

Figure 7.11 Effects of mutations in the three *lac* operators. Müller-Hill and colleagues placed wild-type and mutated *lac* operon fragments on λ phage DNA and allowed these DNAs to lysogenize *E. coli* cells (Chapter 8). This introduced these *lac* fragments, containing the three operators, the *lac* promoter, and the *lacZ* gene, into the cellular genome. The cell contained no other *LacZ* gene, but it had a wild-type *lacI* gene. Then Müller-Hill and coworkers assayed for β-galactosidase produced in the presence and absence of the inducer IPTG. The ratio of activity in the presence and absence of inducer is the repression given at right. For example, the repression observed with all three operators was 1,300-fold. The phage designations at left are the λ phages used to lysogenize the cells. λ Ewt 123 (top) was wild-type in all three operators (green). All the other phages had one or more operators deleted (red X).

shows the spatial arrangement of these operators, the classical (major) operator O_1, centered at position +11, the downstream auxiliary operator O_2, centered at position +412, and the upstream auxiliary operator O_3, centered at position −82. We have already discussed the classical operator, and the role investigators have tradi-

tionally ascribed to it alone. But Benno Müller-Hill and others have more recently investigated the auxiliary operators and have discovered that they are not just trivial copies of the major operator. Instead, they play a significant role in repression. Müller-Hill and colleagues demonstrated this role by showing that removal of either of the auxiliary operators decreased repression only slightly, but removal of both auxiliary operators decreased repression about 50-fold. Figure 7.11 outlines the results of these experiments and shows that all three operators together repress transcription 1,300-fold, two operators together repress from 440- to 700-fold, but the classical operator by itself represses only 18-fold.

In 1996, Mitchell Lewis and coworkers provided a structural basis for this cooperativity among operators. They determined the crystal structure of the *lac* repressor and its complexes with 21-bp DNA fragments containing operator sequences. Figure 7.12 summarizes their findings. We can see that the two dimers in a repressor tetramer are independent DNA-binding entities that interact with the major groove of the DNA. It is also clear that the two dimers within the tetramer are bound to separate operator sequences. It is easy to imagine these two operators as part of a single long piece of DNA.

SUMMARY Two competing hypotheses seek to explain the mechanism of repression of the *lac* operon. One is that the RNA polymerase can bind to the *lac* promoter in the presence of the repressor, but the repressor inhibits the transition from abortive transcription to processive transcription. The other is that the repressor, by binding to the operator, blocks access by the polymerase to the adjacent promoter. The latest evidence supports the latter hypothesis. In addition to the classical (major) *lac* operator adjacent to the promoter, two auxiliary *lac* operators exist: one each upstream and downstream. All three operators are required for optimum repression, two work reasonably well, but the classical operator by itself produces only a modest amount of repression.

Positive Control of the *lac* Operon

As you might predict, positive control is the opposite of negative control; that is, the operon is turned off (actually, turned down to a basal level) unless something intervenes to turn it on. If negative control is like the brake of a car, positive control is like the accelerator pedal. In the case of the *lac* operon, this means that removing the repressor from the operator (releasing the brake) is not enough to activate the operon. An additional, positive factor (stepping on the accelerator) is needed.

(a)   (b)

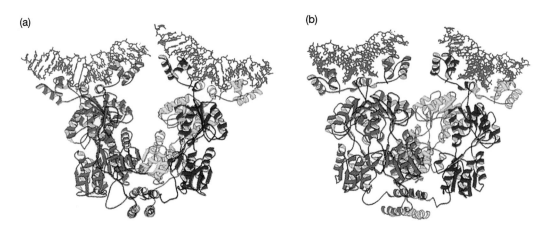

Figure 7.12   Structure of the *lac* repressor tetramer bound to two operator fragments. Lewis Lu and colleagues performed x-ray crystallography on *lac* repressor bound to 21-bp DNA fragments containing the major *lac* operator sequence. The structure presents the four repressor monomers in pink, green, yellow, and red, and the DNA fragments in blue. Two repressor dimers interact with each other at bottom to form tetramers. Each of the dimers contains two DNA-binding domains that can be seen interacting with the DNA major grooves at top. The structure shows clearly that the two dimers can bind independently to separate *lac* operators. Panels (**a**) and (**b**) are "front" and "side" views of the same structure. (*Source:* Lewis et al., Crystal structure of the lactose operon processor and its complexes with DNA and inducer. *Science* 271 (1 Mar 1996), f. 6, p. 1251. © AAAS.)

What is the selective advantage of this positive control system? It keeps the operon turned off when the level of glucose is high. If there were no way to respond to glucose levels, the presence of lactose alone would suffice to activate the operon. But that would be inappropriate if glucose was still available, because *E. coli* cells metabolize glucose more easily than lactose; it would therefore be wasteful for them to activate the *lac* operon in the presence of glucose. In fact, *E. coli* cells keep the *lac* operon in a relatively inactive state as long as glucose is present. This selection in favor of glucose metabolism and against use of other energy sources has long been attributed to the influence of some breakdown product, or *catabolite*, of glucose. It is therefore known as **catabolite repression.**

The ideal positive controller of the *lac* operon would be a substance that sensed the lack of glucose and responded by activating the *lac* promoter so that RNA polymerase could bind and transcribe the structural genes (assuming, of course, that lactose is present and the repressor is therefore not bound to the operator). One substance that responds to glucose concentration is a nucleotide called **cyclic-AMP (cAMP)** (Figure 7.13). As the level of glucose drops, the concentration of cAMP rises.

Catabolite Activator Protein   Ira Pastan and his colleagues demonstrated that cAMP, added to bacteria, could overcome catabolite repression of the *lac* operon and a number of other operons, including the *gal* and *ara* operons. The latter two govern the metabolism of the sugars galactose and arabinose, respectively. In other words, cAMP rendered these genes active, even in the presence of glucose. This finding implicated cAMP strongly in the positive control of the *lac* operon. Does this mean that cAMP is the positive effector? Not exactly. The positive

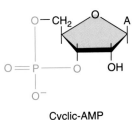

Cyclic-AMP

Figure 7.13   Cyclic-AMP. Note the cyclic 5′-3′ phosphodiester bond (blue).

controller of the *lac* operon is a complex composed of two parts: cAMP and a protein factor.

Geoffrey Zubay and coworkers showed that a crude cell-free extract of *E. coli* would make β-galactosidase if supplied with cAMP. This finding led the way to the discovery of a protein in the extract that was necessary for the stimulation by cAMP. Zubay called this protein **catabolite activator protein,** or **CAP.** Later, Pastan's group found the same protein and named it **cyclic-AMP receptor protein,** or **CRP.** To avoid confusion, we will refer to this protein from now on as CAP, regardless of whose experiments we are discussing. However, the gene encoding this protein has been given the official name *crp.*

Pastan found that the dissociation constant for the CAP–cAMP complex was $1–2 \times 10^{-6}$ M. However, they also isolated a mutant whose CAP bound about 10 times less tightly to cAMP. If CAP–cAMP really is important to positive control of the *lac* operon, we would expect reduced production of β-galactosidase by a cAMP-supplemented cell-free extract of these mutant cells. Figure 7.14 shows that this is indeed the case. To make the point even more strongly, Pastan showed that β-galactosidase synthesis by this mutant extract (plus cAMP) could be stimulated about threefold by the addition of wild-type CAP.

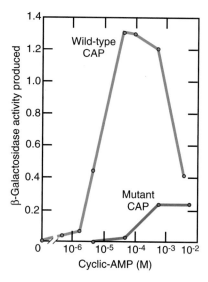

Figure 7.14 Stimulation of β-galactosidase synthesis by cAMP with wild-type and mutant CAP. Pastan and colleagues stimulated cell-free bacterial extracts to make β-galactosidase in the presence of increasing concentrations of cAMP with a wild-type extract (red), or an extract from mutant cells that have a CAP with reduced affinity for cAMP (green). This mutant extract made much less β-galactosidase, which is what we expect if the CAP–cAMP complex is important in *lac* operon transcription. Too much cAMP obviously interfered with β-galactosidase synthesis in the wild-type extract. This is not surprising because cAMP has many effects, and some may indirectly inhibit some step in expression of the *lacZ* gene in vitro. (*Source:* Emmer, M., et al., "Cyclic AMP Receptor Protein of E. coli: Its Role in the Synthesis of Inducible Enzymes," *Proceedings of the National Academy of Sciences* 66(2): 480–487, June 1970.)

SUMMARY Positive control of the *lac* operon, and certain other inducible operons that code for sugar-metabolizing enzymes, is mediated by a factor called catabolite activator protein (CAP), which, in conjunction with cyclic-AMP, stimulates transcription. Because cyclic-AMP concentration is depressed by glucose, this sugar prevents positive control from operating. Thus, the *lac* operon is activated only when glucose concentration is low and a corresponding need arises to metabolize an alternative energy source.

Mechanism of CAP Action

How do CAP and cAMP stimulate *lac* transcription? Zubay and colleagues discovered a class of *lac* mutants in which CAP and cAMP could not stimulate *lac* transcription. These mutations mapped to the *lac* promoter, suggesting that the binding site for the CAP–cAMP complex lies in the promoter. Later molecular biological work, which we will discuss shortly, has shown that the CAP–cAMP binding site (the **activator site**) lies just upstream of the promoter. Pastan and colleagues went on to show that this binding of CAP and cAMP to the activator site helps RNA polymerase to form an open promoter complex.

Figure 7.15 shows how this experiment worked. First, Pastan and colleagues allowed RNA polymerase to bind to the *lac* promoter in the presence or absence of CAP and cAMP. Then they challenged the promoter complex by adding nucleotides and rifampicin simultaneously to see if it was an open or closed promoter complex. If it was closed, it should still be rifampicin-sensitive because the DNA melting step takes so much time that it would allow the antibiotic to inhibit the polymerase before initiation could occur. However, if it was an open promoter complex, it would be primed to polymerize nucleotides. Because nucleotides reach the polymerase before the antibiotic, the polymerase has time to initiate transcription. Once it has initiated an RNA chain, the polymerase becomes resistant to rifampicin until it completes that RNA chain. In fact, Pastan and colleagues found that when the polymerase–promoter complex formed in the absence of CAP and cAMP it was still rifampicin-sensitive. Thus, it had only formed a closed promoter complex. On the other hand, when CAP and cAMP were present when polymerase associated with the promoter, a rifampicin-resistant open promoter complex formed.

Figure 7.15 presents a dimer of CAP–cAMP at the activator site on the left and polymerase at the promoter on the right. How do we know that is the proper order? The first indication came from genetic experiments. Mutations to the left of the promoter prevent stimulation of transcription by CAP and cAMP, but still allow a low level of transcription. An example is a deletion called L1, whose position is shown in Figure 7.16. Because this deletion completely obliterates positive control of the *lac* operon by CAP and cAMP, the CAP-binding site must lie at least partially within the deleted region. On the other hand, since the L1 deletion has no effect on CAP-independent transcription, it has not encroached on the promoter, where RNA polymerase binds. Therefore, the right-hand end of this deletion serves as a rough dividing line between the activator site and the promoter.

The CAP-binding sites in the *lac*, *gal*, and *ara* operons all contain the sequence TGTGA. The conservation of this TGTGA sequence suggests that it is an important part of the CAP-binding site, and we also have direct evidence for this. For example, footprinting studies show that binding of the CAP–cAMP complex protects the G's in this sequence against methylation by dimethyl sulfate, suggesting that the CAP–cAMP complex binds tightly enough to these G's that it hides them from the methylating agent.

The *lac* operon, and other operons activated by CAP and cAMP, have remarkably weak promoters. Their −35 boxes are particularly unlike the consensus sequences; in fact, they are scarcely recognizable. This situation is actually not surprising. If the *lac* operon had a strong promoter, RNA polymerase could form open promoter com-

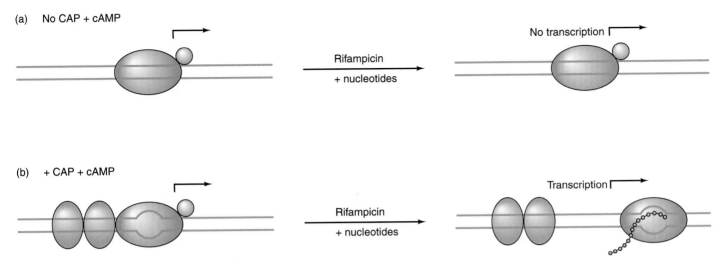

(a)   No CAP + cAMP

No transcription

Rifampicin
+ nucleotides

(b)   + CAP + cAMP

Transcription

Rifampicin
+ nucleotides

Figure 7.15   CAP plus cAMP allow formation of an open promoter complex. (a) When RNA polymerase binds to the *lac* promoter without CAP, it forms a closed promoter complex. This is susceptible to inhibition when rifampicin is added along with nucleotides, so no transcription occurs. **(b)** When RNA polymerase binds to the *lac* promoter in the presence of CAP and cAMP (purple), it forms an open promoter complex. This is not susceptible to inhibition when rifampicin and nucleotides are added at the same time because the open promoter complex is ready to polymerize the nucleotides, which reach the polymerase active site before the antibiotic. Once the first phosphodiester bond forms, the polymerase is resistant to rifampicin inhibition until it reinitiates. Thus, transcription occurs under these conditions, demonstrating that CAP and cAMP facilitate formation of an open promoter complex. The RNA is shown in green.

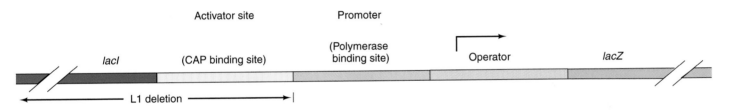

Activator site

Promoter

lacI   (CAP binding site)   (Polymerase binding site)   Operator   *lacZ*

L1 deletion

Figure 7.16   The *lac* control region. The activator–promoter region, just upstream of the operator, contains the activator site, or CAP binding site, on the left (yellow) and the promoter, or polymerase binding site, on the right (pink). These sites have been defined by footprinting experiments and by genetic analysis. An example of the latter approach is the L1 deletion, whose right-hand end is shown. The L1 mutant shows basal transcription of the *lac* operon, but no stimulation by CAP and cAMP. Thus, it still has the promoter, but lacks the activator site.

plexes readily without help from CAP and cAMP, and it would therefore be active even in the presence of glucose. Thus, this promoter has to be weak to be dependent on CAP and cAMP. In fact, strong mutant *lac* promoters are known (e.g., the *lac* UV5 promoter) and they do not depend on CAP and cAMP.

SUMMARY The CAP–cAMP complex stimulates transcription of the *lac* operon by binding to an activator site adjacent to the promoter and helping RNA polymerase to bind to the promoter.

Recruitment   How does CAP–cAMP recruit polymerase to the promoter? Such **recruitment** has two steps: (1) Formation of the closed promoter complex, and (2) conversion of the closed promoter complex to the open pro-

moter complex. William McClure and his colleagues have summarized these two steps in the following equation:

$$R + P \rightleftharpoons RP_C \rightarrow RP_o$$
$$K_B \qquad k_2$$

where R is RNA polymerase, P is the promoter, RP_C is the closed promoter complex, and RP_O is the open promoter complex. McClure and coworkers devised kinetic methods of distinguishing between the two steps and determined that CAP–cAMP acts directly to stimulate the first step by increasing K_B. CAP–cAMP has little if any effect on k_2, so the second step is not accelerated. Nevertheless, by increasing the rate of formation of the closed promoter complex, CAP–cAMP provides more raw material (closed promoter complex) for conversion to the open promoter complex. Thus, the net effect of CAP–cAMP is to increase the rate of open promoter complex formation.

Table 7.2   Activation of *lac* P1 Transcription by CAP–cAMP

| | Transcripts (cpm) | | | | P1/UV5 (%) | | |
| | −cAMP–CAP | | +cAMP–CAP | | | | |
Enzyme	P1	UV5	P1	UV5	−cAMP–CAP	+cAMP–CAP	Activation (fold)
α-WT	46	797	625	748	5.8	83.6	14.4
α-256	53	766	62	723	6.9	8.6	1.2
α-235	51	760	45	643	6.7	7.0	1.0

Philip Malan and McClure also showed that CAP–cAMP works in another way to stimulate transcription of the *lac* operon. William Reznikoff and colleagues had demonstrated in 1978 that the *lac* operon has an alternative promoter 22 bp upstream of the major promoter. McClure named the major promoter P_1 and the alternative promoter P_2. Malan and McClure's in vitro transcription experiments showed that CAP–cAMP reduced initiation at P_2, while stimulating initiation at P_1. Because P_2 appears to be an inefficient promoter, limiting the amount of polymerase binding there should free up more polymerase to bind at the more efficient promoter, P_1, and should therefore result in more *lac* transcription. That is what CAP–cAMP appears to do.

Two hypotheses explain how binding CAP and cAMP to the activator site facilitates binding of polymerase to the promoter to form the closed promoter complex. Because both hypotheses have strong experimental support, both may be part of the answer. One long-standing hypothesis is that CAP and RNA polymerase actually touch when they bind to their respective DNA sites, and therefore they bind cooperatively. The other is that CAP–cAMP causes the DNA to bend.

What evidence supports the first hypothesis? First, CAP and RNA polymerase cosediment on ultracentrifugation in the presence of cAMP, suggesting that they have an affinity for each other. Second, CAP and RNA polymerase, when both are bound to their DNA sites, can be chemically cross-linked to each other, suggesting that they are in close proximity. Third, DNase footprinting experiments (Chapter 5) show that the CAP–cAMP footprint lies adjacent to the polymerase footprint. Thus, the DNA binding sites for these two proteins are close enough that the proteins could interact with each other as they bind to their DNA sites. Fourth, several CAP mutations decrease activation without affecting DNA binding (or bending), and some of these mutations alter amino acids in the region of CAP (**activation region I [ARI]**) that is thought to interact with polymerase. Finally, the polymerase site that is presumed to interact with ARI on CAP is the carboxyl terminal domain of the α-subunit (the **α-CTD**), and deletion of the α-CTD prevents activation by CAP–cAMP.

Kazuhiko Igarashi and Akira Ishihama provided evidence for the importance of the α-CTD of RNA polymerase in activation by CAP–cAMP. They transcribed cloned *lac* operons in vitro with RNA polymerases reconstituted from separated subunits. All the subunits were wild-type, except in some experiments, in which the α-subunit was a truncated version lacking the CTD. One of the truncated α-subunits ended at amino acid 256 (of the normal 329 amino acids); the other ended at amino acid 235. Table 7.2 shows the results of run-off transcription from a CAP–cAMP-dependent *lac* promoter (P1) and a CAP–cAMP-independent *lac* promoter (*lac* UV5) with reconstituted polymerases containing the wild-type or truncated α-subunits in the presence and absence of CAP–cAMP. As expected, CAP–cAMP did not stimulate transcription from the *lac* UV5 promoter because it is a strong promoter that is CAP–cAMP-insensitive. Also as expected, transcription from the *lac* P1 promoter was stimulated over 14-fold by CAP–cAMP. But the most interesting behavior was that of the polymerases reconstituted with truncated α-subunits. These enzymes were just as good as wild-type in transcribing from either promoter in the absence of CAP–cAMP, but they could not be stimulated by CAP–cAMP. Thus, the α-CTD, missing in these truncated enzymes, is not necessary for reconstitution of an active RNA polymerase, but it is necessary for stimulation by CAP–cAMP. Other experiments by Richard Ebright and colleagues have shown that the N-terminal domain of the RNA polymerase α-subunit (α-NTD) is also involved in interaction with CAP–cAMP, at the **activator region II, or ARII**.

Figure 7.17 illustrates one hypothesis of activation, in which the CAP–cAMP dimer binds to its activator site and simultaneously binds to the carboxyl-terminal domain of the polymerase α-subunit (α-CTD), facilitating binding of polymerase to the promoter. This would be the functional equivalent of the α-CTD binding to an UP element in the DNA and thereby enhancing polymerase binding to the promoter.

It is easy to imagine direct interaction between CAP and polymerase in the *lac* operon because the binding sites for these two proteins are so close together. However, the

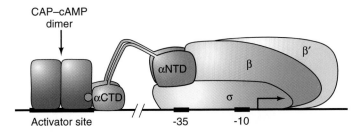

Figure 7.17 Hypothesis for CAP–cAMP activation of *lac* transcription. The CAP–cAMP dimer (purple) binds to its activator site on the DNA, and the α-CTD (red) interacts with a specific site on the CAP protein (brown dot). This strengthens binding between polymerase and promoter. (*Source:* Busby, S. and R.H. Ebright, "Promoter structure, promoter recognition, and transcription activation in prokaryotes," *Cell* 79:742, 1994. Copyright 1994 Cell Press, Cambridge, MA. Reprinted by permission.)

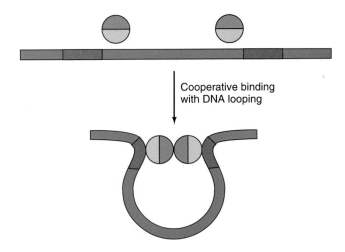

Figure 7.18 DNA looping enables proteins to bind cooperatively to widely separated sites. Cooperative binding between two proteins (blue and yellow) to remote sites (red) is possible if the DNA in between the sites bends (loops out) to allow the two proteins to interact with each other.

ara operon's activator site is far upstream of the promoter, so how can CAP and polymerase interact in that operon? These proteins can interact if the DNA in between their target sites bends, as illustrated in Figure 7.18.

Binding of CAP–cAMP to the activator site does indeed cause the DNA to bend, even in the *lac* operon, and this supports the second hypothesis. This bending can be observed, and even measured, by gel electrophoresis (Figure 7.19). When a piece of DNA is bent, it migrates more slowly during electrophoresis. Furthermore, as Figure 7.19b and c illustrate, the closer the bend is to the middle of the DNA, the more slowly the DNA electrophoreses. Hen-Ming Wu and Donald Crothers took advantage of this phenomenon by preparing DNA fragments of the *lac* operon, all the same length, with the CAP-binding site located at different positions in each. Next, they bound CAP and cAMP to each fragment and electrophoresed the DNA–protein complexes. If CAP binding really did bend the DNA, then the different fragments should have migrated at different rates. If the DNA did not bend, they all should have migrated at the same rate. Figure 7.19d demonstrates that the fragments really did migrate at different rates. Moreover, the more pronounced the DNA bend, the greater the difference in electrophoretic rates should be. In other words, the shape of the curve in Figure 7.19 should give us an estimate of the degree of bending of DNA by CAP and cAMP. In fact, the bending seems to be about 90 degrees.

Thomas Steitz and colleagues have examined the shape of CAP–DNA complexes by x-ray crystallography and have confirmed that the DNA in these complexes bends by about 90 degrees. Figure 7.20 demonstrates that this is a very pronounced bend indeed. How could such bending contribute to tighter binding between RNA polymerase and the *lac* promoter? One way would be to bring the polymerase binding site on CAP into the right position to interact with the polymerase and therefore to attract it to the promoter. Indeed, the CAP in a CAP–*lac* promoter complex lies close to the −35 box of the promoter. Moreover, the side chains of the two amino acids implicated by genetic analysis in polymerase binding are in a position to interact with the nearby polymerase.

If DNA bending is really involved in activation of the *lac* operon by CAP and cAMP, then substitution of a naturally bent segment of DNA for the CAP binding site should cause activation even in the absence of CAP and cAMP. Marc Gartenberg and Crothers reported such an experiment in 1991. They replaced the CAP binding site in the *lac* operon with tracts of $(dA)_n \bullet (dT)_n$, which are intrinsically bent. They found that this bent DNA caused a 10-fold stimulation of transcription from the *lac* promoter. Furthermore, the bent DNA induced a similar increase in the rate of formation of open promoter complexes.

Another way that DNA bending could facilitate polymerase binding to the promoter would be to cause the DNA to wrap around the polymerase, bringing an upstream binding site into contact with the enzyme, as illustrated in Figure 7.21. This could explain how CAP activates the *ara* promoter, even though it binds to an activator site centered at position −103 (103 bp upstream of the transcription start site), far from the polymerase binding site in that operon. In spite of its remote location, it could still cause the DNA to wrap around, so the polymerase could interact with the hypothetical upstream binding site. This model could even be modified to help explain the *gal* promoter, in which the −35 box and CAP binding site overlap. In this case, it is hard to envision RNA polymerase and CAP interacting simultaneously at the −35 box region. However, if the RNA polymerase interacts with the −10, but not the −35 box, which has little affinity for polymerase, CAP could still cause the DNA to

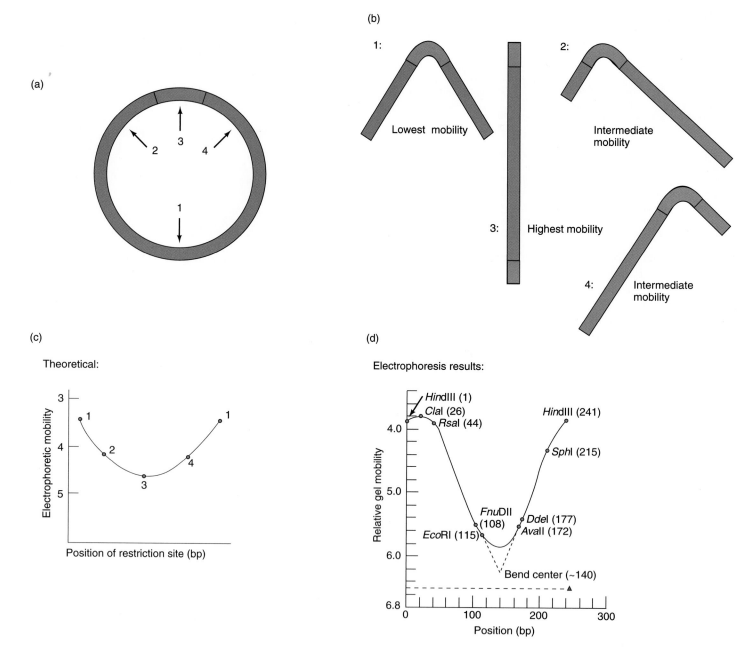

Figure 7.19 Electrophoresis of CAP–cAMP-promoter complexes. (a) Map of a hypothetical DNA circle, showing a protein binding site at center (red), and cutting sites for four different restriction sites (arrows). (b) Results of cutting DNA in panel (a) with each restriction enzyme, then adding a DNA-binding protein, which bends DNA. Restriction enzyme 1 cuts across from the binding site, leaving it in the middle; restriction enzymes 2 and 4 place the binding site off center; and restriction enzyme 3 cuts within the binding site, allowing little if any bending of the DNA. (c) Theoretical curve showing the relationship between electrophoretic mobility and bent DNA, with the bend at various sites along the DNA. Note that the mobility is lowest when the bend is closest to the middle of the DNA fragment (at either end of the curve). Note also that mobility increases in the downward direction on the *y* axis. (d) Actual electrophoresis results with CAP-cAMP and DNA fragments containing the *lac* promoter at various points in the fragment, depending on which restriction enzyme was used to cut the DNA. The name of each restriction enzyme and the position of its cutting site is given on the curve. The symmetrical curve allowed Wu and Crothers to extrapolate to a bend center that corresponds to the CAP–cAMP binding site in the *lac* promoter. (*Source:* Wu, H.M., and D.M. Crothers, " The locus of sequence-directed and protein-induced DNA bending." *Nature* 308:511, 1984. Copyright 1984 Macmillan Magazines Limited. Reprinted by permission.)

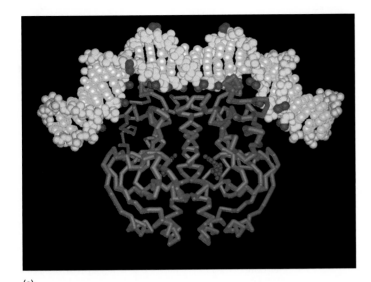

(a)

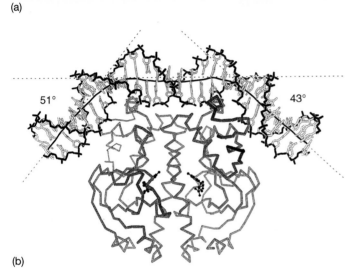

(b)

Figure 7.20 Structure of the CAP–cAMP–DNA complex. (a) The protein backbones of the two CAP monomers are shown below the bent DNA. The DNA-binding domains of CAP are shown in purple. The bases of the DNA are depicted in white and the sugar–phosphate backbones are in yellow. Regions of DNA suspected (on the basis of other evidence) to be involved in CAP binding are shown in red. Phosphodiester bonds rendered hypersensitive to DNAse by CAP binding are shown in blue. These occur at the kinks where the DNA minor groove is opened up by CAP binding and made accessible to DNase. **(b)** Angle of DNA bending in the complex. The helical axis is represented by the solid black line running down the middle of the DNA helix. The bend away from horizontal on the right is 43 degrees and on the left is 51 degrees, adding up to a total bend of 94 degrees. (*Source:* Schultz et al., Crystal structure of a CAP-DNA complex: The DNA is bent by 90 degrees. *Science* 253 (30 Aug 1991) f. 2a-b, p. 1003. © AAAS.)

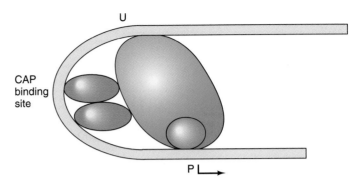

Figure 7.21 A DNA-bending model for the activity of CAP. CAP and cAMP (purple) associate with the activator site and cause a bend in the DNA (yellow). This allows the RNA polymerase (red, with blue σ-factor) to interact with the promoter (P) and with a hypothetical upstream site (U), bending the DNA a full 180 degrees. The double contact between polymerase and DNA presumably stabilizes its binding. (*Source:* Lilley, D.M.J. "When the CAP fits bent DNA," *Nature* 354:359 1991. Copyright 1991 Macmillan Magazines Limited. Reprinted with permission.)

> **SUMMARY** CAP–cAMP binding to the *lac* activator site recruits RNA polymerase to the adjacent *lac* promoter to form a closed promoter complex. This closed complex then converts to an open promoter complex. CAP–cAMP causes recruitment through protein–protein interactions, by bending the DNA, or probably by a combination of these phenomena.

7.2 The *mal* Regulon

Another set of genes stimulated by CAP is the *mal* regulon. These genes code for a set of enzymes involved in metabolizing the sugar maltose, a disaccharide made up of two glucose units. The genes encoding these enzymes are regulated together, but, in contrast to an operon, they are not contiguous and are controlled by multiple promoters. We call such a set of noncontiguous, coordinately controlled genes a **regulon**. In this section, we will consider how regulon genes can be controlled together, even though they are separated in space. We will also learn about another transcription activator, the MalT protein.

The Role of CAP in the *mal* Regulon

Because maltose is an alternative energy source like lactose, it is not surprising that the *mal* regulon is also controlled by CAP. But CAP exerts its influence in a novel way in this case. To begin with, another protein, called **MalT**, also regulates the *mal* promoters. This protein requires ATP and the maltose regulon inducer, *maltotriose*, for its activity. Some *mal* promoters (*malEp* and *malKp*) depend on both MalT and CAP. Others (*pulAp, pulCp,*

bend so as to bring another upstream element close to the RNA polymerase.

CAP is just one of a growing number of bacterial transcription activators. We will examine another example, MalT, when we discuss the *mal* regulon later in this chapter, and still more examples in Chapter 9.

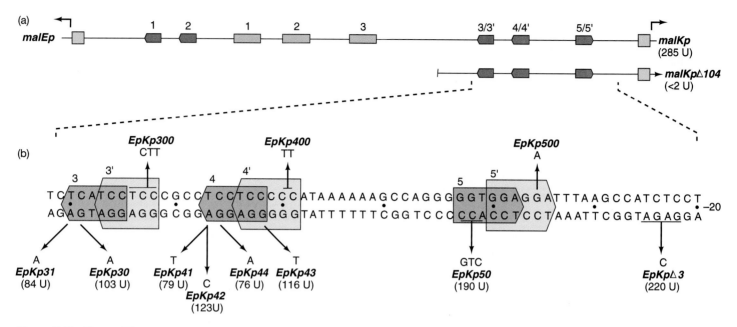

Figure 7.22   The *malEp–malKp* region. (a) The control region between the *malEp* and *malKp* transcription start sites includes a –10 box (yellow) upstream of each start site (bent arrows); five MalT-binding sites (red arrows); and three CAP binding sites (blue). Below the wild-type map is the map of the *malKpΔ104* deletion mutant, which is missing everything to the left of position –104 with respect to the *malKp* transcription start site, including the CAP binding sites. The numbers in parentheses represent the units of amylomaltase produced from *malKp*, or the mutant, in vivo. **(b)** The sequence of part of *malKpΔ104*, showing the changes in sequence and amylomaltase production for each of several mutants. For example, an A → T mutation in the third base from the left has created mutant *EpKp31*, which produces 84 units of amylomaltase, compared with fewer than 2 units produced by the parent, *malKpΔ104*. The large open box-shaped arrows represent the orientations of the MalT-binding sites, but not the direction of transcription resulting from MalT binding at each site. The preferred binding sites (3, 4, and 5) are in dark pink; the alternative sites (3′, 4′, and 5′), are in light pink. (*Source:* Richet et al., "A New Mechanism for Coactivation of Transcription Initiation" *Cell* 66:1186, 1991. Copyright 1991 Cell Press, Cambridge, MA. Reprinted by permission.)

and *malPp*) rely only on MalT, whereas one (*malTp*) is regulated by CAP alone.

The two promoters regulated by both MalT and CAP are especially interesting. Each of these promoters controls a three-gene operon: *malEp* controls the *malEFG* operon; and *malKp* controls the *malK-lamB-malM* operon. Figure 7.22a shows that these promoters are divergent (point away from each other) and are separated by a considerable distance (271 bp). In between the promoters lie the binding sites for the regulatory proteins. In fact, each protein has multiple binding sites. The MalT protein binds to five different binding sites, represented by red arrows. Two of these (sites 1 and 2) lie near the *malEp* promoter; the other three (sites 3, 4, and 5) are close to the *malKp* promoter. The three CAP binding sites, represented by blue boxes, lie between the two groups of MalT binding sites.

The fact that two different proteins cooperate to control the two promoters in this region raises an obvious question: How does this cooperation work? A closer look at the sequence of MalT binding sites 3, 4, and 5 (Figure 7.22b) gives us a clue to the answer: Each of the three binding sites (GGAT/GGA) is actually two sites that overlap by three bases and are therefore offset by three bases. For example, the site 3 region includes this 9-bp sequence: GGAG-GATGA. This includes one binding site stretching from

base 1 to base 6, which reads GGAGGA; and another, from base 4 to base 9, which reads GGATGA. The first of these is site 3, and the other is called site 3′. Sites 4 and 5 have their overlapping counterparts, sites 4′ and 5′, respectively. Olivier Raibaud and his colleagues noticed that sites 3, 4, and 5, though they have stronger affinity for MalT than do sites 3′, 4′, and 5′, are not positioned perfectly with respect to the promoter's –10 box. Instead they are exactly 3 bp too far away. This means that CAP could stimulate transcription from the *malKp* promoter simply by forcing MalT to bind to sites 3′, 4′, and 5′, which are 3 bp closer to the promoter's –10 box and therefore perfectly positioned to stimulate that promoter's activity.

Evidence for the Role of CAP in the *mal* Regulon

This hypothesis allows several predictions, which Raibaud and colleagues have tested and confirmed. First, they made a deletion mutant (*malKpΔ104*), which lacks the CAP binding sites and therefore cannot be stimulated by CAP. We would predict that this mutant would use MalT binding sites 3, 4, and 5, rather than 3, 4′, and 5′. As a consequence, it should have low promoter activity in vivo and therefore make only small amounts of *mal* products, including amylomaltase. Indeed, the mutant cells make

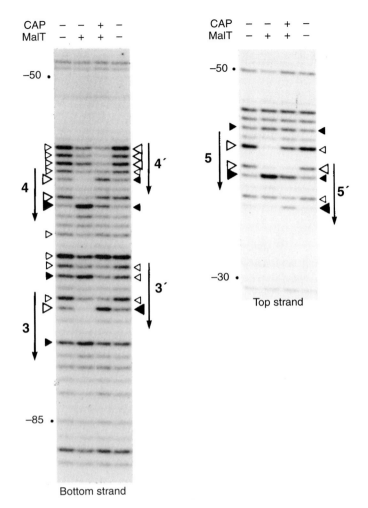

CAP − − + −
MalT − + + −

−50 •

4

3

−85 •

4′

3′

Bottom strand

CAP − − + −
MalT − + + −

−50 •

5

−30 •

Top strand

5′

Figure 7.23   DMS footprint of MalT bound to wild-type *malKp* in the presence and absence of CAP. The presence or absence of CAP and MalT are indicated at top. Open and filled arrows at the side indicate protection from, or enhanced exposure to, DMS, respectively. Note that CAP shifts the MalT footprints by 3 bp, from sites 3, 4, and 5 to sites 3′, 4′, and 5′, respectively. (*Source:* Richet et al., A new mechanism for co-activation of transcription initiation: Repositioning of an activator triggered by the binding of a second activator. *Cell* 66, f. 5, p. 1189. © Cell Press.)

less than 2 units of this enzyme, compared with 285 units made by wild-type cells.

We would also predict that further mutations to the deletion mutant that blocked binding to sites 3, 4, or 5 would force MalT to bind to the alternative sites and restore at least some of the mutant's lost activity. In fact, Raibaud and coworkers isolated six such spontaneous revertants and determined their base sequences in this control region. Figure 7.22b shows these six mutations (all point mutations in sites 3 and 4), along with the revertants' activities. In all cases, great gains in *mal* activity occurred (up to between 76 and 123 units of amylomaltase). Furthermore, even though spontaneous mutations did not occur in MalT binding site 5, the hypothesis predicts that mutations there will cause an increase in promoter activity. In accord with this prediction, when Raibaud and col-

leagues changed site 5 from GGTGGA to GTCGGA, they stimulated enzyme production up to 190 units. Finally, we would predict that deletion of three bases between site 5 and the −10 box would position sites 3, 4, and 5 perfectly with respect to the promoter and would therefore restore activity to the deletion mutant. Again, experiment verified the prediction. A site-directed mutant called *EpKpΔ3,* with 3 bp deleted, had almost full activity (220 units).

So far, all the findings accord with the hypothesis, but we have not seen direct evidence showing which binding sites are occupied in the deletion mutant or its revertants. Nor have we seen that CAP has any effect on this binding. To measure binding directly, Raibaud and coworkers used three different kinds of footprinting: DNase, DMS, and oxygen radical (Chapter 5). The DMS footprinting pattern is easiest to interpret, and Figure 7.23 shows an example, with wild-type DNA in the presence of MalT alone, or MalT plus CAP. Notice that with MalT alone, regions 3 and 4 are protected from methylation by DMS, and one base in each binding site, marked with a closed triangle, is rendered more susceptible to methylation. Now notice how these protected regions and extrasusceptible bases move up three 3 bp, to sites 3′ and 4′, respectively, in the presence of CAP. Figure 7.23b shows footprinting on the opposite strand to elucidate sites 5 and 5′, which are oriented oppositely to sites 3 and 4. Although these results are not as clear as the others, we can see that the extrasusceptible base has moved down 3 bp, from site 5 to site 5′ in the presence of MalT plus CAP. Thus, CAP really does shift binding from the less favorable sites (3, 4, and 5) to the more favorable sites (3′, 4′, and 5′) and thereby stimulates transcription.

SUMMARY CAP apparently stimulates transcription of two promoters in the *mal* regulon by shifting the binding of the transcription factor MalT from a set of binding sites that are not aligned well with the promoters to another set of sites (only 3 bp away) that are better aligned with the promoters.

7.3   The *ara* Operon

We have already mentioned that the *ara* operon of *E. coli* is another catabolite-repressible operon. It has several interesting features to compare with the *lac* operon. First, two *ara* operators exist: *araO*$_1$ and *araO*$_2$. The former regulates transcription of a control gene called *araC.* The other operator is located far upstream of the promoter it controls (P_{BAD}), between positions −265 and −294, yet it still governs transcription. Second, the CAP binding site is about 200 bp upstream of the *ara*

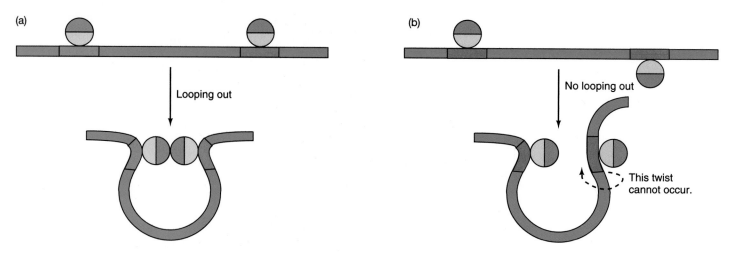

Figure 7.24   Proteins must bind to the same face of the DNA to interact by looping out the DNA. (a) Two proteins with DNA-binding domains (yellow) and protein–protein interaction domains (blue) bind to sites (red) on the same face of the DNA double helix. These proteins can interact because the intervening DNA can loop out without twisting. (b) Two proteins bind to sites on opposite sides of the DNA duplex. These proteins cannot interact because the DNA is not flexible enough to perform the twist needed to bring the protein interaction sites together.

promoter, yet CAP can still stimulate transcription. Third, the operon has another system of negative regulation, mediated by the AraC protein.

The *ara* Operon Repression Loop

How can $araO_2$ control transcription from a promoter over 250 bp downstream? The most reasonable explanation is that the DNA in between these remote sites (the operator and the promoter) loops out as illustrated in Figure 7.24a. Indeed, we have good evidence that DNA looping is occurring. Robert Lobell and Robert Schleif found that if they inserted DNA fragments containing an integral number of double-helical turns (multiples of 10.5 bp) between the operator and the promoter, the operator still functioned. However, if the inserts contained a non-integral number of helical turns (e.g., 5 or 15 bp), the operator did not function. This is consistent with the general notion that a double-stranded DNA can loop out and bring two protein-binding sites together as long as these sites are located on the same face of the double helix. However, the DNA cannot twist through the 180 degrees required to bring binding sites on opposite faces around to the same face so they can interact with each other through looping (see Figure 7.24). In this respect, DNA resembles a piece of stiff coat hanger wire: It can be bent relatively easily, but it resists twisting.

The simple model in Figure 7.24 assumes that proteins bind first to the two remote binding sites, then these proteins interact to cause the DNA looping. However, Lobell and Schleif found that the situation is more subtle than that. In fact, the *ara* control protein (**AraC**) has 3 binding sites, as illustrated in Figure 7.25a. In addition to the far upstream site, $araO_2$, araC can bind to $araO_1$, located between positions −106 and −144, and to

$araI$, which really includes two half-sites: $araI_1$ (−56 to −78) and $araI_2$ (−35 to −51), each of which can bind one monomer of AraC. The *ara* operon is also known as the *araCBAD* operon, for its four genes, *araA–D*. Three of these genes, *araB*, *A*, and *D*, encode the arabinose metabolizing enzymes; they are transcribed rightward from the promoter $araP_{BAD}$. The other gene, *araC*, encodes the control protein AraC and is transcribed leftward from the $araP_C$ promoter.

In the absence of arabinose, when no *araBAD* products are needed, AraC exerts negative control, binding to $araO_2$ and $araI_1$, looping out the DNA in between and repressing the operon (Figure 7.25b). On the other hand, when arabinose is present, it apparently changes the conformation of AraC so that it no longer binds to $araO_2$, but occupies $araI_1$ and $araI_2$ instead. This breaks the repression loop, and the operon is derepressed (Figure 7.25c). As in the *lac* operon, however, derepression isn't the whole story. Positive control mediated by CAP and cAMP also occurs, and Figure 7.25c shows this complex attached to its binding site upstream of the *araBAD* promoter.

Evidence for the *ara* Operon Repression Loop

What is the evidence for the looping model of *ara* operon repression? First, Lobell and Schleif used electrophoresis to show that AraC can cause loop formation in the absence of arabinose. Instead of the entire *E. coli* DNA, they used a small (404-bp) supercoiled circle of DNA, called a **minicircle**, that contained the $araO_2$ and *araI* sites, 160 bp apart. They then added AraC and measured looping by taking advantage of the fact that looped supercoiled DNAs have a higher electrophoretic mobility than

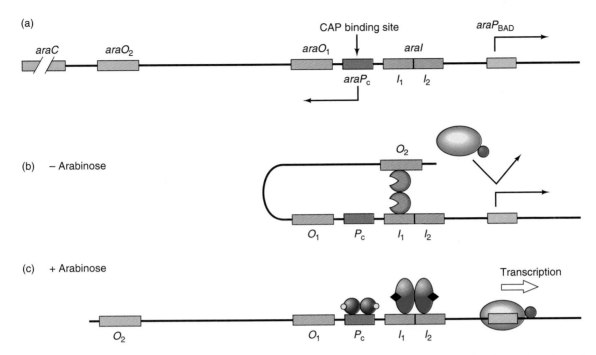

Figure 7.25 Control of the *ara* operon. (a) Map of the *ara* control region. There are four AraC binding sites (*araO₁*, *araO₂*, *araI₁*, and *araI₂*), which all lie upstream of the *ara* promoter, *araP*BAD. (b) Negative control. In the absence of arabinose, monomers of AraC (green) bind to O_2 and I_1, bending the DNA and blocking access to the promoter by RNA polymerase (red and blue). (c) Positive control. Arabinose (black) binds to AraC, changing its shape so it prefers to bind as a dimer to I_1 and I_2, and not to O_2. This opens up the promoter (salmon) to binding by RNA polymerase. If glucose is absent, the CAP–cAMP complex (purple and yellow) is in high enough concentration to occupy the CAP binding site, which stimulates polymerase binding to the promoter. Now active transcription can occur.

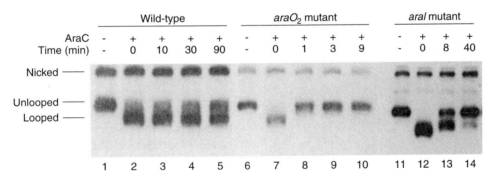

Figure 7.26 Effects of mutations in *araO₂* and *araI* on the stability of looped complexes with AraC. Lobell and Schleif prepared labeled minicircles (small DNA circles) containing either wild-type or mutant AraC binding sites, as indicated at top. Then they added AraC to form a complex with the labeled DNA. Next they added an excess of unlabeled DNA containing an *araI* site as a competitor, for various lengths of time. Finally they electrophoresed the protein–DNA complexes to see whether they were still in looped or unlooped form. The looped DNA was more supercoiled than the unlooped DNA, so it migrated faster. The wild-type DNA remained in a looped complex even after 90 min in the presence of the competitor. By contrast, dissociation of AraC from the mutant DNAs, and therefore loss of the looped complex, occurred much faster. It lasted less than 1 min with the *araO₂* mutant DNA and was half gone in less than 10 min with the *araI* mutant DNA. (*Source:* Lobell and Schleif, DNA looping and unlooping by AraC protein. *Science* 250 (1990), f. 2, p. 529. © AAAS.)

the same DNAs that are unlooped. Figure 7.26 shows one such assay. Comparing lanes 1 and 2, we can see that the addition of AraC causes the appearance of a new, high-mobility band that corresponds to the looped minicircle.

This experiment also shows that the stability of the loop depends on binding of AraC to both *araO₂* and *araI*. Lobell and Schleif made looped complexes with a wild-type minicircle, with a minicircle containing a mutant *araO₂* site, and with a minicircle containing mutations in both *araI* sites. They then added an excess of unlabeled wild-type minicircles and observed the decay of each of the looped complexes. Lanes 3–5 show only about 50% conversion of the looped to unlooped wild-type minicircle in 90 min. Thus, the half-time of dissociation of the wild-type looped complex is about 100 min. In contrast, the *araO₂* mutant minicircle's conversion from

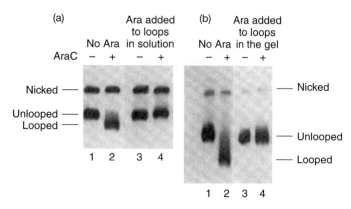

(a)

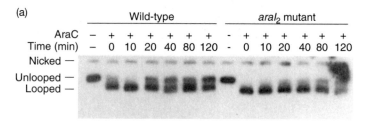

(a)

Figure 7.27   **Arabinose breaks the loop between *araO₂* and *araI*.** (a) Lobell and Schleif added arabinose to preformed loops before electrophoresis. In the absence of arabinose, AraC formed a DNA loop (lane 2). In the presence of arabinose, the loop formed with AraC was broken (lane 4). (b) This time the investigators added arabinose to the gel after electrophoresis started. Again, in the absence of arabinose, looping occurred (lane 2). However, in the presence of arabinose, the loop was broken (lane 4). The designation Ara at top refers to arabinose. (*Source:* Lobell and Schleif, *Science* 250 (1990), f. 4, p. 530. © AAAS.)

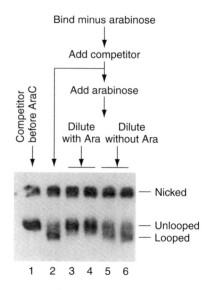

Figure 7.28   **Reversibility of loop breakage by arabinose.** Lobell and Schleif formed loops with AraC and labeled DNA, then added excess unlabeled competitor DNA, then arabinose, then diluted with buffer, or with buffer containing arabinose. The flow diagram at top shows these steps. Lane 1, competitor added before AraC, no looped complex detectable; lane 2, competitor added after AraC, looped complex is stable; lanes 3 and 4, arabinose as well as competitor added, complex is unstable and loops break; lanes 5 and 6, buffer without arabinose added to dilute arabinose, looped complex reforms. (*Source:* Lobell and Schleif, *Science* 250 (1990), f. 5, p. 530. © AAAS.)

Figure 7.29   **Contact with *araI₂* in the unlooped, but not in the looped complex.** (a) Lobell and Schleif assayed the dissociation rates of the looped complex as in Figure 7.26, with wild-type and *araI₂* mutant DNA. No difference between the wild-type and mutant DNA could be detected, showing no contact between AraC and *araI₂* in the looped state. (b) To assay binding of AraC to *araI₂* in the unlooped state, the investigators bound AraC to labeled wild-type or *araI₂* mutant DNA, broke the loop with competitor DNA and arabinose for 3 or 15 min, as indicated at top, linearized the DNA with *Hind*III, and then electrophoresed the products. Free, linear DNA can be distinguished from linear DNA bound to AraC by its faster migration rate. The wild-type DNA remained bound to AraC in the unlooped state, but the *araI₂* mutant DNA did not. Thus, *araI₂* contacts AraC in the unlooped state, but not when the loop forms. (*Source:* Lobell and Schleif, *Science* 250 (1990), f. 7, p. 531. © AAAS.)

that arabinose added to looped minicircles immediately before electrophoresis eliminates the band corresponding to the looped DNA. Figure 7.27 illustrates this phenomenon. Even though the loop was broken, Lobell and Schleif showed that it could re-form if arabinose was removed. They used arabinose to prevent looping, then diluted the DNA into buffer containing excess competitor DNA, either with or without arabinose. The buffer with arabinose maintained the broken loop, but the buffer without arabinose diluted the sugar to such an extent that the loop could re-form. As usual, loop formation was measured by electrophoresis; Figure 7.28 shows the results.

This experiment has an important corollary: A single dimer of AraC is sufficient for loop formation. We know this because the initial AraC–DNA complex was formed in the presence of a large excess of unlabeled competitor DNA and in the presence of arabinose, which ensures that a single dimer of AraC binds to *araI*, but not to *araO₂*. The loops were then formed in the presence of excess competitor DNA, which prevents any free AraC dimers from binding to *araO₂*. Thus, the

looped to unlooped took less than 1 min (compare lanes 7 and 8). The *araI* mutant's half-time of loop breakage is also short—less than 10 min. Thus, both *araO₂* and *araI* are involved in looping by AraC because mutations in either one greatly weaken the DNA loop.

Next, Lobell and Schleif demonstrated that arabinose breaks the repression loop. They did this by showing

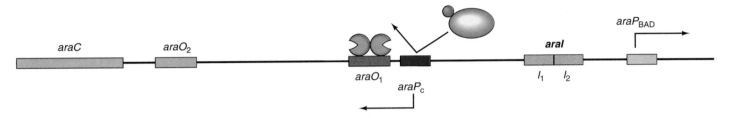

Figure 7.30   Autoregulation of *araC*. AraC (green) binds to *araO₁* and prevents transcription leftward from P_C through the *araC* gene. This can presumably happen whether or not arabinose is bound to AraC, that is, with the control region either unlooped or looped.

original dimer of AraC, still bound to *araI*, must cause looping by binding also to *araO₂*. By labeling AraC with ^{35}S and DNA with ^{32}P, and measuring the content of each radioisotope in the looped complex, Lobell and Schleif confirmed that the stoichiometry of AraC binding is one dimer of AraC per DNA minicircle.

What happens to the AraC monomer bound to *araO₂* when the loop opens up? Apparently it binds to *araI₂*. To demonstrate this, Lobell and Schleif first showed by **methylation interference** that AraC contacts *araI₁*, but not *araI₂*, in the looped state. The strategy was to partially methylate the minicircle DNA, bind AraC to loop the DNA, separate looped from unlooped DNA by electrophoresis, and then break the looped and unlooped DNAs at their methylated sites. Those sites that are important for looping will be unmethylated in the looped DNA, but methylated in the unlooped DNA. Indeed, two *araI₁* bases were heavily methylated in the unlooped DNA, but only lightly methylated in the looped DNA. In contrast, no *araI₂* bases showed this behavior. Thus, it appears that AraC does not contact *araI₂* in the looped state.

Lobell and Schleif confirmed this conclusion by showing that mutations in *araI₂* have no effect on AraC binding in the looped state, but have a strong effect on binding in the unlooped state. Figure 7.29 shows this effect. The investigators bound AraC to wild-type and *araI₂* mutant minicircles to form a loop, then used arabinose to break the loop. Then they linearized the minicircles with *Hin*dIII and electrophoresed them to see whether AraC was still bound. If so, the linear DNA band would be retarded. Thus, this is a version of the gel mobility shift assay (Chapter 5). As Figure 7.29 shows, AraC did bind to the wild-type DNA after loop breaking, but it did not bind under similar conditions to the *araI₂* mutant DNA. We infer that *araI₂* is necessary for AraC binding in the unlooped state and is therefore contacted by AraC under these conditions.

These data suggest the model of AraC–DNA interaction depicted in Figure 7.25b and c. A dimer of AraC causes looping by simultaneously interacting with *araI₁* and *araO₂*. Arabinose breaks the loop by changing the conformation of AraC so the protein loses its affinity for *araO₂* and binds instead to *araI₂*.

Autoregulation of *araC*

So far, we have only mentioned a role for *araO₁*. It does not take part in repression of *araBAD* transcription; instead it allows AraC to regulate its own synthesis. Figure 7.30 shows the relative positions of *araC*, P_C, and *araO₁*. The *araC* gene is transcribed from P_C in the leftward direction, which puts *araO₁* in a position to control this transcription. As the level of AraC rises, it binds to *araO₁* and inhibits leftward transcription, thus preventing an accumulation of too much repressor. This kind of mechanism, where a protein controls its own synthesis, is called **autoregulation.**

SUMMARY The *ara* operon is controlled by the AraC protein. AraC represses the operon by looping out the DNA between two sites, *araO₂* and *araI₁*, that are 210 bp apart. Arabinose can derepress the operon by causing AraC to loosen its attachment to *araO₂* and to bind to *araI₂* instead. This breaks the loop and allows transcription of the operon. CAP and cAMP further stimulate transcription by binding to a site upstream of *araI*. AraC controls its own synthesis by binding to *araO₁* and preventing leftward transcription of the *araC* gene.

7.4   The *trp* Operon

The *E. coli* **trp** (pronounced "trip") operon contains the genes for the enzymes that the bacterium needs to make the amino acid tryptophan. Like the *lac* operon, it is subject to negative control by a repressor. However, there is a fundamental difference. The *lac* operon codes for **catabolic** enzymes—those that break down a substance. Such operons tend to be turned on by the presence of that substance, lactose in this case. The *trp* operon, on the other hand, codes for **anabolic** enzymes—those that build up a substance. Such operons are generally turned off by that substance. When the tryptophan concentration is high, the products of the *trp* operon are not needed any longer, and we would expect the *trp* operon to be repressed. That is what happens. The *trp* operon also exhibits an extra level of control, called attenuation, not seen in the *lac* operon.

(a)   Low tryptophan: no repression

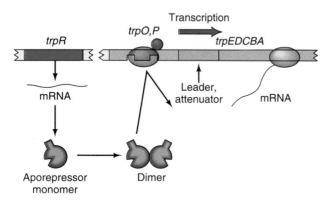

(b)   High tryptophan: repression

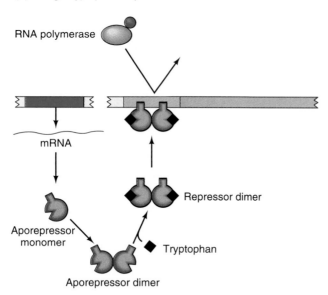

Figure 7.31   Negative control of the *trp* operon. (a) Derepression. RNA polymerase (red and blue) binds to the *trp* promoter and begins transcribing the structural genes (*trpE, D, C, B,* and *A*). Without tryptophan, the aporepressor (green) cannot bind to the operator. (b) Repression. Tryptophan, the corepressor (black), binds to the inactive aporepressor, changing it to repressor, with the proper shape for binding successfully to the *trp* operator. This prevents RNA polymerase from binding to the promoter, so no transcription occurs.

Tryptophan's Role in Negative Control of the *trp* Operon

Figure 7.31 shows an outline of the structure of the *trp* operon. Five genes code for the polypeptides in the enzymes that convert a tryptophan precursor, chorismic acid, to tryptophan. In the *lac* operon, the promoter and operator precede the genes, and the same is true in the *trp* operon. However, the *trp* operator lies wholly within the *trp* promoter, whereas the two loci are merely adjacent in the *lac* operon.

In the negative control of the *lac* operon, the cell senses the presence of lactose by the appearance of tiny amounts of its rearranged product, allolactose. In effect, this inducer causes the repressor to fall off the *lac* operator and derepresses the operon. In the case of the *trp* operon, a plentiful supply of tryptophan means that the cell does not need to spend any more energy making this amino acid. In other words, a high tryptophan concentration is a signal to turn off the operon.

How does the cell sense the presence of tryptophan? In essence, tryptophan helps the *trp* repressor bind to its operator. Here is how that occurs: In the absence of tryptophan, no *trp* repressor exists—only an inactive protein called the **aporepressor**. When the aporepressor binds tryptophan, it changes to a conformation with a much higher affinity for the *trp* operator (Figure 7.31b). This is another allosteric transition like the one we encountered in our discussion of the *lac* repressor. The combination of aporepressor plus tryptophan is the ***trp* repressor;** therefore, tryptophan is called a **corepressor**. When the cellular concentration of tryptophan is high, plenty of corepressor is available to bind and form the active *trp* repressor. Thus, the operon is repressed. When the tryptophan level in the cell falls, the amino acid dissociates from the aporepressor, causing it to shift back to the inactive conformation; the repressor–operator complex is thus broken, and the operon is derepressed. In Chapter 9, we will examine the nature of the conformational shift in the aporepressor that occurs on binding tryptophan and see why this is so important in operator binding.

SUMMARY The negative control of the *trp* operon is, in a sense, the mirror image of the negative control of the *lac* operon. The *lac* operon responds to an inducer that causes the repressor to dissociate from the operator, derepressing the operon. The *trp* operon responds to a repressor that includes a corepressor, tryptophan, which signals the cell that it has made enough of this amino acid. The corepressor binds to the aporepressor, changing its conformation so it can bind better to the *trp* operator, thereby repressing the operon.

Control of the *trp* Operon by Attenuation

In addition to the standard, negative control scheme we have just described, the *trp* operon employs another mechanism of control called **attenuation**. Why is this extra control needed? The answer probably lies in the fact that repression of the *trp* operon is weak—much weaker, for example, than that of the *lac* operon. Thus, considerable transcription of the *trp* operon can occur even in the presence of repressor. In fact, in attenuator mutants where only repression can

(a)   Low tryptophan: transcription of *trp* structural genes

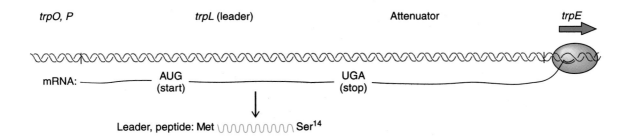

(b)   High tryptophan: attenuation, premature termination

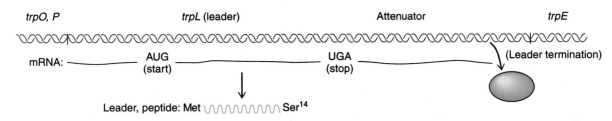

Figure 7.32   Attenuation in the *trp* operon. (a) In the presence of low tryptophan concentration, the RNA polymerase (red) reads through the attenuator, so the structural genes are transcribed. **(b)** In the presence of high tryptophan, the attenuator causes premature termination of transcription, so the structural genes are not transcribed.

operate, the fully repressed level of transcription is only 70-fold lower than the fully derepressed level. The attenuation system permits another 10-fold control over the operon's activity. Thus, the combination of repression and attenuation controls the operon over a 700-fold range, from fully inactive to fully active: (70-fold [repression] × 10-fold [attenuation] = 700-fold). This is valuable because synthesis of tryptophan requires considerable energy.

Here is how attenuation works. Figure 7.31 lists two loci, the *trp* **leader** and the *trp* **attenuator,** in between the operator and the first gene, *trpE*. Figure 7.32 gives a closer view of the leader–attenuator, whose purpose is to attenuate, or weaken, transcription of the operon when tryptophan is relatively abundant. The attenuator operates by causing premature termination of transcription. In other words, transcription that gets started, even though the tryptophan concentration is high, stands a 90% chance of terminating in the attenuator region.

The reason for this premature termination is that the attenuator contains a transcription stop signal (terminator): an inverted repeat followed by a string of eight A-T pairs in a row. Because of the inverted repeat, the transcript of this region would tend to engage in intramolecular base pairing, forming the "hairpin" shown in Figure 7.33. As we learned in Chapter 6, a hairpin followed by a string of U's in a transcript destabilizes the binding between the transcript and the DNA and thus causes termination.

> **SUMMARY** Attenuation imposes an extra level of control on an operon, over and above the repressor–operator system. It operates by causing premature termination of transcription of the operon when the operon's products are abundant.

Defeating Attenuation

When tryptophan is scarce, the *trp* operon must be activated, and that means that the cell must somehow override attenuation. Charles Yanofsky proposed this hypothesis: Something preventing the hairpin from forming would destroy the termination signal, so attenuation would break down and transcription would proceed. A look at Figure 7.34a reveals not just one potential hairpin near the end of the leader transcript, but two. However, the terminator includes only the second hairpin, which is adjacent to the string of U's in the transcript. Furthermore, the two-hairpin arrangement is not the only one available; another, containing only one hairpin, is shown in Figure 7.34c. Note that this alternative hairpin contains elements from each of the two hairpins in the first structure. Figure 7.34b illustrates this concept by labeling the sides of the original two hairpins 1, 2, 3, and 4. If the first of the original hairpins involves elements 1 and 2 and the second involves 3 and 4,

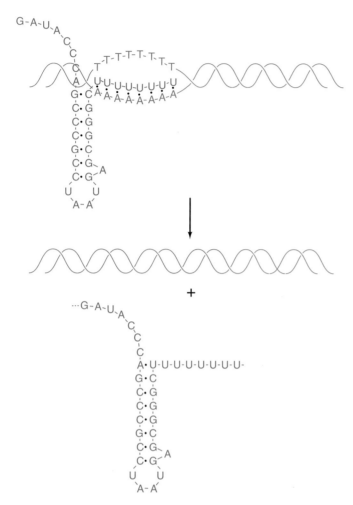

Figure 7.33 Mechanism of attenuation. The transcript of the leader–attenuator region (blue) contains a string of U's and an inverted repeat, so it can form a hairpin structure. When the RNA polymerase pauses at the string of U's, the hairpin forms, then the transcript is released, so termination occurs before transcription can reach the *trp* genes.

then the alternative hairpin in the second structure involves 2 and 3. This means that the formation of the alternative hairpin (Figure 7.34c) precludes formation of the other two hairpins, including the one adjacent to the string of U's, which is a necessary part of the terminator (Figure 7.34a).

The two-hairpin structure involves more base pairs than the alternative, one-hairpin structure; therefore, it is more stable. So why should the less stable structure ever form? A clue comes from the base sequence of the leader region shown in Figure 7.35. One very striking feature of this sequence is that two codons for tryptophan (UGG) occur in a row in element 1 of the first potential hairpin. This may not seem unusual, but tryptophan is a rare amino acid in most proteins; it is found on average only once in every 100 amino acids. So the chance of finding two tryptophan codons in a row *anywhere* is quite small, and the fact that they are found in the *trp* operon is very suspicious.

In bacteria, transcription and translation occur simultaneously. Thus, as soon as the *trp* leader region is transcribed, ribosomes begin translating this emerging mRNA. Think about what would happen to a ribosome trying to translate the *trp* leader under conditions of tryptophan starvation (Figure 7.36a). Tryptophan is in short supply, and here are two demands in a row for that very amino acid. In all likelihood, the ribosome will not be able to satisfy those demands immediately, so it will pause at one of the tryptophan codons. And where does that put the stalled ribosome? Right on element 1, which should be participating in formation of the first hairpin. The bulky ribosome clinging to this RNA site effectively prevents its pairing with element 2, which frees 2 to pair with 3, forming the one-hairpin alternative structure. Because the second hairpin cannot form, transcription does not terminate and attenuation has been defeated. This is desirable, of course, because when tryptophan is scarce, the *trp* operon should be transcribed.

Notice that this mechanism involves a coupling of transcription and translation, where the latter affects the former. It would not work in eukaryotes, where transcription and translation take place in separate compartments. It also depends on transcription and translation occurring at about the same rate. If RNA polymerase outran the ribosome, it might pass through the attenuator region before the ribosome had a chance to stall at the tryptophan codons.

You may be wondering how the polycistronic mRNA made from the *trp* operon can be translated if ribosomes are stalled in the leader at the very beginning. The answer is that each of the genes represented on the mRNA has its own translation start signal (AUG). Ribosomes recognize each of these independently, so translation of the *trp* leader does not affect translation of the *trp* genes. For example, translation of *trpE* can occur, even if translation of the *trp* leader is stalled.

On the other hand, consider a ribosome translating the leader transcript under conditions of abundant tryptophan (Figure 7.36b). Now the dual tryptophan codons present no barrier to translation, so the ribosome continues through element 1 until it reaches the stop signal (UGA) between elements 1 and 2 and falls off. With no ribosome to interfere, the two hairpins can form, completing the transcription termination signal that halts transcription before it reaches the *trp* genes. Thus, the attenuation system responds to the presence of adequate tryptophan and prevents wasteful synthesis of enzymes to make still more tryptophan.

Other *E. coli* operons besides *trp* use the attenuation mechanism. The most dramatic known use of consecutive codons to stall a ribosome occurs in the *E. coli* histidine (*his*) operon, in which the leader region contains seven histidine codons in a row!

Bacillus subtilis uses a different mechanism to establish and override attenuation in the *trp* operon. When tryptophan is plentiful, it binds to a protein called the *trp* **RNA-binding attenuation protein** (**TRAP**). This binding activates TRAP, so it can bind to the leader RNA and cause it to form a terminator. This attenuates transcription of the operon. On the other hand, when tryptophan

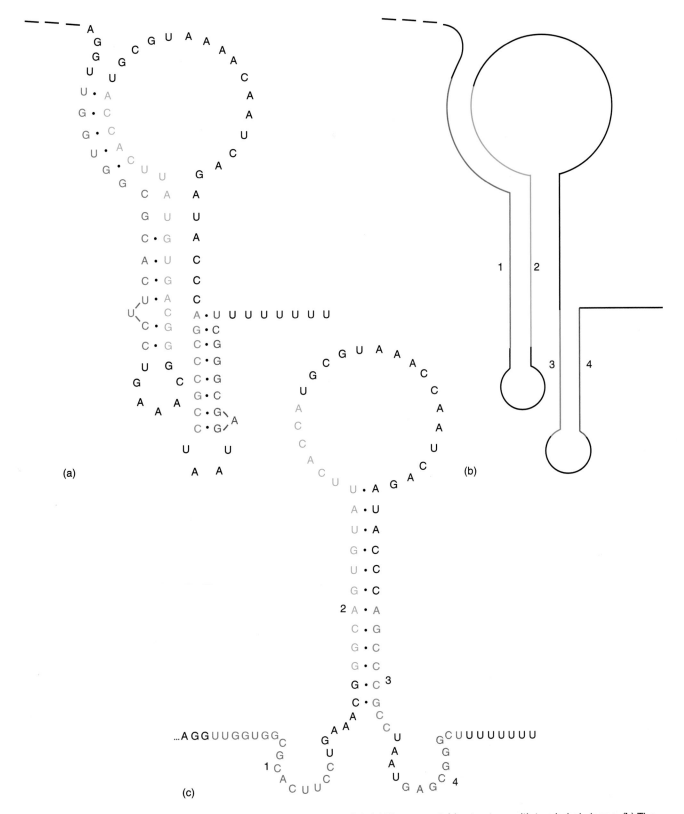

(a)

(b)

(c)

Figure 7.34 Two structures available to the leader–attenuator transcript. (a) The more stable structure, with two hairpin loops. **(b)** The sequences that participate in base pairing to form the hairpins are numbered and colored, so they can be identified in **(c)**, the less stable structure, containing only one hairpin loop. The curved shape of the RNA at the bottom is not meant to suggest a shape for the molecule, it is drawn this way simply to save space.

Met Lys Ala Ile Phe Val Leu Lys Gly Trp Trp Arg Thr Ser Stop
pppA---AUGAAAGCAAUUUUCGUACUGAAAGGUUGGUGGCGCACUUCCUGA

Figure 7.35 Sequence of the leader. The sequence of part of the leader transcript is presented, along with the leader peptide it encodes. Note the two Trp codons in tandem (blue).

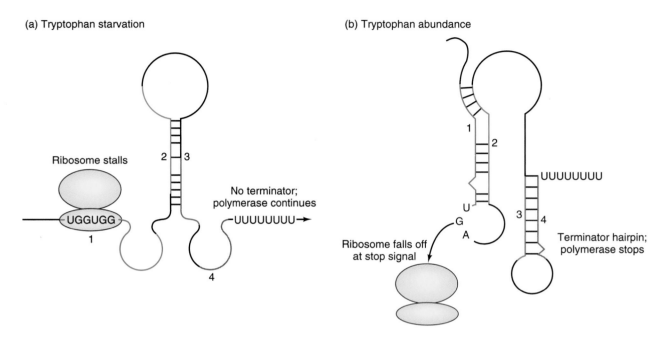

(a) Tryptophan starvation

(b) Tryptophan abundance

Ribosome stalls

No terminator;
polymerase continues
UUUUUUUU→

UGGUGG

2   3

4

1

UUUUUUUU

Ribosome falls off
at stop signal

Terminator hairpin;
polymerase stops

Figure 7.36 Overriding attenuation. (a) Under conditions of tryptophan starvation, the ribosome (yellow) stalls at the Trp codons and prevents element 1 (red) from pairing with element 2 (blue). This forces the one-hairpin structure, which lacks a terminator, to form, so no attenuation should take place. **(b)** Under conditions of tryptophan abundance, the ribosome reads through the two tryptophan codons and falls off at the translation stop signal (UGA), so it cannot interfere with base pairing in the leader transcript. The more stable, two-hairpin structure forms; this structure contains a terminator, so attenuation occurs.

is scarce, it does not bind to TRAP, so TRAP cannot bind to the *trp* leader. Without TRAP, the leader does not form the terminator, so attenuation is defeated.

SUMMARY Attenuation operates in the *E. coli trp* operon as long as tryptophan is plentiful. When the supply of this amino acid is restricted, ribosomes stall at the tandem tryptophan codons in the *trp* leader. Because the *trp* leader is being synthesized just as stalling occurs, the stalled ribosome will influence the way this RNA folds. In particular, it prevents the formation of a hairpin, which is part of the transcription termination signal that causes attenuation. Therefore, when tryptophan is scarce, attenuation is defeated and the operon remains active. This means that the control exerted by attenuation responds to tryptophan levels, just as repression does. *B. subtilis* uses a tryptophan- and RNA-binding protein called TRAP to force the *trp* leader into a terminator structure. When cells are starved for tryptophan, TRAP cannot bind to the *trp* leader, so no terminator forms.

SUMMARY

Lactose metabolism in *E. coli* is carried out by two proteins, β-galactosidase and galactoside permease. The genes for these two, and one additional enzyme, are clustered together and transcribed together from one promoter, yielding a polycistronic message. These functionally related genes are therefore controlled together.

Control of the *lac* operon occurs by both positive and negative control mechanisms. Negative control appears to occur as follows: The operon is turned off as long as repressor binds to the operator, because the repressor prevents RNA polymerase from binding to the promoter to transcribe the three *lac* genes. When the supply of glucose is exhausted and lactose is available, the few molecules of *lac* operon enzymes produce a few molecules of allolactose from the lactose. The allolactose acts as an inducer by binding to the repressor and causing a conformational shift that encourages dissociation from the operator. With the repressor removed, RNA polymerase is free to transcribe the three *lac* genes. A combination of genetic and biochemical experiments revealed the two key elements of negative control of the *lac* operon: the operator and the repressor. DNA sequencing revealed the presence of two auxiliary *lac* operators: one upstream, and one downstream of the major operator. All three are required for optimal repression.

Positive control of the *lac* operon, and certain other inducible operons that code for sugar-metabolizing enzymes, is mediated by a factor called catabolite activator protein (CAP), which, in conjunction with cyclic-AMP (cAMP), stimulates transcription. Because cAMP concentration is depressed by glucose, this sugar prevents positive control from operating. Thus, the *lac* operon is activated only when glucose concentration is low and a corresponding need arises to metabolize an alternative energy source. The CAP–cAMP complex stimulates expression of the *lac* operon by binding to an activator site adjacent to the promoter. CAP–cAMP binding helps RNA polymerase form an open-promoter complex. It does this by recruiting polymerase to form a closed promoter complex, which then converts to an open promoter complex. Recruitment of polymerase occurs through protein–protein interactions, by bending the DNA, or by a combination of these phenomena.

CAP stimulates transcription of two promoters in the *mal* regulon by shifting the binding of the transcription factor MalT from a set of binding sites that are not aligned well with the promoters to another set of sites (only three base pairs away) that are better aligned with the promoters.

The *ara* operon is controlled by the AraC protein. AraC represses the operon by looping out the DNA between two sites, $araO_2$ and $araI_1$, that are 210 bp apart. Arabinose can induce the operon by causing AraC to loosen its attachment to $araO_2$ and to bind to $araI_1$ and $araI_2$ instead. This breaks the loop and allows transcription of the operon. CAP and cAMP further stimulate transcription by binding to a site upstream of *araI*. AraC controls its own synthesis by binding to $araO_1$ and preventing leftward transcription of the *araC* gene.

The *trp* operon responds to a repressor that includes a corepressor, tryptophan, which signals the cell that it has made enough of this amino acid. The corepressor binds to the aporepressor, changing its conformation so it can bind better to the *trp* operator, thereby repressing the operon.

Attenuation operates in the *E. coli trp* operon as long as tryptophan is plentiful. When the supply of this amino acid is restricted, ribosomes stall at the tandem tryptophan codons in the *trp* leader. Because the *trp* leader is being synthesized just as this is taking place, the stalled ribosome will influence the way this RNA folds. In particular, it prevents the formation of a hairpin, which is part of the transcription termination signal that causes attenuation. When tryptophan is scarce, attenuation is therefore defeated and the operon remains active. This means that the control exerted by attenuation responds to

tryptophan levels, just as repression does. *B. subtilis* uses a tryptophan- and RNA-binding protein called TRAP to force the *trp* leader into a terminator structure. When cells are starved for tryptophan, TRAP cannot bind to the *trp* leader, so no terminator forms.

REVIEW QUESTIONS

1. Draw a growth curve of *E. coli* cells growing on a mixture of glucose and lactose. What is happening in each part of the curve?

2. Draw diagrams of the *lac* operon that illustrate (a) negative control and (b) positive control.

3. What are the functions of β-galactosidase and galactoside permease?

4. Why are negative and positive control of the *lac* operon important to the energy efficiency of *E. coli* cells?

5. Describe and give the results of an experiment that shows that the *lac* operator is the site of repressor binding.

6. Describe and give the results of an experiment that shows that polymerase can bind to the *lac* promoter, even if repressor is already bound at the operator.

7. Describe and give the results of an experiment that shows that *lac* repressor prevents RNA polymerase from binding to the *lac* promoter.

8. How do we know that all three *lac* operators are required for full repression? What are the relative effects of removing each or both of the auxiliary operators?

9. Describe and give the results of an experiment that shows the relative levels of stimulation of β-galactosidase synthesis by cAMP, using wild-type and mutant extracts, in which the mutation reduces the affinity of CAP for cAMP.

10. Present a hypothesis for activation of *lac* transcription by CAP–cAMP. Include the C-terminal domain of the polymerase α-subunit in the hypothesis.

11. Describe and give the results of an electrophoresis experiment that shows that binding of CAP–cAMP bends the *lac* promoter region.

12. What other data support DNA bending in response to CAP–cAMP binding?

13. Present a model to explain how DNA bending could stimulate transcription.

14. Present a model for stimulation of the *mal* regulon by CAP.

15. Present evidence for the model in question 14.

16. Explain the fact that insertion of an integral number of DNA helical turns (multiples of 10.5 bp) between the $araO_2$ and *araI* sites in the *araBAD* operon permits repression by AraC, but insertion of a nonintegral number of helical turns prevents repression. Illustrate this phenomenon with diagrams.

17. Use a diagram to illustrate how arabinose can relieve repression of the *araBAD* operon. Show where AraC is located (a) in the absence of arabinose, and (b) in the presence of arabinose.

18. Describe and give the results of an experiment that shows that arabinose can break the repression loop formed by AraC.

19. Describe and give the results of an experiment that shows that both $araO_2$ and $araI$ are involved in forming the repression loop.

20. Describe and give the results of an experiment that shows that removing arabinose from broken repression loops allows them to re-form.

21. Describe and give the results of an experiment that shows that $araI_2$ is important in binding AraC when the DNA is in the unlooped, but not the looped, form.

22. Present a model to explain negative control of the *trp* operon.

23. Present a model to explain attenuation in the *trp* operon.

24. Why does translation of the *trp* leader region not simply continue into the *trp* structural genes (*trpE*, etc.)?

25. How is *trp* attenuation overridden when tryptophan is scarce?

SUGGESTED READINGS

General References and Reviews

Beckwith, J.R. and D. Zipser, eds. 1970. *The Lactose Operon.* Plainview, NY: Cold Spring Harbor Laboratory Press.

Corwin, H.O. and J.B. Jenkins. *Conceptual Foundations of Genetics: Selected Readings.* 1976. Boston: Houghton Mifflin Co.

Jacob, F. 1966. Genetics of the bacterial cell (Nobel lecture). *Science* 152:1470–78.

Lilley, D.M.J. 1991. When the CAP fits bent DNA. *Nature* 354:359–60.

Matthews, K.S. 1996. The whole lactose repressor. *Science* 271:1245–46.

Miller, J.H. and W.S. Reznikoff, eds. 1978. *The operon.* Plainview, NY: Cold Spring Harbor Laboratory Press.

Monod, J. 1966. From enzymatic adaptation to allosteric transitions (Nobel lecture). Science 154:475–83.

Ptashne, M. 1989. How gene activators work. *Scientific American* 260 (January):24–31.

Ptashne, M. 1992. *A Genetic Switch.* Cambridge, MA: Cell Press.

Ptashne, M. and W. Gilbert. 1970. Genetic repressors. *Scientific American* 222 (June):36–44.

Rojo, F. 1999. Repression of transcription initiation in bacteria. *Journal of Bacteriology* 181:2987–91.

Schleif, R. 1987. DNA binding proteins. *Science* 241:1182–87.

Research Articles

Adhya, S. and S. Garges. 1990. Positive control. *Journal of Biological Chemistry* 265:10797–800.

Busby, S. and R.H. Ebright. 1994. Promoter structure, promoter recognition, and transcription activation in prokaryotes. *Cell* 79:743–46.

Chen, B., B. deCrombrugge, W.B. Anderson, M.E. Gottesman, I. Pastan, and R.L. Perlman. 1971. On the mechanism of action of *lac* repressor. *Nature New Biology* 233:67–70.

Chen, Y., Y.W. Ebright, and R.H. Ebright. 1994. Identification of the target of a transcription activator protein by protein–protein photocrosslinking. *Science* 265:90–92.

Emmer, M., B. deCrombrugge, I. Pastan, and R. Perlman. 1970. Cyclic AMP receptor protein of *E. coli:* Its role in the synthesis of inducible enzymes. *Proceedings of the National Academy of Sciences USA* 66:480–87.

Gartenberg, M.R. and D.M. Crothers. 1991. Synthetic DNA bending sequences increase the rate of in vitro transcription initiation at the *Escherichia coli lac* promoter. *Journal of Molecular Biology* 219:217–30.

Gilbert, W. and B. Müller-Hill. 1966. Isolation of the *lac* repressor. *Proceedings of the National Academy of Sciences USA* 56:1891–98.

Igarashi, K. and A. Ishihama. 1991. Bipartite functional map of the *E. coli* RNA polymerase α subunit: Involvement of the C-terminal region in transcription activation by cAMP–CRP. *Cell* 65:1015–22.

Jacob, F. and J. Monod. 1961. Genetic regulatory mechanisms in the synthesis of proteins. *Journal of Molecular Biology* 3:318–56.

Krummel, B. and M.J. Chamberlin. 1989. RNA chain initiation by *Escherichia coli* RNA polymerase. Structural transitions of the enzyme in the early ternary complexes. *Biochemistry* 28:7829–42.

Lee, J. and A. Goldfarb. 1991. *lac* repressor acts by modifying the initial transcribing complex so that it cannot leave the promoter. *Cell* 66:793–98.

Lewis, M., G. Chang, N.C. Horton, M.A. Kercher, H.C. Pace, M.A. Schumacher, R.G. Brennan, and P. Lu. 1996. Crystal structure of the lactose operon repressor and its complexes with DNA and inducer. *Science* 271:1247–54.

Lobell, R.B. and R.F. Schleif. 1991. DNA looping and unlooping by AraC protein. *Science* 250:528–32.

Malan, T.P. and W.R. McClure. 1984. Dual promoter control of the *Escherichia coli* lactose operon. *Cell* 39:173–80.

Oehler, S., E.R. Eismann, H. Krämer, and B. Müller-Hill. 1990. The three operators of the *lac* operon cooperate in repression. *The EMBO Journal* 9:973–79.

Richet, E., D. Vidal-Ingigliardi, and O. Raibaud. 1991. A new mechanism for coactivation of transcription initiation. Repositioning of an activator triggered by the binding of a second activator. *Cell* 66:1185–95.

Schlax, P.J., Capp, M.W., and M.T. Record, Jr. 1995. Inhibition of transcription initiation by *lac* repressor. *Journal of Molecular Biology* 245:331–50.

Schultz, S.C., G.C. Shields, and T.A. Steitz. 1991. Crystal structure of a CAP–DNA complex: The DNA is bent by 90°. *Science* 253:1001–7.

Straney, S. and D.M. Crothers. 1987. Lac repressor is a transient gene-activating protein. *Cell* 51:699–707.

Wu, H.-M., and D.M. Crothers. 1984. The locus of sequence-directed and protein-induced DNA bending. *Nature* 308:509–13.

Yanofsky, C. 1981. Attenuation in the control of expression of bacterial operons. *Nature* 289:751–58.

Zubay, G., D. Schwartz, and J. Beckwith. 1970. Mechanism of activation of catabolite-sensitive genes: A positive control system. *Proceedings of the National Academy of Sciences USA* 66:104–10.

CHAPTER 8

Major Shifts in Prokaryotic Transcription

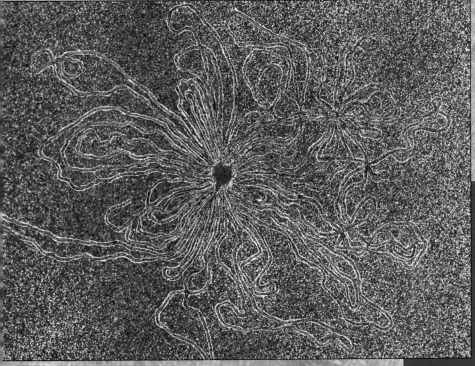

False color transmission electron micrograph of DNA released from an osmotically shocked T₄ bacteriophage (×36,000). © *Biozentrum/ SPL/Photo Researchers, Inc.*

In Chapter 7, we discussed the ways in which bacteria control the transcription of a very limited number of genes at a time. For example, when the *lac* operon is switched on, only three structural genes are activated. At other times in a bacterial cell's life more radical shifts in gene expression take place. When a phage infects a bacterium, it usually subverts the host's transcription machinery to its own use. In the process, it establishes a time-dependent, or temporal, program of transcription. In other words, the early phage genes are transcribed first, then the later genes. By the time phage T4 infection of *E. coli* reaches its late phase, essentially no more transcription of host genes takes place—only transcription of phage genes. This massive shift in specificity would be hard to explain by the operon mechanisms described in Chapter 7. Instead, it is engineered by a fundamental change in the transcription machinery—a change in RNA polymerase itself.

Another profound change in gene expression occurs during sporulation in bacteria such as *Bacillus subtilis*. Here, genes that are needed in the vegetative phase of growth are turned off, and other, sporulation-specific genes are turned on. Again, this switch is accomplished by changes in RNA polymerase. Bacteria also experience stresses such as starvation, heat shock, and lack of nitrogen, and they also respond to these by shifting their patterns of transcription.

Thus, bacteria respond to changes in their environment by global changes in transcription, and these changes in transcription are accomplished by changes in RNA polymerase. This chapter is devoted to this important aspect of gene control. We will see that the most common of these changes in RNA polymerase are actually changes in the σ-factor, which governs the specificity of transcription, but certain phages employ other schemes to alter the RNA polymerase to their own ends. Later chapters will show that eukaryotes do not have σ-factors, so they control their transcription specificity with other proteins called transcription factors.   ■

8.1   Modification of the Host RNA Polymerase

What part of RNA polymerase would be the logical candidate to change the specificity of the enzyme? We have already seen that σ is the key factor in determining specificity of T4 DNA transcription in vitro, so σ is the most reasonable answer to our question, and experiments have confirmed that σ is the correct answer. However, these experiments were not done first with the *E. coli* T4 system, but with *B. subtilis* and its phages, especially phage SPO1.

SPO1, like T4, has a large DNA genome. It has a temporal program of transcription as follows: In the first 5 min or so of infection, the early genes are expressed; next, the middle genes turn on (about 5–10 min after infection); from about the 10-min point until the end of infection, the late genes switch on. Because the phage has a large number of genes, it is not surprising that it uses a fairly elaborate mechanism to control this temporal program. Janice Pero and her colleagues were the leaders in developing the model illustrated in Figure 8.1.

The host RNA polymerase holoenzyme handles transcription of early SPO1 genes, which is analogous to the T4 model, where the earliest genes are transcribed by the host holoenzyme (Chapter 6). This arrangement is necessary because the phage does not carry its own RNA polymerase. When the phage first infects the cell, the host holoenzyme is therefore the only RNA polymerase available. The *B. subtilis* holoenzyme closely resembles the *E. coli* enzyme. Its core consists of two large (β and β′), two small (α), and one very small (ω) polypeptides; its primary σ-factor has a molecular weight of 43,000, somewhat smaller than *E. coli*'s primary σ (70,000).

One of the genes transcribed in the early phase of SPO1 infection is called gene 28. Its product, **gp28**, asso-

ciates with the host core polymerase, displacing the host σ (σ[43]). With this new, phage-encoded polypeptide in place, the RNA polymerase changes specificity. It begins transcribing the phage middle genes instead of the phage early genes and host genes. In other words, gp28 is a novel σ-factor that accomplishes two things: It diverts the host's polymerase from transcribing host genes, and it switches from early to middle phage transcription.

The switch from middle to late transcription occurs in much the same way, except that two polypeptides team up to bind to the polymerase core and change its specificity. These are **gp33** and **gp34**, the products of two

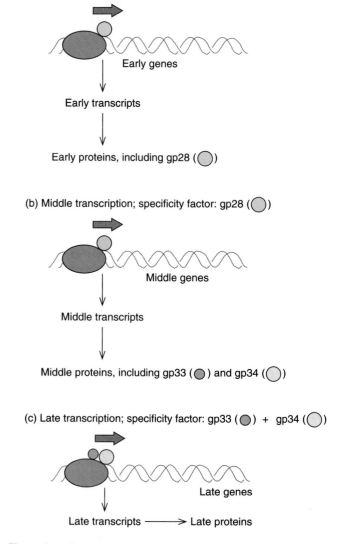

Figure 8.1   Temporal control of transcription in phage SPO1-infected *B. subtilis*. (a) Early transcription is directed by the host RNA polymerase holoenzyme, including the host σ-factor (blue); one of the early phage proteins is gp28 (green), a new σ-factor. **(b)** Middle transcription is directed by gp28, in conjunction with the host core polymerase (red); two middle phage proteins are gp33 and gp34 (purple and yellow, respectively); together, these constitute yet another σ-factor. **(c)** Late transcription depends on the host core polymerase plus gp33 and 34.

phage middle genes (genes 33 and 34, respectively). These proteins replace gp28 and direct the altered polymerase to transcribe the phage late genes in preference to the middle genes. Note that the polypeptides of the host core polymerase remain constant throughout this process; it is the progressive substitution of σ-factors that changes the specificity of the enzyme and thereby directs the transcriptional program. Of course, the changes in transcription specificity also depend on the fact that the early, middle, and late genes have promoters with different sequences. That is how they can be recognized by different σ-factors.

One striking aspect of this process is that the different σ-factors vary quite a bit in size. In particular, host σ, gp28, gp33, and gp34 have molecular weights of 43,000, 26,000, 13,000, and 24,000, respectively. Yet they are capable of associating with the core enzyme and performing a σ-like role. (Of course, gp33 and gp34 must combine forces to play this role.) In fact, even the *E. coli* σ, with a molecular weight of 70,000, can complement the *B. subtilis* core in vitro. The core polymerase apparently has a versatile σ-binding site.

How do we know that the σ-switching model is valid? Two lines of evidence, genetic and biochemical, support it. First, genetic studies have shown that mutations in gene 28 prevent the early-to-middle switch, just as we would predict if the gene 28 product is the σ-factor that turns on the middle genes. Similarly, mutations in either gene 33 or 34 prevent the middle-to-late switch, again in accord with the model.

Pero and colleagues performed the biochemical studies. First, they purified RNA polymerase from SPO1-infected cells. This purification scheme included a phosphocellulose chromatography step, which separated three forms of the polymerase. The first of the separated polymerases, enzyme A, contains the host core polymerase, including δ (Chapter 6), plus all the phage-encoded factors. The other two polymerases, B and C, were missing δ, but B contained

gp28 (called subunit IV in this study) and C contained gp33 and gp34 (called subunits VI and V, respectively, in this study). Figure 8.2 presents the subunit structures of these latter two enzymes, determined by SDS-PAGE. Without δ, these two enzymes were incapable of specific transcription. However, when Pero added δ back and assayed specificity by hybridization-competition, she found that B was specific for the delayed early phage genes, and C was specific for the late genes. Figure 8.3 shows these results.

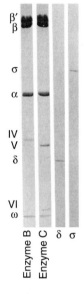

Figure 8.2 Subunit compositions of RNA polymerases in SPO1 phage-infected *B. subtilis* cells. Polymerases were separated by chromatography and subjected to SDS-PAGE to display their subunits. Enzyme B (first lane) contains the core subunits (β′, β, α, and ω), as well as subunit IV (gp28). Enzyme C (second lane) contains the core subunits plus subunits V (gp34) and VI (gp33). The last two lanes contain separated δ and σ subunits, respectively. (*Source:* Pero et al., *In vitro* transcription of a late class of phage SP01 genes. *Nature* 257 (18 Sept 1975): f. 1, p. 249 © Macmillan Magazines Ltd.)

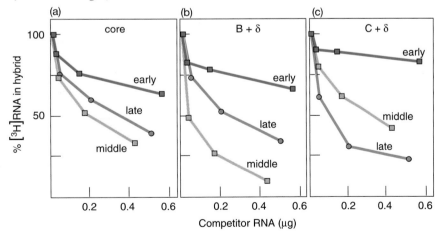

Figure 8.3 Specificities of polymerases B and C. Pero and colleagues measured polymerase specificity by transcribing SPO1 DNA in vitro with (**a**) core polymerase, (**b**) enzyme B , or (**c**) enzyme C, in the presence of [³H]UTP to label the RNA product. Next they hybridized the labeled RNA to SPO1 DNA in the presence of each of the following competitors: early SPO1 RNA (green) made in vivo in the presence of chloramphenicol (CAM); middle RNA (blue) collected from phage-infected cells at 10 min postinfection; and late RNA (red) collected from phage-infected cells 30 min postinfection. The product of the core polymerase is competed roughly equally by all three classes of RNA. On the other hand, competition for the product made by B plus δ is clearly competed best by middle RNA, and the product made by C plus δ is competed best by late RNA. These differences are not as dramatic as one might prefer, but they are easiest to see at low competitor concentration. (*Source:* Reprinted with permission from *Nature* 257:250, 1975. Copyright © 1975 Macmillan Magazines Limited.)

SUMMARY Transcription of phage SPO1 genes in infected *B. subtilis* cells proceeds according to a temporal program in which early genes are transcribed first, then middle genes, and finally late genes. This switching is directed by a set of phage-encoded σ-factors that associate with the host core RNA polymerase and change its specificity of promoter recognition from early to middle to late. The host σ is specific for the phage early genes; the phage gp28 protein switches the specificity to the middle genes; and the phage gp33 and gp34 proteins switch to late specificity.

8.2   The RNA Polymerase Encoded in Phage T7

Phage T7 belongs to a class of relatively simple *E. coli* phages that also includes T3 and φII. These have a considerably smaller genome than SPO1 and, therefore, many fewer genes. In these phages we distinguish three phases of transcription, called classes I, II, and III. (They could just as easily be called early, middle, and late, to conform to SPO1 nomenclature.) One of the five class I genes (gene 1) is necessary for class II and class III gene expression. When it is mutated, only the class I genes are transcribed. Having just learned the SPO1 story, you may be expecting to hear that gene 1 codes for a σ-factor directing the host RNA polymerase to transcribe the later phage genes. In fact, this was the conclusion reached by some workers on T7 transcription, but it was erroneous.

The gene 1 product is actually not a σ-factor but a phage-specific RNA polymerase contained in one polypeptide. This polymerase, as you might expect, transcribes the T7 phage class II and III genes specifically, leaving the class I genes completely alone. Indeed, this enzyme is unusually specific; it will transcribe the class II and III genes of phage T7 and virtually no other natural template. The switching mechanism in this phage is thus quite simple (Figure 8.4). When the phage DNA enters the host cell, the *E. coli* holoenzyme transcribes the five class I genes, including gene 1. The gene 1 product—the phage-specific RNA polymerase—then transcribes the phage class II and class III genes.

A similar polymerase has been isolated from phage T3. It is specific for T3, rather than T7 genes. In fact, T7 and T3 promoters have been engineered into cloning vectors such as pBluescript (Chapter 4). These DNAs can be transcribed in vitro by one phage polymerase or the other to produce strand-specific RNA.

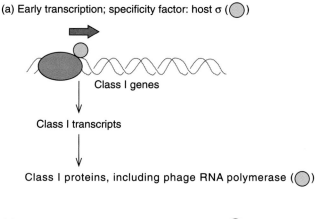

(a) Early transcription; specificity factor: host σ (○)

Class I genes

Class I transcripts

Class I proteins, including phage RNA polymerase (○)

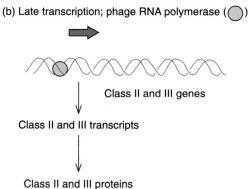

(b) Late transcription; phage RNA polymerase (○)

Class II and III genes

Class II and III transcripts

Class II and III proteins

Figure 8.4   Temporal control of transcription in phage T7-infected *E. coli*. (a) Early (class I) transcription depends on the host RNA polymerase holoenzyme, including the host σ-factor (blue); one of the early phage proteins is the T7 RNA polymerase (green). **(b)** Late (class II and III) transcription depends on the T7 RNA polymerase.

SUMMARY Phage T7, instead of coding for a new σ-factor to change the host polymerase's specificity from early to late, encodes a new RNA polymerase with absolute specificity for the later phage genes. This polymerase, composed of a single polypeptide, is a product of one of the earliest phage genes, gene 1. The temporal program in the infection by this phage is simple. The host polymerase transcribes the earliest (class I) genes, one of whose products is the phage polymerase, which then transcribes the later (class II and class III) genes.

8.3   Control of Transcription during Sporulation

We have already seen how phage SPO1 changes the specificity of its host's RNA polymerase by replacing its σ-factor. In the following section, we will show that the same kind of mechanism applies to changes in gene expression in the host itself during the process of **sporulation**. *B. subtilis* can exist indefinitely in the **vegetative,** or growth, state, as long as nutrients are available and other conditions are appropriate for growth. But, under starvation or other adverse conditions, this organism forms **endospores**—tough, dormant bodies that can survive for years until favorable conditions return (Figure 8.5).

Gene expression must change during sporulation; cells as different in morphology and metabolism as vegetative and sporulating cells must contain at least some different gene products. In fact, when *B. subtilis* cells sporulate, they activate a whole new set of sporulation-specific genes. The switch from the vegetative to the sporulating state is accomplished by a complex σ-switching scheme that turns off transcription of some vegetative genes and turns on sporulation-specific transcription.

Sporulation is a fundamental change, involving large numbers of vegetative genes turning off and sporulation genes turning on. Furthermore, the change in transcription is not absolute; some genes that are active in the vegetative state remain active during sporulation. How is the transcription-switching mechanism able to cope with these complexities?

As you might anticipate, more than one new σ-factor is involved in sporulation. In fact, at least three participate: σ^{29}, σ^{30}, and σ^{32} (also called σ^E, σ^H, and σ^C respectively), in addition to the vegetative σ^{43} (also called σ^A). These polypeptides have molecular masses indicated in their names. Thus, σ^{43} has a molecular mass of 43,000 kD. Each recognizes a different class of promoter. For example, the vegetative σ^{43} recognizes promoters that are very similar to the promoters recognized by the *E. coli* σ-factor, with a −10 box that looks something like TATAAT and a −35 box having the consensus sequence TTGACA. By contrast, the sporulation-specific factors σ^{32} (σ^C) and σ^{29} (σ^E) recognize quite different sequences.

To illustrate the techniques used to demonstrate that these are authentic σ-factors, let us consider some work by Richard Losick and his colleagues on one of them, σ^E (Figure 8.6). First, they showed that this σ-factor confers specificity for a known sporulation gene. To do this, they used polymerases containing either σ^E or σ^A to transcribe a plasmid containing a piece of *B. subtilis* DNA in vitro in the presence of labeled nucleotides. The *B. subtilis* DNA contained promoters for both vegetative and sporulation

genes. The vegetative promoter lay in a restriction fragment 3050 bp long, and the sporulation promoter was in a 770-bp restriction fragment. Losick and coworkers then hybridized the labeled RNA products to Southern blots of the template DNA. This procedure revealed the specificities of the σ-factors: If the vegetative gene was transcribed in vitro, the resulting labeled RNA would hybridize to a 3050-bp band on the Southern blot of the template DNA. On the other hand, if the sporulation gene was transcribed in vitro, the labeled RNA product would hybridize to the 770-bp band. Figure 8.7 shows that when the polymerase contained σ^A, the transcript hybridized

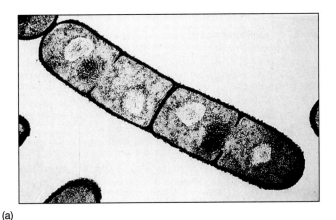

(a)

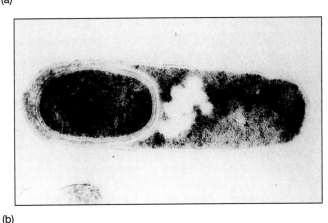

(b)

Figure 8.5 Two types of *B. subtilis* cells. (a) *B. subtilis* vegetative cells and **(b)** a sporulating cell, with an endospore developing at the left end. (*Source:* Courtesy Dr. Kenneth Bott.)

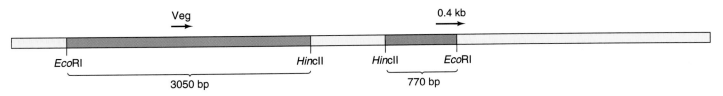

Figure 8.6 Map of part of plasmid p213. This DNA region contains two promoters: a vegetative promoter (Veg) and a sporulation promoter (0.4 kb). The former is located on a 3050-bp *Eco*RI–*Hinc*II fragment (blue); the latter is on a 770-bp fragment (red). (*Source:* Reprinted from Haldenwang et al., *Cell* 23:616, 1981. Copyright 1981, with permission from Elsevier Science.)

Figure 8.7   Specificities of σ^A and σ^E. Losick and colleagues transcribed plasmid p213 in vitro with RNA polymerase containing σ^A (lane 1) or σ^E (lane 2). Next they hybridized the labeled transcripts to Southern blots containing *Eco*RI–*Hinc*II fragments of the plasmid. As shown in Figure 8.6, this plasmid has a vegetative promoter in a 3050-bp *Eco*RI–*Hinc*II fragment, and a sporulation promoter in a 770-bp fragment. Thus, transcripts of the vegetative gene hybridized to the 3050-bp fragment, while transcripts of the sporulation gene hybridized to the 770-bp fragment. The autoradiogram in the figure shows that the σ^A-enzyme transcribed only the vegetative gene, but the σ^E-enzyme transcribed both the vegetative and sporulation genes. (*Source:* Haldenwang et al., A sporulation-induced sigma-like regulatory protein from *B. subtilis. Cell* 23 (Feb 1981), f. 4, p. 618. Reprinted by permission of Elsevier Science.)

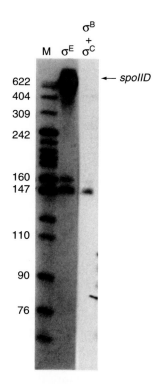

Figure 8.8   Specificity of σ^E determined by run-off transcription from the *spoIID* promoter. Sonenshein and associates prepared a restriction fragment containing the *spoIID* promoter and transcribed it in vitro with *B. subtilis* core RNA polymerase plus σ^E (middle lane) or σ^B plus σ^C (right lane). Lane M contained marker DNA fragments whose sizes are indicated at left. The arrow at the right indicates the position of the expected run-off transcript from the *spoIID* promoter (about 700 nt). Only the enzyme containing σ^E made this transcript. (*Source:* Rong et al., Transcriptional control of the *Bacillus subtilis spoIID* gene. *J. Bacteriology* 165, no. 3 (1986) f. 7, p. 777, by permission of American Society for Microbiology.)

only to the vegetative band (3050 bp). By contrast, when the polymerase contained σ^E, the transcript hybridized to both vegetative and sporulation bands (3050 and 770 bp). Apparently σ^E has some ability to recognize vegetative promoters; however, its main affinity seems to be for sporulation promoters—at least those of the type contained in the 770-bp DNA fragment.

The nature of the sporulation gene contained in the 770-bp fragment was not known, so Abraham Sonenshein and colleagues set out to show that σ^E could transcribe a well-characterized sporulation gene. They chose the *spoIID* gene, which was known to be required for sporulation and had been cloned. They used polymerases containing four different σ-factors, σ^A, σ^B, σ^C, and σ^E, to transcribe a truncated fragment of the gene so as to produce a run-off transcript (Chapter 5). Previous S1 mapping with RNA made in vivo had identified the natural transcription start site. Because the truncation in the *spoIID* gene occurred 700 bp downstream of this start site, transcription from the correct start site in vitro produced a 700-nt run-off transcript. As Figure 8.8 shows, only σ^E could produce this transcript; none of the other σ-factors could direct the RNA polymerase to recognize the *spoIID* promoter.

Losick and his colleagues established that σ^E is itself the product of a sporulation gene, originally called *spoIIG*. Predictably, mutations in this gene block sporula-

tion at an early stage. Without a σ-factor to recognize sporulation genes such as *spoIID*, these genes cannot be expressed, and therefore sporulation cannot occur.

Other sporulation-specific σ-factors come into play later in the sporulation process. One of these is σ^K, which is the product of two separate genes, *spoIVCB* and *spoIIIC* that come together by recombination, an actual rearrangement of DNA sequences, during the sporulation process to form a composite gene called *sig*^{K}. Losick and coworkers have confirmed this rearrangement by sequencing the separate genes at an early stage of sporulation (1h) and the rearranged gene later in sporulation (5h). The sequences of these genes are exactly what we would expect if fusion of the two genes really did occur (Figure 8.9). This rearrangement is irreversible, but that does not matter because it occurs only in the mother cell, not in the endospore. The mother cell is destined to die, but the endospore survives, carrying on the bacterium's unaltered genome. This kind of gene rearrangement during development is unusual, but not unprecedented. We will encounter another example when we study the recombination of antibody genes during vertebrate immune cell development in Chapter 23.

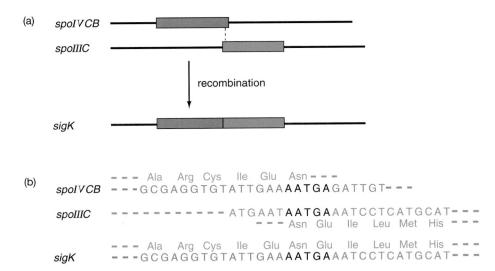

Figure 8.9 The recombination event that forms the *sigK* gene. (a) Schematic outline of the recombination event, showing the two genes, *spoIVCB* (red) and *spoIIIC* (green) before the recombination, followed by the rearranged *sigK* gene. **(b)** Sequences of the *spoIVCB* (red) and *spoIIIC* (green) genes surrounding the breakpoint (black) at which the two genes will be joined, followed by the sequence of the fused gene (*sigK*). The amino acid sequence encoded by each of these gene fragments is shown in blue.

When the original σ was discovered in *E. coli* in 1969, Burgess and Travers speculated that many different σ's would be found and that these would direct transcription of different classes of genes. This principle was first verified in *B. subtilis,* as we have seen, but it has also been confirmed in *E. coli.* For example, a distinct *E. coli* σ (σ^{32}) directs the transcription of a class of genes called **heat shock genes** that are switched on when the bacteria experience high temperature or other environmental insults. During nitrogen starvation, another σ-factor (σ^{54}) directs transcription of genes that encode proteins responsible for nitrogen metabolism. In addition, although gram-negative bacteria such as *E. coli* do not sporulate, they do become relatively resistant to stress on starvation. The genes that confer stress resistance are switched on in stationary phase (nonproliferating) *E. coli* cells by an RNA polymerase bearing the alternative σ factor σ^S. These are all examples of a fundamental coping mechanism: Bacteria tend to deal with changes in their environment with global changes in transcription mediated by shifts in σ-factors. We will consider the roles of some of these alternative σ-factors later in this chapter.

SUMMARY When the bacterium *B. subtilis* sporulates, a whole new set of sporulation-specific genes is turned on, and many, but not all, vegetative genes are turned off. This switch takes place largely at the transcription level. It is accomplished by several new σ-factors that displace the vegetative σ-factor from the core RNA polymerase and direct transcription of sporulation genes instead of vegetative genes. Each σ-factor has its own preferred promoter sequence.

8.4 Genes with Multiple Promoters

The story of *B. subtilis* sporulation is an appropriate introduction to our next topic, multiple promoters, because sporulation genes provided some of the first examples of this phenomenon.

The *B. subtilis* spoVG Gene

One of the sporulation genes with two promoters is *spoVG,* which is transcribed by both $E\sigma^B$ and $E\sigma^E$ (holoenzymes bearing either σ^B or σ^E). Losick and colleagues achieved a partial separation of these holoenzymes by DNA-cellulose chromatography of RNA polymerases from sporulating cells. Then they performed run-off transcription of a cloned, truncated *spoVG* gene with fractions from the peak of polymerase activity. Figure 8.10 shows the results. The fractions on the leading edge of the peak (19 and 20) produced primarily a 110-nt run-off transcript. On the other hand, fractions from the trailing edge of the peak (22 and 23) made predominately a 120-nt run-off transcript. The fraction in the middle (21) made both run-off transcripts.

These workers succeeded in completely separating the two polymerase activities, using another round of DNA-cellulose chromatography. One set of fractions, containing σ^E, synthesized only the 110-nt run-off transcript. Furthermore, the ability to make this transcript paralleled the content of σ^E in the enzyme preparation, suggesting that σ^E was responsible for this transcription activity. To reinforce the point, Losick's group purified σ^E using gel electrophoresis, combined it with core polymerase, and showed that it made only the 110-nt run-off transcript

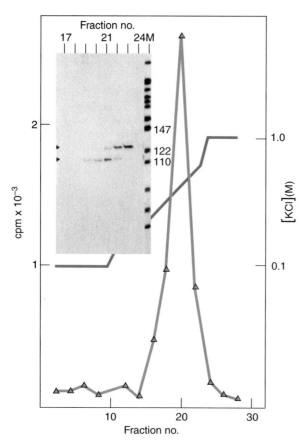

Figure 8.10   Resolution of RNA polymerases that transcribe the *spoVG* gene from two different promoters. Losick and colleagues purified polymerase from *B. subtilis* cells that were running out of nutrients. The last purification step was DNA-cellulose column chromatography. The polymerase activity in each fraction from the column is given by the red line and the scale on the left-hand *y* axis. The salt concentration used to remove the enzyme from the column is given by the green line and the scale on the right-hand *y* axis. The inset shows the results of a run-off transcription assay using a DNA fragment with two *spoVG* promoters spaced 10 bp apart. The fraction numbers at the top of the inset correspond to the fraction numbers from the column at bottom. The last lane (M) contained marker DNA fragments. The two arrowheads at the left of the inset indicate the two run-off transcripts, approximately 110 and 120 nt in length. The column separated a polymerase that transcribed selectively from the downstream promoter and produced the shorter run-off transcript (fractions 19 and 20) from a polymerase that transcribed selectively from the upstream promoter and produced the longer run-off transcript (fractions 22 and 23). (*Source:* Johnson et al., Two RNA polymerase sigma factors from *Bacillus subtilis* discriminate between overlapping promoters for a developmentally regulated gene. *Nature* 302: (28 Apr 1983), f. 2, p. 802. © Macmillan Magazines Ltd.)

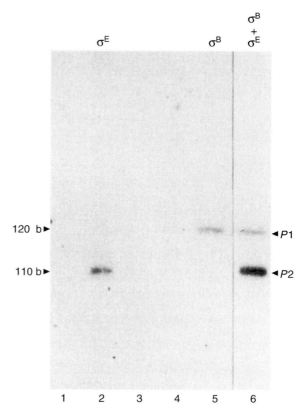

Figure 8.11   Specificities of σ^B and σ^E. Losick and colleagues purified σ-factors σ^B and σ^E by gel electrophoresis and tested them with core polymerase by the same run-off transcription assay used in Figure 8.10. Lane 2, containing σ^E, caused initiation selectively at the downstream promoter (*P2*). Lane 5, containing σ^B, caused initiation selectively at the upstream promoter (*P1*). Lane 6, containing both σ-factors caused initiation at both promoters. The other lanes were the results of experiments with other fractions containing neither σ-factor. (*Source:* Johnson et al., *Nature* 302 (28 Apr 1983), f. 4, p. 803. © Macmillan Magazines Ltd.)

Figure 8.12   Overlapping promoters in *B. subtilis spoVG*. *P1* denotes the upstream promoter, recognized by σ^B; the start of transcription and −10 and −35 boxes for this promoter are indicated in red above the sequence. *P2* denotes the downstream promoter, recognized by σ^E, the start of transcription and −10 and −35 boxes for this promoter are indicated in blue below the sequence.

(Figure 8.11). They also established that σ^B plus core polymerase made only the 120-nt run-off transcript (see Figure 8.11).

These experiments demonstrated that the *spoVG* gene can be transcribed by both $E\sigma^B$ and $E\sigma^E$, and that these two enzymes have transcription start sites that lie 10 bp apart, as shown in Figure 8.12. Knowing the locations of these start sites, we can count the appropriate number of base pairs upstream and find the −10 and −35 boxes of the promoters recognized by each of these σ-factors. Comparing many −10 and −35 boxes recognized by the same σ-factor allowed the identification of consensus sequences, such as those reported in Chapter 6. These are also indicated in Figure 8.12.

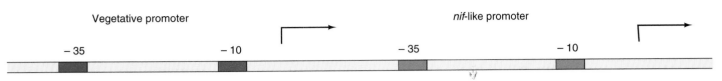

Figure 8.13   Map of the two promoters of the *Anabaena glnA* gene. The downstream, *nif*-like promoter elements are shown in red. The upstream, vegetative promoter elements are shown in green.

The *Anabaena* Glutamine Synthetase Gene

Anabaena 7120 is a *cyanobacterium,* or blue green alga, whose cells are arranged in long chains, or filaments. As long as NH_4^+ is available, the cells can all be vegetative, but when NH_4^+ is in short supply, some of the vegetative cells differentiate into **heterocysts** that can fix nitrogen for all the cells to use. Under nitrogen-fixing conditions, most vegetative genes in the heterocyst are switched off, and the nitrogen-fixing (***nif***) genes are switched on. The main mechanism behind this shift is a now-familiar switch in σ-factors. The vegetative σ, which recognizes the vegetative genes' *E. coli*-like promoters, is replaced by a nitrogen-fixing σ that is specific for the *nif* promoters. However, some vegetative genes must remain active even during nitrogen fixation. Among these is the glutamine synthetase gene (*glnA*). By adding NH_4^+ to glutamate, glutamine synthetase makes the amino acid glutamine, which is required in both vegetative and nitrogen-fixing cells. So how does the *glnA* gene remain active under both conditions? The answer is that the *glnA* gene has two promoters, one for each of the σ-factors. Robert Haselkorn and colleagues used S1 mapping and primer extension analysis to locate the transcription start sites associated with the two promoters. They found that the vegetative start site lies 62 bp upstream of the *nif*-like start site. Interestingly, they also found that *both* promoters are used during the vegetative phase. Figure 8.13 depicts the arrangement of the two promoters for the *glnA* gene.

The *E. coli glnA* Gene

The *glnA* gene of *E. coli* is also regulated in a manner that depends on the availability of fixed nitrogen (NH_4^+). When carbon is the limiting factor, the need for glutamine is small because little protein synthesis can take place. Under these conditions, the gene is transcribed by $E\sigma^{70}$, the major holoenzyme, from a weak promoter called *glnA* P1. On the other hand, when fixed nitrogen is the limiting factor, glutamine synthetase is needed in large quantity to trap as much of the available NH_4^+ as possible in glutamine. The glutamine can then donate amino groups to other cellular compounds. Under these conditions, *glnA* is transcribed by $E\sigma^{54}$ from a strong promoter called *glnA* P2, which lies 115 bp downstream of *glnA* P1. Thus, the switch from the weak to the strong promoter is governed by the replacement of the major σ^{70} by another factor, σ^{54}, also known as σ^N. In Chapter 9, we will learn that σ^{54} is a defective σ that requires ATP to initiate transcrip-

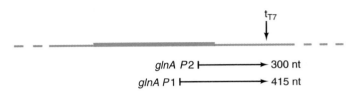

Figure 8.14   Map of the DNA region used to assay the transcription initiation from the *E. coli glnA* gene. The bold red region contains the *glnA* promoters *glnA* P1 and *glnA* P2. Downstream from these promoters, in the blue region of the DNA, lies the T7 phage terminator (t_{T7}). Transcripts initiating at *glnA* P1 or *glnA* P2 and terminating at t_{T7} produce truncated transcripts of 415 and 300 nt, respectively. (*Source:* From Hunt and Magasanik, *Proceedings of the National Academy of Sciences* 82:8454, 1985. Reprinted with permission.)

tion. This σ-factor is apparently able to compete successfully with σ^{70} for the core polymerase, thus forming a new holoenzyme with new promoter specificity. It recognizes not only *glnA* P2, but other promoters involved in the cell's response to nitrogen starvation.

T. P. Hunt and Boris Magasanik mapped the two *glnA* promoters in 1985 by S1 mapping and by a variation of the run-off transcription assay. Instead of using a truncated linear gene, as in normal run-off transcription, these investigators cloned the *glnA* gene into a plasmid upstream of a T7 terminator (Figure 8.14). Transcription of this plasmid in vitro begins at one of the two promoters, depending on the σ-factor employed, then terminates at the T7 terminator, a known distance downstream. The length of the terminated transcript locates the two promoters exactly and also tells which promoter is being used by a given polymerase. Figure 8.15 shows the results obtained with σ^{70}, σ^{54}, and a mixture of the two. It is interesting that when both σ-factors are present, σ^{54} predominates and transcription begins primarily at *glnA* P2. This demonstrates that σ^{54} can compete successfully with σ^{70}, at least in vitro. By the way, the promoter recognized by $E\sigma^{54}$ consists of a –12 box and a –24 box (centered at positions –12 and –24, respectively) instead of the usual –10 and –35 boxes.

SUMMARY Some prokaryotic genes must be transcribed under conditions where two different σ-factors are active. These genes are equipped with two different promoters, each recognized by one of the two σ-factors. This ensures their expression no matter which factor is present and allows for differential control under different conditions.

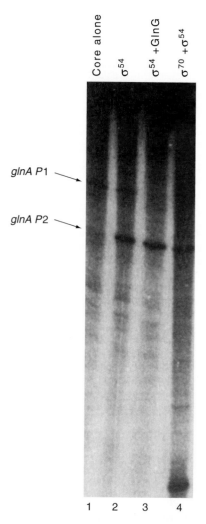

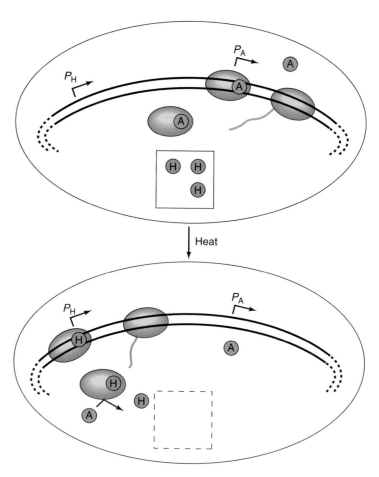

Figure 8.16   Result of heat shock on transcription in _E. coli_. Top: At normal temperature, σ^H (green) is somehow sequestered so it cannot bind to the core polymerase. As a result, the normal σ^A (blue) associates with core, directing it to normal vegetative promoters (P_A). As a result, vegetative transcripts (blue) are produced. Bottom: On heating, the isolated σ^H is released. It competes with σ^A for binding to core and directs the polymerase to heat shock promoters P_H. Consequently, heat shock transcripts (green) are made.

Figure 8.15   Transcription from _glnA P2_ directed by σ^{54}. Hunt and Magasanik transcribed the DNA construct pictured in Figure 8.14 with _E. coli_ core polymerase plus various factors listed at the top. They electrophoresed the transcripts to see which promoter had been used. Lane 1 shows that core polymerase alone gave a small amount of transcription from _glnA P1_. This was probably due to contamination of the core polymerase by σ^{70}. Lane 2 shows that core plus σ^{54} yielded a dark band corresponding to transcription from _glnA P2_. The preference for _glnA P2_ was enhanced by the _glnG_ gene product (lane 3). Lane 4 shows the results of transcription in the presence of both σ^{70} and σ^{54}. Under these conditions, transcription from _glnA P2_ predominated. (_Source:_ Hunt, T. P. and Magasanik, B. Transcription of _glnA_ by purified _Escherichia coli_ components: Core RNA polymerase and the products of _glnF, glnG,_ and _glnL. PNAS_ 82 (Dec 1985) f. 3, p. 8455.)

8.5   The _E. coli_ Heat Shock Genes

When eukaryotic or prokaryotic cells experience an increase in temperature, or a variety of other environmental insults, they mount a defense called the **heat shock response** to minimize damage. They start producing proteins called **molecular chaperones** that bind to proteins partially unfolded by heating and help them fold properly again. They also produce proteases that degrade proteins

that are so badly unfolded that they cannot be refolded, even with the help of chaperones.

Almost immediately after _E. coli_ cells are heated from their normal growth temperature (37°C) to a higher temperature (42°C), normal transcription ceases, or at least decreases, and the synthesis of 17 new, heat shock transcripts begins. These transcripts encode the molecular chaperones and proteases that help the cell survive heat shock. This shift in transcription requires the product of the _htpR_ gene, which encodes a σ-factor with a molecular mass of 32 kD. Hence this factor is called σ^{32}, but it is also known as σ^H, where the H stands for heat shock. In 1984, Grossman and coworkers demonstrated that σ^H really is a σ-factor. They did this by combining σ^H with core polymerase and showing that this mixture could transcribe a variety of heat shock genes in vitro from their natural transcription start sites.

The heat shock response begins in less than 1 min, which is not enough time for synthesis of a new σ-factor. Thus, it is reasonable to suggest that there is some preexisting σ^H, and it is somehow suddenly unmasked so it can compete more successfully with σ^{70} (σ^A) for the core polymerase (Figure 8.16). This notion of competition between σ-factors receives support from the following findings: (1) Mutants that overproduce σ^A, and therefore would have more of this σ-factor to compete with σ^H, have a poor heat shock response. (2) Mutants that underproduce σ^A have an enhanced heat shock response. (3) Mutants with a somewhat defective σ^H have a weak heat shock response when σ^A is normal, but a normal heat shock response when σ^A is underproduced.

This competition among σ-factors for binding to the core polymerase is probably a general phenomenon that explains many of the fundamental switches in gene expression we have encountered in this chapter.

SUMMARY The heat shock response in *E. coli* is governed by an alternative σ-factor, σ^{32} (σ^H) which displaces σ^{70} (σ^A) and directs the RNA polymerase to the heat shock gene promoters.

8.6 Infection of *E. coli* by Phage λ

Many of the phages we have studied so far (T2, T4, T7, and SPO1, for example) are **virulent** phages. When they replicate, they kill their host by **lysing** it, or breaking it open. On the other hand, **lambda** (λ) is a **temperate** phage; when it infects an *E. coli* cell, it does not necessarily kill it. In this respect, λ is more versatile than many phages; it can follow two paths of reproduction (Figure 8.17). The first is the **lytic** mode, in which infection progresses just as it would with a virulent phage. It begins with phage DNA entering the host cell and then serving as the template for transcription by host RNA polymerase. Phage mRNAs are translated to yield phage proteins, the phage DNA replicates, and progeny phages assemble from these DNA and protein components. The infection ends when the host cell lyses to release the progeny phages.

In the **lysogenic** mode, something quite different happens. The phage DNA enters the cell, and its early genes are transcribed and translated, just as in a lytic infection. But then a 27-kD phage protein (the λ **repressor**) appears and binds to the two phage operator regions, ultimately shutting down transcription of all genes except for *cI* (pronounced "c-one," not "c-eye"), the gene for the λ repressor itself. Under these conditions, with only one phage gene active, it is easy to see why no progeny phages can be produced. Furthermore, when lysogeny is established, the phage DNA integrates into the host genome. A bacterium harboring this integrated phage DNA is called a **lysogen**. The integrated DNA is called a **prophage**. The lysogenic state can exist indefinitely and should not be considered a disadvantage for the phage, because the phage DNA in the lysogen replicates right along with the host DNA. In this way, the phage genome multiplies without the necessity of making phage particles; thus, it gets a "free ride." Under certain conditions, such as when the lysogen encounters mutagenic chemicals or radiation, lysogeny can be broken and the phage enters the lytic phase.

SUMMARY Phage λ can replicate in either of two ways: lytic or lysogenic. In the lytic mode, almost all of the phage genes are transcribed and translated, and the phage DNA is replicated, leading to production of progeny phages and lysis of the host cells. In the lysogenic mode, the λ DNA is incorporated into the host genome; after that occurs, only one gene is expressed. The product of this gene, the λ repressor, prevents transcription of all the rest of the phage genes. However, the incorporated phage DNA (the prophage) still replicates, because it has become part of the host DNA.

Lytic Reproduction of Phage λ

The lytic reproduction cycle of phage λ resembles that of the virulent phages we have studied in that it contains three phases of transcription, called **immediate early, delayed early,** and **late.** These three classes of genes are sequentially arranged on the phage DNA, which helps explain how they are regulated, as we will see. Figure 8.18 shows the λ genetic map in two forms: linear, as the DNA exists in the phage particles, and circular, the shape the DNA assumes shortly after infection begins. The circularization is made possible by 12-base overhangs, or "sticky" ends, at either end of the linear genome. These cohesive ends go by the name **cos.** Note that cyclization brings together all the late genes, which had been separated at the two ends of the linear genome.

As usual, the program of gene expression in this phage is controlled by transcriptional switches, but λ uses a switch we have not seen before: **antitermination.** Figure 8.19 outlines this scheme. Of course, the host RNA polymerase holoenzyme transcribes the immediate early genes first. Only two of these genes, *cro* and *N*, lie immediately downstream of the rightward and leftward promoters, P_R and P_L, respectively. At this stage in the lytic cycle, no repressor is bound to the operators that govern these promoters (O_R and O_L, respectively), so transcription proceeds unimpeded. When the polymerase reaches the ends of the immediate early genes, it encounters rho-dependent terminators and stops short of the delayed early genes.

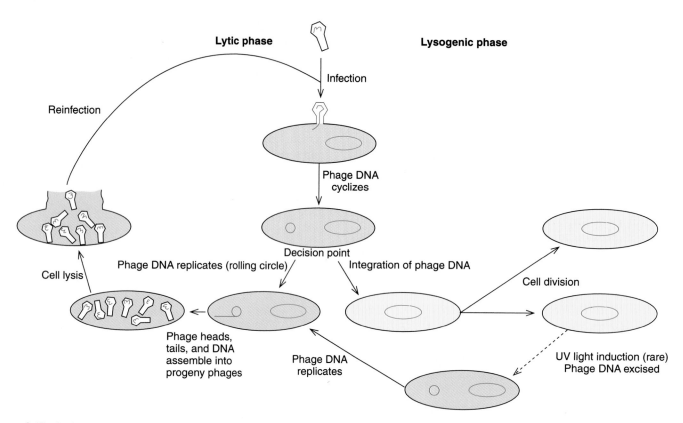

Figure 8.17   Lytic versus lysogenic infection by phage λ. Blue cells are in the lytic phase; yellow cells are in the lysogenic phase; green cells are uncommitted.

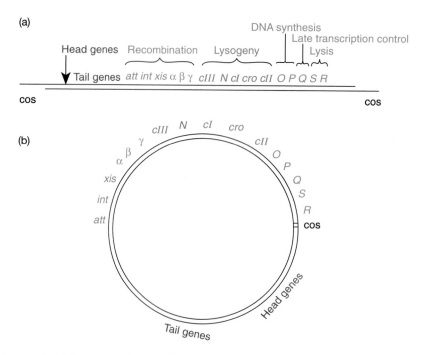

Figure 8.18   Genetic map of phage λ. (**a**) The map is shown in linear form, as the DNA exists in the phage particles; the cohesive ends (cos) are at the ends of the map. The genes are grouped primarily according to function. (**b**) The map is shown in circular form, as it exists in the host cell during a lytic infection after annealing of the cohesive ends.

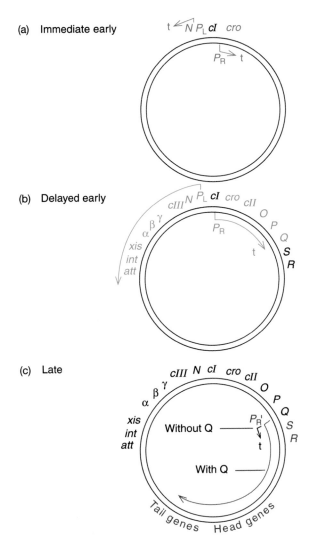

Figure 8.19   Temporal control of transcription during lytic infection by phage λ. **(a)** Immediate early transcription (red) starts at the rightward and leftward promoters (P_R and P_L, respectively) that flank the repressor gene (*cI*); transcription stops at the rho-dependent terminators (t) after the *N* and *cro* genes. **(b)** Delayed early transcription (blue) begins at the same promoters, but bypasses the terminators by virtue of the *N* gene product, N, which is an antiterminator. **(c)** Late transcription (gray) begins at a new promoter ($P_{R'}$); it would stop short at the terminator (t) without the *Q* gene product, Q, another antiterminator. Note that *O* and *P* are protein-encoding delayed early genes, not operator and promoter.

The products of both immediate early genes are crucial to further expression of the λ program. The *cro* gene product is a repressor that blocks transcription of the λ repressor gene, *cI,* and therefore prevents synthesis of λ repressor protein. This is necessary for expression of the other phage genes, which would be blocked by λ repressor. The *N* gene product, **N,** is an **antiterminator** that permits RNA polymerase to ignore the terminators at the ends of the immediate early genes and continue transcribing into the delayed early genes. When this happens, the

delayed early phase begins. Note that the same promoters (P_R and P_L) are used for both immediate early and delayed early transcription. The switch does not involve a new σ-factor or RNA polymerase that recognizes new promoters and starts new transcripts, as we have seen with other phages; instead, it involves an extension of transcripts controlled by the same promoters.

The delayed early genes are important in continuing the lytic cycle and, as we will see in the next section, in establishing lysogeny. Genes *O* and *P* code for proteins that are necessary for phage DNA replication, a key part of lytic growth. The *Q* gene product (**Q**) is another antiterminator, which permits transcription of the late genes.

The late genes are all transcribed in the rightward direction, but not from P_R. The late promoter, $P_{R'}$, lies just downstream of *Q*. Transcription from this promoter terminates after only 194 bases, unless Q intervenes to prevent termination. The *N* gene product cannot substitute for Q; it is specific for antitermination after *cro* and *N*. The late genes code for the proteins that make up the phage head and tail, and for proteins that lyse the host cell so the progeny phages can escape.

> **SUMMARY** The immediate early/delayed early/late transcriptional switching in the lytic cycle of phage λ is controlled by antiterminators. One of the two immediate early genes is *cro*, which codes for a repressor of the *cI* gene that allows the lytic cycle to continue. The other, *N*, codes for an antiterminator, N, that overrides the terminators after the *N* and *cro* genes. Transcription then continues into the delayed early genes. One of the delayed early genes, *Q*, codes for another antiterminator (Q) that permits transcription of the late genes from the late promoter, $P_{R'}$, to continue without premature termination.

Antitermination   How do N and Q perform their antitermination functions? They appear to use two different mechanisms. Let us first consider antitermination by N. Figure 8.20 presents an outline of the process. Panel (a) shows the genetic sites surrounding the *N* gene. On the right is the leftward promoter P_L and its operator O_L. This is where leftward transcription begins. Downstream (left) of the *N* gene is a transcription terminator, where transcription ends in the absence of the *N* gene product (N). Panel (b) shows what happens in the absence of N. The RNA polymerase (pink) begins transcription at P_L and transcribes *N* before reaching the terminator and falling off the DNA, releasing the *N* mRNA. Now that *N* has been transcribed, the N protein appears, and panel (c) shows what happens next. The N protein (purple) binds to the *transcript* of the **N utilization site** (***nut* site,** green)

(a)

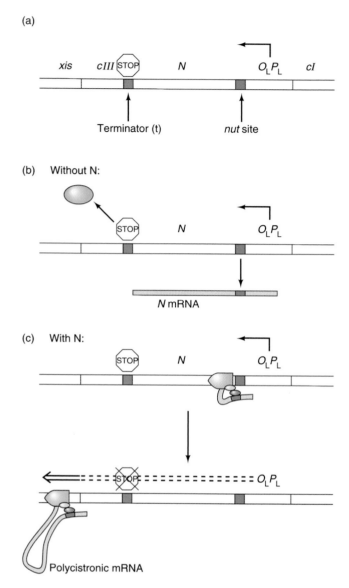

Figure 8.20   Effect of N on leftward transcription. (a) Map of *N* region of λ genome. The genes surrounding *N* are depicted, along with the leftward promoter (*P*L) and operator (*O*L), the terminator (red), and the *nut* site (green). **(b)** Transcription in the absence of N. RNA polymerase (pink) begins transcribing leftward at *P*L and stops at the terminator at the end of *N*. The *N* mRNA is the only product of this transcription. **(c)** Transcription in the presence of N. N (purple) binds to the *nut* region of the transcript, and also to NusA (yellow), which, along with other proteins not shown, has bound to RNA polymerase. This complex of proteins alters the polymerase so it can read through the terminator and continue into the delayed early genes.

and interacts with a complex of host proteins (yellow) bound to the RNA polymerase. This somehow alters the polymerase, turning it into a "juggernaut" that ignores the terminator and keeps on transcribing into the delayed early genes. The same mechanism applies to rightward transcription from *P*R, because a site just to the right of *cro* allows the polymerase to ignore the terminator and enter the delayed early genes beyond *cro*.

How do we know that host proteins are involved in antitermination? Genetic studies have shown that mutations in four host genes interfere with antitermination. These genes encode the proteins NusA, NusB, NusG, and the ribosomal S10 protein. It may seem surprising that host proteins cooperate in a process that leads to host cell death, but this is just one of many examples in which a virus harnesses a cellular process for its own benefit. In this case, the cellular process served by the S10 protein is obvious: protein synthesis. But the Nus proteins also have cellular roles. They allow antitermination in the seven *rrn* operons that encode ribosomal RNAs, as well as some tRNAs.

In vitro studies have shown that two proteins, N and NusA, can cause antitermination if the terminator is close enough to the *nut* site. Figure 8.21a shows the protein complex involved in this short-range antitermination and illustrates the fact that N does not bind by itself to RNA polymerase. It binds to NusA, which in turn binds to polymerase. This figure also introduces the two parts of the *nut* site, known as **boxA** and **boxB**. BoxA is highly conserved among *nut* sites, but boxB varies quite a bit from one *nut* site to another. The transcript of boxB contains an inverted repeat, which presumably forms a stem-loop, as shown in the figure.

Antitermination in vivo is not likely to use this simple antitermination scheme because it occurs at terminators that are at least hundreds of base pairs downstream of the corresponding *nut* sites. We call this kind of antitermination **processive** because the antitermination factors remain associated with the polymerase as it moves a great distance along the DNA. Such processive antitermination requires more than just N and NusA. It also requires the other three host proteins: NusB, NusG, and S10. These proteins presumably help to stabilize the antitermination complex so it persists until it reaches the terminator. Figure 8.21b depicts this stable complex that includes all five antitermination proteins.

Perhaps the most unexpected feature of the antitermination complex depicted in Figure 8.21 is the interaction of the complex with the *transcript* of the *nut* site, rather than with the *nut* site itself. How do we know this is what happens? One line of evidence is that the region of N that is essential for *nut* recognition is an arginine-rich domain that resembles an RNA-binding domain. Asis Das provided more direct evidence, using a gel mobility shift assay to demonstrate binding between N and an RNA fragment containing boxB. Furthermore, when N and NusA have both bound to the complex, they partially protect boxB, but not boxA, from RNase attack. Only when all five proteins have bound is boxA protected from RNase. This is consistent with the model in Figure 8.21.

How do we know the RNA loops as shown in the figure? We don't know for sure, but the easiest way to imagine a signal to the polymerase that persists from the time

(a) Weak, nonprocessive complex

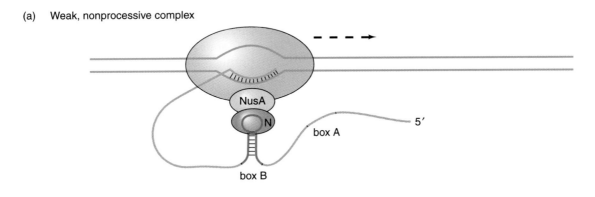

(b) Strong, processive complex

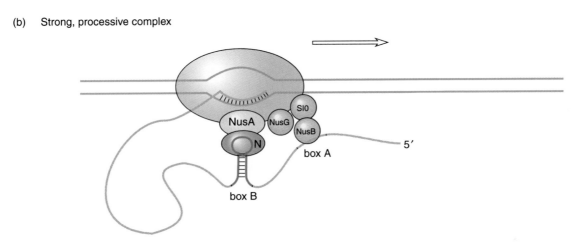

Figure 8.21 Protein complexes involved in N-directed antitermination. (a) Weak, nonprocessive complex. NusA binds to polymerase, and N binds to both NusA and box B of the *nut* site region of the transcript, creating a loop in the growing RNA. This complex is relatively weak and can cause antitermination only at terminators near the *nut* site (dashed arrow). These conditions exist only in vitro. **(b)** Strong, processive complex. NusA tethers N and box B to the polymerase, as in panel **(a)**; in addition, S10 binds to polymerase, and NusB binds to box A of the *nut* site region of the transcript. This provides an additional link between the polymerase and the transcript, strengthening the complex. NusG also contributes to the strength of the complex. This complex is processive and can cause antitermination thousands of base pairs downstream in vivo (open arrow).

N binds to the *nut* site transcript until the polymerase reaches the terminator is to envision N maintaining its association with both the polymerase and the RNA. This would require that the RNA form a loop as shown. In accord with this hypothesis, Jack Greenblatt and colleagues have isolated mutants in the gene encoding the RNA polymerase β-subunit that interfere with N-mediated antitermination. These mutants also fail to protect the *nut* site transcript during transcription in vitro. This suggests that an association exists between the RNA polymerase, N, and the *nut* site transcript during transcription. Again, this is easiest to imagine if the RNA between the *nut* site transcript and the polymerase forms a loop.

How does N prevent termination? One simple hypothesis is that it limits pausing by RNA polymerase. Pausing is a natural part of transcription; RNA polymerase pauses at several DNA sites during a typical elongation process, and pausing is essential for termination of transcription. At a rho-independent terminator, the polymerase pauses long enough for the transcript to be released. At a rho-

dependent terminator, polymerase pauses long enough for rho to catch up with the polymerase and release the transcript. Because N can cause antitermination at both rho-dependent and -independent terminators, it could function at both by restricting the pause time at the terminator. There is evidence that N, NusA, and the *nut* boxB site can indeed limit the pause time at a rho-dependent λ terminator.

The control of late transcription also uses an antitermination mechanism, but with significant differences from the N-antitermination system. Figure 8.22 shows that a Q utilization (*qut*) site overlaps the late promoter ($P_{R'}$). This *qut* site also overlaps a pause site at 16–17 bp downstream of the transcription initiation site. In contrast to the N system, Q binds directly to the *qut* site, not to its transcript.

In the absence of Q, RNA polymerase pauses for several minutes at this site just after transcribing the pause signal. After it finally leaves the pause site, RNA polymerase transcribes to the terminator, where it aborts late

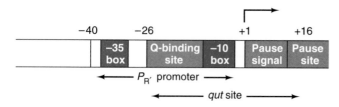

Figure 8.22 Map of the $P_{R'}$ region of the λ genome. The $P_{R'}$ promoter comprises the −10 and −35 boxes. The *qut* site overlaps the promoter and includes the Q binding site upstream of the −10 box, the pause signal downstream of the transcription start site, and the pause site at positions +16 and +17. (*Source:* Reprinted with permission from *Nature* 364:403, 1993. Copyright © 1993 Macmillan Magazines Limited.)

transcription. On the other hand, if Q is present, it recognizes the paused complex and binds to the *qut* site. Q then binds to the polymerase and alters it in such a way that it resumes transcription and ignores the terminator, continuing on into the late genes. The Q-altered polymerase appears to inhibit RNA hairpin formation immediately behind the polymerase, thereby inhibiting terminator activity. Q can cause antitermination by itself in the λ late control region, but NusA makes the process more efficient.

SUMMARY Five proteins (N, NusA, NusB, NusG, and S10) collaborate in antitermination at the λ immediate early terminators. NusA and S10 bind to RNA polymerase, and N and NusB bind to the boxB and boxA regions, respectively, of the *nut* site in the growing transcript. N and NusB bind to NusA and S10, respectively, probably tethering the transcript to the polymerase. This alters the polymerase so it reads through the terminators at the ends of the immediate early genes. Antitermination in the λ late region requires Q, which binds to the Q-binding region of the *qut* site as RNA polymerase is stalled just downstream of the late promoter. This appears to alter the polymerase so it can ignore the terminator and transcribe the late genes.

Establishing Lysogeny

We have mentioned that the delayed early genes are required not only for the lytic cycle but for establishing lysogeny. The delayed early genes help establish lysogeny in two ways: (1) Some of the delayed early gene products are needed for integration of the phage DNA into the host genome, a prerequisite for lysogeny. (2) The products of the *cII* and *cIII* genes allow transcription of the *cI* gene and therefore production of the λ repressor, the central component in lysogeny.

Two promoters control the *cI* gene (Figure 8.23): P_{RM} and P_{RE}. The RM in P_{RM} stands for "repressor mainte-

nance." This is the promoter that is used *during* lysogeny to ensure a continuing supply of repressor to maintain the lysogenic state. It has the peculiar property of requiring its own product—repressor—for activity. We will discuss the basis for this requirement; however, we can see immediately one important implication. This promoter cannot be used to establish lysogeny, because at the start of infection no repressor is present to activate it. Instead, the other promoter, P_{RE}, is used. The RE in P_{RE} stands for "repressor establishment." P_{RE} lies to the right of both P_R and *cro*. It directs transcription leftward through *cro* and then through *cI*. Thus, P_{RE} allows *cI* expression before any repressor is available.

Of course, the natural direction of transcription of *cro* is rightward from P_R, so the leftward transcription from P_{RE} gives an RNA product that is an **antisense** transcript of *cro*, as well as a "sense" transcript of *cI*. The *cI* part of the RNA can be translated to give repressor, but the antisense *cro* part cannot be translated. This antisense RNA also contributes to the establishment of lysogeny because the *cro* antisense transcript binds to *cro* mRNA and interferes with its translation. Because *cro* works against lysogeny, blocking its action promotes lysogeny. Similarly, CII stimulates transcription from a leftward promoter (P_{anti-Q}) within Q. This "backwards" transcription produces Q antisense RNA that blocks production of Q. Because Q is required for late transcription in the lytic phase, obstructing its synthesis favors the alternative pathway—lysogeny.

P_{RE} has some interesting requirements of its own. It has −10 and −35 boxes with no clear resemblance to the consensus sequences ordinarily recognized by *E. coli* RNA polymerase. Indeed, it cannot be transcribed by this polymerase alone in vitro. However, the *cII* gene product, CII, helps RNA polymerase bind to this unusual promoter sequence.

Hiroyuki Shimatake and Martin Rosenberg demonstrated this activity of CII in vitro. Using a filter binding assay, they showed that neither RNA polymerase nor CII protein alone could bind to a fragment of DNA containing P_{RE}, and therefore could not cause this DNA to bind to the nitrocellulose filter. On the other hand, CII protein plus RNA polymerase together could bind the DNA to the filter. Thus, CII must be stimulating RNA polymerase binding to P_{RE}. Furthermore, this binding is specific. CII could stimulate polymerase to bind only to P_{RE} and to one other promoter, P_I. The latter is the promoter for the *int* gene, which is also necessary for establishing lysogeny and which also requires CII. The *int* gene is involved in integration of the λ DNA into the host genome.

We saw in Chapter 7 that CAP–cAMP stimulates RNA polymerase binding to the *lac* promoter by protein–protein interaction. Mark Ptashne and his colleagues have provided evidence that CII may work in a similar way. They used DNase footprinting (Figure 8.24) to map the

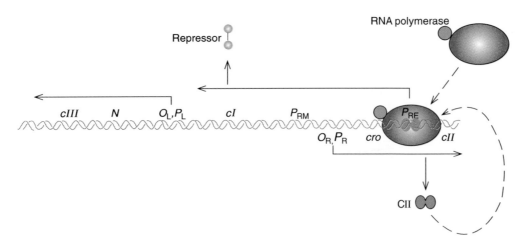

Figure 8.23 Establishing lysogeny. Delayed early transcription from P_R gives *cII* mRNA that is translated to CII (purple). CII allows RNA polymerase (blue and red) to bind to P_{RE} and transcribe the *cI* gene, yielding repressor (green).

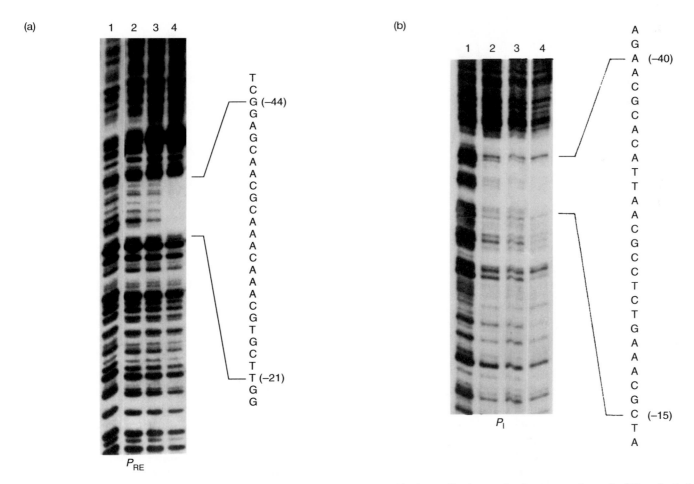

Figure 8.24 Binding of CII at the –35 box of both P_{RE} and P_I promoters of λ phage. Ptashne and colleagues performed a DNase footprint analysis of the interaction between CII and two early λ promoters, P_{RE} (a) and P_I (b). In (a), lanes 1–4 contained the following amounts of CII; lane 1, none; lane 2, 10 pmol; lane 3, 18 pmol; and lane 4, 90 pmol. In (b), lanes 1–4 contained the following amounts of CII: lane 1, none; lane 2, 18 pmol; lane 3, 45 pmol; lane 4,100 pmol. The CII footprint in both promoters includes the –35 box. (*Source:* Ho et al., Bacteriophage lambda protein *cII* binds promoters on the opposite face of the DNA helix from RNA polymerase. *Nature* 304 (1983) f. 3a-b, p. 705. © Macmillan Magazines Ltd.)

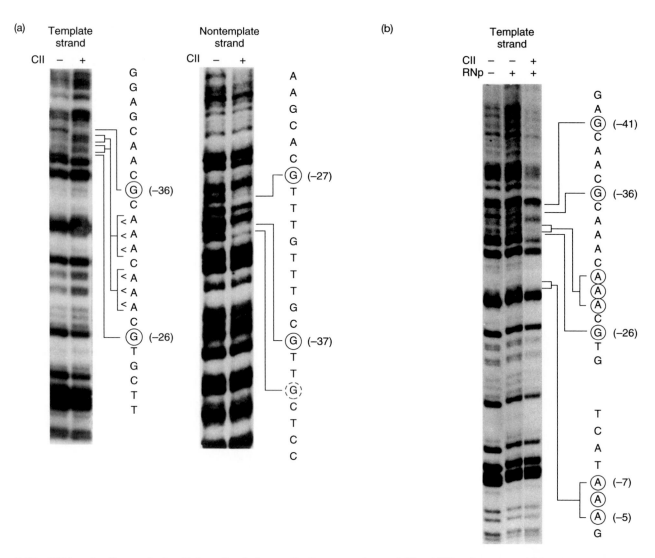

Figure 8.25   DMS protection analysis of interaction between the P_{RE} promoter and CII or RNA polymerase. (a) Ptashne and coworkers labeled the template or nontemplate strand (as indicated at top) of a DNA fragment including P_{RE} and performed DMS footprinting with and without CII, as indicated at top. In this and in all other panels, the circles denote strongly protected bases, the dotted circle represents a weakly protected base, and the < signs denote bases whose methylation is enhanced by protein binding. The perturbations caused by CII were in a region including the −35 box. **(b)** The template strand was labeled and tested for DMS protection with RNA polymerase alone (middle lane), or RNA polymerase plus CII (right lane). Polymerase alone did not perturb the DMS methylation pattern appreciably, indicating no tight binding between polymerase and P_{RE} in the absence of CII. With CII, polymerase caused additional perturbations of the methylation pattern compared with CII alone, especially at the extremes of the perturbed region (−41 and −5 to −7). (*Source:* Ho et al., *Nature* 304 (1983) f. 6, p. 707. © Macmillan Magazines Ltd.)

binding site for CII in the DNA fragment containing P_{RE} and found that CII binds between positions −21 and −44 of the promoter. Of course, this includes the unrecognizable −35 box of the promoter and raises the question; How can two proteins (CII and RNA polymerase) bind to the same site at the same time? The probable answer comes from Ptashne and coworkers' DMS protection experiments with CII (Figure 8.25), which showed that the bases that appear to be contacted by CII (e.g., G's at positions −26 and −36) are on the opposite side of the helix from those thought to be in contact with RNA polymerase (e.g., G at position −41). In other words, these two proteins seem to bind to opposite sides of the DNA helix, and so can bind cooperatively instead of competitively.

Why could CII footprint λ promoter DNA by itself, whereas it could not bind to λ promoter DNA in a filter binding assay? The former is a more sensitive assay. It will detect protein–DNA binding as long as some protein in an equilibrium mixture is bound to DNA and can protect it. By contrast, the latter is a more demanding assay. As soon as a protein dissociates from a DNA, the DNA flows through the filter and is lost, with no opportunity to rebind.

The product of the *cIII* gene, CIII, is also instrumental in establishing lysogeny, but its effect is less direct. It retards destruction of CII by cellular proteases. Thus, the delayed early products CII and CIII cooperate to make possible the establishment of lysogeny by activating P_{RE} and P_I.

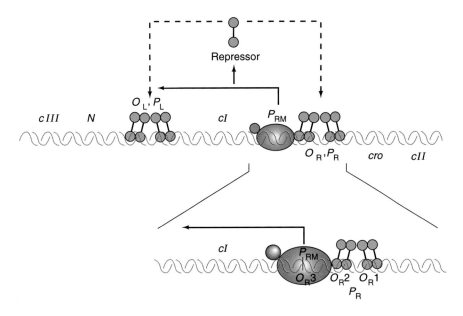

Figure 8.26   Maintaining lysogeny. (top) Transcription (from P_{RM}) and translation of the *cI* mRNA yields a continuous supply of repressor, which binds to O_R and O_L and prevents transcription of any genes aside from *cI*. (bottom) Detail of control region. Repressor (green) forms dimers and binds cooperatively to O_R1 and 2. The protein–protein contact between repressor on O_R2 and RNA polymerase (red and blue) allows polymerase to bind to P_{RM} and transcribe *cI*.

SUMMARY Phage λ establishes lysogeny by causing production of enough repressor to bind to the early operators and prevent further early RNA synthesis. The promoter used for establishment of lysogeny is P_{RE}, which lies to the right of P_R and *cro*. Transcription from this promoter goes leftward through the *cI* gene. The products of the delayed early genes *cII* and *cIII* also participate in this process: CII, by directly stimulating polymerase binding to P_{RE} and P_I; CIII, by slowing degradation of CII.

Autoregulation of the *cI* Gene during Lysogeny

Once the λ repressor appears, it binds to the λ operators, O_R and O_L. This has a double-barreled effect, both tending toward lysogeny. First, the repressor turns off further early transcription, thus interrupting the lytic cycle. The turnoff of *cro* is especially important to lysogeny because the *cro* product (**Cro**) acts to counter repressor activity, as we will see. The second effect of repressor is that it stimulates its own synthesis by activating P_{RM}.

Figure 8.26 illustrates how this self-activation works. The key to this phenomenon is the fact that both O_R and O_L are subdivided into three parts, each of which can bind repressor. The O_R region is more interesting because it controls leftward transcription of *cI*, as well as rightward transcription of *cro*. The three binding sites in the O_R region are called O_R1, O_R2, and O_R3. Their affinities

for repressor are quite different: Repressor binds most tightly to O_R1, next most tightly to O_R2, and least tightly to O_R3. However, binding of repressor to O_R1 and O_R2 is cooperative. This means that as soon as repressor binds to its "favorite" site, O_R1, it facilitates binding of another repressor to O_R2. No cooperative binding to O_R3 normally occurs.

Repressor protein is a dimer of two identical subunits, each of which is represented by a dumbbell shape in Figure 8.26. This shape indicates that each subunit has two domains, one at each end of the molecule. The two domains have distinct roles: One is the DNA-binding end of the molecule; the other is the site for the repressor–repressor interaction that makes dimerization and cooperative binding possible. Once repressor dimers have bound to both O_R1 and O_R2, the repressor occupying O_R2 lies very close to the binding site for RNA polymerase at P_{RM}. So close, in fact, that the two proteins touch each other. Far from being a hindrance, as you might expect, this protein–protein contact is required for RNA polymerase to initiate transcription efficiently at this promoter.

With repressors bound to O_R1 and O_R2, no more transcription from P_{RE} can occur because the repressors block *cII* and *cIII* transcription, and the products of these genes, needed for transcription from P_{RE}, break down very rapidly. However, the disappearance of CII and CIII is not usually a problem, because lysogeny is already established and a small supply of repressor is all that is required to maintain it. That small supply of repressor can be provided as long as O_R3 is left open, because RNA polymerase can transcribe *cI* freely from P_{RM}. Also,

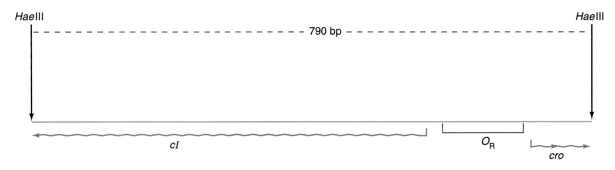

Figure 8.27   Map of the DNA fragment used to assay transcription from *cI* and *cro* promoters. The red arrows denote the in vitro *cI* and *cro* transcripts, some of which ("stutter transcripts") terminate prematurely. (*Source:* From Meyer et al., *Proceedings of the National Academy of Sciences* 72:4787, 1975. Reprinted with permission.)

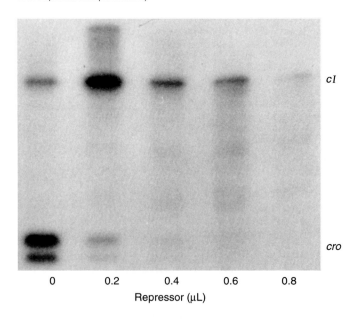

Figure 8.28   Analysis of the effect of λ repressor on *cI* and *cro* transcription in vitro. Ptashne and colleagues performed run-off transcription (which actually produced "stutter" transcripts) using the DNA template depicted in Figure 8.27. They included increasing concentrations of repressor as shown at bottom. Electrophoresis separated the *cI* and *cro* stutter transcripts, which are identified at right. The repressor clearly inhibited *cro* transcription, but it greatly stimulated *cI* transcription at low concentration, then inhibited *cI* transcription at high concentration. (*Source:* Meyer et al. Repressor turns off transcription of its own gene. *PNAS* 72 (Dec 1975), f. 5, p. 4788.)

repressor bound to O_R1 and O_R2 blocks *cro* transcription by interfering with polymerase binding to P_R.

It is conceivable that the concentration of repressor could build up to such a level that it would even fill its weakest binding site, O_R3. In that case, all *cI* transcription would cease because even P_{RM} would be blocked. This halt in *cI* transcription would allow the repressor level to drop, at which time repressor would dissociate first from O_R3, allowing *cI* transcription to begin anew. This mechanism is a means by which repressor can keep its own concentration from rising too high.

Ptashne and colleagues demonstrated with one in vitro experiment three of the preceding assertions: (1) the λ re-

pressor at low concentration can stimulate transcription of its own gene; (2) the repressor at high concentration can inhibit transcription of its own gene; and (3) the repressor can inhibit *cro* transcription. The experiment was a modified run-off assay that used a low concentration of UTP so transcription actually paused before it ran off the end of the DNA template. This produced so-called "stutter" transcripts resulting from the pauses. The template was a 790-bp *Hae*III restriction fragment that included promoters for both the *cI* and *cro* genes (Figure 8.27). Ptashne called the *cro* gene by another name, *tof*, in this paper. The experimenters added RNA polymerase, along with increasing amounts of repressor, to the template and observed the rate of production of the 300-nt stutter transcript from the *cI* gene and the two approximately 110-nt stutter transcripts from the *cro* gene. Figure 8.28 shows the results. At low repressor concentration we can already see some inhibition of *cro* transcription, and we also notice a clear stimulation of *cI* transcription. At higher repressor concentration, *cro* transcription ceases entirely, and at even higher concentration, *cI* transcription is severely inhibited.

SUMMARY The promoter that is used to maintain lysogeny is P_{RM}. It comes into play after transcription from P_{RE} makes possible the burst of repressor synthesis that establishes lysogeny. This repressor binds to O_R1 and O_R2 cooperatively, but leaves O_R3 open. RNA polymerase binds to P_{RM}, which overlaps O_R3 in such a way that it contacts the repressor bound to O_R2. This protein–protein interaction is required for this promoter to work efficiently.

RNA Polymerase/Repressor Interaction   How do we know that λ repressor/RNA polymerase interaction is essential for stimulation at P_{RM}? In 1994, Miriam Susskind and colleagues performed a genetic experiment that provided strong support for this hypothesis. They started with a mutant λ phage in which a key aspartate in the λ

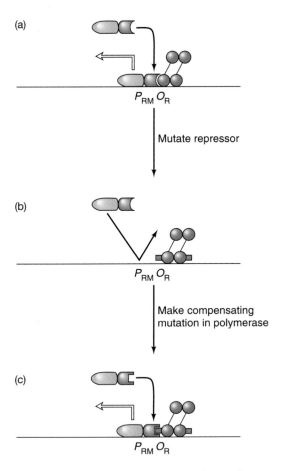

Figure 8.29   Principle of intergenic suppression to detect interaction between λ repressor and RNA polymerase. (a) With wild-type repressor and polymerase, the two proteins interact closely, which stimulates polymerase binding and transcription from P_{RM}. **(b)** The repressor gene has been mutated, yielding repressor with an altered amino acid (red). This prevents binding to polymerase. **(c)** The gene for one polymerase subunit has been mutated, yielding polymerase with an altered amino acid (represented by the square cavity) that restores binding to the mutated repressor. Because polymerase and repressor can now interact, transcription from P_{RM} is restored.

repressor had been changed to an asparagine. This mutant belongs to a class called *pc* for positive control; the mutant repressor is able to bind to the λ operators and repress transcription from P_R and P_L, but it is not able to stimulate transcription from P_{RM} (i.e., positive control of *cI* does not function). Susskind and coworkers reasoned as follows: If direct interaction between repressor and polymerase is necessary for efficient transcription from P_{RM}, then a mutant with a compensating amino acid change in a subunit of RNA polymerase should be able to restore the interaction with the mutant repressor, and therefore restore active transcription from P_{RM}. Figure 8.29 illustrates this concept, which is known as **intergenic suppression,** because a mutation in one gene suppresses a mutation in another.

The search for such intergenic suppressor mutants could be extremely tedious if each one had to be tested separately for the desired activity. It is much more feasible to use a selection (Chapter 4) to eliminate any wild-type polymerase genes, or those with irrelevant mutations, and just keep those with the desired mutations. Susskind and colleagues used such a selection, which is described in Figure 8.30. They used *E. coli* cells with two prophages. One was a λ prophage bearing a *cI* gene with a *pc* mutation indicated by the red X; expression of this *cI* gene was driven by a weak version of the *lac* promoter. The other was a P22 prophage bearing the kanamycin-resistance gene under control of the P_{RM} promoter. Susskind and colleagues grew these cells in the presence of the antibiotic kanamycin, so the cells' survival depended on expression of the kanamycin-resistance gene. With the mutated repressor and the RNA polymerase provided by the cell, these cells could not survive because they could not activate transcription from P_{RM}.

Next, the investigators transformed the cells with plasmids bearing wild-type and mutant versions of a gene (*rpoD*) encoding the RNA polymerase σ-subunit. If the *rpoD* gene was wild-type or contained an irrelevant mutation, the product (σ) could not interact with the mutant repressor, so the cells would not grow in kanamycin. On the other hand, if they contained a suppressor mutation, the σ-factor could join with core polymerase subunits provided by the cell to form a mutated polymerase that could interact with the mutated repressor. This interaction allowed activation of transcription from P_{RM}, and the cells could therefore grow in kanamycin. In principle, these cells with a suppressor mutation in *rpoD* were the *only* cells that could grow, so they were easy to find.

Susskind and associates double-checked their mutated *rpoD* genes by introducing them into cells just like the ones used for selection, but bearing a *lacZ* gene instead of a kanamycin-resistance gene under control of P_{RM}. They assayed for production of the *lacZ* product, β-galactosidase, which is a measure of transcription from P_{RM}, and therefore of interaction between polymerase and repressor. As expected, cells with the mutated repressor that were transformed with the mutated *rpoD* gene gave as many units of β-galactosidase (120) as cells with both wild-type *cI* and *rpoD* genes (100 units). By contrast, cells with the mutated repressor transformed with the wild-type *rpoD* gene gave only 18.5 units of β-galactosidase. When Susskind and colleagues sequenced the *rpoD* genes of all eight suppressor mutants they isolated, they found the same mutation, which resulted in the change of an arginine at position 596 in the σ-factor to a histidine. This is in region 4 of the σ-factor, which is the region that recognizes the −35 box.

These data strongly support the hypothesis that polymerase–repressor interactions are essential for activation of transcription from P_{RM}. They also demonstrate

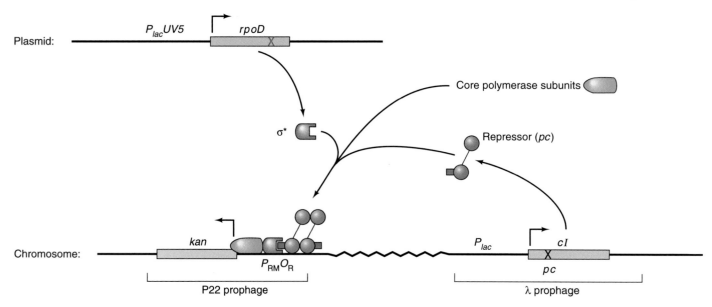

Figure 8.30   **Selection for intergenic suppressor of λ *cI pc* mutation.** Susskind and colleagues used bacteria with the chromosome illustrated (in small part) at bottom. The chromosome included two prophages: (1) a P22 prophage with a kanamycin resistance gene (yellow) driven by a λ P_{RM} promoter with adjacent λ O_R; (2) a λ prophage containing the λ *cI* gene (light green) driven by a weak *lac* promoter. Into these bacteria, Susskind and colleagues placed plasmids bearing mutagenized *rpoD* (σ-factor) genes (light blue) driven by the *lac UV5* promoter. Then they challenged the transformed cells with medium containing kanamycin. Cells transformed with a wild-type *rpoD* gene, or with *rpoD* genes bearing irrelevant mutations, could not grow in kanamycin. However, cells transformed with *rpoD* genes having a mutation (red X) that compensated for the mutation (black X) in the *cI* gene could grow. This mutation suppression is illustrated by the interaction between the mutant σ-factor (blue) and the mutant repressor (green), which permits transcription of the kanamycin resistance gene from P_{RM}.

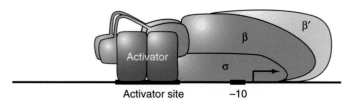

Figure 8.31   **Activation by contacting σ.** The activator (e.g., λ repressor) binds to an activator site that overlaps the weak –35 box of the promoter. This allows interaction between the activator and region 4 of σ, which would otherwise bind weakly, if at all, to the –35 box. This allows the polymerase to bind tightly to a very weak promoter and therefore to transcribe the adjacent gene successfully. (*Source:* Reprinted from Busby and Ebright, *Cell* 79:743, 1994. Copyright 1994, with permission from Elsevier Science.)

that the activating interaction between the two proteins involves the σ-factor, and not the α-subunit, as one might have expected. This provides an example of activation via σ, illustrated in Figure 8.31. Promoters subject to such activation have weak –35 boxes, which are poorly recognized by σ. The activator site, by overlapping the –35 box, places the activator (λ repressor, in this case) in position to interact with region 4 of the σ-factor—in effect substituting for the weakly recognized –35 box.

We saw in Chapter 7 that CAP–cAMP recruits RNA polymerase to the *lac* promoter by stimulating formation of the closed promoter complex. But Diane Hawley and William McClure showed that the λ repressor does not affect this step at P_{RM}. Instead, it stimulates the second step

in recruitment: helping to convert the closed promoter complex at P_{RM} to the open promoter complex.

SUMMARY Intergenic suppressor mutation studies show that the crucial interaction between repressor and RNA polymerase involves region 4 of the σ-subunit of the polymerase. This polypeptide binds near the weak –35 box of P_{RM}, which places the σ-region 4 close to the repressor bound to O_R2. Thus, the repressor can interact with the σ-factor, recruiting RNA polymerase to the weak promoter. In this way, O_R2 serves as an activator site, and λ repressor is an activator of transcription from P_{RM}. It stimulates conversion of the closed promoter complex to the open promoter complex.

Determining the Fate of a λ Infection: Lysis or Lysogeny

What determines whether a given cell infected by λ will enter the lytic cycle or lysogeny? The balance between these two fates is delicate, and we usually cannot predict the actual path taken in a given cell. Support for this assertion comes from a study of the appearance of groups of *E. coli* cells infected by λ phage. When a few phage particles are sprinkled on a lawn of bacteria in a Petri dish, they infect the cells. If a lytic infection takes place, the progeny phages spread to neighboring cells and infect them. After a few hours, we can see a circular hole in the bacterial lawn caused by the death of lytically infected cells. This hole is called a **plaque.** If the infection were 100% lytic, the plaque would be clear, because all the host cells would be killed. But λ plaques are not usually clear. Instead, they are turbid, indicating the presence of live lysogens. This means that even in the local environment of a plaque, some infected cells suffer the lytic cycle, and others are lysogenized.

Let us digress for a moment to ask this question: Why are the lysogens not infected lytically by one of the multitude of phages in the plaque? The answer is that if a new phage DNA enters a lysogen, plenty of repressor is present in the cell to bind to the new phage DNA and prevent its expression. Therefore, we can say that the lysogen is **immune** to **superinfection** by a phage with the same control region, or **immunity region,** as that of the prophage.

Now let us return to the main problem. We have seen that some cells in a plaque can be lytically infected, whereas others are lysogenized. The cells within a plaque are all genetically identical, and so are the phages, so the choice of fate is not genetic. Instead, it seems to represent a race between the products of two genes: *cI* and *cro*. This race is rerun in each infected cell, and the winner determines the pathway of the infection in that cell. If *cI* prevails, lysogeny will be established; if *cro* wins, the infection will be lytic. We can already appreciate the basics of this argument: If the *cI* gene manages to produce enough repressor, this protein will bind to O_R and O_L, prevent further transcription of the early genes, and thereby prevent expression of the late genes that would cause progeny phage production and lysis. On the other hand, if enough Cro is made, this protein can prevent *cI* transcription and thereby block lysogeny (Figure 8.32).

The key to Cro's ability to block *cI* transcription is the nature of its affinity for the λ operators: Cro binds to both O_R and O_L, as does repressor, but its order of binding to the three-part operators is exactly opposite to that of repressor. Instead of binding in the order 1, 2, 3, as repressor does, Cro binds first to O_R3. As soon as that happens, *cI* transcription from P_{RM} stops, because O_R3 overlaps P_{RM}. In other words, Cro acts as a repressor. Furthermore, when Cro fills up all the rightward and leftward operators, it prevents transcription of all the early genes from P_R and P_L, including *cII* and *cIII*. Without the products of these genes, P_{RE} cannot function; so all repressor synthesis ceases. Lytic infection is then ensured. Cro's turning off of early transcription is also required for lytic growth. Continued production of delayed early proteins late in infection aborts the lytic cycle.

But what determines whether *cI* or *cro* wins the race? Surely it is more than just a flip of the coin. Actually, the most important factor seems to be the concentration of

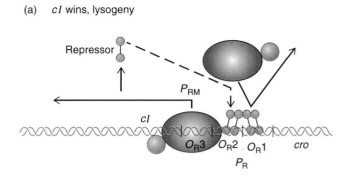

(a)   *cI* wins, lysogeny

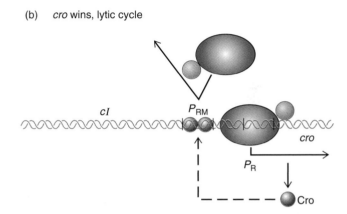

(b)   *cro* wins, lytic cycle

Figure 8.32   The battle between *cI* and *cro*. (a) *cI* wins. Enough repressor (green) is made by transcription of the *cI* gene from P_{RM} that it blocks polymerase (red and blue) from binding to P_R and therefore blocks *cro* transcription. Lysogeny results. **(b)** *cro* wins. Enough Cro (purple) is made by transcription from P_R that it blocks polymerase from binding to P_{RM} and therefore blocks *cI* transcription. The lytic cycle results.

the *cII* gene product, CII. The higher the CII concentration within the cell, the more likely lysogeny becomes. This fits with what we have already learned about CII—it activates P_{RE} and thereby helps turn on the lysogenic program. We have also seen that the activation of P_{RE} by CII works against the lytic program by producing *cro* antisense RNA that can inhibit translation of *cro* sense RNA.

And what controls the concentration of CII? We have seen that CIII protects CII against cellular proteases, but high protease concentrations can overwhelm CIII, destroy CII, and ensure that the infection will be lytic. Such high protease concentrations occur under good environmental conditions—rich medium, for example. By contrast, protease levels are depressed under starvation conditions. Thus, starvation tends to favor lysogeny, whereas rich medium favors the lytic pathway. This is advantageous for the phage because the lytic pathway requires considerable energy to make all the phage DNA, RNAs, and proteins, and this much energy may not be available during starvation. In comparison, lysogeny is cheap; after it is established, it requires only the synthesis of a little repressor.

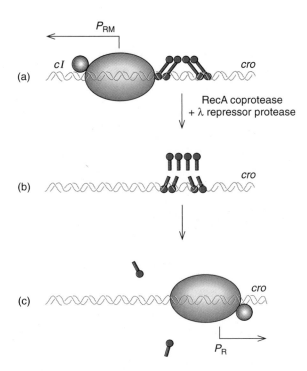

Figure 8.33 Inducing the λ prophage. (a) Lysogeny. Repressor (green) is bound to O_R (and O_L) and *cI* is being actively transcribed from the P_{RM} promoter. **(b)** The RecA coprotease (activated by ultraviolet light or other mutagenic influence) unmasks a protease activity in the repressor, so it can cleave itself. **(c)** The severed repressor falls off the operator, allowing polymerase (red and blue) to bind to P_R and transcribe *cro*. Lysogeny is broken.

SUMMARY Whether a given cell is lytically or lysogenically infected by phage λ depends on the outcome of a race between the products of the *cI* and *cro* genes. The *cI* gene codes for repressor, which blocks O_R1, O_R2, O_L1, and O_L2, turning off all early transcription, including transcription of the *cro* gene. This leads to lysogeny. On the other hand, the *cro* gene codes for Cro, which blocks O_R3 (and O_L3), turning off *cI* transcription. This leads to lytic infection. Whichever gene product appears first in high enough concentration to block its competitor's synthesis wins the race and determines the cell's fate. The winner of this race is determined by the CII concentration, which is determined by the cellular protease concentration, which is in turn determined by environmental factors such as the richness of the medium.

Lysogen Induction

We mentioned that a lysogen can be induced by treatment with mutagenic chemicals or radiation. The mechanism of this induction is as follows: *E. coli* cells respond to DNA-damaging environmental insults, such as mutagens or radiation, by inducing a set of genes whose collective activity is called the **SOS response;** one of these genes, in fact the most important, is *recA*. The *recA* product (**RecA**) participates in recombination repair of DNA damage (which explains part of its usefulness to the SOS response), but environmental insults also induce a new activity in the RecA protein. It becomes a coprotease that stimulates a latent protease, or protein-cleaving activity, in the λ repressor. This protease then cleaves the repressor in half and releases it from the operators, as shown in Figure 8.33. As soon as that happens, transcription begins from P_R and P_L. One of the first genes transcribed is *cro*, whose product shuts down any further transcription of the repressor gene. Lysogeny is broken and lytic phage replication begins.

Surely λ would not have evolved with a repressor that responds to RecA by chopping itself in half unless it provided an advantage to the phage. That advantage seems to be this: The SOS response signals that the lysogen is under some kind of DNA-damaging attack. It is expedient under those circumstances for the prophage to get out by inducing the lytic cycle, rather like rats deserting a sinking ship.

SUMMARY When a lysogen suffers DNA damage, it induces the SOS response. The initial event in this response is the appearance of a coprotease activity in the RecA protein. This causes the repressors to cut themselves in half, removing them from the λ operators and inducing the lytic cycle. In this way, progeny λ phages can escape the potentially lethal damage that is occurring in their host.

SUMMARY

Bacteria experience many different major shifts in transcription pattern (e.g., during phage infection or sporulation), and several mechanisms have evolved to effect these shifts. For example, transcription of phage SPO1 genes in infected *B. subtilis* cells proceeds according to a temporal program in which early genes are transcribed first, then middle genes, and finally late genes. This switching is directed by a set of phage-encoded σ-factors that associate with the host core RNA polymerase and change its specificity from early to middle to late. The host σ is specific for the phage early genes, the phage gp28 switches the specificity to the middle genes; and the phage gp33 and gp34 switch to late specificity.

Phage T7, instead of coding for a new σ-factor to change the host polymerase's specificity from early to late, encodes a new RNA polymerase with absolute specificity for the later phage genes. This polymerase, composed of a single polypeptide, is a product of one of the earliest phage genes, gene 1. The temporal program in the infection by this phage is simple. The host polymerase transcribes the earliest (class I) genes, one of whose products is the phage polymerase, which then transcribes the later genes.

When the bacterium *B. subtilis* sporulates, a whole new set of sporulation-specific genes turns on, and many, but not all, vegetative genes turn off. This switch takes place largely at the transcription level. It is accomplished by several new σ-factors that displace the vegetative σ-factor from the core RNA polymerase and direct transcription of sporulation genes instead of vegetative genes. Each σ-factor has its own preferred promoter sequence.

Some prokaryotic genes must be transcribed under conditions where two different σ-factors are active. These genes are equipped with two different promoters, each recognized by one of the two σ-factors. This ensures their expression no matter which factor is present and allows for differential control under different conditions. The heat shock response in *E. coli* is governed by an alternative σ-factor, σ^{32} (σ^H) which displaces σ^{70} (σ^A) and directs the RNA polymerase to the heat shock gene promoters.

The immediate early/delayed early/late transcriptional switching in the lytic cycle of phage λ is controlled by antiterminators. One of the two immediate early genes is *cro*, which codes for a cI repressor that allows the lytic cycle to continue. The other, *N*, codes for an antiterminator, N, that overrides the terminators after the *N* and *cro* genes. Transcription then continues into the delayed early genes. One of the delayed early genes, *Q*, codes for another antiterminator (Q) that permits transcription of the late genes from the late promoter, $P_{R'}$, to continue without premature termination.

Five proteins (N, NusA, NusB, NusG, and S10) collaborate in antitermination at the λ immediate early terminators. NusA and S10 bind to RNA polymerase, and N and NusB bind to the boxB and boxA regions, respectively, of the *nut* site in the growing transcript. N and NusB bind to NusA and S10, respectively, probably tethering the transcript to the polymerase. This alters the polymerase so it reads through the terminators at the ends of the immediate early genes. Antitermination in the λ late region requires Q, which binds to the Q-binding region of the *qut* site as RNA polymerase is stalled just downstream of the late promoter. This appears to alter the polymerase so it can ignore the terminator and transcribe the late genes.

Phage λ establishes lysogeny by causing production of enough repressor to bind to the early operators and prevent further early RNA synthesis. The promoter used

for establishment of lysogeny is P_{RE}, which lies to the right of P_R and *cro*. Transcription from this promoter goes leftward through the *cI* gene. The products of the delayed early genes *cII* and *cIII* also participate in this process: CII, by directly stimulating polymerase binding to P_{RE}; CIII, by slowing degradation of CII.

The promoter that is used to maintain lysogeny is P_{RM}. It comes into play after transcription from P_{RE} makes possible the burst of repressor synthesis that establishes lysogeny. This repressor binds to O_R1 and O_R2 cooperatively, but leaves O_R3 open. RNA polymerase binds to P_{RM}, which overlaps O_R3 in such a way that it just touches the repressor bound to O_R2. This protein–protein interaction is required for this promoter to work efficiently.

The crucial interaction between the λ repressor and RNA polymerase involves region 4 of the σ-subunit of the polymerase. This polypeptide binds near the weak −35 box of P_{RM}, which places the σ-region 4 close to the repressor bound to O_R2. Thus, the repressor can interact with the σ-factor, attracting RNA polymerase to the weak promoter. In this way, O_R2 serves as an activator site, and λ repressor is an activator of transcription from P_{RM}.

Whether a given cell is lytically or lysogenically infected by phage λ depends on the outcome of a race between the products of the *cI* and *cro* genes. The *cI* gene codes for repressor, which blocks O_R1, O_R2, O_L1, and O_L2, turning off all early transcription, including transcription of the *cro* gene. This leads to lysogeny. On the other hand, the *cro* gene codes for Cro, which blocks O_R3 (and O_L3), turning off *cI* transcription. This leads to lytic infection. Whichever gene product appears first in high enough concentration to block its competitor's synthesis wins the race and determines the cell's fate. The winner of this race is determined by the CII concentration, which is determined by the cellular protease concentration, which is in turn determined by environmental factors such as the richness of the medium.

When a λ lysogen suffers DNA damage, it induces the SOS response. The initial event in this response is the appearance of a coprotease activity in the RecA protein. This causes the repressors to cut themselves in half, removing them from the λ operators and inducing the lytic cycle. In this way, progeny λ phages can escape the potentially lethal damage that is occurring in their host.

REVIEW QUESTIONS

1. Present a model to explain how the phage SPO1 controls its transcription program.

2. Summarize the evidence to support your answer to question 1.

3. Present a model to explain how the phage T7 controls its transcription program.

4. Describe and give the results of an experiment that shows that the *B. subtilis* σ^E recognizes the sporulation-specific 0.4-kb promoter, whereas σ^A recognizes a vegetative promoter.

5. Summarize the mechanism *B. subtilis* cells use to alter their transcription program during sporulation.

6. How do we know that the *B. subtilis* σ^K is the product of two genes that come together into one gene during sporulation?

7. Describe and give the results of an experiment that shows that the *B. subtilis* σ^E recognizes the *spoIID* promoter, but other σ-factors do not.

8. You are studying a gene that you suspect has two different promoters, recognized by two different σ-factors. How would you prove your hypothesis?

9. Why is it an advantage for the *Anabaena* glutamine synthetase gene to have two different promoters?

10. Why is it an advantage for the *E. coli* glutamine synthetase gene to have two different promoters?

11. Present a model for the rapid response of *E. coli* to heat shock.

12. How does λ phage switch from immediate early to delayed early to late transcription during lytic replication.

13. Present a model for N-directed antitermination in λ phage-infected *E. coli* cells. What proteins are involved?

14. The P_{RE} promoter is barely recognizable and is not in itself attractive to RNA polymerase. How can the *cI* gene be transcribed from this promoter?

15. Describe and give the results of an experiment that shows where CII binds on λ DNA.

16. With a diagram, explain DNase footprinting. What kind of information does it provide?

17. How can CII and RNA polymerase bind to the same region of DNA at the same time?

18. Describe and give the results of an experiment that shows that λ repressor regulates its own synthesis both positively and negatively. What does this same experiment show about the effect of repressor on *cro* transcription?

19. Diagram the principle of an intergenic suppression assay to detect interaction between two proteins.

20. Describe and give the results of an intergenic suppression experiment that shows interaction between the λ repressor and the σ-subunit of the *E. coli* RNA polymerase.

21. Present a model to explain the struggle between *cI* and *cro* for lysogenic or lytic infection of *E. coli* by λ phage.

22. Present a model to explain the induction of a λ lysogen by mutagenic insults.

SUGGESTED READINGS

General References and Reviews

Busby, S. and R.H. Ebright. 1994. Promoter structure, promoter recognition, and transcription activation in prokaryotes. *Cell* 79:743–46.

Goodrich, J.A. and W.R. McClure. 1991. Competing promoters in prokaryotic transcription. *Trends in Biochemical Sciences* 15:394–97.

Gralla, J.D. 1991. Transcriptional control—lessons from an *E. coli* data base. *Cell* 66:415–18.

Greenblatt, J., J.R. Nodwell, and S.W. Mason. 1993. Transcriptional antitermination. *Nature* 364:401–6.

Helmann, J.D. and M.J. Chamberlin. 1988. Structure and function of bacterial sigma factors. *Annual Review of Biochemistry* 57:839–72.

Losick, R., and M. Chamberlin. 1976. *RNA Polymerase*. Plainview, NY: Cold Spring Harbor Laboratory Press.

Ptashne, M. 1992. *A Genetic Switch*. Cambridge, MA: Cell Press.

Research Articles

Haldenwang, W.G., N. Lang, and R. Losick. 1981. "A sporulation-induced sigma-like regulatory protein from *B. subtilis.*" *Cell* 23:615–24.

Hawley, D.K. and W.R. McClure. 1982. Mechanism of activation of transcription initiation from the λ P_{RM} promoter. *Journal of Molecular Biology* 157:493–525.

Ho, Y.-S., D.L. Wulff, and M. Rosenberg. 1983. Bacteriophage λ protein cII binds promoters on the opposite face of the DNA helix from RNA polymerase. *Nature* 304:703–8.

Hunt, T.P. and B. Magasanik. 1985. Transcription of *glnA* by purified *E. coli* components: Core RNA polymerase and the products of *glnF, glnG,* and *glnL. Proceedings of the National Academy of Sciences, USA* 82:8453–57.

Johnson, W.C., C.P. Moran, Jr., and R. Losick. 1983. Two RNA polymerase sigma factors from *Bacillus subtilis* discriminate between overlapping promoters for a developmentally regulated gene. *Nature* 302:800–4.

Li, M., H. Moyle, and M.M. Suskind. 1994. Target of the transcriptional activation function of phage λ cI protein. *Science* 263:75–77.

Meyer, B.J., D.G. Kleid, and M. Ptashne. 1975. λ repressor turns off transcription of its own gene. *Proceedings of the National Academy of Science, USA* 72:4785–89.

Pero, J., R. Tjian, J. Nelson, and R. Losick. 1975. In vitro transcription of a late class of phage SPO1 genes. *Nature* 257:248–51.

Rong, S., M.S. Rosenkrantz, and A.L. Sonenshein. 1986. Transcriptional control of the *B. subtilis* SpoIID gene. *Journal of Bacteriology* 165:771–79.

Stragier, P., B. Kunkel, L. Kroos, and R. Losick. 1989. Chromosomal rearrangement generating a composite gene for a developmental transcription factor. *Science* 243:507–12.

Tumer, N.E., S.J. Robinson, and R. Haselkorn. 1983. Different promoters for the *Anabaena* glutamine synthetase gene during growth using molecular or fixed nitrogen. *Nature* 306:337–42.

DNA–Protein Interactions in Prokaryotes

Gene regulation: Computer model of Cro protein bound to DNA.
© Ken Eward/SS/Photo Researchers, Inc.

In Chapters 7 and 8 we discussed several proteins that bind tightly to specific sites on DNA. These include RNA polymerase, *lac* repressor, CAP, *trp* repressor, λ repressor, and Cro. All but the first of these have been studied in detail, and all can locate and bind to one particular short DNA sequence among a vast excess of unrelated sequences. How do these proteins accomplish such specific binding—akin to finding a needle in a haystack? All five proteins have a similar structural motif: two α-helices connected by a short protein "turn." This helix-turn-helix motif (Figure 9.1a) allows the second helix (the recognition helix) to fit snugly into the major groove of the target DNA site (Figure 9.1b). We will see that the configuration of this fit varies considerably from one protein to another, but all the proteins fit their DNA binding sites like a key in a lock. In this chapter we will explore several well-studied examples of specific DNA–protein interactions that occur in prokaryotic cells to see what makes them so specific. In Chapter 12 we will consider several other DNA-binding motifs that occur in eukaryotes.

9.1 The λ Family of Repressors

The repressors of λ and similar phages have recognition helices that lie in the major groove of the appropriate operator as shown in Figure 9.2. The specificity of this binding depends on certain amino acids in the recognition helices that make specific contact with functional groups of certain bases protruding into the major DNA groove, and with phosphate groups in the DNA backbone. Other proteins with helix-turn-helix motifs are not able to bind as well at that same site because they do not have the correct amino acids in their recognition helices. We would like to know which are the important amino acids in these interactions.

Mark Ptashne and his colleagues have provided part of the answer, using repressors from two λ-like phages, **434** and **P22**, and their respective operators. These two phages have very similar molecular genetics, but they have different immunity regions: They make different repressors that recognize different operators. Both repressors resemble the λ repressor in that they contain helix-turn-helix motifs. However, because they recognize operators with different base sequences, we would expect them to have different amino acids in their respective recognition helices, especially those amino acids that are strategically located to contact the bases in the DNA major groove.

Using x-ray diffraction analysis (Box 9.1) of operator–repressor complexes, Stephen Harrison and Ptashne identified the face of the recognition helix of the 434 phage repressor that contacts the bases in the major groove of its operator. By analogy, they could make a similar prediction for the P22 repressor. Figure 9.3 schematically illustrates the amino acids in each repressor that are most likely to be involved in operator binding.

If these are really the important amino acids, one ought to be able to change only these amino acids and thereby alter the specificity of the repressor. In particular, one should be able to employ such changes to alter the 434 repressor so that it recognizes the P22 operator instead of its own. This is exactly what Robin Wharton and Ptashne did. They started with a cloned gene for the 434 repressor and, using mutagenesis techniques (described in Chapter 5), systematically altered the codons for five amino acids in the 434 recognition helix to codons for the five corresponding amino acids in the P22 recognition helix.

Next, they expressed the altered gene in bacteria and tested the product for ability to bind to 434 and P22 operators, both in vivo and in vitro. The in vivo assay was to check for immunity. Recall that an *E. coli* cell lysogenized by λ phage is immune to superinfection by λ because the excess λ repressor in the lysogen immediately binds to the superinfecting λ DNA and prevents its expression. Phages 434 and P22 are λ-like (lambdoid) phages, but they differ in their immunity regions, including the operators. Thus, a 434 lysogen is immune to superinfection by 434, but not by P22. The 434 repressor cannot bind to the P22 operators and therefore cannot prevent superinfection by the P22 phage. The reverse is also true: A P22 lysogen is immune to superinfection by P22, but not by 434.

Instead of creating lysogens, Wharton and Ptashne transformed *E. coli* cells with a plasmid encoding the recombinant 434 repressor, then asked whether the recombinant 434 repressor (with its recognition helix altered to be like the P22 recognition helix) still had its original binding specificity. If so, cells producing the recombinant repressor should have been immune to 434 infection. On the other hand, if the binding specificity had changed, the cells producing the recombinant repressor should have

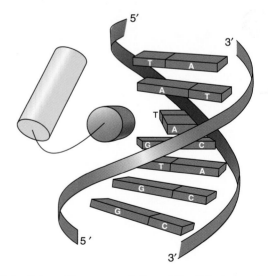

Figure 9.2 Schematic representation of the fit between the recognition helix of a λ repressor monomer and the major groove of the operator region of the DNA. The recognition helix is represented by a red cylinder that lies in the major groove in a position to facilitate hydrogen bonding with the edges of base pairs in the DNA. (*Source:* From Jordan and Pabo, *Science* 242:896, 1988. Copyright © 1988 American Association for the Advancement of Science, Washington, DC. Reprinted by permission.)

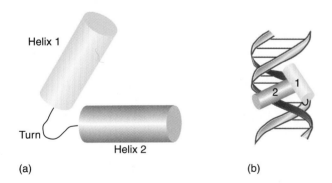

(a)   (b)

Figure 9.1 The helix-turn-helix motif as a DNA-binding element.
(a) The helix-turn-helix motif of the λ repressor. (b) The fit of the helix-turn-helix motif of one repressor monomer with the λ operator. Helix 2 of the motif (red) lies in the major groove of its DNA target; some of the amino acids on the back of this helix (away from the viewer) are available to make contacts with the DNA.

(a)

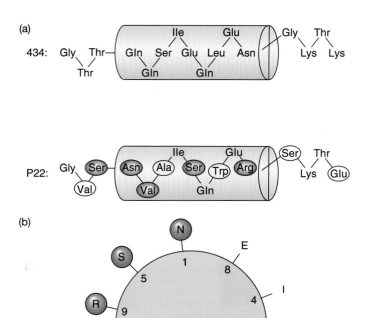

(b)

Figure 9.3 The recognition helices of two λ-like phage repressors. (a) Key amino acids in the recognition helices of two repressors. The amino acid sequences of the recognition helices of the 434 and P22 repressors are shown, along with a few amino acids on either side. Amino acids that differ between these two proteins are circled; these are more likely to contribute to differences in specificity. Furthermore, the amino acids on the side of the helix that faces the DNA are most likely to be involved in DNA binding. These, along with one amino acid in the turn just before the helix (red), were changed to alter the binding specificity of the protein. (b) The recognition helix of the P22 repressor viewed on end. The numbers represent the positions of the amino acids in the protein chain. The left-hand side of the helix faces toward the DNA, so the amino acids on that side are more likely to be important in binding. Those that differ from amino acids in corresponding positions in the 434 repressor are circled in red. (*Source:* (b) Adapted with permission from *Nature* 316:602, 1985. Copyright © 1985 Macmillan Magazines Limited.)

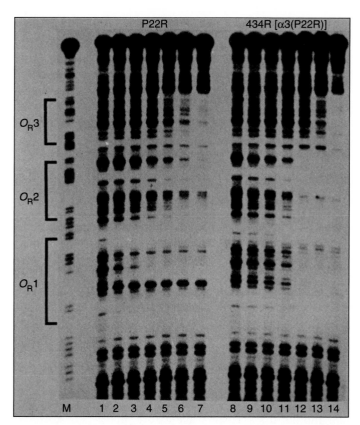

Figure 9.4 DNase footprinting with the recombinant 434 repressor. Wharton and Ptashne performed DNase footprinting with end-labeled P22 phage O_R and either P22 repressor (P22R, lanes 1–7) or the 434 repressor with five amino acids in the recognition helix (α-helix 3) changed to match those in the phage P22 recognition helix (434R[α3(P22R)], lanes 8–14). The two sets of lanes contained increasing concentrations of the respective repressors (0 M in lanes 1 and 8, and ranging from 7.6×10^{-10} M to 1.1×10^{-8} M in lanes 2–7 and from 5.2×10^{-9} M to 5.6×10^{-7} M in lanes 8–14). The marker lane (M) contained the A + G reaction from a Maxam–Gilbert sequencing procedure. The positions of all three rightward operators are indicated with brackets at left. (*Source:* Wharton and Ptashne, Changing the binding specificity of a repressor by redesigning an alpha-helix. *Nature* 316 (15 Aug 1985), f. 3, p. 603. © Macmillan Magazines Ltd.)

been immune to P22 infection. Actually, 434 and P22 do not infect *E. coli* cells, so the investigators used recombinant λ phages with the 434 and P22 immunity regions (λ$_{imm}$434 and λ$_{imm}$P22, respectively) in these tests. They found that the cells producing the altered 434 repressor *were* immune to infection by the λ phage with the P22 immunity region, but not to infection by the λ phage with the 434 immunity region.

To check these results, Wharton and Ptashne measured DNA binding in vitro by DNase footprinting. They found that the purified recombinant repressor could make a "footprint" in the P22 operator, just as the P22 repressor can (Figure 9.4). In control experiments (not shown) they demonstrated that the recombinant repressor could no

longer make a footprint in the 434 operator. Thus, the binding specificity really had been altered by these five amino acid changes. In further experiments, Ptashne and colleagues showed that the first four of these amino acids were necessary and sufficient for either binding activity. That is, if the repressor had TQQE (threonine, glutamine, glutamine, glutamate) in its recognition helix, it would bind to the 434 operator. On the other hand, if it had SNVS (serine, asparagine, valine, serine), it would bind to the P22 operator.

What if Wharton and Ptashne had not tried to *change* the specificity of the repressor, but just to *eliminate* it? They could have identified the amino acids in the repressor that were probably important to specificity, then changed them to other amino acids chosen at random and shown that this recombinant 434 repressor could no longer bind to its operator. If that is all they had done,

BOX 9.1

X-Ray Crystallography

This chapter contains many examples of structures of DNA-binding proteins obtained by the method of **x-ray diffraction analysis,** also called **x-ray crystallography.** This box provides an introduction to this very powerful technique.

X-rays are electromagnetic radiation, just like light rays, but with much shorter wavelengths so they are much more energetic. Thus, it is not surprising that the principle of x-ray diffraction analysis is in some ways similar to the principle of light microscopy. Figure B9.1 illustrates this similarity. In light microscopy (Figure B9.1a), visible light is scattered by an object; then a lens collects the light rays and focuses them to create an image of the object.

In x-ray diffraction, x-rays are scattered by an object (a crystal). But here we encounter a major problem: No lens is capable of focusing x-rays, so we must use a relatively indirect method to create the image. That method is based on the following considerations: When x-rays interact with an electron cloud around an atom, the x-rays scatter in every direction. However, because x-ray beams interact with multiple atoms, most of the scattered x-rays cancel one another due to their wave nature. But x-rays scattered to certain specific directions are amplified in a phenomenon called **diffraction.** Bragg's law, $2d \sin \theta = \lambda$, describes the relationship between the angle (θ) of diffraction and spacing (d) of scattering planes. As you can see in Figure B9.2, x-ray 2 travels $2 \times d \sin \theta$ longer than x-ray 1. Thus, if the wavelength (λ) of x-ray 2 is equal to $2 d \sin \theta$, the resultant rays from the scattered x-ray 1 and x-ray 2 have the same phase and are therefore amplified. On the other hand, the resultant rays are diminished if λ is not equal to $2 d \sin \theta$. These diffracted x-rays are recorded as spots on a collecting device (a detector) placed in the path of the x-rays. This device can be as simple as a sheet of x-ray film, but nowadays much more efficient electronic detectors are available. Figure B9.3 shows a diffraction pattern of a simple protein, lysozyme. Even though the protein is relatively simple (only 129 amino acids), the pattern of spots is complex. To obtain the protein structure in three dimensions, one must rotate the crystal and record diffraction patterns in many different orientations.

The next task is to use the arrays of spots in the diffraction patterns to figure out the structure of the molecule that caused the diffraction. Unfortunately, one cannot reconstruct the **electron-density map** (electron cloud distribution) from the arrays of spots in the diffraction patterns, because information about the physical parameters, called **phase angles,** of individual reflections are not

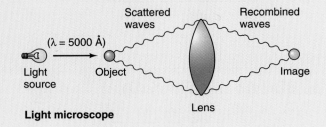

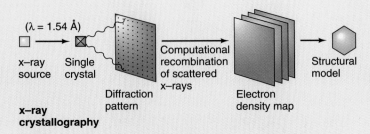

Figure B9.1   Schematic diagram of the procedures followed for image reconstruction in light microscopy (top) and x-ray crystallography (bottom). (*Source:* From Zubay et al., *Principles of Biochemistry.* Copyright © 1993 The McGraw Hill Companies, New York. Reprinted by permission.)

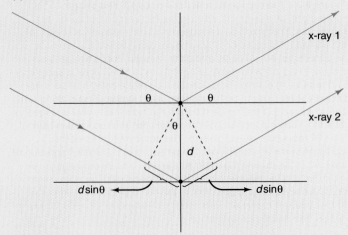

Figure B9.2   Reflection of two x-rays from parallel planes of a crystal. The two x-rays (1 and 2) strike the planes at angle θ and are reflected at the same angle. The planes are separated by distance d. The extra distance traveled by x-ray 2 is $2 d \sin \theta$.

included in the diffraction pattern. To solve this problem, crystallographers make 3–10 different heavy-atom derivative crystals by soaking heavy atom solutions (Hg, Pt, U, etc.) into protein crystals. These heavy atoms tend to bind to reactive amino acid residues, such as cysteine, histidine, and aspartate, without changing the protein structure.

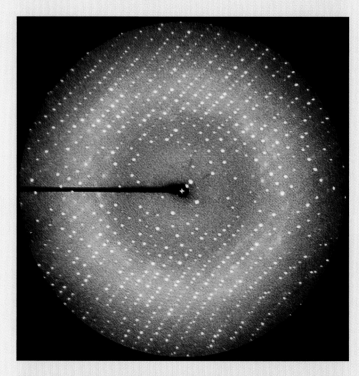

Figure B9.3   Sample diffraction pattern of a crystal of the protein lysozyme. The dark line from the left is the shadow of the arm that holds the beam stop, which protects the detector from the x-ray beam. The location of the crystal is marked by the (+) at the center.

This procedure is called multiple isomorphous replacement (MIR). The phase angles of individual reflections are determined by comparing the diffraction patterns from the native and heavy-atom derivative crystals. Once the phase angles are obtained, the diffraction pattern is mathematically converted to an electron-density map of the diffracting molecule. Then the electron-density map can be used to infer the structure of the diffracting molecule. Using the diffracted rays to create an image of the diffracting object is analogous to using a lens. But this is not accomplished physically, as a lens would; it is done mathematically. Figure B9.4 shows the electron-density map of part of the structure of lysozyme, surrounding a stick diagram representing the molecular structure inferred from the map. Figure B9.5 shows three different representations of the whole lysozyme molecule deduced from the electron-density map of the whole molecule.

Why are single crystals used in x-ray diffraction analysis? It is clearly impractical to place a single molecule of a

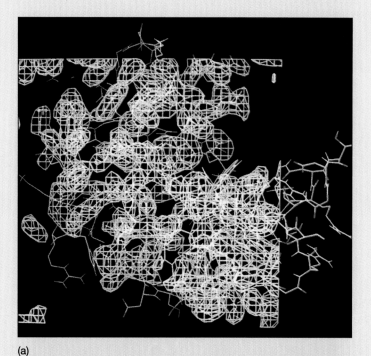

(a)

(b)

Figure B9.4   Electron-density map of part of the lysozyme molecule. (a) Low magnification, showing the electron density map of most of the molecule. The blue cages correspond to regions of high electron density. They surround a stick model of the molecule (red, yellow, and blue) inferred from the pattern of electron density. (b) High magnification, showing the center of the map in panel (a). The resolution of this structure was 2.4 Å so the individual atoms were not resolved. But this resolution is good enough to identify the unique shape of each amino acid. (*Source:* Courtesy Fusao Takusagawa.)

continued

X-ray Crystallography (*continued*)

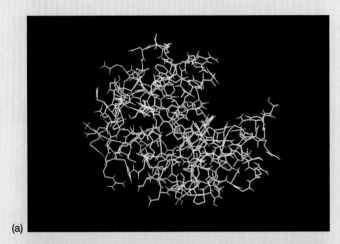

(a)

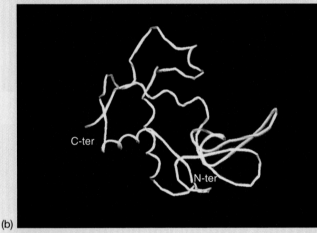

(b)

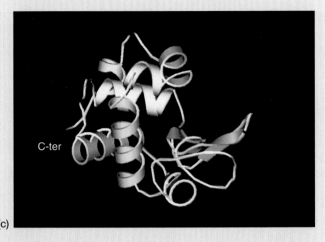

(c)

Figure B9.5   Three representations of the structure of lysozyme calculated from electron density maps such as those in Figure B9.3. (a) Stick diagram as in Figure B9.4a. **(b)** String diagram with α-helices in green, β-sheets in magenta, and random coils in blue. The N-terminus and C-terminus of the protein are marked N-ter and C-ter, respectively. **(c)** Ribbon diagram with same color coding as in panel **(b)**. The helical nature of the α-helices is obvious in this diagram. The cleft at upper right in all three diagrams is the active site of the enzyme. (*Source:* Courtesy Fusao Takusagawa.)

protein in the path of the x-rays; even if it could be done, the diffraction power from a single molecule would be too weak to detect. Therefore, many molecules of protein are placed in the x-ray beam so the signal will be strong enough to detect. Why not just use a protein powder or a solution of protein? The problem with this approach is that the molecules in a powder or solution are randomly oriented, so x-rays diffracted by such a sample would not have an interpretable pattern.

The solution to the problem is to use a crystal of protein. A crystal is composed of many small repeating units (**unit cells**) that are three-dimensionally arranged in a regular way. A unit cell of a protein contains several protein molecules that are usually related by special symmetries. Thus, diffractions by all the molecules in a unit cell in the crystal are the same, and they reinforce one another. To be useful for x-ray diffraction, the smallest dimension of a protein crystal should be at least 0.1 mm. A cubic crystal of this size contains more than 10^{12} molecules (assuming that one protein molecule occupies a $50 \times 50 \times 50$ Å space). Figure B9.6 presents a photograph of crystals of lysozyme suitable for x-ray diffraction analysis. Protein crystals contain not only pure protein but also a large amount of solvent (30–70% of their weight). Thus, their environment in the crystal resembles that in solution, and their three-dimensional structure in the crystal should therefore be

they could have said that the results were consistent with the hypothesis that the altered amino acids are directly involved in binding. But this alternative explanation would remain: These amino acids could simply be important to the overall three-dimensional shape of the repressor protein, and changing them changed this shape and therefore indirectly prevented binding. By contrast, changing specificity by changing amino acids is strong evidence for the direct involvement of these amino acids in binding.

In a related x-ray crystallographic study, Ptashne and coworkers showed that the λ repressor has an amino-terminal arm not found in the repressors of the 434 and P22 phages. This arm contributes to the repressor's binding to the λ operator by embracing the operator. Figure 9.5 shows a computer model of a dimer of λ repressor interacting with λ operator. In the repressor monomer at the top, the helix-turn-helix motif is visible projecting into the major groove of the DNA. At the bottom, we can see the

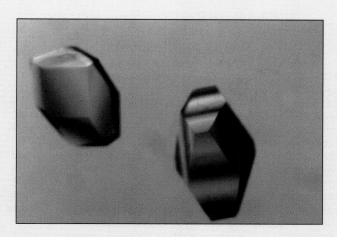

Figure B9.6 Crystals of lysozyme. The photograph was taken using polarizing filters to produce the color in the crystals. The actual size of these crystals is approximately 0.5 × 0.5 × 0.5 mm. (*Source:* Courtesy Fusao Takusagawa.)

atoms have a wavelength of 1.54 Å, which is ideal for high-resolution x-ray diffraction analysis of proteins.

In this chapter we will see protein structures at various levels of resolution. What is the reason for these differences in resolution? A protein crystal in which the protein molecules are relatively well ordered gives many diffraction spots far from the incident beam, that is, from the center of the detector. These spots are produced by x-rays with large diffraction angles (θ, see Figure B9.2). An electron-density map calculated from these diffraction spots from a relatively ordered crystal gives a high-resolution image of the diffracting molecule. On the other hand, a protein crystal whose molecules are relatively poorly arranged gives diffraction spots only near the center of the detector, resulting from x-rays with small diffraction angles. Such data produce a relatively low-resolution image of the molecule.

This relationship between resolution and diffraction angle is another consequence of Bragg's law $2d \sin \theta = \lambda$. Rearranging Bragg's equation, we find $d = \lambda/2 \sin \theta$. So we see that d, the distance between structural elements in the protein, is inversely related to $\sin \theta$. Therefore, the larger the distance between structural elements in the crystal, the smaller the angle of diffraction and the closer to the middle of the pattern the diffracted ray will fall. This is just another way of saying that low-resolution structure (with large distances between elements) gives rise to the pattern of spots near the middle of the diffraction pattern. By the same argument, high-resolution structure gives rise to spots near the periphery of the pattern because they diffract the x-rays at a large angle. When crystallographers can make crystals that are good enough to give this kind of high resolution, they can build a detailed model of the structure of the protein.

The proteins we are considering in this chapter are DNA-binding proteins. In many cases, investigators have prepared cocrystals of the protein and a double-stranded DNA fragment containing the target sequence recognized by the protein. These can reveal not only the shapes of the protein and DNA in the protein–DNA complex, but also the atoms that are involved in the protein–DNA interaction.

close to their structure in solution. In general, then, we can be confident that the protein structures determined by x-ray crystallography are close to their structures in the cell. In fact, most enzyme crystals retain their enzymatic activities.

Why not just use visible light rays to see the structures of proteins and avoid all the trouble involved with x-rays? The problem with this approach lies in **resolution**—the ability to distinguish separate parts of the molecule. The ultimate goal in analyzing the structure of a molecule is to distinguish each atom, so the exact spatial relationship of all the atoms in the molecule is apparent. But atoms have dimensions on the order of angstroms (1 Å = 10^{-10} m), and the maximum resolving power of radiation is one-third of its wavelength ($0.6\lambda/2 \sin \theta$). So we need radiation with a very short wavelength (measured in angstroms) to resolve the atoms in a protein. But visible light has wavelengths averaging about 500 nm (5000 Å). Thus, it is clearly impossible to resolve atoms with visible light. By contrast, x-rays have wavelengths of one to a few angstroms. For example, the characteristic x-rays emitted by excited copper

arm of the other repressor monomer reaching around to embrace the DNA.

Cro also uses a helix-turn-helix DNA binding motif and binds to the same operators as the λ repressor, but it has the exact opposite affinity for the three different operators in a set (Chapter 8). That is, it binds first to O_R3 and last to O_R1, rather than vice versa. Therefore, by changing amino acids in the recognition helices, one ought to be able to identify the amino acids that give Cro

and repressor their different binding specificities. Ptashne and his coworkers accomplished this task and found that amino acids 5 and 6 in the recognition helices are especially important, as is the amino-terminal arm in the λ repressor. When these workers altered base pairs in the operators, they discovered that the base pairs critical to discriminating between O_R1 and O_R3 are at position 3, to which Cro is more sensitive, and at positions 5 and 8, which are selective for repressor binding.

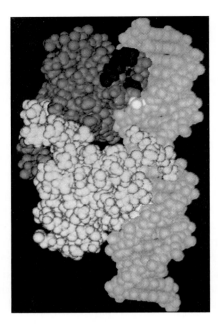

Figure 9.5   Computer model of the λ repressor dimer binding to λ operator (O_R2). The DNA double helix (light blue) is at right. The two monomers of the repressor are in dark blue and yellow. The helix-turn-helix motif of the upper monomer (red and blue) is inserted into the major groove of the DNA. The arm of the lower monomer reaches around to embrace the DNA. (*Source:* Hochschild et al., Repressor structure and the mechanism of positive control. *Cell* 32 (1983) p. 322. Reprinted by permission of Elsevier Science. Photo by Richard Feldman)

SUMMARY The repressors of the λ-like phages have recognition helices that fit sideways into the major groove of the operator DNA. Certain amino acids on the DNA side of the recognition helix make specific contact with bases in the operator, and these contacts determine the specificity of the protein–DNA interactions. In fact, changing these amino acids can change the specificity of the repressor. The λ repressor itself has an extra motif not found in the other repressors, an amino-terminal arm that aids binding by embracing the DNA. The λ repressor and Cro share affinity for the same operators, but they have microspecificities for O_R1 or O_R3, determined by interactions between different amino acids in the recognition helices of the two proteins and different base pairs in the two operators.

High-Resolution Analysis of λ Repressor–Operator Interactions

Steven Jordan and Carl Pabo wished to visualize the λ repressor–operator interaction at higher resolution than previous studies allowed. They were able to achieve a resolution of 2.5 Å by making excellent cocrystals of a repressor fragment and an operator fragment. The repressor fragment encompassed residues 1–92, which included all

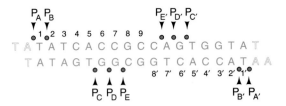

Figure 9.6   The operator fragment used to prepare operator–repressor cocrystals. This 20-mer contains the two λ O_L1 half-sites, each of which binds a monomer of repressor. The half-sites are included within the 17-bp region in boldface; each half-site contains 8 bp, separated by a G-C pair in the middle. The half-site on the left has a consensus sequence; that on the right deviates somewhat from the consensus. The base pairs of the consensus half-site are numbered 1–8; those in the other half-site are numbered 1′–8′. Phosphate groups previously identified by ethylation protection experiments as important in repressor binding are identified by green circles and labeled P_A–P_E in the consensus half-site and P_A′–P_E′ in the nonconsensus half-site. (*Source:* From Jordan and Pabo, *Science* 242:894, 1988. Copyright © 1988 American Association for the Advancement of Science, Washington, DC. Reprinted by permission.)

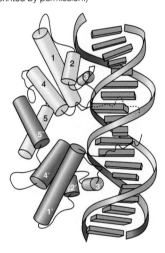

Figure 9.7   Geometry of the λ repressor-operator complex. The DNA (blue) is bound to the repressor dimer, whose monomers are depicted in yellow and purple. The recognition helix of each monomer is shown in red and labeled 3 and 3′.

of the DNA-binding domain of the protein. The operator fragment (Figure 9.6) was 20 bp long and contained one complete site to which the repressor dimer attached. That is, it had two half-sites, each of which bound to a repressor monomer. Such use of partial molecules is a common trick employed by x-ray crystallographers to make better crystals than they can obtain with whole proteins or whole DNAs. In this case, because the primary goal was to elucidate the structure of the interface between the repressor and the operator, the protein and DNA fragments were probably just as useful as the whole protein and DNA since they contained the elements of interest.

General Structural Features   Figure 9.2, used at the beginning of this chapter to illustrate the fit between

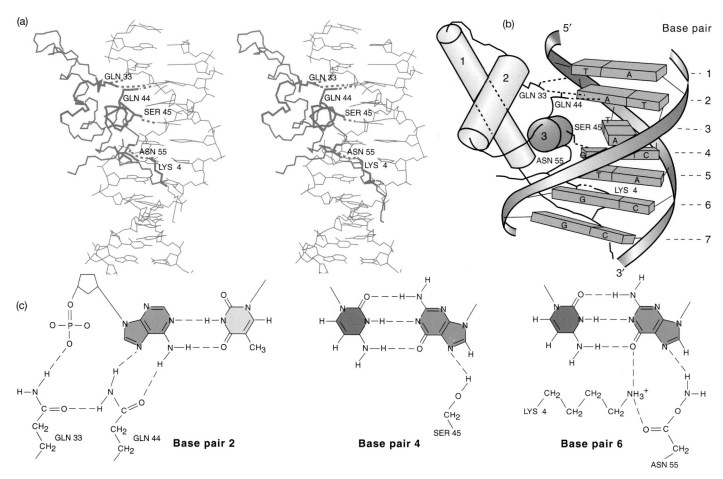

Figure 9.8   Hydrogen bonds between λ repressor and base pairs in the major groove of the operator. (a) Stereo diagram of the complex, with the DNA double helix on the right and the amino terminal part of the repressor monomer on the left. α-Helices 2 and 3 are rendered in bold lines, with the recognition helix almost perpendicular to the plane of the paper. Hydrogen bonds are represented by dashed lines. (b) Schematic diagram of the hydrogen-bonds shown in panel (a). Only the important amino acid side chains are shown. The base pairs are numbered at right. (c) Details of the hydrogen bonds. Structures of the key amino acid side chains and bases are shown, along with the hydrogen bonds in which they participate. (*Source:* From Jordan and Pabo, *Science* 242: 896, 1988. Copyright © 1988 American Association for the Advancement of Science, Washington, DC. Reprinted by permission.)

λ repressor and operator, is based on the high-resolution model from the Jordan and Pabo analysis we are now considering. Figure 9.7, a more detailed representation of the same model, reveals several general aspects of the protein–DNA interaction. First of all, of course, we can see the recognition helices (3 and 3′, red) of each repressor monomer nestled into the DNA major grooves in the two half-sites. We can also see how helices 5 and 5′ approach each other to hold the two monomers together in the repressor dimer. Finally, note that the DNA is similar in shape to the standard B-form of DNA. We can see a bit of bending of the DNA, especially at the two ends of the DNA fragment, as it curves around the repressor dimer, but the rest of the helix is relatively straight.

Interactions with Bases   Figure 9.8 shows the details of the interactions between amino acids in a repressor monomer and bases in one operator half-site. The crucial amino acids participating in these interactions are glutamine 33

(Gln 33), glutamine 44 (Gln 44), serine 45 (Ser 45), lysine 4 (Lys 4), and asparagine 55 (Asn 55). Figure 9.8a is a stereo view of the interactions, where α-helices 2 and 3 are represented by bold lines. The recognition helix (3) is almost perpendicular to the plane of the paper, so the helical polypeptide backbone looks almost like a circle. The key amino acid side chains are shown making hydrogen bonds (dashed lines) to the DNA and to one another.

Figure 9.8b is a schematic diagram of the same amino acid/DNA interactions. It is perhaps easier to see the hydrogen bonds in this diagram. We see that three of the important bonds to DNA bases come from amino acids in the recognition helix. In particular, Gln 44 makes two hydrogen bonds to adenine-2, and Ser 45 makes one hydrogen bond to guanine-4. Figure 9.8c depicts these hydrogen bonds in detail and also clarifies a point made in parts (a) and (b) of the figure: Gln 44 also makes a hydrogen bond to Gln 33, which in turn is hydrogen-bonded to the phosphate preceding base pair number 2. This participation of

Gln 33 is critical. By bridging between the DNA back-bone and Gln 44, it positions Gln 44 and the rest of the recognition helix to interact optimally with the operator. Thus, even though Gln 33 resides at the beginning of helix 2, rather than on the recognition helix, it plays an important role in protein–DNA binding. To underscore the importance of this glutamine, we note that it also appears in the same position in the 434 phage repressor and plays the same role in interactions with the 434 operator, as we will see later in this chapter.

Serine 45 also makes an important hydrogen bond with a base pair, the guanine of base pair number 4. In addition, the methylene (CH_2) group of this serine approaches the methyl group of the thymine of base pair number 5 and participates in a hydrophobic interaction that probably also includes the methyl group of Ala 49. Such hydrophobic interactions involve nonpolar groups like methyl and methylene, which tend to come together to escape the polar environment of the water solvent, much as oil droplets coalesce to minimize their contact with water. Indeed, hydrophobic literally means "water-fearing."

The other hydrogen bonds with base pairs involve two other amino acids that are not part of the recognition helix. In fact, these amino acids are not part of any helix: Asn 55 lies in the linker between helices 3 and 4, and Lys 4 is on the arm that reaches around the DNA. Here again we see an example of a hydrogen bond network, not only between amino acid and base, but between two amino acids. Figure 9.8c makes it particularly clear that these two amino acids each form hydrogen bonds to the guanine of base pair number 6, and also to each other. Such networks add considerably to the stability of the whole complex.

Amino Acid/DNA Backbone Interactions We have already seen one example of an amino acid (Gln 33) that forms a hydrogen bond with the DNA backbone (the phosphate between base pairs 1 and 2). However, this is only one of five such interactions in each half-site. Figure 9.9 portrays these interactions in the consensus half-site, which involve five different amino acids, only one of which (Asn 52) is in the recognition helix. The dashed lines represent hydrogen bonds from the NH groups of the peptide backbone, rather than from the amino acid side chains.

One of these hydrogen bonds, involving the peptide NH at Gln 33, is particularly interesting because of an electrostatic contribution of helix 2 as a whole. To appreciate this, recall from Chapter 3 that all the C=O bonds in a protein α-helix point in one direction. Because each of these bonds is polar, with a partial negative charge on the oxygen and a partial positive charge on the carbon, the whole α-helix has a considerable polarity, with practically a full net positive charge at the amino terminus of the

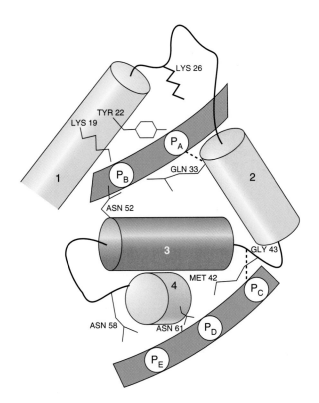

Figure 9.9 Amino acid/DNA backbone interactions. α-Helices 1–4 of the λ repressor are shown, along with the phosphates (P_A–P_E) that are involved in hydrogen bonds with the protein. This diagram is perpendicular to that in Figure 9.8. The side chains of the important amino acids are shown. The two dashed lines denote hydrogen bonds between peptide NH groups and phosphates. (*Source:* From Jordan and Pabo, *Science* 242:897, 1988. Copyright © 1988 American Association for the Advancement of Science, Washington, DC. Reprinted by permission.)

helix. This end of the helix will therefore have a natural affinity for the negatively charged DNA backbone. Now look again at Figure 9.9 and notice that the amino end of helix 2, where Gln 33 is located, points directly at the DNA backbone. This maximizes the electrostatic attraction between the positively charged amino end of the α-helix and the negatively charged DNA and stabilizes the hydrogen bond between the peptide NH of Gln 33 and the phosphate group in the DNA backbone.

Other interactions involve hydrogen bonds between amino acid side chains and DNA backbone phosphates. For example, Lys 19 and Asn 52 both form hydrogen bonds with phosphate P_B. The amino group of Lys 26 carries a full positive charge. Although it may be too far away from the DNA backbone to interact directly with a phosphate, it may contribute to the general affinity between protein and DNA. The large number of amino acid/DNA phosphate contacts suggests that these interactions play a major role in the stabilization of the protein–DNA complex. Figure 9.9 also shows the position of the side chain of Met 42. It probably forms a hydrophobic interaction with three carbon atoms on the deoxyribose between P_C and P_D.

Confirmation of Biochemical and Genetic Data Before the detailed structure of the repressor–operator complex was known, we already had predictions from biochemical and genetic experiments about the importance of certain repressor amino acids and operator bases. In almost all cases, the structure confirms these predictions.

First, ethylation of certain operator phosphates interfered with repressor binding. Hydroxyl radical footprinting had also implicated these phosphates in repressor binding. Now we see that these same phosphates (five per half-site) make important contacts with repressor amino acids in the cocrystal.

Second, methylation protection experiments had predicted that certain guanines in the major groove would be in close contact with repressor. The crystal structure now shows that all of these are indeed involved in repressor binding. One major-groove guanine actually became more sensitive to methylation on repressor binding, and this guanine (G8′) is now seen to have an unusual conformation in the cocrystal. Base pair 8′ is twisted more than any other on its horizontal axis, and the spacing between this base pair and the next is the widest. This unusual conformation could open guanine 8′ up to attack by the methylating agent DMS. Also, adenines were not protected from methylation in previous experiments. This makes sense because adenines are methylated on N3, which resides in the minor groove. Because no contacts between repressor and operator occur in the minor groove, repressor cannot protect adenines from methylation.

Third, DNA sequence data had shown that the A-T base pair at position 2 and the G-C base pair at position 4 were conserved in all 12 half-sites of the operators O_R and O_L. The crystal structure shows why these base pairs are so well conserved: They are involved in important contacts with the repressor.

Fourth, genetic data had shown that mutations in certain amino acids destabilized repressor–operator interaction, whereas other changes in repressor amino acids actually enhanced binding to the operator. Almost all of these mutations can be explained by the cocrystal structure. For example, mutations in Lys 4 and Tyr 22 were particularly damaging, and we now see that both these amino acids make strong contacts with the operator: Lys 4 with guanine–6 (and with Asn 55) and Tyr 22 with P_A. As an example of a mutation with a positive effect, consider the substitution of lysine for Glu 34. This amino acid is not implicated by the crystal structure in any important bonds to the operator, but a lysine in this position could rotate so as to form a salt bridge with the phosphate before P_A and thus enhance protein–DNA binding. This salt bridge would involve the positively charged ε-amino group of the lysine and the negatively charged phosphate.

Figure 9.10 Sequence of the 14-mer used in the 434 phage repressor–operator cocrystal structure. The 14 bp of the nontemplate strand of the 14-mer are numbered 1–14. The surrounding bases are numbered 13–, 14–, +1, and +2. The 14 bases of the template strand are numbered 1′–14′ (right to left), and the surrounding bases are numbered +2′, +1′, 14+′ and 13+′. The central dot indicates the center of symmetry of the 14-mer. The circles between base pairs denote phosphates in contact with the repressor. (*Source:* Adapted with permission from *Nature* 326:847, 1987. Copyright © 1987 Macmillan Magazines Limited.)

SUMMARY The cocrystal structure of a λ repressor fragment with an operator fragment shows many details about how the protein and DNA interact. The most important contacts occur in the major groove, where amino acids make hydrogen bonds with DNA bases and with the DNA backbone. Some of these hydrogen bonds are stabilized by hydrogen-bond networks involving two amino acids and two or more sites on the DNA. The structure derived from the cocrystal is in almost complete agreement with previous biochemical and genetic data.

High-Resolution Analysis of Phage 434 Repressor–Operator Interactions

Harrison, Ptashne, and coworkers used x-ray crystallography to perform a detailed analysis of the interaction between phage 434 repressor and operator. As in the λ cocrystal structure, the crystals they used for this analysis were not composed of full-length repressor and operator, but fragments of each that contained the interaction sites. As a substitute for the repressor, they used a peptide containing the first 69 amino acids of the protein, including the helix-turn-helix DNA-binding motif. For the operator, they used a synthetic 14-bp DNA fragment that contains the repressor-binding site (Figure 9.10). These two fragments presumably bound together as the intact molecules would, and the complex could be crystallized relatively easily. This work provided some important information not available from the earlier x-ray crystallography on the intact repressor and operator.

Repressor–Operator Contacts
Contacts with the Double Helical Backbone. In our discussion of the λ repressor–operator cocrystal, we noted that the amino ends of α-helices carry a partial positive

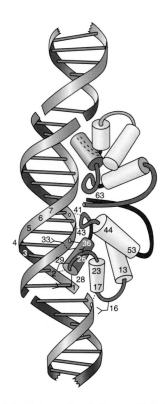

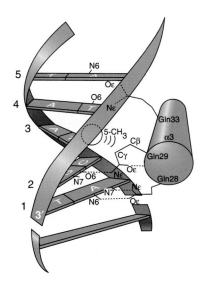

Figure 9.11   Model of interaction between 434 phage repressor dimer and 434 operator. The important amino acid side chains of the bottom repressor monomer are shown making contacts with the DNA, and the key amino acids are numbered. The first four α-helices of the repressor are colored as follows: α-1, α-2, and α-4, yellow; α-3 (recognition helix), red. The seven bases of the bottom half-site (coding strand) are numbered on the DNA backbone (except number 4, where the numeral is beside the backbone). The upper repressor monomer and operator half-site are not labeled. (*Source:* Reprinted with permission from *Nature* 326:847, 1987. Copyright © 1987 Macmillan Magazines Limited.)

Figure 9.12   Detailed model of interaction between recognition helix amino acid side chains and one 434 operator half-site. Hydrogen bonds are represented by dashed lines. The van der Waals interaction between the Gln 29 side chain and the 5-methyl group of the thymine paired to adenine 3 is represented by concentric arcs. (*Source:* Reprinted with permission from *Nature* 326:850, 1987. Copyright © 1987 Macmillan Magazines Limited.)

charge and can therefore interact with the negatively charged DNA backbone. Thus, it is significant that all three of the helices near the DNA-binding motif of the 434 repressor (α2, α3 [the recognition helix], and α4) have their amino termini pointing toward the DNA backbone (recall Figure 9.9). In particular, the crystal structure showed that the peptide backbone NH group of Gln 17, at the amino terminus of α2, approaches phosphate 14- (the phosphate to the 5′ side of nucleoside 14-) closely enough to form a hydrogen bond (Figure 9.11). Given the positive charge on this end of the α-helix, this hydrogen bond should be particularly strong. This is exactly analogous to the situation at Gln 33 at the beginning of the α2 helix in the λ repressor discussed earlier in this chapter. It is also encouraging that phosphate 14- is one that is protected from ethylation when repressor is bound. Thus, these data suggest a close contact between repressor and this phosphate.

Similarly, the backbone NH groups at Arg 43 and Phe 44, which lie at the amino end of α4, make a close approach to phosphate P10′. (The prime indicates the strand opposite the one with P10 or P14.) Two other phosphates (P9′ and P11′) appear to contact the repressor protein as well. All four of these phosphates, P14, P9′, P10′, and P11′, were protected from ethylation by repressor, which further strengthens the hypothesis that they are involved in repressor–operator binding.

In addition to the peptide backbone of the protein, some amino acid side chains also appear to contact the DNA backbone. For example, the Asn 16 side chain contacts P14; the Arg 41 side chain contacts P9′; and the Arg 43 side chain projects into the minor groove near base pair 7. As we will see, the minor groove is narrower than average in this region, so the Arg 41 side chain could make contact with phosphates in the DNA backbone on either side of the groove.

Contacts with Base Pairs. Figure 9.12 summarizes the contacts between the side chains of Gln 28, Gln 29, and Gln 33, all in the recognition helix (α3) of the 434 repressor. Starting at the bottom of the figure, note the two possible hydrogen bonds (represented by dashed lines) between the Oε and Nε of Gln 28 and the N6 and N7 of adenine 1. Next, we see that a possible hydrogen bond between the Oε of Gln 29 and the protein backbone NH of the same amino acid points the Nε of this amino acid directly at the O6 of the guanine in base pair 2 of the operator, which would allow a hydrogen bond between this amino acid and base. Note also the potential van der Waals interactions (represented by concentric arcs) between Cβ and Cγ of Gln 29 and the 5-methyl group of the thymine in base pair 3. Such van der Waals interactions can be explained roughly as follows: Even though all the groups involved are

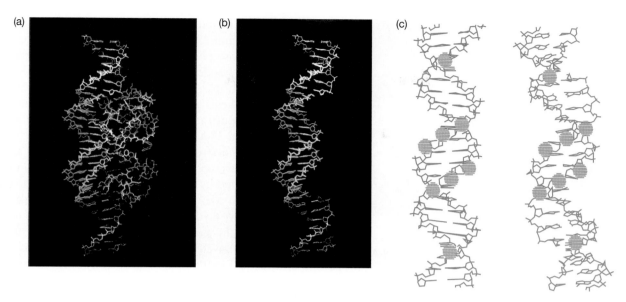

Figure 9.13   Distortion of 434 operator DNA in the repressor–operator complex. (**a**) The operator–repressor complex, showing the central 14-mer in red, and the repressor fragment in light and dark blue. (**b**) The operator alone, to emphasize the distortion in shape. Notice the overall bend in the DNA, and the narrowing of the minor groove at the center of the model. (**c**) The shape of the operator DNA in the operator–repressor complex (right), compared with standard B-DNA (left). Notice again the bend in the DNA, and the narrowing of the minor groove, compared with normal DNA. The phosphates contacted by the repressor are marked with blue circles. (*Source:* Anderson et al., Structure of the repressor-operator complex of bacteriophage 434. *Nature* 326 (30 Apr 1987) f. 1b&c, p. 847. © Macmillan Magazines Ltd.)

nonpolar, at any given instant they have a very small dipole moment due to random fluctuations in their electron clouds. These small dipole moments can cause a corresponding opposite polarity in a very close neighbor. The result is an attraction between the neighboring groups.

> **SUMMARY** X-ray crystallography of a phage 434 repressor-fragment/operator-fragment complex shows close approaches between certain amino acids and certain phosphates in the operator's DNA backbone. This implies hydrogen bonding between the protein and DNA at these sites. This analysis also shows probable hydrogen bonding between three glutamine residues in the recognition helix and three base pairs in the repressor. It also reveals a potential van der Waals contact between one of these glutamines and a base in the operator.

Effects of DNA Conformation   The contacts we have discussed between the repressor and the DNA backbone require that the DNA double helix curve slightly, as illustrated in Figure 9.13. Indeed, crystallography results show that the DNA does curve this way in the DNA–protein complex; we do not know yet whether the DNA bend preexists in this DNA region or whether it is induced by repressor binding. In either case, the base sequence of the operator plays a role by facilitating this bending. That is, some DNA sequences are easier to bend in a given way than others, and the

434 operator sequence is optimal for the bend it must make to fit the repressor. We will discuss this general phenomenon in more detail later in this chapter.

Another notable feature of the conformation of the operator DNA is the compression of the DNA double helix between base pairs 7 and 8, which lie between the two half-sites of the operator. This compression amounts to an overwinding of 3 degrees between base pairs 7 and 8, or 39 degrees, compared with the normal 36 degrees helical twist between base pairs. You can also see this effect in Figure 9.14. Notice the narrowness of the minor groove in the middle of the figure. The major grooves on either side are wider than normal, due to a compensating underwinding of that DNA. Again, the base sequence at this point is optimal for assuming this conformation.

Harrison, Ptashne, and colleagues performed even higher resolution analysis on cocrystals of the N-terminal 434 repressor fragment and a DNA 20-mer containing the 434 operator. This study confirmed the essential elements of the model we have just discussed and provided some additional insights about the shape of the complex. In particular, the newer model, with a longer DNA fragment, shows an even more pronounced bending of the DNA. Figure 9.14 shows the shapes of the standard B-DNA next to the shape of the 434 20–mer as it exists in the cocrystal with the repressor fragment, and the shape of the DNA and protein together. Here again we see the conspicuous bending of the operator DNA to fit the repressor, and the constriction of the DNA minor groove near the middle of the model.

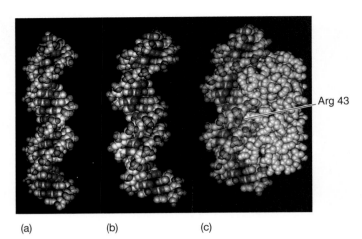

(a)   (b)   (c)

Figure 9.14   Space-filling computer model of distorted DNA in the 434 repressor–operator complex. (a) Standard B-DNA. (b) Shape of the operator-containing 20-mer in the repressor–operator complex. (c) The repressor–operator complex, with the repressor in yellow. Notice how the DNA conforms to the shape of the protein to promote intimate contact between the two. The side chain of Arg 43 can be seen projecting into the minor groove of the DNA near the center of the model. (*Source:* (*a-b*) Aggarwal et al., Recognition of a DNA operator by the repressor of phage 434: A view at high resolution. *Science* 242 (11 Nov 1988) f. 3c, p. 902. © AAAS. (*c*) Aggarwal et al., *Science* 242 (11 Nov 1988) fig. 3b, p. 902. © AAAS.)

SUMMARY The x-ray crystallography analysis of the partial repressor–operator complex shows that the DNA deviates significantly from its normal regular shape. It bends somewhat to accommodate the necessary base/amino acid contacts. Moreover, the central part of the helix, between the two half-sites, is wound extra tightly, and the outer parts are wound more loosely than normal. The base sequence of the operator facilitates these departures from normal DNA shape.

Genetic Tests of the Model   If the apparent contacts we have seen between repressor and operator are important, mutations that change these amino acids or bases should reduce or abolish DNA–protein binding. Alternatively, we might be able to mutate the operator so it does not fit the repressor, then make a compensating mutation in the repressor that restores binding. Also, if the unusual shape assumed by the operator is important, mutations that prevent it from taking that shape should reduce or abolish repressor binding. As we will see, all those conditions have been fulfilled.

First of all, considering all the potential interactions we have described between Gln 28 and 29 and base pairs 1–3, we would predict that mutations in any of those amino acids or bases would reduce promoter–operator binding. In accord with this prediction, we find that base pairs 1–3 are conserved in all 12 half-sites (in O_R1–3 and in O_L1–3) of the 434 operators. Ptashne and colleagues also found that

when they introduced random mutations into the codons for Gln 28 and 29 to produce all possible alternative amino acids in those positions, they obtained no mutant phages capable of lysogeny. In other words, each of these mutations inactivated the repressor. Thus, as predicted, those two amino acids are critical for repressor function.

To demonstrate further the importance of the interaction between Gln 28 and A1, Ptashne and colleagues changed A1 to a T. This destroyed binding between repressor and operator, as we would expect. However, this mutation could be complemented by a mutation at position 28 of the repressor from Gln to Ala. Figure 9.12 reveals the probable explanation: The two hydrogen bonds between Gln 28 and A1 can be replaced by a van der Waals contact between the methyl groups on Ala 28 and T1. The importance of this contact is underscored by the replacement of T1 with a uracil, which does not have a methyl group, or 5-methylcytosine (5MeC), which does. The U-substituted operator does not bind the repressor with Ala 28, but the 5MeC-substituted operator does. Thus, the methyl group is vital to interactions between the mutant operator and mutant repressor, as predicted on the basis of the van der Waals contact.

We strongly suspect that the overwinding of the DNA between base pairs 7 and 8 is important in repressor–operator interaction. If so, substituting G-C or C-G base pairs for the A-T and T-A pairs at positions 6–9 should decrease repressor–operator binding, because G-C pairs do not readily allow the overwinding that is possible with A-T pairs. As expected, repressor did not bind well to operators with G-C or C-G base pairs in this region. This failure to bind well did not prove that overwinding exists, but it was consistent with the overwinding hypothesis.

SUMMARY The contacts between the phage 434 repressor and operator predicted by x-ray crystallography can be confirmed by genetic analysis. When amino acids or bases predicted to be involved in interaction are altered, repressor–operator binding is inhibited.

9.2   The *trp* Repressor

The *trp* repressor is another protein that uses a helix-turn-helix DNA-binding motif. However, recall from Chapter 7 that the aporepressor (the protein without the tryptophan corepressor) is not active. Paul Sigler and colleagues used x-ray crystallography of *trp* repressor and aporepressor to point out the subtle but important difference that tryptophan makes. The crystallography also sheds light on the way the *trp* repressor interacts with its operator, which differs in a fundamental way from the interactions between the λ family of repressors and their operators.

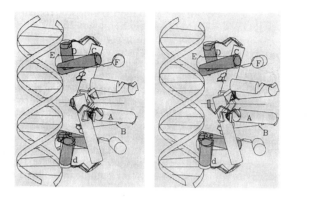
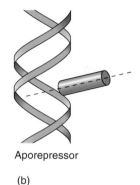
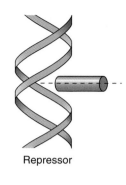

(a)   (b)

Figure 9.15 Comparison of the fit of *trp* repressor and aporepressor with *trp* operator. (a) Stereo diagram. The helix-turn-helix motifs of both monomers are shown in the positions they assume in the repressor (transparent) and aporepressor (dark). The position of tryptophan in the repressor is shown (black polygons). Note that the recognition helix (helix E) in the aporepressor falls back out of ideal position for inserting into the wide groove of the operator DNA. The two almost identical drawings constitute a stereo presentation that allows you to view this picture in three dimensions. To get this 3-D effect, use a stereo viewer, or alternatively, hold the picture 1–2 ft in front of you and let your eyes relax as they would when you are staring into the distance or viewing a "magic eye" picture. After a few seconds, the two images should fuse into one in the center, which appears in three dimensions. This stereo view gives a better appreciation for the fit of the recognition helix and the wide groove of the DNA, but if you cannot get the 3-D effect, just look at one of the two pictures. **(b)** Simplified (nonstereo) diagram comparing the positions of the recognition helix (red) of the aporepressor (left) and the repressor (right) with respect to the DNA major groove. Notice that the recognition helix of the repressor points directly into the major groove, whereas that of the aporepressor points more downward. The dashed line emphasizes the angle of the recognition helix in each drawing.

The Role of Tryptophan

Here is a graphic indication that tryptophan affects the shape of the repressor: When you add tryptophan to crystals of aporepressor, the crystals shatter! When the tryptophan wedges itself into the aporepressor to form the repressor, it changes the shape of the protein enough to break the lattice forces holding the crystal together.

This raises an obvious question: What moves when free tryptophan binds to the aporepressor? To understand the answer, it helps to visualize the repressor as illustrated in Figure 9.15. The protein is actually a dimer of identical subunits, but these subunits fit together to form a three-domain structure. The central domain, or "platform," comprises the A, B, C, and F helices of each monomer, which are grouped together on the right, away from the DNA. The other two domains, found on the left close to the DNA, are the D and E helices of each monomer.

Now back to our question: What moves when we add tryptophan? The platform apparently remains stationary, whereas the other two domains tilt, as shown in Figure 9.15. The recognition helix in each monomer is helix E, and we can see an obvious shift in its position when tryptophan binds. In the top monomer, it shifts from a somewhat downward orientation to a position in which it points directly into the major groove of the operator. In this position, it is ideally situated to make contact with (or "read") the DNA, as we will see.

Sigler refers to these DNA-reading motifs as *reading heads,* likening them to the heads in a tape recorder or the disk drive of a computer. In a computer, the reading heads can assume two positions: engaged and reading the disk, or disengaged and away from the disk. The *trp* repressor works the same way. When tryptophan is present, it inserts itself between the platform and each reading head, as illustrated in Figure 9.15, and forces the reading heads into the best position for fitting into the major groove of the operator. On the other hand, when tryptophan dissociates from the aporepressor, the gap it leaves allows the reading heads to fall back toward the central platform and out of position to fit with the operator.

Figure 9.16a shows a closer view of the environment of the tryptophan in the repressor. It is a hydrophobic pocket that is occupied by the side chain of a hydrophobic amino acid (sometimes tryptophan) in almost all comparable helix-turn-helix proteins, including the λ repressor, Cro, and CAP. However, in these other proteins the hydrophobic amino acid is actually part of the protein chain, not a free amino acid, as in the *trp* repressor. Sigler likened the arrangement of the tryptophan between Arg 84 and Arg 54 to a salami sandwich, in which the flat tryptophan is the salami. When it is removed, as in Figure 9.16b, the two arginines come together as the pieces of bread would when you remove the salami from a sandwich. This model has implications for the rest of the molecule, because Arg 54 is on the surface of the central platform of the repressor dimer, and Arg 84 is on the facing surface of the reading head. Thus, inserting the tryptophan between these two arginines pushes the reading head away from the platform and points it toward the major groove of the operator, as we saw in Figure 9.15.

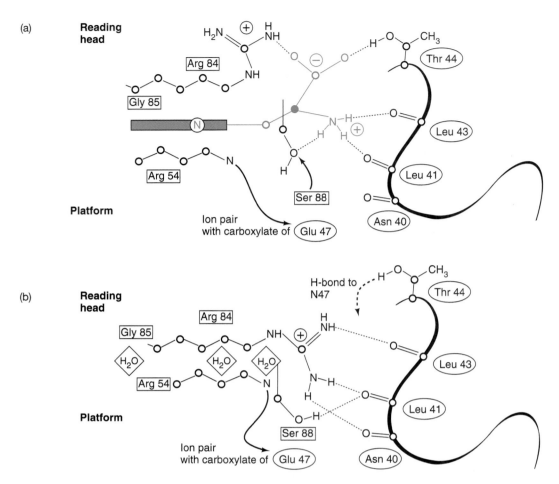

Figure 9.16   Tryptophan binding site in the *trp* repressor. (**a**) Environment surrounding the tryptophan in the *trp* repressor. Notice the positions of Arg 84 above and Arg 54 below the tryptophan side chain (red). (**b**) The same region in the aporepressor, without tryptophan. Notice that the Arg side chains have moved together to fill the gap left by the absent tryptophan. Amino acids with rectangular labels are on one aporepressor subunit; those with oval labels are on the other. (*Source:* Reprinted with permission from *Nature* 327:594, 1987. Copyright © 1987 Macmillan Magazines Limited.)

SUMMARY The *trp* repressor requires tryptophan to force the recognition helices of the repressor dimer into the proper position for interacting with the *trp* operator.

***trp* Repressor–Operator Interactions**   As we have seen, the *trp* repressor points directly into the DNA major groove, rather than lying sideways in it the way the λ repressor does. Thus, it is harder to understand how direct amino acid/DNA base interactions could occur with this repressor. In fact, Sigler suggested that only one such interaction occurs; aside from this one exception, the amino acids of the repressor interact with phosphates in the DNA backbone, as shown in Figure 9.17. But if the DNA backbone is uniform throughout, how can binding be specific? It seems that the backbone in the *trp* operator would look just like the backbone in any other stretch of DNA.

The answer, according to Sigler, is that the *trp* operator is *not* just like any other stretch of DNA. The x-ray crystallography studies on the *trp* repressor–operator complex show that the DNA in the operator bends in a particular way. This bend is just right for interacting with the *trp* repressor, and it allows for an unusually large contact surface (2900 Å^2) between protein and DNA. We know that binding between repressor and operator is still sequence-specific because changes in the base sequence of the operator can create O^c mutant operators that no longer bind repressor. Their altered sequences apparently no longer allow them to bend in the proper way. Sigler terms this "indirect readout," to distinguish it from the direct interactions in other systems—the "reading" of the bases in the λ operator by the amino acids in the λ repressor, for example.

The highest resolution crystallographic studies on this system—to the exceptionally fine level of 1.9 Å—confirm that many of the protein–DNA interactions are indirect. In fact, of the 25 direct contacts, only 1 is an amino

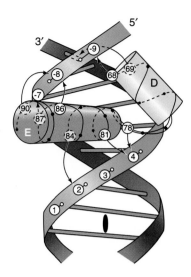

Figure 9.17 Amino acid/DNA backbone interactions in the *trp* repressor–operator complex. This figure reemphasizes the fact that the recognition helix (E, red) in the *trp* repressor points into the major groove of the operator, rather than lying sideways in it. It also shows the contacts between amino acids on the D and E helices (numbered 68–90) and phosphates in the DNA backbone (numbered 1 to 4 on one strand and –7 to –9 on the other). The black oval represents the axis of rotational symmetry between the two operator half-sites.
(*Source:* Reprinted with permission from *Nature* 335:327, 1988. Copyright © 1988 Macmillan Magazines Limited.)

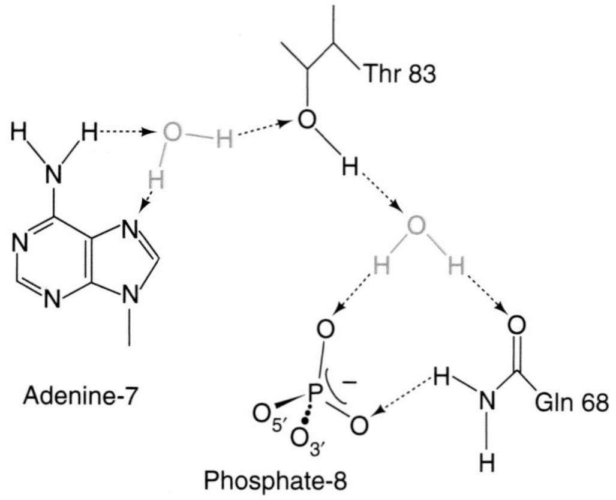

acid/DNA base contact, and of the 18 indirect contacts, 12 involve amino acids and bases. These results also suggest that most of the indirect contacts are mediated by hydrogen bonds through water molecules. That is, water molecules stand between protein and DNA and link the two through hydrogen bonds, as illustrated in Figure 9.18.

It is necessary to understand the distinction between this kind of indirect contact and "indirect readout." Indirect readout is the fit between the amino acids of a protein and the backbone of a DNA helix whose local shape depends on a particular base sequence. No direct amino acid/DNA base contacts participate in this interaction. Indirect contact of protein and DNA through water molecules may still involve amino acid/DNA base interactions, but water molecules bridge the gaps between them. The interaction between *trp* repressor and operator is an example of both. Only one direct contact takes place between a base and an amino acid. All 12 of the other base/amino acid contacts are indirect through water. At the same time, indirect readout of the operator shape by the repressor is also important.

Some have objected that protein–DNA contact through water molecules does not allow for the required specificity, but Sigler argues that the four half-operator/half-repressor complexes in the crystalline repeating unit are all virtually superimposable—water molecules and all. Because they all have the same shape, this shape must be at an energy minimum, and a specific complex, by definition, is at minimum energy.

Figure 9.18 Observed participation of water molecules in interactions between *trp* repressor and operator. Two water molecules (red) appear to form hydrogen bonds with two amino acids (Thr 83 and Gln 68) and two sites on the operator (adenine-7 and phosphate-8). Without the water molecules, the protein and DNA atoms would be too far apart to form hydrogen bonds with one another, at least with the molecules conforming to this shape. (*Source:* Reprinted from *Biochimica et Biophysica Acta*, 1048, B. F. Luisi and P. B. Sigler, "The stereochemistry and biochemistry of the *trp* repressor-operator complex," 113-126, 1990 with permission from Elsevier Science.)

SUMMARY The *trp* repressor binds to the *trp* operator in a less-direct way than the λ family of repressors bind to their operators. The recognition helix of the *trp* repressor points directly into the major groove of the operator, which allows only one direct contact between an amino acid and a base. All the other interactions are either direct contacts between amino acids and DNA backbone phosphates, or indirect contacts through water molecules.

9.3 General Considerations on Protein–DNA Interactions

What contributes to the specificity of binding between a protein and a specific stretch of DNA? The examples we have seen so far suggest two answers: (1) specific interactions between bases and amino acids; and (2) the ability of the DNA to assume a certain shape, which also depends on the DNA's base sequence. These two possibilities are clearly not mutually exclusive, and both apply to many of the same protein–DNA interactions.

Hydrogen Bonding Capabilities of the Four Different Base Pairs

We have seen that different DNA-binding proteins depend to varying extents on contacts with the bases in the DNA. To the extent that they "read" the sequence of bases, one can ask, What exactly do they read? After all, the base pairs do not open up, so the DNA-binding proteins have to sense the differences among the bases in their base-paired condition. And they have to make base-specific contacts with these base pairs, either through hydrogen bonds or van der Waals interactions. Let us examine further the hydrogen-bonding potentials of the four different base pairs.

Consider the DNA double helix in Figure 9.19a. If we were to rotate the DNA 90 degrees so that it is sticking out of the page directly at us, we would be looking straight down the helical axis. Now consider one base pair of the DNA in this orientation, as pictured in Figure 9.19b. The major groove is on top, and the minor groove is below. A DNA-binding protein can approach either of these grooves to interact with the base pair. As it does so, it "sees" four possible contours in each groove, depending on whether the base pair is a T-A, A-T, C-G, or G-C pair.

Figure 9.19c presents two of these contours from both the major and minor groove perspectives. At the very bottom we see line diagrams that summarize what the protein encounters in both grooves for a T-A and a C-G base pair. Hydrogen bond acceptors (oxygen and nitrogen atoms) are represented by A's, and hydrogen bond donors (hydrogen atoms) are represented by D's. The major and minor grooves lie above and below the horizontal lines, respectively. The lengths of the vertical lines represent the relative distances that the donor or acceptor atoms project away from the helical axis toward the outside of the DNA groove. We can see that the T-A and C-G base pairs present very different profiles to the outside world, especially in the major groove. The difference between a purine–pyrimidine pair and the pyrimidine–purine pairs shown here would be even more pronounced.

These hydrogen-bonding profiles assume direct interactions between base pairs and amino acids. However, as we have seen, other possibilities exist. The amino acids can contact the base pairs indirectly through water molecules. They can also "read" the shape of the DNA backbone, either by direct hydrogen bonding or by forming salt bridges, or indirectly through water molecules.

SUMMARY The four different base pairs present four different hydrogen-bonding profiles to amino acids approaching either the major or minor DNA groove.

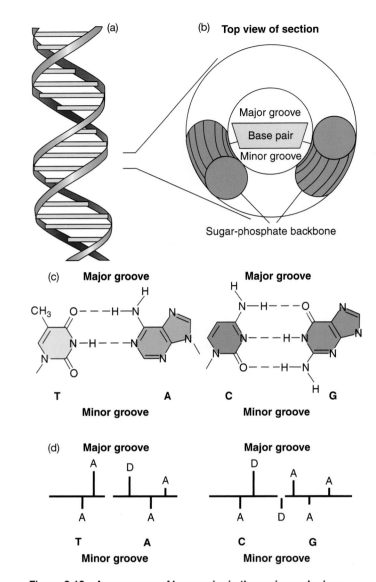

Figure 9.19   Appearance of base pairs in the major and minor grooves of DNA. (a) Standard B-form DNA, with the two backbones in red and blue, and the base pairs in yellow. **(b)** Same DNA molecule seen from the top. Notice the wider opening to the major groove (top), compared with the minor groove (bottom). **(c)** Structural formulas of the two base pairs. Again, the major groove is on top, and the minor groove on the bottom. **(d)** Stick diagrams showing the positions of hydrogen bond acceptors (A) and donors (D) in the major and minor grooves. For example, reading left to right, the major groove of the T-A pair has an acceptor (the C=O of the thymine), then a donor (the NH$_2$) of the adenine), then another acceptor (the N-7 in the ring of the adenine). The relative horizontal positions of these groups are indicated by the point of intersection with the vertical lines. The relative vertical positions are indicated by the lengths of the vertical lines. The two base pairs present different patterns of donors and acceptors in both major and minor grooves, so they are perceived differently by proteins approaching from the outside. By inverting these diagrams left-to-right, you can see that A-T and G-C pairs would present still different patterns. (*Source:* From R. Schleif, *Science* 241:1182–1183, 1988. Copyright © 1988 American Association for the Advancement of Science, Washington, DC. Reprinted by permission.)

The Role of DNA Shape in Specific Binding to Proteins

We have mentioned that when either the λ or phage 434 repressor or the *trp* repressor binds to its respective operator, the DNA in the protein–DNA complex bends somewhat. This distortion may be either preexisting or caused by the protein binding. In either case, it could help explain the specificity of the DNA–protein interaction. If it is preexisting because of a particular base sequence, the protein could have a complementary surface that fits with the bent DNA. If it is induced by the protein, a DNA sequence that readily allows the bending would bind the protein most successfully. The interaction of CAP plus cAMP with the activator-binding site of a catabolite-repressible operon, such as the *lac* operon, provides an extreme example of induced fit between protein and DNA.

As we saw in Chapter 7, CAP–promoter binding results in a bend of about 90 degrees in the DNA. This bend is primarily due to two **kinks,** or abrupt turns in the DNA helix, at which adjacent base pairs "unstack" and no longer lie parallel to each other. Figure 9.20 shows the geometry of one such kink and illustrates the specific protein–DNA interactions. This figure contains only the left-hand side of the CAP-binding site interacting with one monomer of CAP. It shows that this complex is stabilized by some amino acid/DNA base-pair interactions, but by even more amino acid/DNA backbone interactions. The former consist of hydrogen bonds between three amino acid side chains and three base pairs (in the DNA major groove). The latter are hydrogen bonds and ionic interactions between 13 amino acid functional groups and 11 phosphates in the DNA backbone.

Figure 9.20 makes it clear that such extensive interaction between CAP and its DNA target could not occur without the DNA bend. Because the kink between base pairs 5 and 6 is largely responsible for this crucial bend and because the conserved TG sequence (adjacent T–A and G–C base pairs) is supposed to be easier to kink than other sequences, one can propose the following hypothesis: The TG sequence is important, not because of any specific contacts it makes with the protein, but because it allows the all-important kink to occur. Genetic experiments support this interpretation, at least partially. First, changing the T–A base pair at position 6, which does not directly interact with the protein, to a G–C base pair increases the dissociation constant of the CAP–DNA complex by a factor of about 6.7. Mutation of the G–C base pair in position 5 gives results that are harder to interpret because this base pair does participate in direct interactions with the protein.

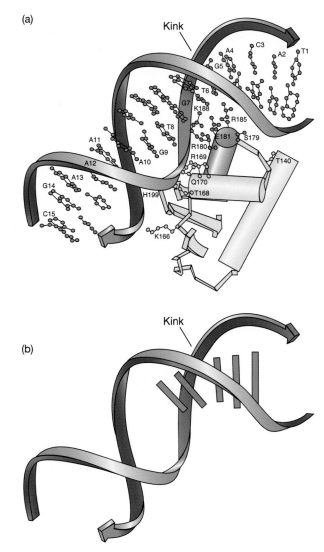

Figure 9.20   One of the DNA kinks in the CAP–DNA complex.
(a) Model showing a monomer of CAP (yellow, with recognition helix in red) bound to one half-site of *lac* DNA (blue). The kink occurs between base pairs 5 and 6 (labeled G5 and T6). Notice that all other base pairs are stacked parallel to their neighbors, but these two have "unstacked," creating the kink. **(b)** Simplified model emphasizing the kink. Here, only the DNA backbone and five base pairs are shown. The two base pairs involved in the kink are highlighted in red. (*Source:* From Schultz et al., *Science* 253: 1004, 1991. Copyright © 1991 American Association for the Advancement of Science, Washington, DC. Reprinted by permission.)

Nevertheless, we again see a significant decrease in affinity between CAP and DNA when we alter this base pair. Thus, the TG sequence is important at least in part because it allows the kink that in turn promotes strong CAP–DNA binding. This same principle probably also applies, though perhaps to a lesser extent, to other specific protein–DNA interactions where distortion of the DNA occurs.

SUMMARY The CAP protein bends its target DNA dramatically when it binds, a bend that is necessary for the specific binding between CAP and DNA. Thus, a DNA sequence that has all or most of the bases with which CAP interacts, but cannot bend, will be relatively unattractive to CAP protein. This is an extreme example of a more general rule that may apply to many protein–DNA associations: It is not always just the amino acid/DNA base-pair interactions that govern the affinity between protein and DNA; the affinity between DNA and protein may also depend on the ability of the DNA to be distorted into a shape that fits the protein.

The Importance of Multimeric DNA-Binding Proteins

Robert Schleif noted that the target sites for DNA-binding proteins are usually symmetric, or repeated, so they can interact with multimeric proteins—those composed of more than one subunit. Most DNA-binding proteins are dimers (some are even tetramers), and this greatly enhances the binding between DNA and protein because the two protein subunits bind cooperatively. Having one there automatically increases the concentration of the other. This boost in concentration is important because DNA-binding proteins are generally present in the cell in very small quantities.

Another way of looking at the advantage of dimeric DNA-binding proteins uses the concept of entropy. **Entropy** can be considered a measure of disorder in the universe. It probably does not come as a surprise to you to learn that entropy, or disorder, naturally tends to increase with time. Think of what happens to the disorder of your room, for example. The disorder increases with time until you expend energy to straighten it up. Thus, it takes energy to push things in the opposite of the natural direction—to create order out of disorder, or make the entropy of a system decrease.

A DNA–protein complex is more ordered than the same DNA and protein independent of each other, so bringing them together causes a decrease in entropy. Binding *two* protein subunits, independently of each other, causes twice the decrease in entropy. But if the two protein subunits are already stuck together in a dimer, orienting one relative to the DNA automatically orients the other, so the entropy change is much less than in independent binding, and therefore requires less energy. Looking at it from the standpoint of the DNA–protein complex, releasing the dimer from the DNA does not provide the same entropy gain as releasing two independently bound proteins would, so the protein and DNA stick together more tightly.

SUMMARY Multimeric DNA-binding proteins have an inherently higher affinity for binding sites on DNA than do multiple monomeric proteins that bind independently of one another.

9.4 DNA-Binding Proteins: Action at a Distance

So far, we have dealt primarily with DNA-binding proteins that govern events that occur very nearby. For example, the *lac* repressor bound to its operator interferes with the activity of RNA polymerase at an adjacent DNA site; or λ repressor stimulates RNA polymerase binding at an adjacent site. However, numerous examples exist in which DNA-binding proteins can influence interactions at remote sites in the DNA. We will see that this phenomenon is common in eukaryotes, but several prokaryotic examples occur as well.

The *gal* Operon

In 1983, S. Adhya and colleagues reported the unexpected finding that the *E. coli gal* operon, which codes for enzymes needed to metabolize the sugar galactose, has two distinct operators, about 97 bp apart. One is located where you would expect to find an operator, adjacent to the *gal* promoter. This one is called O_E, for "external" operator. The other is called O_I, for "internal" operator and is located within the first structural gene, *galE*. The downstream operator was discovered by genetic means: O^c mutations were found that mapped to the *galE* gene instead of to O_E. One way to explain the function of two separated operators is by assuming that they both bind to repressors, and the repressors interact by **looping out** the intervening DNA, as pictured in Figure 9.21. We have already seen an example of this kind of repression by looping out in our discussion of the *ara* operon in Chapter 7.

Duplicated λ Operators

The brief discussion of the *gal* operon just presented strongly suggests that proteins interact over a distance of almost 100 bp, but provided no direct evidence for this contention. Ptashne and colleagues used an artificial system to obtain such evidence. The system was the familiar λ operator–repressor combination, but it was artificial in that the experimenters took the normally adjacent operators and separated them to varying extents. We have seen that repressor dimers normally bind cooperatively to O_R1 and O_R2 when these operators are adjacent. The question is this: Do repressor dimers still bind cooperatively to the operators when they are separated? The answer is that they do, as long as the operators lie on the same face of the DNA double helix. This finding supports the hypothesis that repressors bound to separated *gal* operators probably interact by DNA looping.

Ptashne and coworkers used two lines of evidence to show cooperative binding to the separated λ promoters: DNase footprinting and electron microscopy. If we DNase-footprint two proteins that bind independently to remote DNA sites, we see two separate footprints. However, if we footprint two proteins that bind cooperatively to remote DNA sites through DNA looping, we see two

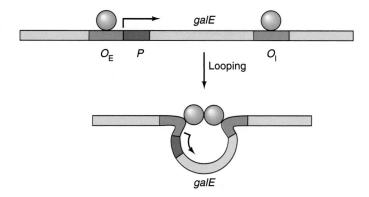

Figure 9.21 Repression of the *gal* operon. The *gal* operon has two operators (red): one external (O$_E$), adjacent to the promoter (green), and one internal (O$_I$), within the *galE* gene (yellow). Repressor molecules (blue) bind to both operators and appear to interact by looping out the intervening DNA (bottom).

separate footprints just as in the previous example, but this time we also see something interesting in between that does not occur when the proteins bind independently. This extra feature is a repeating pattern of insensitivity, then hypersensitivity to DNase. The reason for this pattern is explained in Figure 9.22. When the DNA loops out, the bend in the DNA compresses the base pairs on the inside of the loop, so they are relatively protected from DNase. On the other hand, the base pairs on the outside of the loop are spread apart more than normal, so they become extra sensitive to DNase. This pattern repeats over and over as we go around and around the double helix.

Using this assay for cooperativity, Ptashne and colleagues performed DNase footprinting on repressor bound to DNAs in which the two operators were separated by an integral or nonintegral number of helical turns. Figure 9.23a shows an example of cooperative binding, when the two operators were separated by 63 bp—almost exactly six helical turns. We can see the repeating pattern of lower and higher DNase sensitivity in between the two binding sites. By contrast, Figure 9.23b presents an example of noncooperative binding, in which the two operators were separated by 58 bp—just 5.5 helical turns. Here we see no evidence of a repeating pattern of DNase sensitivity between the two binding sites.

Electron microscopy experiments enabled Ptashne and coworkers to look directly at repressor–operator complexes with integral and nonintegral numbers of helical

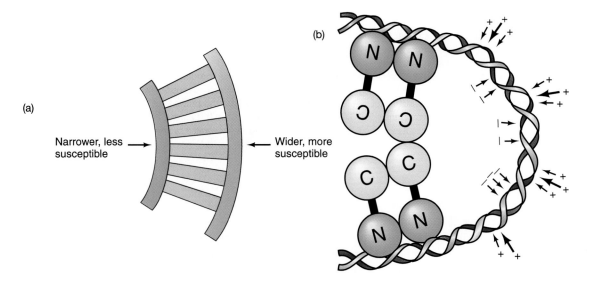

Figure 9.22 Effect of DNA looping on DNase susceptibility. (a) Simplified schematic diagram. The double helix is depicted as a railroad track to simplify the picture. The backbones are in red and blue, and the base pairs are in orange. As the DNA bends, the strand on the inside of the bend is compressed, restricting access to DNase. By the same token, the strand on the outside is stretched, making it easier for DNase to attack. **(b)** In a real helix each strand alternates being on the inside and the outside of the bend. Here, two dimers of a DNA-binding protein (λ repressor in this example) are interacting at separated sites, looping out the DNA in between. This stretches the DNA on the outside of the loop, opening it up to DNase I attack (indicated by + signs). By the same token, looping compresses the DNA on the inside of the loop, obstructing access to DNase I (indicated by the – signs). The result is an alternating pattern of higher and lower sensitivity to DNase in the looped region. Only one strand (red) is considered here, but the same argument applies to the other. (*Source:* (*b*) Reprinted from Hochschild et al., *Cell* 44:685, 1986. Copyright 1986, with permission from Elsevier Science.)

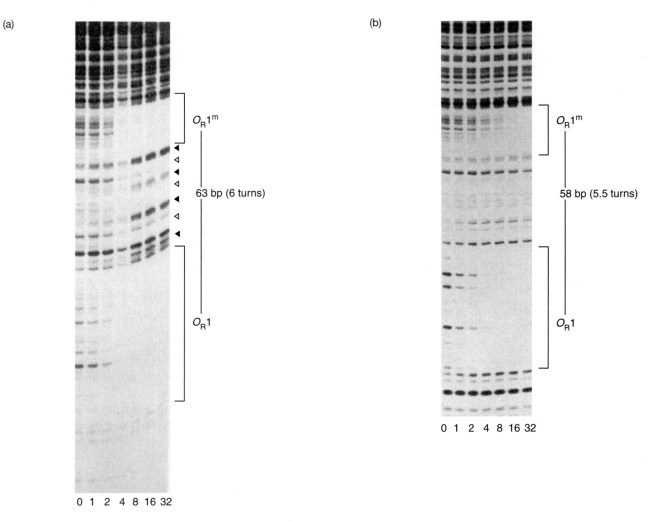

Figure 9.23 DNase footprints of dual operator sites. (a) Cooperative binding. The operators are almost exactly six double-helical turns apart (63 bp), and an alternating pattern of enhanced and reduced cleavage by DNase I appears between the two footprints when increasing amounts of repressor are added. The enhanced cleavage sites are denoted by filled arrowheads, the reduced cleavage sites by open arrowheads. This suggests looping of DNA between the two operators on repressor binding. (b) Noncooperative binding. The operators are separated by a nonintegral number of double-helical turns (58 bp, or 5.5 turns). No alternating pattern of DNase susceptibility appears on repressor binding, so the repressors bind at the two operators independently, without DNA looping. In both (a) and (b), the number at the bottom of each lane gives the amount of repressor monomer added, where 1 corresponds to 13.5 nM repressor monomer in the assay, 2 corresponds to 27 nM repressor monomer, and so on. (*Source:* Hochschild & Ptashne, Cooperative binding of lambda repressors to sites separated by integral turns fo the DNA helix. *Cell* 44 (14 Mar 1986) f. 3a&4, p. 683. Reprinted by permission of Elsevier Science.)

turns between the operators to see if the DNA in the former case really loops out. As Figure 9.24 shows, it does loop out. It is clear when such looping out is occurring, because the DNA is drastically bent. By contrast, Ptashne and colleagues almost never observed bent DNA when the two operators were separated by a nonintegral number of helical turns. Thus, as expected, these DNAs have a hard time looping out. These experiments demonstrate clearly that proteins binding to DNA sites separated by an integral number of helical turns can bind cooperatively by looping out the DNA in between.

SUMMARY When λ operators are separated by an integral number of helical turns, the DNA in between can loop out to allow cooperative binding. When the operators are separated by a nonintegral number of helical turns, the proteins have to bind to opposite faces of the DNA double helix, so no cooperative binding can take place.

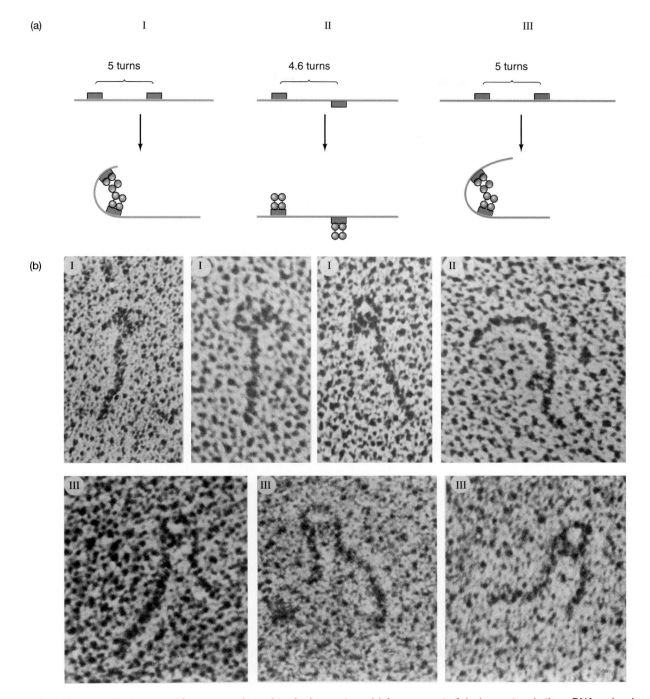

Figure 9.24 Electron microscopy of λ repressor bound to dual operators. (a) Arrangement of dual operators in three DNA molecules. In I, the two operators are five helical turns apart near the end of the DNA; in II, they are 4.6 turns apart near the end; and in III they are five turns apart near the middle. The arrows in each case point to a diagram of the expected shape of the loop due to cooperative binding of repressor to the two operators. In II, no loop should form because the two operators are not separated by an integral number of helical turns and are consequently on opposite sides of the DNA duplex. **(b)** Electron micrographs of the protein–DNA complexes. The DNA types [I, II, or III from panel **(a)**] used in the complexes are given at the upper left of each picture. The complexes really do have the shapes predicted in panel **(a)**. (*Source:* (a) Reprinted with permission from *Nature* 322:751, 1986. Copyright © 1986 Macmillan Magazines Limited. (*b*) From W. Su et al., *Proceedings of the National Academy of Sciences* 87:5507, 1990. Reprinted with permission.)

The *lac* Operon

Can looping out also be so clearly demonstrated between DNA-binding sites that are naturally separated? The answer is yes, and supporting data come from the well-studied *lac* operon, which we once assumed had only one operator. However, as we saw in Chapter 7, the *lac* operon has three operators: one major operator near the transcription start site, and two auxiliary operators, one upstream and one downstream. Moreover, the three operators act cooperatively in repression.

This cooperativity suggests that repressor can bind to two operators at once, looping out the DNA in between. We know this is at least theoretically possible because active dimers of *lac* repressor can bind to a *lac* operator, and a tetramer could therefore be expected to bind to two operators simultaneously. Müller-Hill and colleagues tested the hypothesis that the *lac* repressor can cause looping of DNA containing two separated *lac* operators. They started with a DNA fragment containing two copies of the classical *lac* operator placed 200 bp apart. They used three different assays for looping. The first was elec-

tron microscopy of the sort Ptashne and colleagues used with the λ operators. Again, they observed loops, and the loop sizes were directly related to the spacing between operators. The second was DNase footprinting. Just as with the λ operators, these investigators observed an alternating pattern of DNase sensitivity and resistance when the *lac* operators were an integral number of helical turns apart, but no such pattern when the operators were separated by a nonintegral number of helical turns.

The final assay used gel electrophoresis to detect looping. The cartoons at the left in Figure 9.25 illustrate the DNA–protein complexes resolved by this assay. The DNA fragment by itself has a high electrophoretic mobility, so it can be found near the bottom of the gel after the electrophoresis run. If repressor binds to one of the operators, it will retard the fragment's mobility. If repressor binds to both operators independently, it will retard the fragment further. However, if repressor molecules bind to both operators cooperatively by looping out the DNA, this bent DNA will be retarded even more. (Recall the work on bent DNA–CAP complexes discussed in Chapter 7.)

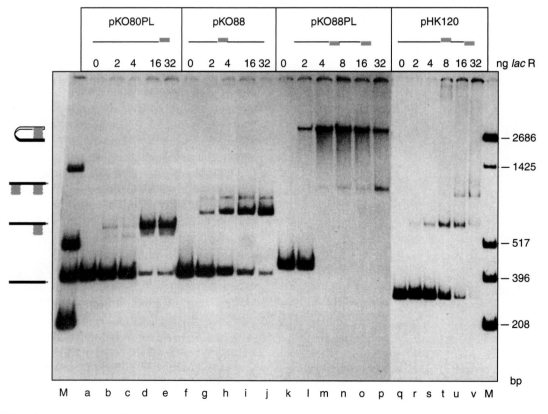

Figure 9.25   DNA looping between two *lac* operators. Müller-Hill and associates prepared radiolabeled DNA constructs with one or two *lac* operators (red), spaced as shown at top. pKO80PL and pKO88 have only one operator; pKO88PL has two operators separated by an integral number of helical turns; and pHK120 has two operators separated by a nonintegral number of helical turns. Next, they bound increasing amounts of *lac* repressor to each of these labeled constructs [amount of repressor in nanograms (ng) shown at the top of each lane] and electrophoresed the protein–DNA complexes. The mobilities of each kind of complex, from bare DNA (bottom) to looped DNA–protein complex (top) are marked at left with cartoons that represent each complex; repressor tetramer is rendered in green. Lanes k–p show that pKO88PL can form a looped complex, whereas lanes q–v demonstrate that pHK120 cannot. The lanes marked M contain DNA markers whose sizes are presented at right. (*Source: (photo)* Krämer et al., *lac* repressor forms loops with linear DNA carrying two suitably paced *lac* operators. *EMBO J.* 6, f. 1, p. 1482.)

Figure 9.25 shows that such looping did indeed occur, but only if the operators were separated by an integral number of helical turns.

As we saw in Chapter 7, Mitchell Lewis and coworkers added more evidence for the looping hypothesis by determining the crystal structure of the *lac* repressor and its complexes with DNA fragments containing operator sequences. The two dimers in a repressor tetramer are independent DNA-binding entities with helix-turn-helix DNA-binding motifs. The two dimers within the tetramer are bound to operator sequences on separate DNA fragments, but one can imagine these two operators as part of a single long piece of DNA, with the intervening DNA looped out.

> **SUMMARY** When an additional operator was placed about 200 bp upstream of the classical *lac* operator on a cloned DNA fragment, the DNA in between could be shown by three methods to loop out on cooperative binding of repressor to the two operators. X-ray crystallography has demonstrated that each dimer in a *lac* repressor tetramer is capable of simultaneous binding to an operator, further supporting the looping hypothesis.

Enhancers

Enhancers are nonpromoter DNA elements that bind protein factors and stimulate transcription. By definition, they can act at a distance. Such elements have been recognized in eukaryotes for two decades, and we will discuss them at length in Chapter 12. More recently, enhancers have also been found in prokaryotes. In 1989, Popham and coworkers described an enhancer that aids in the transcription of genes recognized by an auxiliary σ-factor in *E. coli*—σ^{54}. We encountered this factor before in Chapter 8, in our discussion of the *glnA* gene. It is the σ-factor, also known as σ^N, that comes into play under nitrogen starvation conditions to transcribe the *glnA* gene from the alternative promoter, *glnA*P2.

The σ^{54} factor is defective. DNase footprinting experiments demonstrate that it can cause the $E\sigma^{54}$ holoenzyme to bind stably to the *glnA* promoter (Figure 9.26), but it cannot do one of the important things normal σ-factors do: direct the formation of an open promoter complex. Popham and coworkers assayed this function in two ways: heparin resistance and DNA methylation. When polymerase forms an open promoter complex, it is bound very tightly to DNA. Adding heparin as a DNA competitor does not inhibit the polymerase. On the other hand, when polymerase forms a closed promoter complex, it is relatively loosely bound and will dissociate at a much higher rate. Thus, it is subject to inhibition by an excess

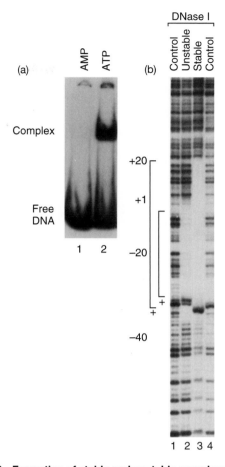

Figure 9.26 Formation of stable and unstable complexes between $E\sigma^{54}$ and the *glnA* promoter. (a) Complex stability measured by heparin sensitivity. Labeled DNA containing the *glnA* promoter was incubated with core polymerase, σ^{54}, and NtrC, in the presence of either AMP or ATP; then heparin was added, and the complexes were electrophoresed. A heparin-resistant (stable) complex formed only in the presence of ATP. Normal σ-factors do not require ATP to form a stable complex between polymerase and promoter. (b) DNase footprinting of the stable and unstable complexes. Complexes were formed with AMP or ATP as in panel (a) and subjected to DNase footprinting. Lanes 1 and 4, control with no added proteins; lane 2, unstable complex formed with AMP; lane 3, stable complex formed with ATP. The unstable complex clearly involves binding at the promoter, but the footprint area is more extensive in the stable complex (see brackets at left). (*Source:* Popham et al., Function of a bacterial activator protein that binds to transcriptional enhancers. *Science* 243 (3 Feb 1989) f. 5a-b & 6, p. 632. © AAAS.)

of the competitor heparin. Furthermore, when polymerase forms an open promoter complex, it exposes the cytosines in the melted DNA to methylation by DMS. Because no melting occurs in the closed promoter complex, no methylation takes place.

By both these criteria—heparin sensitivity and resistance to methylation—$E\sigma^{54}$ fails to form an open promoter complex. Instead, another protein, NtrC (the product of the *ntrC* gene), binds to the enhancer and helps $E\sigma^{54}$ form an open promoter complex (Figure 9.27). The energy

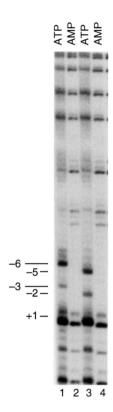

ATP AMP ATP AMP

−6 —
    −5 —
−3 —
    −2 —

+1 —

1 2 3 4

Figure 9.27   The stable complex is an open promoter complex. Stable complexes were formed with ATP in lanes 1 and 3; unstable complexes were formed with AMP in lanes 2 and 4. The end-labeled DNAs were substituted with cytosines in the −6 and −3 positions (lanes 1 and 2), or in the −5 and −2 positions (lanes 3 and 4). The complexes were treated with dimethyl sulfate to methylate any cytosines in melted DNA regions, then chemically cleaved at methylcytosines and electrophoresed to detect cleavage. The newly introduced cytosines were in a single-stranded region in the stable complexes and were methylated, so the DNA was subsequently cleaved at those two points. In the unstable complexes, the DNA remained double-stranded, and so was not methylated and therefore not cleaved. Thus, stable and unstable complexes represent open and closed promoter complexes, respectively. (*Source:* Popham et al., Function of a bacterial activator protein that binds to transcriptional enhancers. *Science* 243 (3 Feb 1989) f. 5a-b & 6, p. 632. © AAAS.)

for the DNA melting comes from the hydrolysis of ATP, performed by an ATPase domain of NtrC. This ATP-dependence is uncommon in prokaryotes. With ordinary, non-enhancer-dependent promoters, the σ-factor can sponsor DNA melting without help from ATP.

How does the enhancer interact with the promoter? The evidence strongly suggests that DNA looping is involved. One clue is that the enhancer has to be at least 70 bp away from the promoter to perform its function. This would allow enough room for the DNA between the promoter and enhancer to loop out. Moreover, the enhancer can still function even if it and the promoter are on separate DNA molecules, as long as the two molecules are linked in a catenane, as shown in Figure 9.28. This would still allow the enhancer and promoter to interact as they would during looping, but it precludes any mechanism (e.g., altering the degree of supercoiling or sliding

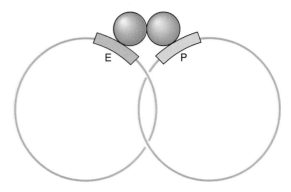

Figure 9.28   Interaction between two sites on separate but linked DNA molecules. An enhancer (E, pink) and a promoter (P, light green) lie on two separate DNA molecules that are topologically linked in a catenane (intertwined circles). Thus, even though the circles are distinct, the enhancer and promoter cannot ever be far apart, so interactions between proteins that bind to them (red and green, respectively) are facilitated.

proteins along the DNA) that requires the two elements to be on the same DNA molecule. We will discuss this phenomenon in more detail in Chapter 12. Finally, and perhaps most tellingly, we can actually observe the predicted DNA loops between NtrC bound to the enhancer and the σ54 holoenzyme bound to the promoter. Figure 9.29 shows the results of electron microscopy experiments performed by Sydney Kustu, Harrison Echols, and colleagues with cloned DNA containing the enhancer–*glnA* region. These workers inserted 350 bp of DNA between the enhancer and promoter to make the loops easier to see. The polymerase holoenzyme stains more darkly than NtrC in most of these electron micrographs, so we can distinguish the two proteins at the bases of the loops, just as we would predict if the two proteins interact by looping out the DNA in between.

Phage T4 provides an example of an unusual, mobile enhancer that is not defined by a set base sequence. Transcription of the late genes of T4 depends on DNA replication; no late transcription occurs until the phage DNA begins to replicate. One reason for this linkage between late transcription and DNA replication is that the late phage σ-factor (σ55), like σ54 of *E. coli,* is defective. It cannot function without an enhancer. But the late T4 enhancer is not a fixed DNA sequence like the NtrC-binding site. Instead, it is the DNA replicating fork. The enhancer-binding protein, encoded by phage genes 44, 45, and 62, is part of the phage DNA replicating machinery. Thus, this protein migrates along with the replicating fork, which keeps it in contact with the moving enhancer.

One can mimic the replicating fork in vitro with a simple nick in the DNA, but the polarity of the nick is important: It works as an enhancer only if it is in the nontemplate strand. This suggests that the T4 late enhancer probably does not act by DNA looping because polarity does not matter in looping. Furthermore, unlike typical enhancers such as the *glnA* enhancer, the T4 late enhancer must be on the same DNA molecule as the promoters it controls. It does not function *in trans* as part of a catenane. This argues against a looping mechanism.

SUMMARY The *E. coli glnA* gene is an example of a prokaryotic gene that depends on an enhancer for its transcription. The enhancer binds the NtrC protein, which interacts with polymerase bound to the promoter at least 70 bp away. Hydrolysis of ATP by NtrC allows the formation of an open promoter complex so transcription can take place. The two proteins appear to interact by looping out the DNA in between. The phage T4 late enhancer is mobile; it is part of the phage DNA-replication apparatus. Because this enhancer must be on the same DNA molecule as the late promoters, it probably does not act by DNA looping.

SUMMARY

The repressors of the λ-like phages have recognition helices that fit sideways into the major groove of the operator DNA. Certain amino acids on the DNA side of the recognition helix make specific contact with bases in the operator, and these contacts determine the specificity of the protein–DNA interactions. Changing these amino acids can change the specificity of the repressor. The λ and Cro protein share affinity for the same operators, but they have microspecificities for O_R1 or O_R3, determined by interactions between different amino acids in the recognition helices of the two proteins and base pairs in the different operators.

The cocrystal structure of a λ repressor fragment with an operator fragment shows many details about how the protein and DNA interact. The most important contacts occur in the major groove, where amino acids on the recognition helix, and other amino acids, make hydrogen bonds with the edges of DNA bases and with the DNA backbone. Some of these hydrogen bonds are stabilized by hydrogen-bond networks involving two amino acids and two or more sites on the DNA. The structure derived from the cocrystal is in almost complete agreement with previous biochemical and genetic data.

X-ray crystallography of a phage 434 repressor-fragment/operator-fragment complex shows close approaches between certain amino acids and certain phosphates in the operator's DNA backbone. This implies hydrogen bonding between the protein and DNA at these sites. This analysis also shows probable hydrogen bonding between amino acid residues in the recognition helix and

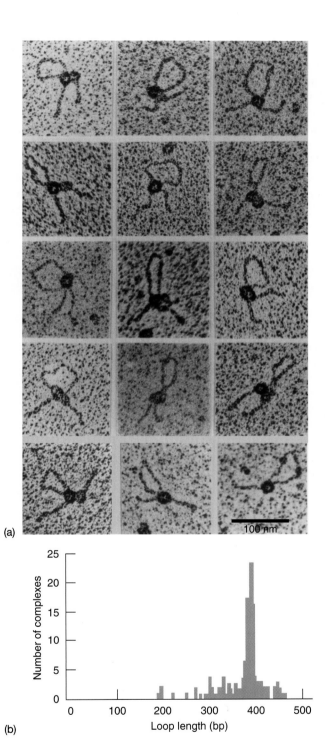

Figure 9.29   Looping the *glnA* promoter–enhancer region. (a) Electron micrographs. Kustu, Echols, and colleagues moved the *glnA* promoter and enhancer apart by inserting a 350-bp DNA segment between them, then allowed the NtrC protein to bind to the enhancer, and RNA polymerase to bind to the promoter. When the two proteins interacted, they looped out the DNA in between, as shown in these electron micrographs. (b) Plot of the loop lengths. The investigators measured the lengths of several loops such as those in part (a) and plotted their lengths versus the number of loops of a given length. The peak occurred at 390 bp, which is exactly the spacing between the enhancer and the promoter. This showed that the loops were not artifacts caused by crossed DNA strands, but were real loops caused by interactions between the two remote proteins. (*Source:* (a) Su et al., DNA-looping and enhancer activity: Association between DNA-bound NtrC activator and RNA polymerase at the bacterial *glnA* promoter. *PNAS* 87 (July 1990) f. 4, p. 5507.)

base pairs in the repressor. It also reveals a potential van der Waals contact between an amino acid in the recognition helix and a base in the operator. The DNA in the complex deviates significantly from its normal regular shape. It bends somewhat to accommodate the necessary base/amino acid contacts. Moreover, the central part of the helix, between the two half-sites, is wound extra tightly, and the outer parts are wound more loosely than normal. The base sequence of the operator facilitates these departures from normal DNA shape.

The *trp* repressor requires tryptophan to force the recognition helices of the repressor dimer into the proper position for interacting with the *trp* operator. The *trp* repressor binds to the *trp* operator in a less-direct way than the λ family of repressors bind to their operators. The recognition helix of the *trp* repressor points directly into the major groove of the operator, which allows only one direct contact between an amino acid and a base. All the other interactions are direct contacts between amino acids and DNA backbone phosphates, or indirect contacts through water molecules.

A DNA-binding protein can interact with the major or minor groove of the DNA (or both). The four different base pairs present four different hydrogen-bonding profiles to amino acids approaching either the major or minor DNA groove, so a DNA-binding protein can recognize base pairs in the DNA even though the two strands do not separate.

The CAP protein bends its target DNA dramatically when it binds, and this bend is necessary for the specific binding between CAP and DNA. Thus, a DNA sequence that has all or most of the bases with which CAP interacts, but cannot bend, will be relatively unattractive to CAP protein. This is an extreme example of a more general rule that may apply to many protein DNA associations. It is not always just the amino acid/DNA base-pair interactions that govern the affinity between protein and DNA; the affinity between DNA and protein may depend on the ability of the DNA to be distorted into a shape that fits the protein.

Multimeric DNA-binding proteins have an inherently higher affinity for binding sites on DNA than do multiple monomeric proteins that bind independently of one another. The advantage of multimeric proteins is that they can bind cooperatively to DNA.

When λ operators are separated by an integral number of helical turns, the DNA in between can loop out to allow cooperative binding. When the operators are separated by a nonintegral number of helical turns, the proteins have to bind to opposite faces of the DNA double helix, so no cooperative binding can take place.

When an additional operator was placed about 200 bp upstream of the major *lac* operator on a cloned DNA fragment, the DNA in between could be shown to loop out on cooperative binding of repressor to the two operators.

The *E. coli glnA* gene is an example of a bacterial gene that depends on an enhancer for its transcription. The enhancer binds the NtrC protein, which interacts with polymerase bound to the promoter at least 70 bp away. Hydrolysis of ATP by NtrC allows the formation of an open promoter complex so transcription can take place. The two proteins appear to interact by looping out the DNA in between. The phage T4 late enhancer is mobile; it is part of the phage DNA-replication apparatus. Because this enhancer must be on the same DNA molecule as the late promoters, it probably does not act by DNA looping.

REVIEW QUESTIONS

1. Draw a rough diagram of a helix-turn-helix domain interacting with a DNA double helix.

2. Describe and give the results of an experiment that shows which amino acids are important in binding between λ-like phage repressors and their operators. Present two methods of assaying the binding between the repressors and operators.

3. In general terms, what accounts for the different preferences of λ repressor and Cro for the three operator sites?

4. Glutamine and asparagine side chains tend to make what kind of bonds with DNA?

5. Methylene and methyl groups on amino acids tend to make what kind of bonds with DNA?

6. A guanine in an operator becomes more sensitive to methylation when the repressor binds. What kind of experiment would lead to this conclusion, and what would the results look like? What kind of alteration in the DNA would explain this finding? How could you find out if this alteration really happens when repressor binds?

7. A glutamine in a DNA-binding protein makes an important hydrogen bond with a cytosine in the DNA. Changing this glutamine to alanine prevents formation of this hydrogen bond and blocks the DNA–protein interaction. Changing the cytosine to thymine restores binding to the mutant protein. Present a plausible hypothesis to explain these findings.

8. What is meant by the term *hydrogen-bond network* in the context of protein–DNA interactions?

9. What is meant by the term *indirect readout* in the context of protein–DNA interactions?

10. You have the following working hypothesis: To bind well to a DNA-binding protein, a DNA target site must twist more tightly and compress the narrow groove between base pairs 4 and 5. Suggest an experiment to test your hypothesis.

11. Draw a rough diagram of the "reading head" model to show the difference in position of the recognition helix of the *trp* repressor and aporepressor, with respect to the *trp* operator.

12. Draw a rough diagram of the "salami sandwich" model to explain how adding tryptophan to the *trp* aporepressor causes a shift in conformation of the protein.

13. In one sentence, contrast the orientations of the λ and *trp* repressors relative to their respective operators.

14. What is Sigler's hypothesis for the nature of most indirect interactions between the *trp* repressor and operator?

15. Consult Figure 2.13 for the structure of an A–T base pair. Based on that structure, draw a line diagram indicating the relative positions of the H-bond acceptor and donor groups in the major and minor grooves. Represent the horizontal axis of the base pair by two segments of a horizontal line, and the relative horizontal positions of the H-bond donors and acceptors by vertical lines. Let the lengths of the vertical lines indicate the relative vertical positions of the acceptors and donors. What relevance does this diagram have for a protein that interacts with this base pair?

16. How is the *lac* promoter able to bend 90 degrees on binding to the CAP–cAMP complex? How does the base sequence of the promoter facilitate this bending?

17. Explain the fact that protein oligomers (dimers or tetramers) bind more successfully to DNA than monomeric proteins do.

18. Use a diagram to explain the alternating pattern of resistance and elevated sensitivity to DNase in the DNA between two separated binding sites when two proteins bind cooperatively to these sites.

19. Describe and give the results of a DNase footprinting experiment that shows that λ repressor dimers bind cooperatively to two operators separated by an integral number of DNA double-helical turns, but noncooperatively to two operators separated by a nonintegral number of turns.

20. Describe and give the results of an electron microscopy experiment that shows the same thing as the experiment in the preceding question.

21. In what way is σ^{54} defective?

22. Describe and give the results of two different assays that demonstrate the defect in σ^{54}.

23. What substances supply the missing function to σ^{54}?

24. Describe and give the results of an experiment that shows that DNA looping is involved in the enhancement of the *E. coli glnA* locus.

25. In what ways is the enhancer for phage T4 σ^{55} different from the enhancer for the *E. coli* σ^{54}?

SUGGESTED READINGS

General References and Reviews

Geiduschek, E.P. 1997. Paths to activation of transcription. *Science* 275:1614–16.

Kustu, S., A.K. North, and D.S. Weiss. 1991. Prokaryotic transcriptional enhancers and enhancer-binding proteins. *Trends in Biochemical Sciences* 16:397–402.

Luisi, B.F. and P.B. Sigler. 1990. The stereochemistry and biochemistry of the *trp* repressor–operator complex. *Biochimica et Biophysica Acta.* 1048:113–26.

Matthews, K.S. 1996. The whole lactose repressor. *Science* 271:1245–46.

Schleif, R. 1988. DNA binding by proteins. *Science* 241:1182–87.

Research Articles

Aggarwal, A.K., D.W. Rodgers, M. Drottar, M. Ptashne, and S.C. Harrison. 1988. Recognition of a DNA operator by the repressor of phage 434: A view at high resolution. *Science* 242:899–907.

Anderson, J.E., M. Ptashne, and S.C. Harrison. 1987. Structure of the repressor–operator complex of bacteriophage 434. *Nature* 326:846–52.

Friedman, A.M., T.D. Fischman, and T.A. Steitz. 1995. Crystal structure of *lac* repressor core tetramer and its implications for DNA looping. *Science* 268:1721–27.

Griffith, J., A. Hochschild, and M. Ptashne. 1986. DNA loops induced by cooperative binding of λ repressor. *Nature* 322:750–52.

Herendeen, D.R., G.A. Kassavetis, J. Barry, B.M. Alberts, and E.P. Geiduschek. 1990. Enhancement of bacteriophage T4 late transcription by components of the T4 DNA replication apparatus. *Science* 245:952–58.

Hochschild, A., N. Irwin, and M. Ptashne. 1983. Repressor structure and the mechanism of positive control. *Cell* 32:319–25.

Hochschild, A. and M. Ptashne. 1986. Cooperative binding of λ repressors to sites separated by integral turns of the DNA helix. *Cell* 44:681–87.

Hochschild, A., J. Douhann III, and M. Ptashne. 1986. How λ repressor and λ cro distinguish between O_R1 and O_R3. *Cell* 47:807–16.

Hunt, T.P. and B. Magasanik. 1985. Transcription of *glnA* by purified *Escherichia coli* components: Core RNA polymerase and the products of *glnF, glnG,* and *glnL. Proceedings of the National Academy of Sciences USA* 82:8453–57.

Jordan, S.R. and C.O. Pabo. 1988. Structure of the lambda complex at 2.5 Å resolution: Details of the repressor–operator interactions. *Science* 242:893–99.

Krämer, H., M. Niemöller, M. Amouyal, B. Revet, B. von Wilcken-Bergmann, and B. Müller-Hill. 1987. *lac* repressor forms loops with linear DNA carrying suitably spaced *lac* operators. *EMBO Journal* 6:1481–91.

Otwinowski, Z., R.W. Schevitz, R.-G. Zhang, C.L. Lawson, A. Joachimiak, R.Q. Marmorstein, B.F. Luisi, and P.B. Sigler. 1988. Crystal structure of *trp* repressor/operator complex at atomic resolution. *Nature* 335:321–29.

Popham, D.L., D. Szeto, J. Keener, and S. Kustu. 1989. Function of a bacterial activator protein that binds to transcriptional enhancers. *Science* 243:629–35.

Sauer, R.T., R.R. Yocum, R.F. Doolittle, M. Lewis, and C.O. Pabo. 1982. Homology among DNA-binding proteins suggests use of a conserved super-secondary structure. *Nature* 298:447–51.

Schevitz, R.W., Z. Otwinowski, A. Joachimiak, C.L. Lawson, and P.B. Sigler. 1985. The three-dimensional structure of *trp* repressor. *Nature* 317:782–86.

Schultz, S.C., G.C. Shields, and T.A. Steitz. 1991. Crystal structure of a CAP–DNA complex: The DNA is bent by 90°. *Science* 253:1001–7.

Su, W., S. Porter, S. Kustu, and H. Echols. 1990. DNA looping and enhancer activity: Association between DNA-bound NtrC activator and RNA polymerase at the bacterial *glnA* promoter. *Proceedings of the National Academy of Sciences USA* 87:5504–8.

Wharton, R.P. and M. Ptashne. 1985. Changing the binding specificity of a repressor by redesigning an α-helix. *Nature* 316:601–5.

Wyman, C., I. Rombel, A.K. North, C. Bustamante, and S. Kustu. 1997. Unusual oligomerization required for activity of NtrC, a bacterial enhancer-binding protein. *Science* 275:1658–61.

Zhang, R.-g., A. Joachimiak, C.L. Lawson, R.W. Schevitz, Z. Otwinowski, and P.B. Sigler. 1987. The crystal structure of *trp* aporepressor at 1.8 Å shows how binding tryptophan enhances DNA affinity. *Nature* 327:591–97.

CHAPTER 10

Eukaryotic RNA Polymerases and Their Promoters

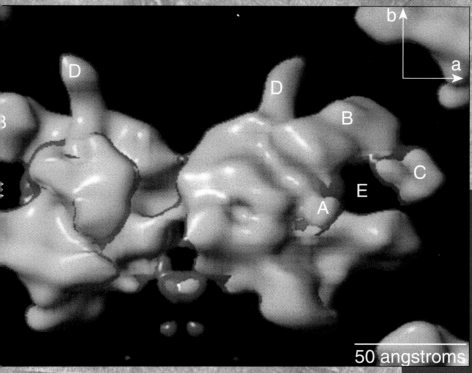

Computer-generated model of two Pol IIΔ4/7 proteins based on electron diffraction data. *Darst et al., Three-dimensional structure of yeast RNA polymerase II at 16Å resolution. Cell 66 (12 July 1991) p. 125, f. 5. Reprinted by permission of Elsevier Science.*

In Chapter 6 we learned that prokaryotes have only one RNA polymerase, which makes all three kinds of RNA: mRNA, rRNA, and tRNA. True, the polymerase can switch σ-factors to meet the demands of a changing environment, but the core enzyme remains essentially the same. Quite a different situation prevails in the eukaryotes. In this chapter we will see that three distinct RNA polymerases occur in the nuclei of eukaryotic cells. Each of these is responsible for transcribing a separate set of genes, and each recognizes a different kind of promoter.

10.1 Multiple Forms of Eukaryotic RNA Polymerase

Several early studies suggested that at least two RNA polymerases operate in eukaryotic nuclei: one to transcribe the major ribosomal RNA genes (those coding for the 28S, 18S, and 5.8S rRNAs in vertebrates), and one or more to transcribe the rest of the nuclear genes. Let us begin with a discussion of these early data.

Indications of Multiple Eukaryotic Polymerases

To begin with, the ribosomal genes are different in several ways from other nuclear genes: (1) They have a different base composition from that of other nuclear genes. For example, rat rRNA genes have a GC content of 60%, but the rest of the DNA has a GC content of only 40%. (2) They are unusually repetitive; depending on the organism, each cell contains from several hundred to over 20,000 copies of the rRNA gene. (3) They are found in a different compartment—the nucleolus—than the rest of the nuclear genes. Several lines of evidence led to this last conclusion. Donald Brown and John Gurdon showed that anucleolate mutant embryos of the South African clawed frog, *Xenopus laevis,* which lacked nucleoli, made no rRNA and had no ribosomes. F.M. Ritossa and Sol Spiegelman analyzed several *Xenopus* mutants with a varying number of nucleoli per cell and found that the amount of rRNA made depended on the number of nucleoli.

C.C. Widnell and J.R. Tata isolated nuclei from rat liver and tested them for RNA synthesis under various conditions. First, they varied the salt concentration (ionic strength) and showed that at high ionic strength the nuclei made an RNA with a low GC content. This base content resembled that of bulk (nonribosomal) DNA. On the other hand, under low ionic strength conditions, the nuclei made RNA with a high, rRNA-like, GC content. Furthermore, transcription at high and low ionic strength responded differently to two different divalent metal ions in the reaction. At high ionic strength, Mn^{2+} stimulated transcription more than Mg^{2+} did. However, at low ionic strength the opposite was true; Mg^{2+} stimulated transcription more than Mn^{2+} did.

Finally, A.O. Pogo and colleagues forced isolated nuclei to synthesize RNA under various conditions in the presence of radiolabeled UTP, and then performed autoradiography to locate the labeled RNA within the nuclei. Figure 10.1 shows the results. In the presence of Mg^{2+} in low ionic strength solution, most of the transcription occurred in the nucleolus. By contrast, in the pres-

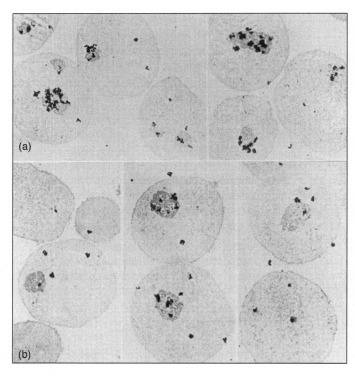

Figure 10.1   Effects of salt concentration and divalent metal ions on location of transcription in isolated nuclei. Pogo and colleagues isolated nuclei from regenerating liver and incubated them with [³H]UTP to radiolabel newly made RNA. They performed the incubations under two conditions: **(a)** low salt concentration plus Mg^{2+}; and **(b)** high salt concentration plus Mn^{2+}. Then they subjected the labeled nuclei to autoradiography to locate the labeled RNA. Under low-salt, Mg^{2+} conditions, the great majority of the RNA synthesis occurred in the nucleoli **(a)**. Under high-salt, Mn^{2+} conditions, the RNA synthesis took place throughout the nuclei **(b)**. (*Source:* Pogo et al., *PNAS* 57, no. 1 (1967) p. 748, f. 1.)

ence of Mn^{2+} in high ionic strength solution, the transcription occurred throughout the nucleus.

All of these results suggested that at least two RNA polymerases were operating in eukaryotic nuclei. One of these synthesized rRNA in the nucleolus and was stimulated by low ionic strength and Mg^{2+}. The other synthesized other RNA in the nucleoplasm (the part of the nucleus outside the nucleolus) and was stimulated by high ionic strength and Mn^{2+}.

Separation of the Three Nuclear Polymerases

Robert Roeder and William Rutter showed in 1969 that eukaryotes have not just two, but three different RNA polymerases. Furthermore, these three enzymes have distinct roles in the cell. These workers separated the three enzymes by DEAE-Sephadex ion-exchange chromatography (Chapter 5).

They named the three peaks of polymerase activity in order of their emergence from the ion-exchange column: **RNA polymerase I, RNA polymerase II,** and **RNA**

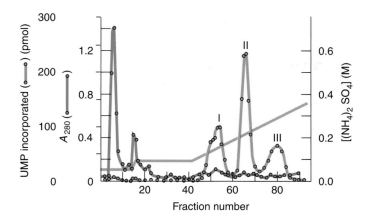

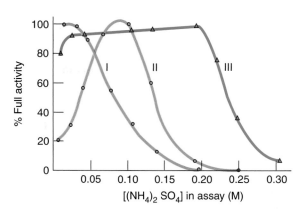

Figure 10.2   **Separation of eukaryotic RNA polymerases.** Roeder and Rutter subjected extracts from sea urchin embryos to DEAE-Sephadex chromatography. Green, protein measured by A_{280}; red, RNA polymerase activity measured by incorporation of labeled UMP into RNA; blue, ammonium sulfate concentration. (*Source:* Reprinted with permission from *Nature* 224:235, 1969. Copyright © 1969 Macmillan Magazines Limited.)

Figure 10.3   **Effect of ionic strength on the three sea urchin RNA polymerases.** The activity of each polymerase, relative to the optimum for each, is given as a function of ammonium sulfate concentration in the enzyme assay. Red, polymerase I; blue, polymerase II; green, polymerase III. (*Source:* Reprinted with permission from *Nature* 224:236, 1969. Copyright © 1969 Macmillan Magazines Limited.)

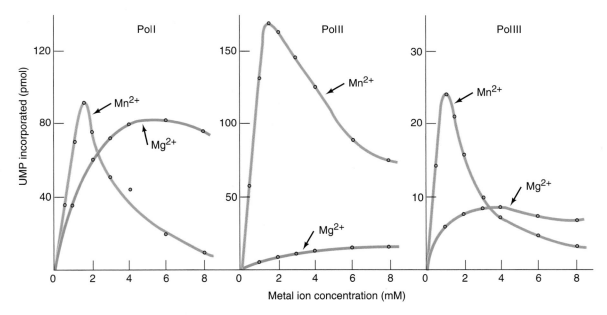

Figure 10.4   **Effect of divalent metal ion on the three sea urchin RNA polymerases.** The activity of each polymerase is given as a function of concentration of Mn^{2+} (blue) or Mg^{2+} (red). (*Source:* Reprinted with permission from *Nature* 224:236, 1969. Copyright © 1969 Macmillan Magazines Limited.)

polymerase III (Figure 10.2). The three enzymes have different properties besides their different behaviors on DEAE-Sephadex chromatography. For example, they have different responses to ionic strength and divalent metals. More importantly, they have distinct roles in transcription: Each makes different kinds of RNA.

After separating the three RNA polymerases, Roeder and Rutter assayed them under different ionic strength and divalent metal ion conditions. Figures 10.3 and 10.4 show the results. Polymerase I is most active at low ionic strength and was about equally active in the presence of Mn^{2+} and Mg^{2+}. Thus, it resembled the low-salt, Mg^{2+}-stimulated en-

zyme thought to be responsible for rRNA synthesis and apparently located in nucleoli. Polymerase II was most active at high ionic strength and had a pronounced preference for Mn^{2+} instead of Mg^{2+}. This enzyme strongly resembled the high-salt, Mn^{2+}-stimulated enzyme thought to be responsible for at least some non-rRNA synthesis and apparently located in the nucleoplasm. Polymerase III was active over a broad range of ionic strengths and had a preference for Mn^{2+}, so it was not readily identifiable with either of the polymerase activities previously detected in isolated nuclei.

Because the ionic strength and metal ion preferences of these polymerases suggested that polymerase I is the

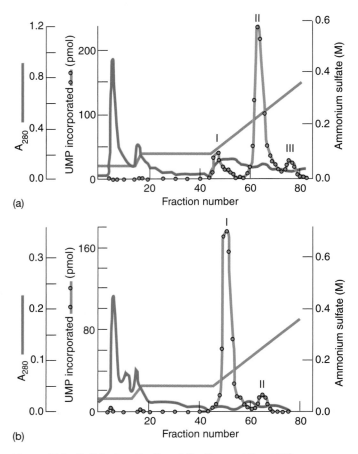

(a)

(b)

Figure 10.5   Cellular localization of the three rat liver RNA polymerases. Roeder and Rutter subjected the polymerases found in the nucleoplasmic fraction (**a**) or nucleolar fraction (**b**) of rat liver to DEAE-Sephadex chromatography as described in Figure 10.2. Colors have the same meanings as in Figure 10.2. (*Source:* From R.G. Roeder and W.J. Rutter, "Specific Nucleolar and Nucleoplasmic RNA Polymerases," *Proceedings of the National Academy of Sciences* 65(3):675–682, March 1970. Reprinted with permission.)

nucleolar rRNA-synthesizing enzyme, and polymerase II is the nucleoplasmic non-rRNA-synthesizing enzyme, Roeder and Rutter next looked in purified nucleoli and nucleoplasm to see if these subnuclear compartments were enriched in the appropriate polymerases. Figure 10.5 shows that polymerase I is indeed located primarily in the nucleolus, and polymerases II and III are found in the nucleoplasm. This made it very likely that polymerase I is the rRNA-synthesizing enzyme, and that polymerases II and III make some other kinds of RNA.

SUMMARY Eukaryotic nuclei contain three RNA polymerases that can be separated by ion-exchange chromatography. These three enzymes have different responses to salt and divalent metal ions, and different locations within the nucleus. RNA polymerase I is found in the nucleolus; the other two polymerases are located in the nucleoplasm. The properties of RNA polymerase I, including its location in the nucleolus, suggest that it transcribes the rRNA genes.

Table 10.1   Roles of Eukaryotic RNA Polymerases

RNA Polymerase	Cellular RNAs Synthesized	Mature RNA (Vertebrate)
I	Large rRNA precursor	28S, 18S, and 5.8S rRNAs
II	hnRNAs	mRNAs
	snRNAs	snRNAs
III	5S rRNA precursor	5S rRNA
	tRNA precursors	tRNAs
	U6 snRNA (precursor?)	U6 snRNA
	7SL RNA (precursor?)	7SL RNA
	7SK RNA (precursor?)	7SK RNA

The Roles of the Three RNA Polymerases

How do we know that the three RNA polymerases have different roles in transcription? The ultimate proof of these roles has come from studies in which the purified polymerases were shown to transcribe certain genes, but not others, in vitro. Such studies have demonstrated that the three RNA polymerases have the following specificities (Table 10.1): Polymerase I makes the large rRNA precursor. In vertebrates, this precursor has a sedimentation coefficient of 45S and is processed to the 28S, 18S, and 5.8S mature rRNAs. Polymerase II makes an ill-defined class of RNA known as **heterogeneous nuclear RNA (hnRNA)** as well as most **small nuclear RNAs (snRNAs)**. We will see in Chapter 14 that most of the hnRNAs are precursors of mRNAs and that the snRNAs participate in the maturation of hnRNAs to mRNAs. Polymerase III makes precursors to the tRNAs, 5S rRNA, and some other small RNAs.

However, even before cloned genes and eukaryotic in vitro transcription systems were available, we had evidence to support most of these transcription assignments. In this section, we will examine the early evidence that RNA polymerase III transcribes the tRNA and 5S rRNA genes.

This work, by Roeder and colleagues in 1974, depended on a toxin called α-**amanitin.** This highly toxic substance is found in several poisonous mushrooms of the genus *Amanita* (Figure 10.6a), including *A. phalloides*, "the death cap," and *A. bisporigera*, which is called "the angel of death" because it is pure white and deadly poisonous. Both species have proven fatal to many inexperienced mushroom hunters. Alpha-amanitin was found to have different effects on the three polymerases. At very low concentrations, it inhibits polymerase II completely while having no effect at all on polymerases I and III. At 1000-fold higher concentrations, the toxin also inhibits polymerase III from most eukaryotes (Figure 10.7).

The plan of the experiment was to incubate mouse cell nuclei in the presence of increasing concentrations of

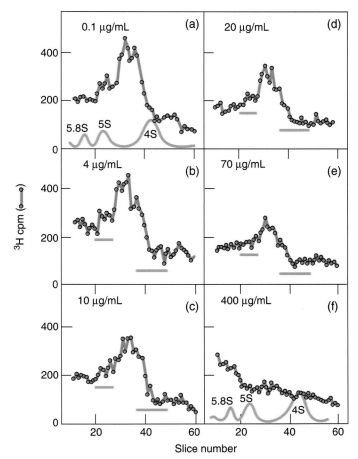

Figure 10.8   Effect of α-amanitin on small RNA synthesis. Weinmann and Roeder synthesized labeled RNA in isolated nuclei in the presence of increasing amounts of α-amanitin (concentration given at the top of each panel). The small labeled RNAs leaked out of the nuclei and were found in the supernatant after centrifugation. The researchers then subjected these RNAs to PAGE, sliced the gel, and determined the radioactivity in each slice (red). They also ran markers (5.8S rRNA, 5S rRNA, and 4S tRNA) in adjacent lanes of the same gel. The positions of all three markers are shown in panels A and F (blue); the positions of the latter two are shown in all other panels by horizontal blue lines. The inhibition of 5S rRNA and 4S tRNA precursor synthesis by α-amanitin closely parallels the effect of the toxin on polymerase III, determined in Figure 10.7. (*Source:* From R. Weinmann and R.G. Roeder, "Role of DNA-Dependent RNA Polymerase III in the Transcription of the tRNA and 5S RNA Genes," *Proceedings of the National Academy of Sciences* 71(5): 1790–1794, May 1974. Reprinted with permission.)

α-amanitin, then to electrophorese the transcripts to observe the effect of the toxin on the synthesis of small RNAs. Figure 10.8 reveals that high concentrations of α-amanitin inhibited the synthesis of both 5S rRNA and 4.5S tRNA precursor. Moreover, this pattern of inhibition of 5S rRNA and tRNA precursor synthesis matched the pattern of inhibition of RNA polymerase III: They both were about half-inhibited at 10 μg/mL of α-amanitin. Therefore, these data support the hypothesis that RNA polymerase III makes these two kinds of RNA. (Actually, polymerase III also synthesizes the 5S rRNA as a slightly larger precursor, but this experiment did not distinguish the precursor from the mature 5S rRNA.) Polymerase III also makes a variety of other small cellular and viral RNAs.

Figure 10.6   Alpha-amanitin. (a) *Amanita phalloides* ("the death cap"), one of the deadly poisonous mushrooms that produces α-amanitin. **(b)** Structure of α-amanitin. (*Source:* (a) Arora, D. *Mushrooms Demystified* 2e, 1986, Plate 50 (Ten Speed Press).)

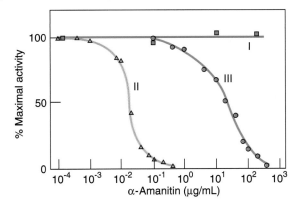

Figure 10.7   Sensitivity of purified RNA polymerases to α-amanitin. Weinmann and Roeder assayed RNA polymerases I (green), II (blue), and III (red) with increasing concentrations of α-amanitin. Polymerase II was 50% inhibited by about 0.02 μg/mL of the toxin, whereas polymerase III reached 50% inhibition only at about 20 μg/mL of toxin. Polymerase I retained full activity even at an α-amanitin concentration of 200 μg/mL. (*Source:* From R. Weinmann and R.G. Roeder, "Role of DNA-Dependent RNA Polymerase III in the Transcription of the tRNA and 5S RNA Genes," *Proceedings of the National Academy of Sciences* 71(5):1790–1794, May 1974. Reprinted with permission.)

These include **U6 snRNA,** a small RNA that participates in RNA splicing (Chapter 14); **7SL RNA,** a small RNA involved in signal peptide recognition in the synthesis of secreted proteins; **7SK RNA,** a small nuclear RNA of unknown function, the adenovirus **VA (virus-associated) RNAs;** and the Epstein–Barr virus **EBER2 RNA.**

Similar experiments were performed to identify the genes transcribed by RNA polymerases I and II. But these studies were not as easy to interpret and they have been confirmed by much more definitive in vitro studies.

SUMMARY The three nuclear RNA polymerases have different roles in transcription. Polymerase I makes the large precursor to the rRNAs (5.8S, 18S, and 28S rRNAs in vertebrates). Polymerase II makes hnRNAs, which are precursors to mRNAs, and most of the snRNAs. Polymerase III makes the precursors to 5S rRNA, the tRNAs, and several other small cellular and viral RNAs.

RNA Polymerase Subunit Structures

The first subunit structures for a eukaryotic RNA polymerase (polymerase II) were reported independently by Pierre Chambon and Rutter and their colleagues in 1971, but they were incomplete. We should note in passing that Chambon named his three polymerases A, B, and C, instead of I, II and III, respectively. However, most investigators use the I, II, III nomenclature of Roeder and Rutter. Roeder and his colleagues presented the first comparison of more or less complete subunit structures for all three nuclear polymerases (from cultured mouse cells) in 1975. We now have very good structural information on all three polymerases from a variety of eukaryotes. Table 10.2 lists the relative molecular masses (in kilodaltons) of the putative subunits known in 1985 for the three yeast polymerases, for which the best data are available. In fact, the genes for all of the subunits of yeast RNA polymerase II have been cloned and sequenced, which means we have detailed information about their primary structures and even some clues about their roles in the polymerase reaction. The names of these polymerase subunits, Rpb1, and so on, derive from the names of the genes that encode them (*RPB1*, and so on). Note the echo of the Chambon nomenclature in the name *RPB*, which stands for RNA polymerase B (or II). The 185-kD polypeptide is not a true subunit of polymerase II. As we will see later in this chapter, it is a degraded form of the 220-kD subunit.

Table 10.2 makes several points. First, all the polymerase structures are complex—even more so than the structure of the bacterial polymerases. Second, all the structures are similar in that each contains two large (greater than 100 kD) subunits, plus a variety of smaller

Table 10.2	Partial Subunit Structures of RNA Polymerases from *Saccharomyces cerevisiae*

Relative molecular masses of RNA polymerase subunits (kD)			
I	II	III	
190	220 (185)	160	
135	150	128	
49		82	
43		53	
40	44.5	40	
		37	
34.5	32	34	
		31	
27	27	27	Rpb5
23	23	23	Rpb6
19	16	19	
14.5	14.5	14.5	Rpb8
14	12.6	22	
12.2			
10	10	10	Rpb10
			Rpb12

Source: Sentenac, A. 1985. "Eukaryotic RNA polymerases." *CRC Critical Reviews in Biochemistry* 18:31–90.

subunits. In this respect, these structures resemble that of the prokaryotic core polymerases, which contain two high-molecular-mass subunits (β and β') plus two low-molecular-mass subunits (α_2). In fact, as we will see later in this chapter, a clear evolutionary relationship is evident between the prokaryotic core polymerase subunits and three of the subunits of all of the eukaryotic polymerases. In other words, the three eukaryotic polymerases are related to the prokaryotic polymerase and to one another.

A third message from Table 10.2 is that the three yeast nuclear polymerases have several subunits in common. In fact, five such **common subunits** exist. In the polymerase II structure, these are called Rpb5, Rpb6, Rpb8, Rpb10, and Rpb12. These are identified on the right in Table 10.2. Rpb12 is a common subunit, but had not been separated from Rpb10 in 1985.

SUMMARY The subunit structures of all three nuclear polymerases from several eukaryotes have been determined. All of these structures contain many subunits, including two large ones, with molecular masses greater than 100 kD. All eukaryotes seem to have at least some common subunits that are found in all three polymerases. Yeast RNA polymerases have five of these common subunits.

Table 10.3 Yeast RNA Polymerase II Subunits

Subunit	SDS-PAGE Mobility (kD)	Protein Mass (kD)	Stoichiometry	Deletion Phenotype
Rpb1	220	190	1.1	Inviable
Rpb2	150	140	1.0	Inviable
Rpb3	45	35	2.1	Inviable
Rpb4	32	25	0.5	Conditional
Rpb5	27	25	2.0	Inviable
Rpb6	23	18	0.9	Inviable
Rpb7	17	19	0.5	Inviable
Rpb8	14	17	0.8	Inviable
Rpb9	13	14	2.0	Conditional
Rpb10	10	8.3	0.9	Inviable
Rpb11	13	14	—	Inviable
Rpb12	10	7.7	—	Inviable

Polymerase II Structure For enzymes as complex as the eukaryotic RNA polymerases it is difficult to tell which polypeptides that copurify with the polymerase activity are really subunits of the enzymes and which are merely contaminants that bind tightly to the enzymes. One way of dealing with this problem would be to separate the putative subunits of a polymerase and then see which polypeptides are really required to reconstitute polymerase activity. Although this strategy worked beautifully for the prokaryotic polymerases, no one has yet been able to reconstitute a eukaryotic nuclear polymerase from its separate subunits. Thus, we must try a different tack.

Another way of approaching this problem is to find the genes for all the putative subunits of a polymerase, mutate them, and determine which are required for activity. This has been accomplished for one enzyme: polymerase II of baker's yeast, *Saccharomyces cerevisiae*. Several investigators used traditional methods to purify yeast polymerase II to homogeneity and identified 10 putative subunits. Later, some of the same scientists discovered two other subunits that had been hidden in the earlier analyses, so the current concept of the structure of yeast polymerase II includes 12 subunits. The genes for all 12 subunits have been cloned and sequenced, which tells us the amino acid sequences of their products. The genes have also been systematically mutated, and the effects of these mutations on polymerase II activity have been observed. As we will see, 10 of these polypeptides are absolutely required for polymerase activity, and 2 are required under certain conditions.

Table 10.3 lists the 12 putative subunits of yeast polymerase II, along with their apparent molecular masses determined by SDS-PAGE (Chapter 5), their actual molecular masses determined from the sequences of their genes, their stoichiometries, and the phenotypes of yeast cells with dele-tion mutations in each gene. Each of these polypeptides is encoded in a single gene in the yeast genome.

Richard Young and his coworkers have corroborated the finding that the 10 originally identified polypeptides are authentic polymerase II subunits, or at least tightly bound contaminants. The method they used is called **epitope tagging** (Figure 10.9), in which they attached a small foreign epitope to one of the yeast polymerase II subunits (Rpb3) by engineering its gene, then introduced this gene into yeast cells lacking a functional Rpb3 gene, and used an antibody directed against this epitope to precipitate the whole enzyme. Before this immunoprecipitation, they labeled the cellular proteins with either ^{35}S or ^{32}P. After immunoprecipitation, they separated the labeled poly-peptides of the precipitated protein by SDS-PAGE and detected them by autoradiography. Figure 10.10a presents the results. This single-step purification method yielded essentially pure polymerase II with 10 apparent subunits. We can also see a few minor polypeptides, but they are also visible in the control in which wild-type enzyme, with no epitope tag, was used. Therefore, they are not polymerase-associated. Figure 10.10b shows a later SDS-PAGE analysis of the same polymerase, performed by Roger Kornberg and colleagues, which distinguished 12 subunits. Rpb11 had coelectrophoresed with Rpb9, and Rpb12 had coelectrophoresed with Rpb10, so both Rpb11 and Rpb12 had been missed in the earlier experiments.

Because Young and colleagues already knew the amino acid compositions of all 10 original subunits, the relative labeling of each polypeptide with ^{35}S-methionine gave them a good estimate of the stoichiometries of sub-units, which are listed in Table 10.3. Figure 10.10a also shows us that two polymerase II subunits are phosphory-lated, because they were labeled by $[\gamma$-^{32}P]ATP. These phos-phoproteins are subunits Rpb1 and Rpb6. Rpb2 is also

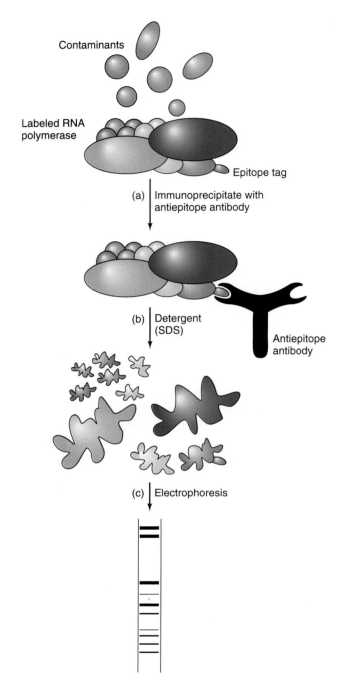

Figure 10.9 Principle of epitope tagging. An extra domain (an epitope tag, red) has been added genetically to one subunit (Rpb3) of the yeast RNA polymerase II. All the other subunits are normal, and assemble with the altered Rpb3 subunit to form an active polymerase. This polymerase has also been labeled by growing cells in labeled amino acids. **(a)** Add an antibody directed against the epitope tag, which immunoprecipitates the whole RNA polymerase, separating it from contaminating proteins (gray). This gives very pure polymerase in just one step. **(b)** Add the strong detergent SDS, which separates and denatures the subunits of the purified polymerase. **(c)** Electrophorese the denatured subunits of the polymerase to yield the electropherogram at bottom.

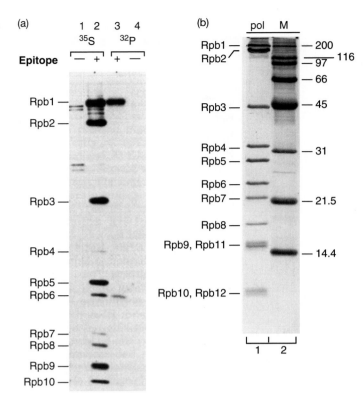

Figure 10.10 Subunit structure of yeast RNA polymerase II.
(a) Apparent 10-subunit structure obtained by epitope tagging. Young and colleagues endowed one of the subunits of yeast polymerase II (Rpb3) with an extra group of amino acids (an epitope tag) by substituting a gene including the codons for this tag for the usual yeast *RPB3* gene. Then they labeled these engineered yeast cells with either [^{35}S]methionine to label all the polymerase subunits, or [γ-^{32}P]ATP to label the phosphorylated subunits only. They immunoprecipitated the labeled protein with an antibody directed against the epitope tag and electrophoresed the products. Lane 1, ^{35}S-labeled protein from wild-type yeast without the epitope tag; lane 2, ^{35}S-labeled protein from yeast having the epitope tag on Rpb3; lane 3, ^{32}P-labeled protein from yeast with the epitope tag; lane 4, ^{32}P-labeled protein from wild-type yeast. The polymerase II subunits are identified at left. **(b)** Apparent 12-subunit structure obtained by multistep purification including immunoprecipitation. Kornberg and colleagues immunoprecipitated yeast RNA polymerase II and subjected it to SDS-PAGE (lane 1), alongside molecular mass markers (lane 2). The marker molecular masses are given at right, and the polymerase II subunits are identified at left. Notice that Rpb9 and Rpb11 almost comigrate, as do Rpb10 and Rpb12. (*Source:* (a) Kolodziej et al., *Mol. Cell Biol.* 10 (May 1990) p. 1917, f. 2. American Society for Microbiology. (b) Sayre et al., *J. Biol. Chem.* 267 (15 Nov 1992) p. 23379, f. 3b. American Society for Biochemistry and Molecular Biology.)

phosphorylated, but at such a low level that Figure 10.10a does not show it.

Young categorized the original 10 polymerase II subunits in three groups: core subunits, which are related in structure and function to the core subunits of bacterial RNA polymerases; common subunits, which are found in all three nuclear RNA polymerases, at least in yeast; and nonessential subunits, which are conditionally dispensable for enzyme activity.

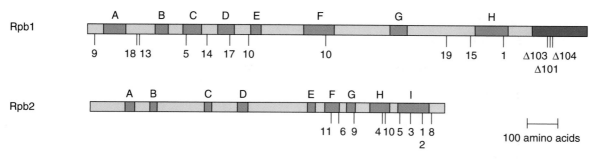

Figure 10.11   Regions of homology between yeast Rpb1 and the *E. coli* polymerase β′ subunit and between Rpb2 and the *E. coli* polymerase β subunit. The regions of homology are shown as red boxes; the carboxyl terminal domain (CTD) in Rpb1 is in green. The positions of conditional mutations that alter polymerase activity or assembly are identified by number. (*Source:* Reprinted from *Trends in Biochemical Sciences*, Vol. 15, N. A. Woychik and R. A. Young, "RNA Polymerase II: subunit structure and function," 347–351. Copyright © 1990 with permission of Elsevier Science.)

Core Subunits   These three polypeptides, Rpb1, Rpb2, and Rpb3, are all absolutely required for enzyme activity. They are homologous to the β′, β, and α subunits, respectively, of *E. coli* RNA polymerase. Figure 10.11 displays the structural similarities between Rpb1 and *E. coli* β′, and between Rpb2 and the *E. coli* polymerase β subunit. The red boxes represent the regions of similarity. The structural similarity between Rpb3 and the *E. coli* α subunit is not as obvious.

How about functional relationships? We have seen (Chapter 6) that the *E. coli* β′ subunit binds DNA, and so does Rpb1. Sean Carroll and David Stollar found a monoclonal antibody (G11) that blocked polymerase II binding to DNA. Moreover, binding polymerase II to DNA blocked interaction with the antibody. This competition between DNA and antibody for binding to the polymerase suggests that the antibody interacts with the enzyme's DNA binding site. Finally, these investigators showed that the antibody binds to two different forms of Rpb1 (215 and 180 kD), but not to Rpb2 (145 kD). Figure 10.12 presents the results of an immunoblot (or Western blot) that illustrates this binding. Lane 1 was probed with an unrelated antibody, which reacted with none of the polymerase subunits. Lane 2 was probed with the G11 antibody. Two bands reacted with the antibody. These are the two dark bands that correspond to the 215- and 180-kD subunits of the polymerase (two versions of Rpb1). Lane 3 is silver-stained RNA polymerase, so all the large polymerase subunits are visible. Altogether, the experimental results obtained by Carroll and Stollar suggested that Rpb1 is the primary site of interaction between polymerase and DNA.

Chapter 6 also showed that the *E. coli* β subunit is at the nucleotide-joining active site of the enzyme. André Sentenac and his colleagues have established that Rpb2 performs the same function. First, they bound the 4-formylphenyl-γ-ester of ATP to polymerases I, II and III, then made the bond covalent by reduction with $NaBH_4$. Then they added [α-^{32}P]UTP to form a phosphodiester bond to the bound ATP analog. Finally, they separated the enzyme subunits by gel electrophoresis and autoradiographed the gel to detect the labeled subunit. In each enzyme, the second largest subunit was labeled. Therefore, the second largest subunit in each seems to be involved in phosphodiester bond formation. In polymerase II, of course, this is Rpb2. The functional similarity among the second largest subunits in all three nuclear RNA polymerases, as well as prokaryotic polymerases, is mirrored by structural similarities among these same subunits, as revealed by the sequences of their genes.

Although, as we have already pointed out, Rpb3 does not closely resemble the *E. coli* α subunit, there is one 20-amino-acid region of great similarity. In addition, the two subunits are about the same size and have the same stoichiometry, two monomers per holoenzyme. Furthermore, the same kinds of polymerase assembly defects are seen in *RPB3* mutants as in *E. coli* α subunit mutants. All of these factors suggest that Rpb3 and *E. coli* α are homologous.

Common Subunits   Five subunits—Rpb5, Rpb6, Rpb8, Rpb10, and Rpb12—are found in all three yeast nuclear polymerases. We know little about the functions of these subunits, but the fact that they are found in all three polymerases suggests that they play roles fundamental to the transcription process, such as localization in the nucleus, or processivity (ability to synthesize long transcripts continuously without falling off the template), or fidelity of transcription.

Nonessential Subunits   Two subunits are not absolutely required for polymerase activity. Mutants in *RPB4* and *RPB9* are viable at normal temperatures, but inviable at both high and low temperatures. Thus, these subunits must not contribute a critical function to the polymerase, at least under normal conditions. Relatively little is known about Rpb9, but quite a few studies have been performed on Rpb4.

In 1989, Woychik and Young isolated a mutant yeast strain they called *rpb-4*, which lacks the gene for subunit Rpb4. As we would expect, the polymerase II from this strain has no Rpb4 protein. Moreover, these workers found that this enzyme also appears to lack Rpb7, even though the gene for this subunit is unaltered. This suggests that Rpb4 plays a role in anchoring Rpb7 to the enzyme. Roger Kornberg and his colleagues have performed in vitro transcription experiments on the polymerase

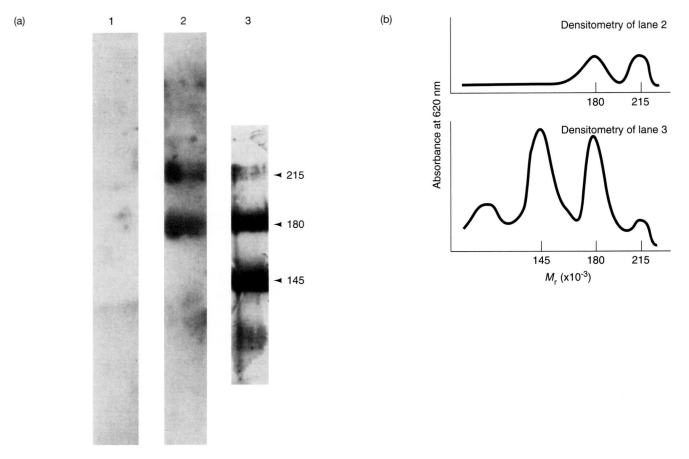

Figure 10.12   Identifying the DNA-binding subunit of polymerase II. Carroll and Stollar found an antibody that could compete with DNA for binding to polymerase II, suggesting that it bound to the DNA binding site of the enzyme. (**a**) They used this antibody (G11) to probe a Western blot of polymerase II subunits. Lane 1, polymerase II subunits probed with an unrelated monoclonal antibody; lane 2, polymerase II subunits probed with G11 antibody; lane 3, silver-stained gel, showing three subunits with relative molecular masses (M_rs) of 215, 180, and 145 kD. The top two of these are really two different forms of the Rpb1 subunit, IIa and IIb, respectively. The third polypeptide is the Rpb2 subunit. (**b**) Densitometer tracings of lanes 2 and 3 of panel (**a**). Because the G11 antibody binds to two forms of the Rpb1 subunit, this is probably the subunit that binds DNA. (*Source:* Carroll and Stollar, Conservation of a DNA-binding site in the largest subunit of eukaryotic RNA polymerase II. *J. Mol. Biol.* 170 (1983) p. 782, f. 2, by permission of Academic Press.)

(pol II Δ4/7) from this strain. They reported that the mutant polymerase works perfectly well when transcribing DNA lacking a promoter. In other words, it carries out elongation and termination normally. We will return to this point later, but it should be emphasized that Rpb7 is an essential subunit, so it must not be completely absent in pol II Δ4/7 mutants.

To measure general transcription efficiency, Kornberg transcribed poly(rC) in vitro with either polymerase II or pol II Δ4/7. Initiation occurs freely and randomly on this single-stranded RNA template, so it provides a measure of promoter-independent initiation and elongation. Both polymerases performed equally well in this assay, as demonstrated by Figure 10.13. To measure termination efficiency, Kornberg used a promoterless DNA with a 5′ extension, and three terminators: TIa, TIb, and TII. Even though such a template lacks a promoter, RNA polymerase can initiate transcription in vitro by starting at the

end with the extension and can then stop at the terminators. Figure 10.14 shows that both polymerases could do this, although pol II Δ4/7 seems to use the TIb terminator less efficiently than ordinary polymerase II does.

On the other hand, and in contrast to wild-type polymerase II, pol II Δ4/7 could not initiate transcription on an intact template with a real promoter. For this assay, Kornberg and colleagues used a template DNA with the yeast CYC1 promoter plus a binding site for the GAL4 transcription factor. The promoter was fused to a DNA cassette containing 377 bp of DNA lacking guanine residues in the nontemplate strand (a G-less cassette, chapter 5). This template can be transcribed in vitro in the presence of ATP, UTP, and CTP (but no GTP) to yield a transcript 377 nt long, which stops at the end of the cassette where the first GTP is needed. Figure 10.15, lane 1, shows that a cell-free extract from the *rpb-4* mutant, by itself, could not transcribe this template. This is because

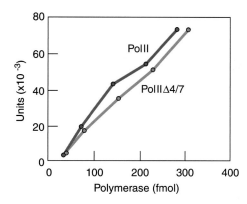

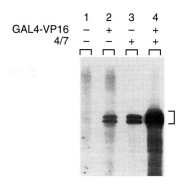

Figure 10.15   Response of pol II Δ4/7 to a yeast promoter. Kornberg and colleagues transcribed a construct containing the yeast *CYC1* promoter and a GAL4 binding site in a cell-free extract from *rpb-4* cells. They supplemented the extracts with subunits 4 and 7 or the composite transcription factor GAL4-VP16 (or both), as indicated at top. Pol IIΔ4/7 by itself was unable to recognize the promoter (lane 1), although with the help of the transcription factor, it could manage a little transcription (lane 2). The mutant polymerase supplemented with subunits 4 and 7 could recognize the promoter even in the absence of the transcription factor (lane 3); in the presence of the factor, transcription was very active (lane 4). (*Source:* Edwards et al., *J. Biol. Chem.* 266 (5 Jan 1991) p. 73, f. 4. American Society for Biochemistry and Molecular Biology.)

Figure 10.13   Promoter-independent transcription by holoenzyme and by pol II Δ4/7. Kornberg and colleagues assayed wild-type holoenzyme and polymerase II lacking Rpb4 and 7, using poly(rC) as template. No promoter is present or required for transcription of this single-stranded RNA template. Both the wild-type (green) and mutant (red) enzymes transcribed this template equally well. (*Source:* From Edwards et al., *Journal of Biological Chemistry* 266:73, 1991. Copyright © 1991 The American Society for Biochemistry & Molecular Biology, Bethesda, MD. Reprinted by permission.)

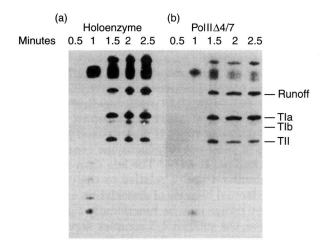

Figure 10.14   Response of wild-type and pol II Δ4/7 to terminators. Kornberg and colleagues made a DNA construct with an extension at the 5′ end to allow initiation, and three termination sites from a human histone H3 gene. They transcribed this construct in vitro with wild-type pol II (**a**) or pol II Δ4/7 (**b**), then subjected the RNA products to gel electrophoresis. Both polymerases could terminate accurately, although the mutant polymerase showed a lower tendency to use the TIb site than the wild-type polymerase did. (*Source:* Edwards et al., Two dissociable subunits of yeast RNA polymerase II stimulate the initiation of transcription at a promoter *in vitro. J. Biol. Chem.* 266 (5 Jan 1991) p. 73, f. 3. American Society for Biochemistry and Molecular Biology.)

the extract contained RNA polymerase II that lacked subunits 4 and 7. Lane 3 demonstrates, as we might anticipate, that addition of subunits 4 and 7 back to the extract reconstituted its activity. Moreover, as we can see in lane 2, the powerful transcription factor GAL4-VP16 could force transcription even in the apparent absence of subunits 4 and 7. Finally, lane 4 shows that GAL4-VP16

together with subunits 4 and 7 provided the cell-free extract with all the components necessary for very strong promoter-dependent transcription.

How do we account for the fact that Rpb4 and Rpb7 are required for promoter-driven transcription in vitro, yet the *rpb-4* mutant, whose polymerase II appears to lack these two subunits, is viable? It may be that a basal level of transcription, undetectable in vitro, may occur in vivo and be sufficient to sustain life. Also, because *RPB7* mutants are nonviable, it is likely that polymerase II in the *rpb-4* mutant really does not lack all Rpb7. It may also be that all vital yeast genes contain one or more binding sites for transcription factors such as GAL4, and therefore benefit from the kind of low-level, but clearly detectable, transcription observed with pol II Δ4/7 in the presence of GAL4-VP16 in vitro (see Figure 10.15, lane 2).

Rpb4 and Rpb7 dissociate relatively easily from polymerase II and are usually found in substoichiometric quantities. Some data even suggest that these two subunits might shuttle from one polymerase II to another. These characteristics are reminiscent of *E. coli* σ-factor, as is the ability to confer promoter-dependent transcription ability on a "core" enzyme. There is even limited sequence similarity between σ and Rpb4. However, Rpb4 and 7 are certainly not exactly analogous to bacterial σ. Unlike σ, they are incapable on their own of stimulating full promoter-dependent transcription activity. Instead, they need the help of one or more transcription factors.

Rpb7 and Rpb11 do not fall into any of the three preceding categories. However, they are absolutely required for polymerase activity, and *RPB7* and *RPB11* mutants are not viable.

SUMMARY The genes encoding all 12 RNA polymerase II subunits in yeast have been cloned and sequenced, and subjected to mutation analysis. Three of the subunits resemble the core subunits of bacterial RNA polymerases in both structure and function, five are found in all three nuclear RNA polymerases, two are not required for activity, at least at normal temperatures, and two fall into none of these three categories. Incorporation of labeled methionine and phosphate into yeast RNA polymerase II allows an estimate of the stoichiometries of the 12 subunits. Two subunits are heavily phosphorylated, and one is lightly phosphorylated.

Heterogeneity of the Rpb1 Subunit   The very earliest studies on RNA polymerase II structure showed some heterogeneity in the largest subunit. Figure 10.16 illustrates this phenomenon in polymerase II from a mouse tumor called a plasmacytoma. We see three polypeptides near the top of the electrophoretic gel, labeled IIo, IIa, and IIb, that are present in smaller quantities than polypeptide IIc. These three polypeptides appear to be related to one another, and indeed two of them seem to derive from the other one. But which is the parent and which are the offspring? Sequencing of the yeast *RPB1* gene predicts a polypeptide product of 210 kD, so the IIa subunit, which has a molecular mass close to 210 kD, seems to be the parent. Furthermore, amino acid sequencing has shown that the IIb subunit lacks a repeating string of seven amino acids with the following consensus sequence: Tyr-Ser-Pro-Thr-Ser-Pro-Ser. Because this sequence is found at the carboxyl terminus of the IIa subunit, it is called the **carboxyl-terminal domain,** or **CTD.** Antibodies against the CTD react readily with the IIa subunit, but not with IIb, reinforcing the conclusion that IIb lacks this domain. A likely explanation for this heterogeneity is that a proteolytic enzyme clips off the CTD, converting IIa to IIb. Because IIb has not been observed in vivo, this clipping seems to be an artifact that occurs during purification of the enzyme. In fact, the sequence of the CTD suggests that it will not fold into a compact structure; instead, it is probably extended and therefore highly accessible to a proteolytic enzyme.

What about the IIo subunit? It is bigger than IIa, so it cannot arise through proteolysis. Instead, it seems to be a phosphorylated version of IIa. This conclusion comes from several lines of evidence. First, proteins are phosphorylated on the hydroxyl groups of serine, threonine, and occasionally tyrosine. The fact that the CTD has such a preponderance of these three amino acids suggests that phosphorylation in this domain is likely. Indeed, subunit IIo can be converted to IIa by incubating it with a phosphatase that removes the phosphate groups. Furthermore, both phosphoserine and phosphothreonine (but not phosphotyrosine) can be found in the CTD of IIo. Can we account for the

Figure 10.16   Partial subunit structure of mouse plasmacytoma RNA polymerase II. The largest subunits are identified by letter on the left, although these subunit designations are not the same as those applied to the yeast polymerase II (see Figure 10.10). Subunits o, a, and b are three forms of the largest subunit, corresponding to yeast Rpb1. Subunit c corresponds to yeast Rpb2. (*Source:* Sklar et al., *PNAS* 72 (Jan 1975) p. 350, f. 2C.)

difference in apparent molecular mass between IIo and IIa simply on the basis of phosphate groups? Apparently not; not enough phosphates are present, so we must devise another explanation for the low electrophoretic mobility of IIo. Perhaps phosphorylation of the CTD induces a conformational change in IIo that makes it electrophorese more slowly and therefore seem larger than it really is. But this conformational change would have to persist even in the denatured protein. Figure 10.17 shows the probable relationships among the subunits IIo, IIa, and IIb, along with some postulated intermediate forms: IIa′, a partially phosphorylated IIa, and IIo′, an overphosphorylated IIo.

The fact that cells contain two forms of the Rpb1 subunit (IIo and IIa) implies that two different forms of RNA polymerase II exist, each of which contains one of these subunits. We call these **RNA polymerase IIO** and **RNA polymerase IIA,** respectively. The nonphysiological form of the enzyme, which contains subunit IIb, is called RNA polymerase IIB.

Do polymerases IIO and IIA have identical or distinct roles in the cell? The evidence strongly suggests that IIA (the unphosphorylated form of the enzyme) is the species that initially binds to the promoter, and that IIO (with its CTD phosphorylated) is the species that carries out elongation. Thus, phosphorylation of the CTD appears to accom-

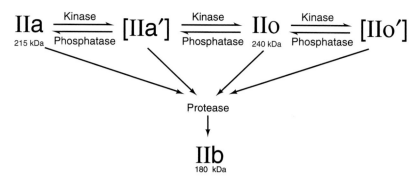

Figure 10.17   Proposed relationships among the different forms of the largest subunit of RNA polymerase II. IIa′ and IIo′ are hypothetical under- and overphosphorylated versions, respectively, of subunit IIo. (*Source:* Reprinted from *Trends in Biochemical Sciences,* vol. 15, J. Corden, "Tails of RNA polymerase II," 383–387. Copyright © 1990 with permission of Elsevier Science.)

pany the transition from initiation to elongation. We will examine the evidence for this hypothesis in Chapter 11.

> **SUMMARY** Subunit IIa is the primary product of the *RPB1* gene in yeast. It can be converted to IIb in vitro by proteolytic removal of the carboxyl-terminal domain (CTD), which is essentially a heptapeptide repeated over and over. Subunit IIa can be converted to IIo by phosphorylating several serines and threonines within the CTD. The enzyme (polymerase IIA) with the IIa subunit is the one that binds to the promoter; the enzyme (polymerase IIO) with the IIo subunit is the one involved in transcript elongation.

The Rpb1 Subunit and α-Amanitin Sensitivity   The toxin α-amanitin has played an important role in the study of eukaryotic RNA polymerases because it allows us to distinguish clearly among the three nuclear RNA polymerases from most eukaryotes. But what makes RNA polymerase II so sensitive to this toxin? Although we do not understand the mechanism by which α-amanitin operates, we know that the *RPB1* product—the largest polymerase II subunit—confers sensitivity to the toxin.

The first evidence that Rpb1 is the subunit confering α-amanitin sensitivity was genetic. Arno Greenleaf and colleagues used ethylmethane sulfonate (EMS, Chapter 20) to mutate genes in the fruit fly (*Drosophila melanogaster*). Some of the mutants were α-amanitin-resistant, which suggested that the mutation had occurred in a gene encoding one of the polymerase II subunits and altered it so its product could no longer interact with α-amanitin. If this was so, then all Greenleaf and colleagues had to do was find the mutated gene and see which subunit it encodes.

To map the mutation, these workers used P element mutagenesis. They introduced the P element (a transposon) into flies, which produced a collection of mutants in which the P element had inserted into, and then inactivated, a variety of genes. Then they mapped these mutations and found two that mapped to the same locus as the

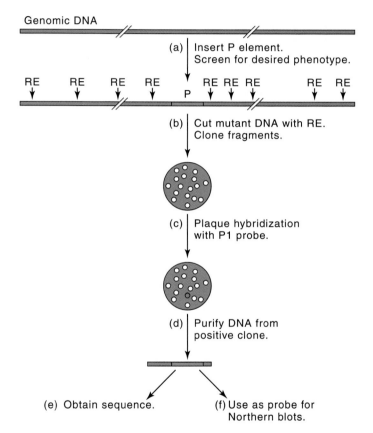

Figure 10.18   A general method for finding a gene mutated with a P element. (a) P elements are inserted at random into the *Drosophila* genome, and mutants with the desired phenotype are picked. **(b)** The mutant DNA is subjected to limited digestion with a restriction endonuclease (RE), and the resulting fragments are cloned into a λ phage or other appropriate vector. **(c)** A labeled P1 probe is used to fish out a positive clone by plaque hybridization (Chapter 4). **(d)** The DNA from a positive clone is purified. This DNA can then be **(e)** sequenced or **(f)** used as a probe for a Northern blot of RNAs from wild-type cells to learn the size of the gene's transcript.

mutation in the α-amanitin-resistant mutant. But where was that locus? The P element greatly simplified the task of answering that question.

Figure 10.18 illustrates the general method of locating a gene mutated by P element insertion. Greenleaf and

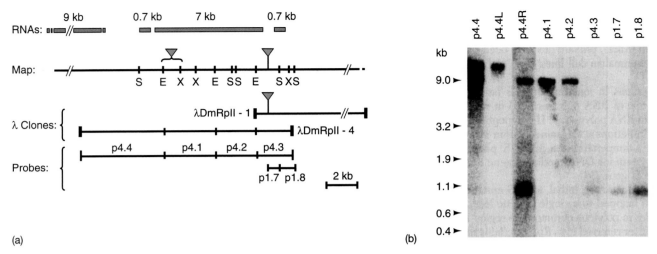

Figure 10.19   Finding the *Drosophila* α-amanitin-resistance gene (*RpII*) (a) Map of the DNA region containing the gene (*RpII*) for a *Drosophila melanogaster* RNA polymerase II subunit. The second line shows the map of the region, with restriction sites as follows: S, *Sph*I; E, *Eco*RI; X, *Xho*I. The red triangles denote sites of P element insertion in different polymerase II mutants. The blue boxes in the top line show the sizes of RNAs that hybridize to DNA probes from the map region directly below. At bottom are the maps of two lambda clones derived from this DNA region, and the extents of probes used in the RNA blot in panel (**b**). (**b**) RNA blot probed with the DNAs shown in panel (**a**). Ingles, Greenleaf, and colleagues electrophoresed and blotted RNA from *D. melanogaster* adults and hybridized the blots to the labeled probes listed at top. The 4.4L and 4.4R probes are single-stranded leftward and rightward (5' → 3') DNA fragments from the 4.4 region. The 4.4R, 4.1, and 4.2 probes all reacted with a 7-kb transcript. (*Source:* Ingles et al., *PNAS* 80 (June 1983) p. 3397, f. 2b.)

colleagues cloned the mutant *Drosophila* DNA into a λ phage vector to create a genomic library. Then they screened the library by plaque hybridization with a P element probe. They identified clones with P elements inserted into them in either of two positions. Greenleaf, James Ingles, and colleagues then found that DNA probes from these clones hybridized to a 7-kb transcript on Northern blots of *Drosophila* RNA (Figure 10.19).

This 7-kb transcript is big enough to code for one of the large polymerase II subunits. Because it reacted with probes from the two P-element-tagged clones, these data were consistent with the hypothesis that the two clones actually encode a large polymerase subunit. Greenleaf and colleagues developed the following strategy to test the hypothesis further. In separate experiments, they inserted two different fragments of one cloned DNA into a λ phage-based expression vector. The resulting recombinant phages produced two different fusion proteins (70 and 65 kD, respectively), which Greenleaf and coworkers then probed with an antibody directed against the largest polymerase II subunit. If the cloned fragments encode parts of the largest polymerase subunit, the fusion proteins should react with this antibody. Figure 10.20 shows that they did. Lanes 3 and 4 contain the 70- and 65-kD fusion proteins, respectively, and the dark bands in both lanes indicate that both fusion proteins reacted with the antibody. This experiment established that the gene that confers α-amanitin resistance (and therefore α-amanitin sensitivity) encodes the largest polymerase II subunit. Thus, this subunit confers α-amanitin sensitivity and resistance.

SUMMARY The largest RNA polymerase II subunit (the product of the *RPB1* gene) governs sensitivity or resistance to α-amanitin.

The Shape of RNA Polymerase II The traditional method for determining the shape of a protein, as we have seen in Chapter 9, is x-ray crystallography. This has been done with RNA polymerases from *E. coli* and phage T7, but, until 1999, it was difficult to produce crystals of RNA polymerase II of high enough quality for x-ray crystallography studies. The problem lay in the heterogeneity of the polymerase caused by the loss of the Rpb4 and Rpb7 subunits from some of the enzymes. Roger Kornberg and colleagues solved this problem by using a mutant yeast polymerase lacking Rbp4 (and therefore lacking Rpb7 as well). This polymerase is capable of transcription elongation, though not initiation at promoters. Thus, is should be adequate for modeling the elongation complex. It produced crystals that were good enough for x-ray crystallography leading to a model with 6-Å resolution.

Figure 10.21 shows a stereo view of this model of yeast polymerase II. Several features stand out. First, a 25-Å channel in the face of the enzyme makes about a 90-degree turn near the middle of the model in Figure 10.21a. A 25-Å channel is wide enough to accommodate a double-stranded DNA, which is the presumed function of this groove. Accordingly, it is marked with a red

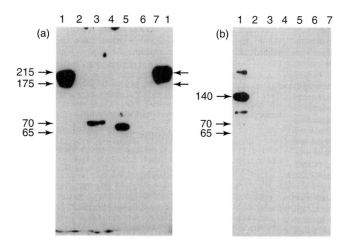

Figure 10.20   The product of the α-amanitin-resistance gene (*RpII*) reacts with an antibody directed against the largest polymerase II subunit. Greenleaf and colleagues inserted fragments of the *RpII* locus (see Figure 10.19) into an expression vector so as to produce fusion proteins of either 70 kD (clone pAG6) or 65 kD (clone pAG13) that reacted with anti-RNA polymerase antibodies. Another clone, pAG4, produced a fusion protein that did not react with these antibodies. Elevated temperature was required to induce fusion protein production in all these clones. **(a)** Immunoblot probed with an antibody directed against the largest polymerase II subunit. Lanes 1, *Drosophila* RNA polymerase II (positive control); lanes 2 and 3, extracts from uninduced and induced clone pAG6; lanes 4 and 5, extracts from uninduced and induced clone pAG13; lanes 6 and 7, extracts from uninduced and induced clone pAG4. **(b)** Immunoblot probed with an antibody directed against the 140-kDa polymerase II subunit. The lanes contained the same substances as in panel **(a)**. The fusion proteins from *RpII* that were antibody-positive reacted with the antibody against the largest subunit (215 kD, and its proteolytic product, 175 kD), but not with antibodies against the second-largest subunit (140 kD). (*Source:* Greenleaf, Amanitin-resistant RNA polymerase II mutations are in the enzyme's largest subunit. *J. Biol. Chem.* 258 (25 Nov 1983) p. 13405, f. 2a-b. American Society for Biochemistry and Molecular Biology.)

line representing DNA in this figure. The second feature is apparent in Figure 10.21b. Here we see that the 25-Å channel is partially blocked near its left end by a column-shaped piece of the enzyme. This blockage leaves two narrower channels: a "residual channel," or hole, whose opening is adjacent to the 3′-end of the RNA, and an open channel between the "arm" and the underlying shelf of the enzyme. Neither of these channels is wide enough to accommodate double-stranded DNA, so the DNA is presumed to thread through the outer, open channel, in single-stranded form. Just beyond this constriction, the outer channel widens again to 25 Å and can accommodate the downstream double-stranded DNA as suggested by the red helix in Figure 10.21a. The DNA would be enclosed in a **sliding clamp** defined on the top side by the tip of the arm and on the bottom by the underlying shelf. This clamp would presumably keep the polymerase from dissociating from the DNA and therefore contribute to the processivity of the enzyme. We observed a similar structure in the *E. coli* polymerase in Chapter 6.

Another obvious feature of this model is the groove extending from the DNA channel diagonally up to the top right. This groove is not wide enough for double-stranded DNA, but it could hold the RNA product, so it is called the **RNA groove**.

The locations of the nucleic acids in this model were based on Kornberg and coworkers' electron crystallography studies of the same enzyme in an elongation complex caught in the act of transcribing. Electrons cannot penetrate through three-dimensional crystals, so investigators typically use two-dimensional crystals (one molecule thick) on the surface of a lipid layer. Also, the resolution of this method is inherently less than that of x-ray crystallography because of the longer wavelength of the electrons, but Kornberg and colleagues surmounted this problem, at least partially, by combining images from two-dimensional crystals in many different orientations.

Figure 10.22 shows a model at 20-Å resolution based on these studies. Panel (a) depicts a dimer of RNA polymerase in the absence of nucleic acids. The columnar piece of the enzyme that divides the 25-Å channel is marked with blue arrows. Panel (b) shows the enzyme in a "paused complex" including a **streptavidin**-tagged DNA template and a 14-nt transcript. The nucleic acids in this paused complex are shown in panel (c). Kornberg and colleagues gave the enzyme a 33-bp length of double-stranded DNA with a 15-nt deoxycytidine tail on one of the DNA strands. Such tails provide transcription start sites about 3–5 nt from the beginning of the double-stranded region. These workers allowed the DNA to grow in the absence of UTP, so transcription paused at the first A in the template strand, where UTP was required for the first time. They attached a biotinylated C to the 5′-end of the nontemplate strand and then attached streptavidin to the biotin, thus labeling the downstream end of the DNA template.

Panel (b) depicts the significant differences between the uncomplexed enzyme in panel (a) and the paused complex containing DNA and RNA. The differences in electron density, shown in yellow, all fit with the predictions in Figure 10.21. The large projection to the left in panel (b) must be the protein streptavidin, so the downstream end of the template must lie to the left. (The right-hand polymerase also shows the streptavidin, but the nucleic acids lie behind this polymerase, so we will not discuss the right-hand part of the figure any further.) We assume that the central electron density labeled "d" is the transcription bubble. Projecting from the transcription bubble, we see three electron densities labeled "a," "b," and "c." The part labeled "a" is connected to the streptavidin, so we assume it is the downstream part of the DNA template. The part labeled "b" is therefore presumably the upstream segment of the template, and the part labeled "c" appears to be the RNA product.

In 2000, Kornberg and colleagues used a soaking procedure that shrank their crystals of the 10-subunit yeast

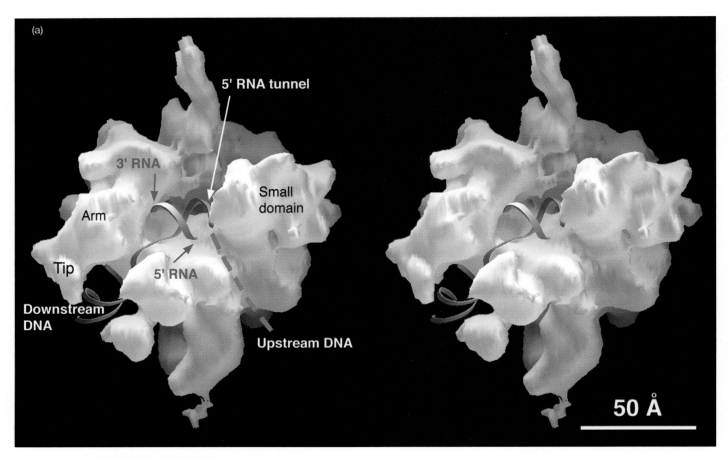

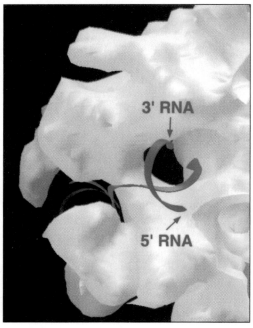

Figure 10.21 Model of yeast RNA polymerase II. (a) Stereo view of the surface of the enzyme based on x-ray diffraction analysis. The red ribbon suggests the presumed path of the DNA template through the 25-Å channel in the enzyme. The blue ribbon indicates the part of the nascent RNA that forms a hybrid with the template DNA strand. **(b)** Closeup view of the "residual channel," showing the proximity of the 3′-end of the nascent RNA to the opening of the residual channel. (*Source:* Fu, J. et al. Yeast RNA Polymerase II at 5 Å Resolution. *Cell* 98 (1999). f. 7, p. 806. Reprinted by permission of Elsevier Science.)

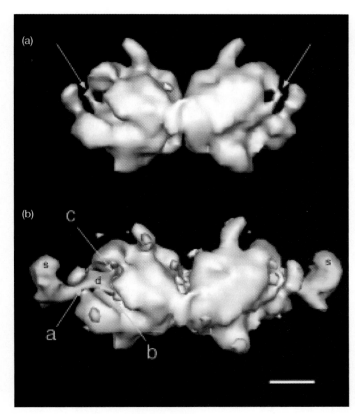

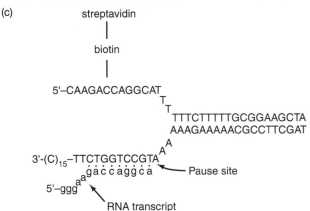

(c)

streptavidin

|

biotin

|

5'–CAAGACCAGGCAT
                         T T
                              T
                         TTTCTTTTTGCGGAAGCTA
                         AAAGAAAAACGCCTTCGAT
                                        A
                                         A
3'-(C)₁₅–TTCTGGTCCGTA
                   gȧċċȧġġċȧ ←— Pause site
               a a
5'–ggg ↖
          RNA transcript

Figure 10.22 Structure of a yeast RNA polymerase II/nucleic acid complex. (a) Computer-generated model of two yeast polymerase II 4/7 proteins based on electron diffraction data. The arrows indicate the regions of electron density in the protein that split the 25-Å channel in two. (b) Computer-generated model to show differences (in yellow) between yeast polymerase II 4/7 and the same protein complexed to the synthetic streptavidin-tagged DNA template/ RNA product shown in panel (c). Labels: s, the streptavidin moiety; a, single-stranded DNA feeding through the constriction in the 25-Å channel; b, extension of the double-stranded DNA template into the upstream part of the 25-Å channel; c, extension of nucleic acid (probably the RNA product) into the RNA groove; d, a polygonal region of nucleic acid, probably containing the RNA–DNA hybrid. (*Source:* (a–b) Poglitsch, C. et al. Electron Crystal Structure of an RNA Polymerase II Transcription Elongation Complex. *Cell* 98 (1999) f. 5, p. 796. Reprinted by permission of Elsevier Science.)

RNA polymerase II to a size that yielded a crystal structure with 3-Å resolution. A major consequence of this enhanced resolution was that the individual subunits of the enzyme are discernible. Figure 10.23 contains a ribbon diagram of this crystal structure, along with a cartoon that identifies each of the subunits (again, Rpb4 and Rpb7 were deliberately omitted from the enzyme).

Several features of this higher resolution structure were not apparent in previous structures:

1. The channel through the enzyme now appears as a deep cleft, with most of Rpb1 on one side, and most of Rpb2 on the other.

2. The opening of the cleft is defined by a set of "jaws" that could grasp the DNA. The upper jaw is composed of Rpb9 and part of Rpb1, and the lower jaw is composed of Rpb5. The jaws are highlighted in Figure 10.24b.

3. The active site of the enzyme can be identified by the single Mg^{2+} ion, which was located by replacing it with Zn^{2+}, Mn^{2+}, and Pb^{2+}. This is also the apparent location of the conserved aspartate motif similar to the one we saw at the active site in the prokaryotic polymerase discussed in Chapter 6.

4. The DNA template was not included in the crystals in this study, but the DNA has been pictured in the cleft according to the position defined by Kornberg and colleagues' two-dimensional crystallographic studies described previously in this section. This is "downstream" DNA, not yet transcribed, and about 20 bp of this double-stranded DNA can fit into the cleft.

5. The sliding clamp is composed of parts of Rpb1, Rpb2, and Rpb6. The position of the clamp is highlighted in Figure 10.24b.

6. A "wall" composed of part of Rpb2 occurs just to the upstream side of the active site (Figure 10.24b). This wall prevents the DNA from extending upstream in the same orientation as the 20 bp of downstream DNA. Therefore, the RNA–DNA hybrid that must lie just upstream of the active site must be at an angle with the downstream DNA. The presumed angle is indicated by the dashed line in Figure 10.24b.

7. Two pores occur in the polymerase ("pore 1" and "pore 2" in Figure 10.24a) that could provide the exit route for the growing RNA. Pore 1 is an attractive candidate because it would place the growing RNA in a groove at the base of the clamp, which would tend to hold the clamp shut on the DNA template. This is consistent with the finding that high affinity binding between polymerase and DNA depends on the presence of an RNA transcript.

This structure sheds some light on the steps that occur near the beginning of transcription. The polymerase must first bind to linear duplex DNA, but the cleft, which ends

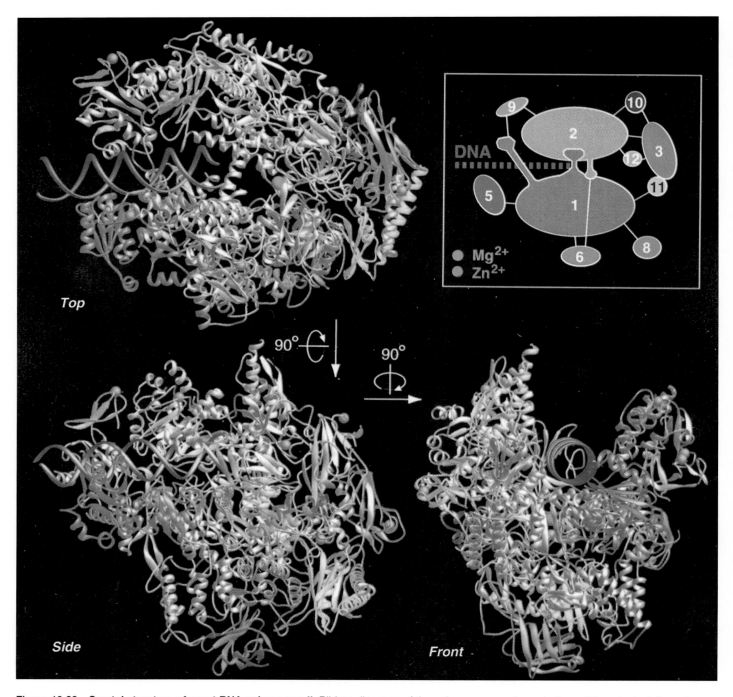

Figure 10.23   Crystal structure of yeast RNA polymerase II. Ribbon diagrams of the polymerase are shown in three different orientations: top, side, and front. The presumed position of the DNA downstream of the active site is shown in dark blue. Eight zinc atoms are depicted as blue spheres, the Mg^{2+} at the active site is shown as a pink sphere. The inset at top right is a cartoon showing the locations of the 10 subunits, and the presumed location of the DNA. The colors of the subunits and DNA are the same as in the ribbon diagrams. The orientation of the cartoon is the same as the ribbon diagram top view at top left. (*Source:* Cramer, P. et al. Architecture of RNA Polymerase II and Implications for the Transcription Mechanism. *Science,* 288 (28 April 2000) f. 3, p. 643 © AAAS.)

just beyond the active site, will not accommodate such a DNA unless it can bend. Melting of the DNA around the active site will allow such bending. We also know that the first products of transcription are short, abortive transcripts. Only when the transcripts reach about 10 nt long does the enzyme move into the elongation phase. And only when the transcripts reach about 20 nt long does the enzyme form a fully stable complex with the template and the RNA product. These numbers are not magic; the architecture of the enzyme and the presumed locations of the template and product rationalize them as follows: When a transcript reaches a length of 10 nt, it would begin to enter

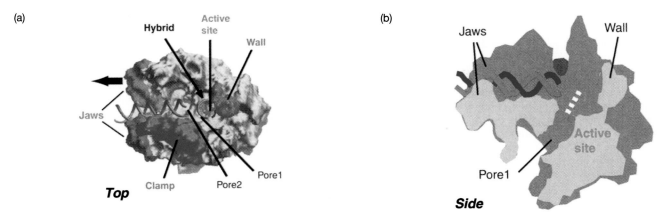

Figure 10.24 Shape of the elongation complex. (a) Major features of the enzyme. The jaws (red), clamp (green), active site, wall (purple), and pores 1 and 2 are indicated. The presumed positions of the downstream DNA (blue helix) and RNA–DNA hybrid (red helix) are also shown. (b) Cutaway view showing the presumed position of the downstream DNA (blue) and the presumed angle of the RNA–DNA hybrid (white dashed line) relative to the active site (red dot corresponding to the Mg^{2+} ion). (*Source:* Cramer, P. et al. Architecture of RNA Polymerase II and Implications for the Transcription Mechanism. *Science* 288 (28 April 2000) f. 6, p. 645. ©AAAS.)

pore 1, which presumably prevents its dissociation from the active site, so it cannot be released as an abortive transcript. When a transcript reaches a length of 20 nt, it would fill the groove at the base of the clamp, serving as a lock to keep the clamp from releasing the DNA template.

SUMMARY X-ray crystallography of yeast RNA polymerase three-dimensional crystals, and electron crystallography of two-dimensional crystals of a paused yeast RNA polymerase on a synthetic DNA template have yielded structural information about the elongation complex. It appears to have a 25-Å channel that makes a right angle turn on the surface of the enzyme. This channel can hold double-stranded DNA for most of its length. A higher resolution crystal structure of the enzyme shows the locations of the 10 essential subunits, and the location of the DNA template can be inferred from the previous studies on two-dimensional crystals. The enzyme contains a cleft that holds the DNA, bounded by a set of jaws on the downstream end, and a wall just beyond the active site at the other end. The DNA must turn to avoid this wall, and DNA melting would make this possible. A predicted RNA exit site places the RNA product in a groove that would hold a sliding clamp shut on the DNA, improving processivity.

10.2 Promoters

We have seen that the three eukaryotic RNA polymerases have different structures and they transcribe different classes of genes. We would therefore expect that the three polymerases would recognize different promoters, and this expectation has been borne out. We will conclude this chapter by looking at the structures of the promoters recognized by all three polymerases.

Class II Promoters

We begin with the promoters recognized by RNA polymerase II (**class II promoters**) because these will be the most familiar to you, resembling to some extent the prokaryotic promoters we have already studied in Chapter 6. Class II promoters can be considered as having two parts: The **core promoter,** and an **upstream element.** The core promoter can have up to three elements: a **TATA box** beginning at approximately position –30, an **initiator** centered on the transcription start site, and a **downstream element** further downstream. Figure 10.25 summarizes the relative positions of all four promoter elements. However, many natural promoters lack recognizable versions of one or more of these sequences.

The TATA Box   By far the most common element in the many class II promoters studied to date is a sequence of bases usually centered about 25 bp upstream of the transcription start site in higher eukaryotes. The consensus sequence of this element is TATAAAA (in the nontemplate strand), although T's frequently replace the A's in the fifth and seventh positions. Its name, *TATA box,* derives from its first four bases. You may have noticed the close similarity between the eukaryotic TATA box and the prokaryotic –10 box. The major difference between the two is position with respect to the transcription start site: –25 versus –10. (TATA boxes in yeast [*Saccharomyces cerevisiae*] have a more variable location, from 50 to 70 bp upstream of their transcription start sites.)

As usual with consensus sequences, exceptions to the rule exist. Indeed, in this case they are plentiful. Sometimes

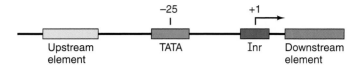

Figure 10.25   A generic polymerase II promoter. This promoter consists of four elements: an initiator region (Inr green) at the site of transcription initiation; a downstream element (blue) downstream of the transcription start site; a TATA box (red) about 25 bp upstream; and an upstream element (yellow) farther upstream. Natural polymerase II promoters may be missing one or more of these elements and still function. In fact, essential initiators and downstream elements are uncommon.

G's and C's creep in, as in the TATA box of the rabbit β-globin gene, which starts with the sequence CATA. Frequently no recognizable TATA box is evident at all. Such TATA-less promoters tend to be found in two classes of genes: (1)The first class comprises the **housekeeping genes** that are constitutively active in virtually all cells because they control common biochemical pathways, such as nucleotide synthesis, needed to sustain cellular life. Thus, we find TATA-less promoters in the cellular genes for adenine deaminase, thymidylate synthetase, and dihydrofolate reductase, all of which encode enzymes necessary for making nucleotides, and in the SV40 region encoding the viral late proteins. These genes tend to have GC boxes that appear to compensate for the lack of a TATA box (Chapter 11). (2)The second class of genes with TATA-less promoters are developmentally regulated genes such as the homeotic genes that control development of the fruit fly or genes that are active during development of the immune system in mammals. We will examine one such gene (the mouse terminal deoxynucleotidyltransferase [TdT] gene) later in this chapter. In general, **specialized genes** (sometimes called "luxury genes"), which encode proteins made only in certain types of cells (e.g., keratin in skin cells and hemoglobin in red blood cells), do have TATA boxes.

What is the function of the TATA box? That seems to depend on the gene. The first experiments to probe this question involved deleting the TATA box and then assaying the deleted DNA for promoter activity by transcription in vitro. In 1980, Rudolph Grosschedl and Max Birnstiel created deletion mutations in a cloned DNA fragment containing two sea urchin histone genes: the histone H2A and H2B genes. They introduced mutations into the H2A gene while leaving the H2B gene alone, then they analyzed the effects of the mutations by injecting the DNA into frog oocytes and measuring the transcription rate of the two genes. That way, the wild-type H2B gene served as a control that should not change from one experiment to another. Any increase or decrease in its rate of transcription could be attributed to variations in experimental technique, and these same fluctuations could be used to correct the results from the H2A gene.

Figure 10.26b shows what happened. As expected, the signals from the H2B gene remained almost constant. By contrast, deletions in the H2A promoter caused easily measurable changes in transcription patterns. A deletion in an upstream sequence that is conserved from one H2A gene to another only served to stimulate transcription of this gene (lane 2), suggesting that this DNA element actually inhibits transcription somewhat. A deletion encompassing the initiator sequence and the region just downstream caused a shorter transcript to be formed, because some of the gene was missing, but it did not change the efficiency of transcription. The payoff came when Grosschedl and Birnstiel deleted the TATA box. This led to little if any decrease in efficiency of transcription, but it did cause a loss of specificity of initiation. Instead of starting at a single initiation site about 25 bp downstream of the TATA box, transcription started at a variety of different sites, giving rise to at least three different transcripts (lane 3). It was as if the polymerase could no longer tell where to start.

Christophe Benoist and Pierre Chambon reached the same conclusion from a study of the SV40 early promoter. As in the experiments we have just discussed, these investigators prepared deletion mutants in the promoter and measured the effects of these deletions. The assays they used were primer extension and S1 mapping. These techniques, described in Chapter 5, produce labeled DNA fragments whose lengths tell us where transcription starts. As Figure 10.27 shows, the P1A, AS, HS0, HS3, and HS4 mutants, which Benoist and Chambon had created by deleting progressively more of the DNA downstream of the TATA box, including the initiation site, simply shortened the S1 signal by an amount equal to the number of base pairs removed by the deletion. This result is consistent with a downstream shift in the transcription start site caused by the deletion. Such a shift is just what we would predict if the TATA box positions transcription initiation approximately 30 bp downstream of the first base of the TATA box. If this is so, what should be the consequences of deleting the TATA box altogether? The H2 deletion extends the H4 deletion through the TATA box and therefore provides the answer to our question: Lane 8 of Figure 10.27b shows that removing the TATA box caused transcription to initiate at a wide variety of sites. Again, it appears that the TATA box is involved in positioning the start of transcription.

Benoist and Chambon reinforced this conclusion by systematically deleting DNA between the TATA box and the initiation site of the SV40 early gene and locating the start of transcription in the resulting shortened DNAs by S1 mapping. Figure 10.28 summarizes the results. Transcription of the wild-type gene begins at three different guanosines, clustered 27–34 bp downstream of the first T of the TATA box. As Benoist and Chambon removed more and more of the DNA between the TATA box and these initiation sites, they noticed that transcription no longer initiated at these sites. Instead, transcription

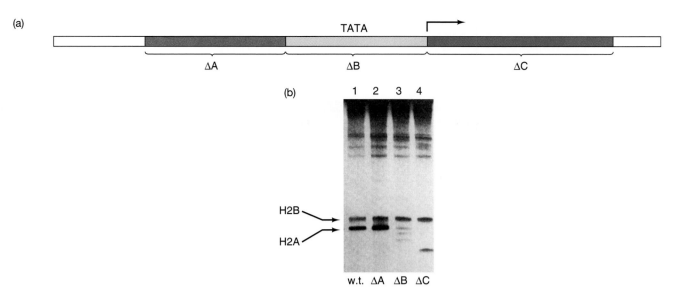

Figure 10.26   Effects of deletions in the sea urchin histone H2A promoter. Grosschedl and Birnstiel injected a plasmid containing the sea urchin histone H2A and H2B genes into frog eggs, along with labeled GTP, then purified the labeled RNA and identified the H2A and H2B transcripts by gel electrophoresis. The H2B gene was always wild-type, so its transcript did not vary. On the other hand, the H2A gene was either wild-type or a deletion mutant. **(a)** Map of the deletions in H2A. Deletion A (red) removed about 60 bp in a conserved upstream region; deletion B (yellow) removed about 60 bp, including the TATA box; and deletion C (blue) removed about 80 bp downstream of the initiation site. **(b)** Results of the assay: Lane 1, wild-type H2A gene; lane 2, deletion A; lane 3, deletion B; lane 4, deletion C. As expected, the H2B signal remained constant. However, the deletions in the H2A control region caused quantitative and qualitative alterations in transcription. Deletion A actually stimulated transcription somewhat. Removing the TATA box (deletion B) caused a decrease in amount of transcription and the appearance of new transcripts that result from heterogeneous start sites. Deletion C had little effect on the amount of transcription, but shifted the start site downstream, so a shorter transcript appeared. (*Source:* (*b*) Grosschedl & Birnstiel, *PNAS* 77 (Mar 1980) p. 1434, f. 3.)

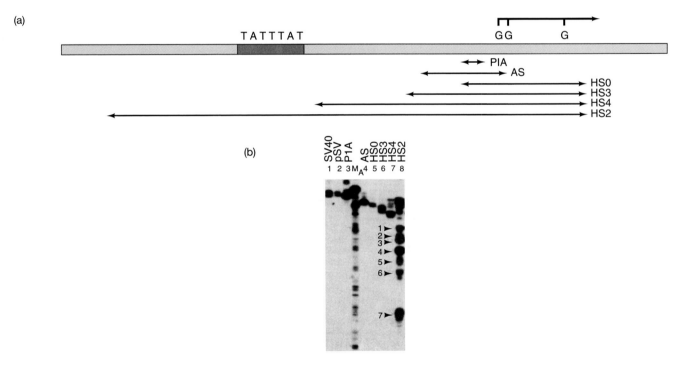

Figure 10.27   Effects of deletions in the SV40 early promoter. (a) Map of the deletions. The names of the mutants are given at the right of each arrow. The arrows indicate the extent of each deletion. The positions of the TATA box (TATTTAT, red) and the three transcription start sites (all Gs) are given at top. **(b)** Locating the transcription start sites in the mutants. Benoist and Chambon transfected cells with either SV40 DNA, or a plasmid containing the wild-type SV40 early region (pSV1), or a derivative of pSV1 containing one of the mutated SV40 early promoters described in panel **(a)**. They located the initiation site (or sites) by S1 mapping. The names of the mutants being tested are given at the top of each lane. The lane denoted M_A contained size markers. The numbers to the left of the bands in the HS2 lane denote novel transcription start sites not detected with the wild-type promoter or with any of the other mutants in this experiment. The heterogeneity in the transcription initiation sites was apparently due to the lack of a TATA box in this mutant. (*Source:* (*b*) Benoist & Chambon, *In vivo* sequence requirements of the SV40 early promoter region. *Nature* 290 (26 Mar 1981) p. 306, f. 3.
© Macmillan Magazines Ltd.)

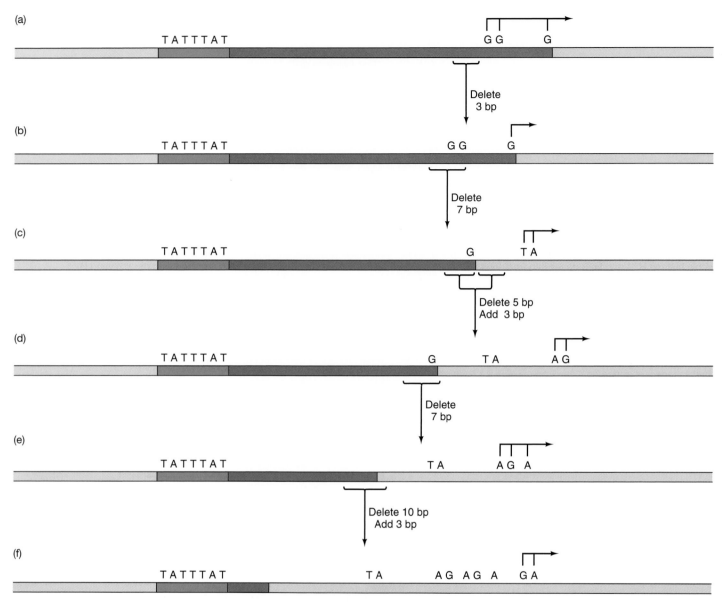

Figure 10.28   Transcription from the SV40 early promoter always starts in approximately the same place(s) relative to the TATA box.
Benoist and Chambon prepared mutants in the SV40 early promoter by deleting progressively more DNA in the region (blue) between the TATA box (red) and the most downstream transcription start site, then located the start of transcription by S1 mapping and primer extension. This figure summarizes the results. (**a**) Wild-type promoter, with three transcription start sites (all at G's) indicated. (**b**) Three base pairs were removed, moving the three original start sites upstream. Transcription started only at the most downstream G, 31 bp downstream of the first T of the TATA box. (**c**) Seven more base pairs of DNA were removed, including the first two original start sites; this moved the third original start site (a G) even farther upstream, but transcription still started about 30 bp downstream of the first T of the TATA box. (**d**) Five more base pairs of DNA were removed, and 3 bp of extra DNA were added, for a net deletion of 2 bp, but transcription still started about 30 bp downstream of the TATA box. (**e**) More DNA (7 bp) was deleted, including the last of the original start sites. Other start sites used in the previous deletion mutants are indicated, and they were also shifted upstream by the deletions. Transcription still occurred about 30 bp downstream of the TATA box. (**f**) Ten more bp of DNA were removed, and 3 bp of extra DNA were added. This left only 2 bp of the region (blue) between the TATA box and the original transcription start sites. Still, transcription began about 30 bp downstream of the TATA box. These results suggest that transcription always starts about 30 bp downstream of the first T of the TATA box, regardless of the sequence in between, and with little regard to the sequence at the start of transcription, except that purines are preferred start sites.

started at other bases, usually purines, that were about 30 bp downstream of the first T of the TATA box. In other words, the distance between the TATA box and the transcription initiation sites remained constant, with little regard to the exact sequence at these initiation sites.

In the two examples considered so far, the TATA box appears to be important for locating the start of transcription, but not for regulating the efficiency of transcription. However, in some other promoters, removal of the TATA box impairs promoter function to such an extent that transcription, even from aberrant start sites, cannot be detected. One such example is the TATA box of the rabbit β-globin gene (which actually begins with the sequence CATA). Grosveld and associates deleted the DNA between positions –20 and +7, which removed the initiator region, but retained the TATA box. When they transfected HeLa cells with this mutated gene and used an S1 assay for location of initiation and strength of transcription, they found that the initiation site moved downstream 27 bp, as we would predict (Figure 10.29). However, transcription remained as strong as with the wild-type promoter. On the other hand, when they deleted the same amount of DNA, but this time between positions –30 and –4 so the TATA box was removed, they detected no transcription. By removing the TATA box, they had destroyed the promoter.

Steven McKnight and Robert Kingsbury provided another example with their studies of the herpes virus thymidine kinase (tk) promoter. They performed **linker scanning mutagenesis,** in which they systematically substituted a 10-bp linker for 10-bp sequences throughout the tk promoter, as illustrated in Figure 10.30. One of the results of this analysis was that mutations within the TATA box destroyed promoter activity (Figure 10.31). In the mutant with the lowest promoter activity (LS –29/–18), the normal sequence in the region of the TATA box had been changed from GCATATTA to CCGGATCC.

Thus, some class II promoters require the TATA box for function, but others need it only to position the transcription start site. And, as we have seen, some class II promoters, most notably the promoters of housekeeping genes, have no TATA box at all, and they still function quite well. How do we account for these differences? As we will see in Chapters 11 and 12, promoter activity depends on assembling a collection of transcription factors and RNA polymerase called a preinitiation complex. This complex forms at the transcription start site and launches the transcription process. In class II promoters, the TATA box serves as the site where this assembly of protein factors begins. The first protein to bind is the TATA-box-binding protein (TBP), which then attracts the other factors. But what about promoters that lack TATA boxes? These still require TBP, but because TBP has no TATA box to which it can bind, it depends on other proteins, which bind to other promoter elements, to hold it in place.

Upstream Elements   McKnight and Kingsbury's linker scanning analysis of the herpes virus tk gene revealed other important promoter elements besides the TATA box. Figure 10.31 shows that mutations in the –47 to –61 and in the –80 to –105 regions caused significant loss of promoter activity. The coding strands of these regions contain the sequences GGGCGG and CCGCCC, respectively. These are so-called **GC boxes,** which are found in a variety of promoters, usually upstream of the TATA box.

Notice that the two GC boxes are in opposite orientations in their two locations in the herpes virus tk promoter.

Chambon and colleagues also found GC boxes in the SV40 early promoter, and not just two copies, but six. Furthermore, mutations in these elements significantly decreased promoter activity (Table 10.4). For example, loss of one GC box decreased transcription to 66% of the wild-type level, but loss of a second GC box decreased transcription all the way down to 13% of the control level. We will see in Chapter 12 that a specific transcription factor called Sp1 binds to the GC boxes and stimulates transcription. Later in this chapter we will discuss DNA elements called enhancers that function like promoters in that they stimulate transcription, but differ from promoters in two important respects: They are position- and orientation-independent. The GC boxes are orientation-independent; they can be flipped 180 degrees and they still function (as occurs naturally in the herpes virus tk promoter). But the GC boxes do not have the position independence of classical enhancers, which can be moved as much as several kilobases away from a promoter, even downstream of a gene's coding region, and still function. If the GC boxes are moved more than a few dozen base pairs away from their own TATA box, they lose the ability to stimulate transcription. Thus, it is probably more proper to consider the GC boxes, at least in these two genes, as promoter elements, rather than enhancers. On the other hand, the distinction is subtle and perhaps borders on semantic.

Another upstream element found in a wide variety of class II promoters is the so-called **CCAAT box** (pronounced "cat box"). In fact, the herpes virus tk promoter has a CCAAT box; the linker scanning study we have discussed failed to detect any loss of activity when this CCAAT box was mutated, but other investigations have clearly shown the importance of the CCAAT box in this and in many other promoters. Just as the GC box has its own transcription factor, so the CCAAT box must bind a transcription factor (the **CCAAT-binding transcription factor [CTF]**, among others) to exert its stimulatory influence.

Initiators and Downstream Elements   Some class II promoters have conserved sequences around their transcription start sites that are required for optimal transcription. These are called **initiators,** and they have the consensus sequence PyPyANT/APyPy, where Py stands for either pyrimidine (C or T), N stands for any base, and the underlined A is the transcription start point. The classic example is the initiator of the adenovirus major late promoter. This, together with the TATA box constitutes a core promoter that can drive transcription of any gene placed behind it, though at a very low level. This promoter is also susceptible to stimulation by upstream elements or enhancers connected to it.

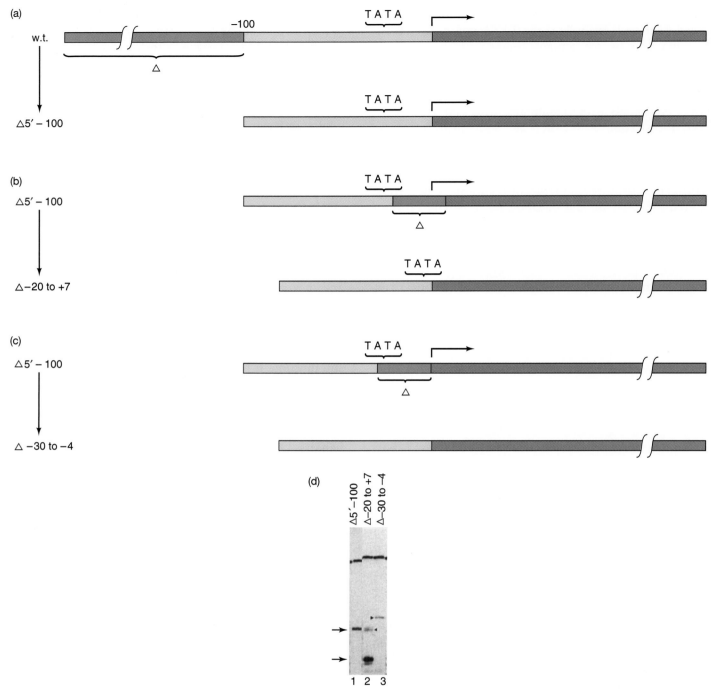

Figure 10.29 Effects of deletions in the rabbit β-globin promoter. (a) Constructing the 5′–100 mutant. At top is the wild-type DNA, showing the positions of the TATA box and start of transcription. The region upstream of position –100, which was deleted to create the Δ5′–100 mutant, is depicted in red. (b) Constructing the Δ–20 to +7 mutant. The starting material was the Δ5′–100 mutant; again, the deleted DNA—27 bp surrounding the transcription start site—is shown in red. (c) Constructing the Δ–30 to –4 mutant. The Δ5′–100 mutant again served as the starting material; this time, 27 bp (red) just upstream from the transcription start site, including the TATA box, were deleted. (d) Assaying the mutant promoters. Grosveld and colleagues tested the effects of deletions by transforming HeLa cells with the DNA constructs containing the mutated promoters and locating the start of transcription and the efficiency of transcription by S1 mapping. Lane 1, transcription of deletion mutant Δ5′–100. This DNA is transcribed from the proper start site with near-wild-type efficiency (arrow). Lane 2, transcription of deletion mutant Δ–20 to +7. This DNA was transcribed actively, but the start site was moved downstream (arrow) because of the deleted DNA. Lane 3, transcription of deletion mutant Δ–30 to –4, in which the TATA box had been deleted. Even though this construct was missing no more DNA than the one in lane 2, no detectable transcription occurred at this promoter. The light bands marked with arrowheads are unrelated transcripts that started at upstream promoters.
(*Source:* (*d*) Grosveld et al., DNA sequences necessary for transcription of the rabbit beta-globin gene *in vivo. Nature* 295 (14 Jan 1982) p. 123, f. 3 (3 lanes). © Macmillan Magazines Ltd.)

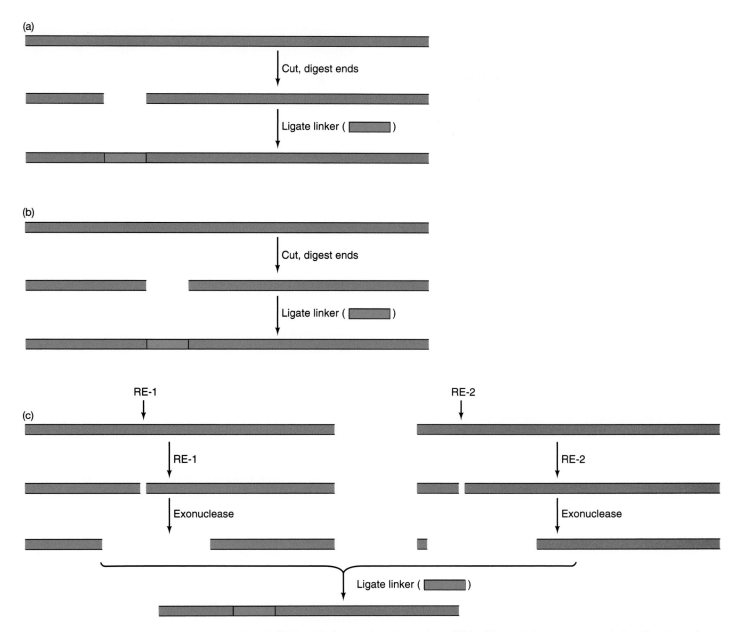

Figure 10.30 Principle of linker scanning. (a) and (b) Simplified procedure. By cutting a DNA with restriction enzymes and digesting the ends, one can create two DNA fragments that almost meet, but are missing about 10 bp, as represented by the gap. Then fill in the gap with a synthetic DNA linker (red). The gap in (b) is farther to the right than the one in (a), so the linker in (b) replaces DNA farther to the right than the one in (a). By repeating this procedure with gaps farther and farther to the right, one can "scan" with 10-bp replacements throughout the DNA region of interest. (c) Procedure for generating the two fragments. Cut the DNA on the left with restriction enzyme 1 (RE-1) and the DNA on the right with restriction enzyme 2 (RE-2). Then treat each pair of DNA fragments with an exonuclease such as Bal 31, which digests away the ends of the fragments. From the digest on the left, obtain the short, left-hand fragment, and from the digest on the right, obtain the long, right-hand fragment. Ligate these together with the linker to generate one of the linker mutants.

Another example of a gene with an important initiator is the mammalian terminal deoxynucleotidyltransferase (TdT) gene, which is activated during development of B and T lymphocytes. Stephen Smale and David Baltimore studied the mouse TdT promoter and found that it contains no TATA box and no apparent upstream promoter elements, but it does contain a 17-bp initiator. This initiator is sufficient to drive basal level transcription of the gene from a single start site located within the initiator sequence. Smale and Baltimore also found that a TATA box or the GC boxes from the SV40 promoter could greatly stimulate transcription starting at the initiator. Thus, this initiator alone constitutes a very simple, but functional, promoter whose efficiency can be enhanced by other promoter elements.

In addition to the initiator, we sometimes find **downstream elements,** downstream of the transcription start site. But no consensus sequence for the downstream elements has been found. These promoter elements have only recently been discovered, and they are not yet well characterized.

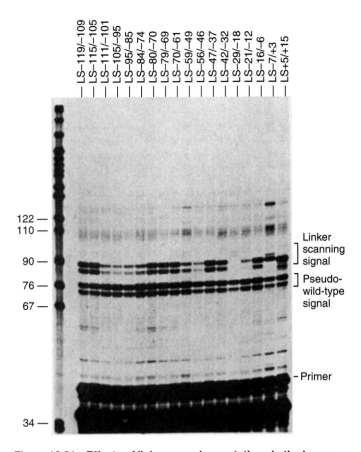

Figure 10.31   Effects of linker scanning mutations in the herpes virus tk promoter. McKnight and Kingsbury made linker scanning mutations throughout the tk promoter, then injected the mutated DNAs into frog oocytes, along with a pseudo-wild-type DNA (mutated at the +21 to +31 position). Transcription from this pseudo-wild-type promoter was just as active as that from the wild-type promoter, so this DNA served as an internal control. The investigators assayed for transcription from the test plasmid and from the control plasmid by primer extension analysis. Transcription from the control plasmid remained relatively constant, as expected, but transcription from the test plasmid varied considerably depending on the locus of the mutations. (*Source:* McKnight and Kingsbury, Transcriptional control signals of a eukaryotic protein-coding gene. *Science* 217 (23 July 1982) p. 322, f. 5. © AAAS.)

SUMMARY Class II promoters may contain any of the following elements: a TATA box; an upstream element (e.g., a GC box or a CCAAT box); an initiator; and a downstream element. At least one of these elements is missing in many promoters. Promoters for highly expressed specialized genes tend to have TATA boxes, but promoters for housekeeping genes tend to lack them. Some promoters have conserved initiators, and some promoters, including the mammalian TdT promoter, contain *only* an initiator, possibly with a downstream element. Most genes have some kind of upstream element, and many have more than one.

Table 10.4   Effects of Mutations in GC Boxes on Expression Directed by the SV40 Early Promoter

Number of Mutations	Number of Intact GC Boxes	Expression (%)
0	6	100
2	5	66
4	4	13
4	3	30
6	2	40
13	1	9
16	0	13

Class I Promoters

What about the promoter recognized by RNA polymerase I? We can refer to this promoter in the singular because each species should have only one kind of gene recognized by polymerase I: the rRNA precursor gene. It is true that this gene is present in hundreds of copies in each cell, but each copy is virtually the same as the others, and they all have the same promoter sequence. However, this sequence is quite variable from one species to another. In particular, the sequences of class I promoters are much more variable than those of the promoters recognized by polymerase II, which tend to have conserved elements, such as TATA boxes and CCATT boxes, in common.

Robert Tjian and colleagues used linker scanning mutagenesis to identify the important regions of the human rRNA promoter. Figure 10.32 shows the results of this analysis: The promoter has two critical regions in which mutations cause a great reduction in promoter strength. One of these, the **core element,** is located at the start of transcription, between positions −45 and +20. The other is the **upstream control element** (UCE), located between positions −156 and −107.

The presence of two promoter elements raises the question of the importance of the spacing between them. In this case, spacing is very important. Figure 10.33 depicts the results of an experiment in which Tjian and colleagues deleted or added DNA fragments of various lengths between the UCE and the core element of the human rRNA promoter. The promoter strength is listed to the right of the diagram of each construct. When Tjian and colleagues removed only 16 bp between the two promoter elements, the promoter strength dropped to 40% of wild-type; by the time they had deleted 44 bp, the promoter strength was only 10%. On the other hand, they could add 28 bp between the elements without affecting the promoter, but adding 49 bp reduced promoter strength by 70%. Thus, the promoter efficiency is more

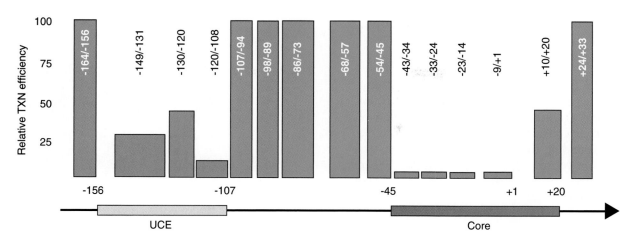

Figure 10.32   Two rRNA promoter elements. Tjian and colleagues used linker scanning to mutate short stretches of DNA throughout the 5′-flanking region of the human rRNA gene. They then tested these mutated DNAs for promoter activity using an in vitro transcription assay. The bar graph illustrates the results, which show that the promoter has two important regions: labeled UCE (upstream control element) and Core. The UCE is necessary for optimal transcription, but basal transcription is possible in its absence. On the other hand, the core element is absolutely required for any transcription to occur. (*Source:* Reprinted from Learned et al., *Cell* 45:848, 1986. Copyright 1986, with permission of Elsevier Science.)

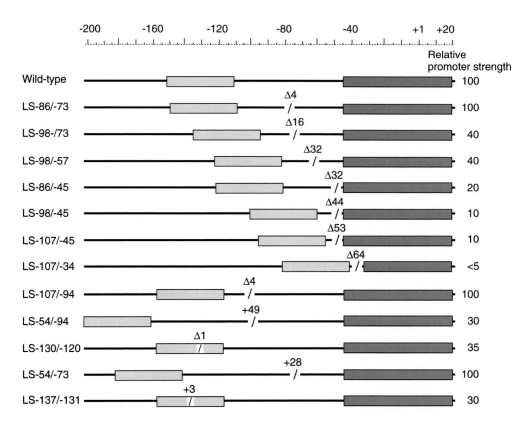

Figure 10.33   Effect of spacing of the two rRNA promoter elements. Tjian and colleagues deleted or added various lengths of DNA between the UCE and the core element of the human rRNA promoter and tested the constructs for promoter strength in a run-off assay. The name of each mutant construct is given at left, a diagram of each is presented in the center, and the relative promoter strength of each is shown at right. (*Source:* From M. M. Haltiner, S. T. Smale, and R. Tjian, *Molecular and Cellular Biology* 6:232, 1986. Copyright © 1986 American Society for Microbiology, Washington, DC. Reprinted by permission.)

sensitive to deletions than to insertions between the two promoter elements.

> **SUMMARY** Class I promoters are not well conserved in sequence from one species to another, but the general architecture of the promoter is well conserved. It consists of two elements, a core element surrounding the transcription start site, and an upstream control element (UCE) about 100 bp farther upstream. The spacing between these two elements is important.

Class III Promoters

As we have seen, RNA polymerase III transcribes a variety of genes that encode small RNAs. These include (1) the "classical" class III genes, including the 5S rRNA and tRNA genes, and the adenovirus VA RNA genes; and (2) some relatively recently discovered class III genes, including the U6 snRNA gene, the 7SL RNA gene, the 7SK RNA gene, and the Epstein–Barr virus EBER2 gene. The latter, "nonclassical" class III genes have promoters that resemble those found in class II genes. By contrast, the "classical" class III genes have promoters located entirely within the genes themselves.

Class III Genes with Internal Promoters   Donald Brown and his colleagues performed the first analysis of a class III promoter, on the gene for the *Xenopus laevis* 5S rRNA. The results they obtained were astonishing. Whereas the promoters recognized by polymerases I and II, as well as by bacterial polymerases, are located mostly in the 5'-flanking region of the gene, the 5S rRNA promoter is located *within* the gene it controls.

The experiments that led to this conclusion worked as follows: First, to identify the 5'-end of the promoter, Brown and colleagues prepared a number of mutant 5S rRNA genes that were missing more and more of their 5'-end and observed the effects of the mutations on transcription in vitro. They scored transcription as correct by measuring the size of the transcript by gel electrophoresis. An RNA of approximately 120 bases (the size of 5S rRNA) was deemed an accurate transcript, even if it did not have the same sequence as real 5S rRNA. They had to allow for incorrect sequence in the transcript because they changed the internal sequence of the gene to disrupt the promoter.

The surprising result (Figure 10.34) was that the entire 5'-flanking region of the gene could be removed without affecting transcription very much. Furthermore, big chunks of the 5'-end of the gene itself could be removed, and a transcript of about 120 nt would still be made. However, deletions beyond about position +50 destroyed promoter function.

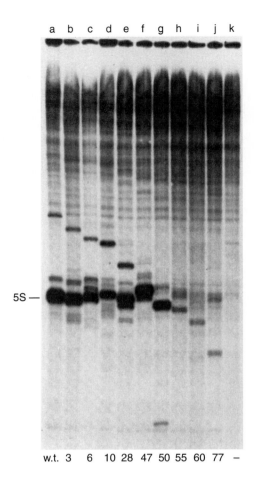

Figure 10.34   Effect of 5'-deletions on 5S rRNA gene transcription. Brown and colleagues prepared a series of deleted *Xenopus laevis* 5S rRNA genes with progressively more DNA deleted from the 5'-end of the gene itself. Then they transcribed these deleted genes in vitro in the presence of labeled substrate and electrophoresed the labeled products. DNA templates: lane a, undeleted positive control; lanes b–j, deleted genes with the position of the remaining 5'-end nucleotide denoted at bottom (e.g., lane b contained the product of a 5S rRNA gene whose 5'-end is at position +3 relative to the wild-type gene); lane k, negative control (pBR322 DNA with no 5S rRNA gene). Strong synthesis of a 5S-size RNA took place with all templates through lane g, in which deletion up to position +50 had occurred. With further deletion into the gene, this synthesis ceased. Lanes h–k also contained a band in this general area, but it is an artifact unrelated to 5S rRNA gene transcription. (*Source:* Sakonju et al., A control region in the center of the 5S RNA gene directs specific initiation of transcription: I. The 5' border of the region. *Cell* 19 (Jan 1980) p. 17, f. 4. Reprinted by permission of Elsevier Science.)

Locating the 3'-end of the promoter was trickier. Brown and colleagues could not simply chew away at the 3'-end of the gene and continue to see a 120-nt transcript because they would be removing the gene's terminator, and transcription would therefore run indefinitely. To get around this problem, Brown and colleagues used the drug cordycepin triphosphate, which is 3'-ATP. When cordycepin is incorporated into a growing RNA it will prevent further elongation because the 3'-end has no hydroxyl group. So when Brown and colleagues transcribed the 5S rRNA gene in the presence of cordycepin triphosphate plus ATP, they obtained a series of shortened transcripts in addition to full-length 5S rRNA. Therefore, they could use these characteristic shortened transcripts as a sign that initiation of 5S rRNA was still occurring and therefore the promoter was still intact, even if the gene's terminator was missing. Figure 10.35 shows that these tell-tale shortened transcripts appeared until the 3′-deletions went further than the +83 position. (Note the difference between lanes f and g of Figure 10.35).

These experiments identified a sensitive region between bases 50 and 83 of the transcribed sequence that could not be encroached on without destroying promoter function. These are the apparent outer boundaries of the internal promoter of the *Xenopus* 5S rRNA gene. Other experiments showed that it is possible to *add* chunks of DNA outside this region without harming the promoter. Roeder and colleagues later performed systematic mutagenesis of bases throughout the promoter region and identified three regions that could not be changed without greatly diminishing promoter function. These sensitive regions are called **box A**, the **intermediate element**, and **box C**. (No box B occurs because a box B had already been discovered in other class III genes, and it had no counterpart in the 5S rRNA promoter.) Figure 10.36a summarizes the results of these experiments on the 5S rRNA promoter. Similar experiments on the other two classical class III genes, the tRNA and VA RNA genes, showed that their promoters contain a **box A** and a **box B** (Figure 10.36b). The sequence of the box A is similar to that of the box A of the 5S rRNA gene. Furthermore, the space in between the two blocks can be altered somewhat without destroying promoter function. Such alteration does have limits, however; if one inserts too much DNA between the two promoter boxes, efficiency of transcription suffers.

> **SUMMARY** RNA polymerase III transcribes a set of short genes. The classical class III genes (5S rRNA, tRNA, and VA RNA genes) have promoters that lie wholly within the genes. The internal promoter of the 5S rRNA gene is split into three regions: box A, a short intermediate element, and box C. The promoters of the latter two classes of genes are split into two parts: box A and box B.

Class III Genes with Polymerase II-like Promoters After Brown and other investigators established the novel idea of internal promoters for class III genes, it was generally assumed that all class III genes worked this way. However, by the mid-1980s some exceptions were discovered. In 1985, Elisabetta Ullu and Alan Weiner conducted in vitro transcription studies on wild-type and mutant 7SL

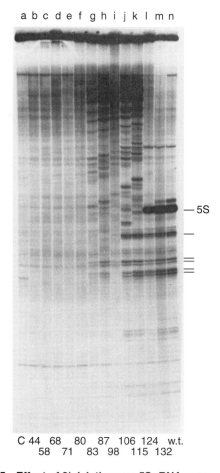

a b c d e f g h i j k l m n

— 5S

—

C 44   68   80   87   106 124   w.t.
   58   71   83   98   115 132

Figure 10.35 Effect of 3′ deletions on 5S rRNA gene transcription. Brown and colleagues prepared a series of 3′-deletions of the *X. laevis* 5S rRNA gene and transcribed these deleted genes in vitro in the presence of labeled substrate and a low concentration of cordycepin triphosphate to terminate transcription at reproducible sites. Templates: lane a, negative control (pBR322 DNA); lanes b–m, deleted genes with the position of the remaining 3′-end nucleotide denoted at bottom (e.g., lane b contained the product of a gene whose 3′-end is at position +44 relative to the wild-type 5S gene); lane n, undeleted positive control. The most prominent bands below the 5S position that are characteristic of proper initiation of 5S rRNA synthesis and termination at sites within the gene are marked at right, as is the 5S product itself. These bands appeared in the products of all genes in which 3′-deletion extended no farther than position +83 (lane g). (*Source:* Bogenhagen et al., A control region in the center of the 5S RNA gene directs specific initiation of transcription: II. The 3′ border of the region. *Cell* 19 (Jan 1980) p. 31, f. 5. Reprinted by permission of Elsevier Science.)

RNA genes that showed that the 5′-flanking region was required for high-level transcription. Without this DNA region, transcription efficiency dropped by 50–100-fold. Ullu and Weiner concluded that the most important DNA element for transcription of this gene lies upstream of the gene. Nevertheless, the fact that transcription still occurred in mutant genes lacking the 5′-flanking region implies that these genes also contain a weak internal promoter. These data help explain why the hundreds of 7SL RNA pseudogenes (nonfunctional copies of the 7SL gene) in the human

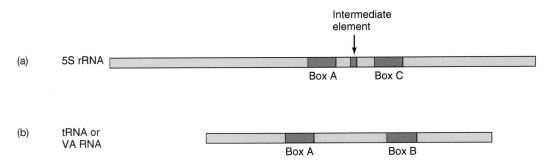

Figure 10.36 Promoters of some polymerase III genes. The promoters of the 5S, tRNA and VA RNA genes are depicted as groups of blue boxes within the genes they control.

genome, as well as the related *Alu* sequences, are relatively poorly transcribed in vivo: They lack the upstream element required for high-level transcription.

Marialuisa Melli and colleagues noticed that the 7SK RNA gene does not have internal sequences that resemble the classic class III promoter. On the other hand, the 7SK RNA gene does have a 5'-flanking region homologous to that of the 7SL RNA gene. On the basis of these observations, they proposed that this gene has a completely external promoter. To prove the point, they made successive deletions in the 5'-flanking region of the gene and tested them for ability to support transcription in vitro. Figure 10.37 shows that deletions up to position –37 still allowed production of high levels of 7SK RNA, but deletions beyond this point were not tolerated. On the other hand, the coding region was not needed for transcription; in vitro transcription analysis of another batch of deletion mutants, this time with deletions within the coding region, showed that transcription still occurred, even when the whole coding region was removed. Thus, this gene lacks an internal promoter.

What is the nature of the promoter located in the region encompassing the 37 bp upstream of the start site? Interestingly enough, a TATA box resides in this region, and changing three of its bases (TAT → GCG) reduced transcription by 97%. Thus the TATA box is required for good promoter function. All this may make you wonder whether polymerase II, not polymerase III, really transcribes this gene after all. If that were the case, low concentrations of α-amanitin should inhibit transcription, but it takes high concentrations of this toxin to block 7SK RNA synthesis. In fact, the profile of inhibition of 7SK RNA synthesis by α-amanitin is exactly what we would expect if polymerase III, not polymerase II, is involved.

Now we know that the other nonclassical class III genes, including the U6 RNA gene and the EBER2 gene, behave the same way. They are transcribed by polymerase III, but they have polymerase II-like promoters. In Chapter 11 we will see that this is not as strange as it seems at first because the TATA-binding protein (TBP) is involved in class III (and class I) transcription, in addition to its well-known role in class II gene transcription.

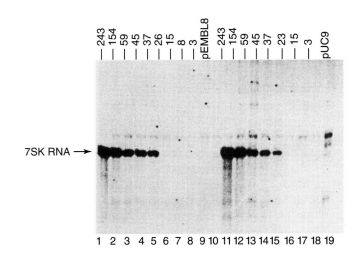

7SK RNA →

Figure 10.37 Effects of 5'-deletion mutations on the 7SK RNA promoter. Melli and colleagues performed deletions in the 5'-flanking region of the human 7SK RNA gene and transcribed the mutated genes in vitro. Then they electrophoresed the products to determine if 7SK RNA was still synthesized. The negative numbers at the top of each lane give the number of base pairs of the 5'-flanking region still remaining in the deleted gene used in that reaction. For example, the template used in lane 9 retained only 3 bp of the 5'-flanking region—up to position –3. Lanes 1–10 contained deleted genes cloned into pEMBL8; lanes 11–19 contained genes cloned into pUC9. The cloning vectors themselves were transcribed in lanes 10 and 19. A comparison of lanes 5 and 6 (or of lanes 15 and 16) shows an abrupt drop in promoter activity when the bases between position –37 and –26 were removed. This suggests that an important promoter element lies in this 11-bp region. (*Source:* Murphy et al., The in vitro transcription of the 7SK RNA gene by RNA polymerase III is dependent only on the presence of an upstream promoter. *Cell* 51 (9) (1987) p. 82, f. 1b. Reprinted by permission of Elsevier Science.)

SUMMARY At least one class III gene, the 7SL RNA gene, contains a weak internal promoter, as well as a sequence in the 5'-flanking region of the gene that is required for high-level transcription. Other class III genes (e.g., 7SK, and U6 RNA genes) lack internal promoters altogether, and contain promoters that strongly resemble class II promoters in that they lie in the 5'-flanking region and contain TATA boxes.

Figure 10.38   Structure of the SV40 virus early control region. As usual, an arrow with a right-angle bend denotes the transcription initiation site, although this is actually a cluster of three sites, as we saw in Figure 10.28. Upstream of the start site we have, in right-to-left order, the TATA box (red), six GC boxes (yellow), and the enhancer (72-bp repeats, blue).

10.3   Enhancers and Silencers

Many eukaryotic genes, especially class II genes, are associated with *cis*-acting DNA elements that are not strictly part of the promoter, yet strongly influence transcription. As we learned in Chapter 9, **enhancers** are elements that stimulate transcription. **Silencers,** by contrast, depress transcription. We will discuss these elements briefly here and expand on their modes of action in Chapters 12 and 13.

Enhancers

Chambon and colleagues discovered the first enhancer in the 5′-flanking region of the SV40 early gene. This DNA region had been noticed before because it contains a conspicuous duplication of a 72-bp sequence, called the *72-bp repeat* (Figure 10.38). When Benoist and Chambon made deletion mutations in this region, they observed profoundly depressed transcription in vivo. This behavior suggested that the 72-bp repeats constituted another upstream promoter element. However, Paul Berg and his colleagues discovered that the 72-bp repeats still stimulated transcription even if they were inverted or moved all the way around to the opposite side of the circular SV40 genome, over 2 kb away from the promoter. The latter behavior, at least, is very un-promoter-like. Thus, such orientation- and position-independent DNA elements are called enhancers to distinguish them from promoter elements.

How do enhancers stimulate transcription? We will see in Chapter 12 that enhancers act through proteins that bind to them. These have several names: **transcription factors, enhancer-binding proteins,** or **activators.** These proteins appear to stimulate transcription by interacting with other proteins called **general transcription factors** at the promoter. This interaction promotes formation of a preinitiation complex, which is necessary for transcription. We will discuss these interactions in much greater detail in Chapters 11 and 12.

We frequently find enhancers upstream of the promoters they control, but this is by no means an absolute rule. In fact, as early as 1983 Susumo Tonegawa and his colleagues found an example of an enhancer *within* a gene. These investigators were studying a gene that encodes the larger subunit of a particular mouse antibody, or immunoglobulin, called γ_{2b}. They introduced this gene into mouse plasmacytoma cells that normally expressed antibody genes, but not this particular gene. To detect efficiency of expression of the transfected cells, they added a labeled amino acid to tag newly made proteins, then electrophoresed the proteins and detected them by autoradiography. The suspected enhancer lay in one of the gene's introns, a region within the gene that is transcribed, but is subsequently cut out of the transcript by a process called splicing (Chapter 14). Tonegawa and colleagues began by deleting two chunks of DNA from this suspected enhancer region, as shown in Figure 10.39a. Then they assayed for expression of the γ_{2b} gene in cells transfected by this mutated DNA. Figure 10.39b shows the results: The deletions within the intron, though they should have no effect on the protein product because they are in a noncoding region of the gene, caused a decrease in the amount of gene product made. This was especially pronounced in the case of the larger deletion ($\Delta2$).

Is this effect due to decreased transcription, or some other cause? Tonegawa's group answered this question by performing Northern blots with RNA from cells transfected with normal and deleted γ_{2b} genes. These blots, shown in Figure 10.39c, again demonstrated a profound loss of function when the suspected enhancer was deleted. But is this really an enhancer? If so, one should be able to move it or invert it and it should retain its activity. Tonegawa and colleagues did this by first inverting the X_2–X_3 fragment, which contained the enhancer, as shown in position/orientation B of Figure 10.40a. Figure 10.40b shows that the enhancer still functioned. Next, they took fragment X_2–X_3 out of the intron and placed it upstream of the promoter (position/orientation C). It still worked. Then they inverted it in its new location (position/orientation D). Still it functioned. Thus, some region within the X_2–X_3 fragment behaved as an enhancer: It stimulated transcription from a nearby promoter, and it was position- and orientation-independent. Finally, these workers compared the expression of this gene when it was transfected into two different types of mouse cells: plasmacytoma cells as before, and fibroblasts. Expression was much more active in plasmacytoma cells. This is also consistent with the presence of an enhancer because fibroblasts do not make antibodies and therefore should not contain enhancer-binding proteins capable of activating the enhancer of an antibody gene. Thus, the antibody gene should not be expressed actively in such cells.

Silencers

Enhancers are not the only DNA elements that can act at a distance to modulate transcription. Silencers also do

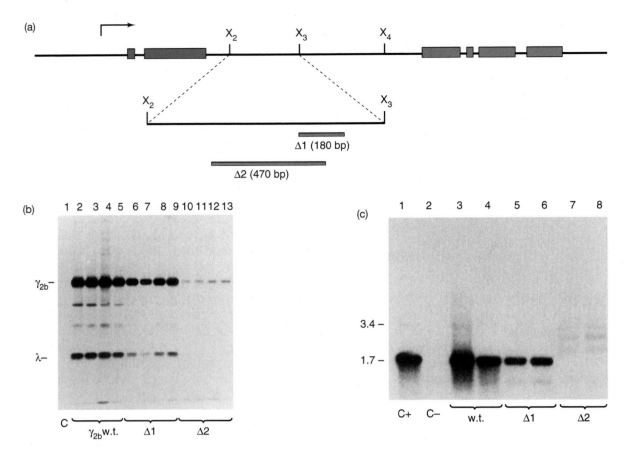

Figure 10.39   Effects of deletions in the immunoglobulin γ$_{2b}$ H-chain enhancer. (a) Map of the cloned γ$_{2b}$ gene. The blue boxes represent the exons of the gene, the parts that are included in the mRNA that comes from this gene. The lines in between boxes are introns, regions of the gene that are transcribed, but then cut out of the mRNA precursor as it is processed to the mature mRNA. X$_2$, X$_3$, and X$_4$ represent cutting sites for the restriction enzyme *Xba*I. Tonegawa and colleagues suspected an enhancer lay in the X$_2$–X$_3$ region, so they made deletions Δ1 and Δ2 as indicated by the red boxes. **(b)** Assay of expression of the γ$_{2b}$ gene at the protein level. Tonegawa and colleagues transfected plasmacytoma cells with the wild-type gene (lanes 2–5), the gene with deletion Δ1 (lanes 6–9), or the gene with deletion Δ2 (lanes 10–13). Lane 1 was a control with untransfected plasmacytoma cells. After transfecting the cells, these investigators added a radioactive amino acid to label any newly made protein, then extracted the protein, immunoprecipitated the γ$_{2b}$ protein, electrophoresed the precipitated protein, and detected the radioactive protein by fluorography (a modified version of autoradiography in which a compound called a fluor is added to the electrophoresis gel). Radioactive emissions excite this fluor to give off photons that are detected by x-ray film. The Δ1 deletion produced only a slight reduction in expression of the gene, but the Δ2 deletion gave a profound reduction. **(c)** Assay of transcription of the γ$_{2b}$ gene. Tonegawa and colleagues electrophoresed and Northern blotted RNA from the following cells: lane 1 (positive control), untransfected plasmacytoma cells (MOPC 141) that expressed the γ$_{2b}$ gene; lane 2 (negative control), untransfected plasmacytoma cells (J558L) that did not express the γ$_{2b}$ gene; lanes 3 and 4, J558L cells transfected with the wild-type γ$_{2b}$ gene; lanes 5 and 6, J558L cells transfected with the gene with the Δ1 deletion; lanes 7 and 8, J558L cells transfected with the gene with the Δ2 deletion. The Δ1 deletion decreased transcription somewhat, but the Δ2 deletion abolished transcription. *(Source: (b-c)* Gillies et al., A tissue-specific transcription enhancer element is located in the major intron of a rearranged immunoglobulin heavy chain gene. *Cell* 33 (July 1983) p. 719, f. 2&3. Reprinted by permission of Elsevier Science.)

this, but—as their name implies—they inhibit rather than stimulate transcription. The mating system (*MAT*) of yeast provides a good example. Yeast chromosome III contains three loci of very similar sequence: *MAT*, *HML*, and *HMR*. Though *MAT* is expressed, the other two loci are not, and silencers located at least 1 kb away seem to be responsible for this genetic inactivity. We know that something besides the inactive genes themselves is at fault, because active yeast genes can be substituted for *HML* or *HMR* and the transplanted genes become inactive. Thus, they seem to be responding to an external negative influence: a silencer. How do silencers work? The available

data indicate that they cause the chromatin to coil up into a condensed, inaccessible, and therefore inactive form, thereby preventing transcription of neighboring genes. We will examine this process in more detail in Chapter 13.

Sometimes the same DNA element can have both enhancer and silencer activity, depending on the protein bound to it. For example, the thyroid hormone response element acts as a silencer when the thyroid hormone receptor binds to it without its ligand, thyroid hormone. But it acts as an enhancer when the thyroid hormone receptor binds along with thyroid hormone. We will revisit this concept in Chapter 12.

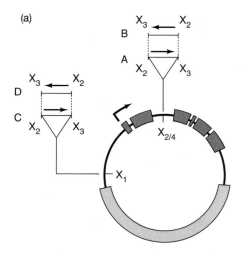
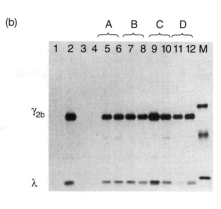

Figure 10.40   The enhancing element in the γ2b gene is orientation- and position-independent. (a) Outline of the mutant plasmids. Tonegawa and colleagues removed the X_2–X_3 region of the parent plasmid containing the γ2b gene (see Figure 10.39a). This deleted the enhancer. Then they reinserted the X_2–X_3 fragment (with the enhancer) in four different ways: plasmids A and B, the fragment was inserted back into the intron in its usual location in the forward (normal) orientation (A), or in the backward orientation (B); plasmids C and D, the fragment was inserted into another *XbaI* site (X_1) hundreds of base pairs upstream of the gene in the forward orientation (C), or in the backward orientation (D). (b) Experimental results. Tonegawa and colleagues tested all four plasmids from (a), as well as the parent, for efficiency of expression as in Figure 10.39b. All functioned equally well. Lane 1, untransfected J558L cells lacking the γ2b gene. Lanes 2–12, J558L cells transfected with the following plasmids: lane 2, the parent plasmid with no deletions; lanes 3 and 4, the parent plasmid with the X_2–X_3 fragment deleted; lanes 5 and 6, plasmid A; lanes 7 and 8, plasmid B; lanes 9 and 10, plasmid C; lanes 11 and 12, plasmid D. Lane M contained protein size markers. (*Source:* Gillies et al., *Cell* 33 (July 1983) p. 721, f. 5. Reprinted by permission of Elsevier Science.)

SUMMARY Enhancers and silencers are position- and orientation-independent DNA elements that stimulate or depress, respectively, the transcription of associated genes. They are also tissue-specific in that they rely on tissue-specific DNA-binding proteins for their activities. Sometimes a DNA element can act as either an enhancer or a silencer depending on what is bound to it.

SUMMARY

Eukaryotic nuclei contain three RNA polymerases that can be separated by ion-exchange chromatography. These three enzymes have different responses to salt and divalent metal ions, and different locations within the nucleus. RNA polymerase I is found in the nucleolus; the other two polymerases are located in the nucleoplasm. The three nuclear RNA polymerases have different roles in transcription. Polymerase I makes a large precursor to the major rRNAs (5.8S, 18S, and 28S rRNAs in vertebrates). Polymerase II synthesizes hnRNAs, which are precursors to mRNAs. It also makes most small nuclear RNAs (snRNAs). Polymerase III makes the precursors to 5S rRNA, the tRNAs, and several other small cellular and viral RNAs.

The subunit structures of all three nuclear polymerases from several eukaryotes have been determined. All of these structures contain many subunits, including two large ones, with molecular masses greater than 100kd. All eukaryotes seem to have at least some common subunits that are found in all three polymerases. The genes encoding all 12 RNA polymerase II subunits in yeast have been cloned and sequenced, and subjected to mutation analysis. Three of the subunits resemble the core subunits of bacterial RNA polymerases in both structure and function, five are found in all three nuclear RNA polymerases, two are not required for activity, at least at normal temperatures, and two fall into none of these three categories.

The largest RNA polymerase II subunit (the product of the RPB1 gene) governs sensitivity or resistance to α-amanitin. Subunit IIa is the primary product of the RPB1 gene in yeast. It can be converted to IIb in vitro by proteolytic removal of the carboxyl-terminal domain (CTD), which is essentially a heptapeptide repeated over and over. Subunit IIa is converted in vivo to IIo by phosphorylating several serines and threonines within the CTD. The enzyme (polymerase IIA) with the IIa subunit is the one that binds to the promoter; the enzyme (polymerase IIO) with the IIo subunit is the one involved in transcript elongation.

Structural studies on yeast RNA polymerase II have revealed a 25 Å wide channel that makes a right angle turn on the surface of the enzyme. This channel can hold double-stranded DNA for most of its length. A higher resolution crystal structure of the enzyme shows the locations of the 10 essential subunits, and the location of the DNA template can be inferred from the previous studies on two-dimensional crystals. The enzyme contains a cleft that holds the DNA, bounded by a set of jaws on the downstream end, and a wall just beyond the active site at the other end. The DNA must turn to avoid this wall, and DNA melting would make this possible. A predicted RNA exit site places the RNA product in a groove that would hold a sliding clamp shut on the DNA, improving processivity.

Class II promoters may contain any of the following four elements: an initiator, a TATA box, an upstream element (e.g., a GC box or a CCAAT box), and a downstream element. At least one of these elements is missing in many promoters. Promoters for highly expressed specialized genes tend to have TATA boxes, but promoters for housekeeping genes tend to lack them. Some promoters have conserved initiators, and some promoters, including the mammalian TdT promoter, contain *only* an initiator. Most genes have some kind of upstream element, and many have more than one.

Class I promoters are not well conserved in sequence from one species to another, but the general architecture of the promoter is well conserved. It consists of two elements: a core element surrounding the transcription start site, and an upstream control element (UCE) about 100 bp farther upstream. The spacing between these two elements is important.

RNA polymerase III transcribes a set of short genes. The classical class III genes (5S rRNA, tRNA, and VA RNA genes), have promoters that lie wholly within the genes. The internal promoter of the 5S rRNA gene is split into three regions: box A, a short intermediate element, and box C. The promoters of the latter two classes of genes are split into two parts: box A and box B. At least one class III gene, the 7SL RNA gene, contains a weak internal promoter, as well as a sequence in the 5′-flanking region of the gene that is required for high-level transcription. Other class III genes (e.g., 7SK, and U6 RNA genes) lack internal promoters altogether and contain promoters that strongly resemble class II promoters in that they lie in the 5′-flanking region and contain TATA boxes.

Enhancers and silencers are position- and orientation-independent DNA elements that stimulate or depress, respectively, the transcription of associated genes. They are also tissue-specific in that they rely on tissue-specific DNA-binding proteins for their activities.

REVIEW QUESTIONS

1. Diagram the elution pattern of the eukaryotic nuclear RNA polymerases from DEAE-Sephadex chromatography. Show what you would expect if you assayed the same fractions in the presence of 1 μg/mL of α-amanitin.

2. Describe and give the results of an experiment that shows that polymerase I is located primarily in the nucleolus of the cell.

3. RNA polymerase I is most active at low ionic strength, responds well to Mg^{2+}, and is located in the nucleolus. What does this suggest about the product of polymerase I?

4. How do we know that polymerase I makes the large rRNA precursor and that polymerase II makes mRNA precursors? Describe a simple in vitro transcription experiment, as well as the experiments described in the text.

5. Describe and give the results of an experiment that shows that polymerase III makes tRNA and 5S rRNA.

6. How many subunits does yeast RNA polymerase II have? Which of these are "core" subunits? How many subunits are common to all three nuclear RNA polymerases?

7. Describe how epitope tagging can be used to purify polymerase II from yeast in one step.

8. Describe and give the results of an experiment that suggests that the largest polymerase II subunit binds to DNA.

9. Describe and give the results of an experiment that shows that the second largest polymerase II subunit contains the catalytic site that forms phosphodiester bonds between nucleotides.

10. What evidence suggests that the largest three subunits of eukaryotic polymerase II are homologous to the β′, β, and α subunits, respectively, of prokaryotic RNA polymerases?

11. What happens to the structure of RNA polymerase II when the RPB4 gene is deleted? What functions of this defective polymerase are normal and abnormal? Present evidence for these conclusions.

12. Some preparations of polymerase II show three different forms of the largest subunit (RPB1). Give the names of these subunits and show their relative positions after SDS-PAGE. What are the differences among these subunits? Present evidence for these conclusions.

13. What is the structure of the CTD of RPB1?

14. Outline a series of experiments that demonstrates that RPB1 confers α-amanitin-sensitivity (or resistance) on polymerase II.

15. Describe the results of electron and x-ray crystallography studies on yeast RNA polymerase II and on an elongation complex involving yeast polymerase II.

16. Draw a diagram of a polymerase II promoter, showing all of the types of elements it could have.

17. What kinds of genes tend to have TATA boxes? What kinds of genes tend not to have them?

18. Transcription of a class II gene starts at a guanosine 30 bp downstream of the first base of the TATA box. You delete 10 bp of DNA between this guanosine and the TATA box and transfect cells with this mutated DNA. Will transcription still start at the same guanosine? If not, where?

19. What are the two most likely effects of removing the TATA box from a class II promoter?

20. Diagram the process of linker scanning. What kind of information does it give?

21. List two common upstream elements of class II promoters.

22. Diagram a typical class I promoter.

23. How were the elements of class I promoters discovered? Present experimental results.

24. Describe and give the results of an experiment that shows the importance of spacing between the elements of a class I promoter.

25. Compare and contrast (with diagrams) the classical and nonclassical class III promoters. Give an example of each.

26. Describe and give the results of experiments that locate the 5′- and 3′-borders of the 5S rRNA gene's promoter.

27. You suspect that a repeated sequence just upstream of a gene is acting as an enhancer. Describe and give the results of an experiment that would test your hypothesis.

28. Explain the fact that enhancer activity is tissue-specific.

SUGGESTED READINGS

General References and Reviews

Corden, J.L. 1990. Tales of RNA polymerase II. *Trends in Biochemical Sciences* 15:383–87.

Sentenac, A. 1985. Eukaryotic RNA Polymerases. *CRC Critical Reviews in Biochemistry* 18:31–90.

Woychik, N.A. and R.A. Young. 1990. RNA polymerase II: Subunit structure and function. *Trends in Biochemical Sciences* 15:347–51.

Research Articles

Benoist, C. and P. Chambon. 1981. In vivo sequence requirements of the SV40 early promoter region. *Nature* 290:304–10.

Bogenhagen, D.F., S. Sakonju, and D.D. Brown. 1980. A control region in the center of the 5S RNA gene directs specific initiation of transcription: II. The 3′ border of the region. *Cell* 19:27–35.

Carroll, S.B. and B.D. Stollar. 1983. Conservation of a DNA-binding site in the largest subunit of eukaryotic RNA polymerase II. *Journal of Molecular Biology* 170:777–90.

Cramer, P., D.A. Bushnell, J. Fu, A.L. Gnatt, B. Maier-Davis, N.E. Thompson, R.R. Burgess, A.M. Edwards, P.R. David, and R.D. Kornberg. 2000. Architecture of RNA polymerase II and implications for the transcription mechanism. *Science* 288:640–49.

Das, G., D. Henning, D. Wright, and R. Reddy. 1988. Upstream regulatory elements are necessary and sufficient for transcription of a U6 RNA gene by RNA polymerase III. *EMBO Journal.* 7:503–12.

Edwards, A.M., C.M. Kane, R.A. Young, and R.D. Kornberg. 1991. Two dissociable subunits of yeast RNA polymerase II stimulate the initiation of transcription at a promoter in vitro. *Journal of Biological Chemistry* 266:71–75.

Fu, J., A.L. Gnatt, D.A. Bushnell, G.J. Jensen, N.E. Thompson, R.R. Burgess, P.R. David, and R.D. Kornberg. 1999. Yeast RNA polymerase II at 5 Å resolution *Cell* 98:799–810.

Gillies, S.D., S.L. Morrison, V.T. Oi, and Susumu Tonegawa. 1983. A tissue-specific transcription enhancer element is located in the major intron of a rearranged immunoglobulin heavy chain gene. *Cell* 33:717–28.

Greenleaf, A.L. 1983. Amanitin-resistant RNA polymerase II mutations are in the enzyme's largest subunit. *Journal of Biological Chemistry* 258:13403–6.

Greenleaf, A.L., L.M. Borsell, P.F. Jiamachello, and D.E. Coulter. 1979. α-Amanitin-resistant *Drosophila melanogaster* with an altered RNA polymerase II. *Cell* 18:613–22.

Grosschedl, B., and M.L. Birnstiel. 1980. Identification of regulatory sequences in the prelude sequences of an H2A histone gene by the study of specific deletion mutations in vivo. *Proceedings of the National Academy of Sciences USA* 77:1432–36.

Grosveld, G.C., E. de Boer, C.K. Shewmaker, and R.A. Flavell. 1982. DNA sequences necessary for transcription of the rabbit β-globin gene in vivo. *Nature* 295:120–26.

Haltiner, M.M., S.T. Smale, and R. Tjian. 1986. Two distinct promoter elements in the human rRNA gene identified by linker scanning mutagenesis. *Molecular and Cellular Biology* 6:227–35.

Ingles, C.J., J. Biggs, J.K.-C. Wong, J.R. Weeks, and A.L. Greenleaf. 1983. Identification of a structural gene for a RNA polymerase II polypeptide in *Drosophila melanogaster* and mammalian species. *Proceedings of the National Academy of Sciences USA* 80:3396–3400.

Kolodziej, P.A., N. Woychik, S.-M. Liao, and R. Young. 1990. RNA polymerase II subunit composition, stoichiometry, and phosphorylation. *Molecular and Cellular Biology* 10:1915–20.

Learned, R.M., T.K. Learned, M.M. Haltiner, and R.T. Tjian. 1986. Human rRNA transcription is modulated by the coordinate binding of two factors to an upstream control element. *Cell* 45:847–57.

McKnight, S.L. and Kingsbury, R. 1982. Transcription control signals of a eukaryotic protein-coding gene. *Science* 217:316–24.

Murphy, S., C. Di Liegro, and M. Melli. 1987. The in vitro transcription of the 7SK RNA gene by RNA polymerase III is dependent only on the presence of an upstream promoter. *Cell* 51:81–87.

Pieler, T., J. Hamm, and R.G. Roeder. 1987. The 5S gene internal control region is composed of three distinct sequence elements, organized as two functional domains with variable spacing. *Cell* 48:91–100.

Poglitsch, C.L., G.D. Meredith, A.L. Gnatt, G.J. Jensen, W-h. Chang, J. Fu, and R.D. Kornberg 1999. Electron crystal structure of an RNA polymerase II transcription elongation complex. *Cell* 98:791–98.

Pogo, A.O., V.C. Littau, V.G. Allfrey, and A.E. Mirsky. 1967. Modification of ribonucleic acid synthesis in nuclei isolated from normal and regenerating liver: Some effects of salt and specific divalent cations. *Proceedings of the National Academy of Sciences USA* 57:743–50.

Roeder, R.G., and W.J. Rutter. 1969. Multiple forms of DNA-dependent RNA polymerase in eukaryotic organisms. *Nature* 224:234–37.

Roeder, R.G., and W.J. Rutter. 1970. Specific nucleolar and nucleoplasmic RNA polymerases. *Proceedings of the National Academy of Sciences USA* 65:675–82.

Sakonju, S., D.F. Bogenhagen, and D.D. Brown. 1980. A control region in the center of the 5S RNA gene directs initiation of transcription: I. The 5′ border of the region. *Cell* 19:13–25.

Sayre, M.H., H. Tschochner, and R.D. Kornberg. 1992. Reconstitution of transcription with five purified initiation factors and RNA polymerase II from *Saccharomyces cerevisiae*. *Journal of Biological Chemistry* 267:23376–82.

Searles, L.L., R.S. Jokerst, P.M. Bingham, R.A. Voelker, and A.L. Greenleaf. 1982. Molecular cloning of sequences from a *Drosophila* RNA polymerase II locus by P element transposon tagging. *Cell* 31:585–92.

Sklar, V.E.F., L.B. Schwartz, and R.G. Roeder. 1975. Distinct molecular structures of nuclear class I, II, and III DNA-dependent RNA polymerases. *Proceedings of the National Academy of Sciences USA* 72:348–52.

Smale, S.T., and D. Baltimore. 1989. The "initiator" as a transcription control element. *Cell* 57:103–13.

Ullu, E., and A.M. Weiner. 1985. Upstream sequences modulate the internal promoter of the human 7SL RNA gene. *Nature* 318:371–74.

Weinman, R., and R.G. Roeder. 1974. "Role of DNA-dependent RNA polymerase III in the transcription of the tRNA and 5S rRNA genes." *Proceedings of the National Academy of Sciences USA* 71:1790–94.

Woychik, N.A., S.M. Liao, P.A. Kolodziej, and R.A Young. 1990. Subunits shared by eukaryotic nuclear RNA polymerases. *Genes and Development* 4:313–23.

Woychik, N.A., et al. 1993. Yeast RNA polymerase II subunit RPB11 is related to a subunit shared by RNA polymerase I and III. *Gene Expression* 3:77–82.

CHAPTER 11

General Transcription Factors in Eukaryotes

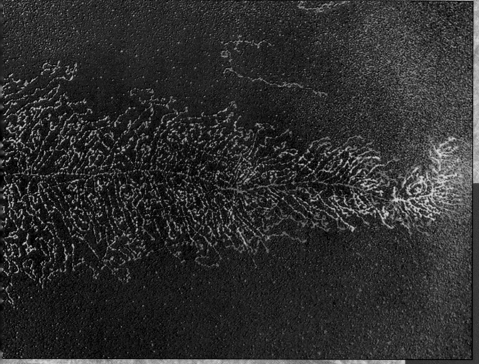

False color transmission electron micrograph of RNAs being synthesized on a DNA template, forming a feather-like structure. The backbone of the feather is a long DNA duplex coated with protein. Numerous RNA molecules extend in clusters from the DNA strand (×14,000). © Oscar L. Miller/SPL/Photo Researchers, Inc.

Eukaryotic RNA polymerases, unlike their prokaryotic counterparts, are incapable of binding by themselves to their respective promoters. Instead, they rely on proteins called *transcription factors* to show them the way. Such factors are grouped into two classes: general transcription factors and gene-specific transcription factors. The general transcription factors by themselves can attract the RNA polymerases to their respective promoters, but only to a weak extent. Therefore, these factors can stimulate only a basal level of transcription. Furthermore, general transcription factors and the three polymerases alone allow for only minimal transcription control, whereas cells really exert exquisitely fine control over transcription. Nevertheless, the task performed by the general transcription factors—getting the RNA polymerases together with their promoters—is not only vital, but also very complex because many polypeptides are required to do the job. In this chapter we will survey the general transcription factors that interact with all three RNA polymerases and their promoters.

11.1 Class II Factors

The general transcription factors combine with RNA polymerase to form a **preinitiation complex** that is competent to initiate transcription as soon as nucleotides are available. This tight binding involves formation of an open promoter complex in which the DNA at the transcription start site has melted to allow polymerase to read it. We will begin with the assembly of preinitiation complexes involving polymerase II. Even though these are by far the most complex, they are also the best studied. Once we see how the class II general transcription factors work, the class I and III mechanisms will be relatively easy to understand.

The Class II Preinitiation Complex

The class II preinitiation complex contains polymerase II and six general transcription factors named **TFIIA, TFIIB, TFIID, TFIIE, TFIIF,** and **TFIIH.** Many studies have shown that the class II general transcription factors and RNA polymerase II bind in a specific order to the growing preinitiation complex, at least in vitro. In particular, Danny Reinberg, as well as Phillip Sharp and their colleagues, performed DNA gel mobility shift and DNase footprinting experiments (Chapter 5) that defined most of the order of factor binding in building the class II preinitiation complex.

Figure 11.1 presents the results of a gel mobility shift experiment (Chapter 5) with a DNA template containing the adenovirus major late promoter (MLP) and transcription factors TFIIA and TFIID. Lane A shows that TFIIA by itself caused no DNA band mobility shift; all the labeled DNA remained in the free DNA band. Lane D contained DNA and TFIID only. A new band appeared, but it appeared not to be a specific factor-DNA complex because it formed on DNA lacking a TATA box. Lane A+D shows that another band (labeled DA) formed in the presence of both factors TFIIA and D. These results suggest that the role of TFIIA is to increase the affinity of TFIID for the TATA box. This is the apparent function of TFIIA in vivo, but other experiments discussed later in this section demonstrate that TFIID can bind independently without the aid of TFIIA, at least under certain conditions in vitro.

Figure 11.2a shows a gel mobility shift assay performed by Danny Reinberg and Jack Greenblatt and their colleagues using two additional transcription factors, TFIIB, and TFIIF, as well as RNA polymerase II. This experiment reveals the existence of four distinct complexes, which are labeled at the left of the figure. When the investigators added TFIID and A alone to DNA containing the adenovirus major late promoter, a **DA complex** formed. When they added TFIIB in addi-

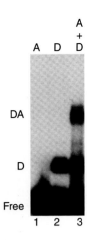

Figure 11.1 Formation of a complex involving TFIID, TFIIA, and a promoter-bearing DNA. Sharp and coworkers mixed a labeled DNA fragment containing the adenovirus major late promoter with TFIIA and TFIID separately and together, then electrophoresed the products. Lane A, with DNA and TFIIA alone, showed only free DNA, which migrated rapidly, almost to the bottom of the gel. Lane D, with DNA and TFIID alone, showed free DNA plus a DNA–protein complex (D). Lane A+D, with both transcription factors, showed a larger complex with both factors (DA). (*Source:* Buratowski et al., Five intermediate complexes in transcription initiation by RNA polymerase II. *Cell* 56 (24 Feb 1989) p. 551, f. 2a-b. Reprinted by permission of Elsevier Science.)

tion to D and A, a new, **DAB complex** formed. The central part of the figure shows what happened when they added various concentrations of RNA polymerase II and TFIIF to the DAB complex. In the lane labeled D+A+B+F, all four of those factors were present, but RNA polymerase was missing. No difference was detectable between the complex formed with these four factors and the DAB complex. Thus, TFIIF does not seem to bind independently to DAB. But when the investigators added increasing amounts of polymerase, two new complexes appeared. These seem to include both polymerase and TFIIF, so the top complex is called the **DABPolF complex.** The other new complex (**DBPolF**) migrates somewhat faster because it is missing TFIIA, as we will see. After they had added enough polymerase to give a maximum amount of DABPolF, the investigators started decreasing the quantity of TFIIF. This reduction in TFIIF concentration decreased the yield of DABPolF, until, with no TFIIF but plenty of polymerase, essentially no DABPolF (or DABPol) complexes formed. These data indicated that RNA polymerase and TFIIF are needed *together* to join the growing preinitiation complex.

Reinberg, Greenblatt, and colleagues assessed the order of addition of proteins by performing the same kind of mobility shift assays, but leaving out one or more factors at a time. In the most extreme example, the lane labeled –D shows what happened when the investigators left out TFIID. No complexes formed, even with all the other factors present. This dependence on TFIID reinforced the hypothesis that TFIID is the first factor to bind;

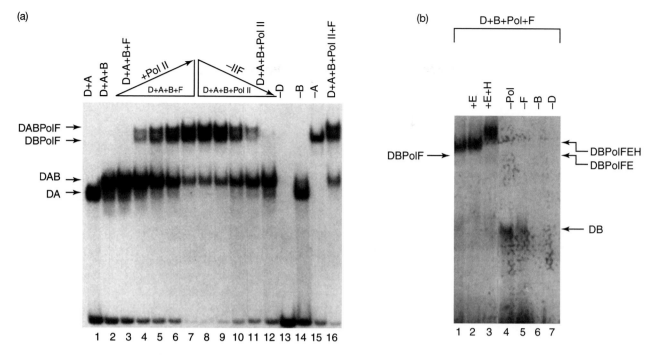

Figure 11.2   Building the preinitiation complex. (a) The DABPolF complex. Reinberg and colleagues performed gel mobility shift assays with TFIID, A, B, and F, and RNA polymerase II, along with labeled DNA containing the adenovirus major late promoter. Lane 1 shows the familiar DA complex, formed with TFIID and A. Lane 2 demonstrates that adding TFIIB caused a new complex, DAB, to form. Lane 3 contained TFIID, A, B, and F, but it looks identical to lane 2. Thus, TFIIF did not seem to bind in the absence of polymerase II. Lanes 4–7 show what happened when the investigators added more and more polymerase II in addition to the four transcription factors: More and more of the large complexes, DABPolF and DBPolF, appeared. Lanes 8–11 contained less and less TFIIF, and we see less and less of the large complexes. Finally, lane 12 shows that essentially no DABPolF or DBPolF complexes formed when TFIIF was absent. Thus, TFIIF appears to bring polymerase II to the complex. The lanes on the right show what happened when Reinberg and colleagues left out one factor at a time. In lane 13, without TFIID, no complexes formed at all. Lane 14 shows that the DA complex, but no others, formed in the absence of TFIIB. Lane 15 demonstrates that DBPolF could still develop without TFIIA. Finally, all the large complexes appeared in the presence of all the factors (lane 16). **(b)** The DBPolFEH complex. Reinberg and colleagues started with the DBPolF complex (lacking TFIIA, lane 1) assembled on a labeled DNA containing the adenovirus major late promoter. Next, they added TFIIE, then TFIIH, in turn, and performed gel mobility shift assays. With each new transcription factor, the complex grew larger and its mobility decreased further. The mobilities of both complexes are indicated at right. Lanes 4–7 show again the result of leaving out various factors, denoted at the top of each lane. At best, only the DB complex forms. At worst, in the absence of TFIID, no complex at all forms. (*Source:* (a) Flores et al., *PNAS* 88 (Nov 1991) p. 10001, f. 2a. (b) Cortes et al., *Mol. & Cell. Biol.* 12 (Jan 1992) p. 418, f. 6b. American Society for Microbiology.)

the binding of all the other factors depends on the presence of TFIID at the TATA box. The lane marked –B shows that TFIIB was needed to add polymerase and TFIIF. In the absence of TFIIB, only the DA complex could form. The lane labeled –A demonstrates that leaving out TFIIA made little difference. Thus, at least in vitro, TFIIA did not seem to be critical. Also, the fact that the band in this lane comigrated with the smaller of the two big complexes suggests that this smaller complex is DB Pol F. Finally, the last lane contained all the proteins and displayed the large complexes as well as some residual DAB complex.

Reinberg and his coworkers extended this study in 1992 with TFIIE and Figure 11.2b demonstrates that they could start with the DBPolF complex and then add TFIIE and TFIIH in turn, producing a larger complex, with reduced mobility, with each added factor. The final preinitiation complex formed in this experiment was DBPolFEH. The last four lanes in this experiment show again that

leaving out any of the early factors (polymerase II, TFIIF, TFIIB, or TFIID) prevents formation of the full preinitiation complex.

Having established the order of addition of most of the general transcription factors (and RNA polymerase) to the preinitiation complex in vitro, we can now turn to the question of where on the DNA each factor binds. Several groups, beginning with Sharp's, have approached this question using footprinting. Figure 11.3 shows the results of a footprinting study on the DA and DAB complexes. Reinberg and colleagues used two different reagents to cut the protein–DNA complexes: 1,10–phenanthroline (OP)-copper ion complex (lanes 1–4 in both panels), and DNase I (lanes 5–8 in both panels). Panel (a) depicts the data on the nontemplate strand, and panel (b) presents the results for the template strand. Panel (a), lanes 3 and 7 show that TFIID and A protect the TATA box. Lanes 3 and 7 in panel (b) show that the DA complex also protects the TATA box region on the template strand. Lanes

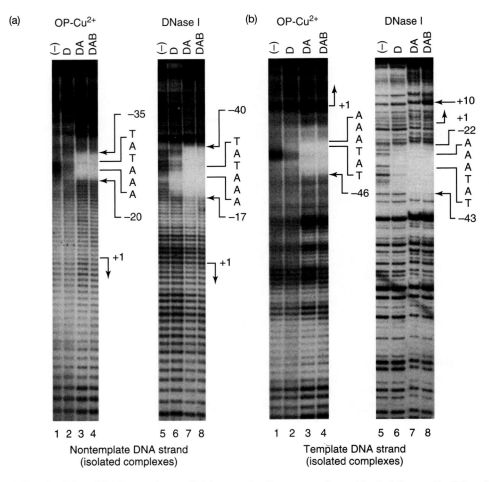

Figure 11.3   Footprinting the DA and DAB complexes. Reinberg and colleagues performed footprinting on the DA and DAB complexes with both DNase and another DNA strand breaker: a 1,10-phenanthroline-copper ion complex (OP-Cu²⁺). **(a)** Footprinting on the nontemplate strand. The DA and DAB complexes formed on the TATA box (TATAAA, indicated at right, top to bottom). **(b)** Footprinting on the template strand. Again, the protected region in both the DA and DAB complexes was centered on the TATA box (TATAAA, indicated at right, bottom to top). The arrow near the top at right denotes a site of enhanced DNA cleavage at position +10. (*Source:* Maldonado et al., *Mol. & Cell. Biol.* 10 (Dec 1990) p. 6344, f. 9. American Society for Microbiology.)

4 and 8 in panel (a) show no change in the nontemplate strand footprint after adding TFIIB to form the DAB complex. Essentially the same results were obtained with the template strand, but one subtle difference is apparent. As lane 8 shows, addition of TFIIB makes the DNA at position +10 even more sensitive to DNase. Thus, TFIIB does not seem to cover a significant expanse of DNA, but it does perturb the DNA structure enough to alter its susceptibility to DNase attack.

RNA polymerase II is a very big protein, so we would expect it to cover a large stretch of DNA and leave a big footprint. Figure 11.4 bears out this prediction. Whereas TFIID, A, and B protected the TATA box region (between positions –17 and –42) in the DAB complex, RNA polymerase II and TFIIF extended this protected region another 34 bases on the nontemplate strand, from position –17 to about position +17. Figure 11.5 summarizes what we have learned about the role of TFIIF in building the

DAB Pol F complex. Polymerase II (red) and TFIIF (green) bind cooperatively, perhaps by forming a binary complex that joins the preformed DAB complex.

SUMMARY Transcription factors bind to class II promoters, including the adenovirus major late promoter, in the following order in vitro: (1) TFIID, apparently with help from TFIIA, binds to the TATA box, forming the DA complex. (2) TFIIB binds next, causing minimal perturbation of the protein–DNA interaction. (3) TFIIF helps RNA polymerase bind to a region extending from at least position –34 to position +17. The remaining factors bind in this order: TFIIE and TFIIH, forming the DABPolFEH preinitiation complex. The participation of TFIIA seems to be optional in vitro.

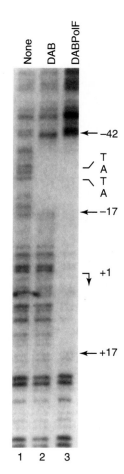

Figure 11.4   Footprinting the DABPolF complex. Reinberg and colleagues performed DNase footprinting with TFIID, A, and B (lane 2) and with TFIID, A, B, and F, and RNA polymerase II (lane 3). When RNA polymerase and TFIIF joined the complex, they caused a large extension of the footprint, to about position +17. This is consistent with the large size of RNA polymerase II. (*Source:* Flores et al., *PNAS* 88 (Nov 1991) p. 10001, f. 2b.)

Structure and Function of TFIID

TFIID is a complex protein containing a **TATA-box-binding protein** (**TBP**) and 8–10 **TBP-associated factors** (**TAFs,** or more specifically, **TAF$_{II}$s**). The subscript "II" is necessary because TBP also participates in transcription of class I and III genes and is associated with different TAFs (TAF$_I$s and TAF$_{III}$s) in class I and III preinitiation complexes, respectively. We will discuss the role of TBP and its TAFs in transcription from class I and III promoters later in this chapter. Let us first discuss the components of TFIID and their activities, beginning with TBP and concluding with the TAF$_{II}$s.

The TATA-Box-Binding Protein   TBP, the first polypeptide in the TFIID complex to be characterized, is highly evolutionarily conserved: Organisms as disparate as yeast, fruit flies, plants, and humans have TATA-box-binding

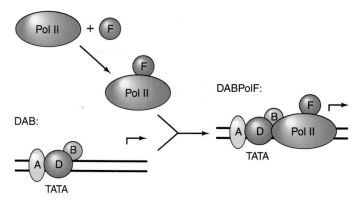

Figure 11.5   Model for formation of the DABPolF complex. TFIIF (green) binds to polymerase II (Pol II, red) and carries it to the DAB complex. The result is the DABPolF complex. This model conveys the conclusion that polymerase II extends the DAB footprint in the downstream direction, and therefore binds to DNA downstream of the binding site for TFIID, A, and B, which centers on the TATA box.

domains that are more than 80% identical in amino acid sequence. These domains encompass the carboxyl-terminal 180 amino acids of each protein and are very rich in basic amino acids. Another indication of evolutionary conservation is the fact that the yeast TBP functions well in a preinitiation complex in which all the other general transcription factors are mammalian.

Tjian's group demonstrated the importance of the carboxyl-terminal 180 amino acids of TBP when they showed that a truncated form of human TBP containing only the carboxyl-terminal 180 amino acids of a human recombinant TBP is enough to bind to the TATA box region of a promoter, just as the native TFIID would. Tjian and colleagues expressed the cloned human TBP gene in both bacteria (bhTBP) and vaccinia virus (vhTBP), then tested these proteins, or the truncated protein expressed in bacteria (bhTBP–180C), using a DNase I footprinting assay. Figure 11.6 shows the results. All three proteins, at approximately the same concentration, protected the same 20-bp region around the TATA box of the adenovirus major late promoter. By contrast, the whole proteins could not protect a mutant promoter in which the TATA box sequence, TATAAAA, was changed to TTATCAT.

How does the TBP in TFIID bind to the TATA box? The original assumption was that it acts like most other DNA-binding proteins (Chapter 9) and makes specific contacts with the base pairs in the major groove of the TATA box DNA. However, this assumption proved to be wrong. Two independent research groups headed by Diane Hawley and Robert Roeder showed convincingly that the TBP in TFIID binds to the minor groove of the TATA box.

Hawley and Roeder and their colleagues performed methylation interference assays with DNA containing the adenovirus major late promoter (MLP) and obtained

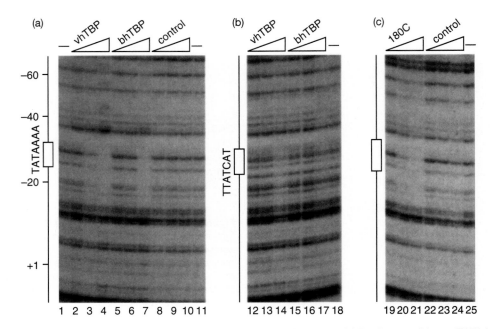

Figure 11.6   Binding of recombinant human TBPs to wild-type and mutant TATA boxes. (a) Binding to wild-type TATA box. Tjian and colleagues added increasing quantities of human TBP produced by cloning in vaccinia virus (vhTBP) or in bacteria (bhTBP) to an adenovirus major late promoter, and then they measured binding efficiency by footprinting. The increasing quantities of protein added are represented by the wedges at top. Both recombinant TBPs created identical footprints in the TATA box, whose position is shown at left. The control contained an equivalent amount of *E. coli* extract lacking TBP. (b) Binding to mutant TATA box. Tjian and colleagues performed the same footprinting experiment, but using a DNA with a mutant TATA box (TTATCAT instead of TATAAAA). Little or no binding occurred. (c) Binding of cloned C-terminal 180 amino acids of human TBP to wild-type TATA box. Tjian and colleagues performed the same experiment with the cloned protein fragment containing the conserved 180 amino acids at the C-terminus of the human TBP (180C). This achieved the same effect as the whole TBP. (*Source:* Peterson et al., Functional domains and upstream activation properties of cloned human TATA binding protein. *Science* 248 (29 June 1990) p. 1628, f. 3. © AAAS.)

essentially the same results. Methylations on A's within the TATA box were the ones that interfered the most with TFIID binding. Figure 11.7 shows the results of Roeder's experiment. These findings were intriguing because adenine methylation occurs at the N-3 position, which is in the DNA minor groove.

Could it be that TFIID binds in the minor, rather than the major, groove? To find out, Roeder and coworkers made a synthetic version of the adenovirus MLP with deoxyuridine substituted for thymidine at selected positions in the TATA box region. Deoxyuridine is missing the 5′-methyl group of thymidine, which occupies the DNA major groove. If TFIID binds in the minor groove, then the substitution of dU for dT should not inhibit binding. In fact, a DNA mobility shift experiment depicted in Figure 11.8 shows that it did not. Substitution of dU for dT in position −30 of the transcribed strand actually enhanced binding to TFIID.

Even more telling was an experiment by Barry Starr and Hawley in which *all* the bases of the TATA box were substituted such that the major groove was changed, but the minor groove was not. This is possible because the hypoxanthine base in inosine (I) looks just like adenine (A) in the minor groove, but much different in the major groove (Figure 11.9a). Similarly, cytosine looks like

thymine in the minor, but not the major, groove. Thus, Starr and Hawley made an adenovirus major late TATA box with all C's instead of T's, and all I's instead of A's (CICIIII instead of TATAAAA, Figure 11.9b). Then they measured TFIID binding to this CICI box and to the standard TATA box by a DNA mobility shift assay. As Figure 11.9c shows, the CICI box worked just as well as the TATA box, but a nonspecific DNA did not bind TFIID at all. Therefore, changing the bases in the TATA box did not affect TFIID binding as long as the minor groove was unaltered. This is strong evidence for binding of TFIID to the minor groove of the TATA box, and for no significant interaction in the major groove.

How does TFIID associate with the TATA box minor groove? Nam-Hai Chua, Roeder, and Stephen Burley and colleagues began to answer this question when they solved the crystal structure of the TBP of a plant, *Arabidopsis thalliana*. The structure they obtained was shaped like a saddle, complete with two "stirrups," which naturally suggested that TBP sits on DNA the way a saddle sits on a horse. The TBP structure has rough two-fold symmetry corresponding to the two sides of the saddle with their stirrups. Then, in 1993, Paul Sigler and colleagues and Stephen Burley and colleagues independently solved the crystal structure of TBP bound to a small

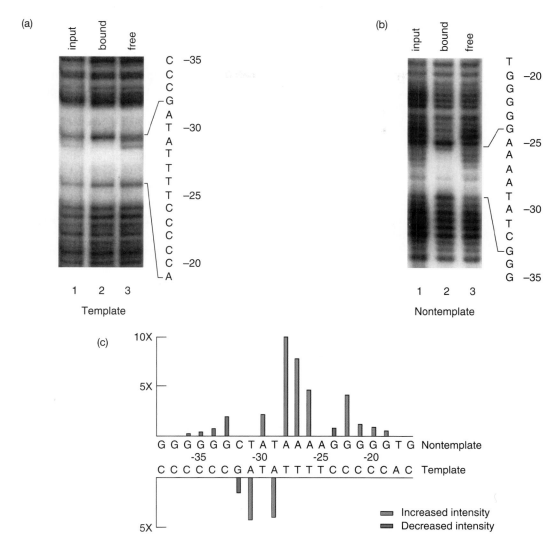

Figure 11.7 Methylation interference at the TATA box. Roeder and coworkers end-labeled DNA containing the adenovirus major late promoter on either nontemplate strand (**a**) or the template (**b**), then methylated the DNA under conditions in which A's were preferentially methylated. Then they added TFIID and filtered the protein–DNA complexes. DNAs that could still bind TFIID were retained, but free DNA flowed through. Finally, they cleaved the filter-bound and free DNAs at methylated sites with NaOH and subjected the fragments to gel electrophoresis. The autoradiographs in panels (**a**) and (**b**) show that the bound DNA did not cleave in the TATA box, so it was not methylated there. On the other hand, the free DNA was cleaved in the TATA box, showing that it had been methylated there. That is why it no longer bound TFIID. (**c**) Summary of methylated bases in the free DNA fractions. The lengths of the bars show the intensities of the bands in the "free" lanes in parts (**a**) and (**b**), which indicate the degree of methylation. The red bars show decreased band intensity and the green bars show increased band intensity in the "free" lanes compared with the "bound" lanes. Most of the methylation occurred on A's, rather than G's. These methyl groups are in the minor groove; because this methylated DNA was incapable of binding TFIID, these results suggest that TFIID binds in the minor groove. In reading the sequences in this and the next figure, remember that the nontemplate strand contains the TATA sequence. (*Source:* Lee et al., Interaction of TFIID in the minor groove of the TATA element. *Cell* 67 (20 Dec 1991) p. 1242, f. 1 b-c. Reprinted by permission of Elsevier Science.)

synthetic piece of double-stranded DNA that contained a TATA box. That allowed them to see how TBP really interacts with the DNA, and it was not nearly as passive as a saddle sitting on a horse.

Figure 11.10 shows this structure. The curved undersurface of the saddle, instead of fitting neatly over the DNA, is roughly aligned with the long axis of the DNA, so its curvature forces the DNA to bend through an angle of 80 degrees. This bending is accomplished by a gross distortion in the DNA helix in which the minor groove is forced open. This opening is most pronounced at the first and last steps of the TATA box (between base pairs 1 and 2 and between base pairs 7 and 8). At each of those sites, two phenylalanine side chains from the stirrups of TBP intercalate, or insert, between base pairs, causing the DNA to kink. This distortion may help explain why the TATA sequence is so well conserved: The T–A step in a DNA double helix is relatively easy to distort, compared with any other dinucleotide step. This argument assumes that distortion of the TATA box is important to transcription

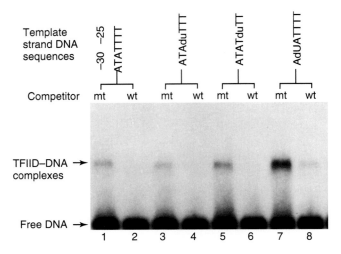

Figure 11.8 Effect of substituting dU for dT on TFIID binding to the TATA box. Roeder and coworkers bound TFIID to labeled DNA containing TATA boxes with the sequences given at top. They did the binding in the presence of excess unlabeled competitor DNA containing either wild-type or mutant TATA boxes (mutant sequence: TAGAGAA). To assay for TFIID–TATA box binding, they electrophoresed the protein–DNA complexes under nondenaturing conditions, which separate free DNA from protein-bound DNA. In all cases, the wild-type TATA box was able to compete, so only free DNA was observed (even-numbered lanes). However, in all cases, the mutant TATA box was unable to compete, even when the labeled TATA box contained a dU instead of a dT. In fact, lane 7 shows that substitution of a dU for a dT in position 2 of the template strand of the TATA box (sequence: AdUATTTT) actually seemed to enhance TFIID–TATA box binding relative to the unsubstitued TATA box (lane 1). Because dU and dT differ in the major groove, but not the minor groove, and the substitution of dU for dT did not inhibit binding, this behavior suggests that TFIID binds in the minor groove. (*Source:* Lee et al., *Cell* 67 (20 Dec 1991) p. 1244, f. 3. Reprinted by permission of Elsevier Science.)

initiation. Indeed, it is easy to imagine that peeling open the DNA minor groove aids the local DNA melting that is part of forming an open promoter complex.

> **SUMMARY** TFIID contains a 38-kD TATA box-binding protein (TBP) plus several other polypeptides known as TBP-associated factors (TAF$_{II}$s). The C-terminal 180 amino acid fragment of the human TBP is the TATA-box-binding domain. The interaction between a TBP and a TATA box is an unusual one that takes place in the DNA minor groove. The saddle-shaped TBP lines up with the DNA, and the underside of the saddle forces open the minor groove and bends the TATA box into an 80-degree curve.

The Versatility of TBP Molecular biology is full of wonderful surprises, and one of these is the versatility of the TATA-binding protein (TBP). This factor functions not only with polymerase II promoters that have a TATA

box, but with TATA-less polymerase II promoters. Astonishingly, it also functions with TATA-less polymerase III promoters, *and* with TATA-less polymerase I promoters. In other words, TBP appears to be a universal eukaryotic transcription factor that operates at all promoters, regardless of their TATA content, and even regardless of the polymerase that recognizes them.

One indication of the widespread utility of TBP came from work by Ronald Reeder and Steven Hahn and colleagues on mutant yeasts with temperature-sensitive TBPs. We would have predicted that elevated temperature would block transcription by polymerase II in these mutants, but it also impaired transcription by polymerases I and III.

Figure 11.11 shows the evidence for this assertion. The investigators prepared cell-free extracts from wild-type and two different temperature-sensitive mutants, with lesions in TBP, as shown in Figure 11.11a. They made extracts from cells grown at 24° C and shocked for 1 h at 37° C, and from cells kept at the lower temperature. Then they added DNAs containing promoters recognized by all three polymerases and assayed transcription by S1 analysis. Figure 11.11b–e depicts the results. The heat shock had no effect on the wild-type extract, as expected (lanes 1 and 2). By contrast, the I143 → N mutant extract could barely support transcription by any of the three polymerases, whether it was heat shocked or not (lanes 3 and 4). Clearly, the mutation in TBP was affecting not only polymerase II transcription, but transcription by the other two polymerases as well. The other mutant, P65 → S, shows an interesting difference between the behavior of polymerase I and the other two polymerases. Whereas this mutant extract could barely support transcription of polymerase II and III genes, whether it had been heat shocked or not, it allowed wild-type levels of transcription by polymerase I if it was not heat-shocked, but heating reduced transcription by polymerase I by about twofold. Finally, wild-type TBP could restore transcription by all three polymerases in mutant extracts.

Not only is TBP universally involved in eukaryotic transcription, it also seems to be involved in transcription in a whole different kingdom of organisms: the **archaea.** Archaea (formerly known as archaebacteria) are single-celled organisms that lack nuclei and usually live in extreme environments, such as hot springs or boiling hot deep ocean vents. They are as different from bacteria as they are from eukaryotes, and in several ways they resemble eukaryotes more than they do prokaryotes. In 1994, Stephen Jackson and colleagues reported that one of the archaea, *Pyrococcus woesei,* produces a protein that is structurally and functionally similar to eukaryotic TBP. This protein is presumably involved in recognizing the TATA boxes that frequently map to the 5'-flanking regions of archaeal genes. Moreover, a TFIIB-like protein has also been found in archaea. Thus, the transcription

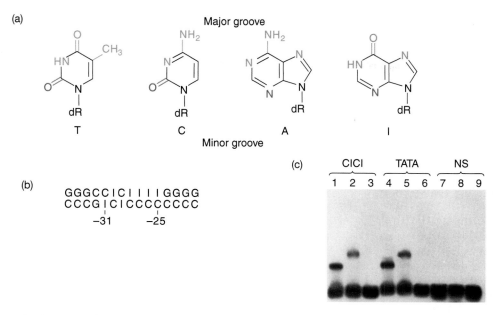

Figure 11.9   Effect of substituting C for T and I for A on TFIID binding to the TATA box. (a) Appearance of nucleosides as viewed from the major and minor grooves. Notice that thymine and cytidine look identical from the minor groove (green, below), but quite different from the major groove (red, above). Similarly, adenosine and inosine look the same from the minor groove, but very different from the major groove. **(b)** Sequence of the adenovirus major late promoter (MLP) TATA box with C's substituted for T's and I's substituted for A's, yielding a CICI box. **(c)** Binding TBP to the CICI box. Starr and Hawley performed gel mobility shift assays using DNA fragments containing the MLP with a CICI box (lanes 1–3) or the normal TATA box (lanes 4–6), or a nonspecific DNA (NS) with no promoter elements (lanes 7–9). The first lane in each set (1, 4, and 7) contained yeast TBP; the second lane in each set (2, 5, and 8) contained human TBP; and the third lane in each set contained just buffer. The yeast and human TBPs gave rise to slightly different size protein–DNA complexes, but substituting a CICI box for the TATA box had little effect on the yield of the complexes. Thus, TBP binding to the TATA box was not significantly diminished by the substitutions. (*Source:* (*b-c*) Starr & Hawley, TFIID binds in the minor groove of the TATA box. *Cell* 67 (20 Dec 1991) p. 1234, f. 2b. Reprinted by permission of Elsevier Science.)

apparatus of the archaea bears at least some resemblance to that in eukaryotes, and suggests that the archaea and the eukaryotes diverged *after* their common ancestor diverged from the bacteria. This evolutionary scheme is also supported by the sequence of archaeal rRNA genes, which bear more resemblance to eukaryotic than to bacterial sequences.

> **SUMMARY** Genetic studies have demonstrated that TBP mutant cell extracts are deficient, not only in transcription of class II genes, but also in transcription of class I and III genes. Thus, TBP is a universal transcription factor required by all three classes of genes. It is also apparently required in transcription of at least some archaeal genes.

The TBP-Associated Factors   Many researchers have contributed to our knowledge of the TBP-associated factors (TAFIIs) in TFIIDs from several organisms. To identify TAFIIs from *Drosophila* cells, Tjian and his colleagues used an antibody specific for TBP to immunoprecipitate TFIID from a crude TFIID preparation. Then they treated the immunoprecipitate with 2.5 M urea to strip the TAFs off of the TFIID–antibody precipitate and displayed the

TAFs by SDS-PAGE. Several such experiments led to the identification of eight TAFs. The genes for all eight polypeptides have been cloned, and Figure 11.12 depicts the pattern of polypeptides in a TFIID assembled in the laboratory from the protein products of these cloned genes. Similar investigations have been performed on the TAFIIs from human and yeast cells. The sizes of the TAFIIs from the three different species do not exactly match, but the similarities in their predicted amino acid sequences allow us to see the homologies, which are represented in Figure 11.13.

Investigators have discovered several functions of the TAFIIs, but the two that have received the most attention are interaction with the promoter and interaction with gene-specific transcription factors. Let us consider the evidence for each of these functions and, where possible, the specific TAFIIs involved in each.

We have already seen the importance of the TBP in binding to the TATA box. But footprinting studies have indicated that the TAFIIs attached to TBP extend the binding of TFIID well beyond the TATA box in some promoters. In particular, Tjian and coworkers showed in 1994 that TBP protected the 20 bp or so around the TATA box in some promoters, but that TFIID protected a region extending to position +35, well beyond the transcription start site. This suggested that the TAFIIs in TFIID were

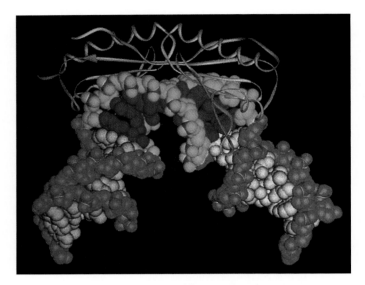

Figure 11.10   Structure of the TBP–TATA box complex. This diagram, based on Sigler and colleagues' crystal structure of the TBP–TATA box complex, shows the backbone of the TBP in olive at top. The long axis of the "saddle" is in the plane of the page. The DNA below the protein is in multiple colors. The backbones in the region that interacts with the protein are in orange, with the base pairs in red. Notice how the protein has opened up the narrow groove and almost straightened the helical twist in that region. One stirrup of the TBP is seen as an olive loop at right center, inserting into the minor groove. The other stirrup performs the same function, but it is out of view in back of the DNA. The two ends of the DNA, which do not interact with the TBP, are in blue and gray: blue for the backbones, and gray for the base pairs. The left end of the DNA sticks about 25 degrees out of the plane of the page, and the right end points inward by the same angle. The overall bend of about 80 degrees in the DNA, caused by TBP, is also apparent. (*Source:* Klug, "Opening the gateway." *Nature* 365 (7 Oct 1993) p. 487, f. 2. © Macmillan Magazines Ltd.)

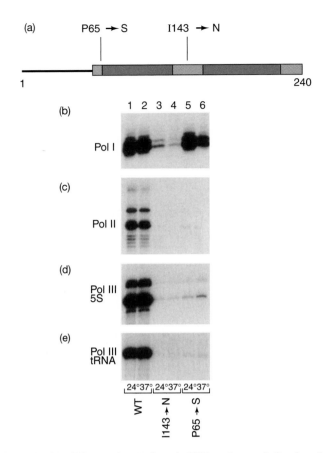

Figure 11.11   Effects of mutations in TBP on transcription by all three RNA polymerases. (a) Locations of the mutations. The boxed region indicates the conserved C-terminal domain of the TBP; red areas denote two repeated elements involved in DNA binding. The two mutations are: P65 → S, in which proline 65 is changed to a serine; and I143 → N, in which isoleucine 143 is changed to asparagine. **(b–e)** Effects of the mutations. Reeder and Hahn made extracts from wild-type or mutant yeasts, as indicated at bottom, and either heat-shocked them at 37 degrees C or left them at 24 degrees C, again as indicated at bottom. Then they tested these extracts by S1 analysis for ability to start transcription at promoters recognized by all three nuclear RNA polymerases: **(b)** the rRNA promoter (polymerase I); **(c)** the CYC1 (polymerase II) promoter; **(d)** the 5S rRNA promoter (polymerase III); and **(e)** the tRNA promoter (also polymerase III). The I143 → N extract was deficient in transcribing from all four promoters even when not heat-shocked. The P65 → S extract was deficient in transcribing from polymerase II and III promoters, but could recognize the polymerase I promoter, even after heat shock. (*Source:* (a) Reprinted from Reeder and Hahn, *Cell* 69:698–699, 1992. Copyright 1992, with permission from Elsevier Science.)

contacting the initiator and downstream elements in these promoters.

To investigate this phenomenon in more detail, Tjian's group tested the abilities of TBP and TFIID to transcribe DNAs bearing two different classes of promoters in vitro. The first class (the adenovirus E1B and E4 promoters) contained a TATA box, but no initiator or downstream element. The second class (the adenovirus major late [AdML] promoter and the *Drosophila* heat shock protein [Hsp70] promoter) contained a TATA box, an initiator, and a downstream element. Figure 11.14 depicts the structures of these promoters, as well as the results of the in vitro transcription experiments. We can see that TBP and TFIID sponsored transcription equally well from the promoters that contained only the TATA box (compare lanes 1 and 2 and lanes 3 and 4). But TFIID had a decided advantage in sponsoring transcription from the promoters that also had an initiatior and downstream element (compare lanes 5 and 6 and lanes 7 and 8). Thus, TAF$_{IIS}$ apparently help TBP facilitate transcription from promoters with initiators and downstream elements.

Which TAF$_{IIS}$ are responsible for recognizing the initiator and downstream element? To find out, Tjian and colleagues performed a photo-cross-linking experiment with *Drosophila* TFIID and a radioactively labeled DNA fragment containing the Hsp70 promoter. They incorporated bromodeoxyuridine (BrdU) into the promoter-containing DNA, then allowed TFIID to bind to the promoter, then irradiated the complexes with UV light to cross-link the protein to the BrdU in the DNA. After washing away unbound protein, the investigators digested the DNA with nuclease to release the proteins, then sub-

(a)

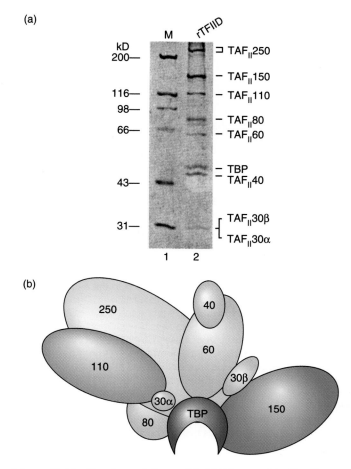

Figure 11.12   Structure of *Drosophila* TFIID assembled in vitro from the products of cloned genes. (a) Gel electrophoresis of the polypeptides. Lane 1 contained molecular mass markers, and lane 2 contained the reconstituted TFIID (rTFIID), with the eight major TAF$_{II}$s. **(b)** Hypothetical structure of the reconstituted *Drosophila* TFIID. (*Source:* Chen et al., Assembly of recombinant TFIID reveals differential co-activator requirements for distinct transcriptional activators. *Cell* 79 (7 Oct 1994) p. 100, f. 7a. Reprinted with permission of Elsevier Science.)

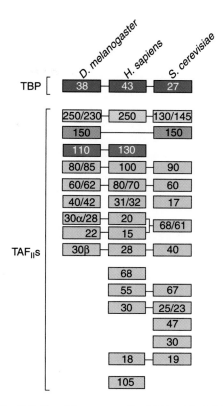

Figure 11.13   Relationships among the TAFs of fruit flies (*D. melanogaster*), humans (*H. sapiens*), and yeast (*S. cerevisiae*). The horizontal lines link homologous proteins. (*Source:* Reprinted from Tansey and Herr, *Cell* 88:729, 1997. Copyright 1997, with permission from Elsevier Science.)

jected the labeled proteins to SDS-PAGE. Figure 11.15, lane 1, shows that two TAF$_{II}$s (TAF$_{II}$250 and TAF$_{II}$150) bound to the Hsp70 promoter and thereby became labeled. When TFIID was omitted (lane 2), no proteins became labeled. Following up on these findings, Tjian and coworkers reconstituted a ternary complex containing only TBP, TAF$_{II}$250, and TAF$_{II}$150 and tested it in the same photo-cross-linking assay. Lane 3 shows that this experiment also yielded labeled TAF$_{II}$250 and TAF$_{II}$150, and lane 4 shows that TBP did not become labeled when it was bound to the DNA by itself. We know that TBP binds to this TATA-box-containing DNA, but it does not become cross-linked to BrdU and therefore does not become labeled. Why not? Probably because this kind of photo-cross-linking works well only with proteins that bind in the major groove, and TBP binds in the minor groove of DNA.

To double-check the binding specificity of the ternary complex (TBP–TAF$_{II}$250–TAF$_{II}$150), Tjian and colleagues performed a DNase footprinting experiment with TBP or the ternary complex. Figure 11.16 shows that TBP caused a footprint only in the TATA box, whereas the ternary complex caused an additional footprint in the initiator and downstream element. This reinforced the hypothesis that the two TAF$_{II}$s bind to the initiator and downstream element.

Further experiments with binary complexes (TBP–TAF$_{II}$250 or TBP–TAF$_{II}$150) showed that these complexes were no better than TBP alone in recognizing initiators and downstream elements. Thus, both TAF$_{II}$s seem to cooperate in enhancing binding to these promoter elements. Furthermore, the ternary complex (TBP–TAF$_{II}$250–TAF$_{II}$150) is almost as effective as TFIID in recognizing a synthetic promoter composed of the AdML TATA box and the TdT initiator. By contrast, neither binary complex functions any better than TBP in recognizing this promoter. These findings support the hypothesis that TAF$_{II}$250 and TAF$_{II}$150 cooperate in binding to the initiator alone, as well as to the initiator plus a downstream element.

The TBP part of TFIID is of course important in recognizing the majority of the well-studied class II promoters, which contain TATA boxes (Figure 11.17a). But what

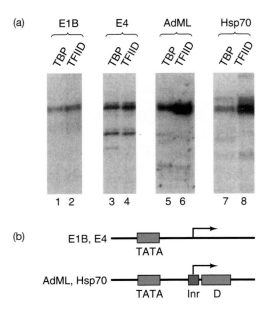

(b)

E1B, E4 ——————■———→————————
                    TATA

AdML, Hsp70 ——————■———■—■———————
                   TATA   Inr D

Figure 11.14   Activities of TBP and TFIID on four different promoters. (**a**) Experimental results. Tjian and colleagues tested a reconstituted *Drosophila* transcription system containing either TBP or TFIID (indicated at top) on templates bearing four different promoters (also as indicated at top). The promoters were of two types diagrammed in panel (**b**). The first type, represented by the adenovirus E1B and E4 promoters, contained a TATA box (red). The second type, represented by the adenovirus major late promoter (AdML) and the *Drosophila* Hsp70 promoter, contained a TATA box plus an initiator (Inr, green) and a downstream element (D, blue). After transcription in vitro, Tjian and coworkers assayed the RNA products by primer extension (top). The autoradiographs show that TBP and TFIID fostered transcription equally well from the first type of promoter (TATA box only), but that TFIID worked much better than TBP in supporting transcription from the second type of promoter (TATA box plus initiator plus downstream element). (*Source:* Verrijzer et al., Promoter recognition by TAFs. *Cell* 81 (30 Jun 1995) p. 1116, f. 1. Reprinted with permission of Elsevier Science.)

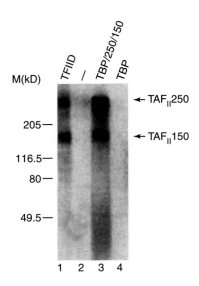

Figure 11.15   Identifying the TAF$_{II}$s that bind to the *hsp70* promoter. Tjian and colleagues photo-crosslinked TFIID to a ^{32}P-labeled template containing the *hsp70* promoter. This template had also been substituted with the photo-sensitive nucleoside bromodeoxyuridine (BrdU). Next, these investigators irradiated the TFIID–DNA complex with UV light to form covalent bonds between the DNA and any proteins in close contact with the major groove of the DNA. Next, they digested the DNA with nuclease and subjected the proteins to SDS-PAGE. Lane 1 of the autoradiograph shows the results when TFIID was the input protein. TAF$_{II}$250 and TAF$_{II}$150 became labeled, implying that these two proteins had been in close contact with the labeled DNA's major groove. Lane 2 is a control with no TFIID. Lane 3 shows the results when a ternary complex containing TBP. TAF$_{II}$250. and TAF$_{II}$150 was the input protein. Again, the two TAF$_{II}$s became labeled, suggesting that they bound to the DNA. Lane 4 shows the results when TBP was the input protein. It did not become labeled, which was expected because it does not bind in the DNA major groove. (*Source:* Verrijzer et al., *Cell* 81 (30 June 1995) p. 1117, f. 2a. Reprinted with permission of Elsevier Science.)

about promoters that lack a TATA box? Even though these promoters cannot bind TBP directly, most still depend on this transcription factor for activity. The key to this apparent paradox is the fact that these TATA-less promoters contain other elements that ensure the binding of TBP. These other elements can be initiators and downstream elements, to which TAF$_{II}$250 and TAF$_{II}$150 can bind and thereby secure the whole TFIID to the promoter (Figure 11.17b). Or they can be upstream elements that bind gene-specific transcription factors, which in turn interact with one or more TAF$_{II}$s to anchor TFIID to the promoter. For example, the gene-specific transcription factor Sp1 binds to upstream elements (GC boxes) and also interacts with at least one TAF$_{II}$(TAF$_{II}$110 in *Drosophila*). This bridging activity apparently helps TFIID bind to the promoter (Figure 11.17c).

The second major activity of TAF$_{II}$s is to participate in the transcription stimulation provided by gene-specific transcription factors, some of which we will study in Chapter 12. Tjian and colleagues demonstrated in 1990

that TFIID is sufficient to participate in such stimulation by the factor Sp1, but TBP is not. These workers mixed general transcription factors, polymerase II, and either TFIID or TBP, with a DNA template containing a TATA box and six GC boxes. Then they measured transcription with a run-off assay. Figure 11.18 shows that Sp1 could stimulate transcription in the presence of TFIID, but not with just TBP. These results suggest that some factors in TFIID are necessary for interaction with upstream-acting factors such as Sp1 and that these factors are missing from TBP. By definition, these factors are TAF$_{II}$s, and they are sometimes called **coactivators.**

We have seen that mixing TBP with subsets of TAF$_{II}$s can produce a complex with the ability to participate in transcription from certain promoters. For example, the TBP–TAF$_{II}$250–TAF$_{II}$150 complex functioned almost as well as the whole TFIID in recognizing a promoter composed of a TATA box and an initiator. Tjian and colleagues used a similar technique to discover which TAF$_{II}$s are involved in activation by Sp1. They built various

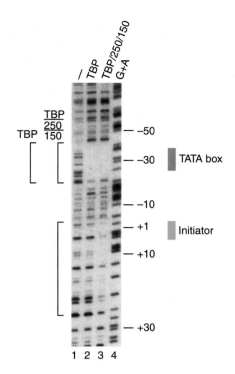

Figure 11.16   DNase I footprinting the *hsp70* promoter with TBP and the ternary complex (TBP, TAF_{II}250, and TAF_{II}150). Lane 1, no protein; lane 2, TBP; lane 3, ternary complex. In both lanes 2 and 3, TFIIA was also added to stabilize the DNA–protein complexes, but separate experiments indicated that it did not affect the extent of the footprints. Lane 4 is a Maxam–Gilbert G+A sequencing lane used as a marker. The extents of the footprints caused by TBP and the ternary complex are indicated by brackets at left. The locations of the TATA box and initiator are indicated by boxes at right. (*Source:* Verrijzer et al., *Cell* 81 (30 June 1995) p. 1117, f. 2c. Reprinted with permission of Elsevier Science.)

complexes involving TBP, TAF_{II}250, TAF_{II}150, and TAF_{II}110, then tested them for the ability to support Sp1-stimulated transcription in TFIID-depleted extracts from either *Drosophila* or human cells. Figure 11.19 shows the results of primer extension assays on these reconstituted extracts, using a template containing three GC boxes upstream of a TATA box. Panel (a) shows that activation by Sp1 in the *Drosophila* extract occurred only when TAF_{II}110 was present. Panel (b) shows the same behavior, but even more pronounced, in the human extract. Thus, TBP and TAF_{II}250 plus TAF_{II}150 were sufficient for basal transcription from this template, but could not support activation by Sp1. Adding TAF_{II}110 in addition to the other two factors and TBP allowed Sp1 to activate transcription.

Tjian and colleages also showed that Sp1 binds directly to TAF_{II}110, but not to TAF_{II}250 or TAF_{II}150. They built an affinity column containing GC boxes and Sp1 and tested it for the ability to retain the three TAF_{II}s. As predicted, only TAF_{II}110 was retained.

Using the same strategy, Tjian and colleagues demonstrated that another gene-specific transcription factor, NTF-1, binds to TAF_{II}150 and requires either TAF_{II}250

and TAF_{II}150 or TAF_{II}250 and TAF_{II}60 to activate transcription in vitro. Thus, different transcription activators work with different combinations of TAF_{II}s to enhance transcription, and all of them seem to have TAF_{II}250 in common. This suggests that TAF_{II}250 serves as an assembly factor around which other TAFs can aggregate. These findings are compatible with the model in Figure 11.20: Each transcription activator interacts with a particular subset of TAF_{II}s, so the holo-TFIID can interact with several activators at once, magnifying their effect and producing strong enhancement of transcription.

In addition to their abilities to interact with promoter elements and gene-specific transcription factors, TAF_{II}s can have enzymatic activities. The best studied of these is TAF_{II}250, which has two known enzymatic activities. It is a histone acetyltransferase (HAT), which attaches acetyl groups to lysine residues of histones. Such acetylation is generally a transcription-activating event. We will study this process in greater detail in Chapter 13. TAF_{II}250 is also a protein kinase that can phosphorylate itself and TFIIF (and TFIIA and TFIIE, though to a lesser extent). These phosphorylation events may modulate the efficiency of assembly of the preinitiation complex.

SUMMARY TFIID contains at least eight TAF_{II}s, in addition to TBP. Most of these TAF_{II}s are evolutionarily conserved in the eukaryotes. The TAF_{II}s serve several functions, but two obvious ones are interacting with core promoter elements and interacting with gene-specific transcription factors. TAF_{II}250 and TAF_{II}150 help TFIID bind to the initiator and downstream elements of promoters and therefore can enable TBP to bind to TATA-less promoters that contain such elements. TAF_{II}250 and TAF_{II}110 help TFIID interact with Sp1 that is bound to GC boxes upstream of the transcription start site. These TAF_{II}s therefore ensure that TBP can bind to TATA-less promoters that have GC boxes. Different combinations of TAF_{II}s are apparently required to respond to various transcription activators, at least in higher eukaryotes. TAF_{II}250 also has two enzymatic activities. It is a histone acetyltransferase and a protein kinase.

Exceptions to the Universality of TAFs and TBP
Genetic studies in yeast call into question the generality of the model in Figure 11.20. Michael Green and Kevin Struhl and their colleagues independently discovered that mutations in yeast TAF genes were lethal, but transcription activation was not affected, at least not in the first genes studied. For example, Green and colleagues made temperature-sensitive mutations in the gene encoding yTAF_{II}145, the yeast homolog of *Drosophila* and human

(a) (b) (c)

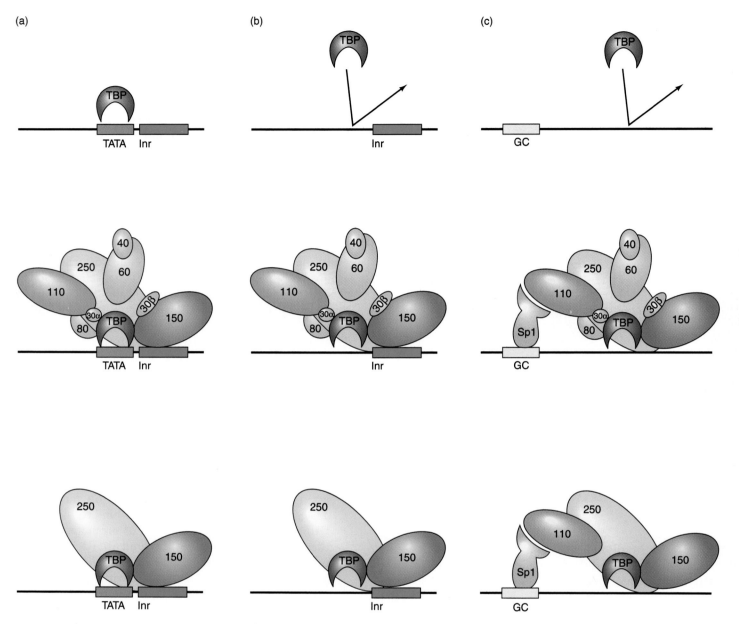

Figure 11.17   Model for the interaction between TBP and TATA-containing or TATA-less promoters. (a) TATA-containing promoter. TBP can bind by itself to the TATA box of this promoter (top). It can also bind in the company of all the TAF$_{II}$s in TFIID (middle). And it can bind with a subset of TAF$_{II}$s (bottom). **(b)** TATA-less promoter with initiator element. TBP cannot bind by itself to this promoter, which contains an initiator but no TATA box (top). The whole TFIID is competent to bind to the TATA-less promoter through interactions between TAF$_{II}$250 (yellow) and TAF$_{II}$150 (brown, middle). TAF$_{II}$250 and TAF$_{II}$150 are sufficient to tether TBP to the initiator (bottom). **(c)** TATA-less promoter with GC boxes. TBP cannot bind to this promoter by itself (top). The whole TFIID can bind to this promoter through interactions with Sp1 bound at the GC boxes (middle). TAF$_{II}$250, TAF$_{II}$150, and TAF$_{II}$110 are sufficient to anchor TBP to the Sp1 bound to the GC boxes. (*Source:* Reprinted from Goodrich et al., *Cell* 84:826, 1996. Copyright 1996, with permission from Elsevier Science.)

TAF$_{II}$250. At the nonpermissive temperature, they found that there was a rapid decrease in the concentration of yTAF$_{II}$145, and at least two other yTAF$_{II}$s. The loss of yTAF$_{II}$145 apparently disrupted the TFIID enough to cause the degradation of other yTAF$_{II}$s. However, in spite of these losses of yTAF$_{II}$s, the in vivo transcription rates of five different yeast genes activated by a variety of gene-specific transcription factors were unaffected at the nonpermissive temperature. These workers obtained the same

results with another mutant in which the yTAF$_{II}$30 gene had been deleted. By contrast, when the genes encoding TBP or an RNA polymerase subunit were mutated, all transcription quickly ceased.

Green, Richard Young, and colleagues followed up these initial studies with a genome-wide analysis of the effects of mutations in two TAF$_{II}$ genes, as well as several other yeast genes. They made temperature-sensitive mutations in TAF$_{II}$145 (the homolog of mammalian TAF$_{II}$250)

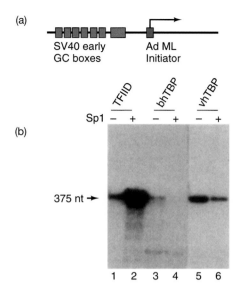

Figure 11.18 Failure of TBP alone to respond to Sp1. (a) Structure of the test promoter. This is a composite Sp1-responsive promoter containing six GC boxes (red) from the SV40 early promoter and the TATA box (blue) and transcription start site (initiator, green) from the adenovirus major late promoter. Accurate initiation from this promoter in the run-off assay described in panel **(b)** should produce a 375-nt transcript. **(b)** In vitro transcription assay. Tjian and colleagues mixed TFIID, or bhTBP, or vhTBP, as shown at top, with TFIIA, TFIIB, TFIIE, TFIIF, and RNA polymerase II, then performed a run-off transcription assay with [α-^{32}P]UTP. Lanes 1 and 2 show that natural TFIID supported a high level of transcription from this promoter, and this transcription was significantly enhanced by the transcription factor Sp1. Lanes 3–6 demonstrate that any transcription due to recombinant human TBP was not stimulated by Sp1 in the absence of TAF$_{II}$s. (*Source:* (a) Reprinted from Goodrich et al., *Cell* 84:1629, 1996. Copyright 1996, with permission from Elsevier Science. (b) Peterson et al., Functional domains and upstream activation properties of cloned human TATA binding protein. *Science* 248 (29 June 1990) p. 1629, f. 5b. © AAAS.)

and in TAF$_{II}$17 (the homolog of mammalian TAF$_{II}$31/32). Then they used high-density oligonucleotide arrays (such as those described in Chapter 24) to determine the extent of expression of each of 5460 yeast genes at an elevated temperature at which the mutant TAF was inactive and at a lower temperature at which the mutant TAF was active. These arrays contain oligonucleotides specific for each gene. Total yeast RNA can then be hybridized to these arrays, and the extent of hybridization to each oligonucleotide is a measure of the extent of expression of the corresponding gene. The investigators compared the hybridization of RNA to each oligonucleotide at low and high temperature and compared the response with the results of a similar analysis of a temperature-sensitive mutation in the largest subunit of RNA polymerase II (Rpb1). Because the latter mutation prevented transcription of *all* class II genes, it provided a baseline with which to compare the effects of mutations in other genes.

Table 11.1 presents the results of this analysis. It is striking that only 16% of the yeast genes analyzed were as dependent on TAF$_{II}$145 as they were on Rpb 1, indicat-

ing that TAF$_{II}$145 is required for transcription of only 16% of yeast genes. This is not what we would expect if the TAFs are essential parts of TFIID, and TFIID is an essential part of the preinitiation complexes formed at all class II genes. Indeed, TAF$_{II}$145, along with TBP, had been regarded as a keystone of TFIID, helping to assemble all the other TAFs in that factor, but this view is clearly not supported by the genome-wide expression analysis. Instead, TAF$_{II}$145 and its homolog in higher organisms appear to be required in the preinitiation complexes formed at only a subset of genes. In yeast, these genes tend to be ones governing progression through the cell cycle.

Mutation of the other yeast TAF (TAF$_{II}$17) had a more pronounced effect. Sixty-seven percent of the yeast genes analyzed were as dependent on this TAF as they were on Rpb1. But that does not mean that TFIID is required for transcription of all these genes, because TAF$_{II}$17 is also part of a transcription adapter complex known as **SAGA** (named for three classes of proteins it contains—SPTs, ADAs, and GCN5—and its enzymatic activity, histone acetyltransferase). Like TFIID, SAGA contains TBP, a number of TAFs, and histone acetyltransferase activity, and appears to mediate the effects of certain transcription activator proteins. So the effect of mutating TAF$_{II}$17 may be due to its role in SAGA or perhaps in other protein complexes yet to be discovered, rather than in TFIID.

Not only are some TAFs not universally required for transcription, the TFIIDs appear to be heterogenous in their TAF compositions. For example, TAF$_{II}$30 is found in only a fraction of human TFIIDs, and its presence correlates with responsiveness to estrogen.

Even more surprisingly, TBP is not universally found in preinitiation complexes in higher eukaryotes. The most celebrated example of an alternative TBP is **TRF1** (**TBP-related factor 1**) in *Drosophila melanogaster*. This protein is expressed in developing neural tissue, binds to TFIIA and TFIIB, and stimulates transcription just as TBP does, and it has its own group of TRF-associated factors called **nTAFs** (for neural TAFs). In 2000, Michael Holmes and Robert Tjian used primer extension analysis in vivo and in vitro to show that TRF1 stimulates transcription of the *Drosophilia tudor* gene. Furthermore, this analysis revealed that the *tudor* gene has two distinct promoters. The first is a downstream promoter with a TATA box recognized by a complex including TBP. The second promoter lies about 77 bp upstream of the first and has a TC box recognized by a complex including TRF1 (Figure 11.21). The TC box extends from position −22 to −33 with respect to the start of transcription and has the sequence ATTGCTTTTCTT in the nontemplate strand. It is protected by a complex of TRF1, TFIIA, and TFIIB in DNase footprinting experiments. However, none of these proteins alone make a footprint in this region, and neither does TBP, or TBP plus TFIIA and TFIIB.

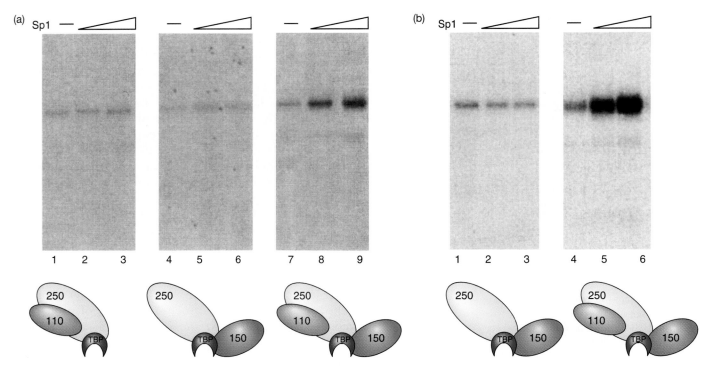

Figure 11.19   Activation by Sp1 requires TAF$_{II}$110. Tjian and colleagues used a primer extension assay to measure transcription from a template containing a TATA box and three upstream GC boxes. They used either a (**a**) *Drosophila* cell extract or (**b**) a human cell extract, each of which had been depleted of TFIID. They replaced the missing TFIID with any of the three different complexes, pictured at bottom, containing combinations of TBP, TAF$_{II}$250, TAF$_{II}$150, and TAF$_{II}$110. For example, lanes 1–3 in panel (**a**) contained TBP, TAF$_{II}$250, and TAF$_{II}$110. They also added no Sp1 (–), or two increasing concentrations of Sp1, represented by the wedges. The autoradiographs show the amount of transcription and therefore the activation achieved by Sp1 with each set of TAF$_{II}$s. Activation was observed in each extract only with all three TAF$_{II}$s. (*Source:* Chen et al., Assembly of recombinant TFIID reveals differential co-activator requirements for distinct transcriptional activators. *Cell* 79 (7 Oct 1994) p. 96, f. 3b-c. Reprinted with permission of Elsevier Science.)

Thus, TRF appears to be a cell type-specific variant of TBP. The presence of alternative TBPs and TAFs raises the possibility that gene expression in higher eukaryotes could be controlled in part by the availability of the appropriate TBP and TAFs, as well as by the activator proteins we will study in Chapter 12. Indeed, the recognition of two different *tudor* promoters by two different TBPs is reminiscent of the recognition of two different prokaryotic promoters for the same gene by RNA polymerases bearing different σ-factors, as we saw in Chapter 8.

The central role of TBP in forming preinitiation complexes has been further challenged by the discovery of a **TBP-free TAF$_{II}$-containing complex (TFTC)** that is able to sponsor preinitiation complex formation without any help from TFIID or TBP. Structural studies by Patrick Schultz and colleagues have provided some insight into how TFTC can substitute for TFIID. They have performed electron microscopy and digital image analysis on both TFTC and TFIID and found that they have strikingly similar three-dimensional structures. Figure 11.22 shows three-dimensional models of the two protein complexes in three different orientations. The most obvious characteristics of both complexes is a groove large enough to accept a double-stranded DNA. In fact, it appears that the protein of both complexes would encircle the DNA and hold

it like a clamp. The only major difference between the two complexes is the projection at the top of TFTC due to domain 5. TFIID lacks both the projection and domain 5.

SUMMARY The TAFs do not appear to be universally required for transcription of class II genes. Even TAF$_{II}$145 (the yeast homolog of human TAF$_{II}$250) is not required for transcription of the great majority of yeast class II genes. Even TBP is not universally required. Some promoters in higher eukaryotes respond to an alternative protein such as TRF1 and not to TBP. Some promoters can be stimulated by a TBP-free TAF$_{II}$-containing complex (TFTC), rather than by TFIID.

Structure and Function of TFIIA and TFIIB

Jeffrey Ranish and coworkers cloned, sequenced, and expressed two genes that encode the two subunits of yeast TFIIA. These two subunits have relative molecular masses of 32 and 13.5 kD. Together, they form a functional unit that has TFIIA activity by several criteria: (1) It cooperates with yeast TBP to cause a gel mobility shift of a DNA

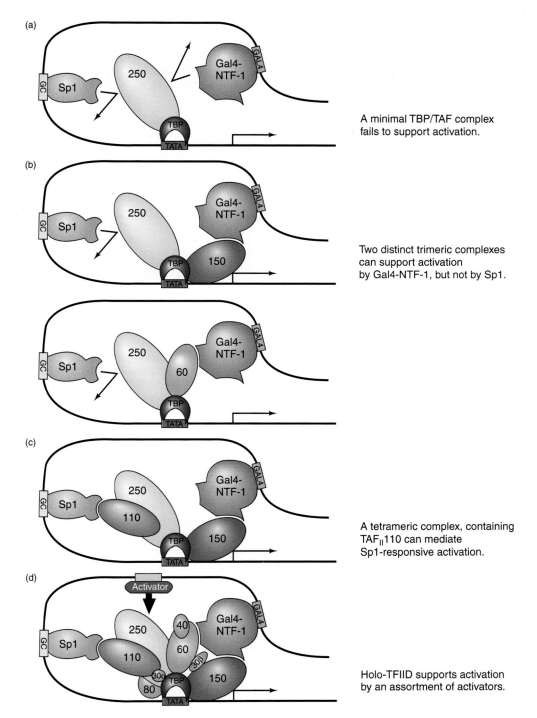

(a) A minimal TBP/TAF complex fails to support activation.

(b) Two distinct trimeric complexes can support activation by Gal4-NTF-1, but not by Sp1.

(c) A tetrameric complex, containing TAF$_{II}$110 can mediate Sp1-responsive activation.

(d) Holo-TFIID supports activation by an assortment of activators.

Figure 11.20 A model for transcription enhancement by activators. (a) TAF$_{II}$250 does not interact with either Sp1 or Gal4-NTF-1 (a hybrid activator with the transcription-activating domain of NTF-1), so no activation takes place. **(b)** Gal4-NTF-1 can interact with either TAF$_{II}$150 or TAF$_{II}$60 and activate transcription; Sp1 cannot interact with either of these TAFs or with TAF$_{II}$250 and does not activate transcription. **(c)** Gal4-NTF-1 interacts with TAF$_{II}$150 and Sp1 interacts with TAF$_{II}$110, so both factors activate transcription. **(d)** Holo-TFIID contains the complete assortment of TAF$_{II}$s, so it can respond to a wide variety of activators, represented here by Sp1, Gal4-NTF-1, and a generic activator at top. (*Source:* Reprinted from Chen et al., *Cell* 79:101, 1994. Copyright 1994, with permission from Elsevier Science.)

313

Table 11.1   Whole Genome Analysis of Transcription Requirements in Yeast

General Transcription Factor (Subunit)	Fraction of Genes Dependent on Subunit Function (%)
TFIID (TAF$_{II}$145)	16
TFIID (TAF$_{II}$17)	67
TFIIE (Tfa1)	54
TFIIH (Kin28)	87

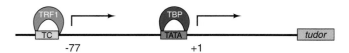

Figure 11.21   The *Drosophila tudor* control region. This gene has two promoters about 77 bp apart. The downstream promoter has a TATA box that attracts a preinitiation complex based on TBP. The upstream promoter has a TC box that attracts a preinitiation complex based on TRF1.

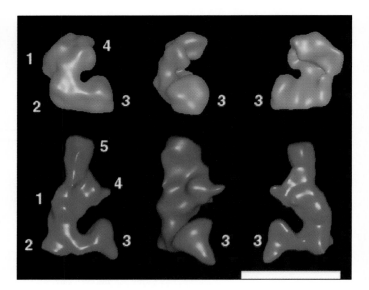

Figure 11.22   Three-dimensional models of TFIID and TFTC. Schultz and colleagues made negatively stained electron micrographs (see Chapter 19, for method) of TFIID and TFTC, then digitally combined images to arrive at an average. Then they tilted the grid in the microscope and analyzed the resulting micrographs to glean three-dimensional information for both proteins. The resulting models for TFIID (green) and TFTC (blue) are shown. (*Source:* Brand, M. et al. Three-Dimensional Structures of the TAF$_{II}$-Containing Complexes TFDIID and TFTC. *Science* 286 (10 Dec 1999) f. 3, p. 2152 © AAAS.)

fragment containing the adenovirus major late promoter. (2) It complements both yeast and mammalian cell-free transcription systems that had been depleted of TFIIA, restoring transcription activity to these particular TFIIA-dependent extracts.

Reinberg and colleagues used TBP affinity chromatography to purify human TFIIA and found that it has three subunits. The *Drosophila* TFIIA has also been purified and found to have three subunits similar to those of the human TFIIA. The molecular masses of the three subunits from these two sources are very similar and are about 30, 20, and 13 kD. The largest two subunits are actually encoded in a single gene, and the resulting fusion protein is processed after translation to yield the two mature polypeptides.

As we have seen, the requirement for TFIIA in formation of a preinitiation complex has been questionable, because it seems to be necessary in some in vitro systems, but not in others. However, Ranish and coworkers showed that mutations in the genes encoding the two subunits of yeast TFIIA are lethal, which suggests that TFIIA really is essential in vivo, at least in yeast. It is now generally accepted that TFIIA is an essential part of the preinitiation complex in eukaryotes in general, but is it a true general transcription factor? Actually, TFIIA can be thought of as a TAF$_{II}$: We have seen that it binds to TBP, and its major function seems to be to stabilize binding between TFIID and the promoter. These are two key characteristics of at least some of the TAF$_{II}$s.

Danny Reinberg and his coworkers cloned and expressed the gene for human TFIIB. This cloned TFIIB product can substitute for the authentic human protein in all in vitro assays, including response to gene-specific transcription factors such as Sp1. This suggests that TFIIB

is a single-subunit factor (M_r = 35 kD) that requires no auxilliary polypeptides such as the TAFs. As we have already discovered, TFIIB is the third general transcription factor to join the preinitiation complex (after TFIID and A), or the second if TFIIA has not yet bound. It is essential for binding RNA polymerase because the polymerase–TFIIF complex will bind to the DAB complex, but not to the DA complex. We will see in Chapter 12 that TFIIB is a probable target of at least some of the common gene-specific transcription factors. In other words, these factors seem to act by facilitating binding of TFIIB to the preinitiation complex. This in turn allows polymerase II and the other factors to bind and, assuming that TFIIB binding is rate-limiting, stimulates transcription.

The position of TFIIB between TFIID and TFIIF/RNA polymerase II in the assembly of the preinitiation complex suggests that TFIIB should have two domains: one to interact with TFIID and the other to interact with TFIIF/polymerase. In fact, Roeder and colleagues showed that TFIIB is indeed composed of two distinct domains, one of which is essential for binding to TFIID, and the other of which is needed for promoting further assembly of the preinitiation complex.

Roeder and Burley and colleagues determined the crystal structure of a heterologous TBP–TFIIB–TATA box complex in 1995. The origin of the TBP was *Arabidopsis*, that of the TFIIB core was human, and the TATA-box-containing DNA was from the the adenovirus major late promoter. Then, in 1996, Timothy Richmond and colleagues solved the crystal structure of a yeast TFIIA–TBP–TATA box complex. Putting these two structures together creates the hypothetical picture of a TFIIA–TFIIB–TBP–TATA box complex shown in Figure 11.23.

This partial preinitiation complex illustrates how TFIIB binds to the downstream portion of TBP (the carboxyl-terminal stirrup) and how TFIIA binds to the upstream portion of TBP (the amino-terminal stirrup). This spatial arrangement is consistent with our understanding of the roles of these proteins. TFIIB's downstream location allows it to interact with TFIIF or RNA polymerase II (or both) so it can position the polymerase at the transcription start site. TFIIA's upstream location allows it to stabilize TBP–TATA box binding and also possibly to interact with transcription activators that bind to upstream elements. The structure is also consistent with Roeder and colleagues' biochemical data showing that TFIIB is composed of two distinct domains, one for binding to TFIID, the other for promoting further assembly of the preinitiation complex.

SUMMARY TFIIA contains two subunits (yeast), or three subunits (fruit flies and humans). This factor may be more properly considered a TAF$_{II}$ because it binds to TBP and stabilizes binding between TFIID and promoters. TFIIB serves as a linker between TFIID and TFIIF/polymerase II. It has two domains, one of which is responsible for binding to TFIID, the other for continuing the assembly of the preinitiation complex. A structure for the TFIIA–TFIIB–TBP–TATA box complex can be imagined, based on the known structures of the TFIIA–TBP–TATA box and TFIIB–TBP–TATA box complexes. This structure shows TFIIA and TFIIB binding to the upstream and downstream stirrups, respectively, of TBP. This puts these two factors in advantageous positions to perform their functions.

Structure and Function of TFIIF

Reinberg and Greenblatt and their colleagues have demonstrated that TFIIF is actually composed of two polypeptides, called RAP30 and RAP70. The RAP stands for RNA polymerase II-associated protein, and the numbers denote the proteins' molecular masses. RAP30 is another protein with weak homology to *E. coli* σ, in particular to the part thought to mediate binding of σ to core. This suggests that RAP30 may have a role in binding TFIIF to polymerase II, and indeed independent evidence supports this idea. Greenblatt has shown that TFIIF binds

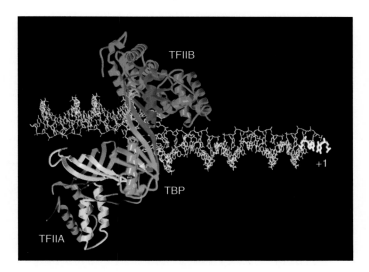

Figure 11.23 Hypothetical structure of a TFIIA–TFIIB–TBP–TATA box complex. This is a combination of two structures: a human core TFIIB-plant TBP-adenovirus TATA box structure, and a yeast TFIIA–TBP–TATA box structure. None of the proteins is complete. They are all core regions that have the key elements needed to do their jobs. The DNA is gray; the two halves of core TBP are light blue (upstream half) and dark blue (downstream half); the amino terminal domain of core TFIIB is red and the carboxyl terminal domain is magenta; the core large subunit of TFIIA is green, and the small subunit is yellow. The transcription start site is at right, denoted "+1." (*Source:* Burley, "Picking up the TAB." *Nature* 381 (9 May 1996) p. 113, f. 1. © Macmillan Magazines Ltd.)

tightly ($K_d = 2 \times 10^{-8}$ M) to *E. coli* RNA polymerase core and can be displaced by *E. coli* σ, which binds even more tightly. He and Reinberg also showed that cloned RAP30 binds to eukaryotic polymerase II in vitro.

TFIIF also plays another σ-like role in eukaryotes: It can reduce nonspecific interactions between polymerase and DNA. Recall from Chapter 6 that σ^{70} of *E. coli* also performs this essential function. TFIIF of rat liver blocks nonspecific binding between polymerase II and DNA and can release polymerase II from a preformed nonspecific complex with DNA. The key to TFIIF's ability to direct polymerase II away from nonspecific DNA interactions may relate to its capacity to interact with both polymerase II and the DAB complex. Thus, it could bind to polymerase and then direct it to promoters where a DAB complex has already assembled. In this way, TFIIF could lead polymerase II to promoters, and away from competing nonspecific DNA sites.

Greenblatt and Reinberg demonstrated that the RAP30 subunit of TFIIF directs the binding of polymerase II to the DAB complex. First, they purified RAP30 and showed that the same fractions that contained RAP30 also had TFIIF transcription activity and could cause the formation of the DABPolF complex (Figure 11.24). They established the presence of RAP30 by probing a Western blot with an anti-RAP30 antibody, measured TFIIF transcription activity by its ability to stimulate specific transcription in the

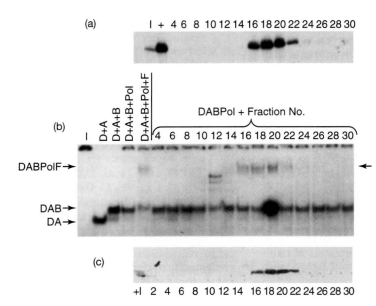

Figure 11.24   Role of TFIIF in binding RNA polymerase II to the preinitiation complex. Greenblatt, Reinberg, and colleagues performed phenyl-Superose micro column chromatography on TFIIF and tested fractions for **(a)** TFIIF transcription factor activity; and **(b)** preinitiation complex formation with RNA polymerase II, using a gel mobility shift assay; and **(c)** content of RAP30, detected by Western blotting and probing with an anti-RAP30 antibody. (RAP70 was also present in these fractions, but no attempt was made to detect it.) **(a)** TFIIF activity assay. Lane I, activity of the protein loaded onto the column (input); lane +, positive control with known TFIIF activity; other lanes are numbered according to their order of elution from the column. The great majority of the TFIIF activity eluted in fractions 16–22. **(b)** Gel mobility shift assay. The lanes on the left show the complexes formed with the TFIIF input fraction alone (I), and with various combinations of highly purified TFIID, A, B, polymerase II, and TFIIF. The numbered lanes show the shifts in the DAB complex produced by addition of polymerase II plus the same column fractions as in part **(a)**. The ability to form the DABPolF complex resided in the same fractions with TFIIF activity (16–22). **(c)** Western blot to detect RAP30. The labeling of the lanes has the same meaning as in panel **(a)**. The fractions with RAP30 (16–22) were the same ones with TFIIF activity and the ability to bring polymerase II into the preinitiation complex. Thus, RAP30 seems to have this activity. (*Source:* Flores et al., *PNAS* 88 (Nov 1991) p. 10000, f. 1.)

presence of all the other factors plus DNA and RNA polymerase II, and used a gel mobility shift assay to confirm the formation of a DABPolF complex. To be sure that a contaminant in the RAP30 preparation was not causing these effects, they obtained RAP30 by expressing its gene in bacteria. This could not possibly be contaminated with other transcription factors, but it could still direct RNA polymerase to join the DABPolF complex.

> **SUMMARY** Binding of the polymerase to the DAB complex requires prior interaction with TFIIF, composed of two polypeptides called RAP30 and RAP70. RAP30 is the protein that ushers polymerase into the growing complex.

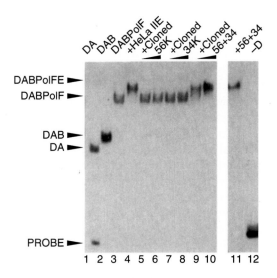

Figure 11.25   Formation of the DABPolFE complex. Tjian, Reinberg, and colleagues performed gel mobility shift assays with various combinations of transcription factors, polymerase II, and a DNA fragment containing the adenovirus major late promoter. The protein components in each lane are given at top, and the complexes formed are indicated at left. Note that TFIID, A, B, F, E, and polymerase II formed the DABPolFE complex, as expected (lane 4). Lanes 5–8 show that increasing quantities of the two subunits of TFIIE, added separately, cannot join the DABPolF complex. However, lanes 9 and 10, demonstrate that the two polypeptides can join the complex if they are added together. Lane 11 is a repeat of lane 10, and lane 12 is identical except that it is missing TFIID. This is a reminder that *everything* depends on TFIID, even with all the other factors present. (*Source:* Peterson et al., Structure and functional properties of human general transcription factor IIE. *Nature* 354 (5 Dec 1991) p. 372, f. 4-5. © Macmillan Magazines Ltd.)

Structure and Function of TFIIE and TFIIH

Tjian and Reinberg and colleagues have cloned the genes for the two subunits of TFIIE: polypeptides of 56 and 34 kD. The sequence of the 56-kD polypeptide contains a region that could form a DNA-binding domain called a zinc finger (Chapter 12). Reinberg had previously determined the native molecular mass of TFIIE as about 200 kD, which suggested that it contains two molecules each of the 56- and 34-kD subunits. The two cloned polypeptides go together to form a factor with activity identical to that of authentic TFIIE. This factor joins the DABPolF complex, causing an extra gel mobility shift, whereas each subunit individually does not (Figure 11.25). It activates transcription from the adenovirus major late promoter in vitro in systems lacking TFIIE (Figure 11.26a). It also supplements a TFIIE-depleted transcription system that is unresponsive to the transcription factor Sp1, thus allowing stimulation by Sp1 (Figure 11.26b). Based on these and other findings, Tjian proposed the model of transcription factor and polymerase II binding to promoters illustrated in Figure 11.27.

TFIIH is the last general transcription factor to join the preinitiation complex. It appears to play multiple roles

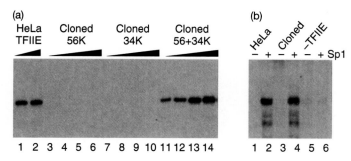

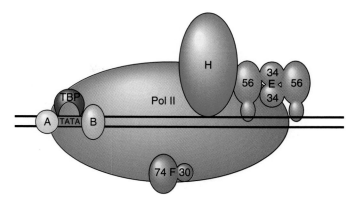

Figure 11.27 The preinitiation complex envisioned by Tjian and Reinberg. This construct contains all of the factors in the DABPolFE complex plus TFIIH (cyan), another general transcription factor we will discuss next. TFIIH is represented by a single symbol, although it actually contains nine subunits. (*Source:* Reprinted with permission from *Nature* 354:373, 1991. Copyright © 1991 Macmillan Magazines Limited.)

Figure 11.26 Dependence of transcription on *both* subunits of TFIIE. (a) Tjian and Reinberg performed run-off transcription of a DNA fragment containing the adenovirus major late promoter in the presence of all transcription factors except TFIIE. They added whole TFIIE or the products of cloned genes encoding the subunits of the transcription factor in increasing concentration, as indicated at top. The wedge shapes illustrate the increase in concentration of each factor from one lane to another. Lanes 1 and 2 show that native TFIIE can reconstitute transcription activity. However, the subunits added separately cannot do this, as portrayed in lanes 3–10. On the other hand, the two subunits together *can* stimulate transcription. (b) The same kind of run-off assays, using the TATA-less G_6I promoter, showed that the TFIIE produced by cloned genes stimulates Sp1-dependent transcription. Lanes 1 and 2 contained native TFIIE purified from HeLa cells. Lanes 3 and 4 contained TFIIE subunits produced by cloned genes. Lanes 5 and 6 had no TFIIE. Clearly, TFIIE is necessary, and the factor made by cloned genes works as well as the native one. Also, as we have seen before, transcription of the TATA-less promoter requires Sp1. (*Source:* Peterson et al., Structure and functional properties of human general transcription factor IIE. *Nature* 354 (5 Dec 1991) p. 372, f. 4-5. © Macmillan Magazines Ltd.)

in transcription initiation; one of these is to phosphorylate RNA polymerase II. As we have already seen in Chapter 10, RNA polymerase II exists in two physiologically meaningful forms: IIA (unphosphorylated) and IIO (with many phosphorylated amino acids in the carboxyl terminal domain [CTD]). The unphosphorylated enzyme, polymerase IIA, is the form that joins the preinitiation complex. But the phosphorylated enzyme, polymerase IIO, carries out RNA chain elongation. This behavior suggests that phosphorylation of the polymerase occurs between the time it joins the preinitiation complex and the time promoter clearance occurs. In other words, phosphorylation of the polymerase could be the trigger that allows the polymerase to shift from initiation to elongation mode. This hypothesis receives support from the fact that the unphosphorylated CTD in polymerase IIA binds much more tightly to TBP than does the phosphorylated form in polymerase IIO. Thus, phosphorylation of the CTD could break the tether that binds the polymerase to the TBP at the promoter and thereby permit transcription elongation to begin. On the other hand, this hypothesis is damaged somewhat by the finding that transcription can sometimes occur in vitro without phosphorylation of the CTD.

Whatever the importance of CTD phosphorylation, Reinberg and his colleagues have demonstrated that TFIIH was a good candidate for the protein kinase that

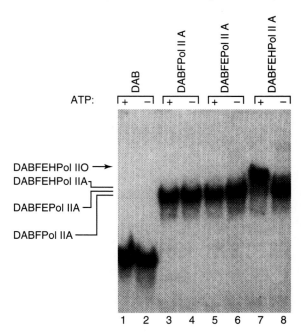

Figure 11.28 Phosphorylation of preinitiation complexes. Reinberg and colleagues performed gel mobility shift assays with preinitiation complexes DAB through DABPolFEH, in the presence and absence of ATP, as indicated at top. Only when TFIIH was present did ATP shift the mobility of the complex (compare lanes 7 and 8). The simplest explanation is that TFIIH promotes phosphorylation of the input polymerase (polymerase IIA) to polymerase IIO. (*Source:* Lu et al., Human general transcription factor IIH phosphorylates the C-terminal domain of RNA polymerase II. *Nature* 358 (20 Aug 1992) p. 642, f.1, 2, 3. © Macmillan Magazines Ltd.)

catalyzes this process. First, these workers showed that the purified transcription factors, by themselves, are capable of phosphorylating the CTD of polymerase II, converting polymerase IIA to IIO. The evidence, shown in Figure 11.28 came from a gel mobility shift assay. Lanes 1–6 demonstrate that adding ATP had no effect on the

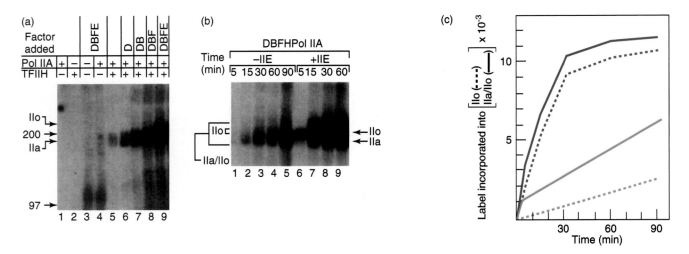

Figure 11.29   TFIIH phosphorylates RNA polymerase II. (a) Reinberg and colleagues incubated polymerase IIA with various mixtures of transcription factors, as shown at top. They included [γ-32P] ATP in all reactions to allow phosphorylation of the polymerase, then electrophoresed the proteins and performed autoradiography to visualize the phosphorylated polymerase. Lane 4 shows that TFIID, B, F, and E, were insufficient to cause phosphorylation. Lanes 5–10 demonstrate that TFIIH alone is sufficient to cause some polymerase phosphorylation, but that the other factors enhance the phosphorylation. TFIIE provides particularly strong stimulation of phosphorylation of the polymerase IIa subunit to IIo. **(b)** Time course of polymerase phosphorylation. Reinberg and colleagues performed the same assay for polymerase phosphorylation with TFIID, B, F, and H in the presence or absence of TFIIE, as indicated at top. They carried out the reactions for 60 or 90 min, sampling at various intermediate times, as shown at top. The small bracket at left indicates the position of the polymerase IIo subunit, and the larger bracket shows the locations of IIa and IIo together (IIa/IIo). Arrows also mark the positions of the two polymerase subunit forms. Note that polymerase phosphorylation is more rapid in the presence of TFIIE. **(c)** Graphic presentation of the data from panel **(b)**. Green and red curves represent phosphorylation in the presence and absence, respectively, of TFIIE. Solid lines and dotted lines correspond to appearance of phosphorylated polymerase subunits IIa and IIo, or just IIo, respectively. (*Source:* Lu et al., Human general transcription factor IIH phosphorylates the C-terminal domain of RNA polymerase II. *Nature* 358 (20 Aug 1992) p. 642, f.1, 2, 3. © Macmillan Magazines Ltd.)

mobility of the DAB, DABPolF, or DABPolFE complexes. On the other hand, after TFIIH was added to form the DABPolFEH complex, ATP produced a change to lower mobility. What accounted for this change? One possibility is that one of the transcription factors in the complex had phosphorylated the polymerase. Indeed, when Reinberg and colleagues isolated the polymerase from the lower mobility complex, it proved to be the phosphorylated form, polymerase IIO. But polymerase IIA had been added to the complex in the first place, so one of the transcription factors had apparently performed the phosphorylation.

Next Reinberg and colleagues demonstrated directly that the TFIIH preparation phosphorylates polymerase IIA. To do this, they incubated purified polymerase IIA and TFIIH together with [γ-32P]ATP under DNA-binding conditions. A small amount of polymerase phosphorylation occurred, as shown in Figure 11.29a. Thus, this TFIIH preparation by itself is capable of carrying out the phosphorylation. By contrast, all the other factors together caused no such phosphorylation. However, these factors could greatly stimulate the phosphorylating capability of TFIIH. Lanes 6–10 show the results with TFIIH plus an increasing set of the other factors. As Reinberg and associates added each new factor, they noticed an increasing efficiency of phosphorylation of the polymerase and accumulation of polymerase IIO. Because the biggest increase in polymerase IIO labeling came with the addi-

tion of TFIIE, these workers performed a time-course study in the presence of TFIIH or TFIIH plus TFIIE. Figure 11.29b shows that the conversion of the IIa subunit to the IIo subunit was much more efficient when TFIIE was present. Figure 11.29c shows the same results graphically.

We know that the carboxyl-terminal domain (CTD) of the polymerase IIa subunit is the site of the phosphorylation because polymerase IIB, which lacks the CTD, is not phosphorylated by the TFIIDBFEH complex, while polymerase IIA, and to a lesser extent, polymerase IIO, are phosphorylated (Figure 11.30a). Also, as we have seen, phosphorylation produces a polypeptide that coelectrophoreses with the IIo subunit, which does have a phosphorylated CTD. To demonstrate directly the phosphorylation of the CTD, Reinberg and colleagues cleaved the phosphorylated enzyme with chymotrypsin, which cuts off the CTD, and electrophoresed the products. The autoradiograph of the chymotrypsin products (Figure 11.30b) shows a labeled CTD fragment, indicating that labeled phosphate has been incorporated into the CTD part of the large polymerase II subunit. The rest of the subunit was not labeled.

To prove that one of the subunits of RNA polymerase II was not helping in the kinase reaction, Reinberg and coworkers cloned a chimeric gene that codes for the CTD as a fusion protein that also includes the DNA-binding domain from the transcription factor GAL4 and the

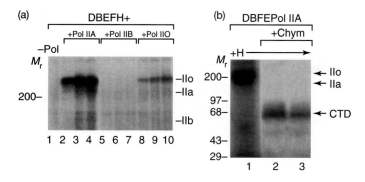

Figure 11.30   TFIIH phosphorylates the CTD of polymerase II.
(a) Reinberg and colleagues phosphorylated increasing amounts of
polymerases IIA, IIB, or IIO, as indicated at top, with TFIID, B, F, E,
and H and radioactive ATP as described in Figure 11.29. Polymerase
IIB, lacking the CTD, could not be phosphorylated. The
unphosphorylated polymerase IIA was a much better phosphorylation
substrate than IIO, as expected. (b) Purification of the phosphorylated
CTD. Reinberg and colleagues cleaved the CTD from the
phosphorylated polymerase IIa subunit with the protease
chymotrypsin (Chym), electrophoresed the products, and visualized
them by autoradiography. Lane 1, reaction products before
chymotrypsin cleavage; lanes 2 and 3, reaction products after
chymotrypsin cleavage. The position of the CTD had been identified in
a separate experiment. (Source: Lu et al., Human general transcription factor
IIH phosphorylates the C-terminal domain of RNA polymerase II. Nature 358
(20 Aug 1992) p. 642, f.1, 2, 3. © Macmillan Magazines Ltd.)

enzyme glutathione-*S*-transferase. It appeared that TFIIH,
all by itself, was capable of phosphorylating the CTD domain of this fusion protein. Thus, this TFIIH preparation
had the appropriate kinase activity, even in the absence of
other polymerase II subunits.

All of the experiments described so far were done
under conditions in which the polymerase (or polymerase
domain) was bound to DNA. Is this important? To find
out, Reinberg's group tried the kinase assay with polymerase II in the presence of DNA that had a complete
promoter, or merely the TATA box or the initiator regions of the promoter, or even no promoter at all. The result was that the TFIIH preparation performed the phosphorylation quite well in the presence of a TATA box, or
an initiator, but did very poorly with a synthetic DNA
(poly [dI-dC]) that contained neither. Thus, TFIIH appears to phosphorylate polymerase II only when it is
bound to DNA. We now know that the kinase activity is
provided by two subunits of TFIIH.

TFIIH is a complex protein, both structurally and
functionally. It contains nine subunits and can be separated into two complexes: a protein kinase complex composed of four subunits, and a five-subunit core TFIIH
complex with two separate DNA helicase/ATPase activities. One of these, contained in the largest subunit of
TFIIH, is essential for viability: When its gene in yeast
(*RAD25*) is mutated, the organism cannot survive. Satya
Prakash and colleagues demonstrated that this helicase is
essential for transcription. First they overproduced the
RAD25 protein in yeast cells, purified it almost to homo-

geneity, and showed that this product had helicase activity. For a helicase substrate, they used a partial duplex
DNA composed of a ^{32}P-labeled synthetic 41-base DNA
hybridized to single-stranded M13 DNA (Figure 11.31a).
They mixed RAD25 with this substrate in the presence
and absence of ATP and electrophoresed the products.
Helicase activity released the short, labeled DNA from its
much longer partner, so it had a much higher electrophoretic mobility and was found at the bottom of the
gel. As Figure 11.31b demonstrates, RAD25 has an ATP-
dependent helicase activity.

Next, Prakash and colleagues showed that transcription
was temperature-sensitive in cells bearing a temperature-
sensitive *RAD25* gene (*rad25*-ts$_{24}$). Figure 11.32 shows the
results of an in vitro transcription assay using a G-less cassette (Chapter 5) as template. This template had a yeast
TATA element upstream of a 400-bp region with no G's in
the nontemplate strand. Transcription in the presence of
ATP, CTP, and UTP (but no GTP) apparently initiated (or
terminated) at two sites within this G-less region and gave
rise to two transcripts, 375 and 350 nt in length, respectively. Transcription must terminate at the end of the G-less
cassette because at that point G's are required to extend the
RNA chain. (The shorter transcript may have come from
premature termination within the G-less cassette, rather
than from a different initiation site.) Panel (a) shows the results of transcription for 0–10 min at the permissive temperature (22° C). It is clear that the *rad25*-ts$_{24}$ mutant
extract gave weaker transcription than the wild-type
(*RAD25*) extract even at low temperature. Panel (b) shows
the results of transcription at the nonpermissive temperature (37° C). The elevated temperature completely inactivated transcription in the *rad25*-ts$_{24}$ mutant extract. Thus,
the *RAD25* product (the TFIIH DNA helicase) is required
for transcription.

What step in transcription requires DNA helicase activity? The chain of evidence leading to the answer begins
with the following consideration: Transcription of class II
genes, unlike transcription of class I and III genes, requires ATP (or dATP) hydrolysis. Of course, the α-β-
bonds of all four nucleotides, including ATP, are hydrolyzed during all transcription, but class II transcription
requires hydrolysis of the β-γ-bond of ATP. The question
arises: What step requires ATP hydrolysis? We would naturally be tempted to look at TFIIH for the answer to this
question because it has two activities (CTD kinase and
DNA helicase) that involve hydrolysis of ATP. The answer appears to be that the helicase activity of TFIIH is
the ATP-requiring step. The main evidence in favor of this
hypothesis is that GTP can substitute for ATP in CTD
phosphorylation, but GTP cannot satisfy the ATP hydrolysis requirement for transcription. Thus, transcription requires ATP hydrolysis for some process besides CTD
phosphorylation, and the best remaining candidate is
DNA helicase.

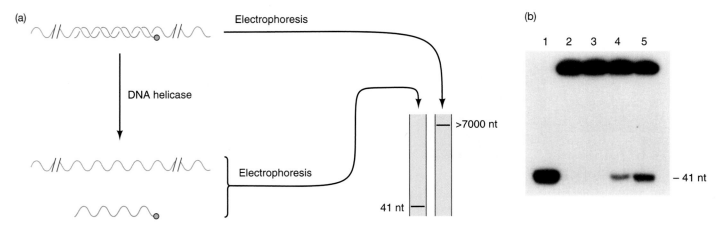

Figure 11.31 Helicase activity of TFIIH. (a) The helicase assay. The substrate consisted of a labeled 41-nt piece of DNA (red) hybridized to its complementary region in a much larger, unlabeled, single-stranded M13 phage DNA (blue). DNA helicase unwinds this short helix and releases the labeled 41-nt DNA from its larger partner. The short DNA is easily distinguished from the hybrid by electrophoresis. **(b)** Results of the helicase assay. Lane 1, heat-denatured substrate; lane 2, no protein; lane 3, 20 ng of RAD25 with no ATP; lane 4, 10 ng of RAD25 plus ATP; lane 5, 20 ng of RAD25 plus ATP. (*Source: (b)* Gudzer et al., RAD25 is a DNA helicase required for DNA repair and RNA polymerase II transcription. *Nature* 369 (16 June 1994) p. 579, f. 2c. © Macmillan Magazines Ltd.)

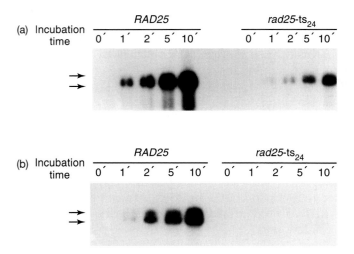

Figure 11.32 The TFIIH DNA helicase gene product (RAD25) is required for transcription in yeast: Prakash and colleagues tested extracts from wild-type (*RAD25*) and temperature-sensitive mutant (*rad25*-ts$_{24}$) cells for transcription of a G-less cassette template at the **(a)** permissive and **(b)** nonpermissive temperatures. After allowing transcription for 0–10 min in the presence of ATP, CTP, and UTP (but no GTP), with one ^{32}P-labeled nucleotide, they electrophoresed the labeled products and detected the bands by autoradiography. The origin of the extract (*RAD25* or *rad25*-ts$_{24}$ cells), as well as the time of incubation in minutes, is given at top. Arrows at left denote the positions of the two G-less transcripts. We can see that transcription is temperature-sensitive when the TFIIH DNA helicase (RAD25) is temperature-sensitive. (*Source:* Gudzer et al., RAD25 is a DNA helicase required for DNA repair and RNA polymerase II transcription. *Nature* 369 (16 June 1994) p. 580, f. 3 b-c. © Macmillan Magazines Ltd.)

Now let us return to the main question: What transcription step requires DNA helicase activity? The most likely answer is promoter clearance. In Chapter 6 we defined transcription initiation to include promoter clearance, but promoter clearance can also be considered a

separate event that serves as the boundary between initiation and elongation. James Goodrich and Tjian asked this question: Are TFIIE and TFIIH required for initiation or for promoter clearance? To find the answer, they devised an assay that measures the production of abortive transcripts (trinucleotides). The appearance of abortive transcripts indicates that a productive transcription initiation complex has formed, including local DNA melting and synthesis of the first phosphodiester bond. Goodrich and Tjian found that TFIIE and TFIIH were not required for production of abortive transcripts, but TBP, TFIIB, TFIIF, and RNA polymerase II were required. Thus, TFIIE and TFIIH are not required for transcription initiation, at least up to the promoter clearance step. However, TFIIH is required for full DNA melting at promoters. If the largest subunit of human TFIIH is mutated, the DNA helicase of that subunit is defective, and the DNA at the promoter does not open completely. This could block promoter clearance, as explained later in this section.

These findings left open the possiblitity that TFIIE and TFIIH are required for either promoter clearance or RNA elongation, or both. To distinguish among these possibilities, Goodrich and Tjian assayed for elongation and measured the effect of TFIIE and TFIIH on that process. By leaving out the nucleotide required in the 17th position, but not before, they allowed transcription to initiate (without TFIIE and TFIIH) on a supercoiled template and proceed to the 16-nt stage. (They used a supercoiled template because transcription on such templates in vitro does not require TFIIE and TFIIH, nor does it require ATP.) Then they linearized the template by cutting it with a restriction enzyme and added ATP to allow transcription to continue in the presence or absence of TFIIE and TFIIH. They found that TFIIE and TFIIH made no difference in this elongation

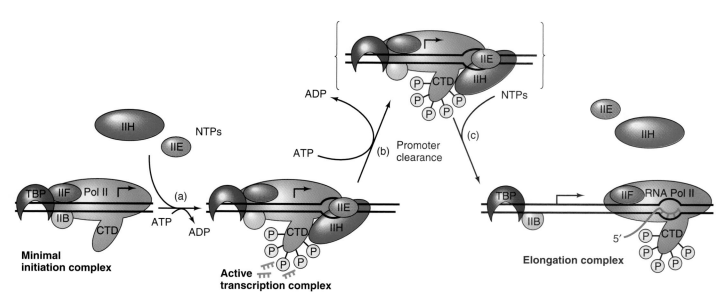

Figure 11.33 A model for the participation of general transcription factors in initiation, promoter clearance, and elongation. (a) TBP (or TFIID), along with TFIIB, TFIIF, and RNA polymerase II form a minimal initiation complex at the initiator. Addition of TFIIH, TFIIE, and ATP allows DNA melting at the initiator region and phosphorylation of the CTD of the largest subunit of RNA polymerase. These events allow production of abortive transcripts (magenta), but the polymerase stalls at position +10 to +12. (b) With energy provided by ATP, the DNA helicase of TFIIH causes further unwinding of the DNA, expanding the transcription bubble. This expansion releases the stalled polymerase and allows it to clear the promoter. (c) With continuous addition of NTPs, the elongation complex continues elongating the RNA. TBP and TFIIB remain at the promoter. TFIIE and TFIIH are not needed for elongation and dissociate from the elongation complex. (*Source:* Reprinted from Goodrich and Tjian, *Cell* 77:153, 1994. Copyright 1994, with permission from Elsevier Science.)

reaction. Thus, because TFIIE and TFIIH appear to have no effect on initiation or elongation, Goodrich and Tjian concluded that TFIIE & TFIIH are required in the promoter clearance step. Figure 11.33 summarizes these findings and more recent data discussed in the next paragraphs.

Tjian and others assumed that the DNA helicase activity of TFIIH acted directly on the DNA at the initiator to melt it. But cross-linking studies performed in 2000 by Tae-Kyung Kim, Richard Ebright, and Danny Reinberg showed that TFIIH (in particular, the subunit bearing the promoter-melting DNA helicase) forms cross-links with DNA between positions +3 and +25, and perhaps farther downstream. This site of interaction for TFIIH is downstream of the site of the first transcription bubble (position −9 to +2). On the other hand, TFIIE cross-links to the transcription bubble region; TFIIB, TFIID, and TFIIF cross-link to the region upstream of the bubble; and RNA polymerase cross-links to the entire region encompassing all the other factors. These findings imply that the DNA helicase of TFIIH is not in contact with the first transcription bubble, and therefore cannot create the bubble by directly unwinding DNA there. Addition of ATP has no effect on the interactions upstream of the transcription bubble, but it does perturb the interactions within and downstream of the bubble.

We know from previous work that the helicase of TFIIH is responsible for creating the transcription bubble, but the cross-linking work described here indicates that it cannot directly unwind the DNA at the transcription bubble. So how does it create the bubble? Kim and associates suggested that

it acts like a molecular "wrench" by untwisting the downstream DNA. Because TFIID and TFIIB (and perhaps other proteins) hold the DNA upstream of the bubble tightly, and this binding persists after addition of ATP, untwisting the downstream DNA would create strain in between and open up the DNA at the transcription bubble. This would allow the polymerase to initiate transcription and move 10–12 bp downstream. But previous work has shown that the polymerase stalls at that point unless it gets further help from TFIIH, which apparently twists the downstream DNA further to lengthen the transcription bubble, releasing the stalled polymerase to clear the promoter.

SUMMARY TFIIE, composed of two molecules each of a 34-kD and a 56-kD polypeptide, binds after polymerase and TFIIF. Both subunits are required for binding and transcription stimulation. A subunit of TFIIH phosphorylates the carboxyl-terminal domain (CTD) of the largest RNA polymerase II subunit. TFIIE greatly stimulates this process in vitro. TFIIE and TFIIH are not essential for formation of an open-promoter complex, or for elongation, but they are required for promoter clearance. TFIIH has a DNA helicase activity that is essential for transcription, presumably because it causes full melting of the DNA at the promoter and thereby facilitates promoter clearance.

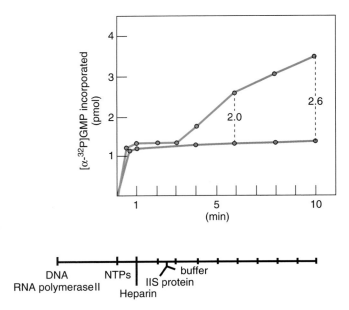

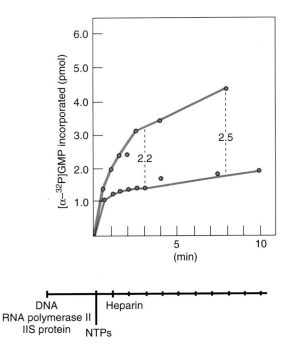

Figure 11.34   Effect of IIS on transcription elongation. Reinberg and Roeder formed elongation complexes as outlined in the time line at bottom. At time –3 min, they added DNA and RNA polymerase, then at time 0 they started the reaction by adding all four NTPs, one of which (GTP) was ^{32}P-labeled. At time +1 min, they added heparin to bind any free RNA polymerase, so all transcription complexes thereafter should be elongation complexes. Finally, at time +2.5 min, they added either IIS (red) or buffer (blue) as a negative control. They allowed labeled GMP incorporation to occur for various lengths of time, then took samples of the reaction mixture and measured the label incorporated into RNA. The dashed vertical lines indicate the fold stimulation of total RNA synthesis by IIS. (*Source:* From D. Reinberg and R. G. Roeder, *Journal of Biological Chemistry* 262:3333, 1987. Copyright © 1987 The American Society for Biochemistry & Molecular Biology, Bethesda, MD. Reprinted by permission.)

Elongation Factors

Eukaryotes control transcription primarily at the initiation step, but they also exert some control during elongation, at least in class II genes. In 1987, Reinberg and Roeder discovered a HeLa cell factor, which they named **TFIIS,** that specifically stimulates transcription elongation in vitro. This factor is homologous to **IIS,** which was originally found by Natori and colleagues in Ehrlich ascites tumor cells.

IIS Stimulates Elongation   Reinberg and Roeder demonstrated that IIS affects elongation, but not initiation, by testing it on preinitiated complexes (Figure 11.34). They incubated polymerase II with a DNA template and nucleotides to allow initiation to occur, then added heparin to bind any free polymerase and block new initiation, then added either IIS or buffer and measured the rate of incorporation of labeled GMP into RNA. Figure 11.34 shows that IIS enhanced the rate of RNA synthesis considerably: the vertical dashed lines show that IIS stimulated GMP incorporation 2.0-fold by the 6-min mark, and 2.6-fold by the 10-min mark.

Figure 11.35   Effect of IIS on transcription initiation and elongation combined. Reinberg and Roeder carried out this experiment in the same manner as in Figure 11.34, except for the order of additions to the reaction. Here, they added IIS (or buffer) at the beginning instead of last (see time line at bottom). Thus, IIS had the opportunity to stimulate both initiation and elongation. The dashed vertical lines show no more stimulation than in Figure 11.34. (*Source:* From D. Reinberg and R. G. Roeder, *Journal of Biological Chemistry* 262:3333, 1987. Copyright © 1987 The American Society for Biochemistry & Molecular Biology, Bethesda, MD. Reprinted by permission.)

It remained possible that IIS also stimulated transcription initiation. To investigate this possibility, Reinberg and Roeder repeated the experiment, but added IIS in the initial incubation, before they added heparin. If IIS really did stimulate initiation as well as elongation, then it should have produced a greater stimulation in this experiment than in the first. But Figure 11.35 shows that the stimulations by IIS in the two experiments were almost identical. Thus, IIS appears to stimulate elongation only.

We can also ask *how* IIS enhances transcription elongation. Reinberg and Roeder performed an experiment that strongly suggested it does so by limiting polymerase pausing. A common characteristic of RNA polymerases is that they do not transcribe at a steady rate. Instead, they pause, sometimes for a long time, before resuming transcription. These pauses tend to occur at certain defined **pause sites,** and we can detect such pausing during in vitro transcription by electrophoresing the in vitro transcripts and finding discrete bands that are shorter than full-length transcripts. Reinberg and Roeder found that IIS minimized the appearance of these short transcripts, indicating that it minimized polymerase pausing. Other workers have since confirmed this conclusion.

It is interesting that an initiation factor (TFIIF) is also reported to play a role in elongation. It apparently does not limit long pauses at defined DNA sites, as IIS does, but limits transient pausing at random DNA sites.

> **SUMMARY** Transcription can be controlled at the elongation level. One factor, IIS, stimulates elongation by limiting long pauses at discrete sites. TFIIF also stimulates elongation, apparently by limiting transient pausing.

IIS Stimulates Proofreading of Transcripts Not only does IIS counteract pausing, it also contributes to proofreading of transcripts, presumably by stimulating an inherent RNase in the RNA polymerase to remove misincorporated nucleotides. Diane Hawley and her colleagues followed the procedure in Figure 11.36a to measure the effect of IIS on proofreading. First, they isolated unlabeled elongation complexes that were stalled at a variety of sites close to the promoter. Next, they walked the complexes to a defined position (Chapter 6) in the presence of radioactive UTP to label the RNA in the complexes. Next, they added ATP or GTP to extend the RNA by one more base, to position +43. The base that is called for at this position is A, but if G is all that is available, the polymerase will incorporate it, though at lower efficiency. Actually, Hawley and colleagues discovered that their ultrapure GTP contained a small amount of ATP, so AMP and GMP were incorporated in about equal quantities at position +43, even though ultrapure GTP was the only nucleotide they added. Next, they either cleaved the products with RNase T1, which cuts after G's, or chased with all four nucleotides to extend the labeled RNA to full length and then cut it with RNase T1. Finally, they subjected all RNase T1 products to electrophoresis and visualized the labeled products by autoradiography.

Simply cleaving the transcript with RNase T1 allowed Hawley and coworkers to measure the relative incorporations of AMP and GMP into position +43 because electrophoresis clearly separated the terminal 7-mers ending in A and G. Figure 11.36b, lane 1 shows the results of an experiment with no chasing. The 7-mer ending in G, the result of misincorporation of G, is about equally represented with the combination of a 7-mer ending in A, and an 8-mer ending in AC, which result from correct incorporation of A (or AC) from nucleotides contaminating the GTP substrate. Lanes 2 and 3 show the effects of chasing in the absence or presence, respectively, of IIS. The chased, full-length transcripts were cleaved with RNase T1, which yielded a 7-mer ending in G_p from a full-length transcript that still contained the misincorporated G, or a 10-mer ending in G_p from a full-length transcript in which proofreading had changed the misincorporated G to an A. When Hawley and colleagues did the chase in the absence of IIS, a significant amount of the misincorporated G

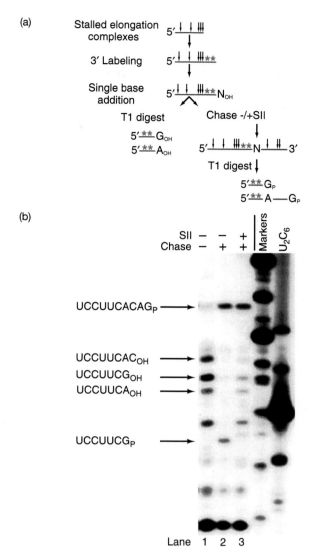

Figure 11.36 TFIIS stimulates proofreading by RNA polymerase II.
(a) Experimental scheme. Hawley and colleagues started with short elongation complexes and 3′-end-labeled the short transcripts by walking the polymerase farther in the presence of [α-32]UTP. Then they added GTP to force misincorporation of G into position +43 where an A was called for. Then they digested the labeled transcripts with RNase T1 to measure the misincorporation of G (left), or chased the transcripts into full length with all four nucleotides, then cleaved the transcripts with RNase T1 to measure the loss of G from position +43 by proofreading. **(b)** Experimental results. Hawley and colleagues electrophoresed the RNase T1 products from part **(a)** and visualized them by autoradiography. Lane 1 contained unchased transcripts. The 7-mer resulting from misincorporation of G ($UCCUUCG_{OH}$), and the 7-mer (UCCUUCA) and 8-mer (UCCUUCAC) resulting from normal incorporation of A (or A and C) are indicated by arrows at left. Lanes 2 and 3 contained RNase T1 products of transcripts chased in the absence (lane 2) or presence (lane 3) of TFIIS. The 7-mer ($UCCUUCG_p$) indicative of the misincorporated G that remained in the chased transcript is denoted by an arrow at left. The 10-mer ($UCCUUCACAG_p$) indicative of incorporation of A in position +43, or G replaced by A at that position by proofreading, is also denoted by an arrow at left. TFIIS allowed removal of all detectable misincorporated G.
(*Source:* (b) Thomas, M. et al. Transcriptional Fidelity and Proofreading by RNA Polymerase II. *Cell* 93 (1988) f. 4, p. 631. Reprinted by permission of Elsevier Science.)

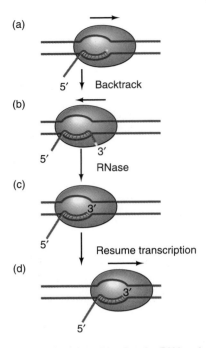

Figure 11.37 A model for proofreading by RNA polymerase II.
(a) The polymerase, transcribing the DNA from left to right, has just incorporated an incorrect nucleotide (yellow). (b) The polymerase backtracks to the left, extruding the 3'-end of the RNA, with its misincorporated nucleotide, out of the active site of the enzyme.
(c) The ribonuclease activity of the polymerase clips off the 3'-end of the RNA, including the incorrect nucleotide. (d) The polymerase resumes transcription.

remained in the RNA (see the band in lane 2 opposite the arrow indicating the 7-mer UCCUUCG$_p$). However, most of the product appeared in the 10-mer (arrow labeled UCCUUCACAG$_p$), which indicates that the polymerase was able to do some proofreading even without IIS). On the other hand, when they include IIS in the chase, Hawley and colleagues discovered that the 7-mer disappeared, and all of the labeled product was in the form of the 10-mer. Thus, IIS stimulates proofreading of the transcript.

One model for proofreading is illustrated in Figure 11.37. The polymerase not only stalls in response to a misincorporated nucleotide, it backtracks, extruding the 3'-end of the RNA out of the polymerase. As a result, the polymerase can cut the RNA near its end, releasing the misincorporated nucleotide, among a few others, then resume transcribing. IIS could aid this process by stimulating the inherent RNase in the RNA polymerase.

SUMMARY IIS stimulates proofreading—the correction of misincorporated nucleotides—presumably by stimulating the RNase activity of the RNA polymerase, allowing it to cleave off a misincorporated nucleotide (and perhaps other nucleotides) and replace it with the correct one.

The Polymerase II Holoenzyme

Our discussion so far has assumed that a preinitiation complex assembles at a class II promoter one protein at a time. This may indeed occur, but mounting evidence suggests that some, and perhaps all, class II preinitiation complexes assemble by binding a preformed **holoenzyme** to the promoter. The holoenzyme contains RNA polymerase and a subset of general transcription factors and other proteins.

Strong evidence for the holoenzyme concept came in 1994 with work from the laboratories of Roger Kornberg and Richard Young. Both groups isolated a complex protein from yeast cells, which contained RNA polymerase II and many other proteins. Kornberg and colleagues used immunoprecipitation with an antibody directed against one component of the holoenzyme to precipitate the whole complex. Figure 11.38 shows the results of SDS-PAGE on the polypeptides contained in this holoenzyme. They recovered the subunits of RNA polymerase II, the subunits of TFIIF, and 17 other polypeptides. They could restore accurate transcription activity to this holoenzyme by adding TBP, TFIIB, E, and H. TFIIF was not required because it was already part of the holoenzyme.

Anthony Koleske and Young used a series of purification steps to isolate a holoenzyme from yeast that contained RNA polymerase II, TFIIB, TFIIF, and TFIIH. All this holenzyme needed for accurate transcription in vitro was TFIIE and TBP, so it contained more of the general transcription factors than the holoenzyme isolated by Kornberg and associates. Koleske and Young also identified some of the other polypeptides in their holoenzyme as **SRB proteins** (**SRB2, SRB4, SRB5,** and **SRB6**). The SRB proteins were discovered by Young and colleagues in a genetic screen whose logic went like this: Deletion of part of the CTD of the largest polymerase II subunit led to ineffective stimulation of transcription by the GAL4 protein, a transcription activator we will study in greater detail in Chapter 12. Young and coworkers then screened for mutants that could suppress this weak stimulation by GAL4. They identified several suppressor mutations in genes they named *SRB*s, for "suppressor of RNA polymerase B." We will discuss the probable basis for this suppression in Chapter 12. For now, it is enough to stress that these SRB proteins are required, at least in yeast, for optimal activation of transcription in vivo, and that they are part of the yeast polymerase II holoenzyme.

Erich Nigg and Ueli Schibler and their colleagues used immunoprecipitation to isolate a holoenzyme from rat liver cells. This complex contained RNA polymerase II, TFIID, B, E, F, and H. It was not clear whether the rat equivalents (if they exist) of the SRB proteins were also present. This holoenzyme was competent to carry out accurate transcription initiation in vitro.

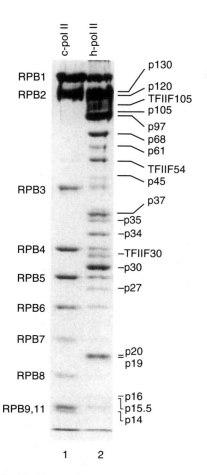

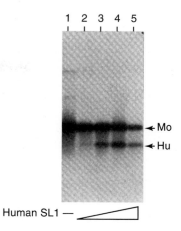

Figure 11.39 SL1 is a species-specific transcription factor. Tjian and colleagues performed a run-off assay with a mouse cell-free extract and two templates, one containing a mouse rRNA promoter, the other containing a human rRNA promoter. The mouse and human templates gave rise to run-off transcripts of 2400 and 1500 nt, respectively. As shown at bottom, lane 1 contained no human SL1, and essentially only the mouse template was transcribed. As Tjian and colleagues added more and more human SL1, they observed more and more transcription of the human template, and less transcription of the mouse template. In lane 5, transcription of both templates seems to be suppressed. (*Source:* Learned et al., *Mol. Cell. Biol* 5, p. 1365, f. 6. American Society for Microbiology.)

Figure 11.38 Purified yeast RNA polymerase II holoenzyme. Kornberg and colleagues used a purification scheme that included immunoprecipitation to isolate a polymerase II holoenzyme from yeast cells, then subjected the polypeptide constituents of this holoenzyme to SDS-PAGE. Lane 2 displays these polypeptides (h-pol II), and lane 1 contains the subunits of the "core RNA polymerase II" (c-pol II) for comparison. (*Source:* Kim et al., A multiprotein mediator of transcriptional activation and its interaction with the C-terminal repeat domain of RNA polymerase II. *Cell* 77 (20 May 1994) p. 602, f. 4b. Reprinted by permission of Elsevier Science.)

> **SUMMARY** Yeast and mammalian cells have an RNA polymerase II holoenzyme that contains many polypeptides in addition to the subunits of the polymerase. The yeast holoenzyme contains a subset of general transcription factors and at least some of the SRB proteins. The rat holoenzyme contains all the general transcription factors except TFIIA.

11.2 Class I Factors

The preinitiation complex that forms at rRNA promoters is much simpler than the polymerase II preinitiation complex we have just discussed. It involves polymerase I, of course, in addition to just two transcription factors, called **SL1** and an **upstream-binding factor, UBF.**

SL1

Tjian and his colleagues discovered SL1 in 1985, when they separated a HeLa cell extract into two functional fractions. One fraction had RNA polymerase I activity, but no ability to initiate accurate transcription of a human rRNA gene in vitro. Another fraction had no polymerase activity of its own, but could direct the polymerase fraction to initiate accurately on a human rRNA template. Furthermore, this transcription factor, SL1, showed species specificity. That is, it could distinguish between the human and mouse rRNA promoter. Figure 11.39 shows this phenomenon. When Tjian and colleagues provided a mouse cell-free extract with two templates, one with the human, and one with the mouse rRNA promoter, and added increasing amounts of human SL1, they observed increasing transcription of the human template at the expense of the mouse template.

Tjian and coworkers went on to show by footprinting that a partially purified polymerase I preparation could bind to the human rRNA promoter (Figure 11.40). In particular, it caused a footprint in a region of the UCE called **site A.** We will see later in this chapter that this binding is not due to polymerase itself, but to a transcription factor called UBF that contaminated the crude polymerase preparation. SL1 by itself caused no footprint, but

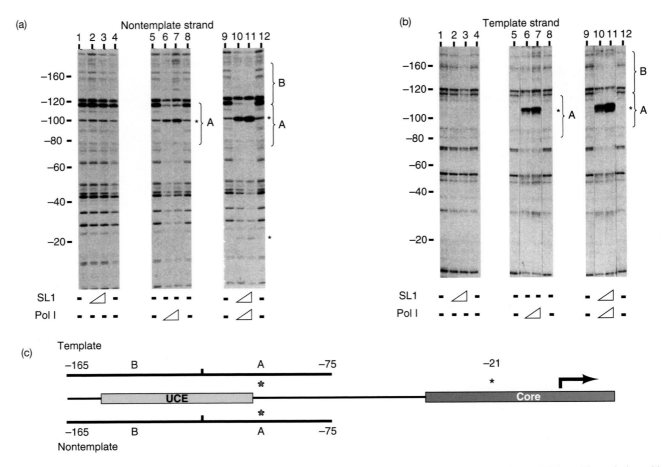

Figure 11.40 Footprinting the rRNA promoter with SL1 and RNA polymerase I. Tjian and colleagues performed DNase I footprinting with either the (**a**) nontemplate strand, or (**b**) the template strand of the human rRNA promoter. They added SL1 or RNA polymerase (or both), as indicated at bottom. Brackets indicate footprint regions, and stars designate sites of enhanced DNase sensitivity. Polymerase I by itself can protect a region (A) of the UCE; polymerase and SL1 together extend the protection into another region (B) of the UCE. Binding of SL1 by itself is not detectable by this assay. (**c**) Summary of footprints. Bars above and below the UCE region represent the footprints on the template and nontemplate strands, respectively, with the A and B sections delineated. Again, stars represent the sites of enhanced cleavage. (*Source:* Learned et al., Human rRNA transcription is modulated by the coordinate binding of two factors to an upstream control element. *Cell* 45 (20 Jun 1986) p. 849, f. 2a-b. Reprinted by permission of Elsevier Science.)

it could combine with the crude polymerase I preparation to extend the UCE footprint upstream into **site B** and cause enhanced DNase cleavage of a site in the core promoter element. Thus, SL1 plays a role in assembling the polymerase I preinitiation complex.

We have already seen that SL1 determines species specificity in rRNA gene transcription. Tjian and colleagues also showed that the core promoter element helps determine species specificity. They constructed hybrid rRNA promoters containing the UCE from mouse and the core element from human, or vice versa, then transcribed these templates in a run-off assay using partially purified human polymerase I and highly purified human SL1. As expected, a completely human promoter supported transcription by the human factors and polymerase (Figure 11.41, lanes 1 and 2), but a wholly mouse promoter did not (lanes 10 and 11). More importantly, the construct with a mouse UCE and a human

core element supported transcription (lanes 7 and 8), but when the promoter elements were switched to human UCE and mouse core element, no transcription occurred (lanes 10 and 11). Thus, the core element is critical, but the UCE is irrelevant to species specificity. Later in this chapter we will see evidence that SL1 is a complex protein composed of TBP and three TAFs. Thus, TBP functions with class I promoters, even though they lack TATA boxes.

SUMMARY SL1 does not bind by itself to the rRNA promoter, but it interacts with crude RNA polymerase I to extend and strengthen the binding to the UCE and perturb the DNA in the core element. SL1 also determines the species specificity of transcription, as does the core promoter element.

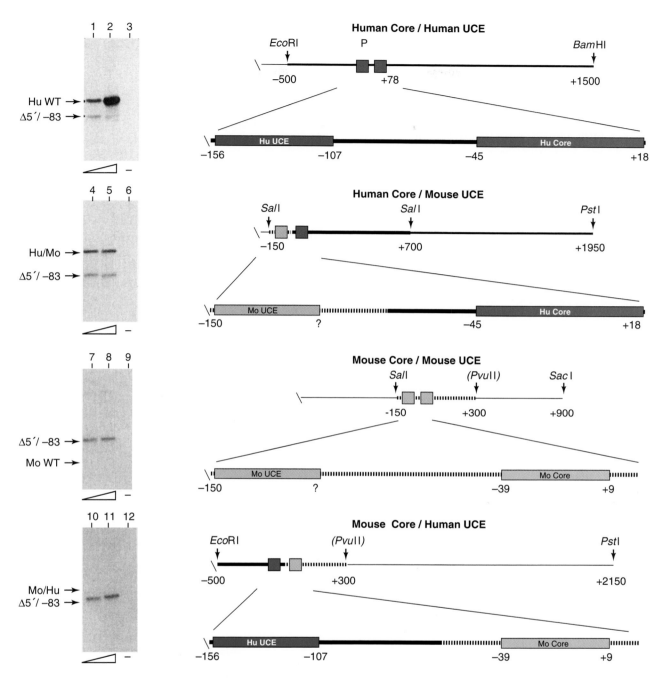

Figure 11.41 The core promoter element determines species specificity. Tjian and colleagues constructed human, mouse, and hybrid human/ mouse rRNA promoters and tested them for promoter activity by a run-off transcription assay with partially purified human RNA polymerase I and highly purified human SL1. All reactions contained a control template, Δ 5'/ –83, which had a human rRNA promoter lacking the UCE. This gave a basal level of transcription in all cases and could be used to normalize the reactions. The expected position of each run-off transcript is indicated at left with an arrow. The first two lanes in each set of three contained increasing quantities of human SL1, as indicated by the "wedges"; the third lane in each set had no SL1. Diagrams of each construct are given at right. Human promoter elements are rendered in green, and mouse elements in pink. Only when the construct contained a human core element did transcription occur. The nature of the UCE was irrelevant. Human SL1 was also required. Thus, the core element determines the species specificity of the rRNA promoter. (*Source:* Learned et al., *Cell* 45 (20 Jun 1986) p. 852, f. 4. Reprinted by permission of Elsevier Science.)

UBF

Because SL1 by itself did not appear to bind directly to the rRNA promoter, but a partially purified RNA polymerase I preparation did, Tjian and his coworkers began a search for DNA-binding proteins in the polymerase preparation. This led to the purification of UBF in 1988. The factor as purified was composed of two polypeptides, of 97 and 94 kD. However, the 97-kD polypeptide alone is sufficient for UBF activity. When Tjian and colleagues performed footprint analysis with this highly purified UBF, they found that it had the same behavior as observed previously with partially purified polymerase I. That is, it gave the same footprint in the core element and site A of the UCE, and SL1 intensified this footprint and extended it to site B. In fact, the footprint was even clearer, especially in the core element, with UBF than with the partially purified polymerase (Figure 11.42). Thus, UBF, not polymerase I, was the agent that bound to the promoter in the previous experiments. These studies did not reveal whether SL1 actually contacts the DNA in a complex with UBF, or whether it merely changes the conformation of UBF so it can contact a longer stretch of DNA that extends into site A.

Tjian and associates also found that UBF stimulates transcription of the rRNA gene in vitro. Figure 11.43 depicts the results of a transcription experiment using the wild-type human rRNA promoter and the mutant promoter ($\Delta 5'-57$) that lacks the UCE, and including various combinations of SL1 and UBF. Polymerase I was present in all reactions, and transcription efficiency was assayed by the S1 technique (Chapter 5). Lane 1 contained UBF, but no SL1, and showed no transcription of either template. This reaffirms that SL1 is absolutely required for transcription. Lane 2 had SL1, but no UBF, and showed a basal level of transcription. This demonstrates again that SL1 by itself is capable of stimulating basal transcription. Moreover, about as much transcription occurred on the mutant template that lacks the UCE as on the wild-type template. Lanes 3 and 4 contained both SL1 and an increasing amount of UBF. Significantly enhanced transcription occurred on both templates, but especially on the template containing the UCE. Tjian and colleagues concluded that UBF is a transcription factor that can stimulate transcription either in the presence or absence of the UCE.

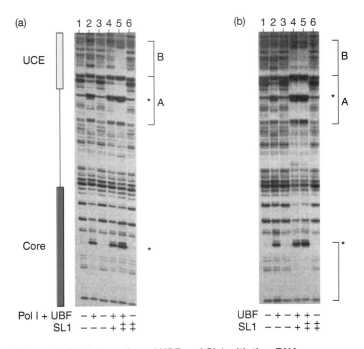

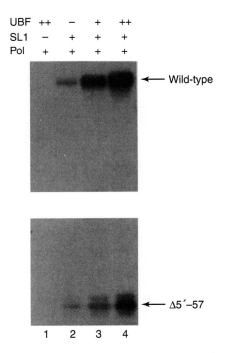

Figure 11.42 Interaction of UBF and SL1 with the rRNA promoter. Tjian and colleagues performed DNase I footprinting with the human rRNA promoter and various combinations of (a) polymerase I + UBF and SL1 or (b) UBF and SL1. The proteins used in each lane are indicated at bottom. The positions of the UCE and core elements are shown at left, and the locations of the A and B sites are illustrated with brackets at right. Stars mark the positions of enhanced DNase sensitivity. SL1 caused no footprint on its own, but enhanced and extended the footprints of UBF in both the UCE and the core element. This enhancement is especially evident in the absence of polymerase I (panel **b**). (*Source:* Bell et al., Functional cooperativity between transcription factors UBF1 and SL1 mediates human ribosomal RNA synthesis. *Science* 241 (2 Sept 1988) p. 1194, f. 3 a-b. © AAAS.)

Figure 11.43 Activation of transcription from the rRNA promoter by SL1 and UBF. Tjian and colleagues used an S1 assay to measure transcription from the human rRNA promoter in the presence of RNA polymerase I and various combinations of UBF and SL1, as indicated at top. The top panel shows transcription from the wild-type promoter; the bottom panel shows transcription from a mutant promoter ($\Delta 5'-57$) lacking UCE function. SL1 was required for at least basal activity, but UBF enhanced this activity on both templates. (*Source:* Bell et al., *Science* 241 (2 Sept 1988) p. 1194, f. 4. © AAAS.)

Two of the findings we have discussed so far suggest that UBF and SL1 interact in stimulating promoter activity via the UCE. First, UBF and SL1 act synergistically to stimulate transcription. Second, the UCE footprint of UBF and SL1 together is greater than the sum of the footprints of the two proteins separately. To test this conclusion, Tjian and coworkers constructed a series of UCE mutants (Figure 11.44a) that could still bind UBF as well as normal (because site A was unaffected), but that prevented the formation of the UBF–SL1 complex (because site B was altered). Transcription from these mutant promoters

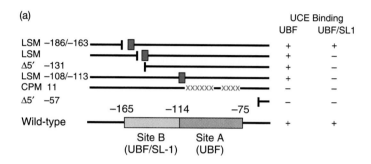

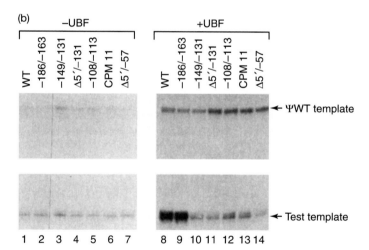

Figure 11.44 Effect of mutations in the UCE on UBF activation. (a) Description of mutants and effects on binding. Inserted linkers are represented by boxes, deletions by spaces, and bases altered following site-directed mutagenesis by X's. The positions of sites A and B of the UCE, relative to the mutations, are given at bottom. Binding of UBF, or UBF/SL1, to each mutant promoter is reported at right. Tjian and colleagues measured the binding by footprinting; the criterion for UBF/SL1 binding was extension of the footprint into site B. (b) Effect on transcription. Tjian and coworkers measured transcription by S1 analysis as in Figure 11.43, in the presence (right panel) or absence (left panel) of UBF. They included SL1 and polymerase I in all cases. They also added a pseudo-wild-type template (ΨWT) as an internal control in all cases. The nature of the test template (wild-type or mutant) is given at the top of each lane. Mutant–186/–163 behaved like the wild-type template in that it supported stimulation by UBF. By contrast, all the other mutant templates were considerably impaired in ability to respond to UBF. (*Source:* (*Source:* Bell et al., *Science* 241 (2 Sept 1988) p. 1195, f. 5a. © AAAS.)

showed very little stimulation by UBF, even in the presence of SL1 (Figure 11.44b). Thus, interaction between UBF and SL1 appears to be necessary for stimulation of transcription mediated by the UCE.

Tjian and colleagues also prepared mutations in the core element and found that one of them blocked both UBF binding and transcription, which is not surprising. Some others allowed UBF binding, but were impaired in transcription. Thus, as we have seen with the UCE, UBF binding to the core element is not sufficient to stimulate transcription.

By 1990, purified UBF and SL1 were available, not only from human but also from mouse. This allowed Tjian and his colleagues to conduct mixing experiments with factors and polymerases from both species that led to the following conclusions: First, in contrast to the human SL1, which does not bind on its own to DNA, the mouse SL1 can bind to the mouse rRNA promoter. Second, human and mouse UBFs are interchangeable. That is, they can substitute for each other in binding or transcription assays regardless of whether the template comes from mouse or human. Third, human and mouse SL1s are *not* interchangeable in transcription reactions. Human templates are active only with human SL1, and mouse templates are active only with mouse SL1. This reinforces the notion that the SL1, not the UBF or polymerase, confers species specificity.

> **SUMMARY** UBF is a transcription factor that stimulates transcription by polymerase I. It can activate the intact promoter, or the core element alone, and it mediates activation by the UCE. UBF and SL1 act synergistically to stimulate transcription.

Structure and Function of SL1

We have been discussing just two factors, UBF and SL1, that are involved in transcription by polymerase I, and one of these, UBF, is probably just a single 97-kD polypeptide. But work presented earlier in this chapter showed that TBP is essential for class I transcription. Where then does TBP fit in? Tjian and coworkers demonstrated in 1992 that SL1 is composed of TBP and three TAFs. First, they purified human (HeLa cell) SL1 by several different procedures. After each step, they used an S1 assay to locate SL1 activity. Then they assayed these same fractions for TBP by Western blotting. Figure 11.45 shows the striking correspondence they found between SL1 activity and TBP content.

If SL1 really does contain TBP, then it should be possible to inhibit SL1 activity with an anti-TBP antibody. Figure 11.46 confirms that this worked as predicted. A nuclear extract was depleted of SL1 activity with an

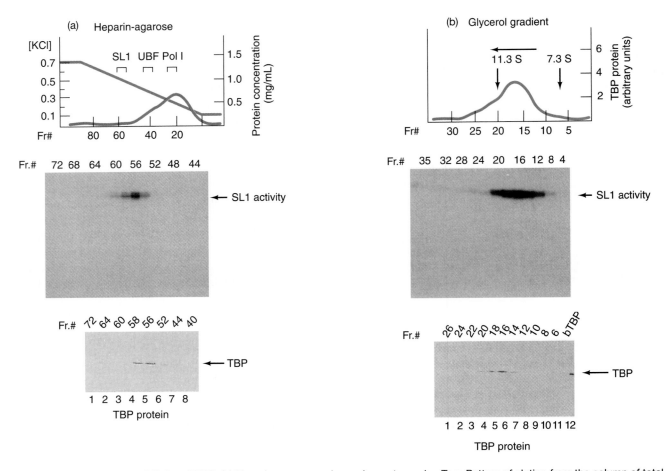

Figure 11.45 Co-purification of SL1 and TBP. (a) Heparin–agarose column chromatography. Top: Pattern of elution from the column of total protein (red) and salt concentration (blue), as well as three specific proteins (brackets). Middle: SL1 activity, measured by S1 analysis, in selected fractions. Bottom: TBP protein, detected by Western blotting, in selected fractions. Both SL1 and TBP were centered on fraction 56. **(b)** Glycerol gradient ultracentrifugation. Top: Sedimentation profile of TBP. Two other proteins, catalase and aldolase, with sedimentation coefficients of 11.3 S and 7.3 S, respectively, were run in a parallel centrifuge tube as markers. Middle and bottom panels, as in panel **(a)**. Both SL1 and TBP sedimented to a position centered around fraction 16. (*Source:* Comai et al., The TATA-binding protein and associated factors are integral components of the RNA polymerase I transcription factor, SL1. *Cell* 68 (6 Mar 1992) p. 968, f. 2a-b. Reprinted by permission of Elsevier Science.)

anti-TBP antibody. Activity could then be restored by adding back SL1, but not just by adding back TBP. Something besides TBP must have been removed.

What other factors are removed along with TBP by immunoprecipitation? To find out, Tjian and colleagues subjected the immunoprecipitate to SDS-PAGE. Figure 11.47 depicts the results. In addition to TBP and antibody (IgG), we see three polypeptides, with molecular masses of 110, 63, and 48 kD (although the 48-kD polypeptide is partially obscured by TBP). Because these were immunoprecipitated along with TBP, they must bind tightly to TBP and are therefore TBP-associated factors, or TAFs, by definition. Hence, Tjian called them **TAF$_I$110, TAF$_I$63**, and **TAF$_I$48**. These are completely different from the TAFs found in TFIID (compare lanes 4 and 5). The TAFs could be stripped off

of the TBP and antibody in the immunoprecipitate by treating the precipitate with 1 M guanidine-HCl and re-precipitating. The antibody and TBP remained together in the precipitate (lane 6) and the TAFs stayed in the supernatant (lane 7). Tjian and colleagues could reconstitute SL1 activity by adding together purified TBP and the three TAFs, and this activity was species-specific, as one would expect. In later work, Tjian and coworkers showed that the TAF$_I$s and TAF$_{II}$s could compete with each other for binding to TBP. This finding suggested that binding of one set of TAFs to TBP is mutually exclusive of binding of the other set.

Thus, both polymerase I and polymerase II rely on transcription factors (SL1 and TFIID, respectively) composed of TBP and several TAFs. The TBP is identical in the two factors but the TAFs are completely different.

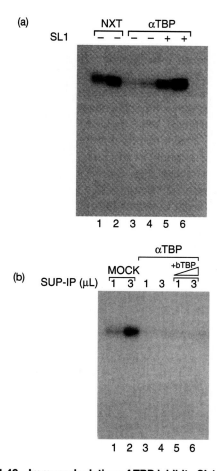

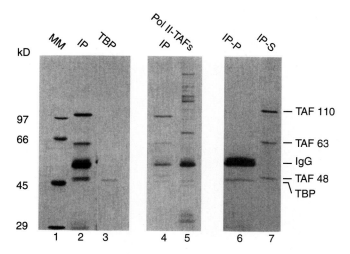

Figure 11.47   The TAFs in SL1. Tjian and colleagues immunoprecipitated SL1 with an anti-TBP antibody and subjected the polypeptides in the immunoprecipitate to SDS-PAGE. Lane 1, molecular weight markers; lane 2, immunoprecipitate (IP); lane 3, purified TBP for comparison; lane 4, another sample of immunoprecipitate; lane 5, TFIID TAFs (PolII-TAFs) for comparison; lane 6, pellet after treating immunoprecipitate with 1 M guanidine–HCl and reprecipitating, showing TBP and antibody; lane 7, supernatant after treating immunoprecipitate with 1 M guanidine–HCl and reprecipitating, showing the three TAFs (labeled at right). (*Source:* Comai et al., The TATA-binding protein and associated factors are integral components of the RNA polymerase I transcription factor, SL1. *Cell* 68 (6 Mar 1992) p. 971, f. 5. Reprinted by permission of Elsevier Science.)

Figure 11.46   Immunodepletion of TBP inhibits SL1 activity.
(**a**) Immunodepletion. Lanes 1 and 2 (positive control), an untreated nuclear extract (NXT) was tested by S1 analysis for ability to transcribe the human rRNA gene. Lanes 3 and 4, a nuclear extract was immunoprecipitated with an anti-TBP antibody (αTBP) and the supernatant was tested for activity. Little activity remained. Lanes 5 and 6, the immunodepleted nuclear extract was supplemented with SL1, which restored activity. Even-numbered lanes contained twice as much nuclear extract as odd-numbered lanes. (**b**) Pure TBP fails to restore SL1 activity to immunodepleted extract. Lanes 1 and 2, a mock-depleted nuclear extract was tested as in part (**a**) for rRNA gene transciption activity. Lanes 3 and 4, a nuclear extract was immunoprecipitated as in part (**a**) and the supernatant (SUP-IP) tested for activity. Little remained. Lanes 5 and 6, the immunodepleted nuclear extract was supplemented with 4 ng or 12 ng, respectively, of recombinant TBP. Little or no restoration of activity occurred. The even-numbered lanes had three times as much nuclear extract as the odd-numbered lanes, as indicated at top. (*Source:* Comai et al., The TATA-binding protein and associated factors are integral components of the RNA polymerase I transcription factor, SL1. *Cell* 68 (6 Mar 1992) p. 970, f. 4. Reprinted by permission of Elsevier Science.)

> **SUMMARY** SL1 is composed of TBP and three TAFs: TAF$_I$110, TAF$_I$63, and TAF$_I$48. Fully functional and species-specific SL1 can be reconstituted from these purified components, and binding of TBP to the TAF$_I$s precludes binding to the TAF$_{II}$s.

11.3 Class III Factors

In 1980, Roeder and his colleagues discovered a factor that bound to the internal promoter of the 5S rRNA gene and stimulated its transcription. They named the factor **TFIIIA.** Since then, two other factors, TFIIIB and C, have been discovered. These two factors participate, not only in 5S rRNA gene transcription, but in all transcription by polymerase III. (Also, as we learned earlier in this chapter, six factors that cooperate with polymerase II have been found and named according to the same system of nomenclature: TFIIA, B, and so on. Unfortunately for students struggling to learn these names, the polymerase I factors do not follow the same pattern. Thus, we have SL1 and UBF, where we might have had TFIA and B.)

Barry Honda and Roeder demonstrated the importance of the TFIIIA factor in 5S rRNA gene transcription when they developed the first eukaryotic in vitro transcription system, from *Xenopus laevis,* and found that it could make no 5S rRNA unless they added TFIIIA. Donald Brown and colleagues went on to show that similar cell-free extracts provided with a 5S rRNA gene and a tRNA gene could make both 5S rRNA and tRNA simultaneously. Furthermore, an antibody against TFIIIA could effectively halt the production of 5S rRNA, but had no effect on tRNA synthesis (Figure 11.48). Thus, TFIIIA is

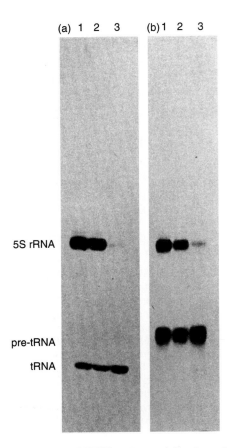

Figure 11.48 Effect of anti-TFIIIA on transcription by polymerase III. Brown and colleagues added cloned 5S rRNA and tRNA genes to (**a**) an oocyte extract, or (**b**) a somatic cell extract in the presence of labeled nucleotide and: no antibody (lanes 1), an irrelevant antibody (lanes 2), or an anti-TFIIIA antibody (lanes 3). After transcription, these workers electrophoresed the labeled RNAs. The anti-TFIIIA antibody blocked 5S rRNA gene transcription in both extracts, but did not inhibit tRNA gene transcription in either extract. The oocyte extract could process the pre-tRNA product to the mature tRNA form, but the somatic cell extract could not. Nevertheless, transcription occurred in both cases. (*Source:* Pelham et al., *PNAS* 78 (Mar 1981) p. 1762, f. 3.)

required for transcription of the 5S rRNA genes, but not the tRNA genes.

If transcription of the tRNA genes does not require TFIIIA, what factors *are* involved? In 1982, Roeder and colleagues separated two new factors they called **TFIIIB** and **TFIIIC** and found that they are necessary and sufficient for transcription of the tRNA genes. We have subsequently learned that these two factors govern transcription of all polymerase III genes, including the 5S rRNA genes. That means that the original extracts that needed to be supplemented only with TFIIIA to make 5S rRNA must have contained TFIIIB and C.

SUMMARY Transcription of all class III genes requires TFIIIB and C, and transcription of the 5S rRNA genes requires these two plus TFIIIA.

TFIIIA

As the very first eukaryotic transcription factor to be discovered, TFIIIA received a considerable amount of attention. It was the first member of a large group of DNA-binding proteins that feature a so-called **zinc finger.** We will discuss the zinc finger proteins in detail in Chapter 12. Here, let us concentrate on the zinc fingers of TFIIIA. The essence of a zinc finger is a roughly finger-shaped protein domain containing four amino acids that bind a single zinc ion. In TFIIIA, and in other typical zinc finger proteins, these four amino acids are two cysteines, followed by two histidines. However, some other zinc finger-like proteins have four cysteines and no histidines. TFIIIA has nine zinc fingers in a row, and these appear to insert into the major groove on either side of the internal promoter of the 5S rRNA gene. This allows specific amino acids to make contact with specific base pairs, forming a tight protein–DNA complex.

TFIIIB and C

TFIIIB and C are both required for transcription of the classical polymerase III genes, and it is difficult to separate the discussion of these two factors because they depend on each other for their activities. Peter Geiduschek and coworkers established in 1989 that a crude transcription factor preparation bound both the internal promoter and an upstream region in a tRNA gene. Figure 11.49 contains DNase footprinting data that led to this conclusion. Lane c is the digestion pattern with no added protein, lane a is the result with factors and polymerase III, and lane b has all this plus three nucleoside triphosphates (ATP, CTP, and UTP), which allowed transcription for just 17 nt, until the first GTP was needed. Notice in lane a that the factors and polymerase strongly protected box B of the internal promoter and the upstream region (U) and weakly protected box A of the internal promoter. Lane b shows that the polymerase shifted downstream and a new region overlapping box A was protected. However, the protection of the upstream region persisted even after the polymerase moved away.

What accounts for the persistent binding to the upstream region? To find out, Geiduschek and colleagues partially purified TFIIIB and C and performed footprinting studies with these separated factors. Figure 11.50 shows the results of one such experiment. Lane b, with TFIIIC alone, reveals that this factor protects the internal promoter, especially box B, but does not bind to the upstream region. When both factors are present, the upstream region is also protected (lane c). Similar DNase footprinting experiments made it clear that TFIIIB by itself does not bind to any of these regions. Its binding is totally dependent on TFIIIC. However, once TFIIIC has sponsored the binding of TFIIIB to the upstream region,

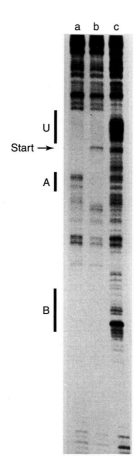

Figure 11.49   Effect of transcription on DNA binding between a tRNA gene and transcription factors. Geiduschek and colleagues performed DNase footprinting with a tRNA gene and an extract containing polymerase III, TFIIIB, and TFIIIC. Lane a contained transcription factors, but no nucleotides. Lane b had factors plus three of the four nucleotides (all but GTP), so transcription could progress for 17 nucleotides, until GTP was needed. Lane c was a control with no added protein. The 17-bp migration of the polymerase in lane b relative to lane a caused a corresponding downstream shift in the footprint around the transcription start site, to a position extending upstream and downstream of the A box. On the other hand, the footprint in the region just upstream of the start of transcription remained unchanged. (*Source:* Kassavetis et al., *Mol. Cell. Biol.* 9, no.171 (June 1989) p. 2555, f. 3. American Society for Microbiology.)

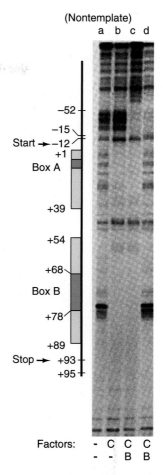

Figure 11.50   Binding of TFIIIB and C to a tRNA gene. Geiduschek and coworkers performed DNase footprinting with a labeled tRNA gene (all lanes), and combinations of purified TFIIIB and C. Lane a, negative control with no factors; lane b, TFIIIC only; lane c, TFIIIB plus TFIIIC; lane d, TFIIIB plus TFIIIC added, then heparin added to strip off any loosely bound protein. Note the added protection in the upstream region afforded by TFIIIB in addition to TFIIIC (lane c). Note also that this upstream protection provided by TFIIIB survives heparin treatment, but the protection of boxes A and B does not. Yellow boxes represent coding regions for mature tRNA. Boxes A and B within these regions are indicated in blue. (*Source:* From Kassavetis et al., *Molecular and Cellular Biology* 9:2558, 1989. Copyright © 1989 American Society for Microbiology, Washington, DC. Reprinted by permission.)

TFIIIB appears to remain there, even after polymerase has moved on (recall Figure 11.49). Moreover, Figure 11.50, lane d, shows that TFIIIB binding persists even after heparin has stripped TFIIIC away from the internal promoter, since the upstream region is still protected from DNase, even though boxes A and B are not.

As we have seen, transcription of the 5S rRNA gene depends on TFIIIA, in addition to TFIIIB and C. And, as we might predict, TFIIIB binding to the 5S rRNA gene depends on prior binding of the other two factors. David Setzer and Brown demonstrated this point by attaching a cloned 5S rRNA gene to a cellulose support, then incubating it with one factor at a time, washing in between

to remove any unbound factor. Finally, after they had exposed the gene to all three factors, they added polymerase III and carried out transcription. To see if any 5S rRNA had been made, they electrophoresed the labeled transcripts. Figure 11.51 shows that the order of addition of the factors matters a great deal. Whenever they added TFIIIB before one of the other factors, no accurate transcription of the 5S rRNA gene occurred. However, if they added TFIIIB last, as in lanes 2 and 5, proper transcription took place. This reinforced the conclusion that TFIIIB cannot bind by itself, but depends on other factors.

The evidence we have seen so far suggests the following model for involvement of transcription factors in

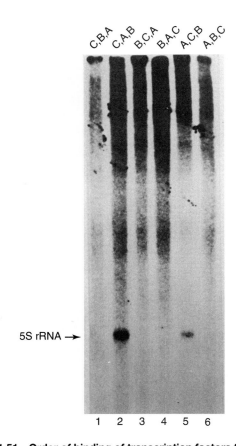

C,B,A C,A,B B,C,A B,A,C A,C,B A,B,C

5S rRNA →

1   2   3   4   5   6

Figure 11.51 Order of binding of transcription factors to a 5S rRNA gene. Setzer and Brown added factors TFIIIA, B, and C, one at a time to a cloned 5S rRNA gene bound to cellulose. After each addition, they washed away any unbound factor before incubation with the next factor. Finally, they added polymerase III and nucleotides, one of which was labeled, and assayed 5S rRNA synthesis by electrophoresing the products. The order of addition of factors is indicated at the top of each lane. Only when TFIIIB was added last did accurate 5S rRNA gene transcription occur. Thus, TFIIIB appears to need the help of the other factors to bind to the gene. (*Source:* Setzer and Brown, Formation and stability of the 5 S RNA transcription complex. *J. Biol Chem.* 260:4 (25 Feb 1985) p. 2488, f. 6. American Society for Biochemistry and Molecular Biology.)

polymerase III transcription (Figure 11.52): First, TFIIIC (or TFIIIA and C, in the case of the 5S rRNA genes) binds to the internal promoter; then these **assembly factors** allow TFIIIB to bind to the upstream region; then TFIIIB helps polymerase III bind at the transcription start site; finally, the polymerase transcribes the gene, probably removing TFIIIC (or A and C) in the process, but TFIIIB remains bound, so it can continue to promote further rounds of transcription.

Geiduschek and colleagues have provided further evidence to bolster this hypothesis. They bound TFIIIC and B to a tRNA gene (or TFIIIA, C, and B to a 5S rRNA gene), then removed the assembly factors, TFIIIC (or A and C) with either heparin or high salt, then separated the remaining TFIIIB–DNA complex from the other factors. Finally, they demonstrated that this TFIIIB–DNA com-

plex was still capable of supporting one round, or even multiple rounds, of transcription by polymerase III (Figure 11.53).

> **SUMMARY** Classical class III genes require two factors, TFIIIB and C, in order to form a preinitiation complex with the polymerase. The 5S rRNA genes also require TFIIIA. TFIIIC and A bind to the internal promoter and help TFIIIB bind to a region just upstream of the transcription start site. TFIIIB then remains bound and can sponsor the initiation of repeated rounds of transcription. TFIIIC (and A, in the case of the 5S rRNA genes) serves one of the same functions as TBP in TATA-containing class II genes. That is, it is the first factor to bind and serves as a center around which the other components of the preinitiation complex can organize.

The Role of TBP

If TFIIIC is necessary for TFIIIB binding in classical class III genes, what about nonclassical genes that have no A or B blocks to which TFIIIC can bind? What stimulates TFIIIB binding to these genes? Because the promoters of these genes have TATA boxes (Chapter 10), and we have already seen that TBP is required for their transcription, it makes sense to propose that the TBP binds to the TATA box and anchors TFIIIB to its upstream binding site.

But what about classical polymerase III genes? These have no TATA box, and TFIIIC clearly plays an organizing role analogous to that played by TBP in TATA-containing polymerase II genes. And yet we have seen that TBP is required for transcription of classical class III genes such as the tRNA and 5S rRNA genes in yeast and human cells. Where does TBP fit into this scheme? It has now become clear that TFIIIB contains TBP along with a small number of TAFs. Geiduschek and coworkers showed that TBP was present even in the purest preparations of TFIIIB, and Nouria Hernandez and Franklin Pugh and their coworkers independently demonstrated the presence of TAFs associated with TBP in TFIIIB. Pugh and colleagues found two TAF$_{III}$s and named them **TAF-172** and **TAF-L.**

Subsequently, Tjian and coworkers have shown by adding factors back to immunodepleted nuclear extracts that TRFI, not TBP, is essential for transcribing *Drosophila* tRNA, 5S rRNA and U6 snRNA genes. Thus, transcription by polymerase III in the fruit fly is another exception to the generality of dependence on TBP.

A unifying principle that emerges from the studies on transcription factors for all three RNA polymerases is that the assembly of a preinitiation complex starts with an assembly factor that recognizes a specific binding site in the

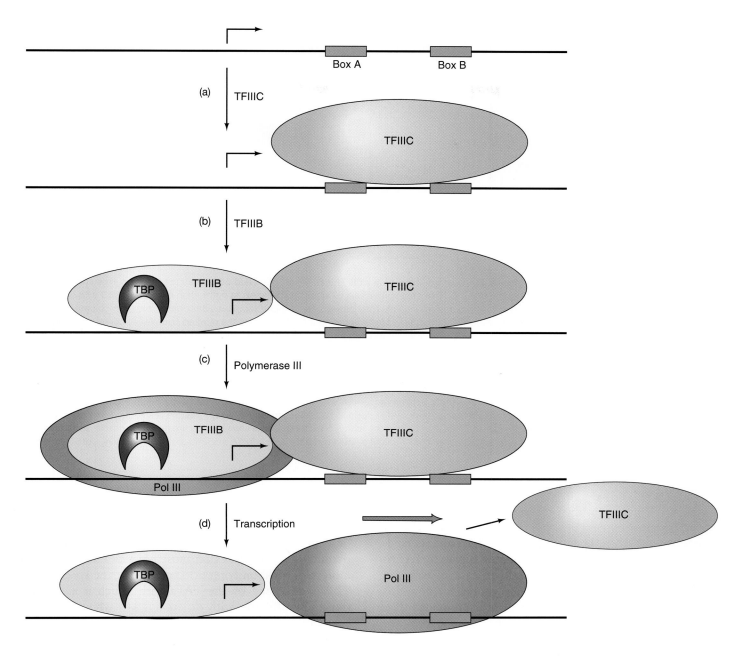

Figure 11.52   Hypothetical scheme for assembly of the preinitiation complex on a classical polymerase III promoter (tRNA), and start of transcription. (**a**) TFIIIC (light green) binds to the internal promoter's A and B blocks (green). (**b**) TFIIIC promotes binding of TFIIIB (yellow), with its TBP (blue) to the region upstream of the transcription start site. (**c**) TFIIIB promotes polymerase III (red) binding at the start site, ready to begin transcribing. (**d**) Transcription begins. As the polymerase moves to the right, making RNA (not shown), it presumably removes TFIIIC from the internal promoter. But TFIIIB remains in place, ready to sponsor a new round of polymerase binding and transcription.

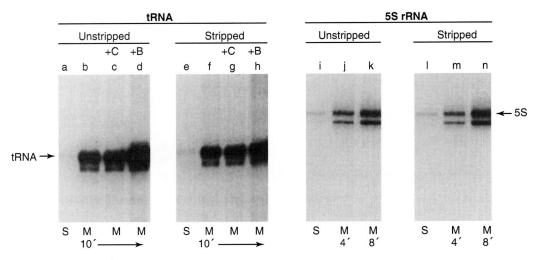

Figure 11.53 Transcription of polymerase III genes complexed only with TFIIIB. Geiduschek and coworkers made complexes containing a tRNA gene and TFIIIB and C (two panels at left), or a 5S rRNA gene and TFIIIA, B, and C (two panels at right), then removed TFIIIC with heparin (lanes e–h), or TFIIIA and C with a high ionic strength buffer (lanes l–n). They passed the stripped templates through gel filtration columns to remove any unbound factors, and demonstrated by gel mobility shift and DNase footprinting (not shown) that the purified complexes contained only TFIIIB bound to the upstream regions of the respective genes. Next, they tested these stripped complexes alongside unstripped complexes for ability to support single-round transcription (S; lanes a, e, i, and l), or multiple-round transcription (M; all other lanes) for the times indicated at bottom. They added extra TFIIIC in lanes c and g, and extra TFIIIB in lanes d and h as indicated at top. They confined transcription to a single round in lanes a, e, i, and l by including a relatively low concentration of heparin, which allowed elongation of RNA to be completed, but then bound up the released polymerase so it could not reinitiate. Notice that the stripped template, containing only TFIIIB, supported just as much transcription as the unstripped template in both single-round and multiple-round experiments, even when the experimenters added extra TFIIIC (compare lanes c and g, and lanes k and n). The only case in which the unstripped template performed better was in lane d, which was the result of adding extra TFIIIB. This presumably resulted from some remaining free TFIIIC that helped the extra TFIIIB bind, thus allowing more preinitiation complexes to form. (*Source:* Kassavetis et al., *S. cerevisiae* TFIIIB is the transcription initiation factor proper of RNA polymerase III, while TFIIIA and TFIIIC are assembly factors. *Cell* 60 (26 Jan 1990) p. 237, f. 3. Reprinted by permission of Elsevier Science.)

promoter. This protein then recruits the other components of the preinitiation complex. For TATA-containing class II promoters, the assembly factor is usually TBP, and its binding site is the TATA box. This presumably applies to TATA-containing class III promoters as well, at least in yeast and human cells. We have already seen a model for how this process begins in TATA-containing class II promoters (Figure 11.5). Figure 11.54 shows, in highly schematic form, the nature of these preinitiation complexes for all kinds of TATA-less promoters. In class I promoters, the assembly factor is UBF, which binds to the UCE and then attracts the TBP-containing SL1 to the core element. TATA-less class II promoters can attract TBP in two ways. TAF$_{II}$s in TFIID can bind to initiators, or they can bind to Sp1 bound to GC boxes. Both methods anchor TFIID to the TATA-less promoter. Classical class III promoters, at least in yeast and human cells, follow the same general scheme. TFIIIC, or in the case of the 5S rRNA genes, TFIIIA plus TFIIIC, play the role of assembly factor, binding to the internal promoter and attracting the TBP-containing TFIIIB to a site upstream of the start point. In *Drosophila* cells, TRFI appears to substitute for TBP in these preinitiation complexes.

Just because TBP does not always bind first, we should not discount its importance in organizing the preinitiation complex on these TATA-less promoters.

Once TBP binds, it helps bring the remaining factors, including RNA polymerase, to the complex. This is a second unifying principle: TBP plays an organizing role in preinitiation complexes on most types of eukaryotic promoters. A third unifying principle is that the specificity of TBP is governed by the TAFs with which it associates; thus, TBP affiliates with different TAFs when it binds to each of the various kinds of promoter.

SUMMARY The assembly of the preinitiation complex on each kind of eukaryotic promoter begins with the binding of an assembly factor to the promoter. With TATA-containing class II (and presumably class III) promoters, this factor is TBP, but other promoters have their own assembly factors. Even if TBP is not the first-bound assembly factor at a given promoter, it becomes part of the growing preinitiation complex on most known promoters and serves an organizing function in building the complex. The specificity of the TBP—which kind of promoter it will bind to—depends on its associated TAFs. TRFI substitutes for TBP, at least in some preinitiation complexes in *Drosophila* class III genes.

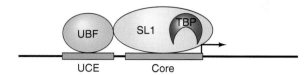

Class I
rRNA

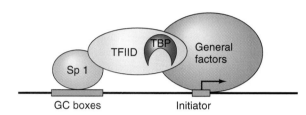

Class II
G₆I

Class III
VA₁

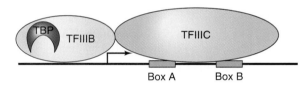

Figure 11.54   Model of preinitiation complexes on TATA-less promoters recognized by all three polymerases. In each case, an assembly factor (green) binds first (UBF, Sp1, and TFIIIC in class I, II, and III promoters, respectively). This in turn attracts another factor (yellow), which contains TBP (blue); this second factor is SL1, TFIID, or TFIIIB in class I, II, or III promoters, respectively. These complexes are sufficient to recruit polymerase for transcription of class I and III promoters, but in class II promoters more general factors (purple) besides polymerase II must bind before transcription can begin. (*Source:* Reprinted from White and Jackson, *Cell* 71:1051, 1992. Copyright 1992, with permission from Elsevier Science.)

SUMMARY

Transcription factors bind to class II promoters in the following order in vitro: (1) TFIID, apparently with help from TFIIA, binds to the TATA box. (2) TFIIB binds next. (3) TFIIF helps RNA polymerase II bind. The remaining factors bind in this order: TFIIE and TFIIH, forming the DABPolFEH preinitiation complex. The participation of TFIIA seems to be optional in vitro.

TFIID contains a TATA-box-binding protein (TBP) plus at least eight other polypeptides known as TBP-associated factors (TAF$_{II}$s). The C-terminal 180 amino acid fragment of the human TBP is the TATA-box-binding domain. The interaction between a TBP and a TATA box is an unusual one that takes place in the DNA minor groove. The saddle-shaped TBP lines up with the DNA, and the underside of the saddle forces open the minor groove and bends the TATA box into an 80-degree curve. TBP is required for transcription of most members of all three classes of genes, not just class II genes.

Most of the TAF$_{II}$s are evolutionarily conserved in the eukaryotes. They serve several functions, but two obvious ones are interacting with core promoter elements and interacting with gene-specific transcription factors. TAF$_{II}$250 and TAF$_{II}$150 help TFIID bind to the initiator and downstream elements of promoters and therefore can enable TBP to bind to certain TATA-less promoters that contain such elements. TAF$_{II}$250 and TAF$_{II}$110 help TFIID interact with Sp1 that is bound to GC boxes upstream of the transcription start site. These TAF$_{II}$s therefore ensure that TBP can bind to TATA-less promoters that have GC boxes. Different combinations of TAF$_{II}$s are apparently required to respond to various transcription activators, at least in higher eukaryotes. TAF$_{II}$250 also has two enzymatic activities. It is a histone acetyltransferase and a protein kinase. TFIID is not universally required, at least in higher eukaryotes. Some promoters in *Drosophila* require an alternative factor, TRF1, and some promoters require a TBP-free TAF$_{II}$-containing complex.

TFIIA contains two subunits (yeast), or three subunits (fruit flies and humans). This factor could be considered a TAF$_{II}$ because it binds to TBP and stabilizes binding between TFIID and promoters. TFIIB serves as a linker between TFIID and TFIIF/polymerase II. It has two domains, one of which is responsible for binding to TFIID, the other for continuing the assembly of the preinitiation complex. A hypothetical structure for the TFIIA–TFIIB–TBP–TATA box complex, based on the known structures of the TFIIA–TBP–TATA box and TFIIB–TBP–TATA box complexes, shows TFIIA and TFIIB binding to the upstream and downstream stirrups, respectively, of TBP. This puts these two factors in advantageous positions to perform their functions.

Binding of the polymerase to the DAB complex requires prior interaction with TFIIF, composed of two polypeptides called RAP30 and RAP70. RAP30 is the protein that ushers polymerase into the growing complex.

TFIIE, composed of two molecules each of a 34-kD and a 56-kD polypeptide, binds after polymerase and TFIIF. Both subunits are required for binding and transcription stimulation. TFIIH phosphorylates the carboxyl-terminal domain (CTD) of the largest RNA polymerase II subunit. TFIIE stimulates this process in vitro. TFIIE and TFIIH are not essential for formation of an open promoter complex, or for elongation, but they are required for promoter clearance. TFIIH has a DNA helicase activity that is essential for transcription, presumably because it facilitates promoter clearance by fully melting the DNA at the promoter.

Transcription can be controlled at the elongation level. One factor, IIS, stimulates elongation by limiting pausing at discrete sites. IIS also participates in proofreading by stimulating an inherent RNase in polymerase II to remove misincorporated nucleotides at the ends of nascent RNAs. TFIIF also stimulates elongation, apparently by limiting random pausing.

Yeast and mammalian cells have been shown to contain an RNA polymerase II holoenzyme with many polypeptides in addition to the subunits of the polymerase. The yeast holoenzyme contains a subset of general transcription factors and at least some of the SRB proteins. The rat holoenzyme contains all the general transcription factors except TFIIA.

SL1 does not bind by itself to the rRNA promoter, but it interacts with UBF to extend and strengthen the binding to the UCE and perturb the DNA in the core element. SL1 also determines the species specificity of transcription, as does the core promoter element. UBF is a transcription factor that stimulates transcription by polymerase I. It can activate the intact promoter, or the core element alone, and it mediates activation by the UCE. UBF and SL1 act synergistically to stimulate transcription.

SL1 is composed of TBP and three TAFs, TAF_I110, TAF_I63, and TAF_I48. Fully functional and species-specific SL1 can be reconstituted from these purified components, and binding of TBP to the TAF_Is precludes binding to the TAF_{II}s.

Classical class III genes require two factors, TFIIIB and C, to form a preinitiation complex with the polymerase. The 5S rRNA genes also require TFIIIA. TFIIIC and A bind to the internal promoter and help TFIIIB bind to a region just upstream of the transcription start site. TFIIIB then remains bound and can sponsor the initiation of repeated rounds of transcription. TFIIIC (and A, in the case of the 5S rRNA genes) serves one of the same functions as TBP in TATA-containing class II genes. That is, it is the first factor to bind and serves as a center around which the other components of the preinitiation complex can organize.

The assembly of the preinitiation complex on each kind of eukaryotic promoter begins with the binding of an assembly factor to the promoter. With TATA-containing class II (and presumably class III) promoters, this factor is usually TBP, but other promoters have their own assembly factors. Even if TBP is not the first-bound assembly factor at a given promoter, it becomes part of the growing preinitiation complex on most known promoters and serves an organizing function in building the complex. The specificity of the TBP—which kind of promoter it will bind to—depends on its associated TAFs, and there are TAFs specific for each of the promoter classes.

REVIEW QUESTIONS

1. List in order the proteins that assemble in vitro to form a class II preinitiation complex.

2. Describe and give the results of an experiment that shows that TFIID is the fundamental building block of the class II preinitiation complex.

3. Describe and give the results of an experiment that shows where TFIID binds.

4. Describe and give the results of an experiment that shows that TFIIF and polymerase II bind together, but neither can bind independently to the preinitiation complex.

5. Show the difference between the footprints caused by the DAB and the DABPolF complexes. What conclusion can you reach, based on this difference?

6. Describe and give the results of an experiment that shows the effect of a mutation in the TATA box on TBP binding.

7. What portion of the TBP is responsible for DNA binding? Describe and give the results of an experiment that demonstrates this.

8. Describe and give the results of an experiment that shows that TFIID can respond to the transcription factor Sp1, but TBP cannot.

9. Present a hypothesis that explains the fact that substitution of dCs for dTs and dIs for dAs, in the TATA box (making a CICI box) has no effect on TFIID binding.

10. What shape does TBP have? What is the geometry of interaction between TBP and the TATA box?

11. Describe and give the results of an experiment that shows TBP is required for transcription from all three classes of promoters.

12. Describe and give the results of an experiment that shows that a class II promoter is more active in vitro with TFIID than with TBP.

13. Describe and give the results of an experiment that identifies the TAFs that bind to a class II promoter containing a TATA box, an initiator, and a downstream element.

14. Describe and give the results of a DNase footprinting experiment that shows how the footprint is expanded by $TAF_{II}250$ and 150 compared with TBP alone.

15. Draw a diagram of a model for the interaction of TBP (and other factors) with a TATA-less class II promoter.

16. Describe and give the results of an experiment that shows that activation of transcription from a class II promoter containing GC boxes depends on Sp1 and $TAF_{II}110$.

17. Whole genome expression analysis indicates that yeast TAF_{II} 145 is required for transcription of only 16% of yeast genes, and TAF_{II} 17 is required for transcription of 67% of yeast genes. Provide a rationale for these results.

18. Present examples of class II preinitiation complexes with:
    a. An alternative TBP
    b. A missing TAF_{II}
    c. No TBP or TBP-like protein

19. What are the apparent roles of TFIIA and TFIIB in transcription?

20. Draw a rough sketch of the TBP–TFIIA–TFIIB ternary complex bound to DNA, showing only the relative positions of the proteins. How do these positions correlate with the apparent roles of the proteins?

21. Describe and give the results of an experiment that shows that TFIIH, but not the other general transcription factors, phosphorylates the IIA form of RNA polymerase II to the IIO form. In addition, include data that show that the other general transcription factors help TFIIH in this task.

22. Describe and give the results of an experiment that shows that TFIIH phosphorylates the CTD of polymerase II.

23. Describe an assay for DNA helicase and show how it can be used to demonstrate that TFIIH is associated with helicase activity.

24. Describe a G-less cassette transcription assay and show how it can be used to demonstrate that the RAD25 DNA helicase activity associated with TFIIH is required for transcription in vitro.

25. Describe and give the results of an experiment that shows that IIS stimulates transcription elongation by RNA polymerase II.

26. Describe and give the results of an experiment that shows that IIS stimulates proofreading by RNA polymerase II.

27. What is the meaning of the term *RNA polymerase II holoenzyme?* How does the holoenzyme differ from the core polymerase II?

28. Describe and give the results of an experiment that shows that SL1 is a species-specific transcription factor for polymerase I.

29. Which general transcription factor is the assembly factor in class I promoters? In other words, which binds first and helps the other bind? Describe a DNase footprinting experiment you would perform to prove this, and show idealized results, not necessarily those that Tjian and colleagues actually obtained. Make sure your diagrams indicate an effect of both transcription factors on the footprints.

30. Fill in the following autoradiograph to show the levels of in vitro transcription from a class I promoter observed under the indicated conditions. Distinguish clearly among these possibilities: no transcription, weak transcription, and strong transcription (see Figure 11.43).

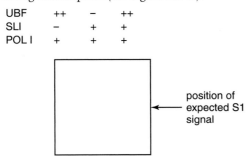

UBF	++	–	++
SLI	–	+	+
POL I	+	+	+

position of expected S1 signal →

31. Describe and give the results of copurification and immunoprecipitation experiments that show that SL1 contains TBP.

32. Describe and give the results of an experiment that identified the TAFs in SL1.

33. How do we know that TFIIIA is necessary for transcription of 5S rRNA, but not tRNA, genes?

34. Geiduschek and colleagues performed DNase footprinting with polymerase III plus TFIIIB and C and a tRNA gene. Show the results they obtained with: No added protein; polymerase and factors; and polymerase, factors and three of the four NTPs. What can you conclude from these results?

35. The classical class III promoters have internal promoters. Nevertheless, TFIIIB and C together cause a footprint in a region upstream of the gene's coding region. Draw a diagram of the binding of these two factors that explains these observations.

36. Draw a diagram of what happens to TFIIIB and C after polymerase III has begun transcribing a classical class III gene such as a tRNA gene. How does this explain how new polymerase III molecules can continue to transcribe the gene, even though factors may not remain bound to the internal promoter?

37. Describe and give the results of a DNase footprint experiment that shows that TFIIIB + C, but not TFIIIC alone, can protect a region upstream of the transcription start site in a tRNA gene. Show also what happens to the footprint when you strip off TFIIIC with heparin.

38. Describe and give the results of an in vitro transcription experiment performed on the 5S rRNA gene that shows that TFIIIB binding to the preinitiation complex depends on prior binding of both TFIIIA and C.

39. Describe and give the results of an experiment that shows the following: Once TFIIIB binds to a classical class III gene, it can support multiple rounds of transcription, even after TFIIIC (or C and A) are stripped off the promoter.

40. Diagram the preinitiation complexes with all three classes of TATA-less promoters. Identify the assembly factors in each case.

SUGGESTED READINGS

General References and Reviews

Buratowski, S. 1997. Multiple TATA-binding factors come back into style. *Cell* 91:13–15.

Burley, S.K. 1996. Picking up the TAB. *Nature* 381:112–13.

Burley, S.K., and R.G. Roeder, 1996. Biochemistry and structural biology of transcription factor IID (TFIID). *Annual Review of Biochemistry* 65:769–99.

Chao, D.M., and R.A. Young, 1996. Activation without a vital ingredient. *Nature* 383:119–20.

Goodrich, J.A., G. Cutler, and R. Tjian. 1996. Contacts in context: Promoter specificity and macromolecular interactions in transcription. *Cell* 84:825–30.

Grant, P. and J.L. Workman. 1998. A lesson in sharing? *Nature* 396:410–11.

Green, M.A. 1992. Transcriptional transgressions. *Nature* 357:364–65.

Hahn, S. 1998. The role of TAFs in RNA polymerase II transcription. *Cell* 95:579–82.

Klug, A. 1993. Opening the gateway. *Nature* 365:486–87.

Sharp, P.A. 1992. TATA-binding protein is a classless factor. *Cell* 68:819–21.

Tansey, W.P., and W. Herr, 1997. TAFs: Guilt by association? *Cell* 88:729–32.

White, R.J., and S.P. Jackson, 1992. The TATA-binding protein: A central role in transcription by RNA polymerases I, II, and III. *Trends in Genetics* 8:284–88.

Research Articles

Brand, M., C. Leurent, V. Mallouh, L. Tora, and P. Schultz. 1999. Three-dimensional structures of the TAF$_{II}$-containing complexes TFIID and TFTC. *Science* 286:2151–53.

Buratowski, S., S. Hahn, L. Guarente, and P. Sharp. (1989). Five intermediate complexes in transcription initiation by RNA polymerase II. *Cell* 56:549–61.

Chen, J.L., L.D. Attardi, C.P. Verrijzer, K. Yokomori, and R. Tjian. 1994. Assembly of recombinant TFIID reveals differential coactivator requirements for distinct transcriptional activators. *Cell* 79:93–105.

Dynlacht, B.D., T. Hoey, and R. Tjian. (1991). Isolation of coactivators associated with the TATA-binding protein that mediate transcriptional activation. *Cell* 66:563–76.

Flores, O., E. Maldonado, and D. Reinberg. 1989. Factors involved in specific transcription by mammalian RNA polymerase II: Factors IIE and IIF independently interact with RNA polymerase II. *Journal of Biological Chemistry* 264:8913–21.

Flores, O., H. Lu, M. Killeen, J. Greenblatt, Z.F. Burton, and D. Reinberg. 1991. The small subunit of transcription factor IIF recruits RNA polymerase II into the preinitiation complex. *Proceedings of the National Academy of Sciences USA* 88:9999–10003.

Guzder, S.N., P. Sung, V. Bailly, L. Prakash, and S. Prakash. 1994. RAD25 is a DNA helicase required for DNA repair and RNA polymerase II transcription. *Nature* 369:578–81.

Ha, I., W.S. Lane, and D. Reinberg. 1991. Cloning of a human gene encoding the general transcription initiation factor IIB. *Nature* 352:689–94.

Hansen, S.K., S. Takada, R.H. Jacobson, J.T. Lis, and R. Tjian. 1997. Transcription properties of a cell type-specific TATA binding protein, TRF. *Cell* 91:71–83.

Holmes, M.C. and Tjian, R. 2000. Promoter-selective properties of the TBP-related factor TRF1. *Science* 288:867–70.

Holstege, F.C.P., E.G. Jennings, J.J. Wyrick, T. I. Lee, C.J. Hengartner, M.R. Green, T.R. Golub, E.S. Lander, and R.A. Young. 1998. Dissecting the regulatory circuitry of a eukaryotic genome. *Cell* 95:717–28.

Honda, B.M., and R.G. Roeder, 1980. Association of a 5S gene transcription factor with 5S RNA and altered levels of the factor during cell differentiation. *Cell* 22:119–26.

Inostroza, J., O. Flores, and D. Reinberg. 1991. Factors involved in specific transcription by mammalian RNA polymerase II:

Purification and functional analysis of general transcription factor IIE. *Journal of Biological Chemistry* 266:9304–08.

Kassavetis, G.A., D.L. Riggs, R. Negri, L.H. Nguyen, and E.P. Geiduschek. 1989. Transcription factor IIIB generates extended DNA interactions in RNA polymerase III transcription complexes on tRNA genes. *Molecular and Cellular Biology* 9:2551–66.

Kassavetis, G.A., B.R. Braun, L.H. Nguyen, and E.P. Geiduschek. 1990. *S. cerevisiae* TFIIIB is the transcription initiation factor proper of RNA polymerase III, while TFIIIA and TFIIIC are assembly factors. *Cell* 60:235–45.

Kim, J.L., D.B. Nikolov, and S.K. Burley. 1993. Co-crystal structure of a TBP recognizing the minor groove of a TATA element. *Nature* 365:520–27.

Kim, T.-K., R.H. Ebright, and D. Reinberg. 2000. Mechanism of ATP-dependent promoter melting by transcription factor IIH. *Science* 288:1418–21.

Kim, Y., J.H. Geiger, S. Hahn, and P.B. Sigler. 1993. Crystal structure of a yeast TBP/TATA-box complex. *Nature* 365:512–20.

Kim, Y. J., S. Björklund, Y. Li, M.H. Sayre, and R.D. Kornberg. 1994. A multiprotein mediator of transcriptional activation and its interaction with the C-terminal repeat domain of RNA polymerase II. *Cell* 77:599–608.

Koleske, A.J., and R.A. Young, 1994. An RNA polymerase II holoenzyme responsive to activators. *Nature* 368:466–69.

Learned, R.M., S. Cordes, and R. Tjian. 1985. Purification and characterization of a transcription factor that confers promoter specificity to human RNA polymerase I. *Molecular and Cellular Biology* 5:1358–69.

Learned, R.M., T.K. Learned, M.M. Haltimer, and R.T. Tjian. 1986. Human rRNA transcription is modulated by the coordinate binding of two factors to an upstream control element. *Cell* 45:847–57.

Lee, D.K., M. Horikoshi, and R.G. Roeder. 1991. Interaction of TFIID in the minor groove of the TATA element. *Cell* 67:1241–50.

Lobo, S.L., M. Tanaka, M.L. Sullivan, and N. Hernandez. 1992. A TBP complex essential for transcription from TATA-less but not TATA-containing RNA polymerase III promoters is part of the TFIIIB fraction. *Cell* 71:1029–40.

Lu, H., L. Zawel, L. Fisher, J.-M. Egly, and D. Reinberg. 1992. Human general transcription factor IIH phosphorylates the C-terminal domain of RNA polymerase II. *Nature* 358:641–45.

Maldonado, E., I. Ha, P. Cortes, L. Weis, and D. Reinberg. 1990. Factors involved in specific transcription by mammalian RNA polymerase II: Role of transcription factors IIA, IID, and IIB during formation of a transcription-competent complex. *Molecular and Cellular Biology* 10:6335–47.

Moqtaderi, Z., Y. Bai, D. Poon, P.A. Weil, and K. Struhl. 1996. TBP-associated factors are not generally required for transcriptional activation in yeast. *Nature* 383:188–91.

Nikolov, D.B., H. Chen, E.D. Halay, A.A. Usheva, K. Hisatake, D.K. Lee, R.G. Roeder, and S.K. Burley. 1995. Crystal structure of a TFIIB-TBP-TATA-element ternary complex. *Nature* 377:119–28.

Ossipow, V., J.-P. Tassan, E.I. Nigg, and U. Schibler. 1995. A mammalian RNA polymerase II holoenzyme containing all components required for promoter-specific transcription initiation. *Cell* 83:137–46.

Pelham, H.B., Wormington, W.M., and D.D. Brown. 1981. Related 5S rRNA transcription factors in *Xenopus* oocytes and somatic cells. *Proceeding of the National Academy of Sciences USA* 78:1760–64.

Peterson, M.G., N. Tanese, B.F. Pugh, and R. Tjian. 1990. Functional domains and upstream activation properties of cloned human TATA binding protein. *Science* 248:1625–30.

Pugh, B.F., and R. Tjian, 1991. Transcription from a TATA-less promoter requires a multisubunit TFIID complex. *Genes and Development* 5:1935–45.

Ranish, J.A., W.S. Lane, and S. Hahn. 1992. Isolation of two genes that encode subunits of the yeast transcription factor IIA. *Science* 255:1127–29.

Rowlands, T., P. Baumann, and S.P. Jackson. 1994. The TATA-binding protein: A general transcription factor in eukaryotes and archaebacteria. *Science* 264:1326–29.

Sauer, F., D.A. Wassarman, G.M. Rubin, and R. Tjian. 1996. TAF$_{II}$s mediate activation of transcription in the Drosophila embryo. *Cell* 87:1271–84.

Schultz, M.C., R.H. Roeder, and S. Hahn. 1992. Variants of the TATA-binding protein can distinguish subsets of RNA polymerase I, II, and III promoters. *Cell* 69:697–702.

Schultz, P., S. Fribourg, A. Poterszman, V. Mallouh, D. Moras, and J.M. Egly. 2000. Molecular structure of human TFIIH. *Cell* 102:599–607.

Setzer, D.R., and D.D. Brown, 1985. Formation and stability of the 5S RNA transcription complex. *Journal of Biological Chemistry* 260:2483–92.

Shastry, B.S., S.-Y. Ng, and R.G. Roeder. 1982. Multiple factors involved in the transcription of class III genes in *Xenopus laevis. Journal of Biological Chemistry* 257:12979–86.

Starr, D.B., and D.K. Hawley, 1991. TFIID binds in the minor groove of the TATA box. *Cell* 67:1231–40.

Taggart, K.P., J.S. Fisher, and B.F. Pugh. 1992. The TATA-binding protein and associated factors are components of Pol III transcription factor TFIIIB. *Cell* 71:1051–28.

Takada, S., J.T. Lis, S. Zhou, and R. Tjian. 2000. A TRF1:BRF complex directs *Drosophila* RNA polymerase III transcription. *Cell* 101:459–69.

Tan, S., Y. Hunziker, D.F. Sargent, and T.J. Richmond. 1996. Crystal structure of a yeast TFIIA/TBP/DNA complex. *Nature* 381:127–34.

Tanese, N. 1991. Coactivators for a proline-rich activator purified from the multisubunit human TFIID complex. *Genes and Development* 5:2212–24.

Thomas, M.J., A.A. Platas, and D.K. Hawley. 1998. Transcriptional fidelity and proofreading by RNA polymerase II. *Cell* 93:627–37.

Verrijzer, C.P., J.-L. Chen, K. Yokomori, and R. Tjian. 1995. Binding of TAFs to core elements directs promoter selectivity by RNA polymerase II. *Cell* 81:1115–25.

Walker, S.S., J.C. Reese, L.M. Apone, and M.R. Green. 1996. Transcription activation in cells lacking TAF$_{II}$s. *Nature* 383:185–88.

Wieczorek, E., M. Brand, X. Jacq, and L. Tora. 1998. Function of TAF$_{II}$-containing complex without TBP in transcription by RNA polymerase II. *Nature* 393:187–91.

CHAPTER 12

Transcription Activators in Eukaryotes

Computer model of the transcription factor p53 interacting with its target DNA site. *Courtesy Nicola P. Pavletich, Sloan-Kettering Cancer Center, Science (15 July 1994) cover. © AAAS.*

In Chapters 10 and 11 we learned about the basic machinery involved in eukaryotic transcription: the three RNA polymerases, their promoters, and the general transcription factors that bring RNA polymerase and promoter together. However, it is clear that this is not the whole story. The general transcription factors by themselves dictate the starting point and direction of transcription, but they are capable of sponsoring only a very low level of transcription (basal level transcription). But transcription of active genes in cells rises above (frequently far above) the basal level. To provide the needed extra boost in transcription, eukaryotic cells have additional, gene-specific transcription factors (activators). The transcription activation provided by these activators also permits cells to control the expression of their genes.

In addition, eukaryotic DNA is complexed with protein in a structure called chromatin. Some chromatin, called heterochromatin, is highly condensed and inaccessible to RNA polymerases, so it cannot be transcribed. Other chromatin (euchromatin) still contains protein, but it is relatively extended. Much of this euchromatin, even though it is relatively open, contains genes that are not transcribed in a given cell because the appropriate activators are not available to turn them on. Instead, other proteins may hide the promoters from RNA polymerase and general transcription factors to ensure that they remain turned off. In this chapter, we will examine the activators that control eukaryotic genes. Then, in Chapter 13, we will look at the crucial relationship among activators, chromatin structure, and gene activity. ■

12.1 Categories of Activators

Activators can either stimulate or inhibit transcription by RNA polymerase II and they have structures composed of at least two functional domains: a **DNA-binding domain** and a **transcription-activation domain.** Many also have a **dimerization domain** that allows the activators to bind to each other, forming homodimers (two identical monomers bound together), heterodimers (two different monomers bound together), or even higher multimers such as tetramers. Some even have binding sites for effector molecules like steroid hormones. Let us consider some examples of these three kinds of structural–functional domains.

DNA-Binding Domains

A protein **domain** is an independently folded region of a protein. Each DNA-binding domain has a **DNA-binding motif**, which is the part of the domain that has a characteristic shape specialized for specific DNA binding. Most DNA-binding motifs fall into the following classes:

1. *Zinc-containing modules.* At least three kinds of zinc-containing modules act as DNA-binding motifs. These all use one or more zinc ions to create the proper shape so an α-helix within the motif can fit into the DNA major groove and make specific contacts there. These zinc-containing modules include:
   a. **Zinc fingers,** such as those found in TFIIIA and Sp1, two transcription factors we have already encountered.
   b. Zinc modules found in the glucocorticoid receptor and other members of this group of nuclear receptors.
   c. Modules containing two zinc ions and six cysteines, found in the yeast activator GAL4 and its relatives.
2. **Homeodomains (HDs).** These contain about 60 amino acids and resemble in structure and function the helix-

turn-helix DNA-binding domains of prokaryotic proteins such as the λ phage repressor. HDs, found in a variety of activators, were originally identified in activators called homeobox proteins that regulate development in the fruit fly *Drosophila.*

3. **bZIP and bHLH motifs.** The CCAAT/enhancer-binding protein (**C/EBP**), the MyoD protein, and many other eukaryotic transcription factors have a highly basic DNA-binding motif linked to one or both of the protein dimerization motifs known as leucine zippers and helix-loop-helix (HLH) motifs.

This list is certainly not exhaustive. In fact, several transcription factors have now been identified that do not fall into any of these categories.

Transcription-Activating Domains

Most activators have one of these domains, and some have more than one. So far, most of these domains fall into three classes, as follows:

1. *Acidic domains.* The yeast activator GAL4 typifies this group. It has a 49-amino-acid domain with 11 acidic amino acids.
2. *Glutamine-rich domains.* The activator Sp1 has two such domains, which are about 25% glutamine. One of these has 39 glutamines in a span of 143 amino acids. In addition, Sp1 has two other activating domains that do not fit into any of these three main categories.
3. *Proline-rich domains.* The activator CTF, for instance, has a domain of 84 amino acids, 19 of which are proline.

Our descriptions of the transcription-activating domains are necessarily nebulous, because the domains themselves are rather ill-defined. The acidic domain, for example, has seemed to require nothing more than a preponderance of acidic residues to make it function, which led to the name "acid blob" to describe this presumably unstructured domain. On the other hand, Stephen Johnston and his colleagues have shown that the acidic activation domain of GAL4 tends to form a defined structure—a β-sheet—in slightly acidic solution. It is possible that the β-sheet also forms under the slightly basic conditions in vivo, but this is not yet clear. These workers also removed all six of the acidic amino acids in the GAL4 acidic domain and showed that it still retained 35% of its normal ability to activate transcription. Thus, not only is the structure of the acidic activating domain unclear, the importance of its acidic nature is even in doubt. With such persistent uncertainty, it has been difficult to draw conclusions about how the structure and function of transcription-activating domains are related.

SUMMARY Eukaryotic activators are composed of at least two domains: a DNA-binding domain and a transcription-activating domain. DNA-binding domains contain motifs such as zinc modules, homeodomains, and bZIP or bHLH motifs. Transcription-activating domains can be acidic, glutamine-rich, or proline-rich.

12.2 Structures of the DNA-Binding Motifs of Activators

By contrast to the transcription-activating domains, most DNA-binding domains have well-defined structures, and x-ray crystallography studies have shown how these structures interact with their DNA targets. Furthermore, these same structural studies have frequently elucidated the dimerization domains responsible for interaction between protein monomers to form a functional dimer, or in some cases, a tetramer. This is crucial, because most classes of DNA-binding proteins are incapable of binding to DNA in monomer form; they must form at least dimers to function. Let us explore the structures of several classes of DNA-binding motifs and see how they mediate interaction with DNA. In the process we will discover the ways some of these proteins can dimerize.

Zinc Fingers

In 1985, Aaron Klug noticed a periodicity in the structure of the general transcription factor TFIIIA. This protein has nine repeats of a 30-residue element. Each element has two closely spaced cysteines followed 12 amino acids later by two closely spaced histidines. Furthermore, the protein is rich in zinc—enough for one zinc ion per repeat. This led Klug to predict that each zinc ion is complexed by the two cysteines and two histidines in each repeat unit to form a finger-shaped domain.

Finger Structure   Michael Pique and Peter Wright used nuclear magnetic resonance spectroscopy to determine the structure in solution of one of the zinc fingers of the *Xenopus laevis* protein Xfin, an activator of certain class II promoters. Note that this structure, depicted in Figure 12.1, really is finger-shaped, as predicted. It is also worth noting that this finger shape by itself does not confer any binding specificity, since there are many different finger proteins, all with the same shape fingers but each binding to its own unique DNA target sequence. Thus, it is the precise amino acid sequences of the fingers, or of neighboring parts of the protein, that determine the DNA

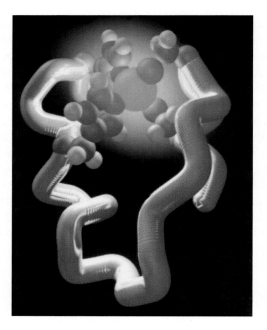

Figure 12.1   Three-dimensional structure of one of the zinc fingers of the *Xenopus* protein Xfin. The zinc is represented by the turquoise sphere at top center. The sulfurs of the two cysteines are represented by yellow spheres. The two histidines are represented by the blue and green structures at upper left. The backbone of the finger is represented by the purple tube. (*Source:* Michael Pique and Peter E. Wright, Dept. of Molecular Biology, Scripps Clinic Research Institute, La Jolla, CA. (cover photo, *Science* 245 (11 Aug 1989)).)

sequence to which the protein will bind. In the Xfin finger, an α-helix (on the left in Figure 12.1) contains several basic amino acids—all on the side that seems to contact the DNA. These and other amino acids in the helix presumably determine the binding specificity of the protein.

Carl Pabo and his colleagues used x-ray crystallography to obtain the structure of the complex between DNA and a member of the TFIIIA class of zinc finger proteins—the mouse protein Zif268. This is a so-called *immediate early protein,* which means that it is one of the first genes to be activated when resting cells are stimulated to divide. The Zif268 protein has three adjacent zinc fingers that fit into the major groove of the DNA double helix. We will see the arrangement of these three fingers a little later in the chapter. For now, let us consider the three-dimensional structure of the fingers themselves. Figure 12.2 presents the structure of finger 1 as an example. The finger shape in this presentation is perhaps not obvious. Still, on close inspection we can see the finger contour, which is indicated by the dashed line. As in the Xfin zinc finger, the left side of each Zif268 finger is an α-helix. This is connected by a short loop at the bottom to the right side of the finger, an antiparallel β-ribbon. Do not confuse this β-ribbon itself with the finger; it is only one half of it. The zinc ion (blue sphere) is in the middle, coordinated by two histidines in the α-helix and by two cysteines in the β-ribbon. All three fingers have almost exactly the same shape.

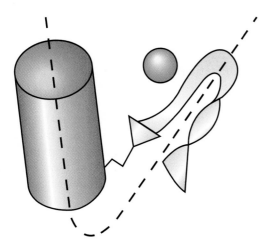

Figure 12.2 Schematic diagram of zinc finger 1 of the Zif268 protein. The right-hand side of the finger is an antiparallel β-strand (yellow), and the left-hand side is an α-helix (red). Two cysteines in the β-strand and two histidines in the α-helix coordinate the zinc ion in the middle (blue). The dashed line traces the outline of the "finger" shape. (*Source:* From Pavletich and Pabo, *Science* 252:812, 1991. Copyright © 1991 American Association for the Advancement of Science, Washington, DC. Reprinted by permission.)

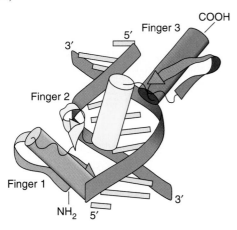

Figure 12.3 Arrangement of the three zinc fingers of Zif268 in a curved shape to fit into the major groove of DNA. As usual, the cylinders and ribbons stand for α-helices and β-strands.

Interaction with DNA How do the fingers interact with their DNA targets? Figure 12.3 shows all three Zif268 fingers lining up in the major groove of the DNA. In fact, the three fingers are arranged in a curve, or C-shape, which matches the curve of the DNA double helix. All the fingers approach the DNA from essentially the same angle, so the geometry of protein–DNA contact is very similar in each case. Binding between each finger and its DNA-binding site relies on direct amino acid–base interactions. The target bases all lie on one strand, within a G-rich motif. The synthetic motif used to form the cocrystals in this experiment had this sequence: 5′-GCGTGGGCG-3′. Furthermore, finger 1 binds to the three bases (GCG) at the 3′-end of the motif, finger 2 binds to the central three bases (TGG), and finger 3 binds to the 5′-triplet (GCG). Because

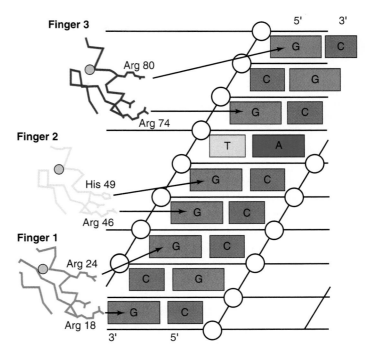

Figure 12.4 Summary of interactions between Zif268 zinc finger amino acids and DNA bases. Each of the three fingers has two amino acids (all but one of the six are arginines) that make specific contact with guanines in the DNA major groove. (*Source:* From Pavletich and Pabo, *Science* 252:813, 1991. Copyright © 1991 American Association for the Advancement of Science, Washington, DC. Reprinted by permission.)

fingers 1 and 3 contact triplets with the same sequence (GCG), we would predict that the three key amino acids in finger 1 that make these contacts would be very similar to those in finger 3 that make the same contacts. In fact they are identical: Arg, Asp, and Arg. Finger 2, which contacts a different triplet (TGG), has a different set of three amino acids in the three key positions: Arg, Asp, and His. The first two amino acids are the same in every case: Arg and Asp. This arginine in each finger makes direct contact with a guanine in each triplet. The aspartate in each finger does not contact DNA; instead, it contacts the preceding arginine to help position it for optimum interaction with its guanine target. Figure 12.4 schematically summarizes the amino acid-base interactions involving all three fingers. For more detailed descriptions of amino acid-base interactions, see Chapter 9.

Each finger also makes two contacts with DNA phosphates through conserved histidine and arginine residues. The histidines that make these contacts are actually the same ones that coordinate the zinc ions in the three fingers. In addition, four other amino acids in the Zif268 peptide contact phosphates in the DNA backbone.

Comparison with Other DNA-Binding Proteins One unifying theme emerging from studies of many, but not all, DNA-binding proteins is the utility of the α-helix in contacting the DNA major groove. We saw many examples of this with the prokaryotic helix-turn-helix domains

(Chapter 9), and we will see several other eukaryotic examples. What about the β-strand in Zif268? It seems to serve the same function as the first α-helix in a helix-turn-helix protein, namely to bind to the DNA backbone and help position the recognition helix for optimal interaction with the DNA major groove.

Zif268 also shows some differences from the helix-turn-helix proteins. Whereas the latter proteins have a single DNA-binding domain per monomer, the finger protein DNA-binding domains have a modular construction, with several fingers making contact with the DNA. This arrangement means that these proteins, in contrast to most DNA-binding proteins, do not need to form dimers or tetramers to bind to DNA. They already have multiple binding domains built in. Also, most of the protein–DNA contacts are with one DNA strand, rather than both, as in the case of the helix-turn-helix proteins. At least with this particular finger protein, most of the contacts are with bases, rather than the DNA backbone. As we saw in Chapter 9, helix-turn-helix proteins, especially the *trp* repressor, do not always behave this way.

In 1991, Nikola Pavletich and Pabo solved the structure of a cocrystal between DNA and a five-zinc-finger human protein called GLI. This provided an interesting contrast with the three-finger Zif268 protein. Again, the major groove is the site of finger–DNA contacts, but in this case one finger (finger 1) does not contact the DNA. Also, the overall geometries of the two finger–DNA complexes are similar, with the fingers wrapping around the DNA major groove, but no simple "code" of recognition between certain bases and amino acids exists. In particular, only one arginine–guanine contact occurs in the GLI complex.

SUMMARY Zinc fingers are composed of an antiparallel β-strand, followed by an α-helix. The β-strand contains two cysteines, and the α-helix two histidines, that are coordinated to a zinc ion. This coordination of amino acids to the metal helps form the finger-shaped structure. The specific recognition between the finger and its DNA target occurs in the major groove.

The GAL4 Protein

The **GAL4** protein is a yeast activator that controls a set of genes responsible for metabolism of galactose. Each of these GAL4-responsive genes contains a GAL4 target site upstream of the transcription start site. These target sites are called **upstream activating sequences**, or **UAS$_G$s**. GAL4 binds to a UAS$_G$ as a dimer. Its DNA-binding motif is located in the first 65 amino acids of the protein, and its dimerization motif is found in residues 65–94. The

DNA-binding motif is similar to the zinc finger in that it contains zinc and cysteine residues, but its structure must be different: Each motif has six cysteines and no histidines, and the ratio of zinc ions to cysteines is 1:3.

Mark Ptashne and Stephen Harrison and their colleagues performed x-ray crystallography on cocrystals of the first 65 amino acids of GAL4 and a synthetic 17-bp piece of DNA. This revealed several important features of the protein–DNA complex, including the shape of the DNA-binding motif and how it interacts with its DNA target, and a weak dimerization motif in residues 50–64.

The DNA-Binding Motif Figure 12.5 depicts the structure of the GAL4 peptide dimer–DNA complex. One end of each monomer contains a DNA-binding motif containing six cysteines that complex two zinc ions (yellow spheres), forming a *bimetal thiolate cluster*. Each of these motifs also features a short α-helix that protrudes into the major groove of the DNA double helix, where it can make specific interactions. The other end of each monomer is an α-helix that serves a dimerization function that we will discuss in the next section.

How does this zinc-containing DNA-binding motif work? Figure 12.6a is a close-up view of one motif, showing the six cysteines and two zinc ions (yellow spheres), and the short recognition helix. Figure 12.6b shows the detailed interaction between amino acids in the recognition module and bases in the DNA.

The Dimerization Motif The GAL4 monomers also take advantage of α-helices in their dimerization, forming a parallel **coiled coil** as illustrated at left in Figure 12.5c. This figure also shows that the dimerizing α-helices point directly at the minor groove of the DNA. Finally, note in Figure 12.5 that the DNA recognition module and the dimerization module in each monomer are joined by an extended linker domain. We will see other examples of coiled coil dimerization motifs when we discuss bZIP and bHLH motifs later in this chapter.

SUMMARY The GAL4 protein is a member of the zinc-containing family of DNA-binding proteins, but it does not have zinc fingers. Instead, each GAL4 monomer contains a DNA-binding motif with six cysteines that coordinate two zinc ions in a bimetal thiolate cluster. The recognition module contains a short α-helix that protrudes into the DNA major groove and makes specific interactions there. The GAL4 monomer also contains an α-helical dimerization motif that forms a parallel coiled coil as it interacts with the α-helix on the other GAL4 monomer.

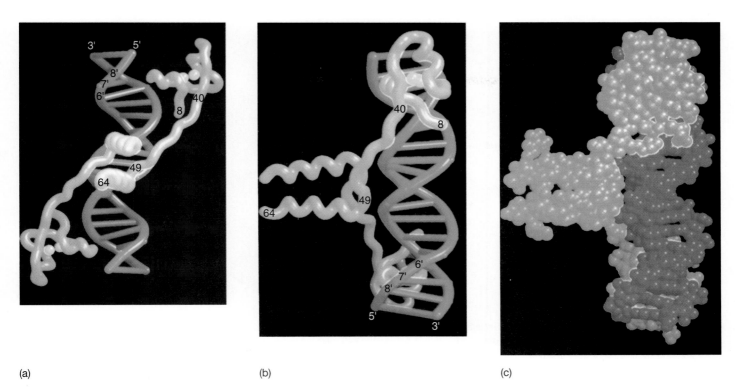

(a)   (b)   (c)

Figure 12.5   Three views of the GAL4–DNA complex. (a) The complex viewed approximately along its two-fold axis of symmetry. The DNA is in red, the protein is in blue, and the zinc ions are represented by yellow spheres. Amino acid residue numbers at the beginnings and ends of the three domains are given on the top monomer: The DNA recognition module extends from residue 8 to 40. The linker, from residue 41 to 49, and the dimerization domain, from residue 50 to 64. **(b)** The complex viewed approximately perpendicular to the view in panel **(a)**. The dimerization elements appear roughly parallel to one another at left center. **(c)** Space-filling model of the complex in the same orientation as in panel **(b)**. Notice that the recognition modules on the two GAL4 monomers make contact with opposite faces of the DNA. Notice also the neat fit between the coiled coil of the dimerization domain and the minor groove of the DNA helix. (*Source:* Marmorstein et al., DNA recognition by GAL4: Structure of a protein-DNA complex. *Nature* 356 (2 April 1992) p. 411, f. 3. © Macmillan Magazines Ltd.)

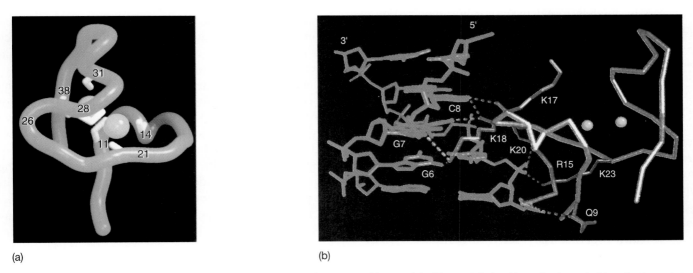

(a)   (b)

Figure 12.6   The DNA recognition module of GAL4. (a) Model of the recognition module. The protein backbone is shown in blue, the bound zinc ions are represented by yellow spheres, and the six cysteine side chains (residues 11, 14, 21, 28, 31, and 38) that coordinate the two zinc ions are depicted as white sticks. The short recognition helix is at right and includes residue 14. **(b)** Specific protein–DNA contacts involving the recognition module. Again, the protein backbone is in blue, with the zinc ions in yellow. The amino acid side chains that make specific contacts with DNA bases or backbone are rendered in green and identified by residue number. The lysine 17 side chain is shown because it could make a hydrogen bond contact with a phosphate if the DNA were longer than the oligomer used in this study. Hydrogen bonds are represented by dashed green lines. (*Source:* Marmorstein et al., *Nature* 356 (2 April 1992) p. 412, f. 4-5. © Macmillan Magazines Ltd.)

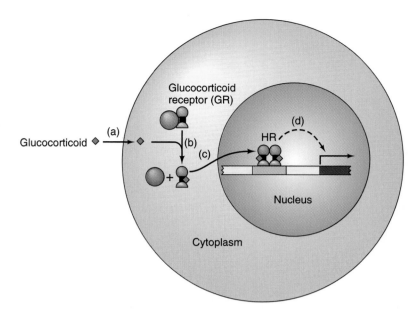

Figure 12.7 Glucocorticoid action. The glucocorticoid receptor (GR) exists in an inactive form in the cytoplasm complexed with heat shock protein 90 (hsp90). **(a)** The glucocorticoid (blue diamond) diffuses across the cell membrane and enters the cytoplasm. **(b)** The glucocorticoid binds to its receptor (GR, red and green), which changes conformation and dissociates from hsp90 (orange). **(c)** The hormone–receptor complex (HR) enters the nucleus, dimerizes with another HR, and binds to a hormone-response element, or enhancer (pink), upstream of a hormone-activated gene (brown). **(d)** Binding of the HR dimer to the enhancer activates (dashed arrow) the associated gene, so transcription occurs (bent arrow).

The Nuclear Receptors

A third class of zinc module is found in the **nuclear receptors.** These proteins interact with a variety of steroid and other hormones that diffuse through the cell membrane. They form hormone-receptor complexes that function as activators by binding to enhancers, or **hormone response elements,** and stimulating transcription of their associated genes. Thus, these activators differ from the others we have studied in that they must bind to an effector (a hormone) in order to function as activators. This implies that they must have an extra important domain—a hormone-binding domain—and indeed they do. Some of the hormones that work this way are the sex hormones (androgens and estrogens); progesterone, the hormone of pregnancy and principal ingredient of common birth control pills; the glucocorticoids, such as cortisol; vitamin D, which regulates calcium metabolism; and thyroid hormone and retinoic acid, which regulate gene expression during development. Each hormone binds to its specific receptor, and together they activate their own set of genes. Some nuclear receptors (e.g., the glucocorticoid receptor) exist in the cytoplasm complexed with a protein called heat shock protein 90 (hsp90). When they meet their ligands, or hormones, in the cytoplasm, they dissociate from hsp90, change conformation, and bind to the hormones. These hormone–receptor complexes can then move into the nucleus to perform their functions (Figure 12.7). On the other hand, the majority of nuclear receptors (e.g., the thyroid hormone receptor) spend all

their time in the nucleus. In the absence of hormone, they bind to their respective enhancers and repress transcription. In the presence of hormone, they form hormone–receptor complexes in the nucleus and function as activators by binding to the same enhancers. This is an example of a protein acting as either a repressor or activator, depending on environmental conditions. Thus, the environment also determines whether the DNA element to which this protein binds, the thyroid hormone response element, serves as an enhancer or as a silencer.

Sigler and colleagues performed x-ray crystallography on cocrystals of the glucocorticoid receptor and either of two oligonucleotides. One oligonucleotide contained two target half-sites separated by the usual 3 bp. The other had the same two half-sites, but separated by 4 instead of 3 bp. This 1-base difference in separation of the binding sites on the DNA made a big difference in the nature of the protein–DNA interaction, as we will see.

Figure 12.8a shows a stereoscopic view of the protein–DNA interaction with the oligonucleotide having an extra base pair between the half-sites. Several features are apparent. (1) The binding domain dimerizes, with one monomer (red) making specific contacts with the DNA, and the other (blue) making nonspecific contacts. In a natural DNA, with the normal 3-bp distance between target half-sites, the fit between protein and DNA would allow specific contacts with both monomers, but the single extra base pair prevents this ideal fit. (2) Each binding motif is a zinc module that contains two zinc ions (yellow

(a)

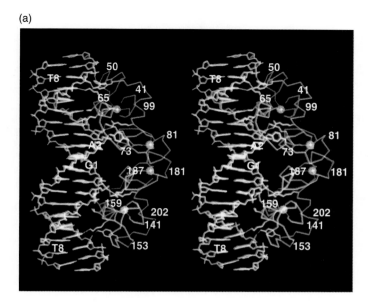

(b)

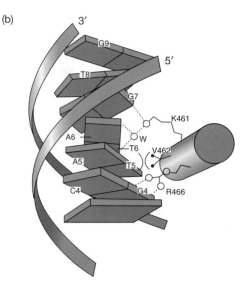

Figure 12.8   Association between the glucocorticoid receptor DNA-binding domain and its DNA target. (a) Stereodiagram showing amino acid/DNA interactions and protein–protein interactions. The DNA (yellow) is on the left and the β-carbon chain of the protein is shown on the right. The protein dimerizes, with one monomer (red) making specific contacts with the DNA, and one monomer (blue) making nonspecific contacts because the spacing between target half-sites is one base pair too much. The zinc ions are represented by yellow spheres and the cysteine side chains coordinating them are colored dark green. Amino acids making contact with the DNA are rendered in light green. DNA bases are identified with letters and numbers, and amino acids are numbered. Note the recognition helices just to the left of the uppermost and lowermost zinc ions. Note also the dimerization domains interacting between the middle two zinc ions. **(b)** The specific amino acid/base interactions involving the recognition helix. A water molecule (W) mediates some of the H-bonding between lysine 461 and the DNA. (*Source:* Luisi et al., Crystallographic analysis of the interaction of the glucocorticoid receptor with DNA. *Nature* 352 (8 Aug 1991) p. 500, f. 4a. © Macmillan Magazines Ltd.)

spheres), rather than the one found in a classical zinc finger. (3) Each zinc ion is complexed to four cysteines to form a finger-like shape (not apparent in the figure). (4) The amino-terminal finger in each binding domain engages in most of the interactions with the DNA target. Most of these interactions involve an α-helix, which is seen here almost from the end, so it looks like a series of squares close together. This α-helix lies just to the left of the upper zinc ion in the specific (red) domain, and just to the left of the lower zinc ion in the nonspecific (blue) domain. Figure 12.8b illustrates the specific amino-acid–base associations between this recognition helix and the DNA target site. Some amino acids outside this helix also make contact with the DNA through its backbone phosphates.

Figure 12.9 contains stereoscopic views of the interactions between the protein dimer and (a) the DNA with the normal 3-bp separation between target half-sites, or (b) 4 bp between these sites. As in the previous figure, red indicates specific, and blue indicates nonspecific, DNA–protein association. All four α-helices are viewed almost from their ends, which makes it easier to locate them by their boxy shape, to see how the amino-terminal helices fit into the major grooves of the DNA, and how the match with the bases is not the same with an extra base pair between half-sites. In particular, notice that the positions of the recognition helices relative to the DNA

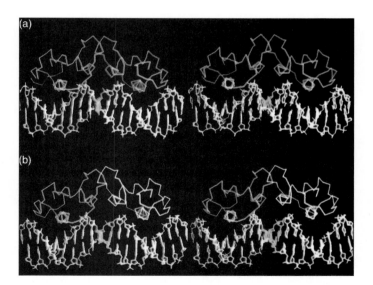

Figure 12.9   Glucocorticoid receptor–DNA interactions involving: (a) DNA with the normal 3-bp spacing between target half-sites; and (b) DNA with an extra base pair between target half-sites. These stereodiagrams show **(a)** the perfect fit between both monomers (red) when the spacing between half-sites is normal, and **(b)** the out-of-register fit with one of the monomers (blue) when the spacing between half-sites is expanded by 1 bp. (*Source:* Luisi et al., *Nature* 352 (8 Aug 1991) p. 500, f. 4e. © Macmillan Magazines Ltd.)

double helix are the same in both specifically interacting domains. On the other hand, when 4 bp separate the target half-sites, the nonspecifically interacting domain's recognition helix is not as deeply engaged in the DNA major groove because it does not encounter the proper bases in the proper places.

Finally, note that all three classes of zinc-containing DNA-binding modules use a common motif—an α-helix—for most of the interactions with their DNA targets.

> **SUMMARY** Some members of the steroid hormone family of receptors bind to their respective ligands and bind to enhancers, and thereby act as activators. The glucocorticoid receptor is representative of this group. It has a DNA-binding domain with two zinc-containing modules. One module contains most of the DNA-binding residues (in a recognition α-helix), and the other module provides the surface for protein–protein interaction to form a dimer. These modules use four cysteine residues to complex the zinc ion, instead of two cysteines and two histidines as seen in classical zinc fingers. One extra base pair between half-sites in the DNA target prevents specific interactions with one of the glucocorticoid receptor monomers.

Homeodomains

Homeodomains are DNA-binding domains found in a large family of activators. Their name comes from the gene regions, called **homeoboxes,** in which they are encoded. Homeoboxes were first discovered in regulatory genes of the fruit fly *Drosophila,* called homeotic genes. Mutations in these genes cause strange transformations of body parts in the fruit fly. For example, a mutation called *Antennapedia* causes legs to grow where antennae would normally be (Figure 12.10).

Homeodomain proteins are members of the helix-turn-helix family of DNA-binding proteins (Chapter 9). Each homeodomain contains three α-helices; the second and third of these form the helix-turn-helix motif, with the third serving as the recognition helix. But most homeodomains have another element, not found in helix-turn-helix motifs: The N-terminus of the protein forms an arm that inserts into the the minor groove of the DNA. Figure 12.11 shows the interaction between a typical homeodomain, from the *Drosophila* homeotic gene *engrailed,* and its DNA target. This view of the protein–DNA complex comes from Thomas Kornberg's and Carl Pabo's x-ray diffraction analysis of cocrystals of the engrailed homeodomain and an oligonucleotide containing the engrailed binding site. Most homeodomain proteins have weak DNA-binding specificity on their own. As a

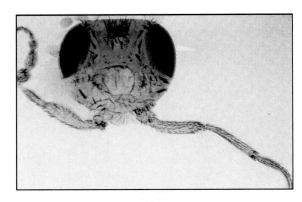

Figure 12.10   The *Antennapedia* phenotype. Legs appear on the head where antennae would normally be. (*Source:* Courtesy Walter J. Gehring, University of Basel, Switzerland.)

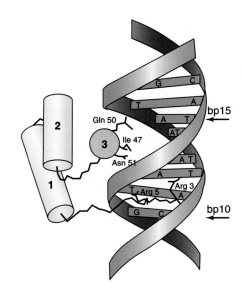

Figure 12.11   Representation of the homeodomain–DNA complex. Schematic model with the three helices numbered on the left, and a ribbon diagram of the DNA target on the right. The recognition helix (labeled 3) is shown on end, resting in the major groove of the DNA. Key amino acid side chains are shown interacting with DNA.

result, they rely on other proteins to help them bind specifically and efficiently to their DNA targets.

> **SUMMARY** The homeodomains in eukaryotic activators contain a DNA-binding motif that functions in much the same way as prokaryotic helix-turn-helix motifs in which a recognition helix fits into the DNA major groove and makes specific contacts there. In addition, the N-terminal arm nestles in the adjacent minor groove.

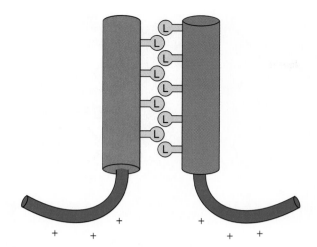

Figure 12.12   Model of leucine zipper. Two protein molecules contain α-helices, represented by the red and blue cylinders. The leucines (circled L's, yellow) on the surfaces of these α-helices are thought to interact like two halves of a zipper. The basic regions (green) outside the leucine zipper (+ charges) participate in DNA binding, but only when the zipper is "zipped shut," that is, when the protein is in dimer form. This model is only hypothetical. It is not based on three-dimensional structural studies.

The bZIP and bHLH Domains

As with several of the other DNA-binding domains we have studied, the **bZIP** and **bHLH** domains combine two functions: DNA binding and dimerization. The *ZIP* and *HLH* parts of the names refer to the **leucine zipper** and **helix-loop-helix parts,** respectively, of the domains, which are the dimerization motifs. The *b* in the names refers to a basic region in each domain that forms the majority of the DNA-binding motif.

Let us consider the structures of these combined dimerization/DNA-binding domains, beginning with the bZIP domain. Figure 12.12 is a schematic diagram to illustrate the concept that each protein monomer contains half of the zipper: an α-helix with leucine (or other hydrophobic amino acid) residues spaced seven amino acids apart, so they are all on one face of the helix. The spacing of the hydrophobic amino acids puts them in position to interact with a similar string of amino acids on the other protein monomer. In this way, the two helices act like the two halves of a zipper.

Figure 12.12 is highly schematic and does not tell us the real three-dimensional shape of the leucine zipper. To get a better idea of the structure of the zipper, Peter Kim and Tom Alber and their colleagues crystallized a synthetic peptide corresponding to the bZIP domain of **GCN4,** a yeast activator that regulates amino acid metabolism. The x-ray diffraction pattern shows that the dimerized bZIP domain assumes a parallel coiled coil structure (Fig-ure 12.13). The α-helices are parallel in that their amino to carboxyl orientations are the same (left to right

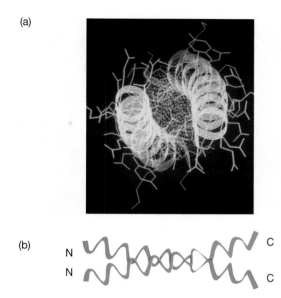

Figure 12.13   Structure of a leucine zipper. (a) Kim and Alber and colleagues crystallized a 33-amino-acid peptide containing the leucine zipper motif of the transcription factor GCN4. X-ray crystallography on this peptide yielded this view along the axis of the zipper with the coiled coil pointed out of the plane of the paper. (b) A side view of the coiled coil with the two α-helices colored red and blue. Notice that the amino ends of both peptides are on the left. Thus, this is a parallel coiled coil. (*Source:* (a) O'Shea et al., X-ray structure of the GCN4 leucine zipper, a two-stranded, parallel coiled coil. *Science* 254 (25 Oct 1991) p. 541, f. 3. © AAAS.)

in panel b). Figure 12.13a in which the coiled coil extends directly out at the reader, gives a good feel for the extent of supercoiling in the coiled coil. Notice the similarity between this and the coiled coil dimerization motif in GAL4 (see Figure 12.5).

This crystallographic study, which focused on the zipper in the absence of DNA, did not shed light on the mechanism of DNA binding. However, Kevin Struhl and Stephen Harrison and their colleagues performed x-ray crystallography on the bZIP domain of GCN4, bound to its DNA target. Figure 12.14 shows that the leucine zipper not only brings the two monomers together, it also places the two basic parts of the domain in position to grasp the DNA like a pair of forceps, or fireplace tongs, with the basic motifs fitting into the DNA major groove.

Harold Weintraub and Carl Pabo and colleagues solved the crystal structure of the bHLH domain of the activator **MyoD** bound to its DNA target. The structure (Figure 12.15) is remarkably similar to that of the bZIP domain–DNA complex we just considered. The helix-loop-helix part is the dimerization motif, but the long helix (helix 1) in each helix-loop-helix domain contains the basic region of the domain, which grips the DNA target via its major groove, just as the bZIP domain does.

Some proteins, such as the oncogene products Myc and Max, have **bHLH-ZIP** domains with both HLH and

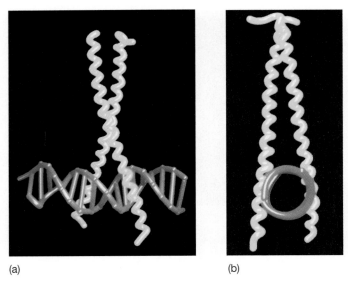

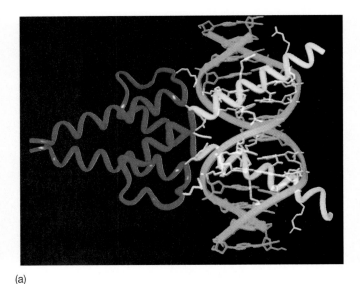

(a)   (b)

(a)

Figure 12.14 Crystal structure of the bZIP motif of GCN4 bound to its DNA target. The DNA (red) contains a DNA target for the bZIP motif (yellow). Notice the coiled coil nature of the interaction between the protein monomers, and the tong-like appearance of the protein grasping the DNA. (*Source:* Ellenberger et al., The GCN4 basic region leucine zipper binds DNA as a dimer of uninterrupted alpha helices: Crystal structure of the protein-DNA complex. *Cell* 71 (24 Dec 1992) p. 1227, f. 3a-b. Reprinted by permission of Elsevier Science.)

(b)

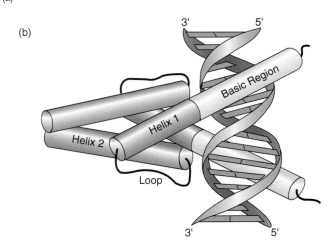

Figure 12.15 Crystal structure of the complex between the bHLH domain of MyoD and its DNA target. Schematic diagram with cylinders representing α-helices. (*Source:* Ma et al., Crystal structure of MyoD bHLH domain-DNA complex: Perspectives on DNA recognition and implications for transcriptional activation. *Cell* 77 (6 May 1994) p. 453, f. 2a. Reprinted by permission of Elsevier Science.)

ZIP motifs adjacent to a basic motif. The bHLH-ZIP domains interact with DNA in a manner very similar to that employed by the bHLH domains. The main difference between bHLH and bHLH-ZIP domains is that the latter may require the extra interaction of the leucine zippers to ensure dimerization of the protein monomers.

SUMMARY The bZIP proteins dimerize through a leucine zipper, which puts the adjacent basic regions of each monomer in position to embrace the DNA target site like a pair of tongs. Similarly, the bHLH proteins dimerize through a helix-loop-helix motif, which allows the basic parts of each long helix to grasp the DNA target site, much as the bZIP proteins do. bHLH and bHLH-ZIP domains bind to DNA in the same way, but the latter have extra dimerization potential due to their leucine zippers.

12.3 Independence of the Domains of Activators

We have now seen several examples of DNA-binding domains in activators, and one detailed example of a transcription-activation domain (the acidic domain).

These domains are separated physically on the proteins, they fold independently of each other to form distinct three-dimensional structures, and they operate independently of each other. Roger Brent and Mark Ptashne demonstrated this independence by creating a chimeric factor with the DNA-binding domain of one protein and the transcription-activating domain of the other. This hybrid protein functioned as an activator, with its specificity dictated by its DNA-binding domain.

Brent and Ptashne started with the genes for two proteins: GAL4 and LexA. We have already studied the DNA-binding and transcription-activating domains of GAL4; LexA is a prokaryotic repressor that binds to *lexA* operators and represses downstream genes in *E. coli* cells. It does not normally have a transcription-activating

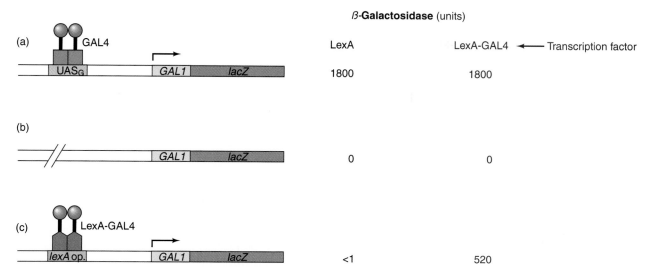

Figure 12.16   Activity of a chimeric transcription factor. Brent and Ptashne introduced two plasmids into yeast cells: (1) a plasmid encoding LexA-GAL4, a hybrid protein containing the transcription-activating domain of GAL4 (green) and the DNA-binding domain of LexA (blue); (2) one of the test plasmid constructs shown in panels a–c. Each of the test plasmids had the *GAL1* promoter linked to a reporter gene (the *E. coli lacZ* gene). The production of β-galactosidase (given at right) is a measure of promoter activity. **(a)** With a UAS$_G$ element, transcription was very active and did not depend on the added transcription factor, because endogenous GAL4 could activate via UAS$_G$. **(b)** With no DNA target site, LexA-GAL4 could not activate, because it could not bind to the DNA near the *GAL1* promoter. **(c)** With the *lexA* operator, transcription was greatly stimulated by the LexA-GAL4 chimeric factor. The LexA DNA-binding domain could bind to the *lexA* operator, and the GAL4 transcription-activating domain could enhance transcription from the *GAL1* promoter.

domain, because that is not its function. By cutting and recombining fragments of the two genes, Brent and Ptashne created a chimeric gene containing the coding regions for the transcription-activating domain of GAL4 and the DNA-binding domain of LexA. To assay the activity of the protein product of this gene, they introduced two plasmids into yeast cells. The first plasmid had the chimeric gene, which produced its hybrid product. The second contained a promoter responsive to GAL4 (either the *GAL1* or the *CYC1* promoter), linked to the *E. coli* β-galactosidase gene, which served as a reporter. The more transcription from the GAL4-responsive promoter, the more β-galactosidase was produced. Therefore, by assaying for β-galactosidase, Brent and Ptashne could determine the transcription rate.

One more element was necessary to make this assay work: a binding site for the chimeric protein. The normal binding site for GAL4 is an upstream enhancer called UAS$_G$. However, this site would not be recognized by the chimeric protein, which has a LexA DNA-binding domain. To make the *GAL1* promoter responsive to activation, the investigators had to introduce a DNA target for the LexA DNA-binding domain. Therefore, they inserted a *lexA* operator in place of UAS$_G$. It is important to note that a *lexA* operator would not normally be found in a yeast cell; it was placed there just for the purpose of this experiment. Now the question is: Did the chimeric protein activate the *GAL1* gene?

The answer is yes, as Figure 12.16 demonstrates. The three test plasmids contained UAS$_G$, no target site, or the *lexA* operator. The activator was either LexA-GAL4, as we have discussed, or LexA (a negative control). With UAS$_G$ present (a), a great deal of β-galactosidase was made, regardless of which activator was present. This is because the yeast cells themselves make GAL4, which can activate via UAS$_G$. When no DNA target site was present (b), no β-galactosidase could be made. Finally, when the *lexA* operator replaced UAS$_G$ (c), the LexA-GAL4 chimeric protein could activate β-galactosidase production over 500-fold, relative to the amount observed with LexA. Thus, one can replace the DNA-binding domain of GAL4 with the DNA-binding domain of a completely unrelated protein, and produce an active activator. This demonstrates that the transcription-activating and DNA-binding domains of GAL4 can operate quite independently.

SUMMARY The DNA-binding and transcription-activating domains of activator proteins are independent modules. We can make hybrid proteins with the DNA-binding domain of one protein and the transcription-activating domain of another, and show that the hybrid protein still functions as an activator.

12.4 Functions of Activators

In prokaryotes, the core RNA polymerase is incapable of initiating meaningful transcription, but the RNA polymerase holoenzyme can catalyze basal level transcription. Basal level transcription is frequently insufficient at weak promoters, so cells have activators to boost this basal transcription to higher levels by a process called **recruitment.** Recruitment takes two forms in prokaryotes: stimulation of formation of a closed promoter complex, or stimulation of isomerization of the closed promoter complex to an open promoter complex. In either case, recruitment leads to the tight binding of RNA polymerase holoenzyme to a promoter.

Eukaryotic activators also recruit RNA polymerase to promoters, but not as directly as prokaryotic activators. The eukaryotic activators stimulate binding of general transcription factors and RNA polymerase to a promoter. Figure 12.17 presents two hypotheses to explain this recruitment: (1) the general transcription factors cause a stepwise build-up of a preinitiation complex; or (2) the general transcription factors and other proteins are already bound to the polymerase in a complex called the **holoenzyme,** and the factors and polymerase are recruited together to the promoter. In either event, it appears that direct contacts between general transcription factors and activators are necessary for recruitment. Which factors do the activators contact? The answer seems to be that many factors can be targets, but the two that were discovered first were TFIID and TFIIB.

Recruitment of TFIID

In 1990, Keith Stringer, James Ingles, and Jack Greenblatt performed a series of experiments to identify the factor that binds to the acidic transcription-activating domain of the herpesvirus transcription factor **VP16.** These workers expressed the VP16 transcription-activating domain as a fusion protein with the *Staphylococcus aureus* protein A, which binds tightly and specifically to immunoglobulin IgG. They immobilized the fusion protein (or protein A by itself) on an agarose IgG column and used these as affinity columns to fish out proteins that interact with the VP16-activating domain. To find out what proteins bind to the VP16-activating domain, they poured HeLa cell nuclear extracts through the columns containing either protein A by itself or the protein A/VP16-activating domain fusion protein. Then they used run-off transcription (Chapter 5) to assay various fractions for ability to transcribe the adenovirus major late locus accurately in vitro. They found that the flow-through from the protein A column still had abundant ability to support transcription, indicating no nonspecific binding of any essential factors to protein A. However, when they tested the flow-through from the

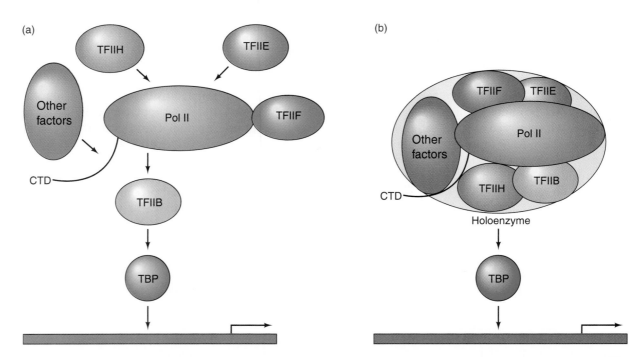

Figure 12.17   Two models for recruitment of yeast preinitiation complex components. (a) Traditional view of recruitment. This scheme calls for stepwise addition of components of the preinitiation complex, as occurs in vitro. CTD refers to the carboxyl-terminal domain of the largest polymerase II subunit (see Chapter 10). **(b)** Recruitment of holoenzyme. Here, TBP binds first, then the holoenzyme (enclosed in beige oval) binds to form the preinitiation complex. (*Source:* Reprinted with permission from *Nature* 368:466, 1994. Copyright © 1994 Macmillan Magazines Limited.)

protein A/VP16-activating domain column they found no transcription activity until they added back the proteins that bound to the column. Thus, some factor or factors essential for in vitro transcription bound to the VP16-activating domain.

Stringer and colleagues knew that TFIID was rate-limiting for transcription in their in vitro system, so they suspected that TFIID was the factor that bound to the affinity column. To find out, they depleted a nuclear extract of TFIID by heating it, then added back the material that bound to either the protein A column or the column containing the protein A/VP16-activating domain. Figure 12.18 shows that the material that bound to protein A by itself could not reconstitute the activity of a TFIID-depleted extract, but the material that bound to the protein A/VP16-activating domain could. This strongly suggested that TFIID binds to the VP16-activating domain.

To check this conclusion, Stringer and colleagues first showed that the material that bound to the VP16-activating domain column behaved just like TFIID on DEAE-cellulose ion-exchange chromatography. Then they assayed the material that bound to the VP16-activating domain column for the ability to substitute for TFIID in a template commitment experiment. In this experiment,

they formed preinitiation complexes on one template, then added a second template to see whether it could also be transcribed. Under these experimental conditions, the commitment to transcribe the second template depended on TFIID. These workers found that the material that bound to the VP16-activating domain column could shift commitment to the second template, but the material that bound to the protein A column could not. These, and similar experiments performed with yeast nuclear extracts, provided convincing evidence that TFIID is the important target of the VP16 transcription-activating domain in this experimental system.

SUMMARY The acidic transcription-activating domain of the herpesvirus transcription factor VP16 binds to TFIID under affinity chromatography conditions.

Recruitment of TFIIB

Young-Sun Lin and Michael Green attacked the problem in a somewhat different way and reached a different conclusion: The acidic activating domain of GAL4 (and several other activating domains, including that of VP16) bind to TFIIB. They showed that an archetypal acidic activator, GAL4, does indeed act by aiding assembly of the general transcription factors on a promoter. Moreover, they showed that TFIID can bind to the promoter without any help from GAL4, but that the other general factors require the assistance of GAL4 to bind.

Lin and Green performed their first set of experiments using a two-step incubation scheme as follows (Figure 12.19). The DNA template they used was a fragment of the adenovirus E4 gene containing the natural TATA box and an added yeast GAL4-binding site, as well as part of the transcribed region of the E4 gene. They attached this double-stranded DNA template to agarose beads. Then, in the first incubation, they added one or more general transcription factors to the template in the presence or absence of GAL4-AH, a hybrid protein containing the DNA-binding domain of GAL4 and a synthetic acidic domain that activates transcription. Because this protein behaves just like GAL4, we will call it GAL4 for convenience. GAL4 was a good choice for this assay because it is not found in mammals, so the experimenters could be sure that any effects apparently due to GAL4 were really caused by that protein, and not by a contaminant in the mammalian extract. After washing the beads to remove any unbound factors, they performed the second incubation by adding nucleotides, one of which was labeled, along with one or more general transcription factors. They measured accurate transcription during this second incubation by a primer extension assay. This

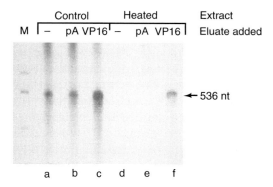

Figure 12.18 Evidence that an acidic activation domain binds TFIID. Stringer and colleagues fractionated a HeLa cell extract by affinity chromatography with a resin containing a fusion protein composed of protein A fused to the VP16-activating domain, or a resin containing just protein A. Then they eluted the proteins bound to each affinity column and tested them for ability to restore in vitro run-off transcription activity to an extract that had been heated to destroy TFIID specifically. Lanes a–c are controls in which the extract had not been heated. Because TFIID was still active, all lanes showed activity. Lanes d–f contained heated extract supplemented with: nothing (–) the eluate from the protein A column (pA), or the eluate from the column that contained the fusion protein composed of protein A and the transcription-activating domain of the VP16 protein (VP16). Only the eluate from the column containing the VP16 fusion protein could replace the missing TFIID and give an accurately initiated run-off transcript with the expected length (536 nt, denoted at right). Thus, TFIID must have bound to the VP16 transcription-activating domain in the affinity column. (*Source:* Stringer, K. et al. Direct and selective binding of an acidic transcriptional activation domain to the TATA-box factor TFIID. *Nature* 345 (1990) f. 2, p. 784. © Macmillan Magazines Ltd.)

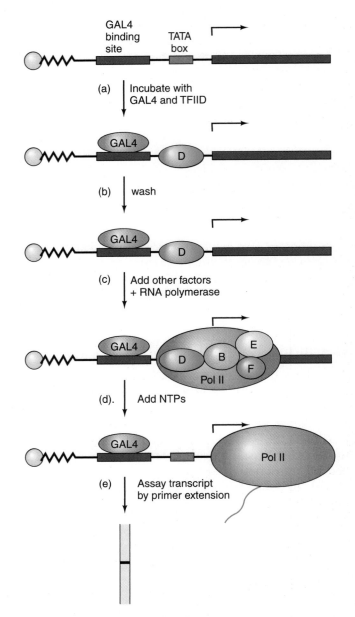

Figure 12.19   Affinity chromatography assay for effect of transcription factors. Illustrated at top is a gene fragment linked to a chromatography bead (yellow). The DNA fragment contains a GAL4-binding site (purple), a TATA box (red), and a transcribed region (dark blue). The assay is performed as follows: **(a)** Incubate with GAL4 (light purple) and TFIID (blue), which bind to their respective binding sites in the 5′-flanking region of the gene. **(b)** Wash the DNA beads to remove any unbound factors. **(c)** Add all the other general transcription factors plus RNA polymerase. If GAL4 and TFIID have bound tightly in the first incubation (part **a**), they will still be available to cooperate with the remaining factors in this step, thus assembling a preinitiation complex, as shown. **(d)** Add nucleoside triphosphates (NTPs); if the preinitiation complex has formed in step **(c)** this will allow RNA polymerase to begin making RNA. **(e)** Assay for transcription by primer extension. If accurate transcription has initiated at the gene's transcription start site, a band of predictable size will be observed. This example shows a positive result. However, a different order of adding factors could give a negative result. For example, if RNA polymerase were given along with GAL4 in the first incubation, it would not bind tightly, and would wash off in step **(b)**. Then, because RNA polymerase would not be added again in step **(c)**, it would not be part of the preinitiation complex, and no transcription would occur.

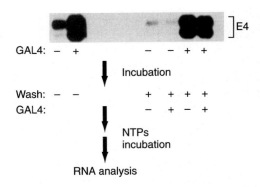

Figure 12.20   GAL4 stimulates preinitiation complex formation. Lin and Green incubated DNA containing the E4 promoter with a crude preparation of transcription factors in the presence or absence of GAL4, as indicated near the top. After washing, they incubated the complexes again either in the presence or absence of GAL4 and assayed for E4 transcription by primer extension. The results of this assay are shown at the very top. The lanes on the left show the great stimulatory effect of GAL4 added in the first incubation. The four lanes on the right show that if GAL4 was missing from the first incubation, transcription was weak, and adding GAL4 in the second incubation—without the other transcription factors—did not help. On the other hand, if GAL4 was present during the first incubation, transcription was equally strong whether or not GAL4 was present in the second incubation. (*Source:* Lin and Green, Mechanism of action of GAL4-AH *in vitro. Cell* 64 (8 Mar 1991) p. 972, f. 1. © Cell Press.)

experimental system allowed them to see what general factors were needed in the second incubation. These must be the factors that did not bind tightly to the preinitiation complex during the first incubation.

If activators, including GAL4, facilitate assembly of preinitiation complexes, we would not expect them to stimulate transcription by preinitiation complexes that have already assembled. In their first experiment, Lin and Green confirmed this expectation. Figure 12.20 shows the results: Complexes formed with crude nuclear extracts in the first incubation (in the presence or absence of GAL4), were not stimulated by GAL4 added in the second incubation, after the wash. Preinitiation complexes had either already formed in the first incubation or had failed to form; in either case, it was too late for GAL4 to offer any help after the first wash. On the other hand, it was clear that GAL4 was a big help if it was present during the first incubation when all the general transcription factors were there. These results suggested that GAL4 acts by stimulating assembly of a preinitiation complex, presumably by recruiting one or more general transcription factors to the promoter.

Next, Lin and Green demonstrated that TFIIB is the limiting factor in forming preinitiation complexes in the presence of GAL4. In Chapter 11 we learned that TFIID is the first factor to bind to the promoter in vitro, and TFIIB comes next. Figure 12.21a shows that this second step, binding of TFIIB, is the one that depended on GAL4 in this assay. In lanes 1 and 2, Lin and Green added

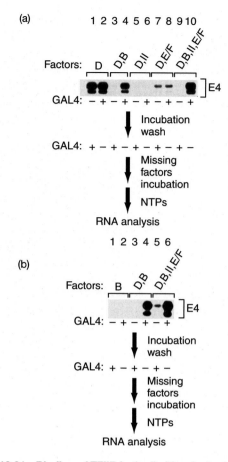

Figure 12.21   Binding of TFIIB is the limiting factor in preinitiation complex formation. Lin and Green performed the two-step assay in the presence of various mixtures of factors in the first incubation, then the other factors in the second incubation, after the wash. The factors used in the first incubation are listed at top, just above the results of the transcription assay. The other factors, missing in the first incubation, were present in the second, labeled "missing factors incubation." **(a)** Lanes 1 and 2 show that GAL4 had no effect if TFIID was the only factor present in the first incubation. This is compatible with the hypothesis that TFIID can bind by itself, without help from GAL4. On the other hand, lanes 3 and 4 demonstrate that GAL4 was needed to help TFIIB bind in the first incubation. **(b)** Lanes 1 and 2 show that TFIIB could not bind in the first incubation in the absence of TFIID, even with help from GAL4. This confirms what we have already learned: Binding of TFIIB can occur only if TFIID is already bound. (*Source:* Lin and Green, *Cell* 64 (8 Mar 1991) p. 974, f. 4. Reprinted by permission of Elsevier Science.)

TFIID in the first incubation, then washed, then added all the other factors, but not TFIID, in the second incubation. Notice that strong transcription occurred under these conditions, regardless of the timing of addition of GAL4. If GAL4 was added during the first incubation, it bound tightly to its binding site on the DNA and was still present during the second incubation with TFIIB. Alternatively, if GAL4 was added during the second incubation, it was also available along with TFIIB. On the other hand, lanes 3 and 4 show that when Lin and Green added TFIID and B to the first incubation, then washed,

then added all the other factors, but not D and B, transcription depended heavily on the presence of GAL4 in the first incubation. This behavior indicated that binding of TFIIB to the complex requires GAL4. Similarly, lanes 5–8 show that if polymerase II or TFIIE and F were included in the first incubation, without TFIIB, but then left out of the second incubation after the wash, little or no transcription occurred. This lack of activity is what we would expect: polymerase II, TFIIE, and TFIIF can bind only to a complex that already includes TFIIB, and TFIIB was not present until after polymerase and TFIIE and F were washed away. The small amount of transcription seen in lanes 7 and 8 is probably due to contamination of the TFIIE/F fraction with TFIIB. Finally, lanes 9 and 10 show that, as expected, all factors plus GAL4 in the first incubation were sufficient to form an active complex. Figure 12.21b contains one more piece of information: Adding TFIIB without TFIID in the first incubation gave no active complex, even in the presence of GAL4. Again, this behavior is expected: TFIIB needs TFIID, in addition to GAL4, to bind to the complex. The reverse order, TFIIB then TFIID, does not work.

The experiments shown so far are consistent with interaction between GAL4 and either TFIIB or TFIID. That is, GAL4 could be contacting TFIIB and promoting its binding to the complex, or it could be contacting TFIID and causing it to recruit TFIIB to the complex. To probe this question, Lin and Green sought to find out which factor can interact directly with an acidic transcription-activation domain such as the one found on GAL4 (or GAL4-AH). For these experiments, they used the very potent acidic activation domain of the herpes virus (HSV-1) activator VP16. They made an affinity chromatography column with this VP16 acidic domain and showed that TFIIB, but not the other factors, could bind to it under high ionic strength conditions. At lower ionic strength, TFIID could also bind, but not quantitatively as TFIIB could. Next, they showed that TFIIB could not bind to a mutated VP16 acidic activation region that was not capable of activating. Thus, the interaction and the activation seem to go hand in hand. In a later publication, Green's group reported that a cloned TFIIB gene product could bind directly to the VP16 acidic activating domain. This demonstrated unequivocally that the binding to TFIIB was direct, not mediated by some other transcription factor such as TFIID. Lin and Green concluded that GAL4, and presumably other activators with acidic activation domains, act by recruiting TFIIB to the preinitiation complex.

How can we reconcile this conclusion with that of others, including Stringer and colleagues, that TFIID is the target of acidic transcription-activating domains? The answer is probably that many proteins can be targets of transcription-activating domains, and the results one observes in any given experiment depend on the experimental conditions. In fact, the next section will show that

GAL4 can recruit other members of the preinitiation complex in addition to TFIIB. Furthermore, if the holoenzyme hypothesis described later in this chapter is valid, most or all of the components of the preinitiation complex are recruited to the promoter as a unit. That would mean that an activator could recruit the preinitiation complex by targeting *any* component of the holoenyzme, and multiple activators bound to multiple enhancers could target many parts of the holoenzyme simultaneously.

> **SUMMARY** Some activators with acidic activation domains, such as GAL4, appear to stimulate transcription by facilitating the binding of TFIIB to the preinitiation complex.

Recruitment of Other General Transcription Factors

Green recognized that his earlier, transcription assay suffered from several problems. For one thing, looking at transcription rather than at factor binding reveals only the first rate-limiting step in preinitiation complex formation. Therefore, this method might miss important downstream steps. It is also possible that GAL4 could stimulate transcription by converting an inactive preinitiation complex to an active one, rather than by causing the formation of more complexes. Thus, it is conceivable that GAL4 merely rearranged the general transcription factors in a complex and did not really recruit TFIIB to the preinitiation complex after all.

To solve these problems, Bob Choy and Green added an assay for factor binding to their transcription assay. They formed complexes with various mixtures of factors and then separated the complexes from the factors by either of two methods: (1) gel exclusion chromatography or (2) forming the complexes on DNA templates linked to metal beads, which could then be purified magnetically. The latter technique is very similar to the agarose bead method they used in their previous studies, but it eliminated the high background binding of factors to agarose beads that would have confused their factor-binding assay. They analyzed their preinitiation complexes for the presence of factors by immunoblotting, or for transcription by a primer extension assay as before. **Immunoblotting (Western blotting)** involves electrophoresing proteins, blotting them to a membrane, and probing the blot with a specific antibody or antiserum to detect a particular protein. The antibody is detected with a labeled secondary antibody or protein A (Chapter 4).

Choy and Green found that GAL4-AH stimulated transcription in the presence of TFIID, which includes TBP and TAFs (Chapter 11), but not in the presence of TBP alone. Does this mean that TBP is incapable of coop-

erating with GAL4-AH to recruit TFIIB? No, an immunoblot assay for TFIIB in the complex showed just as much TFIIB bound in the presence of TBP as with TFIID.

If the problem is not with TFIIB binding, perhaps it concerns the binding of another general transcription factor, later in the assembly of the preinitiation complex. To test this idea, Choy and Green assembled complexes with either TBP or TFIID and examined these complexes for polymerase II, TFIIE, and TFIIF by immunoblotting. Figure 12.22 shows the results. RNA polymerase II, TFIIE, and TFIIF were all recruited to the preinitiation complex, but only in the presence of GAL4-AH and TFIID. TBP could not substitute for TFIID in recruiting these other factors to the complex. This leads to two conclusions: First, GAL4-AH does more than just recruit TFIIB; it also encourages the binding of a downstream factor such as TFIIE or TFIIF. Second, this downstream recruitment requires one or more TAFs, in addition to TBP.

A recurring theme in enhancer action is the multiple target sites found within most enhancers. For example, the classic enhancer, the SV40 70-bp repeat obviously has two copies of the 72-bp DNA target site, but each of these repeats also contains multiple subsites that are the minimal units that can interact with enhancer-binding proteins. These enhancer modules, sometimes called **enhansons**, allow for cooperativity by activators such as GAL4. For instance, we would expect that the enhancement

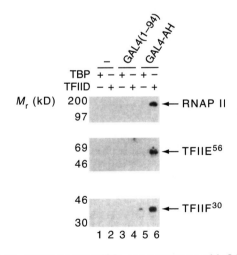

Figure 12.22   TFIID, but not TBP, can cooperate with GAL4-AH to recruit downstream general transcription factors. Choy and Green assayed for recruitment of RNA polymerase II (RNAP II), TFIIE, and TFIIF by immunoblotting preinitiation complexes assembled in the presence of the factors listed at top, and purified by gel exclusion chromatography. They probed the immunoblots with antibodies against the polymerase II large subunit, the 56-kD subunit of TFIIE, and the 30-kD subunit of TFIIF, respectively. In all cases, TFIID could help GAL4-AH recruit the downstream factors, but TBP could not. A polypeptide [GAL4 (1–94)] containing the first 94 amino acids of GAL4, but lacking a transcription-activation domain, was ineffective with either TBP or TFIID. (*Source:* Choy and Green, *Nature* 366 (9 Dec 1993) p. 534, f. 3a. © Macmillan Magazines Ltd.)

provided by five GAL4-binding sites in a row would be much more than five times the enhancement of one GAL4 site. Figure 12.23a shows that this is so. The enhancement by one GAL4-binding site is barely detectable (about twofold), whereas the enhancement by five GAL4-binding sites is 50-fold.

Is this cooperativity exerted at the TFIIB-binding step, or in a downstream factor-binding step? Figure 12.23b demonstrates that it is in a downstream step. The template with the fivefold repeat of the binding site gave only

a twofold boost to TFIIB recruitment, but a very large increase in TFIIE binding.

All of the experiments described so far involve the acidic activation domain of GAL4. Do the same rules apply to other acidic domains, and even to nonacidic activation domains? To find out, Choy and Green ran preinitiation complex assembly experiments with hybrid activators. All had the GAL4 DNA-binding domain, but each had a different transcription-activating domain. The first had the acidic activation domain of VP16. It recruited TFIIB to the preinitiation complex just as well as GAL4-AH. The other factors had either the glutamine-rich transcription-activation domain of Sp1 (GAL4-GLN) or the proline-rich activation domain of CTF (GAL4-PRO). Both of these factors were just as good as GAL4-AH at recruiting both TFIIB and TFIIE to the preinitiation complex.

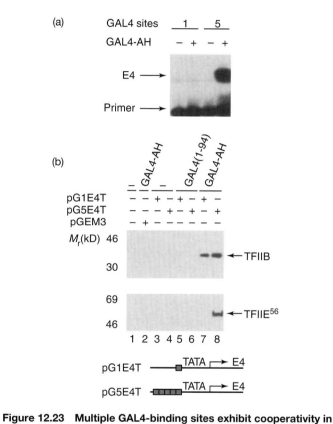

Figure 12.23   Multiple GAL4-binding sites exhibit cooperativity in recruiting TFIIE, but not TFIIB. Shown at bottom are the two different plasmids used in primer extension assays for transcription and immunoblotting assays for preinitiation complex assembly. The first, pG1E4T, has a minimal E4 promoter and one GAL4-binding site. The second, pG5E4T is the same, except it has five GAL4-binding sites. Preinitiation complexes were assembled on metal beads and purified magnetically. **(a)** Transcription assay. Transcription was measured by primer extension on preinitiation complexes assembled with nuclear extract and templates containing one or five GAL4-binding sites, and with or without GAL4-AH, as indicated at top. Clearly, five sites allowed for cooperativity in activation of transcription by GAL4. **(b)** Factor-binding assay. Choy and Green used the immunoblotting assay to measure the amount of TFIIB and TFIIE in preinitiation complexes assembled with nuclear extract with and without activator (GAL4-AH) or mock activator with no transcription–activating domain [GAL4 (1–94)], as indicated at top. The template used is also indicated at top; pGEM3 is a negative control with no eukaryotic promoter. GAL4-AH did not act cooperatively in recruiting TFIIB, but it did in recruiting TFIIE (compare lanes 7 and 8 in both immunoblots). (*Source:* Choy and Green, *Nature* 366 (9 Dec 1993) p. 534, f. 4a-b. © Macmillan Magazines Ltd.)

> **SUMMARY** Chimeric activators bearing acidic, glutamine-rich, and proline-rich transcription-activation domains appear to stimulate transcription by recruiting TFIIB, then one or more other general transcription factors—TFIIF and RNA polymerase II, or TFIIE—to the preinitiation complex.

Recruitment of the Holoenzyme

In Chapter 11 we learned that RNA polymerase II can be isolated from eukaryotic cells as a holoenzyme—a complex containing a subset of general transcription factors and other polypeptides. Much of our discussion so far has been based on the assumption that activators recruit general transcription factors one at a time to assemble the preinitiation complex. But it is also possible that activators recruit the holoenzyme as a unit, leaving only a few other proteins to be assembled at the promoter. In fact, we have good evidence that recruitment of the holoenzyme really does occur.

In 1994, Anthony Koleske and Richard Young isolated from yeast cells a holoenzyme that contained polymerase II, TFIIB, F, and H, and SRB2, 4, 5, and 6. They went on to demonstrate that this holoenzyme, when supplemented with TBP and TFIIE, could accurately transcribe a template bearing a *CYC1* promoter in vitro. Finally, they showed that the activator GAL4-VP16 could activate this transcription. Because the holoenzyme was provided intact, this last finding suggested that the activator recruited the intact holoenzyme to the promoter rather than building it up step by step on the promoter (recall Figure 12.17).

By 1998, investigators had purified holoenzymes from many different organisms, with varying protein compositions. Some contained most or all of the general transcription factors and many other proteins. Vic Meyer and

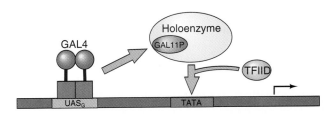

Figure 12.24   Model for recruitment of the GAL11P-containing holoenzyme by the dimerization domain of GAL4. The dimerization domain of GAL4 binds (orange arrow) to GAL11P (purple) in the holoenzyme. This causes the holoenzyme, along with TFIID, to bind to the promoter, activating the gene.

Young suggested the simplifying assumption that the yeast holoenzyme contains all of the general transcription factors except TFIID, and many other proteins, including: 9 Srbs; 6 Meds; 11 members of the Swi/Snf complex; and 5 miscellaneous proteins. Adding all these proteins to the 12 subunits of RNA polymerase II and the 15 subunits of TFIIB, F, E, and H, yielded a yeast holoenzyme composed of 58 polypeptides.

In 1995, Mark Ptashne and colleagues added another strong argument for the holoenzyme recruitment model. They reasoned as follows: If the holoenzyme is recuited as a unit, then interaction between *any* part of an activator (bound near a promoter) and *any* part of the holoenzyme should serve to recruit the holoenzyme to the promoter. This protein–protein interaction need not involve the normal transcription-activating domain of the activator, nor the activator's normal target on a general transcription factor. Instead, any contact between the activator and the holoenzyme should cause activation. On the other hand, if the preinitiation complex must be built up protein by protein, then an abnormal interaction between an activator and a seemingly unimportant member of the holoenzyme should not activate transcription.

Ptashne and colleagues took advantage of a chance observation to test these predictions. They had previously isolated a yeast mutant with a point mutation that changed a single amino acid in a holoenzyme protein (GAL11). They named this altered protein GAL11P (for potentiator) because it responded strongly to weak mutant versions of the activator GAL4. Using a combination of biochemical and genetic analysis, they found the source of the potentiation by GAL11P: The alteration in GAL11 caused this protein to bind to a region of the dimerization domain of GAL4, between amino acids 58 and 97. Because GAL11 (or GAL11P) is part of the holoenzyme, this novel association between GAL11P and GAL4 could recruit the holoenzyme to GAL4-responsive promoters, as illustrated in Figure 12.24. We call the association between GAL11P and GAL4 novel because the part of GAL11P involved is normally functionally inactive, and the part of GAL4 involved is in the dimerization domain, not the acti-

vation domain. It is highly unlikely that any association between these two protein regions occurs normally.

To test the hypothesis that the region of GAL4 between amino acids 58 and 97 is responsible for activation by GAL11P Ptashne and colleagues performed the following experiment. Using gene-cloning techniques, they made a plasmid encoding a fusion protein containing the region between amino acids 58 and 97 of GAL4 and the LexA DNA-binding domain. They introduced this plasmid into yeast cells along with a plasmid encoding either GAL11 or GAL11P, and a plasmid bearing two binding sites for LexA upstream of a GAL1 promoter driving transcription of the *E. coli lacZ* reporter gene. Figure 12.25 summarizes this experiment and shows the results. The LexA-GAL4(58–97) protein is ineffective as an activator when wild-type GAL11 is in the holoenzyme (Figure 12.25a), but works well as an activator when GAL11P is in the holoenzyme (Figure 12.25b).

If activation is really due to interaction between LexA-GAL4(58–97) and GAL11P, we would predict that fusing the LexA DNA-binding domain to GAL11 would also cause activation, as illustrated in Figure 12.25c. In fact, this construct did cause activation, in accord with the hypothesis. Here, no novel interaction between LexA-GAL4 and GAL11P was required because LexA and GAL11 were already covalently joined.

The simplest explanation for these data is that activation, at least in this system, operates by recruitment of the holoenzyme, rather than by recruitment of individual general transcription factors. It is possible, but not likely, that GAL11 is a special protein whose recruitment causes the stepwise assembly of a preinitiation complex. It is much more likely that association between an activator and any component of the holoenzyme can recruit the holoenzyme and thereby cause activation. Ptashne and colleagues conceded that TFIID is an essential part of the preinitiation complex, but is apparently not part of the yeast holoenzyme. They proposed that TFIID might have bound to the promoter cooperatively with the holoenzyme in their experiments.

SUMMARY Activation, at least in certain promoters in yeast, appears to function by recruitment of the holoenzyme, rather than by recruitment of individual components of the holoenzyme one at a time.

12.5 Interaction among Activators

We have seen several examples of crucial interactions among different types of transcription factors. Obviously, the general transcription factors must interact to form the

(a)   **WT Cells**

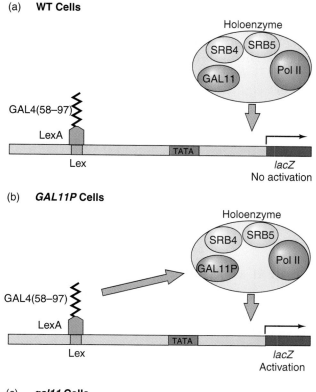

(b)   **GAL11P Cells**

(c)   **gal11 Cells**

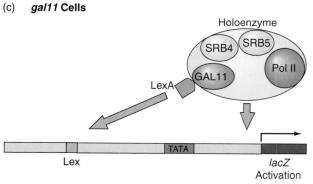

Figure 12.25   Activation by GAL11P and GAL11-LexA. Ptashne and colleagues transformed cells with a plasmid containing a *lexA* operator 50 bp upstream of a promoter driving transcription of a *lacZ* reporter gene, plus the following plasmids: (**a**) a plasmid encoding amino acids 58–97 of GAL4 coupled to the DNA-binding domain of LexA plus a plasmid encoding wild-type GAL11; (**b**) a plasmid encoding amino acids 58–97 of GAL4 coupled to the DNA-binding domain of LexA plus a plasmid encoding GAL11P; (**c**) a plasmid encoding GAL11 coupled to the DNA-binding domain of LexA. They assayed for production of the *lacZ* product, β-galactosidase. Results: (**a**) The GAL4(58–97) region did not interact with GAL11, so no activation occurred. (**b**) The GAL4(58–97) region bound to GAL11P, recruiting the holoenzyme to the promoter, so activation occurred. (**c**) The LexA–GAL11 fusion protein could bind to the *lexA* operator, recruiting the holoenzyme to the promoter, so activation occurred. (*Source:* Reprinted from Barberis et al., *Cell* 81:365, 1995. Copyright 1995, with permission from Elsevier Science.)

preinitiation complex. But activators and general transcription factors also interact. For example, we have just learned that GAL4 and other activators interact with TFIIB and other general transcription factor(s). In addition, activators usually interact with one another in activating a gene. This

can occur in two ways: Individual factors can interact to form a protein dimer to facilitate binding to a single DNA target site. Alternatively, specific factors bound to different DNA target sites can collaborate in activating a gene.

Dimerization

We have already mentioned a number of different means of interaction between protein monomers in DNA-binding proteins. In Chapter 9 we discussed the helix-turn-helix proteins such as the λ repressor and observed that the interaction between the monomers of this protein place the recognition helices of the two monomers in just the right position to interact with two major grooves exactly one helical turn apart. The recognition helices are antiparallel to each other so they can recognize the two parts of a palindromic DNA target. Earlier in this chapter we discussed the coiled coil dimerization domains of the GAL4 protein and the similar leucine zippers of the bZIP proteins, and the looped dimerization domain of the glucocorticoid receptor.

In Chapter 9 we discussed the advantage that a protein dimer has over a monomer in binding to DNA. This advantage can be summarized as follows: The affinity of binding between a protein and DNA varies with the square of the free energy of binding. Because the free energy depends on the number of protein–DNA contacts, doubling the contacts by using a protein dimer instead of a monomer quadruples the affinity between the protein and the DNA. This is significant because most activators have to operate at very low concentrations. The fact that the great majority of DNA-binding proteins are dimers is a testament to the advantage to this arrangement. Let us consider one celebrated example of a heterodimer, a dimer composed of two different monomers: Jun and Fos.

The Jun–Fos Dimer   Jun and **Fos** are the products of the *jun* and *fos* genes, which are proto-oncogenes that play a role in the development of some tumors. The Jun and Fos proteins are DNA-binding proteins of the bZIP family. Jun dimers are one form of the activator **AP1**, but Jun and Fos cooperate to bind DNA more tightly than Jun dimers (Fos dimers do not bind at all), so the most abundant natural form of AP1 is probably a Jun–Fos heterodimer.

T. Kouzarides and E. Ziff demonstrated that Jun and Fos bind to their DNA target better together than either one does separately. They also showed that the bZIP domains of each protein are crucial for binding. They expressed each protein, or the bZIP domain of each protein, by transcribing the respective genes in vitro and then translating the mRNAs in vitro. They mixed these proteins in various combinations with an oligonucleotide containing the AP1 target, TGACTCA. This is called the **TPA response element, TRE,** because it responds to a class of compounds including TPA (12-O-tetradecanoylphorbol-13-acetate). These **phorbol esters** stimulate cell division by

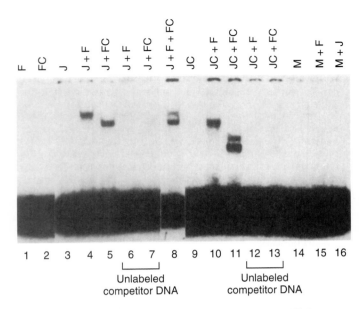

Figure 12.26   Binding of Fos and Jun to DNA with an AP-1 binding site. Kouzarides and Ziff mixed Fos, Jun, and the bZIP domains of Fos and Jun (Fos-core and Jun-core, respectively) with a labeled DNA that contains an AP-1 binding site, then measured DNA–protein binding with a gel mobility shift assay. Letters at the top of each lane indicate the proteins added (F = Fos; J = Jun; FC = Fos-core; JC = Jun-core; M = Myc, another bZIP protein). Lanes 6, 7, 12, and 13 contained excess unlabeled DNA competitor, which prevented interaction with the labeled DNA. Note that combinations of Fos or Fos-core with Jun or Jun-core bound to the DNA, but each of these proteins did not bind when added alone. The interaction was also specific, because it did not occur with Myc. Thus, the Fos–Jun heterodimer bound better to AP-1 binding sites than Fos or Jun homodimers. (*Source:* Kouzarides and Ziff, The role of the leucine zipper in the fos-jun interaction. *Nature* 336 (15 Dec 1988) p. 649, f. 3a. © Macmillan Magazines Ltd.)

activating genes with TREs. Finally, they used a gel mobility shift assay to measure protein–DNA binding.

Figure 12.26 shows the results of the assay. At the low concentrations used in this study, neither Fos nor Jun alone was capable of binding to the DNA, but the two proteins together could bind. Similarly, fragments of Fos or Jun containing the bZIP domains (the Fos-core and Jun-core, respectively) could bind only in combination, not alone. (Other investigators had shown that Jun could bind by itself, but that was at a higher protein concentration).

This experiment also confirmed that the Fos–Jun complex is a dimer and not a tetramer. Lane 8 contains Jun plus Fos plus Fos-core. We already know from lanes 4 and 5 that Jun–Fos and Jun–Fos-core give bands of different mobilities. Therefore, we expect that a mixture of Jun, Fos, and Fos-core would give both these bands, if they form dimers, but a third band (Jun$_2$–Fos–Fos-core) if they form tetramers. No such third band appeared, suggesting that these proteins form only dimers. This does not mean that activators never form anything but dimers. The tumor suppressor p53 is an activator that acts as a tetramer.

Would the same dimerization and DNA binding occur with Fos or Jun plus other bZIP proteins, such as Myc? Apparently not, because neither Fos plus Myc nor Jun plus Myc bound to the TRE-containing DNA (lanes 15 and 16). Thus, in spite of the structural similarity of the bZIP proteins, their interactions with their DNA targets are specific.

SUMMARY Dimerization is a great advantage to an activator because it increases the affinity between the activator and its DNA target. Some activators form homodimers, but others, such as Jun and Fos, function better as heterodimers.

Action at a Distance

We have seen that both prokaryotic and eukaryotic enhancers can stimulate transcription, even though they are located some distance away from the promoters they control. How does this action at a distance occur? In Chapter 9 we learned that the evidence favors looping out of DNA in between the two remote sites to allow the DNA-binding proteins to interact. We will see that this same scheme also seems to apply to eukaryotic enhancers.

Among the most reasonable hypotheses to explain the ability of enhancers to act at a distance are the following (Figure 12.27): (a) An activator binds to an enhancer and changes the topology, or shape, of the whole DNA molecule, perhaps by causing supercoiling. This in turn opens the promoter up to general transcription factors. (b) An activator binds to an enhancer and then slides along the DNA until it encounters the promoter, where it can activate transcription by virtue of its direct contact with the promoter DNA. (c) An activator binds to an enhancer and, by looping out DNA in between, interacts with proteins at the promoter, stimulating transcription. (d) An activator binds to an enhancer and a downstream segment of DNA to form a DNA loop. By enlarging this loop, the protein tracks toward the promoter. When it reaches the promoter, it interacts with proteins there to stimulate transcription.

Notice that the first two of these models demand that the two elements, enhancer and promoter, be on the same DNA molecule. A change in topology of one DNA molecule cannot influence transcription on a second, and an activator cannot bind to an enhancer on one DNA and slide onto a second molecule that contained the promoter. On the other hand, the third model simply requires that the enhancer and promoter be relatively near each other, not necessarily on the same molecule. This is because the essence of the looping model is not the looping itself, but the interaction between the proteins bound to remote sites. In principle, this would work just as well if the proteins were bound to two sites on different DNA

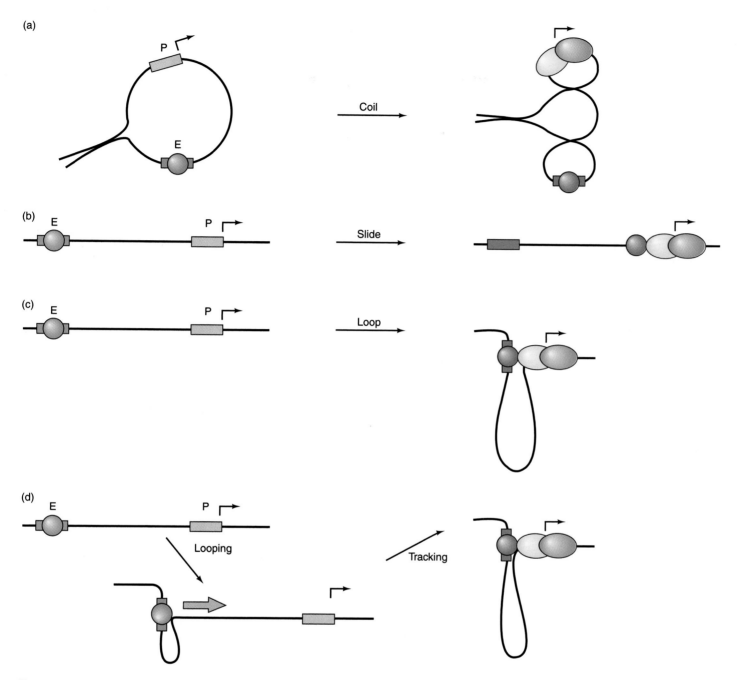

Figure 12.27 Four hypotheses of enhancer action. (a) Change in topology. The enhancer (E, blue) and promoter (P, orange) are both located on a loop of DNA. Binding of a gene-specific transcription factor (green) to the enhancer causes supercoiling that facilitates binding of general transcription factors (yellow) and polymerase (red) to the promoter. (b) Sliding. A transcription factor binds to the enhancer and slides down the DNA to the promoter, where it facilitates binding of general transcription factors and polymerase. (c) Looping. A transcription factor binds to the enhancer and, by looping out the DNA in between, binds to and facilitates the binding of general transcription factors and polymerase to the promoter. (d) Facilitated tracking. A transcription factor binds to the enhancer and causes a short DNA segment to loop out downstream. Increasing the size of this loop allows the factor to track along the DNA until it reaches the promoter, where it can facilitate the binding of general transcription factors and RNA polymerase.

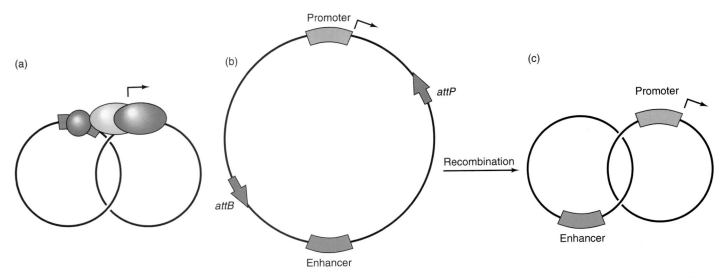

Figure 12.28   Isolating enhancer and promoter on two plasmids linked in a cantenane. (a) Hypothetical interaction between a gene-specific transcription factor (green) bound to an enhancer (blue) on one plasmid, and general transcription factors (yellow) and polymerase (red) bound to the promoter (not visible) in the other plasmid of the catenane. **(b)** Creation of the cantenane. The promoter and enhancer are on opposite sides of a plasmid that also contains the λ phage *attB* and *attP* sites. **(c)** In vitro recombination between the *att* sites creates the catenane with enhancer and promoter on two linked plasmids.

molecules, as long as the molecules were tethered together somehow so they would not float apart and prevent interactions between the bound proteins. Figure 12.28a shows how this might happen.

Thus, if we could arrange to put an enhancer on one DNA molecule and a promoter on another, and get the two molecules to link together in a catenane, we could test the hypotheses. If the enhancer still functioned, we could eliminate the first two. Marietta Dunaway and Peter Dröge did just that. They constructed a plasmid with the *Xenopus laevis* rRNA promoter plus an rRNA minigene on one side and the rRNA enhancer on the other, with the λ phage integration sites, *attP* and *attB*, in between, as shown in Figure 12.28b. These are targets of site-specific recombination (Chapter 23), so placing them on the same molecule and allowing recombination produces a cantenane, as illustrated in Figure 12.28c.

Finally, these workers injected combinations of plasmids into *Xenopus* oocytes and measured their transcription by quantitative S1 mapping. The injected plasmids were the catenane, the unrecombined plasmid containing both enhancer and promoter, or two separate plasmids, each containing either the enhancer or promoter. In quantitative S1 mapping, a reference plasmid is needed to correct for the variations among oocytes. In this case, the reference plasmid contained an rRNA minigene (called ψ52) with a 52-bp insert, whereas the rRNA minigenes of the test plasmids (called ψ40) all contained a 40-bp insert. Dunaway and Dröge included probes for both these minigenes in their assay, so we expect to see two signals, 12 nt apart, if both genes are transcribed. We are most interested in the *ratio* of these two signals, which tells us how well each test plasmid is transcribed relative to the reference plasmid, which should behave the same in each case.

Figure 12.29a shows the test plasmid results in the lanes marked "a" and the reference plasmid results in the lanes marked "b." The plasmids used to produce the transcripts in each lane are pictured in panel (b). Note that the same plasmids were used in both lane a and lane b of each set in panel (a). Only the probes were different. These were the results: Lanes 1 show that when the plasmid contained the promoter alone, the test plasmid signal was weaker than the reference plasmid signal. That is because the test probe had a lower specific radioactivity than the reference probe. Lanes 2 demonstrate that the enhancer right next to the promoter (its normal position) greatly enhanced transcription in the test plasmid—its signal was much stronger than the reference plasmid signal. Lanes 3 show that the enhancer still worked, though not quite as well, when placed opposite the promoter on the plasmid. Lanes 4 are the most important. They show that the enhancer still worked when it was on a separate plasmid that formed a catenane with the plasmid containing the promoter. Lanes 5 verify that the enhancer did not work if it was on a separate plasmid *not* linked in a catenane with the promoter plasmid. Finally, lanes 6 show that the enhancement observed in lanes 4 was not due to a small amount of contamination by unrecombined plasmid. In lanes 6, the investigators added 5% of such a plasmid and observed no significant increase in the test plasmid signal.

These results lead to the following conclusion about enhancer function: The enhancer does not need to be on

(a)

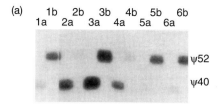

(b)

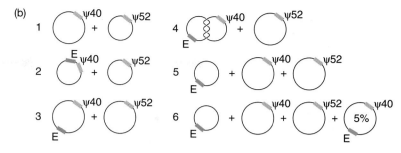

Figure 12.29   Results of the catenane experiment. Dunaway and Dröge injected mixtures of plasmids into *Xenopus* oocytes and measured transcription rates by quantitative S1 mapping. They injected a test plasmid and a reference plasmid in each experiment and assayed for transcription of each with separate probes. The results of the test and reference assays are given in lanes a and b, respectively, of each experiment. The plasmids injected in each experiment are given in panel (b). For example, the plasmids used in the experiments in lanes 1a and 1b are labeled 1. The plasmids on the left, labeled ψ40 (or ψ40 plus another plasmid), are the test plasmids. The ones on the right, labeled ψ52, are the reference plasmids. The 40 and 52 in these names denote the size inserts each has to distinguish it from the other. Both plasmids were injected and then assayed with the test plasmid (lane 1a) or the reference plasmid (lane 1b). Lanes 4a and 4b demonstrate that transcription of the catenane with the enhancer on one plasmid and the promoter on the other is enhanced relative to transcription of the plasmid containing just the promoter (lanes 1a and 1b). This is evident in the much higher ratio of the signals in lanes 4a and 4b relative to the ratio of the signals in lanes 1a and 1b. (*Source:* Dunaway and Dröge, Transactivation of the *Xenopus* rRNA gene promoter by its enhancer. *Nature* 341 (19 Oct 1989) p. 658, f. 2a. © Macmillan Magazines Ltd.)

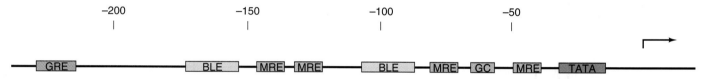

Figure 12.30   Control region of the human metallothionine gene. Upstream of the transcription start site at position +1 we find, in 3′–5′ order: the TATA box; a metal response element (MRE) that allows the gene to be stimulated in response to heavy metals; a GC box that responds to the activator Sp1; another MRE; a basal level enhancer (BLE) that responds to the activator AP-1; two more MREs; another BLE; and a glucocorticoid response element (GRE) that allows the gene to be stimulated by an activator composed of a glucocorticoid hormone and its receptor.

the same DNA with the promoter, but it does need to be relatively nearby, presumably so the proteins bound to enhancer and promoter can interact. This is difficult to reconcile with models involving supercoiling or sliding (Figures 12.27a and b), but is consistent with the DNA looping and facilitated tracking models (Figure 12.27c and d). In the catenane, no looping or tracking is required because the enhancer and promoter are on different DNA molecules; instead, protein–protein interactions can occur without looping, as illustrated in Figure 12.28a.

SUMMARY The essence of enhancer function—protein–protein interaction between activators bound to the enhancers, and general transcription factors and RNA polymerase bound to the promoter—seems in many cases to be mediated by looping out the DNA in between. This can also account for the effects of multiple enhancers on gene transcription, at least in theory. DNA looping could bring the activators bound to each enhancer close to the promoter where they could stimulate transcription, perhaps in a cooperative way.

Multiple Enhancers

Many genes have more than one enhancer, so they can respond to multiple stimuli. For example, the metallothionine gene, which codes for a protein that apparently helps eukaryotes cope with poisoning by heavy metals, can be turned on by several different agents, as illustrated in Figure 12.30. Thus, each of the activators that bind at these enhancers must be able to interact with the preinitiation complex assembling at the promoter, presumably by looping out any intervening DNA.

This scheme allows for very fine control over the expression of genes. Different combinations of activators produce different levels of expression of a given gene in different cells. In fact, the presence or absence of various enhancers near a gene reminds one of a binary code, where the presence is an "on" switch, and the absence is an "off" switch. Of course, the activators also have to be present to throw the switches. It may not be a simple additive arrangement, however, since multiple enhancers are known to act cooperatively.

Eric Davidson and colleagues provided a beautiful example of multiple enhancers in the *Endo 16* gene of a sea urchin. This gene is active in the early embryo's vegetal plate—a group of cells that produces the endodermal

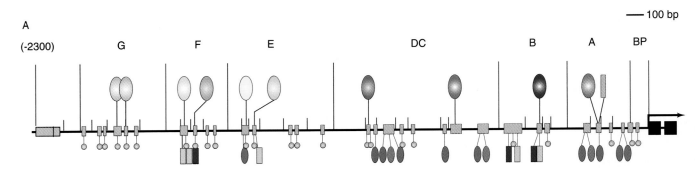

Figure 12.31 **Modular arrangement of enhancers at the sea urchin *Endo 16* gene.** The large and medium-size colored ovals and rectangles represent activators, and the small orange ovals represent architectural transcription factors, bound to enhancer elements (pink boxes). The enhancers are arranged in clusters, or modules, as indicated by the regions labeled G, F, E, DC, B, and A. Long vertical lines denote restriction sites that define the modules. BP stands for "basal promoter." (*Source:* Reprinted with permission from Yuh et al, *Science,* 279:1897, 1998. Copyright 1998 American Association for the Advancement of Science.)

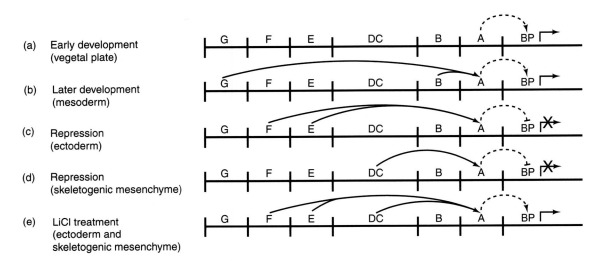

Figure 12.32 **The integrative role of module A in the sea urchin *Endo 16* gene.** (**a**) In early development, A is the only module operating and it stimulates transcription from the basal promoter (BP), as indicated by the dashed arrow. (**b**) In later development in the mesoderm, module A integrates inputs from modules G and B and stimulates transcription. (**c**) In later development in the ectoderm, module A integrates inputs from modules F and E and inhibits transcription, as indicated by the dashed "barrier" symbol. (**d**) In later development in skeletogenic mesenchyme, module A receives input from module DC and inhibits transcription. (**e**) LiCl treatment reverses the signal from module A in ectoderm and skeletogenic mesenchyme, causing stimulation, rather than inhibition, of transcription. (*Source:* Reprinted with permission from Yuh et al, *Science,* 279:1897, 1998. Copyright 1998 American Association for the Advancement of Science.)

tissues, including the gut. Davidson and colleagues began by testing DNA in the *Endo 16* 5′-flanking region for the ability to bind nuclear proteins. They found dozens of such regions, arranged into six modules, as illustrated in Figure 12.31.

How do we know that all these modules that bind nuclear proteins are actually involved in gene activation? Chiou-Hwa-Yuh and Davidson tested them by linking them alone and in combinations to the CAT reporter gene (Chapter 5), reintroducing these constructs into sea urchin eggs, and observing the patterns of expression of the reporter gene in the resulting developing embryo. They found that the reporter gene was switched on in different parts of the embryo and at different times, depending on the exact combination of modules attached. Thus, the modules were responding to activators that were distributed non-uniformly in the developing embryo.

Although all the elements may be able to function independently in vitro, the situation is more organized in vivo, as illustrated in Figure 12.32. Module A appears to be the only one that interacts directly with the basal transcription apparatus; all the other modules work through A. Some of the upstream modules (B and G) act synergistically through A to stimulate *Endo 16* transcription in endoderm cells. The other modules (DC, E, and F) act synergistically through A to block *Endo 16* transcription in nonendoderm cells (modules E and F play this role in ectoderm cells, and module DC plays this role in skeletogenic mesenchyme cells). The effect of LiCl is very interesting in this context. LiCl is a **teratogen** (an agent that causes abnormal development). In sea urchin embryos, it causes too much endoderm to form at the expense of the surrounding ectoderm in developing embryos. And Davidson and colleagues have demonstrated that LiCl causes modules DC, E, and F to act through module A to stimulate, rather than repress *Endo 16* activity. Thus, by turning the gene on in preectoderm cells where it is supposed to be turned off, LiCl converts these cells to endoderm, accounting for the overproduction of endoderm.

> **SUMMARY** Multiple enhancers enable a gene to respond differently to different combinations of activators. This arrangement gives cells exquisitely fine control over their genes in different tissues, or at different times in a developing organism.

Architectural Transcription Factors

The looping mechanism we have discussed for bringing together activators and general transcription factors is quite feasible for proteins bound to DNA elements that are separated by at least a few hundred base pairs because DNA is flexible enough to allow such bending. On the other hand, many enhancers are located much closer to the promoters they control, and that presents a problem: DNA looping over such short distances will not occur spontaneously, because short DNAs behave more like rigid rods than like flexible strings.

How then do activators and general transcription factors bound close together on a stretch of DNA interact to stimulate transcription? They can still approach each other if something else intervenes to bend the DNA more than the DNA itself would normally permit. We now have several examples of **architectural transcription factors** whose sole (or main) purpose seems to be to change the shape of a DNA control region so that other proteins can interact successfully to stimulate transcription. Rudolf Grosschedl and his colleagues provided the first example of a eukaryotic architectural transcription factor. They used the human T-cell receptor α-chain (TCRα) gene control region, which contains three enhancers, binding sites for the activators Ets-1, LEF-1, and CREB within just 112 bp of the transcription start site (Figure 12.33).

LEF-1 is the lymphoid enhancer-binding factor, which binds to the middle enhancer pictured in Figure 12.33 and helps activate the TCRα gene. However, previous work by Grosschedl and others had shown that LEF-1 by itself cannot activate TCRα gene transcription. So what is its role? Grosschedl and coworkers established that it acts by binding primarily to the minor groove of the enhancer and bending the DNA by 130 degrees.

These workers demonstrated minor groove binding by two methods. First, they showed that methylating six enhancer adenines on N3 (in the minor groove) interfered with enhancer function. Then they substituted these six A–T pairs with I–C pairs, which look the same in the minor groove, but not the major groove, and found no

Figure 12.33 Control region of the human T-cell receptor α-chain (TCRα) gene. Within 112 bp upstream of the start of transcription lie three enhancer elements, which bind Ets-1, LEF-1 and CREB. These three enhancers are identified here by the transcription factors they bind, not by their own names.

loss of enhancer activity. This is the same strategy Stark and Hawley used to demonstrate that TBP binds to the minor groove of the TATA box (Chapter 11).

Next, using the same electrophoretic assay Wu and Crothers used to show that CAP bends *lac* operon DNA (Chapter 7), Grosschedl and coworkers showed that LEF-1 bends DNA. They placed the LEF-1 binding site at different positions on linear DNA fragments, bound LEF-1, and measured the electrophoretic mobilities. The mobility was greatly retarded when the binding site was in the middle of the fragment, suggesting significant bending.

They also showed that the DNA bending is due to a so-called **HMG domain** on LEF-1. **HMG proteins** are small nuclear proteins that have a high electrophoretic mobility (hence, **h**igh **m**obility **g**roup, or HMG). To show the importance of the HMG domain of LEF-1, they prepared a purified peptide containing just the HMG domain and showed that it caused the same degree of bending (130 degrees) as the full-length protein (Figure 12.34). Extrapolation of the mobility curve to the point of maximum mobility (where the bend-inducing element should be right at the end of the DNA fragment) indicates that the bend occurs at the LEF-1 binding site. Because LEF-1 does not enhance transcription by itself, it seems likely that it acts indirectly by bending the DNA. This presumably allows the other activators to contact the basal transcription machinery at the promoter and thereby enhance transcription.

Tom Maniatis and colleagues observed a similar phenomenon in the human interferon-β (IFNβ) control region, which also has an enhancer close to the promoter. Figure 12.35 depicts this complex enhancer with all the activators that bind to it. Among these proteins are four copies of HMG I(Y), of which only three are shown. Unlike LEF-1 (and despite its name), HMG I(Y) does not have an HMG domain, but gel electrophoresis experiments performed by Maniatis and colleagues showed that it does bend its A–T-rich DNA target. In fact, the target is pre-bent, but HMG I(Y) and the activator nuclear factor-kappa B (NF-κB), actually unbend it as they bind cooperatively to the enhancer.

HMG I(Y) is necessary for virus-inducible expression of the IFNβ gene, but does not function as an activator in the absence of other activators. However, Maniatis and

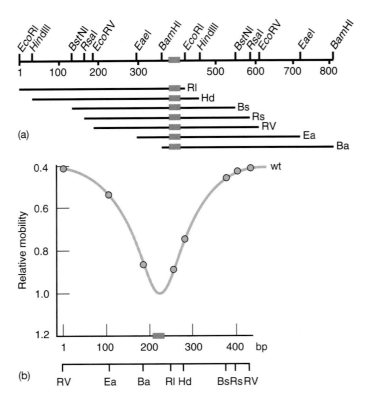

(a)

(b)

Figure 12.34   The activator LEF-1 bends its DNA target. (a) DNAs used in the bending assay. Grosschedl and coworkers prepared a set of DNA fragments containing the LEF-1 binding site (red box) at various points, from one end of the fragment to the other. Identifying letters at the right end of each fragment indicate the restriction enzyme used in their preparation: RI, *Eco*RI; Hd, *Hind*III; Bs, *Bst*NI; Rs, *Rsa*I; RV, *Eco*RV; Ea, *Eae*I; Ba, *Bam*HI. The restriction map at top shows where each of these enzymes cuts the DNA. **(b)** The bending assay. These workers bound LEF-1 to each fragment and measured its electrophoretic mobility. This is a plot of these relative mobilities. Fragments with the binding site near the ends showed the highest mobilities (near center of plot). Extrapolation of the descending lines of the plot should point to the center of bending, which is right at the LEF-binding site (red box). Fragments with the binding site near the middle showed the lowest mobilities (both ends of plot). The steepness of the plot indicates a great deal of DNA bending (about 130 degrees) on LEF-1 binding. (*Source:* Reprinted from Giese et al., *Cell* 69:188, 1992. Copyright 1992, with permission from Elsevier Science.)

his colleagues demonstrated that HMG I(Y) is necessary for binding two other activators, ATF-2 (activation-transcription factor 2) and NF-κB, to the enhancer. It appears that the role of HMG 1(Y) is to remodel the enhancer DNA so as to bring the other activators into contact with one another in a so-called **enhanceosome,** as illustrated in Figure 12.36. This would explain an apparent paradox: IFNβ is an antiviral protein whose gene is induced by virus infection of a cell, yet the IFNβ gene has an enhancer that binds NF-κB, which is produced in response to many signals and stressful conditions besides virus infection. Why don't these other signals and conditions activate the IFNβ gene? The answer could be that activation of this gene requires the assembly of the whole enhanceosome in the proper shape, and this requires *all* of the activators to be present at once—a condition that occurs primarily during virus infection.

A key feature of this scheme is cooperativity. All of the factors must cooperate to form the enhanceosome. Maniatis and colleagues have shown that HMG 1(Y), in particular, is required for cooperativity. Without it, some stimulation of the IFNβ gene is possible, but cooperativity is greatly enhanced by the presence of HMG 1(Y). This means that it takes a much lower concentration of the other activators to give maximum stimulation of the gene in the presence of HMG 1(Y) than in its absence (Figure 12.37).

Does the three-dimensional structure of the enhanceosome play a role in the cooperativity we have seen? Maniatis and colleagues noted that all the HMG 1(Y) binding sites are located on the same face of the DNA helix. This arrangement suggests that this protein must bind on one side of the DNA in order to bend the DNA in the proper shape for cooperativity to occur. To address this issue, Maniatis and coworkers inserted either 6 bp (one-half of a helical turn) or 10 bp (a full helical turn) between HMG 1(Y) binding sites in the IFNβ enhances. They found that inserting 6 bp, which would place the two flanking HMG 1(Y) binding sites on opposite sides of the

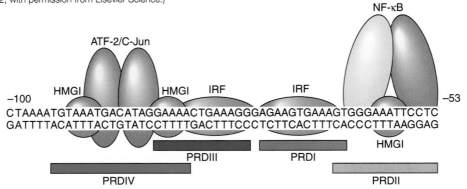

Figure 12.35   The human IFNβ (interferon-β) enhancer, showing the activators and their binding sites. ATF-2 is a dimeric activator (activation transcription factor 2) that binds to PRDIV (positive regulatory domain IV), along with two molecules of HMG (high-mobility group) I(Y). The former binds in the major groove, the latter in the minor groove. IRF (interferon regulatory factor) transcription factors bind to PRDIII and I. NF-κB (nuclear factor-kappa B), an activator of immune system genes, binds to PRDII, along with a molecule of HMG I(Y). Again, the former factor binds in the major groove, the latter in the minor groove. This accounts for the fact that the two factors seem to bind simultaneously to the same DNA region. (*Source:* Reprinted from Du et al., *Cell* 74:888, 1993. Copyright 1993, with permission from Elsevier Science.)

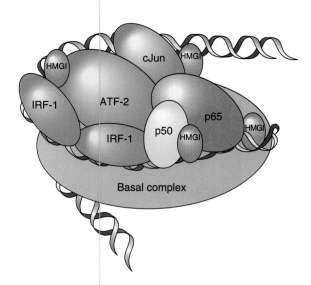

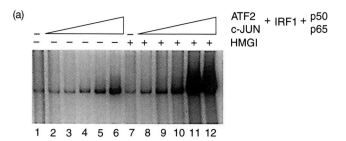

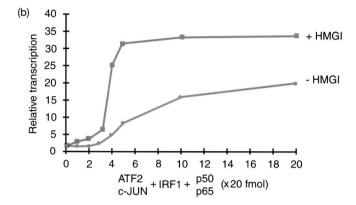

Figure 12.36   A model of the enhanceosome assembled at the human IFNβ gene. The activators IRF-1, ATF-2/c-Jun, HMG 1(Y), and NF-κB (represented by its subunits, p50 and p65) are shown as colored ovals bending the DNA and interacting at multiple sites with the basal complex (orange). The basal complex contains at least the general transcription factors and RNA polymerase II, and probably other holoenzyme components. The relative sizes of the proteins and DNA are not drawn to scale. (*Source:* Reprinted from Thanos and Maniatis, *Cell* 83:1098, 1995. Copyright 1995, with permission from Elsevier Science.)

Figure 12.37   Effect of HMG 1(Y) on cooperativity of enhanceosome formation on the IFNβ enhancer.
(a) Phosphorimager results. Maniatis and colleagues performed primer extension in vitro transcription reactions after assembling an IFNβ enhanceosome from purified components including increasing amounts of the activators ATF-2/c-JUN, IRF-1, and p50/65, in the absence (lanes 2–6) or presence (lanes 8–12) of HMG 1(Y). The wedges at top show the amount of each activator [except HMG 1(Y)] increasing from 20 to 100 fmol. Much greater cooperativity occurred in the presence of HMG 1(Y). **(b)** Graph of results. Maniatis and colleagues plotted the results of the experiment in panel **(a)**, but also including activator amounts of 200 and 400 fmol. The steep sigmoidal (S-shaped) nature of the curve with HMG 1(Y) shows highly cooperative activation. The more gentle slope of the curve without HMG 1(Y) shows weak cooperativity. (*Source:* Kim, T. K. and Maniatis, T. The Mechanism of Transcriptional Synergy of an In Vitro Assembled Interferon-β Enhanceosome. *Molecular Cell* 1 (1997) f. 5, p. 124. Reprinted by permission of Elsevier Science.)

DNA helix, reduced cooperativity, but inserting 10 bp, which would leave the binding sites on the same side of the DNA, had little effect on cooperativity.

> **SUMMARY** The activator LEF-1 binds to the minor groove of its DNA target through its HMG domain and induces strong bending in the DNA. LEF-1 does not enhance transcription by itself, but the bending it induces probably helps other activators bind and interact with other activators and the general transcription factors to stimulate transcription. Another activator, HMG 1(Y), probably plays a similar role in the human interferon-β control region. These proteins are examples of architectural transcription factors. In the case of the IFNβ enhancer, activation seems to require the cooperative binding of several activators, including HMG 1(Y), to form an enhanceosome with a specific shape.

Insulators

We know that enhancers can act at a great distance from the promoters they activate. For example, the wing margin enhancer in the *Drosophila cut* locus is separated by 85 kb from the promoter. With a range that large, some enhancers will likely be close enough to other, unrelated genes to activate them as well. How does the cell prevent such inappropriate activation? Higher organisms, includ-

ing at least *Drosophila* and mammals, use DNA elements called **insulators** to block activation of unrelated genes by nearby enhancers. Gary Felsenfeld has defined an insulator as a "neutral barrier to the influence of neighboring elements." Thus, insulators can protect a gene from both activation and repression by nearby enhancers and silencers.

How do insulators work? The details are not clear yet, but we do know that insulators define boundaries between DNA domains. Thus, an insulator placed between an enhancer and the promoter it usually activates abolishes that activation. Similarly, an insulator placed between a silencer and a gene it usually represses abolishes that repression. It appears that the insulator creates a boundary between the domain of the gene and that of the enhancer (or silencer) so the gene can no longer feel the activating (or repressing) effects (Figure 12.38).

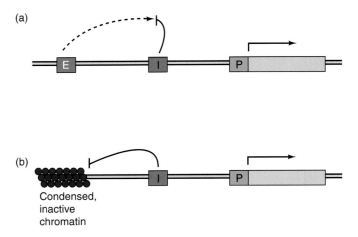

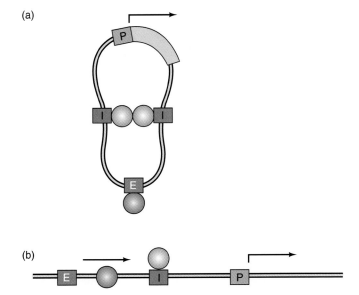

Figure 12.38   Insulator function. (a) Insulating against enhancer activity. The insulator between a promoter and an enhancer prevents the promoter from feeling the activating effect of the enhancer. **(b)** Insulating against silencer activity. The insulator between a promoter and condensed, repressive chromatin (induced by a silencer) prevents the promoter from feeling the repressive effect of the condensed chromatin (indeed, prevents the condensed chromatin from engulfing the promoter).

Figure 12.39   Two hypotheses for the mechanism of insulator activity. (a) Looping model. Two insulators (red) flank an enhancer (blue). When proteins (purple) bind to these insulators, they interact with one another, isolating the enhancer on a loop so it cannot stimulate transcription from the nearby promoter (orange). **(b)** Sliding model. An activator has bound to an enhancer and a stimulatory signal (green), perhaps the activator itself, is sliding along the DNA from the enhancer toward the promoter. But the insulator (red), perhaps with a protein or proteins attached, stands in the way and prevents the signal from reaching the promoter.

We also know that insulator function depends on protein binding. For example, certain *Drosophila* insulators contain the sequence GAGA and are known as **GAGA boxes.** These require the GAGA-binding protein **Trl** for insulator activity. Genetic experiments have shown that insulator activity can be abolished by mutations in either the GAGA box itself, or in the *trl* gene, which encodes Trl.

One can imagine many mechanisms for insulator function. We can easily eliminate one of these: a model in which the insulator induces a silenced, condensed chromatin domain upstream of the insulator. If that were the case, then a gene placed upstream of an insulator would always be silenced. But experiments with *Drosophila* have shown that such upstream genes are still potentially active and can be activated by their own enhancers. Figure 12.39 illustrates two more models of insulator action. The first requires interaction between insulators on either side of an enhancer, which isolates the enhancer on a loop so it cannot interact with the promoter. The second involves a signal that somehow moves progressively from the enhancer to the promoter, and the insulator blocks the progression of this signal. However, both of these hypotheses have experimental evidence weighing against them.

The first hypothesis cannot explain how isolating an enhancer in a loop would keep it from interacting with a promoter many thousands of base pairs away. A long stretch of DNA in between would seem to be flexible enough to permit such interaction. Besides, insulators are not always found in pairs flanking an enhancer. The second hypothesis is hard to reconcile with an experiment performed by J. Krebs and Dunaway similar in concept to the one by Dunaway and Dröge we discussed earlier in

this chapter. In that earlier experiment, Dunaway and Dröge placed a promoter and an enhancer on separate DNA circles linked in a catenane and showed that the enhancer still worked. In the later experiment, Krebs and Dunaway used the same catenane construct, but this time they surrounded either the enhancer or promoter with two *Drosophila* insulators: *scs* and *scs'*. They found that in both cases, the insulators blocked enhancer activity. On the other hand, a single insulator in either circle had little effect on enhancement. Both experiments from Dunaway's group are compatible with a signal propagating from the enhancer to the promoter only if the signal can jump from one DNA circle to another. Also, the finding that a single insulator cannot block enhancement activity implies that the signal, whatever it is, would have to be able to move in either direction from the enhancer.

We know from the Dunaway and Dröge experiment (see Figure 12.29) that the signal cannot be a change in the topology of the DNA. Only two other candidates for signals come readily to mind. One is a protein that slides down the DNA from the enhancer to the promoter. The other is a protein that polymerizes with the same or other proteins to build a solid chain of proteins along the DNA from the enhancer to the promoter. Again, both these concepts are difficult to fit into a model that calls for bidirectional propagation of the signal, and for a signal that can jump from one DNA to another. Thus, with no hypothesis that fits all the data, we are still looking for a good explanation for insulator action.

> **SUMMARY** Insulators are DNA elements that shield genes from activation or repression by enhancers or silencers. They typically block such activation or repression on one side of a DNA element, but not on the other. Thus, they establish boundaries between DNA regions in a chromosome. No one has proposed a model of insulator action that can account for all the data available.

Mediators

Some class II activators may be capable of recruiting the basal transcription complex all by themselves, possibly by contacting one or more general transcription factors or RNA polymerase. But many, if not most, cannot. Roger Kornberg and colleagues provided the first evidence that something else must be involved when they studied **activator interference**, or **squelching**, in 1989 and 1990. Squelching occurs when increasing the concentration of one activator inhibits the activity of another activator in an in vitro transcription experiment, presumably by competing for a scarce factor required by both activators. A reasonable candidate for such a limiting factor would be a general transcription factor, but Kornberg and coworkers discovered that adding very large quantities of the general transcription factors did not relieve squelching. This finding suggested that some other factor must be required by both activators.

What was this other factor? In 1990, Kornberg and colleagues partially purified a yeast protein that could relieve squelching. Then, in 1991, they purified this factor further and demonstrated directly that it could stimulate activated transcription, but not basal transcription in vitro. They called it a **mediator** because it appeared to mediate the effect of an activator. Their assay for transcription used a G-less cassette (Chapter 5) driven by the yeast *CYC1* promoter and a GAL4-binding site. They added increasing concentrations of the mediator in the absence and presence of the activator GAL4-VP16. Figure 12.40 shows the results: The mediator had no effect on transcription in the absence of the activator (lanes 3–6), but it greatly stimulated transcription in the presence of the activator (lanes 7–10). A similar experiment with the yeast activator GCN4 yielded comparable results, showing that the mediator could cooperate with more than one activator having an acidic activation domain.

In Chapter 7 we learned that cyclic-AMP (cAMP) stimulates transcription of bacterial operons by binding to an activator (CAP) and causing it to bind to activator

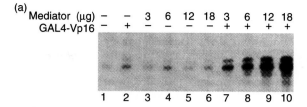

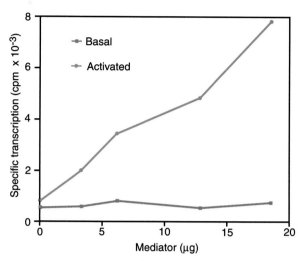

(b)

Figure 12.40   Discovery of a mediator. Kornberg and colleagues placed the yeast *CYC1* promoter downstream of a GAL4-binding site and upstream of a G-less cassette, so transcription of the G-less cassette depended on both the *CYC1* promoter and GAL4. Then they transcribed this construct in vitro in the absence of GTP and in the presence of the amounts of mediator shown at the top of panel (**a**), and in the absence (–) or presence (+) of the activator GAL4- VP16 as indicated at the top of panel (**a**). They included a labeled nucleotide to label the products of the in vitro transcription reactions and electrophoresed the labeled RNAs. (**a**) Phosphorimager scan of the electropherogram. (**b**) Graphical presentation of the results in panel (**a**). Note that the mediator greatly stimulates transcription in the presence of the activator, but has no effect on unactivated (basal) transcription. (*Source:* Flanigan, P. et al. A mediator required for activation of RNA polymerase II transcription *in vitro. Nature* 350 (4 Apr 1991) f. 2, p. 437. © Macmillan Magazines Ltd.)

target sites in the operon control regions. Cyclic-AMP also participates in transcription activation in eukaryotes, but it does so in a less direct way, through a series of steps called a **signal transduction pathway**. When the level of cAMP rises in a eukaryotic cell, it stimulates the activity of **protein kinase A (PKA)** and causes this enzyme to move into the cell nucleus. In the nucleus, PKA phosphorylates an activator called the **cAMP response element-binding protein (CREB)**, which binds to the **cAMP response element** (CRE) and activates associated genes.

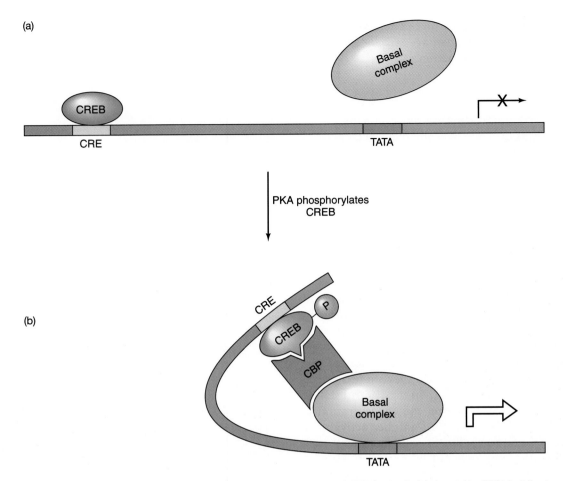

Figure 12.41 A model for activation of a CRE-linked gene. (a) Unphosphorylated CREB (turquoise) is bound to CRE, but the basal complex (RNA polymerase plus general transcription factors, orange) is not bound to the promoter in significant quantity and may not even have assembled yet. Thus, the gene is not activated. **(b)** PKA has phosphorylated CREB, which causes CREB to associate with CBP (red). CBP, in turn, associates with at least one component of the basal transcription complex, recruiting it to the promoter. Now transcription is activated.

Because phosphorylation of CREB is necessary for activation of transcription, one would expect this phosphorylation to cause CREB to move into the nucleus or to bind more strongly to CREs, but neither of these things actually seems to happen—CREB localizes to the nucleus and binds to CREs very well even without being phosphorylated. How, then, does phosphorylation of CREB effect activation? The key to the answer appeared in 1993 with the discovery of the **CREB-binding protein** (CBP). CBP binds to CREB much more avidly after CREB has been phosphorylated by protein kinase A. Then, CBP can contact and recruit elements of the basal transcription apparatus, or it could recruit the holoenzyme as a unit. By coupling CREB to the transcription apparatus, CBP acts as a **mediator** of activation (Figure 12.41).

Since 1993 when CBP was discovered, many mediators, sometimes also called **coactivators,** have been identified. In 1999, Tjian and colleagues isolated a mediator required for activation of transcription in vitro by the transcription factor Sp1. When they purified this mediator, which they called cofactor required for Sp1 activation (**CRSP**), they discovered that it had nine putative subunits. They separated these subunits by SDS-PAGE, transferred them to a nitrocellulose membrane, then cleaved each polypeptide with a protease to generate peptides that could be sequenced. The sequences revealed that some of the subunits of CRSP are unique, but many of them are identical, or at least homologous, to other known mediators—the yeast mediator, for example. Thus, different mediators seem to be assembled by "mixing and matching" subunits from a variety of other mediators.

The mediator role of CBP is not limited to cAMP-responsive genes. It also serves as a mediator in genes responsive to the nuclear receptors. This explains why no one could detect direct interaction between the transcription-activation domains of the nuclear receptors and any of the general transcription factors. The reason seems to be that the nuclear receptors do not contact the

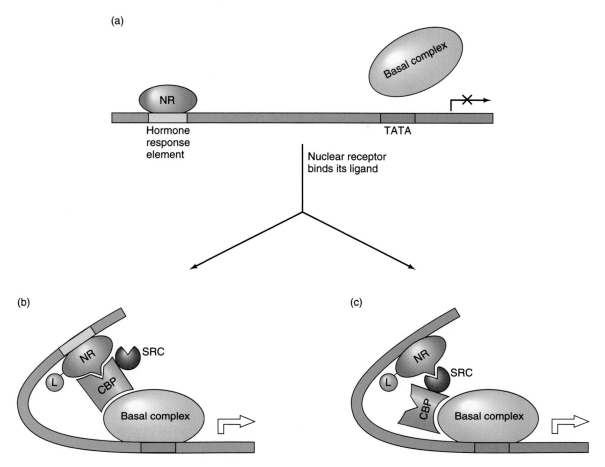

Figure 12.42   Models for activation of a nuclear receptor-activated gene. (a) A nuclear receptor (without its ligand) is bound to its hormone response element, but it cannot contact the basal transcription complex, so the linked gene is not activated. Depending on the type, the nuclear receptor could also be dissociated from its DNA target in the absence of its ligand. The nuclear receptor bound to its DNA target without its ligand may also actively inhibit transcription. (b) The nuclear receptor has bound to its ligand (purple) and is now able to interact with CBP, which binds to at least one component of the basal transcription apparatus, recruiting it to the promoter and activating transcription. (c) The nuclear receptor, bound to its ligand, can also interact with the steroid receptor coactivator (SRC, green), which is bound to CBP. CBP, in turn, contacts the basal complex, recruiting it to the promoter and activating transcription.

basal transcription apparatus directly. Instead, CBP, or its homologue, **P300,** acts as a mediator between the nuclear receptors and the basal transcription apparatus. In principle, CBP can play this mediator role in two ways (Figure 12.42). First, it can interact directly with the activation domains of the nuclear receptors. Second, CBP is bound to other proteins called **steroid receptor coactivators (SRCs),** and these can interact with the nuclear receptors while CBP contacts the basal transcription apparatus. Either mechanism would recruit the basal apparatus to the promoter and activate transcription.

Still another important class of activators use CBP as a mediator. A variety of growth factors and cellular stresses initiate a cascade of events (another signal transduction pathway) that results in the phosphorylation and activation of a protein kinase called **mitogen-activated protein kinase (MAPK).** The activated MAPK enters the nucleus

and phosphorylates activators such as Sap-1a and the Jun monomers in AP-1. These activators then use CBP to mediate activation of their target genes, which finally stimulate cell division. Figure 12.43 illustrates the three important pathways that converge on CBP.

Besides recruiting the basal transcription apparatus to the promoter, CBP may play another role in gene activation. CBP has histone acetyl transferase activity, which adds acetyl groups to histones. As we will see in Chapter 13, histones are general repressors of gene activity. Moreover, acetylation of histones tends to loosen their grip on DNA and relax their repression of transcription. Thus, the association between activators and CBP at an enhancer brings the histone acetyltransferase to the enhancer, where it can acetylate histones and activate the nearby gene. We will discuss this phenomenon in greater detail in Chapter 13.

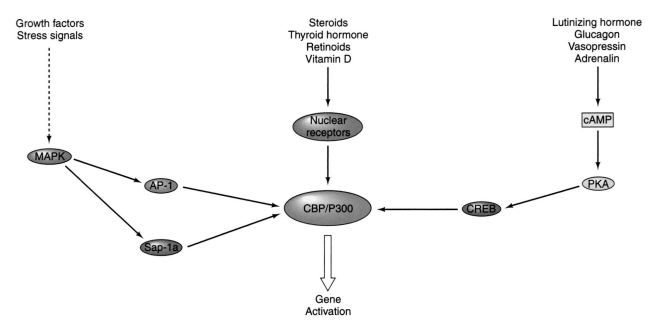

Figure 12.43   Multiple roles of CBP/P300. Three pathways that use CBP/P300 to mediate transcription activation are shown converging on CBP/P300 (red) at center. The arrows between pathway members simply indicate position within the pathway (e.g., MAPK acts on AP-1), without indicating the nature of the action (e.g., phosphorylation). This scheme has also been simplified by omitting branches in the pathways. For example MAPK and PKA also phosphorylate nuclear receptors, although the importance of this phosphorylation is unclear. (*Source:* Reprinted with permission from *Nature* 383:23, 1996. Copyright © 1996 Macmillan Magazines Limited.)

SUMMARY Several different activators, including CREB, the nuclear receptors, and AP-1, do not activate transcription by contacting the basal transcription apparatus directly. Instead, they contact a mediator called CBP (or its homolog P300), which in turn contacts the basal transcription apparatus and recruits it to promoters.

12.6 Regulation of Transcription Factors

Transcription factors regulate gene activity, but what regulates the regulators? We have encountered several examples of such regulation in this chapter and list them here, followed by some other examples we have not discussed:

1. Binding between certain nuclear receptors (e.g., the thyroid hormone receptor) and their ligands changes the receptors from transcription repressors to transcription activators.

2. Binding between certain nuclear receptors (e.g., the glucocorticoid receptor) and their ligands causes the receptor to dissociate from an inhibitory protein in the cytoplasm, translocate to the nucleus, and activate transcription.

3. Phosphorylation of certain activators (e.g., CREB and Jun) allows them to interact with mediators (e.g., CBP) that in turn stimulate transcription.

4. Phosphorylation of certain inhibitor proteins (e.g., IκB) marks them for destruction by proteolysis, which releases them from their target activators (e.g., NF-κB). This allows the liberated activators to stimulate transcription.

5. Phosphorylation of certain activators (e.g., β-catenin) marks them for destruction by proteolysis. Inactivation of the relevant protein kinases (e.g., GSK-3β) prevents phosphorylation of the activators, so the activators avoid destruction and can stimulate transcription.

SUMMARY Activators are subject to a variety of regulation mechanisms. Some are regulated by ligand binding, others by phosphorylation, others by binding to inhibitory proteins, and others by proteolysis.

Signal Transduction Pathways

The phosphorylations of CREB, Jun, IκB, and β-catenin, mentioned in the preceding section, are all the results of signal transduction pathways. So signal transduction pathways play a major role in the control of transcription. Let us explore the concept of signal transduction further

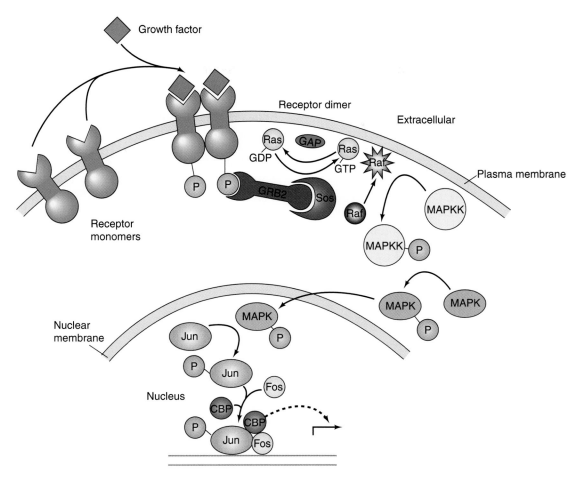

Figure 12.44 A signal transduction pathway involving MAP kinase. Signal transduction begins (top) when a growth factor or other extracellular signaling molecule (red) binds to its receptor (blue). In this case, the receptor dimerizes on binding its ligand. The intracellular protein tyrosine kinase domain of each receptor monomer then phosphorylates its partner. The new phosphotyrosines can then be recognized by an adapter molecule called GRB2 (dark green), which in turn binds to the Ras exchanger Sos. Sos (gray) is activated to replace GDP on Ras with GTP, thus activating Ras (yellow). Ras delivers Raf (purple) to the cell membrane, where Raf becomes activated. The protein serine/threonine kinase domain of Raf is activated at the membrane, so it phosphorylates MAP kinase kinase (MAPKK, gold), which phosphorylates MAP kinase (MAPK, pink), which enters the nucleus and phosphorylates the transcription factor Jun (light green). This activates Jun, which associates with another transcription factor (Fos, light purple). Together, activated Jun and Fos constitute the activator AP-1, which binds CBP, which in turn can stimulate transcription of certain genes. The end result is more rapid cell division.

and examine a well-studied example. Cells are surrounded by a semipermeable membrane that keeps the cell contents from escaping and provides some protection from noxious substances in the cell's environment. This barrier between the interior of a cell and its environment means that mechanisms had to evolve to allow cells to sense the conditions in their surroundings and to respond accordingly. Signal transduction pathways provide these mechanisms. Because the responses a cell makes to its environment usually require changes in gene expression, signal transduction pathways usually end with activation of a transcription factor that activates a gene or set of genes.

Figure 12.43 already outlined two signal transduction pathways, the protein kinase A pathway, and the MAP kinase pathway. Both rely heavily on protein phosphorylation cascades to activate members of the pathway and ul-

timately to activate transcription. Let us explore the MAP kinase pathway in more detail and see how aberrant members of the pathway can lead a cell to lose control over its growth and become a cancer cell.

Figure 12.44 presents a MAP kinase pathway with mammalian names for the proteins. The same pathway operates in other organisms (famously in *Drosophila*) where the proteins have different names. The pathway begins when an extracellular agent, such as a growth factor, interacts with a receptor in the cell membrane. The agent (**epidermal growth factor** [EGF], for example) binds to the extracellular domain of its receptor. This binding stimulates two adjacent receptors to come together to form a dimer, causing the intracellular domains, which have protein tyrosine kinase activity, to phosphorylate each other. Notice how the transmembrane receptor has transduced the signal across the cell membrane into the cell (Latin, *transducere,* meaning "to lead across"). Once the intracellular domains of the receptors are phosphorylated, the new phosphotyrosines attract adapter proteins such as **GRB2** (pronounced "grab two") that have specialized phosphotyrosine binding sites called **SH2 domains.** These are named for similar sites on an oncoprotein called pp60^{src}, which can transform cells from normal to tumor-like behavior; SH stands for "Src homology." GRB2 has another domain called **SH3** (also found in pp60^{src}) that attracts proteins with a particular kind of hydrophobic α-helix, such as **Sos.** Sos is a **Ras exchanger** that can replace GDP on the protein **Ras** with GTP, thereby activating the Ras protein. Ras contains an endogenous GTPase activity that can hydrolyze the GTP to GDP, inactivating the Ras protein. This GTPase activity is very weak by itself, but it can be strongly stimulated by another protein called **GTPase activator protein** (**GAP**). Thus, GAP is an inhibitor of this signal transduction pathway.

Once activated, Ras attracts another protein, **Raf,** to the inner surface of the cell membrane, where Raf is activated. Raf is another protein kinase, but it adds phosphate groups to serines and threonines rather than to tyrosines. Its target is another protein serine kinase called **MAPKK** (**MAP kinase kinase**). In turn, MAPKK phosphorylates another group of proteins lumped together under the name **MAP kinase** (**MAPK**). MAPs are activated by mitogens like epidermal growth factor, which induce cells to divide. Two such MAPs are Jun and Fos, which are also oncoproteins. Phosphorylation activates Jun, which binds to Fos, creating the transcription factor known as AP-1. This factor, in conjunction with CBP, then stimulates transcription of genes whose products promote cell division.

Thus, one signal transduction pathway that begins with a growth factor interacting with the surface of a cell and ends with enhanced transcription of growth-promoting genes, can be pictured as follows:

Growth factor → receptor → GRB2 → Sos → Ras → Raf →
MAPKK → MAPK → MAP → enhanced transcription → more
cell division

It is not surprising that the genes encoding many of the carriers in this pathway, including at least the EGF receptor, Sos, Ras, Raf, Jun, and Fos are **oncogenes,** whose mutation can lead to runaway cell growth and cancer. If these genes overproduce their products, or make products that are hyperactive, the whole pathway can speed up, leading to abnormally enhanced cell growth and, ultimately, to cancer.

Notice the amplifying power of this pathway. One molecule of EGF can lead to the activation of many molecules of Ras, each of which can activate many molecules of Raf. And, because Raf and the kinases that follow it in the pathway are all enzymes, each can activate many molecules of the next member of the pathway. By the end, one molecule of EGF can yield a great number of activated transcription factors, leading to a burst of new transcription. We should also note that this is only one pathway leading through Ras. In reality, the pathway branches at several points, rather like a web.

SUMMARY Signal transduction pathways begin with a signaling molecule that interacts with a receptor on the cell surface, which sends the signal into the cell, and frequently leads to altered gene expression. Many signal transduction pathways, including the MAP kinase pathway, rely on protein phosphorylation to pass the signal from one protein to another. This amplifies the signal at each step.

SUMMARY

Eukaryotic activators are composed of at least two domains: a DNA-binding domain and a transcription-activating domain. DNA-binding domains include motifs such as a zinc module, homeodomain, bZIP, or bHLH motif. Transcription-activating domains can be acidic, glutamine-rich, or proline-rich.

Zinc fingers are composed of an antiparallel β-strand, followed by an α-helix. The β-strand contains two cysteines, and the α-helix two histidines, that are coordinated to a zinc ion. This coordination of amino acids to the metal helps form the finger-shaped structure. The specific recognition between the finger and its DNA target occurs in the major groove.

The DNA-binding motif of the GAL4 protein contains six cysteines that coordinate two zinc ions in a bimetal thiolate cluster. This DNA-binding motif contains a short α-helix that protrudes into the DNA major groove and makes specific interactions there. The GAL4 monomer also contains an α-helical dimerization motif that forms a parallel coiled coil with the α-helix on the other GAL4 monomer.

Some members of the steroid hormone family of receptors bind to their respective ligand and move to the nucleus (if they are not there already), bind to enhancers, and thereby act as activators. The glucocorticoid receptor is representative of this group. It has a DNA-binding domain containing two zinc modules. One module contains most of the DNA-binding residues (in a recognition α-helix), and the other module provides the surface for protein–protein interaction to form a dimer. These zinc modules use four cysteine residues to complex the zinc ion, instead of two cysteines and two histidines as seen in classical zinc fingers. One extra base pair between half-sites in the DNA target prevents specific interactions with one of the glucocorticoid receptor monomers.

The homeodomains in eukaryotic activators contain a DNA-binding motif that functions in much the same way as prokaryotic helix-turn-helix motifs, where a recognition helix fits into the DNA major groove and contacts specific residues there.

The bZIP proteins dimerize through a leucine zipper, which puts the adjacent basic regions of each monomer in position to embrace the DNA target site like a pair of tongs. Similarly, the bHLH proteins dimerize through a helix-loop-helix motif, which allows the basic parts of each long helix to grasp the DNA target site, much as the bZIP proteins do. bHLH and bHLH-ZIP domains bind to DNA in the same way, but the latter have extra dimerization potential due to their leucine zippers.

The DNA-binding and transcription-activation domains of activator proteins are independent modules. Hybrid proteins with the DNA-binding domain of one protein and the transcription-activation domain of another still function as activators.

Activators function by contacting general transcription factors and stimulating the assembly of preinitiation complexes at promoters. For class II promoters, this assembly may occur by stepwise buildup of the general transcription factors and RNA polymerase II, as observed in vitro, or it may occur by recruitment of a large holoenzyme that includes RNA polymerase and most of the general transcription factors. Additional factors (perhaps just TBP or TFIID) may be recruited independently of the holoenzyme. To do their job, activators can interact with a variety of factors, including TFIID and TFIIB.

Dimerization is a great advantage to an activator because it increases the affinity between the activator and its DNA target. Some activators form homodimers, but others, such as Jun and Fos, function better as heterodimers.

The essence of enhancer function—protein–protein interaction between activators bound to the enhancers, and general transcription factors and RNA polymerase bound to the promoter—seems in many cases to be mediated by looping out the DNA in between. At least in theory, this can also account for the effects of multiple enhancers on gene transcription. DNA looping could bring the activators bound to each enhancer close to the promoter where they could stimulate transcription, perhaps in a cooperative way.

Multiple enhancers enable a gene to respond differently to different combinations of activators. This arrangement gives cells exquisitely fine control over their genes in different tissues, or at different times in a developing organism.

The activator LEF-1 binds to the minor groove of its DNA target through its HMG domain and induces strong bending in the DNA. LEF-1 does not enhance transcription by itself, but the bending it induces probably helps other activators bind and interact with other activators and the general transcription factors to stimulate transcription. Another activator, HMG 1(Y), probably plays a similar role in the human interferon-β control region. These proteins are examples of architectural transcription factors. In the case of the IFNβ enhancer, activation seems to require the cooperative binding of several activators, including HMG 1(Y), to form an enhanceosome with a specific shape.

Insulators are DNA elements that shield genes from activation or repression by enhancers or silencers. They usually operate on one side of a DNA element but not on the other. Thus, by an incompletely understood mechanism, they establish boundaries between chromosome regions.

Several different activators, including CREB, the nuclear receptors, and AP-1, do not activate transcription by contacting the basal transcription apparatus directly. Instead, they contact a mediator called CBP (or its homolog P300), which in turn contacts the basal transcription apparatus and recruits it to promoters.

Signal transduction pathways begin with a signaling molecule that interacts with a receptor on the cell surface, which sends the signal into the cell, and frequently leads to altered gene expression. Many signal transduction pathways, including the MAP kinase pathway, rely on protein phosphorylation to pass the signal from one protein to another. This enzymatic action amplifies the signal at each step.

REVIEW QUESTIONS

1. List three different classes of DNA-binding domains found in eukaryotic transcription factors.

2. List three different classes of transcription-activation domains in eukaryotic transcription factors.

3. Draw a detailed diagram of a zinc finger. Point out the DNA-binding motif of the finger.

4. List one important similarity and three differences between a typical prokaryotic helix-turn-helix domain and the Zif268 zinc finger domain.

5. Draw a diagram of the dimer composed of two molecules of the N-terminal 65 amino acids of the GAL4 protein, interacting with DNA. Your diagram should show clearly the dimerization domains and the motifs in the two DNA-binding domains interacting with their DNA-binding sites. What metal ions and coordinating amino acids, and how many of each, are present in each DNA-binding domain?

6. In general terms, what is the function of a nuclear receptor? Give three examples.

7. What metal ions and coordinating amino acids, and how many of each, are present in each DNA-binding domain of a nuclear receptor? What part of the DNA-binding domain contacts the DNA bases?

8. What is the effect of adding one extra base pair between the half-sites recognized by the glucocorticoid receptor?

9. What is the nature of the homeodomain? What other DNA-binding domain does it most resemble?

10. Draw a schematic diagram of a leucine zipper seen from the side. Draw a more realistic diagram of a leucine zipper seen from the end. How do these diagrams illustrate the relationship between the structure and function of the leucine zipper?

11. Draw a diagram of a bZIP protein interacting with its DNA-binding site.

12. Describe and show the results of an experiment that illustrates the independence of the DNA-binding and transcription-activating domains of a gene-specific transcription factor.

13. Present two models of recruitment of the class II preinitiation complex, one involving a holoenzyme, the other not.

14. Describe and give the results of an experiment that shows that an acidic transcription-activating domain binds to TFIID.

15. Describe and give the results of an experiment that shows that GAL4-AH stimulates transcription only if added before the preinitiation complex forms.

16. Describe and give the results of an experiment that shows that a function of GAL4 is recruitment of TFIIB to the preinitiation complex.

17. Design an experiment to show that TFIIB binds directly to an acidic activating domain. Show sample positive results.

18. TFIID, but not TBP, can cooperate with GAL4-AH in activating transcription from the adenovirus E4 promoter. However, TBP and TFIID are equally effective in helping GAL4 recruit TFIIB to the preinitiation complex. Present a hypothesis to explain these data. Describe and show the results of a further experiment that supports this hypothesis.

19. Describe and show the results of an experiment that demonstrates that multiple GAL4-binding sites are much more effective in enhancing transcription than a single GAL4-binding site.

20. Why are multiple GAL4-binding sites more effective in transcription enhancement? Describe and show the results of an experiment that answers this question.

21. Present a model (in diagram form) that explains the interactions of a transcription activator such as GAL4 with general transcription factors to enhance transcription. Which step(s) are TAF-dependent and -independent?

22. Present evidence that favors the holoenzyme recruitment model.

23. Why is a protein dimer (or tetramer) so much more effective than a monomer in DNA binding? Why is it important for a transcription activator to have a high affinity for DNA?

24. Present three models to explain how an enhancer can act on a promoter hundreds of base pairs away.

25. Describe and give the results of an experiment that shows the effect of isolating an enhancer on a separate circle of DNA intertwined with another circle of DNA that contains the promoter. Which model(s) of enhancer activity does this experiment favor? Why?

26. What advantage do multiple enhancers confer on a gene?

27. Explain how LiCl causes too much endoderm to form in developing sea urchin embryos.

28. LEF-1 is an activator of the human T-cell receptor α-chain, yet LEF-1 by itself does not activate this gene. How does LEF-1 act? Describe and show the results of an experiment that supports your answer.

29. Does LEF-1 bind in the major or minor groove of its DNA target? Present data to support your answer.

30. What do insulators do?

31. Describe and give the results of an experiment that shows the effects of a mediator.

32. Draw diagrams to illustrate the action of CBP as a mediator of (a) phosphorylated CREB; (b) a nuclear receptor.

33. How do signal transduction pathways amplify their signals? Present an example.

SUGGESTED READINGS

General References and Reviews

Bell, A.C. and G. Felsenfeld. 1999. Stopped at the border: boundaries and insulators. *Current Opinion in Genetics and Development.* 9:191–98.

Blackwood, E.M. and J.T. Kadonaga. 1998. Going the distance: A current view of enhancer action. *Science* 281:60–63.

Carey, M. 1994. Simplifying the complex. *Nature* 368:402–3.

Goodrich, J.A., G. Cutler, and R. Tjian. 1996. Contacts in context: Promoter specificity and macromolecular interactions in transcription. *Cell* 84:825–30.

Hampsey, M. and D. Reinberg. 1999. RNA polymerase II as a control panel for multiple coactivator complexes. *Current Opinion in Genetics and Development* 9:132–39.

Janknecht, R. and T. Hunter. 1996. A growing coactivator network. *Nature* 383:22–23.

Liusi, B. 1992. Zinc standard for economy. *Nature* 397:379–80.

Montminy, M. 1997. Something new to hang your hat on. *Nature* 387:654–55.

Myer, V.E. and R.A. Young. 1998. RNA polymerase II holoenzymes and subcomplexes. *Journal of Biological Chemistry* 273:27757–60.

Nordheim, A. 1994. CREB takes CBP to tango. *Nature* 370:177–78.

Ptashne, M. and A. Gann. 1997. Transcriptional activation by recruitment. *Nature* 386:569–77.

Rhodes, D. and A. Klug. 1993. Zinc fingers. *Scientific American* 268 (Feb.):56–65.

Roush, W. 1996. "Smart" genes use many cues to set cell fate. *Science* 272:652–53.

Sauer, F. and R. Tjian. 1997. Mechanisms of transcription activation: Differences and similarities between yeast, *Drosophila,* and man. *Current Opinion in Genetics and Development* 7:176–81.

Tjian, R. and T. Maniatis. 1994. Transcriptional activation: A complex puzzle with few easy pieces. *Cell* 77:5–8.

Werner, M.H. and S.K. Burley. 1997. Architectural transcription factors: Proteins that remodel DNA. *Cell* 88:733–36.

Wolffe, A.P. 1994. Architectural transcription factors. *Science* 264:1100–1.

Research Articles

Barberis, A., J. Pearlberg, N. Simkovich, S. Farrell, P. Reinagle, C. Bamdad, G. Sigal, and M. Ptashne. 1995. Contact with a component of the polymerase II holoenzyme suffices for gene activation. *Cell* 81:359–68.

Brent, R. and M. Ptashne. 1985. A eukaryotic transcriptional activator bearing the DNA specificity of a prokaryotic repressor. *Cell* 43:729–36.

Choy, B. and M.R. Green. 1993. Eukaryotic activators function during multiple steps of preinitiation complex assembly. *Nature* 366:531–36.

Du, W., Thanos, D., and T. Maniatis. 1993. Mechanisms of transcriptional synergism between distinct virus-inducible enhancer elements. *Cell* 74:887–98.

Dunaway, M. and P. Dröge. 1989. Transactivation of the *Xenopus* rRNA gene promoter by its enhancer. *Nature* 341:657–59.

Ellenberger, T.E., C.J. Brandl, K. Struhl, and S.C. Harrison. 1992. The GCN4 basic region leucine zipper binds DNA as a dimer of uninterrupted α helices: Crystal structure of the protein–DNA complex. *Cell* 71:1223–37.

Flanagan, P.M., et al. 1991. A mediator required for activation of RNA polymerase II transcription in vitro. *Nature* 350:436–38.

Geise, K., J. Cox, and R. Grosschedl. 1992. The HMG domain of lymphoid enhancer factor 1 bends DNA and facilitates assembly of functional nucleoprotein structures. *Cell* 69:185–95.

Hegde, R.S., S.R. Grossman, L.A. Laimins, and P.B. Sigler. 1992. Crystal structure at 1.7 Å of the bovine papillomavirus-1 E2 DNA-binding domain bound to its DNA target. *Nature* 359:505–12.

Kim, T.K. and T. Maniatis. 1997. The mechanism of transcriptional synergy of an in vitro assembled interferon-β enhanceosome. *Molecular Cell* 1:119–29.

Kissinger, C.R., B. Liu, E. Martin-Blanco, T.B. Kornberg, and C.O. Pabo. 1990. Crystal structure of an engrailed homeodomain–DNA complex at 2.8 Å resolution: A framework for understanding homeodomain–DNA interactions. *Cell* 63:579–90.

Koleske, A.J. and R.A. Young. 1994. An RNA polymerase II holoenzyme responsive to activators. *Nature* 368:466–69.

Kouzarides, T. and E. Ziff. 1988. The role of the leucine zipper in the fos–jun interaction. *Nature* 336:646–51.

Krebs, J.E. and Dunaway, M. 1998. The scs and scs' insulator elements impart a *cis* requirement on enhancer–promoter interactions. *Molecular Cell* 1:301–08.

Lee, M.S. 1989. Three-dimensional solution structure of a single zinc finger DNA-binding domain. *Science* 245:635–37.

Leuther, K.K., J.M. Salmeron, and S.A. Johnston. 1993. Genetic evidence that an activation domain of GAL4 does not require acidity and may form a β sheet. *Cell* 72:575–85.

Lin, Y.-S., and M.R. Green. 1991. Mechanism of action of an acidic transcriptional activator in vitro. *Cell* 64:971–81.

Lin, Y.-S., I. Ha, E. Maldonado, D. Reinberg, and M.R. Green. 1991. Binding of general transcription factor TFIIB to an acidic activating region. *Nature* 353:569–71.

Luisi, B.F., W.X. Xu, Z. Otwinowski, L.P. Freedman, K.R. Yamamoto, and P.B. Sigler. 1991. Crystallographic analysis of the interaction of the glucocorticoid receptor with DNA. *Nature* 352:497–505.

Ma, P.C.M., M.A. Rould, H. Weintraub, and C.O. Pabo. 1994. Crystal structure of MyoD bHLH domain–DNA complex: Perspectives on DNA recognition and implications for transcription activation. *Cell* 77:451–59.

Marmorstein, R., M. Carey, M. Ptashne, and S.C. Harrison. 1992. DNA recognition by GAL4: Structure of a protein–DNA complex. *Nature* 356:408–14.

O'Shea, E.K., J.D. Klemm, P.S. Kim, and T. Alber. 1991. X-ray structure of the GCN4 leucine zipper, a two-stranded, parallel coiled coil. *Science* 254:539–44.

Pavletich, N.P., and C.O. Pabo. 1991. Zinc finger–DNA recognition: Crystal structure of a Zif 268–DNA complex at 2.1 Å. *Science* 252:809–17.

Ryu, L., L. Zhou, A.G. Ladurner, and R. Tjian. 1999. The transcriptional cofactor complex CRSP is required for activity of the enhancer-binding protein Sp1. *Nature* 397:446–50.

Stringer, K.F., C.J. Ingles, and J. Greenblatt. 1990. Direct and selective binding of an acidic transcriptional activation domain to the TATA-box factor TFIID. *Nature* 345:783–86.

Thanos, D. and T. Maniatis. 1995. Virus induction of human IFNβ gene expression requires the assembly of an enhanceosome. *Cell* 83:1091–1100.

Yuh, C.-H. and E.H. Davidson. 1996. Modular *cis*-regulatory organization of *Endo 16,* a gut-specific gene of the sea urchin embryo. *Development* 122:1069–82.

Yuh, C.-H., B. Hamid, and E.H. Davidson. 1998. Genomic *cis*-regulatory logic: Experimental and computational analysis of a sea urchin gene. *Science* 279:1896–1902.

Chromatin Structure and Its Effects on Transcription

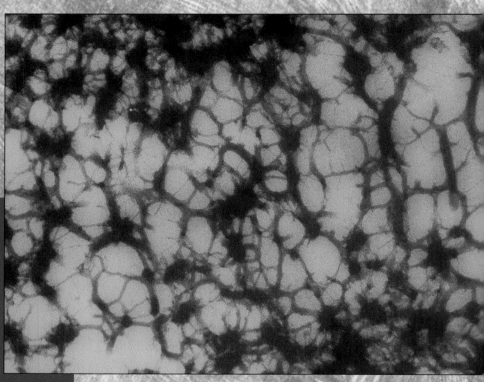

Chromatin in developing human spermatid. (×300,000). © David M. Phillips/Visuals Unlimited.

I n our discussion of transcription of eukaryotic genes, we have so far been ignoring an important point: Eukaryotic genes do not exist naturally as naked DNA molecules, or even as DNA molecules bound only to transcription factors. Instead, they are complexed with other proteins, especially basic proteins called histones, to form a substance known as chromatin. As we will see, histones can repress gene transcription. In fact, for years it was assumed that they were rather uninteresting impediments to gene expression. Now we realize that they interact in a dynamic way with DNA and transcription factors to help control gene expression. Let us examine first the structure of histones and chromatin and then look at the way histones can cooperate with other chromatin proteins to control transcription.

13.1 Histones

Most eukaryotic cells contain five different kinds of histones: **H1, H2A, H2B, H3,** and **H4.** Figure 13.1 shows that these five molecules can be separated by gel electrophoresis. These are extremely abundant proteins; the mass of histones in eukaryotic nuclei is equal to the mass of DNA. They are also unusually basic—at least 20% of their amino acids are arginine or lysine—and have a pronounced positive charge at neutral pH. For this reason, they can be extracted from cells with strong acids, such as 1.5 N HCl—conditions that would destroy most proteins. Also because of their basic nature, the histones migrate toward the cathode during nondenaturing electrophoresis, unlike most other proteins, which are acidic and therefore move toward the anode. Most of the histones are also well conserved from one organism to another. The most extreme example of this is histone H4. Cow histone H4 differs from pea H4 in only two amino acids out of a total of 102, and these are conservative changes—one basic amino acid (lysine) substituted for another (arginine), and one hydrophobic amino acid (valine) substituted for another (isoleucine). In other words, in all the eons since the cow and pea lines have diverged from a common ancestor, only two amino acids in histone H4 have changed. Histone H3 is also extremely well conserved; histones H2A and H2B are moderately well conserved; but histone H1 varies considerably among organisms. Table 13.1 lists some of the characteristics of histones.

Figure 13.1 gives the impression that each histone is a homogeneous species. However, higher resolution separations of the histones have revealed much greater variety. This variety stems from two sources: gene reiteration and posttranslational modification. The histone genes are not

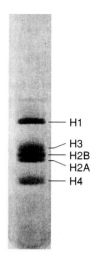

Figure 13.1 Separation of histones by electrophoresis. The histones of calf thymus were separated by polyacrylamide gel electrophoresis. (*Source:* Panyim and Chalkley. *Archives of Biochem. & Biophys.* 130, 1969, f. 6A, p. 343.)

single-copy genes like most protein-encoding genes in eukaryotes. Instead, they are repeated many times: 10–20 times in the mouse, and about 100 times in *Drosophila.* Many of these copies are identical, but some are quite different. Histone H1 (the lysine-rich histone) shows the greatest variation, with at least six subspecies in the mouse. One H1 variant is called H1°. Birds, fish, amphibians, and reptiles have another lysine-rich histone that could be an extreme variant of H1, but it is so different from H1 that it is generally called by a distinct name, **H5.** Histone H4 shows the least variation; only two variant species have ever been reported, and these are rare. It is assumed that the variant species of a given histone all play

Table 13.1 General Properties of the Histones

Histone Type	Histone	Molecular Weight (M_r)	Degree of Conservation
Core histones— can assemble DNA to particles	H3	15,400	Highly conserved
	H4	11,340	Highly conserved
Core histones— more peripheral	H2A	14,000	Moderate variation among tissues and species
	H2B	13,770	Moderate variation among tissues and species
H1-like histones	H1	21,500	Varies markedly among tissues and species
	H1°	~21,500	Variable, mostly present in nonreplicating cells
	H5	21,500	Very variable, present only in transcriptionally inactive cells of some species

essentially the same role, but each may influence the properties of chromatin somewhat differently.

The second cause of histone heterogeneity is such a rich source of variation that Kensal van Holde has described it as "baroque." The most common histone modification is acetylation, which can occur on N-terminal amino groups and on lysine ε-amino groups. Other modifications are lysine ε-amino methylation and phosphorylation, including serine and threonine O-phosphorylation and lysine and histidine N-phosphorylation. These modifications are dynamic processes, so modifying groups can be removed as well as added. At least some of the histone modifications influence chromatin structure and function. Histone acetylation, in particular, plays an important role in governing gene activity. We will discuss this phenomenon later in this chapter.

13.2 Nucleosomes

The length-to-width ratio of a typical human chromosome is more than 10 million to one. Such a long, thin molecule would tend to get tangled if it were not folded somehow. Another way of considering the folding problem is that the total length of human DNA, if stretched out, would be about 2 m, and this all has to fit into a nucleus only about 10 μm in diameter. In fact, if you laid all the DNA molecules in your body end to end, they would reach to the sun and back many times. Obviously, a great deal of DNA folding must occur in your body and in all other living things. We will see that eukaryotic chromatin is indeed folded in several ways. The first order of folding involves structures called **nucleosomes,** which have a core of histones, around which the DNA winds.

Maurice Wilkins showed as early as 1956 that x-ray diffraction patterns of DNA in intact nuclei exhibited sharp bands, indicating a repeating structure larger than the double helix itself. Subsequent x-ray diffraction work by Aaron Klug, Roger Kornberg, Francis Crick, and others showed a strong repeat at intervals of approximately 100 Å. This corresponds to a string of nucleosomes, which are about 110 Å in diameter. Kornberg found in 1974 that he could chemically cross-link histones H3 and H4, or histones H2A and H2B in solution. Moreover, he found that H3 and H4 exist as a tetramer (H3–H4)$_2$ in solution. He also noted that chromatin is composed of roughly equal masses of histones and DNA. In addition the concentration of histone H1 is about half that of the other histones. This corresponds to one histone octamer (two molecules each of H2A, H2B, H3, and H4) plus one molecule of histone H1 per 200 bp of DNA. Finally, he reconstituted chromatin from H3–H4 tetramers, H2A–H2B oligomers, and DNA and found that this reconstituted chromatin produced the same x-ray diffraction pattern as natural

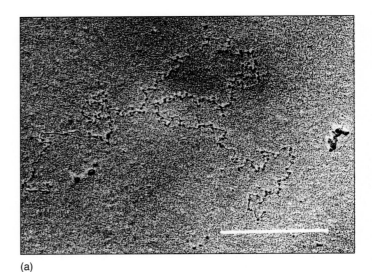

(a)

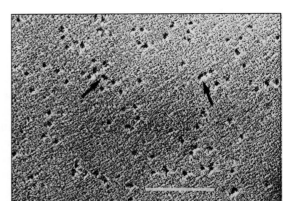

(b)

Figure 13.2   Early electron micrographs of nucleosomes.
(**a**) Nucleosome strings. Chambon and colleagues used trypsin to remove histone H1 from chromatin isolated from chicken red blood cells, revealing a beads-on-a-string structure. The bar represents 500 nm. (**b**) Isolated nucleosomes. Chambon's group used micrococcal nuclease to cut between nucleosomes, then isolated these particles by ultracentrifugation. The arrows point to two representative nucleosomes. The bar represents 250 nm. (*Source:* Oudet et al., Electron microscopic and biochemical evidence that chromatin structure is a repeating unit. *Cell* 4 (1975), f. 4b & 5, pp. 286–87. Reprinted by permission of Elsevier Science.)

chromatin. Several workers, including Gary Felsenfeld and L.A. Burgoyne, had already shown that chromatin cut with a variety of nucleases yielded DNA fragments about 200 bp long. Based on all these data, Kornberg proposed a repeating structure of chromatin composed of the histone octamer plus one molecule of histone H1 complexed with about 200 bp of DNA.

G.P. Georgiev and coworkers discovered that histone H1 is much easier than the other four histones to remove from chromatin. In 1975, Pierre Chambon and colleagues took advantage of this phenomenon to selectively remove histone H1 from chromatin with trypsin or high salt buffers, and found that this procedure yielded chromatin with a "beads-on-a-string" appearance (Figure 13.2a).

(a)

(b)

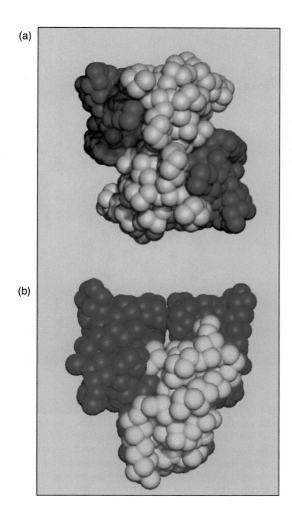

(c)

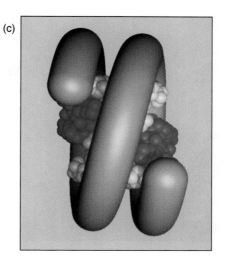

Figure 13.3   Two views of the histone octamer based on x-ray crystallography and a hypothetical path for the nucleosomal DNA. The H2A–H2B dimers are dark blue; the (H3–H4)$_2$ tetramer is light blue. The octamer in panel (**b**) is rotated 90 degrees downward relative to the octamer in panel (**a**). The thin edge of the wedge is pointing toward the viewer in panel (**a**) and downward in panel (**b**), where it is clear that the narrowing of the wedge occurs primarily in the H3–H4 tetramer. (**c**) Hypothetical path of the DNA around the histone octamer. The 20 Å-diameter DNA (blue-gray tube) nearly obscures the octamer, which is shown in the same orientation as in part (**a**). (*Source:* (*a-b*) Arents et al., The nucleosomal core histone octamer at 3.1Å resolution: A tripartite protein assembly and a left-handed superhelix. *PNAS* 88 (Nov 1991), f. 3, p. 10150. (*c*) Arents et al., Topography of the histone octamer surface: Repeating structural motifs utilized in the docking of nucleosomal DNA. *PNAS* 90 (Nov 1993), f. 3a, 1 & 4, pp. 10490-91. © National Academy of Sciences, USA.)

They named the beads nucleosomes. Figure 13.2b shows some of the nucleosomes that Chambon and coworkers purified from chicken red blood cells, using micrococcal nuclease to cut the DNA string between the beads.

J.P. Baldwin and colleagues subjected chromatin to neutron-scattering analysis, which is similar to x-ray diffraction, but uses a beam of neutrons instead of x-rays. The pattern of scattering of the neutrons by the sample gives clues to the three-dimensional structure of the molecules in the sample. These investigators found a ring of scattered neutrons corresponding to a repeat distance of about 105 Å, which agreed with the x-ray diffraction analysis. Moreover, the overall pattern suggested that the protein and DNA occupied separate regions within the nucleosomes. Based on these data, Baldwin and coworkers proposed that the core histones (H2A, H2B, H3, and H4) form a ball, with the DNA wrapped around the outside. Having the DNA on the outside also has the advantage that it minimizes the amount of bending the DNA would have to do. In fact, DNA is such a stiff molecule that it could not bend tightly enough to fit inside a nucleosome. These workers also placed histone H1 on the outside, in accord with its ease of removal from chromatin.

Several research groups have used x-ray crystallography to determine a structure for the histone octamer. According to the work of Evangelos Moudrianakis and his colleagues in 1991, the octamer takes on different shapes when viewed from different directions, but most viewpoints reveal a three-part architecture. This tripartite structure contains a central (H3–H4)$_2$ core attached to two H2A–H2B dimers, as shown in Figure 13.3. The overall structure is shaped roughly like a disc, or hockey puck, that has been worn down to a wedge shape. Notice that this structure is consistent with Kornberg's data on the association between histones in solution and with the fact that the histone octamer dissociates into an (H3–H4)$_2$ tetramer and two H2A–H2B dimers.

Where does the DNA fit in? It was not possible to tell from these data, because the crystals did not include DNA. However, grooves on the surface of the proposed octamer defined a left-handed helical ramp that could provide a path for the DNA (Figure 13.3c). In 1997, Timothy Richmond and colleagues succeeded in crystallizing a nucleosomal core particle that did include DNA. The nucleosome, as originally defined, contained about 200 bp of DNA. This is the length of DNA released by subjecting

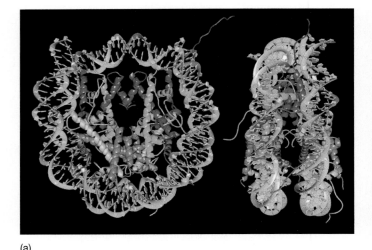

(a)

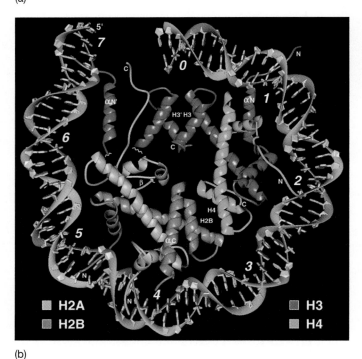

(b)

(c)

Figure 13.4 Crystal structure of a nucleosomal core particle. Richmond and colleagues crystallized a core particle composed of a 146-bp DNA and cloned core histones, then determined its crystal structure. (**a**) Two views of the core particle, seen face-on (left) and edge-on (right). The DNA on the outside is rendered in tan and green. The core histones are rendered as follows: H2A, yellow; H2B, red; H3, blue; and H4, green. Note the H3 tail extending through a cleft between the minor grooves of the two adjacent turns of the DNA around the core (the blue tail visible at the top of both parts of the panel). (**b**) Half of the core particle, showing 73 bp of DNA plus at least one molecule each of the core histones. (**c**) Core particle with DNA removed. (*Source:* (a–b) Luger, K. et al. Crystal Structure of the Nucleosome Core Particle at 2.8Å Resolution. *Nature* 389 (18 Sep 1997) f. 1, p. 252. © Macmillan Magazines Ltd. (c) Rhodes, Daniela. Chromatin structure: The nucleosome core all wrapped up. *Nature* 389 (18 Sep 1997) f. 2, p. 233. © Macmillan Magazines Ltd.)

chromatin to a mild nuclease treatment. However, exhaustive digestion with nuclease gives a **core particle** with 146 bp of DNA and the histone octamer containing all four **core histones** (H2A, H2B, H3, and H4), but no histone H1, which is relatively easily removed.

Figure 13.4 depicts the structure determined by Richmond and colleagues. We can see the DNA winding almost twice around the core histones. We can also see the H3–H4 tetramer near the top and the two H2A–H2B dimers near the bottom. This arrangement is particularly obvious on the right in panel a. The architecture of the histones themselves is interesting. All of the core histones contain the same fundamental **histone fold,** which consists of three α-helices linked by two loops. All of them also contain extended tails that make up about 28% of the mass of the core histones. Because the tails are relatively unstructured, the crystal structure does not include most of their length. The tails are especially evident with the DNA removed in panel c. The tails of H2B and H3 pass out of the core particle through a cleft formed from two adjacent DNA minor grooves (see the long blue tail at the top of the left part of panel a). One of the H4 tails is exposed to the side of the core particle (see the right part of panel a). This tail is rich in basic residues and can interact strongly with an acidic region of an H2A–H2B dimer in an adjacent nucleosome. Such interactions may play a role in nucleosome cross-linking, which we will discuss later in this chapter.

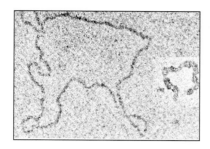

Figure 13.5   Condensation of DNA in nucleosomes. Deproteinized SV40 DNA is shown next to an SV40 minichromosome (inset) in electron micrographs enlarged to the same scale. The condensation of DNA afforded by nucleosome formation is apparent. (*Source:* Griffith, Jack. Chromatin structure: Deduced from a minichromosome. *Science* 187:1202 (28 March 1975). © AAAS.)

This and other models of the nucleosome indicate that the DNA winds almost twice around the core, condensing the length of the DNA by a factor of 6 to 7. Jack Griffith also observed this magnitude of condensation in his 1975 study of the SV40 minichromosome. Because SV40 DNA replicates in mammalian nuclei, it is exposed to mammalian histones, and therefore forms typical nucleosomes. Figure 13.5 shows two views of the SV40 DNA. The main panel shows the DNA after all protein has been stripped off. The inset shows the minichromosome with all its protein—at the same scale. The reason the minichromosome looks so much smaller is that the DNA is condensed by winding around the histone cores in the nucleosomes.

The Nucleosome Filament

Where exactly can we find histone H1 in the nucleosome? Several lines of evidence place it at the point of entry and exit of DNA from the core particle, as shown in Figure 13.6. One piece of evidence for this placement is the behavior of chromatin with and without histone H1 at low ionic strength. For example, consider Figure 13.7. Notice that the chromatin containing histone H1 (panels a and b) contains nucleosomes with a relatively regular zigzag orientation, whereas the chromatin lacking histone H1 (panels c and d) exhibits no such regular structure. The zigzag structure is just what we would predict if the DNA entered and exited the nucleosome near the same point, as pictured in Figure 13.8. The fact that histone H1 is necessary for chromatin to take on this zigzag appearance suggests that this protein lies near the point where DNA enters and exits the nucleosome. Without histone H1, the DNA is not forced to enter and exit on the same side of the nucleosome, so the zigzag conformation disappears.

We should not necessarily conclude from Figure 13.8 that chromatin actually exists as a planar zigzag. An alternative explanation is that the nucleosome string consists of an irregular open helix that becomes flattened on the

Figure 13.6   A model for the location of histone H1 in the nucleosome. Klug and colleagues postulated that histone H1 (yellow) was found outside the core nucleosome, near the point where DNA (red) enters and exits the core particle. The striations on the histone octamer (blue) show the shape of the octamer according to the best resolution structural studies available in 1980. (*Source:* Reprinted with permission from Butler, *CRC Critical Reviews in Biochemistry,* 15:61. Copyright CRC Press, Boca Raton, Florida.)

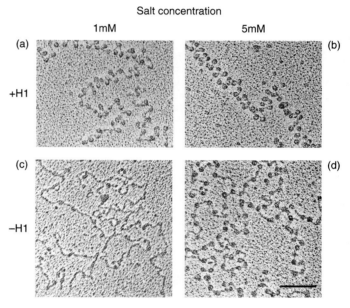

Figure 13.7   Appearance of chromatin at low salt concentrations with and without histone H1. The salt concentrations are listed at top, and the presence or absence of histone H1 is indicated at left. Note the ordered zigzag appearance of the nucleosomes in the presence of histone H1, and the loss of this structure without histone H1. At 5 mM NaCl in the absence of H1, the nucleosomes look like beads on an extended string, rather than a zigzag. At 1 mM NaCl in the absence of H1, it is even difficult to see the nucleosomes at all. The bar represents 100 nm. (*Source:* Thoma, Koller and Klug. Involvement of histone H1 in the organization of the nucleosome and of the salt-dependent superstructures of chromatin. *J. Cell Biology* 83 (Nov 1979) f. 4&6, p. 410. © Rockefeller University Press.)

electron microscope grid. Figure 13.9a presents a diagram of such an irregular open helix, and Figure 13.9b shows a planar projection of the same diagram. Notice the planar zigzag appearance, which is an artifact of flattening the helix.

(a)

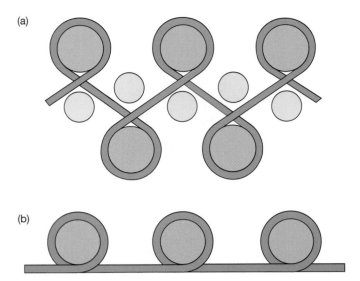

(b)

Figure 13.8   A model for the role of histone H1 in organizing chromatin. (**a**) Chromatin containing histone H1. Molecules of histone H1 (yellow) lie outside the nucleosome core and ensure that the DNA makes almost two turns around the core, entering and exiting near the same spot. This gives rise to a zigzag appearance to the chromatin. (**b**) Chromatin lacking histone H1. Here, histone H1 has been removed, so the DNA is no longer constrained to form a zigzag. Instead, the chromatin can be stretched into the linear "string-of-beads" conformation. (*Source:* From K. E. van Holde, *Chromatin.* Copyright © 1989 Springer-Verlag New York Inc., New York, NY. Reprinted by permission.)

SUMMARY Eukaryotic DNA combines with basic protein molecules called histones to form structures known as nucleosomes. These structures contain four pairs of core histones (H2A, H2B, H3, and H4) in a wedge-shaped disc, around which is wrapped a stretch of about 146 bp of DNA. Histone H1, apparently bound to DNA outside the disc, completes the nucleosome structure and may constrain the DNA to enter and exit near the same point. The first order of chromatin folding is represented by a string of nucleosomes, which condenses the DNA by a factor of 6 to 7.

The 30-nm Fiber

After the string of nucleosomes, the next order of chromatin folding produces a fiber about 30 nm in diameter. Unfortunately, it has not been possible to crystallize any component of chromatin larger than the nucleosome core, so researchers have had to rely on lower resolution methods such as electron microscopy (EM). Figure 13.10 depicts the results of an EM study that shows how the string of nucleosomes condenses to form the 30-nm fiber at increasing ionic strength. The degree of this condensation

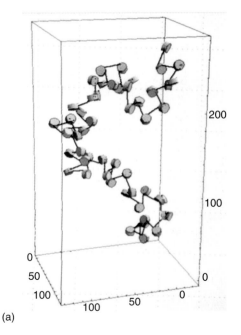

(a)

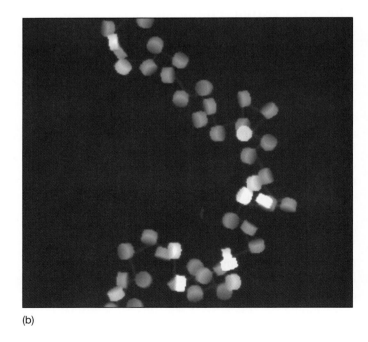

(b)

Figure 13.9   An open helical model of the extended nucleosome filament. (**a**) The model in three dimensions. (**b**) A planar projection of the model in part (**a**). Notice the apparent flat zigzag structure that actually derives from flattening an open helix. (*Source:* Courtesy S. Leuba, in van Holde and Zlatanova, Chromatin higher order structure: Chasing a mirage? *J. Biol. Chem.* 270 (14 Apr 1995), f. 1, p. 8375. American Society for Biochemistry and Molecular Biology.)

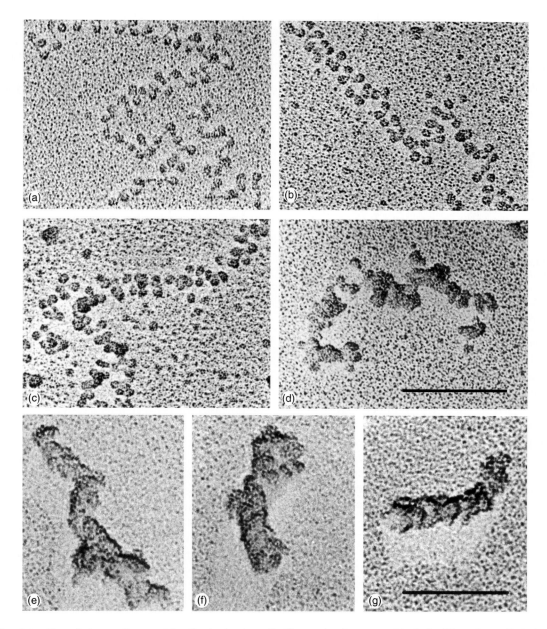

Figure 13.10 Condensation of chromatin on raising the ionic strength. Klug and colleagues subjected rat liver chromatin to buffers of increasing ionic strength, during fixation for electron microscopy. Panels (a)–(c) were at low ionic strength, panel (d) at moderate ionic strength, and panels (e)–(g) at high ionic strength. More specifically, the fixing conditions in each panel were the following, plus 0.2 mM EDTA in each case: (a) 1 mM triethylamine hydrochloride (TEACl); (b and c) 5 mM TEACl; (d) 40 mM NaCl, 5mM TEACl; (e)–(g) 100 mM NaCl, 5 mM TEACl. The bars represent 100 nm. (*Source:* Thoma, Koller and Klug. Involvement of histone H1 in the organization of the nucleosome and of the salt-dependent superstructures of chromatin. *J. Cell Biology* 83 (1979) f. 4, p. 408. © Rockefeller University Press.)

is another six- to sevenfold in addition to the approximately six- to sevenfold condensation in the nucleosome itself.

Aaron Klug and colleagues used electron microscopy and x-ray diffraction analysis of oriented fibers to determine the structure of the 30-nm fiber. They concluded that this fiber is a **solenoid,** or hollow contact helix of nucleosomes, as illustrated in Figure 13.11. The term "solenoid" ordinarily refers to a tube-shaped electromagnet

and comes from the Greek *solen,* which means "pipe." The term *contact helix* means that the nucleosomes in one turn of the helix are in contact with those of the next. This would account for the 100–110-Å spacing observed by both electron microscopy and x-ray diffraction analysis of the 30-nm fiber. Because the nucleosomes are in contact, the distance between turns of the helix is simply the diameter of a nucleosome, or about 110 Å. The number of nucleosomes per helical turn in the solenoid is

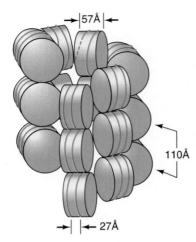

Figure 13.11   The solenoid model of chromatin folding. A string of nucleosomes coils into a hollow tube, or solenoid. Each nucleosome is represented by a blue cylinder with a wire-like DNA (red) coiled around it. For simplicity, the solenoid is drawn with six nucleosomes per turn and with the nucleosomes parallel to the solenoid axis.
(*Source:* Reprinted from Widom and Klug, *Cell* 43:210, 1985. Copyright 1985, with permission from Elsevier Press.)

about six, but this does not have to be an integral number. Also, adjacent nucleosomes in the nucleosome fiber are in contact with each other in the solenoid, in contrast to their spacing in the simple 110-Å nucleosome fiber. The helix in the solenoid appears sometimes to be right-handed, as shown in Figure 13.11, but other times it appears to be left-handed. It is possible that the helical handedness could shift at discontinuities in the solenoid. The solenoid model in Figure 13.11 shows the disc-shaped nucleosomes in a vertical orientation, but they could be tilted.

C. L. Woodcock and colleagues have proposed a variable zigzag ribbon structure for the 30-nm fiber (Figure 13.12). The most likely version of this structure features alternating regular and irregular arrangements of a zigzag string of nucleosomes (Figure 13.12b). Some investigators, including Manfred Renz and Wolf Strätling, have invoked a **superbead** structure, rather than a solenoid, for this fiber. A superbead is an aggregate of several nucleosomes (an oligonucleosome), and these superbeads could be strung together in a relatively disordered way to generate the 30-nm fiber. This would explain the irregularity of the 30-nm fiber in many electron micrographs, but it fails to explain fully the prominent 110-Å spacing revealed by x-ray diffraction analysis of the 30-nm fiber.

Van Holde and Jordanka Zlatanova have argued that the 30-nm fiber has *no* regular structure. They find the evidence for a regular helix inconclusive. Indeed, it is only a minority of chromatin images in electron micrographs that show clear helical structure, and left-handed and right-handed helices can coexist in the same preparation. These workers envisioned an uncondensed, irregular, open helical chromatin, as illustrated in Figure 13.9a, condensing to form a structure with no significant regular helical content.

Because the distance between nucleosomes is irregular in natural chromatin, they found it difficult to imagine contraction of an irregular open helical structure into a regular condensed helix. Because van Holde and Zlatanova's models of the open helix and condensed structure both have diameters of about 30 nm, they prefer to call the latter the "condensed fiber," rather than the 30-nm fiber.

The Role of Histone H1 in Chromatin Folding

Early evidence suggested that histone H1 must play a critical role in forming the 30-nm fiber because no such fiber occurred in the absence of this histone. In fact, some evidence suggests that H1–H1 interactions are critical in forming this structure. In one such experiment, Jean Thomas and Roger Kornberg used a chemical cross-linking reagent to covalently link any two proteins that are close together in chromatin. They found that histone H1 is linked to other molecules of H1 much more frequently than to molecules of core histones. This suggests that molecules of H1 are indeed close together in chromatin, but other studies have revealed that this occurs even in the absence of the 30-nm fiber. Another argument against the importance of H1–H1 interactions in 30-nm fiber formation is that H1 can be largely (but not completely) destroyed with proteases and the fiber will still form.

Thus, it is possible, but not likely, that histone H1 plays no part in forming the 30-nm fiber. A compromise hypothesis seems to account for most of the data. It stipulates that the role of histone H1 is to space the nucleosomes properly on chromatin. This proper spacing then favors 30-nm fiber formation. Presumably, the chromatin digested with proteases leaves enough of each histone H1 molecule to carry out this nucleosome spacing function.

SUMMARY The nucleosome fiber is further folded into a 30-nm, or condensed, fiber. One candidate for the structure of this fiber is a solenoid. Another is a variable zigzag ribbon. Another possibility is that the nucleosomes form superbeads that are arranged in an irregular order. Still another is that an open, irregular helical fiber folds into an irregular condensed fiber with significant contents of neither helix nor superbead. Histone H1 appears to play a part in formation of the 30-nm fiber, but its exact role is unclear.

Higher Order Chromatin Folding

The 30-nm fiber probably accounts for most of the chromatin in a typical interphase nucleus, but further orders of folding are clearly needed, especially in mitotic

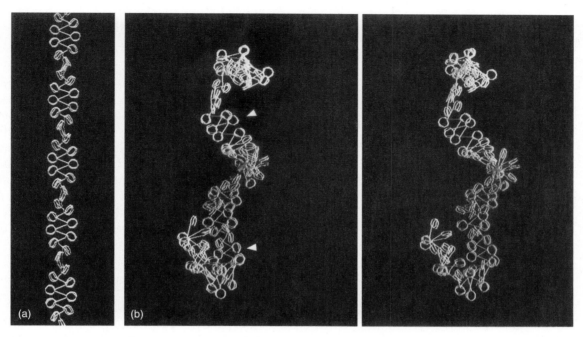

Figure 13.12 Zigzag models of Woodcock and colleagues. (a) Model based on constant linker length and angle at which the linker enters and exits the nucleosome. **(b)** Stereoscopic model based on variable linker length (±2 bp) and exit–entry angle (±15 degrees). The arrowheads denote short regions of constant length and exit–entry angle that resemble the model in part **(a)**. (*Source:* Horowitz et al., The three-dimensional architecture of chromatin *in situ*: Electron tomography reveals fibers composed of a continuously variable zig-zag nucleosomal ribbon. *J. Cell Biology* 125:1 (April 1994) p. 8, f. 10. © Rockefeller University Press.)

chromosomes, which have condensed so much that they become visible with a light microscope. The favorite model for the next order of condensation is a series of radial loops, as pictured in Figure 13.13. Cheeptip Benyajati and Abraham Worcel produced the first evidence in support of this model in 1976 when they subjected *Drosophila* chromatin to mild digestion with DNase I, then measured the sedimentation coefficients of the digested chromatin. They found that the coefficients decreased gradually with digestion, then reached a plateau value. Worcel had previously shown that the *E. coli* nucleoid (the DNA-containing complex) exhibited similar behavior, which was caused by the introduction of nicks into more and more superhelical loops of the bacterial DNA. As each loop was nicked once, it relaxed to an open circular form and slightly decreased the sedimentation coefficient of the whole complex. But eukaryotic chromosomes are linear, so how can the DNA in them be supercoiled? If the chromatin fiber is looped as it is in *E. coli* and held fast at the base of each loop, then each loop would be the functional equivalent of a circle and could be supercoiled. Indeed, the winding of DNA in the nucleosomes would provide the strain necessary for supercoiling. Figure 13.14 illustrates this concept and shows how relaxation of a supercoiled loop gives much less compact chromatin in that region, which would reduce the sedimentation coefficient.

How big are the loops? Worcel calculated that each loop in a *Drosophila* chromosome contains about 85 kb,

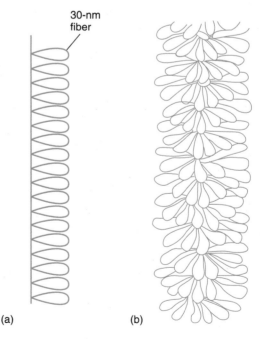

Figure 13.13 Radial loop models of chromatin folding. (a) This is only a partial model, showing some of the loops of chromatin attached to a central scaffold; of course, all the loops are part of the same continuous 30-nm fiber. **(b)** A more complete model, showing how the loops are arranged in three dimensions around the central scaffold. (*Source:* Reprinted from Marsden and Laemmli *Cell* 17:856, 1979. Copyright 1979, with permission from Elsevier Press.)

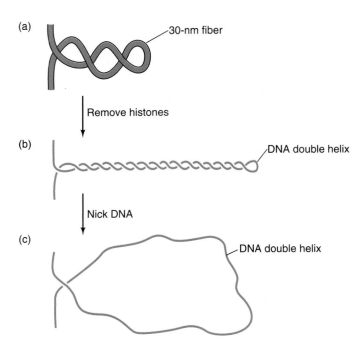

Figure 13.14 Relaxing supercoiling in chromatin loops. (a) A hypothetical chromatin loop composed of the 30-nm fiber, with some superhelical turns. (b) The chromatin loop with histones removed. Without histones, the nucleosomes and 30-nm fiber have disappeared, leaving a supercoiled DNA duplex. Note that the helical turns here are superhelices, not ordinary turns in a DNA double helix. (c) A relaxed chromatin loop. The DNA has been nicked to relax the superhelix. Now we see a relaxed DNA double helix that forms a loop. With each step from (a) to (c), the apparent length of the loop increases, but these increases are not drawn to scale.

but other investigators, working with vertebrate species and using a variety of techniques, have made estimates ranging from 35 to 83 kb.

The images of chromosomes in Figure 13.15 also support the loop idea. Figure 13.15a shows the edge of a human metaphase chromosome, with loops clearly visible. These loops are about 50 nm thick, so they may be twisted or folded versions of the 30-nm fiber. Figure 13.15b depicts a cross section of a swollen human chromosome in which the 50-nm folding or twisting is gone, but the 30-nm fiber is preserved. Radial loops are clearly visible. Figure 13.15c shows part of a deproteinized human chromosome. Loops of DNA are anchored to a central scaffold in the skeleton of the chromosome. All these pictures strongly support the notion of a radially looped fiber in chromosomes.

SUMMARY Sedimentation and EM studies have revealed a radial loop structure in eukaryotic chromosomes. The 30-nm fiber seems to undergo some kind of folding back or coiling to form a 50-nm fiber, which then forms loops between 35 and 85 kb long, anchored to the central matrix of the chromosome.

13.3 Chromatin Structure and Gene Activity

Enthusiasm for histones as important regulators of gene activity has been inconsistent. When it first became clear that histones could turn off transcription when added to DNA in vitro, molecular biologists got excited. Then, when the role of histones in chromatin structure was elucidated, most investigators tended to focus on this structural role and forget about histones as regulators of genetic activity. Histones were then viewed as mere scaffolding for the DNA. Now we have come full circle and are rediscovering the regulatory functions of histones.

The Effects of Histones on 5S rRNA Gene Transcription

As early as 1984, Donald Brown and his colleagues had shown that histones, particularly histone H1, have a repressive effect on gene activity in vitro. They focused their attention on the *Xenopus laevis* 5S rRNA genes, which are transcribed by RNA polymerase III, with the help of TFIIIA, TFIIIB, and TFIIIC.

The haploid number of *Xenopus* 5S rRNA genes is quite large, about 20,000, and these genes compose two different families. The first family includes approximately 98% of the genes; because their transcription occurs only in oocytes, they are called the **oocyte 5S rRNA genes.** The smaller family contains only about 400 genes, which are transcribed in both oocytes and somatic cells and are therefore called the **somatic 5S rRNA genes.** These observations raised a crucial question: What causes the oocyte genes to be active in oocytes but inactive in somatic cells?

A role for nucleosomes could be postulated as follows: The oocyte gene, or at least its promoter, is complexed with nucleosomes in somatic cells and is therefore repressed, but is relatively free of nucleosomes in oocytes and is therefore active. In fact, Brown's group found evidence to support this hypothesis.

First, Brown and colleagues transcribed purified DNA from oocyte and somatic 5S genes and found that RNA polymerase III plus the three transcription factors could transcribe both genes very well. Next, the investigators gently extracted chromatin from oocytes and from somatic cells and tried to transcribe both in vitro. Here they obtained a much different result: The oocyte 5S gene in oocyte chromatin was active in vitro, but the same gene in somatic cell chromatin was not. Something in the somatic cell chromatin was repressing the oocyte gene.

As we have just suggested, that "something" involves nucleosomes and, therefore, histones. Brown and coworkers found that the oocyte 5S gene unexpectedly remained active in some preparations of somatic cell chromatin. To find out the basis for this surprising activity, they looked

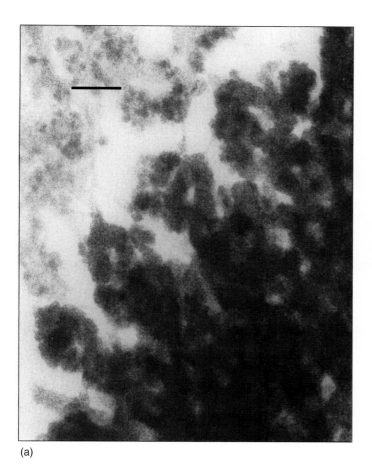

(a)

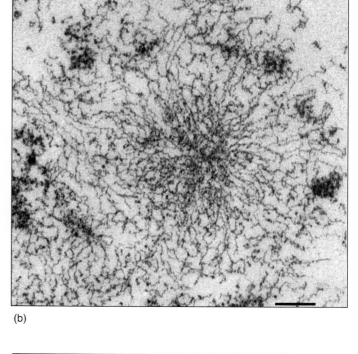

(b)

Figure 13.15   Three views of loops in human chromosomes.
(a) Scanning transmission electron micrograph of the edge of a human chromosome isolated with hexylene glycol. Bar represents 100 nm.
(b) Transmission electron micrograph of cross sections of human chromosomes swollen with EDTA. The chromatin fiber visible here is the 30-nm nucleosome fiber, not the 50-nm fiber apparent in panel (a). Bar represents 200 nm. (c) Transmission electron micrograph of a deproteinized human chromosome showing DNA loops emanating from a central scaffold. Bar represents 2 μm (2000 nm). *(Source:*
(a) Marsden and Laemmli, Metaphase chromosome structure: Evidence for a radial loop model. Cell 17 (Aug 1979) f. 5, p. 855. Reprinted by permission of Elsevier Science. (b) Marsden and Laemmli, Cell 17 (Aug 1979) f. 1, p. 851. Reprinted by permission of Elsevier Science. (c) Paulson and Laemmli, The structure of histone-depleted metaphase chromosomes. Cell 12 (1977) f. 5, p. 823. Reprinted by permission of Elsevier Science.)

(c)

at the pattern of histones in the active and inactive chromatin. The inactive chromatin contained all five histones, but the active chromatin was missing histone H1 (Figure 13.16a). Salt treatment or ion-exchange column chromatography also reactivated the oocyte 5S rRNA gene in somatic cell chromatin, and these treatments also selectively removed histone H1 (Figure 13.16b).

What if histone H1 is added back to somatic cell chromatin? Brown and colleagues did this, and Figure 13.17

shows the results. When the level of histone H1 reached one molecule per 200 bp of DNA, its natural level in chromatin, 5S rRNA synthesis fell dramatically. To find out whether the remaining 5S rRNA synthesis was oocyte or somatic, these workers performed an RNase protection assay as follows: They hybridized the labeled RNA products to a complementary DNA probe representing the 3′-part of the oocyte gene, then digested the hybrid with RNase. This probe protects a set of oocyte 5S rRNA

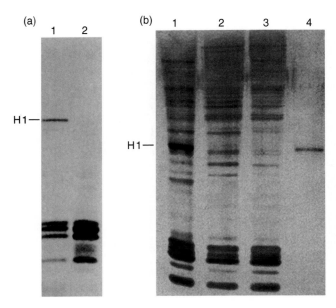

Figure 13.16 Histone content of active and inactive chromatin.
(a) Histones in "accidentally" derepressed chromatin. Brown and colleagues examined the histone contents of somatic cell chromatin by gel electrophoresis. Lane 1 contains histones from repressed chromatin, and lane 2 contains histones from chromatin that was accidentally derepressed during purification, perhaps because of partial proteolysis of histone. The latter chromatin lacks histone H1. (b) Histones in treated chromatin. Brown and colleagues electrophoresed histones from somatic cell chromatin treated as follows: lane 1, untreated; lane 2, washed with 0.6 M KCl; lane 3, subjected to ion-exchange chromatography. Lane 4 contains a histone H1 marker. Histone H1 is absent in both of the treated chromatins. (*Source:* Schliessel and Brown, The transcriptional regulation of *Xenopus* 5S RNA genes in chromatin: The roles of active stable transcription complexes and histone H1. *Cell* 37 (July 1984) f. 3b, 4. pp. 905–06. Reprinted by permission of Elsevier Science.)

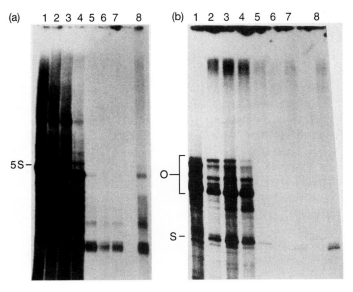

Figure 13.17 Effect of histone H1 on 5S rRNA gene transcription.
Brown and colleagues washed somatic cell chromatin with 0.6 M KCl to derepress it, then added back increasing concentrations of histone H1. (a) Transcription assay. The investigators used an in vitro assay to measure the rate of transcription of the 5S genes in chromatin reconstituted with the following ratios of histone H1 per 200 bp of DNA: lane 2, 0; lane 3, 0.5; lane 4, 0.75; lane 5, 1.0; lane 6, 1.25; lane 7, 1.5. Lane 1 contained RNA made from purified DNA instead of from chromatin. Lane 8 contained RNA made from unwashed chromatin. (b) RNase protection assay for oocyte versus somatic 5S rRNA synthesis. Brown and colleagues used this assay on RNAs from each of the lanes in panel (a). The signals corresponding to oocyte and somatic 5S rRNAs are marked O and S, respectively, at left. Adding one molecule of histone H1 per 200 bp of DNA leaves only somatic 5S rRNA synthesis (lane 5). (*Source:* Schliessel and Brown, *Cell* 37 (July 1984) f. 6, p. 907. Reprinted by permission of Elsevier Science.)

fragments, the largest of which is 89 nt long. Mismatches between the probe and somatic 5S rRNA result in a labeled RNA fragment only 64 nt long, which is easily distinguished from the 89-nt oocyte signal. Figure 13.17b demonstrates that the small amount of remaining 5S rRNA synthesized was all from somatic 5S rRNA genes.

An attractive interpretation of these results is that somatic cells contain transcription factors (TFIIIA, B, and C) that are able to form stable preinitiation complexes with the somatic 5S rRNA genes, but are less successful in forming these complexes with the oocyte 5S genes. Thus, nucleosomes form on the oocyte genes, including their promoters, and histone H1 cross-links these nucleosomes in an ordered array (perhaps in a solenoid) and keeps them repressed. By contrast, the transcription factors engaged on the somatic genes prevent nucleosomes from forming, or at least from forming an ordered, cross-linked structure. Therefore, these genes remain active. In other words, a kind of race occurs between factors and histones, as pictured in Figure 13.18. If the factors win, they keep the gene active, but if histones win, they tie up the gene's control region in cross-linked nucleosomes and repress the gene.

If this model is valid, then one should be able to remove histone H1 from somatic chromatin with salt, then add transcription factors from oocytes, and finally add back histone H1. Adding the factors first guarantees that they will win the race and the gene will be active. Brown and colleagues did this, and the oocyte gene was indeed activated, as Figure 13.19 shows.

SUMMARY The two families of 5S rRNA genes in the frog *Xenopus laevis* are oocyte genes and somatic genes. The former are expressed only in oocytes; the latter are turned on in both oocytes and somatic cells. The apparent reason for this difference is that the somatic genes form more stable complexes with transcription factors. The transcription factors seem to keep the somatic genes active by preventing nucleosomes from forming a stable complex with the internal control region. This stable complex requires the participation of histone H1. Once the complex forms, transcription factors are excluded and the gene is repressed.

(a) Transcription factors win:

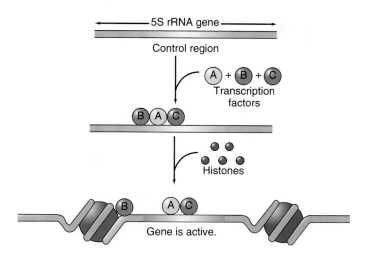

(b) Histones win:

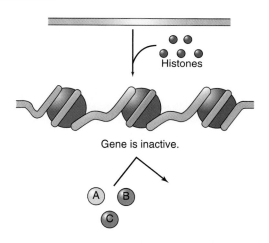

Figure 13.18 The race between transcription factors and histones for the 5S rRNA control region. (a) Transcription factors TFIIIA, B, and C (yellow, red, and green respectively) win the race and associate with the control region (purple). Histones may still form outside the control region, but they are not cross-linked. (b) Histones (blue) win the race, associate with the control region, and tie it up in a cross-linked nucleosome. The gene is unavailable to transcription factors and remains inactive.

The Effects of Histones on Transcription of Class II Genes

James Kadonaga and his colleagues showed that the same principles concerning the interactions between histones and class III genes also apply to histones and class II genes.

Core Histones In 1991, Paul Laybourne and Kadonaga performed a detailed study to distinguish between the effects of the core histones and of histone H1 on transcrip-

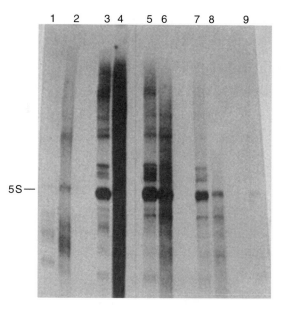

Figure 13.19 Transcription factors protect 5S rRNA genes from repression. Brown and colleagues removed histone H1 from somatic cell chromatin by incubating with 0.6 M KCl, then used gel exclusion chromatography to separate free histone H1 from the derepressed chromatin. Then they added back oocyte extracts containing transcription factors (lanes 1, 3, 5, and 7), or purified RNA polymerase III (lanes 2, 4, 6, and 8), then added back histone H1 (lanes 5–9). Finally, they assayed for 5S rRNA synthesis with the same transcription assay as in Figure 13.17. Lanes 1 and 2 contained untreated chromatin; lanes 3 and 4 contained no histone H1; lanes 5 and 6 contained histone H1 at 1.0 molecule per 200 bp of DNA; lanes 7 and 8 contained histone H1 at 1.5 molecule per 200 bp of DNA; lane 9 contained chromatin to which histone H1 at 1.0 molecule per 200 bp of DNA had been added *prior* to adding the cell extract. Clearly, adding cell extracts to the chromatin before adding histone H1 keeps the 5S rRNA genes active. Adding RNA polymerase III to the chromatin before histone HI has a similar, but less pronounced effect. (*Source:* Schliessel and Brown, *Cell* 37 (July 1984) f. 9, p. 909. Reprinted by permission of Elsevier Science.)

tion by RNA polymerase II in vitro. They found that the core histones (H2A, H2B, H3, and H4) formed core nucleosomes with cloned DNA and caused a mild repression (about fourfold) of genetic activity. Transcription factors had no effect on this repression. When they added histone H1, in addition to the core histones, the repression became much more profound: 25- to 100-fold. This repression *could* be blocked by activators. In this respect, these factors resembled the factors (presumably TFIIIA, B, and C), which could compete with histone H1 for the control region of the *Xenopus* 5S rRNA gene.

Laybourne and Kadonaga's experimental strategy was to reconstitute chromatin from plasmid DNA containing a well-defined cloned gene, and histones in the presence or absence of activators that were known to affect transcription of the cloned gene in question. They also added topoisomerase I to keep the DNA relaxed. Then they used a primer extension assay to test whether the reconstituted chromatin could be transcribed by a nuclear extract. In the first studies, these workers used only the core

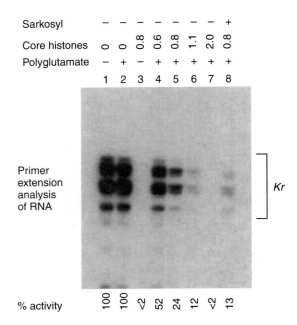

Sarkosyl	–	–	–	–	–	–	–	+
Core histones	0	0	0.8	0.6	0.8	1.1	2.0	0.8
Polyglutamate	–	+	–	+	+	+	+	+
	1	2	3	4	5	6	7	8

Primer extension analysis of RNA — *Kr*

| % activity | 100 | 100 | <2 | 52 | 24 | 12 | <2 | 13 |

Figure 13.20   In vitro transcription of reconstituted chromatin. Laybourne and Kadonaga reconstituted chromatin with plasmid DNA containing the *Drosophila Krüppel* gene and core histones in varying ratios of protein to DNA, as indicated at top. Then they performed primer extension analysis to measure efficiency of transcription. Diverse signals corresponding to *Krüppel* gene transcription are indicated by the bracket at right. Lane 1, naked DNA; lane 2, naked DNA plus polyglutamate (used as a vehicle to help histones deposit onto DNA); lanes 3–7, chromatin at various core histone–DNA ratios; lane 8, sarkosyl was included to prevent reinitiation, so only one round of transcription occurred. Core histones can inhibit transcription of the *Krüppel* gene in a dose-dependent manner. (*Source:* Laybourn and Kadonaga, Role of nucleosomal cores and histone H1 in regulation of transcription by RNA polymerase II. *Science* 254 (11 Oct 1991) f. 2B, p. 239. © AAAS.)

histones, not histone H1. They added a mass ratio of histones to DNA of 0.8 to 1.0, which is enough to form an average of one nucleosome per 200 bp of DNA.

Using such reconstituted chromatin that contained the *Drosophila Krüppel* gene, Laybourne and Kadonaga showed that a *Drosophila* nuclear extract could transcribe the *Krüppel* gene (Figure 13.20). However, core histones in quantities that produced nucleosomes at a density of one nucleosome per 200 bp, which is the physiological density, caused partial repression of transcription (down to 25% of the control value; compare lanes 2 and 5). Notice that the transcription start sites as detected by this method are quite heterogeneous in this gene, so we see a cluster of primer extension products.

The authors pointed to two possible explanations for the 75% repression observed with the core histones. First, the nucleosomes could slow the progress of all RNA polymerases by about 75%, but not stop any of them. Second, 75% of the polymerases could be blocked entirely by nucleosomes, but 25% of the promoters might have been left free of nucleosomes and thus could remain available

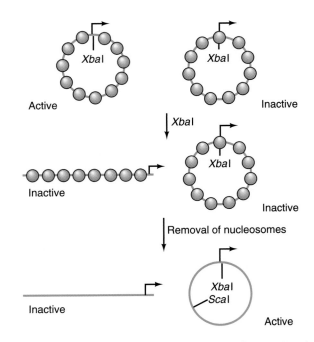

Figure 13.21   Scheme for distinguishing engaged promoters from those not engaged with nucleosomes. Because the *Xba*I site of the plasmid lies very close to the transcription start site of the *Krüppel* gene, it will be available for cutting if the gene is active, but not available if the gene is inactive and the start site is covered by a nucleosome. Thus, *Xba*I will cut the active gene and render it inactive. Now all templates are inactive. If we remove the nucleosomes, the previously inactive templates are rendered active, but the linearized templates remain inactive. (*Source:* From Laybourn and Kadonaga, *Science* 254:240, 1991. Copyright © 1991 American Association for the Advancement of Science, Washington, DC. Reprinted by permission.)

to RNA polymerase. How can one distinguish between these two possibilities? Laybourne and Kadonaga suggested a clever method, outlined in Figure 13.21. If a promoter is not blocked with nucleosomes and is available to RNA polymerase, it should also be available to a restriction enzyme. In particular, because the *Krüppel* gene contains an *Xba*I restriction site just downstream of the transcription start site, this enzyme should cut any such genes not protected by nucleosomes. This would render these promoters inactive in subsequent transcription assays, because a gene would no longer be present to transcribe. Therefore, if hypothesis number 2 is correct, the 25% of the promoters that are not tied up in nucleosomes should be digested with *Xba*I, and therefore inactivated. This is exactly what happened. Lanes 9 and 10 of Figure 13.22 demonstrate that cutting with *Xba*I completely inactivated the remaining active chromatin, which suggests that this active chromatin did not have nucleosomes over the promoter region of the *Krüppel* gene. Furthermore, the activator Sp1 could not restore the lost activity. Lanes 11 and 12 show the same behavior for dialyzed chromatin, and lanes 7 and 8 show that naked DNA was completely inactivated by the restriction enzyme, as expected. Another restriction enzyme, *Sca*I, which cuts 1700 bp upstream of the transcription start site, had no effect, as

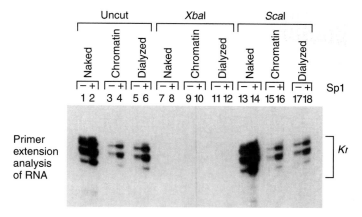

Figure 13.22   Assay for state of chromatin at the *Krüppel* promoter. Laybourne and Kadonaga reconstituted chromatin using core histones and plasmid DNA containing the *Krüppel* gene. Then they cleaved this chromatin with *Xba*I as in Figure 13.21. Then they assayed for *Krüppel* transcription by primer extension. The range of *Krüppel* transcripts is indicated by the bracket at right. The state of the chromatin assayed in each lane is indicated at top. Lanes 1–6 show the results of transcribing uncut templates of three types: naked DNA, chromatin, or dialyzed chromatin, as indicated. Lanes 7–12 and 13–18 show the results of the same assay on the same three templates cut with *Xba*I, or *Sca*I, respectively. All even lanes represent assays done in the presence of Sp1; all odd lanes represent assays done in the absence of Sp1. *Xba*I, which cuts near the promoter, destroyed all remaining *Krüppel* transcription. Thus, the promoter activity remaining in *Krüppel* genes complexed with core histones was due to promoters not blocked by nucleosomes. (*Source:* Laybourn and Kadonaga, *Science* 254 (11 Oct 1991) f. 4, p. 241. © AAAS.)

illustrated in lanes 13–18. Thus, simply cutting the DNA was not what caused the damage. Rather, cutting just after the transcription start site was the destructive event. Another conclusion we can draw from Figure 13.22 is that the activator Sp1 could stimulate transcription of chromatin containing nucleosome cores by a factor of 2.5 to 3, the same amount of transcription stimulation obtained with naked DNA. Thus, Sp1 cannot reverse the repression caused by these nucleosome cores after they have assembled.

SUMMARY The core histones (H2A, H2B, H3, and H4) assemble nucleosome cores on naked DNA. Transcription of reconstituted chromatin with an average of one nucleosome core per 200 bp of DNA exhibits about 75% repression relative to naked DNA. The remaining 25% is due to promoter sites not covered by nucleosome cores. Activators such as Sp1 cannot counteract the repression caused by nucleosome core formation.

Histone H1   Based on its postulated role as a nucleosome cross-linker, we would expect that histone H1 would add to the inhibition of transcription caused by the core histones in reconstituted chromatin. This is indeed

the case, as Laybourne and Kadonaga demonstrated. They reconstituted chromatin with DNA containing two enhancer–promoter constructs: (1) pG$_5$E4 (five GAL4-binding sites coupled to the adenovirus E4 minimal promoter); and (2) pSV-Kr (six GC boxes from the SV40 early promoter coupled to the *Drosophila Krüppel* minimal promoter). In this experiment, they added not only the core histones, but histone H1 in various quantities, from 0 to 1.5 molecules per core nucleosome. Then they transcribed the reconstituted chromatin in vitro.

The odd lanes in Figure 13.23 show that increasing amounts of histone H1 caused a progressive loss of template activity, until transcription was barely detectable. However, at moderate histone H1 levels (0.5 molecules per core histone), activators could prevent much of the repression. For example, on chromatin reconstituted from the pG$_5$E4 plasmid, the hybrid activator GAL4-VP16, which interacts with GAL4-binding sites, caused a 200-fold greater template activity. Part of this (eightfold) is due to the stimulatory activity of the activator, observed even on naked DNA. The remaining 25-fold stimulation is apparently due to **antirepression,** the prevention of repression by histone H1. Similarly, when the reconstituted chromatin contained the pSV-Kr promoter, the activator Sp1, which binds to the GC boxes in the promoter, caused a 92-fold increase in template activity. Because true activation by Sp1 on naked DNA was only 2.8-fold, 33-fold of the 92-fold stimulation was antirepression. The true activation component is what we studied in Chapter 12, in which the experimenters used naked DNAs as the templates in their transcription assays.

These data are consistent with the model in Figure 13.24. Histone H1 can cause repression in the cases studied here by binding to the linker DNA between nucleosomes that happens to contain a transcription start site. The activators Sp1 and GAL4-VP16 (represented by the green oval) can prevent this effect if added at the same time as histone H1. But these factors cannot reverse the effects of preformed nucleosome cores, even without histone H1. Some activators, represented by the purple rectangle, are more powerful than this and seem to be able to shoulder nucleosomes aside, at least if the nucleosomes are not cross-linked by histone H1. For example, the glucocorticoid hormone–receptor complex seems to be able to bind to a site occupied by a nucleosome.

Kadonaga and colleagues have also studied another protein, called **GAGA factor,** which binds to several GA-rich sequences in the *Krüppel* promoter and to other *Drosophila* promoters. It has no transcription-stimulating activity of its own; in fact it slightly inhibits transcription. But GAGA factor prevents repression by histone H1 when added to DNA before the histone and can therefore cause a significant net increase in transcription rate. Thus, the GAGA factor seems to be a pure antirepressor, unlike the more typical activators we have been studying, which have both antirepression and transcription stimulation activities.

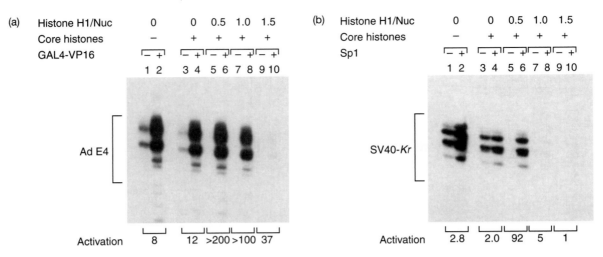

Figure 13.23 Competing effects of histones and activators on transcription. Laybourne and Kadonaga reconstituted chromatin in the presence and absence of core histones and histone H1 as indicated at top. Then they assayed for transcription by primer extension in the presence or absence of an activator as indicated. Apparent degrees of activation by each activator are given below each pair of lanes. The true activation by each activator is seen in lanes 1 and 2 of each panel, where naked DNA was the template. Any higher levels of apparent activation in the other lanes, where chromatin served as the template, were due to antirepression. (**a**) Effect of GAL4–VP16. Chromatin contained the adenovirus E4 promoter with five GAL4 binding sites. The signals corresponding to E4 transcription are indicated by the bracket at left. (**b**) Effect of Sp1. Chromatin contained the *Krüppel* promoter plus the SV40 promoter, which is responsive to Sp1. The signals corresponding to *Krüppel* transcription are indicated at left. (*Source:* Laybourn and Kadonaga, *Science* 254 (11 Oct 1991) f. 7, p. 243. © AAAS.)

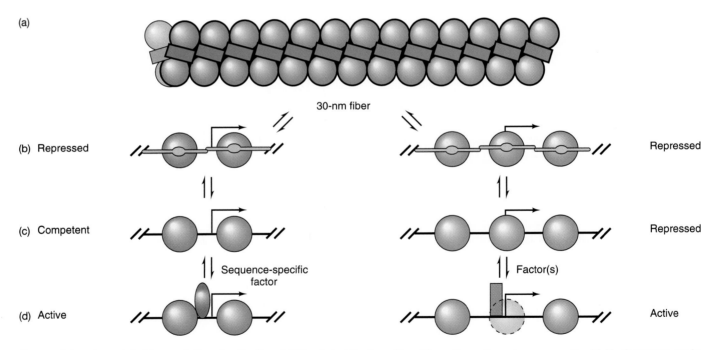

Figure 13.24 A model of transcriptional activation. (**a**) We start at the top with a 30-nm fiber, shown here as a solenoid. Nucleosomes are in blue. (**b**) The 30-nm fiber can open up to give two kinds of repressed chromatin. On the right, a cross-linked nucleosome (blue) covers the promoter, keeping it repressed. On the left, no nucleosomes cover the promoter, but histone H1 (yellow) does, so the gene is still repressed. (**c**) When we remove histone H1, we can get two chromatin states: On the left the promoter is uncovered, so the gene is competent to be transcribed. On the right, a nucleosome still covers the promoter, so it remains repressed. (**d**) Antirepression. Most activators, for example GAL4 and Sp1 (green), act as shown at left, by binding to the chromatin and preventing nucleosomes from binding to the promoter. Others, the glucocorticoid hormone–receptor complex (purple), for example, seem to play a more active role by binding to repressed chromatin, removing the repressing nucleosomes from the promoter, and thus turning it on. (*Source:* From Laybourn and Kadonaga, *Science* 254:243, 1991. Copyright © 1991 American Association for the Advancement of Science, Washington, DC. Reprinted by permission.)

SUMMARY Histone H1 causes a further repression of template activity, in addition to that produced by core nucleosomes. This repression can be counteracted by transcription factors. Some, like Sp1 and GAL4, act as both antirepressors (preventing repression by histone H1) and as transcription activators. Others, like GAGA factor, are just antirepressors. The antirepressors presumably compete with histone H1 for binding sites on the DNA template.

Nucleosome Positioning

The model of activation and antirepression in Figure 13.24 asserts that transcription factors can cause antirepression by removing nucleosomes that obscure a promoter or by preventing their binding to the promoter in the first place. Both these scenarios embody the idea of **nucleosome positioning**, in which activators force the nucleosomes to take up positions around, but not within, the promoter.

Nucleosome-Free Zones Several lines of evidence demonstrate nucleosome-free zones in the control regions of active genes. M. Yaniv and colleagues performed a particularly graphic experiment on the control region of SV40 virus DNA. SV40 DNA in an infected mammalian cell exists as a minichromosome, as described earlier in this chapter. Yaniv noticed that some actively transcribed SV40 minichromosomes have a conspicuous nucleosome-free zone late in infection (Figure 13.25). We would expect this nucleosome-free region to include at least one late promoter. In fact, the SV40 early and late promoters lie very close to each other, with the 72-bp repeat

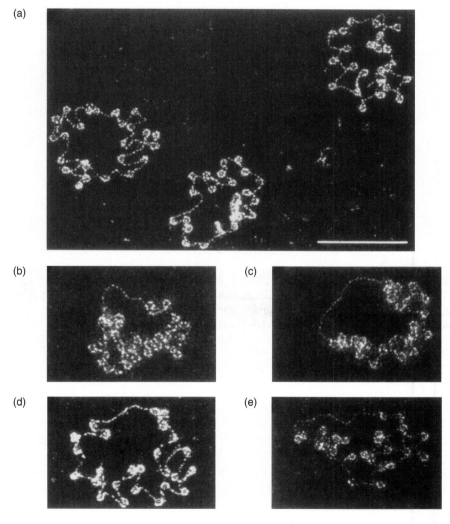

(a)

(b)　　(c)

(d)　　(e)

Figure 13.25 Nucleosome-free zones in SV40 minichromosomes. (a) Three examples of minichromosomes with no extensive nucleosome-free zones. **(b–e)** Four examples of SV40 minichromosomes with easily detectable nucleosome-free regions. The bar represents 100 nm. (*Source:* Saragosti et al., Absence of nucleosomes in a fraction of SV40 chromatin between the origin of replication and the region coding for the late leader RNA. *Cell* 20 (May 1980) f. 2, p. 67. Reprinted by permission of Elsevier Science.)

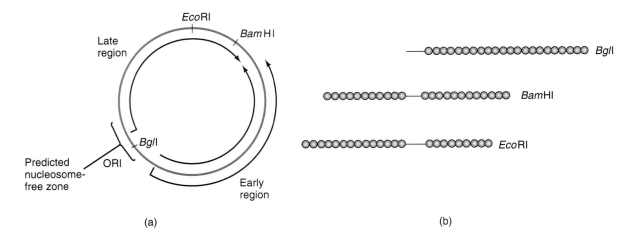

(a)   (b)

Figure 13.26   Experimental scheme to locate the nucleosome-free zone in the SV40 minichromosome. (a) Map of SV40 genome showing the cutting sites for three restriction enzymes *Bgl*I, *Bam*HI, and *Eco*RI. The control region surrounds the origin of replication (ORI), with the late control region on the clockwise side. **(b)** Expected results of cleavage of minichromosome from late infected cells with three restriction enzymes, assuming that the late control region is nucleosome-free. All three enzymes should cut once to linearize the minichromosome. *Bgl*I is predicted to cut at one end of the nucleosome-free zone and should therefore produce a minichromosome with a nucleosome-free zone at one end. *Bam*HI is predicted to cut at a site diametrically opposed to the nucleosome-free zone and should therefore produce a minichromosome with the zone in the middle. In the same way, *Eco*RI should yield a minichromosome with the zone somewhat asymmetrically located.

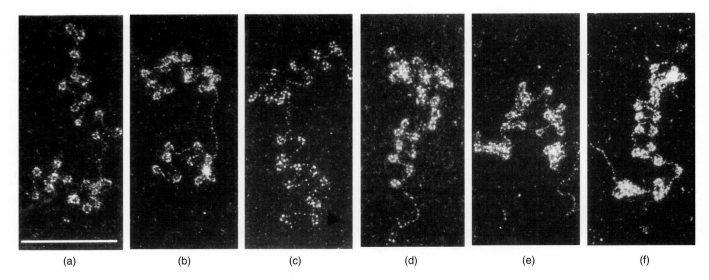

(a)   (b)   (c)   (d)   (e)   (f)

Figure 13.27   Locating the nucleosome-free zone on the SV40 minichromosome. Yaniv and colleagues cut SV40 minichromosomes from late infected cells with either *Bam*HI (panels **a–c**) or *Bgl*I (panels **d–f**). Just as predicted in Figure 13.26, *Bam*HI produced a centrally located nucleosome-free zone, and *Bgl*I yielded a nucleosome-free zone at the end of the minichromosome. The bar represents 100 nm. (*Source:* Saragosti et al., *Cell* 20 (May 1980) f. 4, p. 69. Reprinted by permission of Elsevier Science.)

enhancer in between. Is this the nucleosome-free zone? The problem with a circular chromosome is that it has no beginning and no end, so we cannot tell what part of the circle we are looking at without a marker of some kind. Yaniv and colleagues used restriction sites as markers. A *Bgl*I restriction site occurs close to one end of the control region, and *Bam*HI and *Eco*RI sites occur around on the other side of the circle, as illustrated in Figure 13.26a. Therefore, if the nucleosome-free region includes the control region, *Bgl*I will cut within that zone, and the other two restriction enzymes will cut at remote sites, as illustrated in Figure 13.26b. Fig-

ure 13.27 shows that cutting with *Bam*HI or *Bgl*I produced exactly the expected results. Cutting with *Eco*RI (not shown) also fulfilled the prediction.

We can even tell that *Bgl*I cut asymmetrically within the nucleosome-free region, because it left a long nucleosome-free tail at one end of the linearized minichromosome, but not at the other. This is what we would expect if the nucleosome-free zone corresponds to one of the SV40 promoters, which are asymmetrically arranged relative to the *Bgl*I site. On the other hand, it is not what we would expect if the nucleosome-free zone corresponds to the viral origin of replication, which almost coincides with the *BgI*I site.

DNase Hypersensitivity   Another sign of a nucleosome-free DNA region is hypersensitivity to DNase. Chromatin regions that are actively transcribed are **DNase-sensitive** (~10-fold more sensitive than bulk chromatin). But the control regions of active genes are **DNase-hypersensitive** (~100-fold more sensitive than bulk chromatin). For example, the control region of SV40 DNA is DNase-hypersensitive, as we would expect. Yaniv demonstrated this by isolating chromatin from SV40 virus-infected monkey cells, mildly digesting this chromatin with DNase I, then purifying the SV40 DNA, cutting it with *Eco*RI, electrophoresing the fragments, Southern blotting, and probing the blot with radioactive SV40 DNA. Figure 13.26a shows that the *Eco*RI and *BgI*I sites lie 67% (and 33%) apart on the circle. Therefore, if the nucleosome-free region near the *BgI*I site is really DNase-hypersensitive, then DNase will cut there and *Eco*RI will cut at its unique site, yielding two fragments containing about 67% and 33% of the total SV40 genome. In fact, as Figure 13.28 demonstrates, experiments carried out 24 h, 34 h, and 44 h after virus infection all produced a large amount of the 67% product, and lesser amounts of the 33% product and shorter fragments. This suggests that DNase I is really cutting the chromatin in a relatively small region around the *BgI*I site. Thus, the nucleosome-free region and the DNase-hypersensitive region coincide.

DNase hypersensitivity of the control regions of active genes is a general phenomenon. For example, the 5′-flanking region of the β-globin gene in red blood cells is DNase-hypersensitive. In fact, the DNase hypersensitivity of the globin genes gives a good indication of the activity of those genes at any given time. Harold Weintraub and colleagues examined the DNase hypersensitivity of the embryonic and adult β-globin genes in embryonic and adult chick red blood cells, and found that the embryonic gene is active—and DNase-hypersensitive—in embryonic cells, whereas the adult gene is inactive and not DNase-hypersensitive. In adult cells, the reverse is true.

Figure 13.29 illustrates the principle involved in detecting a DNase-hypersensitive gene by Southern blotting. We see at the top of panels a and b the arrangement of nucleosomes on an active and an inactive gene, and the positions of three recognition sites for a restriction endonuclease (RE). If DNase I is used to lightly digest nuclei containing the inactive gene, nothing happens because no DNase-hypersensitive sites are present. On the other hand, if the same thing is done to nuclei containing the active gene, the DNase will attack the hypersensitive site near the promoter. Now the protein is removed from both DNAs, which are then cut with the RE. All three RE sites remain in the inactive DNA, so it yields two fragments, 6 and 4 kb long, respectively, just as untreated DNA would.

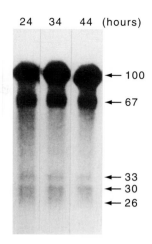

24   34   44   (hours)

← 100

← 67

← 33
← 30
← 26

Figure 13.28   Locating the region of DNase hypersensitivity in the SV40 minichromosome. Yaniv and colleagues isolated nuclei from SV40 virus-infected monkey cells at 24, 34, and 44 h after infection and treated them with DNase I. Then they cleaved the treated minichromosomes with *Eco*RI and analyzed the DNA products by electrophoresis, Southern blotting, and probing with radioactive SV40 DNA. Because *Eco*RI cuts 33% of the way clockwise around the circle from the nucleosome-free zone, we would expect to see two fragments, corresponding to 33% and 67% of the whole length of the SV40 genome, assuming that the nucleosome-free zone and the DNase-hypersensitive region coincide. Actually, the 67% fragment is very prevalent, but the 33% fragment is partially degraded into smaller fragments. Thus, the DNase hypersensitive region does correspond to the nucleosome-free zone, which is large enough to produce a range of degradation products. (*Source:* Saragosti et al., *Cell* 20 (May 1980) f. 7, p. 71. Reprinted by permission of Elsevier Science.)

But one RE site has been destroyed by DNase in the active DNA, so cleaving it with RE produces smaller products. Sometimes the irregular cutting by DNase I yields products with "ragged" ends, so no one product stands out. In such a case, the normal bands seem to disappear altogether, not just shift to smaller size. We observed an indication of this phenomenon in Figure 13.28 in which the recovery of one of the products (the 67% product) was much greater than that of the other (the 33% product) when SV40 minichromosomes were analyzed for DNase hypersensitivity. Also note that this assay would work just as well if no RE site was present in the DNase-hypersensitive region. In that case, the inactive gene would yield a 10-kb product, and the active gene would yield two smaller products (5 and 3 kb in this example).

Figure 13.30 shows the results of just such an experiment performed by Weintraub to probe the DNase hypersensitivity of the α-globin and ovalbumin genes in chick red blood cells and in cultured chick lymphoid cells (MSB cells) that do not express the globin genes. DNase completely removed the bands corresponding to the two chick α-globin genes in red blood cells, but had no effect on the bands in MSB cells. For comparison, DNase had no effect on the ovalbumin gene in either red blood cells or MSB cells. Weintraub speculated that the extra ovalbumin

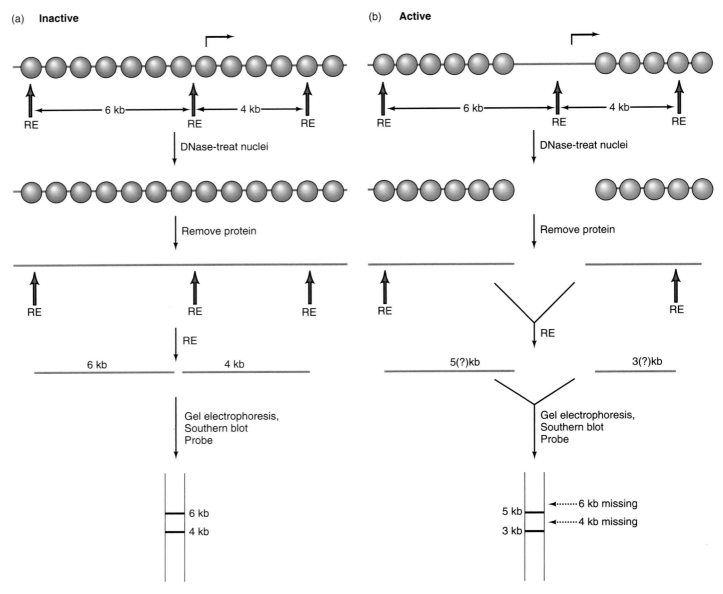

Figure 13.29 Experimental scheme to detect DNase-hypersensitive regions. (a) Inactive gene, no DNase hypersensitivity. The gene and its promoter are complexed with nucleosomes. Therefore, no DNA will be degraded when nuclei containing this gene are subjected to mild treatment with DNase I. Next, isolate the DNA from these nuclei, removing all the protein and digest with a restriction endonuclease (RE). This creates two fragments, 6 kb and 4 kb, in the neighborhood of the promoter. These fragments can be detected by electrophoresis, Southern blotting, and hybridizing with a labeled probe for the gene in question. (b) Active gene, DNase hypersensitivity. An active gene has a nucleosome-free zone around its promoter. Thus, when nuclei containing this active gene are subjected to mild DNase I treatment, that zone will be digested. Next, isolate the DNA from these nuclei, removing all the protein, and digest with an RE. A considerable amount of DNA has been removed by the DNase, including an RE site, so the fragments created by RE are different from those observed with the inactive gene (or with naked DNA). In this hypothetical example, they are 5 kb and 3 kb, but in practice one may not be able to see them at all because variability in the sites of cleavage by DNase can give a variety of products that appear as a smear, if at all.

band in MSB cells was due to a gene polymorphism in those cells; that is, it was an allele of the ovalbumin gene with a different pattern of *Bam*HI sites from that seen in the animal that gave the red blood cells. It seems not to be due to degradation of the upper ovalbumin band because that band was not decreased in intensity.

SUMMARY Active genes tend to have DNase-hypersensitive control regions. At least part of this hypersensitivity is due to the absence of nucleosomes.

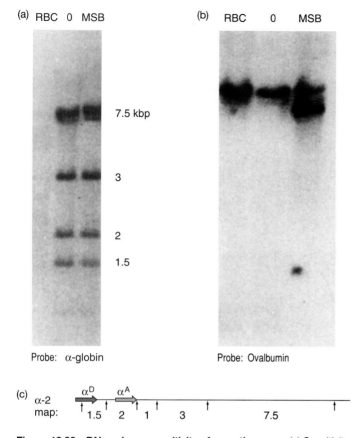

(a) RBC 0 MSB

7.5 kbp

3

2

1.5

Probe: α-globin

(b) RBC 0 MSB

Probe: Ovalbumin

(c)
α-2
map:

αD   αA

1.5   2   1   3   7.5

Figure 13.30 DNase hypersensitivity of an active gene. (a) Sensitivity of α-globin genes. Weintraub and colleagues treated nuclei from either 14-day chick red blood cells (RBC) or cultured MSB cells with DNase, then isolated the DNA, cleaved it with *Bam*HI, electrophoresed, Southern blotted the fragments, and probed the blot with a DNA from the α-2 globin region, whose map is shown in panel **(c)**. The middle lane, marked 0, was DNA purified from untreated nuclei, then cleaved with *Bam*HI. The *Bam*HI restriction map of the two α-2 genes (αD and αA), panel **(c)**, reveals that fragments of 1.5, 2, 1, 3, and 7.5 kb should be generated by *Bam*HI cleavage of this DNA. All but the 1-kb band are clearly visible in the blot. All the bands in this region are sensitive to DNase in red blood cells, but not in MSB (lymphoid) cells. **(b)** Sensitivity of the ovalbumin gene. Weintraub and colleagues repeated the same experiment as in panel **(a)**, but hybridized the Southern blot to an ovalbumin probe instead of an α-globin probe. The ovalbumin gene is not DNase-sensitive in either red blood cells or lymphoid cells. **(c)** *Bam*HI restriction map of the α-2 region. The two α-globin genes are represented by the colored open arrows at left. The *Bam*HI restriction sites are located by the vertical arrows below. The sizes of the *Bam*HI fragments are indicated. (*Source:* Stalder et al., Tissue-specific DNA cleavages in the globin chromatin domain introduced by DNase I. Cell 20 (June 1980) f. 2, p. 453. Reprinted by permission of Elsevier Science.)

Detecting Positioned Nucleosomes Kenneth Zaret and colleagues performed finer analysis of the state of the chromatin in active and inactive genes and found a precisely positioned array of nucleosomes around the enhancer of an active gene, but not around the enhancer of the same gene when it is inactive. They studied the rat serum albumin gene, which is active in liver, but not in other tissues. First, they treated nuclei from various rat

tissues with **microccocal nuclease, (MNase)** which cuts the linker DNA between nucleosomes. Then they isolated the DNA, cut it with *Eco*RI, which generates a 2-kb restriction fragment containing the albumin enhancer. Finally, they electrophoresed the DNA fragments, Southern blotted them, and hybridized the blot to a probe for DNA at the left end of the 2-kb *Eco*RI fragment, as illustrated in Figure 13.31c.

Figure 13.31a depicts an electrophoresis gel stained with ethidium bromide to show the major DNA bands. High MNase concentrations yielded a ladder with a spacing of about 200 bp. This is what we expect of a nuclease that cuts between nucleosomes, which contain 200 bp of DNA. Of course, the digestion was incomplete, or most of the DNA would have been in a single 200 bp fragment. Figure 13.31b shows the Southern blot. Lanes 5–10 demonstrate that treatment of liver nuclei with high concentrations of MNase created a ladder of bands in the albumin enhancer region. This suggests that nucleosomes are precisely positioned in the enhancer region. If they were randomly positioned, the cutting sites for MNase would also have been random, and we would see nothing but perhaps a faint smear. In fact, lanes 12 and 13 show that such a smear resulted when Zaret and colleagues digested spleen nuclei with increasing concentrations of MNase. Thus, the precise positioning of nucleosomes in the enhancer is tissue-specific. It occurs only in the tissue (liver) where the gene is active, not in spleen (and other tissues) where it is inactive.

The sizes of the bands in the ladder in panel b give us an idea of the locations of the nuclease cutting sites and, therefore, the locations of the positioned nucleosomes in the enhancer. Further experiments refined these locations to the single-base-pair level. Panel c shows these cutting sites, as well as the inferred positions of the nucleosomes on a map of the albumin enhancer region. The positioned nucleosomes are numbered N3, N2, and N1, upstream to downstream. This panel also shows the binding sites within the enhancer for several activators (green boxes).

Panel b, lane 15, also shows what happened when Zaret and coworkers treated liver nuclei with DNase I instead of MNase. This located the DNase-hypersensitive site right over the strongest nucleosome signal (N1). This seems paradoxical because DNase-hypersensitive sites are assumed to be nucleosome-free. In this case, a nucleosome seems to be present, but it is presumably perturbed enough to allow access to DNase I (but not MNase). Alternatively, there could be no nucleosome at that site, but transcription factors bound there could protect an unusually long, 200-bp region from MNase but not from DNase I.

Zaret's group also used in vivo footprinting to locate the in vivo binding sites for some of the activators that bind to this promoter. They found that three of the enhancer sites are occupied by their respective activators:

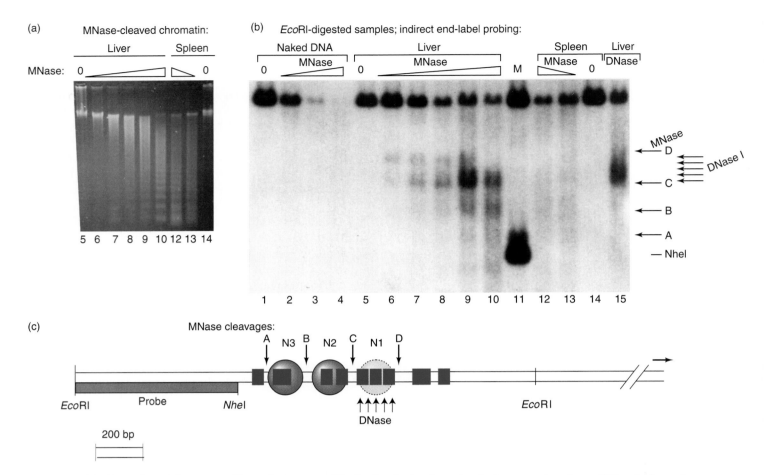

Figure 13.31   Detection of tissue-specific nucleosome positioning by electrophoresis of micrococcal nuclease (MNase)-cleaved chromatin. (a) Zaret and colleagues cleaved liver and spleen chromatin with MNase, electrophoresed the fragments, then stained them with ethidium bromide. We can see the appearance of a ladder with 200-bp spacing between bands in both liver and spleen samples at high MNase concentration, indicating that the nuclease is cutting between nucleosomes in chromatin from both tissues. **(b)** Southern blot to show positioned nucleosomes at the serum albumin enhancer. Zaret and coworkers treated chromatin from liver and spleen, as well as naked DNA, in the same way as in panel **(a)**, except that they Southern blotted the fragments after electrophoresis and then hybridized the blot to the probe illustrated in panel **(c)**. This probe hybidizes upstream of the enhancer, so if nucleosomes are positioned in the locations indicated by N3, N2, and N1 (see panel [**c**]), and if MNase digestion is incomplete, discreet bands of increasing size should be visible in the blot. In fact, such bands are apparent in the blot of liver chromatin (lanes 6–10), but not the blot of spleen chromatin (lanes 12 and 13) or of naked DNA (lanes 2–4). Zaret also cleaved liver chromatin with DNase I before purifying the DNA and cutting with *Eco*RI. This located a DNase-hypersensitive site that corresponded to the position of the N1 nucleosome (lane 15). Lanes 1, 5, and 14 show the results of using no nuclease. Lane 11 shows DNA cut with *Nhe*I as well as with *Eco*RI. The fragment predicted on the left side of panel **(c)** is seen in this lane and marked at right. Positions of bands resulting from cleavage at each of the MNase sites (A–D, see panel [**c**]) and at the cluster of DNase I sites, are shown at right. **(c)** Map of the enhancer region showing location of positioned nucleosomes (N1–N3, blue). Because N1 coincides with a DNase-hypersensitive region, this nucleosome is probably perturbed to allow access to the DNase. Therefore, this nucleosome is shown dotted and colored light blue. Vertical arrows show the MNase cleavage sites (A–D) between positioned nucleosomes. The green boxes show the locations of binding of known activators. (*Source:* McPherson et al., An active tissue-specific enhancer and bound transcription factors existing in a precisely positioned nucleosomal array. *Cell* 75 (22 Oct 1993) f. 1, p. 388. Reprinted by permission of Elsevier Science.)

C/EBP; hepatocyte nuclear factor 3 (HNF3); and HNF3 and NF1 (a ubiquitous nuclear factor). Remarkably, these proteins all bind to areas (N2 and N3) that seem to be occupied by positioned nucleosomes. Presumably, they bind to the DNA on the outsides of the nucleosomes. It is also highly likely that at least some of these activators help position the nucleosomes. The tissue specificity of some of the activators would explain the tissue specificity of the nucleosome positioning.

SUMMARY Some active genes have an array of precisely positioned nucleosomes in their control regions. The same control regions, in tissues where these genes are not active, do not have positioned nucleosomes. The positioned nucleosomes can overlap the binding sites for activators. Thus, activators and nucleosomes can coexist in the same place. In fact, the activators almost certainly have a role in positioning the nucleosomes.

Histone Acetylation

Vincent Allfrey discovered in 1964 that histones are found in both acetylated and unacetylated forms. Acetylation occurs on the amino groups on lysine side chains. Allfrey also showed that acetylation of histones correlates with gene activity. That is, unacetylated histones, added to DNA, tend to repress transcription, but acetylated histones are weaker repressors of transcription. These findings implied that enzymes in nuclei acetylate and deacetylate histones and thereby influence gene activity. To investigate this hypothesis, one needs to identify these enzymes, yet they remained elusive for over 30 years, in part because they are present in low quantities in cells.

Finally, in 1996, James Brownell and David Allis succeeded in identifying and purifying a **histone acetyltransferase** (**HAT**), an enzyme that transfers acetyl groups from a donor (acetyl-CoA) to core histones. These investigators used a creative strategy to isolate the enzyme: They started with *Tetrahymena* (ciliated protozoan) cells because this organism has histones that are heavily acetylated, which suggests that the cells contain relatively high concentrations of HAT. They prepared extracts from macronuclei (the large *Tetrahymena* nuclei that contain the active genes) and subjected them to gel electrophoresis in an SDS gel impregnated with histones. To detect HAT activity, they soaked the gel in a solution of acetyl-CoA with a radioactive label in the acetyl group. If the gel contained a band with HAT activity, the HAT would transfer labeled acetyl groups from acetyl-CoA to the histones. This would create a labeled band of acetylated histones in the gel at the position of the HAT activity. To detect the labeled histones, they washed away the unreacted acetyl-CoA, then subjected the gel to fluorography. Figure 13.32 shows the result: a band of HAT activity corresponding to a protein 55 kD in size. Accordingly, Brownell and Allis named this protein **p55**.

Allis and colleagues followed this initial identification of the HAT activity with a classic molecular cloning scheme to learn more about p55 and its gene. They began by purifying the HAT activity further, using standard biochemical techniques. Once they had purified the HAT activity essentially to homogeneity, they isolated enough of it to obtain a partial amino acid sequence. Using this sequence, they designed a set of degenerate oligonucleotides (Chapter 4) that coded for parts of the amino acid sequence and therefore hybridized to the macronuclear genomic DNA (or to cellular RNA). Using these oligonucleotides as primers, and total cellular RNA as template, they performed RT-PCR as explained in Chapter 4, then cloned the PCR products. They obtained the base sequences of some of the cloned PCR products and checked them to verify that the internal parts also coded for known HAT amino acid sequences. None of the PCR clones contained complete cDNAs, so these workers extended them in both the 5′- and 3′-directions, using

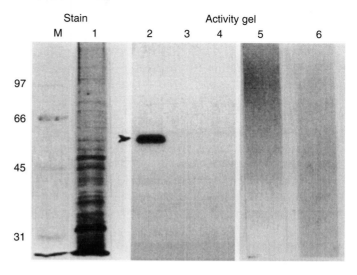

Figure 13.32   Activity gel assay for histone acetyltransferase (HAT) activity. Brownell and Allis electrophoresed a *Tetrahymena* macronuclear extract in an SDS-polyacrylamide gel containing histones (lanes 2–4), bovine serum albumin (BSA, lane 5), or no protein (lanes 1 and 6). After electrophoresis, they either silver-stained the gel to detect protein (lanes M and 1), or treated it with acetyl-CoA labeled in its acetyl group with ³H to detect HAT activity. After washing to remove unreacted acetyl-CoA, they subjected the gel to fluorography to detect ³H-acetyl groups. Lane 2 showed a clear band of ³H-acetylated histones, which indicated the presence of HAT activity. Lanes 3 and 4 failed to show activity because the HAT in the nuclear extracts was inactivated by heating (lane 3) or by treatment with *N*-ethylmaleimide (lane 4) prior to electrophoresis. Lane 5, with BSA instead of histones, also showed no activity, as did lane 6, with no protein substrate. Lane M contained molecular mass marker proteins. (*Source:* Brownell and Allis, *PNAS* 92 (July 1995) f. 1, p. 6365. © National Academy of Sciences, USA.)

rapid amplification of cDNA ends (RACE, Chapter 5). Finally, they obtained a cDNA clone that encoded the full 421-amino-acid p55 protein.

The amino acid sequence inferred from the base sequence of the p55 cDNA was very similar to the amino acid sequence of a yeast protein called **Gcn5p** (Figure 13.33). Gcn5p had been identified as a mediator of acidic transcription activators such as Gcn4p, so the amino acid sequence similarity suggested that both p55 and Gcn5p are HATs that are involved in gene activation. To verify that Gcn5p has HAT activity, Allis and colleagues expressed its gene in *E. coli,* then subjected it and p55 to the SDS-PAGE activity gel assay. Both proteins showed clear HAT activity. Thus, at least one HAT (Gcn5p) has both HAT and transcription coactivator activities. It appears to play a direct role in gene activation by acetylating histones.

It is important to note that p55 and Gcn5p are type A HATs (**HAT A's**) that exist in the nucleus and are apparently involved in gene regulation. They acetylate the lysine-rich N-terminal tails of core histones. Fully acetylated histone H3 has acetyl groups on lysines 9, 14, and 18, and fully acetylated histone H4 has acetyl groups on

Tetrahymena p55/yeast Gcn5p

```
Gcn5p    1 M V T K H Q I E E D H L D G A T T D P E V K R V K L E N N V E E I Q P E Q A E T N K Q E G T D K E N   50

 p55     1 . M A D Q E K S A Q D A Q N A A P Q E T A F V G M N G E E T G L G F A T R D Q G A K V E E D Q G L L   49
Gcn5p   51 K G K F E K E T E R I G G S E V V T D V E K G I V K F E F D G V E Y T F K E R P S V V E E N E G K I   100
                                                              I                          *
 p55    50 D F D I L T N D G T H R N M K L L I D L K N I F S R Q L P K M P K E Y I V K L V L D R H H E S M V I   99
Gcn5p  101 E F R V V N N D N T K E N M M V L T G L K N I F Q K Q L P K M P K E Y I A R L V Y D R S H L S M A V   150
                                                      II
 p55   100 L K N K Q K V I G G I C F R Q Y K P Q R F A E V A F L A V T A N E Q V R G Y G T R L M N K F K D H M   149
Gcn5p  151 I R K P L T V V G G I T Y R P F D K R E F A E I V F C A I S S T E Q V R G Y G A H L M N H L K D Y V   200
               III                                                                       IV
 p55   150 Q K . Q N I E Y L L T Y A D N F A I G Y F K K Q G F T K E H R M P Q E K W K G Y I K D Y D G G T L M   198
Gcn5p  201 R N T S N I K Y F L T Y A D N Y A I G Y F K K Q G F T K E I T L D K S I W M G Y I K D Y E G G T L M   250
                      *
 p55   199 E C Y I H P Y V D Y G N I S Q I I K R Q K E L L I E R I K K L S L N E K V F S G K E Y A A L I Q N S   248
Gcn5p  251 Q C S M L P R I R Y L D A G K I L L L Q E A A L R R K I R T I S K S H I V R P G L E Q F K D L N N I   300

 p55   249 M D N E D P E N P K V N P S D I P G V A F S G W E W K D Y H E L K K S K E R S F N L Q C A N V I G N   298
Gcn5p  301 K . . . . . . . . P I D P M T I P G L K E A G W T P E M D A L A Q R P K R G P H D A A I Q N I L T E   342
                                                        V-bromodomain
 p55   299 M K R H K Q S W P F L D P V N K D D V P D Y Y D V I T D P I D I K A I E K K L Q N N Q Y V D K D Q F   348
Gcn5p  343 L Q N H A A A W P F L Q P V N K E E V P D Y Y D F I K E P M D L S T M E I K L E S N K Y Q K M E D F   392

 p55   349 I K D V K R I F T N A K I Y N Q P D T I Y Y K A A K E L E D F V E P Y L T K L K D T K E S H T P S N   398
Gcn5p  393 I Y D A R L V F N N C R M Y N G E N T S Y Y K Y A N R L E K F F N N K V K E I P E Y S H L I D . . .   439

 p55   399 N N S A H G S K K P L P V S Q K K V Q K R N E
```

Figure 13.33   Comparison of the inferred amino acid sequences of p55 and Gcn5p. Vertical lines indicate identical residues; dots indicate similar residues; and double dots indicate very similar residues. Colors denote domains that are particularly well conserved. Asterisks in domains I and IV denote residues that are invariant in all HATs that have been sequenced. These amino acids may participate in the HAT reaction. (*Source:* Reprinted from Brownell et al., *Cell* 84:846, 1996. Copyright 1996, with permission from Elsevier Science.)

lysines 5, 8, 12, and 16. Lysines 9 and 14 of histone H3 and Lysines 5, 8, and 16 of histone H4 are acetylated in active chromatin and deacetylated in inactive chromatin. Type B HATs (**HAT B's**) are found in the cytoplasm and acetylate newly synthesized histones H3 and H4 so they can be assembled properly into nucleosomes. The acetyl groups added by HAT B's are later removed in the nucleus by histone deacetylases. All known HAT A's, including p55 and Gcn5p, contain a **bromodomain** (domain V in Figure 13.33), while all known HAT B's lack a bromodomain. Bromodomains allow proteins to bind to acetylated lysines. This would be useful to HAT A's, which must recognize partially acetylated histone tails and add acetyl groups to the other lysine residues. But HAT B's would have no use for a bromodomain, because they must recognize newly synthesized core histones that are unacetylated.

Since Allis's group's initial discovery of p55, at least two coactivators besides Gcn5p have been found to have HAT A activity. These are CBP/p300 (Chapter 12) and TAF$_{II}$250 (Chapter 11). All three of these coactivators cooperate with activators to enhance transcription. The fact that they have HAT A activity suggests a mechanism for part of this transcription enhancement: By binding near the transcription start site, they could acetylate core histones in the nucleosomes in the neighborhood, neutralizing some of their positive charge and thereby loosening their hold on the DNA (and perhaps on neighboring nucleosomes). This would allow remodeling of the chromatin to make it more accessible to the transcription apparatus, thus stimulating transcription.

It is interesting in this context that TAF$_{II}$250 has a double bromodomain module capable of recognizing two neighboring acetylated lysines, such as we would find on hypoacetylated core histones in inactive chromatin. In fact, Tjian and colleagues have determined that a protein fragment containing the double bromodomain module binds tightest to two acetylated lysines spaced seven amino acids apart, such as we would find in partially acetylated histone H4, with acetylated lysines 5 and 12. Thus, another role of TAF$_{II}$250 may be to recognize partially acetylated histones in inactive chromatin and to usher its partners, TBP and the other TAF$_{II}$s, into such chromatin to begin the activation process.

SUMMARY Histone acetylation occurs in both the cytoplasm and nucleus. Cytoplasmic acetylation is carried out by a HAT B and prepares histones for incorporation into nucleosomes. The acetyl groups are later removed in the nucleus. Nuclear acetylation is catalyzed by a HAT A and correlates with transcription activation. A variety of coactivators have HAT A activity, which may allow them to loosen the association of nucleosomes with a gene's control region, thereby enhancing transcription.

Histone Deacetylation

If core histone acetylation is a transcription-activating event, we would predict that core histone deacetylation would be a repressing event. In accord with this hypothesis, chromatin with underacetylated core histones is less transcriptionally active than average chromatin. Figure 13.34 outlines the apparent mechanism behind this repression: Known transcription repressors, such as nuclear receptors without their ligands, interact with corepressors, which in turn interact with histone deacetylases. These deacetylases then remove acetyl groups from the basic tails of core histones in nearby nucleosomes, tightening the grip of the histones on the DNA, thus stabilizing the nucleosomes and keeping transcription repressed. This repression can be considered **silencing,** although it is less severe than the silencing seen in heterochromatic regions of chromosomes, such as the ends, or telomeres. Some of the best studied corepressors are **Sin3** (yeast), **Sin3A** and **Sin3B** (mammals), and **NCoR/SMRT** (mammals). NCoR stands for "nuclear receptor **co**repressor" and SMRT stands for "silencing **m**ediator for **r**etinoid and **t**hyroid hormone receptors." These proteins interact with unliganded retinoic acid receptor (RAR-RXR), a heterodimeric nuclear receptor.

How do we know a physical association exists among transcription factors, corepressors, and histone deacetylases? One way to answer this question has been to add epitope tags to one of the components, then to immunoprecipitate the whole complex with an antibody against the tag. For example, Robert Eisenman and coworkers used epitope tagging to demonstrate a ternary complex among a transcription factor **Mad-Max,** a mammalian Sin3 corepressor (Sin3A), and a histone deacetylase (HDAC2). **Max** is a transcription factor that can serve as an activator or a repressor, depending on its partner in the heterodimer. If it associates with Myc to form a Myc-Max dimer, it acts as a transcription activator. On the other hand, if it associates with Mad to form a Mad-Max dimer, it acts as a repressor.

Part of the repression caused by Mad-Max comes from histone deacetylation, which suggests some kind of

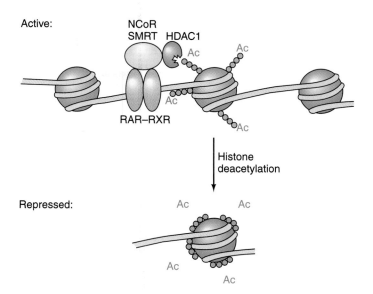

Figure 13.34   Model for participation of histone deacetylase in transcription repression. A heterodimer of retinoic acid receptor (RAR) and retinoic acid receptor X (RXR) binds to an enhancer (top). In the absence of the ligand, retinoic acid, the receptor dimer binds to the corepressor NCoR, which binds to the histone deacetylase HDAC1. The deacetylase then removes acetyl groups (red) from the lysine side chains (gray) on core histones of nearby nucleosomes. This deacetylation allows the lysine side chains to associate more closely with DNA (bottom), stabilizing the nucleosomes, and thereby inhibiting transcription.

interaction between a histone deacetylase and Mad. By analogy to the RAR-RXR–NCoR/SMRT–HDAC1 interaction illustrated in Figure 13.34, we might expect some corepressor like NCoR/SMRT to mediate this interaction between Mad and a histone deacetylase. To show that this interaction really does occur in vivo, and that it is mediated by a corepressor (Sin3A), Eisenman and coworkers used the following epitope-tagging strategy. They transfected mammalian cells with two plasmids. The first plasmid encoded epitope-tagged histone deacetylase (**HDAC2** tagged with a small peptide called the FLAG epitope [FLAG-HDAC2]). The second plasmid encoded Mad1, or a mutant Mad1 (Mad1Pro) having a proline substitution that blocked both interaction with Sin3A and repression of transcription. Then Eisenman and coworkers prepared extracts from these transfected cells and immunoprecipitated complexes using an anti-FLAG antibody. After electrophoresis, they blotted the proteins and first probed the blots with antibodies against Sin3A, then stripped the blots and probed them with antibodies against Mad1.

Figure 13.35 depicts the results. Lanes 1–3 are negative controls from cells containing a FLAG-encoding plasmid, rather than a FLAG-HDAC2-encoding plasmid. Immunoprecipitation of these lysates with an anti-FLAG antibody should not have precipitated HDAC2 or any

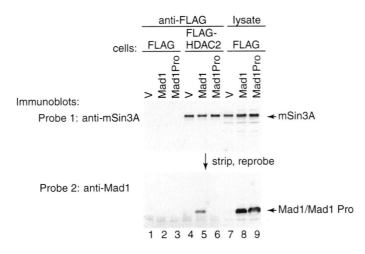

Figure 13.35   Evidence for a ternary complex involving HDAC2, Sin3A, and Mad1. Eisenman and coworkers transfected cells with a plasmid encoding either the FLAG epitope alone, or FLAG–HDAC2, as indicated at the top beside the designation "cells"; and a plasmid encoding either no Mad1 (V), Mad1, or Mad1Pro, also as indicated at top. They immunoprecipitated cell lysates with an anti-FLAG antibody (lanes 1–6, designated "anti-FLAG" at top) or just collected lysates (lanes 7–9, designated "lysate" at top) and electrophoresed the immunoprecipitates or lysates. After electrophoresis, they blotted the proteins to a membrane and probed the immunoblots, first with an anti-Sin3A antibody (top blot). Then, after stripping the first blot, they probed with an anti-Mad1 antibody that reacts with both Mad1 and Mad1Pro (bottom blot). Finally, they detected antibodies bound to proteins on the blot with a secondary antibody conjugated to horseradish peroxidase. They detected the presence of this enzyme with a substrate that becomes chemiluminescent on reaction with peroxidase. The positions of Sin3A and Mad1/Mad1Pro are indicated beside the blots at right. (*Source:* Laherty et al., Histone deacetylases associated with the mSin3 co-repressor mediate Mad transcriptional repression. *Cell* 89 (2 May 1997) f. 3, p. 352. Reprinted by permission of Elsevier Science.)

proteins associated with it. Accordingly, no Sin3A or Mad1 were found in the blots. Lanes 4–6 contained extracts from cells transfected with a plasmid encoding FLAG–HDAC2, and plasmids encoding: no Mad1 (lane 4); Mad1 (lane 5); and Mad1Pro (lane 6). All three lanes contained Sin3A, which indicated that this protein coprecipitated with FLAG-HDAC2. However, only lane 5 contained Mad1. It was expected that lane 4 would not contain Mad1 because no Mad1 plasmid was provided. It is significant that lane 6 did not contain Mad1Pro, even though a plasmid encoding this protein was included in the transfection. Because Mad1Pro cannot bind to Sin3A, it would not be expected to coprecipitate with FLAG-HDAC2 unless it interacted directly with HDAC2. The fact that it did not coprecipitate supports the hypothesis that Mad1 must bind to Sin3A, and not to HDAC2. This is another way of saying that the corepressor Sin3A mediates the interaction between the transcription factor Mad1 and the histone deacetylase HDAC2. Lanes 7–9

show the results of simply electrophoresing whole-cell lysates without any immunoprecipitation. The two blots show that these lysates contained plenty of Sin3A and abundant Mad1, if a Mad1-encoding plasmid was given (lane 8), or Mad1Pro, if that was the protein present (lane 9). Thus, the lack of Mad1Pro in lane 6 could not be explained by the failure of the plasmid encoding Mad1Pro to produce Mad1Pro protein.

We have now seen two examples of proteins that can be either activators or repressors, depending on other molecules bound to them. Some nuclear receptors behave this way depending on whether or not they are bound to their ligands. Mad proteins behave this way depending on whether they are bound to Myc or Max proteins. Figure 13.36 illustrates this phenomenon for a nuclear receptor, **thyroid hormone receptor (TR)**. TR forms heterodimers with RXR and binds to the enhancer known as the **thyroid hormone response element (TRE)**. In the absence of **thyroid hormone**, it serves as a repressor. Part of this repression is due to its interaction with NCoR, Sin3, and a histone deacetylase known as mRPD3, which deacetylates core histones in neighboring nucleosomes. This deacetylation stabilizes the nucleosomes and therefore represses transcription.

In the presence of thyroid hormone, the TR–RXR dimer serves as an activator. Part of the activation is due to binding to p300/CBP, P/CAF, and TAF$_{II}$250, all three of which are histone acetyltransferases that acetylate histones in neighboring nucleosomes. This acetylation destabilizes the nucleosomes and therefore stimulates transcription. Notice that the significant targets of the histone acetyltransferases and the histone deacetylases are core histones, not histone H1. Thus, the core histones, as well as H1, play important roles in nucleosome stabilization and destabilization.

Acetylation of core histone tails apparently does more than just inhibit binding of these tails to DNA. As we saw earlier in this chapter (see Figure 13.4), Timothy Richmond and colleagues' x-ray crystallography of core nucleosome particles revealed an interaction between histone H4 in one nucleosome core and the histone H2A–H2B dimer in the adjacent nucleosome core in the crystal lattice. In particular, the very basic region of the N-terminal tail of histone H4 (residues 16–25) interacts with an acidic pocket in the H2A–H2B dimer of the adjoining nucleosome. This interaction could help explain the cross-linking of nucleosomes that blocks access to transcription factors and therefore represses transcription. This hypothesis would also help explain why acetylating the tails of the core histones has an activating effect: Neutralizing the positive charge of the N-terminal tail of histone H4 by acetylation would help prevent nucleosome cross-linking and therefore help deter repression of transcription.

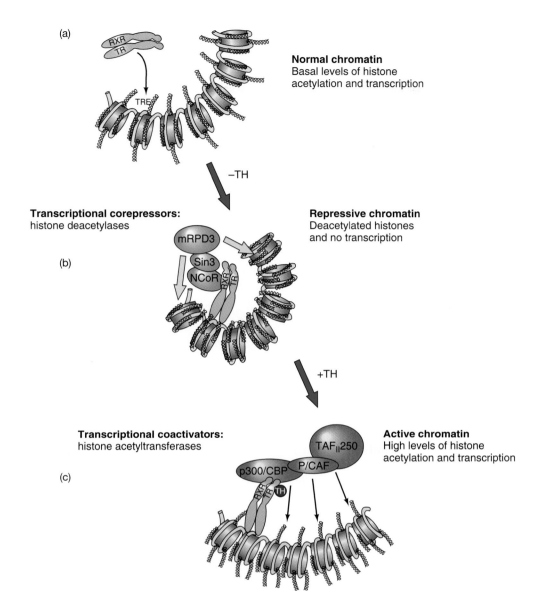

(a)

RXR
TR

TRE

Normal chromatin
Basal levels of histone
acetylation and transcription

−TH

Transcriptional corepressors:
histone deacetylases

(b)

mRPD3

Sin3

NCoR

RXR
TR

Repressive chromatin
Deacetylated histones
and no transcription

+TH

Transcriptional coactivators:
histone acetyltransferases

(c)

TAF$_{II}$250

P/CAF

p300/CBP

RXR
TR
TH

Active chromatin
High levels of histone
acetylation and transcription

Figure 13.36   A model for activation and repression by the same nuclear receptor. (a) Normal chromatin. No nuclear receptor (TR–RXR dimer) is bound to the thyroid hormone response element (TRE). Core histone tails are moderately acetylated. Transcription occurs at a basal level. **(b)** Repressive chromatin. The nuclear receptor is bound to the TRE in the absence of thyroid hormone (TH). The nuclear receptor interacts with either of the corepressors Sin3 and NCoR, which interact with the histone deacetylase mRPD3. The deacetylase cleaves acetyl groups off of the tails of core histones in surrounding nucleosomes, tightening the binding between histones and DNA, and helping to repress transcription. **(c)** Active chromatin. Thyroid hormone (purple) binds to the TR part of the nuclear receptor dimer, changing its conformation so it binds to one or more of the coactivators p300/CBP, P/CAF, and TAF$_{II}$250. These coactivators are all HAT A's that acetylate the tails of core histones in nearby nucleosomes, loosening the binding between histones and DNA and helping to activate transcription. (*Source:* Reprinted with permission from *Nature* 387:17, 1997. Copyright © 1997 Macmillan Magazines Limited.)

SUMMARY Transcription repressors such as unliganded nuclear receptors and Mad-Max bind to DNA sites and interact with corepressors such as NCoR/SMRT and Sin3, which in turn bind to histone deacetylases such as HDAC1 and 2. This assembly of ternary protein complexes brings the histone deacetylases close to nucleosomes in the neighborhood. The deacetylation of core histones allows the basic tails of the histones to bind strongly to DNA and to histones in neighboring nucleosomes, stabilizing and cross-linking the nucleosomes, and thereby inhibiting transcription.

Chromatin Remodeling

Histone acetylation is frequently essential for gene derepression but it is not sufficient because it deals only with the tails of the core histones, which lie outside the nucleosome core. Acetylation of these core histone tails can disrupt nucleosome cross-linking, as we will see in the next section, but it leaves the nucleosomes intact. Something else is needed to "remodel" the nucleosome cores to permit access of transcription factors. Two classes of proteins participate in this **chromatin remodeling,** and both require ATP for activity. The first is the **Swi/Snf** family (pronounced "switch-sniff") and the second is the **ISwi** family, which stands for "imitation switch." The Swi/Snf proteins alter the structure of nucleosome cores to make the DNA more accessible, not only to transcription factors, but also to nucleases. The ISwi proteins actually move the nucleosomes on the DNA.

Both families of proteins may yield the nucleosome-free regions around enhancers and promoters that are characteristic of active genes. In fact, we would predict that a nucleosome-free enhancer would be an important early requirement for gene activation. Thus, it is not surprising that Swi/Snf appears to be one of the first coactivators to arrive on the scene when a yeast gene is activated.

Kim Nasmyth and colleagues studied protein association with the *HO* gene of yeast, which plays a key role in switching the mating type. The expression of *HO* depends on a series of protein factors that appear at different phases of the cell cycle. Nasmyth and colleagues used a technique called **chromatin immunoprecipitation (CHIP)** as follows (Figure 13.37): First, they fused DNA fragments encoding short regions (epitopes) of a protein (Myc) to the ends of genes encoding the proteins known to associate with the *HO* gene. This led to the production of fusion proteins with the Myc epitopes at their C-termini. Then they synchronized the yeast cells, so most of them went through the cell cycle together. They obtained cells in various phases of the cell cycle and added formaldehyde to form covalent bonds between DNA and

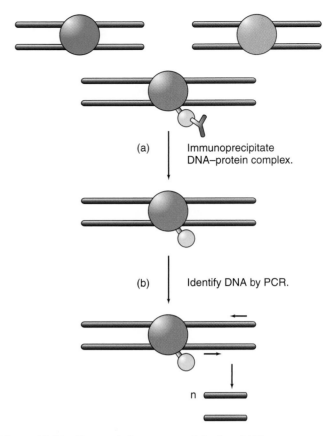

Figure 13.37 Chromatin immunoprecipitation. (a) The immunoprecipitation step. An antibody (red) has bound to an epitope (yellow) attached to a protein of interest (purple), which in turn is bound to a specific site on a double-stranded DNA (blue). The antibody does not bind to the other proteins (green and orange) to which the epitope is not attached. The antibody will actually bind to two DNA–protein complexes, which will also be bound to other molecules of the antibody, forming a complex that can easily be purified by centrifugation. **(b)** Identifying the DNA in the immunoprecipitate. Primers specific for the DNA of interest are used in a PCR reaction to amplify a portion of the DNA. Production of a DNA fragment of the correct predicted size indicates that the protein did indeed bind to the DNA of interest.

any proteins bound to it. Then they made cell extracts and immunoprecipitated the protein–DNA complexes with antibodies directed against the Myc epitopes. Recall that the Myc epitopes were attached to the proteins known to associate with *HO,* so the immunoprecipitated protein–DNA complexes should contain both these fusion proteins and the *HO* gene. To verify that these complexes contained the *HO* gene, they performed PCR with *HO*-specific primers. The PCR product should be a band of predictable size if the *HO* gene is really present.

The experimental results showed that a protein known as Swi5 bound first to the control region of *HO.* Next, Swi/Snf bound, followed by the SAGA complex (Chapter 11), which contains the HAT Gcn5, which then recruited the activator SBF. Other proteins, including general transcription factors and RNA polymerase II bound in turn after SBF. Both

Swi/Snf and SAGA are absolutely required for activation of *HO,* and they could act in concert to remodel the chromatin around the *HO* promoter. For example, Swi/Snf could disrupt the core histones around the gene's control region, and SAGA, by acetylating the tails of the core histones, could enhance the disruption and possibly make it permanent.

> **SUMMARY** Nucleosome remodeling agents such as Swi/Snf and ISwi are required for activation of some genes. These proteins disrupt the core histones in nucleosomes, and move nucleosomes, respectively. Such remodeling, combined with core histone acetylation, appears to create nucleosome-free enhancers that can bind readily to transcription activators.

Heterochromatin and Silencing

Most of the chromatin we have discussed in this chapter is in a class known as **euchromatin.** This chromatin is relatively extended and open and at least potentially active. By contrast, **heterochromatin** is very condensed and its DNA is inaccessible. In higher eukaryotes it even appears as clumps when viewed microscopically (Figure 13.38). In the yeast *Saccharomyces cerevisiae,* the chromosomes are too small to produce such clumps, but heterochromatin still exists, and it has the same repressive character as in higher eukaryotes. In fact, it can silence gene activity up to 3 kb away. Yeast heterochromatin is found at the telomeres, or tips of the chromosomes, and in the permanently repressed mating loci mentioned at the end of Chapter 10. In the fission yeast, *Schizosaccharomyces pombe,* heterochromatin is also found at the centromeres of chromosomes.

It is particularly convenient to do genetic and biochemical experiments in yeast, so molecular biologists have exploited this organism to learn about the structure of heterochromatin and the way in which it silences genes, not only within the heterochromatin, but in neighboring regions of the chromosome. The silencing of genes near the telomere is called the **telomere position effect** (**TPE**) because the silencing of a gene is dependent on its position in the chromosome: If it is within about 3 kb of the telomere, it is silenced; if it is farther away, it is not.

Studies on yeast telomeric heterochromatin have shown that several proteins bind to the telomeres and are presumably involved in forming heterochromatin. These are **RAP1, SIR2, SIR3, SIR4,** and histones H3 and H4. (SIR stands for silencing information regulator.) Yeast telomeres consist of many repeats of this sequence: $C_{2-3}A(CA)_{1-5}$. (Of course, the opposite strand of the telomere has the complementary sequence.) This sequence, commonly called $C_{1-3}A$, is the binding site for the

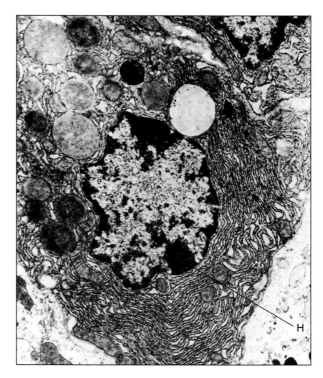

Figure 13.38 Interphase nucleus showing heterochromatin. Bat stomach lining cell with nucleus at center. Dark areas around periphery of nucleus are heterochromatin (H). (*Source:* Courtesy Dr. Keith Porter.)

RAP1 protein, the only telomeric protein that binds to a specific site in DNA. RAP1 then recruits the SIR proteins to the telomere in this order: SIR3-SIR4-SIR2. As we have already seen, histones H3 and H4 are core histones of the nucleosome. Both SIR3 and SIR4 bind directly to the N-terminal tails of these two histones at residues 4–20 of histone H3 and residues 16–29 of histone H4.

Because RAP1 binds only to telomeric DNA, we might expect to find it associated only with the telomere, but we find it in the "subtelomeric" region adjacent to the telomere, along with the SIR proteins. To explain this finding, Michael Grunstein and his colleagues have proposed the model in Figure 13.39: RAP1 binds to the telomeric DNA, the SIR proteins bind to RAP1 and to histones in the nucleosomes of the subtelomeric region. Then, protein–protein interactions cause the telomere to fold back on the subtelomeric region.

Earlier in this chapter, we learned that removing acetyl groups from core histones has a repressive effect on gene activity. Thus, we would predict that core histones in silenced chromatin would be poor in acetyl groups, or hypoacetylated. Indeed, whereas histone H4 in euchromatin is acetylated on lysines 5, 8, 12, and 16, histone H4 in yeast heterochromatin is acetylated only on lysine 12. What role might this hypoacetylation play in silencing? We know that lysine 16 of histone H4 is part of the domain (residues 16–29) that interacts with the SIR proteins (SIR3

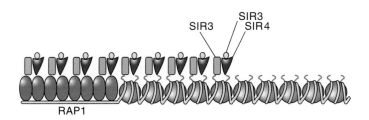

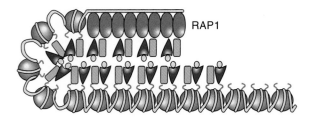

Figure 13.39   Model of telomere structure. RAP1 (green) binds to the telomere, and recruits SIR3 (red) and SIR4 (dark blue), which in turn attract SIR2 (yellow). SIR3 and SIR4 also bind to the N-terminal tails of histones H3 and H4 (thin blue lines). Interaction among the SIR proteins then causes the end of the chromosome to fold back on itself, so RAP1 is associated with the subtelomeric part of the chromosome. (*Source:* Grunstein, M. 1998. "Yeast heterochromatin: Regulation of its assembly and inheritance by histones." *Cell* 93: 325–28. Cell Press, Cambridge, MA. Reprinted by permission.)

in particular). Thus, acetylation of lysine 16 of histone H4 may block its interaction with SIR3, averting the formation of heterochromatin, and therefore preventing silencing.

Genetic experiments in yeast provide support for this hypothesis. Changing lysine 16 of histone H4 to a glutamine mimics the acetylation of this residue by removing its positive charge. This mutation also mimics acetylation in blocking the silencing of genes placed close to yeast telomeres and mating loci. On the other hand, changing lysine 16 to an arginine preserves the positive charge of the amino acid and thus mimics to some extent the deacetylated form of lysine. This mutation has less of an effect on silencing.

SUMMARY Euchromatin is relatively extended and potentially active, whereas heterochromatin is condensed and genetically inactive. Heterochromatin can also silence genes as much as 3 kb away. Formation of heterochromatin at the tips of yeast chromosomes (telomeres) depends on binding of the protein RAP1 to telomeric DNA, followed by recruitment of the proteins SIR3, SIR4, and SIR2, in that order. Heterochromatin at other locations in the chromosome also depends on the SIR proteins. SIR3 and SIR4 also interact directly with histones H3 and H4 in nucleosomes. Acetylation of lysine 16 of histone H4 in nucleosomes prevents its interaction with SIR3 and therefore blocks heterochromatin formation. This is another way in which histone acetylation promotes gene activity.

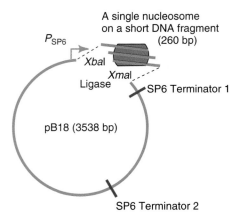

Figure 13.40   Construction of a plasmid with a single positioned nucleosome. Clark and Felsenfeld started with a plasmid containing an SP6 promoter and two terminators. They cleaved this plasmid with *Xba*I and *Xma*I and ligated it to a 260-bp *Xba*I–*Xma*I fragment containing a single positioned nucleosome on a 5S rRNA gene. The investigators then carefully purified the recombinant plasmid, which contained a single positioned nucleosome immediately downstream of the promoter, and therefore directly in the path of an advancing RNA polymerase. (*Source:* Reprinted from Clark and Felsenfeld, *Cell* 71:12, 1992. Copyright 1992, with permission from Elsevier Science.)

Nucleosomes and Transcription Elongation

We have seen that nucleosomes must be absent from a gene's control region, or at least nudged aside as activators and general transcription factors bind to their respective DNA sites. But how does RNA polymerase deal with the nucleosomes that lie within the transcribed region of a gene? Two hypotheses deal with this important question. First, the polymerase could somehow loosen the association between DNA and histones enough to make its way through a nucleosome. Second, the polymerase could actually displace the nucleosomes it encounters as it transcribes a gene.

Gary Felsenfeld has presented evidence for the second hypothesis. He began with the supposition that RNA polymerase causes positive supercoiling ahead of itself, and negative supercoiling behind, due to the DNA unwinding and rewinding that is constantly happening during transcription (Chapter 6). He then found that nucleosomes could form on both positively and negatively supercoiled DNA, but that they preferred negative supercoils. This led him to propose that RNA polymerase displaces nucleosomes as it transcribes a gene, and these nucleosomes then re-form on the negatively supercoiled DNA in the wake of transcription.

Like any good hypothesis, this one is testable. David Clark and Felsenfeld made a plasmid with one nucleosome at a well-defined site downstream of a promoter recognized by phage SP6 RNA polymerase. Then they transcribed this construct with SP6 polymerase and looked to see what happened to the nucleosome after transcription. As anticipated, the nucleosome was displaced to a position behind the polymerase.

Figure 13.40 shows the method Clark and Felsenfeld used to create the plasmid with one positioned

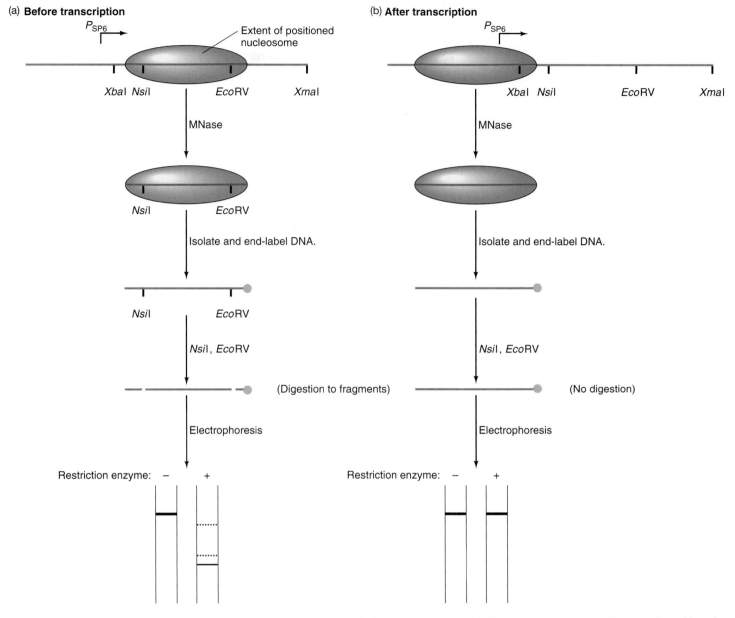

Figure 13.41   Scheme to detect the repositioning of a nucleosome during transcription. (a) Before transcription, the nucleosome is positioned in any of several locations within the XbaI–XmaI fragment. One such position is shown. Note that the blue oval merely indicates the extent of the positioned nucleosome. It is not intended to represent a real nucleosome, which would not be so extended, and would have the DNA wrapped around it. Clark and Felsenfeld digested the plasmid with micrococcal nuclease (MNase) to produce nucleosome cores, then isolated and end-labeled the nucleosome core DNA. Next they digested the DNA with NsiI or EcoRV (or both). Finally, they electrophoresed the fragments and detected the labeled fragments with a phosphorimager (Chapter 5). Most of the nucleosomes were located over one or both of the restriction sites, so most of the nucleosome core DNA should be cut by these restriction enzymes. The smallest fragment will be labeled, but its size heterogeneity will probably make it hard to detect. (b) After transcription, many of the nucleosomes have moved away from their original position to new locations that do not have NsiI and EcoRV sites. Therefore, the same treatment as in panel (a) should leave this labeled nucleosome core DNA intact.

nucleosome. They started with a 260-bp DNA fragment, added histones to assemble a nucleosome on this fragment, then ligated the fragment, along with its nucleosome, into a plasmid just downstream of an SP6 promoter and upstream of two SP6 terminators. Then they isolated intact circles by gel filtration and sucrose density gradient ultracentrifugation. Finally, they transcribed this con-

struct with SP6 RNA polymerase and then checked to see whether the nucleosome had moved.

Figure 13.41 explains the assay for nucleosome movement. Clark and Felsenfeld took plasmids before (panel a) and after (panel b) transcription and treated them with micrococcal nuclease to degrade all the DNA except that which is involved in a nucleosome core. Then they

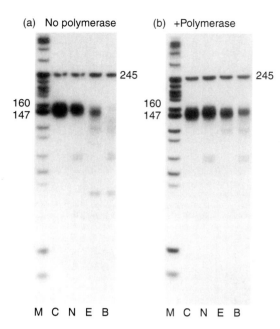

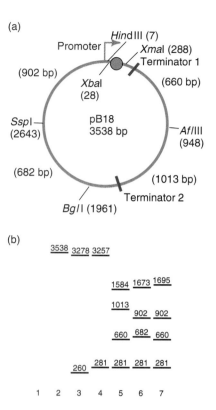

Figure 13.42 Experimental results showing relocation of nucleosomes. Electrophoresis lanes in both panels: M, marker DNAs; C, control DNA not treated with restriction enzymes; N, *Nsi*I-treated DNA; E, *Eco*RV-treated DNA; B, DNA treated with both enzymes. **(a)** Before transcription (no polymerase), the nucleosome core DNA fragment that migrates between the 147- and 160-kb markers is partially sensitive to each enzyme, especially *Eco*RV, and almost completely sensitive to both enzymes together. This indicates that most of the positioned nucleosomes lie over one or both of these sites before transcription. **(b)** After transcription (+polymerase), the nucleosome core DNA is much less sensitive to *Nsi*I and *Eco*RV, indicating that many of the nucleosomes have moved to other regions that lack these restriction sites. (*Source:* Clark and Felsenfeld, A nucleosome core is transferred out of the path of a transcribing polymerase. *Cell* 71 (2 Oct 1992) f. 3b-c, p. 13. Reprinted by permission of Elsevier Science.)

Figure 13.43 Restriction map of pB18. **(a)** Circular map of pB18, showing the location of the positioned nucleosome (blue), the promoter (orange), the terminators (violet), and cutting sites for various restriction enzymes. **(b)** Predicted electrophoretic patterns of fragments generated in the following restriction digests: 2, linearization with *Sph*I, whose cutting site is not shown; 3, cutting with *Xba*I and *Xma*I; 4, cutting with *Hind*III and *Xma*I; 5, cutting with *Hind*III, *Xma*I, *Afl*III, and *Bgl*II; 6, cutting with *Hind*III, *Xma*I, *Bgl*II, and *Ssp*I; 7, cutting with *Hind*III, *Xma*I, *Afl*III, and *Ssp*I. The numbers over the bands represent the fragment sizes in base pairs. (*Source:* Reprinted from Clark and Felsenfeld, *Cell* 71:15, 1992. Copyright 1992, with permission from Elsevier Science.)

removed the protein from the nucleosomal DNA, end-labeled this DNA, cut it with one or both of the restriction enzymes *Nsi*I and *Eco*RV, then electrophoresed the resulting DNA fragments. Here is the trick to the assay: The plasmid contains only one cutting site for each of these two restriction enzymes, and these sites lie in the region where the positioned nucleosome initially forms. Therefore, before transcription, the nucleosome core would cover at least one of these two restriction sites, and the DNA isolated from the nucleosome cores before transcription will be sensitive to one or both of the enzymes. On the other hand, if the nucleosome moves to another site during transcription, it will cover DNA that contains neither restriction site (Figure 13.41b). The DNA isolated from these nucleosome cores will not be sensitive to either *Nsi*I or *Eco*RV.

Figure 13.42 shows the results of this assay. Before transcription (no polymerase) the nucleosome core DNA was somewhat sensitive to each restriction enzyme and almost completely sensitive to both together (lane B). By

contrast, after transcription (+polymerase), the nucleosome core DNA was much less sensitive to the two enzymes, even when they were added together. The residual sensitivity was probably due at least in part to the following two factors. First, some templates (about 23%) were not transcribed because the nucleosome took up a position too close to the promoter. Second, some nucleosomes (about 13%) shifted away from the original *Nsi*I–*Eco*RV region on their own, before transcription began.

Thus, transcription seems to move the nucleosome, but where? To answer this question, Clark and Felsenfeld used micrococcal nuclease to isolate nucleosome core DNA before and after transcription, then labeled this DNA and used it to probe Southern blots of the plasmid DNA cut with various restriction enzymes as shown in Figure 13.43. The fragments that hybridized after transcription should be the ones in which nucleosome cores migrated during transcription.

Figure 13.44 contains the results of the blots. Before transcription (Figure 13.44b), nucleosome core DNA

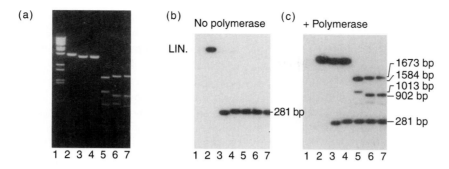

Figure 13.44   Results of Southern blot to locate repositioned nucleosomes. (a) Ethidium bromide-stained gel with fragments generated by the restriction digests listed in Figure 13.43b. Lane 1 contained size markers. **(b)** Southern blot hybridized to labeled nucleosome core DNA before transcription. Nucleosomal core DNA hybridized almost exclusively to the 260-bp *Xba*I–*Xma*I fragment (lane 3) or the 281-bp *Hin*dIII–*Xma*I fragment (lanes 4–7). **(c)** Southern blot hybridized to labeled nucleosome core DNA after transcription. The nucleosome core DNA hybridized to other fragments, especially those such as the 1584-bp fragment (lane 5) and the 902-bp fragment (lanes 6 and 7), which lie in the untranscribed region. (*Source:* Clark and Felsenfeld, *Cell* 71 (2 Oct 1992) f. 6, p. 15. Reprinted by permission of Elsevier Science.)

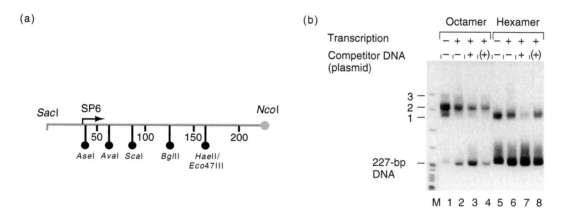

Figure 13.45   Assay to distinguish repositioning of a nucleosome core *in cis* or *in trans*. (a) Restriction map of the DNA fragment containing the original positioned nucleosome. This 227-bp DNA fragment was labeled at its *Nco*I end as indicated by the orange circle. Note the SP6 promoter near the left end. **(b)** Experimental results. Felsenfeld and colleagues positioned a nucleosome on the end-labeled 227-bp DNA fragment from panel **(a)**. Then they mixed this construct with a 170-fold excess (in moles of base pairs) of competitor DNA in the presence or absence of RNA polymerase. Finally, they electrophoresed the mixture in a nucleoprotein gel that preserved DNA–protein complexes, including nucleosomes (octamers) and partial nucleosomes lacking one histone H2A–H2B pair (hexamers). Nucleoprotein complexes migrated slowly and were found near the top of the gel. Complex 1 was composed of hexamers; complexes 2 and 3 were complete nucleosomes, with the nucleosome asymmetrically located in complex 2 and centrally positioned on the DNA fragment in complex 3. The free 227-bp fragment migrated much more rapidly. Lane M, markers; lanes 1–8 contained polymerase or competitor DNA (or both) as indicated at top, and nucleosome octamers (lanes 2–4) or hexamers (lanes 5–8). In lanes 4 and 8, the competitor was added 30 sec after the polymerase instead of simultaneously with it. (*Source:* Studitsky et al., A histone octamer can step around a transcribing polymerase without leaving the template. *Cell* 76 (28 Jan 1994) f. 4a, p. 375. Reprinted by permission of Elsevier Science.)

hybridized to the whole linearized plasmid (LIN.) in lane 1, or to the 281-bp *Hin*dIII–*Xma*I fragment that contained most of the original nucleosome cores (lanes 2–7). The fact that no other discrete band is visible means that the 13% of nucleosome cores that spontaneously moved elsewhere do not move to one or a few defined positions, but are randomly positioned.

Figure 13.44c shows what happened after transcription. Nucleosome core DNA hybridized to other fragments besides the one containing the original nucleosome. In particular, the bands that hybridized best were the ones that lie outside the transcribed region. The best example of this is lane 5, where the 1584-bp *Bgl*II–*Hin*dIII band, which lies outside the transcribed region, was clearly fa-

vored over the 1013-bp *Afl*III–*Bgl*II band, which includes mostly transcribed DNA. However, this preference was not overwhelming. For example, in lane 6, hybridization to the 1673-bp *Xma*I–*Bgl*II (transcribed) band was as strong as hybridization to the 902-bp *Ssp*I–*Hin*dIII (nontranscribed) band. Taken together, these data suggest that transcription forces the nucleosome into a new location, which tends to lie behind the polymerase.

Does transcription force the nucleosome core histones to dissociate from the DNA during repositioning, or do the histones move without dissociating from the DNA? To find out, Felsenfeld and colleagues transcribed an end-labeled 227-bp DNA fragment with a single positioned nucleosome (Figure 13.45a), but they also added a great

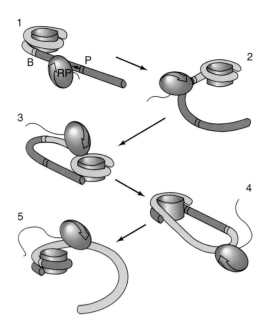

Figure 13.46 Felsenfeld's spooling model for nucleosome repositioning. 1. We begin with an RNA polymerase (RP, red) bound to a promoter (P) just upstream of a nucleosome. The border (B) between nucleosomal core DNA (pink) and upstream DNA (purple) is marked with a yellow band. 2. As the polymerase advances toward the border, the DNA begins to unspool from the nucleosome core. 3. The upstream end of the DNA contacts the protein core where the original DNA has unspooled, forming a loop that contains the polymerase. 4. As the original core DNA unspools, the upstream DNA spools onto the core; the polymerase continues to advance around the loop. 5. Upstream DNA has replaced the original DNA, releasing the downstream end of the DNA and allowing the polymerase to pass. (*Source:* Reprinted from Studitsky et al., *Cell* 76:379, 1994. Copyright 1994, with permission from Elsevier Science.)

excess of a negatively supercoiled competitor plasmid with its own site for nucleosome positioning. Then they electrophoresed the resulting nucleoprotein complexes to see whether the nucleosomes had jumped from the template DNA. Figure 13.45b, lane 3, demonstrates that this competitor did indeed cause some loss of nucleosomes from the template DNA, leaving more free 227-bp DNA fragments. However, even under these conditions, with a great excess of competitor, fully 30% of the cores on transcribable templates were transferred *in cis* (to a site on the same DNA molecule), rather than *in trans* (to a site on the competitor DNA). Further experiments showed that transfer *in cis* is about four times more likely than transfer *in trans* if original and competitor DNAs are present in equal quantities. Figure 13.45b also shows that the competitor could no longer work if it was added only 30 sec after transcription began (lane 4). By that time, one round of transcription had occurred, and the cores had moved to a site in the promoter, preventing further transcription.

The nucleosome cores that transferred *in cis* in the experiment depicted in Figure 13.45b must never have lost contact with the original DNA. If they had, they would have been picked up by the great excess of competitor

DNA. How can a nucleosome core transfer to a site *behind* the RNA polymerase without ever losing contact with the DNA? It seems that the core would somehow have to negotiate its way around the advancing polymerase to do this. Felsenfeld has proposed a spooling mechanism that could solve this problem (Figure 13.46). We begin with an RNA polymerase bound to a promoter just upstream of a nucleosome. As the polymerase approaches the nucleosome, the core DNA begins to uncoil. The upstream end of the DNA then approaches the core nucleosome to form a loop. Now, as the old core DNA uncoils, upstream DNA coils around the core histones to form a new nucleosome upstream of the original location. Further experimentation will be needed to test this hypothesis.

SUMMARY Felsenfeld and colleagues have performed in vitro experiments with plasmids containing a single positioned nucleosome to examine the fate of nucleosomes in the path of RNA polymerase. These studies suggest the following scenario: When RNA polymerase moves through nucleosomal core DNA, it displaces the nucleosome cores, most to a new location behind the advancing polymerase on the same DNA molecule. This may be accomplished by a spooling mechanism in which upstream DNA spools around core histones as downstream DNA unspools, with the RNA polymerase transcribing through a loop between these spooling and unspooling DNAs.

SUMMARY

Eukaryotic DNA combines with basic protein molecules called histones to form structures known as nucleosomes. These structures contain four pairs of histones (H2A, H2B, H3, and H4) in a wedge-shaped disc, around which is wrapped a stretch of 146 bp of DNA. Histone H1, apparently bound to DNA outside the disc, completes the nucleosome structure, and may constrain the DNA to enter and exit near the same point. The first order of chromatin folding is represented by a string of nucleosomes, which condenses the DNA by a factor of 6 to 7. The second order of chromatin condensation occurs when the nucleosome fiber is further folded into a 30-nm, or condensed, fiber. One candidate for the structure of this fiber is a solenoid. Another is a variable zigzag ribbon. Another possibility is that the nucleosomes form superbeads that are arranged in an irregular order. Still another is that an open, irregular helical fiber folds into an irregular condensed fiber with significant contents of neither helix nor superbead. Histone H1 appears to play a part in formation of the 30-nm fiber, but its exact role is unclear. The third order of chromatin condensation appears to involve formation of a radial loop structure in eukaryotic chromosomes. The 30-nm fiber seems to undergo some kind of folding back or coiling to form a 50-nm fiber, which then forms loops between 35 and 85 kb long, anchored to the central matrix of the chromosome.

The apparent reason for the difference in expression of *Xenopus laevis* oocyte and somatic 5S rRNA genes in somatic cells is that the somatic genes form more stable complexes with transcription factors. The transcription factors seem to keep the somatic genes active by preventing nucleosomes from forming a stable complex with the internal control region. This stable complex requires the participation of histone H1. Once it forms, transcription factors are excluded and the gene is repressed.

The core histones (H2A, H2B, H3, and H4) assemble nucleosome cores on naked DNA. Transcription of a class II gene in reconstituted chromatin with an average of one nucleosome core per 200 bp of DNA exhibits about 75% repression relative to naked DNA. The remaining 25% is due to promoter sites not covered by nucleosome cores. Activators such as Sp1 cannot counteract the repression caused by nucleosome core formation. Histone H1 causes a further repression of template activity, in addition to that produced by core nucleosomes. This repression can be counteracted by transcription factors. Some, like Sp1 and GAL4, act as both antirepressors (preventing repression by histone H1) and as transcription activators. Others, like GAGA factor, are just antirepressors. The antirepressors presumably compete with histone H1 for binding sites on the DNA template.

Active genes tend to have DNase-hypersensitive control regions. At least part of this hypersensitivity is due to the absence of nucleosomes. Some active genes have an array of precisely positioned nucleosomes over their control regions. The same control regions, in tissues where these genes are not active, do not have positioned nucleosomes. The positioned nucleosomes can overlap the binding sites for activators. Thus, activators and nucleosomes can coexist in the same place. In fact, the activators almost certainly have a role in positioning the nucleosomes.

Histone acetylation occurs in both the cytoplasm and nucleus. Cytoplasmic acetylation is carried out by a HAT B and prepares histones for incorporation into nucleosomes. The acetyl groups are later removed in the nucleus. Nuclear acetylation is catalyzed by a HAT A and correlates with transcription activation. A variety of coactivators have HAT A activity, which may allow them to loosen the association of nucleosomes with each other and with a gene's control region, thereby enhancing transcription.

Transcription repressors such as unliganded nuclear receptors and Mad-Max bind to DNA sites and interact with corepressors such as NCoR/SMRT and Sin3, which in turn bind to histone deacetylases such as HDAC1 and 2. This assembly of ternary protein complexes brings the histone deacetylases close to nucleosomes in the neighborhood. The deacetylation of core histones allows the basic tails of the histones to bind strongly to DNA, stabilizing the nucleosomes and inhibiting transcription.

Nucleosome remodeling agents such as Swi/Snf and ISwi are required for activation of some genes. These proteins disrupt the core histones in nucleosomes, and move nucleosomes, respectively. Such remodeling, combined with core histone acetylation appears to create nucleosome-free enhancers that can bind readily to transcription activators.

Euchromatin is relatively extended and potentially active, whereas heterochromatin is condensed and genetically inactive. Heterochromatin can also silence genes as much as 3 kb away. Formation of heterochromatin at the tips of yeast chromosomes (telomeres) depends on binding of the protein RAP1 to telomeric DNA, followed by recruitment of the proteins SIR3, SIR4, and SIR2, in that order. Heterochromatin at other locations in the chromosome also depends on the SIR proteins. SIR3 and SIR4 also interact directly with histones H3 and H4 in nucleosomes. Acetylation of lysine 16 of histone H4 in nucleosomes prevents its interaction with SIR3 and therefore prevents heterochromatin formation. This is another way in which histone acetylation promotes gene activity.

When RNA polymerase moves through nucleosomal core DNA, it displaces the nucleosome cores, most to a new location, apparently behind the advancing polymerase on the same DNA molecule. This may be accomplished by a spooling mechanism in which upstream DNA spools around core histones as downstream DNA unspools, with the RNA polymerase transcribing through a loop between these spooling and unspooling DNAs.

REVIEW QUESTIONS

1. Diagram a nucleosome as follows: (a) On a drawing of the histones without the DNA, show the rough positions of all the histones. (b) On a separate drawing, show the path of DNA around the histones.

2. Diagram a 110-Å fiber composed of a string of nucleosomes.

3. Cite electron microscopic evidence for a six- to sevenfold condensation of DNA in nucleosomes.

4. Draw two models to explain the zigzag appearance of a chromatin fiber in the presence of histone H1.

5. Cite electron microscopic evidence for formation of a condensed fiber (30-nm fiber) at high ionic strength.

6. Diagram the solenoid model of the condensed chromatin fiber (30-nm fiber).

7. Draw a model to explain the next order of chromatin folding after the 30-nm fiber. Cite biochemical and microscopic evidence to support the model.

8. Describe and give the results of an experiment that shows the effect of histone H1 on 5S rRNA transcription in vitro.

9. Draw a model to explain the competing effects of transcription factors and histone H1 on transcription of the 5S rRNA genes in reconstituted chromatin.

10. Describe and give the results of an experiment that shows the competing effects of histone H1 and the activator GAL4-VP16 on transcription of the adenovirus E4 gene in reconstituted chromatin.

11. Present two models for antirepression by transcription activators.

12. Describe and give the results of an experiment that shows that the nucleosome-free zone in active SV40 chromatin lies at the viral late gene control region.

13. Describe and give the results of an experiment that shows that the zone of DNase hypersensitivity in SV40 chromatin lies at the viral late gene control region.

14. Diagram a general technique for detecting a DNase-hypersensitive DNA region.

15. Describe and give the results of an experiment that shows the presence of a positioned nucleosome.

16. Describe and give the results of an activity gel assay that shows the existence of a histone acetyltransferase (HAT) activity.

17. Present a model for the involvement of a corepressor and histone deacetylase in transcription repression.

18. Describe and give the results of an epitope-tagging experiment that shows interaction among the following three proteins: the repressor Mad1, the corepressor Sin3A, and the histone deacetylase HDAC2.

19. Present a model for activation and repression by the same protein, depending on the presence or absence of that protein's ligand.

20. Describe how you could use a chromatin immunoprecipitation procedure to detect the proteins associated with a particular gene at various points in the cell cycle.

21. Present a model to explain why lysine 16 in histone H4 is thought to be critical for silencing. What evidence supports this hypothesis?

22. Describe a scheme to detect repositioning of a positioned nucleosome during transcription.

23. Describe and give the results of an experiment that shows where a nucleosome moves after RNA polymerase has passed.

24. Describe and give the results of an experiment that shows that a nucleosome repositions at least partly to the same DNA (repositioning *in cis*).

SUGGESTED READINGS

General References and Reviews

Brownell, J.E. and C.D. Allis. 1996. Special HATs for special occasions: Linking histone acetylation to chromatin assembly and gene activation. *Current Opinion in Genetics and Development* 6:176–84.

Butler, P.J.G. 1983. The folding of chromatin. *CRC Critical Reviews in Biochemistry* 15:57–91.

Felsenfeld, G. 1996. Chromatin unfolds. *Cell* 86:13–19.

Grunstein, M. 1998. Yeast heterochromatin: Regulation of its assembly and inheritance by histones. *Cell* 93:325–28.

Pazin, M.J. and J.T. Kadonaga. 1997. What's up and down with histone deacetylation and transcription? *Cell* 89:325–28.

Pennisi, E. 1996. Linker histones, DNA's protein custodians, gain new respect. *Science* 274:503–4.

Pennisi, E. 1997. Opening the way to gene activity. *Science* 275:155–57.

Pennisi, E. 2000. Matching the transcription machinery to the right DNA. *Science* 288:1372–73.

Perlman, T., and B. Vennström. 1995. The sound of silence. *Nature* 377:387–88.

Rhodes, D. 1997. The nucleosome core all wrapped up. *Nature* 389:231–33.

Roth, S.Y., and C.D. Allis. 1996. Histone acetylation and chromatin assembly: A single escort, multiple dances? *Cell* 87:5–8.

van Holde, K., and J. Zlatanova. 1995. Chromatin higher order structure: Chasing a mirage? *Journal of Biological Chemistry* 270:8373–76.

Wolffe, A.P. 1996. Histone deacetylase: A regulator of transcription. *Science* 272:371–72.

Wolffe, A.P. 1997. Sinful repression. *Nature* 387:16–17.

Wolffe, A.P. and D. Pruss. 1996. Targeting chromatin disruption: Transcription regulators that acetylate histones. *Cell* 84:817–19.

Xu, L., C.K. Glass, and M.G. Rosenfeld. 1999. Coactivator and corepressor complexes in nuclear receptor function. *Current Opinion in Genetics and Development* 9:140–47.

Research Articles

Arents, G. and E.N. Moudrianakis. 1993. Topography of the histone octamer surface: Repeating structural motifs utilized in the docking of nucleosomal DNA. *Proceedings of the National Academy of Sciences USA* 90:10489–93.

Arents, G., R.W. Burlingame, B.-C. Wang, W.E. Love, and E.N. Moudrianakis. 1991. The nucleosomal core histone octamer at 3.1 Å resolution: A tripartite protein assembly and a left-handed superhelix. *Proceedings of the National Academy of Sciences USA* 88:10148–52.

Benyajati, C., and A. Worcel. 1976. Isolation, characterization, and structure of the folded interphase genome of *Drosophila melanogaster*. *Cell* 9:393–407.

Brownell, J.E. and C.D. Allis. 1996. An activity gel assay detects a single, catalytically active histone acetyltransferase subunit in *Tetrahymena* macronuclei. *Proceedings of the National Academy of Sciences USA* 92:6364–68.

Brownell, J.E., J. Zhou, T. Ranalli, R. Kobayashi, D.G. Edmondson, S.Y. Roth, and C.D. Allis. 1996. Tetrahymena histone acetyltransferase A: A homolog of yeast Gcn5p linking histone acetylation to gene activation. *Cell* 84:843–51.

Clark, D.J. and G. Felsenfeld. 1992. A nucleosome core is transferred out of the path of a transcribing polymerase. *Cell* 71:11–22.

Cosma, M.P., T. Tanaka, and K. Nasmyth. 1999. Ordered recruitment of transcription and chromatin remodeling factors to a cell cycle- and developmentally regulated promoter. *Cell* 97:299–311.

Griffith, J. 1975. Chromatin structure: Deduced from a minichromosome. *Science* 187:1202–3.

Jacobson, R.H., A.G. Ladurner, D.S. King, and R. Tjian. 2000. Structure and function of a human TAF$_{II}$250 double bromodomain module. *Science* 288:1422–28.

Laherty, C.D., W.-M. Yang, J.-M. Sun, J.R. Davie, E. Seto, and R.N. Eisenman. 1997. Histone deacetylases associated with the mSin3 corepressor mediate Mad transcriptional repression. *Cell* 89:349–56.

Laybourn, P.J. and J.T. Kadonaga. 1991. Role of nucleosomal cores and histone H1 in regulation of transcription by RNA polymerase II. *Science* 254:238–45.

Luger, K., A.W. Mäder, R.K. Richmond, D.F. Sargent, and T.J. Richmond. 1997. Crystal structure of the nucleosome core particle at 2.8 Å resolution. *Nature* 389:251–60.

Marsden, M.P.F. and U.K. Laemmli. 1979. Metaphase chromosome structure: Evidence for a radial loop model. *Cell* 17:849–58.

McPherson, C.E., E.-Y. Shim, D.S. Friedman, and K.S. Zaret. 1993. An active tissue-specific enhancer and bound transcription factors existing in a precisely positioned nucleosomal array. *Cell* 75:387–98.

Ogryzko, V.V., R.L. Schiltz, V. Russanova, B.H. Howard, and Y. Nakatani. 1996. The transcriptional coactivators p300 and CBP are histone acetyltransferases. *Cell* 87:953–59.

Panyim, S. and R. Chalkley. 1969. High resolution acrylamide gel electrophoresis of histones. *Archives of Biochemistry and Biophysics* 130:337–46.

Paulson, J.R. and U.K. Laemmli. 1977. The structure of histone-depleted interphase chromosomes. *Cell* 12:817–28.

Saragosti, S., G. Moyne, and M. Yaniv. 1980. Absence of nucleosomes in a fraction of SV40 chromatin between the origin of replication and the region coding for the late leader RNA. *Cell* 20:65–73.

Schlissel, M.S. and D.D. Brown. 1984. The transcriptional regulation of *Xenopus* 5S RNA genes in chromatin: The roles of active stable transcription complexes and histone H1. *Cell* 37:903–13.

Stalder, J., A. Larsen, J.D. Engel, M. Dolan, M. Groudine, and H. Weintraub. 1980. Tissue-specific DNA cleavages in the globin chromatin domain introduced by DNAase I. *Cell* 20:451–60.

Studitsky, V.M., D.J. Clark, and G. Felsenfeld. 1994. A histone octamer can step around a transcribing polymerase without leaving the template. *Cell* 76:371–82.

Taunton, J., C.A. Hassig, and S.L. Schreiber. 1996. A mammalian histone deacetylase related to the yeast transcriptional regulator Rpd3p. *Science* 272:408–11.

Thoma, F., T. Koller, and A. Klug. 1979. Involvement of histone H1 in the organization of the nucleosome and of the salt-dependent superstructure of chromatin. *Journal of Cell Biology* 83:403–27.

Thomas, J.O. and R.D. Kornberg. 1975. Cleavable cross-links in the analysis of histone-histone interactions. *FEBS Letters* 58:353–58.

Widom, J. and A. Klug. 1985. Structure of the 300 Å chromatin filament: x-ray diffraction from oriented samples. *Cell* 43:207–13.

CHAPTER 14

Posttranscriptional Events I: Splicing

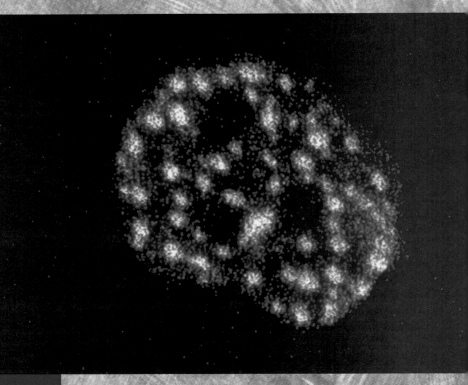

Electron micrograph of human pre-mRNA splicing factor in a cell nucleus. (×14,500). © *Dr. David Spector/Peter Arnold, Inc.*

Prokaryotic gene expression can be summarized very briefly as follows: First, RNA polymerase transcribes a gene, or set of genes, in an operon. Then, even while transcription is still occurring, ribosomes bind to the mRNA and translate it to make protein. We have already studied the transcription part of this scheme in Chapters 6–8, and it may seem to be quite complex. However, the situation in eukaryotes is much more intricate. In Chapter 13 we discussed the complex chromatin structure that distinguishes eukaryotes, but the complexities do not stop there.

In eukaryotes, the compartments in which transcription and translation occur are different. Transcription takes place in the nucleus, whereas translation takes place in the cytoplasm. This means that transcription and translation cannot occur simultaneously as they do in prokaryotes. Instead, transcription has to finish, then the transcript has to make its way into the cytoplasm before translation can begin. This allows an interval between transcription and translation known as the posttranscriptional phase.

Many events peculiar to eukaryotes occur in this period. In this and the next two chapters, we will deal with these posttranscriptional events.

We will see that most eukaryotic genes, in contrast to typical prokaryotic genes, are interrupted by nonsense DNA. RNA polymerase cannot distinguish the coding region of the gene from the nonsense, so it transcribes everything. Thus, one posttranscriptional task is to remove the nonsense DNA from the original transcript, in a process called splicing. Eukaryotes also tack special structures onto the 5′- and 3′-ends of their mRNAs. The 5′-structure is called a cap, and the 3′-structure is a string of AMPs called poly(A). All three of these events occur in the nucleus before the mRNA emigrates to the cytoplasm. The first of these, splicing, will be the subject of this chapter. ∎

14.1 Genes in Pieces

If we expressed the sequence of the human β-globin gene as a sentence, here is how it might look:

This is *bhgty* the human β-globin *qwtzptlrbn* gene.

Two regions (italicized) within the gene obviously make no sense: they contain sequences totally unrelated to the globin coding sequences surrounding them. These are sometimes called **intervening sequences, or IVSs**, but they usually go by the name Walter Gilbert gave them: **introns**. Similarly, the parts of the gene that make sense are sometimes called coding regions, or expressed regions, but Gilbert's name for them is more popular: **exons**. Some genes, especially in lower eukaryotes, have no introns at all; others have an abundance. The current record (at least 60 introns) is held by the dystrophin gene, whose failure causes Duchenne muscular dystrophy.

Evidence for Split Genes

Consider the major late locus of adenovirus—the first place introns were found, by Phillip Sharp and his colleagues in 1977. Several lines of evidence converged at that time to show that the genes of the adenovirus major late locus are interrupted, but perhaps the easiest to understand comes from studies using a technique called **R-looping**.

In R-looping experiments, RNA is hybridized to its DNA template. In other words, the DNA template strands are separated to allow a double-stranded hybrid to form between one of these strands and the RNA product. Such a hybrid double-stranded polynucleotide is actually a bit more stable than a double-stranded DNA under the conditions of the experiment. After the hybrid forms, it is examined by electron microscopy. These experiments can be done in two basic ways: (1) using DNA whose two

strands are separated only enough to let the RNA hybridize or (2) completely separating the two DNA strands before hybridization. Sharp and colleagues used the latter method, hybridizing single-stranded adenovirus DNA to mature mRNA for one of the viral coat proteins: the hexon protein. Figure 14.1 shows the results. (Do not be confused by the similarity between the terms exon and hexon. They are not related; hexon is just the name of a viral coat protein.)

If the hexon gene had no introns, a smooth, linear hybrid would occur where the mRNA lined up with its DNA template. But what if introns *do* occur in this gene? Clearly, no introns are present in the mature mRNA, or they would code for nonsense that would appear in the protein product. Therefore, introns are sequences that occur in the DNA but are missing from mRNA. That means the hexon DNA and hexon mRNA will not be able to form a smooth hybrid. Instead, the intron regions of the DNA will not find counterparts in the mRNA and so will form unhybridized loops. Of course, that is exactly what happened in the experiment shown in Figure 14.1. The loops there are made of DNA, but we still call them R loops because hybridization with RNA caused them to form.

The electron micrograph shows an RNA–DNA hybrid interrupted by three single-stranded DNA loops (labeled A, B, and C). These loops represent the introns in the hexon gene. Each loop is preceded by a short hybrid region, and the last loop is followed by a long hybrid region. Thus, the gene has four exons: three short ones near the beginning, followed by one large one. The three short exons are transcribed into a leader region that appears at the 5′-end of the hexon mRNA before the coding region; the long exon contains the coding region of the gene. In fact, the major late genes have different coding regions, but all share the same leader region encoded in the same three short exons.

When we discover something as surprising as introns in a virus, we wonder whether it is just a bizarre viral phenomenon that has no relationship to eukaryotic cellular processes. Thus, it was important to determine whether eukaryotic cellular genes also have introns. One of the first such demonstrations was an R-looping experiment done by Pierre Chambon and colleagues, using the chicken ovalbumin gene. They observed six DNA loops of various sizes that could not hybridize to the mRNA, so this gene contains six introns spaced among seven exons. It is also interesting that most of the introns were considerably longer than most of the exons. This preponderance of introns is typical of higher eukaryotic genes. Introns in lower eukaryotes such as yeast tend to be shorter and much rarer.

So far we have discussed introns only in mRNA genes, but some tRNA genes also have introns, and even rRNA genes sometimes do. The introns in both these latter types

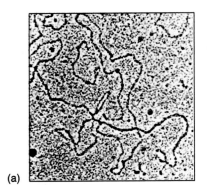

(a)

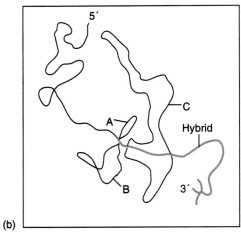

(b)

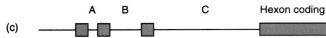

(c)

Figure 14.1   R-looping experiments reveal introns in adenovirus.
(a) Electron micrograph of a cloned fragment of adenovirus DNA
containing the 5′-part of the late hexon gene, hybridized to mature
hexon mRNA. The loops represent introns in the gene that cannot
hybridize to mRNA. **(b)** Interpretation of the electron micrograph,
showing the three intron loops (labeled A, B, and C), the hybrid (heavy
red line), and the unhybridized region of DNA upstream of the gene
(upper left). The fork at the lower right is due to the 3′-end of the
mRNA, which cannot hybridize because the 3′-end of the gene is not
included. Therefore, the mRNA forms intramolecular double-stranded
structures that have a forked appearance. **(c)** Linear arrangements of
the hexon gene, showing the three short leader exons, the two introns
separating them (A and B), and the long intron (C) separating the
leaders from the coding exon of the hexon gene. All exons are
represented by red boxes. (*Source:* (a) Berget, Susan M., Moore, Claire, and
Sharp, Phillip. *PNAS* 74: 3173, 1977.)

of genes are a bit different from those in mRNA genes.
For example, tRNA introns are relatively small, ranging
in size from 4 to about 50 bases long. Not all tRNA genes
have introns; those that do have only one, and it is adjacent to the DNA bases corresponding to the anticodon of
the tRNA. Genes in mitochondria and chloroplasts can
also have introns. Indeed, these introns are some of the
most interesting, as we will see.

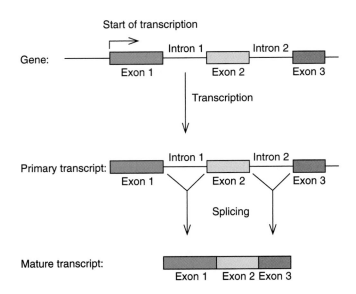

Figure 14.2   Outline of splicing. The introns in a gene are
transcribed along with the exons (colored boxes) in the primary
transcript. Then they are removed as the exons are spliced together.

SUMMARY Most higher eukaryotic genes coding
for mRNA and tRNA, and a few coding for rRNA,
are interrupted by unrelated regions called introns.
The other parts of the gene, surrounding the introns, are called exons; the exons contain the sequences that finally appear in the mature RNA
product. Genes for mRNAs have been found with
anywhere from zero to 60 introns. Transfer RNA
genes have either zero or one.

RNA Splicing

Consider the problem introns pose. They are present in
genes but not in mature RNA. How is it that the information in introns does not find its way into the mature RNA
products of the genes? The two main possibilities are:
(1) The introns are never transcribed; the polymerase
somehow jumps from one exon to the next and ignores
the introns in between. (2) The introns are transcribed,
yielding a primary transcript, an overlarge gene product
that is cut down to size by removing the introns. As
wasteful as it seems, the latter possibility is the correct
one. The process of cutting introns out of immature
RNAs and stitching together the exons to form the final
product is called **RNA splicing**. The splicing process is
outlined in Figure 14.2, although, as we will see later in
the chapter, this picture is considerably oversimplified.

How do we know splicing takes place? Actually, at the
time introns were discovered, circumstantial evidence to
support splicing already existed. A class of large nuclear
RNAs called heterogeneous nuclear RNA (hnRNA),

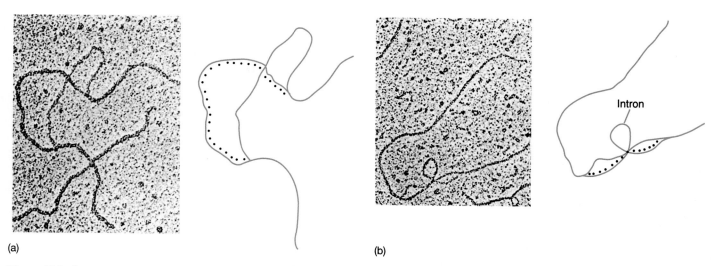

(a)

(b)

Figure 14.3   Introns are transcribed. (a) R-looping experiment in which mouse globin mRNA precursor was hybridized to a cloned mouse β-globin gene. A smooth hybrid formed, demonstrating that the introns are represented in the mRNA precursor. **(b)** Similar R-looping experiment in which mature mouse globin mRNA was used. Here, the large intron in the gene looped out, showing that this intron was no longer present in the mRNA. The small intron was not detected in this experiment. In the interpretive drawings, the dotted black lines represent RNA and the solid red lines represent DNA. (*Source:* Tilghman, S., Curtis, P. Tiemeier, P., and Weissmann, C. *PNAS* 75: 1312, 1978.)

widely believed to be precursors to mRNA, had been found (Chapter 10). These hnRNAs are the right size (larger than mRNAs) and have the right location (nuclear) to be unspliced mRNA precursors. Furthermore, hnRNA turns over very rapidly, which means it is made and converted to smaller RNAs quickly. This too suggested that these RNAs are merely intermediates in the formation of more stable RNAs. However, no direct evidence existed to show that hnRNA could be spliced to yield mRNA.

The mouse β-globin mRNA and its precursor provided an ideal place to look for such evidence. The mouse globin mRNA precursor is a member of the hnRNA population. It is found only in the nucleus, turns over very rapidly, and is about twice as large (1500 bases) as mature globin mRNA (750 bases). Also, mouse immature red blood cells make so much globin (about 90% of their protein) that α- and β-globin mRNAs are abundant and can be purified relatively easily; even their precursors exist in appreciable quantities. This abundance made experiments feasible. Furthermore, the β-globin precursor is the right size to contain both exons and introns. Charles Weissmann and Philip Leder and their coworkers used R-looping to test the hypothesis that the precursor still contained the introns.

The experimental plan was to hybridize mature globin mRNA, or its precursor, to the cloned globin gene, then observe the resulting R-loops (Figure 14.3). We know what the results with the mature mRNA should be. Because this RNA has no intron sequences, the introns in the gene will loop out. On the other hand, if the precursor RNA still has all the intron sequences, no such loops will form. That is what happened. You may have a little difficulty recognizing the structures in Figure 14.3 because

this R-looping was done with double-, instead of single-stranded, DNA. Thus, the RNA hybridized to one of the DNA strands, displacing the other. The precursor RNA gave a smooth, uninterrupted R-loop; the mature mRNA gave an R-loop interrupted by an obvious loop of double-stranded DNA, which represents the large intron. The small intron was not visible in this experiment. Notice that the term *intron* can be used for intervening sequences in either DNA or RNA.

SUMMARY Messenger RNA synthesis in eukaryotes occurs in stages. The first stage is synthesis of the primary transcription product, an mRNA precursor that still contains introns copied from the gene, if any were present. This precursor is part of a pool of large nuclear RNAs called hnRNAs. The second stage is mRNA maturation. Part of the maturation of an mRNA precursor is the removal of its introns in a process called splicing. This yields the mature-sized mRNA.

Splicing Signals

Consider the importance of accurate splicing. If too little RNA is removed from an mRNA precursor, the mature RNA will be interrupted by nonsense regions. If too much is removed, important sequences may be left out.

Given the importance of accurate splicing, signals must occur in the mRNA precursor that tell the splicing machinery exactly where to "cut and paste." What are these signals? One way to find out is to look at the base sequences of a number of different genes, locate the intron boundaries, and see what sequences are common to all of them. In principle, these common sequences could be part of the signal for splicing. The most striking observation, first made by Chambon, is that almost all introns in nuclear mRNA precursors begin and end the same way:

exon/GU–intron–AG/exon

In other words, the first two bases in the intron of a transcript are GU and the last two are AG. This kind of conservation does not occur by accident; surely the GU–AG motif is part of the signal that says, "Splice here." However, GU–AG motifs occur too frequently. A typical intron will contain several GU's and AG's within it. Why are these not used as splice sites? The answer is that GU–AG is not the whole story; splicing signals are more complex than that. The sequences of many intron–exon boundaries have now been examined and a consensus sequence has emerged:

5′-AG/GUAAGU–intron–YNCUR$\underline{A}$C–Y$_n$NAG/G-3′

where the slashes denote the exon–intron borders, Y is either pyrimidine (U or C), Y$_n$ denotes a string of about nine pyrimidines, R is either purine (A or G), $\underline{A}$ is a special A that participates in forming a branched splicing intermediate we will discuss later in this chapter, and N is any base. The consensus sequences at exon–intron boundaries in yeast mRNA precursors are also well studied, and a little different from those in mammals:

5′-/GUAUGU–intron–UACUA$\underline{A}$C–YAG/-3′

Finding consensus sequences is one thing; showing that they are really important is another. Several research groups have found ample evidence supporting the importance of these splice junction consensus sequences. Their experiments were of two basic types. In one, they mutated the consensus sequences at the splice junctions in cloned genes, then checked whether proper splicing still occurred. In the other, they collected defective genes from human patients with presumed splicing problems and examined the genes for mutations near the splice junctions. Both approaches gave the same answer: Disturbing the consensus sequences usually inhibits normal splicing.

SUMMARY The splicing signals in nuclear mRNA precursors are remarkably uniform. The first two bases of the intron are almost always GU, and the last two are almost always AG. The 5′- and 3′-splice sites have consensus sequences that extend beyond the GU and AG motifs. The whole consensus sequences are important to proper splicing; when they are mutated, abnormal splicing can occur.

14.2 The Mechanism of Splicing of Nuclear mRNA Precursors

The splicing scheme in Figure 14.2 gave only the precursor and the product, with no indication about the mechanism cells use to get from one to the other. Let us now explore the intricate and quite unexpected mechanism of nuclear mRNA precursor splicing.

A Branched Intermediate

One of the essential details missing from Figure 14.2 is that the intermediate in nuclear mRNA precursor splicing is branched, so it looks like a **lariat,** or cowboy's lasso. Figure 14.4 outlines the two-step lariat model of splicing. The first step is the formation of the lariat-shaped intermediate. This occurs when the 2′-hydroxyl group of an adenosine nucleotide in the middle of the intron attacks the phosphodiester bond between the first exon and the G at the beginning of the intron, forming the loop of the lariat and simultaneously separating the first exon from the

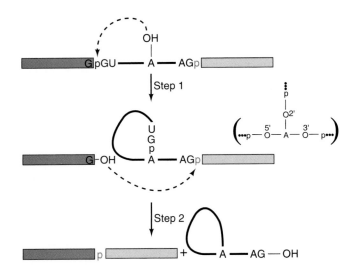

Figure 14.4   Simplified mechanism of nuclear mRNA precursor splicing. In step 1, the 2′-hydroxyl group of an adenine nucleotide within the intron attacks the phosphodiester bond linking the first exon (blue) to the intron. This attack, indicated by the dashed arrow at top, breaks the bond between exon 1 and intron, yielding the free exon 1 and the lariat-shaped intron–exon 2 intermediate, with the GU at the 5′-end of the intron linked through a phosphodiester bond to the branchpoint A. The lariat is a consequence of the internal attack of one part of the RNA precursor on another part of the same molecule. At right in parentheses is the branchpoint showing that the adenine nucleotide is involved in phosophdiester bonds through its 2′-, 3′-, and 5′-hydroxyl groups. In step 2, the free 3′-hydroxyl group on exon 1 attacks the phosphodiester bond between the intron and exon 2. This yields the spliced exon 1/exon 2 product and the lariat-shaped intron. Note that the phosphate (red) at the 5′-end of exon 2 becomes the phosphate linking the two exons in the spliced product.

intron. The second step completes the splicing process: The 3'-hydroxyl group left at the end of the first exon attacks the phosphodiester bond linking the intron to the second exon. This forms the exon–exon phosphodiester bond and releases the intron, in lariat form, at the same time.

This mechanism seemed unlikely enough that rigorous proof had to be presented for it to be accepted. In fact, very good evidence supports the existence of all the intermediates shown in Figure 14.4, much of it collected by Sharp and his research group.

First and foremost, what is the evidence for the branched intermediate? The first indication of a strangely shaped RNA created during splicing came in 1984, when Sharp and colleagues made a cell-free splicing extract and used it to splice an RNA with an intron. This splicing substrate was a radioactive transcript of the first few hundred base pairs of the adenovirus major late region. This transcript contained the first two leader exons, with a 231-nt intron in between. After allowing some time for splicing, these workers electrophoresed the RNAs and found the unspliced precursor plus a novel band with unusual behavior on gel electrophoresis (Figure 14.5). It migrated faster than the precursor on a 4% polyacrylamide gel, but slower than the precursor on a 10% polyacrylamide gel. This kind of behavior is characteristic of circular (or lariat-shaped) RNAs.

Was this strange RNA a splicing product? Yes; another experiment by Sharp's group (Figure 14.6) showed that it accumulated more and more as splicing progressed. It turned out to be the lariat-shaped intron that had been removed from the precursor. This experiment also showed the existence of another RNA with anomalous electrophoretic behavior. Its concentration rose during the first part of the splicing process, then fell later on, suggesting that it was a splicing intermediate. It is actually exon 2 with the lariat-shaped intron still attached. Both this RNA and the intron have anomalous electrophoretic behavior because they are lariat-shaped.

The two-step mechanism in Figure 14.4 allows the following predictions, each of which Sharp and colleagues verified.

1. The excised intron has a 3'-hydroxyl group. This is required if exon 1 attacks the phosphodiester bond as shown at the beginning of step 2, because this will remove the phosphate attached to the 3'-end of the intron, leaving just a hydroxyl group.

2. The phosphorus atom between the 2 exons in the spliced product comes from the 3'- (downstream) splice site.

3. The intermediate (exon 2 plus intron) and the spliced intron contain a branched nucleotide that has both its 2'- and 3'-hydroxyl groups bonded to other nucleotides.

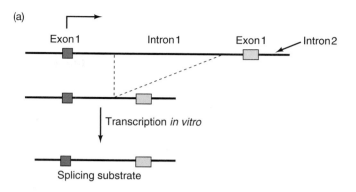

(a)

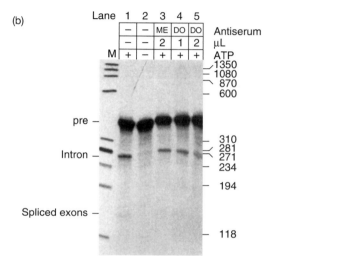

(b)

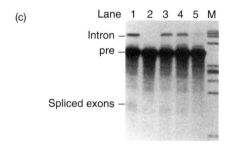

(c)

Figure 14.5 Anomalous electrophoretic behavior of spliced-out intron. (a) Splicing substrate. Sharp and colleagues started with a cloned fragment from the adenovirus major late region containing leader exons 1 (blue) and 2 (yellow), all of intron 1 and part of intron 2. They made a deletion in intron 1, yielding a shortened intron only 231 bp long. Finally, they transcribed this construct in vitro to produce a labeled splicing substrate. **(b)** Electrophoresis. Sharp and colleagues incubated the labeled splicing substrate from panel **(a)** with HeLa cell nuclear extract, then electrophoresed the resulting RNAs on **(b)** a 4%, or **(c)** a 10% polyacrylamide gel. Lane 1 and lanes 3–5 contained ATP: lane 2 did not. Lane 3 contained 2 μL of control serum; lanes 4 and 5 contained 1 and 2 μL, respectively, of an antiserum directed against an RNA–protein component required for splicing (see page 430, on snurps). Under splicing conditions (lanes 1,3, and 4), a band corresponding to the liberated intron was observed below the precursor band in the 4% gel, but above the precursor band in the 10% gel. (*Source:* (b-c) Grabowski et al., Messenger RNA splicing in vitro: an excised intervening sequence and a potential intermediate. *Cell* 37 (June 1984) f. 2, p. 417. Reprinted by permission of Elsevier Science.)

(a)

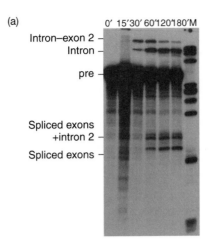

(b)

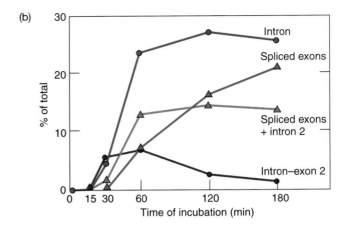

Figure 14.6   Time course of intermediate and liberated intron appearance. (a) Electrophoresis. Sharp and colleagues carried out splicing reactions as described in Figure 14.5 and electrophoresed the products after various times, indicated at top, on a 10% polyacrylamide gel. The products are identified at left. The top band contained the intron–exon 2 intermediate. The next band contained the intron. Both these RNAs were lariat-shaped, as suggested by their anomalously low electrophoretic mobilities. The next band contained the precursor. The bottom two bands contained two forms of the spliced exons: the upper one was still attached to the piece of intron 2, and the lower one seemed to lack that extra RNA. (b) Graphic presentation. Sharp and colleagues plotted the intensities of each band from panel (a) to show the accumulation of each RNA species as a function of time. (*Source:* Grabowski et al., *Cell* 37 (June 1984) f. 4, p. 419. Reprinted by permission of Elsevier Science.)

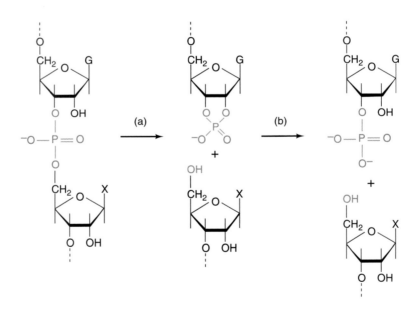

Figure 14.7   Mechanism of RNase T1. RNase T1 cuts RNA as follows: (a) The RNase cleaves the bond between the phosphate attached to the 3′-hydroxyl of a guanine nucleotide and the 5′-hydroxyl of the next nucleotide, generating a cyclic 2′, 3′-phosphate intermediate. (b) The cyclic intermediate opens up, yielding an oligonucleotide ending in a guanosine 3′-phosphate.

4. The branch involves the 5′-end of the intron binding to a site within the intron.

    Let us look at the evidence for each.

1. The excised intron has a 3′-hydroxyl group. Sharp and colleagues digested the precursor and the excised intron with ribonuclease T1. This enzyme cuts RNA after every G according to the scheme in Figure 14.7, leaving a phosphate attached to the 3′-terminal G.

This generates a set of oligonucleotide products ending in Gp. If we know the sequence of the precursor, we can predict what the RNase T1 products should be. In particular, because the intron ends in a string of (mostly) pyrimidines followed by the sequence AG, we would predict that one of the RNase T1 products (called oligonucleotide 48) would have this sequence:

UCCCUUUUUUUUCCACAGp

And, indeed, these workers obtained this oligonucleotide (among many others) when they digested the precursor. But when they digested the excised intron, they got a slightly different oligonucleotide (called oligo 48′) that migrated slightly more slowly on gel electrophoresis. Could oligo 48′ migrate more slowly because it was missing its 3′-terminal phosphate? If so, one should be able to remove this phosphate from oligo 48 and make it just like oligo 48′. Sharp and his coworkers demonstrated this by treating oligo 48 and oligo 48′ with alkaline phosphatase to remove any 3′-terminal phosphates. After this treatment, oligo 48 behaved exactly like oligo 48′. Therefore, oligo 48′ (and its precursor, the excised intron) really was missing its 3′-terminal phosphate, just as the lariat model predicts.

2. The phosphorus atom between the two exons in the spliced product comes from the second exon. Sharp and colleagues approached this problem as follows: After splicing, the sequence at the exon–exon junction is:

$$GpGpGp*Cp$$

where the starred phosphate is at the junction of exons 1 and 2. The question is, where did this junction phosphate come from? If we can show it came from the cytosine nucleotide, we have proven our case. Sharp and coworkers used a variation on an old technique called **nearest neighbor analysis** to demonstrate this.

In nearest neighbor analysis (Figure 14.8a), RNA is synthesized in the presence of a precursor (say CTP) with a ^{32}P-labeled α-phosphate. This labeled phosphate is incorporated into the phosphodiester bond on the 5′-side of this cytosine and of all other cytosines incorporated into the RNA. Next, the labeled RNA is hydrolyzed with base. This breaks all the phosphodiester bonds in the RNA, yielding mononucleotides with the phosphates attached to their 3′-hydroxyl groups. Notice that the labeled phosphate that came into the RNA on a cytosine nucleotide has been passed to its nearest neighbor on the left, a uracil nucleotide in this case. Of course, cytosine will be incorporated throughout the RNA, so each of the four nucleotides (A, C, G, and U) will be its nearest neighbor with a certain frequency and will therefore become labeled, just as UMP is labeled in this example. By separating all four labeled nucleotides and determining the radioactivity in each, one can determine how often each is C's nearest neighbor.

Figure 14.8b shows how Sharp and colleagues modified the nearest neighbor method to show where the phosphate joining the two exons comes from. They labeled the precursor RNA with [α-^{32}P]CTP to introduce a labeled phosphate just at the 5′-end of exon 2 (because the first base in exon 2 is a cytosine). Then they allowed splicing to occur and cleaved the spliced product with RNase A. This enzyme cuts after pyrimidine nucleotides, leaving a 3′-phosphate. This means that RNase A will generate the tetranucleotide shown above (GpGpGpCp). Now we can ask the question: Is this tetranucleotide labeled? The answer is yes, and the label is in the starred phosphate: GpGpGp*Cp, which must have come from the C, which was originally labeled. Thus, the phosphate at the junction of the two exons in the spliced product comes from the second exon.

3. The intermediate (exon 2 plus intron) and the spliced intron contain a branched nucleotide that has both its 2′- and 3′-hydroxyl groups bonded to other nucleotides. Sharp and coworkers cut the splicing intermediate with either **RNase T2** or **RNase P1.** Both enzymes cut after every nucleotide in an RNA, but RNase T2 leaves nucleoside-3′-phosphates just as RNase T1 does, whereas RNase P1 generates nucleoside-5′-phosphates. Both enzymes yielded novel products among the normal nucleoside monophosphates. Thin-layer chromatography allowed the charges of these two products to be determined. The T2 product had a charge of –6, whereas the P1 product had a charge of –4. An ordinary mononucleotide would have a charge of –2.

What are these unusual products? Their charges are consistent with the structures shown in Figure 14.9, given that each phosphodiester bond has one negative charge and each terminal phosphate has two negative charges. To prove that these structures were correct, Sharp and colleagues treated the RNase P1 product with periodate and aniline to remove the 2′- and 3′-nucleosides by β-elimination. The product of this reaction should be a nucleoside 2′,3′,5′-trisphosphate. To verify this assignment, these workers subjected the product to two-dimensional thin-layer chromatography and found that it comigrated with adenosine 2′,3′, 5′-trisphosphate. Thus, a branched nucleotide occurs, and it is an adenine nucleotide.

4. The branch involves the 5′-end of the intron binding to a site within the intron. Figure 14.10 shows the strategy Sharp and colleagues used to demonstrate the joining of the 5′-end of the intron to the branchpoint of the lariat. They hybridized the lariat-shaped splicing intermediate or spliced intron to a single-stranded DNA representing the 5′-end of the intron, then digested the hybrid with RNase T1. The sequence of the intron predicts that one of the RNase T1 products of the unspliced intron will be a 14-mer called oligo 46. However, previous experiments had shown that the spliced intron yields an alternative version called

(a) Nearest neighbor analysis:

1. Incorporation of (α - ^{32}P) nucleotides:

$$pppApGpU - OH + pp\overset{*}{p}C \longrightarrow pppApGpU\overset{*}{p}C - OH$$

2. Base hydrolysis:

$$pppApGpU\overset{*}{p}C - OH \xrightarrow{OH^-} pppAp + Gp + U\overset{*}{p} + C - OH \longrightarrow \text{analyze labeled products}$$

(b) Identifying origin of phosphate between spliced exons:

1. Incorporation of (α - ^{32}P) nucleotides:

$$\underline{\text{exon 1}}\ GpGpGp \vdots GpU \xrightarrow{\text{intron}} ApG - OH + pp\overset{*}{p}C \longrightarrow \underline{\text{exon 1}}\ GpGpGp \vdots GpU \xrightarrow{\text{intron}} ApG\overset{*}{p} \vdots Cp\ \underline{\text{exon 2}}$$

2. Splice:

$$\underline{\text{exon 1}}\ GpGpG\overset{*}{p} \vdots Cp\ \underline{\text{exon 2}}$$

3. RNase A:

$$GpGpG\overset{*}{p}Cp$$

4. Base hydrolysis:

$$Gp + Gp + G\overset{*}{p} + Cp$$

Figure 14.8 Strategy of an experiment to trace the origin of the phosphate between the two exons in a spliced transcript. (a) Nearest neighbor analysis. This technique is not the one Sharp and colleagues used, but it embodies the same principle. First, label RNA by synthesizing it in the presence of an α-^{32}P-labeled nucleotide, CTP in this case. The labeled phosphate (red) is incorporated between the C and its nearest neighbor on the 5′-side, a U in this case. Next, hydrolyze the labeled RNA with alkali, which cleaves on the 3′-side of every base, yielding mononucleotides. Notice that this transfers the labeled phosphate to the nearest neighbor on the 5′-side, so the uracil nucleotide is now labeled instead of the cytosine nucleotide. Finally, separate all four nucleotides and determine the radioactivity in each. This tells how frequently each nucleotide is C's nearest neighbor on the 5′-side. **(b)** Identifying the origin of the phosphate between spliced exons. Step 1: Sharp and his colleagues labeled the splicing precursor with [α-^{32}P] CTP, which labels the phosphate (red) between the intron and the second exon. Step 2: Our hypothesis of splicing places the labeled phosphate in the bond between the two spliced exons. The alternative scheme would involve the phosphate originally attached to the end of the first exon (blue in step 1), which would not be labeled because it entered the RNA on a GTP (the first G in the intron). Step 3: Next, Sharp and coworkers cleaved the spliced RNA with RNase A, which cuts after the pyrimidine C, yielding the oligonucleotide GpGpGpCp, in which the phosphate between the last G and the C is labeled (red). Finally, they cleaved this oligonucleotide with alkali, cutting it into individual nucleotides. They analyzed these nucleotides for radioactivity and found that GMP was labeled. This indicated that the labeled phosphate at the end of the intron really did form the bridge between spliced exons in the final RNA.

oligo 46′, which has the branched nucleotide. This branch is what distinguishes oligo 46′ from 46. Sharp's group wanted to know whether oligo 46′ was attached to the 5′-end of the intron. If so, it should hybridize to the DNA probe complementary to the 5′-end of the intron. To find out if it did, they isolated the hybrid containing the 5′-end of the intron (after the RNase cleavage in step [b]), denatured it to remove the hybridized DNA probe, then subjected the released RNA to another round of RNase T1 digestion to identify the region of RNA attached to the 5′-end of the intron. As anticipated, they found oligo 46′, demonstrating that the 5′-end of the intron is attached to the branched oligonucleotide and is therefore almost certainly involved in the branch itself.

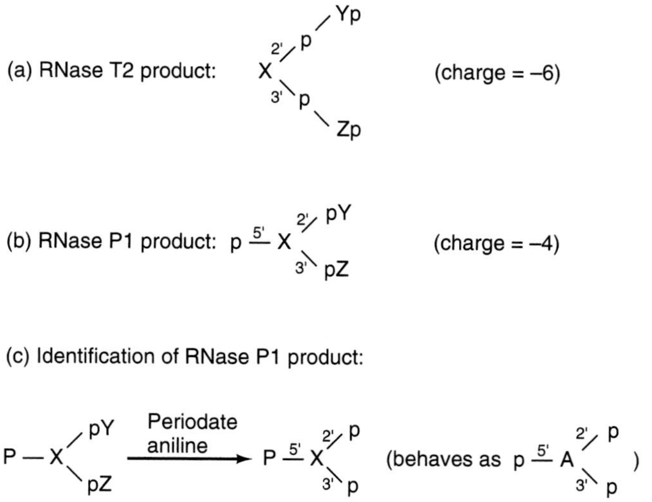

(a) RNase T2 product: (charge = –6)

(b) RNase P1 product: (charge = –4)

(c) Identification of RNase P1 product:

$$P - X \begin{smallmatrix} pY \\ pZ \end{smallmatrix} \xrightarrow[\text{aniline}]{\text{Periodate}} P \underset{3'}{\overset{5'}{-}} X \begin{smallmatrix} 2' \\ p \\ 3' \\ p \end{smallmatrix} \quad (\text{behaves as } p \underset{3'}{\overset{5'}{-}} A \begin{smallmatrix} 2' \\ p \\ 3' \\ p \end{smallmatrix})$$

Figure 14.9 Direct evidence for a branched nucleotide. (a) Sharp and colleagues digested the splicing intermediate with RNase T2. This yielded a product with a charge of –6. This is consistent with the branched structure pictured here. **(b)** Digestion with RNase P1 gave a product with a charge of –4, consistent with this branched structure. **(c)** Sharp and colleagues treated the P1 product with periodate and aniline to eliminate the nucleosides bound to the 2′- and 3′-phosphates of the branched nucleotide. The resulting product copurified with adenosine-2′,3′,5′-trisphosphate, verifying the presence of a branch and demonstrating that the branch occurs at an adenine nucleotide.

A Signal at the Branch

Is there something special about the adenine nucleotide that participates in the branch, or can any A in the intron serve this function? Study of many different introns has revealed the existence of a consensus sequence, and the fact that this sequence, and no other, can form the branch.

The first hint of a special region within the intron came from experiments with the yeast actin gene performed by Christopher Langford and Dieter Gallwitz in 1983. These workers cloned the actin gene, made numerous mutations in it, and reintroduced these mutant genes into normal yeast cells. Then they assayed for splicing by S1 mapping. Figure 14.11 shows the results: First, when they removed a region between 35 and 70 bp upstream of the intron's 3′-splice site (mutant number 1), they blocked splicing. This suggested that this 35-bp region contains a sequence, represented in the figure by a small red box,

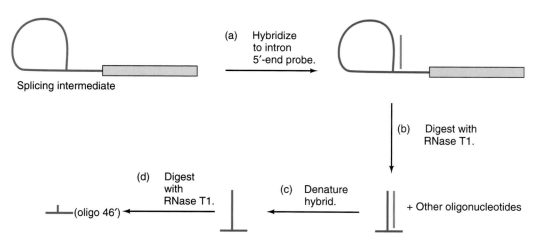

Figure 14.10 Demonstration that the 5′-end of the intron is involved in the branch. (a) Sharp and colleagues hybridized a DNA probe for the 5′-end of the intron to the lariat-shaped splicing intermediate. Because the 5′-end of the intron is involved in the branch, that is where the probe hybridized. **(b)** Next, these workers digested the hybrid with RNase T1. The branch DNA probe protected the 5′-end of the intron from digestion. **(c)** Next, they denatured the hybrid to release the protected RNA fragment. **(d)** Finally, they digested again with RNase T1 and analyzed the oligonucleotides to see what was attached to the 5′-end of the intron. They found oligo 46′, the branched oligonucleotide, demonstrating that the 5′-end of the intron is involved in the branch.

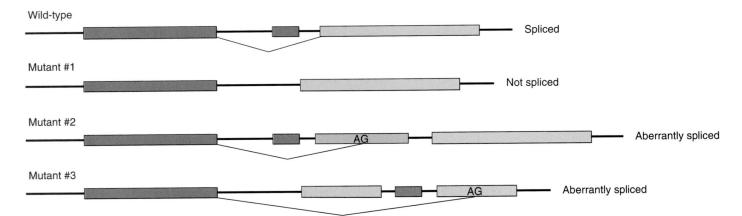

Figure 14.11   Demonstration of a critical signal within a yeast intron. Langford and Gallwitz made mutant yeast actin genes in vitro, reintroduced them into yeast cells, and tested them for splicing there. The wild-type gene contained two exons (blue and yellow). The intron contained a conserved sequence (red) found in all yeast introns. Yeast cells spliced this gene properly. To make mutant number 1, Langford and Gallwitz deleted the conserved intron sequence, which destroyed the ability of this gene's transcript to be spliced. Mutant number 2 had extra, nonintron DNA (pink) inserted into the intron downstream of the conserved intron sequence. The transcript of this gene was aberrantly spliced to the first AG within the insert. To construct mutant number 3, Langford and Gallwitz moved the conserved intron sequence downstream into the second exon. The transcript of this gene was aberrantly spliced to the first AG downstream of the relocated conserved sequence. These experiments suggested that the conserved sequence is critical for splicing and that it designates a downstream AG as the 3'-splice site.

that is important for splicing. When they inserted an extra DNA segment between this "special sequence" and the second exon (mutant number 2), splicing occurred from the usual 5'-splice site, but not to the correct 3'-splice site. Instead, the aberrant 3'-splice site was the first AG downstream of the special intron sequence. This AG lay within the inserted segment of DNA. This result suggested that the special intron sequence tells the splicing machinery to splice to an AG at some appropriate distance downstream. If one inserts a new AG in front of the usual one, splicing may go to the new site. Finally, mutant number 3 contained the special intron sequence within the second exon. Again in this case, the 3'-splice site became the first AG downstream of the special sequence in its new location, which happened to be in the second exon.

The special intron sequence is so important because it contains the branchpoint adenine nucleotide: the final A in the sequence UACUAAC. In fact, this is the nearly invariant sequence around the branchpoint in all yeast nuclear introns. Higher eukaryotes have a more variable consensus sequence surrounding the branchpoint A: $U_{47}NC_{63}U_{53}R_{72}\underline{A}_{91}C_{47}$, where R is either purine (A or G), and N is any base. The subscripts indicate the frequency with which a base is found in that position. For example, the branchpoint A (underlined) is found in this position 91% of the time. The first U is frequently replaced by a C, so this position usually contains a pyrimidine.

SUMMARY In addition to the consensus sequences at the 5'- and 3'-ends of nuclear introns, branchpoint consensus sequences also occur. In yeast, this sequence is almost invariant: UACUAAC. In higher eukaryotes, the consensus sequence is more variable. In all cases, the branched nucleotide is the final A in the sequence. The yeast branchpoint sequence also appears to tell the splicing machinery which downstream AG to select as the 3'-splice site.

Spliceosomes

Edward Brody and John Abelson discovered in 1985 that the lariat-shaped splicing intermediates in yeast are not free in solution, but bound to 40S particles they called **spliceosomes.** These workers added labeled pre-mRNAs to cell-free extracts and used a glycerol gradient ultracentrifugation procedure to purify the spliceosomes. Figure 14.12 shows a prominent 40S peak containing labeled RNAs. Analysis of these RNAs by electrophoresis revealed the presence of lariats: the splicing intermediate and the spliced-out intron. To further demonstrate the importance of these spliceosomes to the splicing process, Brody and Abelson tried to form spliceosomes with a mutant pre-mRNA that had an A→C mutation at the branchpoint that rendered it unspliceable. This RNA was severely impaired in its ability to form spliceosomes.

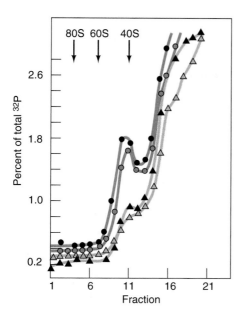

Figure 14.12   Yeast spliceosomes. Brody and Abelson incubated a labeled yeast pre-mRNA with a yeast splicing extract, then subjected the mixture to glycerol gradient ultracentrifugation. Finally, they determined the radioactivity in each gradient fraction by scintillation counting. Two different experiments with a wild-type pre-mRNA (red) and two different experiments with a mutant pre-mRNA with a base alteration at the 5′-splice site (blue) are shown. The wild-type pre-mRNA and, to a lesser extent, the mutant pre-mRNA show a clear association with a 40S aggregate. (*Source:* Reprinted from Brody and Abelson, *Science* 228:965, 1985. Copyright © 1985 American Association for the Advancement of Science, Washington, DC. Reprinted by permission.)

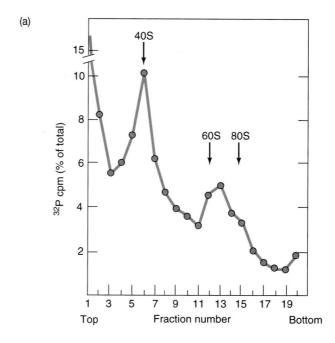

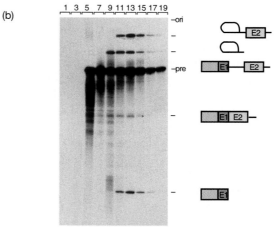

Figure 14.13   Human spliceosomes. (a) Glycerol gradient. Sharp and colleagues started with the same labeled adenovirus major late splicing substrate described in Figure 14.5. They placed this RNA into a HeLa cell nuclear extract under splicing conditions for 30 min, then subjected the labeled RNAs to ultracentrifugation on a glycerol gradient. Radioactivity is plotted against the fraction number from the gradient, with the bottom of the centrifuge tube at right. Two major peaks, at 40S and 60S, are apparent. **(b)** Electrophoresis. To look for splicing intermediates and products in the two peaks. Sharp and colleagues electrophoresed individual fractions and subjected the gel to autoradiography. The RNA bands are identified by name and diagram of structure at right. The lariat represents the intron; E1 (blue) and E2 (yellow) are exons 1 and 2, respectively; the pink box represents vector sequences included in the splicing substrate by in vitro transcription. Clearly, the 60S peak contains splicing intermediates and products. (*Source:* Grabowski et al., A multicomponent complex is involved in the splicing of messenger RNA precursors. *Cell* 42 (August 1985) f. 1, p. 346. Reprinted by permission of Elsevier Science.)

Sharp and his colleagues isolated spliceosomes from human (HeLa) cells, also in 1985, and showed that they sedimented at 60S. First, these investigators added a labeled adenovirus splicing substrate to a HeLa cell nuclear extract, then subjected the complexes to glycerol gradient ultracentrifugation. Figure 14.13a depicts the profile of radioactivity they observed, which includes two major peaks, at 40S and 60S. To determine which of these peaks had splicing activity, Sharp and coworkers extracted RNAs from fractions across the gradient and electrophoresed them. Figure 14.13b shows the results. The splicing intermediate (intron–exon 2) was found exclusively in the 60S peak, as was an RNA containing only the first exon. The splicing products, intron and spliced exons, were not as obviously confined to the 60S peak, but they were not detectable in the 40S peak. The only labeled RNA found in the 40S peak was the unspliced substrate. It is clear that the 60S particles, and not the 40S particles, are authentic spliceosomes because they do not form in the cold or in the absence of ATP, whereas the 40S particles do. Furthermore, the 40S particles form even with an RNA that lacks splice sites, whereas the 60S particles do not.

SUMMARY Splicing takes place on a particle called a spliceosome. Yeast spliceosomes and mammalian spliceosomes have sedimentation coefficients of about 40S, and about 60S, respectively.

Snurps

In principle, consensus sequences at the ends and branch-point of an intron could be recognized by either proteins or nucleic acids. We now have excellent evidence that small RNAs called **small nuclear RNAs (snRNAs)** are the agents that recognize these critical splicing signals. Joan Steitz and her coworkers first focused attention on snRNAs as potential participants in splicing. These snRNAs exist in the cell coupled with proteins in complexes called **small nuclear ribonuclear proteins (snRNPs,** pronounced "snurps"). They can be resolved by gel electrophoresis into individual species designated **U1, U2, U4, U5,** and **U6.** All these snRNPs join the spliceosome and play crucial roles in splicing.

U1 snRNP   Joan Steitz and, independently, J. Rogers and R. Wall, noticed in 1980 that U1 snRNA has a region whose sequence is almost perfectly complementary to both 5′- and 3′-splice site consensus sequences. They proposed that U1 snRNA base-paired with these splice sites, bringing them together for splicing. We now know that splicing involves a branch within the intron, which rules out such a simple mechanism. Nevertheless, base pairing between U1 snRNA and the 5′-splice site not only occurs, it is essential for splicing.

We know that this base pairing with U1 is essential because of genetic experiments performed by Yuan Zhuang and Alan Weiner in 1986. They introduced alterations into one of the three alternative 5′-splice sites of the adenovirus E1A gene. Splicing of this gene normally occurs from each of these 5′-sites to a common 3′-site to yield three different mature mRNAs, called 9S, 12S, and 13S (Figure 14.14). The mutations (at the 12S 5′-splice site) disturbed the potential base pairing with U1. To measure the effects of these mutations on splicing, Zhuang and Weiner performed an RNase protection assay on RNA from cells transfected with plasmids bearing the 5′-splice site mutations in the E1A gene. Figure 14.14 shows the length in nucleotides (nt) of the signals expected from splicing at each of the three sites.

The first mutation Zhuang and Weiner tested was actually a double mutation. The fifth and sixth bases (+5 and +6) of the intron were changed from GG to AU (Figure 14.15a). This disrupted a GC base pair between the G(+5) of the intron and a C in U1, but introduced a new potential base pair between U(+6) of the intron and an A in U1. In spite of this new potential base pair, the overall

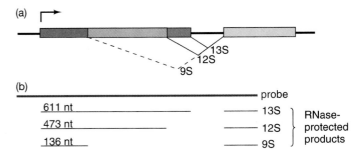

Figure 14.14   Splicing scheme of adenovirus E1A gene and RNase protection assay to detect each spliced product. (**a**) Splicing scheme. Three alternative 5′-splice sites (at the borders of the red, orange, and blue blocks and at the end of the blue block) combine with one 3′-splice site at the beginning of the yellow block to produce three different spliced mRNAs: the 9S, 12S, and 13S mRNAs, respectively. (**b**) RNase protection assay. The labeled riboprobe is represented by the purple line at top. Each alternative splicing product protects different-size fragments of this probe from digestion by RNase. These sizes in nucleotides (nt) are given above each fragment. (*Source:* Reprinted from Zhuang and Weiner, *Cell* 46:829, 1986. Copyright 1986, with permission from Elsevier Science.)

base pairing between mutant splice site and U1 should have been weakened because the number of *contiguous* base pairs was lower. Was splicing affected? Figure 14.16 (lane 4) shows that the mutation essentially abolished splicing at the 12S site and caused a concomitant increase in splicing at the 13S and 9S sites. Next, these workers made a compensating mutation in the U1 that restored base pairing with the mutant splice site. They introduced the mutant U1 gene into HeLa cells on the same plasmid that bore the mutant E1A gene. Figure 14.16 (lane 5) shows that this mutant U1 not only restored base pairing, it also restored splicing at the 12S site.

Thus, base pairing between the splice site and U1 is required for splicing. But is it sufficient? If one could make a mutant splice site with weakened base pairing to U1 whose splicing could not be suppressed by a compensating mutation in U1, one could prove that this base pairing is not enough to ensure splicing. Figures 14.15b and 14.16 show how Zhuang and Weiner demonstrated just this. This time, they mutated the 13S 5′-splice site, changing an A to a U in the +3 position, which interrupted a string of six base pairs. This abolished 13S splicing, while stimulating 12S and, to a lesser degree, 9S splicing (Figure 14.16, lane 6). A compensating mutation in the U1 gene restored the six base pairs, but failed to restore splicing at the 13S site (lane 7). Thus, base pairing between the 5′-splice site and U1 is not sufficient for splicing.

SUMMARY Genetic experiments have shown that base pairing between U1 snRNA and the 5′-splice site of an mRNA precursor is necessary, but not sufficient, for splicing.

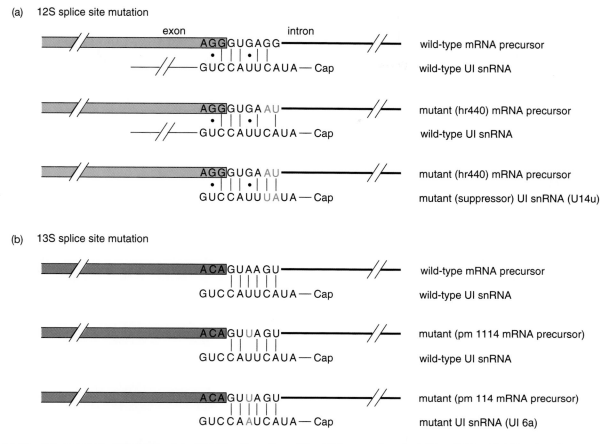

Figure 14.15   Alignment of wild-type and mutant 5′-splice sites with wild-type and mutant U1 snRNAs. (a) 12S splice site mutation. The wild-type and mutant sequences are identified at right. Watson–Crick base pairs between the mRNA precursor and U1 RNA are represented by vertical lines; wobble base pairs, by dots. Mutated bases are represented by red letters. The end of the exon is represented by an orange box as in Figure 14.14. **(b)** 13S splice site mutation. All symbols as in panel **(a)** except that the end of the exon is represented by a blue box as in Figure 14.14.

U6 snRNP   Why do base changes in U1 sometimes fail to compensate for base changes in the 5′-splice site? We can imagine a variety of answers to this question, including the possibility that some protein or proteins must also recognize the sequence at the 5′-splice site. In that case, changes in U1 might not be enough to restore recognition of this site by the spliceosome. It is also possible that *another* snRNA must base-pair with the 5′-splice site. Altering the U1 sequence to match a mutant splice site would not restore its pairing to this other snRNA, so splicing could still be prevented.

Two research groups, led by Christine Guthrie and Joan Steitz, have shown that another snRNA does indeed base-pair with the 5′-splice site. This is U6 snRNA. Steitz first demonstrated that U6 might be involved in events near the 5′-splice site when she showed that U6 could be chemically cross-linked to intron position +5. Based on this finding, she postulated that the ACA in the invariant sequence ACAGAG in U6 base-pairs with the conserved UGU in positions +4 to +6 of 5′-splice sites (Figure 14.17).

Erik Sontheimer and Joan Steitz used cross-linking studies to show that U6 binds to a site very close to the 5′-end of the intron in the spliceosome. Their experimental strategy went like this: First they made a model splicing precursor with a single intron, flanked by two exons (the first two exons and the first intron of the adenovirus major late locus). Then they substituted 4-thiouridine (4-thioU) for the nucleotides at either of two positions: the last nucleotide in the first exon, or the second nucleotide of the intron. The 4-thioU residue is photosensitive; when it is activated by ultraviolet light, it forms covalent cross-links to other RNAs with which it is in contact. By isolating these cross-linked structures, they could discover the RNAs that base-pair with the nucleotides at the 5′-splice site.

Figure 14.18 shows the cross-links that occurred when Sontheimer and Steitz placed the 4-thioU in the second position of the intron. In addition to a link with U2, which we will discuss later in this chapter, this experiment showed a linkage between U6 and the second nucleotide in the intron. Moreover, this analysis revealed the timing

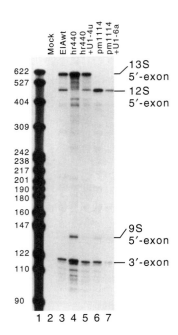

Figure 14.16   Results of RNase protection assay. Zhuang and Weiner tested the wild-type and mutant 5′-splice sites and wild-type and mutant U1 snRNAs pictured in Figure 14.15 by transfecting HeLa cells with plasmids containing these genes, then detected splicing by RNase protection as illustrated in Figure 14.14. Lane 1, size markers, with lengths in base pairs indicated at left. Lane 2, mock-transfected cells (negative control). Lane 3, wild-type E1A gene with wild-type U1 snRNA. Signals were visible for the 13S and 12S products, but not for the 9S product, which normally does not appear until late in infection. Lane 4, mutant hr440 with an altered 12S 5′-splice site. No 12S signal was apparent. Lane 5, mutant hr440 plus mutant U1 snRNA (U1–4u). Splicing at the 12S 5′-site was restored. Lane 6, mutant pm1114 with an altered 13S 5′-splice site. No 13S signal was apparent. Lane 7, mutant pm1114 plus mutant U1 snRNA (U1–6a). Even though base pairing between 5′-splice site and U1 snRNA was restored, no 13S splicing occurred. (*Source:* Zhuang and Weiner, A compensatory base change in U1 snRNA suppresses a 5′ splice site mutation. *Cell* 46 (12 Sept 1986) f. 1a, p. 829. Reprinted by permission of Elsevier Science.)

of formation of this U6–intron linkage. It did not appear in significant amounts until 20 min into the splicing reaction. By comparison, the intron–exon2 (intron–E2) lariat became prevalent at 15 min. This finding suggested that the first step in splicing, formation of the lariat, occurs before U6 associates with the 5′-end of the intron. However, as we will see, there is good evidence that U6 has already associated with the 5′-splice site before the first step in splicing. Thus, it is probably just the association between U6 and this particular nucleotide in the intron that occurs after the initial step.

How do we know that U6 associates with the splicing substrate before the formation of the lariat intermediate? David Wassarman and Steitz used another kind of cross-linking and Northern blotting to demonstrate this point. Instead of incorporating the cross-linking agent into the splicing substrate itself, they added the agent (psoralen) to the solvent. This has the disadvantage that it might be ex-

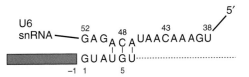

Figure 14.17   A model for interaction between a yeast 5′-splice site and U6 snRNA. The invariant ACA (nt 47–49) of yeast U6 base-pairs with the UGU (nt 4–6) of the intron. (*Source:* From Lesser and Guthrie, *Science* 262:1983, 1993. Copyright © 1993 American Association for the Advancement of Science, Washington, DC. Reprinted by permission.)

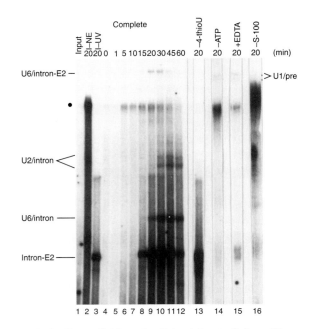

Figure 14.18   Cross-linking of splicing intermediates with 4-thioU. Sontheimer and Steitz incubated a labeled adenovirus splicing precursor containing 4-thioU in the second position of the intron with a nuclear splicing extract. After various periods of time (indicated at top), they removed samples, UV-irradiated them to cross-link the 4-thioU to any RNAs in the neighborhood, and electrophoresed the products. Lane 1, input RNA with no incubation; lane 2, 20-min incubation with no nuclear extract (NE); lane 3, 20-min incubation followed by no UV-irradiation; lanes 4–12, incubation for the times indicated at top; lane 13, no 4-thioU labeling; lane 14, no ATP; lane 15, EDTA added to chelate magnesium and block splicing; lane 16, a fraction clarified by high-speed ultracentrifugation was used instead of nuclear extract. The dot at left denotes a band resulting from an intramolecular cross-link. "E2" stands for "exon 2." (*Source:* Sontheimer and Steitz, The U5 and U6 small nuclear RNAs as active site components of the spliceosome. *Science* 262 (24 Dec 1993) f. 3, p. 1991. © AAAS.)

cluded from the active site of the spliceosome and therefore fail to reveal an RNA–RNA complex, but it has the advantage that it can be reversed to reveal the RNA species that associate at any given time in the splicing process.

Figure 14.19 shows the results of a psoralen cross-linking experiment that demonstrated association between U6 and the splicing precursor (as well as other RNA–RNA associations). Wassarman and Steitz followed these steps in performing the experiment:

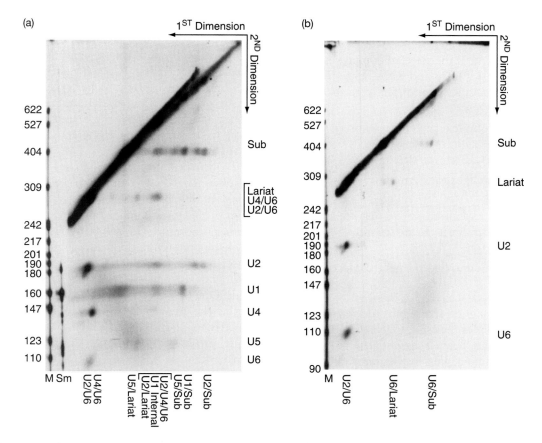

Figure 14.19   Two-dimensional electrophoresis to detect splicing complexes. (a) Complexes involving U1, U2, U4, and U5. Wassarman and Steitz formed splicing complexes using a biotin-labeled splicing precursor, cross-linked RNAs in the complexes with psoralen, and isolated snRNP-containing complexes by first precipitating with an anti-m$_3$G antibody (which recognizes all snRNPs except U6), then precipitating with streptavidin beads, which recognize the splicing precursor. Next they end-labeled the RNAs in the complexes with [^{32}P]pCp and RNA ligase, then electrophoresed these complexes in the horizontal dimension as shown at top right. They dissociated the complexes with ultraviolet light (254 nm) and electrophoresed the RNAs in the vertical dimension to identify each RNA in each complex. The marker positions at right identify the individual RNA species in each complex, and the mixtures of RNAs in the complexes are suggested at bottom. Sometimes, two or more complexes copurified. This was the case with U2/U4/U6, a U1 internally linked molecule, and the U2–lariat complex. **(b)** Complexes containing U6. Because U6 does not have an m$_3$G cap, Wassarman and Steitz isolated complexes containing this snRNP using a biotin-labeled U6-specific oligonucleotide and avidin beads. Other conditions were as in panel **(a)**. The marker positions at right identify both splicing substrate and lariat, as well as U2, as components of U6-containing complexes. U6 itself is not detectable in the first two of these complexes, but we know it is there because these complexes were selected by a U6-specific probe. Sm refers to the snRNAs that can be immunoprecipitated with anti-snRNP antiserum from systemic lupus erythematosis patients. This antiserum reacts with a set of proteins (Sm proteins) found in U1, U2, U4, U5, and U6 snRNPs. Sub refers to the splicing substrate. (*Source:* Wasserman and Steitz, Interactions of small nuclear RNA's with precursor messenger RNA during *in vitro* splicing. *Science* 257 (25 Sept 1992) f. 2, p. 1919. © AAAS.)

1. They incubated a HeLa cell nuclear extract for 30 min with an adenovirus major late splicing substrate that was labeled with biotin so it could be easily purified later.

2. They added psoralen and exposed the mixture to 365-nm light during a further 60-min incubation to cross-link RNAs within splicing complexes.

3. They did a double selection to collect all splicing complexes. By definition, these complexes should contain two things: (a) snRNPs and (b) the splicing precursor or lariat intermediate. This two-stage selection for snRNPs started with a step to isolate everything containing a spliceosomal snRNP (U1, U2,

U4, U5, or U6). All of these except U6 have a cap containing trimethylguanine (m$_3$G), so they were selected by immunoprecipitation with an antibody directed against m$_3$G). To find complexes containing U6, they used a U6-specific oligonucleotide tagged with biotin. Any complexes containing U6 formed hybrids with this oligonucleotide and thus became precipitable with streptavidin beads, because streptavidin binds tightly to biotin. In the second step, Wassarman and Steitz began with the snRNP-containing complexes isolated in the first stage and selected complexes that also contained the biotin-labeled splicing substrate, using streptavidin beads.

This was not done with the U6-complexes because they contained non-biotin-labeled splicing substrate.

4. They labeled the RNAs in the selected complexes at their 3′-ends with [5′-^{32}P]pCp and RNA ligase. This enzyme treats pCp as an RNA and joins it to the 3′-end of the other RNA.

5. They subjected the complexes to electrophoresis in one dimension.

6. They exposed the electrophoretic gel containing the complexes to 254-nm light for 20 min to reverse the cross-links between RNAs in the complexes.

7. They turned the electrophoretic gel 90 degrees and subjected it to electrophoresis in the second dimension.

8. Finally, they autoradiographed the gel to detect labeled RNAs.

The beauty of reversing the cross-links becomes apparent at this point. Complexes that stay together, or individual RNAs, do not change their electrophoretic mobilities, so they line up in a diagonal line in the autoradiograph. However, complexes that fall apart because of the reversal of cross-links give rise to individual RNAs that migrate more rapidly in the second dimension and are found below the diagonal line. These can be identified by their mobilities. Figure 14.19a shows complexes involving U1, U2, U4, U5, the splicing substrate, and the lariat intermediate. Figure 14.19b shows that U6 is found in complexes with both the substrate and lariat. U6 is hard to see in this picture, but the substrate and lariat are clearly present, and we infer that U6 is present because the complexes were selected with a U6-specific oligonucleotide. Thus, U6 binds to the splicing substrate both before and after the initial step in splicing. This experiment also demonstrated the presence of a U2–U6 complex, which can be predicted based on sequence complementarity between these two RNAs. Later in this chapter we will see how base pairing between U2 and U6 helps to form a structure that constitutes the active site of the spliceosome. The identities of the snRNPs in each complex were confirmed by Northern blotting, using labeled antisense snRNAs as probes. When these workers used U6 antisense RNA as the probe, they detected the U6–substrate and U6–lariat complexes and showed that the latter complex formed within 15 min.

SUMMARY The U6 snRNP associates with the 5′-end of the intron by base pairing. This association first occurs prior to formation of the lariat intermediate, but its character may change after this first step in splicing. The association between U6 and the splicing substrate is essential for the splicing process. U6 also associates with U2 during splicing.

U2 snRNP The consensus branchpoint sequence in yeast is complementary to a sequence in U2 snRNP, as shown in Figure 14.20, and genetic analysis has shown that base pairing between these two sequences is essential for splicing. Christine Guthrie and her colleagues provided such genetic evidence when they mutated the branchpoint sequence and showed that the defective splicing this caused could be reversed by a complementary mutation in the yeast U2 gene.

To do these experiments, these workers provided a histidine-dependent yeast mutant with a fused actin-*HIS4* gene containing an intron in the actin portion. If the transcript of this gene is spliced properly, the *HIS4* part of the fusion protein product will be active, and the cells can live on media containing the histidine precursor histidinol, because the *HIS4* product converts histidinol to histidine. Next, they introduced mutations into the splicing branchpoint. One of these, a U to A change in position 257, converted the nearly invariant sequence UACUAAC to UACAAAC and inhibited splicing by 95%. This also prevented growth on histidinol. Another mutation, a C to A transversion in position 256, converted the branch sequence UACUAAC to UAAUAAC and inhibited splicing by 50%.

To test for suppression of these mutations by mutant U2s, Guthrie and colleagues introduced a plasmid bearing the mutant U2s into yeast. They made sure the plasmid was retained by endowing it with a selectable marker: the *LEU2* gene. (The host cells were *LEU*⁻.) It was necessary to provide an *extra* copy of the U2 gene because making a mutation in the cell's only copy of the U2 gene could cause the splicing of all other genes to fail. Figure 14.21 shows that the U2s that restored complementary binding to the mutant branch sites really did restore splicing. This was especially apparent in the case of the A257 mutant, where no growth was observed with the wild-type U2, but abundant growth occurred with the U2 that had the mutation that restored base pairing with the mutant branch site.

Besides base-pairing with the branchpoint, U2 also base-pairs with U6. This association can be predicted on the basis of the sequences of the two RNAs, and genetic analysis by Guthrie and her colleagues provided direct evidence for the base pairing. First, Guthrie and colleagues discovered lethal mutations in the ACG sequence of yeast U6, which base-pairs to another snRNA, U4 (see Figure 14.25). These workers showed in two ways that the ability of these mutants to disrupt base pairing with U4 was not the problem. First, they introduced corresponding mutations into U4 that would cause the same disruption of the U4–U6 interaction and showed that these did not affect cell growth. Second, they introduced compensating mutations into U4 that would restore base pairing with the mutant U6 and showed that these did not suppress the lethal U6 mutations.

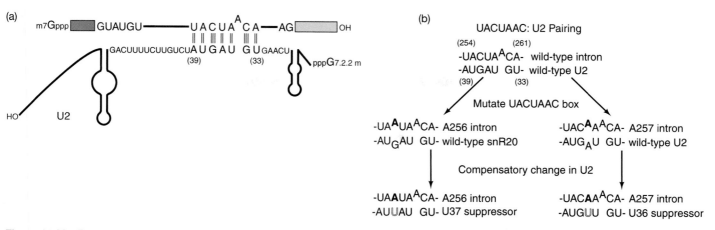

Figure 14.20 Base pairing between yeast U2 and yeast branchpoint sequences. (a) Proposed base pairing between wild-type yeast U2 and the invariant yeast branchpoint sequence. Note that the A at the branch site bulges out (top) and does not participate in the base pairing. (b) Proposed base pairing between wild-type and mutant yeast U2s and branchpoints. The bold letters indicate mutations (A's) introduced into the branchpoint at positions 256 and 257; the outlined letters represent compensating mutations (U's) introduced into U2. (*Source:* Reprinted from Parker et al., *Cell* 49:230, 1987. Copyright 1987, with permission from Elsevier Science.)

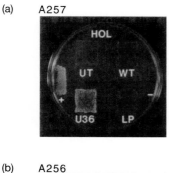

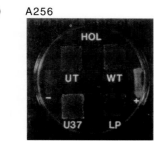

Figure 14.21 Demonstration of U2 snRNP-branchpoint base pairing by mutation suppression. Growth of A257 (panel a) and A256 (panel b) mutants on HOL medium was measured in the presence of wild-type and suppressor mutant U2. The abbreviations under each patch of cells denote the nature of the U2 added, if any: UT, untransformed (no U2 added); WT, wild-type U2; U36, U2 with mutation that restores base pairing with A257; U37, U2 with mutation that restores base pairing with A256; LP, a colony that lost its U2 plasmid. The positive control in each plate (+) contained a wild-type fusion gene and no extra U2. The negative control in each plate contained no fusion gene. (*Source:* Parker et al., Recognition of the TACTAAC box during mRNA splicing in yeast involves base pairing to the U2-like snRNA. *Cell* 49 (24 Apr 1987) f. 3, p. 232. Reprinted by permission of Elsevier Science.)

Apparently, U6 interacts with something else besides U4, and the lethal U6 mutations interfere with this interaction. Hiten Madhani and Christine Guthrie demonstrated that U2 is the other molecule with which U6 interacts. They introduced lethal mutations into residues

56–59 of U6 and found that these mutations could be suppressed by compensating mutations in residues 23 and 26–28 of U2, which restored base pairing with the mutant U6 molecules. This crucial base pairing between U2 and U6 forms a region called helix I, which will be summarized later in Figure 14.26.

Other workers (Jian Wu and James Manley, and Banshidar Datta and Alan Weiner) have used similar genetic analysis of splicing efficiency in mammalian cells to demonstrate interaction between the 5′-end of U2 and the 3′-end of U6, to form another base-paired domain called helix II. Mutations in U2 could be suppressed by compensating mutations in U6 that restored base pairing. This interaction is nonessential in yeast, but necessary in mammals, at least for high splicing efficiency.

SUMMARY The U2 snRNP base-pairs with the conserved sequence at the splicing branchpoint. This base pairing is essential for splicing. U2 also forms vital base pairs with U6, forming a region called helix I, that apparently helps orient these snRNPs for splicing. In addition, the 5′-end of U2 interacts with the 3′-end of U6, forming a region called helix II, that is important for splicing in mammalian cells, but not in yeast cells.

U5 snRNP We have now seen evidence for the participation of U1, U2, and U6 snRNPs in splicing. What about U5? It has no obvious complementarity with any snRNP or conserved region of a splicing substrate, yet it does seem to associate with both exons, perhaps positioning them for the second splicing step.

Sontheimer and Steitz have provided evidence for the involvement of U5 with the ends of the exons during

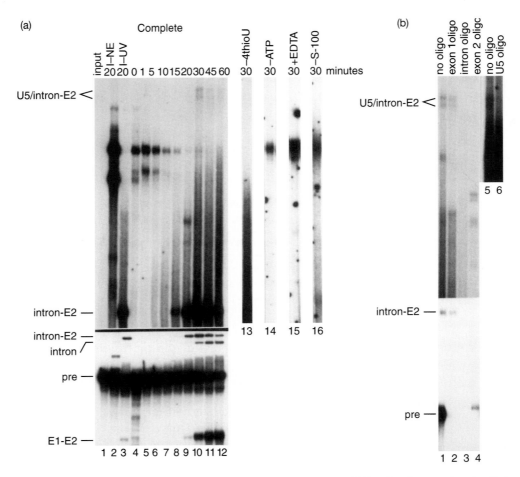

Figure 14.22 Detection of a complex between U5 and the 5′-end of the second exon. (a) Forming the complex. Sontheimer and Steitz placed 4-thioU in the first position of the second exon of a labeled splicing substrate and cross-linked it to whatever RNAs were nearby at various times during splicing. Then they electrophoresed the products and detected them by autoradiography. The U5/intron–E2 doublet appears near the top late in the splicing process (after 30 min). Lane designations are as in Figure 14.18. **(b)** Identification of the RNAs in the complex. Sontheimer and Steitz irradiated the splicing mix after 30 min of splicing to form cross-links, then incubated it with DNA oligonucleotides complementary to U5 and other RNAs, then added RNase H to degrade any RNAs hybridized to the oligonucleotides. Finally, they electrophoresed and autoradiographed the products. The oligonucleotides (oligos) used were as follows: lanes 1 and 5, no oligo; lane 2, anti-exon-1 oligo; lane 3, anti-intron oligo; Lane 4, anti-exon-2 oligo; lane 6, anti-U5 oligo. The anti-intron, anti-exon-2, and anti-U5 oligos all destroyed the complex, indicating that the complex is composed of the intron, second exon, and U5. (*Source:* Sontheimer and Steitz, The U5 and U6 small nuclear RNAs as active site components of the spliceosome. *Science* 262 (24 Dec 1993) f. 4, p. 1992. © AAAS.)

splicing, again using 4-thioU-substituted splicing substrates. In one such experiment, they substituted 4-thioU for the normal C in the first position of the second exon of an adenovirus major late splicing substrate. This change still allowed normal splicing to occur. When they cross-linked the 4-thioU to whatever snRNP was near the 5′-end of the second exon, they created a doublet complex (U5/intron–E2) that appeared at 30 min after the onset of splicing (Figure 14.22a). This was late enough that the first splicing step had already occurred. Many other complexes also formed, but we will not discuss them here.

To show that this doublet complex really does include U5, the intron, and exon 2, Sontheimer and Steitz hybridized the complex to DNA oligonucleotides complementary to these RNAs, then treated the complex with RNase H, which degrades the RNA part of an RNA–

DNA hybrid. Figure 14.22b shows that oligonucleotides complementary to U5, the intron, and the second exon, but not the first exon, cooperated with RNase H to degrade the complex. Thus, the complex appears to include U5 and the intron–exon-2 splicing intermediate. The interaction between U5 and the second exon is base-specific because substitution of 4-thioU for the second base in the second exon did not result in formation of any bimolecular RNA complexes.

To identify the bases in U5 or U6 involved in the 4-thioU cross-links to the splicing intermediates, Sontheimer and Steitz exploited primer extension blockage. They used oligonucleotides complementary to sequences in the snRNPs as primers for reverse transcription of the snRNPs in the complexes. Wherever reverse transcriptase encounters a cross-link, it will stop, yielding a DNA of

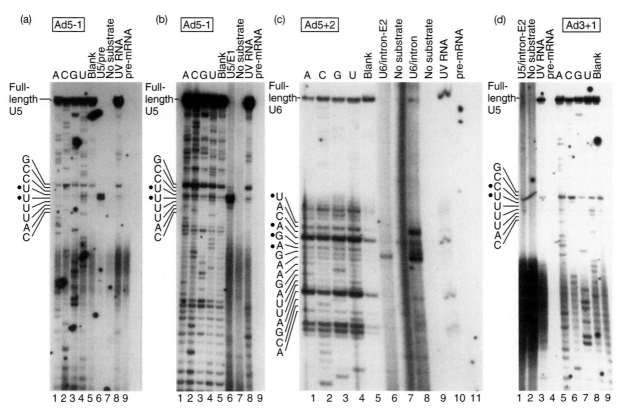

Figure 14.23   Identification of snRNP bases cross-linked to 4-thioU in various positions in the splicing substrate. Sontheimer and Steitz used primer extension to map the bases in U5 and U6 cross-linked to 4-thioU in the following positions: the last base in the first exon (Ad5-1, panels **a** and **b**); the second base in the intron (Ad5+2, panel **c**); or the first base in the second exon (Ad3+1, panel **d**). They formed cross-linked complexes with these RNAs, then excised the complexes from the electrophoresis gels and added primers specific for either U5 or U6, and performed primer extension analysis. The first four lanes in panels (**a–c**) and lanes 5–8 in panel (**d**) are sequencing lanes using the same primer as in the primer extension assays. The lanes marked "blank" are control sequencing lanes with no template. The experimental lanes are lanes 6 in panels (**a** and **b**), lanes 6 and 8 in panel (**c**), and lane 1 in panel (**d**). These are the results of primer extension with: the U5/splicing precursor complex (U5/pre, panel **a**); the U5/exon 1 complex (U5/E1, panel **b**); the U6/intron–exon 2 complex (U6/intron-E2, panel **c**), and the U6/intron complex, panel (**c**); and the U5/intron-exon 2 complex (U5/intron–E2, panel **d**). The other lanes are controls as follows: "no substrate," substrate was omitted from the reaction mix, then a slice of gel was cut out from the position where complex would be if substrate were included; "UV RNA," total RNA from an extract lacking substrate; "pre-mRNA," uncross-linked substrate. The cross-linked bases in the snRNPs are marked with dots at the left of each panel. (*Source:* Sontheimer and Steitz, *Science* 262 (24 Dec 1993) f. 5, p. 1993. © AAAS.)

defined length. This length corresponds to the distance between the primer binding site and the cross-link, and therefore the exact postion of the cross-link. Figure 14.23 shows the results. Panels (a) and (b) demonstrate that two adjacent U's in U5 cross-link to the last base in the first exon, when either the intact splicing substrate or just the first exon was used. Skipping panel (c) for a moment, panel (d) demonstrates that one of the same U's that were involved in cross-links to the end of the first exon is also involved, along with an adjacent C, in cross-links to the first base in the second exon. Panel (c) shows that four bases in U6 cross-link to the second base in the intron. The sum of the results with U5 suggest that this snRNP is involved in binding to the 3′-end of the first exon and the 5′-end of the second exon, as illustrated in Figure 14.24. This would allow it to position the two exons for splicing.

SUMMARY The U5 snRNP associates with the last nucleotide in one exon and the first nucleotide of the next. This presumably lines the two exons up for splicing.

U4 snRNP   Most of what we know about U4 concerns its association with U6. We have known for some time that the sequences of U4 and U6 suggest an association to form two base-paired stems, called stem I and stem II (Figure 14.25). We have also seen evidence for association between U4 and U6. In fact, Figure 14.19 showed a particularly strong signal corresponding to a U4–U6 complex. Does U4 have any direct role to play in splicing? Apparently not. U4 dissociates from U6 after splicing is underway and can then be removed from the spliceosome

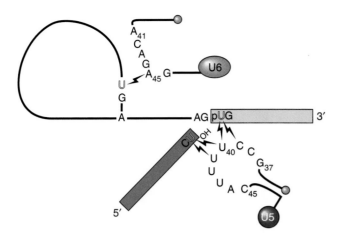

Figure 14.24 Summary of U5 and U6 interactions with the splicing substrates revealed by 4-thioU cross-linking. The outlined U's represent 4-thioUs introduced into the splicing substrate. The lightning bolts illustrate cross-links between snRNP bases and 4-thioUs in the splicing substrates. Exon 1 is blue and exon 2 is yellow. The small purple dots are caps at the 5'-ends of the snRNAs. Note the role U5 can play in positioning the two exons for the second step in splicing. (*Source:* From Sontheimer and Steitz, *Science* 262:1995, 1993. Copyright © 1993 American Association for the Advancement of Science, Washington, DC. Reprinted by permission.)

using gentle procedures. Thus, its role may be to bind and sequester U6 until it is time for U6 to participate in splicing by binding to the 5'-splice site. It is worth noting that some U6 bases that participate in base pairing with U4 to form stem I are also involved in the essential base pairing to U2 that we discussed earlier in this chapter. This underscores the importance of removing U4, so U6 can base-pair to U2 and help form an active spliceosome.

Figure 14.25 makes another interesting point: The gene for U6 is interrupted by an mRNA-type intron in at least two species of yeast. The positions of the introns in *Schizosaccharomyces pombe* and *Rhodosporidium dacryoidum* are shown.

> **SUMMARY** U4 base-pairs with U6, and its role seems to be to bind U6 until U6 is needed in the splicing reaction. The U6 gene is split by an mRNA-type intron in at least two yeast species.

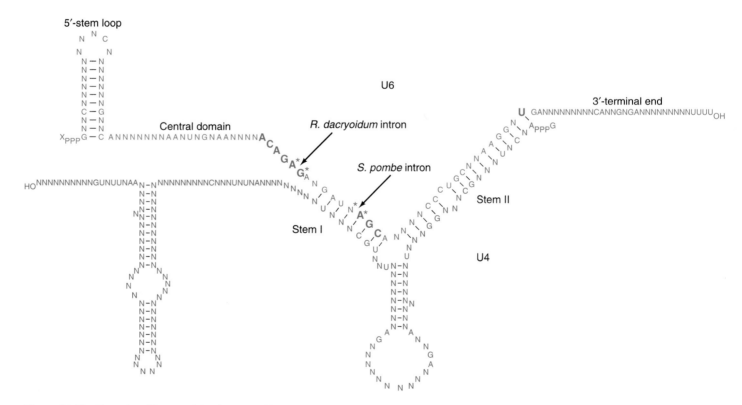

Figure 14.25 Postulated base pairing between U4 and U6 RNAs. The conserved nucleotides in the two RNAs are shown. All others are simply represented as N. Nucleotides in bold are in positions where mutations strongly impair growth. Stars designate bases in positions where mutations cause at least partial inhibition of the second splicing step in vitro. Arrows mark the positions of introns in the gene for U6 in two yeast species.
(*Source:* From Guthrie, *Science* 253:160, 1991. Copyright © 1991 American Association for the Advancement of Science, Washington, DC. Reprinted by permission.)

(a) Spliceosomal
     pre-mRNA

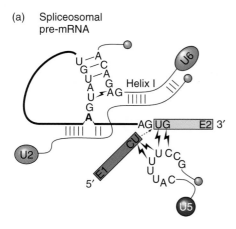

(b) Group II intron

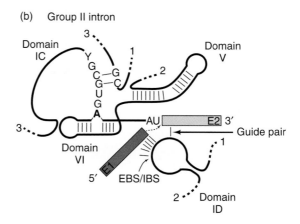

Figure 14.26 A model to compare the active center of a spliceosome to the active center of a group II intron. (a) Spliceosome. This is a variation on Figure 14.24, but including U2 (reddish brown). All other colors have the same significance as in Figure 14.24. The branchpoint A is bold, and the intron is rendered with a thick line. Dashed arrow represents the attack by exon 1 on the intron/exon 2 bond that is about to occur. (b) Group II intron. The intron is drawn in the same shape as the proposed spliceosomal structure in panel (a), to illustrate the similarities. Only parts of the intron are shown; the missing parts are suggested by dotted lines with numbers to indicate connections between parts. The exons are colored and the branchpoint A is bold. Dashed arrow represents the attack by exon 1 on the intron/exon 2 bond that is about to occur. (*Source:* (a) From Wise, *Science* 262:1978, 1993. Copyright © 1993 American Association for the Advancement of Science, Washington, DC. Reprinted by permission. (b) From Sontheimer and Steitz, *Science* 262:1995, 1993. Copyright © 1993 American Association for the Advancement of Science, Washington, DC. Reprinted by permission.)

A Summary of snRNP Involvement in mRNA Splicing   We will see later in this chapter that some other types of introns are self-splicing. That is, they do not rely on a spliceosome, but have all the catalytic activity they need to splice themselves. These self-splicing introns fall into two classes. One class, group II introns, use a lariat intermediate just like the lariat intermediates in nuclear mRNA splicing. Thus, it is tempting to speculate that the spliceosomal snRNPs substitute for parts of the group II intron in forming a similar structure that juxtaposes exons 1 and 2 for splicing.

Figure 14.26 depicts models for splicing both spliceosomal and group II introns. Panel (a) shows a variation on the model for the second step in nuclear mRNA splicing presented in Figure 14.24; panel (b) shows an equivalent model for a group II intron. Several features are noteworthy. First, the U5 loop, by contacting exons 1 and 2 and positioning them for splicing, substitutes for domain ID of a group II intron. Such RNA regions are called **internal guide sequences** because of their function in guiding other RNA regions into the proper position for catalysis. Second, the U6 region that base-pairs with the 5′-splice site substitutes for domain IC of a group II intron. Third, the U2–U6 helix I resembles domain V of a group II intron. Finally, the U2–branchpoint helix substitutes for domain VI of a group II intron. In both cases, base pairing around the branchpoint A causes this key nucleotide to bulge out, presumably helping it in its task of forming the branch. Because group II introns are catalytic RNAs (ribozymes), the similarities presented in Figure 14.26 strongly imply that the snRNPs, which substitute for group II intron elements at the center of splicing activity, also catalyze the splicing reactions.

SUMMARY The spliceosomal complex (substrate, U2, U5, and U6) poised for the second step in splicing can be drawn in the same way as a group II intron at the same stage of splicing. Thus, the spliceosomal snRNPs seem to substitute for elements at the center of catalytic activity of the group II introns and probably have the spliceosome's catalytic activity.

Spliceosome Assembly and Function

The spliceosome is composed of many components, proteins as well as RNAs. The components of the spliceosome assemble in a stepwise manner, and part of the order of assembly has been discovered. We call the assembly, function, and disassembly of the spliceosome the **spliceosome cycle**. In this section, we will discuss this cycle. We will see that by controlling the assembly of the spliceosome, a cell can regulate the quality and quantity of splicing and thereby regulate gene expression.

The Spliceosome Cycle   When various research groups first isolated spliceosomes, they did not find U1 snRNP. This was surprising because U1 is clearly involved in base pairing to the 5′-splice site and is essential for splicing. The fact is that U1 *is* part of the spliceosome, but the methods used in the first spliceosome purifications were probably too harsh to retain U1. To emphasize the importance of this snRNP, Stephanie Ruby and John Abelson discovered in 1988 that U1 is the first snRNP to bind to the splicing precursor. These workers used a clever technique to

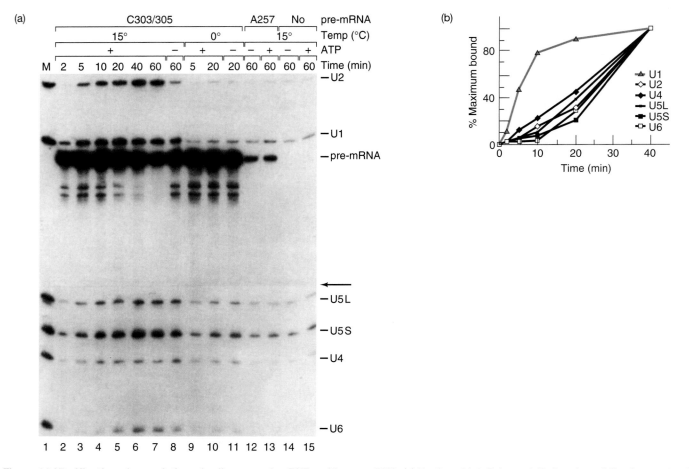

Figure 14.27   Kinetics of association of spliceosomal snRNPs with pre-mRNA. (a) Northern blot. Ruby and Abelson immobilized a yeast actin pre-mRNA to agarose beads by hybridizing it to an RNA (the anchor RNA) tethered through biotin–avidin links to the beads. They incubated this RNA–bead construct with yeast nuclear extract at either 15 or 0°C, in the presence or absence of ATP, and for 2–60 min, as indicated at top. The pre-mRNA was mutated in the 3′-splice site (C303/305), or in the conserved branchpoint (A257). The former would assemble a spliceosome, but the latter would not. The lanes marked "No" contained no pre-mRNA, only anchor RNA. After the incubation step, these workers washed away unbound material, extracted RNAs from the complexes, electrophoresed and blotted the RNAs, and hybridized the blots to probes for U1, U2, U4, U5, and U6. Two forms of U5 (U5 L and U5 S) were recognized. Lane 15, with no pre-mRNA, showed background binding of most snRNAs and served as a control for the other lanes. U1 bound first, then the other snRNPs bound. None of the snRNPs bound in significant amounts to the A257 mutant RNA. All snRNPs, including U1 and U4, remained bound after 60 min. **(b)** Graphic representation of amount of each snRNA bound to the complex as a function of time. U1 (red) clearly bound first, with all the others following later. (*Source:* Ruby and Abelson, An early hierarchic role of U1 small nuclear ribonucleoprotein in spliceosome assembly. *Science* 242 (18 Nov 1988). © AAAS.)

measure spliceosome assembly. They immobilized a yeast pre-mRNA on agarose beads by hybridizing it to an "anchor RNA" joined to the beads through a biotin–avidin linkage. Then they added yeast nuclear extract for varying periods of time. They washed away unbound material, then eluted the RNAs, which they electrophoresed, blotted, and probed with radioactive probes for all spliceosomal snRNAs.

Figure 14.27 contains the results, which show that U1 was the first snRNP to bind to the splicing substrate. At the 2-min time point, it was the only snRNP whose association with the pre-mRNA was above background; compare lane 2 with lane 15 in panel (a). Panel (a) also demonstrates that ATP was required for optimum binding of all snRNPs except U1. Figure 14.27b is a graph of the

time course of association of all spliceosomal snRNPs with the substrate. U1 stands out from all the others as the first snRNP to join the spliceosome.

To probe more deeply into the order of spliceosome assembly, these workers inactivated either U1 or U2 by incubating extracts with DNA oligonucleotides complementary to key parts of these two snRNAs plus RNase H, then used the same spliceosome assembly assay as before. RNase H degrades the RNA part of an RNA–DNA hybrid, so the parts of the snRNAs in a hybrid with the DNA oligomers were degraded. The parts that hybridized to the pre-mRNA (the 5′-splice site and the branchpoint, respectively) were selected for degradation. The results in Figure 14.28 make two main points: (1) Inactivating U1 prevented U1 binding, as expected, and also prevented

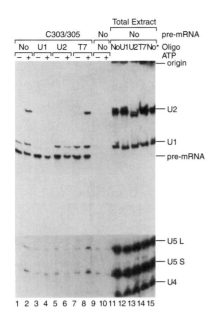

Figure 14.28   Effect of inactivation of U1 or U2 on assembly of the spliceosome. Ruby and Abelson inactivated either U1 or U2 by incubation with RNase H and a DNA oligonucleotide complementary to a key part of either snRNA. Lanes 11–15 show the patterns of labeled snRNAs in an extract after treating with RNase H and: no oligonucleotide (No); an anti-U1 oligonucleotide (U1); an anti-U2 oligonucleotide (U2); or an anti-T7 oligonucleotide (T7). The latter served as a second negative control. Treatment with RNase H and anti-U1 led to essentially complete conversion to a truncated form that electrophoresed slightly faster than the parent RNA. Treatment with RNase H and anti-U2 led to near-elimination of full-size U2, and appearance of a small amount of truncated U2. Lanes 1–10 show the results of spliceosome assembly experiments, as described in Figure 14.27, under the following conditions, as indicated at top: C303/305 pre-mRNA, or no pre-mRNA; extracts treated with RNase H and no oligonucleotide, anti-U1, anti-U2, or anti-T7 oligonucleotides; and with or without ATP. Inactivating U1 prevented binding of U1, U2, and U5. Inactivating U2 prevented binding of U2 and U5. (*Source:* Ruby and Abelson, *Science* 242 (18 Nov 1988) f. 7, p. 1032. © AAAS.)

binding of all other snRNPs (compare lanes 2 and 4). (2) Inactivating U2 prevented U2 binding, as expected, and also prevented U5 binding. However, it did not prevent U1 binding (compare lanes 2 and 6). Taken together, these results indicate that U1 binds first, then U2 binds with the help of ATP, and then the rest of the snRNPs join the spliceosome.

As we will discuss later in this chapter, U6, once freed from association with U4, displaces U1 from its binding site at the 5′-splice site. We know from other experiments that, when U1 is displaced, it exits the spliceosome along with U4. This leaves an active spliceosome containing only U2, U5, and U6. Indeed, the replacement of U1 by U6 seems to be the event that activates the spliceosome to carry out the splicing reaction. Jonathan Staley and Christine Guthrie demonstrated in 1999 that activation can be blocked by changing the base sequence of the 5′-splice site so that it base-pairs even better with U1. This presumably

made it harder for U6 to compete with U1 for binding to the 5′-splice site, and as a result, release of U1 and U4, as well as splicing, was inhibited. Conversely, with binding between U1 and the 5′-splice site held constant, enhancing the base-pairing between U6 and the 5′-splice site allowed more activation (release of U1 and U4) and therefore more splicing. Staley and Guthrie went on to show that a protein known as **Prp28**, one of the proteins in U5 snRNP, appears to be required, along with ATP, for exchange of U1 for U6 at the 5′-splice site.

Figure 14.29 illustrates the yeast spliceosome cycle. The first complex to form, composed of splicing substrate plus U1 and perhaps other substances, is called the **commitment complex (CC)**. As its name implies, the commitment complex is committed to splicing out the intron at which it assembles. Next, U2 joins, with help from ATP, to form the **A complex**. Next, U4–U6 and U5 join to form the **B1 complex**. U4 then dissociates from U6 to allow: (1) U6 to displace U1 from the 5′-splice site in an ATP-dependent reaction that activates the spliceosome, (2) U1 and U4 to exit the spliceosome, and (3) U6 to base-pair with U2. The activated spliceosome is also known as the **B2 complex**. ATP then provides the energy for the first splicing step, which separates the two exons and forms the lariat splicing intermediate, both held in the **C1 complex**. With energy from a second molecule of ATP, the second splicing step occurs, joining the two exons and removing the lariat-shaped intron, all held in the **C2 complex**. In the next step, the spliced, mature mRNA exits the complex, leaving the intron bound to the **I complex**. Finally, the I complex dissociates into its component snRNPs, which can be recycled into another splicing complex, and the lariat intermediate, which is debranched and degraded.

SUMMARY The spliceosome cycle includes the assembly, splicing activity, and disassembly of the spliceosome. Assembly begins with the binding of U1 to the splicing substrate to form a commitment complex. U2 is the next snRNP to join the complex, followed by the others. The binding of U2 requires ATP. When U6 dissociates from U4, it displaces U1 at the U5 splice site. This ATP-dependent step activates the spliceosome and allows U1 and U4 to be released.

3′-Splice Site Selection   During step 2 of the splicing process, the 3′-hydroxyl group of exon 1 attacks the phosphodiester bond linking an AG at the end of the intron to the first nucleotide of exon 2. This AG is ideally between 18 and 40 nt downstream of the branchpoint. AG's that are closer to the branchpoint are usually skipped. What determines which AG is used? Robin Reed and colleagues

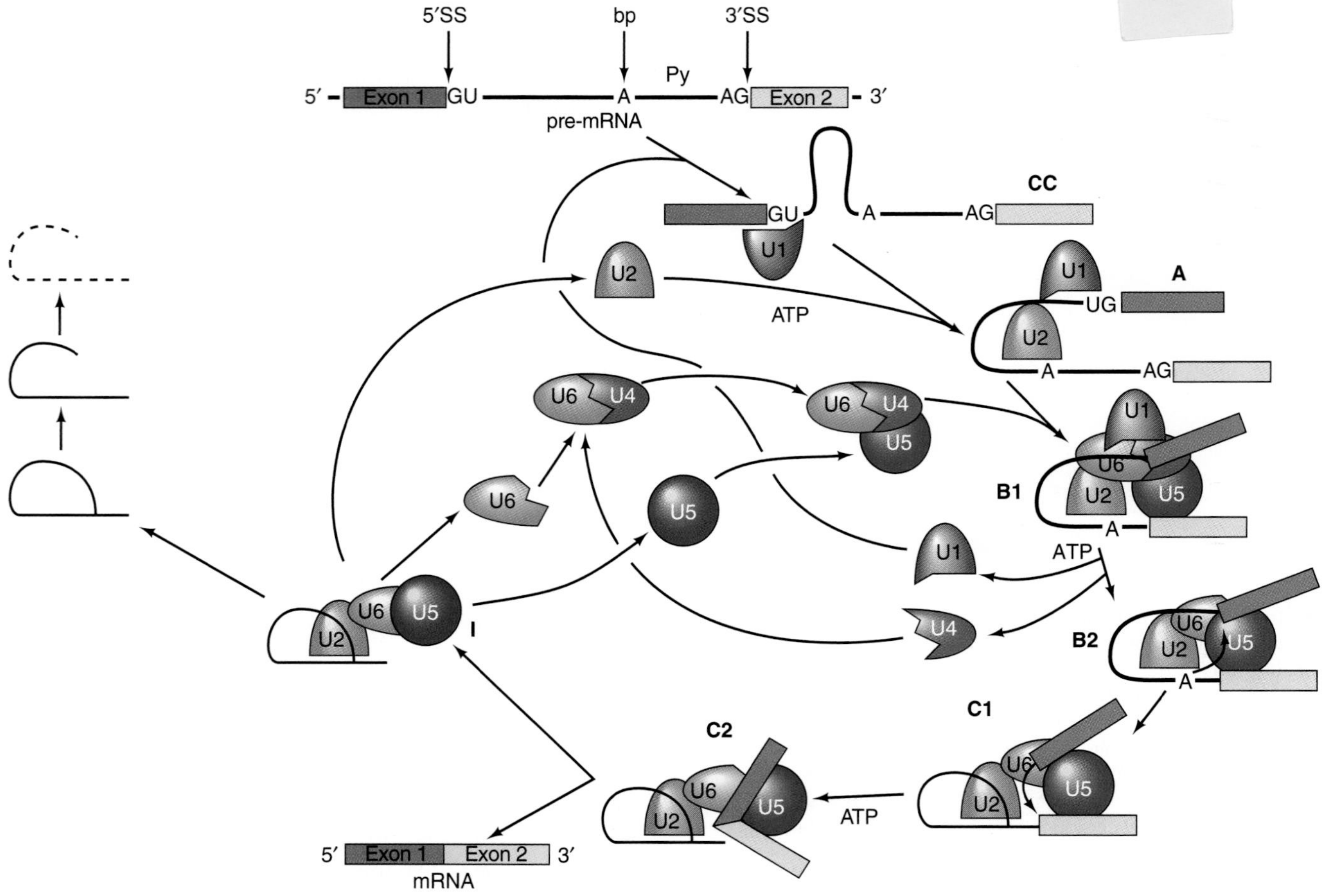

Figure 14.29   The spliceosome cycle. The text gives a description of the events in the cycle. (*Source:* After Sharp, *Cell* 77:811, 1994. Copyright 1994 Cell Press/The Nobel Foundation. Reprinted by permission.)

have found that a **splicing factor** known as **Slu7** is required for selection of the proper AG. Without Slu7, the correct AG is not used, but an incorrect AG may come into play.

Katrin Chua and Reed immunodepleted a HeLa cell extract of Slu7 by treating the extract with an anti-Slu7 antiserum linked to Sepharose beads. Separation of the extract from the beads leaves an extract depleted of Slu7. They also prepared a mock-depleted extract by treating the extract with Sepharose beads linked to preimmune serum, which contained no anti-Slu7 antibodies. Then they tested these extracts for ability to splice a labeled model pre-mRNA made from part of the adenovirus major late transcript that was modified so it contained a single AG located 23 nt downstream of the branchpoint sequence (Figure 14.30a). After incubating the model splicing substrate with an extract, Chua and Reed tested for splicing by electrophoresing the products.

Figure 14.30a shows the results with the "natural" substrate. The mock-depleted extract completed steps 1 and 2 of the splicing reaction, yielding mature mRNA, intron, and relatively little of the unspliced exons. On the

other hand, the extract depleted of human Slu7 (ΔhSlu7) yielded almost no mature mRNA or intron, but abundant exon 1 and lariat-exon 2. Thus, step 2 of splicing was blocked. This could mean that Slu7 is necessary for recognizing the normal AG at the 3'-splice site.

Chua and Reed next asked what would happen if they inserted an extra AG only 11 nt downstream of the branchpoint sequence. Figure 14.30b shows that the mock-depleted extract yielded mRNA spliced at the natural AG 23 nt downstream of the branchpoint sequence, but very little mRNA spliced at the AG unnaturally close to the branchpoint. By contrast, the extract depleted in Slu7 spliced most of the mRNA at the unnatural AG and very little at the natural AG. In further experiments, the depleted extract exhibited the same aberrant behavior when the two AGs were at 11 and 18 nt or 9 and 23 nt downstream of the branchpoint. Furthermore, it spliced to an incorrect AG placed downstream, as well as upstream, of the proper one, but not to the proper one itself. (In all cases, the incorrect AG had to be within about 30 nt of the branchpoint to be a target for aberrant splicing.) Thus, not only is Slu7 needed to recognize the

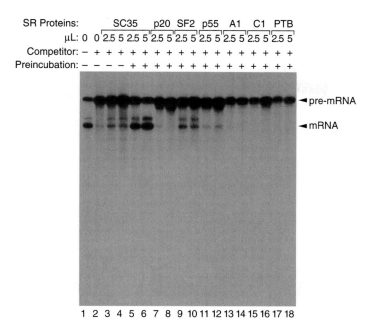

Figure 14.32   Commitment activities of several RNA-binding proteins with the human β-globin pre-mRNA. Fu ran the commitment assay with two different concentrations of the seven different RNA-binding SR proteins listed at top. Lane 1 contained no competitor and no SR protein; lane 2 contained competitor, but no SR protein; lanes 3 and 4 included competitor, but no preincubation with SC35. All other lanes included competitor and preincubation with an SR protein as usual. The positions of the pre-mRNA and mature mRNA are shown at right. (*Source: Fu, Nature* 365 (2 Sept 1993) f. 2b, p. 84. © Macmillan Magazines Ltd.)

snRNA is a prominent and early participant in commitment, they decided to look for genes encoding proteins that interacted with U1 snRNA. To find these genes, they employed a **synthetic lethal screen** as follows: First, they introduced a temperature-sensitive mutation into the gene encoding U1 snRNA. The mutant U1 snRNA functioned at low temperature (30° C) but not at high temperature (37° C). They reasoned that the strain carrying this altered U1 snRNA would be especially sensitive to mutations in proteins that interact with snRNA. These second mutations could render the yeast strain inviable, even at the low temperature, so such mutations were called "mutant-u-die," abbreviated Mud. Thus, the second mutations may not have been lethal in wild-type cells, but they became lethal in cells bearing the first mutation. In this sense, their lethality was "synthetic"—it depended on a conditional lethal mutation already created in the cell. One mutation discovered this way mapped to the *MUD2* gene, which encodes the protein **Mud2p**.

Subsequent work showed that the function of Mud2p depended on a natural sequence at the lariat branchpoint, near the 3′-end of the intron. This suggested that Mud2p interacted not only with U1 snRNA at the 5′-end of the intron, but with some other substance near the 3′-end of the intron. A major question remained: Does Mud2p by itself make these interactions with the 5′- and 3′-ends of the intron, or does it rely on other factors? In 1997, Nadja Abovich and Rosbash used another synthetic lethal

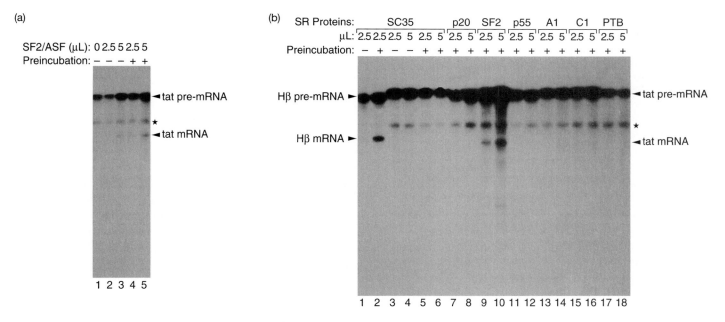

Figure 14.33   Commitment activities of several RNA-binding proteins with either the β-globin pre-mRNA or the HIV tat pre-mRNA. (a) Effect of SF2/ASF on commitment with tat pre-mRNA. Fu ran the commitment assay with the concentrations of the SF2/ASF shown at top, and either without (lanes 1–3) or with (lanes 4 and 5) preincubation with the splicing factor. Comparing lanes 5 and 3 gives the clearest view of the effect of SF2/ASF. (b) Specificity of SR proteins. Fu tested two different concentrations of the seven different SR proteins listed at top for commitment with the tat pre-mRNA. Lanes 1 and 2 contained human β-globin pre-mRNA with and without preincubation with SC35. All other lanes contained tat pre-mRNA. The positions of the β-globin pre-/and mature mRNA are indicated at left; the positions of the tat pre-/and mature mRNA are indicated at right. The stars denote bands resulting from artifactual tat pre-mRNA degradation. (*Source: Fu, Nature* 365 (2 Sept 1993) f. 3, p. 84. © Macmillan Magazines Ltd.)

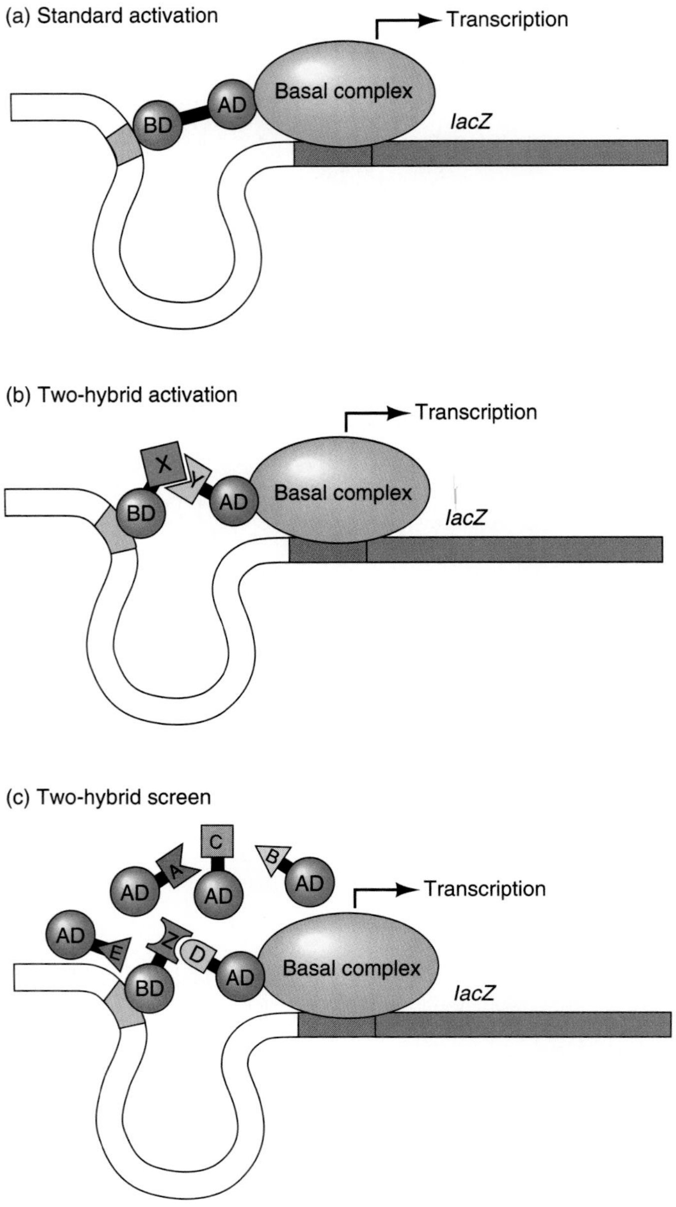

Figure 14.34 Principle of the yeast two-hybrid assay. (a) Standard model of transcription activation. The DNA-binding domain (BD, red) of an activator binds to an enhancer (pink), and the activating domain (AD green) interacts with the basal complex (orange), recruiting it to the promoter (reddish brown). This stimulates transcription. **(b)** Two-hybrid assay for protein–protein interaction. Link one protein (X, turquoise) to a DNA-binding domain to form one hybrid protein; link another protein (Y, yellow) to a transcription-activating domain to form a second hybrid protein. When plasmids encoding these two hybrid proteins are introduced into yeast cells bearing the appropriate promoter, enhancer, and reporter gene (*lacZ*, purple, in this case), the two hybrid proteins can get together as shown to serve as an activator. Activated transcription produces abundant reporter gene product, which can be detected with a colorimetric assay, using X-gal, for example. One hybrid protein contributes a DNA-binding domain, and the other contributes a transcription-activating domain. The two parts of the activator are held together by the interaction between proteins X and Y. If X and Y do not interact, no activator will form, and no activation of the reporter gene will occur. The GAL4 DNA-binding domain and transcription-activating domain are traditionally used in this assay, but other possibilities exist, such as the ones used by Abovich and Rosbash. **(c)** Two-hybrid screen for a protein that interacts with protein Z. Yeast cells are transformed with two plasmids: one encoding a binding domain (red) coupled to a "bait" protein (Z, turquoise). The other is a set of plasmids containing many cDNAs coupled to the coding region for a transcription activating domain. Each of these encodes a fusion protein containing the activating domain (green) fused to an unknown cDNA product (the "prey"). Each yeast cell is transformed with just one of these prey-encoding plasmids, but several of their products are shown together here for convenience. One prey protein (D, yellow) interacts with the bait protein, Z. This brings together the DNA-binding domain and the transcription-activating domain so they can activate the reporter gene. Now the experimenter can purify the prey plasmid from this positive clone and thereby get an idea about the nature of the prey protein.

screen to answer this question. They introduced a mutation into the *MUD2* gene, then looked for second mutations that would kill the *MUD2* mutant cells, but not wild-type cells. One gene identified by this screen is called *MSL-5* (Mud synthetic lethal-5). It encodes a protein originally named Msl5p, but renamed **BBP** (**branchpoint bridging protein**) once its binding properties were clarified.

Abovich and Rosbash suspected that BBP forms a bridge between the 5′- and 3′-ends of an intron, by binding to U1 snRNP at the 5′-end and to Mud2p at the 3′-end. To test this hypothesis, they used a combination of methods, including a **yeast two-hybrid assay.** Figure 14.34 describes a generic version of this very sensitive technique, which is designed to demonstrate binding—even transient binding—between two proteins. The yeast two-hybrid

assay takes advantage of two facts, discussed in Chapter 12: (1) that transcription activators typically have a DNA-binding domain and a transcription-activating domain; and (2) that these two domains have self-contained activities. To assay for binding between two proteins, X and Y, one can arrange for yeast cells to express these two proteins as fusion proteins, pictured in Figure 14.34b. Protein X is fused to a DNA-binding domain, and protein Y is fused to a transcription-activating domain. Now if proteins X and Y interact, that brings the DNA-binding domain and the transcription-activating domain together, and activates transcription of a reporter gene (typically *lacZ*).

One can even use the yeast two-hybrid system to fish for unknown proteins that interact with a known protein

(Z). In such a screen (Figure 14.32c), one would prepare a library of cDNAs linked to the coding region for a transcription-activating domain and express these hybrid genes, along with a gene encoding the DNA-binding domain—Z hybrid gene, in yeast cells. In practice, each yeast cell would make a different fusion protein (AD-A, AD-B, AD-C, etc.), along with the BD–Z fusion protein, but they are all pictured here together for simplicity. We can see that AD–D binds to BD–Z and activates transcription, but none of the other fusion proteins can do this because they cannot interact with BD–Z. Once clones that activate transcription are found, the plasmid bearing the AD–D hybrid gene is isolated and the D portion is sequenced to find out what it codes for. Because the yeast two-hybrid assay is indirect, it is subject to artifacts. Thus, the protein–protein interactions suggested by such an assay should be verified with a direct assay.

Abovich and Rosbash already knew which proteins were likely to interact, so they made plasmids expressing these proteins as fusion proteins containing the protein of interest plus either a DNA-binding domain or a transcription-activating domain. They transfected yeast cells with various pairs of these plasmids. In one experiment, for example, one plasmid encoded a hybrid protein containing the LexA DNA-binding domain linked to BBP; the other plasmid encoded a hybrid protein containing the B42 transcription-activating domain linked to Mud2p. Figure 14.35a (first column, first row) shows that cells bearing these two plasmids experienced activation of the *lacZ* gene, as demonstrated by the dark stain on the X-gal indicator plate. Thus, BBP bound to Mud2p in this assay. Figure 14.35a (first column, second row) shows that BBP also bound to Prp40p, a polypeptide component of U1 snRNP. On the other hand, Mud2p did not bind to Prp40p. Thus, BBP serves as a bridge between Mud2p, presumably bound at the branchpoint near the 3′-end of the intron, and to U1 snRNP at the 5′-end of the intron. In this way, BBP could help define the intron and help bring the two ends of the intron together for splicing. Abovich and Rosbash included Prp8p in this experiment as a positive control because they already knew it bound to Prp40p. Figure 14.35b summarizes the protein–protein interactions suggested by this yeast two-hybrid assay. These workers confirmed these interactions by showing that BBP tethered to Sepharose beads coprecipitated both Prp40p and Mud2p.

Abovich and Rosbash noted that the yeast Mud2p and BBP proteins resemble two mammalian proteins called U2AF65 and **SF1**, respectively. If these two mammalian proteins behave like their yeast counterparts, they should bind to each other. To test this hypothesis, these workers used the same yeast two-hybrid assay and coprecipitation procedure and found that U2AF65 and SF1 do indeed interact. Figure 14.36 illustrates the bridging functions of BBP and its mammalian counterpart, SF1, which could also be called mBBP (for mammalian BBP). We already en-

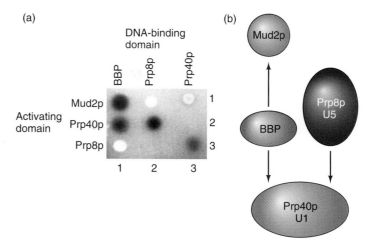

Figure 14.35   Yeast two-hybrid assays for interactions between BBP and other proteins. (a) Results of the assays. The proteins linked to the DNA-binding domain are listed at top, and the proteins linked to the transcription-activating domain are listed at left. Abovich and Rosbash spotted cells bearing the indicated pairs of plasmids on an indicator plate containing X-gal to measure the activation of the *lacZ* reporter gene. A dark stain indicates activation. For example, the darkly stained yeast cells in column 1, rows 1 and 2, indicated interaction between BBP and Mud2p, and between BBP and Prp40p (a component of U1 snRNP). The other positive reactions indicated interactions between Prp40p and Prp8p (a component of U5 snRNP). (b) Summary of results. This schematic shows the protein–protein interactions revealed by the yeast two-hybrid assay results in panel (a). (*Source:* Abovich and Rosbash, Cross-intron bridging interactions in the yeast commitment complex are conserved in mammals. *Cell* 89 (2 May 1997) f. 5, p. 406. Reprinted by permission of Elsevier Science.)

countered U2AF in our discussion of 3′-splice site selection and learned that U2AF65 binds to the polypyrimidine tract just upstream of the 3′-splice site and that U2AF35 binds to the AG at the 3′-splice site. In Figure 14.36, these two proteins are pictured together at the 3′-splice site and designated U2AF.

Further work by Rosbash's group demonstrated that BBP also recognizes the branchpoint UACUACC sequence and binds at (or very close to) this sequence in the commitment complex. Thus, BBP is also an RNA-binding protein, and the BBP now also stands for "branchpoint binding protein."

SUMMARY In the yeast commitment complex, the branchpoint bridging protein (BBP) binds to a U1 snRNP protein at the 5′-end of the intron, and to Mud2p near the 3′-end of the intron. It also binds to the RNA near the 3′-end of the intron. Thus, it bridges the intron and could play a role in defining the intron prior to splicing. The mammalian BBP counterpart, SF1 (mBBP), might serve the same bridging function in the mammalian commitment complex.

Yeast

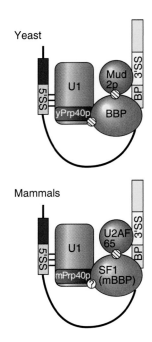

Mammals

Figure 14.36 Summary of intron-bridging protein–protein interactions in yeast and mammals. 5′SS is the 5′-splicing signal; BP is the branchpoint, and 3′SS is the 3′-splicing signal. The cross-hatched circles denote interactions demonstrated by Abovich and Rosbash. (*Source:* Reprinted from Abovich and Rosbash, *Cell* 89:409, 1997. Copyright 1997, with permission from Elsevier Science.)

Alternative Splicing

Our previous discussion of commitment leads naturally to another important topic: **alternative splicing.** About 1 in 20 eukaryotic pre-mRNAs can be spliced in more than one way, leading to two or more alternative mRNAs that encode different proteins. In humans, at least 40% of transcripts are subject to alternative splicing. The switch from one alternative splicing pattern to another undoubtedly involves commitment, and we will return to this theme at the end of this section.

David Baltimore and colleagues discovered the first example of alternative splicing, the mouse immunoglobulin μ heavy-chain gene, in 1980. The μ heavy chain exists in two forms, a secreted form (μ_s), and a membrane-bound form (μ_m). The difference in the two proteins lies at the carboxyl terminus, where the membrane-bound form has a hydrophobic region that anchors it to the membrane, and the secreted form lacks this membrane anchor. Using hybridization, Baltimore and colleagues found that the two proteins are encoded in two separate mRNAs that are identical at their 5′-ends, but differ at their 3′-ends. When these workers cloned the germline gene for the constant region of the μ heavy chain (the C_μ gene), they noticed that it encoded both the secreted and membrane-bound 3′-regions, and each of these was contained in a separate exon. Thus, two different modes of splicing of a common pre-mRNA could give two alterna-

tive mature mRNAs encoding μ_s and μ_m, as illustrated in Figure 14.37. In this way, alternative splicing can determine the nature of the protein product of a gene and therefore control gene expression.

Alternative splicing can have profound biological effects. Perhaps the best example is the sex determination system in *Drosophila.* Sex in the fruit fly is determined by a pathway that includes alternative splicing of the pre-mRNAs from three different genes: *Sex lethal* (*Sxl*); *transformer* (*tra*); and *doublesex* (*dsx*). Figure 14.38 illustrates this alternative splicing pattern. Males splice the transcripts of these genes in one way, which leads to male development; females splice them in a different way, which leads to development of a female.

Moreover, these genes function in a cascade as follows: Female-specific splicing of *Sxl* transcripts gives an active product that reinforces female-specific splicing of *Sxl* transcripts and also causes female-specific splicing of *tra* transcripts, which leads to an active *tra* product. (Actually, about half the *tra* transcripts are spliced according to the male pattern even in females, but this simply yields inactive product, so the female pattern is dominant.) The active *tra* product, together with the product of another gene, *tra-2,* causes female-specific splicing of transcripts of the *dsx* gene. This female-specific *dsx* product inactivates male-specific genes and therefore leads to female development.

By contrast, male-specific splicing of *Sxl* transcripts gives an inactive product because it includes an exon with a stop codon. This permits default (male-specific) splicing of *tra* transcripts, which again leads to an inactive product because of the inclusion of an exon with a stop codon. With no *tra* product, the developing cells splice the *dsx* transcripts according to the default, male-specific pattern, yielding a product that inactivates female-specific genes and therefore leads to development of a male.

How is this alternative splicing controlled? Knowing what we do about splicing commitment, we might guess that RNA-binding splicing factors would be involved. Indeed, because the products of *Sxl* and *tra* can determine which splice sites will be used in *tra* and *dsx* transcripts, respectively, we would predict that these proteins are splicing factors that cause commitment to the female-specific pattern of splicing. In accord with this hypothesis, the products of both *Sxl* and *tra* are SR proteins.

To further elucidate the mechanism of splice site selection, Tom Maniatis and his colleagues focused on the female-specific splicing of *dsx* pre-mRNA by Tra and Tra-2 (the products of *tra* and *tra-2,* respectively). They discovered that these two proteins act by binding to a regulatory region about 300 nt downstream of the female-specific 3′-splice site in the *dsx* pre-mRNA. This region contains six repeats of a 13-nt sequence, so it is known as the **repeat element.**

Tra and Tra-2 are necessary for commitment to female-specific splicing of *dsx* pre-mRNA, but are they

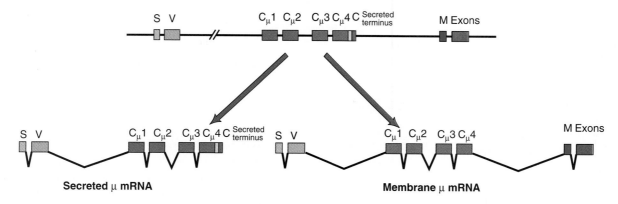

Figure 14.37   Alternative splicing pattern in the mouse immunoglobulin μ heavy-chain gene. The structure of the gene is shown at top. The boxes represent exons: The S exon (pink) encodes the signal peptide that allows the protein product to be exported to the plasma membrane, or secreted from the cell. The V exons (orange) encode the variable region of the protein. The C exons (blue) encode the constant region of the protein. Near the end of the fourth constant exon ($C_\mu 4$) lies the coding region (yellow) for the secreted terminus of the μ_s protein. This is followed by a short untranslated region (red), then by a long intron, then by two exons. The first of these (green) encodes the membrane anchor region of the μ_m mRNA. The second (red) is the untranslated region found at the end of the μ_m mRNA. The arrows pointing left and right indicate the splicing patterns that produce the secreted and membrane versions of the μ heavy chain (μ_s and μ_m, respectively). (*Source:* Reprinted from Early et al., *Cell* 20:318, 1980. Copyright 1980, with permission from Elsevier Science.)

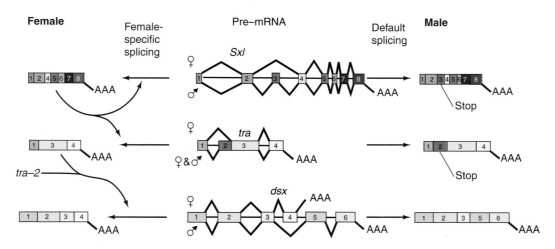

Figure 14.38   Alternative splicing cascade in *Drosophila* sex determination. The structures of the *Sxl*, *tra,* and *dsx* pre-mRNAs common to both males and females are shown at center, with the female-specific splicing pattern indicated above each, and the male-specific pattern below. Thus, female-specific splicing of the *Sxl* pre-mRNA includes exons 1, 2, and 4–8, whereas male-specific (default) splicing of the same transcript includes all exons (1–8), including exon 3, which has a stop codon. This means that male-specific splicing of this transcript gives a shortened, inactive protein product. Similarly, female-specific splicing of the *tra* pre-mRNA includes exons 1, 3, and 4, leading to an active protein product, whereas male-specific splicing of the same transcript includes all four exons, including exon 2 with a stop codon. Again, the male protein is inactive. The curved arrows indicate the positive effects of gene products on splicing. That is, the female *Sxl* product causes female-specific splicing of both *Sxl* and *tra* pre-mRNAs, and the female *tra* product, together with the *tra-2* product, causes female-specific splicing of *dsx* transcripts. (*Source:* Reprinted with permission from *Nature* 340:523, 1989. Copyright © 1989 Macmillan Magazines Limited.)

sufficient? To find out, Ming Tian and Maniatis developed a commitment assay that worked as follows: They began with a labeled, shortened *dsx* pre-mRNA containing only exons 3 and 4, with the intron in between. This model pre-mRNA can be spliced in vitro. Then they added Tra, Tra-2, and a **micrococcal nuclease (MNase)**-treated nuclear extract to supply any proteins, besides Tra and Tra-2, that might be needed for commitment. The MNase degrades snRNAs, but leaves proteins intact. Then the experimenters added an untreated nuclear ex-

tract, along with an excess of competitor RNA. If commitment occurred during the preincubation, the labeled pre-mRNA would be spliced. If not, the competitor RNA would block splicing. To assay for splicing, Tian and Maniatis electrophoresed the RNAs and detected RNA species by autoradiography. They found that Tra and Tra-2 alone, without the MNase-treated extract, were not enough to cause commitment. However, something in the extract could complement these proteins, resulting in commitment.

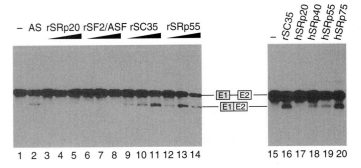

Figure 14.39   Commitment assay for female-specific splicing of *dsx* pre-mRNA. Tian and Maniatis assayed for the ability of various SR proteins to complement Tra and Tra-2 in an in vitro *dsx* splicing assay. Lane 1 contained no complementing protein. Lane 2 contained a mixture of SR proteins precipitated by ammonium sulfate (AS). Lanes 3–14 contained various amounts of the SR proteins indicated at the top of the lanes. Lane 15 is another negative control identical to lane 1. Lane 16 contained the highest amount of recombinant SC35, as in lane 11. Lanes 17–20 contained the purified nonrecombinant SR proteins indicated at the top of each lane. The electrophoretic mobilities of the splicing substrate (top band) and the spliced product (bottom band) are indicated between the two autoradiographs. (*Source:* Tian and Maniatis, A splicing enhancer complex controls alternative splicing of *doublesex* pre-mRNA. *Cell* 74 (16 July 1993) f. 5, p. 108. Reprinted by permission of Elsevier Science.)

To identify the other required factors, Tian and Maniatis first did a bulk purification of SR proteins and found that this SR protein mixture could complement Tra and Tra-2. Next, they obtained four pure recombinant SR proteins, and highly purified, nonrecombinant preparations of two others and tested them in the commitment assay with Tra and Tra-2. In this assay, the purified proteins took the place of the MNase-treated nuclear extract in the previous experiment. Figure 14.39, lane 1, shows that no splicing occurred with Tra and Tra-2 alone, in the absence of any other SR proteins. Lane 2 shows that a mixture of SR proteins prepared by ammonium sulfate (AS) precipitation could complement Tra and Tra-2. The other lanes show the effects of recombinant and highly purified SR proteins. Among these, some worked, and some did not. In particular, SC35, SRp40, SRp55, and SRp75 could complement Tra and Tra-2, but SRp20 and SF2/ASF could not. Thus, Tra, Tra-2, plus any one of the active proteins was enough to cause commitment to female-specific splicing of the *dsx* pre-mRNA.

We assume that commitment involves binding of SR proteins to the pre-mRNA, and we already know that Tra and Tra-2 bind to the repeat element, but do the other SR proteins also bind there? To find out, Tian and Maniatis performed affinity chromatography with a resin linked to an RNA containing the repeat element. After eluting the proteins from this RNA, they electrophoresed and immunoblotted (Western blotted) them. Finally, they probed the immunoblot in three separate experiments with antibodies against Tra, Tra-2, and SR

proteins in general. They detected Tra and Tra-2 as expected, and also found large amounts of SRp40 and a band that could contain either SF2/ASF or SC35. Because SC35, but not SF2/ASF, could complement Tra and Tra-2 in the commitment assay, we assume that this latter band corresponds to SC35. No significant amounts of any SR proteins bound to the RNA in the absence of Tra and Tra-2. This experiment demonstrates only that two SR proteins bind well to repeat-element-containing RNA in the presence of Tra and Tra-2. It does not necessarily mean a relationship exists between this binding and commitment. However, the fact that the two SR proteins that bind are also ones that complement Tra and Tra-2 in commitment is suggestive.

SUMMARY The transcripts of many eukaryotic genes are subject to alternative splicing. This can have profound effects on the protein products of a gene. For example, it can make the difference between a secreted or a membrane-bound protein; it can even make the difference between activity and inactivity. In the fruit fly, the products of three genes in the sex determination pathway are subject to alternative splicing. Female-specific splicing of the *tra* transcript gives an active product that causes female-specific splicing of the *dsx* pre-mRNA, which produces a female fly. Male-specific splicing of the *tra* transcript gives an inactive product that allows default, or male-specific, splicing of the *dsx* pre-mRNA, producing a male fly. Tra and its partner Tra-2 act in conjuction with one or more other SR proteins to commit splicing at the female-specific splice site on the *dsx* pre-mRNA. Such commitment is probably the basis of most, if not all, alternative splicing schemes.

14.3   Self-Splicing RNAs

One of the most stunning discoveries in molecular biology in the 1980s was that some RNAs could splice themselves without aid from a spliceosome or any other proteins. Thomas Cech (pronounced "Check") and his coworkers made this discovery in their study of the 26S rRNA gene of the ciliated protozoan, *Tetrahymena*. This rRNA gene is a bit unusual in that it has an intron, but the thing that really attracted attention when this work was published in 1982 was that the purified 26S rRNA precursor spliced itself in vitro. In fact, this was just the first example of self-splicing RNAs containing introns called **group I introns.** Subsequent work revealed another class of RNAs containing introns called **group II introns,** some of whose members are self-splicing.

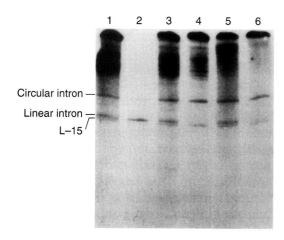

Figure 14.40 In vitro transcription of plasmids containing the 26S rRNA intron yields excised introns. Cech and coworkers used *E. coli* RNA polymerase to transcribe two different plasmids containing the rRNA intron. The reaction mix contained labeled nucleotides to label the products. These workers electrophoresed the labeled RNA products and visualized them by autoradiography. Lane 1, transcripts from *Eco*RI-cleaved pIVS13; lane 2, linear and circular intron markers; lane 3, transcripts from *Eco*RI-cleaved pIVS11; lane 4, transcripts from lane 3 incubated for 30 min under conditions that favor circularization of the intron (higher temperature, MgCl₂, and ammonium sulfate concentration); lane 5, transcripts from supercoiled pIVS11; lane 6, transcripts from lane 5 treated as in lane 4. The large RNAs in lane 5 are indistinct because the template was not linearized, so no definite stop-point for transcription exists. At left are the positions of the circular and linear introns, as well as the linear intron after losing 15 nt from its 5′-end. (*Source:* Kruger et al., Self-splicing RNA: Autoexcision and autocyclization of the ribosomal RNA intervening sequence of *Tetrahymena. Cell* 31 (Nov 1982) f. 2, p. 149. Reprinted by permission of Elsevier Science.)

Group I Introns

To make the self-splicing RNA, Cech and coworkers cloned part of the 26S rRNA gene containing the intron, forming two different recombinant plasmids, and transcribed them in vitro with *E. coli* RNA polymerase. When they electrophoresed the labeled products of these transcription reactions (Figure 14.40), they observed four large RNA products, plus three smaller RNAs corresponding in size to the linear and circular introns (plus a linear intron missing 15 nt) they had observed in previous studies. This suggested that the RNA was being spliced, and that the excised intron was circularizing.

Was this splicing carried out by the RNA itself, or was the RNA polymerase somehow involved? To answer this question, Cech and coworkers ran the RNA polymerase reaction in the presence of polyamines (spermine, spermidine, and putrescine) that inhibit splicing. Then they electrophoresed the products, excised all four RNA bands plus the material that remained at the origin, and purified the RNAs. Next they incubated these RNAs under splicing conditions (no polyamines) and reelectrophoresed them. When they autoradiographed the electrophoretic gel, they could see the intron in the lanes containing RNA

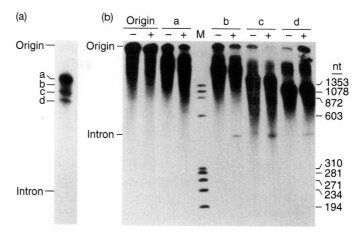

Figure 14.41 The splicing reaction can be separated from the transcription reaction. (a) Transcription. Cech and coworkers transcribed plasmid pIVS11 under conditions that inhibited splicing (presence of polyamines), then electrophoresed the labeled products. Four bands (designated a–d at left) appeared; a and b were difficult to distinguish in this experiment. (b) Splicing. Cech and coworkers took RNA from the origin and bands a–d from panel (a) and either electrophoresed them directly (lanes –) or incubated them under splicing conditions (no polyamines) before electrophoresis (lanes marked +). The nature of the RNA (origin or band a–d) is indicated at top. The RNA at the origin and in band a of panel (a) yielded no intron, whereas the RNAs from bands b–d did yield intron, but only under splicing conditions. (*Source:* Kruger et al., *Cell* 31 (Nov 1982) f. 3, p. 150. Reprinted by permission of Elsevier Science.)

from bands b–d, but not from the origin or band a (Figure 14.41). Thus, bands b–d appear to be 26S rRNA precursors that can splice themselves without any protein, even RNA polymerase.

The band we are calling the intron is the right size, but is it really what we think it is? Cech and coworkers sequenced the first 39 nt of this RNA and showed that they corresponded exactly to the first 39 nt of the intron. Therefore, it seemed clear that this RNA really was the intron.

Cech's group also discovered that the linear intron—the RNA we have been discussing so far—can cyclize by itself. All they had to do was raise the temperature and the Mg²⁺ and salt concentrations, and at least some of the purified linear intron would convert to circular intron, as shown in Figure 14.42.

So far, we have seen that the rRNA precursor can remove its intron, but can it splice its exons together? Cech and coworkers used a model splicing reaction to show that it can (Figure 14.43a). They began by cloning a part of the *Tetrahymena* 26S rRNA gene including 303 bp of the first exon, the whole intron, and 624 bp of the second exon into a vector with a promoter for phage SP6 polymerase. To generate the labeled splicing substrate, they transcribed this DNA in vitro with SP6 polymerase in the presence of [α-³²P]ATP. Then they incubated this RNA under splicing conditions with and without GTP and electrophoresed the products. Lane 1 displays the products of

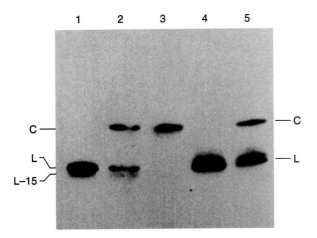

Figure 14.42 Autocyclization of the linear intron. Cech and coworkers isolated labeled linear intron, either from *Tetrahymena* nuclei or from an in vitro splicing reaction, by electrophoresis, treated it in various ways, then electrophoresed the products a second time. Lane 1, untreated linear intron from nuclei; lane 2, linear intron from nuclei incubated under cyclization conditions (high temperature and high MgCl₂ and salt concentrations); lane 3, circular intron marker; lane 4, linear intron produced by transcription of pIVS11 and splicing the RNA in vitro; lane 5, linear intron from lane 4 incubated under splicing conditions. The minor band just above the linear intron in lanes 4 and 5 is an unrelated transcript from the plasmid. Positions of circular (C), linear (L) introns are indicated at left and right. The position of the linear intron lacking 15 nt at its 5′-end (L-15) is indicated at left. (*Source:* Kruger et al., *Cell* 31 (Nov 1982) f. 5, p. 152. Reprinted by permission of Elsevier Science.)

the reaction with GTP. The familiar linear intron is present, as well as a small amount of circular intron. In addition, a prominent band representing the ligated exons appeared. By contrast, lane 2 shows that no such products appeared in the absence of GTP; only the substrate was present. This is what we expect because splicing is dependent on GTP, and it reinforces the conclusion that these products are all the result of splicing. In summary, these data argue strongly for true splicing, including the joining of exons.

Cech's group had already shown that splicing of the 26S rRNA precursor involved addition of a guanine nucleotide at the 5′-end of the intron. To verify that self-splicing in the absence of protein used the same mechanism, they performed a two-part experiment. In the first part, they incubated the splicing precursor with [α-³²P]GTP under splicing and nonsplicing conditions, then electrophoresed the products to see if the intron had become labeled. Figure 14.44a shows that it had, and a similar experiment with [γ-³²P]GTP gave the same results. In the second part, these workers 5′-end-labeled the intron with [α-³²P]GTP in the same way and sequenced the product. It gave exactly the sequence expected for the linear intron, with an extra G at the 5′-end (Figure 14.44b). This G could be removed by RNase T1, demonstrating that it

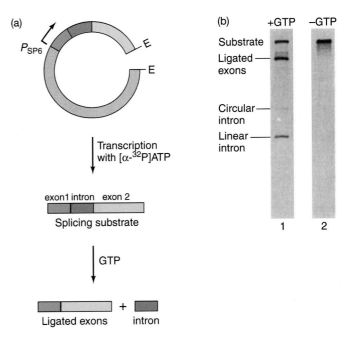

Figure 14.43 Demonstration of exon ligation. (a) Experimental scheme. Cech and coworkers constructed a plasmid containing part of the *Tetrahymena* 26S rRNA gene: 303 bp of exon 1 (blue); the 413-bp intron (red); and 624 bp of exon 2 (yellow). They linearized the plasmid by cutting it with *Eco*RI, creating *Eco*RI ends (E), then transcribed the plasmid in vitro with phage SP6 RNA polymerase and [α-³²P]ATP. This yielded the labeled splicing substrate. They incubated this substrate under splicing conditions in the presence or absence of GTP, then electrophoresed the splicing reactions and detected the labeled RNAs by autoradiography. **(b)** Experimental results. In the presence of GTP (lane 1), a prominent band representing the ligated exons appeared, in addition to bands representing the linear and circular intron. In the absence of GTP (lane 2), only the substrate band appeared. Thus, exon ligation appears to be a part of the self-splicing reaction catalyzed by this RNA. (*Source:* (b) Kruger et al., Intermolecular exon ligation of the rRNA precursor of *Tetrahymena*: Oligonucleotides can function as 5′ exons. *Cell* 43 (Dec 1985) f. 1a, p. 432. Reprinted by permission by Elsevier Science.)

is attached to the end of the intron by a normal 5′-3′-phosphodiester bond. Figure 14.45 presents a model for the splicing of the *Tetrahymena* 26S rRNA precursor, up to the point of ligating the two exons together and formation of the linear intron.

We have seen that the excised intron can cyclize itself. Cech and his coworkers showed that this cyclization actually involves the loss of 15 nt from the 5′-end of the linear intron. Three lines of evidence led to this conclusion: (1) When the 5′-end of the linear intron is labeled, none of this label appears in the circularized intron. (2) At least two RNase T1 products (actually three) found at the 5′-end of the linear intron are missing from the circular intron. (3) Cyclization of the intron is accompanied by the accumulation of an RNA 15-mer that contains the missing RNase T1 products.

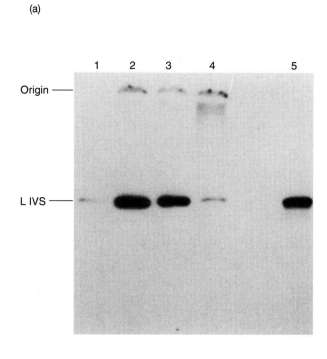

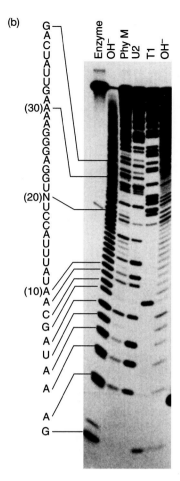

Figure 14.44   Addition of GMP to the 5′-end of the excised intron. (a) Radioactive GTP labels the intron during splicing. Cech and coworkers transcribed plasmid pIVS11 under nonsplicing conditions with no labeled nucleotides. They isolated this unlabeled 26S rRNA precursor and incubated it under splicing conditions in the presence of [α-^{32}P]GTP. Then they chromatographed the products on Sephadex G-50, electrophoresed the column fractions, and autoradiographed the gel. Lanes 1–4 are successive fractions from the Sephadex column. Lane 5 is a linear intron marker. Lanes 2 and 3 contain the bulk of the linear intron, and it is labeled, indicating that it had incorporated a labeled guanine nucleotide. **(b)** Sequence of the labeled intron. Cech and coworkers used an enzymatic method to sequence the 5′-end of the RNA. They cut it with base (OH⁻), which cuts after every base; RNase Phy M, which cuts after A and U; RNase U2, which cuts after A; and RNase T1, which cuts after G. Treatment of each RNA sample is indicated at top. The deduced sequence is given at left. Note the 5′-G at bottom. (*Source:* Kruger et al., Self-splicing RNA: Autoexcision and autocyclization of the ribosomal RNA intervening sequence of *Tetrahymena. Cell* 31 (Nov 1982) f. 4, p. 151. Reprinted by permission of Elsevier Science.)

But this is not the end of the process. After cyclization, the circular intron opens up again at the very same phosphodiester bond that formed the circle in the first place. Then the intron recyclizes by removing four more nucleotides from the 5′-end. Finally, the intron opens up at the same bond that just formed, yielding a shortened linear intron.

Figure 14.46 presents a detailed mechanism of the cyclization and relinearization of the excised intron. Notice that throughout the splicing process, for every phosphodiester bond that breaks a new one forms. Thus, the free energy change of each step is near zero, so no exogenous source of energy, such as ATP, is required. Another gen-

eral feature of the process is that the bonds that form to make the circular introns are the same ones that break when the circle opens up again. This tells us that these bonds are special; the three-dimensional shape of the RNA must strain these bonds to make them easiest to break during relinearization. This strain would help to explain the catalytic power of the RNA.

At first glance, there appears to be a major difference between the splicing mechanisms of spliceosomal introns and group I introns: Whereas the group I introns use an exogenous nucleotide in the first step of splicing, spliceosomal introns use a nucleotide that is integral to the intron itself. However, on closer examination we see that

(a)

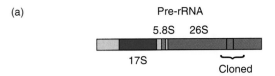

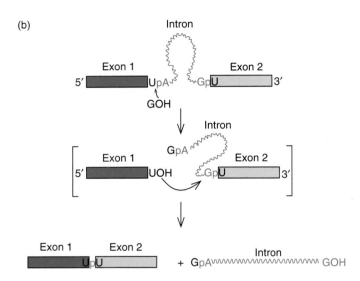

Figure 14.45 Self-splicing of *Tetrahymena* rRNA precursor.
(a) Structure of the rRNA precursor, containing the 17S, 5.8S, and 26S sequences. Note the intron within the 26S region (red). The cloned segment used in subsequent experiments is indicated by a bracket. (b) Self-splicing scheme. In the first step (top), a guanine nucleotide attacks the adenine nucleotide at the 5′-end of the intron, releasing exon 1 (blue) from the rest of the molecule and generating the hypothetical intermediates shown in brackets. In the second step, exon 1 (blue) attacks exon 2 (yellow), performing the splicing reaction that releases a linear intron (red), and joins the two exons together. Finally, in a series of reactions not shown here, the linear intron loses 19 nt from its 5′-end.

the difference might not be as great as it seems. Michael Yarus and his colleagues used molecular modeling techniques to predict the lowest energy conformation of the *Tetrahymena* 26S rRNA intron as it associates with GMP. They proposed that part of the intron folds into a double helix with a pocket that holds the guanine nucleotide through hydrogen bonds (Figure 14.47). This guanine, held fast to the intron, behaves in essentially the same way as the adenine in spliceosomal introns. Of course, it cannot form a lariat because it is not covalently linked to the intron.

Until the discovery of self-splicing RNAs, biochemists thought that the catalytic parts of enzymes were made only of protein. Sidney Altman had shown a few years

earlier that **RNase P,** which cleaves extra nucleotides off the 5′-ends of tRNA precursors, has an RNA component called **M1.** But RNase P also has a protein component, which could have held the catalytic activity of the enzyme. In 1983, Altman confirmed that the M1 RNA is the catalytic component of RNase P. This enzyme and self-splicing RNAs are examples of catalytic RNAs, which we call **ribozymes.** Actually, the reactions we have seen so far, in which group I introns participate, are not enzymatic in the strict sense, because the RNA itself changes. A true enzyme is supposed to emerge unchanged at the end of the reaction. But the final linearized group I intron from the *Tetrahymena* 26S rRNA precursor can act as a true enzyme by adding nucleotides to, and subtracting them from, an oligonucleotide.

SUMMARY Group I introns, such as the one in the *Tetrahymena* 26S rRNA precursor, can be removed in vitro with no help from protein. The reaction begins with an attack by a guanine nucleotide on the 5′-splice site, adding the G to the 5′-end of the intron, and releasing the first exon. In the second step, the first exon attacks the 3′-splice site, ligating the two exons together, and releasing the linear intron. The intron cyclizes twice, losing nucleotides each time, then linearizes for the last time.

Group II Introns

The introns of fungal mitochondrial genes were originally classified as group I or group II according to certain conserved sequences they contained. Later, it became clear that mitochondrial and chloroplast genes from many species contained group I and II introns, and that RNAs containing both classes of intron have members that are self-splicing. However, the mechanisms of splicing used by RNAs with group I and group II introns are different. Whereas the initiating event in group I splicing is attack by a guanine nucleotide, the initiating event in group II splicing involves intramolecular attack by an A residue in the intron to form a lariat.

The lariat formation by group II introns sounds very similar to the situation in spliceosomal splicing of nuclear mRNA precursors, and the similarity extends to the overall shapes of the RNAs in the spliceosomal complex and of the group II introns, as we saw in Figure 14.26. This implies a similarity in function between the spliceosomal snRNPs and the catalytic part of the group II introns. It may even point to a common evolutionary origin of these

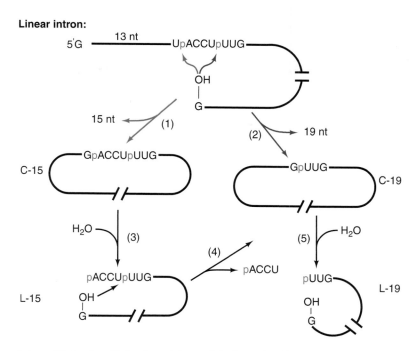

Linear intron:

Figure 14.46 Fate of the linear intron. We begin with the linear intron originally excised from the 26S rRNA precursor. This can be cyclized in two ways: In reaction 1 (green arrows), the 3′-terminal G attacks the bond between U-15 and A-16, removing a 15-nt fragment and giving a circular intron (C-15). In the alternative reaction (2, blue arrows), the terminal G attacks 4 nt farther into the intron, removing a 19-nt fragment and leaving a smaller circular intron (C-19). Reaction 3, C-15 can open up at the same bond that closed the circle, yielding a linear intron (L-15). Reaction 4, the terminal G of L-15 can attack the bond between the first two U's, yielding the circular intron C-19. Reaction 5, C-19 opens up to yield the linear intron L-19.

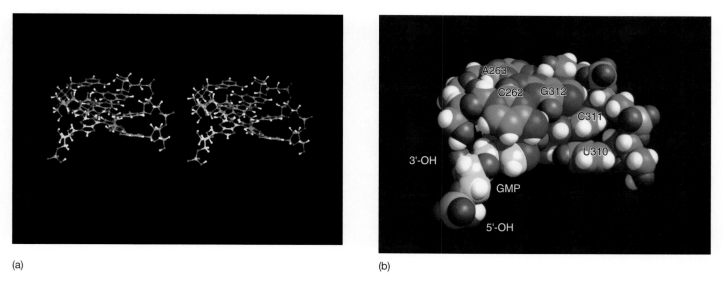

(a)   (b)

Figure 14.47 Two views of GMP held in a pocket of the 26S rRNA intron. (a) A cross-eyed stereogram that can be viewed in three dimensions by crossing the eyes until the two images merge. Carbon atoms of RNA, green; carbon atoms of G, yellow; phosphorus, lavender. Other atoms are standard colors. **(b)** Space-filling model. Colors are as in part (**a**). (*Source:* Yarus et al., An axial binding site in the *Tetrahymena* precursor RNA. *J. Mol. Biol.* 222 (1991) f. 7c-d, p. 1005, by permission of Academic Press.)

RNA species. In fact, it has been proposed that nuclear pre-mRNA introns descended from bacterial group II introns. These bacterial introns presumably got into eukaryotic cells because they inhabited the bacteria that invaded the precursors of modern eukaryotic cells and evolved into mitochondria. This hypothesis has become even more attractive since the discovery of group II introns in archaea, as well as in two classes of bacteria: cyanobacteria and purple bacteria. If we assume that the group II introns are older than the common ancestor of these two bacterial lineages, then they are old enough to have inhabited the bacteria that were the ancestors of modern eukaryotic organelles. Nevertheless, convergent evolution to a common mechanism also remains a possibility.

SUMMARY RNAs containing group II introns self-splice by a pathway that uses an A-branched lariat intermediate, just like the spliceosomal lariats. The secondary structures of the splicing complexes involving spliceosomal systems and group II introns are also strikingly similar.

14.4 tRNA Splicing

Nuclear mRNA introns have no special secondary structure to show where splicing should occur. Instead, the spliceosomal snRNPs bend the mRNA precursor into the proper shape for splicing. As we have seen, this shape resembles the secondary structure naturally assumed by group II self-splicing introns. In this respect, splicing of tRNA introns resembles splicing of group II introns; all unspliced tRNA precursors appear to have their introns in the same place—one nucleotide to the 3′-side of the anticodon—and all have essentially the same shape. This presumably makes it easier for the splicing machinery to find the intron and remove it. Figure 14.48 is a schematic diagram of a typical yeast tRNA precursor, showing the shape of the molecule and the position of the intron.

The mechanism of splicing of tRNA precursors is also different from what we have encountered so far. All the examples of splicing we have studied have two steps: (1) an attack on the 5′-splice site by a nucleotide in the intron or by a free nucleotide; and (2) an attack by the first exon on the second exon, which splices the two together. In both steps, the number of phosphodiester bonds is conserved. For every bond that breaks, a new one forms. Splicing of tRNA also proceeds in two steps, but they are different from the ones described so far. In the first step, a **tRNA endonuclease** cuts out the

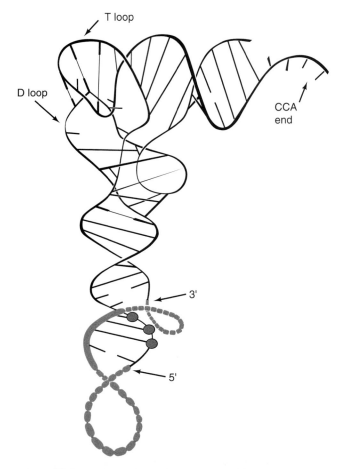

Figure 14.48   Consensus tertiary structure of yeast tRNA precursors. The molecule keeps its well-known L-shape, even with the intron (red) still in place, one nucleotide to the 3′-side of the anticodon (green). (*Source:* From Lee and Knapp, *Journal of Biological Chemistry* 260:3113, 1985. Copyright © 1985 The American Society for Biochemistry & Molecular Biology, Bethesda, MD. Reprinted by permission.)

intron. In the second step, an **RNA ligase** stitches the two half-tRNAs together.

John Abelson and colleagues clearly demonstrated the autonomy of these two reactions in yeast by separating the two enzymes and showing they could operate independently. In one experiment, they labeled six different pre-tRNAs, then incubated them with a membrane fraction containing the endonuclease, with or without the ligase. Figure 14.49 shows that the endonuclease alone yielded half-tRNAs and introns, but no spliced tRNAs. Adding either ATP or ligase alone to the endonuclease reaction made no difference, but adding ATP *plus* ligase yielded spliced tRNAs.

The detailed mechanism of the ligation step in tRNA splicing varies from one organism to another. Aaron

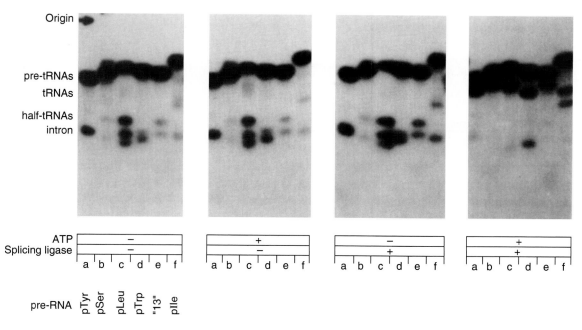

Figure 14.49   Independence of the endonuclease and ligation reactions in yeast tRNA splicing. Abelson and colleagues labeled six different yeast pre-tRNAs, as indicated at bottom (a–f). Then they incubated these molecules with a membrane fraction containing the endonuclease activity in the presence or absence of ATP and ligase, as indicated at bottom. The half-tRNAs and introns were generally in the positions indicated at left, except that the short tRNA^{Tyr} half-molecules migrated along with the introns, and the large tRNA^{Ile} intron migrated midway between the half-tRNAs and the tRNAs. Only when endonuclease, ligase, and ATP were all included did Abelson and colleagues see disappearance of the half-molecules and appearance of mature ligated tRNAs. (*Source:* Peebles et al., Precise excision of intervening sequences from precursor tRNAs by a membrane-associated yeast endonuclease. *Cell* 32 (Feb 1983) f. 3, p. 527. Reprinted by permission of Elsevier Science.)

Shatkin and coworkers showed that ligation in HeLa cells works as we probably would predict: The 3′-phosphate on the 5′-half-molecule forms the phosphodiester bond between half-molecules. However, Abelson and colleagues found something more complex in yeast.

Figure 14.50 depicts the splicing mechanism proposed for yeast tRNAs. The first product after the endonuclease step is a 2′,3′-cyclic phosphate. Next, a cyclic phosphodiesterase cleaves the cyclic bond, leaving a 2′-phosphomonoester. At the same time, a polynucleotide kinase phosphorylates the free hydroxyl group at the 5′-end of the 3′-half-molecule, and the RNA ligase extracts an AMP from ATP and donates it to the 5′-end of the 3′-half-molecule, leaving an activated RNA with a 5′,5′-phosphoanhydride bond. During the ligation step, the 3′-half-molecule loses its AMP, and the remaining 5′-phosphate forms the phosphodiester bond between the two half-molecules. The 2′-phosphate remains for a while after the phosphodiester bond forms, yielding an unusual 2′-phosphomonoester, 3′,5′-phosphodiester linkage. Finally, a phosphatase removes the 2′-phosphate, leaving an ordinary 3′,5′-phosphodiester bond between the two half-molecules.

> **SUMMARY** If a tRNA precursor has an intron, it is always in the same place, one nucleotide to the 3′-side of the anticodon. These tRNA precursors are spliced in two steps. First the intron is cut out by a membrane-bound endonuclease, and then the two half-molecules are ligated together. The ligation step can be a simple formation of a 3′,5′-phosphodiester bond, as in HeLa cells, or it can go through a 2′-phosphomonoester, 3′,5′-phosphodiester intermediate, as in yeast.

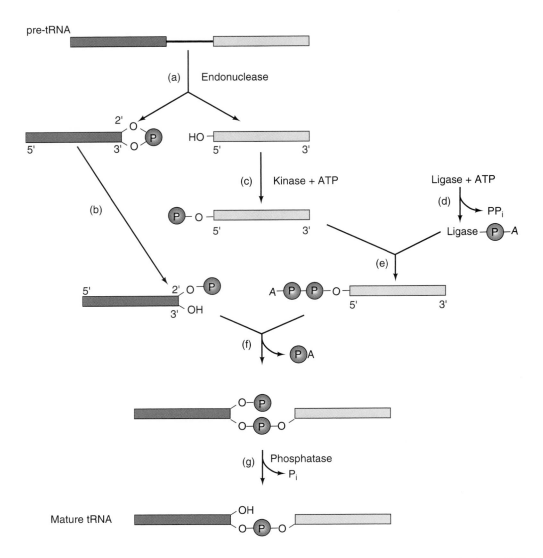

Figure 14.50   Mechanism of tRNA splicing. (a) An endonuclease removes the intron from the tRNA precursor, leaving a 2′,3′-cyclic phosphate at the end of the 5′-half-molecule (blue). **(b)** The cyclic phosphate opens, leaving the phosphate on the 2′-hydroxyl group. **(c)** A kinase adds a phosphate to the 5′-end of the 3′-half-molecule (yellow). **(d)** A ligase accepts an AMP from ATP. **(e)** The ligase donates the adenylate group to the 5′-end of the 3′-half-molecule. **(f)** The two half-molecules are joined together, with adenylate as the by-product. **(g)** Finally, the phosphate is removed from the 2′-hydroxyl group of the 5′-half-molecule, yielding the mature tRNA product.

SUMMARY

Nuclear mRNA precursors are spliced via a lariat-shaped, or branched, intermediate. In addition to the consensus sequences at the 5′- and 3′-ends of nuclear introns, branchpoint consensus sequences also occur. In yeast, this sequence is nearly invariant: UACUAAC. In higher eukaryotes, the consensus sequence is more variable: YNCURAC. In all cases, the branched nucleotide is the final A in the sequence. The yeast branchpoint sequence also determines which downstream AG is the 3′-splice site.

Splicing appears to take place on a particle called a spliceosome. Yeast and mammalian spliceosomes have sedimentation coefficients of about 40S and 60S, respectively. Genetic experiments have shown that base pairing between U1 snRNA and the 5′-splice site of an mRNA precursor is necessary, but not sufficient, for splicing. The U6 snRNP also associates with the 5′-end of the intron by base pairing. This association first occurs prior to formation of the lariat intermediate, but its character may change after this first step in splicing. The association between U6 and the splicing substrate is essential for the splicing process. U6 also associates with U2 during splicing.

The U2 snRNA base-pairs with the conserved sequence at the splicing branchpoint. This base pairing is essential for splicing. U2 also forms vital base pairs with U6, forming a region called helix I, that apparently helps orient these snRNPs for splicing. The U4 snRNA base-pairs with U6, and its role seems to be to bind U6 until U6 is needed in the splicing reaction. The U5 snRNP associates with the last nucleotide in one exon and the first nucleotide of the next. This presumably lines the 5′- and 3′-splice sites up for splicing.

The spliceosomal complex (substrate, U2, U5, and U6) poised for the second step in splicing can be drawn in the same way as a group II intron at the same stage of splicing. Thus, the spliceosomal snRNPs seem to substitute for elements at the center of catalytic activity of the group II introns, and probably have the spliceosome's catalytic activity.

The spliceosome cycle includes the assembly, splicing activity, and disassembly of the spliceosome. Assembly begins with the binding of U1 to the splicing substrate to form a commitment complex. U2 is the next snRNP to join the complex, followed by the others. The binding of U2 requires ATP. U6 dissociates from U4, then displaces U1 at the 5′-splice site. This ATP-dependent step activates the spliceosome and allows release of U1 and U4. The splicing factor Slu7 is required for correct 3′-splice site selection. In its absence, splicing to the correct 3′-splice site AG is specifically suppressed and splicing to aberrant AG's within about 30 nt of the branchpoint is activated. U2AF is also required for 3′-splice site recognition. The 65-kD U2AF subunit binds to the polypyrimidine tract upstream of the 3′-splice site, and the 35-kD subunit binds to the 3′-splice site AG.

Commitment to splice at a given site is determined by an RNA-binding protein, which presumably binds to the splicing substrate and recruits other spliceosomal components, starting with U1. For example, the SR proteins SC35 and SF2/ASF commit splicing on human β-globin pre-mRNA and HIV tat pre-mRNA, respectively. In the yeast commitment complex, the branchpoint bridging protein (BBP) binds to a U1 snRNP protein at the 5′-end of the intron, and to Mud2p near the 3′-end of the intron. It also binds to the RNA near the 3′-end of the intron. Thus, it bridges the intron and could play a role in defining the intron prior to splicing. The mammalian counterpart of BBP, SF1 (mBBP), may serve the same function in the mammalian commitment complex.

The transcripts of many eukaryotic genes are subject to alternative splicing. This can have profound effects on the protein products of a gene. For example, it can make the difference between a secreted or a membrane-bound protein; it can even make the difference between activity and inactivity. In the fruit fly, the products of three genes in the sex determination pathway are subject to alternative splicing. Female-specific splicing of the *tra* transcript gives an active product that causes female-specific splicing of the *dsx* pre-mRNA, which produces a female fly. Male-specific splicing of the *tra* transcript gives

an inactive product that allows default, or male-specific, splicing of the *dsx* pre-mRNA, producing a male fly. Tra and its partner Tra-2 act in conjuction with one or more other SR proteins to commit splicing at the female-specific splice site on the *dsx* pre-mRNA. Such commitment is undoubtedly the basis of most, if not all, alternative splicing schemes.

Group I introns, such as the one in the *Tetrahymena* 26S rRNA precursor, can be removed with no help from protein in vitro. The reaction begins with an attack by a guanine nucleotide on the 5′-splice site, adding the G to the 5′-end of the intron and releasing the first exon. In the second step, the first exon attacks the 3′-splice site, ligating the two exons together and releasing the linear intron. The intron cyclizes twice, losing nucleotides each time, then linearizes for the last time.

RNAs containing group II introns self-splice by a pathway that uses an A-branched lariat intermediate, just like the spliceosomal lariats. The secondary structures of the splicing complexes involving spliceosomal systems and group II introns are also strikingly similar.

If a tRNA precursor has an intron, it is always in the same place, one nucleotide 3′ of the anticodon. These tRNA precursors are spliced in two steps. First the intron is cut out by a membrane-bound endonuclease, and then the two half-molecules are ligated together. The ligation step can be a simple formation of a 3′,5′-phosphodiester bond, as in HeLa cells, or it can go through a 2′-phosphomonoester, 3′,5′-phosphodiester intermediate, as in yeast.

REVIEW QUESTIONS

1. You are investigating a gene with one large intron and two short exons. Show the results of R-looping experiments done under the following conditions:
   a. Using mRNA and single-stranded DNA
   b. Using mRNA and double-stranded DNA
   c. Using mRNA precursor and single-stranded DNA
   d. Using mRNA precursor and double-stranded DNA

2. Describe and show the results of an R-looping experiment that demonstrates that an intron is transcribed.

3. Diagram the lariat mechanism of splicing.

4. Present gel electrophoretic data that suggest that the excised intron is circular, or lariat-shaped.

5. Present gel electrophoretic data that distinguish between a lariat-shaped splicing intermediate (the intron–exon-2 intermediate) and a lariat-shaped product (the excised intron).

6. The lariat model makes four predictions: (a) the excised intron has a 3′-hydroxyl group; (b) the phosphorus atom between the two spliced exons comes from the 3′-end of the intron; (c) there is a branched nucleotide; and (d) the 5′-end of the intron is involved at the branch. Describe and show the results of experiments that confirm each of these predictions.

7. Diagram the mechanism of RNase T1 action. Because this is the same mechanism used in base hydrolysis of RNA, how does this explain why DNA is not subject to base hydrolysis?

8. Describe and give the results of an experiment that shows that a sequence (UACUAAC) within a yeast intron is required for splicing.

9. Describe and show the results of an experiment that demonstrates that this UACUAAC sequence within a yeast intron dictates splicing to an AG downstream.

10. What relationship does the UACUAAC sequence have to the lariat model of splicing?

11. Describe and show the results of an experiment that demonstrates that human spliceosomes have a sedimentation coefficient of 60S.

12. Describe and show the results of an experiment that demonstrates that base pairing between U1 snRNP and the 5′-exon–intron boundary is required for splicing.

13. Describe and show the results of an experiment that demonstrates that base pairing between U1 and the 5′-exon–intron boundary is not sufficient for splicing.

14. What snRNP besides U1 and U5 must bind near the 5′-exon–intron boundary in order for splicing to occur? Present two kinds of cross-linking data to support this conclusion.

15. What are two major differences between RNA cross-linking experiments using 4-thioU and those using psoralen?

16. Describe and show the results of an experiment that demonstrates that base pairing between U2 snRNP and the branchpoint sequence is required for splicing. In this experiment, why was it not possible to mutate the cell's only copy of the U2 gene?

17. Besides base pairing with the pre-mRNA, U6 base-pairs with two snRNAs. What are they?

18. Describe and show the results of an experiment that demonstrates that U5 contacts the 3′-end of the upstream exon and the 5′-end of the downstream exon during splicing. Make sure your experiment(s) provide positive identification of the RNA species involved, not just electrophoretic mobilities.

19. Describe and show the results of an experiment that demonstrates which bases in U5 can be cross-linked to bases in the pre-mRNA.

20. Draw a diagram of a pre-mRNA as it exists in a spliceosome just before the second step in splicing. Show the interactions with U2, U5, and U6 snRNPs. This scheme resembles the intermediate stage for splicing of what kind of self-splicing RNA?

21. Describe and show the results of an experiment that demonstrates that U1 is the first snRNP to bind to the splicing substrate.

22. Describe and show the results of an experiment that demonstrates that binding of all other snRNPs to the spliceosome depends on U1, and that binding of U2 requires ATP.

23. Describe and show the results of an experiment that demonstrates that Slu7 is required for selection of the proper AG at the 3′-splice site.

24. Describe a splicing commitment assay to screen for splicing factors involved in commitment. Show sample results.

25. Describe a yeast two-hybrid assay for interaction between two known proteins.

26. Describe a yeast two-hybrid screen for finding an unknown protein that interacts with a known protein.

27. Describe and give the results of a yeast two-hybrid assay that shows interaction between yeast branchpoint bridging protein (BBP) and two other proteins. What are the two other proteins, and where are they found with respect to the ends of the intron in the commitment complex?

28. Diagram the alternative splicing of the immunoglobulin μ heavy-chain transcript. Focus on the exons that are involved in one or the other of the alternative pathways, rather than the ones that are involved in both. What difference in the protein products is caused by the two pathways of splicing?

29. Describe and show the results of an experiment that demonstrates self-splicing by a group I intron.

30. Describe and show the results of an experiment that demonstrates autocyclization by the linear group I intron.

31. Describe and show the results of an experiment that demonstrates that an RNA containing a group I intron can splice together its two exons.

32. Describe and show the results of an experiment that demonstrates that a guanine nucleotide is added to the end of a spliced-out group I intron.

33. Draw a diagram of the steps involved in autosplicing of an RNA containing a group I intron. You do not need to show cyclization of the intron.

34. Diagram the steps involved in forming the L-19 intron from the original excised linear intron product of the *Tetrahymena* 26S pre-rRNA. Do not go through the C-15 intermediate.

35. Diagram the process of splicing human pre-tRNA.

SUGGESTED READINGS

General References and Reviews

Baker, B.S. 1989. "Sex in flies: The splice of life." *Nature* 340:521–24.

Guthrie, C. 1991. Messenger RNA splicing in yeast: Clues to why the spliceosome is a ribonucleoprotein. *Science* 253:157–63.

Lamm, G.M. and A.I. Lamond. 1993. Non-snRNP protein splicing factors. *Biochimica et Biophysica Acta* 1173:247–65.

Manley, J. 1993. Question of commitment. *Nature* 365:14.

Murray, H.L. and K.A. Jarrell. 1999. Flipping the switch to an active spliceosome. *Cell* 96:599–602.

Patrusky, B. 1992. The intron story. *Mosaic* 23:22–33.

Sharp, P.A. 1994. Split genes and RNA splicing. (Nobel Lecture.) *Cell* 77:805–15.

Weiner, A.M. 1993. mRNA splicing and autocatalytic introns: Distant cousins or the products of chemical determinism? *Cell* 72:161–64.

Wise, J.A. 1993. Guides to the heart of the spliceosome. *Science* 262:1978–79.

Research Articles

Abovich, N. and M. Rosbash. 1997. Cross-intron bridging interactions in the yeast commitment complex are conserved in mammals. *Cell* 89:403–12.

Berget, S.M., C. Moore, and P. Sharp. 1977. Spliced segments at the 5′ terminus of adenovirus 2 late mRNA. *Proceedings of the National Academy of Sciences USA* 74:3171–75.

Berglund, J.A., K. Chua, N. Abovich, R. Reed, and M. Rosbash. 1997. The splicing factor BBP interacts specifically with the pre-mRNA branchpoint sequence UACUAAC. *Cell* 89:781–87.

Brody, E. and J. Abelson. 1985. The spliceosome: Yeast pre-messenger RNA associated with a 40S complex in a splicing-dependent reaction. *Science* 228:963–67.

Chua, K. and R. Reed. 1999. The RNA splicing factor hSlu7 is required for correct 3′ splice-site choice. *Nature* 402:207–10.

Early, P., J. Rogers, M. Davis, K. Calame, M. Bond, R. Wall, and L. Hood. 1980. Two mRNAs can be produced from a single immunoglobulin γ gene by alternative RNA processing pathways. *Cell* 20:313–19.

Fu, X.-D. 1993. Specific commitment of different pre-mRNAs to splicing by single SR proteins. *Nature* 365:82–85.

Grabowski, P., R.A. Padgett, and P.A. Sharp. 1984. Messenger RNA splicing in vitro: An excised intervening sequence and a potential intermediate. *Cell* 37:415–27.

Grabowski, P., S.R. Seiler, and P.A. Sharp. 1985. A multicomponent complex is involved in the splicing of messenger RNA precursors. *Cell* 42:345–53.

Kruger, K., P.J. Grabowski, A.J. Zaug, J. Sands, D.E. Gottschling, and T.R. Cech. 1982. Self-splicing RNA: Autoexcision and autocylization of the ribosomal RNA intervening sequence of *Tetrahymena*. *Cell* 31:147–57.

Langford, C. and D. Gallwitz. 1983. Evidence for an intron-contained sequence required for the splicing of yeast RNA polymerase II transcripts. *Cell* 33:519–27.

Lee, M.-C. and G. Knapp. 1985. Transfer RNA splicing in *Saccharomyces cerevisiae*. *Journal of Biological Chemistry* 260:3108–15.

Lesser, C.F. and C. Guthrie. 1993. Mutations in U6 snRNA that alter splice site specificity: Implications for the active site. *Science* 262:1982–88.

Liao, X.C., J. Tang, and M. Rosbash. 1993. An enhancer screen identifies a gene that encodes the yeast U1 snRNP A protein: Implications for snRNP protein function in pre-mRNA splicing. *Genes and Development* 7:419–28.

Parker, R., P.G. Siliciano, and C. Guthrie. 1987. Recognition of the TACTAAC box during mRNA splicing in yeast involves base pairing to the U2-like snRNA. *Cell* 49:229–39.

Peebles, C.L., P. Gegenheimer, and J. Abelson. 1983. Precise excision of intervening sequences from precursor tRNAs by a membrane-associated yeast endonuclease. *Cell* 32:525–36.

Ruby, S.W. and J. Abelson. 1988. An early hierarchic role of U1 small nuclear ribonucleoprotein in spliceosome assembly. *Science* 242:1028–35.

Sontheimer, E.J. and J.A. Steitz. 1993. The U5 and U6 small nuclear RNAs as active site components of the spliceosome. *Science* 262:1989–96.

Staley, J.P. and C. Guthrie. 1999. An RNA switch at the 5′ splice site requires ATP and the DEAD box protein Prp28p. *Molecular Cell* 3:55–64.

Tian, M. and T. Maniatis. 1993. A splicing enhancer complex controls alternative splicing of *doublesex* pre-mRNA. *Cell* 74:105–14.

Tilghman, S.M., P. Curtis, D. Tiemeier, P. Leder, and C. Weissmann. 1978. The intervening sequence of a mouse β-globin gene is transcribed within the 15S β-globin mRNA precursor. *Proceedings of the National Academy of Sciences USA* 75:1309–13.

Wassarman, D.A. and J.A. Steitz. 1992. Interactions of small nuclear RNA's with precursor messenger RNA during in vitro splicing. *Science* 257:1918–25.

Yarus, M., I. Illangesekare, and E. Christian. 1991. An axial binding site in the *Tetrahymena* precursor RNA. *Journal of Molecular Biology* 222:995–1012.

Zhuang, Y. and A.M. Weiner. 1986. A compensatory base change in U1 snRNA suppresses a 5′-splice site mutation. *Cell* 46:827–35.

CHAPTER 15

Posttranscriptional Events II: Capping and Polyadenylation

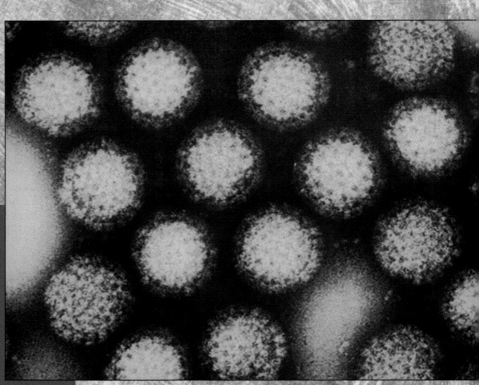

False color electron micrograph of human reovirus (×1,000,000).
© K. G. Murti/Visuals Unlimited.

Besides splicing, eukaryotic cells perform several other posttranscriptional modifications of their RNAs. Messenger RNAs are subject to two kinds of posttranscriptional modification, or processing, known as capping and polyadenylation. In capping, a special blocking nucleotide (a cap) is added to the 5′-end of a pre-mRNA. In polyadenylation, a string of AMPs (poly[A]) is added to the 3′-end of the pre-mRNA. These steps are essential for the proper function of mRNAs and will be our topics in this chapter.

15.1 Capping

By 1974, several investigators had discovered that mRNA from a variety of eukaryotic species and viruses was methylated. Moreover, a significant amount of this methylation was clustered at the 5'-end of mRNAs, in structures we call **caps**. In this section we will examine the structure and synthesis of these caps.

Cap Structure

Before gene cloning became routine, viral mRNAs were much easier to purify and investigate than cellular mRNAs. Thus, the first caps to be characterized came from viral RNAs. Bernard Moss and his colleagues produced vaccinia virus mRNAs in vitro and isolated their caps as follows: They labeled the methyl groups in the RNA with ^{3}H-S-adenosyl methionine (AdoMet, a methyl donor), or with ^{32}P-nucleotides, then subjected the labeled RNA to base hydrolysis. The major products of this hydrolysis were mononucleotides, but the cap could be separated from these by DEAE-cellulose chromatography. Figure 15.1 shows the chromatographic behavior of the vaccinia virus caps. They behaved as a substance with a net charge near −5. Furthermore, the red and blue curves

in Figure 15.1b show that the ^{3}H(methyl) and ^{32}P labels practically coincided, demonstrating that the caps were methylated. Aaron Shatkin and his coworkers obtained very similar results with reovirus caps.

To determine the exact structure of the reovirus cap, Yasuhiro Furuichi and Kin-Ichiro Miura performed the following series of experiments. They found that they could label the cap with [β,γ-^{32}P]ATP (but not with [γ-^{32}P]ATP). This result indicated that the β-phosphate, but not the γ-phosphate, was retained in the cap. Because the β-phosphate of a nucleoside triposphate remains only in the first nucleotide in an RNA, this finding reinforced the notion that the cap was at the 5'-terminus of the RNA. But the β-phosphate must be protected, or blocked, by some substance (X), because it cannot be removed with alkaline phosphatase.

This raised the next question: What is X? The blocking agent could be removed with phosphodiesterase, which cuts both phosphodiester and phosphoanhydride bonds (e.g., the bond between the α- and β-phosphates in a nucleotide). This enzyme released a charged substance likely to be Xp. Next, Furuichi and Miura removed the phosphate from Xp with phosphomonoesterase, leaving just X, and subjected this substance to paper electrophoresis. Figure 15.2 shows that X coelectrophoresed

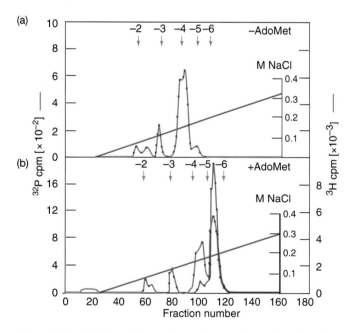

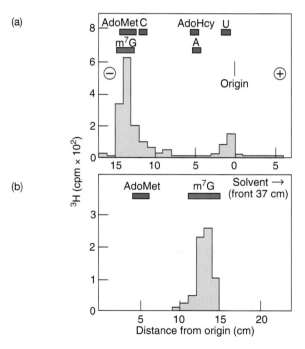

Figure 15.1 DEAE-cellulose chromatographic purification of vaccinia virus caps. Wei and Moss allowed vaccinia virus particles to synthesize caps in the presence of [β,γ-^{32}P]GTP and **(a)** in the absence and **(b)** presence of S-adenosyl [methyl-^{3}H] methionine. Then they digested the labeled, capped RNAs with KOH and separated the products by DEAE-cellulose column chromatography. ^{3}H (blue) and ^{32}P (red) radioactivities (in counts per minute) are plotted versus column fraction number. Salt concentrations (green) of each fraction are also plotted. The positions and net charges of markers are shown at the top of each panel. (*Source:* From C. M. Wei and B. Moss, "Methylated Nucleotides Block 5'-Terminus of Vaccinia Virus Messenger RNA," *Proceedings of the National Academy of Sciences* 72(1):318–322, January 1975.)

Figure 15.2 Identification of the capping substance (X) as 7-methyl-guanosine. Miura and Furuichi used phosphomonoesterase to digest the ^{3}H-labeled capping substance (Xp) to yield X. They electrophoresed this digest **(a)** along with a series of markers (S-adenosylmethionine, AdoMet; m^{7}G; S-adenosylhomocysteine, AdoHcy; adenosine, A; and uridine, U). Because electrophoresis did not resolve AdoMet and m^{7}G, these workers subjected the digest to paper chromatography **(b)** along with markers for AdoMet and m^{7}G. The radioactivity in X cochromatographed with the m^{7}G marker.
(*Source:* Reprinted with permission from *Nature* 253:375, 1975. Copyright © 1975 Macmillan Magazines Limited.)

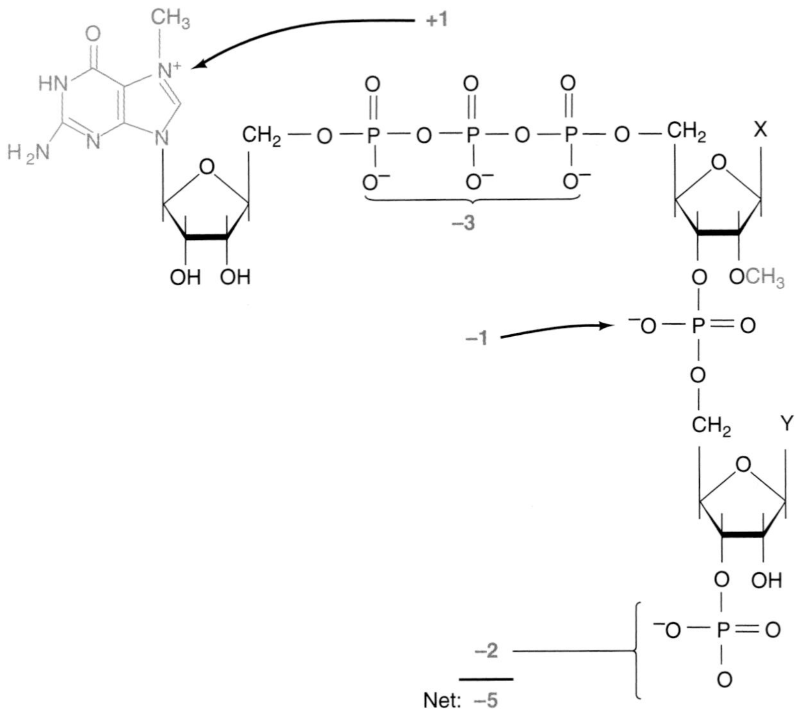

Figure 15.3   Reovirus cap structure, highlighting the charges. The m^7G (blue guanine with red methyl group) contributes a positive charge, the triphosphate linkage contributes three negative charges, the phosphodiester bond contributes one negative charge, and the terminal phosphate contributes two negative charges. The net charge is therefore about –5.

with **7-methylguanosine (m^7G)**. Thus, the capping substance is m^7G.

Another product of phosphodiesterase cleavage of the cap was pAm (2′-O-methyl-AMP). Thus, m^7G is linked to pAm in the cap. What is the nature of the linkage? The following two considerations tell us that it is a triphosphate: (1) The α-phosphate, but not the β- or γ-phosphate, of GTP was retained in the cap. (2) The β- and α-phosphates of ATP are retained in the cap. Thus, because one phosphate comes from the capping GTP, and two come from the nucleotide (ATP) that initiated RNA synthesis, there are three phosphates (a triphosphate linkage) between the capping nucleotide (m^7G) and the next nucleotide. Furthermore, because both ATP and GTP have their phosphates in the 5′-position, the linkage is very likely to be 5′ to 5′.

How do we explain the charge of the reovirus cap; about –5? Figure 15.3 provides a rationale. Three negative charges come from the triphosphate linkage between the m^7G and the penultimate (next-to-end) nucleotide. One negative charge comes from the phosphodiester bond between the penultimate nucleotide and the next nucleotide. (This bond is not broken by alkali because the 2′-hydroxyl group is methylated.) Two more negative charges come from the terminal phosphate in the cap. This makes a total of six negative charges, but the m^7G provides a positive charge, which gives the purified reovirus cap a charge of about –5.

Other viral and cellular mRNAs have similar caps, although the extent of 2′-O-methylation can vary. Figure 15.4 shows three forms of cap found in eukaryotic viruses and cells. **Cap 1** is the same as the cap shown in Figure 15.3. **Cap 2** has another 2′-O-methylated nucleotide (two in a row). And **cap 0** has no 2′-O-methylated nucleotides. Cap 2 is found only in eukaryotic cells, cap 1 is found in both cellular and viral RNAs, and cap 0 is found only in certain viral RNAs. As we saw in Chapter 14, some snRNAs have another kind of cap, which contains a trimethylated guanosine. We will discuss these caps later in this chapter.

Cap Synthesis

To determine how caps are made, Moss and his colleagues, and Furuichi and Shatkin and their colleagues, studied capping of model substrates in vitro. These investigators used cores from vaccinia virus and reovirus, respectively, to provide the capping enzymes. Both these human viruses replicate in the cytoplasm of their host cells, so they do not have access to the host nuclear machinery. Therefore, they must carry their own transcription and capping systems right in their virus cores. In both viruses, we observe the same sequence of events, as illustrated in Figure 15.5. (a) A nucleotide phosphohydrolase (also called **RNA triphosphatase**) clips the γ-phosphate off the triphosphate at the 5′-end of the growing RNA (or model substrate), leaving a diphosphate. (b) A guanylyl transferase attaches GMP from GTP to the diphosphate at the end of the RNA, forming the 5′–5′-triphosphate linkage. (c) A methyl transferase transfers the methyl group from *S*-adenosylmethionine (AdoMet) to the 7-nitrogen of the capping guanine. (d) Another methyl transferase uses

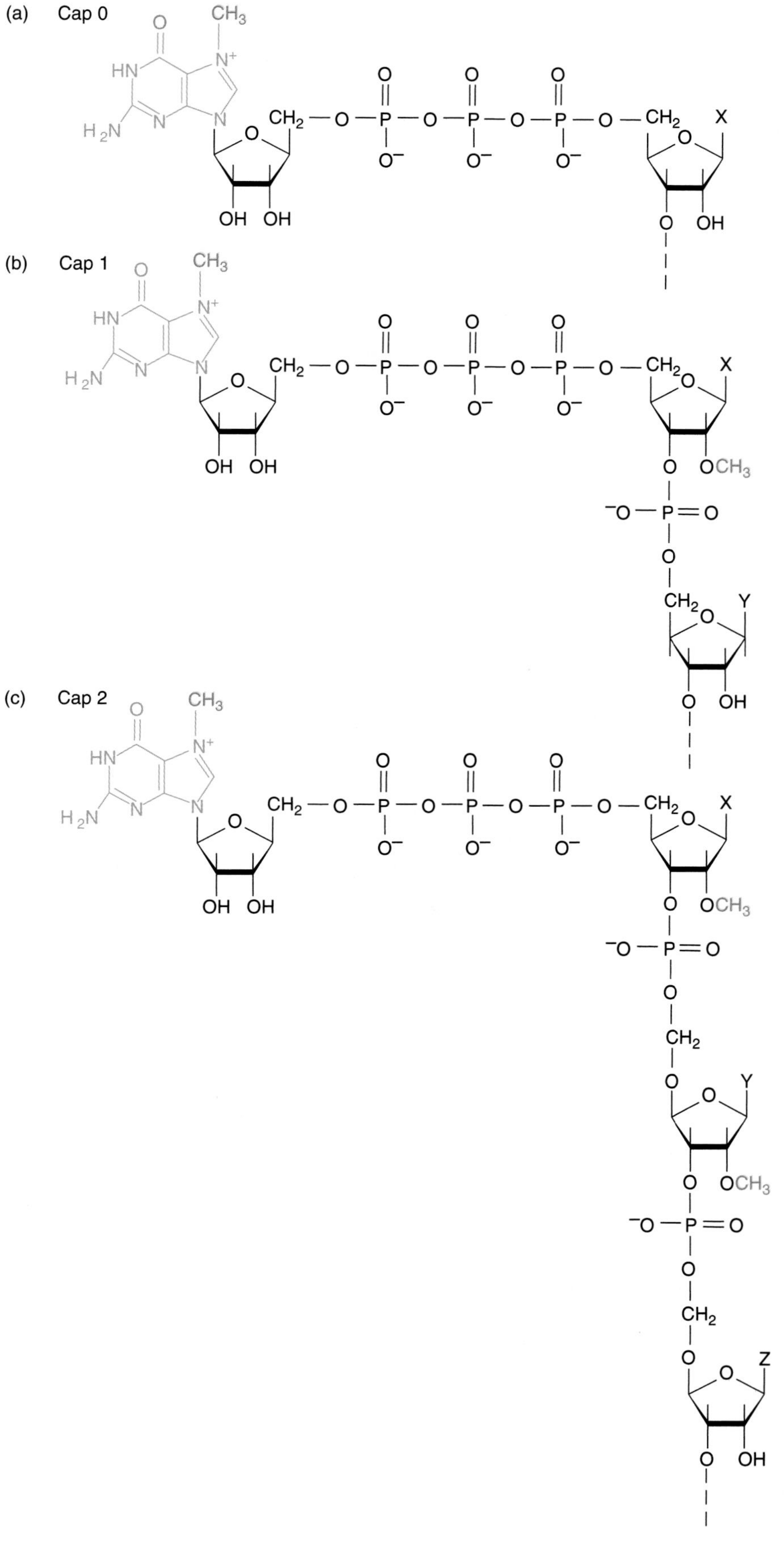

Figure 15.4 Three cap structures. The only differences among caps 0, 1, and 2 are the methyl groups, highlighted in red. Each 2'-O-methyl group prevents hydrolysis of the corresponding phosphodiester bond and therefore adds one more nucleotide (one more negative charge) to the isolated cap.

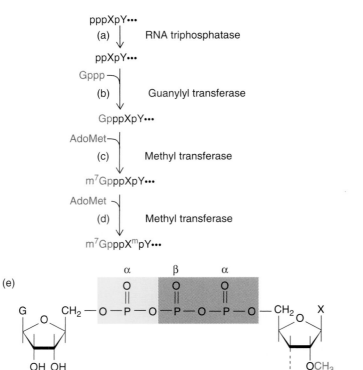

Figure 15.5 Sequence of events in capping. (a) RNA triphosphatase cleaves the γ-phosphate from the 5′-end of the growing RNA. **(b)** Guanylyl transferase adds the GMP part of GTP (blue) to form a triphosphate linkage, blocking the 5′-end of the RNA. **(c)** A methyl transferase adds a methyl group (red) from AdoMet to the N^7 of the blocking guanine. **(d)** Another methyl transferase adds a methyl group (red) from AdoMet to the 2′-hydroxyl group of the penultimate nucleotide. The product is cap 1. To form a cap 2, the next nucleotide (Y) would be methylated in a repeat of step **(d)**. **(e)** The origin of the phosphates in the triphosphate linkage. The α- and β-phosphates from the initiating nucleotide (XTP) are highlighted in green, and the α-phosphate from the capping GTP is highlighted in yellow.

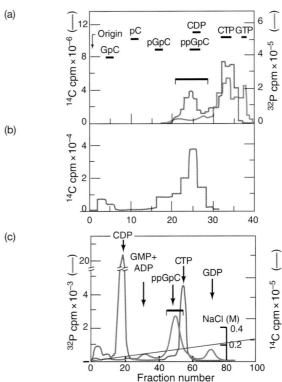

Figure 15.6 Identification of ppGpC as an intermediate in reovirus cap synthesis. (a) First purification step. Furuichi and colleagues added [^{14}C]CTP and [^{32}P]GTP to reovirus cores to label caps and capping intermediates. Then they analyzed the mixture by paper electrophoresis with the markers listed at top. One radioactive intermediate (bracket) coelectrophoresed with the ppGpC and CDP markers. **(b)** Conversion of ppGpC to GpC. Furuichi and colleagues treated the bracketed radioactive material from panel **(a)** with alkaline phosphatase, which should convert ppGpC to GpC, then electrophoresed the products. This time, a significant peak coelectrophoresed with the GpC marker. **(c)** Positive identification of ppGpC. Furuichi and colleagues subjected the bracketed material in **(a)** to ion-exchange chromatography on Dowex resin with the markers indicated at top. The major ^{32}P peak (red) coincided with the ppGpC marker. *(Source: From Furuichi et al., Journal of Biological Chemistry 251:5051, 1976. Copyright © 1976 The American Society for Biochemistry & Molecular Biology, Bethesda, MD. Reprinted by permission.)*

another molecule of AdoMet to methylate the 2′-hydroxyl of the penultimate nucleotide.

To verify that this really is the correct pathway, the investigators isolated each of the enzymes we have listed and all of the intermediates. For example, Furuichi and colleagues started with the labeled model substrate pppGpC, which resembles the 5′-end of a newly initiated reovirus mRNA. How do we know that the virus cores can remove a terminal phosphate and convert this starting material to ppGpC? These workers blocked the guanylyl transferase reaction with an excess of by-product (PP$_i$), which should cause ppGpC to build up, if it exists. They looked directly for this intermediate by the scheme in Figure 15.6. First, they performed paper electrophoresis with markers and showed that a significant labeled product coelectrophoresed with the ppGpC marker. Unfortunately, CDP also electrophoresed to this position, so the product could not be clearly identified. Next, they treated the

product with alkaline phosphatase to convert any ppGpC to GpC and reelectrophoresed it. Now a peak of radioactivity appeared in the GpC position. This was encouraging, but to positively identify ppGpC, these workers subjected the putative ppGpC peak from panel (a) to ion-exchange chromatography on a Dowex resin and obtained a radioactive peak that comigrated uniquely with the ppGpC marker. Thus, ppGpC is a real intermediate in the capping scheme.

When is the cap added? In some viruses, such as cytoplasmic polyhedrosis virus (CPV), lack of AdoMet completely inhibits transcription, suggesting that transcription depends on capping. This implies that capping in this virus is a very early event and presumably occurs soon after the first phosphodiester bond forms in the pre-mRNA. In other viruses, such as vaccinia virus, transcription occurs normally in the absence of AdoMet, so transcription and capping may not be so tightly coupled in that virus.

Unlike CPV and vaccinia virus, adenovirus replicates in the nucleus and therefore presumably takes advantage of the host cell's capping system. Adenovirus should therefore tell us more about when capping of eukaryotic pre-mRNAs occurs. James Darnell and colleagues performed an experiment that showed that adenovirus capping occurs early in the transcription process. These workers measured the incorporation of [³H]adenosine into the cap and the first dozen or so adenylate residues of the adenovirus major late transcripts (pre-mRNAs). First, they added [³H]adenosine to label the cap (the bold A in m⁷GpppA) and other adenosines in adenovirus pre-mRNAs during the late phase of infection. Then they separated large from small mRNA precursors by gradient centrifugation. Then they hybridized the small RNAs to a small restriction fragment that included the major late transcription start site. Any short RNAs that hybridized to this fragment were likely to be newly initiated RNAs, not just degradation products of mature RNAs. They eluted these nascent fragments from the hybrids and looked to see whether they were capped. Indeed they were, and no pppA, which would have been present on uncapped RNA, could be detected. This experiment demonstrated that caps are added to adenovirus major late pre-mRNA before the chain length reaches about 70 nt. It is now generally accepted that capping in eukaryotic cells occurs even earlier than that: before the pre-mRNA chain length reaches 30 nt.

> **SUMMARY** Caps are made in steps: First, an RNA triphosphatase removes the terminal phosphate from a pre-mRNA; next, a guanylyl transferase adds the capping GMP (from GTP). Next, two methyl transferases methylate the N⁷ of the capping guanosine and the 2'-O-methyl group of the penultimate nucleotide. These events occur early in the transcription process, before the chain length reaches 30 nt.

Functions of Caps

Caps appear to serve at least four functions. (1) They protect mRNAs from degradation. (2) They enhance the translatability of mRNAs. (3) They enhance the transport of mRNAs from the nucleus into the cytoplasm. (4) They enhance the efficiency of splicing of mRNAs. In this section we will discuss the first three of these functions, then deal with the fourth later in the chapter.

Protection The cap is joined to the rest of the mRNA through a triphosphate linkage found nowhere else in the RNA. The cap might therefore be expected to protect the

mRNA from attack by RNases that begin at the 5'-end of their substrates and that cannot cleave triphosphate linkages. In fact, good evidence supports the notion that caps protect mRNAs from degradation.

Furuichi, Shatkin, and colleagues showed in 1977 that capped reovirus RNAs are much more stable than uncapped RNAs. They synthesized newly labeled reovirus RNA that was either capped with m⁷GpppG, "blocked" with GpppG, or uncapped. Then they injected each of the three kinds of RNA into *Xenopus* oocytes, left them there for 8 h, then purified them and analyzed them by glycerol gradient ultracentrifugation. Reovirus RNAs exist in three size classes, termed large (l), medium (m), and small (s). Figure 15.7a shows a glycerol gradient ultracentrifugation separation of these three RNA classes. Furuichi and colleagues included RNAs with all three kinds of 5'-ends in this experiment, and no significant differences could be seen. All three size classes are clearly visible. Figure 15.7b shows what happened to these RNAs after 8 h in *Xenopus* oocytes. RNAs with all three kinds of 5'-ends had suffered degradation, but this degradation was much more pronounced for the uncapped RNAs. Thus, the *Xenopus* oocytes contain nucleases that degrade the viral RNAs, but the caps appear to provide protection from these nucleases.

Figure 15.7c underscores the fact that the protection was provided by the cap or block. In this experiment, Furuichi and colleagues chemically treated the 5'-ends of capped and blocked reovirus RNAs to remove the cap or block and yield pppGᵐ or pppG, respectively, at the 5'-ends of the RNAs. Then they injected the modified RNAs into *Xenopus* oocytes for 8 h and subjected them to the same sedimentation analysis as before. The decapped and deblocked RNAs obviously suffered substantial degradation in the oocytes.

Finally, Furuichi and colleagues performed the same experiment as in panel (b), but using wheat germ extracts instead of *Xenopus* oocytes. Again, uncapped RNA was much less stable than capped or blocked RNA (Figure 15.7d).

Translatability Another important function of the cap is to provide translatability. We will see in Chapter 17 that a eukaryotic mRNA gains access to the ribosome for translation via a cap-binding protein that recognizes the cap. If there is no cap, the cap-binding protein cannot bind and the mRNA is very poorly translated. Using an in vivo assay, Daniel Gallie has documented the stimulatory effect of the cap on translation. In this procedure, Gallie introduced the firefly luciferase mRNA, with and without a cap, and with and without poly(A), into tobacco cells. Luciferase is an easy product to detect because of the light it generates in the presence of luciferin and ATP. Table 15.1 illustrates that the poly(A) at the 3'-end and the cap at the 5'-end act synergistically to stabilize and,

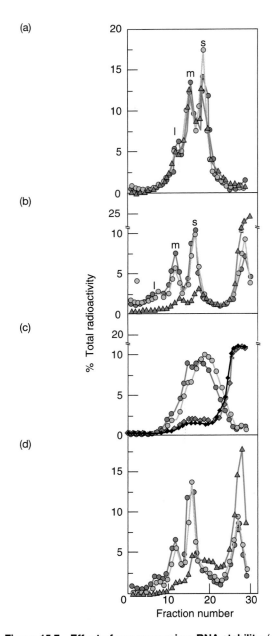

Figure 15.7 Effect of cap on reovirus RNA stability. (a) Appearance of newly synthesized RNAs. Furuichi and colleagues made labeled reovirus RNAs with capped (green), blocked (blue), or uncapped (red) 5′-ends, then subjected these RNAs to glycerol gradient ultracentrifugation. The three size classes of RNA are labeled l, m, and s. **(b)** Effect of incubation in *Xenopus* oocytes. Furuichi and colleagues injected the RNAs with the three different 5′-ends into *Xenopus* oocytes. After 8 h they purified the RNAs and performed the same sedimentation analysis as in panel **(a)**. Colors have the same meaning as in panel **(a)**. **(c)** Effect of decapping or deblocking on RNA stability. Furuichi and colleagues removed caps and blocks from labeled reovirus RNAs by β-elimination, yielding pppGᵐ and pppG ends, respectively. Then they incubated these RNAs in *Xenopus* oocytes for 8 h and performed sedimentation analysis. Green, pppGᵐ-RNA before incubation; blue, pppG-RNA before incubation; red, pppGᵐ-RNA after incubation; black, pppG-RNA after incubation. **(d)** Effect of incubation in wheat germ extract. Conditions were as in panel **(b)**, except that RNAs were incubated in wheat germ extracts instead of *Xenopus* oocytes. Colors as in panel **(a)**. (*Source:* Reprinted with permission from *Nature* 266:236, 1977. Copyright © 1977 Macmillan Magazines Limited.)

especially, to enhance the translation of luciferase mRNA. Poly(A) provided a 21-fold boost in translation of a capped mRNA, but that was a minor effect compared with the 297-fold stimulation of translation that the cap conferred on a polyadenylated mRNA.

Transport of mRNA The cap also appears to facilitate the transport of a mature RNA out of the nucleus. Jörg Hamm and Iain Mattaj studied the behavior of U1 snRNA to reach this conclusion. Most of the snRNA genes, including the U1 snRNA gene, are normally transcribed by RNA polymerase II, and the transcripts receive monomethylated (m^7G) caps in the nucleus. They migrate briefly to the cytoplasm, where they bind to proteins to form snRNPs, and their caps are modified to trimethylated ($m^{2,2,7}G$) structures. Then they reenter the nucleus, where they participate in splicing and other activities. The U6 snRNA is exceptional. It is made by polymerase III and is not capped. It retains its terminal triphosphate and remains in the nucleus. Hamm and Mattaj wondered what would happen if they arranged for the U1 snRNA gene to be transcribed by polymerase III instead of polymerase II. If it failed to be capped and remained in the nucleus, that would suggest that capping is important for transporting an RNA out of the nucleus.

Thus, Hamm and Mattaj placed the *Xenopus* U1 snRNA gene under the control of the human U6 snRNA promoter, so it would be transcribed by polymerase III. Then they injected this construct into *Xenopus* oocyte nuclei, along with a labeled nucleotide and a *Xenopus* 5S rRNA gene, which acted as an internal control. They also included 1 μg/mL of α-amanitin to inhibit RNA polymerase II and therefore ensure that no transcripts of the U1 gene would be made by polymerase II. In addition to the wild-type U1 gene, these workers also used several mutant U1 genes, with lesions in the regions coding for protein-binding sites. Loss of ability to associate with the proper proteins in the cytoplasm rendered the products of these mutant genes unable to return to the nucleus once they had been transported to the cytoplasm. Twelve hours after injection, Hamm and Mattaj dissected the oocytes into nuclear and cytoplasmic fractions and electrophoresed the labeled products in each. They compared the cellular locations of capped U1 snRNAs made by RNA polymerase II and uncapped U1 snRNA made by polymerase III.

Figure 15.8a shows the results. Virtually all the uncapped U1 snRNA made by polymerase III remained in the nucleus (see the top row of the autoradiograph). The material in the cytoplasm that migrated slightly slower than U1 is probably 5.8S rRNA. At any rate, we can tell it is not U1 because it is present in the oocytes that were not injected with a U1 snRNA gene (see the panel labeled "no U1").

On the other hand, the U1 snRNAs made by polymerase II were transported to the cytoplasm (see the third

Table 15.1 Synergism between Poly(A) and Cap during Translation of Luciferase mRNA in Tobacco Protoplasts

mRNA	Luciferase mRNA Half-Life (min)	Luciferase Activity (light units/mg protein)	Relative Effect of Poly(A) on Activity	Relative Effect of Cap on Activity
Uncapped				
Poly(A)⁻	31	2,941	1	1
Poly(A)⁺	44	4,480	1.5	1
Capped				
Poly(A)⁻	53	62,595	1	21
Poly(A)⁺	100	1,331,917	21	297

Source: From D. R. Gallie, "The cap and poly(A) tail function synergistically to regulate mRNA translational efficiency," *Genes & Development* 5:2108–2116, 1991. Copyright © 1991 Cold Spring Harbor Laboratory Press, Cold Spring Harbor, NY. Reprinted by permission.

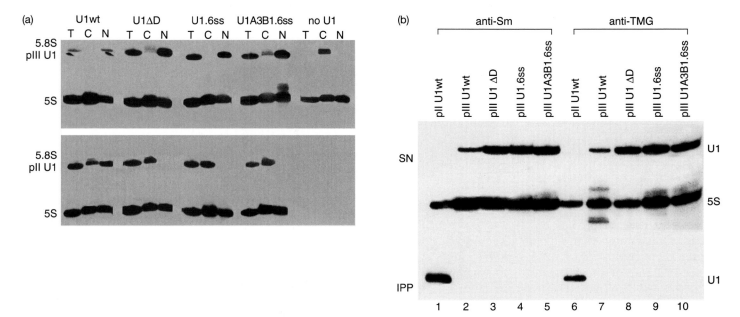

Figure 15.8 Capping of U1 snRNA is necessary for its transport to the cytoplasm. (a) Assay for transport. Hamm and Mattaj placed the *Xenopus* U1 snRNA gene, or various U1 mutant genes, under control of either a polymerase II or a polymerase III promoter. They injected these constructs into the nuclei of *Xenopus* oocytes along with [α-³²P]GTP and a 5S rRNA gene as an internal control. Constructs with the polymerase III promoter were also accompanied by α-amanitin to inhibit any transcription by polymerase II. After 12 h, these workers extracted RNA from nucleus and cytoplasm and electrophoresed it. The U1 gene being analyzed in each panel is identified at top: U1wt is the wild-type gene; the others, except for "no U1" are mutant U1 genes whose products are defective in protein binding. Total cell RNA, cytoplasmic RNA, and nuclear RNA are identified at top as T, C, and N, respectively. The bands corresponding to U1, 5.8S rRNA, and 5S rRNA are identified at left. The top group of gels contained the U1 gene under control of the U6 polymerase III promoter, as indicated by "pIII U1" at left. The bottom group of gels contained the U1 gene under control of its own polymerase II promoter, as indicated by "pII U1" at left. **(b)** Immunoprecipitation. To verify that the polymerase III constructs remained uncapped and in the nucleus, Hamm and Mattaj isolated RNAs from cells injected with constructs containing the polymerase II or polymerase III promoter (as indicated by pII and pIII, respectively) and immunoprecipitated the labeled RNAs with either an anti-Sm antibody or an anti-trimethyl guanosine (TMG) antibody, as indicated at top. Then they electrophoresed the RNAs from the immunoprecipitate (IPP, bottom) and also those that remained in the supernatant (SN, top). (*Source:* Hamm and Mattaj, Monomethylated cap structures facilitate RNA export from the nucleus. *Cell* 63 (5 Oct 1990) p. 111, f. 2. Reprinted by permission of Elsevier Science.)

row of the autoradiograph). A considerable amount of the wild-type RNA appeared in the nucleus, but this seems to be RNA that had migrated to the cytoplasm, complexed with proteins, and then returned to the nucleus. The behavior of the mutant RNAs supports this conclusion, because all three of them could not complex with the appropriate proteins, and all three remained in the cytoplasm.

An objection could be raised to this experiment. The U1 snRNAs made by polymerase III may actually have entered the cytoplasm, but returned to the nucleus so rapidly that they were undetectable in the cytoplasm. If so, they should have complexed with proteins and have been capped with a trimethyl cap, both of which are required for reentry into the nucleus. To test this possibility,

Hamm and Mattaj immunoprecipitated the polymerase II and polymerase III products from the injected cells with antibodies directed against the snRNP proteins (anti-Sm antibodies) or the trimethylated guanosine cap (anti-TMG). Then they electrophoresed the supernatant and immunoprecipitate fractions. As Figure 15.8b demonstrates, the only RNA precipitable by either antibody was the wild-type U1 snRNA made by polymerase II (lanes 1 and 6, bands labeled "IPP," for immunoprecipitate). None of the polymerase III products could be precipitated by either antibody, so they remained in the supernatant, labeled "SN." Therefore, none of the polymerase III products appeared to have spent any time in the cytoplasm, where at least the wild-type U1 snRNA should have been complexed with protein that would have bound to the anti-Sm antibody.

Finally, as we will see later in this chapter, the cap is essential for proper splicing of a pre-mRNA.

> **SUMMARY** The cap provides: (1) protection of the mRNA from degradation; (2) enhancement of the mRNA's translatability; (3) transport of the mRNA out of the nucleus; and (4) proper splicing of the pre-mRNA.

15.2 Polyadenylation

We have already seen that hnRNA is a precursor to mRNA. One finding that suggested such a relationship between these two types of RNA was that they shared a unique structure at their 3′-ends: a long chain of AMP residues called **poly(A)**. Neither rRNA nor tRNA has a poly(A) tail. The process of adding poly(A) to RNA is called **polyadenylation.** Let us examine first the nature of poly(A) and then the polyadenylation process.

Poly(A)

James Darnell and his coworkers performed much of the early work on poly(A) and polyadenylation. To purify HeLa cell poly(A) from the rest of the mRNA molecule, Diana Sheiness and Darnell released it with two enzymes: RNase A, which cuts after the pyrimidine nucleotides C and U, and RNase T1, which cuts after G nucleotides. In other words, they cut the RNA after every nucleotide except the A's, preserving only pure runs of A's. Next, Sheiness and Darnell electrophoresed the poly(A)s from nuclei and from cytoplasm to determine their sizes. Figure 15.9 shows the results, which demonstrate that both poly(A)s have major peaks that electrophoresed more slowly than 5S rRNA, at about 7S. Sheiness and Darnell estimated that this corresponded to about 150–200 nt. The poly(A) species observed in this experiment were labeled for only

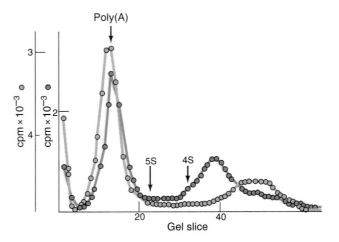

Figure 15.9   Size of poly(A). Sheiness and Darnell isolated radioactively labeled hnRNA from the nuclei (blue), and mRNA from the cytoplasm (red) of HeLa cells, then released poly(A) from these RNAs by RNase A and RNase T1 treatment. They electrophoresed the poly(A)s, collected fractions, and determined their radioactivities by scintillation counting (Chapter 5). They included 4S tRNA and 5S rRNA as size markers. Both poly(A)s electrophoresed more slowly than the 5S marker, corresponding to molecules about 200 nt long. (*Source:* Reprinted with permission from *Nature New Biology* 241:267, 1973. Copyright © 1973 Macmillan Magazines Limited.)

12 min, so they were newly synthesized. Little difference in size between these fresh nuclear and cytoplasmic poly(A)s is noticeable. However, cytoplasmic poly(A) is subject to shortening, as we will see later in this chapter. Now that poly(A)s from many different organisms have been analyzed, we see an average size of fresh poly(A) of about 250 nt.

It is apparent that the poly(A) goes on the 3′-end of the mRNA or hnRNA because it can be released very quickly with an enzyme that degrades RNAs from the 3′-end inward. Furthermore, complete RNase digestion of poly(A) yielded one molecule of adenosine and about 200 molecules of AMP. Figure 15.10 demonstrates that this requires poly(A) to be at the 3′-end of the molecule. This experiment also reinforced the conclusion that poly(A) is about 200 nt long. Finally, RNase T1 digestion of hnRNA or mRNA yields poly(A) with no G's, which means that no G's are downstream of poly(A), and RNase A yields poly(A) with no pyrimidines, which means that no pyrimidines are downstream of poly(A). Thus, poly(A) must be at the very end of the RNA molecule.

We also know that poly(A) is not made by transcribing DNA because typical genomes contain no runs of T's long enough to encode it. In particular, we find no runs of T's at the ends of any of the thousands of eukaryotic genes that have been sequenced. Furthermore, actinomycin D, which inhibits DNA-directed transcription, does not inhibit polyadenylation. Thus, poly(A) must be added posttranscriptionally. In fact, there is an enzyme in nuclei called **poly(A) polymerase (PAP)** that adds AMP residues one at a time to mRNA precursors.

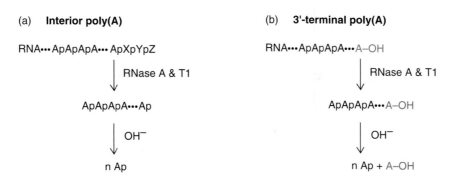

Figure 15.10 Finding poly(A) at the 3′-end of hnRNA and mRNA. (a) Interior poly(A). If poly(A) were located in the interior of an RNA molecule, RNase A and RNase T1 digestion would yield poly(A) with a phosphate at the 3′-end, then base hydrolysis would give only AMP. **(b)** Poly(A) at the 3′-end of hnRNA and mRNA. Because poly(A) is located at the 3′-end of these RNA molecules, RNase A and T1 digestion yields poly(A) with an unphosphorylated adenosine at the 3′-end. Base hydrolysis gives AMP plus one molecule of adenosine. In fact, the ratio of AMP to adenosine was 200, suggesting a poly(A) length of about 200 nt.

We know that poly(A) is added to mRNA precursors because it is found on hnRNA. Even specific unspliced mRNA precursors (the 15S mouse globin mRNA precursor, for example) contain poly(A). However, as we will see later in this chapter, splicing of some introns in a pre-mRNA can occur before polyadenylation. Once an mRNA enters the cytoplasm, its poly(A) turns over; in other words, it is constantly being broken down by RNases and rebuilt by a cytoplasmic poly(A) polymerase.

> **SUMMARY** Most eukaryotic mRNAs and their precursors have a chain of AMP residues about 250 nt long at their 3′-ends. This poly(A) is added post-transcriptionally by poly(A) polymerase.

Function of Poly(A)

Most mRNAs contain poly(A). One noteworthy exception is histone mRNA, which manages to perform its function without a detectable poly(A) tail. This exception notwithstanding, the near universality of poly(A) in eukaryotes raises the question: What is the purpose of poly(A)? One line of evidence suggests that it helps protect mRNAs from degradation. Another indicates that it stimulates translation of mRNAs to which it is attached. Let us consider this evidence.

Protection of mRNA To examine the stabilizing effect of poly(A), Michel Revel and colleagues injected globin mRNA, with and without poly(A) attached, into *Xenopus laevis* oocytes and measured the rate of globin synthesis at various intervals over a 2-day period. They found that there was little difference at first. However, after only 6 h, the mRNA without poly(A) [**poly(A)⁻ RNA**] could no longer support translation, while the mRNA with poly(A) [**poly(A)⁺ RNA**] was still quite actively translated (Figures

15.11 and 15.12). The simplest explanation for this behavior is that the poly(A)⁺ RNA has a longer lifetime than the poly(A)⁻ RNA, and that poly(A) is therefore the protective agent. However, as we will see, poly(A) plays an even bigger role in efficiency of translation of mRNA.

Translatability of mRNA Several lines of evidence indicate that poly(A) also enhances the translatability of an mRNA. One of the proteins that binds to a eukaryotic mRNA during translation is **poly(A)-binding protein I, (PAB I)**. Binding to this protein seems to boost the efficiency with which an mRNA is translated. One line of evidence in favor of this hypothesis is that excess poly(A) added to an in vitro reaction inhibited translation of a capped, polyadenylated mRNA. This finding suggested that the excess poly(A) was competing with the poly(A) on the mRNA for an essential factor, presumably for PAB I. Without this factor, the mRNA could not be translated well. Carrying this argument one step further leads to the conclusion that poly(A)⁻ RNA, because it cannot bind PAB I, cannot be translated efficiently.

To test the hypothesis that poly(A)⁻ RNA is not translated efficiently, David Munroe and Allan Jacobson compared the rates of translation of two synthetic mRNAs, with and without poly(A), in rabbit reticulocyte extracts. They made the mRNAs (rabbit β-globin [RBG] mRNA and vesicular stomatitis virus N gene [VSV.N] mRNA) by cloning their respective genes into plasmids under the control of the phage SP6 promoter, then transcribing these genes in vitro with SP6 RNA polymerase. They endowed the synthetic mRNAs with various length poly(A) tails by adding poly(T) to their respective genes with terminal transferase and dTTP for varying lengths of time before cloning and transcription.

Munroe and Jacobson tested the poly(A)⁺ and poly(A)⁻ mRNAs for both translatability and stability in the reticulocyte extract. Figure 15.13a shows the effects of

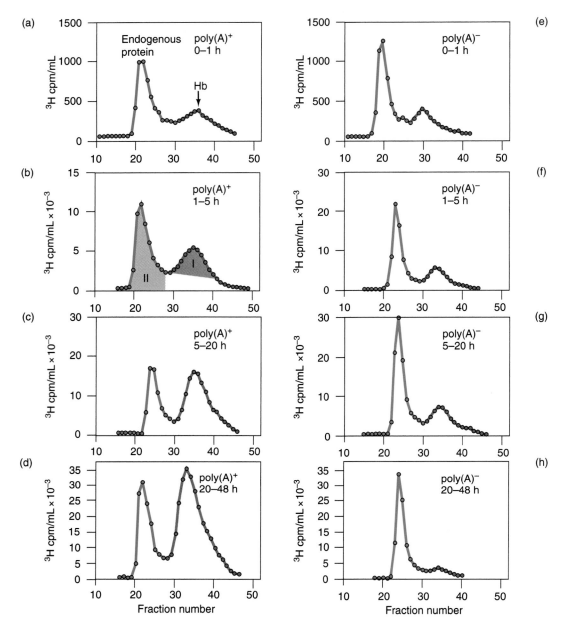

Figure 15.11 **Effect of poly(A) on translation of globin mRNA in oocytes.** Revel and colleagues injected poly(A)+ and poly(A)- globin mRNA (as indicated) into *Xenopus laevis* oocytes. Then, at the time intervals indicated at upper right of each graph, they added [³H]histidine to label newly synthesized globin. After the labeling period, they subjected the labeled proteins to gel filtration on Sephadex G–100 to separate the globin from larger proteins made from endogenous mRNAs. Finally, they plotted the radioactivity in each fraction versus the fraction number. Panel (**a**) shows the positions of the endogenous proteins and of globin (Hb). Panel (**b**) shows the method of quantifying the area of each peak. The poly(A)+ mRNA remained active much longer than the poly(A)- mRNA. (*Source:* From G. Huez et al., "Role of the Polyadenylate Segment in the Transition of Globin Messenger RNA in *Xenopus* Oocytes," *Proceedings of the National Academy of Sciences* 71(8):3143–3146, August 1974.)

both capping and polyadenylation on translatability of the VSV.N mRNA. Both capped and uncapped mRNAs were translated better with poly(A) than without. Figure 15.13b shows the effect of poly(A) on mRNA lifetime in the extract. Polyadenylation made no difference in the stability of either mRNA. Munroe and Jacobson interpreted these results to mean that the extra translatability conferred by poly(A) was not due to stabilization of the mRNAs, but to enhanced translation per se. If so, what aspect of transla-

tion is enhanced by poly(A)? These studies suggested that it is a step at the very beginning of the translation process: association between mRNA and ribosomes. We will see in Chapter 17 that many ribosomes bind sequentially at the beginning of eukaryotic mRNAs and read the message in tandem. An mRNA with more than one ribosome translating it at once is called a polysome. Munroe and Jacobson contended that poly(A)+ mRNA forms polysomes more successfully than poly(A)- mRNA.

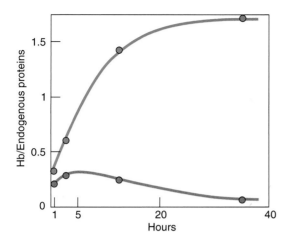

Figure 15.12   Time course of translation of poly(A)⁺ (green) and poly(A)⁻ (red) globin mRNA. Revel and colleagues plotted the ratio of radioactivity incorporated into globin and endogenous protein versus the midpoint of the labeling time. Radioactivity in each peak was determined by integration as indicated in Figure 15.11b. (*Source:* From G. Huez et al., "Role of the Polyadenylate Segment in the Transition of Globin Messenger RNA in *Xenopus* Oocytes," *Proceedings of the National Academy of Sciences* 71(8):3143–3146, August 1974.)

These workers measured the incorporation of labeled mRNAs into polysomes as follows. They labeled poly(A)⁺ mRNA with ^{32}P and poly(A)⁻ mRNA with ^{3}H, then incubated these RNAs together in a reticulocyte extract. Then they separated polysomes from monosomes by sucrose gradient ultracentrifugation. Figure 15.14a indicates that the poly(A)⁺ VSV.N mRNA was significantly more associated with polysomes than was poly(A)⁻ mRNA. In parallel experiments, the RBG mRNA exhibited the same behavior. Figure 15.14b shows the effect of *length* of poly(A) attached to RBG mRNA on the extent of polysome formation. We see the greatest increase as the poly(A) grows from 15 to 30 nt, and a more gradual increase as more A residues are added.

Munroe and Jacobson's finding that poly(A) did not affect the stability of mRNAs seems to contradict the earlier work by Revel and colleagues. Perhaps the discrepancy arises from the fact that the early work was done in intact frog eggs, whereas the later work used a cell-free system. Earlier in this chapter, Table 15.1 showed that poly(A)

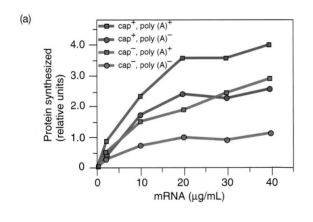

(a)

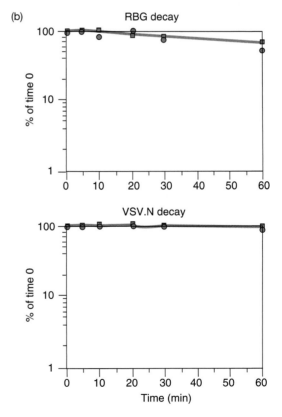

(b)

Figure 15.13   Effect of polyadenylation on translatability and stability of mRNAs. (a) Translatability. Munroe and Jacobson incubated VSV.N mRNAs with ^{35}S-methionine in rabbit reticulocyte extracts. The mRNAs were capped (green) or uncapped (red), and poly(A)⁺ (68 As; squares) or poly(A)⁻ (circles). After allowing 30 min for protein synthesis, these workers electrophoresed the labeled products and measured the radioactivity of the newly made protein by quantitative fluorography. Poly(A) enhanced the translatability of both capped and uncapped mRNAs. **(b)** mRNA stability. Munroe and Jacobson labeled either RBG or VSV.N mRNAs, as indicated at top of each graph, with ^{32}P and incubated the RNAs in rabbit reticulocyte extracts for the times indicated. Then they electrophoresed the labeled RNAs, autoradiographed the gels, and quantified the amount of RNA in the bands by densitometry (Chapter 5). Squares, poly(A)⁺; circles, poly(A)⁻. Poly(A) made no difference in the stability of either mRNA.

(*Source:* From Munroe and Jacobson, *Molecular and Cellular Biology* 10:3445, 1990. Copyright © 1990 American Society for Microbiology, Washington, DC. Reprinted by permission.)

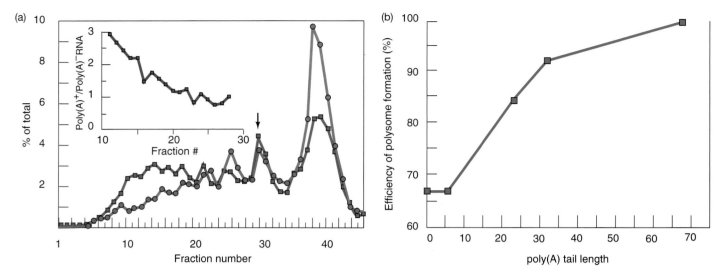

Figure 15.14 Effect of polyadenylation on recruitment of mRNA to polysomes. (a) Polysome profiles. Munroe and Jacobson mixed ^{33}P-labeled poly(A)+ (blue) and ^{3}H-labeled poly(A)− (red) mRNA with a rabbit reticulocyte extract, then separated polysomes from monosomes by sucrose gradient ultracentrifugation. The arrow denotes the monosome peak; fractions to the left of this peak are polysomes, and one can see the disome, trisome, and even higher polysome peaks. The poly(A)+ mRNA is clearly better at associating with polysomes, especially the higher polysomes. The inset shows the ratio of poly(A)+ to poly(A)− RNA in fractions 11–28. Again, this demonstrates a preferential association of poly(A)+ mRNA with polysomes (the lower fraction numbers). **(b)** Efficiency of polysome formation as a function of poly(A) length on VSV.N mRNA. The efficiency at a tail length of 68 is taken as 100%. (*Source:* From Munroe and Jacobson, *Molecular and Cellular Biology* 10:3447–8, 1990. Copyright © 1990 American Society for Microbiology, Washington, DC. Reprinted by permission.)

stimulated transcription of luciferase mRNA. The stabilizing effect of poly(A) on this mRNA was two-fold at most, whereas the overall increase in luciferase production caused by poly(A) was up to 20-fold. Thus, this system also suggested that enhancement of translatability by poly(A) seems to be more important than mRNA stabilization.

> **SUMMARY** Poly(A) enhances both the lifetime and translatability of mRNA. The relative importance of these two effects seems to vary from one system to another. At least in rabbit reticulocyte extracts, poly(A) seems to enhance translatability by helping to recruit mRNA to polysomes, thereby promoting initiation of translation.

Basic Mechanism of Polyadenylation

It would be logical to assume that poly(A) polymerase simply waits for a transcript to be finished, then adds poly(A) to the 3′-end of the RNA. However, this is not what ordinarily happens. Instead, the mechanism of polyadenylation usually involves clipping an mRNA precursor, even before transcription has terminated, and then adding poly(A) to the newly exposed 3′-end (Figure 15.15). Thus, contrary to expectations, polyadenylation and transcription termination are not necessarily linked. RNA polymerase can still be elongating an RNA chain, while the polyadenylation apparatus has already

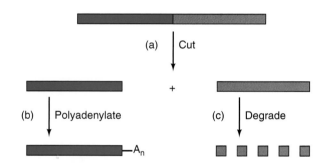

Figure 15.15 Overview of the polyadenylation process. (a) Cutting. The first step is cleaving the transcript, which may actually still be in the process of being made. The cut occurs at the end of the RNA region (green) that will be included in the mature mRNA. **(b)** Polyadenylation. The poly(A) polymerase adds poly(A) to the 3′-end of the mRNA. **(c)** Degradation of the extra RNA. All RNA (red) lying beyond the polyadenylation site is superfluous and is destroyed.

located a polyadenylation signal somewhere upstream, cut the growing RNA, and polyadenylated it.

Joseph Nevins and Darnell provided some of the first evidence for this model of polyadenylation. They chose to study the adenovirus major late transcription unit because it serves as the template for several different overlapping mRNAs, each of which is polyadenylated at one of five separate sites. Figure 15.16a is a schematic diagram of this transcription unit, illustrating the use of the five different polyadenylation sites to produce five different mRNAs. (This diagram is conceptual only and is not intended to be drawn to scale.) Recall from Chapter 14 that

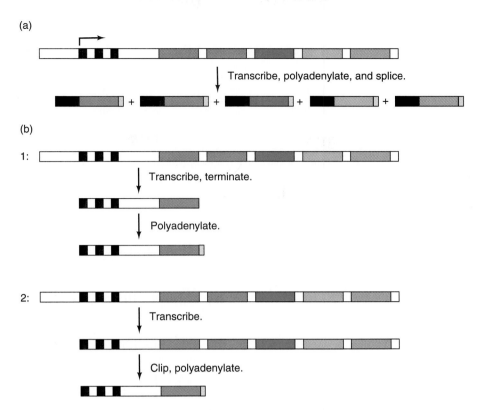

Figure 15.16   Adenovirus major late transcripts are clipped and polyadenylated while transcription is still in progress. (a) Schematic diagram of the adenovirus major late transcription unit. The leader exons are in black, and the coding regions are in different colors. After transcription, polyadenylation, and splicing, we have multiple mRNAs, each with a poly(A) (yellow) at the end of its coding region. **(b)** Two hypotheses for polyadenylation of the major late mRNAs. In each case we consider the polyadenylation of the first gene only. (1) Transcription terminates at the end of each coding region and polyadenylation occurs at those sites. (2) Transcription continues to the end of the major late region, and the primary transcript is clipped and polyadenylated at each clipping site.

each of these mature mRNAs has the same three leader exons spliced to a different coding region. The poly(A) of each is attached to the 3′-end of the coding region. Figure 15.16b presents two alternative hypotheses for the relationship between transcription termination and polyadenylation in this system. (1) Transcription terminates immediately downstream of a polyadenylation site, and then polyadenylation occurs. For example, if gene A is being expressed, transcription will proceed only to the end of coding region A, then terminate, and then polyadenylation will occur at the 3′-end left by that termination event. (2) Transcription goes at least to the end of the last coding exon, and polyadenylation can occur at any polyadenylation site, presumably even before transcription of the whole major late region is complete.

The first hypothesis, that transcription does not always go clear to the end, was easy to eliminate. Nevins and Darnell hybridized radioactive RNA made in cells late in infection to DNA fragments from various positions throughout the major late region. If primary transcripts of the first gene stopped after the first polyadenylation site, and only transcripts of the last gene made it all the way to the end, then much more RNA would hybridize to frag-

ments near the 5′-end of the major late region than to fragments near the 3′-end. But RNA hybridized to all the fragments equally well—to fragments near the 3′-end of the region just as well as to fragments near the 5′-end. Therefore, once a transcript of the major late region is begun, it is elongated all the way to the end of the region before it terminates. In other words, the major late region contains only one transcription terminator, and it lies at the end of the region. Thus, this whole region can be called a **transcription unit** to denote the fact that it is transcribed as a whole, even though it contains multiple genes.

This behavior of transcribing far past a polyadenylation site before clipping and polyadenylating the transcript seems wasteful because all the RNA past the polyadenylation site will be destroyed without being used. So the question naturally arises: Is this method of polyadenylation unique to viruses, or does it also occur in ordinary cellular transcripts? To find out, Erhard Hofer and Darnell isolated labeled RNA from Friend mouse erythroleukemia cells that had been induced with dimethyl sulfoxide (DMSO) to synthesize large quantities of globin, and therefore to transcribe the globin genes at a

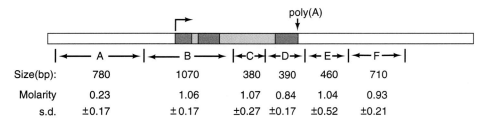

Figure 15.17   Transcription beyond the polyadenylation site. Hofer and Darnell isolated nuclei from DMSO-stimulated Friend erythroleukemia cells and incubated them with [^{32}P]UTP to label run-on RNA—mostly globin pre-mRNA. Then they hybridized this labeled RNA to DNA fragments A–F, whose locations and sizes are given in the diagram at top. The molarities of RNA hybridization to each fragment are given beneath each, with their standard deviations (s.d.). In the physical map at top, the exons are in red and the introns are in yellow. (*Source:* Reprinted from Hofer and Darnell, *Cell* 23:586, 1981. Copyright 1981, with permission from Elsevier Science.)

high rate. They hybridized the labeled transcripts to cloned fragments representing various parts of the mouse β-globin gene, and regions downstream of the gene (Figure 15.17). They observed just as much hybridization to fragments lying over 500 bp downstream of the polyadenylation site as to fragments within the globin gene. This demonstrated that transcription continues at least 500 bp downstream of this polyadenylation site. In further studies, these workers found that transcription finally terminated in regions lying even further downstream. Thus, transcription significantly beyond the polyadenylation site occurs in cellular, as well as viral, transcripts.

> **SUMMARY** Transcription of eukaryotic genes extends beyond the polyadenylation site. Then the transcript is cleaved and polyadenylated at the 3′-end created by the cleavage.

Polyadenylation Signals

If the polyadenylation apparatus does not recognize the ends of transcripts, but binds somewhere in the middle to cleave and polyadenylate, what is it about a polyadenylation site that attracts this apparatus? The answer to this question depends on what kind of eukaryote or virus we are discussing. Let us first consider mammalian **polyadenylation signals.** By 1981, molecular biologists had examined the sequences of dozens of mammalian genes and had found that the only obvious common feature they had was the sequence AATAAA about 20–30 bp before the polyadenylation site. At the RNA level, the sequence AAUAAA occurs in most mammalian mRNAs about 20 nt upstream of their poly(A). Molly Fitzgerald and Thomas Shenk tested the importance of the **AAUAAA sequence** in two ways. First, they deleted nucleotides between this sequence and the polyadenylation site and sequenced the 3′-ends of the resulting RNAs. They found that the deletions simply shifted the polyadenylation site downstream by roughly the number of nucleotides deleted.

This result suggested that the AAUAAA sequence is at least part of a signal that causes polyadenylation approximately 20 nt downstream. If so, then deleting this sequence should abolish polyadenylation altogether. These workers used an S1 assay as follows to show that it did. They created a recombinant SV40 virus (mutant 1471) with duplicate polyadenylation signals 240 bp apart, at the end of the late region. S1 analysis of the 3′-ends of the late transcripts (Chapter 5) revealed two signals 240 bp apart (Figure 15.18). [We can ignore the poly(A) in this kind of experiment because it does not hybridize to the probe.] Thus, both polyadenylation sites worked, implying some readthrough of the first site. Then Fitzgerald and Shenk deleted either the first AATAAA (mutant 1474) or the second AATAAA (mutant 1475) and reran the S1 assay. This time, the polyadenylation site just downstream of the deleted AATAAA did not function, demonstrating that AATAAA is necessary for polyadenylation. We shall see shortly, however, that this is only part of the mammalian polyadenylation signal.

Is the AATAAA invariant, or is some variation tolerated? Early experiments with manipulated signals (AATACA, AATTAA, AACAAA, and AATGAA) suggested that no deviation from AATAAA could occur without destroying polyadenylation. But by 1990, a compilation of polyadenylation signals from 269 vertebrate cDNAs showed some variation in these natural signals, especially in the second nucleotide. Marvin Wickens compiled these data, which defined a consensus sequence (Figure 15.19). The most common sequence, at the RNA level, is AAUAAA, and it is the most efficient in promoting polyadenylation. The most common variant is AUUAAA, and it is about 80% as efficient as AAUAAA. The other variants are much less common, and also much less efficient.

By now it has also become clear that AAUAAA by itself is not sufficient for polyadenylation. If it were, then polyadenylation would occur downstream of the many AAUAAA sequences found in introns, but it does not. Several investigators found that polyadenylation can be disrupted by deleting sequences immediately downstream of the polyadenylation site. This raised the suspicion that

(a)

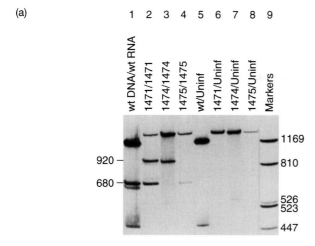

(b)

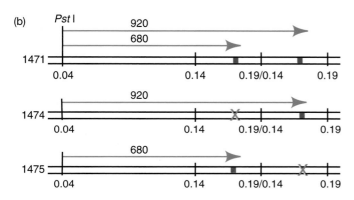

Figure 15.18   Importance of the AATAAA sequence to polyadenylation. Fitzgerald and Shenk created recombinant SV40 viruses with the following characteristics (bottom panel): Mutant 1471 contained duplicate late polyadenylation sites (green) 240 bp apart within the duplicated region, which extends from 0.14 to 0.19 map units. Mutant 1474 contained a 16-bp deletion (red) at the AATAAA in the upstream site, and mutant 1475 contained the same kind of 16-bp deletion (red) in the downstream site. Then they performed S1 analysis with a probe that should yield a 680-nt signal if the upstream polyadenylation signal works, and a 920-nt signal if the downstream polyadenylation signal works (blue arrows). **(a)** The lanes are marked at the top with the probe designation, followed by the RNA (or template) designation. Lane 1, using only wild-type probe and template, showed the wild-type signal at 680 nt, as well as an artifactual signal not usually seen. Lanes 5–8 are uninfected negative controls. The top band in each lane represents reannealed S1 probe and can be ignored. The results, diagrammed in the bottom panel **(b)**, show that deletion of an AATAAA prevents polyadenylation at that site. (*Source:* Fitzgerald and Shenk, The sequence 5′-AAUAAA-3′ forms part of the recognition site for polyadenylation of late SV40 mRNAs. *Cell* 24 (April 1981) p. 257, f. 7. Reprinted by permission of Elsevier Science.)

the region just downstream of the polyadenylation site contains another element of the polyadenylation signal. The problem was that that region is not highly conserved among vertebrates. Instead, there is simply a tendency for it to be GT- or T-rich (at the DNA level).

These considerations suggested that the minimum efficient polyadenylation signal (at the DNA level) is the se-

Figure 15.19   Summary of data on 369 vertebrate polyadenylation signals. The consensus sequence (in RNA form) appears at top, with the frequency of appearance of each base. The substitution of U for A in the second position is frequent enough (12%) that it is listed separately, below the main consensus sequence. Below, the polyadenylation efficiency is plotted for each variant polyadenylation signal. The base that deviates from normal is printed larger than the others in blue. The standard AAUAAA is given at the bottom, with the next most frequent (and active) variant (AUUAAA) just above it. (*Source:* Reprinted from *Trends in Biochemical Sciences*, vol. 15, M. Wickens, "How the messenger got its tail: addition of poly(A) in the nucleus," 278. Copyright © 1990 with permission of Elsevier Science.)

quence AATAAA followed about 20 bp later by a GT- or T-rich sequence. Anna Gil and Nicholas Proudfoot tested this hypothesis by examining the very efficient rabbit β-globin polyadenylation signal, which contains an AATAAA, followed 24 bp later by a GT-rich region, immediately followed by a T-rich region. They began by inserting an extra copy of the whole polyadenylation signal upstream of the natural one, then testing for polyadenylation at the two sites of this mutant clone (clone 3) by S1 analysis. This DNA supported polyadenylation at the new site at a rate 90% as high as at the original site. Thus, the inserted polyadenylation site is active. Next, they created a new mutant clone [clone 2(v)] by deleting a 35-bp fragment containing the GT- and T-rich region (GT/T) in the new (upstream) polyadenylation signal. This abolished polyadenylation at the new site, reaffirming that this 35-bp fragment is a vital part of the polyadenylation signal.

To define the minimum efficient polyadenylation site, these workers added back various sequences to clone 2 and tested for polyadenylation. Figure 15.20 summarizes the results, which show that neither the GT-rich nor the T-rich sequence by itself could reconstitute an efficient polyadenylation signal: Clone GT had the GT-rich region, but was only 30% as active as the wild-type signal; clone A–T had the T-rich region, but had only 30% of the

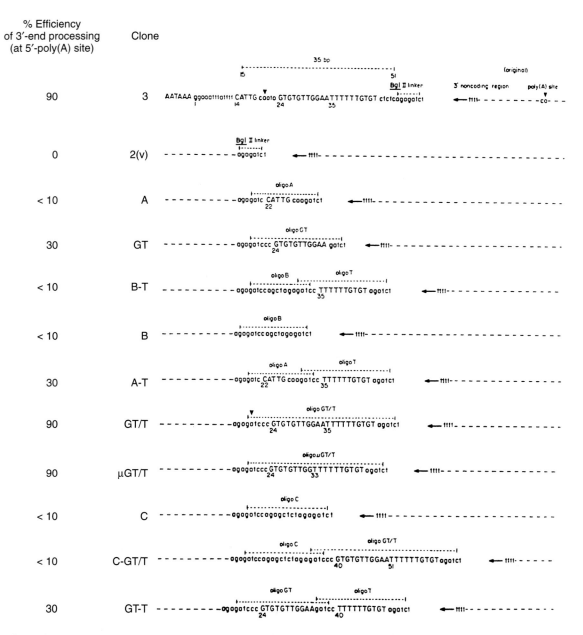

Figure 15.20   Summary of activities of mutant polyadenylation signals. Gil and Proudfoot created the mutant polyadenylation signals whose sequences are shown here and tested them by S1 analysis. The polyadenylation sites are indicated by vertical arrowheads. Natural sequences are in capital letters; sequences that came from linkers are in lowercase. Numbers below the sequences indicate the spacing between the numbered base and the AATAAA sequence. (*Source:* Reprinted from Gil and Proudfoot, *Cell* 49:401, 1987. Copyright 1987, with permission from Elsevier Science.)

normal activity. Furthermore, the position of the GT/T region is important. In clone C–GT/T it is shifted 16 bp further downstream of the AATAAA element, and this clone has less than 10% of normal activity. Moreover, the spacing between the GT-rich and T-rich sequences within GT/T is important. Clone GT–T had both, but they were separated by an extra 5 bp, and this mutant signal had only 30% of the normal activity. Thus, an efficient polyadenylation signal has an AATAAA motif followed 23–24 bp later by a GT-rich motif, followed immediately by a T-rich motif.

To examine further the nature of the polyadenylation signal, Proudfoot and colleagues made a synthetic polyadenylation signal (SPA) as illustrated in Figure 15.21. It was equipped with two *Bgl*II linkers so an unrelated sequence could be inserted in place of the natural sequence between the AATAAA and GT/T motifs. S1 analysis showed that this substitution made no difference in efficiency of polyadenylation.

Next, Proudfoot and colleagues wondered what would happen if they placed the SPA in competition with a

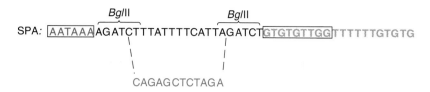

Figure 15.21   Structure of the synthetic polyadenylation site (SPA). Proudfoot and colleagues synthesized this oligonucleotide with AATAAA, GT-rich, and T-rich motifs. They included two *Bgl*II sites so novel sequences could be inserted between AATAAA and GT/T, as shown here. (*Source:* From N. Levitt et al., "Definition of an efficient synthetic poly(A) site," *Genes & Development* 3:1019–1025, 1989. Copyright © 1989 Cold Spring Harbor Laboratory Press, Cold Spring Harbor, NY. Reprinted by permission.)

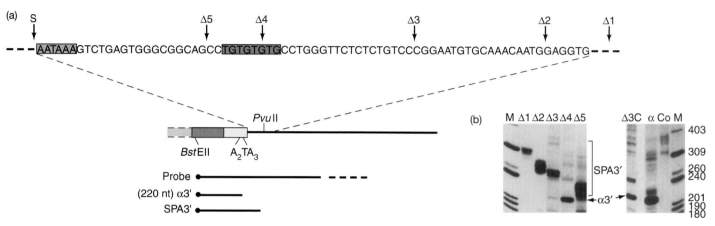

Figure 15.22   Competition between a natural polyadenylation signal and the SPA. (a) Construction of mutant clones. Proudfoot and colleagues made deletions at the 3'-end of a human α-globin gene cloned into pSVed. The deletions extended up to the points indicated (Δ1–Δ5). Then they added the SPA to these new ends, transfected HeLa cells with the mutant constructs, and assayed for polyadenylation by S1 mapping with the 3'-labeled probe illustrated at bottom. The sizes of the S1 signals expected for polyadenylation at the natural site (α3') and at the SPA site (SPA3') are indicated. The α3'-signal should be 220 nt, but the size of the SPA signal should vary from one clone to the next because of its varying position. **(b)** Results. The S1 signals for each mutant clone are shown. Lane α shows the results with the wild-type gene with no deletion or SPA. Lane Co is a control with tRNA instead of transfected cell RNA. Lanes M are size markers. (*Source:* Levitt et al., Definition of an efficient synthetic poly(A) site. *Genes & Development* 3 (1989) p. 1022, f. 3.)

natural polyadenylation signal. To find out, they made deletions at the 3'-end of a cloned human α-globin gene and attached the SPA to these new 3'-ends. Then they placed these mutant genes in a plasmid that could deliver them to HeLa cells for transient expression. Figure 15.22a shows the extents of these deletions (Δ1–Δ5) and the results of S1 assays for efficiency of polyadenylation. Deletions Δ1–Δ3 simply involve adding the SPA downstream of the α-globin gene's natural polyadenylation signal. One might have expected these mutant genes to use the natural polyadenylation signal, but instead they used the SPA exclusively (Figure 15.22b). Δ3C is a control that has the same deletion as Δ3, but no SPA. The fact that polyadenylation occurs with this clone at the natural site indicates that the site itself is functional, but can simply be out-competed by a SPA added in clones Δ1–Δ3. Why is the SPA able to compete so well? Presumably, it is because the SPA contains both a GT-rich and a T-rich motif, whereas the α-globin polyadenylation signal contains only a GT-rich motif. Deletion Δ4 placed the SPA immediately adjacent to the α-globin polyadenylation signal; for some reason, the polyadenylation machinery preferred the natural signal in this case. Fi-

nally, Δ5 deleted the natural GT-rich motif, so it was not surprising that the SPA was used.

Proudfoot and colleagues also tried placing their SPA in an exon or an intron of the rabbit β-globin gene, as diagrammed in Figure 15.23a. They found that it would function very well—better, in fact, than the natural polyadenylation site—if placed in exon 3 of the gene (Figure 15.23b, lane SPAE). However, the SPA would not function at all if placed in intron 2 of the gene (Figure 15.23b, lane SPAI). The loss of function of the SPA in an intron could indicate that this intron, with the SPA in it, is removed before polyadenylation occurs. Alternatively, it could simply indicate that the SPA had been inserted into a region of the intron that somehow disrupted its activity.

To distinguish between these two possibilities, Proudfoot and colleagues deleted the 5'-splice site of the intron, creating clone SPAIΔD, then tested for polyadenylation using an RNase protection assay. This deletion prevented splicing and therefore prevented premature removal of the SPA from the pre-mRNA. It also restored polyadenylation at the SPA, as shown in Figure 15.23c, lane SPAIΔD. In other words, the polyadenylation signal would function if it was located in an exon, but not in an intron. These

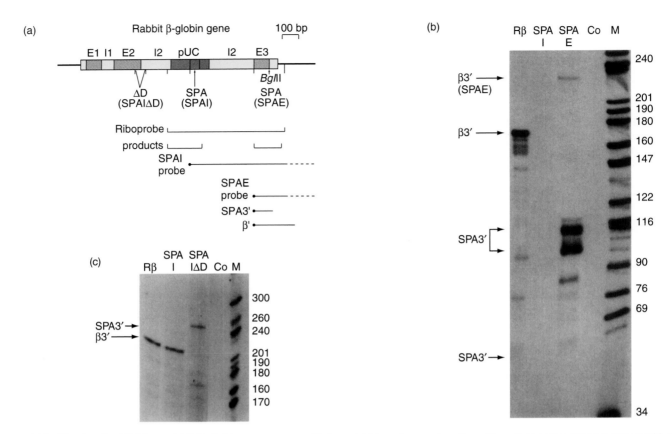

Figure 15.23   Testing the SPA in a rabbit β-globin gene exon and intron. (a) Map of gene alterations. Proudfoot and colleagues began with a β-globin clone whose intron 2 contained pUC vector sequences (green), including the pUC multiple cloning site (MCS, red). They placed the SPA into the *Bgl*II site of exon 3 (E3) or into the MCS in intron 2 (I2) at the positions shown (SPAE and SPAI, respectively). ΔD shows the splice site removed to create clone SPAIΔD. (b) S1 results. Proudfoot and colleagues transfected cells with the normal and mutated constructs and analyzed the 3′-ends of the resulting β-globin RNAs by S1 mapping. Lane Rβ shows the position of the normal 3′-polyadenylation site with the parent gene that contains no inserted SPA. Lanes SPAE and SPAI show the results with those two constructs. Note that SPAE showed strong polyadenylation at the SPA, but SPAI did not. (c) RNase protection assay of SPAI and SPAIΔD using the riboprobe shown in panel (a). Lane designations have the same meaning as in panel (b), except for the addition of SPAIΔD. Note that the SPA was not used in Rβ or SPAI, but was used in SPAIΔD. (*Source:* Levitt et al., *Genes & Development* 3 (1989) p. 1023, f. 4.)

findings suggest that the transcript of this gene is at least partially spliced before it is polyadenylated. We have already seen that many pre-mRNAs, including adenovirus pre-mRNAs, are polyadenylated. In those cases, polyadenylation appears to precede splicing. Thus, the order of splicing and polyadenylation may vary from one transcript to another.

Plants and yeast mRNAs are also polyadenylated, but their polyadenylation signals are different from those of mammals. Yeast genes usually lack an AATAAA sequence near their polyadenylation sites. In fact, it is difficult to discern a pattern in the yeast polyadenylation signals, other than a general AT-richness upstream of the polyadenylation site. Plant genes may have an AATAAA in the appropriate position, and deletion of this sequence prevents polyadenylation. But plant and animal polyadenylation signals are not the same: Single-base substitutions within the AATAAA of the cauliflower mosaic virus do not have near the negative effect they have in ver-

tebrate polyadenylation signals. Furthermore, animal signals do not function when placed at the ends of plant genes.

SUMMARY An efficient mammalian polyadenylation signal consists of an AAUAAA motif about 20 nt upstream of a polyadenylation site in a pre-mRNA, followed 23 or 24 bp later by a GU-rich motif, followed immediately by a U-rich motif. Many variations on this theme occur in nature, which results in variations in efficiency of polyadenylation. Plant polyadenylation signals also usually contain an AAUAAA motif, but more variation is allowed in this region than in an animal AAUAAA. Yeast polyadenylation signals are more different yet, and rarely contain an AAUAAA motif.

Cleavage and Polyadenylation of a Pre-mRNA

The process commonly known as polyadenylation really involves both RNA cleavage and polyadenylation. In this section we will briefly discuss the factors involved in the cleavage reaction, then discuss the polyadenylation reaction in more detail.

Pre-mRNA Cleavage Several proteins are necessary for cleavage of mammalian pre-mRNAs prior to polyadenylation. One of these proteins is also required for polyadenylation, so it was initially called "cleavage and polyadenylation factor," or "CPF," but it is now known as **cleavage and polyadenylation specificity factor (CPSF)**. Cross-linking experiments have demonstrated that this protein binds to the AAUAAA signal. Shenk and colleagues reported in 1994 that another factor participates in recognizing the polyadenylation site. This is the **cleavage stimulation factor (CstF)**, which, according to cross-linking data, binds to the G/U region. Thus, CPSF and CstF bind to sites flanking the cleavage and polyadenylation site. Binding of either CPSF or CstF alone is unstable, but together the two factors bind cooperatively and stably.

Still another pair of RNA-binding proteins required for cleavage are the **cleavage factors I and II (CF I and CF II)**. It is also likely that poly(A) polymerase itself is required for cleavage because cleavage is followed immediately by polyadenylation. In fact, the coupling between cleavage and polyadenylation is so strong that no cleaved, unpolyadenylated RNAs can be detected.

Another protein that is intimately involved in cleavage is RNA polymerase II. The first hint of this involvement was the discovery that RNAs made in vitro by RNA polymerase II were capped, spliced, and polyadenylated properly, but those made by polymerases I and III were not. In fact, even RNAs made by RNA polymerase II lacking the carboxyl-terminal domain (CTD) of the largest subunit were not efficiently spliced and polyadenylated. These data suggested that the CTD was involved somehow in splicing and polyadenylation.

In 1998, Yutaka Hirose and James Manley showed that the CTD stimulates the cleavage reaction, and this stimulation is not dependent on transcription. First, these workers tested the phosphorylated and unphosphorylated forms of polymerase II (IIO and IIA, Chapter 10) for ability to stimulate cleavage in the presence of all the other cleavage and polyadenylation factors. They incubated ^{32}P-labeled adenovirus L3 pre-mRNA with CPSF, CstF, CFI, CFII, poly(A) polymerase, and either RNA polymerase IIA or IIO. After the incubation period, they electrophoresed the products and autoradiographed the gel to see if the pre-mRNA had been cleaved in the right place. Figure 15.24 depicts the results. Both polymerases

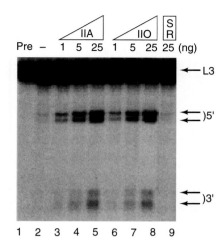

Figure 15.24 Effect of RNA polymerases IIA and IIO on prepolyadenylation mRNA cleavage in vitro. Hirose and Manley prepared a ^{32}P-labeled adenovirus L3 pre-mRNA and incubated it with all the cleavage and polyadenylation factors [CPSF, CstF, CFI, CFII, and poly(A) polymerase] plus polymerase IIA, IIO, no protein (–), or purified HeLa cell SR proteins, as indicated at top. (The amounts of the various proteins are given in nanograms.) Then the investigators electrophoresed the RNA products and detected them by autoradiography. The positions of the 5'- and 3'-cleavage fragments, and the pre-mRNA are indicated at right. Lane 1 contained precursor alone. Both IIA and IIO stimulated cleavage of the pre-mRNA to the appropriate 5'- and 3'-fragments. (Source: Hirose, Y. and Manley, J. RNA polymerase II is an essential mRNA polyadenylation factor. *Nature* 395 (3 Sep 1998) f. 2, p. 94. © Macmillan Magazines Ltd.)

IIA and IIO stimulated correct cleavage of the pre-mRNA, yielding 5'- and 3'-fragments of the expected sizes.

To verify that the CTD is the important part of polymerase II in stimulating cleavage, Hirose and Manley expressed the CTD as a fusion protein with glutathione-S-transferase (Chapter 4), then purified the fusion protein by glutathione affinity chromatography. They phosphorylated part of the fusion protein preparation on its CTD component and tested the phosphorylated and unphosphorylated fusion proteins in the cleavage assay with the adenovirus L3 pre-mRNA. Figure 15.25a shows that both the phosphorylated and unphosphorylated CTDs stimulated cleavage, but the phosphorylated form worked about five times better than the unphosphorylated one. That makes sense because the CTD is phosphorylated in polymerase IIO, which is the form that carries out transcription. It is unclear why phosphorylation made no difference when whole polymerase II was used in Figure 15.24.

If the CTD is the key to stimulating cleavage of the pre-mRNA, then polymerase IIB, the proteolytic product of polymerase IIA that lacks the CTD, should not stimulate, and Figure 15.25b shows that it does not. Thus, RNA polymerase II, and the CTD in particular, appears to be required for efficient cleavage of a pre-mRNA prior to polyadenylation. Figure 15.26 summarizes our knowledge about the complex of proteins that assembles on a pre-mRNA just before cleavage.

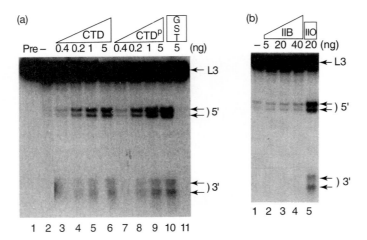

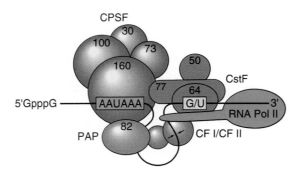

Figure 15.25   Effect of the RPB1 CTD on prepolyadenylation mRNA cleavage in vitro. Hirose and Manley incubated a labeled pre-mRNA with cleavage and polyadenylation factors and assayed for cleavage as in Figure 15.24. (a) They included phosphorylated or unphosphorylated GST–CTD fusion proteins or GST alone, as indicated at top, in the cleavage reaction. (b) They included RNA polymerase IIB or IIO, as indicated at top, in the cleavage reaction. The phosphorylated CTD stimulated cleavage more than the unphosphorylated CTD, and polymerase IIB, which lacks the CTD did not stimulate cleavage at all. (*Source:* Hirose, Y and Manley, J. RNA polymerase II is an essential mRNA polyadenylation factor. *Nature* 395 (3 Sep 1998) f. 3, p. 94. © Macmillan Magazines Ltd.)

Figure 15.26   A model for the precleavage complex. This partly hypothetical model shows the apparent positions of all the proteins presumed to be involved in cleavage, with respect to the two parts of the polyadenylation signal (green and yellow). (*Source:* After *Trends in Biochemical Sciences,* vol. 21, E. Wahle and W. Keller, "The biochemistry of polyadenylation," 247–250. Copyright © 1996 with permission of Elsevier Science.)

SUMMARY Polyadenylation requires both cleavage of the pre-mRNA and polyadenylation at the cleavage site. Cleavage in mammals requires several proteins: CPSF, CstF, CFI, CFII, poly(A) polymerase, and RNA polymerase II (in particular, the CTD of RPB1).

Initiation of Polyadenylation   Once a pre-mRNA has been cleaved downstream of its AAUAAA motif, it is ready to be polyadenylated. The polyadenylation of a cleaved RNA occurs in two phases. The first, initiation, depends on the AAUAAA signal and involves slow addition of at least 10 A's to the pre-mRNA. The second phase, elongation, is independent of the AAUAAA motif, but depends on the oligo(A) added in the first phase. This second phase involves the rapid addition of 200 or more A's to the RNA. Let us begin with the initiation phase.

Strictly speaking, the entity we have been calling "the polyadenylation signal" is really the cleavage signal. It is what attracts the cleavage enzyme to cut the RNA 15–25 nt downstream of the AAUAAA motif. Polyadenylation itself, that is, the addition of poly(A) to the 3′-end created by the cleavage enzyme, cannot use the same signal. This must be true because the cleavage enzyme has already removed the downstream part of the signal (the GU-rich and U-rich elements).

What is the signal that causes polyadenylation per se? It seems to be AAUAAA, followed by at least 8 nt at the end of the RNA. We know this because short synthetic oligonucleotides (as short as 11 nt) containing AAUAAA can be polyadenylated in vitro. The optimal length between the AAUAAA and the end of the RNA is 8 nt.

To study the process of polyadenylation by itself in vitro, it is necessary to divorce it from the cleavage reaction. Molecular biologists accomplish this by using labeled, short RNAs that have an AAUAAA sequence at least 8 nt from the 3′-end. These substrates mimic pre-mRNAs that have just been cleaved and are ready to be polyadenylated. The assay for polyadenylation is electrophoresis of the labeled RNA. If poly(A) has been added, the RNA will be much bigger and will therefore electrophorese much more slowly. It will also be less discrete in size, because the poly(A) tail varies somewhat in length from molecule to molecule. In this section, we will use the term *polyadenylation* to refer to the addition of poly(A) to the 3′-end of such a model RNA substrate.

Figure 15.27 shows how Marvin Wickens and his colleagues used this assay to demonstrate that two fractions are needed for polyadenylation: poly(A) polymerase and a specificity factor. We now know that this specificity factor is CPSF. At high substrate concentrations, the poly(A) polymerase can catalyze the addition of poly(A) to the 3′-end of any RNA, but at low substrate concentrations it cannot polyadenylate by itself (lane 1). Neither can CPSF, which recognizes the AAUAAA signal (lane 2). But together, these two substances can polyadenylate the synthetic substrate (lane 3). Lane 4 demonstrates that both fractions together will not polyadenylate a substrate with an aberrant signal (AAUCAA).

Michael Sheets and Wickens have also shown that polyadenylation itself consists of two phases. The first is dependent on the AAUAAA signal and lasts until about

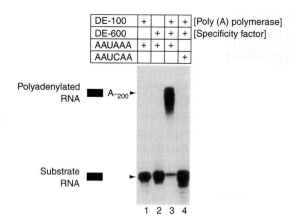

Figure 15.27 Separation of poly(A) polymerase and specificity factor activities. Wickens and colleagues separated HeLa cell poly(A) polymerase and specificity factor activities by DEAE-Sepharose chromatography. The polymerase eluted at 100 mM salt, so it is called the DE-100 fraction; the specificity factor eluted at 600 mM salt, so it is designated the DE-600 fraction. These workers tested the separated activities on a synthetic substrate consisting of nucleotides –58 to +1 of SV40 late mRNA, whose 3′-end is at the normal polyadenylation site. After they incubated the two fractions, separately or together, with the substrate and labeled ATP, they electrophoresed the labeled RNA and autoradiographed the gel. The components in the reactions in each lane are listed at top. The positions of substrate and polyadenylated product are listed at left. (*Source:* Bardwell et al., The enzyme that adds poly(A) to mRNAs is a classical poly(A) polymerase. *Mol. Cell. Biol.* 10 (Feb 1990) p. 847, f. 1. American Society for Microbiology.)

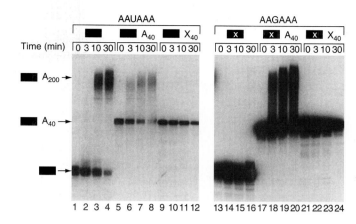

Figure 15.28 Demonstration of two phases in polyadenylation. Sheets and Wickens performed polyadenylation reactions in HeLa nuclear extracts with the following labeled substrates: 1. The standard 58-nt substrate containing the 3′-end of an SV40 late mRNA, represented by a black box; 2. The same RNA with a 40-nt poly(A), represented by a black box followed by A_{40}; 3. The same RNA with a 40-nt 3′-tag containing vector sequence, represented by a black box followed by X_{40}; substrates 1–3 containing an aberrant AAGAAA instead of AAUAAA, are represented with white X's within the black boxes. Sheets and Wickens used four different reaction times with each substrate, and the substrate in each set of lanes is indicated by its symbol at top. The electrophoretic mobility of substrates and products are indicated at left. (*Source:* Sheets and Wickens, Two phases in the addition of a poly(A) tail. *Genes & Development* 3 (1989) p. 1402, f. 1.)

10 A's have been added; the second is AAUAAA-independent and lasts until the poly(A) chain has grown to its full length—about 250 A's. To demonstrate this point, Sheets and Wickens used several different model RNA substrates. The first is simply the same terminal 58 nt of the SV40 late mRNA, including the AAUAAA, used in Figure 15.27. The second is the same RNA with 40 A's [a short poly(A)] at the 3′-end. The third is the same RNA with 40 nt from the vector instead of a short poly(A) at the 3′-end. They also used an analogous set of three substrates that had an AAGAAA signal instead of AAUAAA.

Sheets and Wickens used each of these substrates in standard polyadenylation reactions with HeLa cell nuclear extracts. Figure 15.28, lanes 1–4, shows that the extract could polyadenylate the usual model substrate with an AAUAAA signal. Lanes 5–8 show that polyadenylation also occurred with the model substrate that already had 40 A's at its end (A_{40}). The polyadenylated signal was weaker in this case, but the radioactivity of the substrate was also lower. On the other hand, the extract could not polyadenylate the model substrate with 40 non-poly(A) nucleotides at its end (X_{40}). Lanes 13–16 demonstrate that the extract could not polyadenylate the substrate with an aberrant AAGAAA signal and no poly(A) pre-added. However, lanes 17–20 make the most telling point: The extract is able to polyadenylate the substrate with an aberrant AAGAAA signal and 40 A's already

added to the end. Thus, by the time 40 A's have been added, polyadenylation is independent of the AAUAAA signal. But these extra nucleotides must be A's; the X_{40} substrate with an aberrant AAGAAA signal could not be polyadenylated (lanes 21–24).

Sheets and Wickens went on to show that the shortest poly(A) that could override the effect of a mutation in AAUAAA is 9 A's, but 10 A's work even better. These findings suggest the following hypothesis: After cleavage of the pre-mRNA, polyadenylation depends on the AAUAAA signal and CPSF until the poly(A) reaches about 10 A's in length. From then on, polyadenylation is independent of the AAUAAA and CPSF, but dependent on the poly(A) at the 3′-end of the RNA.

If CPSF recognizes the poladenylation signal AAUAAA, we would predict that CPSF binds to this signal in the pre-mRNA. Walter Keller and colleagues have demonstrated this directly, using gel mobility shift and RNA–protein cross-linking procedures. Figure 15.29 illustrates the results of both kinds of experiments. Panel (a) shows that CPSF binds to a labeled RNA containing an AAUAAA signal, but not to the same RNA with a U → G mutation in the AAUAAA motif. Panel (b) demonstrates that an oligonucleotide bearing an AAUAAA motif, but not an AAGAAA motif, can be cross-linked to two polypeptides (35 and 160 kD) in a CPSF preparation. Furthermore, these complexes will not form in the presence of unlabeled competitor RNAs containing AAUAAA;

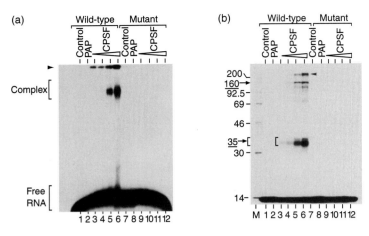

Figure 15.29 CPSF binds to the AAUAAA motif. (a) Gel mobility shift assay. Keller and colleagues mixed a labeled oligoribonucleotide with poly(A) polymerase (PAP), or CPSF in various concentrations, then electrophoresed the mixture. The wild-type oligo contained the AAUAAA motif, and the mutant oligo contained an AAGAAA motif. The controls contained no added proteins. CPSF could form a complex with the wild-type but not the mutant oligo. The band at the top (arrowhead) is material that remained at the top of the gel, rather than a specific band. **(b)** SDS-PAGE of proteins cross-linked to oligoribonucleotides. Keller and colleagues illuminated each of the mixtures from panel (a) with ultraviolet light to cross-link proteins to the oligo. Then they electrophoresed the complexes on an SDS polyacrylamide gel. Major bands appeared at 35 and 160 kD (arrows). (*Source:* Keller et al., Cleavage and polyadenylation factor CPF specifically interacts with the pre-mRNA 3′ processing signal AAUAAA. *EMBO Journal* 10 (1991) p. 4243, f. 2.)

competitor RNAs containing AAGAAA cannot compete. All of these findings bolster the conclusion that CPSF binds directly to the AAUAAA motif.

SUMMARY Short RNAs that mimic a newly created mRNA 3′-end can be polyadenylated. The optimal signal for initiation of such polyadenylation of a cleaved substrate is AAUAAA followed by at least 8 nt. Once the poly(A) reaches about 10 nt in length, further polyadenylation becomes independent of the AAUAAA signal and depends on the poly(A) itself. Two proteins participate in the initiation process: poly(A) polymerase and CPSF, which binds to the AAUAAA motif.

Elongation of Poly(A) We have seen that elongation of an initiated poly(A) chain 10 nt or more in length is independent of CPSF. However, purified poly(A) polymerase binds to and elongates poly(A) only very poorly by itself. This implies that another specificity factor can recognize an initiated poly(A) and direct poly(A) polymerase to elongate it. Elmar Wahle has purified a poly(A)-binding protein that has these characteristics.

Figure 15.30b shows the results of PAGE on fractions from the last step in purification of the poly(A)-binding protein. A major 49-kD polypeptide is visible, as well as a minor polypeptide with a lower molecular mass. Because the latter band varied in abundance, and was even invisible in some preparations, Wahle concluded that it was not related to the poly(A)-binding protein. Wahle tested the fractions containing the 49-kD protein for poly(A) binding by a nitrocellulose filter binding assay [panel (a)], and found that the peak of poly(A)-binding activity coincided with the peak of abundance of the 49-kD polypeptide. Next, he tested the same fractions for ability to stimulate polyadenylation of a model RNA substrate in the presence of poly(A) polymerase and CPSF [panel (c)]. Again, he found that the peak of activity coincided with the abundance of the 49-kD polypeptide. Thus, the 49-kD polypeptide is a poly(A)-binding protein, but differs from the major, 70-kD poly(A)-binding protein, (**PAB I**) found earlier in the cytoplasm, so Wahle named it **poly(A)-binding protein II (PAB II)**.

PAB II can stimulate polyadenylation of a model substrate, just as CPSF can, but it binds to poly(A) rather than to the AAUAAA motif. This suggests that PAB II is active in elongation, rather than initiation, of polyadenylation. If so, then its substrate preference should be different from that of CPSF. In particular, it should stimulate polyadenylation of RNAs that already have an oligo(A) attached, but not RNAs with no oligo(A). The results in Figure 15.31 confirm this prediction. Panel (a) shows that an RNA lacking oligo(A) (L3 pre) could be polyadenylated by poly(A) polymerase (PAP) plus CPSF, but not by PAP plus PAB II. However, PAP plus CPSF plus PAB II polyadenylated this substrate best of all. Presumably, CPSF serves as the initiation factor, then PAB II directs the polyadenylation of the substrate once an oligo(A) has been added, and does this better than CPSF can. Predictably, an L3 pre substrate with a mutant AAUAAA signal (AAGAAA) could not be polyadenylated by any combination of factors, because it depends on CPSF for initiation, and CPSF depends on an AAUAAA signal.

Figure 15.31b shows that the same RNA with an oligo(A) at the end behaved differently. It could be polyadenylated by PAP in conjunction with *either* CPSF or PAB II. This makes sense because this substrate has an oligo(A) that PAB II can recognize. It is interesting that both factors together produced even better polyadenylation of this substrate. This suggests that PAP might interact with both factors, directly or indirectly, during the elongation phase. Finally, panel (b) demonstrates that PAB II, in the absence of CPSF, could direct efficient polyadenylation of the mutant RNA with an AAGAAA motif, as long as the RNA had an oligo(A) to begin with. Again, this makes sense because the oligo(A) provides a recognition site for PAB II and therefore makes it independent of CPSF and the AAUAAA motif.

Figure 15.32 presents a model of initiation and elongation of polyadenylation. Optimal activity during the ini-

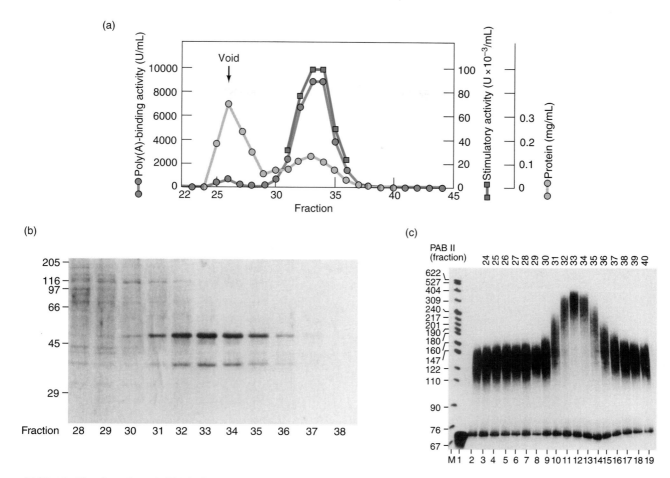

Figure 15.30   Purification of a poly(A)-binding protein. (a) Summary of results. Wahle subjected the poly(A)-binding protein to a final gel filtration chromatographic purification step on Sephadex G-100. In this panel, he plotted three parameters against fraction number from the G-100 column. Red, poly(A)-binding activity determined by a filter binding assay; green, polyadenylation-stimulating activity [see panel **(c)**]; blue, protein concentration. "Void" indicates proteins that eluted in the void volume. These large proteins were not included in the gel spaces on the column. **(b)** SDS-PAGE analysis. Wahle subjected aliquots of each fraction from the G-100 column in panel **(a)** to SDS-PAGE and stained the proteins in the gel with Coomassie Blue. Sizes of marker polypeptides are given at left. A 49-kD polypeptide reached maximum concentration in the fractions (32–35) that had peak poly(A)-binding activity and polyadenylation-stimulatory activity. **(c)** Assay for polyadenylation stimulatory activity. Wahle added aliquots of each fraction from the G-100 column to standard polyadenylation reactions containing labeled L3pre RNA substrate. Lane 1 contained only substrate, with no poly(A) polymerase. The increase in size of poly(A) indicates stimulatory activity, which peaked in fractions 32–35. (*Source:* Wahle, A novel poly(A)-binding protein acts as a specificity factor in the second phase of messenger RNA polyadenylation. *Cell* 66 (23 Aug 1991) p. 761, f. 1. Reprinted by permission of Elsevier Science.)

tiation phase requires PAP, CPSF, CstF, CF I, CF II and the two-part polyadenylation signal (the AAUAAA and G/U motifs flanking the polyadenylation site). The elongation phase requires PAP, PAB II, and an oligo(A) at least 10 nt long. It is enhanced by CPSF. Table 15.2 lists all these protein factors, their structures, and their roles.

> **SUMMARY** Elongation of poly(A) in mammals requires a specificity factor called poly(A)-binding protein II (PAB II). This protein binds to a preinitiated oligo(A) and aids poly(A) polymerase in elongating the poly(A) to 250 nt or more. PAB II acts independently of the AAUAAA motif. It depends only on poly(A), but its activity is enhanced by CPSF.

Poly(A) Polymerase

In 1991, James Manley and colleagues cloned cDNAs encoding bovine poly(A) polymerase (PAP). Sequencing of these clones revealed two different cDNAs that differed at their 3′-ends, apparently because of two alternative splicing schemes. This in turn should give rise to two different PAPs (**PAP I** and **PAP II**) that differ in their carboxyl termini. Figure 15.33 shows the putative architecture of **PAP II**. This protein has several regions whose sequences match (more or less) the consensus sequences of known functional domains of other proteins. These are, in order from N-terminus to C-terminus: an RNA-binding domain (RBD); a **polymerase module (PM)**; two nuclear localization signals (NLS 1 and 2); and several serine/threonine-rich regions (S/T). By 1996, four additional PAP cDNAs

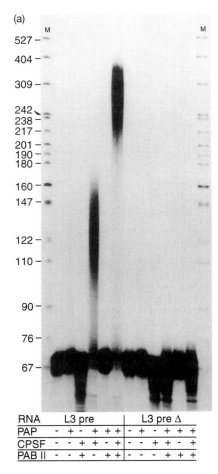
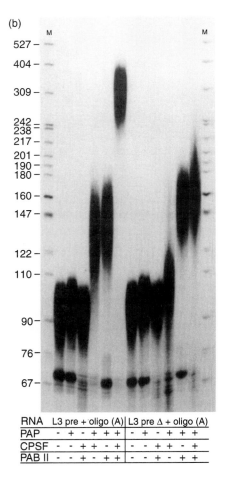

Figure 15.31   Effect of CPSF and PAB II on polyadenylation of model substrates. (a) Polyadenylation of RNAs lacking oligo(A). Wahle carried out polyadenylation reactions in the presence of the RNAs and proteins listed at bottom. L3 pre was the standard substrate RNA with an AAUAAA motif; L3 preΔ was the same, except that AAUAAA was mutated to AAGAAA. PAB II could not direct polyadenylation of L3 pre without help from CPSF. **(b)** Polyadenylation of RNAs containing oligo(A). All conditions were the same as in panel **(a)** except that the substrates contained oligo(A) at their 3′-ends. This allowed PAB II to work in the absence of CPSF and to work on the substrate with a mutant AAUAAA motif. The first and last lanes in both panels contained markers. (*Source:* Wahle, *Cell* 66 (23 Aug 1991) p. 764, f. 5. Reprinted by permission of Elsevier Science.)

had been discovered. Two of these were short and could arise from polyadenylation within the pre-mRNA. Another was long and could come from a pseudogene. The most important PAP in most tissues is probably PAP II.

Because the polymerase module, which presumably catalyzes the polyadenylation reaction, lies near the amino terminus of the protein, it would be interesting to know how much of the carboxyl end of the protein is required for activity. To examine the importance of the carboxyl end, Manley and colleagues expressed full-length and 3′-deleted versions of the PAP I cDNA by transcribing them in vitro with SP6 RNA polymerase, then translating these transcripts in cell-free reticulocyte extracts. This generated a full-length protein of 689 amino acids, and truncated proteins of 538, 379, and 308 amino acids. Then they tested each of these proteins for specific polyadenylation activity in the presence of calf thymus CPSF (called specificity factor [SF] by Manley). Figure 15.34 shows the results. The full-length and 538-

amino-acid proteins have activity, but the smaller proteins do not. (The small amount of activity seen with these proteins is also seen with SF alone.) Thus, the S/T domain is not necessary for activity, but sequences extending at least 150 amino acids toward the carboxyl terminus from the polymerase module are essential, at least in vitro.

SUMMARY Cloning and sequencing cDNAs encoding calf thymus poly(A) polymerase reveal a mixture of 5 cDNAs derived from alternative splicing and alternative polyadenylation. The structures of the enzymes predicted from the longest of these sequences include an RNA-binding domain, a polymerase module, two nuclear localization signals, and a serine/threonine-rich region. The latter region, but none of the rest, is dispensable for activity in vitro.

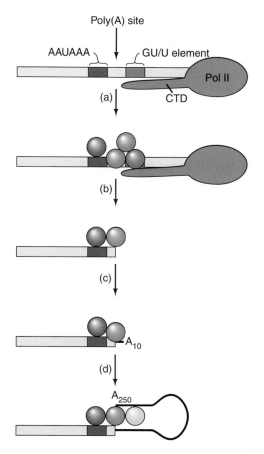

Figure 15.32 Model for polyadenylation. (a) CPSF (blue), CstF (brown), and CF I and II (gray) assemble on the pre-mRNA, guided by the AAUAAA and GU/U motifs. **(b)** Cleavage occurs, stimulated by the CTP of RNA polymerase II; CstF and CF I and II leave the complex; and poly(A) polymerase (red) enters. **(c)** poly(A) polymerase (orange), aided by CPSF, initiates poly(A) synthesis, yielding an oligo(A) at least 10 nt long. **(d)** PAB II (yellow) enters the complex and allows the rapid extension of the oligo(A) to a full-length poly(A). At this point, the complex presumably dissociates.

Turnover of Poly(A)

Figure 15.9 showed some evidence of a slight difference in size between nuclear and cytoplasmic poly(A). However, that experiment involved newly labeled RNA, so the poly(A) had not had much time to break down. Sheiness and Darnell performed another study on RNA from cells that were continuously labeled with RNA precursors for 48 h. This procedure gave a population of poly(A)s at their "steady-state" sizes; that is, the natural sizes one would observe by peeking into a cell at any given time. Figure 15.35 shows an apparent difference in the sizes of nuclear and cytoplasmic poly(A)s. The major peak of nuclear poly(A) was 210 ± 20 nt, whereas the major peak of cytoplasmic poly(A) was 190 ± 20 nt. Furthermore, the cytoplasmic poly(A) peak showed a much broader skew toward smaller species than the nuclear poly(A) peak. This broad peak encompassed RNAs at least as small as

50 nt. Thus, poly(A) seems to undergo considerable shortening in the cytoplasm.

In 1970, Maurice Sussman proposed a "ticketing" hypothesis that held that each mRNA has a "ticket" that allows it entry to the ribosome for translation. Each time it is translated, the mRNA gets its "ticket punched." When it accumulates enough "punches," it can no longer be translated. Poly(A) would make an ideal ticket; the punches would then be progressive shortening of the poly(A) every time it is translated. To test this idea, Sheiness and Darnell tested the rate of shortening of poly(A) in the cytoplasm under normal conditions, and in the presence of emetine, which inhibits translation. They observed no difference in the size of cytoplasmic poly(A), whether or not translation was occurring. Thus, the shortening of poly(A) does not depend on translation, and the ticket, if it exists at all, seems not to be poly(A).

Poly(A) is not just shortened in the cytoplasm; it turns over. That is, it is constantly being shortened by RNases and lengthened by a cytoplasmic poly(A) polymerase. The general trend, however, is toward shortening, and ultimately an mRNA will lose all or almost all of its poly(A). By that time, its demise is near.

SUMMARY Poly(A) turns over in the cytoplasm. RNases tear it down, and poly(A) polymerase builds it back up. When the poly(A) is gone, the mRNA is slated for destruction.

Cytoplasmic Polyadenylation The best studied cases of cytoplasmic polyadenylation are those that occur during oocyte maturation. Maturation of *Xenopus* oocytes, for example, occurs in vitro on stimulation by progesterone. The immature oocyte cytoplasm contains a large store of mRNAs called **maternal messages**, or **maternal mRNAs**, many of which are almost fully deadenylated. During maturation, some maternal mRNAs are polyadenylated, and others are deadenylated.

To find out what controls this maturation-specific cytoplasmic polyadenylation, Wickens and colleagues injected two mRNAs into *Xenopus* oocyte cytoplasm. The first was a synthetic 3′-fragment of D7 mRNA, a *Xenopus* mRNA known to undergo maturation-specific polyadenylation. The second was a synthetic 3′-fragment of an SV40 mRNA. As Figure 15.36 shows, the D7 RNA was polyadenylated, but the SV40 RNA was not. This implied that the D7 RNA contained a sequence or sequences that are required for maturation-specific polyadenylation, and that these are lacking in the SV40 RNA.

Wickens and colleagues noted that *Xenopus* RNAs that were known to undergo polyadenylation during oocyte maturation all contained the sequence UUUUUAU, or a close relative, upstream of the AAUAAA signal. Is this the

Table 15.2   Mammalian Factors Required for 3′-cleavage and Polyadenylation

Factor	Polypeptides (kD)	Properties
Poly(A) polymerase (PAP)	82	Required for cleavage and polyadenylation; catalyzes poly(A) synthesis
Cleavage and polyadenylation specificity factor (CPSF)	160 100 70 30	Required for cleavage and polyadenylation; binds AAUAAA and interacts with PAP and CstF
Cleavage stimulation factor (CstF)	77 64 50	Required only for cleavage; binds the downstream element and interacts with CPSF
Cleavage factor I (CF I)	68 59 25	Required only for cleavage; binds RNA
Cleavage factor II (CF II)	unknown	Required only for cleavage
RNA polymerase II (especially CTD)	many	Required only for cleavage
Poly(A)-binding protein II (PAB II)	33	Stimulates poly(A) elongation; binds growing poly(A) tail; essential for poly(A) tail length control

Source: Reprinted from *Trends in Biochemical Sciences,* vol. 21, E. Wahle and W. Keller, "The biochemistry of polyadenylation," 247–250. Copyright © 1996 with permission of Elsevier Science.

Figure 15.33   Schematic representation of calf thymus poly(A) polymerase. The putative functional domains are the following: the RNA-binding domain (RBD, blue); the polymerase module (PM, red); the nuclear localization signals (NLS 1 and 2, green); and the serine/threonine-rich domains (S/T, yellow). (*Source:* Reprinted with permission from *Nature* 353:232, 1991. Copyright © 1991 Macmillan Magazines Limited.)

key? To find out, these workers inserted this sequence upstream of the AAUAAA in the SV40 RNA and retested it. Figure 15.37 demonstrates that addition of this sequence caused polyadenylation of the SV40 RNA. In light of this character, the UUUUUAU sequence has been dubbed the **cytoplasmic polyadenylation element (CPE).**

Is the AAUAAA also required for cytoplasmic polyadenylation? To answer this question, Wickens and colleagues made point mutations in the AAUAAA motif and injected the mutated RNAs into oocyte cytoplasm. As Figure 15.38 shows, alteration of AAUAAA to either AAUAUA or AAGAAA completely abolished polyadenylation. Thus, this motif is required for both nuclear and cytoplasmic polyadenylation.

SUMMARY Maturation-specific polyadenylation of *Xenopus* maternal mRNAs in the cytoplasm depends on two sequence motifs: the AAUAAA motif near the end of the mRNA and an upstream motif called the cytoplasmic polyadenylation element (CPE), which is UUUUUAU or a closely related sequence.

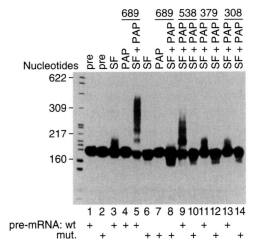

Figure 15.34   Specific polyadenylation carried out by full-length and C-terminally truncated poly(A) polymerases. Manley and colleagues expressed full-length and 3′-truncated versions of PAP I cDNA and tested them for specific polyadenylation activity in vitro with purified calf thymus CPSF (SF). All lanes contained pre-mRNAs with AAUAAA (wild-type) or AAAAAA (mutant) motifs, as indicated at bottom. Other components are listed at top; the numbers (689, 538, 379, and 308) indicate the number of amino acids in the PAPs synthesized in vitro. Wild-type RNA and SF were required for any of the PAPs to operate. Only the full-length and the 538-amino-acid versions of PAP could cooperate with these components to cause polyadenylation. (*Source:* Roabe et al., *Nature* 353 (19 Sept 1991) p. 233, f. 5. © Macmillan Magazines Ltd.)

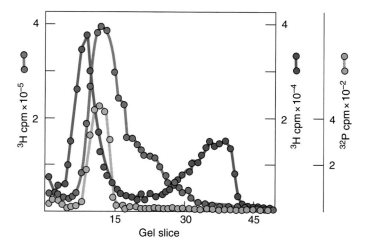

Figure 15.35 Shortening of cytoplasmic poly(A). Sheiness and Darnell labeled HeLa cells with [3H]-adenine for 48 h, then isolated nuclear (green) and cytoplasmic (red) poly(A)+ RNA and analyzed it by gel electrophoresis. They also included a [32P]5S rRNA as a marker (blue). (*Source:* Reprinted with permission from *Nature New Biology* 241:266, 1973. Copyright © 1973 Macmillan Magazines Limited.)

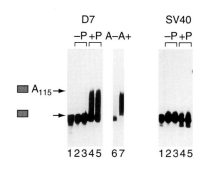

Figure 15.36 Maturation-specific polyadenylation of two RNAs. Wickens and colleagues injected labeled RNAs into *Xenopus* oocyte cytoplasm and stimulated maturation-specific polyadenylation with progesterone. After a 12-h incubation, they isolated the labeled RNA products, electrophoresed them, and visualized them by autoradiography. The two RNAs, as indicated at top, were synthetic 3′-fragments of either the *Xenopus* mRNA (D7), which normally undergoes maturation-specific polyadenylation, or an SV40 mRNA, which does not. The mobilities of unpolyadenylated RNA and RNA with a 115-nt poly(A) are indicated by the red boxes at left. The presence or absence of progesterone during the incubation is indicated at top by +P and −P, respectively. Lanes 6 and 7 contained RNA that was fractionated by oligo(dT)-cellulose chromatography. RNA that did not bind to the resin is designated A−, and RNA that did bind is designated A+. (*Source:* Fox et al., Poly(A) addition during maturation of frog oocytes: Distinct nuclear and cytoplasmic activities and regulation by the sequence UUUUUAU. *Genes & Development* 3 (1989) p. 2154, f. 3.)

UAAUUUUUAUAAGCUGCAAUAAACAAGUUAACAACCUCUAG_OH

UAACCAUUAUAAGCUGCAAUAAACAAGUUAACAACCUCUAG_OH

(a)

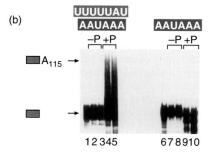

(b)

Figure 15.37 Demonstration that UUUUUAU confers maturation-specific polyadenylation. Wickens and colleagues performed the same experiment as described in Figure 15.36, using the same SV40 3′-mRNA fragment with and without an added UUUUUAU motif upstream of the AAUAAA motif. (a) Sequences of the two injected RNAs, with the UUUUUAU and AAUAAA motifs highlighted. (b) Results. Lanes 2–5 contained RNA from oocytes injected with the RNA having both a UUUUUAU and an AAUAAA sequence, as shown at top. Lanes 7–10 contained RNA from oocytes injected with the RNA having only an AAUAAA sequence. Presence or absence of progesterone during the incubation is indicated at top as in Figure 15.36. Lanes 1 and 6 had uninjected RNA. Markers at left as in Figure 15.36. The UUUUUAU motif was essential for polyadenylation. (*Source:* Fox et al., *Genes & Development* 3 (1989) p. 2155, f. 5.)

(a) UACUUUUUAUACUAGUCAAUAAACAAGUUAACAACCUCUAG_OH

UACUUUUUAUACUAGUCAAUAUACAAGUUAACAACCUCUAG_OH

UACUUUUUAUACUAGUCAAGAAACAAGUUAACAACCUCUAG_OH

(b)

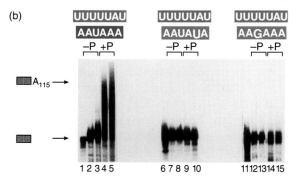

Figure 15.38 Abolition of maturation-specific polyadenylation by mutations in the AAUAAA motif. Wickens and colleagues started with the labeled synthetic SV40 RNA with an added UUUUUAU motif, described in Figure 15.37, made two single-base changes in the AAUAAA motif and assayed these RNAs for maturation-specific polyadenylation as in Figures 15.36 and 15.37 (a) Sequences of the RNAs, with UUUUUAU and AAUAAA motifs highlighted. The top AAUAAA sequence is normal; the two lower sequences have a change to AAUAUA and AAGAAA, respectively, where the change is boldfaced. (b) Results. The sequences of the UUUUUAU and AAUAAA motifs are given above the appropriate lanes. Other designations have the same meaning as in Figure 15.37. Lanes 1, 6, and 11 represent uninjected RNAs. Both changes in AAUAAA abolished polyadenylation. (*Source:* Fox et al., *Genes & Development* 3 (1989) p. 2155, f. 6.)

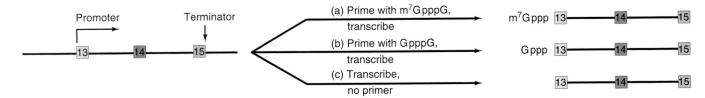

Figure 15.39   Production of capped and uncapped splicing substrates. Shimura and colleagues cloned exons 13, 14, and 15, with their natural introns, into a vector between an *E. coli* promoter and terminator. They transcribed this gene fragment under three conditions to produce three different splicing substrates: (**a**) They primed with m⁷GpppG and transcribed to yield an RNA with a methylated cap. (**b**) They primed with GpppG and transcribed to yield an RNA with an unmethylated cap. (**c**) They transcribed with no primer to yield an uncapped RNA.

15.3   The Effects of the Cap and Poly(A) on Splicing

Now that we have studied capping, polyadenylation, and splicing, we can appreciate that these processes are related. In particular, the cap can be essential for splicing, but only for splicing out the first intron. Similarly, the poly(A) can be essential for splicing out the last intron. Let us consider first the effect of capping on splicing.

Dependence of Splicing on the Cap

In 1987, Yoshiro Shimura and colleagues made a series of labeled splicing substrates by transcribing fragments of the chicken δ-crystallin gene in vitro. One such fragment (Figure 15.39) contained exons 13, 14, and 15, separated by their natural introns. To cap the RNAs, Shimura and colleagues transcribed the gene fragment in the presence of m⁷GpppG or GpppG, which primed in vitro transcription, adding the methylated or unmethylated caps in the process. To make uncapped RNAs, they performed the in vitro transcription without a primer. Next, these workers placed the labeled, capped and uncapped splicing substrates in HeLa nuclear extracts and allowed splicing to occur for various times. Finally, they electrophoresed the mixtures to separate products and intermediates from precursors.

Figure 15.40 depicts the results. With the capped precursor (panel a), intermediates appeared with one or the other intron spliced out, but the most prominent species at the end was the fully spliced product. Essentially the same thing happened with the precursor with the unmethylated cap. But the uncapped precursor gave a much different pattern. A little bit of intermediate with the first intron missing appeared, along with a predominant amount of intermediate with the second intron missing. No fully spliced product was detectable. Thus, it appears that the absence of cap selectively inhibits splicing of the first intron.

Is the unmethylated cap really just as good as the methylated one in promoting splicing? Probably not, be-

cause separate experiments showed that splicing of a substrate with an unmethylated cap could be inhibited by S-adenosyl homocysteine, an inhibitor of cap methylation. In other words, it appears that cap methylation was occurring in the splicing extract, even though a methylating agent was not added to the extract. This conclusion was bolstered by inhibition studies with cap analogs, in which methylated caps were more effective inhibitors of splicing than were unmethylated caps.

Does capping stimulate splicing of the first intron because it is closest, or is there something special about the sequence of this intron that makes it dependent on the cap? To find out, Shimura and colleagues made another splicing substrate with the order of introns reversed. That is, the 14–15 intron came first, and the 13–14 intron second. Again, the cap stimulated splicing out of the first intron (the 14–15 exon in this case), but not the second (the 13–14 intron). Therefore, the sequence of the intron and its bordering exons is irrelevant. What matters is proximity to the cap. The cap stimulates removal of whichever intron is closest.

In 1994, Mattaj and his colleagues characterized a **cap-binding complex (CBC)** that may help explain the effect of cap on splicing. This CBC is composed of two **cap-binding proteins (CBPs)**, with molecular masses of 80 and 20 kD (CBP80 and CBP20, respectively). An antibody against CBP80 immunoprecipitates the whole complex, containing both proteins. Figure 15.41 depicts the results of experiments that demonstrate the importance of the CBC. Panel (a) shows that immunodepletion of a HeLa nuclear extract with anti-CBP80 severely inhibited splicing of a model substrate with one intron. (A further experiment demonstrated that this splicing activity could be restored by adding back the CBC). Panels (b) and (c) verify that the antibody depleted the extract of both CBP80 and CBP20. Panel (d) provides evidence that the same antibody also inhibited formation of prespliceosomes (complexes A, B, and C). Thus, an early step in spliceosome formation (at least on the first intron) seems to be recognition of the cap. The CBC seems to be important in this recognition.

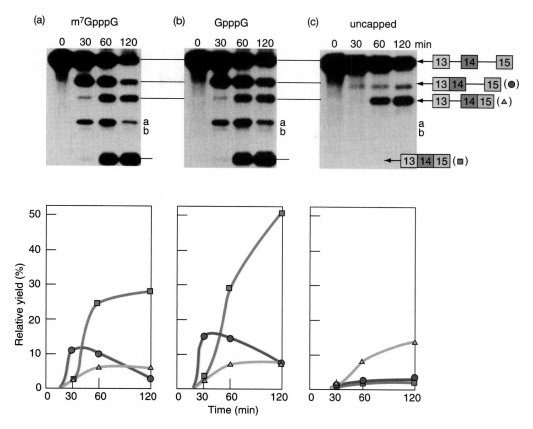

Figure 15.40   Effect of cap on splicing a substrate with two introns. Shimura and colleagues made ^{32}P-labeled model capped and uncapped splicing substrates as shown in Figure 15.39. Then they incubated these substrates in HeLa nuclear extracts for the lengths of time indicated at top. Finally, they electrophoresed the mixtures and autoradiographed the gels to detect splicing precursors, intermediates, and products. The positions of the major RNA species are indicated schematically at right, along with symbols that relate to the graphs below. Band a is the lariat form of the first intron; band b is a linear RNA containing exons 13 and 14 with the first intron in between. Panels (a), (b), and (c) show the results with the capped and methylated, capped, and uncapped precursors, respectively. Below each autoradiograph, graphs show the appearance and disappearance of the RNA species identified by the symbols to the right of the autoradiographs. (*Source:* Ohno et al., Preferential excision of the 5' proximal intron from mRNA precursors with two introns as mediated by the cap structure. *Proc. Natl. Acad. Sci.* 84 (1987) p. 5189, f 3.)

> **SUMMARY** Removal of the first intron from model pre-mRNAs in vitro is dependent on the cap. This effect may be mediated by a cap-binding complex that is involved in spliceosome formation.

Effect of Poly(A) on Splicing

Polyadenylation has been reported to occur both before and after splicing, depending on the system in question. In fact, in some pre-mRNAs, certain introns are spliced out before polyadenylation, and other introns are spliced out after. Maho Niwa and Susan Berget used model splicing substrates to illustrate this point, and also to show that polyadenylation stimulates removal of the last intron in a pre-mRNA, but not upstream introns. They started with three labeled model substrates containing an intron from the adenovirus major late transcription unit (Fig-

ure 15.42a). One of these (WT) contained a normal polyadenylation signal (AAUAAA), so it could be polyadenylated. The second was the same, except for a mutation to AAGAAA in the signal, so it could not be polyadenylated. The third (SB 2) had unrelated RNA at its 3'-end, with no AAUAAA signal, so it could not be polyadenylated. Niwa and Berget added each of these substrates to splicing extracts for various lengths of time, then electrophoresed the products, and visualized them by autoradiography. Figure 15.42b shows the results. The wild-type substrate exhibited a great deal of splicing, as measured by the appearance of spliced, nonpolyadenylated RNA (S+) and spliced, polyadenylated RNA (A+S+). Panel (c) depicts a quantification of the appearance of spliced products in panel (b). The nonpolyadenylated substrates were spliced 5- to 10-fold more slowly than the polyadenylated substrate.

We have just seen that the cap affects splicing of only the closest intron—the first one. Based on this, we might

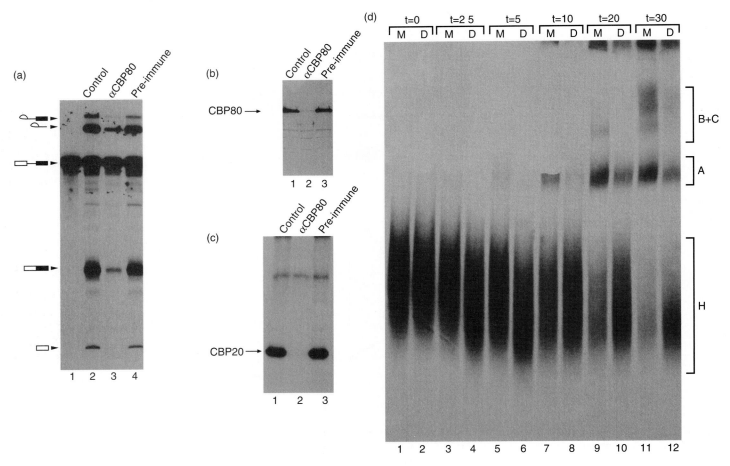

Figure 15.41   Effect of CBC on splicing and prespliceosome formation. (a) Effect on splicing. Mattaj and colleagues immunodepleted HeLa nuclear extracts of CBC using an anti-CBP80 antibody, then tested the extract for ability to splice a labeled model substrate with one intron. They displayed the products of the splicing reaction by gel electrophoresis and visualized them by autoradiography. Lane 1 contains the labeled substrate alone. Lane 2 is a positive control with no antiserum. Lane 3 contains an extract that was immunodepleted with anti-CBP80 antiserum. Lane 4 contains an extract that was immunodepleted with a preimmune (nonspecific) serum. **(b)** Removal of CBP80 by the anti-CBP80 antiserum. Equivalent amounts of nuclear extract were treated with no serum (lane 1), anti-CBP80 (lane 2), or preimmune serum (lane 3), then subjected to gel electrophoresis and immunoblotting with a CBP80 antibody probe. **(c)** Removal of CBP20 by the anti-CBP80 antiserum. Conditions were identical to those in panel **(b)** except that the immunoblot was probed with an anti-CBP antibody. **(d)** Effect on prespliceosome assembly. Mattaj and colleagues allowed CBC-immunodepleted (D) or mock-depleted (M) nuclear extracts to form prespliceosomes with a labeled model substrate for the times (in minutes) indicated at top. Then they displayed the complexes formed by gel electrophoresis and visualized them by autoradiography. The positions of the substrate (H) and pre-spliceosomes (A, B, and C) are indicated at right. (*Source:* Izaurralde et al., A nuclear cap binding protein complex involved in pre-mRNA splicing. *Cell* 78 (26 Aug 1994) p. 663, f. 7. Reprinted by permission of Elsevier Science.)

expect the poly(A) to affect splicing of only the nearest intron—the last one. To find out, Niwa and Berget made a pair of model substrates with two introns that were identical except for their lengths. One of these substrates (WT) had a wild-type AAUAAA signal; the other (AAGAAA) had a mutant signal. These workers subjected these two substrates to the same treatment as in the previous experiment. Figure 15.43 depicts the results. Panel (b) shows that the first intron of the unpolyadenylated substrate is spliced very well, but the second one is not. By contrast, both introns are rapidly spliced out of the polyadenylated substrate. This suggests that the poly(A) is

required for active removal of the proximal (nearby) intron, but not for removal of the distal (distant) intron.

This conclusion was reinforced by the fact that one of the prominent products derived from the wild-type substrate, labeled S1+, is partly spliced, but not polyadenylated. Thus, removal of the first intron from this substrate occurred quite well in the absence of polyadenylation. The idea that polyadenylation affects splicing of only the last intron also received support from a further experiment, in which the splicing substrates had three identical introns (except for size). Again, lack of polyadenylation inhibited removal of the third intron, but not the two upstream introns.

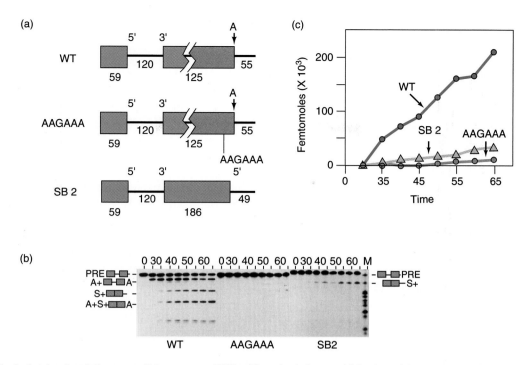

Figure 15.42 Effect of polyadenylation on splicing a pre-mRNA with a single intron. (a) Design of the substrates. Each substrate had the same intron, derived from the adenovirus major late region. The first substrate (WT) contained a wild-type polyadenylation signal (AAUAAA). The second (AAGAAA) had a mutant polyadenylation signal with that sequence. The third (SB 2) had unrelated RNA at its 3′-end, and no AAUAAA signal. (b) Splicing reactions. Niwa and Berget incubated each labeled substrate in a nuclear extract for the times indicated at top, then electrophoresed the products and visualized them by autoradiography. The substrates in each experiment are indicated at bottom. Precursor, intermediate, and product RNAs are identified by schematic diagrams at left and right. Pre, precursor RNA; A+, polyadenylated precursor; S+, spliced, unpolyadenylated product; A+S+, spliced, polyadenylated product. (c) Quantification of splicing. Niwa and Berget determined the amount of label in the bands corresponding to each spliced product in panel (b) with a blot analyzer and plotted the results. Green, splicing of WT RNA; blue, splicing of SB 2 RNA; red, splicing of AAGAAA RNA. (*Source:* Niwa & Berget, Mutation of the AAUAAA polyadenylation signal depresses *in vitro* splicing of proximal but not distal introns. *Genes & Development* 5 (1991) p. 2088, f. 1.)

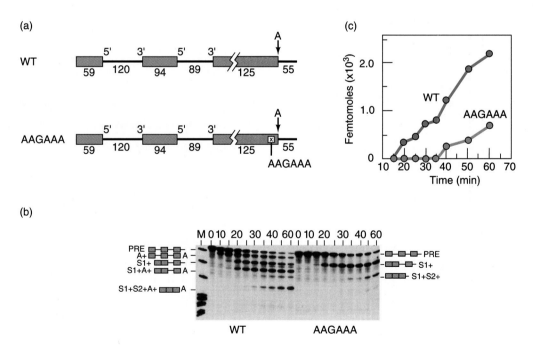

Figure 15.43 Effect of polyadenylation on splicing a two-intron substrate. All colors and names have the same meaning as in Figure 15.42, except for these new ones: S1+, nonpolyadenylated intermediate with first intron spliced out; S1+A+, polyadenylated intermediate with first intron spliced out; S1+S2+, nonpolyadenylated, fully spliced product; S1+S2+A+ polyadenylated, fully spliced product. (*Source:* Niwa & Berget, *Genes & Development* 5 (1991) p. 2091, f. 4.)

SUMMARY Polyadenylation of model substrates in vitro is required for active removal of the intron closest to the poly(A). However, splicing any other introns out of these substrates occurs at a normal rate even without polyadenylation.

SUMMARY

Caps are made in steps: First, an RNA triphosphatase removes the terminal phosphate from a pre-mRNA. Next, a guanylyl transferase adds the capping GMP (from GTP). Next, two methyl transferases methylate the N^7 of the capping guanosine and the 2′-O-methyl group of the penultimate nucleotide. These events occur early in the transcription process, before the chain length reaches 30. The cap ensures proper splicing of at least some pre-mRNAs, facilitates transport of the mature mRNA out of the nucleus, protects the mRNA from degradation, and enhances the mRNA's translatability.

Most eukaryotic mRNAs and their precursors have a poly(A) about 250 nt long at their 3′-ends. This poly(A) is added posttranscriptionally by poly(A) polymerase. Poly(A) enhances both the lifetime and translatability of mRNA. The relative importance of these two effects seems to vary from one system to another.

Transcription of eukaryotic genes extends beyond the polyadenylation site. Then the transcript is cleaved and polyadenylated at the 3′-end created by the cleavage. An efficient mammalian polyadenylation signal consists of an AAUAAA motif about 20 nt upstream of a polyadenylation site in a pre-mRNA, followed 23 or 24 bp later by a GU-rich motif, followed immediately by a U-rich motif. Many variations on this theme occur in nature, which results in variations in efficiency of polyadenylation. Plant polyadenylation signals also usually contain an AAUAAA motif, but more variation is allowed in this region than in an animal AAUAAA. Yeast polyadenylation signals are more different yet and rarely contain an AAUAAA motif.

Polyadenylation requires both cleavage of the pre-mRNA and polyadenylation at the cleavage site. Cleavage requires several proteins: CPSF, CstF, CF I, CF II, poly(A) polymerase, and the CTD of the RNA polymerase II largest subunit. Short RNAs that mimic a newly created mRNA 3′-end can be polyadenylated. The optimal signal for initiation of such polyadenylation of a cleaved substrate is AAUAAA, followed by at least 8 nt. Once the poly(A) reaches about 10 nt in length, further polyadenylation becomes independent of the AAUAAA signal, and depends on the poly(A) itself. Two proteins participate in the initiation process: poly(A) polymerase and CPSF, which binds to the AAUAAA motif.

Elongation requires a specificity factor called poly(A)-binding protein II (PAB II). This protein binds to a preinitiated oligo(A) and aids poly(A) polymerase in elongating the poly(A) up to 250 nt or more. PAB II acts independently of the AAUAAA motif. It depends only on poly(A), but its activity is enhanced by CPSF.

Calf thymus poly(A) polymerase is probably a mixture of at least three proteins derived from alternative RNA processing. The structures of the enzymes predicted from these sequences include an RNA-binding domain, a polymerase module, two nuclear localization signals, and a serine/threonine-rich region. The latter region, but none of the rest, is dispensable for activity in vitro.

Poly(A) turns over in the cytoplasm. RNases tear it down, and poly(A) polymerase builds it back up. When the poly(A) is gone, the mRNA is slated for destruction. Maturation-specific polyadenylation of maternal mRNAs in the cytoplasm depends on two sequence motifs: the AAUAAA motif near the end of the mRNA, and an upstream motif called the cytoplasmic polyadenylation element (CPE), which is UUUUUAU or a closely related sequence.

Caps and poly(A) play a role in splicing, at least in removal of the introns closest to the two ends of the pre-mRNA. Removal of the first intron can depend on the cap. This effect may be mediated by a cap-binding complex that is involved in spliceosome formation. Polyadenylation of model substrates in vitro is required for active removal of the intron closest to the poly(A). However, splicing any other introns out of these substrates occurs at a normal rate even without polyadenylation.

REVIEW QUESTIONS

1. You label a capped eukaryotic mRNA with ^{3}H-AdoMet and ^{32}P, then digest it with base and subject the products to DEAE-cellulose chromatography. Show the elution of cap 1 with respect to oligonucleotide markers of known charge. Draw the structure of cap 1 and account for its apparent charge.

2. How do we know that the cap contains 7-methylguanosine?

3. Outline the steps in capping.

4. Describe and show the results of an experiment that demonstrates the effect of capping on RNA stability.

5. Describe and give the results of an experiment that shows the synergistic effects of capping and polyadenylation on translation.

6. Describe and give the results of an experiment that shows the effect of capping on mRNA transport into the cytoplasm.

7. Describe and give the results of an experiment that shows the size of poly(A).

8. How do we know that poly(A) is at the 3′-end of mRNAs?

9. How do we know that poly(A) is added posttranscriptionally?

10. Describe and give the results of experiments that show the effects of poly(A) on mRNA translatability, mRNA stability, and recruitment of mRNA into polysomes.

11. Draw a diagram of the polyadenylation process, beginning with an RNA that is being elongated past the polyadenylation site.

12. Describe and give the results of an experiment that shows that transcription does not stop at the polyadenylation site.

13. Describe and give the results of an experiment that shows the importance of the AAUAAA polyadenylation motif. What other motif is frequently found in place of AAUAAA? Where are these motifs found with respect to the polyadenylation site?

14. Describe and give the results of an experiment that shows the importance of the G/U-rich and U-rich polyadenylation motifs. Where are these motifs with respect to the polyadenylation site?

15. Describe and give the results of an experiment that shows the effect of the RPB1 CTD on pre-mRNA cleavage prior to polyadenylation.

16. What happens to polyadenylation when you place a synthetic polyadenylation site into the large intron of the rabbit β-globin gene? What happens to polyadenylation when you impair splicing of this intron? How do you interpret these results?

17. Describe and give the results of an experiment that shows the importance of poly(A) polymerase and the specificity factor CPSF.

18. Describe and give the results of an experiment that shows the effect on polyadenylation of adding 40 A's to the end of a polyadenylation substrate that has an altered AAUAAA motif.

19. Describe and give the results of an experiment that shows that CPSF binds to AAUAAA, but not AAGAAA.

20. Describe and give the results of an experiment that shows the effects of CPSF and PAB II on polyadenylation of substrates with AAUAAA or AAGAAA motifs, with and without oligo(A) added. How do you interpret these results?

21. Present a diagram of polyadenylation that illustrates the roles of CPSF, the cleavage enzyme, poly(A) polymerase, RNA polymerase II, (PAP), and PAB II.

22. What part of the poly(A) polymerase PAP I is required for polyadenylation activity? Cite evidence.

23. Describe and give the results of an experiment that identifies the cytoplasmic polyadenylation element (CPE) that is necessary for cytoplasmic polyadenylation.

24. Describe and give the results of an experiment that shows that the cap plays a role in splicing of the first intron in a pre-mRNA.

25. Describe and give the results of an experiment that shows that the poly(A) influences splicing of the last intron in a pre-mRNA.

SUGGESTED READINGS

General References and Reviews

Barabino, S.M.L. and W. Keller. 1999. Last but not least: Regulated poly(A) tail formation. *Cell* 99:9–11.

Bentley, D. 1998. A tale of two tails. *Nature* 395:21–22.

Colgan, D.F. and Manley, J.L. 1997. Mechanism and regulation of mRNA polyadenylation. *Genes and Development* 11:2755–66.

Manley, J.L. and Y. Takagaki. 1996. The end of the message— Another link between yeast and mammals. *Science* 274:1481–82.

Proudfoot, N. 1996. Ending the message is not so simple. *Cell* 87:779–81.

Wahle, E. and W. Keller. 1996. The biochemistry of polyadenylation. *Trends in Biochemical Sciences* 21:247–51.

Wickens, M. 1990. How the messenger got its tail: Addition of poly(A) in the nucleus. *Trends in Biochemical Sciences* 15:277–81.

Research Articles

Bardwell, V.J., D. Zarkower, M. Edmonds, and M. Wickens. 1990. The enzyme that adds poly(A) to mRNA is a classical poly(A) polymerase. *Molecular and Cellular Biology* 10:846–49.

Fitzgerald, M. and T. Shenk. 1981. The sequence 5′-AAUAAA-3′ forms part of the recognition site for polyadenylation of late SV40 mRNAs. *Cell* 24:251–60.

Fox, C.A., M.D. Sheets, and M.P. Wickens. 1989. Poly(A) addition during maturation of frog oocytes: Distinct nuclear and cytoplasmic activities and regulation by the sequence UUUUUAU. *Genes and Development* 3:2151–56.

Furuichi, Y. and K-I. Miura. 1975. A blocked structure at the 5′ terminus of mRNA from cytoplasmic polyhedrosis virus. *Nature* 253:374–75.

Furuichi, Y., A. LaFiandra, and A.J. Shatkin. 1977. 5′-terminal structure and mRNA stability. *Nature* 266:235–39.

Furuichi, Y., S. Muthukrishnan, J. Tomasz, and A.J. Shatkin. 1976. Mechanism of formation of reovirus mRNA 5′-terminal blocked and methylated sequence m^{7}GpppG^{m}pC. *Journal of Biological Chemistry* 251:5043–53.

Gallie, D.R. 1991. The cap and poly(A) tail function synergistically to regulate mRNA translational efficiency. *Genes and Development* 5:2108–16.

Gil, A. and N.J. Proudfoot. 1987. Position-dependent sequence elements downstream of AAUAAA are required for efficient rabbit β-globin mRNA 3′-end formation. *Cell* 49:399–406.

Hamm, J. and I.W. Mattaj. 1990. Monomethylated cap structures facilitate RNA export from the nucleus. *Cell* 63:109–18.

Hirose, Y. and J.L. Manley. 1998. RNA polymerase II is an essential mRNA polyadenylation factor. *Nature* 395:93–96.

Hofer, E. and J.E. Darnell. 1981. The primary transcription unit of the mouse β-major globin gene. *Cell* 23:585–93.

Huez, G., G. Marbaix, E. Hubert, M. Leclercq, U. Nudel, H. Soreq, R. Salomon, B. Lebleu, M. Revel, and U.Z. Littauer. 1974. Role of the polyadenylate segment in the translation of globin messenger RNA in *Xenopus* oocytes. *Proceedings of the National Academy of Sciences USA* 71:3143–46.

Izaurralde, E., J. Lewis, C. McGuigan, M. Jankowska, E. Darzynkiewicz, and I.W. Mattaj. 1994. A nuclear cap binding protein complex involved in pre-mRNA splicing. *Cell* 78:657–68.

Keller, W., S. Bienroth, K.M. Lang, and G. Christofori. 1991. Cleavage and polyadenylation factor CPF specifically interacts with the pre-mRNA 3′ processing signal AAUAAA. *EMBO Journal* 10:4241–49.

Levitt, N., D. Briggs, A. Gil, and N.J. Proudfoot. 1989. Definition of an efficient synthetic poly(A) site. *Genes and Development* 3:1019–25.

McDonald, C.C., J. Wilusz, and T. Shenk. 1994. The 64-kilodalton subunit of the CstF polyadenylation factor binds to pre-mRNAs downstream of the cleavage site and influences cleavage site location. *Molecular and Cellular Biology* 14:6647–54.

Munroe, D. and A. Jacobson. 1990. mRNA poly(A) tail, a 3′ enhancer of a translational initiation. *Molecular and Cellular Biology* 10:3441–55.

Niwa, M. and S. Berget. 1991. Mutation of the AAUAAA polyadenylation signal depresses in vitro splicing of proximal but not distal introns. *Genes and Development* 5:2086–95.

Ohno, M., H. Sakamoto, and Y. Shimura. 1987. Preferential excision of the 5′ proximal intron from mRNA precursors with two introns as mediated by the cap structure. *Proceedings of the National Academy of Sciences USA* 84:5187–91.

Raabe, T., F.J. Bollum, and J.L. Manley. 1991. Primary structure and expression of bovine poly(A) polymerase. *Nature* 353:229–34.

Sheets, M.D. and M. Wickens. 1989. Two phases in the addition of a poly(A) tail. *Genes and Development* 3:1401–12.

Sheiness, D. and J.E. Darnell. 1973. Polyadenylic acid segment in mRNA becomes shorter with age. *Nature New Biology* 241:265–68.

Wahle, E. 1991. A novel poly(A)-binding protein acts as a specificity factor in the second phase of messenger RNA polyadenylation. *Cell* 66:759–68.

Wei, C.M. and Moss, B. 1975. Methylated nucleotides block 5′-terminus of vaccinia virus mRNA. *Proceedings of the National Academy of Sciences, USA* 72:318–22.

CHAPTER 16

Posttranscriptional Events III: Other Events

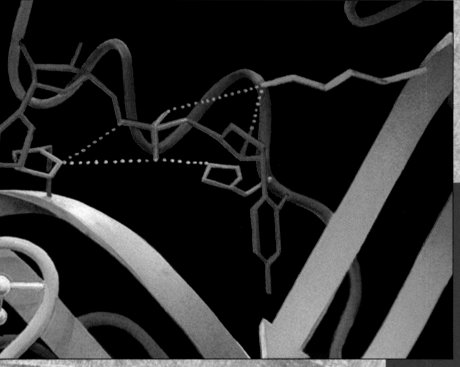

Computer model of Ribonuclease S. © *Irving Geis/Peter Arnold, Inc.*

In the previous two chapters, we examined splicing, capping, and polyadenylation, which covers most of what happens to pre-mRNAs in eukaryotic cells. However, in a few organisms, other specialized pre-mRNA processing events occur. For example, parasitic protozoa called trypanosomes, as well as some parasitic worms and the free-living protist *Euglena*, carry out *trans*-splicing of pre-mRNAs. This involves splicing together two independent transcripts. Trypanosomes also have mitochondria, called kinetoplasts, that edit their mRNAs by adding or deleting nucleotides after transcription. In contrast to these rather esoteric processing events, most organisms process their rRNAs and tRNAs by more conventional mechanisms. Finally, eukaryotes control some of their gene expression by regulating posttranscriptional processes, primarily mRNA degradation. All of these posttranscriptional events will be our subjects in this chapter.

16.1 Ribosomal RNA Processing

The rRNA genes of both eukaryotes and prokaryotes are transcribed as larger precursors that must be processed (cut into pieces) to yield rRNAs of mature size. However, this is not just a matter of removing unwanted material at either end of an overly long molecule. Instead, several different rRNA molecules are imbedded in a long precursor, and each of these must be cut out. Let us consider rRNA processing, first in eukaryotes, then in prokaryotes.

Eukaryotic rRNA Processing

The rRNA genes in eukaryotes are repeated several hundred times and clustered together in the nucleolus of the cell. Their arrangement in amphibians has been especially well studied, and, as Figure 16.1a shows, they are separated by regions called **nontranscribed spacers (NTSs)**. NTSs are distinguished from **transcribed spacers,** regions of the gene that are transcribed as part of the rRNA precursor and then removed in the processing of the precursor to mature rRNA species.

This clustering of the reiterated rRNA genes in the nucleolus made them easy to find and therefore provided Oscar Miller and his colleagues with an excellent opportunity to observe genes in action. These workers looked at amphibian nuclei with the electron microscope and uncovered a visually appealing phenomenon, shown in Figure 16.1b. The DNA containing the rRNA genes can be seen winding through the picture, but the most obvious feature of the micrograph is a series of "Christmas tree" structures. These include the rRNA genes (the trunk of the tree) and growing rRNA transcripts (the branches of the tree). We will see shortly that these transcripts are actually rRNA precursors, not mature rRNA molecules. The spaces between "Christmas trees" are the nontranscribed spacers. You can even tell the direction of transcription from the lengths of the transcripts within a given gene; the shorter RNAs are at the beginning of the gene and the longer ones are at the end.

We have seen that mRNA precursors frequently require splicing but no other trimming. On the other hand, rRNAs and tRNAs first appear as precursors that sometimes need splicing, but they also have excess nucleotides at their ends, or even between regions that will become separate mature RNA sequences. These excess regions must also be removed. This trimming of excess regions from an RNA precursor is called **processing**. RNA processing is similar to splicing in that unnecessary RNA is removed, but it differs from splicing in that no RNAs are stitched together during processing.

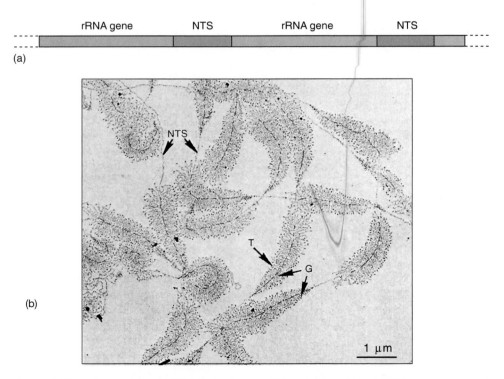

(a)

(b)

Figure 16.1   Transcription of rRNA precursor genes. (a) Map of a portion of the newt (amphibian) rRNA precursor gene cluster, showing the alternating rRNA genes (orange) and nontranscribed spacers (NTS, green). **(b)** Electron micrograph of part of a newt nucleolus, showing rRNA precursor transcripts (T) being synthesized in a "Christmas tree" pattern on the tandemly duplicated rRNA precursor genes (G). At the base of each transcript is an RNA polymerase I, not visible in this picture. The genes are separated by nontranscribed spacer DNA (NTS). (*Source:* (b) O. L. Miller et al., *CSH Symposia on Quantitative Biology* 35 (1970) p. 506.)

For example, mammalian RNA polymerase I makes a **45S rRNA precursor,** which contains the **28S, 18S,** and **5.8S rRNAs,** imbedded between transcribed spacer RNA regions. The processing of the precursor (Figure 16.2) takes place in the **nucleolus,** the nuclear compartment where rRNAs are made and ribosomes are assembled. The first step is to cut off the spacer at the 5′-end, leaving a 41S intermediate. The next step involves cleaving the 41S RNA into two pieces, 32S and 20S, that contain the 28S and 18S sequences, respectively. The 32S precursor also retains the 5.8S sequence. Finally, the 32S intermediate is split to yield the mature 28S and 5.8S RNAs, which base-pair with each other, and the 20S intermediate is trimmed to mature 18S size.

What is the evidence for this sequence of events? As long ago as 1964, Robert Perry used a **pulse-chase** procedure to establish a precursor–product relationship between the 45S precursor and the 18S and 28S mature rRNAs. He labeled mouse L cells for a short time (a short pulse) with [3H]uridine and found that the labeled RNA sedimented as a broad peak centered at about 45S. Then he "chased" the label in this RNA into 18S and 28S rRNAs. That is, he added excess unlabeled uridine to dilute the labeled nucleoside and observed that the amount

of label in the 45S precursor decreased as the amount of label in the mature 18S and 28S rRNAs increased. This suggested that one or more RNA species in the 45S peak was a precursor to 18S and 28S rRNAs. In 1970, Robert Weinberg and Sheldon Penman found the key intermediates by labeling poliovirus-infected HeLa cells with [3H]uridine and separating the labeled RNAs by gel electrophoresis. Ordinarily, processing intermediates are too short-lived to accumulate to detectable levels, but poliovirus infection slowed processing down enough that the intermediates could be seen. The major species observed were 45S, 41S, 32S, 28S, 20S, and 18S (Figure 16.3).

In 1973, Peter Wellauer and Igor Dawid visualized the precursor, intermediates, and products of human rRNA processing by electron microscopy (Figure 16.4). Each RNA species had its own capacity for intramolecular base pairing, so each had its own secondary structure that could be detected by electron microscopy. Once David and Wellauer had identified these "signatures" of all the RNA species, they could recognize them in the 45S precursor and thereby locate the 28S and 18S species in the precursor. Although they originally got the order backwards, we now know that the arrangement is: 5′-18S-5.8S-28S-3′. The details of this processing scheme are not

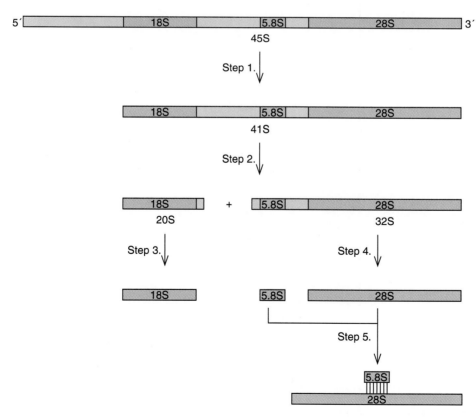

Figure 16.2 Processing scheme of 45S human (HeLa) rRNA precursor. Step 1: The 5′-end of the 45S precursor RNA is removed, yielding the 41S precursor. Step 2: The 41S precursor is cut into two parts, the 20S precursor of the 18S rRNA, and the 32S precursor of the 5.8S and 28S rRNAs. Step 3: The 3′-end of the 20S precursor is removed, yielding the mature 18S rRNA. Step 4: The 32S precursor is cut to liberate the 5.8S and 28S rRNAs. Step 5: The 5.8S and 28S rRNAs associate by base pairing.

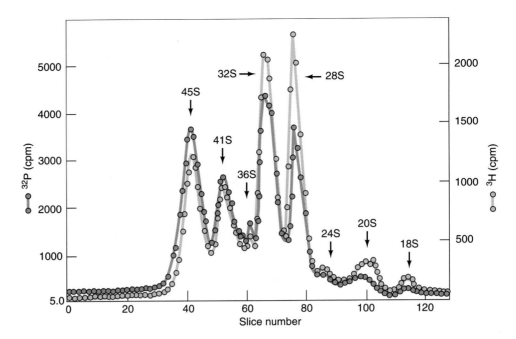

Figure 16.3   Isolation of 45S rRNA processing intermediates from poliovirus-infected HeLa cells. Penman and colleagues labeled RNA in virus-infected cells with [^{32}P]phosphate and with [^{3}H]methionine, which labeled the many methyl groups in rRNAs and their precursors. They isolated nucleolar RNA (mostly rRNA) from these cells, subjected it to gel electrophoresis, sliced the gel, determined the ^{32}P and ^{3}H radioactivity in each slice, then plotted these radioactivity values in cpm versus slice number. (*Source:* Reprinted from *Journal of Molecular Biology* 47:169 (1970) by permission of Academic Press Limited, London.)

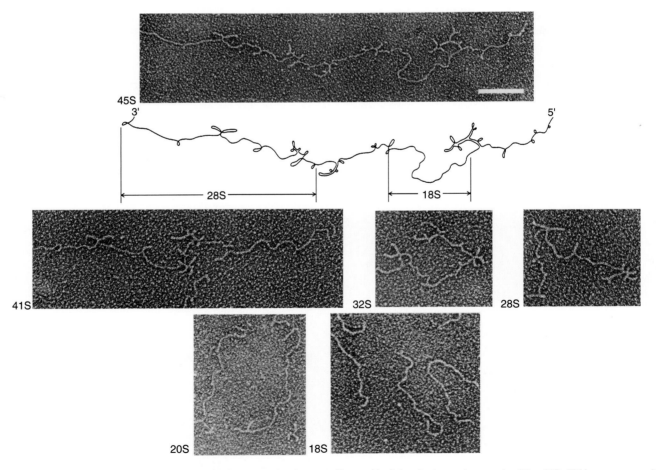

Figure 16.4   Electron micrographs of human rRNA processing intermediates. (Top) An electron micrograph of the 45S rRNA precursor, with interpretive diagram below. Bar represents 200 nm; magnification is the same in all micrographs. (Middle and bottom) The characteristic structures of the smaller intermediates and final products allow us to find their positions on the 45S precursor. The positions of the 18S and 28S final products are indicated on the diagram. Note that this RNA is written 3′ → 5′, left to right, which is contrary to convention. (*Source:* Wellauer and Dawid, Secondary structure maps of RNA: Processing of HeLa ribosomal RNA. *PNAS* 70, no. 10 (Oct 1973) p. 2828.)

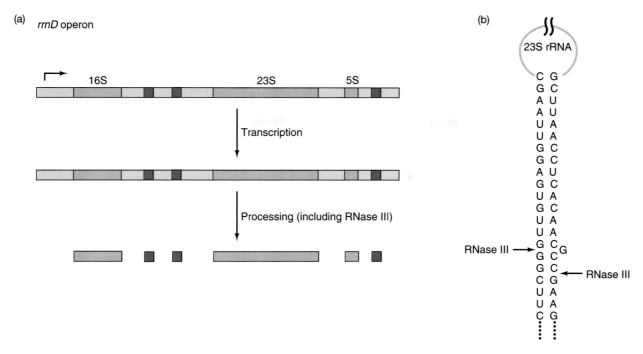

Figure 16.5 Processing bacterial rRNA precursors. (a) Structure of the *E. coli rrnD* operon. This operon is typical of the rRNA-encoding operons of *E. coli* in that it includes regions that code for tRNAs (red), as well as rRNA-coding regions (orange), embedded in transcribed spacers (yellow). As usual with bacterial operons, this one is transcribed to produce a long composite RNA. This RNA is then processed by enzymes, including RNase III, to yield mature products. **(b)** Sequence analysis has shown that the spacers surrounding the 23S rRNA gene are complementary, so they can form an extended hairpin with the 23S rRNA region at the top. The observed cleavage sites for RNase III are in the stem, offset by 2 bp. The regions surrounding the 16S rRNA gene can also form a hairpin stem, with a somewhat more complex structure.

universal; even the mouse does things a little differently, and the frog precursor is only 40S, which is quite a bit smaller than 45S. Still, the basic mechanism of rRNA processing, including the order of mature sequences in the precursor, is preserved throughout the eukaryotic kingdom.

SUMMARY Ribosomal RNAs are made in eukaryotes as precursors that must be processed to release the mature rRNAs. The order of RNAs in the precursor is 18S, 5.8S, 28S in all eukaryotes, although the exact sizes of the mature rRNAs vary from one species to another. In human cells, the precursor is 45S, and the processing scheme creates 41S, 32S, and 20S intermediates.

Prokaryotic rRNA Processing

The bacterium *E. coli* has seven *rrn* operons that contain rRNA genes. Figure 16.5a presents an example, *rrnD*, which has three tRNA genes in addition to the three rRNA genes. Transcription of the operon yields a 30S precursor, which must be cut up to release the three rRNAs and three tRNAs.

RNase III is the enzyme that performs at least the initial cleavages that separate the individual large rRNAs. One type of evidence leading to this conclusion is genetic: A mutant with a defective RNase III gene accumulates 30S rRNA precursors. In 1980, Joan Steitz and her colleagues compared the sequences of the spacers between the rRNAs in two different precursors (from the *rrnX* and *rrnD* operons) and found considerable similarity. These sequences revealed complementary sequences flanking both 16S and 23S rRNA regions of the precursors. This complementarity predicts two extended hairpins (Figure 16.5b) involving stems created by base pairing between two spacers, with the rRNA regions looping out in between. The RNase III cleavage sites in this model are in the stems. Another ribonuclease, **RNase E,** is responsible for removing the 5S rRNA from the precursor.

SUMMARY Prokaryotic rRNA precursors contain tRNAs as well as all three rRNAs. The rRNAs are released from their precursors by RNase III and RNase E.

16.2   Transfer RNA Processing

Transfer RNAs are made in all cells as overly long precursors that must be processed by removing RNA at both ends. In the nuclei of eukaryotes, these precursors contain a single tRNA; in prokaryotes, a precursor may contain one or more tRNAs, and sometimes a mixture of rRNAs and tRNAs, as we saw in Figure 16.5. Because the tRNA processing schemes in eukaryotes and prokaryotes are so similar, we will consider them together.

Cutting Apart Polycistronic Precursors

The first step in processing prokaryotic RNAs that contain more than one tRNA is to cut the precursor up into fragments with just one tRNA gene each. This means cutting between tRNAs in precursors that have two or more tRNAs, or cutting between tRNAs and rRNAs in precursors, such as the one in Figure 16.5, that have both tRNAs and rRNAs. The enzyme that performs both these chores seems to be RNase III.

Forming Mature 5′-Ends

After RNase III has cut the tRNA precursor into pieces, the tRNA still contains extra nucleotides at both 5′- and 3′-ends. As such, it resembles the primary transcripts of eukaryotic tRNA genes, which are monocistronic (single-gene) precursors with extended 5′- and 3′-ends. Maturation of the 5′-end of a prokaryotic or eukaryotic tRNA involves a single cut just at the point that will be the 5′-end of the mature tRNA, as shown in Figure 16.6. The enzyme that catalyzes this cleavage is **RNase P.**

RNase P from both bacteria and eukaryotic nuclei is a fascinating enzyme. It contains two subunits, but unlike other dimeric enzymes we have studied, one of these subunits is made of RNA, not protein. In fact, the majority of the enzyme is RNA because the RNA (the **M1 RNA**) has a molecular mass of about 125 kD, and the protein has a mass of only about 14 kD. When Sidney Altman and his colleagues first isolated this enzyme and discovered that it is a ribonucleoprotein, they faced a critical question: Which part has the catalytic activity, the RNA or the protein? The heavy betting at that time was on the protein because all enzymes that had ever been studied were made of protein, not RNA. In fact, early studies on RNase P showed that the enzyme lost all activity when the RNA and protein parts were separated.

Then, in 1982, Thomas Cech and colleagues found autocatalytic activity in a self-splicing intron (Chapter 14). Shortly thereafter, Altman and Norman Pace and their colleagues demonstrated the catalytic activity of the M1 part of RNase P in 1983. As Figure 16.7 illustrates, the trick was magnesium concentration. The early studies had been performed with 5–10 mM Mg^{2+}, under these condi-

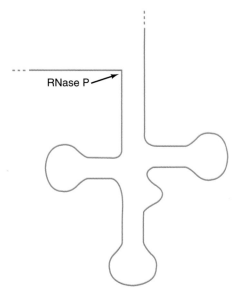

Figure 16.6   RNase P action. RNase P makes a cut at the site that will become the mature 5′-end of a tRNA. Thus, this enzyme is all that is needed to form mature 5′-ends.

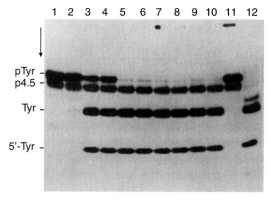

Figure 16.7   The M1 RNA of *E. coli* RNase P has enzymatic activity. Altman and Pace and colleagues purified the M1 RNA from RNase P and incubated it with ^{32}P-labeled pre-tRNA^{Tyr} (pTyr) and p4.5S RNA from *E. coli* (p4.5) for 15 min at the Mg^{2+} and NH$_4$Cl concentrations indicated at top. Then they electrophoresed the RNAs and visualized them by autoradiography. Lane 11, no additions; lane 12, crude *E. coli* RNase P. At the higher Mg^{2+} concentrations, the M1 RNA by itself cleaved the pTyr to form mature 5′-ends, but had no effect on the p4.5 substrate under any of the conditions used. (*Source:* Guerrier-Takada et al., The RNA moiety of ribonuclease P is the catalytic subunit of the enzyme. *Cell* 35 (Dec 1983) p. 851, f. 4A. Reprinted by permission of Elsevier Science.)

tions, both the protein and RNA parts of RNase P are required for activity. Figure 16.7 shows the effect of Mg^{2+} concentration over the range 5 mM to 50 mM using M1 RNA alone. Altman, Pace, and colleagues used two different substrates: pre-tRNA^{Tyr} and pre-4.5S RNA from *E. coli*. Figure 16.7, lanes 1–3 show the difference among 5, 10 and 20 mM Mg^{2+}, respectively. At 5 mM Mg^{2+}

neither substrate showed any maturation by cleavage of the extra nucleotides from the 5′-end. Even at 10 mM Mg^{2+}, the cleavage of pre-tRNA was barely detectable. By contrast, at 20 mM Mg^+, approximately half the pre-tRNA was cleaved to mature form, releasing the extra nucleotides as a single fragment, labeled "5′-Tyr" in the figure. Increasing the Mg^{2+} concentration to 30, 40, and 50 mM Mg^{2+} (lanes 5, 7, and 9, respectively) further enhanced 5′-processing of the pre-tRNA, but did not cause any pre-4.5S processing. Lane 12 demonstrates that crude RNase P (the dimeric form of the enzyme that contains both the RNA and protein subunits) can cleave both substrates at 10 mM Mg^{2+}.

Eukaryotic nuclear RNase P is very much like the prokaryotic enzyme. For example, the yeast nuclear RNase P contains a protein and an RNA part, and the RNA has the catalytic activity. However, Peter Gegenheimer and his colleagues have found that spinach chloroplast RNase P appears not to have an RNA at all. This enzyme is not inhibited by micrococcal nuclease, as it should be if it contains a catalytic RNA, and it has the density expected of pure protein, not a ribonucleoprotein that is mostly RNA. It is still conceivable that the chloroplast RNase P harbors a very small RNA, but all efforts to detect such an RNA have failed.

SUMMARY Extra nucleotides are removed from the 5′-ends of pre-tRNAs in one step by an endonucleolytic cleavage catalyzed by RNase P. RNase Ps from bacteria and eukaryotic nuclei have a catalytic RNA subunit called M1 RNA. Spinach chloroplast RNase P appears to lack an RNA subunit.

Forming Mature 3′-Ends

Transfer RNA 3′-end maturation is considerably more complex than 5′-maturation because not one, but six RNases take part. Murray Deutscher and other investigators have shown that the following RNases can remove nucleotides from the 3′-ends of tRNAs in vitro: **RNase D, RNase BN, RNase T, RNase PH, RNase II**, and **polynucleotide phosphorylase (PNPase)**. Genetic experiments by Deutscher and colleagues have also demonstrated that each of these enzymes is necessary for the most efficient 3′-end processing. If the genes encoding any of these enzymes were inactivated, the efficiency of tRNA processing suffered. Inactivation of all of the genes at once was lethal to bacterial cells. On the other hand, the presence of any one of the enzymes was sufficient to ensure viability and tRNA maturation, although the efficiency varied depending on the active RNase.

Zhongwei Li and Deutscher assayed 3′-end maturation activity in vitro using labeled tRNA substrates and ex-

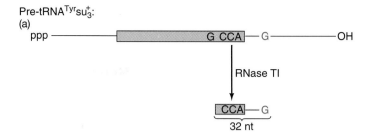

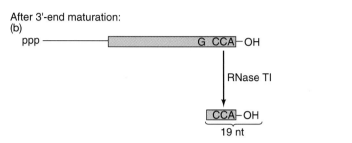

Figure 16.8   Substrate for in vitro assay of tRNA 3′-end maturation. (a) Digestion of the precursor for tRNA^{Tyr} su$_3^+$ by RNase T1 yields a 32-nt oligonucleotide that bridges the CCA that will appear at the mature 3′-end. **(b)** Digestion of the mature tRNA^{Tyr} su$_3^+$ with RNase T1 produces a 19-nt oligonucleotide that ends at the mature 3′-end of the tRNA.

tracts from wild-type or RNase-deficient cells. One substrate (Figure 16.8) was a precursor to a tRNA called tRNA^{Tyr}su$^+_3$. This is a so-called suppressor tRNA that can insert tyrosine at a stop codon and therefore suppress termination of translation. We will have more to say about such tRNAs in Chapter 18. For the purposes of this experiment, we can treat it just like any other tRNA. This tRNA precursor has the advantage of two G's flanking the mature 3′-end, these G's are 32 nt apart and no other G's occur in between. This means that cleavage of the precursor with RNase T1, which cuts only after G's, will yield a unique oligonucleotide 32 nt long. After correct processing of the 3′-end, the CCA is the new 3′-terminus, and RNase T1 digestion will yield a new 19-nt fragment instead of the 32 nt fragment. (Note that RNase T1 is different from RNase T.)

Figure 16.9 illustrates the results of in vitro processing of the tRNA^{Tyr}su$^+_3$ substrate with wild-type and RNase-deficient mutant extracts. Lanes 2–5 in panel (a) show that the wild-type extract can form mature 3′-ends and even cut the terminal A off the mature 3′-end. Lane 5 shows that ATP provides the AMP to replace this terminal A. By contrast, panel (b) shows the same experiment with a mutant lacking RNases I, II, D, BN, and T. The only RNases remaining in this mutant that can cleave tRNA precursors were RNase PH and PNPase, and these both depend on inorganic phosphate for activity. Thus, Li and Deutscher could minimize their activity by dialyzing the mutant extract to remove phosphate. Lane 2 shows

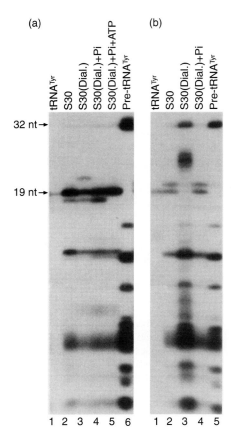

Figure 16.9   Assay for maturation of tRNA^TyrsuX. (a) Assay with wild-type extract. Li and Deutscher incubated a ^{32}P-labeled tRNA^TyrsuX with a wild-type *E. coli* extract for 30 min, then boiled the mixture and reacted it with RNase T1 for 1 h. Then they electrophoresed the reaction products and visualized them by autoradiography. Lanes 1 and 6 are markers with tRNA^TyrsuX and pre-tRNA^TyrsuX, respectively. Lane 2 represents an incubation with an S30 extract (a supernatant from an ultracentrifugation at 30,000 × g for 30 min) with no other treatment. Lane 3 used an extract dialyzed to remove inorganic phosphate. Lane 4 used a dialyzed extract with phosphate added back. Lane 5 was like lane 4, with ATP added, so the terminal A of the tRNA could be added back. All reactions could produce the mature tRNA 3'-end. (b) Assay with mutant extract. All conditions were as in panel (a) except that the S30 extract came from a mutant lacking RNases-I, II, D, BN, and T. No lane was added ATP, so the marker with pre-tRNA^TyrsuX was in lane 5. (*Source:* Li and Deutscher, The role of individual exoribonucleases in processing at the 3' end of *Escherichia coli* tRNA precursors. *J. Biol. Chem.* 269 (1994) pp. 6066–7, figs. 2, 3, 4. American Society for Biochemistry and Molecular Biology.)

that some mature 3'-ends still appeared when the undialyzed extract was used. However, lane 3, which used a dialyzed extract, showed no mature 3'-ends. Lane 4 demonstrates that phosphate was the key ingredient that was missing in the dialyzed extract. When the investigators added it back, they restored correct processing.

Genetic studies had already suggested that RNase T and RNase PH were important in correct maturation of tRNA 3'-ends. To check this conclusion, Li and Deutscher performed in vitro processing studies (like those described in Figure 16.9) with the tRNA^TyrsuX substrate and mutant

extracts lacking one or both of these enzymes. An extract from the RNase PH mutant tended to stop 2 bases short of the mature 3'-end. Thus RNase PH seems to be involved in removing the last two extra bases from the 3'-end. A mutant extract lacking RNase T activity, but having RNase PH activity, yielded some mature 3'-ends, as well as 3'-ends with one extra base. Thus, RNase PH, in the absence of RNase T, can readily cut the first of the two extra bases off, but has more difficulty removing the last extra base. Removal of the last base appears to be the main function of RNase T. Extracts from double mutants in RNases PH and T formed almost no mature 3'-ends and left a preponderance of tRNAs with two extra bases. This reinforces the conclusion that these two enzymes are important partners in final 3'-end maturation. By contrast, a mutant extract lacking RNases D, II, and BN was just as capable as wild-type extracts of forming mature 3'-ends. Thus, in contrast to RNases PH and T, these three RNases are relatively unimportant in final 3'-end maturation.

If RNases PH and T remove the final two nucleotides, what enzyme(s) remove the other 75 extra nucleotides at the 3'-end of the pre-tRNA^TyrsuX? Candidates include an endonuclease, or a 3'-exonuclease such as RNAse II or PNPase. To test the importance of these two enzymes, Li and Deutscher performed the 3'-processing reaction with an extract from RNase II, PNPase double mutants. Mature 3'-ends could be made with all single-mutant extracts, at least in the presence of phosphate, but no mature, and very little near-mature, 3'-ends were formed with the double mutant extract at high temperature. Thus it seems that RNase II and PNPase cooperate to remove the bulk of the 3'-trailer from pre-tRNA. This opens the way for RNases PH and T to complete the job (Figure 16.10).

Li and Deutscher followed up these studies with three other types of experiments. When they complemented mutant extracts with the missing enzymes, normal activity was restored. They also reconstituted processing activity from purified RNases and showed that leaving out one or more RNases had essentially the same effect as mutations in the genes for those enzymes. Finally, they performed the same processing assays with another substrate, pre-tRNA^Arg, and obtained results that were qualitatively similar to those with pre-tRNA^TyrsuX although some quantitative differences did occur.

SUMMARY RNase II and polynucleotide phosphorylase cooperate to remove most of the extra nucleotides at the end of a tRNA precursor, but stop at the +2 stage, with two extra nucleotides remaining. RNases PH and T are most active in removing the last two nucleotides from the RNA, with RNase T being the major participant in removing the very last nucleotide.

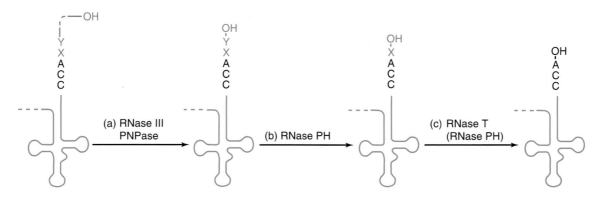

Figure 16.10   Model for processing the 3′-end of an *E. coli* tRNA. (a) RNase III and PNPase cooperate to remove all but two of the extra bases (red) at the 3′-end of the tRNA precursor. (b) RNase PH removes the next-to-last extra base (Y, red). (c) RNase T, perhaps with help from RNase PH removes the last extra base (X, red), leaving a mature tRNA.

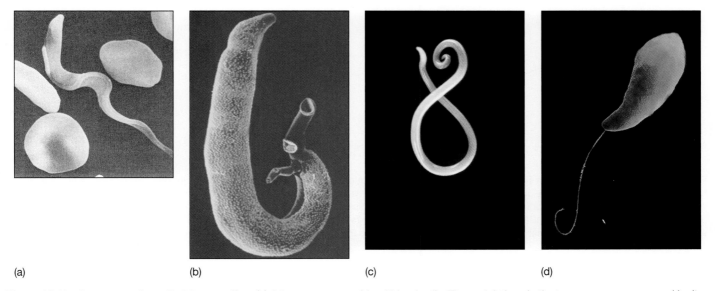

(a)   (b)   (c)   (d)

Figure 16.11   Some organisms that *trans*-splice. (a) A trypanosome amid red blood cells. The undulations in the trypanosome are caused by its single flagellum, which extends down one side of the cell. **(b)** *Schistosoma mansoni,* a parasitic worm. **(c)** *Ascaris lumbricoides,* another parasitic worm. **(d)** *Euglena,* a free-living protist. (*Source:* (a) Donelson and Turner, How the trypanosome changes its coat. *Scientific American* 45 (Feb 1985). (b) © G. Shih-R. Kessel/Visuals Unlimited. (c) © Arthur M. Siegelman/Visuals Unlimited. (d) © David M. Phillips/Visuals Unlimited.)

16.3   *Trans*-Splicing

In Chapter 14 we considered the sort of splicing that occurs in almost all eukaryotic species. This splicing can be called *cis*-splicing, because it involves two or more exons that exist together in the same gene. As unlikely as it may seem, in another alternative, *trans*-splicing, the exons are not part of the same gene at all and may not even be found on the same chromosome.

The Mechanism of *Trans*-Splicing

Trans-splicing occurs in several oganisms (Figure 16.11), including parasitic and free-living worms (e.g., *Caenorhabditis elegans*), but it was first discovered in **trypanosomes,** a group of parasitic flagellated protozoa, one

species of which causes African sleeping sickness. The genes of trypanosomes are expressed in a manner we would never have predicted based on what we have discussed in this book so far. Piet Borst and his colleagues laid the groundwork for these surprising discoveries in 1982 when they sequenced the 5′-end of an mRNA encoding a trypanosome surface coat protein and the 5′-end of the gene that encodes this same protein and discovered that they did not match. The mRNA had 35 extra nucleotides that were missing from the gene. As molecular biologists sequenced more and more trypanosome mRNAs, they discovered that they all had the same 35-nt leader, called the **spliced leader (SL),** but none of the genes encoded the SL. Instead, the SL is encoded by a gene that is repeated about 200 times in the trypanosome genome. This gene encodes only the SL, plus a 100-nt

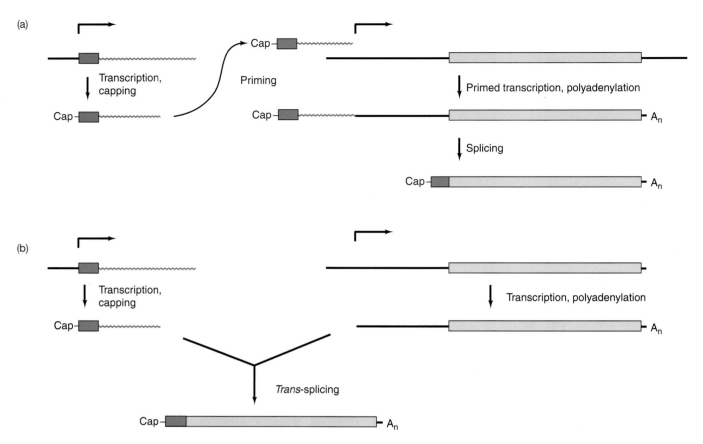

Figure 16.12   Two hypotheses for joining the SL to the coding region of an mRNA. (**a**) Priming by the SL intron. The SL (blue), with its attached half-intron (red), is transcribed to yield a 135-nt RNA. This RNA then serves as a primer for transcription of a coding region (yellow), including its attached half-intron (black). This produces a transcript including the SL plus the coding region, with a whole intron in between. The intron can then be spliced out to yield the mature mRNA. (**b**) *Trans*-splicing. The SL with its attached half-intron is transcribed; independently, the coding region with its half-intron is transcribed. Then these two separate RNAs undergo *trans*-splicing to produce the mature mRNA.

sequence that is joined to the leader through a consensus 5′-splice sequence. Thus, this minigene is composed of a short SL exon, followed by what looks like the 5′-part of an intron.

How can we explain the production of an mRNA derived from two widely separated DNA regions that are sometimes even found on separate chromosomes? Two classes of explanations are plausible. First (Figure 16.12a), the SL (with or without its intron) could be transcribed, and this transcript could then serve as a primer for transcription of any one of the coding regions elsewhere in the genome. Alternatively (Figure 16.12b), RNA polymerases could transcribe an SL and a coding region separately, and these two independent transcripts could then be spliced together.

If such *trans*-splicing really occurs, then we would not expect to see lariat-shaped intermediates. Instead, we should find Y-shaped intermediates that form when the branchpoint in the intron attacks the 5′-end of the intron attached to the short leader exon, as illustrated in Figure 16.13. Finding the Y-shaped intermediate would go a long way toward proving that *trans*-splicing really takes place.

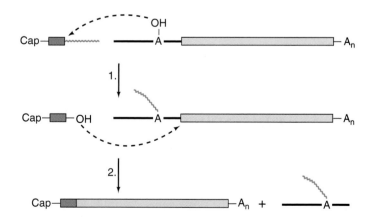

Figure 16.13   Detailed *trans*-splicing scheme for a trypanosome mRNA. Step 1: The branchpoint adenosine within the half-intron (black) attached to the coding exon (yellow) attacks the junction between the leader exon (blue) and its half-intron (red). This creates a Y-shaped intron–exon intermediate analogous to the lariat intermediate created by *cis*-splicing. Step 2: The leader exon attacks the splice site between the branched intron and the coding exon. This produces the spliced, mature mRNA plus the Y-shaped intron.

Nina Agabian and colleagues reported evidence for the intermediate in 1986.

First, they demonstrated that the SL intron becomes attached to poly(A)⁺ RNA, as it should in the Y-shaped intermediate (or in a lariat-shaped intermediate). To do this, they used a primer that hybridizes to the SL intron 25 nt from the splice junction and used it to sequence poly(A)⁺ RNA. As Figure 16.14 shows, a strong stop occurred in all four lanes where the reverse transcriptase reached the branchpoint and stopped after adding the last nucleotide. The enzyme can go no farther when it encounters such a branch. The sequence up to that point matched the sequence of the SL intron. Thus, the SL intron is associated with an RNA that is polyadenylated, which must be the main body of an mRNA. The other strong stop 12 nt earlier appears to be a sequencing artifact, as the sequence continued just as strongly after the stop.

The association of the SL intron with an mRNA coding region is compatible with either *cis-* or *trans-*splicing, so it does not rule out either of the hypotheses in Figure 16.13. However, the unique feature of the Y-shaped structure, which distinguishes it from a normal, lariat intermediate, is that the 3′-end of the SL intron in the Y-shaped structure is free (see Figure 16.13). This means that treatment of the Y-shaped *splicing* intermediate with debranching enzyme, which breaks the 2′–5′ phosphodiester bond at the branchpoint, should yield a 100-nt fragment as a by-product (Figure 16.15). This contrasts with the results we expect from a lariat-shaped intermediate, which would simply be linearized. Figure 16.16 shows the results of gel electrophoresis of labeled total RNA and poly(A)⁺ RNA after treatment with debranching enzyme. In both cases, the expected 100-nt fragment appeared, thus corroborating the *trans-*splicing hypothesis.

> **SUMMARY** Trypanosome mRNAs are formed by *trans*-splicing between a short leader exon and any one of many independent coding exons.

Polycistronic Arrangement of Coding Regions in Trypanosomes

In considering the expression of trypanosome genes, we have been tacitly assuming that their coding regions are transcribed as other eukaryotic genes are—individually, one at a time. However, this is one more way in which trypanosomes show us something unexpected. Borst and colleagues discovered in 1987 that the coding regions in trypanosomes are located in long, polycistronic transcription units, so many can be transcribed together as part of a long precursor RNA. *Trans*-splicing then joins each of these coding regions to the SL.

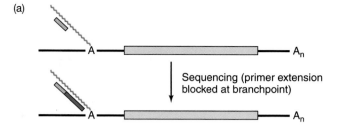

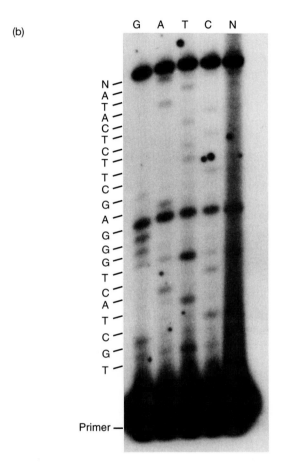

Figure 16.14 Demonstration that the SL half-intron is associated with poly(A)⁺ RNA. (a) Experimental plan. Agabian and colleagues isolated polyadenylated RNA from trypanosomes and subjected it to sequencing using reverse transcriptase and a primer (orange) that hybridizes to the 100-nt half-intron (red) originally attached to the SL. Sequencing extended the primer, but the maximum extension could reach only to the branchpoint, where it had to stop. (b) Experimental results. The four sequencing lanes are shown, along with a lane including no dideoxy nucleotides (N). The sequence, which corresponds to the 25 nt in the half-intron between the primer binding site and the branchpoint, is shown at left. The strong stop in all lanes, labeled N at left, corresponds to the G attached to the branchpoint A. That is where the reverse transcriptase had to stop. Thus, the SL half-intron is linked to a polyadenylated mRNA-coding region. (*Source:* (b) Murphy et al., Identification of a novel Y branch structure as an intermediate in trypanosome mRNA processing: Evidence for *Trans* splicing. *Cell* 47 (21 Nov 1986) p. 520, f. 4. Reprinted by permission of Elsevier Science.)

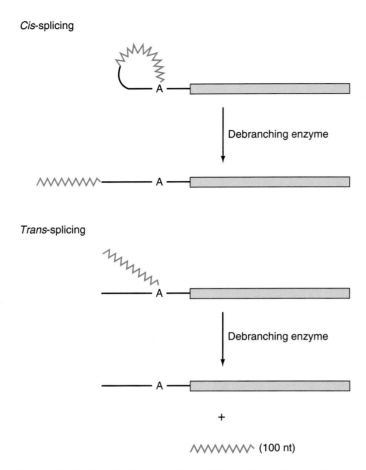

Figure 16.15 Treating hypothetical splicing intermediates with debranching enzyme. (a) *Cis*-splicing. The debranching enzyme simply opens the lariat up to a linear form. **(b)** *Trans*-splicing. Because the 100-nt half-intron (red) is open at its 3′-end instead of being involved in a lariat, debranching enzyme releases it as an independent RNA.

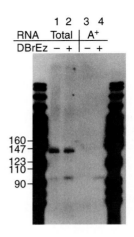

Figure 16.16 Release of the SL half-intron from a larger RNA by debranching enzyme. Agabian and colleagues labeled trypanosome RNA with ^{32}P and treated total RNA, or poly(A)$^+$ RNA, with debranching enzyme (DBrEz) as indicated at top. Then they electrophoresed the products and visualized them by autoradiography. The 100-nt SL half-intron is clearly detectable in both enzyme-treated RNA samples. (Source: Murphy et al., *Cell* 47 (21 Nov 1986) p. 521, f. 5. Reprinted by permission of Elsevier Science.)

What is the evidence for this polycistronic arrangement of coding regions? Borst and coworkers focused on a region of the genome containing the coding region for a surface coat protein called VSG 221, as well as at least eight other coding regions. **VSG** stands for **variable surface glycoprotein.** (Trypanosomes evade their host's immune defenses by periodically changing their surface coat protein, just as immunity to the old coat threatens their survival. That is why we call the surface proteins "variable.") Borst and colleagues collected nuclei from trypanosomes in the bloodstream of rats and allowed them to extend and label preinitiated RNAs by run-on transcription. Then they hybridized this labeled trypanosome RNA to DNA fragments representing the VSG 221-coding region and 57 kb of DNA upstream. The run-on RNA hybridized to all DNA fragments over this 57 kb region about equally, suggesting that the whole region is transcribed as a unit.

This run-on experiment left open the possibility that the 57-kb region is transcribed in small sections, rather than as one long unit. To clarify this point, Borst and colleagues used a technique called **inactivation of transcription by ultraviolet irradiation.** They began by irradiating trypanosomes from rat blood with increasing doses of UV radiation to introduce pyrimidine dimers (chapter 20) into the parasite's DNA. Then they allowed at least 1.5 h for any RNA polymerases still transcribing the damaged DNA to terminate. Then they isolated nuclei from the irradiated cells and allowed RNA synthesis to occur in the presence of a labeled nucleotide. Because RNA polymerase has difficulty traversing pyrimidine dimers, the efficiency of transcription will drop off noticeably as the polymerase transcribes farther and farther downstream of the promoter. Thus, transcription of the 3′-end of a long transcription unit will be much more sensitive to UV irradiation than transcription of the 5′-end. This can be detected by dot-blotting or Southern blotting DNA fragments along the length of the transcription unit and hybridizing them to the labeled transcripts of the UV-irradiated DNA.

Borst and colleagues performed this analysis on the 57-kb transcription unit that has the VSG 121 gene at its 3′-end. Figure 16.17a is a physical map of this transcription unit, showing the locations of cloned DNA fragments used as probes. Figure 16.17b shows the results of the inactivation study. The farther downstream of the promoter, the more transcription was inactivated by UV irradiation. This is strong evidence for the existence of a long transcription unit governed by a single promoter.

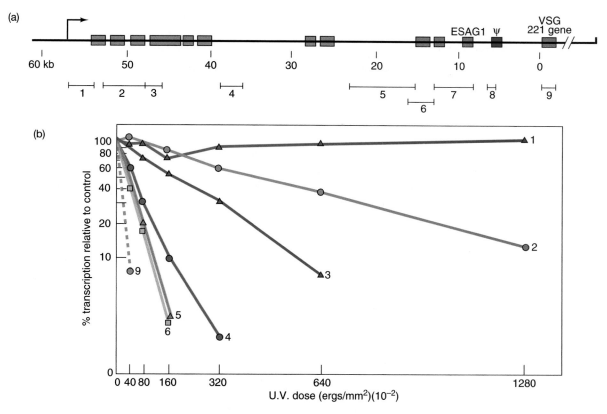

Figure 16.17 **Demonstration of a polycistronic transcription unit in a trypanosome.** (a) Physical map of the transcription unit. Red boxes represent open reading frames (ORFs). The green box represents a pseudo-gene (ψ). Blue boxes encode stable RNAs. The locations of the cloned DNA fragments used in the ultra (UV) inactivation analysis are indicated by number at bottom. (b) Ultraviolet inactivation analysis. The relative transcription rates of RNAs hybridizing to the cloned fragments from panel (a) are plotted as a function of UV dose. Transcription of regions farthest from the promoter (e.g., clones 6 and 9) is most sensitive to irradiation. In fact, the line corresponding to clone 9 is dashed because no transcription was detected even at the lowest UV dose. (*Source:* Reprinted from Johnson et al., *Cell* 51:275 & 277, 1987. Copyright 1987, with permission from Elsevier Science.)

SUMMARY Trypanosome coding regions, including genes encoding rRNAs and tRNAs, are arranged in long, polycistronic transcription units governed by a single promoter.

16.4 RNA Editing

Trans-splicing is not the only bizarre occurrence in trypanosomatids. These organisms also have unusual mitochondria called **kinetoplasts,** which contain two types of circular DNA linked together into large networks (Figure 16.18). There are 25–50 identical **maxicircles,** 20–40 kb in size, which contain the mitochondrial genes, and about 10,000 1–3 kb-**minicircles,** which have a role in mitochondrial gene expression. In 1986, Rob Benne and his colleagues discovered that the sequence of the cytochrome oxidase (COII) mRNA from trypanosomes does not match the sequence of the COII gene; the mRNA contains four nucleotides that are missing from the gene (Figure 16.19). Furthermore, these missing nucleotides cause a

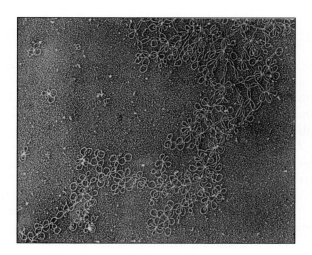

Figure 16.18 **Part of the network of kinetoplast minicircles and maxicircles from *Leishmania tarentolae.*** (*Source: Cell* 61 (1 June 1990) cover (acc. Sturm & Simpson, pp. 871-884). Reprinted by permission of Elsevier Science.)

frameshift (a shift in the frame in which a ribosome reads the mRNA; see Chapter 18) that should seemingly inactivate the gene. But somehow the mRNA has been supplied with these four nucleotides, averting the frameshift.

```
COX II DNA: •••GTATAAAAGTAGA   G A ACCTGG•••

COX II RNA: •••GUAUAAAAGUAGAUUGUAUACCUGG•••
```

Figure 16.19 Comparison of the sequence of part of the COII gene of a trypanosome with its mRNA product. Four U's in the mRNA are not represented by T's in the gene. These four U's are presumably added to the RNA by editing.

Of course, one possibility is that the gene Benne and colleagues sequenced did not actually code for the mRNA, but was a **pseudo-gene,** a duplicate copy of a gene that has been mutated so it does not function and is no longer used. The active gene could reside elsewhere, and these workers could have missed it. The problem with this explanation is that, try as they might, Benne and his coworkers could find no other COII gene in either the kinetoplast or the nucleus. Furthermore, they found the same missing nucleotides in the COII genes of two other trypanosomatids. For these and other reasons, Benne and coworkers concluded that the mRNAs of trypanosomatids are copied from incomplete genes called **cryptogenes** and then edited by adding the missing nucleotides, which are all UMPs.

By 1988, a number of trypanosomatid kinetoplast genes and corresponding mRNAs had been sequenced, revealing editing as a common phenomenon in these organisms. In fact, some RNAs are very extensively edited (**panedited**). For example, a 731-nt stretch of the COIII mRNA of *Trypanosoma brucei* contains 407 UMPs added by editing; editing also deletes 19 encoded UMPs from this stretch of the COIII mRNA. Part of this sequence is presented in Figure 16.20.

```
•••uAuAuGuuuuGuuGuuuAuuAuGuGAuuAuGGuuuuGuuuuuuA
    T
uuGGUAuuuuuuAGAuuuAuuuAAuuuGuuGAuAAAuACAuuuu

AUUUGuuUGuuAGuGGuuuAuuuGuuAAuuuuuuuGuuuuGuGU
                          TT
UUUUGGuuuAGGuuuuuuuGuuGUUGuuGuuuuGuAuuAuGAuu
              TTTT
GAGuuuGuuGuuuGGuuuuuuGuuuuuGuGAAACCAGuuAUGAG
 TT                                  TTTT
AGUUUGCAuuGuuAuuuAuuACAuuAAGuuG GGUGuuuuuGGu
                                   TT
uCuAuuuuAuuuuuAuuGGAuuuAuUACAuuuuAUGCAuGuuuu

uuuAGGuGuuuuGuuGuuGuuuAuuuGuuuuAGCGuuuGuuuA

AuuuuuuGuGuAuGGAuACACGuuuuGuuuuuuuGuAuuGuGuu

uGuuuAuAuuGACAuuuuGuuGAUUUAGuuuGAuuuuuuuuAuu

GCGAuuuGuuuAuuuuGAuGuuuuAuGGuuAuGu  uuuGuGu
                                      T
GuGuAAuuuuAuuGGuGuuuuUUUAGUUGuuGAAGuuA•••
```

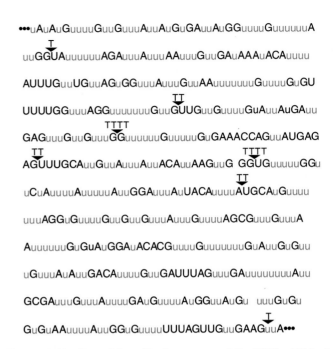

Figure 16.20 Part of the edited sequence of the COIII mRNA of *T. brucei*. The U's added by editing are shown in red lowercase; the T's present in the gene, but absent (as U's) in the mRNA are shown in black above the sequence. (*Source:* Reprinted from *Cell* 53:cover, 1988. Copyright 1988, with permission from Elsevier Science.)

Mechanism of Editing

We have been assuming that editing is a posttranscriptional event. This seems like a good bet because unedited transcripts can be found along with edited versions of the same mRNAs. Moreover, editing occurs in the poly(A) tails of mRNAs, which are added posttranscriptionally.

One important clue about the mechanism of editing is that partially edited transcripts have been isolated, and these are always edited at their 3′-ends but not at their 5′-ends. This suggests strongly that editing proceeds in a 3′→5′ direction. Kenneth Stuart and colleagues first reported this phenomenon in 1988. Their experimental tool was RT-PCR, starting with reverse transcriptase to make the first DNA strand from an RNA template, followed by standard PCR (see Chapter 4).

In one experiment, Stuart and coworkers used pairs of PCR primers in which both were edited primers, both unedited primers, or one of each. A completely edited RNA will hybridize only to edited primers and give a PCR signal, whereas it will not hybridize to unedited primers, so any PCR protocol including at least one unedited primer will not give a signal from this RNA. By contrast, a completely unedited RNA will react only with unedited primers. But the real test is to use an unedited 5′-primer and an edited 3′-primer to detect 3′-edited transcripts, or an edited 5′-primer and an unedited 3′-primer to detect 5′-edited transcripts. If editing goes from 3′ to 5′ in the transcript, then 3′-edited transcripts, but not 5′-edited transcripts, should be detected. The advantage of the PCR method is that it amplifies very small amounts of RNA, such as partially edited RNAs, to easily detectable bands of DNA. Figure 16.21 depicts the results of this analysis. Lanes 1–4 show the PCR products of *Trypanosoma brucei* kinetoplast RNA with different combinations of primers. We see signals only when both primers were edited, or the 3′-primer was edited. We see no signal when only the 5′-primer was edited. Thus, 3′-editing occurred in the absence of 5′-editing, but 5′-editing did not occur without 3′-editing. This is consistent with editing in the 3′→5′ direction. Lanes 5–6 and 7–10 are positive and negative controls, respectively.

What determines where the editing system should add or delete UMPs? Larry Simpson and colleagues found the answer in 1990 when they discovered **guide RNAs (gRNAs)** encoded in *Leishmania* maxicircles. They began with a computer search of the 21-kb part of the maxicircle DNA sequence that was known at that time. This search revealed seven short sequences that could produce short RNAs (gRNAs) complementary to parts of five different edited mitochondrial mRNAs. In principle, such

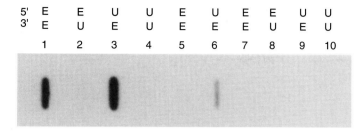

Figure 16.21 PCR analysis of direction of editing. Stuart and colleagues performed RT-PCR with kinetoplast RNA and edited (E) or unedited (U) 5′- and 3′-primers, as indicated at top. Then they slot-blotted the PCR products and hybridized them to a labeled probe and detected hybridization by autoradiography. PCR templates: lanes 1–4, RNA from wild-type cells; lanes 5–6, a 3′-edited cDNA (positive control); lanes 7–10, RNA from a mutant that lacks mitochondrial DNA (negative control). (*Source:* Abraham et al., Characterization of cytochrome *c* oxidase III transcripts that are edited only in the 3′ region. *Cell* 55 (21 Oct 1988) p. 269, f. 2a. Reprinted by permission of Elsevier Science.)

gRNAs could direct the insertion and deletion of UMPs over a stretch of several dozen nucleotides in the mRNA, as illustrated in Figure 16.22a and b. Once that editing is done, another gRNA could hybridize near the 5′-end of the newly edited region and direct editing of a new segment, as Figure 16.22c and d demonstrate. Working in this way from the 3′-end of the mRNA toward the 5′-end, successive gRNAs bind to regions edited by their predecessor gRNAs and direct further editing until they have finished the whole editing job.

One notable feature of the base pairing between gRNAs and mRNA is the existence of G–U base pairs, as well as standard Watson–Crick base pairs. In Chapter 18 we will learn that G–U base pairs are also common during codon–anticodon pairing in translation, and one of the two bases can accommodate these nonstandard base pairs by wobbling slightly from the position it would occupy in Watson–Crick base pairs. The importance of these G–U base pairs in editing probably derives from the fact that they are weaker than Watson–Crick base pairs. This means that the 5′-end of a new gRNA, by forming Watson–Crick base pairs with the newly edited region of an mRNA, can displace the 3′-end of an old gRNA, whose base pairing with the mRNA includes weak G–U pairs (Figure 16.23).

Later in 1990, Nancy Sturm and Simpson found that minicircles also encode gRNAs. But besides the coding potential, Simpson and colleagues found direct evidence for the existence of gRNAs. They electrophoresed kinetoplastid RNA, Northern blotted it, and hybridized it to labeled oligonucleotide probes designed to detect gRNAs, according to the sequences of putative gRNA genes in maxicircles. Figure 16.24 shows that this procedure detected small RNAs, most of which appeared to be shorter than 80 nt.

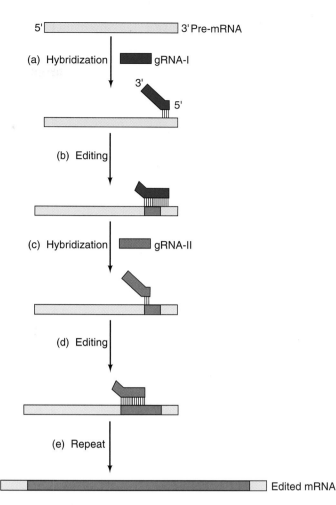

Figure 16.22 Model for the role of gRNAs in editing. (**a**) In the first step, gRNA-I (dark blue) hybridizes through its 5′-end to a region of the pre-mRNA that requires no editing. (**b**) Most of the rest of gRNA-I directs editing of part of the pre-mRNA. The edited portion is shown in red, and the pre-mRNA has grown in length, due to the inserted UMPs. (**c**) A new gRNA, gRNA-II (light blue), displaces gRNA-I by hybridizing to the 5′-end of the newly edited region of the pre-mRNA. (**d**) gRNA-II directs editing of a new part of the pre-mRNA. (**e**) The previous steps are repeated with additional gRNAs until the RNA is completely edited.

The precise mechanism of editing, the cutting and pasting required to insert and delete UMPs, remained unclear for several years, but the enzyme activities found in kinetoplasts provided some hints. For example, kinetoplasts have a **terminal uridylyl transferase (TUTase)** that could add extra UMPs (uridylates) to the mRNA during editing. Because the mRNA has to be cut to accept these new UMPs, it must also be ligated together again, and kinetoplasts also contain an **RNA ligase.** The major remaining question concerned the source of uridylates for editing. UTP could provide them. On the other hand, uridylates at the ends of gRNAs could be transferred to the pre-mRNA by transesterification. That is, the uridylates could be plucked off of the ends of gRNAs and transferred directly to the pre-mRNA.

Then, in 1994, Scott Seiwert and Stuart used a mitochondrial extract and a gRNA to edit a synthetic pre-mRNA. They found that deletion of UMPs required three enzymatic activities (Figure 16.25a): (1) an endonuclease that follows directions from the gRNA and cuts the pre-mRNA at the site where a UMP needs to be removed; (2) a 3'-exonuclease that is specific for terminal uridines; and (3) an RNA ligase. In 1996, using a similar in vitro system, Stuart and colleagues demonstrated that UMP insertion follows a similar three-step pathway (Figure 16.25b):

(1) a gRNA-directed endonuclease cuts at the site where UMP insertion is required; (2) an enzyme (probably TU-Tase) transfers UMPs from UTP, as directed by the gRNA; and (3) an RNA ligase puts the two pieces of RNA back together. Figure 16.25c shows the formation

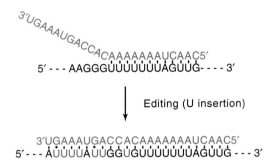

Figure 16.23 Editing of part of a hypothetical RNA. The gRNA (blue) binds via Watson–Crick base pairs to an edited portion of a pre-mRNA. The 3'-end of the gRNA then serves as the template for insertion of U residues (red). Most of the base pairs between newly inserted U's and the gRNA are Watson–Crick A–U pairs, but some are wobble G–U pairs, denoted by dots.

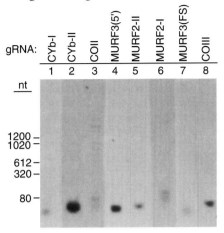

Figure 16.24 Evidence for gRNAs. Simpson and colleagues Northern blotted RNA from the mitochondria of *Leishmania tarentolae* and probed the blots with labeled oligonucleotides that would hybridize to gRNAs. The gRNAs are identified at top. (*Source:* Blum et al., A model for RNA editing in kinetoplastid mitochondria: "Guide" RNA molecules transcribed from maxicircle DNA provide the edited information. *Cell* 60 (26 Jan 1990) p. 191, f. 3a. Reprinted by permission of Elsevier Science.)

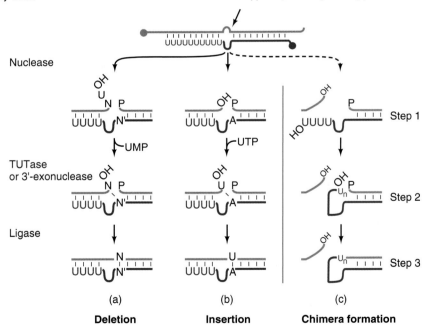

Figure 16.25 Mechanism of RNA editing. The mechanisms of (a) U deletion, (b) U insertion, and (c) chimera formation are shown, starting with a hybrid between a pre-mRNA (magenta) and a gRNA (dark blue) at top. The arrow indicates the position at which the nuclease cuts the pre-mRNA for editing. (a) U deletion. Step 1: A nuclease clips the pre-mRNA just to the 3'-side of the U to be deleted. Step 2: An exonuclease removes the UMP at the end of the left-hand RNA fragment. Base pairing occurs between base N in the pre-mRNA and base N' in the gRNA. Step 3: RNA ligase puts the two halves of the pre-mRNA back together. (b) U insertion. Step 1: A nuclease clips the pre-mRNA at the site where the gRNA dictates that a U should be inserted. Step 2: TUTase transfers a UMP from UTP to the 3'-end of the left-hand RNA fragment. This U base-pairs with an A in the gRNA. Step 3: RNA ligase puts the pre-mRNA back together. (c) Chimera formation. Step 1: A nuclease clips the pre-mRNA, and base pairing between the left-hand pre-mRNA fragment and the gRNA fails. Step 2: The free 3'-end of the gRNA approaches the 5'-end of the right-hand pre-mRNA fragment. Step 3: RNA ligase joins the gRNA to the right-hand pre-mRNA fragment, forming a chimera. This is a side reaction that does not appear to be a significant part of the editing process. (*Source:* From Seiwert, *Science* 274:1637, 1996. Copyright © 1996 American Association for the Advancement of Science, Washington, DC. Reprinted by permission.)

of chimeras between the broken RNA substrate and gRNA. This appears to be a mere side reaction, with no implications for editing itself.

SUMMARY Trypanosomatid mitochondria encode incomplete mRNAs that must be edited before they can be translated. Editing occurs in the 3'→5' direction by successive action of one or more guide RNAs. These gRNAs hybridize to the unedited region of the mRNA and provide A's and G's as templates for the incorporation of U's missing from the mRNA. Sometimes the gRNA is missing an A or G to pair with a U in the mRNA, in which case the U is removed. The mechanism of removing U's involves: (1) cutting the pre-mRNA just beyond the U to be removed; (2) removal of the U by an exonuclease; and (3) ligating the two pieces of pre-mRNA together. The mechanism of adding U's uses the same first and last step, but the middle step (step 2) involves addition of one or more U's from UTP by TUTase instead of removing U's.

16.5 Posttranscriptional Control of Gene Expression

In our discussions of the mechanisms of prokaryotic and eukaryotic transcription, we saw many examples of transcriptional control. It makes sense to control gene expression by blocking the first step—transcription. That is the least wasteful method because the cell expends no energy making an mRNA for a protein that is not needed.

Although transcriptional control is the most prevalent form of control of gene expression, it is by no means the only way. We have already seen in Chapter 15 that poly(A) stabilizes and confers translatability on an mRNA, and special sequences in the 3'-untranslated region of an mRNA, called cytoplasmic polyadenylation elements (CPEs), govern the efficiency of polyadenylation of maternal messages during oocyte maturation. In this way, these CPEs serve as controllers of gene expression.

But an even more important posttranscriptional control of gene expression is control of mRNA stability. In fact, Joe Harford has pointed out that "cellular mRNA levels often correlate more closely with transcript stability than with transcription rate." We will devote the remainder of this chapter to a discussion of this topic.

Casein mRNA Stability

The response of mammary gland tissue to the hormone prolactin provides a good example of control of mRNA stability. When cultured mammary gland tissue is stimulated with prolactin, it responds by producing the milk

Table 16.1 Effect of Prolaction on Half-Life of Casein mRNA

	RNA half-life (h)	
Species of RNA	− Prolactin	+ Prolactin
rRNA	>790	>790
Poly(A)$^+$RNA(short-lived)	3.3	12.8
Poly(A)$^+$RNA(long-lived)	29	39
Casein mRNA	1.1	28.5

Source: Reprinted from Guyette et al., *Cell* 17:1013, 1979. Copyright 1979, with permission from Elsevier Science.

protein casein. One would expect an increase in casein mRNA concentration to accompany this casein buildup, and it does. The number of casein mRNA molecules increases about 20-fold in 24 h following the hormone treatment. But this does not mean the rate of casein mRNA synthesis has increased 20-fold. In fact it only increases about two- to threefold. The rest of the increase in casein mRNA level depends on an approximately 20-fold increase in stability of the casein mRNA.

Pierre Chambon and his colleagues performed a pulse-chase experiment to measure the **half-life** of casein mRNA. The half-life is the time it takes for half the RNA molecules to be degraded. Chambon and colleagues radioactively labeled casein mRNA for a short time in vivo in the presence or absence of prolactin. In other words, they gave the cells a pulse of radioactive nucleotides, which the cells incorporated into their RNAs. Then they transferred the cells to medium lacking radioactivity. This chased the radioactivity out of the RNA, as labeled RNAs broke down and were replaced by unlabeled ones. After various chase times, the experimenters measured the level of labeled casein mRNA by hybridizing it to a cloned casein gene. The faster the labeled casein mRNA disappeared, the shorter its half-life. The conclusion, shown in Table 16.1, was that the half-life of casein mRNA increased dramatically, from 1.1 h to 28.5 h, in the presence of prolactin. At the same time, the half-life of total polyadenylated mRNA increased only 1.3- to 4-fold in response to the hormone. It appears prolactin causes a selective stabilization of casein mRNA that is largely responsible for the enhanced expression of the casein gene.

SUMMARY A common form of posttranscriptional control of gene expression is control of mRNA stability. For example, when mammary gland tissue is stimulated by prolactin, the synthesis of casein protein increases dramatically. However, most of this increase in casein is not due to an increase in the rate of transcription of the casein gene. Instead, it is caused by an increase in the half-life of casein mRNA.

Transferrin Receptor mRNA Stability

One of the best studied examples of posttranscriptional control concerns iron homeostasis (control of iron concentration) in mammalian cells. Iron is an essential mineral for all eukaryotic cells, yet it is toxic in high concentrations. Consequently, cells have to regulate the intracellular iron concentration carefully. Mammalian cells do this by regulating the amounts of two proteins: an iron import protein called the **transferrin receptor,** and an iron storage protein called **ferritin.** Transferrin is an iron-bearing protein that can get into a cell via the transferrin receptor on the cell surface. Once the cell imports transferrin, it passes the iron to cellular proteins, such as cytochromes, that need iron. Alternatively, if too much iron is in the cell, it stores the iron in the form of ferritin. Thus, when a cell needs more iron, it increases the concentration of transferrin receptors to get more iron into the cell and decreases the concentration of ferritin, so not as much iron will be stored and more will be available. On the other hand, if a cell has too much iron, it decreases the concentration of transferrin receptors and increases the concentration of ferritin. It employs posttranscriptional strategies to do both these things: It regulates the rate of translation of ferritin mRNA, and it regulates the stability of the transferrin receptor mRNA. We will deal with the regulation of ferritin mRNA translation in Chapter 17. Here we are concerned with the latter process: controlling the stability of the mRNA encoding the transferrin receptor.

Joe Harford and his colleagues reported in 1986 that depleting intracellular iron by chelation resulted in an increase in transferrin receptor (TfR) mRNA concentration. On the other hand, increasing the intracellular iron concentration by adding hemin or iron salts decreased the TfR mRNA concentration. The changes in TfR mRNA concentrations with fluctuating intracellular iron concentration are not caused primarily by changes in the rate of synthesis of TfR mRNA. Instead, these alterations in TfR mRNA concentration largely depend on changes in the TfR mRNA half-life. In particular, the TfR mRNA half-life increases from about 45 min when iron is plentiful to many hours when iron is in short supply. We will examine the data on mRNA half-life a little later in this chapter. First, however, we need to inspect the structure of the mRNA, which makes possible the modulation in its lifetime.

Iron Response Elements   Lukas Kühn and his colleagues cloned a TfR cDNA in 1985 and found that it encoded an mRNA with a 96-nt 5′-untranslated region (**5′-UTR**), a 2280-nt coding region, and a 2.6-kb 3′-untranslated region (**3′-UTR**). To test the effect of this long 3′-UTR, Dianne Owen and Kühn deleted 2.3 kb of the 3′-UTR and transfected mouse L cells with this shortened construct. They also made similar constructs with the normal TfR

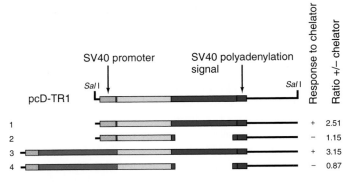

Figure 16.26   Effect of the 3′-UTR on the iron-responsiveness of cell surface concentration of TfR. Owen and Kühn made the TfR gene constructs diagrammed here. The DNA regions within the boxes are color-coded as follows: SV40 promoter, orange; TfR promoter, blue; TfR 5′-UTR, black; TfR-coding region, yellow; TfR 3′-UTR, green; SV40 polyadenylation signal, purple. These workers then transfected cells with each construct and assayed for concentration of TfR on the cell surface, using fluorescent antibodies. The ratio of cell surface TfR in the presence and absence of the iron chelator (desferrioxamine) is given at right, along with a qualitative index of response to chelator (+ or −). (*Source:* From Owen and Kühn, *The EMBO Journal* 6:1288, 1987. Copyright © 1987 Oxford University Press, Oxford, UK. Reprinted by permission.)

promoter replaced by an SV40 viral promoter. Then they used fluorescent antibodies to detect TfR on the cell surfaces. Figure 16.26 summarizes the results. With the wild-type gene, the cells responded to an iron chelator by increasing the surface concentration of TfR about threefold. Owen and Kühn observed the same behavior when the TfR gene was controlled by the SV40 promoter, demonstrating that the TfR promoter was not responsible for iron responsiveness. On the other hand, the gene with the deleted 3′ UTR did not respond to iron; the same concentration of TfR appeared on the cell surface in the presence or in the absence of the iron chelator. Thus, the part of the 3′-UTR deleted in this experiment apparently included the iron response element.

Of course, the appearance of TfR receptor on the cell surface does not necessarily reflect the concentration of TfR mRNA. To check directly for an effect of iron on TfR mRNA concentration, Owen and Kühn performed S1 analysis (Chapter 5) of TfR mRNA in cells treated and untreated with iron chelator. As expected, the iron chelator increased the concentration of TfR mRNA considerably. But this response to iron disappeared when the gene had a deleted 3′-UTR.

What about the 3′-UTR confers responsiveness to iron? Harford and colleagues narrowed the search when they discovered that deletion of just 678 nt from the middle of the 3′-UTR eliminated most of the iron responsiveness.

Computer analysis of the critical 678-nt region of the 3′-UTR revealed that its most probable structure includes five hairpins, or stem loops, as illustrated in Figure 16.27. Even more interesting is the fact that the overall structures of these stem loops, including the base sequences in the

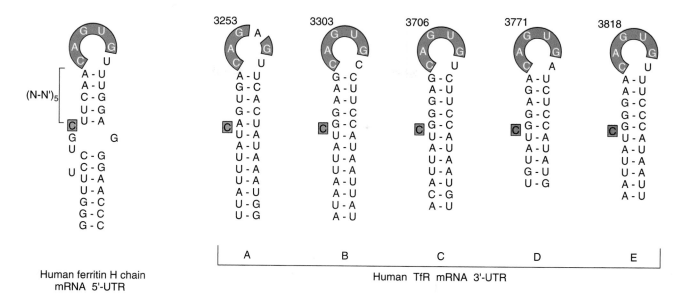

Figure 16.27 **Comparison of five stem loop structures in the 3′-UTR of the TfR mRNA with the IRE in the 5′-UTR of the ferritin mRNA.** The conserved, looped-out C and the conserved bases in the loop are highlighted in blue and red, respectively. (*Source:* Reprinted from Casey et al., *Science* 240:926, 1988. Copyright © 1988 American Association for the Advancement of Science, Washington, DC. Reprinted by permission.)

loops, bear a strong resemblance to a stem loop found in the 5′-UTR of the ferritin mRNA. This stem loop, called an **iron-response element** (**IRE**), is responsible for the ability of iron to stimulate translation of the ferritin mRNA. The implication is that these TfR IREs are the mediators of the responsiveness of TfR expression to iron.

Harford and colleagues went on to show by gel mobility shift assays (Chapter 5) that human cells contain a protein or proteins that bind specifically to the human TfR IREs (Figure 16.28). This binding could be competed with excess TfR mRNA or ferritin mRNA, which also has an IRE, but it could not be competed by β-globin mRNA, which has no IRE. Thus, the binding is IRE-specific. This finding underscores the similarity between the ferritin and TfR IREs and suggests that they may even bind the same protein(s). However, binding of the protein(s) to the two mRNAs has different effects, as we have seen.

> **SUMMARY** The 3′-UTR of the TfR mRNA contains five stem loops called iron response elements (IREs). Deletion of two or more of these IREs in tandem causes loss of response to iron.

The Rapid Turnover Determinant Knowing that iron regulates the TfR gene by controlling mRNA stability, and knowing that a protein binds to one or more IREs in the 3′-UTR of TfR mRNA, we assume that the IRE-binding protein protects the mRNA from degradation. This kind of regulation demands that the TfR mRNA be inherently unstable. If it were a stable mRNA, relatively little would be gained by stabilizing it further. In fact, the

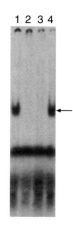

Figure 16.28 **Gel mobility shift assay for IRE-binding proteins.** Harford and colleagues prepared a labeled 1059-nt transcript corresponding to the region of the human TfR mRNA 3′-UTR that contains the five IREs. They mixed this labeled RNA with a cytoplasmic extract from human cells (with or without competitor RNA), electrophoresed the complexes, and visualized them by autoradiography. Lane 1, no competitor; lane 2, TfR mRNA competitor; lane 3, ferritin mRNA competitor; lane 4, β-globin mRNA competitor. The arrow points to a specific protein–RNA complex, presumably involving one or more IRE-binding proteins. (*Source:* Koeller et al., *PNAS* 86 (1989) p. 3576, f. 3.)

mRNA *is* unstable, and Harford and coworkers have demonstrated that this instability is caused by a **rapid turnover determinant** that also lies in the 3′-UTR.

What is this rapid turnover determinant? Because the human and chicken TfR genes are controlled in the same manner, they probably have the same kind of rapid turnover determinant. Therefore, a comparison of the 3′-UTRs of these two mRNAs might reveal common

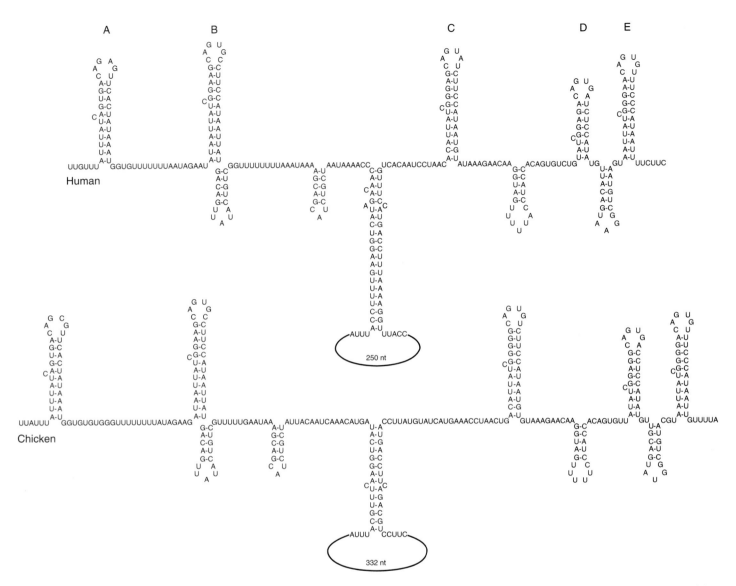

Figure 16.29   Structures of human and chicken IRE regions in the 3′-UTR of the TfR mRNAs. The IREs are labeled A–E. The sequences of the long loops in the middle are not given. (*Source:* From D. M. Koeller et al., "A cytosolic protein binds to structural elements within the iron regulatory region of the transferrin receptor mRNA," *Proceedings of the National Academy of Sciences* 86:3574–3578, May 1989. Reprinted by permission.)

features that would suggest where to start the search. Harford and colleagues compared the 678-nt region of the TfR mRNA from human with the corresponding region of the chicken TfR mRNA and found a great deal of similarity in the region containing the IREs. Figure 16.29 presents these two structures. Both have two IREs in the 5′-part of the region, then a stem with a large loop (250 nt in human and 332 nt in chicken), then the other three IREs. The 5′ and 3′ IRE-containing regions in the human mRNA are very similar to the corresponding regions in the chicken mRNA, but the loop region in between and the regions farther upstream and downstream have no detectable similarity. This suggested that the rapid turnover determinant should be somewhere among the IREs. Harford and coworkers identified some of its elements by mu-

tagenizing the TfR mRNA 3′- UTR and observing which mutations stabilized the mRNA.

The first mutants they looked at were simple 5′- or 3′-deletions, as illustrated in Figure 16.30. They transfected cells with these constructs and assayed for iron regulation by comparing the TfR mRNA and protein levels after treatment with either hemin or the iron chelator desferrioxamine. They measured mRNA levels by Northern blotting and protein levels by immunoprecipitation. They found that deletion of the 250-nt central loop (construct 2) or deletion of IRE A (construct 3) had no effect on iron regulation. Both of these constructs gave responses that were similar to that of the wild-type construct (construct 1), that is, a big increase in TfR mRNA and protein when cells were treated with iron chelator

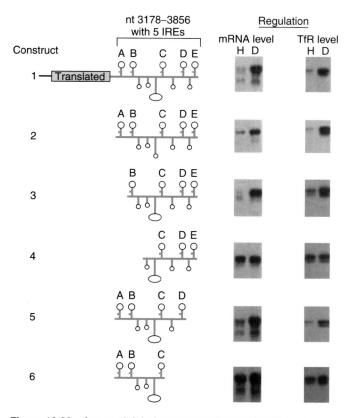

Figure 16.30   Assay of deletion mutants in the 3′-UTR for iron-responsiveness. Harford and colleagues made deletions in the 678-nt IRE-containing region of the 3′-UTR of the human TfR gene and transfected cells with these constructs (1–6), whose structures are shown schematically. They assayed for mRNA concentration after treatment with hemin (H) or desferrioxamine (D) by Northern blotting. They also assayed for TfR protein after the same treatments by immunoprecipitation. These Northern blotting and immunoprecipitation results are shown at right. (*Source:* Casey et al., Iron regulation of transferrin receptor. *EMBO Journal* 8 (8 Jul 1989) p. 3694, f. 1.)

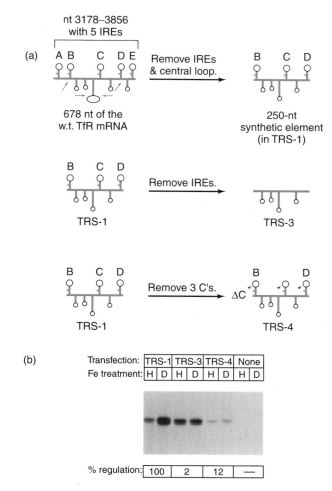

Figure 16.31   Effects on iron-responsiveness of deletions in the IRE region of the TfR 3′-UTR. (a) Creation of deletion mutants. Harford and colleagues generated the TRS-1 mutant by removing IREs A and E, and the large central loop, as shown by the arrows. From TRS-1, they generated TRS-3 by removing the remaining three IREs, and TRS-4 by deleting a single C at the 5′-end of each IRE loop. **(b)** Testing mutants for iron response. These workers transfected cells with each construct, treated half the cells with hemin (H) and the other half with desferrioxamine (D), and assayed for TfR biosynthesis by immunoprecipitation. The autoradiograph is shown, with transfected construct and iron treatment shown at top. A summary of the percentage regulation by iron is given at bottom. TRS-3 shows no regulation and a constitutively high level of TfR synthesis, suggesting a stable mRNA. TRS-4 shows little regulation and a low level of TfR synthesis, suggesting an unstable mRNA. (*Source:* Casey et al., *EMBO Journal* 8 (8 Jul 1989) p. 3695, f. 3B.)

rather than with hemin. However, deletion of both IREs A and B (construct 4) eliminated iron regulation: The levels of TfR mRNA and protein were the same (and high) with both treatments. Thus, the TfR mRNA is stable when IRE B is removed, so this IRE seems to be part of the rapid turnover determinant. The 3′-deletions gave a similar result. Deletion of IRE E had little effect on iron regulation, but deletion of both IREs D and E stabilized the TfR mRNA, even in the presence of hemin. Thus, IRE D appears to be part of the rapid turnover determinant.

Based on these findings, we would predict that IRE A, IRE E, and the central loop could be deleted without altering iron regulation. Accordingly, Harford and colleagues made a synthetic element they called TRS-1 that was missing these three parts as illustrated in Figure 16.31a. As expected, mRNAs containing this element retained full iron responsiveness. Next, these workers made two alterations to TRS-1 (Figure 16.31a). The first, TRS-3, had lost all three of its IREs. All that remained

were the other stem loops, pictured pointing downward in Figure 16.31. The other, TRS-4, had lost only three bases, the C's at the 5′-end of the loop in each IRE. Figure 16.31b shows the effects of these two alterations. TRS-3, with no IREs, had lost virtually all iron responsiveness, and the TfR RNA was much more stable than the wild-type mRNA. TRS-4, with a C missing from each IRE, had lost most of its iron responsiveness, but the mRNA remained unstable. Thus, this mRNA retained its rapid

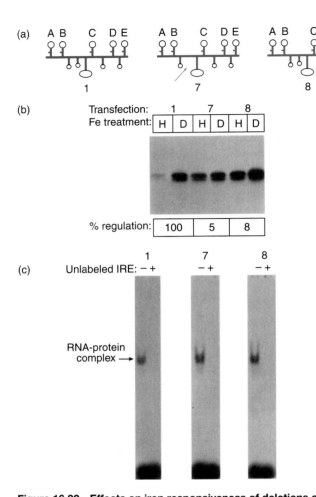

Figure 16.32   Effects on iron responsiveness of deletions of non-IRE stem loops from the TfR 3′-UTR. (a) The deletions. Harford and coworkers started with construct 1, containing all five IREs, and made constructs 7 and 8 by deleting one non-IRE stem loop as shown by the arrows. (b) Assay for iron response. These workers transfected cells with each construct and applied the usual iron treatments, as denoted by H and D, at top. They assayed for TfR biosynthesis by immunoprecipitation. The percentage regulation by iron is given at bottom. (c) Interaction of mutant constructs with IRE-binding protein. These workers labeled constructs 1, 7, and 8, as indicated at top, and mixed them with cell extracts in the absence (−) or presence (+) of unlabeled competitor IRE-containing RNA. All constructs bound equally well to the IRE-binding protein. (*Source:* Casey et al., *EMBO Journal* 8 (8 Jul 1989) p. 3696, f. 4.)

turnover determinant, but had lost the ability to be stabilized by the IRE-binding protein. In fact, as we would expect, gel mobility shift assays showed that TRS-4 could not bind the IRE-binding protein.

To pin down the rapid turnover determinant still further, Harford and colleagues made two more deletion mutants, this time starting with construct 1, which has the full 678-nt element from the TfR 3′-UTR. They made constructs 7 and 8 by deleting a non-IRE stem loop, as illustrated in Figure 16.32a. Then they tested these constructs by transfection and immunoprecipitation as before. Figure 16.32b shows the results. Both constructs 7 and 8 show almost total loss of iron responsiveness and a constitutively high level of TfR expression. Thus, both of the

deleted stem loops appear to be essential to confer rapid turnover of the mRNA. To demonstrate that this effect was not due to an inability of the mRNAs to interact with the IRE-binding protein, these workers assayed protein–RNA binding as before by gel mobility shift. Figure 16.32c shows that both constructs 7 and 8 are just as capable of binding to the IRE-binding protein as is the wild-type mRNA, and this binding in every case is competed by unlabeled IRE.

> **SUMMARY** IREs A and E, and the large central loop of the TfR 3′-UTR can be deleted without altering the response to iron. However, removing all of the IREs, or either one of two non-IRE stem loops renders the TfR mRNA constitutively stable and therefore destroys the rapid turnover determinant. Thus, each of these non-IRE stem loops, and at least one of IREs B–D, are all part of this determinant.

TfR mRNA Stability   So far we have seen that the level of TfR mRNA responds to intracellular iron concentration and that this responsiveness is determined by the 3′-UTR, rather than by the promoter. This strongly suggests that iron regulates the TfR mRNA half-life, rather than the rate of mRNA synthesis, but we have not presented explicit evidence for this hypothesis. To provide such evidence, Ernst Müllner and Lukas Kühn measured the rate of TfR mRNA decay in the presence and absence of the iron chelator desferrioxamine. They found that the TfR mRNA was very stable when the iron concentration was low. On the other hand, at high iron concentration the TfR mRNA decayed much faster. These two half-lives were 30 and 1.5 hours, respectively, so iron appears to destabilize the TfR mRNA by approximately 30/1.5, or 20-fold.

Harford and colleagues reached a similar conclusion by testing constructs having the 3′-UTRs previously described in Figure 16.31: TRS-1, -3, and -4. They transfected cells with these constructs and tested for TfR mRNA half-life after hemin or iron chelator treatment as follows: They blocked further mRNA synthesis with actinomycin D, then isolated RNA from the cells at various times after actinomycin D treatment, and measured TfR mRNA concentration by Northern blotting. Figure 16.33a shows the results for TRS-1 mRNA, which we have already seen is fully iron responsive. In the presence of hemin, the mRNA level had dropped dramatically by 180 min after transcription was blocked by actinomycin D. The mRNA half-life under these conditions was about 45 min. By contrast, the iron chelator desferrioxamine strongly stabilized the TfR mRNA. Under these conditions, the mRNA level had not dropped significantly even by 180 min after drug treatment. Thus, the mRNA half-life when iron is removed is much longer than 3 h.

Next, Harford and coworkers tested the mutant mRNAs, TRS-3 and TRS-4, as well as TRS-1, for half-life after

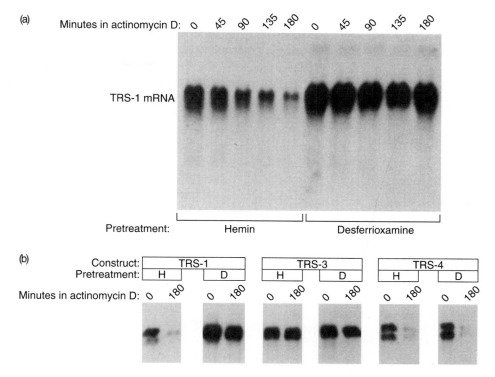

Figure 16.33   Effect of iron on stability of wild-type and mutant TfR mRNAs. (a) Wild-type TfR mRNA. Harford and colleagues examined the rate of degradation of TRS-1 mRNA (Figure 16.31) on inhibition of further RNA synthesis by actinomycin D, under two sets of conditions: (1) Pretreatment with hemin to raise the intracellular iron concentration in mouse fibroblasts (left), and (2) Pretreatment with desferrioxamine to lower the intracellular iron concentration (right). The times (in minutes) after actinomycin D addition are given at top. **(b)** Comparison of behavior of TRS-1, -3, and -4 TfR mRNAs in response to iron. Harford and colleagues treated cells as in panel **(a)**, except that they used only one period of actinomycin D treatment (180 min). The constructs, pretreatments with hemin (H), or desferrioxamine (D), and time in actinomycin D are listed at top. TRS-1 could be stabilized in the absence of iron, but TRS-3 was constitutively stable (could not be destabilized) and TRS-4 was constitutively unstable (could not be stabilized). (*Source:* Koeller et al., *PNAS* 88 (1991) p. 7780, f. 1.)

treatment of cells with either hemin or desferrioxamine. Figure 16.33b illustrates what happened. As before, TRS-1 was unstable when iron concentration was high, but stabilized when iron was removed by chelation. TRS-3, with its missing IREs and rapid turnover determinant, was stable under both conditions. By contrast, TRS-4, with mutated IREs, was incapable of binding the IRE-binding protein. However, its rapid turnover determinant was still intact, so it was unstable regardless of the intracellular iron concentration.

SUMMARY When the iron concentration is high, the TfR mRNA decays rapidly. When the iron concentration is low, the TfR mRNA decays much more slowly. This difference in mRNA stability is about 20-fold, and it is much more important in determining the rate of accumulation of TfR mRNA than the difference in mRNA synthesis (about twofold) at high and low iron concentrations. The TRS-3 mRNA, with no rapid turnover determinant, is constitutively stable (stable in both high and low iron concentrations). The TRS-4 mRNA, with no ability to bind the IRE-binding protein, is constitutively unstable.

The TfR mRNA Degradation Pathway Harford and colleagues have investigated the mechanism by which TfR mRNA is degraded and have found that the first event appears to be an endonucleolytic cut within the IRE region. Unlike the degradation of many other mRNAs, there seems to be no requirement for deadenylation before TfR degradation can begin.

These workers began their study by treating human plasmacytoma cells (ARH-77 cells) with hemin and showing by Northern blotting that the level of TfR mRNA dropped precipitously in 8 h. When they exposed the blot for a longer time, they found that a new RNA species, about 1000–1500 nt shorter than full-length TfR mRNA, appeared during the period in which the TfR mRNA was breaking down (Figure 16.34a). This RNA was also found in the poly(A)⁻ fraction (Figure 16.34b), suggesting that it had lost its poly(A). But the size of this shortened RNA suggested that it had lost much more than just its poly(A). The simplest explanation was that it had been cut by an endonuclease within its 3′-UTR, which removed over 1000 3′-terminal nucleotides, including the poly(A).

To test this hypothesis, Harford's group treated cells containing either the TRS-1 or TRS-4 construct with desferrioxamine to remove iron, then isolated RNA, cut the

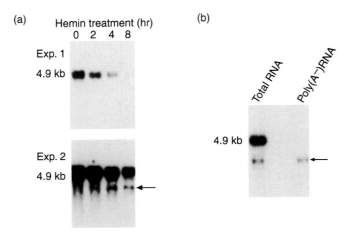

(a)

Hemin treatment (hr)
0 2 4 8

Exp. 1

4.9 kb

Exp. 2

4.9 kb

(b)

Total RNA   Poly(A⁻)RNA

4.9 kb

Figure 16.34 Appearance of a shortened TfR mRNA on hemin treatment. (a) Detection of the shortened product in human cells. Harford and colleagues isolated RNAs from ARH-77 cells treated for various times with hemin (as indicated at top), electrophoresed them, Northern blotted the RNAs, and hybridized the blot to a TfR probe. Top (Exp. 1), short autoradiographic exposure to detect the full-length, 4.9-kb transcript only; bottom (Exp. 2), long exposure to detect a shorter RNA (arrow) in addition to the full-length transcript. (b) The shorter RNA is found in the poly(A)⁻ fraction. Harford and colleagues subjected RNA from a 6-h hemin exposure as in panel (a) to oligo(dT)-cellulose chromatography to separate poly(A)⁺ and poly(A)⁻ RNAs, then subjected total RNA and poly(A)⁻ RNAs to the same analysis as in panel (a). The poly(A)⁻ fraction contains only the shortened RNA fragment. (*Source:* Binder et al., Evidence that the pathway of transferrin receptor mRNA degradation involves an endonucleolytic cleavage within the 3′ UTR and does not involve poly(A) tail shortening. *EMBO Journal* 13 (1994) p. 1970, f. 1.)

TfR mRNA near the translation termination codon with an oligonucleotide and RNase H, as described in Chapter 14, then performed primer extension with a labeled primer that hybridized to the SV40 sequence beyond the end of the 3′-UTR of the TfR coding region. Thus, this primer extension analysis detected the 3′-part of the mRNA, whereas the Northern blot analysis detected the 5′-part. Figure 16.35 shows the structure of the TRS-1 or –4 3′-UTR, with the RNase H cleavage site at one end and the primer-binding site at the other. The purpose of the RNase H cleavage was to shorten the RNA enough that it would yield a primer extension product that would be small enough to electrophorese a measurable distance into the gel.

Figure 16.35 demonstrates a clear full-length signal for both TRS-1 and –4, as well as a shorter fragment from each RNA. This shorter fragment is much more prominent in the lane containing TRS-4 products than in the lane with TRS-1 products. This is exactly what we expect if this fragment represents a degradation product of the TfR mRNA, because the TRS-4 RNA is much more unstable in the presence of the iron chelator than the TRS-1 RNA. The size of the shorter primer extension product places its 5′-end between IREs C and D, as illustrated in Figure 16.35. This is consistent with the hypothesis that

the first step in degradation of the TfR mRNA is an endonucleolytic cut more than 1000 nt from the 3′-end of the mRNA.

All the data we have considered are consistent with this hypothesis (Figure 16.36): When iron concentrations are low, an IRE-binding protein binds to the rapid turnover determinant in the 3′-UTR of the TfR mRNA. This protects the mRNA from degradation. When iron concentrations are high, iron binds to the IRE-binding protein, causing it to dissociate from the rapid turnover determinant, opening it up to attack by a specific endonuclease that clips off a 1-kb fragment from the 3′-end of the TfR mRNA. This destabilizes the mRNA and leads to its rapid degradation.

The protein that binds to the IREs in both the transferrin receptor mRNA and the ferritin mRNA (Chapter 17) has now been identified as a form of **aconitase,** an enzyme that converts citrate to isocitrate in the citric acid cycle. The enzymatically active form of aconitase is an iron-binding protein that does not bind to the IREs. However, the apoprotein form of aconitase, which lacks iron, binds to the IREs in mRNAs.

SUMMARY The initiating event in TfR mRNA degradation seems to be an endonucleolytic cleavage of the mRNA more than 1000 nt from its 3′-end, within the IRE region. This cleavage does not require prior deadenylation of the mRNA.

Posttranscriptional Gene Silencing (RNA Interference)

For years, molecular biologists have been using antisense RNA to inhibit expression of selected genes in living cells. At first, the rationale was that the antisense RNA, which is complementary to mRNA, would base-pair to the mRNA and inhibit its translation. The strategy usually worked, but the rationale was flawed. As Su Guo and Kenneth Kenphues established in 1995, injecting sense RNA into cells worked just as well as antisense RNA in blocking expression of a particular gene. Then, in 1998, Andrew Fire and colleagues showed that double-stranded RNA (**dsRNA**) worked much better than either sense or antisense RNA. In fact, the reason sense and antisense RNAs worked appears to be that they were contaminated with small amounts of dsRNA, and the dsRNA actually blocked gene expression.

Also, beginning about 1990, molecular biologists began noticing that placing transgenes into various organisms sometimes had the opposite of the desired effect. Instead of turning on the transgene, organisms sometimes turned off, not only the transgene, but the normal cellular copy of the gene as well. This phenomenon was called by

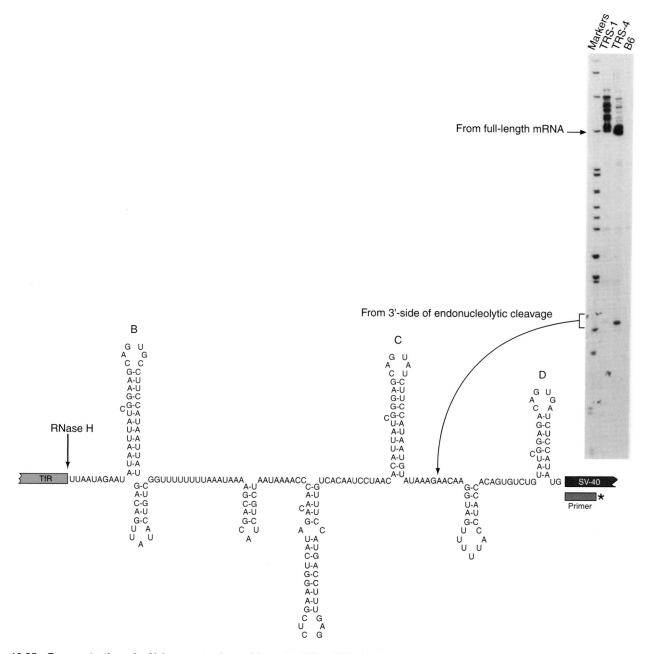

Figure 16.35   Demonstration of a 3′-fragment released from the TfR mRNA. Harford and colleagues transfected cells with TRS-1 or -4 constructs, as indicated at top, isolated RNAs, and subjected them to primer extension with a primer (red) that hybridized as shown. Before primer extension analysis, RNAs were cleaved at the 5′-end (see arrow at left) by oligonucleotide-mediated RNase H degradation. This was done to produce a primer extension product of the full-length transcript that was short enough to analyze. Both the TRS-1 and TRS-4 cells yielded a strong signal corresponding to the full-length transcript (arrow near top of autoradiograph). Both also yielded a signal corresponding to a 3′-RNA fragment cut at the point between IREs C and D indicated by the curved arrow. The bracket at the end of the arrow indicates the signal; the arrow points to the cleavage site. The lane marked B6 contained RNA from untransfected B6 cells. (*Source:* Binder et al., *EMBO Journal* 13 (1994) p. 1973, f. 5.)

several names: cosuppression and **posttranscriptional gene silencing (PTGS)** in plants, **RNA interference (RNAi)** in animals such as nematodes (*Caenorhabditis elegans*) and fruit flies, and quelling in fungi. To avoid confusion, we will refer to this phenomenon as RNAi from now on, regardless of the species under study.

Fire and colleagues showed that injecting *C. elegans* gonads with dsRNA caused RNAi in the resulting embryos. Furthermore they detected a lack of the corresponding mRNA in embryos undergoing RNAi (Figure 16.37). However, the dsRNA had to include exon regions; dsRNA corresponding to introns and promoter sequences

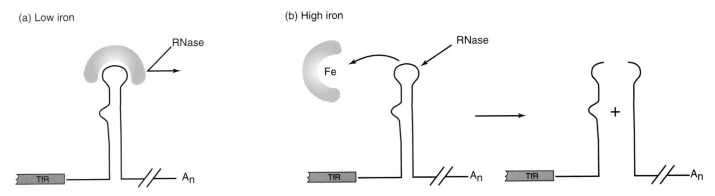

Figure 16.36   Model for destabilization of TfR mRNA by iron. (a) Under low-iron conditions, the aconitase apoprotein (orange) binds to the IREs in the 3′-UTR of the TfR mRNA. This protects the RNA from degradation by RNases. **(b)** Under high-iron conditions, iron binds to the aconitase apoprotein, removing it from the IREs, and opening the IREs up to attack by RNase. The RNase clips the mRNA at least once, exposing its 3′-end to further degradation.

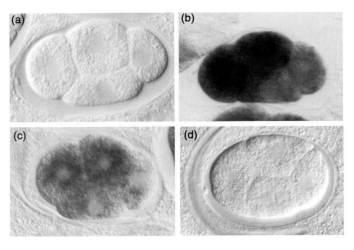

Figure 16.37   Double-stranded RNA-induced RNA interference causes destruction of a specific mRNA. Fire and colleagues injected antisense or dsRNA corresponding to the *C. elegans mex-3* mRNA into *C. elegans* ovaries. After 24 h, they fixed the embryos in the treated ovaries and subjected them to in situ hybridization (Chapter 5) with a probe for *mex-3* mRNA. **(a)** Embryo from a negative control parent with no hybridization probe. **(b)** Embryo from a positive control parent that was not injected with RNA. **(c)** Embryo from a parent that was injected with *mex-3* antisense RNA. A considerable amount of *mex-3* mRNA remained. **(d)** Embryo from a parent that was injected with dsRNA corresponding to part of the *mex-3* mRNA. No detectable *mex-3* mRNA remained. (*Source:* Fire, A. et al. Potent and specific genetic interference by double-stranded RNA in Caenorhabditis elegans. *Nature* 391 (1998) f. 3, p. 809 © Macmillan Magazines Ltd.)

did not cause RNAi. Finally, these workers demonstrated that the effect of the dsRNA crossed cell boundaries, at least in *C. elegans.*

Is this loss of a particular mRNA in response to the corresponding dsRNA caused by repression of transcription of the gene or destruction of the mRNA? In 1998, Fire and colleagues, as well as others, demonstrated that RNAi is a posttranscriptional process that involves mRNA degradation. Several investigators reported the presence of short pieces of dsRNA in cells undergoing RNAi. In 2000,

Scott Hammond and collaborators purified a nuclease from *Drosophila* embryos undergoing RNAi that digests the targeted mRNA. The partially purified preparation that contained this nuclease activity also contained a 25-nt RNA fraction that could be detected on Northern blots with probes for either the sense or antisense strand of the targeted mRNA. Degradation of the 25-nt RNA with micrococcal nuclease destroyed the ability of the preparation to digest the mRNA. These data suggested that a nuclease digests the dsRNA that initiates RNAi into fragments about 25 nt long, and these fragments then associate with the nuclease and provide guide sequences that allow the nuclease to target the corresponding mRNA.

Phillip Zamore, Thomas Tuschl, Phillip Sharp, and David Bartel developed a system based on *Drosophila* embryo lysates that carried out RNAi in vitro. This system allowed these workers to look at individual steps in the RNAi process. The embryos had been injected with dsRNA corresponding to luciferase mRNA, so they targeted that mRNA for destruction. First, Zamore and collaborators showed that RNAi requires ATP. They depleted their extract of ATP by incubating it with hexokinase and glucose, which converts ATP to ADP and transfers the lost phosphate group to glucose. The ATP-depleted extract no longer carried out the degradation of the target, luciferase mRNA.

Next, these workers performed experiments in which they labeled one strand of the dsRNA at a time (or both) and showed that labeled short RNAs of 21–23 nt appeared, no matter which strand was labeled (Figure 16.38). The appearance of the 21–23-nt RNAs did not require the presence of mRNA (e.g., compare lanes 2 and 3), so these short RNAs apparently derived from dsRNA, not mRNA. When capped antisense luciferase RNA was labeled (lanes 11 and 12), a small amount of 21–23-nt RNAs appeared, and that amount increased in the presence of mRNA (lane 12). This result suggested that the labeled antisense RNA was hybridizing to the

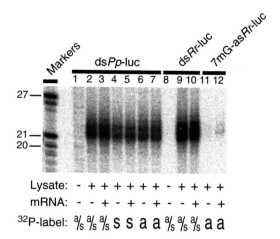

Figure 16.38 Generation of 21–23-nt RNA fragments in an RNAi-competent *Drosophila* embryo extract. Zamore and collaborators added ds luciferase RNA from *Photinus pyralis* (*Pp*-luc RNA) or from *Renilla reniformis* (*Rr*-luc RNA), as indicated at top, to lysates in the presence or absence of the corresponding mRNA, as indicated at bottom. The dsRNAs were labeled in the sense strand (s), in the antisense strand (a), or in both strands (a/s), as indicated at bottom. RNA markers from 17- to 27-nt long were included in the lane at left. Lanes 11 and 12 contained labeled, capped antisense *Rr*-luc RNA in the absence and presence of mRNA, respectively. (*Source:* Zamore, D. et al. RNAi: Double-Stranded RNA Directs the ATP-Dependent Cleavage of mRNA at 21 to 23 Nucleotide Intervals. *Cell* 101 (2000) f. 3, p. 28. Reprinted by permission of Elsevier Science.)

added mRNA to generate a dsRNA that could be degraded to the short RNA pieces. In summary, all these results suggest that a nuclease degrades dsRNA into short pieces.

Next, Zamore and collaborators showed that the dsRNA dictated where the corresponding mRNA would be cleaved. They added three different dsRNAs, whose ends differed by about 100 nt, to their RNAi extracts, then added 5′-labeled mRNA, allowed RNA cleavage to occur, and electrophoresed the products. Figure 16.39 shows the results: The dsRNA (C) whose 5′-end was closest to the 5′-end of the mRNA yielded the shortest fragments; the next dsRNA(B), whose 5′-end was about 100 nt further downstream, yielded mRNA fragments about 100 nt longer; and the third dsRNA, whose 5′-end was about another 100 nt further downstream, yielded mRNA fragments about another 100 nt longer. This close relationship between the position of the dsRNA relative to the mRNA, and the position at which cleavage began, strongly suggests that the dsRNA determined the sites of cleavage of the mRNA.

Next, Zamore and collaborators performed high-resolution gel electrophoresis of the mRNA degradation products from Figure 16.39. The results, presented in Figure 16.40, are striking. The major cleavage sites in the mRNA are mostly at 21–23-nt intervals, producing a set of RNA fragments whose lengths differ by multiples of 21–23 nt. The one obvious exception is the site marked

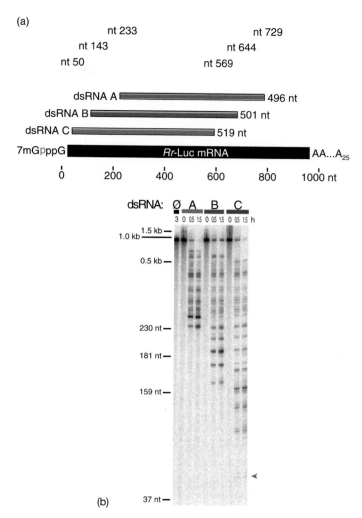

Figure 16.39 The dsRNA dictates the boundaries of cleavage of mRNA in RNAi. Zamore and collaborators added the three dsRNAs pictured in panel (**a**) to an embryo extract along with an *Rr*-luc mRNA, 5′-labeled in one of the phosphates of the cap. (**b**) Experimental results. The 5′-end-labeled mRNA degradation products were electrophoresed. The dsRNAs included in the reactions are indicated and color-coded at top. The first lane, marked 0, contained no dsRNA. Reactions were incubated for the times (in h) indicated at top. The arrowhead indicates a faint cleavage site that lies outside the position of RNA C. Otherwise, the sites cleaved lie within the positions of the three dsRNAs on the mRNA. (*Source:* Zamore, D. et al. RNAi: Double-Stranded RNA Directs the ATP-Dependent Cleavage of mRNA at 21 to 23 Nucleotide Intervals. *Cell* 101 (2000) f. 5, p. 30. Reprinted by permission of Elsevier Science.)

by an arrowhead, which lies only 9 nt from the previous cleavage site. This exceptional site lies within a run of seven uracil residues, which is interesting in light of the fact that 14 of 16 cleavage sites mapped were at uracils. After this exceptional site, the 21–23-nt interval resumed for the rest of the mapped cleavage sites. These results support the hypothesis that the 21–23-nt dsRNAs determine where the mRNA will be cut and suggest that cleavage takes place preferentially at uracils.

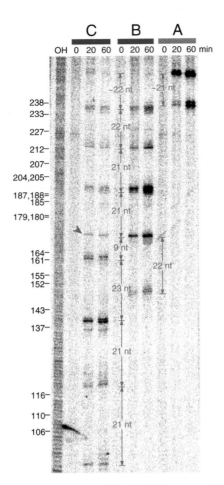

Figure 16.40 Cleavages of mRNA in RNAi occur at 21–23-nt intervals. Zamore and collaborators performed high-resolution denaturing polyacrylamide gel electrophoresis on the products of RNAi in the presence of all three of the dsRNAs from Figure 16.39. The cleavages, with one notable exception (arrowhead), occured at 21–23-nt intervals. The exceptional band indicates a cleavage at only a 9-nt interval, but cleavages thereafter were at 21–23-nt intervals. (*Source:* Zamore, D, et al. RNAi: Double-Stranded RNA Directs the ATP-Dependent Cleavage of mRNA at 21 to 23 Nucleotide Intervals. *Cell* 101 (2000) f. 6, p. 31. Reprinted by permission of Elsevier Science.)

Zamore and collaborators summarized their data with the model for RNAi in Figure 16.41. The process begins with dsRNA, which initiates RNAi. The dsRNA may be introduced into cells experimentally or by transcription of both strands of a transgene. The next step is degradation of the dsRNA into short pieces, about 21–23 nt long. The RNase that clips dsRNA may be a member of the RNase III family discussed earlier in this chapter. RNase III is the only well-studied nuclease that cuts dsRNA specifically. The RNase that created the short double-stranded pieces of RNA presumably remains bound to an RNA piece and uses it as a template to find and degrade the corresponding mRNA. One way it could do this is by employing an ATP-dependent RNA helicase to unwind the dsRNA (which would explain the ATP-dependence of the process). Then it could remain bound to the antisense strand, which could hybridize to mRNA, bringing the RNase to its target.

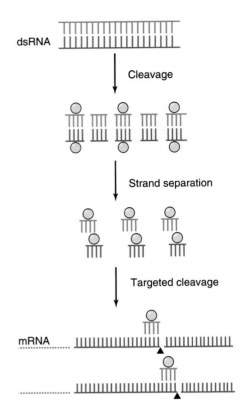

Figure 16.41 A model for RNAi. The process begins with dsRNA, which a cellular nuclease cleaves to fragments 21–23 nt long. The nuclease remains bound to the fragments and uses an ATP-dependent RNA helicase to denature the dsRNAs. The nucleases that are bound to the antisense 21–23-nt RNAs can hybridize to sites in the mRNA and dictate cleavage of the mRNA at or near their ends, usually at a uracil residue. (*Source:* Reprinted from Zamore et al., *Cell* 101:32, 2000. Copyright 2000, with permission from Elsevier Science.)

What is the physiological significance of RNAi? Double-stranded RNA does not normally occur in eukaryotic cells, but it does occur during infection by certain RNA viruses that replicate through dsRNA intermediates. So one important function of RNAi may be to inhibit the replication of viruses by degrading their mRNAs. But Fire and other investigators have also found that some of the genes required for RNAi are also required to prevent certain transposons from transposing within the genome. Thus, RNAi may have utility even in cells that are not infected by a virus.

SUMMARY: Posttranscriptional gene silencing, or RNA interference, occurs when a cell encounters dsRNA or an added transgene. The added dsRNA, or one derived from the transgene, is degraded into 21–23-nt fragments by a nuclease. The nuclease then presumably associates with the dsRNA, perhaps denaturing it with a helicase activity. The 21–23-nt fragment can then dictate the sites to attack on the corresponding mRNA.

SUMMARY

Ribosomal RNAs are made in eukaryotes as precursors that must be processed to release the mature rRNAs. The order of RNAs in the precursor is 18S, 5.8S, 28S in all eukaryotes, although the exact sizes of the mature rRNAs vary from one species to another. In human cells, the precursor is 45S, and the processing scheme creates 41S, 32S, and 20S intermediates.

Extra nucleotides are removed from the 5′-ends of pre-tRNAs in one step by an endonucleolytic cleavage catalyzed by RNase P. RNase P's from bacteria and eukaryotic nuclei have a catalytic RNA subunit called M1 RNA. Spinach chloroplast RNase P appears to lack RNA. RNase II and polynucleotide phosphorylase cooperate to remove most of the extra nucleotides at the 3′-end of a tRNA precursor, but stop at the +2 stage, with two extra nucleotides remaining. RNases PH and T are most active in removing the last two nucleotides from the RNA, with RNase T being the major participant in removing the very last nucleotide.

Trypanosome mRNAs are formed by *trans*-splicing between a short leader exon and any one of many independent coding exons. Trypanosome coding regions, including genes encoding rRNAs and tRNAs, are arranged in long, polycistronic transcription units governed by a single promoter.

Trypanosomatid mitochondria (kinetoplastids) encode incomplete mRNAs that must be edited before they can be translated. Editing occurs in the 3′ → 5′ direction by successive action of one or more guide RNAs. These gRNAs hybridize to the unedited region of the mRNA and provide A's and G's as templates for the incorporation of U's missing from the mRNA. Sometimes the gRNA is missing an A or G to pair with a U in the mRNA, in which case the U will be removed. The mechanism of removing U's involves: (1) cutting the pre-mRNA just beyond the U to be removed; (2) removal of the U by an exonuclease; and (3) ligating the two pieces of pre-mRNA together. The mechanism of adding U's uses the same first and last step, but the middle step (step 2) involves addition of one or more U's by TUTase instead of removing U's.

A common form of posttranscriptional control of gene expression is control of mRNA stability. For example, when mammary gland tissue is stimulated by prolactin, the synthesis of casein protein increases dramatically. However, most of this increase in casein is not due to an increase in the rate of transcription of the casein gene. Instead, it is caused by an increase in the half-life of casein mRNA.

Another well-studied example of control of mRNA stability is the control of the stability of the transferrin receptor mRNA. When cells have abundant iron, the level of tranferrin receptor is reduced to avoid accumulation of too much iron in cells. Conversely, when cells are starved for iron, they increase the concentration of transferrin receptor to transport as much iron as possible into the cells. The expression of the transferrin receptor (Tfr) gene is controlled by controlling the mRNA stability as follows: The 3′-UTR of the TfR mRNA contains five stem loops called iron response elements (IREs). Deletion of two or more of these IREs in tandem causes loss of response to iron. IREs A and E, and the large central loop of the TfR 3′-UTR, can be deleted without altering the response to iron. However, removing all of the IREs, or either one of two non-IRE stem loops, renders the TfR mRNA constitutively stable and therefore destroys the rapid turnover determinant. Thus, each of these non-IRE stem loops and at least one of IREs B–D are all part of this determinant. When the iron concentration is high, the TfR mRNA decays rapidly. When the iron concentration is low, the TfR mRNA decays much more slowly. This difference in mRNA stability is about 20-fold, and it is much more important in determining the rate of accumulation of TfR mRNA than the difference in mRNA synthesis (about twofold) at high and low iron concentrations. The TRS-3 mRNA, with no rapid turnover determinant, is constitutively stable (stable in both high and low iron concentrations). The TRS-4 mRNA, with no ability to bind the IRE-binding protein, is constitutively unstable. The initiating event in TfR mRNA degradation seems to be an endonucleolytic cleavage of the mRNA more than 1000 nt from its 3′-end within the IRE region. This cleavage does not require prior deadenylation of the mRNA.

Posttranscriptional gene silencing, or RNA interference, occurs when a cell encounters dsRNA or an added transgene. The added dsRNA, or one derived from the transgene, is degraded into 21–23-nt fragments by a nuclease. The nuclease then presumably associates with the dsRNA, perhaps denaturing it with a helicase activity. The 21–23-nt fragment can then dictate the sites to attack on the corresponding mRNA.

REVIEW QUESTIONS

1. Draw the structure of a mammalian rRNA precursor, showing the locations of all three mature rRNAs.

2. What is the function of RNase P? What is unusual about this enzyme (at least the bacterial and eukaryotic nuclear forms of the enzyme)?

3. Illustrate the difference between *cis*- and *trans*-splicing.

4. Describe sequencing data that suggest that the trypanosome spliced leader is attached to a poly(A)+ RNA.

5. Describe and give the results of an experiment that shows that a Y-shaped intermediate exists in the splicing of a trypanosome pre-mRNA. Show how this result is compatible with *trans*-splicing, but not with *cis*-splicing.

6. Describe what we mean by RNA editing. What is a cryptogene?

7. Describe and give the results of an experiment that shows that editing of kinetoplast mRNA goes in the 3'- to 5'-direction.

8. Draw a diagram of a model of RNA editing that fits the data at hand. What enzymes are involved?

9. Present direct evidence for guide RNAs.

10. Describe and give the results of an experiment that shows that prolactin controls the casein gene primarily at the post-transcriptional level.

11. What two proteins are most directly involved in iron homeostasis in mammalian cells? How do their levels respond to changes in iron concentration?

12. How do we know that a protein binds to the iron response elements (IREs) of the TfR mRNA?

13. Describe and give the results of an experiment that shows that one kind of mutation in the TfR IRE region results in an iron unresponsive and stable mRNA, and another kind of mutation results in an iron unresponsive and unstable mRNA. Interpret these results in terms of the rapid turnover determinant and interaction with IRE-binding protein(s).

14. Describe and give the results of an experiment that shows that iron-responsiveness really depends on instability of TfR mRNA.

15. Present a model for the involvement of aconitase in determining the stability of TfR mRNA.

16. What evidence suggests that RNA interference depends on mRNA degradation?

17. Present a model for the mechanism of RNA interference.

SUGGESTED READINGS

General References and Reviews

Bass, B.L. 2000. Double-stranded RNA as a template for gene silencing. *Cell* 101:235–38.

Nilsen, T.W. 1994. Unusual strategies of gene expression and control in parasites. *Science* 264:1868–69.

Seiwert, S.D. 1996. RNA editing hints of a remarkable diversity in gene expression pathways. *Science* 274:1636–37.

Simpson, L. and D.A. Maslov. 1994. RNA editing and the evolution of parasites. *Science* 264:1870–71.

Solner-Webb, B. 1996. Trypanosome RNA editing: Resolved. *Science* 273:1182–83.

Research Articles

Abraham, J.M., J.E. Feagin, and K. Stuart. 1988. Characterization of cytochrome c oxidase III transcripts that are edited only in the 3' region. *Cell* 55:267–72.

Binder, R., J.A. Horowitz, J.P. Basilion, D.M. Koeller, R.D. Klausner, and J.B. Harford. 1994. Evidence that the pathway of transferrin receptor mRNA degradation involves an endonucleolytic cleavage within the 3' UTR and does not involve poly(A) tail shortening. *EMBO Journal* 13:1969–80.

Blum, B., N. Bakalara, and L. Simpson. 1990. A model for RNA editing in kinetoplastid mitochondria: "Guide" RNA molecules transcribed from maxicircle DNA provide the edited information. *Cell* 60:189–98.

Casey, J.L., M.W. Hentze, D.M. Koeller, S.W. Caughman, T.A. Rovault, R.D. Klausner, and J.B. Harford. 1988. Iron-responsive elements: Regulatory RNA sequences that control mRNA levels and translation. *Science* 240:924–28.

Casey, J.L., D.M. Koeller, V.C. Ramin, R.D. Klausner, and J.B. Harford. 1989. Iron regulation of transferrin receptor mRNA levels requires iron-responsive elements and a rapid turnover determinant in the 3' untranslated region of the mRNA. *EMBO Journal* 8:3693–99.

Feagin, J.E., J.M. Abraham, and K. Stuart. 1988. Extensive editing of the cytochrome c oxidase III transcript in *Trypanosoma brucei*. *Cell* 53:413–22.

Fire, A., S. Xu, M.K. Montgomery, S.A. Kostas, S.E. Driver, and C.C. Mello. 1998. Potent and specific genetic interference by double-stranded RNA in *Caenorhabditis elegans*. *Nature* 391:806–11.

Guerrier-Takada, C., K. Gardiner, T. Marsh, N. Pace, and S. Altman. 1983. The RNA moiety of ribonuclease P is the catalytic subunit of the enzyme. *Cell* 35:849–57.

Guyette, W.A., R.J. Matusik, and J.M. Rosen. 1979. Prolactin-mediated transcriptional and post-transcriptional control of casein gene expression. *Cell* 17:1013–23.

Hammond, S.M., E. Bernstein, D. Beach, and G.J. Hannon. 2000. An RNA-directed nuclease mediates post-transcriptional gene silencing in *Drosophila* cells. *Nature* 404:293–96.

Johnson, P.J., J.M. Kooter, and P. Borst. 1987. Inactivation of transcription by UV irradiation of *T. brucei* provides evidence for a multicistronic transcription unit including a VSG gene. *Cell* 51:273–81.

Kable, M.L., S.D. Seiwart, S. Heidmann, and K. Stuart. 1996. RNA editing: A mechanism for gRNA-specified uridylate insertion into precursor mRNA. *Science* 273:1189–95.

Koeller, D.M., J.L. Casey, M.W. Hentze, E.M. Gerhardt, L.-N.L. Chan, R.D. Klausner, and J.B. Harford. 1989. A cytosolic protein binds to structural elements within the iron regulatory region of the transferrin receptor mRNA. *Proceedings of the National Academy of Sciences USA* 86:3574–78.

Koeller, D.M., J.A. Horowitz, J.L. Casey, R.D. Klausner, and J.B. Harford. 1991. Translation and the stability of mRNAs encoding the transferrin receptor and *c-fos*. *Proceedings of the National Academy of Sciences USA* 88:7778–82.

LeBowitz, J.H., H.Q. Smith, L. Rusche, and S.M. Beverley. 1993. Coupling of poly(A) site selection and *trans* splicing in *Leishmania*. *Genes and Development* 7:996–1007.

Li, Z. and M.P. Deutscher. 1994. The role of individual exoribonucleases in processing at the 3′ end of *Escherichia. coli* tRNA precursors. *Journal of Biological Chemistry* 269:6064–71.

Miller, O.L., Jr., B.R. Beatty, B.A. Hamkalo, and C.A. Thomas, Jr. 1970. Electron microscopic visualization of transcription. *Cold Spring Harbor Symposia on Quantitative Biology.* 35:505–12.

Müllner, E.W. and L.C. Kühn. 1988. A stem-loop in the 3′ untranslated region mediates iron-dependent regulation of transferrin receptor mRNA stability in the cytoplasm. *Cell* 53:815–25.

Murphy, W.J., K.P. Watkins, and N. Agabian. 1986. Identification of a novel Y branch structure as an intermediate in trypanosome mRNA processing: Evidence for *trans* splicing. *Cell* 47:517–25.

Owen, D. and L.C. Kühn. 1987. Noncoding 3′ sequences of the transferrin receptor gene are required for mRNA regulation by iron. *EMBO Journal* 6:1287–93.

Seiwert, S.D. and K. Stuart. 1994. RNA editing: Transfer of genetic information from gRNA to precursor mRNA in vitro. *Science* 266:114–17.

Wellauer, P.K. and I.B. David. 1973. Secondary structure map of RNA: Processing of HeLa ribosomal RNA. *Proceedings of the National Academy of Sciences, USA* 70:2827–31.

Weinberg, R.A., and S. Penman. 1970. Processing of 45S nucleolar RNA. *Journal of Molecular Biology* 47:169–78.

Zamore, P.D., T. Tuschl, P.A. Sharp, and D.P. Bartel. 2000. RNAi: Double-stranded RNA directs the ATP-dependent cleavage of mRNA at 21 to 23 nucleotide intervals. *Cell* 101:25–33.

CHAPTER 17

The Mechanism of Translation I: Initiation

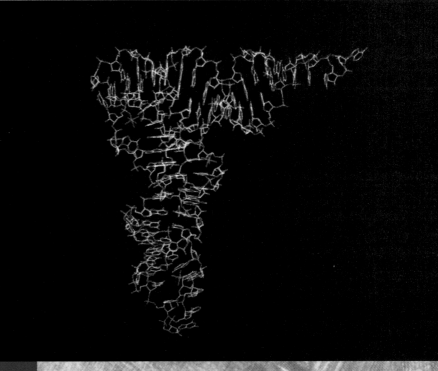

Computer model of a tRNA. © *Tripos Associates/Peter Arnold, Inc.*

Translation is the process by which ribosomes read the genetic message in mRNA and produce a protein product according to the message's instructions. Ribosomes therefore serve as protein factories. Transfer RNAs (tRNAs) play an equally important role as adaptors that can bind an amino acid at one end and interact with the mRNA at the other. Chapter 3 presented an outline of the translation process. In this chapter we will begin to fill in some of the details.

We can conveniently divide the mechanism of translation into three phases: initiation, elongation, and termination. In the initiation phase, the ribosome binds to the mRNA, and the first amino acid, attached to its tRNA, also binds. During the elongation phase, the ribosome adds one amino acid at a time to the growing polypeptide chain. Finally, in the termination phase, the ribosome releases the mRNA and the finished polypeptide. The overall scheme is similar in prokaryotes and eukaryotes, but there are significant differences, especially in the added complexity of the eukaryotic translation initiation system.

This chapter concerns the initiation of translation in eukaryotes and prokaryotes. Because the nomenclatures of the two systems are different, it is easier to consider them separately. Therefore, let us begin with a discussion of the simpler system, initiation in prokaryotes. Then we will move on to the more complex eukaryotic scheme. ∎

17.1 Initiation of Translation in Prokaryotes

Two important events must occur even before translation initiation can take place. One of these prerequisites is to generate a supply of **aminoacyl-tRNAs** (tRNAs with their cognate amino acids attached). In other words, amino acids must be covalently bound to tRNAs. This process is called **tRNA charging**; the tRNA is said to be "charged" with an amino acid. Another preinitiation event is the dissociation of ribosomes into their two subunits. This is necessary because the cell assembles the initiation complex on the small ribosomal subunit, so the two subunits must separate to make this assembly possible.

tRNA Charging

All tRNAs have the same three bases (CCA) at their 3′-ends, and the terminal adenosine is the target for charging. An amino acid is attached by an ester bond between its carboxyl group and the 2′- or 3′-hydroxyl group of the terminal adenosine of the tRNA, as shown in Figure 17.1. Charging takes place in two steps (Figure 17.2), both catalyzed by the enzyme **aminoacyl-tRNA synthetase**. In the first reaction, the amino acid is activated, using energy from ATP; the product of the reaction is aminoacyl-AMP. The pyrophosphate by-product is simply the two end phosphate groups (the β- and γ-phosphates), which the ATP lost in forming AMP.

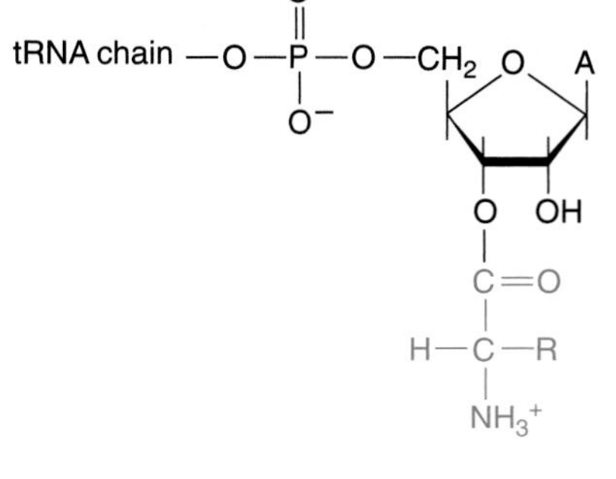

Figure 17.1 Linkage between tRNA and an amino acid. Some amino acids are bound initially by an ester linkage to the 3′-hydroxyl group of the terminal adenosine of the tRNA as shown, but some bind initially to the 2′-hydroxyl group. In any event, the amino acid is transferred to the 3′-hydroxyl group before it is incorporated into a protein.

(1) amino acid + ATP → aminoacyl-AMP + pyrophosphate (PP$_i$)

The bonds between phosphate groups in ATP (and the other nucleoside triphosphates) are high-energy bonds. When they are broken, this energy is released. In this case, the energy is trapped in the **aminoacyl-AMP**, which is why we call this an **activated amino acid**. In the second reaction of charging, the energy in the aminoacyl-AMP is used to transfer the amino acid to a tRNA, forming **aminoacyl-tRNA**.

(2) aminoacyl-AMP + tRNA → aminoacyl-tRNA + AMP

The sum of reactions 1 and 2 is this:

(3) amino acid + ATP + tRNA → aminoacyl-tRNA + AMP + PP$_i$.

Just like other enzymes, an aminoacyl-tRNA synthetase plays a dual role. Not only does it catalyze the reaction leading to an aminoacyl-tRNA, but it determines the specificity of this reaction. Only 20 synthetases exist, one for each amino acid, and they are very specific. Each will almost always place an amino acid on the right kind of tRNA. This is essential to life: If the aminoacyl-tRNA synthetases made many mistakes, proteins would be put together with a correspondingly large number of incorrect amino acids and could not function properly. We will return to this theme and see how the synthetases select the proper tRNAs and amino acids in Chapter 19.

SUMMARY Aminoacyl-tRNA synthetases join amino acids to their cognate tRNAs. They do this very specifically in a two-step reaction that begins with activation of the amino acid with AMP, derived from ATP.

Dissociation of Ribosomes

We learned in Chapter 3 that ribosomes consist of two subunits. The 70S ribosomes of *E. coli*, for example, contain one 30S and one 50S subunit. Each subunit has one or two ribosomal RNAs and a large collection of ribosomal proteins. We will see shortly that both prokaryotic and eukaryotic cells build translation initiation complexes on the small ribosomal subunit. This implies that the two ribosomal subunits must dissociate after each round of translation for a new initiation complex to form. But as early as 1968, Matthew Meselson and colleagues provided direct evidence for the dissociation of ribosomes, using an experiment outlined in Figure 17.3. These workers labeled *E. coli* ribosomes with heavy isotopes of nitrogen (^{15}N), carbon (^{13}C), and hydrogen (deuterium), plus a little ^{3}H as a radioactive tracer. The ribosomes so labeled became much denser than their normal counterparts grown in ^{14}N, ^{12}C, and hydrogen, as illustrated in Figure 17.4a. Next, the investigators placed cells with

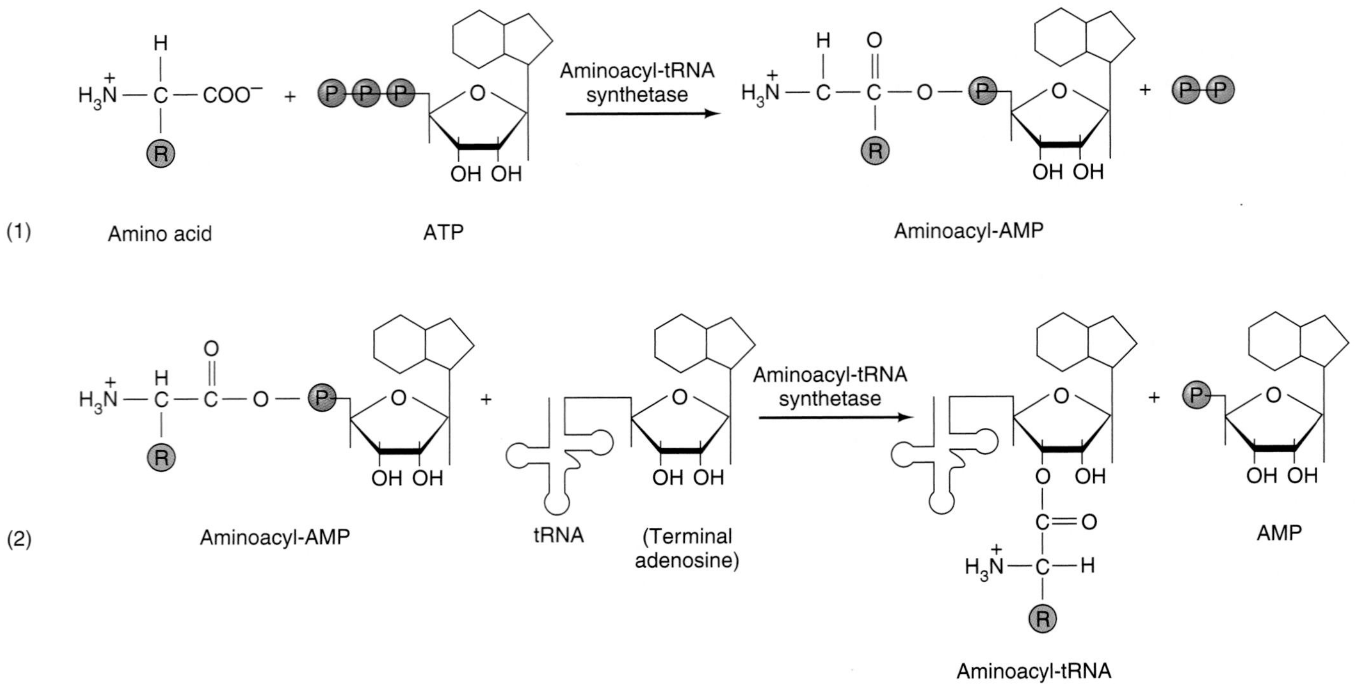

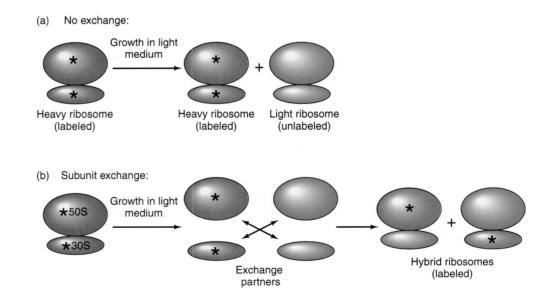

Figure 17.2 Aminoacyl-tRNA synthetase activity. Reaction 1: The aminoacyl-tRNA synthetase couples an amino acid to AMP, derived from ATP, to form an aminoacyl-AMP, with pyrophosphate (P-P) as a by-product. Reaction 2: The synthetase replaces the AMP in the aminoacyl-AMP with tRNA, to form an aminoacyl-tRNA, with AMP as a by-product. The amino acid is joined to the 3′-hydroxyl group of the terminal adenosine of the tRNA. For simplicity, the adenine base is merely indicated in outline.

Figure 17.3 Experimental plan to demonstrate ribosomal subunit exchange. Meselson and colleagues made ribosomes heavy (red) by growing *E. coli* in the presence of heavy isotopes of nitrogen, carbon, and hydrogen, and made them radioactive (asterisks) by including some ^{3}H. Then they shifted the cells with labeled, heavy ribosomes to light medium containing the standard isotopes of nitrogen, carbon, and hydrogen. (a) No exchange. If no ribosome subunit exchange occurs, the heavy ribosomal subunits will stay together, and the only labeled ribosomes observed will be heavy. The light ribosomes made in the light medium will not be detected because they are not radioactive. (b) Subunit exchange. If the ribosomes dissociate into 50S and 30S subunits, heavy subunits can associate with light ones to form labeled hybrid ribosomes.

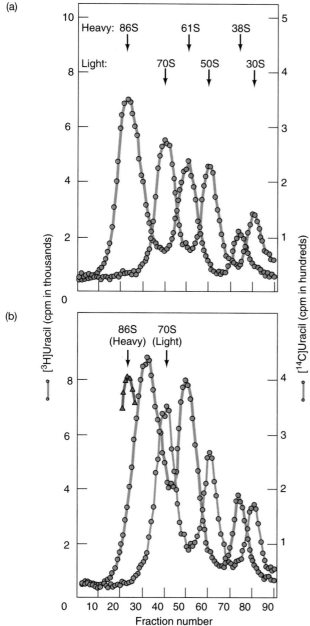

Figure 17.4   Demonstration of ribosomal subunit exchange.
(a) Sedimentation behavior of heavy and light ribosomes. Meselson and coworkers made heavy ribosomes labeled with [³H]uracil as described in Figure 17.3, and light (ordinary) ribosomes labeled with [¹⁴C]uracil. Then they subjected these ribosomes to sucrose gradient centrifugation, collected fractions from the gradient, and detected the two radioisotopes by liquid scintillation counting. The positions of the light ribosomes and subunits (70S, 50S, and 30S; blue) and of the heavy ribosomes and subunits (86S, 61S, and 38S; red) are indicated at top. **(b)** Experimental results. Meselson and colleagues cultured *E. coli* cells with ³H-labeled heavy ribosomes as in panel (a) and shifted these cells to light medium for 3.5 generations. Then they extracted the ribosomes, added ¹⁴C-labeled light ribosomes as a reference, and subjected the mixture of ribosomes to sucrose gradient ultracentrifugation. They collected fractions and determined their radioactivity as in panel **(a)**: ³H, red; ¹⁴C, blue. The position of the 86S heavy ribosomes (green) was determined from heavy ribosomes centrifuged in a parallel gradient. The ³H-labeled ribosomes (leftmost red peak) were hybrids that sedimented midway between the light (70S) and heavy (86S) ribosomes. (*Source:* Reprinted from *Journal of Molecular Biology,* vol. 31, R. O. R. Kaempfer, M. Meselson and H. J. Raskas, "Cyclic Dissociation into Stable Subunits and Re-formation of Ribosomes during Bacterial Growth," 277–289, 1968, by permission of Academic Press Limited, London.)

labeled, heavy ribosomes in medium with ordinary light isotopes of nitrogen, carbon, and hydrogen. After 3.5 generations, they isolated the ribosomes and measured their masses by sucrose density gradient centrifugation with ¹⁴C-labeled light ribosomes for comparison. Figure 17.4b shows the results. As expected, they observed heavy radioactively labeled ribosomal subunits (38S and 61S instead of the standard 30S and 50S). But the labeled whole ribosomes had a hybrid sedimentation coefficient, in between the standard 70S and the 86S they would have had if both subunits were heavy. This indicated that subunit exchange had occurred. Heavy ribosomes had dissociated into subunits and taken new, light partners.

More precise resolution of the ribosomes on CsCl gradients demonstrated two species: one with a heavy large subunit and a light small subunit, and one with a light large subunit and a heavy small subunit, as predicted in Figure 17.3. These workers performed the same experiments on yeast cells and obtained the same results, so eukaryotic ribosomes also cycle between intact ribosomes (80S) and ribosomal subunits (40S and 60S).

Because few biological reactions occur spontaneously, we would predict that the dissociation of bacterial ribosomes into subunits would require participation by outside agents. Indeed, *E. coli* cells have three **initiation factors: IF-1, IF-2,** and **IF-3,** two of which (IF-1 and IF-3) take part in ribosome dissociation. Severo Ochoa and his colleagues isolated the initiation factors in 1966 by washing ribosomes with buffers of progressively higher salt concentration. They extracted IF-1 and -2 from ribosomes with a wash of 0.5 M NaCl, and IF-3 with a wash of 1.0 M NaCl. In 1970, Ochoa and his coworkers used sucrose gradient centrifugation to measure the relative amounts of ribosomes and ribosomal subunits in the presence and absence of IF-3 or IF-1. The results in Figure 17.5 demonstrate that a highly purified preparation of IF-3 could apparently cause dissociation of 70S *E. coli* ribosomes into 30S and 50S subunits, but that IF-1 could not. A similar experiment showed that IF-2 was also ineffective as a ribosome dissociation factor.

Steven Sabol and Ochoa went on to show that IF-3 binds to 30S ribosomal particles but does not remain bound when 30S and 50S particles get together to form intact ribosomes. They labeled IF-3 with ³⁵S, then added various amounts of this labeled factor to ribosomes from *E. coli* strain MRE600, which has an unusually high concentration of free, single ribosomes under certain conditions. Then they separated ribosomes and subunits by sucrose gradient centrifugation. Figure 17.6 shows that the more IF-3 these workers added to the ribosomes, the more dissociation occurred. Moreover, the labeled IF-3 was associated exclusively with the 30S particles, not with the intact ribosomes. Then they allowed the 30S particles, with labeled IF-3 attached, to combine with 50S particles to form ribosomes. Sucrose gradient centrifugation

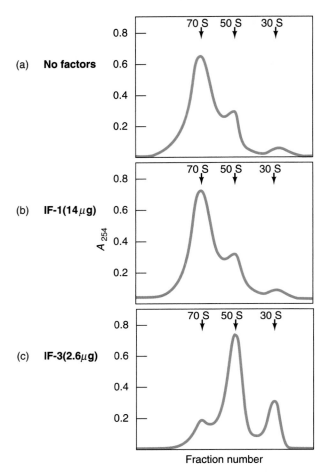

Figure 17.5 Effect of IF-3 in dissociating ribosomes. Ochoa and colleagues mixed *E. coli* ribosomes with (a) no factors, (b) IF-1 , or (c) IF-3 , then separated intact ribosomes from ribosomal subunits by sucrose gradient ultracentrifugation. They assayed the concentration of ribosomes or subunits in each fraction by measuring the absorbance at 254 nm (A_{254}), the wavelength at which RNA, including ribosomal RNA, absorbs maximally. Only IF-3 caused a significant dissociation. (*Source:* Reprinted with permission from *Nature* 228:1271, 1970. Copyright © 1970 Macmillan Magazines Limited.)

revealed that the IF-3 remained attached to free 30S particles, but was utterly absent from intact ribosomes. Thus, IF-3 dissociates from 30S particles before they are incorporated into ribosomes.

These experiments leave the impression that IF-3 is actively promoting the dissociation of ribosomes and that the other two factors have no role to play. But the truth is somewhat different: IF-1 actually promotes dissociation, and IF-3 binds to the free 30S subunits and prevents their reassociation with 50S subunits. The experiments we have just examined did not really measure the rate of dissociation of ribosomes. Instead, they measured the equilibrium concentrations of intact ribosomes and ribosomal subunits. Those concentrations reflect both dissociation and association rates.

To assay dissociation itself, Markus Noll and Hans Noll added ^{14}C-labeled 50S subunits to intact ribosomes and measured the rate of incorporation of ^{14}C into intact ribosomes. This sounds at first like an assay for association, not dissociation. However, because the rate of association is much faster than the rate of dissociation, the incorporation of labeled 50S subunits into 70S ribosomes depends on the slow dissociation of unlabeled ribosomes. That is because disassociation of ribosomes provides unlabeled 30S subunits with which the labeled 50S subunits can combine (Figure 17.7). Figure 17.8 shows the result of this experiment, in which all three initiation factors were used. These factors accelerated the incorporation of labeled 50S subunits into 70S ribosomes. Later experiments with individual initiation factors showed that IF-1, but not IF-2 or IF-3, gave the same amount of stimulation as all three factors together. The modest magnitude of this acceleration—about threefold—explains why Ochoa and colleagues did not observe an effect of IF-1 in the experiments that measured the equilibrium concentrations of subunits and ribosomes: The equilibrium lies so far in the direction of intact ribosomes that a threefold boost in dissociation rate still leaves the vast majority of the ribosomes in the associated state.

If IF-1 is actually the factor that promotes ribosome dissociation, what is the role of IF-3? We know that IF-3 has the strongest effect on the equilibrium of ribosomes and subunits, and that it binds to the 30S subunit and remains bound to it until a new 50S subunit attaches. These findings are all compatible with the scheme pictured in Figure 17.9: IF-1 stimulates ribosome dissociation, and IF-3 binds to the free 30S subunits and prevents them from reassociating with 50S subunits to re-form ribosomes. This interference with the reassociation of subunits has a profound effect on the equilibrium between ribosomes and free subunits, favoring the build-up of subunits.

Raymond Kaempfer and Meselson devised a cell-free translation system in which ribosome subunit cycling still occurred. Inhibitors of RNA and DNA synthesis had no effect on this ribosome subunit exchange, but four different inhibitors of protein synthesis blocked it. Thus, translation seems to be necessary for ribosomal subunit cycling, but is cycling necessary for translation? If so, it would suggest that ribosomes cycle at the end of each round of translation.

Christine Guthrie and Masayasu Nomura provided the needed evidence in 1968. To understand their first experiment we need to give a preview of some information about a special initiating aminoacyl-tRNA that we will discuss in more detail later in this chapter: The first codon in most messages is AUG. An AUG in the interior of a message codes for methionine, but an AUG at the beginning of a message is the initiation codon and attracts a special aminoacyl-tRNA bearing an N-formyl methionine (fMet).

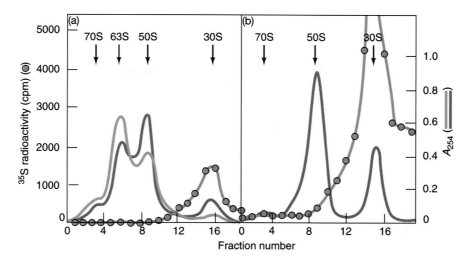

Figure 17.6 Binding of IF-3 to 30S ribosomal subunits. Sabol and Ochoa mixed *E. coli* strain MRE600 ribosomes with varying concentrations of 35S-labeled IF-3, then subjected the ribosomes to sucrose gradient ultracentrifugation and measured the concentration of ribosomes and subunits (A254) and IF-3 (35S radioactivity) in each fraction. (a) Profiles of ribosomes and subunits with no IF-3 (blue) and with 22 pmol of IF-3 (brown); profile of 35S-IF-3 (red). (b) Profile of ribosomes and subunits with 109 pmol of IF-3 (brown); profile of IF-3 (red). Note that IF-3 is associated only with the 30S subunit. (*Source:* Reprinted with permission from *Nature New Biology* 234:234, 1971. Copyright © 1971 Macmillan Magazines Limited.)

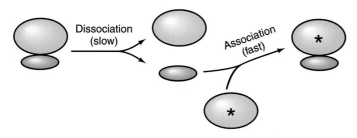

Figure 17.7 Assay for ribosome dissociation. Noll and Noll assayed ribosome dissociation indirectly by adding 14C-50S ribosomal subunits (asterisk) to intact ribosomes and measuring the rate of uptake of the labeled subunits into whole ribosomes. This rate depends on the slow rate of dissociation of ribosomes to provide 30S partners for the labeled 50S subunits.

Guthrie and Nomura made ribosomes heavy with 15N and D2O, then mixed them with some light 50S subunits, a synthetic mRNA containing AUG and GUG codons, initiation factors, and two aminoacyl-tRNAs ([3H]fMet-tRNA and [14C]Val-tRNA). The idea was that the fMet-tRNA would bind to ribosomes that were initiating translation, but the Val-tRNA would bind to ribosomes that mimicked the elongation mode. Because the ready-made heavy/heavy ribosomes could bind to interior valine codons (GUGs) in this synthetic system, they would seem to be in the elongation phase, and the Val-tRNA would be expected to bind to them. However, if ribosome cycling must occur before true initiation, the fMet-tRNA should bind only to ribosomes that have cycled. These ribosomes should be hybrids formed from a heavy 30S subunit and a light 50S subunit, as pictured in Figure 17.10a. In fact, this is just what Nomura observed, as we see in

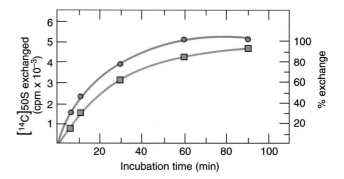

Figure 17.8 Initiation factors stimulate the rate of ribosomal subunit exchange. Noll and Noll measured the extent of ribosomal subunit exchange as a function of time, using 14C-labeled 50S subunits, by sucrose gradient ultracentrifugation. The exchanges were performed without (blue squares) or with (red circles) initiation factors. The blue line represents the theoretical curve for spontaneous exchange. Factors caused a threefold increase in the initial rate of exchange. Further experiments showed that the only factor that had this effect was IF-1. (*Source:* Reprinted with permission from *Nature New Biology* 238:226, 1972. Copyright © 1972 Macmillan Magazines Limited.)

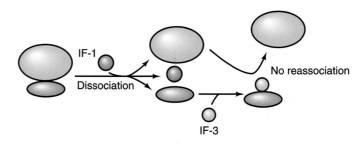

Figure 17.9 Roles of IF-1 and IF-3 in ribosome dissociation. IF-1 (green) promotes dissociation of the ribosomal subunits (pink and blue). IF-3 (tan) then binds to the free 30S subunit and prevents its reassociation with the free 50S subunit.

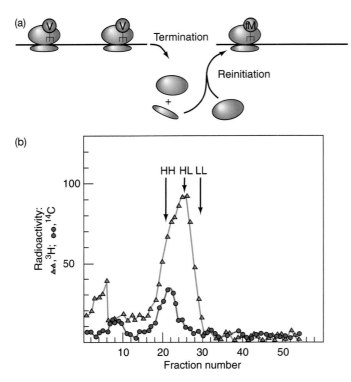

Figure 17.10 Demonstration that ribosome cycling is necessary for reinitiation of translation. (a) Experimental plan. Guthrie and Nomura made heavy ribosomes (red) by growing *E. coli* cells in ^{15}N medium. They added light 50S ribosomal subunits (blue), along with [^{14}C]Valine-tRNA (green) and [^{3}H]fMet-tRNA (magenta), then used gradient ultracentrifugation to measure the densities of ribosomes with which the two labeled aminoacyl-tRNAs associated. The valine-tRNA is involved in elongation and should therefore associate with preformed, heavy ribosomes, as pictured at left. On the other hand, if ribosome cycling is required for reinitiation, a significant amount of fMet-tRNA should associate with hybrid heavy/light ribosomes (red and blue) as pictured at right. **(b)** Experimental results. Guthrie and Nomura fixed the ribosomes with formaldehyde, separated the heavy, light, and hybrid ribosomes by CsCl gradient ultracentrifugation, and plotted the positions of the [^{3}H]fMet-tRNA (magenta) and [^{14}C]Val-tRNA (green). The positions of heavy/heavy (HH), heavy/light (HL), and light/light (LL) ribosomes are given at top. The Val-tRNA associated with the uncycled, HH ribosomes, but the fMet-tRNA associated primarily with cycled, hybrid (HL) ribosomes. (*Source:* (*b*) Reprinted with permission from *Nature* 219:234, 1968. Copyright © 1968 Macmillan Magazines Limited.)

Figure 17.10b. It is not surprising that some fMet-tRNA is also bound to heavy/heavy 70S ribosomes: Because the ratio of light 50S particles to heavy 70S ribosomes was only 2/3, there was a good chance that a dissociated heavy 30S subunit would reassociate with a heavy 50S subunit, forming a heavy/heavy 70S ribosome.

SUMMARY *E. coli* ribosomes dissociate into subunits at the end of each round of translation. IF-1 actively promotes this dissociation, and IF-3 binds to the free 30S subunit and prevents its reassociation with a 50S subunit to form a whole ribosome.

Formation of the 30S Initiation Complex

Once IF-1 and -3 have done their job in dissociating ribosomes into 50S and 30S subunits, the cell builds a complex on the 30S ribosomal subunit, including mRNA, aminoacyl-tRNA, and initiation factors. This is known as the **30S initiation complex.** We have already seen that IF-3 is capable of binding by itself to 30S subunits, but IF-1 and IF-2 stabilize this binding. Similarly, IF-2 can bind to 30S particles, but achieves much more stable binding with the help of IF-1 and -3. IF-1 does not bind by itself, but does so with the assistance of the other two factors. All three factors bind close together at a site on the 30S subunit near the 3′-end of the 16S rRNA. Once the three initiation factors have bound, they attract two other key players to the complex: mRNA and the first aminoacyl-tRNA. The order of binding of these two substances appears to be random.

The First Codon and the First Aminoacyl tRNA In 1964, Fritz Lipmann showed that digestion of leucyl-tRNA with RNase yielded the adenosyl ester of leucine (Figure 17.11a). This is what we expect, because we know that the amino acid is bound to the 3′-hydroxyl group of the terminal adenosine of the tRNA. However, when K.A. Marcker and Frederick Sanger tried the same procedure with methionyl-tRNA, they found not only the expected adenosyl-methionine ester, but also an adenosyl-N-formyl methionine ester (Figure 17.11b). This demonstrated that the tRNA with which they started was esterified, not only to methionine, but also to a methionine derivative, **N-formyl-methionine,** which is abbreviated **fMet.** Figure 17.11c compares the structures of methionine and N-formyl methionine.

Next, B.F.C. Clark and Marcker showed that *E. coli* cells contain two different tRNAs that can be charged with methionine. They separated these two tRNAs by an old preparatory method called countercurrent distribution. The faster moving tRNA, now called **tRNA$_m^{Met}$** could be charged with methionine, but the methionine could not be formylated. That is, it could not accept a formyl group onto its amino group. The slower moving tRNA was called **tRNA$_f^{Met}$**, to denote the fact that the methionine attached to it could be formylated. Notice that the methionine formylation takes place *on* the tRNA. The tRNA cannot be charged directly with formyl-methionine. Clark and Marcker went on to test the two tRNAs for two properties: (1) the codons they respond to, and (2) the positions within the protein into which they placed methionine.

The assay for codon specificity used a method introduced by Marshall Nirenberg, which we will describe more fully later in this chapter. The strategy is to make a labeled aminoacyl-tRNA, mix it with ribosomes and a variety of trinucleotides, such as AUG. A trinucleotide that

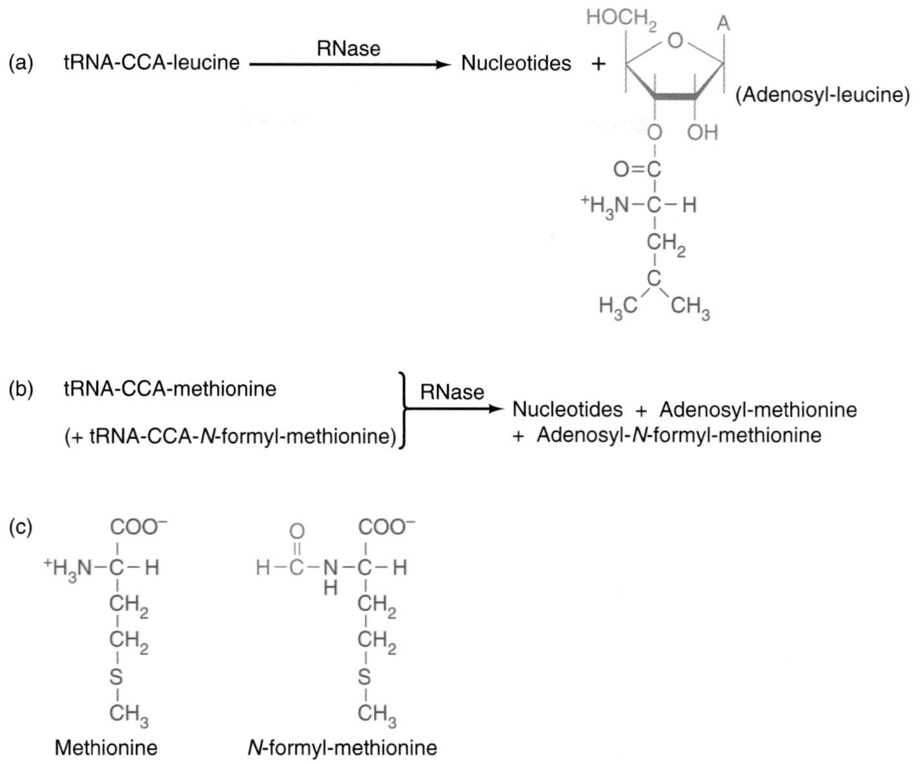

(a) tRNA-CCA-leucine $\xrightarrow{\text{RNase}}$ Nucleotides + (Adenosyl-leucine)

(b) tRNA-CCA-methionine

(+ tRNA-CCA-*N*-formyl-methionine) $\Big] \xrightarrow{\text{RNase}}$ Nucleotides + Adenosyl-methionine + Adenosyl-*N*-formyl-methionine

(c)

Methionine   *N*-formyl-methionine

Figure 17.11 Discovery of *N*-formyl-methionine. (a) Lipmann and colleagues degraded leucyl-tRNA with RNase to yield nucleotides plus adenosyl-leucine. The leucine was attached to the terminal A of the ubiquitous CCA sequence at the 3′-end of the tRNA. **(b)** Marcker and Sanger performed the same experiment with what they assumed was pure methionyl-tRNA. However, they obtained a mixture of adenosyl-amino acids: adenosyl-methionine and adenosyl-*N*-formyl-methionine, demonstrating that the aminoacyl-tRNA with which they started was a mixture of methionyl-tRNA and *N*-formyl-methionyl-tRNA. **(c)** Structures of methionine and *N*-formyl-methionine, with the formyl group of fMet highlighted in red.

codes for a given amino acid will usually cause the appropriate aminoacyl-tRNA to bind to the ribosomes. In the case at hand, tRNA$_m^{\text{Met}}$ responded to the codon AUG, whereas tRNA$_f^{\text{Met}}$ responded to AUG, GUG, and UUG. As we have already indicated, tRNA$_f^{\text{Met}}$ is involved in initiation, which suggests that all three of these codons, AUG, GUG, and UUG, can serve as initiation codons. Indeed, sequencing of many *E. coli* genes has confirmed that AUG is the initiating codon in more than 90% of the genes, whereas GUG and UUG are initiating codons in about 8% and 1% of the genes, respectively.

Next, Clark and Marcker determined the positions in the protein chain in which the two tRNAs placed methionines. To do this, they used an in vitro translation system with a synthetic mRNA that had AUG codons scattered throughout it. When they used tRNA$_m^{\text{Met}}$, methionines were incorporated primarily into the interior of the protein product. By contrast, when they used tRNA$_f^{\text{Met}}$, methionines went only into the first position of the polypeptide. Thus, tRNA$_f^{\text{Met}}$ appears to serve as the initiating aminoacyl-tRNA. Is this due to the formylation

of the amino acid, or to some characteristic of the tRNA? To find out, Clark and Marcker tried their experiment with formylated and unformylated methionyl-tRNA$_f^{\text{Met}}$. They found that formylation made no difference; in both cases, this tRNA directed incorporation of the first amino acid. Thus, the tRNA part of formyl-methionyl-tRNA$_f^{\text{Met}}$ is what makes it the initiating aminoacyl-tRNA.

Martin Weigert and Alan Garen reinforced the conclusion that tRNA$_f^{\text{Met}}$ is the initiating aminoacyl-tRNA with an in vivo experiment. When they infected *E. coli* with R17 phage and isolated newly synthesized phage coat protein, they found fMet in the N-terminal position, as it should be if it is the initiating amino acid. Alanine was the second amino acid in this new coat protein. On the other hand, mature phage R17 coat protein has alanine in the N-terminal position, so maturation of this protein must involve removal of the N-terminal fMet. Examination of many different bacterial and phage proteins has shown that the fMet is frequently removed. In some cases the methionine remains, but the formyl group is always removed.

SUMMARY The initiation codon in prokaryotes is usually AUG, but it can also be GUG, or more rarely, UUG. The initiating aminoacyl-tRNA is N-formyl-methionyl-tRNA$_f^{Met}$. N-formyl-methionine (fMet) is therefore the first amino acid incorporated into a polypeptide, but it is frequently removed from the protein during maturation.

Binding mRNA and fMet-tRNA$_f^{Met}$ to the 30S Ribosomal Subunit We have seen that the initiating codon is AUG, or sometimes GUG or UUG. But these codons also occur in the interior of a message. An interior AUG codes for ordinary methionine, and GUG and UUG code for valine and leucine, respectively. How does the cell detect the difference between an initiation codon and an ordinary codon with the same sequence? Two explanations come readily to mind: Either a special primary structure (RNA sequence) or a special secondary RNA structure occurs near the initiation codon that identifies it as an initiation codon and allows the ribosome to bind there. In 1969, Joan Argetsinger Steitz searched for such distinguishing characteristics in the mRNA from an *E. coli* phage called R17. This phage belongs to a group of small spherical RNA phages, which also includes phages f2 and MS2. These are **positive strand phages,** which means that their genomes are also their mRNAs. Thus, these phages provide a convenient source of pure mRNA. These phages are also very simple; each has only three genes, which encode the A protein (or maturation protein), the coat protein, and the replicase. Steitz searched the neighborhoods of the three initiation codons in phage R17 mRNA for distinguishing primary or secondary structures. She began by binding ribosomes to R17 mRNA under conditions in which the ribosomes would remain at the initiation sites. Then she digested the RNA not protected by ribosomes with RNase A. Finally, she sequenced the initiation regions protected by the ribosomes. She found no obvious sequence or secondary structure similarities around the start sites.

In fact, subsequent work has shown that the secondary structures at all three start sites are inhibitory; relaxing these secondary structures actually enhances initiation. This is particularly true of the replicase gene, in which the initiation codon is buried in a double-stranded structure that also involves part of the coat gene, as illustrated for MS2 phage RNA in Figure 17.12a. This explains why the replicase gene of these phages cannot be translated until the coat protein is translated: The ribosomes moving through the coat gene open up the secondary structure hiding the initiation codon of the replicase gene (Figure 17.12b).

We have seen that secondary structure does not identify the start codons, and the first start site sequences did not reveal any obvious similarities, so what does constitute a ribosome binding site? The answer is that it *is* a special sequence, but sometimes, as in the case of the R17 coat protein gene, it diverges so far from the consensus sequence that it is hard to recognize. Richard Lodish and his colleagues laid some of the groundwork for the discovery of this sequence in their work on the translation of the f2 coat mRNA by ribosomes from different bacteria. They found that *E. coli* ribosomes could translate all three f2 genes in vitro, but that ribosomes from the bacterium *Bacillus stearothermophilus* could translate only the A protein gene. The real problem was in translating the coat gene; as we have seen, the translation of the replicase gene depends on translating the coat gene, so the inability of *B. stearothermophilus* ribosomes to translate the f2 replicase gene was simply an indirect effect of its inability to translate the coat gene. With mixing experiments, Lodish and coworkers demonstrated that the *B. stearothermophilus* ribosomes, not the initiation factors, were at fault.

Next, Nomura and his colleagues performed more detailed mixing experiments using R17 phage RNA. They found that the important element lay in the 30S ribosomal subunit. If the 30S subunit came from *E. coli,* the R17 coat gene could be translated. If it came from *B. stearothermophilus,* this gene could not be translated. Finally, they dissociated the 30S subunit into its RNA and protein components and tried them in mixing experiments. This time, two components stood out: one of the ribosomal proteins, called S12, and the 16S ribosomal RNA. If either of these components came from *E. coli,* translation of the coat gene was active. If either came from *B. stearothermophilus,* translation was depressed (though not as much as if the whole ribosomal subunit came from *B. stearothermophilus*).

These findings stimulated J. Shine and L. Dalgarno to look for possible interactions between the 16S rRNA and sequences around the start sites of the R17 genes. They noted that all binding sites contained, just upstream of the initiation codon, all or part of this sequence: AGGAGGU, which is complementary to the boldfaced part of the following sequence, found at the very 3'-end of *E. coli* 16S rRNA: 3'HO-AUUCCUCCAC5'. Note that the hydroxyl group denotes the 3'-end of the 16S rRNA, and that this sequence is written 3' → 5', so its complementarity to the AGGAGGU sequence is obvious. This relationship is very provocative, especially considering that the complementarity between the coat protein sequence and the 16S rRNA is the weakest of the three genes, and therefore would be likely to be the most sensitive to alterations in the sequence of the 16S rRNA.

The story gets even more intriguing when we compare the sequences of the *E. coli* and *B. stearothermophilus* 16S rRNAs and find an even poorer match between the R17 coat ribosome binding site and the *Bacillus* 16S rRNA. Table 17.1 presents these comparisons, along with

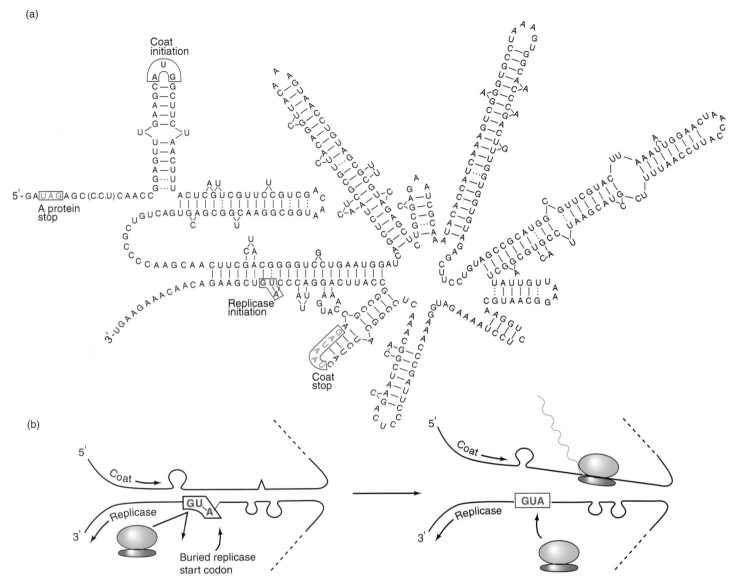

Figure 17.12 Potential secondary structure in MS2 phage RNA and its effect on translation. (a) The sequence of the coat gene and surrounding regions in the MS2 RNA, and potential secondary structure. Independent evidence suggests that the "petals" in this "flower model" really do form. Initiation and termination codons are boxed and labeled. **(b)** Effect of translation of coat gene on replicase translation. At left, the coat gene is not being translated, and the replicase initiation codon (AUG, green, written right to left here) is buried in a stem that is base-paired to part of the coat gene. Thus, the replicase gene cannot be translated. At right, a ribosome is translating the coat gene. This disrupts the base pairing around the replicase initiation codon and opens it up to ribosomes that can now translate the replicase gene. (*Source:* (a) Reprinted with permission from *Nature* 237:84, 1972. Copyright © 1972 Macmillan Magazines Limited.)

the results of a ribosome-binding assay with unfolded R17 RNA, in which secondary structure did not play a role. Note that the *Bacillus* 16S rRNA can make four Watson–Crick base pairs with the A protein and replicase ribosome-binding sites, but only two such base pairs with the coat protein gene. The *E. coli* 16S rRNA can make at least three base pairs with the ribosome-binding sites of all three genes. Could the base pairing between 16S rRNA and the region upstream of the translation initiation site be vital to ribosome binding? If so, it would explain the

inability of the *Bacillus* ribosomes to bind to the R17 coat protein initiation site, and it would also identify the AGGAGGU sequence as the ribosome-binding site. As we will see, other evidence shows that this really is the ribosome-binding site, and it has come to be called the **Shine–Dalgarno sequence,** or **SD sequence,** in honor of its discoverers.

To bolster their hypothesis, Shine and Dalgarno isolated ribosomes from two other bacterial species, *Pseudomonas aeruginosa* and *Caulobacter crescentus*, sequenced

Table 17.1 R17 Ribosome-Binding Site Sequences and Proposed Pairing with the 3'-Terminus of 16S RNA*

Ribosome source	R17 cistron	Possible pairing of 3'-terminus with appropriate region of ribosome-binding site sequence	Number of base pairs possible	Ribosomes binding to unfolded R17, MS2, or f2 RNA
E. coli B		_{OH}AUUCCUCCACPy ⟵ 16S RNA		
	A protein	(5') CU<u>AGGAGGU</u>UU(3')	7	+
	Replicase	ACAU<u>GAGG</u>AUU	4	+
	Coat protein	ACC<u>GGGG</u>UUUG	3(4)‡	+
		↓		
		[A]†		
B. stearothermophilus		_{OH}AUCUUUCCUPy ⟵ 16S RNA		
	A protein	(5')AUCCU<u>AGGA</u>G(3')	>4	+
	Replicase	ACAUG<u>AGGA</u>U	4	+
	Coat protein	AACCGG<u>GG</u>UU	2(1)‡	−
		↓		
		A†		
P. aeruginosa		_{OH}AUUCCUCUCPy ⟵ 16S RNA		
	A protein	(5') CU<u>AGGAG</u>GUU(3')	5	+
	Replicase	UG<u>AGGA</u>UUAC	4	+
	Coat protein	ACC<u>GGGG</u>UUU	3(4)‡	+
		↓		
		[A]†		
C. crescentus		_{OH}UCUUUCCUPy ⟵ 16S RNA		
	A protein	(5')UCCU<u>AGGA</u>G(3')	>4	Not done
	Replicase	CAUG<u>AGGA</u>U	4	Not done
	Coat protein	ACCGG<u>GG</u>UU	2(1)‡	Not done
		↓		
		A†		

Source: Reprinted with permission from *Nature* 254:37, 1975. Copyright © 1975 Macmillan Magazines Limited.

*The 3'-terminal sequences of the 16S rRNAs from each bacterial species are given above the sequences of the ribosome-binding regions of each cistron of R17 (or f2 or MS2) phage.

†The third G in the GGGG sequence of the coat cistron ribosome-binding site was mutated to A in some of the R17 phages used in this experiment. This caused a difference in the number of base pairs that could form between the R17 initiation site and the 16S rRNAs.

‡The number of base pairs that form with the GGAG variant is given in parentheses.

the 3'-ends of their 16S rRNAs, and tested the ribosomes for the ability to bind to the three R17 initiation sites. In accord with their other results, they found that whenever three or more base pairs were possible between the 16S rRNA and the sequence upstream of the initiation codon, ribosome binding occurred. Whenever fewer than 3 bp were possible, no ribosome binding occurred.

Steitz and Karen Jakes added strong evidence in favor of the Shine–Dalgarno hypothesis. They bound *E. coli* ribosomes to the R17 A protein gene's initiation region, then treated the complexes with a sequence-specific RNase called colicin E3, which cuts near the 3'-end of the 16S rRNA of *E. coli*. Next, they fingerprinted the RNA and found a double-stranded RNA fragment, as pictured in Figure 17.13. One strand of this RNA was an oligonucleotide from the A protein gene initiation site, including

the Shine–Dalgarno sequence. Base-paired to it was an oligonucleotide from the 3'-end of the 16S rRNA. This demonstrated directly that the Shine–Dalgarno sequence base-paired to the 3'-end of the 16S rRNA and left little doubt that this was indeed the ribosome binding site. It is also important to remember that prokaryotic mRNAs are usually polycistronic. That is, they contain information from more than one cistron, or gene. Each cistron represented on the mRNA has its own initiation codon and its own ribosome-binding site. Thus, ribosomes bind independently to each initiation site, and this provides a means for controlling gene expression, by making some initiation sites more attractive to ribosomes than others.

Anna Hui and Herman De Boer produced the best evidence for the importance of base pairing between the Shine–Dalgarno sequence and the 3'-end of the 16S rRNA

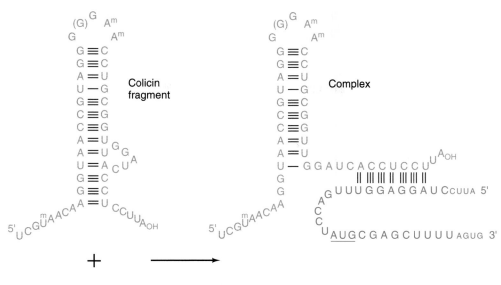

5' AUUcCUAGGAGGUUUGACCUAUGCGAGCUUUUAGUG 3'

R17 A protein initiator region

Figure 17.13 **Potential structure of the colicin fragment from the 3′-end of *E. coli* 16S rRNA and the initiator region of the R17 phage A protein cistron.** The initiation codon (AUG) is underlined. An "m" on the colicin fragment denotes a methylated base. The lines between bases represent relative strengths of base pairs: G-C > A-U > G-U. (*Source:* From J. A. Steitz and K. Jakes, "How ribosomes select initiator regions in mRNA," *Proceedings of the National Academy of Sciences* 72(12):4734–4738, December 1975.)

in 1987. They cloned a mutant human growth hormone gene into an *E. coli* expression vector bearing a wild-type Shine–Dalgarno (SD) sequence (GGAGG), which is complementary to the wild-type 16S rRNA anti-SD sequence (CCUCC). This gave high levels of human growth hormone protein. Then they mutated the SD sequence to either CCUCC or GUGUG, which would not base-pair with the anti-SD sequence on the 16S rRNA. Neither of these constructs produced very much human growth hormone. But the clincher came when they mutated the anti-SD sequence in the 16S rRNA gene to either GGAGG or CACAC, which restored the base pairing with CCUCC and GUGUG, respectively. Now the mRNA with the mutant CCUCC SD sequence was translated very well by the mutant cells with the 16S rRNA having the GGAGG anti-SD sequence, and the mRNA with the mutant GUGUG SD sequence was translated very well in cells with the 16S rRNA having the CACAC anti-SD sequence. This kind of complementation is very hard to explain unless base pairing occurs between these sequences.

What factors are involved in binding mRNA to the 30S ribosomal subunit? In 1969, Albert Wahba and colleagues showed that all three initiation factors are required for optimum binding, but that IF-3 is the most important of the three. They mixed ^{32}P-labeled mRNAs from two *E. coli* phages, R17 and MS2, and from tobacco mosaic virus (TMV), with ribosomal subunits and initiation factors, either singly or in combinations. These viruses all have RNA genomes that serve as mRNAs, so they are convenient sources of mRNAs for experiments like this. Table 17.2, experiment 1, shows the results. IF-2 or IF-2 + IF-1 showed little ability to cause R17 mRNA to bind to ribosomes, but IF-3 by itself could cause significant binding. IF-1 stimulated this binding further, and all three factors worked best of all. Thus, IF-3 seems to be the primary factor involved in mRNA binding to ribosomes, but the other two factors also assist in this task. We have seen that IF-3 is already bound to the 30S subunit, by virtue of its role in keeping 50S subunits from associating with the free 30S particles. The other two initiation factors also bind near the IF-3 binding site on the 30S subunit, where they can participate in assembling the 30S initiation complex.

SUMMARY The 30S initiation complex is formed from a free 30S ribosomal subunit plus mRNA and fMet-tRNA$_f^{Met}$. Binding between the 30S prokaryotic ribosomal subunit and the initiation site of a message depends on base pairing between a short RNA sequence called the Shine–Dalgarno sequence just upstream of the initiation codon, and a complementary sequence at the 3′-end of the 16S rRNA. This binding is mediated by IF-3, with help from IF-1 and IF-2. All three initiation factors have bound to the 30S subunit by this time.

Table 17.2   Role of Initiation Factors in Formation of the 30S Initiation Complex with Natural mRNA

Experiments	Ribosomes	mRNA	Factor additions	Ribosomal binding (pmol)	
				mRNA	fMet-tRNA$_f^{Met}$
1	30S + 50S	R17	IF-3	2.7	0.1
			IF-1 + IF-3	4.8	0.2
			IF-2	0.3	0.3
			IF-1 + IF-2	0.4	0.4
			IF-2 + IF-3	2.5	1.3
			IF-1 + IF-2 + IF-3	6.2	6.6
2	30S	MS2	IF-1 + IF-3		0.0
			IF-2		1.8
			IF-1 + IF-2		3.7
			IF-2 + IF-3		2.7
			IF-1 + IF-2 + IF-3		7.3
3	30S + 50S	TMV	IF-1 + IF-3		0.5
			IF-2		1.7
			IF-1 + IF-2		3.1
			IF-2 + IF-3		8.3
			IF-1 + IF-2 + IF-3		16.9

Source: From A. Wahba et al., "Initiation of Protein Synthesis in Escherichia coli, II," *Cold Spring Harbor Symp. Quant. Biology* 34:292, 1969. Copyright © 1969 Cold Spring Harbor Laboratory Press, Cold Spring Harbor, NY. Reprinted by permission.

Binding fMet-tRNA$_f^{Met}$ to the 30S Initiation Complex   If IF-3 bears the primary responsibility for binding mRNA to the 30S ribosome, which initiation factor plays this role for fMet-tRNA$_f^{Met}$? Table 17.2 shows that the answer is IF-2. This is especially apparent in experiments 2 and 3, using MS2 and TMV mRNA, respectively. In both cases, IF-1 and IF-3 together yielded little or no fMet-tRNA$_f^{Met}$ binding, whereas IF-2 by itself could cause significant binding. Again, as is the case with mRNA binding, all three factors together yielded optimum fMet-tRNA$_f^{Met}$ binding.

In 1971, Sigrid and Robert Thach showed that one mole of GTP binds to the 30S ribosomal subunit along with every mole of fMet-tRNA$_f^{Met}$, but the GTP is not hydrolyzed until the 50S ribosomal subunit joins the complex and IF-2 departs. Their experiment used three different radioactive labels: GTP was labeled with ^{3}H in the guanosine part, and with ^{32}P in the γ-phosphate. fMet-tRNA$_f^{Met}$ was labeled with ^{14}C. After mixing the labeled components with 30S ribosomal subunits and other factors, these workers passed the mixtures through filters to catch the ribosomes and let the unbound nucleotides and RNAs flow through. Then they determined the amount of each radioisotope bound to the filter. Table 17.3 shows the results. Experiment 1 demonstrated that equimolar amounts of GTP and fMet-tRNA$_f^{Met}$ bound to the 30S

particle, and that IF-2 was required for both to bind. Experiment 2 showed that the binding of GTP to the 30S subunit depended on the trinucleotide AUG, which served as an mRNA substitute, and on fMet-tRNA$_f^{Met}$. Experiment 3 corroborated the findings from experiments 1 and 2, and also had a negative control that confirmed that 30S ribosomal subunits were also required. In both experiments 2 and 3, the amounts of ^{3}H- and ^{32}P-labeled GTP bound to the complex in the complete reactions were the same, showing that no GTP hydrolysis had occurred (the ^{32}P label was in the γ-phosphate, which would have been released as phosphate if hydrolysis took place). Finally, experiment 4 showed that addition of 50S ribosomal subunits released the GTP, but not the fMet-tRNA$_f^{Met}$, which stayed to initiate protein synthesis. We will discuss this matter further later in this chapter.

In 1973, John Fakunding and John Hershey performed in vitro experiments with labeled IF-2 and fMet-tRNA$_f^{Met}$ to show the binding of both to the 30S ribosomal subunit, and the lack of necessity for GTP hydrolysis for such binding to occur. They labeled fMet-tRNA$_f^{Met}$ with ^{3}H, and IF-2 by phosphorylating it with [^{32}P]ATP. This phosphorylated IF-2 retained full activity. Then they mixed these components with 30S ribosomal subunits in the presence of either GTP or an unhydrolyzable analog of GTP, **GDPCP**. This analog has a methylene

Table 17.3	Presence of GTP in the 30S Initiation Complex			
		pmol bound		
Experiment	Conditions	[^{3}H]GTP	[^{32}P]GTP	[^{14}C]fMet-tRNA
1	Complete	2.4		2.2
	Minus IF-2	0.7		0.2
2	Complete	2.1	2.0	2.3
	Minus AUG	0.6	0.2	0.1
	Minus fMet-tRNA	0.7	0.8	0.0
3	Complete	2.0	1.6	
	Minus AUG	1.4	0.5	
	Minus fMet-tRNA	0.7	0.4	
	Minus IF-2	0.5	0.5	
	Minus 30S ribosomes	0.5	0.4	
4	Complete	2.3	2.4	2.5
	Minus IF-2	0.3	0.5	0.5
	Complete, plus 50S ribosomes	0.6	0.6	2.5
	Minus IF-2, plus 50S ribosomes	0.3	0.5	0.4

Source: Reprinted with permission from *Nature New Biology* 229:219, 1971. Copyright © 1971 Macmillan Magazines Limited.

linkage (-CH$_2$-) between the β- and γ-phosphates where ordinary GTP would have an oxygen atom, which explains why it cannot be hydrolyzed to GDP and phosphate. After mixing all these components together, Fakunding and Hershey displayed the initiation complexes by sucrose gradient ultracentrifugation. Figure 17.14 shows the results. All of the labeled IF-2 and a significant amount of the fMet-tRNA$_f^{Met}$ comigrated with the 30S ribosomal subunit, indicating the formation of an initiation complex. The same results were seen in the presence of either authentic GTP or GDPCP, demonstrating that GTP hydrolysis is not required for binding of either IF-2 or fMet-tRNA$_f^{Met}$ to the complex. Indeed, IF-2 can bind to 30S subunits in the absence of GTP, but only at unnaturally high concentrations of IF-2.

This kind of experiment also allowed Fakunding and Hershey to estimate the stoichiometry of binding between 30S subunit, IF-2, and fMet-tRNA$_f^{Met}$. They added more and more IF-2 to generate a saturation curve, as illustrated in Figure 17.15. The curve leveled off at 0.7 moles of IF-2 bound per 30S subunit. Because some of the 30S subunits were probably not competent to bind IF-2, this number seems close enough to 1.0 to conclude that the real stoichiometry is 1:1. Furthermore, at saturating IF-2 concentration, 0.69 mol of fMet-tRNA$_f^{Met}$ bound to the 30S subunits. This is almost exactly the amount of IF-2 that bound, so the stoichiometry of fMet-tRNA$_f^{Met}$ also appears to be 1:1. However, as we will see, IF-2 is ultimately released from the initiation complex, so it can recycle and bind another fMet-tRNA$_f^{Met}$ to another complex. In this way, it really acts catalytically.

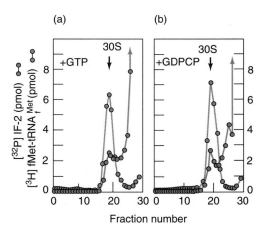

Figure 17.14   Formation of 30S initiation complex with GTP or GDPCP. Fakunding and Hershey mixed [^{32}P]IF-2, [^{3}H]fMet-tRNA$_f^{Met}$ and AUG, an mRNA substitute, with 30S ribosomal subunits and either (**a**) GTP or (**b**) the unhydrolyzable GTP analog GDPCP. Then they centrifuged the mixtures in sucrose gradients and assayed each gradient fraction for radioactive IF-2 (blue) and fMet-tRNA$_f^{Met}$ (red). Both substances bound to 30S ribosomes equally well with GTP and GDPCP. (*Source:* From Fakunding and Hershey, *Journal of Biological Chemistry* 248:4208, 1973. Copyright © 1973 The American Society for Biochemistry & Molecular Biology, Bethesda, MD. Reprinted by permission.)

Next, Fakunding and Hershey investigated the effect of IF-1 and IF-3 on binding of IF-2 to the 30S ribosomal subunit in the absence of fMet-tRNA$_f^{Met}$ and AUG. They mixed labeled IF-2 with different combinations of factors and 30S ribosomes and then centrifuged the mixtures to see whether or not IF-2 had bound. Without fMet-tRNA$_f^{Met}$ and AUG, IF-2 binding to 30S particles is less stable, and Figure 17.16 shows that IF-2 could not bind

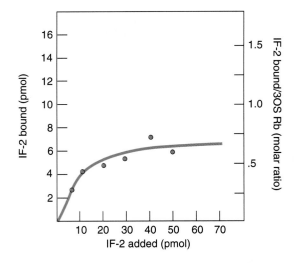

Figure 17.15 Stoichiometry of binding between IF-2 and 30S subunit. Fakunding and Hershey measured [^{32}P]IF-2 binding to 30S ribosomes by sucrose gradient ultracentrifugation, under the conditions describe in Figure 17.14, then plotted IF-2 bound in pmol (left y axis) or as the ratio of IF-2 bound per 30S subunit (right y axis). The binding leveled off at a molar ratio of about 0.7. (*Source:* From Fakunding and Hershey, *Journal of Biological Chemistry* 248:4209, 1973. Copyright © 1973 The American Society for Biochemistry & Molecular Biology, Bethesda, MD. Reprinted by permission.)

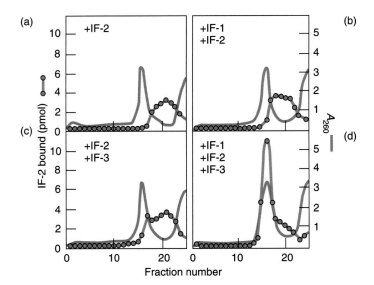

Figure 17.16 Cooperative binding of initiation factors to the 30S ribosomal subunit. Fakunding and Hershey mixed 30S ribosomes with GTP, [^{32}P]IF-2 and various combinations of the other two initiation factors, then tested for binding of IF-2 by sucrose gradient ultracentrifugation. Factors used: (**a**) IF-2 only; (**b**) IF-2 plus IF-1; (**c**) IF-2 plus IF-3; (**d**) all three initiation factors. Blue, 30S ribosomes, as detected by A_{260}; red, [^{32}P]IF-2. Only when all three factors were present could IF-2 bind under these in vitro conditions. (*Source:* From Fakunding and Hershey, *Journal of Biological Chemistry* 248:4208, 1973. Copyright © 1973 The American Society for Biochemistry & Molecular Biology, Bethesda, MD. Reprinted by permission.)

to 30S particles by itself, in the absence of other factors and GTP. Similarly, no binding occurred when IF-2 was accompanied by either IF-1 or IF-3. However, when all three factors were present, IF-2 bound. This suggests that all three factors can bind cooperatively to the 30S subunit. Indeed, the binding of all three factors seems to be the first step in formation of the 30S initiation complex. Once bound, the factors can direct the binding of mRNA and fMet-tRNA$_f^{Met}$ yielding a complete 30S initiation complex, which consists of a 30S ribosomal subunit plus one molecule each of mRNA, fMet-tRNA$_f^{Met}$ GTP, IF-1, IF-2, and IF-3.

> **SUMMARY** IF-2 is the major factor promoting binding of fMet-tRNA$_f^{Met}$ to the 30S initiation complex. The other two initiation factors play important supporting roles. GTP is also required for IF-2 binding at physiological IF-2 concentrations, but it is not hydrolyzed in the process. The complete 30S initiation complex contains one 30S ribosomal subunit plus one molecule each of mRNA, fMet-tRNA$_f^{Met}$ GTP, IF-1, IF-2, and IF-3.

Formation of the 70S Initiation Complex

For elongation to occur, the 50S ribosomal subunit must join the 30S initiation complex to form the **70S initiation complex.** In this process, IF-1 and IF-3 dissociate from the

complex. Then GTP is hydrolyzed to GDP and phosphate, as IF-2 leaves the complex. We will see that GTP hydrolysis does not drive the binding of the 50S ribosomal subunit. Instead, it drives the release of IF-2, which would otherwise interfere with formation of an active 70S initiation complex.

We have already seen that GTP is part of the 30S initiation complex, and that it is removed when the 50S ribosomal subunit joins the complex. But how is it removed? Jerry Dubnoff and Umadas Maitra demonstrated in 1972 that IF-2 contains a ribosome-dependent GTPase activity that hydrolyzes the GTP to GDP and inorganic phosphate (P_i). They mixed [γ-^{32}P]GTP with salt-washed ribosomes (devoid of initiation factors), or with IF-2, or with both, and plotted the $^{32}P_i$ released. Figure 17.17 shows that ribosomes or IF-2 separately could not hydrolyze the GTP, but together they could. Thus, IF-2 and ribosomes together constitute a GTPase. Our examination of the 30S initiation complex showed that the 30S ribosomal subunit cannot complement IF-2 this way because GTP is not hydrolyzed until the 50S particle joins the complex.

What is the function of GTP hydrolysis? Fakunding and Hershey's experiments with labeled IF-2 also shed light on this question: They showed that GTP hydrolysis is necessary for removal of IF-2 from the ribosome. These workers formed 30S initiation complexes with labeled

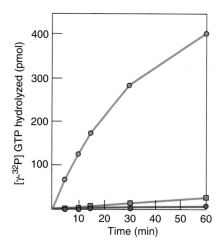

Figure 17.17   Ribosome-dependent GTPase activity of IF-2.
Dubnoff and Maitra measured the release of labeled inorganic
phosphate from [γ-^{32}P]GTP in the presence of IF-2 (green), ribosomes
(blue), and IF-2 plus ribosomes (red). Together, ribosomes and IF-2
could hydrolyze the GTP. (*Source:* From Dubnoff and Maitra, *Journal of
Biological Chemistry* 247:2878, 1972. Copyright © 1972 The American Society for
Biochemistry & Molecular Biology, Bethesda, MD. Reprinted by permission.)

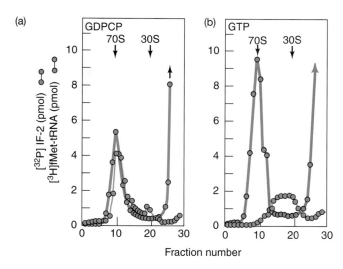

**Figure 17.18   Effect of GTP hydrolysis on release of IF-2 from the
ribosome.** Fakunding and Hershey mixed [^{32}P]IF-2 (blue) and
[^{3}H]fMet-tRNA$_f^{Met}$ (red) with 30S ribosomal subunits to form 30S
initiation complexes. Then they added 50S ribosomal subunits in the
presence of either (**a**) GDPCP, or (**b**) GTP, and then analyzed the
complexes by sucrose gradient ultracentrifugation as in Figure 17.14.
(*Source:* From Fakunding and Hershey, *Journal of Biological Chemistry* 248:4210,
1973. Copyright © 1973 The American Society for Biochemistry & Molecular
Biology, Bethesda, MD. Reprinted by permission.)

IF-2 and fMet-tRNA$_f^{Met}$ and either GDPCP or GTP,
added 50S subunits and then ultracentrifuged the mixtures to see which components remained associated with
the 70S initiation complexes. Figure 17.18 shows the results. With GDPCP, both IF-2 and fMet-tRNA$_f^{Met}$ remained associated with the 70S complex. By contrast,
GTP allowed IF-2 to dissociate, while fMet-tRNA$_f^{Met}$ remained with the 70S complex. This demonstrated that
GTP hydrolysis is required for IF-2 to leave the ribosome.

Another feature of Figure 17.18 is that much *more*
fMet-tRNA$_f^{Met}$ bound to the 70S initiation complex in the
presence of GTP than in the presence of GDPCP. This
hints at the catalytic function of IF-2: Hydrolysis of GTP
is necessary to release IF-2 from the 70S initiation complex so it can bind another molecule of fMet-tRNA$_f^{Met}$ to
another 30S initiation complex. This recycling constitutes
catalytic activity. However, if the factor remains stuck to
the 70S complex because of failure of GTP to be hydrolyzed, it cannot recycle and therefore acts only stoichiometrically. Maitra and his colleagues showed the
same thing by measuring [^{3}H]fMet-tRNA$_f^{Met}$ binding to
ribosomes in the presence of increasing concentrations of
IF-2. Table 17.4 shows that more and more fMet-
tRNA$_f^{Met}$ bound to 30S ribosomes as the concentration of
IF-2 rose. However, these workers achieved the same effect by adding 50S, as well as 30S ribosomal subunits.
This allowed GTP hydrolysis and release of IF-2 so it
could act catalytically, rather than stoichiometrically.

Is GTP hydrolysis also required to prime the ribosome
for translation? Apparently not, since Maitra and colleagues removed GTP from 30S initiation complexes by
gel filtration and found that these complexes were compe-

Table 17.4	Role of Initiation Factors in Binding fMet-tRNA$_f^{Met}$ to Ribosomes	
	[^{3}H]fMet-tRNA$_f^{Met}$ (pmol) bound to:	
Factor additions	**30S**	**30S + 50S**
None	0.8	0.5
IF-3(0.5 pmol)	0.8	0.7
IF-1(0.15 µg) + IF-3(0.15 µg) + IF-2(0.55 pmol)	1.0	5.7
IF-2(1.65 pmol)	1.0	
IF-2(8.8 pmol)	2.2	
IF-2(16.5 pmol)	4.2	

Source: From Dubnoff et al., *Journal of Biological Chemistry* 247:2886, 1972.
Copyright © 1972 The American Society for Biochemistry & Molecular Biology,
Bethesda, MD. Reprinted by permission.

tent to accept 50S subunits and then carry out peptide
bond formation. The GTP was not hydrolyzed in this procedure, and a similar procedure with GDPCP gave the
same results, so GTP hydrolysis is not a prerequisite for an
active 70S initiation complex, at least under these experimental conditions. This reinforces the notion that the real
function of GTP hydrolysis is to remove IF-2 and GTP itself from the 70S initiation complex so it can go about its
business of linking together amino acids to make proteins.

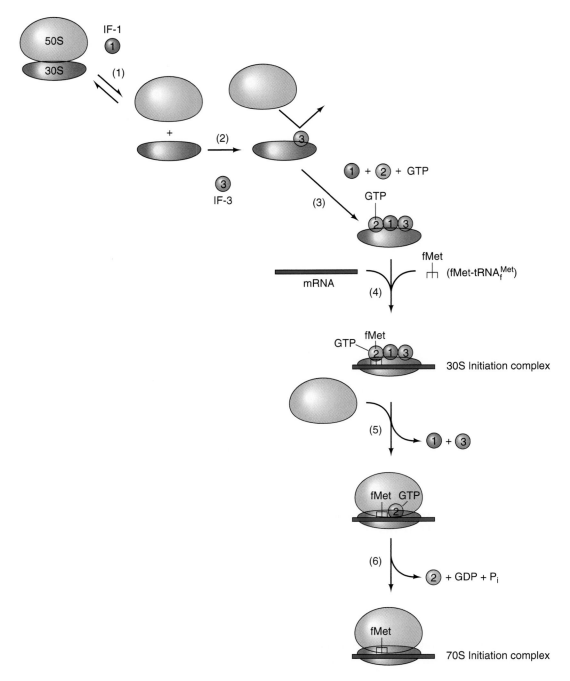

Figure 17.19   Summary of prokaryotic translation initiation. See the text for a description of steps 1–6.

SUMMARY GTP is hydrolyzed after the 50S subunit joins the 30S complex to form the 70S initiation complex. This GTP hydrolysis is carried out by IF-2 in conjunction with the 50S ribosomal subunit. The purpose of this hydrolysis is to release IF-2 and GTP from the complex so polypeptide chain elongation can begin.

Summary of Initiation in Prokaryotes

Figure 17.19 summarizes what we have learned about translation initiation in prokaryotes. It includes the following features:

1. Dissociation of the 70S ribosome into 50S and 30S subunits, under the influence of IF-1.
2. Binding of IF-3 to the 30S subunit, which prevents reassociation between the ribosomal subunits.

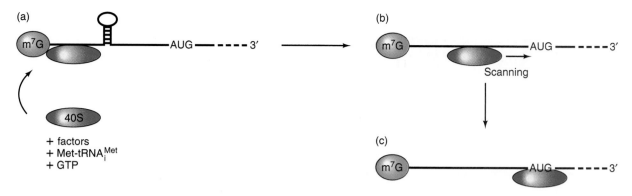

Figure 17.20   A simplified version of the scanning model for translation initiation. (a) The 40S ribosomal subunit, along with initiation factors, Met-tRNA$_i^{Met}$, and GTP, recognize the m^{7}G cap (red) at the 5′-end of an mRNA and allow the ribosomal subunit to bind at the end of the mRNA. All the other components (factors, etc.) are omitted for simplicity. **(b)** The 40S subunit is scanning the mRNA toward the 3′-end, searching for an initiation codon. It has melted a stem loop structure in its way. **(c)** The ribosomal subunit has located an AUG initiation codon and has stopped scanning. Now the 60S ribosomal subunit can join the complex and initiation can occur.

3. Binding of IF-1, IF-2, and GTP alongside IF-3.

4. Binding of mRNA and fMet-tRNA$_f^{Met}$ to form the 30S initiation complex. These two components can apparently bind in either order, but IF-2 sponsors fMet-tRNA$_f^{Met}$ binding, and IF-3 sponsors mRNA binding. In each case, the other initiation factors also help.

5. Binding of the 50S subunit, with loss of IF-1 and IF-3.

6. Dissociation of IF-2 from the complex, with simultaneous hydrolysis of GTP. The product is the 70S initiation complex, ready to begin elongation.

17.2 Initiation in Eukaryotes

Several features distinguish eukaryotic translation initiation from prokaryotic. First, eukaryotic initiation begins with methionine, not *N*-formyl-methionine. But the initiating tRNA is different from the one that adds methionines to the interiors of polypeptides (tRNA$_m^{Met}$). The initiating tRNA bears an unformylated methionine, so it seems improper to call it tRNA$_f^{Met}$. Accordingly, it is frequently called **tRNA$_i^{Met}$** or **tRNA$_i$**. A second major difference distinguishing eukaryotic translation initiation from prokaryotic is that eukaryotic mRNAs contain no Shine–Dalgarno sequence to show the ribosomes where to start translating. Instead, most eukaryotic mRNAs have caps (Chapter 15) at their 5′-ends, which direct initiation factors to bind and begin searching for an initiation codon. This different mechanism of initiation and the initiation factors it requires will be our topics in this section.

The Scanning Model of Initiation

Most prokaryotic mRNAs are polycistronic. They contain information from multiple genes, or cistrons, and each cistron has its own initiation codon and ribosome-binding site. But polycistronic mRNAs are rare in eukaryotes, except for the transcripts of certain viruses. Thus, eukaryotic cells are usually faced with the task of finding a start codon near the 5′-end of a transcript. They accomplish this by recognizing the cap at the 5′-end, then **scanning** the mRNA in the 5′→3′ direction until they encounter a start codon, as illustrated in Figure 17.20.

Marilyn Kozak first developed this scanning model in 1978, based on four considerations: (1) In no known instance was eukaryotic translation initiated at an internal AUG, as in a polycistronic mRNA. (2) Initiation did not occur at a fixed distance from the 5′-end of an mRNA. (3) In all of the first 22 eukaryotic mRNAs examined, the first AUG downstream of the cap was used for initiation. (4) As we saw in Chapter 15, the cap at the 5′-end of the mRNA facilitates initiation. We will see more definitive evidence for the scanning model later in this chapter.

The simplest version of the scanning model has the ribosome recognizing the first AUG it encounters and initiating translation there. However, a survey of 699 eukaryotic mRNAs revealed that the first AUG is not the initiation site in 5–10% of the cases. Instead, in those cases, the ribosome skips over one or more AUGs before encountering the right one and initiating translation. This raises the question: What sets the right AUG apart from the wrong ones? To find out, Kozak examined the sequences surrounding initiating AUGs and found that the consensus sequence was CCRCC<u>AUG</u>G, where R is a purine (A or G), and the initiation codon is underlined.

If this is really the optimum sequence, then mutations should reduce its efficiency. To check this hypothesis, Kozak systematically mutated nucleotides around the initiation codon in a cloned rat preproinsulin gene. She substituted a synthetic ATG-containing oligonucleotide for the normal initiating ATG, then introduced mutations into this initiation region, placed the mutated genes under control of the SV40 virus promoter, introduced them into

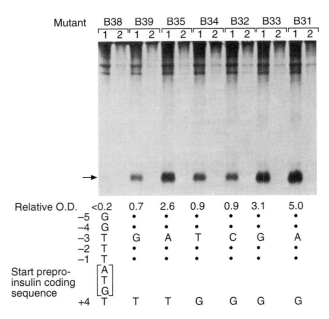

Mutant	B38	B39	B35	B34	B32	B33	B31
	1 2	1 2	1 2	1 2	1 2	1 2	1 2
Relative O.D.	<0.2	0.7	2.6	0.9	0.9	3.1	5.0
−5	G	•	•	•	•	•	•
−4	G	•	•	•	•	•	•
−3	T	G	A	T	C	G	A
−2	T	•	•	•	•	•	•
−1	T	•	•	•	•	•	•
Start preproinsulin coding sequence	A T G						
+4	T	T	T	G	G	G	G

Figure 17.21   Effects of single base changes in positions −3 and +4 surrounding the initiating ATG. Starting with a cloned rat preproinsulin gene under the control of an SV40 viral promoter, Kozak replaced the natural initiation codon with a synthetic oligonucleotide containing an ATG. She then mutagenized the nucleotides at positions −3 and +4 as shown at bottom, introduced the manipulated genes into COS cells growing in medium containing [³⁵S]methionine to label any proinsulin produced. She purified the proinsulin by immuno-precipitation, then electrophoresed it and detected the labeled protein by fluorography. This is a technique similar to autoradiography in which the electrophoresis gel is impregnated with a fluorescent compound to amplify the relatively weak radioactive emissions from an isotope such as ³⁵S. The arrow at left indicates the position of the proinsulin product. Kozak subjected the proinsulin bands in the fluorograph to densitometry to quantify their intensities. These are listed as relative O. D., or optical density, beneath each band. Optimal initiation occurred with a purine in position −3 and a G in position +4. Proinsulin is the product of the preproinsulin gene because the "signal peptide" at the amino terminus of preproinsulin is removed during translation, yielding proinsulin. The signal peptide directs the growing polypeptide, along with the ribosome and mRNA, to the endoplasmic reticulum (ER). This ensures that the polypeptide enters the ER and can therefore be secreted from the cell. (*Source:* Kozak, Point mutations define a sequence flanking the AUG initiator codon that modulates translation by eukaryotic ribosomes. *Cell* 44 (31 Jan 1986) p. 286, f. 2. Reprinted by permission of Elsevier Science.)

monkey (COS) cells, then labeled newly synthesized proteins with ³⁵S-methionine, immunoprecipitated the proinsulin, electrophoresed it, and detected it by fluorography, a technique akin to autoradiography. Finally, she scanned the fluorograph with a densitometer to quantify the production of proinsulin. The better the translation initiation, the more proinsulin was made. Throughout this discussion we will refer to the initiation codon as ATG, rather than AUG, because the mutations were done at the DNA level.

Figure 17.21 shows some of the results, which include alterations in positions −3 and +4, where the A in ATG is position +1. The best initiation occurred with a G or an A

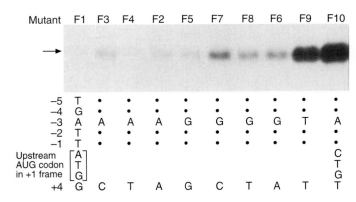

Mutant	F1	F3	F4	F2	F5	F7	F8	F6	F9	F10
−5	T	•	•	•	•	•	•	•	•	•
−4	G	•	•	•	•	•	•	•	•	•
−3	A	A	A	A	G	G	G	G	T	A
−2	T	•	•	•	•	•	•	•	•	•
−1	T	•	•	•	•	•	•	•	•	C
Upstream AUG codon in +1 frame	A T G									T G
+4	G	C	T	A	G	C	T	A	T	T

Figure 17.22   Influence of the context of an upstream "barrier" ATG. Kozak made a construct having the normal ATG initiation codon of the rat preproinsulin gene preceded by an out-of-frame ATG, then made mutations in the −3 and +4 positions surrounding the upstream ATG (shown at bottom) and assayed the effect on proinsulin synthesis as in Figure 17.21. The arrow at left indicates the position of correctly initiated proinsulin. The more favorable the context of the upstream ATG, the better it serves as a barrier to correct downstream initiation. (*Source:* Kozak, *Cell* 44 (31 Jan 1986) p. 288, f. 6. Reprinted by permission of Elsevier Science.)

in position −3 and a G in position +4. Similar experiments showed that the best initiation of all occurred with the sequence ACC<u>ATG</u>G.

If this really is the optimum sequence for translation initiation, introducing it out of frame and upstream of the normal initiation codon should provide a barrier to scanning ribosomes and force them to initiate out of frame. The more this occurs, the less proinsulin should be produced. Kozak performed this experiment with the As of the two ATGs 8 nt apart as follows: <u>ATG</u>NCACC<u>ATG</u>G. Note that the downstream ATG is in an optimal neighborhood, so initiation should start there readily if the ribosome can reach it without initiating upstream first. Figure 17.22 shows the results. Mutant F10 had no upstream ATG, and initiation from the normal ATG was predictably strong. Mutant F9 had the upstream ATG in a very weak context, with Ts in both −3 and +4 positions. Again, this did not interfere much with initiation at the downstream ATG. But all the other mutants exhibited strong interference with normal initiation, and the strength of this interference was related to the context of the upstream ATG. The closer it resembled the optimal sequence, the more it interfered with initiation at the downstream ATG. This is just what the scanning model predicts.

What about natural mRNAs that have an upstream ATG in a favorable context, yet still manage to initiate from a downstream ATG? Kozak noted that these mRNAs have in-frame stop codons between the two ATGs, and she argued that initiation at the downstream ATG actually represents reinitiation by ribosomes that have initiated at the upstream start codon, terminated at the stop codon, then continued scanning for another start

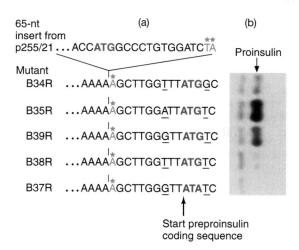

Figure 17.23 An upstream ATG is an ineffective barrier if followed by an in-frame stop codon. Kozak placed a stop codon in frame with a barrier ATG upstream of the normal initiation codon of the rat preproinsulin gene, made mutations in the −3 and +4 positions (underlined) around the normal ATG, as shown in part (**a**), and (**b**) assayed for proinsulin synthesis as in Figure 17.21. Even though the upstream ATG is in a favorable context, the stop codon downstream makes it an ineffective barrier. (*Source:* Kozak, *Cell* 44 (31 Jan 1986) p. 289, f. 7. Reprinted by permission of Elsevier Science.)

codon. To illustrate the effect of a stop codon between the two ATGs, Kozak made another set of constructs with such a stop codon and tested them by the same assay. Figure 17.23 shows that abundant initiation occurred at the downstream ATG in this case, as long as the downstream ATG was in a good environment.

Note that an initiation codon and a downstream termination codon in the same reading frame define the boundaries of an **open reading frame (ORF).** Such an ORF potentially encodes a protein; whether it is actually translated in vivo is another matter. Further experiments have revealed another requirement for efficient reinitiation at a downstream ORF: The upstream ORF must be short. In every case in which a bicistronic mRNA with a full-sized upstream ORF has been examined, reinitiation at the downstream ORF has been extremely inefficient. Perhaps by the time a ribosome finishes translating a long ORF, the initiation factors needed for reinitiation have diffused away, so it ignores the second ORF.

To check rigorously the hypothesis that an upstream AUG is favored over downstream AUGs, Kozak created mRNAs with exact repeats of the initiation region of the rat preproinsulin cistron. She then tested these for the actual translation initiation site by isolating the resulting proteins and electrophoresing them to determine their sizes, which tell us which initiation site the ribosomes used in making them. Figure 17.24 shows that in each case, the farthest upstream AUG was used, which is again consistent with the scanning model.

What is the effect of mRNA secondary structure on efficiency of initiation? Hairpins in the mRNA can affect initiation both positively and negatively. Kozak showed that a stem loop 12–15 nt downstream of an AUG in a weak context could prevent ribosomes from skipping that initiation site. The hairpin presumably stalled the ribosome at the AUG long enough for initiation to occur. Secondary structure can also have a negative effect. Kozak tested the effects of two different stem loop structures in the leader of an mRNA. One was relatively short and had a free energy of formation (or stability) of −30 kcal/mol; the other was much longer, with a higher stability of −62 kcal/mol. She introduced these stem-loops into various positions in the leader of the chloramphenicol acetyl transferase (CAT) gene, then transcribed the altered genes and translated their transcripts in vitro in the presence of [^{35}S]methionine. Finally, she electrophoresed the CAT proteins and detected them by fluorography. The results in Figure 17.25a show that a −30-kcal stem loop 52 nt downstream of the cap does not interfere with translation, even if it includes the initiating ATG. However, a −30-kcal stem loop only 12 nt downstream of the cap strongly inhibits translation, presumably because it interferes with binding of the 40S ribosome and factors at the cap. Furthermore, a −62-kcal stem loop placed 71 nt downstream of the cap completely blocked appearance of the CAT protein.

Why was the construct with the stable hairpin not translated? The simplest explanation is that the very stable stem loop blocked the scanning 40S ribosomal subunit and would not let it through to the initiation codon. This effect was observed only *in cis* (on the same molecule). When construct 3 and 4 (or 3 and 1) were tested together, translation occurred on the linear mRNA made from construct 3 (lanes 4 and 6). This indicates that the untranslatable constructs were not poisoning the translation system somehow.

The fact that construct 2 is translated well, even though its initiation codon lies buried in a hairpin, suggests that the ribosome and initiation factors can unwind a certain amount of double-stranded RNA, as predicted by Kozak in her original scanning model (see Figure 17.20). However, as we have just seen, this unwinding ability has limits; the long hairpin in construct 4 effectively blocks the ribosomes from reaching the initiation codon.

To conclude that the long hairpin in construct 4 really trapped the 40S ribosomal subunit in the mRNA leader and would not let it pass through to the initiation codon, one would like direct evidence that the 40S subunit is positioned just upstream of the hairpin. Kozak produced such evidence using RNase T1 fingerprinting. Figure 17.26 shows a detailed picture of construct 4, as well as the two constructs, 4-L and 4-R, used to create it. Construct 4-L contains the left-hand part of the hairpin, up to a *Pst*I site, which is marked. Construct 4-R contains the

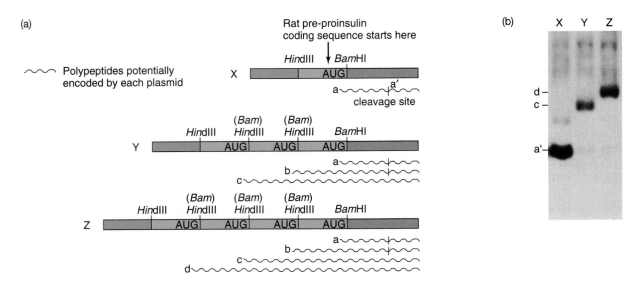

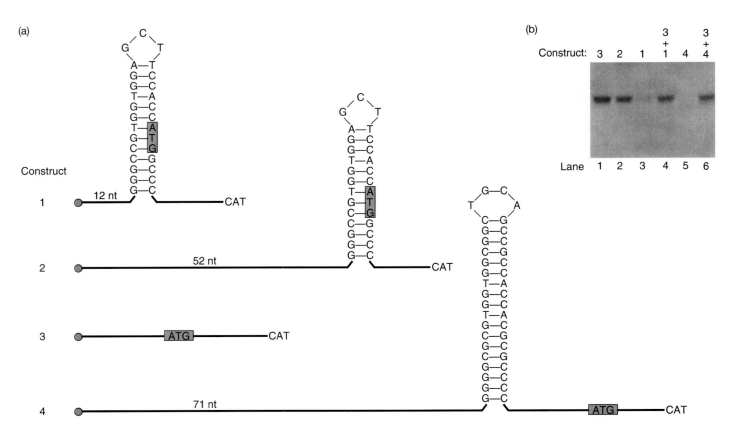

Figure 17.24 Effect of multiple replicas of the translation initiation site. (a) Constructs. Kozak created three constructs, X, Y, and Z, with one, three, and four copies of the rat preproinsulin initiation site. The site was included on a HindIII-BamHI DNA fragment that could be linked head to tail as shown in constructs Y and Z to give tandem repeated copies of the initiation site. The wavy lines beneath each construct denote the polypeptides potentially encoded by each. (b) Results. Kozak introduced these constructs into mammalian cells and allowed them to be transcribed and the mRNAs to be translated. Then she immunoprecipitated proinsulin-containing polypeptides, electrophoresed them, and detected them by fluorography. Authentic proinsulin was the product of construct X, because it had only the natural initiation site, and the signal peptide could be cleaved. However, construct Y, with three initiation codons, produced only product c, which initiated at the farthest upstream codon. Similarly, construct Z, with four initiation codons, produced only product d, which initiated at the farthest upstream codon. Note that the signal peptide cannot be removed from product c or d because it is internal to these proteins. *Source: Kozak, Translation of insulin-related polypeptides from messenger RNAs with tandemly reiterated copies of the ribosome binding site. Cell 34 (Oct 1983) p. 975, f. 4. Reprinted by permission of Elsevier Science.)*

Figure 17.25 Effect of secondary structure in an mRNA leader on translation efficiency. (a) mRNA constructs. Kozak made the synthetic leader constructs pictured here, with the cap in red and the initiation codon highlighted in green, and attached the CAT ORF to the 3′-end of each. In this illustration, T replaces U, because the constructs were made of DNA. (b) Results of in vitro translation. Kozak transcribed each construct in vitro and then translated each mRNA in vitro in a rabbit reticulocyte extract with [³⁵S]methionine. She electrophoresed the labeled proteins and detected them by fluorography. The short hairpin near the cap (construct 1) interfered, as did the long hairpin between the cap and the initiation codon (construct 4). *(Source: (b) Kozak, Mol. Cell. Biol. 9 (1989) p. 5136, f. 3. American Society for Microbiology.)*

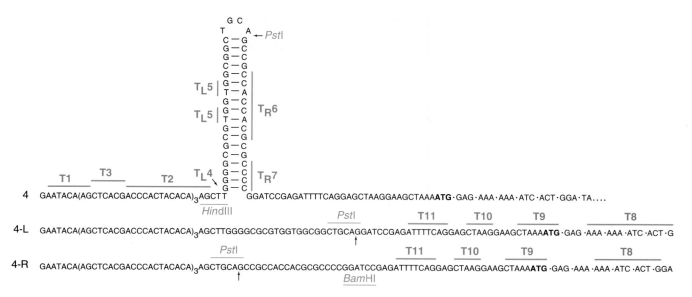

Figure 17.26   Summary of RNase protection experiment with constructs 4, 4-L, and 4-R. The DNA sequences of the templates used to make each RNA construct are presented, including the hairpin structure of construct 4. The RNase T1 products detected by fingerprinting are overlined and identified by number. Note that each oligo ends in G because RNase T1 cleaves only after Gs. The T_L4 oligonucleotide derives from the CTTG sequence that overlaps the *Hind*III site. (*Source:* From M. Kozak, *Molecular and Cellular Biology* 9:5137, 1989. Copyright © 1989 American Society for Microbiology, Washington, DC. Reprinted by permission.)

right-hand part of the hairpin, beginning with a *Pst*I site. Construct 4 was created by cutting 4-L and 4-R with *Pst*I and ligating them together.

Kozak labeled all three constructs with ^{32}P, then incubated them with wheat germ ribosomes in the presence of the antibiotic sparsomycin to block initiation. Then she treated the complexes with pancreatic RNase to degrade any RNA not protected by ribosomes. She purified the protected RNA regions by centrifugation, then cleaved them with RNase T1 and fingerprinted the fragments to identify them. Figure 17.26 shows the positions of the T1 fragments, which are labeled T1–T11. Note that when ribosomes were incubated with constructs 4-L and 4-R, neither of which has a significant hairpin, the protected region includes fragments T8–T11, which cluster around the AUG start codon. Thus, with these constructs, the ribosome made it to the start codon. By contrast, when construct 4 was used, the region upstream of the hairpin was protected (fragments T1–T3), but the region around the start codon was not protected. Thus, the ribosome was not able to move past the hairpin. (Fragments within the hairpin were also protected because RNase T1 cannot degrade double-stranded RNA.)

How do ribosomes recognize an AUG start codon? Thomas Donahue and colleagues have shown that the initiator tRNA (tRNA$_i^{Met}$) plays a critical role. They changed the anticodon of one of the four yeast tRNA$_i^{Met}$s to 3′-UCC-5′ so it would recognize the codon AGG instead

of AUG. Then they placed *his4* genes with various mutant initiation codons into a *his4*⁻ yeast strain. Figure 17.27a shows that the *his4* gene bearing an AGG codon in place of the initiation codon could support yeast growth. None of the other substitute initiation codons worked, presumably because they could not pair with the UCC anticodon in the altered initiator tRNA. In another experiment, these workers placed a second AGG 28 nt upstream of the AGG in the initiation site and out of frame with it. This construct could not support growth. This result is in accord with the scanning model, as illustrated in Figure 17.27b. The initiator tRNA, with a UCC anticodon in this case, binds to the 40S ribosomal subunit and the complex scans the mRNA searching for the first initiation codon (AGG in this case). Since the first AGG is out-of-frame with the *his4* coding region, translation will occur in the wrong reading frame and will soon encounter a stop codon and terminate prematurely.

The scanning model has some apparent exceptions. The best documented of these concern the polycistronic mRNAs of the picornaviruses such as poliovirus, which lack caps. In these cases, ribosomes can apparently enter at internal initiation codons without scanning. A few capped cellular mRNAs also have **internal ribosome entry sequences (IRESs)** that can attract ribosomes directly without help from the cap. We will discuss this phenomenon in more detail later in this chapter.

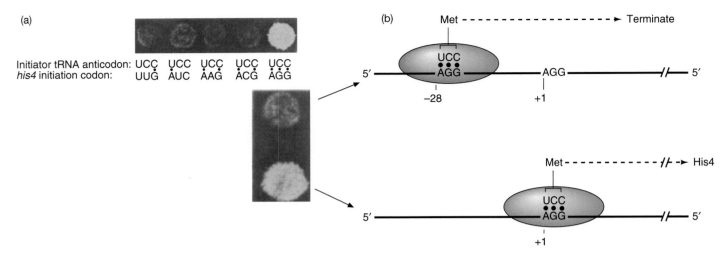

Figure 17.27   Role of initiator tRNA in scanning. (a) An initiator tRNA with an altered anticodon can recognize a complementary initiation codon. Donahue and colleagues mutated the anticodon of one of the initiator tRNAs in the yeast *Saccharomyces cerevisiae* to 3'-UCC-5' and introduced the gene encoding this altered tRNA into *his4⁻* cells, using a high-copy yeast vector. Then they changed the initiation codon of the *his4* gene to any of the five versions listed at the bottom and tested the mutant yeast cells for growth in the absence of histidine. When the initiation codon was AGG, it could base-pair with the UCC anticodon on the initiator tRNA, so the mutant mRNA could be translated and growth occurred. (b) Effect of an extra AGG upstream and out of frame. Donahue and colleagues made a *his4* construct with an extra AGG in good context beginning at position −28 (top), placed it in cells bearing the initiator tRNA with the UCC anticodon, and tested these cells for ability to grow in the absence of histidine. Growth was much reduced compared with cells with no upstream AGG (bottom). The scanning 40S ribosomal subunit apparently encountered the first AGG and initiated there, producing a shortened *his4* product. (*Source:* (a) Cigan et al., tRNA$_i^{Met}$ functions in directing the scanning ribosomes to the start site of translation. *Science* 242 (7 Oct 1988) p. 94, f. 1B & C (left). © AAAS.)

SUMMARY Eukaryotic ribosomes, together with the initiator tRNA (tRNA$_i^{Met}$), generally locate the appropriate start codon by binding to the 5'-cap of an mRNA and scanning downstream until they find the first AUG in a favorable context. The best context is a purine in the −3 position and a G in the +4 position where the A of the AUG is +1. In 5–10% of the cases, the ribosomes will bypass the first ATG and continue to scan for a more favorable one. Sometimes they apparently initiate at an upstream AUG, translate a short ORF, then continue scanning and reinitiate at a downstream AUG. This mechanism works only with short upstream ORFs. Secondary structure near the 5'-end of an mRNA can have positive or negative effects. A hairpin just past an AUG can force a ribosome to pause at the AUG and thus stimulate initiation. A very stable stem loop between the cap and an initiation site can block ribosome scanning and thus inhibit initiation. Some viral and cellular mRNAs contain IRESs that attract ribosomes directly to the interior of the RNA.

Eukaryotic Initiation Factors

We have seen that prokaryotic translation initiation requires initiation factors and so does initiation in eukary-otes. As you might expect, though, the eukaryotic system is more complex than the prokaryotic. One level of extra complexity we have already seen is the scanning process. Factors are needed to recognize the cap at the 5'-end of an mRNA and bind the 40S ribosomal subunit nearby. In this section we will examine the factors involved at the various stages of initiation in eukaryotes. We will also see that some of these steps are natural sites for regulation of the translation process.

Overview of Translation Initiation in Eukaryotes

Figure 17.28 provides an outline of the initiation process in eukaryotes, showing the major classes of initiation factors involved. Some of the initiation factors are recognizable as eukaryotic counterparts of prokaryotic factors we have already studied. The most obvious of these is **eIF2,** which, like IF-2 is responsible for binding the initiating aminoacyl-tRNA (Met-tRNA$_i^{Met}$) to the ribosome. Notice also that the eukaryotic initiation factor names all begin with e, which stands for "eukaryotic." Like IF-2, eIF2 requires GTP to do its job, and this GTP is hydrolyzed to GDP when the factor dissociates from the ribosome. Then GTP must replace GDP on the factor for it to function again. This requires an exchange factor, **eIF2B,** which exchanges GTP for GDP on eIF2. This factor is also called **GEF,** for guanine nucleotide exchange factor. Notice that all of the factors acting at a given step are given the same

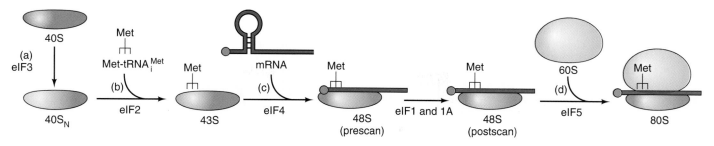

Figure 17.28   Summary of translation initiation in eukaryotes. (a) The eIF3 factor converts the 40S ribosomal subunit to 40S$_N$, which resists association with the 60S ribosomal particle and is ready to accept the initiator aminoacyl-tRNA. **(b)** With the help of eIF2, Met-tRNA$_i^{Met}$ binds to the 40S$_N$ particle, forming the 43S complex. **(c)** Aided by eIF4, the mRNA binds to the 43S complex, forming the 48S complex. **(d)** The eIF5 factor helps the 60S ribosomal particle bind to the 48S complex, yielding the 80S complex that is ready to begin translating the mRNA.

number. For example, we have seen that at least two factors (eIF2 and eIF2B) are required for initiator aminoacyl-tRNA binding, and both of these share the number 2. Another eukaryotic factor whose function bears at least some resemblance to that of a prokaryotic factor is **eIF3**, which binds to the 40S (small) ribosomal subunit and discourages its reassociation with the 60S (large) subunit. In this way, it resembles IF-3. **eIF4** is a complex cap-binding protein that allows the 40S ribosomal particle to bind to the 5′-end of an mRNA. Once the 40S particle has bound at the cap, it requires **eIF1** (and **eIF1A**) to scan to the initiation codon. **eIF5** is a factor with no known prokaryotic counterpart. It stimulates association between the 60S ribosomal subunit and the 40S initiation complex, which is actually called the 48S complex because it includes mRNA and many factors in addition to the 40S ribosomal subunit. **eIF6** is another antiassociation factor, like eIF3. It binds to the 60S ribosomal subunit and discourages premature association with the 40S subunit.

SUMMARY The functions of several of the eukaryotic initiation factors parallel those of the prokaryotic factors. For example, eIF2 is involved in binding Met-tRNA$_i^{Met}$ to the ribosome. eIF2B activates eIF2 by replacing its GDP with GTP. eIF1 and eIF1A aid in scanning to the initiation codon. eIF3 binds to the 40S ribosomal subunit and inhibits its reassociation with the 60S subunit. eIF4 is a cap-binding protein that allows the 40S ribosomal subunit to bind to the 5′-end of an mRNA. eIF5 encourages association between the 60S ribosomal subunit and the 48S complex (40S subunit plus mRNA and Met-tRNA$_i^{Met}$). eIF6 binds to the 60S subunit and blocks its reassociation with the 40S subunit.

Function of eIF4F   Now we come to a major novelty of eukaryotic translation initiation: the role of the cap. We have seen in Chapter 15 that the cap greatly stimulates the efficiency of translation of an mRNA. That implies that some factor can recognize the cap at the 5′-end of an mRNA and aid in the translation of that mRNA. Nahum Sonenberg, William Merrick, Aaron Shatkin, and colleagues identified a cap-binding protein in 1978 by cross-linking it to a modified cap as follows: First they oxidized the ribose of the capping nucleotide on a ^{3}H-reovirus mRNA to convert its 2′- and 3′-hydroxyl groups to a reactive dialdehyde. Then they incubated this altered mRNA with initiation factors. Free amino groups of any factor that binds to the modified cap should bind covalently to one of the reactive aldehydes. This bond can be made permanent by reduction. After cross-linking, the investigators digested all of the RNA but the cap with RNase, then electrophoresed the products to measure the sizes of any proteins cross-linked to the labeled cap. Figure 17.29 shows that a polypeptide with a M_r of about 24 kD bound, even at low temperature. At higher temperature, another pair of polypeptides of higher molecular mass (50–55 kD) bound. However, the binding of these high M_r polypeptides was not competed by unlabeled m^{7}GDP, whereas binding to the 24-kD polypeptide was competed by the cap analog. This suggested that the 24-kD polypeptide bound specifically to the cap, but the 50–55-kD-polypeptides did not. On the other hand, GDP competed with the 50–55-kD polypeptides for binding to the mRNA, but it did not compete with the 24-kD polypeptide. This may mean that the larger polypeptides are GDP-binding proteins, rather than cap-binding proteins.

Sonenberg, Shatkin, and colleagues followed up their discovery of the cap-binding protein by purifying it by affinity chromatography on a m^{7}GDP-Sepharose column. Then they added this protein to HeLa cell-free extracts and demonstrated that it stimulated transcription of capped, but not uncapped, mRNAs (Figure 17.30). They used viral mRNAs in both experiments: Sindbis virus mRNA for capped mRNA, and encephalomyocarditis virus mRNA for uncapped mRNA. (Encephalomyocarditis virus is a picornavirus similar to poliovirus.)

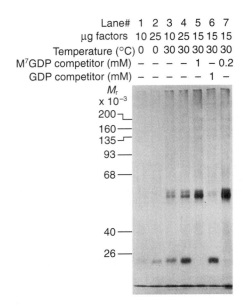

Lane#	1	2	3	4	5	6	7
µg factors	10	25	10	25	15	15	15
Temperature (°C)	0	0	30	30	30	30	30
M⁷GDP competitor (mM)	–	–	–	–	1	–	0.2
GDP competitor (mM)	–	–	–	–	–	1	–

Figure 17.29   Identifying a cap-binding protein by chemical crosslinking. Sonenberg and colleagues placed a reactive dialdehyde in the ribose of the capping nucleotide of a ^{3}H-reovirus mRNA. Then they mixed initiation factors with this mRNA to cross-link any cap-binding protein via a Schiff base between an aldehyde on the cap and a free amino group on the protein. They made this covalent bond permanent by reduction with NaBH$_3$CN. Then they digested these complexes with RNase to remove everything but the cap, and electrophoresed the labeled cap–protein complexes to detect the sizes of any polypeptides that bound to the cap. The conditions in each lane were as listed at top. Note that the binding of the 24-kD band is competed by m⁷GDP, but that of the 50–55-kD bands is not. (*Source:* Sonenberg et al., *PNAS* 75 (1978) p. 4844, f. 1.)

As we have seen, picornavirus mRNAs are not capped. Nevertheless, these viruses have mechanisms for ensuring that their mRNAs are translated. Moreover, they take advantage of the cap-free nature of their mRNAs to eliminate competition from capped host mRNAs. They do this by inactivating the host cap-binding protein, thus blocking translation of capped host mRNAs. Molecular biologists have taken advantage of this situation by using poliovirus-infected cell extracts as an assay system for the cap-binding protein. Any protein that can restore translation of capped mRNAs to such extracts must contain the cap-binding protein. This assay revealed that the 24-kD protein by itself was quite labile, but a higher molecular mass complex was much more stable. Sonenberg and collaborators have refined this analysis to demonstrate that the active purified complex contains three polypeptides: The original 24-kD cap-binding protein, and two other polypeptides with M_rs of 50 kD and 220 kD (Figure 17.31). These polypeptides have now been given new names: The 24-kD cap-binding protein is **eIF4E**; the 50-kD polypeptide is **eIF4A,** and the 220 kDa polypeptide is **eIF4G.** The whole three-polypeptide complex is called **eIF4F.**

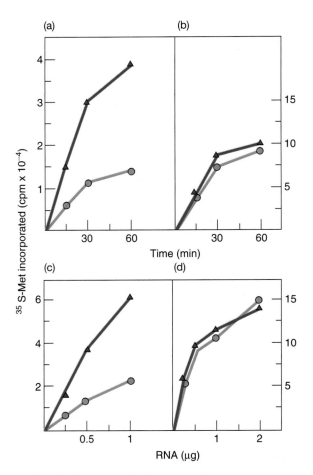

Figure 17.30   Cap-binding protein stimulates translation of capped, but not uncapped, mRNA. Shatkin and collaborators used HeLa cell-free extracts to translate capped and uncapped mRNAs in the presence of [^{35}S]methionine. Panels **(a)** and **(c)**: translation of capped Sindbis virus mRNA with (green) or without (red) cap-binding protein. Panels **(b)** and **(d)**: translation of uncapped encephalomyocarditis virus (EMC) with (green) or without (red) cap-binding protein. (*Source:* Reprinted with permission from *Nature* 285:331, 1980. Copyright © 1980 Macmillan Magazines Limited.)

> **SUMMARY** eIF4F is a cap-binding protein composed of three parts: eIF4E has the actual cap-binding activity; it is stabilized by the two other subunits: eIF4A and eIF4G.

Functions of eIF4A and eIF4B   The eIF4A polypeptide is a subunit of eIF4F, but it also has an independent existence and an independent function: It is a member of the so-called **DEAD protein** family, which has the consensus amino acid sequence Asp (D), Glu (E), Ala (A), Asp (D), and has **RNA helicase** activity. It can therefore unwind the hairpins that are frequently found in the 5′-leaders of eukaryotic mRNAs. To do this job effectively, eIF4A needs the help of **eIF4B**, which has an RNA-binding domain and can stimulate the binding of eIF4A to mRNA.

Arnim Pause and Sonenberg used a well-defined in vitro system to demonstrate the activities of both eIF4A and 4B. They started with the products of the eIF4A and 4B genes cloned in bacteria, so there was no possibility of contamination by other eukaryotic proteins. Then they added the labeled RNA helicase substrate pictured on the right in Figure 17.32. This is actually two 40-nt RNAs with complementary 5′-ends, which form a 10-bp RNA double helix. If an RNA helicase unwinds this 10-bp structure, it separates the two 40-nt monomers. Elec-

trophoresis then easily discriminates between monomers and dimer. The more monomers form, the greater is the RNA helicase activity.

Figure 17.32 depicts the results. A small amount of eIF4A (with ATP) caused a very modest amount of unwinding (lane 3), suggesting that this factor has some RNA helicase activity of its own. However, this helicase activity was stimulated by eIF4B (lane 5), and this activity depended on ATP (compare lanes 4 and 5). Greater amounts of eIF4A produced even greater RNA helicase activity (lanes 6 and 7). To show that eIF4B has no helicase activity of its own, Pause and Sonenberg added eIF4B and ATP without eIF4A and observed no helicase activity (lane 8). Thus, these two factors cooperate to unwind RNA helices, including hairpins, and this activity depends on ATP.

> **SUMMARY** eIF4A has RNA helicase activity that can unwind hairpins found in the 5′-leaders of eukaryotic mRNAs. It is aided in this task by another factor, eIF4B, and requires ATP for activity.

Functions of eIF4G We have seen that most eukaryotic mRNAs are capped, and the cap serves to help the ribosome bind. But some viral mRNAs are uncapped; these mRNAs, as well as a few cellular mRNAs, have IRESs that can help ribosomes bind. Furthermore, we know that the poly(A) tail at the 5′-end of mRNAs stimulates translation. At least in yeast, this latter process involves recruitment of ribosomes to the message via a poly(A)-binding protein called **Pab1p**. The eIF4G protein participates in all of these kinds of initiations by serving as an adaptor that can interact with a variety of different proteins.

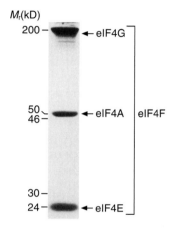

Figure 17.31 Components of eIF4F (complete cap-binding protein). Sonenberg and colleagues purified the cap-binding protein using a series of steps, including m⁷GTP affinity chromatography. Then they displayed the subunits of the purified protein by SDS-PAGE. The relative molecular masses (in kilodaltons) of the subunits and markers (200, 46, and 30 kD) are given at left. The whole complex, composed of three polypeptides, is called eIF4F. (*Source:* Edery et al., Involvement of eukaryotic initiation factor 4A in the cap recognition process. *J. Biol. Chem.* 258 (25 Sept 1983) p. 11400, f. 2. American Society for Biochemistry and Molecular Biology.)

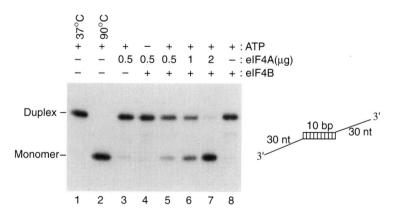

Figure 17.32 RNA helicase activity of eIF4A. Pause and Sonenberg tested combinations of ATP, eIF4A, and eIF4B (as indicated at top) on the radioactive helicase substrate shown at right. RNA helicase unwinds the 10-bp double-stranded region of the substrate, converting the dimer to two monomers. The dimer and monomers are then easily separated by gel electrophoresis, as indicated at left, and detected by autoradiography. The first two lanes are just substrate at low and high temperatures. The high temperature melts the double-stranded region of the substrate, yielding monomers. Lanes 3–8 show that ATP and eIF4A are required for helicase activity, and eIF4B stimulates this activity. (*Source:* Pause & Sonenberg, DEAD box RNA helicase. *EMBO Journal* 11 (1992) p. 2644, f. 1.)

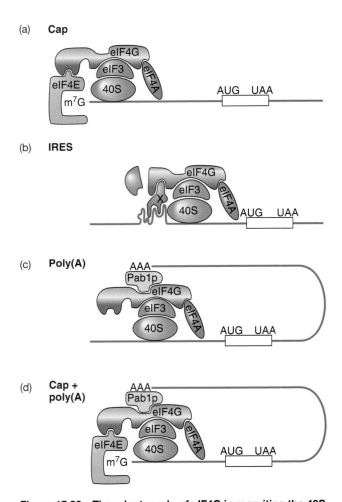

(a) **Cap**

(b) **IRES**

(c) **Poly(A)**

(d) **Cap + poly(A)**

Figure 17.33   The adapter role of eIF4G in recruiting the 40S ribosomal particle in four different situations. (a) Capped mRNA. eIF4G (green) serves as an adapter between eIF4E (orange), bound to the cap, and eIF3 (purple), bound to the 40S ribosomal particle (blue). The formation of this chain of molecules recruits the 40S particle to a site on the mRNA (dark green) near the cap, where it can begin scanning. eIF4A (brown) is also bound to eIF4G, but does not play a role in the interactions illustrated here. **(b)** An mRNA with an IRES. The IRES presumably interacts with an RNA-binding protein (X, red), which binds to eIF4G, ensuring recruitment of the 40S particle. This interaction happens even after removal of the N-terminal part of eIF4G by a picornaviral protease, which blocks binding to capped cellular mRNAs. **(c)** Participation of poly(A). Poly(A) at the 3′-ends of yeast mRNAs binds to Pab1p (yellow), which binds to eIF4G, thus helping to recruit the 40S particle. **(d)** Synergism between cap and poly(A). eIF4E bound to the cap and Pab1p bound to the poly(A) both bind to eIF4G and act synergistically in recruiting the 40S particle. (*Source:* From Hentze, *Science* 275:501, 1997. Copyright © 1997 American Association for the Advancement of Science, Washington, DC. Reprinted by permission.)

Figure 17.33 illustrates four different ways in which eIF4G can participate in translation initiation. In panel (a) we see the function eIF4G performs in initiating on ordinary, capped mRNAs. The amino terminus of eIF4G binds to eIF4E, which in turn binds to the cap. The central portion of eIF4G binds to eIF3, which in turn binds to the 40S ribosomal particle. Thus, by tethering together

eIF4E and eIF3, eIF4G can bring the 40S particle close to the 5′-end of the mRNA, where it can begin scanning. Panel (b) depicts the corruption of translation initiation by a picornavirus such as poliovirus. A viral protease cleaves off the amino terminal domain from eIF4G so it can no longer interact with eIF4E in recognizing caps. Thus, capped cellular mRNAs go untranslated. However, an alternative protein (X) can bind to eIF4G and also to an IRES in the viral mRNA, so 40S particles are still recruited to the viral mRNA. Panel (c) shows how Pab1p bound to the poly(A) tail on an mRNA can interact with eIF4G and thereby recruit 40S particles to the mRNA. Finally, panel (d) illustrates the simultaneous interactions between eIF4G and eIF4E bound to the cap and between eIF4G and Pab1p bound to the poly(A) tail of the mRNA. This dual binding of eIF4G would presumably recruit 40S particles very strongly and would help explain the synergistic stimulation of translation by the cap and the poly(A).

SUMMARY eIF4G is an adaptor protein that is capable of binding to a variety of other proteins, including eIF4E (the cap-binding protein), eIF-3 (the 40S ribosomal particle-binding protein), and Pab1p (a poly[A]-binding protein). By interacting with these proteins, eIF4G can recruit 40S ribosomal particles to the mRNA and thereby stimulate translation initiation.

Functions of eIF1 and eIF1A   eIF1 causes only a modest (about 20%) stimulation of translation activity in vitro. Thus, it was long thought to be dispensable. However, the genes encoding both eIF1 and eIF1A are essential for yeast viability, so their products are hardly dispensable. But what roles do they play? Tatyana Pestova and colleagues found the answer. Without eIF1 and eIF1A, the 40S particle scans only a few nucleotides, if at all, and remains only loosely bound to the mRNA. With these factors, the 40S particle scans to the initiation codon and forms a stable 48S complex.

Pestova and coworkers used a **toeprint assay** based on the primer extension technique (Chapter 5) to locate the leading edge of the 40S ribosomal particle as it bound to an mRNA. They isolated complexes between the 40S particle and a mammalian β-globin mRNA, then mixed them with a primer that binds downstream of the initiation codon on the mRNA. Then they extended the primer with nucleotides and reverse transcriptase. When the reverse transcriptase hits the leading edge of the 40S subunit, it stops, so the length of the extended primer shows where that leading edge lies. If you think of the 40S particle as a foot, its leading edge would be the toe, which is why we call this a toeprint assay. Finally, Pestova and colleagues electrophoresed the primer extension products to measure

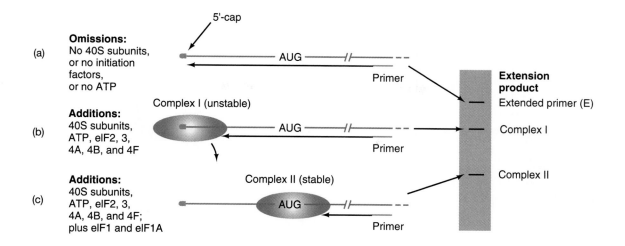

Figure 17.34   Principle of toeprint assay. (a) Negative control. Leave out an essential ingredient, so no complex can form between 40S ribosomal subunits and mRNA. With no 40S particle to block the reverse transcriptase, the primer is extended to the 5′-end of the mRNA. This yields a long extended primer (E) corresponding to naked mRNA. **(b)** Complex formed in the absence of eIF1 and eIF1A. Add all the components listed at left, but omit eIF1 and 1A. Complex I forms at the cap, but does not progress far, if at all. Thus, the primer is extended a long distance to the leading edge of the 40S particle. **(c)** Complex formed in the presence of eIF1 and eIF1A. the 40S ribosomal particle has scanned downstream to the initiation codon (AUG) and formed a stable complex (complex II). Thus the primer is extended only a short distance before it is blocked by the leading edge of the 40S particle in the 48S complex. (*Source:* Adapted from Jackson, *Nature* 394:830, 1998. Copyright © 1998 Macmillan Magazines Ltd. Reprinted by permission.)

their sizes. Figure 17.34 presents a schematic view of this procedure.

The actual results are presented in Figure 17.35. Lanes 1 and 2 contained only mRNA or mRNA and 40S subunits, with no factors, so it is not surprising that no complex formed. Lane 3 contained mRNA, 40S subunits, and eIF2, 3, 4A, 4B, and 4F. These factors promoted formation of complex I only, with no trace of complex II. The leading edge of the 40S particle under these circumstances was between positions +21 and +24 relative to the cap of the mRNA, about where we would expect it if the 40S particle bound at the cap and did not begin scanning or scanned at most a short distance. Lane 4 contains all the factors in lane 3, plus a mixture of initiation factors obtained by washing ribosomes with a saline solution, then collecting those proteins that could be precipitated by ammonium sulfate concentrations between 50 and 70%. Clearly, this mixture of factors, along with others, could promote the formation of complex II, whose leading edge was between positions +15 and +17 relative to the A of the AUG initiation codon, about where we would expect it if the 40S particle was centered on the initiation codon.

Next, Pestova and colleagues purified the important proteins in the 50–70% ammonium sulfate fraction to homogeneity and obtained partial amino acid sequences to identify them. They turned out to be eIF1 and eIF1A. Figure 17.35, lanes 5 and 6 show that each of these factors individually had little or no ability to stimulate complex II formation. On the other hand, lane 7 demonstrates that these two factors together caused complex II to be formed almost exclusively. Thus, these two factors

act synergistically to promote complex II formation. In lane 8, complex I was allowed to form for 5 min, then eIF1 and eIF1A were added. Under these conditions, only complex II formed. Thus, complex I was not a dead end; initiation factors could convert it to complex II.

Did eIF1 and eIF1A convert complex I to complex II by simply causing the 40S particle to scan further on the same mRNA, or did these factors cause the 40S particle to dissociate from the mRNA and bind again to scan to the initiation codon? To find out, Pestova and colleagues formed complex I on a radiolabeled mRNA, then added eIF1 and eIF1A with and without a 15-fold excess of unlabeled competitor mRNA. They purified 48S complexes (presumably equivalent to complex II) by sucrose gradient ultracentrifugation and checked these complexes for radioactivity by scintillation counting (Chapter 5).

As expected (Figure 17.36), they found a clear radioactive peak of 48S complexes in the absense of competitor mRNA. However, they found no radioactive peak of 48S complexes when they added the competitor mRNA at the beginning of the incubation *or* when they added the competitor mRNA after complex I had formed for 5 min. Thus, eIF1 and eIF1A did not simply allow 40S particles in complex I to scan downstream and form complex II on the same, labeled mRNA. If they did, labeled 48S complexes would have been seen when these factors and the competitor mRNA were added after 5 min, when complex I had already formed on the labeled mRNA. Instead, these factors disrupted complex I on the labeled mRNA and forced a new complex to form on the excess, unlabeled mRNA. Presumably, the 40S particles abandoned

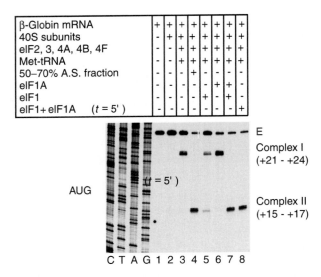

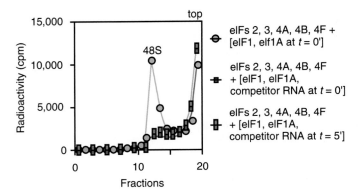

Figure 17.36   Effect of competitor RNA on formation of 48S complex. Pestova and colleagues incubated [^{32}P]β-globin mRNA with 40S ribosomal particles plus the initiation factors and unlabeled competitor RNA combinations indicated at right: blue, no competitor; green, competitor, along with eIF1 and eIF1A, added at time zero; red, competitor, along with eIF1 and eIF1A, added after 5 min of incubation (by which time complex I had formed). After the incubations, the investigators subjected the mixtures to sucrose gradient ultracentrifugation to detect the formation of stable 48S complexes involving 40S particles, [^{32}P]mRNA, and Met-tRNA$_i^{Met}$. They plotted the radioactivity in counts per minute (cpm) detected in each fraction by scintillation counting. The top of the gradient was in fraction 19, as indicated at top right. (*Source:* Reprinted by permission from Pestova et al., *Nature* 394:856, 1998. Copyright © 1998 Macmillan Magazines Ltd.)

Figure 17.35   Results of toeprint assay. Pestova and colleagues carried out a toeprint assay as described in Figure 17.34, using mammalian β-globin mRNA. The components added to each assay are listed at the top of lanes 1–8 "50–70% A.S. fraction" (lane 4) refers to the factors obtained by precipitating proteins from a ribosome salt wash with ammomium sulfate concentrations between 50 and 70% saturated. "eIF1 + eIF1A (t = 5′)" refers to eIF1 and eIF1A added 5 min after adding the other components of the assay. Lanes C, T, A, and G were the results of sequencing a DNA corresponding to the β-globin mRNA. These sequencing lanes were included as markers to determine the exact positions of the leading edges (toeprints) of the 40S ribosomal particle in the complexes. The position of the initiation codon (AUG) is given at left. The bands corresponding to full-length extended primer (E) and complexes I and II are given at right, with the leading edge of the 40S particle relative to the cap and the initiation codon, respectively. eIF1 and eIF1a were required for complex II formation. (*Source:* Pestova, T. et al. Eukaryotic ribosomes require initiation factors 1 and 1A to locate initiation codons. *Nature* 394 (27 Aug 1998) f. 2, p. 855. © Macmillan Magazines Ltd.)

the labeled mRNA, bound to the caps of (mostly) unlabeled mRNAs, and scanned to the initiation codons of these unlabeled mRNAs, forming complex II. Thus, eIF1 and eIF1A are not only essential for proper 48S complex formation, they appear to disrupt improper complexes between 40S ribosomal particles and mRNA.

SUMMARY eIF1 and eIF1A act synergistically to promote formation of a stable 48S complex, involving initiation factors, Met-tRNA$_i^{Met}$, and 40S ribosomal particles bound at the initiation codon of an mRNA. eIF1 and eIF1A appear to act by dissociating improper complexes between 40S particles and mRNA and encouraging the formation of stable 48S complexes.

17.3 Control of Initiation

We have already examined control of eukaryotic gene expression at the transcriptional and post-transcriptional levels. But control also occurs at the translational level. Naturally enough, most of this control happens at the initiation step. We have also learned that most of the control of prokaryotic gene expression occurs at the transcription level. The very short lifetime (only 1–3 min) of the great majority of prokaryotic mRNAs is consistent with this scheme, because it allows bacteria to respond quickly to changing circumstances. It is true that different cistrons on a polycistronic transcript can be translated better than others. For example, the *lacZ, Y,* and *A* cistrons yield protein products in a molar ratio of 10:5:2. However, this ratio is constant under a variety of conditions, so it seems to reflect the relative efficiencies of the Shine–Dalgarno sequences of the three cistrons as well as differential degradation of parts of the polycistronic mRNA. However, some examples of real control of prokaryotic translation do occur. Let us consider two of them and then make a fuller investigation of control of eukaryotic translation.

Prokaryotic Translational Control

Messenger RNA secondary structure can play a role in translation efficiency, as we observed in Figure 17.12 earlier in this chapter. We learned that the initiation codon of the replicase cistron of the MS2 family of RNA phages is buried in a double-stranded structure that also involves part of the coat gene. This explains why the replicase gene of these phages cannot be translated until the coat protein is translated: The ribosomes moving through the coat gene open up the secondary structure that hides the initiation codon of the replicase gene.

Prokaryotic translation can also be controlled by feedback repression. The best studied example is translation of the cistrons on the polycistronic mRNAs encoding the *E. coli* ribosomal protein genes. One such mRNA includes cistrons encoding both the L11 and L1 ribosomal proteins, (Figure 17.37). The L1 protein, when made in moderate quantities, binds tightly to a hairpin loop structure in 23S rRNA. However, when L1 (and L11) are more abundant than 23S rRNA, L1 binds to a similar stem loop structure near the translation start site of the L11 cistron. This represses translation of both the L11 and L1 cistrons, because their translation is coupled.

> **SUMMARY** Prokaryotic mRNAs are very short-lived, so control of translation is not common in these organisms. However, some translational control does occur. Messenger RNA secondary structure can govern translation initiation, as in the replicase gene of the MS2 class of phages, or ribosomal proteins can feedback inhibit the translation of their own mRNAs. L1 does this by binding to a hairpin loop at the beginning of the mRNA that encodes L11 and L1.

Eukaryotic Translational Control

Eukaryotic mRNAs are much longer-lived than prokaryotic ones, so there is more opportunity for translational control. The rate-limiting factor in translation is usually initiation, so we would expect to find most control exerted at this level. In fact, the most common mechanism of such control is phosphorylation of initiation factors, and we know of cases where such phosphorylation can be inhibitory, and others where it can be stimulatory. We also know of at least one case in which an initiation factor is bound and inhibited by another protein until that protein is phosphorylated. This phosphorylation releases the initiation factor so initiation can occur. Finally, there is an example of a protein binding directly to the 5′-untranslated region of an mRNA and preventing its translation. Removal of this protein activates translation.

Phosphorylation of Initiation Factor eIF2α   The best known example of inhibitory phosphorylation occurs in reticulocytes, which make one protein, hemoglobin, to the exclusion of almost everything else. But sometimes reticulocytes are starved for heme, the iron-containing part of he-

(a)

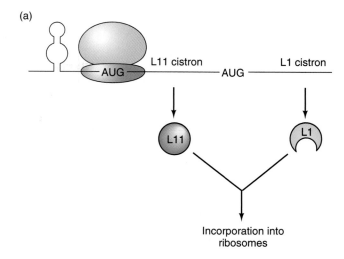

(b)

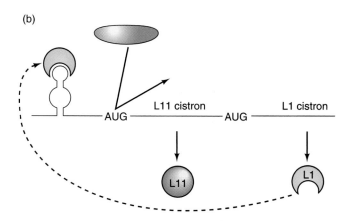

Figure 17.37   Translational repression of the L11 operon.
(a) Moderate amounts of the operon's products, the L11 and L1 ribosomal proteins, are made. All the products can be incorporated into ribosomes, where L1 binds to a stem loop structure on 23S rRNA similar to the one pictured on the mRNA near the initiation codon for the L11 cistron. (b) Too much L11 and L1 proteins are made. The excess L1 finds no 23S rRNA to which to bind, so it binds to the stem loop near the initiation codon for the L11 cistron. This blocks the access of the ribosomes to the L11 cistron and, because translation of the two cistrons is linked, it blocks access to the L1 cistron as well. Thus, L11 and L1 protein synthesis is blocked until the excess L1 disappears.

moglobin, so it would be wasteful to go on producing α- and β-globins, the protein parts. Instead of stopping the production of the globin mRNAs, reticulocytes block their translation as follows (Figure 17.38). The absence of heme unmasks the activity of a protein kinase called the **heme-controlled repressor,** or **HCR.** This enzyme phosphorylates one of the subunits of eIF2, known as **eIF2α.** The phosphorylated form of eIF2 binds more tightly than usual to eIF2B, which is an initiation factor whose job is to exchange GTP for GDP on eIF2. When eIF2B is stuck fast to phosphorylated eIF2, it cannot get free to exchange GTP for GDP on other molecules of eIF2, so eIF2 remains in the inactive

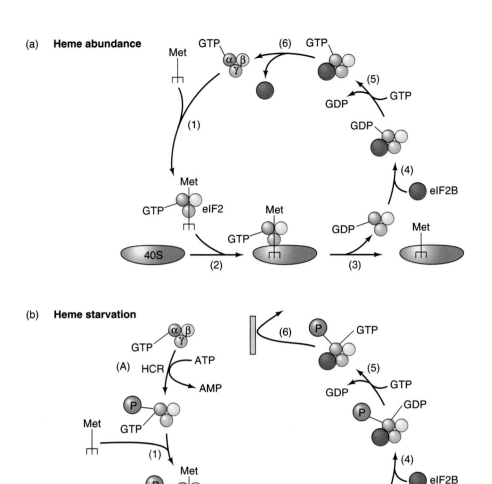

Figure 17.38   Repression of translation by phosphorylation of eIF2α. (a) Heme abundance, no repression. Step 1, Met-tRNA$_i^{Met}$ binds to the eIF2-GTP complex, forming the ternary Met-tRNA$_i^{Met}$ GTP-eIF2 complex. The eIF2 factor is a trimer of nonidentical subunits (α [green], β [yellow], and γ [orange]). Step 2, the ternary complex binds to the 40S ribosomal particle (blue). Step 3, GTP is hydrolyzed to GDP and phosphate, allowing the GDP–eIF2 complex to dissociate from the 40S ribosome, leaving Met-tRNA$_i^{Met}$ attached. Step 4, eIF2B (red) binds to the eIF2–GDP complex. Step 5, eIF2B exchanges GTP for GDP on the complex. Step 6, eIF2B dissociates from the complex. Now eIF2–GTP and Met-tRNA$_i^{Met}$ can get together to form a new complex to start a new round of initiation. (b) Heme starvation leads to translational repression. Step A, HCR (activated by heme starvation) attaches a phosphate group (purple) to the α-subunit of eIF2. Then, steps 1–5 are identical to those in panel (a), but step 6 is blocked because the high affinity of eIF2B for the phosphorylated eIF2α prevents its dissociation. Now eIF2B will be tied up in such complexes, and translation initiation will be repressed.

GDP-bound form and cannot attach Met-tRNA$_i^{Met}$ to 40S ribosomes. Thus, translation initiation grinds to a halt.

The antiviral proteins known as **interferons** follow this same pathway. In the presence of interferon and double-stranded RNA, which appears in many viral infections, but not in normal cellular life, another eIF2α kinase is activated. This one is called **DAI**, for **double-stranded RNA-activated inhibitor of protein synthesis**. The effect of DAI is the same as that of HCR—blocking translation initiation. This is useful in a virus-infected cell because the virus has taken over the cell, and blocking translation will block production of progeny viruses, thus short-circuiting the infection.

SUMMARY Eukaryotic mRNA lifetimes are relatively long, so there is more opportunity for translation control than in prokaryotes. The α-subunit of eIF2 is a favorite target for translation control. In heme-starved reticulocytes, HCR is activated, so it can phosphorylate eIF2α and inhibit initiation. In virus-infected cells, another kinase, DAI, is activated; it also phosphorylates eIF2α and inhibits translation initiation.

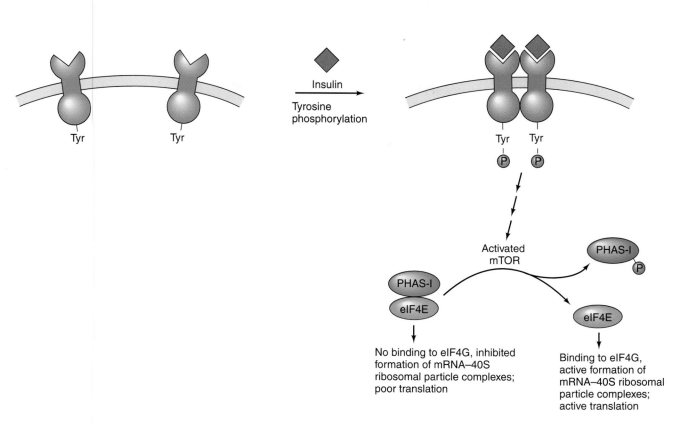

Figure 17.39 Stimulation of translation by phosphorylation of PHAS-I. Insulin, or a growth factor such as EGF, binds to its receptor at the cell surface. Through a series of steps, this activates the protein kinase mTOR. One of the targets of mTOR is PHAS-I. When PHAS-I is phosphorylated by mTOR, it dissociates from eIF4E, releasing it to bind to eIF4G and therefore to participate in active translation initiation.

Phosphorylation of an eIF4E-Binding Protein The rate-limiting step in translation initiation is cap binding by the cap-binding factor eIF4E. Thus, it is intriguing that eIF4E is also subject to phosphorylation, and that this stimulates, rather than represses, translation initiation. Phosphorylated eIF4E binds the cap with about four times the affinity of unphosphorylated eIF4E, which explains the stimulation of translation. We saw that the conditions that favor eIF2α phosphorylation and translation repression are unfavorable for cell growth, (e.g., heme starvation and virus infection). This suggests that the conditions that favor eIF4E phosphorylation and translation stimulation should be favorable for cell growth, and this is generally true. Indeed, stimulation of cell division with insulin or mitogens leads to an increase in eIF4E phosphorylation.

Insulin and various growth factors, such as platelet-derived growth factor (PDGF), also stimulate translation in mammals by an alternative pathway that involves eIF4E. We have known for many years that insulin and many growth factors interact with specific receptors at the cell surface (Figure 17.39). These receptors have intracellular domains with protein tyrosine kinase activity. When they interact with their ligands, these receptors can dimerize and autophosphorylate. In other words, the tyrosine kinase domain of one monomer phosphorylates a tyrosine on the other monomer. This triggers several signal trans-

duction pathways, as described in Chapter 12. One of these activates a protein called mTOR. One of the targets of mTOR is a protein called PHAS-I, which binds to eIF4E and inhibits its activity. In particular, PHAS-I inhibits binding between eIF4E and eIF4G. But once phosphorylated by mTOR, PHAS-I dissociates from eIF4E, which is then free to bind eIF4G and promote formation of active complexes between mRNA and 40S ribosomal particles (Figures 17.39 and 17.33). Thus, translation is stimulated.

Sonenberg and John Lawrence and colleagues discovered human PHAS-I in 1994 in a **far-Western** screen for proteins that bind to eIF4E. A far-Western screen is similar to a screen of an expression library with an antibody (Chapter 4), except that the probe is a labeled ordinary protein instead of an antibody. Thus, we are looking for the interaction between two non-antibody proteins instead of the recognition of a protein by an antibody. In this case, the investigators probed a human expression library (in λgt11) with a derivative of eIF4E, looking for eIF4E-binding proteins. The probe was eIF4E, coupled to the phosphorylation site of heart muscle kinase (HMK), which was then phosphorylated with [γ-^{32}P]ATP to label it. Of about one million plaques screened, nine contained proteins that bound the eIF4E probe. Three of these contained at least part of the gene that codes for the eIF4G subunit of eIF4F, so it is not surprising that these bound

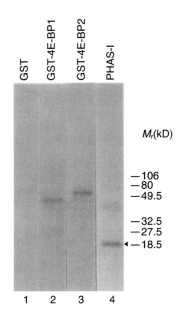

Figure 17.40 Binding between eIF4E and 4E-BP1, 4E-BP2, and PHAS-I. GST, GST-4E-BP1, GST-4E-BP2, and oligohistidine-PHAS-I, as indicated at top, were subjected to SDS-PAGE, then blotted to nitrocellulose, and the blot probed with ^{32}P-labeled HMK-eIF4E. GST did not bind to the probe, but all the other proteins did, demonstrating that they all bind to eIF4E. (*Source:* Pause et al., Insulin-dependent stimulation of protein synthesis by phosphorylation of a regulator of 5′-cap function. *Nature* 371 (27 Oct 1994) p. 763, f. 2. © Macmillan Magazines Ltd.)

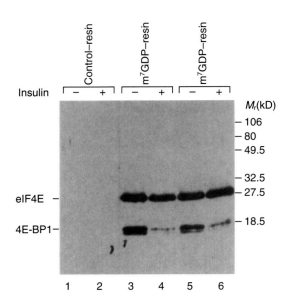

Figure 17.41 Effect of insulin on binding between eIF4E and 4E-BP1. Lawrence, Sonenberg, and collaborators incubated rat adipose tissue with and without insulin, then homogenized the tissue and purified cap-binding proteins in the extract by binding them to m^{7}GDP-agarose. Next, they removed the cap-binding proteins from the resin with SDS, displayed them by SDS-PAGE, and then blotted them to nitrocellulose. Then they probed the blot with antibodies directed against both eIF4E and 4E-BP1, and detected the antibodies with a labeled secondary antibody. Lanes 1 and 2 show that nothing bound to control resin lacking m^{7}GDP. Lanes 3–6 contain proteins bound to m^{7}GDP-agarose; these proteins were from insulin-treated (+) or untreated (–) tissue, as indicated at top. The positions of eIF4E and 4E-BP1 are indicated at left. The fact that insulin treatment greatly reduced the amount of 4E-BP1 that bound to the cap analogue resin demonstrated that insulin treatment reduces the interaction between eIF4E and 4E-BP1. (*Source:* Pause et al., *Nature* 371 (27 Oct 1994) p. 765, f. 4c. © Macmillan Magazines Ltd.)

to eIF4E. The other six positive clones coded for two related proteins, named 4E-BP1 and 4E-BP2 (for eIF4E-binding proteins 1 and 2). The 4E-BP1 protein turned out to be homologous to a previously discovered rat protein called PHAS-I, which has very interesting properties, as we will discuss shortly.

To verify that 4E-BP1 and 4E-BP2, and PHAS-I, really bind to eIF4E, these workers expressed 4E-BP1 and -BP2 as fusion proteins with glutathione-*S*-transferase (GST) and expressed PHAS-I as a fusion protein with oligohistidine (Chapter 4). Then they performed a far-Western blot in which they electrophoresed the GST- and oligohistidine-fusion proteins by SDS-PAGE, then blotted the proteins to nitrocellulose and probed the blot with radiolabeled HMK-eIF4E, as before. Figure 17.40 shows that GST by itself did not bind the eIF4E probe, but all the other proteins did. Thus, these three proteins really do bind to eIF4E.

Because PHAS-I was known to be phosphorylated in response to insulin, it is important to know whether this phosphorylation changes the interaction between 4E-BP1 and eIF4E. To find out, the investigators made extracts of insulin-treated and -untreated adipose (fat) tissue, then performed affinity chromatography on m^{7}GDP-agarose. They removed the proteins bound to the m^{7}GDP-resin with SDS and subjected them to SDS-PAGE. Then they blotted the proteins to nitrocellulose and probed the blot

with antibodies against both eIF4E and 4E-BP1. Figure 17.41 shows that eIF4E from insulin-treated cells binds just as tightly to the cap analog resin as does eIF4E from untreated cells. However, the amount of 4E-BP1 binding to the cap analog resin was much reduced in insulin-treated cells. Because 4E-BP1 binds to the resin through its interaction with eIF4E, this result strongly suggests that the binding between 4E-BP1 and eIF4E is weakened in insulin-treated cells.

Next, these workers showed that both 4E-BP1 and 4E-BP2 can inhibit cap-dependent, but not cap-independent, translation in vitro and in vivo. They used a capped bicistronic mRNA containing the transcripts of two reporter genes (Figure 17.42). The first of these, at the 5′-end of the mRNA, was chloramphenicol acetyl transferase (CAT), whose translation is cap-dependent; the second was luciferase, whose translation was driven by the internal ribosome entry site (IRES) from a picornavirus. CAT and luciferase are both good choices for reporter genes (see Chapter 5) because we can detect their products with convenient, very sensitive assays. Luciferase

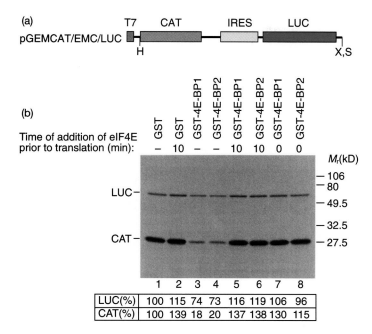

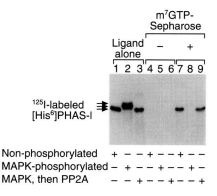

Figure 17.42 Effect of 4E-BP1 and 4E-BP2 on cap-dependent and cap-independent translation in vitro. (a) Bicistronic construct. Lawrence, Sonenberg, and coworkers made a bicistronic RNA by transcribing the plasmid (pictured here) in vitro with T7 RNA polymerase, then capping the transcript at its 5′-end. Translation of the CAT cistron (red) was cap-dependent, but translation of the luciferase cistron (green) could occur from the internal ribosome entry site (IRES, yellow). (b) Experimental results. These workers added the capped RNA to rabbit reticulocyte extracts along with the other substances indicated at top: lane 1, GST; lane 2, GST added 10 min after eIF4E; lane 3, 4E-BP1; lane 4, 4E-BP2; lane 5, 4E-BP1 added 10 min after eIF4E; lane 6, 4E-BP2 added 10 minutes after eIF4E; lane 7, 4E-BP1 added at the same time as eIF4E; lane 8, 4E-BP2 added at the same time as eIF4E. Lanes 3 and 4 show that the 4E-BP fusion proteins inhibited translation of the cap-dependent cistron (CAT), but had little effect on translation of the cap-independent cistron (luciferase, LUC). Lanes 5–8 show that eIF4E could reverse this inhibition when added either 10 min before, or at the same time as, the fusion proteins. (*Source:* Pause et al., *Nature* 371 (27 Oct 1994) p. 766, f. 5a (bottom). © Macmillan Magazines Ltd.)

Figure 17.43 Effect of phosphorylation of PHAS-I on binding to eIF4E. Sonenberg and Lawrence and colleagues prepared recombinant oligohistidine-tagged rat PHAS-I ([His$_6$]PHAS-I) and treated it in three different ways: (1) no modifications; (2) phosphorylation in vitro with MAP kinase; and (3) phosphorylation, followed by dephosphorylation with the protein kinase PP2A. Then they labeled each of these PHAS-I preparations with ^{125}I and tested them for ability to bind to m^{7}GTP-Sepharose resin that had (+) or had not (−) been incubated with extract from insulin-treated cells. Proteins eluted from the resin were subjected to SDS-PAGE. Lanes 1–3 (ligand alone) show the three forms of [(His$_6$)]PHAS-I before passing through the column. Lanes 4–6 show that none of the forms of [(His$_6$)]PHAS-I bound to the column in the absence of cellular extract. Thus, none of the PHAS-I samples had any appreciable affinity on its own for the m^{7}GTP resin. Lanes 7 and 9 show that the two unphosphorylated forms of PHAS-I bound to the resin that had been incubated with extracts from insulin-treated cells. The basis for this binding is presumably that PHAS-I bound to eIF4E from the extract, which in turn bound to the cap analog (m^{7}GTP) on the resin. On the other hand, lane 8 demonstrates that the PHAS-I sample that had been phosphorylated in vitro could not bind to the resin, presumably because it could not bind to the eIF4E immobilized on the resin.
(*Source:* Lin et al., PHAS-I as a link between mitogen-activated protein kinase and translation initiation. *Science* 266 (28 Oct 1994) p. 655, f. 3b. © AAAS.)

synthesis served as an internal control in this experiment because it should occur whether or not cap-dependent translation is working. Figure 17.42a shows the results of in vitro assays of CAT and luciferase translation in rabbit reticulocyte extracts. Sonnenberg and Lawrence added 4E-BP1 and 4E-BP2 (as GST-fusion proteins) to the cell-free extracts at various times. As expected, luciferase production was relatively constant throughout, but when they added either GST–4E-BP1 or 4E–BP2, CAT production decreased considerably. If they added eIF4E at the same time as GST–4E-BP1 or –4E-BP2, or 10 minutes earlier, they observed no inhibition of CAT translation. This suggested that the extra eIF4E was enough to sop up the GST–4E-BP1 or –4E-BP2 and prevent it from inhibiting initiation of translation. They observed similar effects in vivo by introducing plasmids bearing the two reporter genes, along with plasmids encoding 4E-BP1 and 4E-BP2, into cells and assaying for the two reporter gene products. That is, CAT synthesis was inhibited about fivefold in the presence of 4E-BP1 or 4E-BP2, whereas luciferase synthesis was not affected.

In a separate study, Lawrence and Sonenberg and coworkers examined directly the effect of phosphorylation of PHAS-I. These workers used an enzyme called MAP kinase to phosphorylate PHAS-I (actually [His$_6$]PHAS-I) in vitro, then tested its ability to bind to eIF4E. Figure 17.43 shows that unphosphorylated PHAS-I bound to eIF4E immobilized on m^{7}GTP-Sepharose (lane 7), but that PHAS-I that had been phosphorylated did not bind (lane 8). Finally, PHAS-I that was phosphorylated and then dephosphorylated with the protein phosphatase PP2A, bound to eIF4E as well as unphosphorylated PHAS-I did (lane 9). Thus, phosphorylation of PHAS-I really does weaken the interaction between PHAS-I and eIF4E. MAP kinase was used in these experiments, although it is now thought that mTOR is the enzyme that phosphorylates PHAS-I in vivo.

SUMMARY Insulin and a number of growth factors stimulate a pathway involving a protein kinase known as mTOR. One of the targets for mTOR kinase is a protein called PHAS-I (rat) or 4E-BP1 (human). Upon phosphorylation by mTOR, this protein dissociates from eIF4E and releases it to participate in more active translation initiation.

Stimulation by an mRNA-Binding Protein We have seen that mRNA secondary structure can influence translation of prokaryotic genes. This is also true in eukaryotes. Let us consider a well-studied example of repression of translation of an mRNA by interaction between an RNA secondary structure element (a stem loop) and an RNA-binding protein. In Chapter 16 we learned that the concentrations of two iron-associated proteins, the transferrin receptor and ferritin, are regulated by iron concentration. When the serum concentration of iron is high, the synthesis of the transferrin receptor slows down due to destabilization of the mRNA encoding this protein. At the same time, the synthesis of **ferritin**, an intracellular iron storage protein, increases. Ferritin consists of two polypeptide chains, L and H. Iron causes an increased level of translation of the mRNAs encoding both ferritin chains.

What causes this increased efficiency of translation? Two groups arrived at the same conclusion almost simultaneously. The first, led by Hamish Munro, examined translation of the rat ferritin mRNAs; the second, led by Richard Klausner, studied translation of the human ferritin mRNAs. Recall from Chapter 16 that the 3'-untranslated region (3'-UTR) of the transferrin receptor mRNA contains several stem loop structures called iron response elements (IREs) that can bind proteins. We also saw that the ferritin mRNAs have a very similar IRE in their 5'-UTRs. Furthermore, the ferritin IREs are highly conserved among vertebrates, much more so than the coding regions of the genes themselves. These observations strongly suggest that the ferritin IREs play a role in ferritin mRNA translation.

To test this prediction, Munro and colleagues made DNA constructs containing the CAT reporter gene flanked by the 5'- and 3'-UTRs from the rat ferritin L gene. In one construct (pLJ5CAT3), CAT transcription was driven by a very strong retroviral promoter–enhancer. In the other (pWE5CAT3), CAT transcription was under the control of the weak β-actin promoter. Next, they introduced these DNAs into mammalian cells and tested for CAT production in the presence of an iron source (hemin), an iron chelator (desferal), or no additions. Figure 17.44 shows the results. When cells carried the CAT gene in the WE5CAT3 plasmid, CAT mRNA was relatively scarce. Under these circumstances, CAT

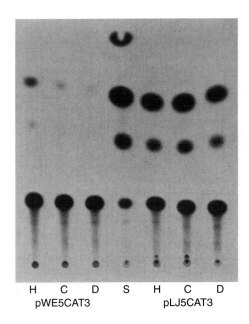

Figure 17.44 Relief of repression of recombinant 5CAT3 translation by iron. Munro and colleagues prepared two recombinant genes with the CAT reporter gene flanked by the 5'- and 3'-UTRs of the rat ferritin L gene. They introduced this construct into cells under control of a weak promoter (the β-actin promoter in the plasmid pWE5CAT3) or a strong promoter (a retrovirus promoter–enhancer in the plasmid pLJ5CAT3). They treated the cells in lanes H with hemin, and those in lanes D with the iron chelator desferal to remove iron. The cells in lanes C were untreated. They assayed CAT activity in each group of cells as described in Chapter 5. Lane S was a standard CAT reaction showing the positions of the chloramphenicol substrate and the acetylated forms of the antibiotic. The lanes on the left show that when the CAT mRNA is not abundant, its translation is inducible by iron. By contrast, the lanes on the right show that when the mRNA is abundant, its translation is not inducible by iron. (*Source:* Aziz & Munro, Iron regulates ferritin mRNA translation through a segment of its 5' untranslated region. *PNAS* 84 (1997) p. 8481, f. 6.)

production was low, but inducible by iron (compare left-hand lanes C and H) and inhibited by the iron chelator (compare left-hand lanes C and D). By contrast, when cells carried the pLJ5CAT3, the CAT mRNA was relatively abundant, and CAT production was high and noninducible. The simplest explanation for these results is that a repressor binds to the IRE in the ferritin 5'-UTR and blocks translation of the associated CAT cistron. Iron somehow removes the repressor and allows translation to occur. CAT production was not inducible when the CAT mRNA was abundant because the mRNA molecules greatly outnumbered the repressor molecules. With little repression happening, induction cannot be observed.

How do we know that the IRE is involved in repression? In fact, how do we even know that the 5'-UTR, and not the 3'-UTR, is important? Munro and colleagues answered these questions by preparing two new constructs, one containing the 5'-UTR, but lacking the 3'-UTR, and

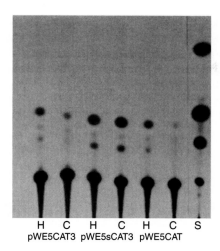

Figure 17.45   Importance of the IRE in the 5′-UTR of pWE5CAT3 for iron inducibility. Munro and colleagues transfected cells with the parent plasmid pWE5CAT3, as described in Figure 17.44, and with two derivatives: pWE5sCAT3, which lacked the first 67 nt of the ferritin 5′-UTR, including the IRE; and pWE5CAT, which lacked the ferritin 3′-UTR. These cells were either treated (H) or not treated (C) with hemin. Then the experimenters assayed each batch of cells for CAT activity. Loss of the IRE caused a loss of iron inducibility. (*Source:* Aziz & Munro, *PNAS* 84 (1987) p. 8482, f. 7.)

one containing both UTRs, but lacking the first 67 nt, including the IRE. Figure 17.45 shows that pWE5CAT, the plasmid lacking the ferritin mRNA's 3′-UTR, still supported iron induction of CAT. On the other hand, pWE5sCAT3, which lacked the IRE, was expressed at a high level with or without added iron. This result not only indicates that the IRE is responsible for induction, it also reinforces the conclusion that the IRE mediates repression because loss of the IRE leads to high CAT production even without iron.

We can conclude that some repressor protein(s) must bind to the IRE in the ferritin mRNA 5′-UTR and cause repression until removed somehow by iron. Because such great conservation of the IREs occurs in the ferritin mRNAs and the transferrin receptor mRNAs, we suspect that at least some of these proteins might operate in both cases. In fact, as we learned in Chapter 16, the aconitase apoprotein is the IRE-binding protein. When it binds to iron, it dissociates from the IRE, relieving repression.

SUMMARY Ferritin mRNA translation is subject to induction by iron. This induction seems to work as follows: A repressor protein (aconitase apoprotein), binds to a stem loop iron response element (IRE) near the 5′-end of the 5′-UTR of the ferritin mRNA. Iron removes this repressor and allows translation of the mRNA to proceed.

SUMMARY

Two events must occur as a prelude to protein synthesis: First, aminoacyl-tRNA synthetases join amino acids to their cognate tRNAs. They do this very specifically in a two-step reaction that begins with activation of the amino acid with AMP, derived from ATP. Second, ribosomes must dissociate into subunits at the end of each round of translation. In bacteria, IF-1 actively promotes this dissociation, whereas IF-3 binds to the free 30S subunit and prevents its reassociation with a 50S subunit to form a whole ribosome.

The initiation codon in prokaryotes is usually AUG, but it can also be GUG, or more rarely, UUG. The initiating aminoacyl-tRNA is N-formyl-methionyl-tRNA$_f^{Met}$. N-formyl-methionine (fMet) is therefore the first amino acid incorporated into a polypeptide, but it is frequently removed from the protein during maturation.

The 30S initiation complex is formed from a free 30S ribosomal subunit plus mRNA and fMet-tRNA$_f^{Met}$. Binding between the 30S prokaryotic ribosomal subunit and the initiation site of an mRNA depends on base pairing between a short RNA sequence called the Shine–Dalgarno sequence just upstream of the initiation codon, and a complementary sequence at the 3′-end of the 16S rRNA. This binding is mediated by IF-3, with help from IF-1 and IF-2. All three initiation factors have bound to the 30S subunit by this time.

IF-2 is the major factor promoting binding of fMet-tRNA$_f^{Met}$ to the 30S initiation complex. The other two initiation factors play important supporting roles. GTP is also required for IF-2 binding at physiological IF-2 concentrations, but it is not hydrolyzed in the process. The complete 30S initiation complex contains one 30S ribosomal subunit plus one molecule each of mRNA, fMet-tRNA$_f^{Met}$, GTP, IF-1, IF-2, and IF-3. GTP is hydrolyzed after the 50S subunit joins the 30S complex to form the 70S initiation complex. This GTP hydrolysis is carried out by IF-2 in conjunction with the 50S ribosomal subunit. The purpose of this hydrolysis is to release IF-2 and GTP from the complex so polypeptide chain elongation can begin.

Eukaryotic ribosomes, together with the initiator tRNA (tRNA$_i^{Met}$), generally locate the appropriate start codon by binding to the 5′-cap of an mRNA and scanning downstream until they find the first AUG in a favorable context. The best context contains a purine at position −3 and a G at position + 4. In 5–10% of the cases, the ribosomes will bypass the first ATG and continue to scan for a more favorable one. Sometimes they apparently initiate at an upstream AUG, translate a short ORF, then continue scanning and reinitiate at a downstream AUG. This mechanism works only with short upstream ORFs. Secondary structure near the 5′-end of an mRNA can have positive or negative effects. A hairpin just past an AUG can force a ribosome to pause at the AUG and thus stimulate initiation. A very stable stem loop between the cap and an initiation site can block ribosome scanning and thus inhibit initiation.

The functions of several of the eukaryotic initiation factors parallel those of the prokaryotic factors. For example, eIF2 is involved in binding Met-tRNA$_i^{Met}$ to the ribosome. eIF2A activates eIF2 by replacing its GDP with GTP. eIF3 binds to the 40S ribosomal subunit and inhibits its reassociation with the 60S subunit. eIF5 encourages association between the 43S complex (40S subunit plus mRNA and Met-tRNA$_i^{Met}$). eIF6 binds to the 60S subunit and blocks its reassociation with the 40S subunit.

eIF4F is a cap-binding protein composed of three parts: eIF4E has the actual cap-binding activity; it is stabilized by the two other subunits: eIF4A and eIF4G. eIF4A has RNA helicase activity that can unwind hairpins found in the 5′-leaders of eukaryotic mRNAs. It is aided in this task by another factor, eIF4B, and requires ATP for activity. eIF4G is an adaptor protein that is capable of binding to a variety of other proteins, including eIF4E (the cap-binding protein), eIF3 (the 40S ribosomal particle-binding protein), and Pab1p (a poly[A]-binding protein). By interacting with these proteins, eIF4G can recruit 40S ribosomal particles to the mRNA and thereby stimulate translation initiation.

eIF1 and eIF1A act synergistically to promote formation of a stable 48S complex, involving initiation factors, Met-tRNA$_i^{Met}$, and 40S ribosomal particles bound at the initiation codon of an mRNA. eIF1 and eIF1A appear to act by dissociating improper complexes between 40S particles and mRNA and encouraging the formation of stable 48S complexes that scan to the initiation codon.

Prokaryotic mRNAs are very short-lived, so control of translation is not common in these organisms. However, some translational control does occur. Messenger RNA secondary structure can govern translation initiation, as in the replicase gene of the MS2 class of phages, or ribosomal proteins can feedback inhibit the translation of their own mRNAs. L1 does this by binding to a hairpin loop at the beginning of the mRNA that encodes L11 and L1.

Eukaryotic mRNA lifetimes are relatively long, so there is more opportunity for translation control than in prokaryotes. The α-subunit of eIF2 is a favorite target for translation control. In heme-starved reticulocytes, HCR is activated, so it can phosphorylate eIF2α and inhibit initiation. In virus-infected cells, another kinase, DAI is activated; it also phosphorylates eIF2α and inhibits translation initiation. Insulin and a number of growth factors stimulate a pathway involving a protein kinase called mTOR. One of the targets for mTOR is a protein called PHAS-I (rat) or 4E-BP1 (human). On phosphorylation by mTOR, this protein dissociates from eIF4E and releases it to participate in more active translation initiation. Ferritin mRNA translation is subject to induction by iron. This induction seems to work as follows: A repressor protein, or proteins, binds to a stem loop iron response element (IRE) near the 5′-end of the 5′-UTR of the ferritin mRNA. Iron somehow removes this repressor and allows translation of the mRNA to proceed.

REVIEW QUESTIONS

1. Describe and give the results of an experiment that shows that ribosomes dissociate and reassociate.

2. How do IF-1 and IF-3 participate in ribosome dissociation? Draw a diagram and present evidence.

3. How do we know that IF-3 binds to ribosomal 30S particles, but does not remain bound when 50S subunits join to form 70S ribosomes?

4. Describe and give the results of an experiment that shows that ribosome cycling is necessary for reinitiation of translation.

5. What are the two methionyl-tRNAs called? What are their roles?

6. Why does translation of the MS2 phage replicase cistron depend on translation of the coat cistron?

7. Present data (exact base sequences are not necessary) to support the importance of the Shine–Dalgarno sequence in translation initiation. Select the most convincing data.

8. Present data to show the effects of the three initiation factors in mRNA-ribosome binding.

9. Describe and give the results of an experiment that shows the role (if any) of GTP hydrolysis in forming the 30S initiation complex.

10. Describe and give the results of an experiment that shows the role of GTP hydrolysis in release of IF-2 from the ribosome.

11. What is the stoichiometry of binding between IF-2 and the 30S particle? How do we know?

12. Describe and give the results of an experiment that shows cooperative binding of initiation factors to the 30S subunit.

13. Present data to show the effects of the three initiation factors in fMet-tRNA$_i^{Met}$ binding to the ribosome.

14. Draw a diagram to summarize the initiation process in *E. coli*.

15. Write the sequence of an ideal eukaryotic translation initiation site. Aside from the AUG, what are the most important positions?

16. Draw a diagram of the scanning model of translation initiation.

17. Present evidence that a scanning ribosome can bypass an AUG and initiate at a downstream AUG.

18. Under what circumstances is an upstream AUG in good context not a barrier to initiation at a downstream AUG? Present evidence.

19. Describe and give the results of an experiment that shows the effects of secondary structure in an mRNA leader on scanning.

20. Draw a diagram of the steps in translation initiation in eukaryotes, showing the effects of each class of initiation factor.

21. Describe and give the results of an experiment that identified the cap-binding protein.

22. Describe and give the results of an experiment that shows that cap-binding protein stimulates translation of capped, but not uncapped, mRNAs.

23. What is the subunit structure of eIF4F? Molecular masses are not required.

24. Describe and give the results of an experiment that shows the roles of eIF4A and eIF4B in translation.

25. Describe a toeprint assay involving 30S ribosomal particles and a fictitious prokaryotic mRNA. What results would you expect?

26. How do we know that eIF1 and eIF1A do not cause conversion of complex I to complex II by stimulating scanning on the same mRNA?

27. Present a model for feedback repression of translation of the L1 and L11 ribosomal protein cistrons.

28. Present a model for repression of translation by phosphorylation of eIF2α.

29. Present a model to explain the effect of PHAS-I phosphorylation on translation efficiency.

30. Describe and give the results of an experiment that shows the importance of the IRE in the ferritin mRNA to iron inducibility of ferritin production.

31. Present a hypothesis for iron inducibility of ferritin production in mammalian cells. Make sure your hypothesis explains why ferritin production is not inducible in cells in which the ferritin gene is driven by a strong promoter.

SUGGESTED READINGS

General References and Reviews

Hentze, M.W. 1997. eIF4G: A multipurpose ribosome adapter? *Science* 275:500–1.

Jackson, R. J. 1998. Cinderella factors have a ball. *Nature* 394:829–31.

Kozak, M. 1989. The scanning model for translation: An update. *Journal of Cell Biology* 108:229–41.

Kozak, M. 1991. Structural features in eukaryotic mRNAs that modulate the initiation of translation. *Journal of Biological Chemistry* 266:19867–70.

Lawrence, J.C. and Abraham, R.T. 1997. PHAS/4E-BPs as regulators of mRNA translation and cell proliferation. *Trends in Biochemical Sciences*. 22:345–49.

Proud, C.G. 1994. Turned on by insulin. *Nature* 371:747–48.

Rhoads, R.E. 1993. Regulation of eukaryotic protein synthesis by initiation factors. *Journal of Biological Chemistry* 268:3017–20.

Sachs, A.B. 1997. Starting at the beginning, middle, and end: Translation initiation in eukaryotes. *Cell* 89:831–38.

Thach, R.E. 1992. Cap recap: The involvement of eIF-4F in regulating gene expression. *Cell* 68:177–80.

Research Articles

Aziz, N., and H.N. Munro. 1987. Iron regulates ferritin mRNA translation through a segment of its 5′-untranslated region. *Proceedings of the National Academy of Sciences USA* 84:8478–82.

Baughman, G., and M. Nomura. 1983. Localization of the target site for translational regulation of the L11 operon and direct evidence for translational coupling in *Escherichia coli*. *Cell* 34:979–88.

Cigan, A.M., L. Feng, and T.F. Donahue. 1988. tRNA$_f^{Met}$ functions in directing the scanning ribosome to the start site of translation. *Science* 242:93–96.

Dubnoff, J.S., A. H. Lockwood, and U. Maitra. 1972. Studies on the role of guanosine triphosphate in polypeptide chain initiation in *Escherichia coli*. *Journal of Biological Chemistry* 247:2884–94.

Edery, I., M. Hümbelin, A. Darveau, K.A.W. Lee, S. Milburn, J.W.B. Hershey, H. Trachsel, and N. Sonenberg. 1983. Involvement of eukaryotic initiation factor 4A in the cap recognition process. *Journal of Biological Chemistry* 258:11398–403.

Fakunding, J.L., and J.W.B. Hershey. 1973. The interaction of radioactive initiation factor IF-2 with ribosomes during initiation of protein synthesis. *Journal of Biological Chemistry* 248:4206–12.

Guthrie, C. and M. Nomura. 1968. Initiation of protein synthesis: A critical test of the 30S subunit model. *Nature* 219:232–35.

Kaempfer, R.O.R., M. Meselson, and H.J. Raskas. 1968. Cyclic dissociation into stable subunits and reformation of ribosomes during bacterial growth. *Journal of Molecular Biology* 31:277–89.

Kozak, M. 1986. Point mutations define a sequence flanking the AUG initiator codon that modulates translation by eukaryotic ribosomes. *Cell* 44:283–92.

Kozak, M. 1989. Circumstances and mechanisms of inhibition of translation by secondary structure in eucaryotic mRNAs. *Molecular and Cellular Biology* 9:5134–42.

Lin, T.-A. 1994. PHAS-I as a link between mitogen-activated protein kinase and translation initiation. *Science* 266:653–56.

Min Jou, W., G. Haegeman, M. Ysebaert, and W. Fiers. 1972. Nucleotide sequence of the gene coding for the bacteriophage MS2 coat protein. *Nature* 237:82–88.

Noll, M. and H. Noll. 1972. Mechanism and control of initiation in the translation of R17 RNA. *Nature New Biology* 238:225–28.

Pause, A. and N. Sonenberg. 1992. Mutational analysis of a DEAD box RNA helicase: The mammalian translation initiation factor eIF-4A. *EMBO Journal* 11:2643–54.

Pause, A., G.J. Belsham, A.-C. Gingras, O. Donzé, T.-A. Lin, J.C. Lawrence, and N. Sonenberg. 1994. Insulin-dependent stimulation of protein synthesis by phosphorylation of a regulator of 5′-cap function. *Nature* 371:762–67.

Pestova, T.V., S.I. Borukhov, and C.V.T. Hellen. 1998. Eukaryotic ribosomes require initiation factors 1 and 1A to locate initiation codons. *Nature* 394:854—59.

Sabol, S., M.A.G. Sillero, K. Iwasaki, and S. Ochoa. 1970. Purification and properties of initiation factor F$_3$. *Nature* 228:1269–75.

Sabol, S. and S. Ochoa. 1971. Ribosomal binding of labeled initiation factor F$_3$. *Nature New Biology* 234:233–36.

Sonenberg, N., M.A. Morgan, W.C. Merrick, and A.J. Shatkin. 1978. A polypeptide in eukaryotic initiation factors that crosslinks specifically to the 5′-terminal cap in mRNA. *Proceedings of the National Academy of Sciences USA* 75:4843–47.

Sonenberg, N., H. Trachsel, S. Hecht, and A.J. Shatkin. 1980. Differential stimulation of capped mRNA translation in vitro by cap binding protein. *Nature* 285:331–33.

Steitz, J.A. and K. Jakes. 1975. How ribosomes select initiator regions in mRNA: Base pair formation between the 3′-terminus of 16S rRNA and the mRNA during initiation of protein synthesis in *Escherichia coli. Proceedings of the National Academy of Sciences USA* 72:4734–38.

Wahba, A.J., K. Iwasaki, M.J. Miller, S. Sabol, M.A.G. Sillero, and C. Vasquez. 1969. Initiation of protein synthesis in *Escherichia coli,* II. Role of the initiation factors in polypeptide synthesis. *Cold Spring Harbor Symposia* 34:291–99.

CHAPTER 18

The Mechanism of Translation II: Elongation and Termination

False color transmission electron micrograph of a gene from the bacterium *Escherichia coli* showing coupled transcription and translation. The DNA fiber (in yellow) runs across the image from bottom left, with numerous ribosomes (in red) attached to each mRNA chain (×32,600). © Oscar Miller/SPL/Photo Researchers, Inc.

The elongation processes in prokaryotes and eukaryotes are very similar. Accordingly, we will consider the processes together, discussing the prokaryotic system first, then noting some differences in the eukaryotic system.

As we learned in Chapter 17, the initiation process in bacteria creates a ribosome primed with an mRNA and the initiating aminoacyl-tRNA, fMet-tRNA$_f^{Met}$, ready to begin elongating a polypeptide chain. Before we look at the steps involved in this elongation process, let us consider some fundamental questions about the nature of elongation: (1) In what direction is a polypeptide synthesized? (2) In what direction does the ribosome read the mRNA? (3) What is the nature of the genetic code that dictates which amino acids will be incorporated in response to the mRNA?

18.1 The Direction of Polypeptide Synthesis and of mRNA Translation

Proteins are made one amino acid at a time, but where does synthesis begin? Do protein chains grow in the amino-to-carboxyl direction, or the reverse? In other words, which amino acid is inserted first into a growing polypeptide—the amino-terminal amino acid, or the carboxyl-terminal one? Howard Dintzis provided definitive proof of the amino → carboxyl direction in 1961 with a study of α- and β-globin synthesis in isolated rabbit reticulocytes (immature red blood cells). He labeled the growing globin chains for various short lengths of time with [³H]leucine, and for a long time with

[¹⁴C]leucine. Then he separated the α- and β-globins, cut them into peptides with trypsin, and separated the peptides. He then plotted the relative amounts of [³H]leucine incorporated into the peptides versus the positions of the peptides, from N-terminus to C-terminus, in the proteins. The long labeling with [¹⁴C]leucine should have labeled all peptides equally, so it could be used as an internal control for losses of certain peptides during purification, and for differences in leucine content from one peptide to another.

Figure 18.1 shows how this procedure can tell us the direction of translation. It is important to notice that the protein chains are in all stages of completion when the labeled amino acid is added. Thus, some are just starting, some are partly finished, and some are almost finished. This means that label will be incorporated into the first peptide only in those proteins whose synthesis had just begun when the label was added. The others will be

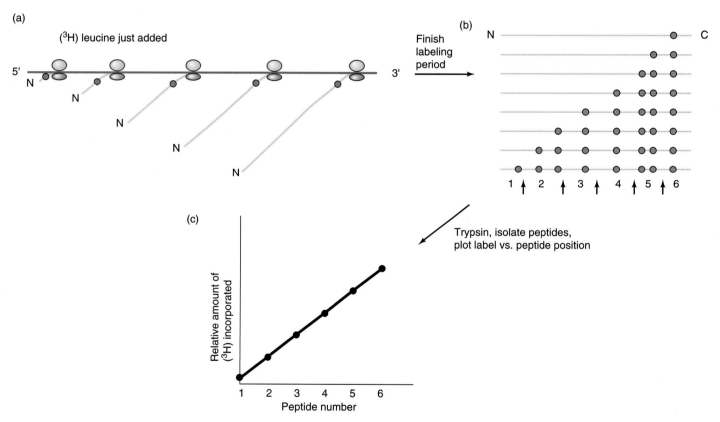

Figure 18.1   Experimental strategy to determine the direction of translation. (a) Labeling the protein. Consider an mRNA (green) being translated by several ribosomes (pink and blue), assuming that the mRNA is translated in the 5′ → 3′ direction and the proteins are made in the amino (N) to carboxyl (C) direction. A labeled amino acid (³H-leucine) has just been added to the system, so it has begun to be incorporated into the growing protein chains (blue), as indicated by the red dots. It is incorporated near the N-terminus in the polypeptides on the left, where protein synthesis has just begun, but only near the C-terminus in the polypeptides on the right, which are almost completed. (b) Distribution of label in completed proteins after a moderate labeling period. The proteins near the top, with label only near the C-terminus correspond to the nearly completed proteins near the right in panel (a). Those near the bottom, with label distributed toward the N-terminus, correspond to the growing proteins near the left in panel (a). These have had time to incorporate label throughout a greater length of the protein. Cutting sites for trypsin within the protein are indicated by arrows at bottom, and the resulting peptides are numbered 1–6 according to their positions in the protein. (c) Model experimental results. We plot the relative amount of ³H labeling in each of the peptides, 1–6, and find that the C-terminal peptides are the most highly labeled. This is what we expect if translation started at the N-terminus. If it had started at the C-terminus (opposite to the picture in panel [a]), then the N-terminal peptides would be the most highly labeled.

labeled in downstream peptides, but not in the first one. By contrast, the end of the protein where protein synthesis ends will be relatively rich in label after a short labeling time. Intermediate peptides will show intermediate levels of labeling. Thus, if translation starts at the amino terminus, labeling will be strongest in carboxyl-terminal peptides. Figure 18.2 shows the results. Labeling of the peptides of both α- and β-globins increased from the amino terminus to the carboxyl terminus, and this disparity was especially noticeable with short labeling times. Therefore, protein synthesis starts at the amino terminus of the protein.

Is the mRNA read in the 5′-to-3′-direction or the reverse? Knowing that proteins grow in the amino-to-carboxyl direction, it is easy to show that mRNAs are read in the 5′-to-3′-direction. When molecular biologists first started using synthetic mRNAs as templates for protein synthesis in the 1960s, some of these messages held the answer to our question. For example, when Ochoa and his colleagues translated the mRNA: $5'\text{-AUGUUU}_n\text{-}3'$, they obtained fMet-Phe_n, where the fMet was at the amino terminus. We know that AUG codes for fMet and UUU codes for phenylalanine (Phe). We see that fMet is incorporated into the amino terminal position of the protein, which means it was added first, before any of the phenylalanines.

Therefore the mRNA must have been read from the 5′-end, because that is where the fMet codon is.

> **SUMMARY** Messenger RNAs are read in the 5′-to-3′-direction, the same direction in which they are synthesized. Proteins are made in the amino-to-carboxyl direction, which means that the amino terminal amino acid is added first.

18.2 The Genetic Code

The term **genetic code** refers to the set of three-base code words (**codons**) in mRNAs that stand for the 20 amino acids in proteins. Like any code, this one had to be broken before we knew what the codons stood for. Indeed, before 1960, other more basic questions about the code were still unanswered. These included: Do the codons overlap? Are there gaps, or "commas," in the code? How many bases make up a codon? These questions were answered in the 1960s by a series of imaginative experiments, which we will examine here.

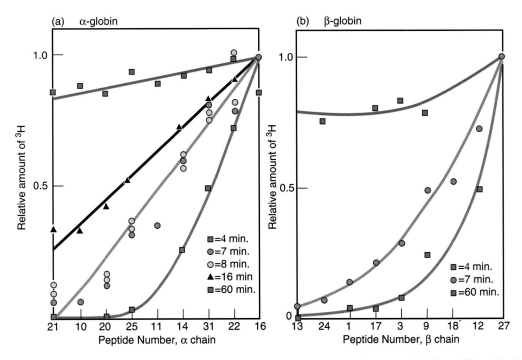

Figure 18.2 Determining the direction of translation. Dintzis carried out the experimental plan outlined in Figure 18.1 with rabbit reticulocytes, which make almost nothing but α- and β-globins. He labeled the reticulocytes with [³H]leucine for various lengths of time, then separated the α- and β-globins, cut each protein into peptides with trypsin, and determined the label in each peptide. He plotted the relative amount of ³H label against the peptide number, with the N-terminal peptide on the left, and the C-terminal peptide on the right. The curves for **(a)** α- and **(b)** β-globin showed the most label in the C-terminal peptides, especially after short labeling times. This is what we expect if translation starts at the N-terminus of a protein. Note that the peptide numbers are not related to their position in the protein, as they are in the example in Figure 18.1. Labeling times are given at lower right of each panel. The blue line in panel **(a)** represents both 7- and 8-min labeling times (blue and yellow symbols, respectively).
(*Source:* From H. M. Dintzis, "Assembly of the Peptide Chains of Hemoglobin," *Proceedings of the National Academy of Sciences* 47:255, 1961.)

Nonoverlapping Codons

In a nonoverlapping code, each base is part of at most one codon. In an overlapping code, one base may be part of two or even three codons. Consider the following micromessage:

AUGUUC

Assuming that the code is triplet (three bases per codon) and this message is read from the beginning, the codons will be AUG and UUC if the code is nonoverlapping. On the other hand, an overlapping code might yield four codons: AUG, UGU, GUU, and UUC. As early as 1957, Sydney Brenner concluded on theoretical grounds that a fully overlapping triplet code like this would be impossible.

However, given the data available in 1957, a *partially* overlapping code remained possible, but A. Tsugita and H. Frankel-Conrat laid it to rest with the following line of reasoning: If the code is nonoverlapping, a change of one base in an mRNA (a missense mutation) would change no more than one amino acid in the resulting protein. For example, consider another micromessage:

AUGCUA

Assuming that the code is triplet (three bases per codon) and this message is read from the beginning, the codons will be AUG and CUA if the code is nonoverlapping. A change in the third base (G) would change only one codon (AUG) and therefore at most only one amino acid. On the other hand, if the code were overlapping, base G could be part of three adjacent codons (AUG, UGC, and GCU). Therefore, if the G were changed, up to three adjacent amino acids could be changed in the resulting protein. But when the investigators introduced one-base alterations into mRNA from tobacco mosaic virus (TMV), they found that these never caused changes in more than one amino acid. Hence, the code must be nonoverlapping.

No Gaps in the Code

If the code contained untranslated gaps, or "commas," mutations that add or subtract a base from a message might change a few codons, but we would expect the ribosome to get back on track after the next comma. In other words, these mutations might frequently be lethal, but in many cases the mutation should occur just before a comma in the message and therefore have little, if any, effect. If no commas were present to get the ribosome back on track, these mutations would be lethal except when they occur right at the end of a message.

Such mutations do occur, and they are called **frameshift mutations**; they work as follows. Consider another tiny message:

AUGCAGCCAACG

If translation starts at the beginning, the codons will be AUG, CAG, CCA, and ACG. If we insert an extra base (X) right after base U, we get:

AUXGCAGCCAACG

Now this would be translated from the beginning as AUX, GCA, GCC, AAC. Notice that the extra base changes not only the codon (AUX) in which it appears, but every codon from that point on. The **reading frame** has shifted one base to the left; whereas C was originally the first base of the second codon, G is now in that position. This could be compared to a filmstrip viewing machine that slips a cog, thus showing the last one-third of one picture together with the first two-thirds of the next.

On the other hand, a code with commas would be one in which each codon is flanked by one or more untranslated bases, represented by Z's in the following message. The commas would serve to set off each codon so the ribosome could recognize it:

AUGZCAGZCCAZACGZ

Deletion or insertion of a base anywhere in this message would change only a single codon. The comma (Z) at the end of the damaged codon would then put the ribosome back on the right track. Thus, addition of an extra base (X) to the first codon would give the message:

AUXGZCAGZCCAZACGZ

The first codon (AUXG) is now wrong, but all the others, still neatly set off by Z's, would be translated normally.

When Francis Crick and his colleagues treated bacteria with acridine dyes that usually cause single-base insertions or deletions, they found that such mutations were very severe; the mutant genes gave no functional product. This is what we would expect of a "comma-less" code with no gaps; base insertions or deletions cause a shift in the reading frame of the message that persists until the end of the message.

Moreover, Crick found that adding a base could cancel the effect of deleting a base, and vice versa. This phenomenon is illustrated in Figure 18.3, where we start with an artificial gene composed of the same codon, CAT, repeated over and over. When we add a base, G, in the third position, we change the reading frame so that all codons thereafter read TCA. When we start with the wild-type gene and delete the fifth base, A, we change the reading frame in the other direction, so that all subsequent codons read ATC. Crossing these two mutants sometimes gives a recombined "pseudo-wild-type" gene like the one on line 4 of the figure. Its first two codons, CAG and TCT, are wrong, but thereafter the insertion and deletion cancel, and the original reading frame is restored. All codons from that point on read CAT.

1. Wild-type: CAT ¦CAT ¦CAT ¦CAT ¦CAT
2. Add a base: CAG¦TCA ¦TCA ¦TCA ¦TCA
3. Delete a base: CAT ¦CTC ¦ATC ¦ATC ¦ATC
4. Cross #2 and #3: CAG¦TCT ¦CAT ¦CAT ¦CAT
5. Add 3 bases: CAG¦GGT¦CAT ¦CAT ¦CAT

Figure 18.3   Frameshift mutations. Line 1: An imaginary gene has the same codon, CAT, repeated over and over. The vertical dashed lines show the reading frame, starting from the beginning. Line 2: Adding a base, G (red), in the third position changes the first codon to CAG and shifts the reading frame one base to the left so that every subsequent codon reads TCA. Line 3: Deleting the fifth base, A (marked by the triangle), from the wild-type gene changes the second codon to CTC and shifts the reading frame one base to the right so that every subsequent codon reads ATC. Line 4: Crossing the mutants in lines 2 and 3 sometimes gives "pseudo-wild-type" revertants with an insertion and a deletion close together. The end result is a DNA with its first two codons altered, but all the other ones put back into the correct reading frame. Line 5: Adding three bases, GGG (red), after the first two bases disrupts the first two codons, but leaves the reading frame unchanged. The same would be true of deleting three bases.

The Triplet Code

Crick and Leslie Barnett discovered that a presumed set of three insertions or deletions could produce a pseudo-wild-type gene (Figure 18.3, line 5). This of course demands that a codon consist of three bases. As Crick remarked to Barnett when he saw the experimental result, "We're the only two [who] know it's a triplet code!" Actually, Crick and Bartlett were inferring that their pseudo-wild-type genes contained three insertions or deletions. They had no way of sequencing the genes to make sure, so more experiments were needed.

In 1961, Marshall Nirenberg and Johann Heinrich Matthaei performed a groundbreaking experiment that laid the foundation for confirming the triplet nature of the code and for breaking the genetic code itself. The experiment was deceptively simple; it showed that synthetic RNA could be translated in vitro. In particular, when Nirenberg and Matthaei translated poly(U), a synthetic RNA composed only of U's, they made polyphenylalanine. Of course, that told them that a codon for phenylalanine contains only U's. This finding by itself was important, but the long-range implication was that one could design synthetic mRNAs of defined sequence and analyze the protein products to shed light on the nature of the code. Gobind Khorana and his colleagues were the chief practitioners of this strategy.

Here is how Khorana's synthetic messenger experiments confirmed that the codons contain three bases: First, if the codons contain an odd number of bases, then a repeating dinucleotide poly[UC] or UCUCUCUC . . . should contain two alternating codons (UCU and CUC, in this case), no matter where translation starts. The result-

ing protein would be a repeating dipeptide—two amino acids alternating with each other. If codons have an even number of bases, only one codon (UCUC, for example) should be repeated over and over. Of course, if translation started at the second base, the single repeated codon would be different (CUCU). In either case, the resulting protein would be a homopolypeptide, containing only one amino acid repeated over and over. Khorana found that poly(UC) translated to a repeating dipeptide, poly(serine-leucine) (Figure 18.4a), proving that the codons contained an odd number of bases.

Repeating triplets were translated to homopolypeptides, as had been expected if the number of bases in a codon was three or a multiple of three. For example, poly(UUC) translated to polyphenylalanine plus polyserine plus polyleucine (Figure 18.4b). The reason for three different products is that translation can start at any point in the synthetic message. Therefore, poly(UUC) can be read as UUC, UUC, and so on, UCU, UCU, and so on, or CUU, CUU, and so on, depending on where translation starts. In all cases, once translation begins, only one codon is encountered, as long as the number of bases in a codon is divisible by 3.

Repeating tetranucleotides were translated to repeating tetrapeptides. For example, poly(UAUC) yielded poly(tyrosine-leucine-serine-isoleucine) (Figure 18.4c). As an exercise, you can write out the sequence of such a message and satisfy yourself that it is compatible with codons having three bases, or nine, or even more, but not six. (We already know six cannot be right because it is not an odd number.) Because codons are not likely to be as cumbersome as nine bases long, three is the best choice. Look at the problem another way: Three is the lowest number that gives enough different codons to specify all 20 amino acids. (The number of permutations of four different bases taken 3 at a time is 4^3, or 64.) There would be only 16 two-base codons ($4^2 = 16$), not quite enough. But there would be over 200,000 ($4^9 = 262,144$) nine-base codons. Nature is usually more economical than that.

SUMMARY The genetic code is a set of three-base code words, or codons, in mRNA that instruct the ribosome to incorporate specific amino acids into a polypeptide. The code is nonoverlapping: that is, each base is part of only one codon. It is also devoid of gaps, or commas; that is, each base in the coding region of an mRNA is part of a codon.

Breaking the Code

Obviously, Khorana's synthetic mRNAs gave strong hints about some of the codons. For example, because poly(UC) yields poly(serine-leucine), we know that one of

(a)   UCUCUCUCUCUC
       Ser Leu Ser Leu

(b)   UUCUUCUUCUUC   or   UUCUUCUUCUUC   or   UUCUUCUUCUUC
       Phe Phe Phe Phe   Ser Ser Ser   Leu Leu Leu

(c)   UAUCUAUCUAUC
       Tyr Leu Ser Ile

Figure 18.4 Coding properties of several synthetic mRNAs. (a) Poly(UC) contains two alternating codons, UCU and CUC, which code for serine (Ser) and leucine (Leu), respectively. Thus, the product is poly(Ser-Leu). **(b)** Poly(UUC) contains three codons, UUC, UCU, and CUU, which code for phenylalanine (Phe), serine (Ser), and leucine (Leu), respectively. The product is therefore poly(Phe), or poly(Ser), or poly(Leu), depending on which of the three reading frames the ribosome uses. **(c)** Poly(UAUC) contains four codons in a repeating sequence: UAU, CUA, UCU, and AUC, which code for tyrosine (Tyr), leucine (Leu), serine (Ser), and isoleucine (Ile), respectively. The product is therefore poly(Tyr-Leu-Ser-Ile).

the codons (UCU or CUC) codes for serine and other codes for leucine. The question remains: Which is which? Nirenberg developed a powerful assay to answer this question. He found that a trinucleotide was usually enough like an mRNA to cause a specific aminoacyl-tRNA to bind to ribosomes. For example, the triplet UUU will cause phenylalanyl-tRNA to bind, but not lysyl-tRNA or any other aminoacyl-tRNA. Therefore, UUU is a codon for phenylalanine. This method was not perfect; some codons did not cause any aminoacyl-tRNA to bind, even though they were authentic codons for amino acids. But it provided a nice complement to Khorana's method, which by itself would not have given all the answers either, at least not easily.

Here is an example of how the two methods could be used together: Translation of the polynucleotide poly(AAG) yielded polylysine plus polyglutamate plus polyarginine. There are three different codons in that synthetic message: AAG, AGA, and GAA. Which one codes for lysine? All three were tested by Nirenberg's assay, yielding the results shown in Figure 18.5. Clearly, AGA and GAA caused no binding of [^{14}C]lysyl-tRNA to ribosomes, but AAG did. Therefore, AAG is the lysine codon in poly(AAG). Something else to notice about this experiment is that the triplet AAA also caused lysyl-tRNA to bind. Therefore, AAA is another lysine codon. This illustrates a general feature of the code: In most cases, more than one triplet codes for a given amino acid. In other words, the code is **degenerate**.

Figure 18.6 shows the entire genetic code. As predicted, there are 64 different codons and only 20 different amino acids, yet all of the codons are used. Three are "stop" codons found at the ends of messages, but all the others specify amino acids, which means that the code is highly degenerate. Leucine, serine, and arginine have six different codons; several others, including proline, threo-

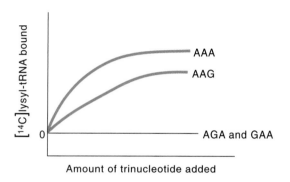

Figure 18.5 Binding of lysyl-tRNA to ribosomes in response to various codons. Lysyl-tRNA was labeled with radioactive carbon (^{14}C) and mixed with *E. coli* ribosomes in the presence of the following trinucleotides: AAA, AAG, AGA, and GAA. Lysyl-tRNA-ribosome complex formation was measured by binding to nitrocellulose filters. (Unbound lysyl-tRNA does not stick to these filters, but a lysyl-tRNA–ribosome complex does.) AAA was a known lysine codon, so binding was expected with this trinucleotide. (*Source:* From Khorana, *Biochemical Journal* 109:715, 1968. Copyright © 1968 The Biochemical Society, London, UK. Reprinted by permission.)

nine, and alanine, have four; isoleucine has three; and many others have two. Just two amino acids, methionine and tryptophan, have only one codon.

SUMMARY The genetic code was broken by using either synthetic messengers or synthetic trinucleotides and observing the polypeptides synthesized or aminoacyl-tRNAs bound to ribosomes, respectively. There are 64 codons in all. Three are stop signals, and the rest code for amino acids. This means that the code is highly degenerate.

Second position

		U	C	A	G	
First position (5'-end)	U	UUU } Phe UUC } UUA } Leu UUG }	UCU } UCC } Ser UCA } UCG }	UAU } Tyr UAC } UAA } STOP UAG }	UGU } Cys UGC } UGA STOP UGG Trp	U C A G
	C	CUU } CUC } Leu CUA } CUG }	CCU } CCC } Pro CCA } CCG }	CAU } His CAC } CAA } Gln CAG }	CGU } CGC } Arg CGA } CGG }	U C A G
	A	AUU } AUC } Ile AUA } AUG Met	ACU } ACC } Thr ACA } ACG }	AAU } Asn AAC } AAA } Lys AAG }	AGU } Ser AGC } AGA } Arg AGG }	U C A G
	G	GUU } GUC } Val GUA } GUG }	GCU } GCC } Ala GCA } GCG }	GAU } Asp GAC } GAA } Glu GAG }	GGU } GGC } Gly GGA } GGG }	U C A G

Third position (3'-end)

Figure 18.6 The genetic code. All 64 codons are listed, along with the amino acid for which each codes. To find a given codon—ACU, for example—we start with the wide horizontal row labeled with the name of the first base of the codon (A) on the left border. Then we move across to the vertical column corresponding to the second base (C). This brings us to a box containing all four codons beginning with AC. It is now a simple matter to find the one among these four we are seeking, ACU. We see that this triplet codes for threonine (Thr), as do all the other codons in the box: ACC, ACA, and ACG. This is an example of the degeneracy of the code. Notice that three codons (magenta) do not code for amino acids; instead, they are stop signals.

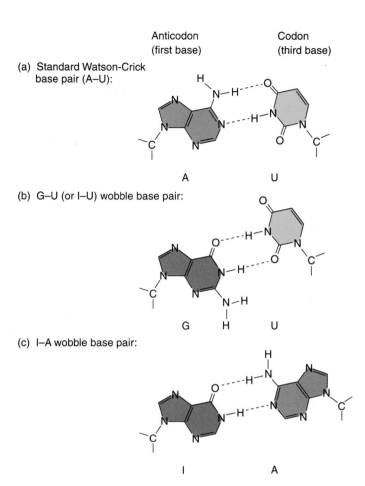

Anticodon
(first base)

Codon
(third base)

(a) Standard Watson-Crick base pair (A–U):

A   U

(b) G–U (or I–U) wobble base pair:

G   H   U

(c) I–A wobble base pair:

I   A

Figure 18.7 Wobble base pairs. (a) Relative positions of bases in a standard (A–U) base pair. The base on the left here and in the wobble base pairs (b) and (c) is the first base in the anticodon. The base on the right is the third base in the codon. (b) Relative positions of bases in a G–U (or I–U) wobble base pair. Notice that U has to "wobble" upward to pair with the G (or I). (c) Relative positions of bases in an I–A wobble base pair. The A has to rotate (wobble) counterclockwise in order to form this pair.

Unusual Base Pairs between Codon and Anticodon

How does an organism cope with multiple codons for the same amino acid? One way would be to have multiple tRNAs (**isoaccepting species**) for the same amino acid, each one specific for a different codon. This is part of the answer, but there is more to it than that; we can get along with considerably fewer tRNAs than that simple hypothesis would predict. Again Francis Crick anticipated experimental results with insightful theory. In this case, Crick hypothesized that the first two bases of a codon must pair correctly with the anticodon according to Watson–Crick base-pairing rules (Figure 18.7a), but the last base of the codon can "wobble" from its normal position to form unusual base pairs with the anticodon. This proposal was called the **wobble hypothesis.** In particular, Crick proposed that a G in an anticodon can pair not only with a C

in the third position of a codon (the **wobble position**), but also with a U. This would give the **wobble base pair** shown in Figure 18.7b. Notice how the U has wobbled from its normal position to form this base pair.

Furthermore, Crick noted that one of the unusual nucleosides found in tRNA is **inosine (I)**, which has a structure similar to that of guanosine. This nucleoside can ordinarily pair like G, so we would expect it to pair with C (Watson–Crick base pair) or U (wobble base pair) in the third position (the wobble position) of a codon. But Crick proposed that inosine could form still another kind of wobble pair, this time with A in the third position of a codon (Figure 18.7c). That means an anticodon with I in the first position can potentially pair with three different codons ending with C, U, or A.

The wobble phenomenon reduces the number of tRNAs required to translate the genetic code. For example,

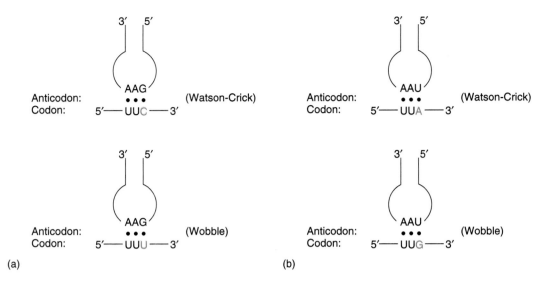

Figure 18.8 The wobble position. (a) An abbreviated tRNA with anticodon 3′-AAG-5′ is shown base-pairing with two different codons for phenylalanine: UUC and UUU. The wobble position (the third base of the codon) is highlighted in red. The base-pairing with the UUC codon (top) uses only Watson–Crick pairs; the base-pairing with the UUU codon (bottom) uses two Watson–Crick pairs in the first two positions of the codon, but requires a wobble pair (G–U) in the wobble position. **(b)** A similar situation, in which a tRNA with anticodon AAU base-pairs with two different codons for leucine: UUA and UUG. Pairing with the UUG codon requires a G–U wobble pair in the wobble position.

consider the two codons for phenylalanine, UUU and UUC, listed at the top left of Figure 18.6. According to the wobble hypothesis, they can both be recognized by an anticodon that reads 3′-AAG-5′ (Figure 18.8a). The G in the 5′-position of the anticodon could form a Watson–Crick G–C base pair with the C in the UUC, or a G–U wobble base pair with the U in UUU. Similarly, the two leucine codons in the same box, UUA and UUG, can both be recognized by the anticodon 3′-AAU-5′ (Figure 18.8b). The U can form a Watson–Crick pair with the A in UUA, or a wobble pair with the G in UUG.

> **SUMMARY** Part of the degeneracy of the genetic code is accommodated by isoaccepting species of tRNA that bind the same amino acid but recognize different codons. The rest is handled by wobble, in which the third base of a codon is allowed to move slightly from its normal position to form a non-Watson–Crick base pair with the anticodon. This allows the same aminoacyl-tRNA to pair with more than one codon. The wobble pairs are G–U (or I–U) and I–A.

The (Almost) Universal Code

In the years after the genetic code was broken, all organisms examined, from bacteria to humans, were shown to share the same code. Therefore it was generally assumed (incorrectly, as we will see) that the code was universal, with no deviations whatsoever. This apparent universality

led in turn to the notion of a single origin of present life on earth.

The reasoning for this idea goes like this: Nothing is inherently advantageous about each specific codon assignment we see. There is no obvious reason, for example, why UUC should make a good codon for phenylalanine, whereas AAG is a good one for lysine. Rather, the genetic code may be an "accident"; it just happened to evolve that way. However, once these codons were established, there was a very good reason why they did not change: A change that fundamental would almost certainly be lethal.

Consider, for instance, a tRNA for the amino acid cysteine and the codon it recognizes, UGU. For that relationship to change, the anticodon of the cysteinyl-tRNA would have to change so it can recognize a different codon, say UCU, which is a serine codon. At the same time, all the UCU codons in that organism's genome that code for important serines would have to change to alternate serine codons so they would not be recognized as cysteine codons. The chances of all these things happening together, even over vast evolutionary time, are negligible. That is why the genetic code is sometimes called a "frozen accident"; once it was established, for whatever reasons, it had to stay that way. So a universal code would be powerful evidence for a single origin of life. After all, if life started independently in two places, we would hardly expect the two lines to evolve the same genetic code by accident!

In light of all this, it is remarkable that the genetic code is not absolutely universal; there are some exceptions to the rule. The first of these to be discovered were in the genomes of mitochondria. In mitochondria of the fruit fly *D. melanogaster*, UGA is a codon for tryptophan rather

Table 18.1 Deviations from the "Universal" Genetic Code

Source	Codon	Usual Meaning	New Meaning
Fruit fly mitochondria	UGA	Stop	Tryptophan
	AGA & AGG	Arginine	Serine
	AUA	Isoleucine	Methionine
Mammalian mitochondria	AGA & AGG	Arginine	Stop
	AUA	Isoleucine	Methionine
	UGA	Stop	Tryptophan
Yeast mitochondria	CUN*	Leucine	Threonine
	AUA	Isoleucine	Methionine
	UGA	Stop	Tryptophan
Higher plant mitochondria	UGA	Stop	Tryptophan
	CGG	Arginine	Tryptophan
Candida albicans nuclei	CTG	Leucine	Serine
Protozoa nuclei	UAA & UAG	Stop	Glutamine
Mycoplasma	UGA	Stop	Tryptophan

*N = Any base

than for "stop." Even more remarkably, AGA in these mitochondria codes for serine, whereas it is an arginine codon in the universal code. Mammalian mitochondria show some deviations, too. Both AGA and AGG, though they are arginine codons in the universal code, have a different meaning in human and bovine mitochondria; there they code for "stop." Furthermore, AUA, ordinarily an isoleucine codon, codes for methionine in these mitochondria.

These aberrations might be dismissed as relatively unimportant, occurring as they do in mitochondria, which have very small genomes coding for only a few proteins and therefore more latitude to change than nuclear genomes. But exceptional codons also occur in nuclear genomes and prokaryotic genomes. In at least three ciliated protozoa, including *Paramecium*, UAA and UAG, which are normally stop codons, code for glutamine. In the prokaryote *Mycoplasma capricolum*, UGA, normally a stop codon, codes for tryptophan. In the pathogenic yeast, *Candida albicans*, CTG, usually a leucine codon, codes for serine. Deviations from the standard genetic code are summarized in Table 18.1.

Clearly, the so-called universal code is not really universal. Does this mean that the evidence now favors more than one origin of present life on earth? If the deviant codes were radically different from the standard one, this might be an attractive possibility, but they are not. In many cases, the novel codons are stop codons that have been recruited to code for an amino acid: glutamine or tryptophan. There is a well-established mechanism for this sort of occurrence, as we will see later in this chapter. The vast majority of known examples of codons that have switched their meaning from one amino acid to another occur in mitochondria.

Again, mitochondrial genomes, because they code for far fewer proteins than nuclear genomes or even prokaryotic genomes, might be expected to change a codon safely every now and then. In summary, even if the code is not universal, a standard code does exist from which the deviant ones almost certainly evolved. Therefore, the evidence still strongly favors a single origin of life.

What about the argument that the code is random: that the existing codons have no inherent advantage? Actually, when we consider the code's effectiveness in dealing with mutations, we find that it is an excellent code indeed. First, consider the fact that single-base changes in the code are likely to result in a shift to a chemically similar amino acid. For example, leucine, isoleucine, and valine all have very similar hydrophobic side chains. And their codons are also very similar, differing only in the first base. So, to pick a particularly advantageous example, a mutation in the first base of the isoleucine codon AUA, could yield UUA, CUA, or GUA. The first two are leucine codons, and the last is a valine codon. Thus, none of these mutations would cause much change in the corresponding amino acid, which minimizes the chance of causing serious damage to the protein product of the mutated gene.

When we consider two other factors, the code looks even better: First, **transitions** (the change of one purine to another, or one pyrimidine to another, are much more common mutations than **transversions**, the change of a purine to a pyrimidine, or vice versa. Second, the ribosome is much more likely to misread the first and third bases in a codon than the second. Considering these things, we can calculate the probability that a single base change will result in no change or just a modest change in the encoded amino acid, for all the possible three-base codes. Then we can see how our natural code stacks up against the others. Figure 18.9 presents a result of this mathematical analysis, which shows that our code is literally one in a million. Only one in a million other possible codes would work better than ours in minimizing the effects of mutations. Given those odds, it seems less likely that our code is just an accident, and not the result of honing by evolution.

SUMMARY The genetic code is not strictly universal. In certain eukaryotic nuclei and mitochondria and in at least one bacterium, codons that cause termination in the standard genetic code can code for amino acids such as tryptophan and glutamine. In several mitochondrial genomes, and in the nuclei of at least one yeast, the sense of a codon is changed from one amino acid to another. These deviant codes are still closely related to the standard one from which they probably evolved. It is not clear whether the genetic code is a frozen accident or the product of evolution, but its ability to cope with mutations suggests that it has been subject to evolution.

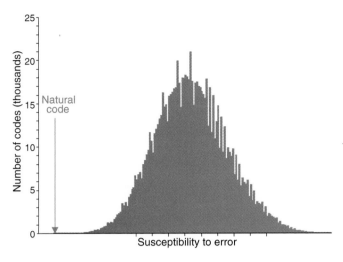

Figure 18.9   Susceptibility of genetic codes to error. The susceptibility to error of all possible triplet genetic codes with four bases is plotted against the number of codes (in thousands) having each susceptibility value. Our own natural code lies far outside the normal distribution, with a very low susceptibility to error. In fact, only one code in a million has a lower susceptibility. (*Source:* Vogel, G. Tracking the History of the Genetic Code. *Science* 281 (17 Jul 1998) 329–331 © AAAS. Source: Freeland and Hurst.)

18.3   The Elongation Mechanism

Elongation of a polypeptide chain occurs in three steps that are repeated over and over. We will survey these steps first, then come back and fill in the details, along with experimental evidence.

Overview of Elongation

Figure 18.10 schematically depicts the three elongation steps through two rounds of elongation (adding two amino acids to a growing polypeptide chain) in *E. coli*. We start with mRNA and fMet-tRNA$_f^{Met}$ bound to a ribosome. There are three binding sites for aminoacyl-tRNAs on the ribosome. Two of these are called the **P (peptidyl) site** and the **A (aminoacyl) site.** In our schematic diagram, the P site is on the left and the A site is on the right. The fMet-tRNA$_f^{Met}$ is in the P site. A binding site for deacylated tRNA called the **E (exit) site** is omitted from Figure 18.10 for simplicity. We will discuss the E site later in this chapter. Detailed below are the elongation events as shown in Figure 18.10:

a. To begin elongation, we need another amino acid to join with the first. This second amino acid arrives bound to a tRNA, and the nature of this aminoacyl-tRNA is dictated by the second codon in the message. The second codon is in the A site, which is otherwise empty, so our second aminoacyl-tRNA will bind to

this site. Such binding requires a protein **elongation factor** known as **EF-Tu** and GTP.

b. Next, the first peptide bond forms. An enzyme called **peptidyl transferase**—an integral part of the large ribosomal subunit—transfers the fMet from its tRNA in the P site to the aminoacyl-tRNA in the A site. This forms a two-amino acid unit called a dipeptide linked to the tRNA in the A site. This whole assembly in the A site is a dipeptidyl-tRNA. What remains in the P site is a deacylated tRNA—a tRNA without its amino acid.

The peptidyl transferase step in prokaryotes is inhibited by an important antibiotic called **chloramphenicol,** whose structure is given in Figure 18.11, along with those of some other common antibiotics. This drug has no effect on most eukaryotic ribosomes, which makes it selective for bacterial invaders in higher organisms. However, the mitochondria of eukaryotes have their own ribosomes, and chloramphenicol does inhibit their peptidyl transferase. Thus, chloramphenicol's selectivity for bacteria is not absolute.

c. In the next step, called **translocation,** the mRNA with its peptidyl-tRNA attached in the A site moves one codon's length to the left. This has the following results: (1) The deacylated tRNA in the P site (the one that lost its amino acid during the peptidyl transferase step when the peptide bond formed) leaves the ribosome. (2) The dipeptidyl-tRNA in the A site, along with its corresponding codon, moves into the P site.

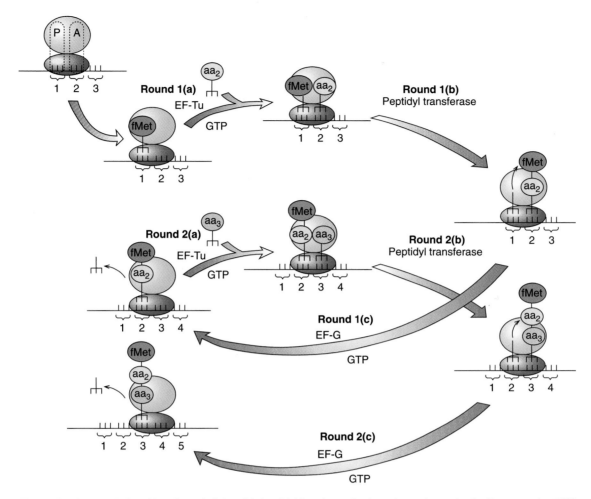

Figure 18.10 Elongation in translation. Note first of all that this is a highly schematic view of protein synthesis. For example, tRNAs are represented by fork-like structures that show only the two business ends of the molecule. Upper left: A ribosome with an mRNA attached is shown to illustrate two sites, P and A, roughly delineated with dotted lines. Round I: **(a)** EF-Tu brings in the second aminoacyl-tRNA (yellow) to the A site on the ribosome. The P site is already occupied by fMet-tRNA (magenta). **(b)** Peptidyl transferase forms a peptide bond between fMet-tRNA and the second aminoacyl-tRNA. **(c)** In the translocation step, EF-G shifts the message and the tRNAs one codon's width to the left. This moves the dipeptidyl-tRNA into the P site, bumps the deacylated tRNA aside, and opens up the A site for a new aminoacyl-tRNA. In round 2, these steps are repeated to add one more amino acid (green) to the growing polypeptide.

(3) The codon that was "waiting in the wings" to the right moves into the A site, ready to interact with an aminoacyl-tRNA. Translocation requires an elongation factor called **EF-G** plus GTP.

The process then repeats itself to add another amino acid: (a) EF-Tu, in conjunction with GTP, brings the appropriate aminoacyl-tRNA to match the new codon in the A site. (b) Peptidyl transferase brings the dipeptide from the P site and joins it to the aminoacyl-tRNA in the A site, forming a tripeptidyl-tRNA. (c) EF-G translocates the tripeptidyl-tRNA, together with its mRNA codon, to the P site.

We have now completed two rounds of peptide chain elongation. We started with an aminoacyl-tRNA (fMet-tRNA$_f^{Met}$) in the P site, and we have lengthened the chain by two amino acids to a tripeptidyl-tRNA. This process continues over and over until the ribosome reaches the last codon in the message. The protein is now complete; it is time for chain termination.

SUMMARY Elongation takes place in three steps: (1) EF-Tu, with GTP, binds an aminoacyl-tRNA to the ribosomal A site. (2) Peptidyl transferase forms a peptide bond between the peptide in the P site and the newly arrived aminoacyl-tRNA in the A site. This lengthens the peptide by one amino acid and shifts it to the A site. (3) EF-G, with GTP, translocates the growing peptidyl-tRNA, with its mRNA codon, to the P site.

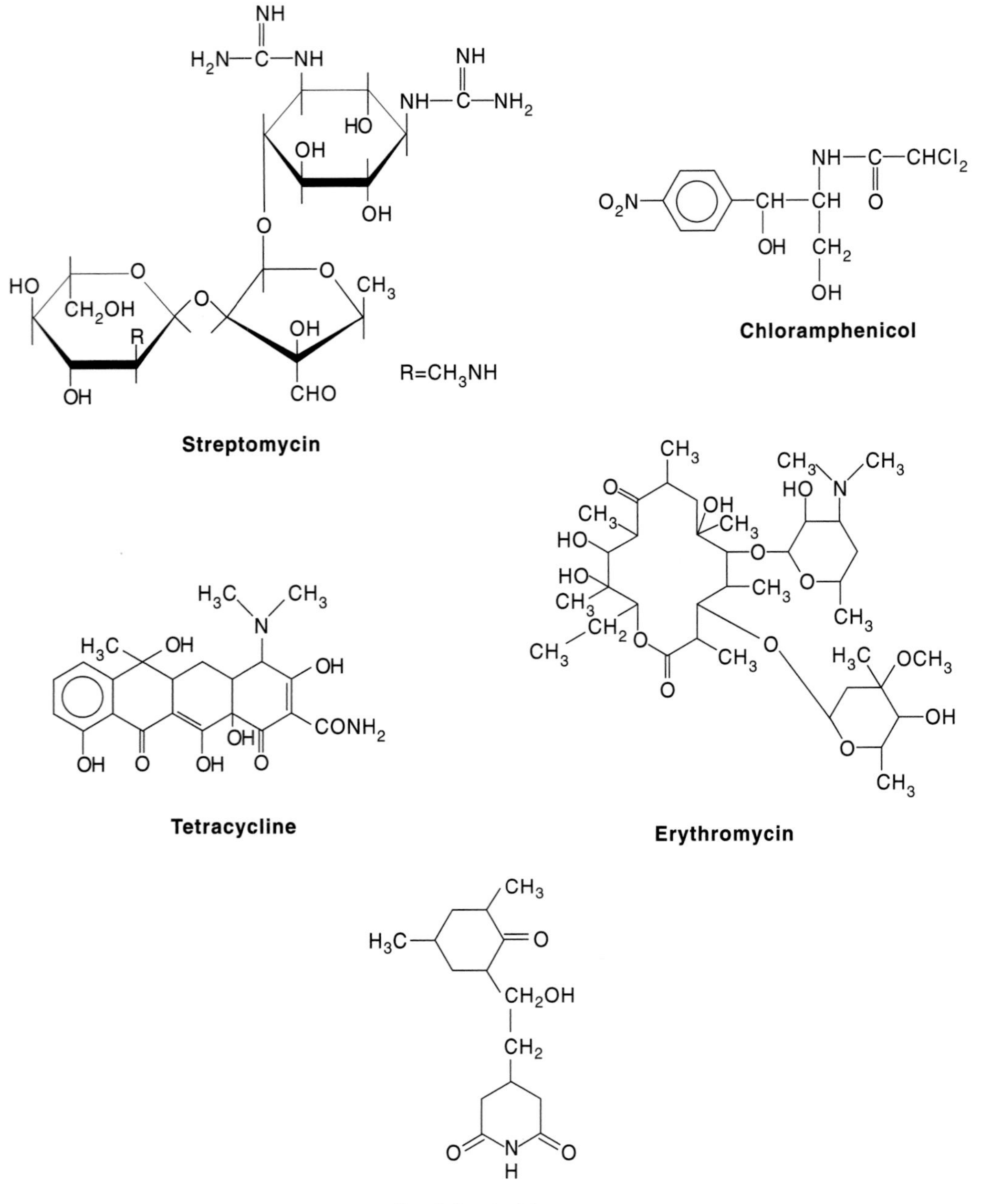

Figure 18.11 The structures of some antibiotics that inhibit protein synthesis by binding to ribosomes.

A Three-Site Model of the Ribosome

The ribosome depicted in the preceding section was simplified for ease of understanding. One simplification was the inclusion of just two binding sites for tRNA—the A and P sites—though there is good evidence for a third site (the E site). We will begin our discussion with the evidence for the A and P sites, and then examine the evidence for the E site. The existence of the A and P sites is based on experiments with the antibiotic **puromycin** (Figure 18.12). This drug is an amino acid coupled to an

adenosine analog. Thus, it resembles the aminoacyl adenosine at the end of an aminoacyl-tRNA. In fact, it looks enough like an aminoacyl-tRNA that it binds to the A site of a ribosome. Then it can form a peptide bond with the peptide in the P site, yielding a peptidyl puromycin. At this point the ruse is over. The peptidyl puromycin is not tightly bound to the ribosome and so is soon released, aborting translation prematurely. This is why puromycin kills bacteria and other cells.

The link between puromycin and the two-site model is this: Before translocation, because the A site is occupied

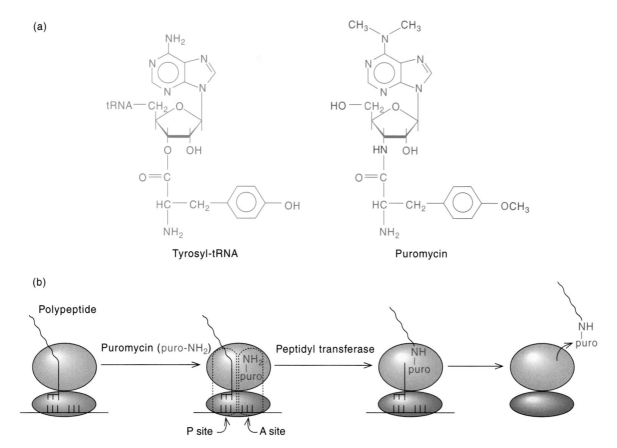

Figure 18.12 Puromycin structure and activity. (a) Comparison of structures of tyrosyl-tRNA and puromycin. Note the rest of the tRNA attached to the 5′-carbon in the aminoacyl-tRNA, where there is only a hydroxyl group in puromycin. The differences between puromycin and tyrosyl-tRNA are highlighted in magenta. **(b)** Mode of action of puromycin. First, puromycin (puro-NH_2) binds to the open A site on the ribosome. (The A site must be open for puromycin to bind.) Next, peptidyl transferase joins the peptide in the P site to the amino group of puromycin in the A site. Finally, the peptidyl-puromycin dissociates from the ribosome, terminating translation prematurely.

by a peptidyl-tRNA, puromycin cannot bind and release the peptide; after translocation, the peptidyl-tRNA has moved to the P site, and the A site is open. At this point puromycin can bind and release the peptide. We therefore see two states the ribosome can assume: puromycin reactive and puromycin unreactive. Those two states require at least two binding sites on the ribosome for the peptidyl-tRNA.

Puromycin can be used to show whether an aminoacyl-tRNA is in the A or the P site. If it is in the P site, it can form a peptide bond with puromycin and be released. However, if it is in the A site, it prevents puromycin from binding to the ribosome and is not released.

This same procedure can be used to show that fMet-tRNA goes to the P site in the 70s initiation complex. In our discussion of initiation in Chapter 17, we assumed that the fMet-tRNA$_f^{Met}$ goes to the P site. This certainly makes sense, because it would leave the A site open for the second aminoacyl-tRNA. Using the puromycin assay, M. S. Bretscher and Marcker showed in 1966 that it does indeed go to the P site. They mixed [^{35}S]fMet-tRNA$_f^{Met}$

with ribosomes, the trinucleotide AUG, and puromycin. If AUG attracted fMet-tRNA$_f^{Met}$ to the P site, then the labeled fMet should have been able to react with puromycin, releasing labeled fMet-puromycin. On the other hand, if the fMet-tRNA$_f^{Met}$ went to the A site, puromycin should not have been able to bind, so no release of labeled amino acid should have occurred. Figure 18.13 shows that the fMet attached to tRNA$_f^{Met}$ was indeed released by puromycin, whereas the methionine attached to tRNA$_m^{Met}$ was not. Thus, fMet-tRNA$_f^{Met}$ goes to the P site, but methionyl-tRNA$_m^{Met}$ goes to the A site. One could argue that it was the fMet, not the tRNA$_f^{Met}$ that made the difference in this experiment. To eliminate that possibility, Bretscher and Marcker performed the same experiment with Met-tRNA$_f^{Met}$ and found that its methionine was also released by puromycin (Figure 18.13c). Thus, the tRNA, not the formyl group on the methionine, is what targets the aminoacyl-tRNA to the P site.

In 1981, Knud Nierhaus and coworkers presented evidence for a third ribosomal site called the E site, where

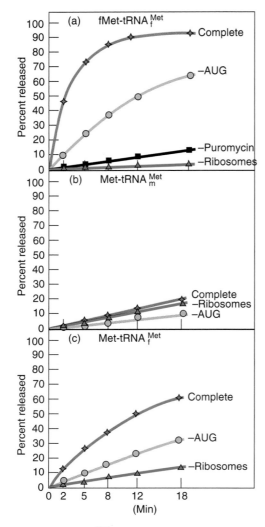

Figure 18.13   fMet-tRNA$_f^{Met}$ occupies the ribosomal P site.
Bretscher and Marcker used a puromycin-release assay to determine the location of fMet-tRNA$_f^{Met}$ on the ribosome. They mixed ^{35}S-labeled fMet-tRNA$_f^{Met}$ (a), Met-tRNA$_m^{Met}$ (b), or Met-tRNA$_f^{Met}$ (c) with ribosomes, AUG, and puromycin, and tested for release of labeled fMet- or Met-puromycin by precipitating tRNA and protein with perchloric acid. Aminoacyl-puromycin released from the ribosome is acid-soluble, whereas aminoacyl-tRNA bound to the ribosome is acid-insoluble. The complete reactions contained all ingredients; control reactions lacked one ingredient, as indicated beside each curve. Met or fMet attached to tRNA$_f^{Met}$ went to the P site and was released. Met attached to tRNA$_m^{Met}$ stayed in the A site and was not released by puromycin. (*Source:* Reprinted with permission from *Nature* 211:382–3, 1966. Copyright © 1966 Macmillan Magazines Limited.)

Table 18.2   Binding of tRNAs and Aminoacyl-tRNAs to *E. coli* Ribosomes

mRNA	tRNA Species	Binding sites No.	Location
Poly(U)	Acetyl-Phe-tRNA^{Phe}	1	P or A
Poly(U)	Phe-tRNA^{Phe}	2	P and A
Poly(U)	tRNA^{Phe}	3	P, E, and A
None	tRNA^{Phe}	1	P
None	Phe-tRNA^{Phe}	0	—
None	Acetyl-Phe-tRNA^{Phe}	1	P

Source: From H.-J. Rheinberger, H. Sternbach, and K. H. Nierhaus, "Three tRNA binding sites on *Escherichia coli* ribosomes," *Proceedings of the National Academy of Sciences* 78(9):5310–5314, September 1981.

binding site could be either the A site or P site. On the other hand, two molecules of Phe-tRNA^{Phe} could bind, one to the A site, and the other to the P site. Finally, three molecules of deacylated tRNA^{Phe} could bind. We can explain these results most easily by postulating a third site that presumably binds deacylated tRNA on its way out of the ribosome. Hence the E, for exit. In the absence of mRNA, only one tRNA can bind. This can be either deacylated tRNA^{Phe} or acetyl-Phe-tRNA^{Phe}. Nierhaus and colleagues speculated that the binding site was the P site, and subsequent work has confirmed this suspicion. We will discuss the E site in greater detail in Chapter 19.

SUMMARY Puromycin resembles an aminoacyl-tRNA and so can bind to the A site, couple with the peptide in the P site, and release it as peptidyl puromycin. On the other hand, if the peptidyl-tRNA is in the A site, puromycin will not bind to the ribosome, and the peptide will not be released. This defines two sites on the ribosome: a puromycin-reactive site (P), and a puromycin unreactive site (A). fMet-tRNA$_f^{Met}$ is puromycin reactive in the 70S initiation complex, so it is in the P site. Binding studies have identified a third binding site (the E site) for deacylated tRNA. Such tRNAs presumably bind to the E site as they exit the ribosome.

E stands for "exit." Their experimental strategy was to bind radioactive deacylated tRNA^{Phe} (tRNA^{Phe} lacking phenylalanine), or Phe-tRNA^{Phe}, or acetyl-Phe-tRNA^{Phe} to *E. coli* ribosomes and to measure the number of molecules bound per 70S ribosome. Table 18.2 shows the results of binding experiments carried out in the presence or absence of poly(U) mRNA. Only one molecule of acetyl-Phe-tRNA^{Phe} could bind at a time to a ribosome, and the

Elongation Step 1: Binding an Aminoacyl-tRNA to the A Site of the Ribosome

Our detailed understanding of the elongation process began in 1965 when Yasutomi Nishizuka and Fritz Lipmann used anion exchange chromatography to separate two protein factors required for peptide bond formation in *E. coli*. They named one factor T, for transfer, because

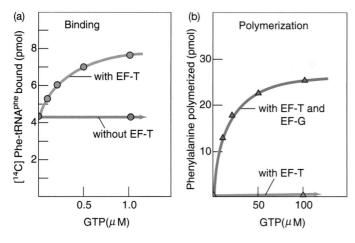

Figure 18.14   Effects of EF-T and GTP on Phe-tRNA^Phe binding to ribosomes and on poly-Phe synthesis. (a) Binding Phe-tRNA^Phe to ribosomes. Ravel mixed ^14C-Phe-tRNA^Phe with washed ribosomes and various concentrations of GTP in the presence or absence of EF-T. She measured Phe-tRNA^Phe–ribosome binding by filtering the mixture and determining the labeled Phe bound to the ribosomes on the filter. Considerable nonenzymatic binding occurred in the absence of EF-T and GTP, but the EF-T-dependent binding required GTP.
(b) Polymerization of phenylalanine. Ravel mixed labeled Phe-tRNA^Phe with ribosomes, EF-T, and various concentrations of GTP in the presence and absence of EF-G. She measured polymerization of Phe by acid precipitation as follows: She precipitated the poly(Phe) with trichloroacetic acid (TCA), heated the precipitate in the presence of TCA to hydrolyze any phe-tRNA^Phe, and trapped the precipitated poly(Phe) on filters. Polymerization required both EF-T and EF-G and a high concentration of GTP. (*Source:* From J. M. Ravel, "Demonstration of a Guanosine Triphosphate-Dependent Enzymatic Binding of Aminoacyl-Ribonucleic Acid to *Escherichia coli* Ribosomes," *Proceedings of the National Academy of Sciences* 57:1815, 1967.)

it transfers aminoacyl-tRNAs to the ribosome. The second factor they called G because of its GTPase activity. (T also has GTPase activity, as we will see.) Then Jean Lucas-Lenard and Lipmann showed that T is actually composed of two different proteins, which they called Tu (where the **u** stands for unstable) and Ts (where the **s** stands for stable). These three factors, which we now call **EF-Tu, EF-Ts,** and **EF-G,** participate in the first and third steps in elongation. Let us consider first the activities of EF-Tu and -Ts because they are involved in the first elongation step.

Joanne Ravel showed in 1967 that unfractionated EF-T (Tu plus Ts) had GTPase activity. Furthermore, EF-T required GTP to bind an aminoacyl-tRNA to the ribosome. To demonstrate this phenomenon, she made [^14C]Phe-tRNA^Phe and added it to washed ribosomes along with EF-T and an increasing concentration of GTP. Then she filtered the ribosomes through nitrocellulose. Labeled Phe-tRNA^Phe that bound to ribosomes stuck to the filter, but unbound Phe-tRNA^Phe washed through. Figure 18.14a depicts the results. Background nonenzymatic binding of the Phe-tRNA^Phe to the ribosomes was rather

high in the absence of EF-T and GTP, but this was not physiologically significant. Ignoring that background, we can see that GTP was necessary for EF-T-dependent binding of Phe-tRNA^Phe to the ribosomes.

When Ravel added both EF-T and EF-G to washed ribosomes in the presence of poly(U) and labeled Phe-tRNA^Phe she found that the ribosomes made labeled polyphenylalanine. And this polymerization of amino acids required an even higher concentration of GTP than the aminoacyl-tRNA-binding reaction did.

When we examined initiation of translation we learned that IF-2-mediated binding of fMet-tRNA_f^Met to ribosomes also required GTP, but that GTP hydrolysis was not required. Could the same be true of EF-T and binding of ordinary aminoacyl-tRNAs to ribosomes? Anne-Lise Haenni and Lucas-Lenard showed in 1968 that this is indeed the case. They labeled N-acetyl-Phe-tRNA with ^14C and Phe-tRNA with ^3H. Then they mixed these labeled aminoacyl-tRNAs with EF-T and either GTP or the unhydrolyzable analog, GDPCP. Under the non-physiological conditions of this experiment, the N-acetyl-Phe-tRNA^Phe went to the P site. These workers measured binding of aminoacyl-tRNAs to ribosomes by filter binding, as described in Figure 18.14. They also measured peptide bond formation between the N-acetyl-Phe in the P site and the Phe-tRNA^Phe in the A site by extracting the dipeptide product and identifying it by paper electrophoresis. Table 18.3 shows that N-acetyl-Phe-tRNA^Phe could bind to the P site, and that Phe-tRNA^Phe could bind to the A site, with the help of EF-T and either GTP or GDPCP. (In fact, N-acetyl-Phe-tRNA^Phe did not even need EF-T to bind to the P site.) Thus, GTP hydrolysis is not needed for EF-T to promote aminoacyl-tRNA binding to the ribosomal A site. In marked contrast, formation of the peptide bond between N-Acetyl-Phe and Phe-tRNA^Phe required GTP hydrolysis. This is analogous to the situation in initiation, where IF-2 can bind fMet-tRNA_f^Met to the P site without GTP hydrolysis, but subsequent events are blocked until GTP is broken down.

These same workers also demonstrated that *both* EF-Tu and EF-Ts are required for Phe-tRNA^Phe binding to the ribosome. The assay was the same as in Table 18.3, except that no GDPCP was used and that EF-Tu and EF-Ts were separated from each other (except for some residual contamination of the EF-Tu fraction with EF-Ts), and added separately. Table 18.4 shows that both EF-Tu and -Ts are required for Phe-tRNA^Phe-ribosome binding. The small amount of binding seen with EF-Tu alone resulted from contamination of the factor by EF-Ts.

Figure 18.15 presents a model for the detailed mechanism by which EF-Tu and EF-Ts cooperate to cause transfer of aminoacyl-tRNAs to the ribosome. First, EF-Tu and GTP form a binary (two-part) complex. Then aminoacyl-tRNA joins the complex, forming a ternary (three-part) complex composed of EF-Tu, GTP, and aminoacyl-tRNA.

Table 18.3   Effect of GTP and GDPCP on Aminoacyl-tRNA Binding to Ribosomes and on Binding Plus Peptide Bond Formation

Additions	N-acetyl-Phe-tRNA^{Phe} bound (^{14}C) (pmol)	N-acetyl diPhe-tRNA formed (^{14}C or ^{3}H) (pmol)	Phe-tRNA bound (^{3}H) (pmol)
None	7.6	0.4	0.1
EF-T + GTP	3.0	4.5	2.8
EF-T + GDPCP	7.0	0.5	4.8

Source: From A.-L. Haenni and J. Lucas-Lenard, "Stepwise Synthesis of a Tripeptide," *Proceedings of the National Academy of Sciences* 61:1365, 1968.

Table 18.4   Requirement for Both EF-Ts and EF-Tu to Bind [^{3}H]Phe-tRNA to Ribosomes Carrying Prebound N-acetyl-[^{14}C]Phe-tRNA

Additions	[^{3}H]Phe-tRNA bound (pmol)
None	2.8
EF-Ts + GTP	2.8
EF-Tu + GTP	5.2
EF-Ts + EF-Tu + GTP	11.6

Source: From A.-L. Haenni and J. Lucas-Lenard, "Stepwise Synthesis of a Tripeptide," *Proceedings of the National Academy of Sciences* 61:1365, 1968.

This ternary complex then delivers its aminoacyl-tRNA to the ribosome's A site. EF-Tu and GTP remain bound to the ribosome. Next, GTP is hydrolyzed and an EF-Tu·GDP complex dissociates from the ribosome. Finally, EF-Ts exchanges GTP for GDP on the complex, yielding an EF-Tu·GTP complex.

What is the evidence for this scheme? Herbert Weissbach and colleagues found in 1967 that an EF-T preparation and GTP could form a complex that was retained by a nitrocellulose filter. They labeled GTP, mixed it with EF-T, and found that the labeled nucleotide bound to the filter. This meant that GTP had bound to a protein in the EF-T preparation, presumably EF-T itself, to form a complex. Julian Gordon then discovered that adding an aminoacyl-tRNA to the EF-Tu·GTP complex caused the complex to be released from the filter. One interpretation of this behavior is that the aminoacyl-tRNA joined the EF-Tu·GTP complex to form a ternary complex that could no longer bind to the filter. Two lines of evidence support this hypothesis. First, deacylated tRNA (tRNA lacking an amino acid) did not release the EF-T·GTP complex from the filter (Figure 18.16). This fits the hypothesis because deacylated tRNA would not be expected to enter into a complex whose job is to deliver an aminoacyl-

tRNA to the ribosome. Furthermore, other investigators found that fMet-tRNA$_f^{Met}$, Met-tRNA$_f^{Met}$, and N-acetyl-Phe-tRNA^{Phe}, like deacylated tRNA, failed to release the EF-T·GTP complex from the filter. Thus, only aminoacyl-tRNAs that are destined to bind to the A site could complex with EF-T and GTP to form the ternary complex. The second line of evidence supporting the existence of the ternary complex is the fact that EF-T activity passes through the filter into the filtrate when aminoacyl-tRNA is added (Figure 18.17). This means that EF-T was part of the binary complex, and we already know from the labeling experiment that GTP was part of this complex. Since EF-T by itself, or bound just to GTP, would bind to the filter, we conclude that it flows through because it is in a complex with both GTP and aminoacyl-tRNA.

Ravel and her collaborators gave us additional evidence for the formation of the ternary complex with the following experiment. They labeled GTP with ^{3}H, and Phe-tRNA^{Phe} with ^{14}C, and mixed them with EF-T, then subjected the mixture to gel filtration on Sephadex G100 (Chapter 5). This gel filtration resin excludes relatively large proteins, such as EF-T, so they flow through rapidly in a fraction called the void volume. By contrast, relatively small substances like GTP, and even Phe-tRNA^{Phe}, enter the pores in the resin and are thereby retarded; they emerge later from the column, after the void volume. In fact, the smaller the molecule, the longer it takes to elute from the column. Figure 18.18 shows the results of this gel filtration experiment. A fraction of both labeled substances, GTP and Phe-tRNA^{Phe}, emerged relatively late, in their usual positions. These fractions represented free GTP and Phe-tRNA^{Phe}. However, significant fractions of both substances eluted much earlier, around fraction 20, demonstrating that they must be complexed to something larger. The predominant larger substance in this experiment was EF-T, and the experiments we have already discussed implicate EF-T in this complex, so we infer that a ternary complex, involving Phe-tRNA^{Phe}, GTP, and EF-T has formed.

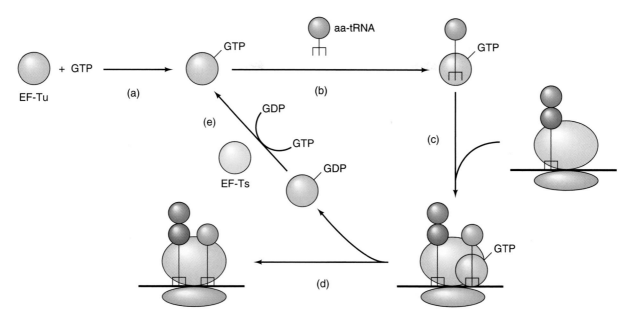

Figure 18.15 Model of binding aminoacyl-tRNAs to the ribosome A site. (**a**) EF-Tu couples with GTP to form a two-part complex. (**b**) This complex associates with an aminoacyl-tRNA to form a ternary complex. (**c**) The ternary complex binds to a ribosome with a peptidyl-tRNA in its P site and an empty A site. (**d**) GTP is hydrolyzed and the resulting EF-Tu·GDP complex dissociates from the ribosome, leaving the new aminoacyl-tRNA in the A site. (**e**) EF-Ts exchanges GTP for GDP on EF-Tu, regenerating the EF-Tu·GTP complex.

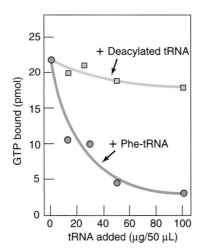

Figure 18.16 Effect of aminoacyl-tRNA and deacylated tRNA on the EF-T·GTP complex. Gordon mixed either Phe-tRNA or deacylated tRNA with the EF-Tu·GTP complex containing [³H]GTP, then filtered the mixture through nitrocellulose and plotted the amount of GTP in the complex still bound to the filter versus the amount of tRNA added. Phe-tRNA^{Phe} caused a big decrease in the amount of complex bound, whereas deacylated tRNA^{Phe} had little effect. (*Source:* From J. Gordon, "A Stepwise Reaction Yielding a Complex Between a Supernatant Fraction From *E. coli,* Guanosine, 5'-Triphosphate, and Aminoacyl-SRNA," *Proceedings of the National Academy of Sciences* 59:180, 1968.)

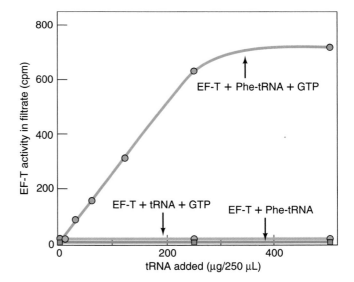

Figure 18.17 Appearance of EF-T in nitrocellulose filtrate after addition of aminoacyl-tRNA. Gordon mixed the following ingredients, filtered them through nitrocellulose, and then tested the filtrate for EF-T activity using a phenylalanine polymerization assay. Red, EF-T + Phe-tRNA^{Phe} + GTP; blue, EF-T + deacylated tRNA + GTP; green, EF-T + Phe-tRNA. Only when EF-T, Phe-tRNA^{Phe}, and GTP were all present did an EF-T complex pass through the filter. (*Source:* From J. Gordon, "A Stepwise Reaction Yielding a Complex Between a Supernatant Fraction From *E. coli,* Guanosine, 5'-Triphosphate, and Aminoacyl-SRNA," *Proceedings of the National Academy of Sciences* 59:182, 1968.)

So far, we have not distinguished between EF-Ts and EF-Tu in these experiments. Herbert Weissbach and his collaborators did this by separating the two proteins and testing them separately. They found that EF-Tu is the factor that binds GTP in the two-part complex. What then is the role of EF-Ts? These investigators demonstrated that this factor is essential for conversion of the EF-Tu·GDP

complex to the EF-Tu·GTP complex. However, EF-Ts has little, if any, effect when it is presented with the pre-formed EF-Tu·GTP complex or with EF-Tu itself (Figure 18.19). Thus, it seems that EF-Ts does not form a complex directly from EF-Tu and GTP. Instead, it converts EF-Tu·GDP to EF-Tu·GTP by exchanging the guanine nucleotide.

How does EF-Ts perform its exchange duty? David Miller and Weissbach showed that EF-Ts can displace GDP from EF-Tu·GDP (Figure 18.20) by forming an EF-Ts·EF-Tu complex. How does this displacement work?

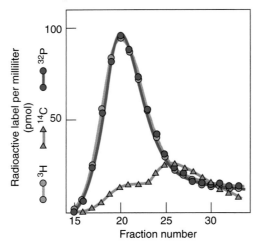

Figure 18.18   Formation of a ternary complex among EF-T, aminoacyl-tRNA, and GTP. Ravel and colleagues mixed ^{14}C-Phe-tRNA^{Phe} with GTP (labeled in the guanine part with ^{3}H and in the γ-phosphate with ^{32}P), and EF-T. Then they passed the mixture through a Sephadex G100 gel filtration column to separate large molecules, such as EF-T, from relatively small molecules such as GTP and Phe-tRNA^{Phe}. They assayed each fraction for the three radioisotopes to detect GTP and Phe-tRNA^{Phe}. Both of these substances were found at least partly in a large-molecule fraction (around fraction 20), so they were bound to the EF-T in a complex. (*Source:* From Ravel et al., *Biochemical and Biophysical Research Communications* 32:12, 1968. Copyright © 1968 Academic Press, reproduced with permission of the publisher.)

X-ray crystallography studies on EF-Tu·EF-Ts complexes by Reuben Leberman and colleagues have shown that one of the main consequences of EF-Ts binding to EF-Tu·GDP is disruption of the Mg^{2+}-binding center of EF-Tu. The weakened binding between EF-Tu and Mg^{2+} leads to dissociation of GDP, which opens the way for binding of GTP to EF-Tu.

Why is EF-Tu needed to escort aminoacyl-tRNAs to the ribosome? The ester bond joining the amino acid to its cognate tRNA is easily broken, and sequestering the aminoacyl-tRNA within the EF-Tu protein protects this labile compound from hydrolysis. But the concentration of aminoacyl-tRNAs in the cell is quite high. Is there enough EF-Tu to go around? Yes, because EF-Tu is one of the most abundant proteins in the cell. For example, EF-Tu constitutes 5% of the total protein in *E. coli* cells, and the reason for this abundance appears to be the important protective role that EF-Tu plays.

SUMMARY A ternary complex formed from EF-Tu, aminoacyl-tRNA, and GTP delivers an aminoacyl-tRNA to the ribosome's A site, without hydrolysis of the GTP. In the next step, EF-Tu hydrolyzes GTP with its ribosome-dependent GTPase activity, and an EF-Tu·GDP complex dissociates from the ribosome. EF-Ts regenerates an EF-Tu·GTP complex by exchanging GTP for GDP attached to EF-Tu. Addition of aminoacyl-tRNA then reconstitutes the ternary complex for another round of translation elongation.

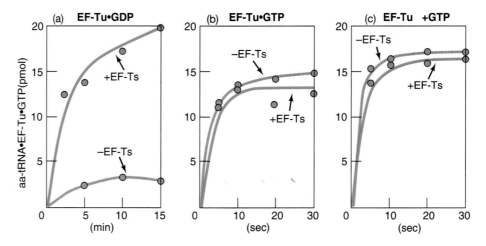

Figure 18.19   Effect of EF-Ts on ternary complex formation. Weissbach and colleagues attempted to form the ternary complex with [^{14}C]Phe-tRNA, [^{3}H]GTP, and the EF-Tu preparations listed at top, with (red) and without (blue) EF-Ts. They measured ternary complex formation by loss of radioactivity trapped by nitrocellulose filtration. EF-Ts stimulated complex formation only when EF-Tu·GDP was the substrate. EF-Tu·GTP (panel **b**) or EF-Tu+GTP (panel **c**) could form the complex spontaneously, with no help from EF-Ts. (aa-tRNA = aminoacyl-tRNA). (*Source:* From Weissbach et al., *Archives of Biochemistry and Biophysics,* 137:267, 1970. Copyright © 1970 Academic Press, reproduced with permission of the publisher.)

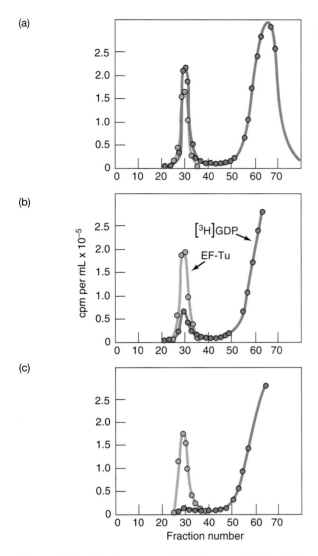

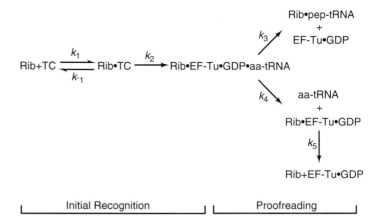

Figure 18.21   Probable mechanism of incorporation of amino acids into protein. Each rate constant denotes the rate of the reaction represented by the arrow with which it is associated. For example, k_1 represents the rate of binding of a ternary complex (TC) to a ribosome (Rib); k_{-1} represents the rate of the reverse reaction. (*Source:* Reprinted from *Trends in Biochemical Sciences,* vol. 13, R. C. Thompson, "EFTu provides an internal kinetic standard for translational accuracy," 91–93. Copyright © 1988 with permission of Elsevier Science.)

Figure 18.20   Displacement of GDP from an EF-Tu·GDP complex by EF-Ts. Miller and Weissbach mixed an EF-Tu·[^{3}H]GDP complex with three different amounts of EF-Ts, then detected the amount of GDP remaining in the complex by gel filtration through Sephadex G-25. The three panels contained the following amounts of EF-Ts: (**a**), 500 units; (**b**), 14,000 units; (**c**), 25,000 units. Red, ^{3}H-GDP; blue, EF-Tu. (*Source:* From Miller and Weissbach, *Biochemical and Biophysical Research Communications* 38:1019, 1970. Copyright © 1970 Academic Press, reproduced with permission of the publisher.)

Proofreading   As we will see in Chapter 19, part of the accuracy of protein synthesis comes from charging of tRNAs with the correct amino acids. But part also comes in elongation step 1: The ribosome usually binds the aminoacyl-tRNA called for by the codon in the A site; However, if it makes a mistake in this initial recognition step, it still has a chance to correct it by rejecting an incorrect aminoacyl-tRNA before it can donate its amino acid to the growing polypeptide. This process is called **proofreading.**

Robert Thompson summarized these concepts with the scheme in Figure 18.21. The *k* values are rate con-

stants that indicate the rates of each of the reactions. These rate constants vary, depending on whether the ribosome has bound a ternary complex with the correct or incorrect aminoacyl-tRNA. The most variation from correct to incorrect occurs in k_{-1} and k_4. The first of these represents the rate of dissociation of the ternary complex from the ribosome just after it has bound. This should not happen if the aminoacyl-tRNA in the ternary complex is correct, but it does happen at a low rate. It happens at a much higher rate if the aminoacyl-tRNA is incorrect because the base pairing between codon and anticodon is imperfect and does not hold the incorrect ternary complex as tightly as perfect base pairs do. In general, this dissociation of incorrect ternary complexes occurs before GTP has a chance to be hydrolyzed, so the aminoacyl-tRNA from those ternary complexes does not usually remain bound to the ribosome.

The second rate constant k_4 represents the rate of dissociation of an aminoacyl-tRNA (derived from the ternary complex) from the ribosome. Again, this happens at a much higher rate if the aminoacyl-tRNA is incorrect than it does when it is correct, because of the weakness of the imperfect codon–anticodon base pairing. This is generally fast enough that an incorrect aminoacyl-tRNA dissociates from the ribosome before it has a chance to be incorporated into the nascent polypeptide.

A general principle that emerges from the analysis of accuracy in translation is that a high degree of accuracy and a high rate of translation are incompatible. In fact, accuracy and speed are inversely related: The faster translation goes, the less accurate it becomes. This is because the ribosome must allow enough time for incorrect ternary complexes and aminoacyl-tRNAs to leave before the incorrect amino acid is irreversibly incorporated into the growing polypeptide. If translation goes faster, more incorrect amino acids will be incorporated. Conversely, if translation goes more slowly, accuracy will be higher, but then proteins may not be made fast enough to sustain life. So there is a delicate balance between speed and accuracy of translation.

One of the most important factors in this balance is the rate of hydrolysis of GTP by EF-Tu. If the rate were higher, less time would be available for the first proofreading step: EF-Tu would hydrolyze GTP to GDP quickly without giving sufficient time for ternary complexes bearing improper aminoacyl-tRNAs to dissociate from the ribosome. On the other hand, if the rate were lower, there would be ample time for proofreading, but translation would be too slow. What is the proper rate? In *E. coli*, the average time between binding of the ternary complex and hydrolysis of GTP is several milliseconds. Then it takes several milliseconds more for EF-Tu·GDP to dissociate from the ribosome. Proofreading takes place during both of these pauses, and shortening one or both could be devastating to accuracy of translation.

How large an error rate in translation can a cell tolerate? What if it were 1%, for example? Ninety-nine percent accuracy sounds pretty good until you consider that the lengths of most polypeptides are much more than 100 amino acids. They average about 300 amino acids long, and some are more than 1000 amino acids long. The probability p of producing an error-free polypeptide, given an error rate per amino acid (ε) and a polypeptide length (n) is given by the following expression:

$$p = (1-\varepsilon)^n$$

For example, with an error rate of 1%, an average size polypeptide would be produced error-free only about 5% of the time, and a 1000 amino acid polypeptide would almost never be error-free. On the other hand, if the error rate were 0.1%, an average size polypeptide would be produced error-free about 74% of the time, but a 1000 amino acid polypeptide would be made error-free only about 37% of the time. This would still pose a problem for large polypeptides. What if the error rate were only 0.01%? At that rate, about 97% of average size polypeptides, and about 91% of 1000 amino acid polypeptides would be produced error-free. That seems like an acceptable rate of production of defective proteins, and the observed error rate per amino acid added, at least in *E. coli*, is in fact close to 0.01%.

An important antibiotic known as **streptomycin** interferes with proofreading so the ribosome makes more mistakes. For example, normal ribosomes incorporate phenylalanine almost exclusively in response to the synthetic message poly(U). But streptomycin greatly stimulates the incorporation of isoleucine, and to a lesser extent, the incorporation of serine and leucine in response to poly(U).

Certain natural conditions allow us to see what happens when the rate of translation is either faster or slower than normal. For example, mutants in ribosomal proteins, such as *ram,* or in EF-Tu, such as *tufAr,* double k_3, the rate of peptide bond formation. In these mutants, accuracy of translation suffers because not enough time is available for incorrect aminoacyl-tRNAs to dissociate from the ribosome.

By contrast, in streptomycin-resistant mutants such as *strA,* k_3 is only half the normal value, so peptide bond formation occurs at half the normal rate. This allows extra time for incorrect aminoacyl-tRNAs to leave the ribosome, so translation is extra accurate. Another example of decreased rate of translation is seen when *E. coli* cells are starved for amino acids. Under these conditions, the concentration of a GTP derivative, **ppGpp** (guanosine 3′-diphosphate, 5′-diphosphate), increases. This nucleotide, sometimes known as "magic spot" because it was first discovered as a novel spot on a chromatogram, slows down the rate of translation, presumably to save the ribosomes from making mistakes when the concentrations of aminoacyl-tRNAs are low. This would happen if the ribosomes had to wait for scarce aminoacyl-tRNAs and accepted the wrong one by mistake.

SUMMARY The protein-synthesizing machinery achieves accuracy during elongation in a two-step process. First, it gets rid of ternary complexes bearing the wrong aminoacyl-tRNA before GTP hydrolysis occurs. If this screen fails, it can still eliminate the incorrect aminoacyl-tRNA in the proofreading step before the wrong amino acid can be incorporated into the growing protein chain. According to Thompson's hypothesis, both these screens rely on the weakness of incorrect codon–anticodon base pairing to ensure that dissociation will occur more rapidly than either GTP hydrolysis or peptide bond formation. The balance between speed and accuracy of translation is delicate. If peptide bond formation goes too fast, incorrect aminoacyl-tRNAs do not have enough time to leave the ribosome, so their amino acids are incorporated into protein. But if translation goes too slowly, proteins are not made fast enough for the organism to grow successfully. The actual error rate, about 0.01% per amino acid added, strikes a good balance between speed and accurary.

Elongation Step 2: Peptide Bond Formation

After the initiation factors and EF-Tu have done their jobs, the ribosome has fMet-tRNA$_f^{Met}$ in the P site and an aminoacyl-tRNA in the A site. Now it is time to form the

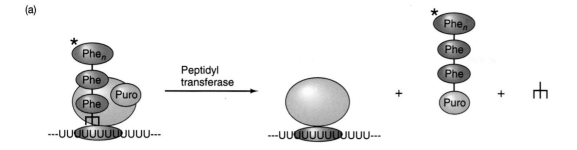

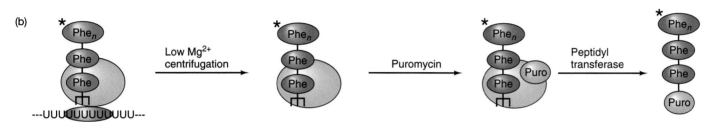

Figure 18.22   The puromycin reaction as an assay for peptidyl transferase. (a) Standard puromycin reaction. Load labeled poly(Phe)-tRNA into the P site by running a translation reaction with poly(U) as messenger. Then add puromycin. When a peptide bond forms between the labeled poly(Phe) and puromycin, the labeled peptidyl-puromycin is released. (b) Reaction with 50S subunits only. Again, load labeled poly(Phe)-tRNA into the ribosome's P site, incubate in a low Mg^{2+} buffer, and then centrifuge to separate the 50S-poly(Phe)–tRNA complex from the 30S subunit and the message. Then add puromycin and detect peptidyl transferase by the release of labeled peptidyl puromycin. The by-products of the reaction (50S subunits and tRNA) are not pictured. The asterisks denote the label in the poly(Phe).

first peptide bond. You might be expecting a new group of elongation factors to participate in this event, but there are none. Instead, the ribosome itself contains the enzymatic activity, called **peptidyl transferase,** that forms peptide bonds. No soluble factors are needed.

The classic assay for peptidyl transferase was invented by Robert Traut and Robert Monro and uses a labeled aminoacyl-tRNA or peptidyl-tRNA bound to the ribosomal P site, and puromycin. The release of labeled aminoacyl- or peptidyl-puromycin depends on forming a peptide bond between the amino acid or peptide in the P site, and puromycin in the A site, as depicted in Figure 18.22a. Traut and Monro also discovered that this system could be modified somewhat to show that the 50S ribosomal subunit, without any help from the 30S subunit or soluble factors, could carry out the peptidyl transferase reaction (Figure 18.22b). First, they allowed ribosomes to carry out poly(Phe) synthesis, using poly(U) as mRNA. This placed poly(Phe)-tRNA in the P site. Then they removed the 30S subunits by incubation with buffer having a low Mg^{2+} concentration, followed by ultracentrifugation. Then they washed away any remaining initiation or elongation factors with salt solutions, leaving the 50S subunits bound to poly(Phe)-tRNA. Ordinarily, such primed 50S subunits would be unreactive with puromycin, but these workers found that they could elicit puromycin re-

activity with 33% methanol (ethanol also worked). In both assays, one must distinguish the released peptidyl-tRNA from the peptidyl-tRNA still bound to ribosomes. Traut and Monro originally accomplished this by sucrose gradient centrifugation, as shown in Figure 18.23. Later, a more convenient filter-binding assay was developed. Figure 18.23a is a negative control with no puromycin, and the poly(Phe) remained bound to the 50S subunit, as expected. Figure 18.23b is a positive control in which the poly(Phe) was released by destroying the ribosomes with urea and RNase. Figure 18.23c and 18.23d show the experimental results with puromycin plus and minus GTP, respectively. Peptidyl transferase appeared to be working, since puromycin could release the poly(Phe). This reaction occurred even in the absence of GTP, as the peptidyl transferase reaction should.

The puromycin reaction with 50S subunits seems to demonstrate that the 50S subunit contains the peptidyl transferase activity, but could the rather unphysiological conditions (33% methanol, puromycin) be distorting the picture? One encouraging sign is that the reaction of a peptide with puromycin seems to follow the same mechanism as normal peptide synthesis. Also, M.A. Gottesmann substituted poly(A) for poly(U), and therefore poly(Lys) for poly(Phe), and also substituted lysyl-tRNA for puromycin, and found the same kind of reaction, demonstrating that

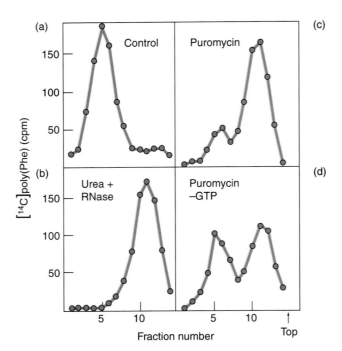

Figure 18.23   Puromycin assay for peptide bond formation. Traut and Monro loaded ribosomes with [^{14}C]polyphenylalanine, incubated them with or without puromycin, then subjected the products to sucrose gradient centrifugation to separate ribosome-bound poly(Phe) from free poly(Phe) that had been released from the ribosomes. The poly(Phe)-loaded ribosomes were treated as follows: (**a**) no treatment; (**b**) treated with urea and RNase; (**c**) treated with puromycin; (**d**) treated with puromycin in the absence of GTP. (*Source:* Reprinted from *Journal of Molecular Biology*, vol. 10, R. R. Traut and R. E. Monro, "The Puromycin Reaction and Its Relation to Protein Synthesis," 63–72, 1964, by permission of Academic Press Limited, London.)

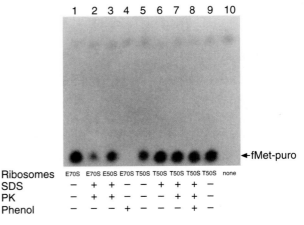

Figure 18.24   Effects of protein-removing reagents on peptidyl transferase activities of *E. coli* and *Thermus aquaticus* ribosomes. Noller and collaborators treated ribosomes with SDS, proteinase K (PK), or phenol, or combinations of these treatments, as indicated at bottom. Then they tested the treated ribosomes for peptidyl transferase by the fragment reaction using CAACCA-f[^{35}S]Met. They isolated f[^{35}S]Met-puromycin by high-voltage paper electrophoresis and detected it by autoradiography. The ribosome source is listed at bottom: E70S and E50S are 70S ribosomes and 50S ribosomal subunits, respectively, from *E. coli*; T50S refers to 50S ribosomal subunits from *Thermus aquaticus*. The position of fMet-puromycin is indicated at right. (*Source:* Noller et al., Unusual resistance of peptidyl transferase to protein extraction procedures. *Science* 256 (1992) p. 1417, f. 2. © AAAS.)

the puromycin reaction is a valid model for peptide bond formation. Furthermore, these reactions are all blocked by chloramphenicol and other antibiotics that inhibit the normal peptidyl transferase reaction, suggesting that the model reactions use the same pathway as the normal one.

For decades, no one knew what part of the 50S subunit had the peptidyl transferase activity. However, as soon as Thomas Cech and coworkers demonstrated in the early 1980s that some RNAs have catalytic activity, some molecular biologists began to suspect that the 23S rRNA might actually catalyze the peptidyl transferase reaction. In 1992, Harry Noller and his coworkers presented strong evidence that this is so. As their assay for peptidyl transferase, they used a modification of the puromycin reaction called the **fragment reaction.** This procedure, pioneered by Monro in the 1960s, uses a fragment of labeled fMet-tRNA$_f^{Met}$ in the P site and puromycin in the A site. The fragment can be CCA-fMet, or CAACCA-fMet. Either one resembles the whole fMet-tRNA$_f^{Met}$ enough that it can bind to the P site. Then the labeled fMet can react with puromycin to release labeled fMet-puromycin.

The task facing Noller and collaborators was to show that they could remove all the protein from 50S particles,

leaving only the rRNA, and that this rRNA could catalyze the fragment reaction. To remove the protein from the rRNA, these workers treated 50S subunits with three harsh agents known for their ability to denature or degrade protein: phenol, SDS, and proteinase K (PK). Figure 18.24, lanes 1–4, shows that the peptidyl transferase activity of *E. coli* 50S subunits survived SDS and proteinase K treatment, but not extraction with phenol. The ability to withstand SDS and PK was impressive, but it leaves us wondering why phenol extraction would disrupt the peptidyl transferase any more than the other two agents.

Noller and colleagues reasoned that phenol might be disrupting some higher-order RNA structure that is essential for peptidyl transferase activity. If so, they postulated that the rRNA from a thermophilic bacterium might be more sturdy and therefore might keep its native structure even after phenol extraction. To test this hypothesis, they tried the same experiment with 50S subunits from a thermophilic bacterium, *Thermus aquaticus*, that inhabits scalding hot springs. Lanes 5–9 of Figure 18.24 demonstrate that the peptidyl transferase activity of *T. aquaticus* 50S subunits survives treatment with all three of these agents.

If the fragment activity really represents peptidyl transferase, it should be blocked by peptidyl transferase inhibitors like chloramphenicol and carbomycin. Furthermore, if rRNA is a key factor in peptidyl transferase, then

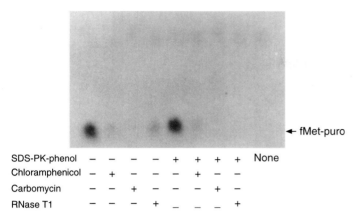

SDS-PK-phenol	−	−	−	−	+	+	+	+	None
Chloramphenicol	−	+	−	−	−	+	−	−	
Carbomycin	−	−	+	−	−	−	+	−	
RNase T1	−	−	−	+	−	−	−	+	

← fMet-puro

Figure 18.25 Sensitivity of *Thermus aquaticus* fragment reaction to peptidyl transferase inhibitors and RNase. Noller and coworkers treated *Thermus aquaticus* 50S ribosomal subunits with protein denaturants, peptidyl transferase inhibitors, or RNase T1 as indicated at bottom. Then they assayed for peptidyl transferase by the fragment reaction as described in the text and in Figure 18.24. (*Source:* Noller et al., *Science* 256 (1992) p. 1418, f. 4. © AAAS.)

the fragment reaction should be inhibited by RNase. Figure 18.25 shows that Noller and colleagues satisfied both of these sets of conditions. The fragment reactions carried out by either intact or treated *T. aquaticus* 50S subunits are inhibited by carbomycin, chloramphenicol, and RNase, just as they should be.

Do these experiments show that ribosomal RNA is the only component of peptidyl transferase? Noller and coworkers stopped short of that conclusion, in part because they could not eliminate all protein from their preparations, even after vigorous treatment with protein-destroying agents. In fact, their subsequent work in collaboration with Alexander Mankin demonstrated that eight ribosomal proteins remained associated with rRNA even after such vigorous treatment.

Mankin, Noller, and colleagues subjected *T. aquaticus* 50S ribosomal particles to the same protein-destroying agents used in Noller's original experiments. Then they performed sucrose gradient ultracentrifugation on the remaining material, and found that the material retaining peptidyl transferase activity sedimented as 50S and 80S particles, which they called *KSP50* and *KSP80* particles. The K, S, and P stand for proteinase K, SDS, and Phenol, respectively. Next, they examined intact 50S particles, as well as KSP50 and KSP80 particles, to see which RNAs and proteins they contained. They identified 23S and 5S rRNAs by gel electrophoresis. To separate and identify the protein in these particles, they used two-dimensional electrophoresis (Chapter 5). Amazingly, eight proteins remained more-or-less intact, and four of them (L2, L3, L13, and L22) were present in near-stoichiometric quantities. The other four (L15, L17, L18, and L21) were reduced in quantity. Mankin, Noller, and colleagues

double-checked the identities of these eight proteins by sequencing N-terminal peptides derived from each one. Because identical proteins and RNAs appeared in both particles, it is likely that the KSP80 particles are simply dimers of KSP50 particles.

Earlier studies on reconstitution of peptidyl transferase from purified components had shown that peptidyl transferase activity could be reconstituted from just 23S rRNA and proteins L2, L3, and L4. Of these, only L4 was missing from the KSP particles. Thus, considering the reconstitution data together with the KSP particle data, we can conclude that the minimum components necessary for peptidyl transferase activity are 23S rRNA and proteins L2 and L3.

What role does 23S rRNA play in the peptidyl transferase activity? It is tempting to speculate that it has a catalytic role, but we cannot reach that conclusion based on the data presented so far. However, in 2000 Thomas Steitz and his colleagues performed x-ray crystallography studies on 50S ribosomal particles and they found no proteins—only 23S rRNA—near the peptidyl transferase active center. So it appears that 23S rRNA really does have the peptidyl transferase catalytic activity. We will examine this subject in detail in Chapter 19.

SUMMARY Peptide bonds are formed by a ribosomal enzyme called peptidyl transferase, which resides on the 50S ribosomal particle. The minimum components necessary for peptidyl transferase activity in vitro are 23S rRNA and proteins L2 and L3. X-ray crystallography studies show that 23S rRNA is at the catalytic center of peptidyl transferase and therefore appears to have peptidyl transferase activity in vivo.

Elongation Step 3: Translocation

Once the peptidyl transferase has done its job, we have a peptidyl-tRNA in the A site and a deacylated tRNA in the P site. The next step, translocation, moves the mRNA and peptidyl-tRNA one codon's length through the ribosome. This places the peptidyl-tRNA in the P site and ejects the deacylated tRNA. The translocation process requires the elongation factor EF-G, which hydrolyzes GTP after translocation is complete. In this section we will examine the translocation process in more detail.

Three-nucleotide Movement of mRNA during Translocation
First of all, it certainly makes sense that translocation should move the mRNA exactly 3 nt (one codon's length) through the ribosome; any other length of movement would tend to shift the ribosome into a different reading frame, yielding aberrant protein products. But what is the

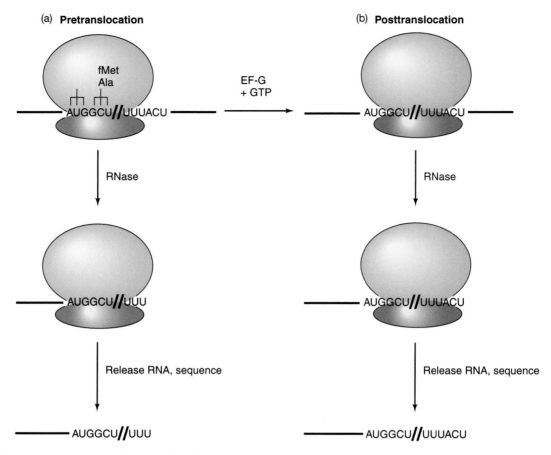

Figure 18.26   Movement of mRNA 3 nt through the ribosome on translocation. Lengyel and collaborators made a pretranslocation complex (**a**) with phage f2 mRNA, fMet-tRNA$_f^{Met}$, Ala-tRNA, and ribosomes. They converted this to a posttranslocation complex (**b**) by adding EF-G and GTP from *Bacillus stearothermophilus*. Then they treated both complexes with RNase to digest any mRNA not protected by the ribosome, then released the remaining, protected fragment and sequenced its 3′-end. The protected RNA from the posttranslocation complex was 3 nt longer (ACU) at its 3′-end, showing that translocation moves the mRNA 3 nt. The discontinuity (//) in the mRNA indicates that 6 nt have been omitted from these drawings for the sake of simplicity.

evidence? Peter Lengyel and colleagues provided data in support of the 3-nt hypothesis in 1971. As outlined in Figure 18.26, they created a pretranslocation complex with a phage mRNA, ribosomes, and aminoacyl-tRNAs, but left out EF-G and GTP to prevent translocation. Then they made a posttranslocation complex by adding EF-G and GTP. They treated each of these complexes with pancreatic ribonuclease to digest any mRNA not protected by the ribosome, then released the protected RNA fragment and sequenced it. They found that the sequence of the 3′-end of the fragment was UUU in the pretranslocation complex, and UUUACU in the posttranslocation complex. This indicated that translocation moved the mRNA 3 nt to the left, so 3 additional nucleotides (ACU) entered the ribosome and became protected. As an added check on the 3′-terminal sequences of the protected RNAs, these workers finished translating them before they released them for sequencing. They found that the protected mRNA fragment in the pretranslocation complex produced a peptide ending in phenylalanine, encoded by

UUU, but the protected mRNA fragment in the post-translocation complex produced a peptide ending in threonine, encoded by ACU. Thus, translocation had moved the mRNA exactly 3 nt, one codon's worth, through the ribosome.

SUMMARY Each translocation event moves the mRNA one codon's length, 3 nt, through the ribosome.

Role of GTP and EF-G   Translocation in *E. coli* depends on GTP and a GTP-binding protein called EF-G, as we learned earlier in this chapter. In eukaryotes, a homologous protein known as **EF-2** carries out the same process. Yoshito Kaziro and colleagues demonstrated this dependence on GTP and EF-G in 1970. Then, in 1974, they amplified their findings by showing *when* during the translocation process GTP is required. First, they created the

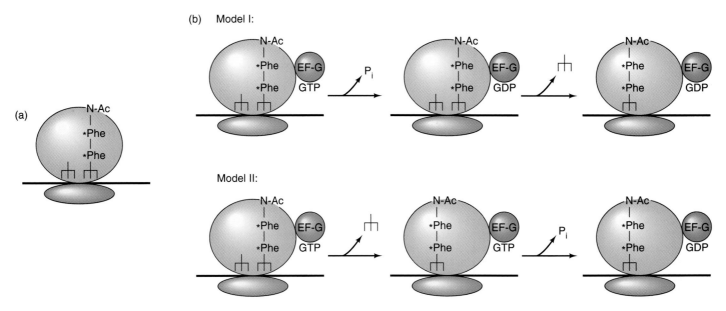

Figure 18.27   Dependence of translocation on EF-G and GTP. (**a**) Translocation substrate. Kaziro and colleagues created a translocation substrate by loading ribosomes with *N*-acetyl-di-Phe-tRNA in the A site and a deacylated tRNA in the P site as follows: First, they mixed ribosomes and poly(U) RNA with *N*-acetyl-Phe-tRNA, which went to the P site. Then they added ordinary Phe-tRNA, which went to the A site. Peptidyl transferase then formed a peptide bond, yielding *N*-acetyl-diPhe-tRNA in the A site, and a deacylated tRNA in the P site. (**b**) Two models for the timing of GTP hydrolysis. In model I, GTP hydrolysis occurs before or during translocation. In model II, it occurs after translocation.

translocation substrate pictured in Figure 18.27a, with ^{14}C-labeled *N*-acetyl-diPhe-tRNA in the A site and a deacylated tRNA in the P site. This substrate is poised to undergo translocation, which can be measured in two ways: The first assay was the release of the deacylated tRNA from the ribosome. Kaziro and coworkers assayed the release of tRNA by filtering the reaction mix and testing the filtrate for the presence of tRNA by attempting to charge it with labeled Phe. The second assay for translocation was puromycin reactivity. As soon as translocation occurs, the labeled dipeptide in the P site can join with puromycin and be released. Table 18.5 shows that neither GTP nor EF-G alone caused significant translocation, but that both together did promote translocation, measured by the release of deacylated tRNA.

At what point in this process is GTP hydrolyzed? There are two main possibilities, illustrated in Figure 18.27b: Model I calls for GTP hydrolysis pretranslocation. Model II allows for GTP hydrolysis after translocation has occurred. As unlikely as it may sound, model II was once the preferred hypothesis. Kaziro and colleagues performed experiments with an unhydrolyzable analog of GTP, GDPCP. If GTP is not needed until after translocation, then this GTP analog should promote translocation, as natural GTP does. Table 18.5 shows that GDPCP does yield a significant amount of translocation, though not quite as much as GTP does. However, when they used GDPCP, the investigators found that they had to add stoichiometric quantities of EF-G (equimolar with the ribosomes). Ordinarily, translation requires only catalytic amounts of EF-G, because EF-G

Table 18.5   Roles of EF-G and GTP in Translocation

Additions	tRNA Released	
	(pmol)	Δ
Experiment 1		
None	0.8	
GTP	1.8	1.0
EF-G	2.4	1.6
EF-G, GTP	12.6	11.8
EF-G, GDPCP	7.5	6.7
Experiment 2		
None	1.6	
EF-G	1.5	0
EF-G, GTP	5.1	3.5
EF-G, GTP, fusidic acid	6.7	5.1
EF-G, GDPCP	4.3	2.7
EF-G, GDPCP, fusidic acid	4.7	3.1

Source: From Inoue-Yokosawa, *Journal of Biological Chemistry* 249:4322, 1974. Copyright © 1974 The American Society for Biochemistry & Molecular Biology, Bethesda, MD. Reprinted by permission.

can be recycled over and over. But when GTP hydrolysis is not possible, as with GDPCP, recycling cannot occur. This suggested a function for GTP hydrolysis: release of EF-G from the ribosome, so both EF-G and ribosome can participate in another round of elongation.

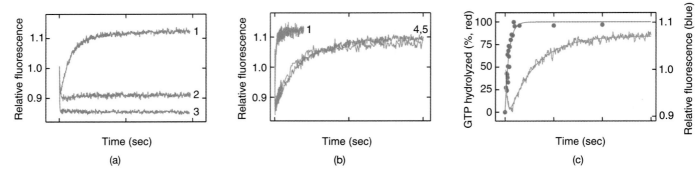

Figure 18.28   Kinetics of translocation. Wintermeyer and colleagues used stopped-flow kinetic experiments to measure translocation. They plotted the relative fluorescence of a fluorescent derivative of fMet-Phe-tRNA[Phe] bound to the A site as a function of time in seconds. The rise in fluorescence was taken as a measure of translocation. (**a**) Effect of antitranslocation antibiotics as follows: trace 1, no antibiotics; trace 2, viomycin; trace 3, thiostrepton. (**b**) Effect of GTP analogues. The following GTP analogues were added to the translocation reaction: trace 1, GTP; trace 4, an unhydrolyzable GTP analogue (caged GTP); trace 5, GDP. (**c**) Timing of GTP hydrolysis and translocation. Wintermeyer and colleagues measured translocation by stopped-flow kinetics as in panels (**a**) and (**b**) and GTP hydrolysis by release of $^{32}P_i$ from [^{32}P]GTP in a stopped-flow device. GTP hydrolysis occurs first, and about five times faster than translocation. (*Source:* (a) Rodnina, M. et al. Hydrolysis of GTP by elongation factor G drives tRNA movement on the ribosome. *Nature* 385 (2 Jan 1997) f. 1, p. 37. © Macmillan Magazines Ltd. (b) Rodnina, M. et al. Hydrolysis of GTP by elongation factor G drives tRNA movement on the ribosome. *Nature* 385 (2 Jan 1997) f. 1, p. 7. © Macmillan Magazines Ltd. (c) Rodnina, M. et al. Hydrolysis of GTP by elongation factor G drives tRNA movement on the ribosome. *Nature* 385 (2 Jan 1997) f. 2, p. 38. © Macmillan Magazines Ltd.)

Experiment 2, reported in the bottom part of Table 18.5, includes data on the effect of the antibiotic **fusidic acid.** This substance blocks the release of EF-G from the ribosome after GTP hydrolysis. This would normally greatly inhibit translation because it would halt the process after only one round of translocation. In this experiment, however, one round of translocation is all that can occur in any event, so fusidic acid has no effect. Kaziro and colleagues repeated these same experiments, using puromycin reactivity as their assay for translocation, and obtained essentially the same results. They also tried GDP in place of GTP and found that it could not support translocation.

Kaziro and colleagues concluded that GTP hydrolysis is not absolutely required for translocation (although it did help). Therefore, they reasoned, GTP hydrolysis must follow translocation. But their assays took several minutes, much longer than the millisecond time scale at which translation reactions take place. So they could not measure GTP hydrolysis and translocation and really tell which one happened first. To answer the question rigorously, we need a **kinetic experiment** that can measure events from one millisecond to the next. in 1997, Wolfgang Wintermeyer and colleagues performed such kinetic experiments and showed conclusively that GTP hydrolysis is very rapid and occurs before translocation.

Part of these workers' experimental plan was to load pretranslocation ribosomes in vitro with a fluorescent peptidyl-tRNA in the A site and a deacylated tRNA in the P site. Then they added EF-G·GTP and instantly began measuring the fluorescence of the complex. Such kinetic experiments on millisecond (ms) time scales are possible using a **stopped-flow** apparatus in which two or more solutions are forced simultaneously into a mixing chamber, and then immediately into another chamber for analysis. The mixing time in these experiments is of the order of only 2 ms. After an initial drop, the fluorescence increased significantly, as shown in Figure 18.28a, trace 1. This increase in fluorescence must be related to translocation, because it is prevented by two antibiotics that block translocation, viomycin and thiostrepton (Figure 18.28a, traces 2 and 3). Translocation worked much better with GTP (Figure 18.28b, trace 1) than with an unhydrolyzable GTP analog (a "caged" GTP, Figure 18.28b, trace 4) or with GDP (Figure 18.28b, trace 5).

Next, Wintermeyer and colleagues compared the timing and speed of GTP hydrolysis and translocation. They measured GTP hydrolysis with [γ-^{32}P]GTP, again with a stopped-flow device. This time, they rapidly mixed the radioactive GTP with the other components and then, after only milliseconds, forced the mixture into another chamber where the reaction was stopped with perchlorate solution. They measured $^{32}P_i$ released by liquid scintillation counting. Again, they assayed translocation by fluorescence increase. Figure 18.28c shows that GTP hydrolysis occurred first, and about five times faster than translocation. Thus, Wintermeyer and colleagues concluded that GTP hydrolysis precedes and drives translocation.

SUMMARY GTP and EF-G are necessary for translocation. GTP hydrolysis precedes translocation and significantly accelerates it. For a new round of elongation to occur, EF-G must be released from the ribosome, and that release depends on GTP hydrolysis.

The Structures of EF-Tu and EF-G

Notice the similarity between the activities of the two elongation factors, EF-Tu and EF-G. Both are ribosome-dependent GTPases. Both require GTP to carry out their functions. GTP hydrolysis in both cases is not required for the elongation reaction itself, but is required to release the factor from the ribosome. One difference in mechanism is that EF-Tu depends on an exchange factor, EF-Ts, to replace GDP with GTP, whereas EF-G can carry out this exchange on its own.

These similarities led to the suggestion that both elongation factors bind to the same site on the ribosome, namely the site that stimulates their GTPase activities. If this is true, then the two factors should have similar structures, just as two keys that fit the same lock must have similar shapes. X-ray crystallography studies on the two proteins have shown that this is true, with one qualification: It is actually the EF-Tu·tRNA·GTP ternary complex that has a shape very similar to that of the EF-G·GDP binary complex. This makes sense because EF-Tu binds to the ribosome as a ternary complex with tRNA and GTP, whereas EF-G binds as a binary complex with GTP only. To avoid GTP hydrolysis, the experimenters used unhydrolyzable GTP analogs, GDP in the case of EF-G, and GDPNP in the case of EF-Tu·tRNA. Figure 18.29 depicts the three-dimensional structures of the two complexes. We can see that the lower part of the EF-G protein (red, right) mimics the shape of the tRNA portion of the tRNA (red, left) in the EF-Tu ternary complex. This presumably allows both complexes to bind at or close to the same site on the ribosome.

Two other translation factors also have ribosome-dependent GTPase activities: the prokaryotic initiation factor IF-2 (Chapter 17) and the termination factor RF-3 (see later in this chapter). Because they also seem to rely on the same GTPase-activating center on the ribosome, it is reasonable to predict that they are structurally similar to at least parts of the two complexes depicted in Figure 18.29.

SUMMARY The three-dimensional shapes of the EF-Tu·tRNA·GDPNP ternary complex and the EF-G·GDP binary complex have been determined by x-ray crystallography. As predicted, they are very similar.

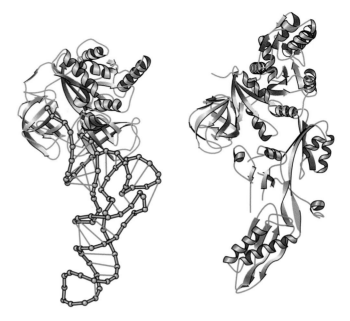

Figure 18.29   Comparison of the three-dimensional shapes of the EF-Tu·tRNA·GDPNP ternary complex (left) and the EF-G·GDP binary complex (right). The tRNA part of the ternary complex and the corresponding protein part of the binary complex are highlighted in red.

GTPases and Translation

We have now seen two examples of proteins that use hydrolysis of GTP to drive important steps in the elongation phase of translation: EF-Tu and EF-G. Recall from Chapter 17 that IF-2 plays a similar role in the initiation phase. Finally, at the end of this chapter we will discover that another factor (RF3) plays the same role in translation termination.

What do all of these processes have in common? All use energy from GTP to drive molecular movements essential for translation. IF-2 and EF-Tu both bring aminoacyl-tRNAs to the ribosome (IF-2 transports the initiating aminoacyl-tRNA (fMet-tRNA$_f^{Met}$) to the P site of the ribosome, while EF-Tu transports the elongating aminoacyl-tRNAs to the A site of the ribosome). EF-G sponsors translocation, in which the mRNA and the peptidyl-tRNA move from the A site to the P site of the ribosome. And RF3 helps catalyze termination, in which the bond linking the finished polypeptide to the tRNA is broken and the polypeptide exits the ribosome.

All of these factors belong to a large class of proteins known as **G proteins** that perform a wide variety of cellular functions. Most of the G proteins share the following features, illustrated in Figure 18.30:

1. They are GDP- and GTP-binding proteins. In fact the "G" in "G protein" comes from "guanine nucleotide."

2. They cycle among three conformational states, depending on whether they are bound to GDP, GTP,

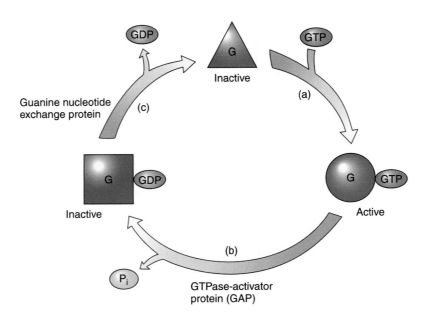

Figure 18.30 Generalized G protein cycle. The G protein at top (red triangle) is in the unbound state with neither GDP nor GTP bound. This state is normally short-lived. **(a)** GTP binds to the unbound G protein, changing its conformation (represented by the change from triangular to circular shape), and thereby activating it. **(b)** A GTPase-activator protein (GAP) stimulates the intrinsic GTPase activity of the G protein, causing it to hydrolyze its GTP to GDP. This results in another conformational change, represented by the change to square shape, which inactivates the G protein. **(c)** A guanine nucleotide exchange protein removes the GDP from the G protein, changing it back to the original unbound state, which is ready to accept another GTP.

or neither nucleotide, and these conformational states determine their activities.

3. When they are bound to GTP they are activated to carry out their functions.

4. They have intrinsic GTPase activity.

5. Their GTPase activity is stimulated by another agent called a **GTPase activator protein** (**GAP**).

6. When a GAP stimulates their GTPase activity, they cleave their bound GTP to GDP, inactivating themselves.

7. They are reactivated by another protein called a **guanine nucleotide exchange protein.** This factor removes GDP from the inactive G protein and allows another molecule of GTP to bind. One guanine nucleotide exchange protein comes immediately to mind: EF-Ts. We have seen that EF-Ts is essential for replacing GDP with GTP on EF-Tu.

Recall that the GTPases of all the G proteins involved in translation are stimulated by the ribosome. Thus, we might predict that the GAP for all these G proteins would be a protein or proteins at some site(s) on the ribosome. In fact, a set of ribosomal proteins and parts of a ribosomal RNA, collectively known as the **GTPase-associated site** has been discovered on the ribosome. It is really two sites, and consists of the ribosomal protein L11, a complex of the ribosomal proteins L10 and L12, and the 23S rRNA. We will discuss this topic in more detail in Chapter 19 and see how interaction between one of the translation factors and the GTPase-associated site stimulates GTPase activity.

SUMMARY Several translation factors harness the energy of GTP to catalyze molecular motions. These factors belong to a large class of G proteins that are activated by GTP, have intrinsic GTPase activity that is activated by an external factor (GAP), are inactivated when they cleave their own GTP to GDP, and are reactivated by another external factor (a guanine nucleotide exchange protein) that replaces GDP with GTP.

18.4 Termination

The elongation cycle repeats over and over, adding amino acids one at a time to the growing polypeptide product. Finally, the ribosome encounters a stop codon, signaling that it is time for the last step in translation: termination.

Termination Codons

The first termination codon (the **amber codon**) was discovered by Seymour Benzer and Sewell Champe in 1962 as a conditional mutation in a T4 phage. The amber

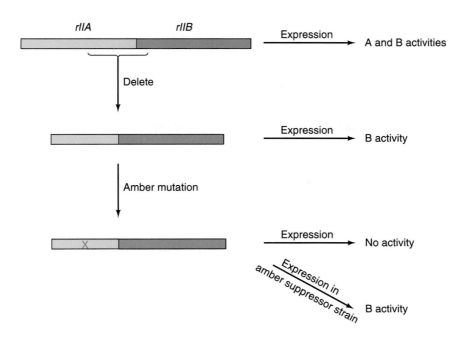

Figure 18.31   Effects of an amber mutation in a fused gene. Benzer and Champe deleted the DNA shown by the bracket, fusing the *rIIA* and *B* cistrons together. Expression of this fused gene yielded *B* activity, but no *A* activity. An amber mutation in the *A* cistron inactivated *B* activity, which could be restored by transferring the gene to an amber suppressor strain (*E. coli* CR63). The amber mutation caused premature translation termination in the *A* cistron, and the amber suppressor prevented this termination, allowing production of the B part of the fusion protein.

mutation was conditional in that the mutant phage was unable to replicate in wild-type *E. coli* cells, but could replicate in a mutant, **suppressor** strain. Certain mutations in the *E. coli* alkaline phosphatase gene were also suppressed by the same suppressor strain, so it appeared that they were also amber mutations. We now know that amber mutations create termination codons that cause translation to stop prematurely in the middle of an mRNA, and therefore give rise to incomplete proteins. What was the evidence for this conclusion?

First of all, amber mutations have severe effects. Ordinary missense mutations change at most one amino acid in a protein, which may or may not affect the function of the protein, but even if the protein is inactive, it can usually be detected with an antibody. By contrast, *E. coli* strains with amber mutations in the alkaline phosphatase gene produce no detectable alkaline phosphatase activity or protein. This fits the hypothesis that the amber mutations caused premature termination of the alkaline phosphatase, so no full-size protein could be found.

A genetic experiment by Benzer and Champe further strengthened this hypothesis. They introduced a deletion into the adjacent *rIIA* and *B* genes of phage T4 which fused the two genes together, as shown in Figure 18.31. The fused gene gave a fusion protein with B activity, but no A activity. Then they introduced an amber mutation into the *rIIA* part of the fused gene. This mutation blocked *rIIB* activity, and this block was removed by an amber suppressor. How could a mutation in the *A* cistron

block the expression of the *B* cistron, which lies downstream? Translation termination at the amber mutation is an obvious explanation. If translation stops at the amber codon, it would never reach the *B* cistron. Moreover, according to this logic, the amber suppressor overrides the translation termination at the amber mutation and allows translation to continue on into the *B* cistron.

More direct evidence for the amber mutation as a translation terminator came from studies by Brenner and colleagues on the head protein gene of phage T4. When this phage infects *E. coli* B, head protein accounts for more than 50% of the protein made late in infection, which makes it easy to purify. When these investigators introduced amber mutations into the head protein gene, they were unable to isolate intact head protein from infected cells, but they could isolate fragments of head protein. And tryptic digestion of these fragments yielded peptides that could be identified as amino-terminal peptides. Thus, the products of head protein genes with amber mutations were all amino-terminal protein fragments, as shown in Figure 18.32. Because translation starts at a protein's amino terminus, this experiment demonstrated that the amber mutations caused termination of translation before it had a chance to reach the carboxyl terminus.

The amber mutation defined one translation stop codon, but two others have similarly colorful names, **ochre** and **opal.** Ochre mutations were originally distinguished by the fact that they were not suppressed by **amber suppressors.** Instead, they have their own class of

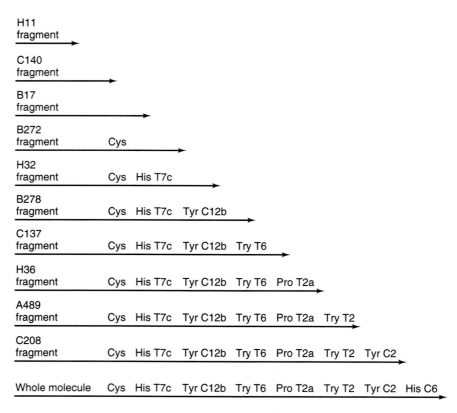

Figure 18.32   Amino-terminal protein fragments produced by T4 head protein amber mutants. Benzer and colleagues replicated T4 phage mutants with amber mutations in the head gene, then extracted the shortened proteins and analyzed them by cleaving them with trypsin (or chymotrypsin) and ordering the peptides released by this cleavage. The deduced head protein fragments are represented by horizontal arrows and displayed in order of increasing length. The name of each amber mutant protein fragment is given at left, and the names of the tryptic or chymotryptic peptides are given above each arrow. Note that all protein fragments appear to start at the normal amino terminus and stop somewhere before the normal carboxyl terminus. The bottom arrow represents the whole protein. (*Source:* Reprinted with permission from *Nature* 201:16, 1964. Copyright © 1964 Macmillan Magazines Limited.)

ochre suppressors. Similarly, opal mutations are suppressed by **opal suppressors.**

Since amber mutations are caused by mutagens that give rise to missense mutations, we suspect that these mutations come from the conversion of an ordinary codon to a stop codon by a one-base change. We know that only three unassigned "nonsense" codons occur in the genetic code: UAG, UAA, and UGA. We assume these are stop codons, so the simplest explanation for the results we have seen so far is that one of these is the **amber codon,** one is the **ochre codon,** and one is the **opal codon.** But which is which?

Martin Weigert and Alan Garen answered this question in 1965, not by sequencing DNA or RNA, but by sequencing protein. They studied an amber mutation at one position in the alkaline phosphatase gene of *E. coli.* The amino acid at this position in wild-type cells was tryptophan, whose sole codon is UGG. Because the amber mutation originated with a one-base change, we already know that the amber codon is related to UGG by a one-base change. To find out what that change was, Weigert and Garen determined the amino acids inserted in this po-

sition by several different revertants. The revertants presumably arose by one-base changes from the amber codon. Some of these had tryptophan in the key position, but most had other amino acids: serine, tyrosine, leucine, glutamate, glutamine, and lysine. These other amino acids could substitute for tryptophan well enough to give at least some alkaline phosphatase activity. The puzzle is to deduce the one codon that is related by one-base changes to at least one codon for each of these amino acids, including tryptophan. Figure 18.33 demonstrates that UAG is the solution to this puzzle and therefore must be the amber codon.

By the same logic, including the fact that amber mutants can mutate by single-base changes to ochre mutants, Sydney Brenner and collaborators reasoned that the ochre codon must be UAA. Severo Ochoa and colleagues verified that UAA is a stop signal when they showed that the synthetic message AUGUUUUAAA$_n$ directed the synthesis and release of the dipeptide fMet-Phe. (AUG codes for fMet; UUU codes for Phe; and UAA codes for stop.) With UAG and UAA assigned to the amber and ochre codons, respectively, UGA must be the opal codon, by

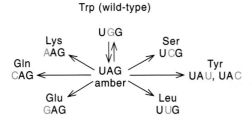

Figure 18.33 The amber codon is UAG. The amber codon (middle) came via a one-base change from the tryptophan codon (UGG), and the gene reverts to a functional condition in which one of the following amino acids replaces tryptophan: serine, tyrosine, leucine, glutamate, glutamine, or lysine. The magenta represents the single base that is changed in all these revertants, including the wild-type revertant that codes for tryptophan.

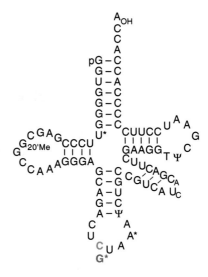

Figure 18.34 Comparison of sequence of wild-type *E. coli* tRNA^Tyr and *E. coli* amber suppressor tRNA. The G* (green) present in the wild-type tRNA^Tyr is replaced by a C (red) in the suppressor tRNA. (*Source:* Reprinted with permission from *Nature* 217:1021, 1968. Copyright © 1968 Macmillan Magazines Limited.)

elimination. Now that we have the base sequences of thousands of genes, it is abundantly clear that these three codons really do serve as stop signals. Sometimes we even find two stop codons in a row (e.g., UAAUAG), which provides a fail-safe stop signal even if termination at one codon is suppressed.

> **SUMMARY** Amber, ochre, and opal mutations create termination codons (UAG, UAA, and UGA, respectively) in the middle of a message and thereby cause premature termination of translation. These three codons are also the natural stop signals at the ends of coding regions in mRNAs.

Stop Codon Suppression

How do suppressors overcome the lethal effects of premature termination signals? Mario Capecchi and Gary Gussin showed in 1965 that tRNA from a suppressor strain of *E. coli* could suppress an amber mutation in the coat cistron of phage R17 mRNA. This identified tRNA as the suppressor molecule, but how does it work? Brenner and collaborators found the answer when they sequenced a suppressor tRNA. They placed the gene for a suppressor tRNA on a ϕ80 phage and used this recombinant phage to infect *E. coli* cells bearing an amber mutation in the *lacZ* gene. Because of this suppressor tRNA, infected cells were able to suppress the amber mutation by inserting a tyrosine instead of terminating. When Brenner and colleagues sequenced this suppressor tRNA they found only one difference from the sequence of the wild-type tRNA^Tyr: a change from C to G in the first base of the anticodon, as shown in Figure 18.34.

Figure 18.35 illustrates how this altered tRNA can suppress an amber codon. We start with a codon, CAG, which encodes glutamine (Gln). It pairs with the anticodon 3′-GUC-5′ on a tRNA^Gln. Assume that the CAG

codon is mutated to UAG. Now it can no longer pair with the tRNA^Gln; instead, it attracts the termination machinery to stop translation. Now a second mutation occurs in the anticodon of a tRNA^Tyr, changing it from AUG to AUC (again reading 3′→5′). This new tRNA is a suppressor tRNA because it has an anticodon complementary to the amber codon UAG. Thus, it can pair with the UAG stop codon and insert tyrosine into the growing polypeptide, allowing the ribosome to get past the stop codon without terminating translation.

> **SUMMARY** Most suppressor tRNAs have altered anticodons that can recognize stop codons and prevent termination by inserting an amino acid and allowing the ribosome to move on to the next codon.

Release Factors

Because the stop codons are triplets, just like ordinary codons, one might expect that these stop codons would be decoded by tRNAs, just as other codons are. However, work begun by Mario Capecchi in 1967 proved that tRNAs do not ordinarily recognize stop codons. Instead, proteins called **release factors** (**RFs**) do. Capecchi devised the following scheme (Figure 18.36) to identify the release factors: He began with *E. coli* ribosomes plus an R17 mRNA that was mutated in the seventh codon of the coat cistron to UAG (amber). The codon preceding this amber codon was ACC, which codes for threonine. He incubated the ribosomes with this mRNA in the absence of threonine

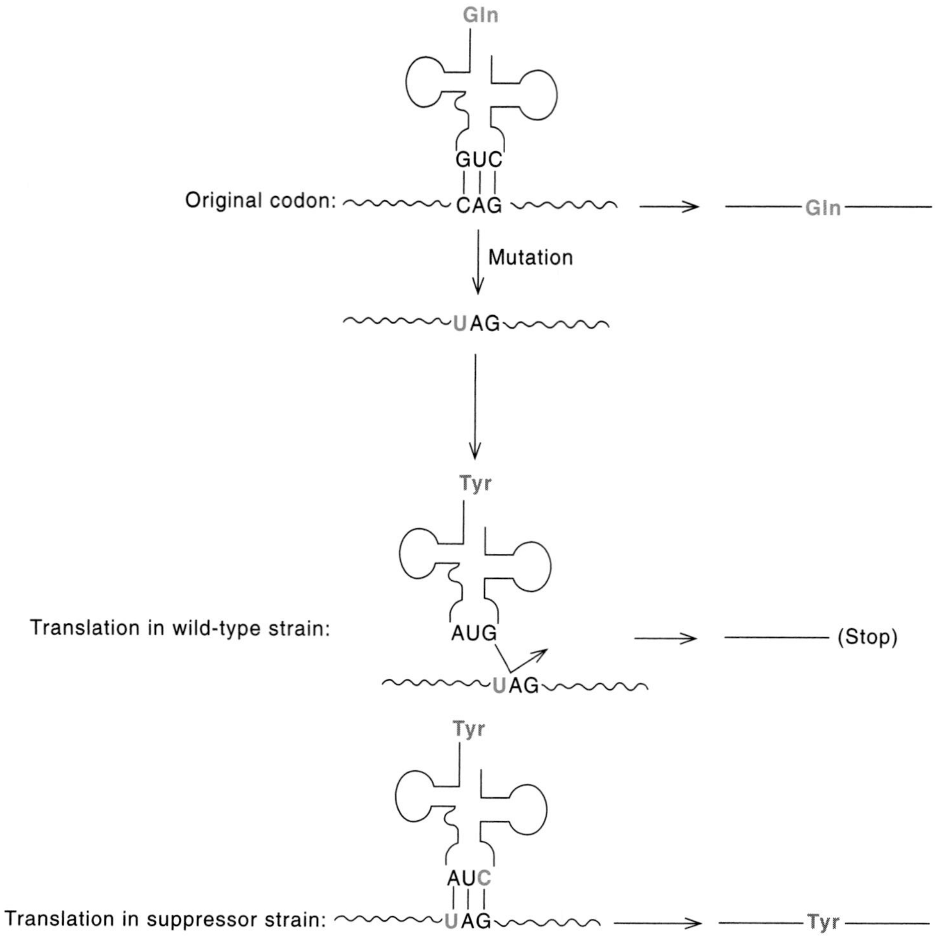

Figure 18.35 Mechanism of suppression. Top: The original codon in the wild-type *E. coli* gene was CAG, which was recognized by a glutamine tRNA. Middle: This codon mutated to UAG, which was translated as a stop codon by a wild-type strain of *E. coli*. Notice the tyrosine tRNA, whose anticodon (AUG) cannot translate the amber codon. Bottom: A suppressor strain contains a mutant tyrosine tRNA with the anticodon AUC instead of AUG. This altered anticodon recognizes the amber codon and causes the insertion of tyrosine (gray) instead of allowing termination.

so they would make a pentapeptide and then stall at the threonine codon. Then he isolated the ribosomes with the pentapeptide attached and placed them in a system containing only EF-Tu, EF-G (attached to the ribosomes) and [^{14}C]threonyl-tRNA. The ribosomes incorporated the labeled threonine into the peptide, producing a labeled hexapeptide in the P site, poised on the brink of release. To find the release factor, Capecchi added ribosomal supernatant fractions until one released the labeled peptide. He discovered that this factor, which he called release factor (RF), was not a tRNA, but a protein.

Nirenberg and colleagues devised a simpler technique (Figure 18.37), which was a takeoff on their assay for identifying codons, examined earlier in this chapter. They formed a ternary complex with ribosomes, the triplet AUG, and [^{3}H]fMet-tRNA$_f^{Met}$. The initiation codon and aminoacyl-tRNA went to the P site in the complex, and the labeled amino acid was therefore eligible for release. Incubation of this complex with a crude release factor

preparation and any of the three termination codons (UAG, UAA, or UGA) caused release of the labeled fMet. In this assay, the termination trinucleotide went to the A site and dictated release if the appropriate release factor was present. Table 18.6 shows that one factor (**RF-1**) cooperated with the stop codons UAA and UAG to cause release of the fMet, while another factor (**RF-2**) cooperated with UAA and UGA. Subsequent studies showed that UAA or UAG could direct the binding of purified RF-1 to the ribosome, while UAA or UGA could direct RF-2 binding. This reinforced the idea that the RFs could recognize specific translation stop signals. A third release factor, (**RF-3**), a ribosome-dependent GTPase, binds GTP and helps the other two release factors bind to the ribosome. Based on EF-G's mimicry of the shape of EF-Tu bound to a tRNA, Jens Nyborg and colleagues have predicted that RF-3 will have a structure resembling the protein part of the EF-Tu·tRNA·GTP ternary complex, and that RF-1 and RF-2 will mimic the structure of tRNA. The facts that

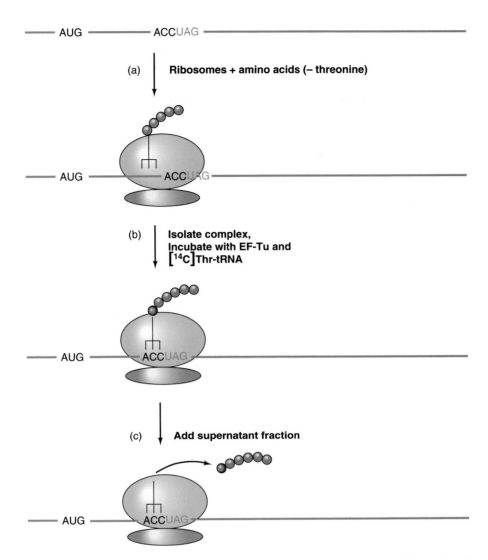

Figure 18.36   An assay for release factor. Capecchi began with a mutant R17 mRNA with a UAG termination codon (blue) following the sixth codon, ACC, which codes for threonine. **(a)** He added a translation system containing all the amino acids except threonine. This allowed the synthesis of a pentapeptide (green). **(b)** He isolated the complex and incubated it with Ef-Tu and [^{14}C]Thr-tRNA, which formed a hexapeptide, including [^{14}C]Thr (red), which was ready for release. **(c)** Finally, he added a supernatant fraction containing the release factor, which released the labeled hexapeptide.

RF-1 and -2 compete with tRNA for binding to the ribosome, recognize codons as tRNAs do, and are about the same size as tRNAs are consistent with this hypothesis.

We now know how the bacterial release factors mimic tRNAs in recognizing stop codons specifically. Yoshikazu Nakamura and colleagues discovered through genetic selection and biochemical studies that a tripeptide determinant in each RF protein discriminates among the three stop codons. These workers had already discovered that RF1 and RF2 have very similar architectures, with seven domains in equivalent locations. If these release factors really contain the peptide equivalent of an anticodon, it probably resides in one of these seven domains. So they set about to generate recombinant genes encoding hybrid RFs with one domain at a time switched. For example,

one such hybrid protein would have six domains from RF1 and one from RF2. Then they tested each of these hybrid proteins in two ways: First, they examined their ability to complement a mutation in bacteria with a mutant RF1 or RF2 gene. Second, they looked at their ability to recognize the stop codons in an in vitro release assay. The key difference was in the ability to recognize UAG or UGA. RF1 recognizes the former, and RF2 recognizes the latter, but both factors recognize UAA. They found that domain D was the important one. If domain D came from RF1, the hybrid protein had RF1 activity, even if all the other domains came from RF2. Similarly, if domain D came from RF2, the hybrid protein had RF2 activity.

Next, Nakamura and coworkers divided domain D into three subdomains, swapped these between factors,

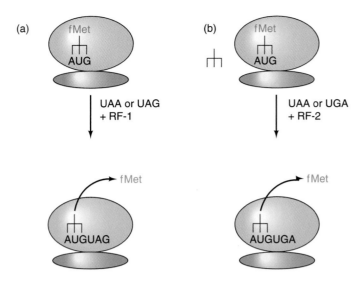

Figure 18.37   Nirenberg's assay for release factors. Nirenberg loaded the P site of ribosomes with the initiation codon AUG and [³H]fMet-tRNA$_f^{Met}$. Then he added one of the termination codons plus a release factor, which released the labeled fMet. (**a**) RF-1 is active with UAA or UAG. (**b**) RF-2 is active with UAA or UGA.

Table 18.6	**Response of RF-1 and -2 to Stop Codons**		
		pmol [³H]fMet Released in Presence of:	
Additions			
Release factor	**Stop codon**	**0.012 M Mg²⁺**	**0.030 M Mg²⁺**
RF-1	None	0.12	0.15
RF-1	UAA	0.47	0.86
RF-1	UAG	0.53	1.20
RF-1	UGA	0.08	0.10
RF-2	None	0.02	0.14
RF-2	UAA	0.22	0.77
RF-2	UAG	0.02	0.14
RF-2	UGA	0.33	1.08

Source: From E. Scolnick et al., "Release Factors Differing in Specificity for Terminator Codons," *Proceedings of the National Academy of Sciences* 61:772, 1968.

and tested for RF specificity. They identified a 15-amino-acid region that could change the specificity of the factor. In this region, five amino acids are perfectly conserved among RFs of a given class, but differ between RF1 and RF2. So these five amino acids almost certainly include the crucial ones. By changing one amino acid at a time and testing the products, they were able to identify a tripeptide that recognizes the stop codons. In particular, a Pro-Ala-Thr sequence in RF1 binds to the last two bases in the stop codon UAG, but cannot bind to the last two bases in the stop codon UGA. Conversely, a Ser-Pro-Phe sequence in RF2 binds to the last two bases in UGA, but cannot bind to the last two bases in UAG. Both tripeptides can recognize the third stop codon, UAA.

Figure 18.38 shows how this works. Only the first and third amino acids of each tripeptide are involved. The first recognizes the second base of the codon. Thr or Ser in the first position can recognize either the ring hydrogen of A or the amino group of G in the second base of the stop codon, so these amino acids do not discriminate between UAG and UGA. Pro or Phe in the first position can interact only with the ring hydrogen of A, not the amino group of G, so the tripeptide on RF1, with Pro in the first position, can recognize only UAG, not UGA Similarly, Thr or Ser in the third position of the peptide can recognize either G or A, but Pro or Phe in the third position can recognize only A, not G. Thus, the tripeptide on RF2, with Phe in the third position, can recognize only UGA, not UAG. Because both tripeptides are permissive for A in the second and third positions, both RFs can recognize UAA.

What about eukaryotic release factors? The first such factor (**eRF**) was discovered by a technique similar to Nirenberg's in 1971. Then, in 1994, a collaborative group led by Lev Kisselev finally purified eRF, still using an assay based on Nirenberg's, and succeeded in cloning and sequencing the eRF gene. Their approach to cloning and sequencing the gene was a widely used one: Using an fMet release assay similar to Nirenberg's to detect eRF, they purified the eRF activity until it gave one major band on SDS-PAGE, then subjected this protein to two-dimensional gel electrophoresis to purify it away from all other proteins. They cut out the eRF spot from this electrophoresis step, cleaved the protein with trypsin, and subjected four of the tryptic peptides to microsequencing. The sequences strongly resembled those of proteins from humans, *Xenopus laevis,* yeast, and the small flowering plant *Arabidopsis thaliana.* Thus, they were able to use the *Xenopus* gene (C11), which had already been cloned, as a probe to find the corresponding human gene in a human cDNA library. To verify that the products of the cloned *Xenopus* and human genes (C11 and TB3-1, respectively) had eRF activity, Kisselev and colleagues expressed these genes in bacteria or yeast and tested them in the fMet release assay with tetra-nucleotides, some of which contained stop codons. Table 18.7 shows that both proteins released fMet from loaded ribosomes, but only in the presence of a stop codon. The *Xenopus* protein was expressed with an oligohistidine (His) tag, so Kisselev and colleagues included unrelated His-tagged proteins as negative controls. They also showed that an antibody against C11 blocked its release factor activity, but an irrelevant antibody (anti-Eg5) did not.

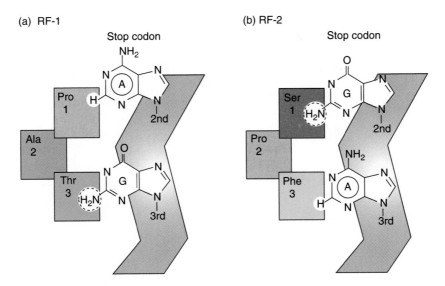

Figure 18.38 Role of the tripeptide determinant in RF-1 and RF-2 in recognizing stop codons. (a) RF-1 contains the tripeptide Pro-Ala-Thr. The first amino acid, proline, can interact with the ring hydrogen of A, but not the amino group of G, whereas the third amino acid, threonine, can interact with either A or G. Thus, this tripeptide allows RF-1 to bind to UAG (or UAA), but not to UGA. (b) RF-2 contains the tripeptide Ser-Pro-Phe. The first amino acid, serine, like threonine, can interact with either A or G. But the third amino acid, phenylalanine, can interact with the ring hydrogen of A, but not the amino group of G. Thus, this tripeptide allows RF-2 to bind to UGA (or UAA), but not to UAG. (*Source:* Reprinted from Nakamura et al., *Cell* 101:349, 2000. Copyright 2000, with permission from Elsevier Science.)

Table 18.7 Polypeptide Chain Release Activity of Purified *X. laevis* C11 (expressed with a His tag in *E. coli*) and Human TB3-1 (expressed in yeast) Proteins

Protein	Tetranucleotide added	Antibodies added	f[^{35}S]Met released (pmol)
His-tagged C11	UAAA	—	0.38
	UGAA	—	0.39
	UAGA	—	0.35
	poly (A, G, U)	—	0.28
	UGGA	—	0
	UAAA	Anti-C11	0
	UAAA	Anti-Eg5	0.35
His-tagged human protymosine	UAAA	—	0
His-tagged HIV reverse transcriptase	UAAA	—	0
TB3-1	UAAA	—	0.20
	UGAA	—	0.23
	UAGA	—	0.17
	UGGA	—	0

Source: Reprinted with permission from *Nature* 372:701, 1994. Copyright © 1994 Macmillan Magazines Limited.

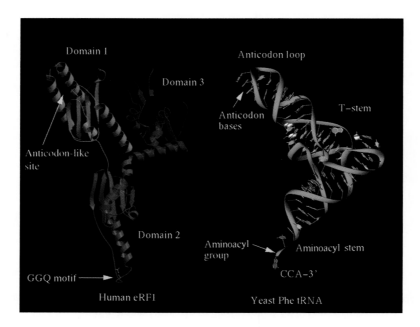

Figure 18.39   Molecular mimicry by eRF1. The crystal structures of human eRF1 and yeast tRNA^{Phe} are shown side by side for comparison. The disposition of domains 1, 2, and 3 of the eRF1 are very similar to those of the anticodon loop, the aminoacyl stem (acceptor stem) and the T-stem, respectively, of the tRNA. The GGQ motif of eRF1 is in an analogous location to the aminoacyl group of an aminoacyl-tRNA and therefore in a position to participate in hydrolysis of the ester bond between the peptide and the tRNA in the neighboring peptidyl-tRNA in the P site of the ribosome. (*Source:* Song, H. et al. The crystal structure of human eukaryotic release factor eRF1-mechanism of stop codon recognition and peptidyl-tRNA hydrolysis. *Cell* 100 (4 Feb 2000) f. 6, p. 318. Reprinted by permission of Elsevier Science.)

Table 18.7 demonstrates that eRF can recognize all three stop codons, unlike either of the two prokaryotic release factors, which can recognize only two. Does eRF collaborate with a G protein as prokaryotic RF1 and RF2 do? Michel Philippe and colleagues found that the answer is yes when they discovered a eukaryotic counterpart (eRF3) of prokaryotic RF3 in *X. laevis* cells in 1995. Another member of the eRF3 family, a yeast protein known as Sup35, has a guanine nucleotide-binding domain and is essential for yeast growth. Thus, eukaryotes contain two release factors, eRF (renamed **eRF1**) and eRF3. The eRF1 factor performs the same role as both RF1 and RF2, and eRF3 performs the role of bacterial RF3.

David Barford and colleagues have obtained a crystal structure of eRF1, which reinforces the idea of the release factors as molecular mimics of tRNAs. As Figure 18.39 suggests, the structure of eRF1 bears a striking resemblance to the two-ended structure of a tRNA. On one end, it has an anticodon-like region that presumably recognizes all three termination codons. On the other end, where an amino acid or polypeptide would be attached in a tRNA, lies the active site of the release factor. This region is characterized by an invariant Gly-Gly-Gln motif that appears to participate in hydrolysis of the bond between the finished polypeptide and its tRNA. Consistent with this hypothesis is the fact that substitution of any amino acid in the Gly-Gly-Gln motif completely inactivates the protein.

It is worth noting that the anticodon-like motif in eRF1 is quite different from the corresponding motifs in

RF1 and RF2 identified by Nakamura and coworkers. This is consistent with the fact that the sequence of eRF1 does not resemble the sequences of either RF1 or RF2 (except for the Gly-Gly-Gln motif), though the bacterial release factors strongly resemble each other. Thus, the eukaryotic (and archaeal) and prokaryotic release factors may not share common ancestry and may therefore have evolved different means of recognizing stop codons.

SUMMARY Prokaryotic translation termination is mediated by three factors: RF-1, -2, and -3. RF-1 recognizes the termination codons UAA and UAG; RF-2 recognizes UAA and UGA. RF-3 is a GTP-binding protein that facilitates binding of RF-1 and -2 to the ribosome. Eukaryotes have two release factors: eRF1, which recognizes all three termination codons, and eRF3, a ribosome-dependent GTPase that help eRF1 release the finished polypeptide.

SUMMARY

Messenger RNAs are read in the 5′ → 3′ direction, the same direction in which they are synthesized. Proteins are made in the amino to carboxyl direction, which means that the amino-terminal amino acid is added first.

The genetic code is a set of three-base code words, or codons, in mRNA that instruct the ribosome to incorporate specific amino acids into a polypeptide. The code is nonoverlapping: that is, each base is part of only one codon. It is also devoid of gaps, or commas; that is, each base in the coding region of an mRNA is part of a codon. There are 64 codons in all. Three are stop signals, and the rest code for amino acids. This means that the code is highly degenerate.

Part of the degeneracy of the genetic code is accommodated by isoaccepting species of tRNA that bind the same amino acid but recognize different codons. The rest is handled by wobble, in which the third base of a codon is allowed to move slightly from its normal position to form a non-Watson–Crick base pair with the anticodon. This allows the same aminoacyl-tRNA to pair with more than one codon. The wobble pairs are G–U (or I–U) and I–A.

The genetic code is not strictly universal. In certain eukaryotic nuclei and mitochondria and in at least one bacterium, codons that cause termination in the standard genetic code can code for amino acids such as tryptophan and glutamine. In several mitochondrial genomes, the sense of a codon is changed from one amino acid to another. These deviant codes are still closely related to the standard one from which they probably evolved.

Elongation takes place in three steps: (1) EF-Tu, with GTP, binds an aminoacyl-tRNA to the ribosomal A site. (2) Peptidyl transferase forms a peptide bond between the peptide in the P site and the newly arrived aminoacyl-tRNA in the A site. This lengthens the peptide by one amino acid and shifts it to the A site. (3) EF-G, with GTP, translocates the growing peptidyl-tRNA, with its mRNA codon, to the P site.

Puromycin resembles an aminoacyl-tRNA, and so can bind to the A site, couple with the peptide in the P site, and release it as peptidyl puromycin. On the other hand, if the peptidyl-tRNA is in the A site, puromycin will not bind to the ribosome, and the peptide will not be released. This defines two sites on the ribosome: the P site, in which the peptide in a peptidyl-tRNA is puromycin reactive, and the A site, in which the peptide in a peptidyl-tRNA is puromycin unreactive. fMet-tRNA$_f^{Met}$ is puromycin reactive in the 70S initiation complex, so it is in the P site. Binding studies have identified a third binding site (the E site) for deacylated tRNA. Such tRNAs bind to the E site as they exit the ribosome.

A ternary complex formed from EF-Tu, aminoacyl-tRNA, and GTP delivers an aminoacyl-tRNA to the ribosome's A site, without hydrolysis of the GTP. In the next step, GTP is hydrolyzed by a ribosome-dependent GTPase activity of EF-Tu, and an EF·Tu-GDP complex dissociates from the ribosome. EF-Ts regenerates an EF-Tu·GTP complex by exchanging GTP for GDP attached to EF-Tu. Addition of aminoacyl-tRNA then reconstitutes the ternary complex for another round of translation elongation.

The protein-synthesizing machinery achieves accuracy during elongation in a two-step process. First, it gets rid of ternary complexes bearing the wrong aminoacyl-tRNA before GTP hydrolysis occurs. If this screen fails, it can still eliminate the incorrect aminoacyl-tRNA in the proofreading step before the wrong amino acid can be incorporated into the growing protein chain. Both these screens may rely on the weakness of incorrect codon–anticodon base pairing to ensure that dissociation will occur more rapidly than either GTP hydrolysis or peptide bond formation. The balance between speed and accuracy of translation is delicate. If peptide bond formation goes too fast, incorrect aminoacyl-tRNAs do not have enough time to leave the ribosome, so their amino acids are incorporated into protein. But if translation goes too slowly, proteins are not made fast enough for the organism to grow successfully.

Peptide bonds are formed by a ribosomal enzyme called peptidyl transferase. This activity resides on the 50S subunit. The 23S rRNA contains the catalytic center of the peptidyl transferase.

Each translocation event moves the mRNA one codon's length, 3 nt, through the ribosome. GTP and EF-G are necessary for translocation, but the GTP is not hydrolyzed until after the translocation event. GTP hydrolysis releases EF-G from the ribosome, so a new round of elongation can occur. The three-dimensional shapes of the EF-Tu·tRNA·GDPNP ternary complex and the EF-G·GDP binary complex have been determined by x-ray crystallography. As predicted, they are very similar.

Amber, ochre, and opal mutations create termination codons (UAG, UAA, and UGA, respectively) in the middle of a message and thereby cause premature termination of translation. These three codons are also the natural stop signals at the ends of coding regions in mRNAs. Most suppressor tRNAs have altered anticodons that can recognize stop codons and prevent termination by inserting an amino acid and allowing the ribosome to move on to the next codon.

Prokaryotic translation termination is mediated by three factors: RF-1, -2, and -3. RF-1 recognizes the termination codons UAA and UAG; RF-2 recognizes UAA and UGA. RF-3 is a GTP-binding protein that facilitates binding of RF-1 and -2 to the ribosome. Eukaryotes have two release factors: eRF1, which recognizes all three termination codons, and eRF3, a ribosome-dependent GTPase that helps eRF1 release the finished polypeptide.

REVIEW QUESTIONS

1. Describe and give the results of an experiment that shows that translation starts at the amino terminus of a protein.

2. How do we know that mRNAs are read in the $5' \rightarrow 3'$ direction?

3. How do we know that the genetic code is: (a) nonoverlapping; (b) commaless; (c) triplet; (d) degenerate?

4. Describe and give the results of an experiment that reveals two of the codons for an amino acid.

5. Diagram a wobble base pair. You do not have to show the positions of all the atoms, just the shape of the base pair. Contrast this with the shape of a Watson–Crick base pair. What is the importance of wobble in translation?

6. Diagram the translation elongation process in prokaryotes.

7. Diagram the mode of action of puromycin.

8. Describe and give the results of an experiment that shows that fMet-tRNA$_f^{Met}$ occupies the P site of the ribosome.

9. Describe and give the results of an experiment that shows that EF-Ts releases GDP from EF-Tu.

10. What step in translation does chloramphenicol block?

11. Diagram the roles of EF-Tu and EF-Ts in translation.

12. Present evidence for the formation of a ternary complex among EF-Tu, GTP, and aminoacyl-tRNA.

13. Describe and give the results of an experiment that shows that ribosomal RNA is likely to be the catalytic agent in peptidyl transferase.

14. What are the initial recognition and proofreading steps in protein synthesis?

15. Describe and give the results of an experiment that shows that the mRNA moves in 3-nt units in the translocation step.

16. Describe and give the results of an experiment that shows that EF-G and GTP are both required for translocation. What are the effects of (a) substituting GDPCP for GTP, and (b) adding fusidic acid in this single-translocation event assay?

17. Describe an experiment that shows that GTP hydrolysis precede translocation.

18. What would be the effect on a G protein's activity of: (a) inhibiting its GAP? (b) inhibiting its guanine nucleotide exchange protein?

19. Present direct evidence that the amber codon is a translation terminator.

20. Present evidence that the amber codon is UAG.

21. Explain how an amber suppressor works.

22. Present evidence that the amber suppressor is a tRNA.

23. Describe an assay for a release factor.

24. What are the roles of RF-1, RF-2, and RF-3?

25. How do we know which amino acids in RF1 and RF2 recognize the termination codons?

26. What are the roles of eRF1 and eRF3?

SUGGESTED READINGS

General References and Reviews

Kaziro, Y. 1978. The role of guanosine 5′-triphosphate in polypeptide chain elongation. *Biochimica et Biophysica Acta* 505:95–127.

Khorana, H.G. 1968. Synthesis in the study of nucleic acids. *Biochemistry Journal* 109:709–25.

Nakamura, Y., K. Ito, and L.A. Isaksson. 1996. Emerging understanding of translation termination. *Cell* 87:147–50.

Nakamura, Y., K. Ito, and M. Ehrenberg. 2000. Mimicry grasps reality in translation termination. *Cell* 101:349—52.

Nierhaus, K.H. 1996. An elongation factor turn-on. *Nature* 379:491–92.

Thompson, R.C. 1988. EFTu provides an internal kinetic standard for translational accuracy. *Trends in Biochemical Sciences* 13:91–93.

Tuite, M.F. and I. Stansfield. 1994. Knowing when to stop. *Nature* 614–15.

Research Articles

Benzer, S. and S.P. Champe. 1962. A change from nonsense to sense in the genetic code. *Proceedings of the National Academy of Sciences, USA* 48:1114–21.

Brenner, S., A.O.W. Stretton, and S. Kaplan. 1965. Genetic code: The "nonsense" triplets for chain termination and their suppression. *Nature* 206:994–98.

Bretscher, M.S., and K.A. Marcker. 1966. Peptidyl-sRibonucleic acid and amino-acyl-sRibonucleic acid binding sites on ribosomes. *Nature* 21:380–84.

Dintzis, H.M. 1961. Assembly of the peptide chains of hemoglobin. *Proceedings of the National Academy of Sciences USA* 47:247–61.

Frovola, L., X. Le Goff, H.H. Rasmussen, S. Cheperegin, G. Drugeon, M. Kress, I. Arman, A.-L. Haenni, J.E. Celis, M. Philippe, J. Justesen, and L. Kisselev. 1994. A highly conserved eukaryotic protein family possessing properties of polypeptide chain release factor. *Nature* 372:701–3.

Goodman, H.M., J. Abelson, A. Landy, S. Brenner, and J. D. Smith. 1968. Amber suppression: A nucleotide change in the anticodon of a tyrosine transfer RNA. *Nature* 217:1019–24.

Gordon, J. 1968. A stepwise reaction yielding a complex between a supernatant fraction from *E. coli*, guanosine 5′-triphosphate, and aminoacyl-sRNA. *Proceedings of the National Academy of Sciences USA* 59:179–83.

Haenni, A.-L. and J. Lucas-Lenard. 1968. Stepwise synthesis of a tripeptide. *Proceedings of the National Academy of Sciences USA* 61:1363–69.

Inoue-Yokosawa, N., C. Ishikawa, and Y. Kaziro. 1974. The role of guanosine triphosphate in translocation reaction catalyzed by elongation factor G. *Journal of Biological Chemistry* 249:4321–23.

Ito, K., M. Uno, and Y. Nakamura. 2000. A tripeptide "anticodon" deciphers stop codons in messenger RNA. *Nature* 403:680–84.

Khaitovich, P., A.S. Mankin, R. Green, L. Lancaster, and H.F. Noller. 1999. Characterization of functionally active subribosomal particles from *Thermus aquaticus*. *Proceedings of the National Academy of Sciences USA* 96:85–90.

Last, J.A., W.M. Stanley, Jr., M. Salas, M.B. Hille, A.J. Wahba, and S. Ochoa. 1967. Translation of the genetic message, IV. UAA as a chain termination codon. *Proceedings of the National Academy of Sciences USA* 57:1062–67.

Miller, D.L. and H. Weissbach. Interactions between the elongation factors: The displacement of GDP from the Tu-GDP complex by factor Ts. *Biochemical and Biophysical Research Communications* 38:1016–22.

Nirenberg, M. and P. Leder. 1964. RNA codewords and protein synthesis: The effect of trinucleotides upon binding of sRNA to ribosomes. *Science* 145:1399–1407.

Nissen, P., M. Kjeldgaard, S. Thirup, G. Polekhina, L. Reshetnikova, B.F.C. Clark, and J. Nyborg. 1995. Crystal structure of the ternary complex of Phe-tRNA[Phe], EF-Tu, and a GTP analog. *Science* 270:1464–71.

Noller, H.F., V. Hoffarth, and L. Zimniak. 1992. Unusual resistance of peptidyl transferase to protein extraction procedures. *Science* 256:1416–19.

Ravel, J.M. 1967. Demonstration of a guanine triphosphate-dependent enzymatic binding of aminoacyl-ribonucleic acid to *Escherichia coli* ribosomes. *Proceedings of the National Academy of Sciences USA* 57:1811–16.

Ravel, J.M., R.L. Shorey, and W. Shire. 1968. The composition of the active intermediate in the transfer of aminoacyl-RNA to ribosomes. *Biochemical and Biophysical Research Communications* 32:9–14.

Rheinberger, H.-J., H. Sternbach, and K.H. Nierhaus. 1981. Three tRNA binding sites on *Escherichia coli* ribosomes. *Proceedings of the National Academy of Sciences USA* 78:5310–14.

Rodnina, M.V., A. Savelsbergh, V.I. Katunin, and W. Wintermeyer. 1997. Hydrolysis of GTP by elongation factor G drives tRNA movement on the ribosome. *Nature* 385:37–41.

Sarabhai, A.S., A.O.W. Stretton, S. Brenner, and A. Bolle. 1964. Co-linearity of the gene with the polypeptide chain. *Nature* 201:13–17.

Scolnick, E., R. Tompkins, T. Caskey, and M. Nirenberg. 1968. Release factors differing in specificity for terminator codons. *Proceedings of the National Academy of Sciences USA* 61:768–74.

Song, H., P. Mugnier, A.K. Das, H.M. Webb, D.R. Evans, M.F. Tuite, B.A. Hemmings, and D. Barford. 2000. The crystal structure of human eukaryotic release factor eRF1—Mechanism of stop codon recognition and peptidyl-tRNA hydrolysis. *Cell* 100:311–21.

Thach, S.S. and R.E. Thach. 1971. Translocation of messenger RNA and "accommodation" of fMet-tRNA. *Proceedings of the National Academy of Sciences USA* 68:1791–95.

Traut, R.R. and R.E. Monro. 1964. The puromycin reaction and its relation to protein synthesis. *Journal of Molecular Biology* 10:63–72.

Weigert, M.G. and A. Garen. 1965. Base composition of nonsense codons in *E. coli*. *Nature* 206:992–94.

Weissbach, H., D.L. Miller and J. Hachmann. 1970. Studies on the role of factor Ts in polypeptide synthesis. *Archives of Biochemistry and Biophysics* 137:262–69.

Zhouravleva, G., L. Frolova, X. Le Goff, R. Le Guellec, S. Inge-Vechtomov, L. Kisselev, and M. Philippe. 1995. Termination of translation in eukaryotes is governed by two interacting polypeptide chain release factors, eRF1 and eRF3. *EMBO Journal* 14:4065–72.

Ribosomes and Transfer RNA

I n Chapter 3 we examined a few aspects of translation. We learned that ribosomes are the cell's protein factories and that transfer RNA plays a crucial adaptor role, binding an amino acid at one end and an mRNA codon at the other. Chapters 17 and 18 expanded on the mechanisms of translation initiation, elongation, and termination, without dealing in depth with ribosomes and tRNA. Let us continue our discussion of translation with a closer look at these two essential substances.

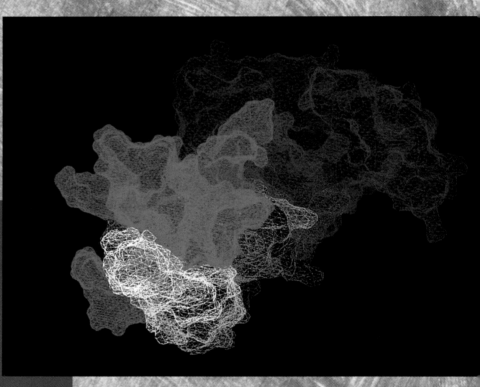

Computer generated model of transfer RNA (magenta) bound to aminoacyl-tRNA synthetase. © Kenneth Eward/BioGrafx/SS/Photo Researchers, Inc.

19.1 Ribosomes

Chapter 3 introduced the *E. coli* ribosome as a two-part structure with a sedimentation coefficient of 70S. The two subunits of this structure are the 30S and 50S ribosomal particles. In this section we will focus on the prokaryotic ribosome, its overall structure, composition, assembly, and function.

Gross Ribosomal Structure

The diameter of a ribosome is only about one-tenth the wavelength of visible light. For this reason, an individual ribosome cannot reflect visible light and will be invisible when viewed with a light microscope. On the other hand, ribosomes are big enough to see with an electron microscope, although it takes some practice to recognize what you are looking at, as we will see.

The earliest electron micrographs of *E. coli* ribosomes showed two subunits, one larger than the other, but did not reveal anything about the shapes of the subunits. In the 1970s, James Lake used negative staining techniques to produce electron micrographs of high enough resolution to yield information about the shapes of the subunits and how they fit together in the intact ribosome. To perform negative staining, the experimenter first places a drop containing a suspension of the sample (ribosomes, in this case) on a microscope grid, then wicks away the excess liquid with paper, then adds a drop of a solution of a heavy metal salt (commonly uranyl acetate, a uranium salt), then wicks away the excess and allows the grid to dry. The uranium does not penetrate a ribosome, but tends to build up in a "shell" surrounding it. This allows the electrons in the microscope to pass relatively freely through the ribosome, but they are stopped by the uranium atoms around each ribosome. The result is that we see a light area corresponding to the ribosome, outlined by a darker zone of uranium. This is why we call this technique *negative* staining.

Figure 19.1 shows the results of Lake's electron microscopy of negatively stained *E. coli* 30S ribosomal subunits. The images are arranged in rows, each showing the subunit in a different orientation. Each image by itself may look ambiguous, but once we look at enough of them we get the impression of an asymmetrical particle, as illustrated on the right side of Figure 19.1. Figure 19.2 shows a model of the 30S ribosome with names attached to its various features. The top is called the **head,** and the bottom is called the **base.** Protruding from the base to the right is the **platform,** which is separated from the head by the **cleft.**

Figure 19.3 shows a gallery of electron micrographs of negatively stained 50S ribosomal subunits. As in Figure 19.1, Lake interpreted the overall shapes and orienta-

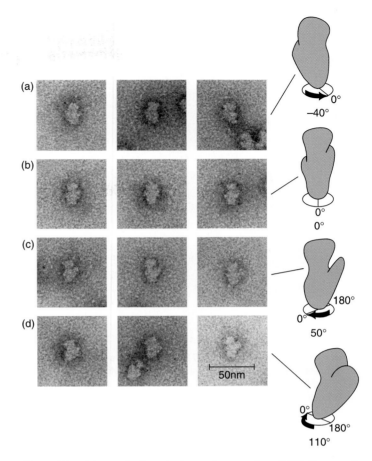

Figure 19.1 Transmission electron micrographs of 30S ribosomal subunits. Lake negatively stained purified 30S ribosomal subunits with uranyl acetate then viewed them with a transmission electron microscope. The subunit images in rows **a–d** are in four different orientations, as indicated by the cartoons at right. The 0-degree orientation (row **b**) is looking directly at the "back" of the 30S subunit. (*Source:* Lake, Ribosome structure determined by electron microscopy of *Escherichia coli* small subunits, large subunits, and monomeric ribosomes. *J. Mol. Biol.* 105 (1976) f. 2, p. 136, by permission of Academic Press.)

tions of the particles with cartoons to the right of each row in the gallery. Figure 19.4 contains another model of the 50S subunit in two orientations, with the obvious structural features identified. The large projection in the center is called the **central protuberance;** the long projection on the right (bottom picture) is called the **stalk;** the projection to the left is called the **ridge;** the space between the ridge and the central protuberance is called the **valley.**

Lake also examined the whole 70S ribosome by the same technique and generated a model for the way the 30S and 50S subunits fit together (Figure 19.5). Notice that the 30S and 50S subunits nestle together, with the 30S platform extending into the valley of the 50S subunit.

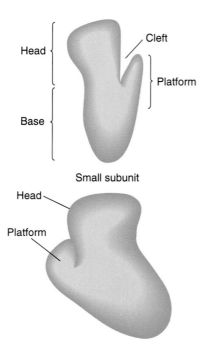

Figure 19.2   Model of the *E. coli* 30S ribosomal subunit in two orientations. The structural features are given the names head, base, cleft, and platform. (*Source:* From J. A. Lake, *Scientific American* 245:84, 1981. All rights reserved. Reprinted by permission. Copyright © George V. Kelvin/ *Scientific American*.)

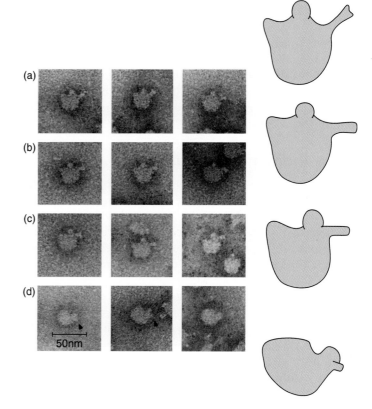

Figure 19.3   Electron micrographs of *E. coli* 50S ribosomal subunits, prepared as in Figure 19.1. The subunit images in rows **a–d** are in four different orientations, as indicated by the cartoons at right. The arrowheads in the first two frames of row (**d**) point to the stalk (see Figure 19.4). (*Source:* Lake, *J. Mol. Biol.* 105 (1976) f. 5, p. 139, by permission of Academic Press.)

In 1995, Joachim Frank and colleagues used a higher resolution technique called cryoelectron microscopy to give a more detailed view of the *E. coli* 70S ribosome. They froze a suspension of ribosomes, which immobilizes the ribosomes, but keeps them in a hydrated state. Then they took electron micrographs of the ice-imbedded ribosomes and used a computer to combine 4300 such images to produce a three-dimensional map. All of the features previously identified by Lake and others were evident; in addition, several new features could be discerned. These include a spur on the 30S particle, a channel through the neck of the 30S subunit, and a tunnel that begins in the interface canyon between the two subunits and leads into the 50S subunit.

Fine Structure of the 70S Ribosome

Harry Noller and colleagues succeeded in obtaining crystals of ribosomes from the bacterium *Thermus thermophilus* that were suitable for x-ray crystallography. By 1999, they had obtained crystal structures of these ribosomes. These studies have provided the most detailed structure to date of the intact ribosome, at a resolution as great as 7.8 Å.

Figure 19.6 shows the crystal structure of the 70S ribosome. We can see the features of the ribosome that Lake had seen decades earlier, but in much greater detail. In fact, the detail is a little confusing at first, and tends to hide the overall shapes. But viewing Figure 19.6 in 3-D with a viewer or with the "magic eye" technique helps make the ribosomal features recognizable.

One of the remarkable features of the ribosome crystal structure is the small amount of contact between the two subunits. Contrary to the traditional view of the 30S subunit nestling snugly in the 50S subunit, the actual points of contact between subunits are small and discrete. Figure 19.7 illustrates the areas of contact on both subunits, separated from each other. The limited contact between the two subunits leaves considerable space for other molecules to bind. Indeed, the cavity between the two subunits contains the active centers of the ribosome: the A site, P site, and E site. The floor of this cavity consists of the bridges between the body and platform of the 30S subunit

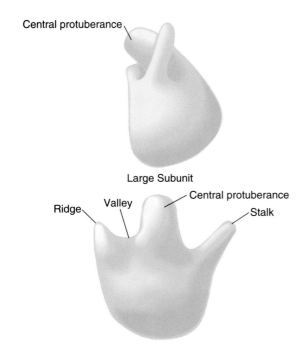

Figure 19.4 Model of the *E. coli* 50 ribosomal subunit in two orientations. The structural features, from left to right, are called the ridge, valley, central protuberance, and stalk. (*Source:* From J. A. Lake, *Scientific American* 245:85, 1981. All rights reserved. Reprinted by permission. Copyright © George V. Kelvin/ Scientific American.)

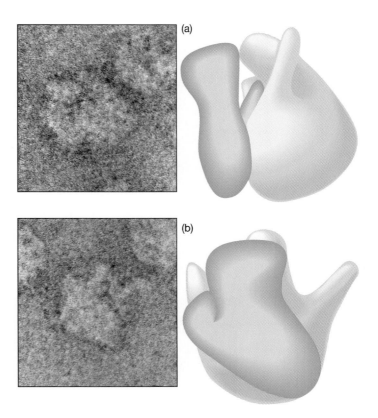

Figure 19.5 Electron micrographs of *E. coli* 70S ribosomes in two different orientations, with models of each at right. (*Source:* Lake, J. The ribosome. *Scientific American* 245 (Aug. 1981) p. 85.)

and the face of the 50S subunit (including B7, B2b, and B2a in Figure 19.7c).

To probe the binding sites for mRNA and tRNAs in the ribosome, Noller and colleagues prepared crystals of ribosomes complexed with: (a) an mRNA fragment and a tRNA fragment (a portion of the anticodon stem loop [ASL] containing 6 bp of the stem loop) bound in the ribosomal P site; (b) an mRNA fragment, plus the same ASL in the P site as in panel (a), and a deacylated tRNA bound to the ribosomal A site; (c) an mRNA fragment, plus a $tRNA_f^{Met}$ bound to the ribosomal P site. Differences in electron densities among these complexes allowed Noller and colleagues to model the locations of tRNAs in the A, P, and E sites, as shown in Figure 19.8. We can see that the tRNAs in the A and P sites lie parallel to one another, as we would expect because they must participate together in the peptidyl transferase reaction. On the other hand, the tRNA in the E site lies at a considerable angle to the other two tRNAs. Because the deacylated tRNA in the E site is on the way out of the ribosome, it need not assume a particular orientation relative to the other two tRNAs.

Figure 19.9 is a more schematic view of the ribosome that emphasizes three important points: First, a large cavity exists between the two ribosomal subunits that can ac-

commodate the three tRNAs. Second, the tRNAs interact with the 30S subunit through their anticodon ends, which bind to the mRNA that is also bound to the 30S subunit. Third, the tRNAs interact with the 50S subunit through their acceptor stems. This makes sense because the acceptor stems must come together during the peptidyl transferase reaction, which takes place on the 50S subunit. During this reaction, the peptide, linked to the acceptor stem of the peptidyl-tRNA in the P site, joins the amino acid, linked to the acceptor stem of the aminoacyl-tRNA in the A site.

Eukaryotic cytoplasmic ribosomes are more complex than prokaryotic ones. In mammals, the whole ribosome has a sedimentation coefficient of 80S and is composed of a 40S and a 60S subunit. Eukaryotic organelles also have their own ribosomes, but these are less complex. In fact, they are even simpler than prokaryotic ribosomes.

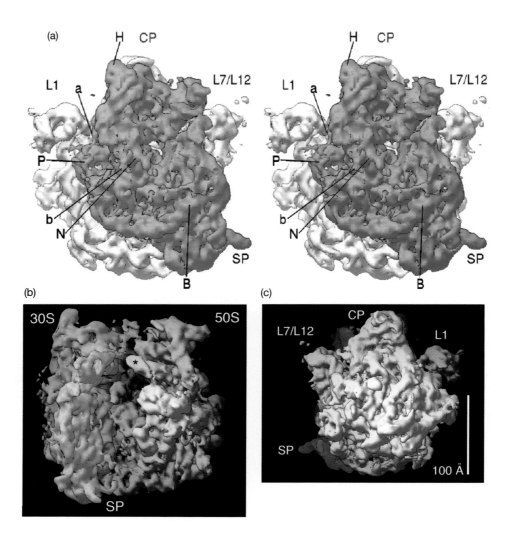

Figure 19.6   Crystal structure of the 70S ribosome. (a) Stereo diagram of the electron density of the 70S ribosome at 7.8 Å resolution. The 30S subunit is in purple and the 50S subunit is in gray. The major features of each subunit are labeled as follows: 30S particle: head, H; neck, N; body, B, platform, P; spur, SP; and contacts between head and platform, a and b. 50S particle: central protuberance, CP; stalk (which contains proteins L7 and L12), L7/L12; ridge (which contains protein L1), L1. (b) View of 70S ribosome rotated 90 degree relative to the view in panel (a). The asterisk marks a finger of the 50S subunit at the ribosomal A site. (c) View of the 70S ribosome rotated 180 degree relative to the view in panel (a). (*Source:* Cate, J. et al. X-ray crystal structures of 70S ribosome functional complexes. *Science* 285 (24 Sep 1999) f. 2, p. 2097. © AAAS.)

SUMMARY The *E. coli* 70S ribosome is composed of two parts, a 50S and a 30S subunit. Electron microscopy has revealed the overall shapes and several structural features of these two subunits, (the head, platform, and cleft of the 30S particle, and the ridge, valley, central protuberance, and stalk of the 50S particle). EM work has also revealed how the two subunits fit together to make the whole ribosome. Cryoelectron microscopy has elucidated more details of the ribosomal structure, especially a canyon at the interface between the two ribosomal particles. X-ray crystallography of 70S ribo-somes with and without tRNAs has provided an even more detailed look at the shapes of the two subunits, their interactions with each other, and their interactions with mRNA and tRNAs. The tRNAs occupy the canyon between the two subunits. They interact with the 30S subunit through their anticodon ends, and with the 50S subunit through their acceptor stems. Eukaryotic cytoplasmic ribosomes are larger and more complex than their prokaryotic counterparts, but eukaryotic organellar ribosomes are smaller than prokaryotic ones.

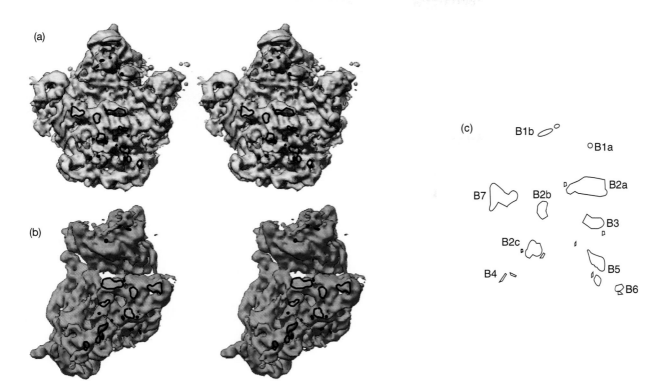

Figure 19.7   Contacts between ribosomal subunits. (a) Stereo view of 50S ribosomal subunit, showing points of contact with the 30S subunit outlined with heavy black lines. (b) Stereo view of 30S subunit, showing points of contact with the 50S subunit outlined with heavy black lines. These are the mirror image of those in panel (a). (c) Summary of points of contact between the two subunits, as seen on the front of the 50S subunit. (*Source:* (a-b) Cate, J. et al. X-ray crystal structures of 70S ribosome functional complexes. *Science* 285 (24 Sep 1999) f. 3, p. 2098. © AAAS.)

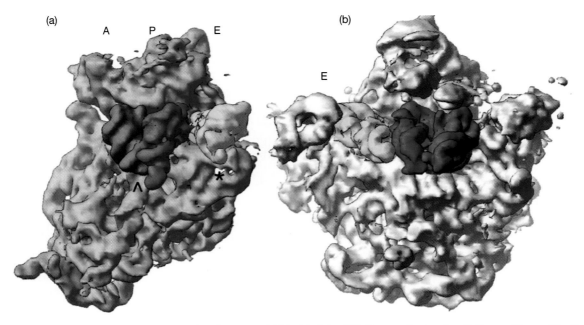

Figure 19.8   Positions of the three tRNAs bound to the ribosome. (a) Model of the 30S ribosomal subunit, with the positions of the three tRNAs indicated as follows: green, A site; blue, P site; yellow, E site. (b) Model of the 50S ribosomal subunit with the positions of the three tRNAs indicated and color-coded as in panel (a). The position of a particular tRNA was determined by electron density differences between structures that contained the tRNA and those that did not. The carat in the A site indicates that the CCA end of the A site tRNA could not be determined. The asterisk in the E site indicates that the CCA end of the E site tRNA could not be determined. (*Source:* Cate, J. et al. X-ray crystal structures of 70S ribosome functional complexes. *Science* 285 (24 Sep 1999) f. 5c, p. 2100. © AAAS.)

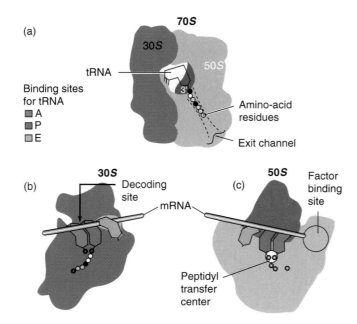

(a)

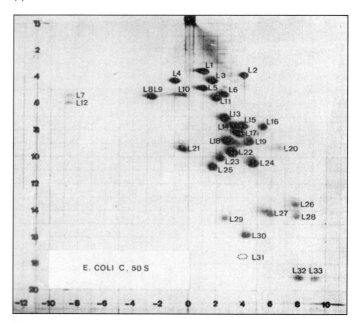

(a)

Figure 19.9 Schematic representation of the ribosome. (a) 70S ribosome, showing the large cavern between subunits, which can accommodate three tRNAs at a time. The peptidyl tRNA in the P site is shown, with the nascent polypeptide feeding through an exit channel in the 50S subunit. Notice that the interaction between the tRNA and the 30S subunit is through the tRNA's anticodon end, but the interaction between the tRNA and the 50S subunit is through the tRNA's acceptor stem. **(b)** The 30S subunit, with an mRNA and all three tRNAs bound. Colors of tRNAs are as in Figure 19.8. **(c)** The 50S subunit with an mRNA and all three tRNAs bound. Again the tRNAs are colored as in Figure 19.8. (*Source:* Reprinted with permission from Liljas, *Science* 285:2078, 1999. Copyright © 1999 American Association for the Advancement of Science.)

Ribosome Composition

We learned in Chapter 3 that the *E. coli* 30S ribosomal sub-unit is composed of a molecule of 16S rRNA and 21 ribosomal proteins, whereas the 50S particle contains two rRNAs (5S and 23S) and 34 ribosomal proteins. The rRNAs were relatively easy to purify by phenol extracting ribosomes to remove the proteins, leaving rRNA in solution. Then the sizes of the rRNAs could be determined by ultracentrifugation.

On the other hand, the ribosomal proteins are much more complex mixtures and had to be resolved by finer methods. The 30S ribosomal proteins can be displayed by one-dimensional SDS-PAGE to reveal a number of different bands ranging in mass from about 60 down to about 8 kD, but some of the proteins are incompletely resolved by this method. In 1970, E. Kaldschmidt and H. G. Wittman used two-dimensional gel electrophoresis to give almost complete resolution of the ribosomal proteins from both subunits. In this version of the technique, the two steps were simply native PAGE (no SDS) performed at two different pH values and acrylamide concentrations.

Figure 19.10 depicts the results of two-dimensional electrophoresis on *E. coli* 30S and 50S proteins. Each

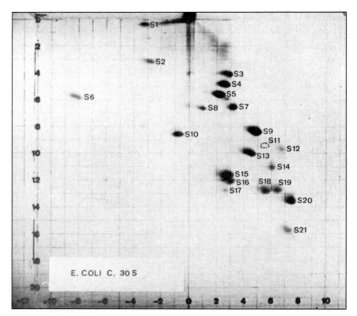

(b)

Figure 19.10 Two-dimensional gel electrophoresis of proteins from (a) *E. coli* **30S subunits and (b)** *E. coli* **50S subunits.** Proteins are identified by number, with S designating the small ribosomal subunit, and L, the large subunit. Electrophoresis in the first dimension (horizontal) was run at pH 8.6 and 8% acrylamide; electrophoresis in the second dimension (vertical) was run at pH 4.6 and 18% acrylamide. Proteins S11 and L31 were not visible on these gels, but their positions from other experiments are marked with dotted circles. (*Source:* Kaltschmidt & Wittmann, Ribosomal proteins XII. Number of proteins in small and large ribosomal subunits of *Escherichia coli* as determined by two-dimensional gel electrophoresis. *PNAS* 67 (1970) f. 1-2, pp. 1277–78.)

spot contains a protein, identified as S1–S21 for the 30S proteins, and L1–L33 (L34 is not visible) for the 50S proteins. The S and L stand for small and large ribosomal subunits. The numbering starts with the largest protein and ends with the smallest. Thus, S1 is about 60kD and S21 is about 8kD. You can see almost all the proteins, and almost all of them are resolved from their neighbors.

Eukaryotic ribosomes are more complex. The mammalian 40S subunit contains an 18S rRNA and about 30 proteins. The mammalian 60S subunit holds three rRNAs (5S, 5.8S, and 28S) and about 40 proteins. As we learned in Chapters 10 and 16, the 5.8S, 18S, and 28S rRNAs all come from the same transcript, made by RNA polymerase I, but the 5S rRNA is made as a separate transcript by RNA polymerase III. Eukaryotic organellar rRNAs are even smaller than their prokaryotic counterparts. For example, the mammalian mitochondrial small ribosomal subunit has an rRNA with a sedimentation coefficient of only 12S.

> **SUMMARY** The *E. coli* 30S subunit contains a 16S rRNA and 21 proteins (S1–S21). The 50S subunit contains a 5S rRNA, a 23S rRNA, and 34 proteins (L1–L34). Eukaryotic cytoplasmic ribosomes are larger and contain more RNAs and proteins than their prokaryotic counterparts.

Ribosome Assembly

Once the ribosomal proteins were all separated, they could be added back to rRNA to produce fully active ribosomes, at least in principle. Masayasu Nomura and his colleagues achieved the first such reconstitution with the 30S ribosomal subunit of *E. coli*. These workers added the purified proteins back to 16S rRNA and regenerated an active 30S particle. The test for activity was to combine the reconstituted 30S particles with 50S particles to give 70S ribosomes that could carry out translation in vitro.

This in vitro reconstitution demonstrated that the 30S subunit is a self-assembling entity. No extraneous material was needed for the components of the small subunit to assemble themselves into an active unit. This allowed Nomura to investigate an interesting question: In what order do the components assemble? He answered this question in several ways. In one set of experiments, Nomura and colleagues radioactively labeled all the proteins, then added each, separately, to 16S rRNA and measured its binding by scintillation counting. (Without 16S rRNA, no particle forms at all.) Two proteins, S4 and S8, bound very well to the 16S rRNA (80–90% as much as in native 30S particles); four others bound less well; and the rest, not at all. This suggested that S4 and S8 were among the

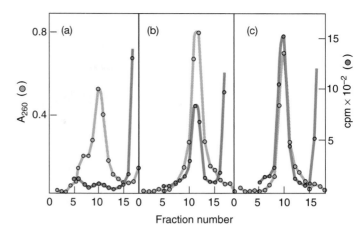

Figure 19.11   Effects of 30S ribosomal proteins on S12 binding to 16S rRNA–protein complexes. Nomura and colleagues added [³H]S12 to 16S rRNA plus the following proteins: (**a**) S4, S7, S8, S13, S16, and S20; (**b**) S4, S8, S16, and S17; (**c**) all proteins except S12. After incubation, these workers subjected the complexes to sucrose gradient ultracentrifugation, collected samples, and determined their radioactivities (red) by scintillation counting and their absorbances at 260 nm (blue) spectrophotometrically. They observed the most binding in the control (**c**), but significant binding also occurred in (**b**), with only four other proteins, and no binding occurred in (**a**), which lacked S17. (*Source:* From Held et al., *Journal of Biological Chemistry* 249:3109, 1974. Copyright © 1974 The American Society for Biochemistry & Molecular Biology, Bethesda, MD. Reprinted by permission.)

first proteins to bind. Indeed, once these two bound, a new set of proteins bound, and after these, yet another set could bind.

Figure 19.11 shows an example of this kind of experiment, in which Nomura's group was attempting to identify the proteins required for S12 binding. Panel (a) demonstrates that S4, S7, S8, S13, S16, and S20 are not sufficient to promote S12 binding. By contrast, panel (b) shows that S4, S8, S16, and S17 are sufficient. The only new protein in panel (b) that is not present in panel (a) is S17. It follows that S17 is important in binding S12, although this experiment does not tell us whether S12 binds directly to S17 or not.

Following this type of strategy, Nomura built up the **assembly map** depicted in Figure 19.12. The arrows in the map show that one protein facilitates the binding of another. For example, S17 facilitates binding of S12. The thicker the arrow, the stronger the effect. For example, S17 has a stronger effect on S12 binding than on S5 binding.

Does this assembly map have any relationship to the assembly process in vivo? The evidence suggests that it does. For example, the proteins added last in the map would be expected to be on the outside of the ribosome. Indeed, these proteins are most susceptible to attack by proteolytic enzymes and protein-binding reagents. Furthermore, cold-sensitive mutants of *E. coli* make defective 30S particles with only nine proteins at low temperatures. These are presumably the first proteins added

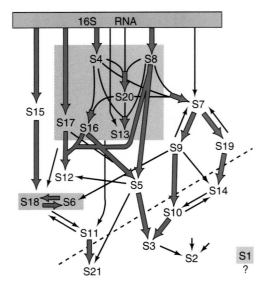

Figure 19.12   Assembly map of *E. coli* 30S ribosomal subunit. The dark arrows indicate strong facilitating effects, the thin arrows, weaker ones. The arrow from the large shaded box, leading to S11 indicates that some of the proteins within the box facilitate S11 binding, but it is not clear which. Proteins above the dashed line are found in a reconstitution intermediate particle (RI). Those below the line are missing from this intermediate. (*Source:* From Held et al., *Journal of Biological Chemistry* 249:3109, 1974. Copyright © 1974 The American Society for Biochemistry & Molecular Biology, Bethesda, MD. Reprinted by permission.)

to 16S rRNA in vivo, and all but one of them are also in the early part of the assembly map.

Reassembling the 50S subunits proved to be more challenging. At first, in fact, the 50S rRNAs and proteins from *E. coli* would not self-assemble in vitro. The problem was that the 50S assembly process was so temperature-dependent that the ribosomal proteins denatured at the high temperatures needed for assembly. To get around this problem, experimenters turned to the thermophilic bacterium *Bacillus stearothermophilus*. They reasoned that bacteria that live in very hot environments should have heat-stable proteins that could stand up to the temperatures needed to reconstitute the 50S subunit. They were right. Active *B. stearothermophilus* 50S particles can be reconstituted in vitro, and the optimum temperature for such reconstitution is 60° C. By 1974, even *E. coli* 50S subunits were successfully reassembled in vitro, using a two-stage approach that avoided extremely high temperatures. In spite of this early success, no comprehensive assembly map of the 50S ribosomal subunit from any organism is yet available.

SUMMARY Assembly of the 30S ribosomal subunit in vitro begins with 16S rRNA. Proteins then join sequentially and cooperatively, with proteins added early in the process helping later proteins to bind to the growing particle.

Fine Structure of the 30S Subunit

The reassembly studies we have just discussed shed some light on the locations of proteins within the 30S subunit: The proteins that bind to 16S rRNA first seem to be most protected against external agents and therefore tend to be buried deeper in the ribosome than proteins that join later and are available to external agents. But studies of ribosome fine structure have been aided most by three more sensitive methods: **immunoelectron microscopy, neutron diffraction,** and **x-ray diffraction studies.**

As we have seen, antibodies are Y-shaped proteins with specific antigen-binding sites at the ends of the two arms of the Y. Thus, they can bind to two molecules of antigen at the same time. Here is how Lake and colleagues used immunoelectron microscopy of ribosomes to take advantage of this bifunctional character of antibodies. From Lawrence Kahan's laboratory, they obtained antibodies directed against each of the ribosomal proteins, then reacted them one-by-one with whole ribosomes or ribosomal subunits. Then they viewed the antibody–ribosome complexes by electron microscopy. Figure 19.13 shows the result of such an experiment with an antibody directed against the ribosomal protein S13. The Y-shaped antibody molecules are even visible in some of these micrographs, and, as usual, the interpretive diagrams are a great aid to seeing what is actually going on. It appears that S13 is located near the top of the 30S subunit head, on the side near the platform. Clearly, this method gives direct clues about the positions of proteins exposed to the surface of ribosomes, but it cannot tell us the positions of buried proteins. Marina Stöffler-Meilicke and Georg Stöffler used immunoelectron microscopy to produce a partial map of the proteins in both the 30S and 50S ribosomal subunits of *E. coli*, which is presented in Figure 19.14. The designation 7/12 in panels (e) and (f) refers to proteins L7 and L12, which are identical in amino acid sequence and differ only in whether they are acetylated. The only other example of two *E. coli* ribosomal proteins that derive from the same gene is S20 and L26. Thus, in this one case, the same protein seems to be involved in both ribosomal subunits.

To get a more complete picture of the arrangement of ribosomal proteins, Peter Moore and his colleagues exploited the neutron diffraction method. They systematically assembled 30S ribosomal subunits with pairs of proteins labeled with deuterium, a heavy isotope of hydrogen. All the other proteins were unlabeled. Then they bombarded these 30S particles with a beam of neutrons. The neutrons are sensitive to the deuterium atoms and their diffraction depends on the spacing between the two proteins that contain deuterium. Repeating this experiment over and over with different pairs of labeled proteins enabled Moore and coworkers to deduce the spacings between all pairs of proteins, and this in turn

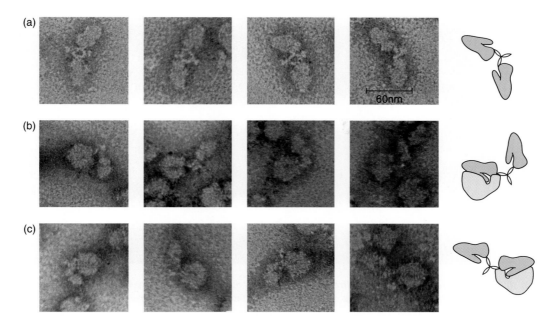

Figure 19.13   Immunoelectron microscopy of *E. coli* ribosomes with anti-S13 antibody. Lake reacted an antibody directed against S13 with pure **(a)** 30S particles or **(b)** and **(c)** mixtures of 30S particles and 70S ribosomes. Then he subjected the complexes to electron microscopy. The last image in each row is interpreted with a diagram at right. The Y-shaped protein linking the particles is the antibody. (*Source:* Lake, Ribosome structure determined by electron microscopy of *Escherichia coli* small subunits, large subunits, and monomeric ribosomes. *J. Mol. Biol.* 105 (1976) f. 11b–d, p. 147, by permission of Academic Press.)

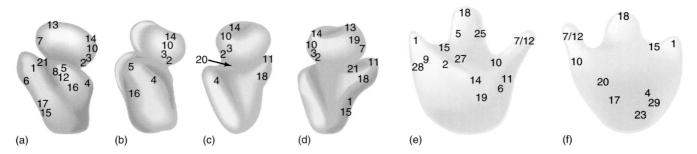

Figure 19.14   Partial map of *E. coli* 30S and 50S proteins based on immunoelectron microscopy. Panels **(a)**–**(d)** show the 30S subunit in four different orientations created by rotating through 180 degrees around the vertical axis. Panels **(e)** and **(f)** show the 50S subunit from the front and back. (*Source:* From Stöffler and Stöffler-Meilicke in Hill et al., eds., *The Ribosome: Structure, Function & Evolution.* Copyright © American Society for Microbiology.)

allowed them to build the map of the 30S ribosomal proteins shown in Figure 19.15. It is satisfying that many of the near neighbors in this map are also proteins that are close together in the assembly map. This suggests that in those cases, the cooperativity of binding is due to direct binding between the proteins.

As soon as the sequence of the *E. coli* rRNAs became known, molecular biologists began proposing models for their secondary structures. The idea is to find the most stable molecule—the one with the most intramolecular base pairing. Figure 19.16 depicts a consensus secondary structure for the 16S rRNA that has been verified by x-ray crystallograph of 30S ribosomal subunits. Note the

extensive base pairing proposed for this molecule. Note also how the molecule can be divided into three almost independently folded domains (one of which has two subdomains), highlighted in different colors.

How does the three-dimensional arrangement of the 16S rRNA relate to the positions of the ribosomal proteins in the intact ribosomal subunit? The best way to obtain such information is to perform x-ray crystallography, but that is a difficult task with an asymmetric object as large as a ribosomal subunit. Despite the difficulty, V. Ramakrishnan and colleagues succeeded in 2000 in solving the crystal structure for the *Thermus thermophilus* 30S subunit to a resolution of 3.0 Å. At almost the same time, a

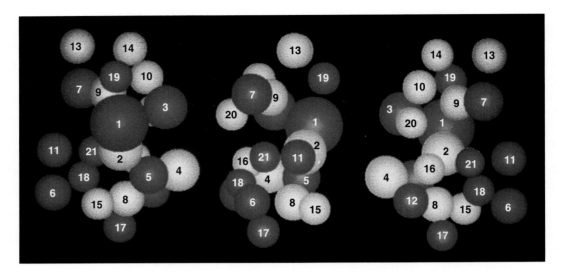

Figure 19.15 Map of 30S proteins based on neutron diffraction. Three views of the map are shown, with the middle view rotated 90 degrees around a vertical axis, with respect to the two flanking views, which are thus rotated 180 degrees with respect to each other. The sizes of the balls are related to the sizes of the corresponding proteins. (*Source:* Capel et al., Positions of S2, S13, S16, S17, S19, and S21 in the 30S ribosomal subunit of *Escherichia coli. J. Mol. Biol.* 200 (1988) f. 13, p. 78, by permission of Academic Press.)

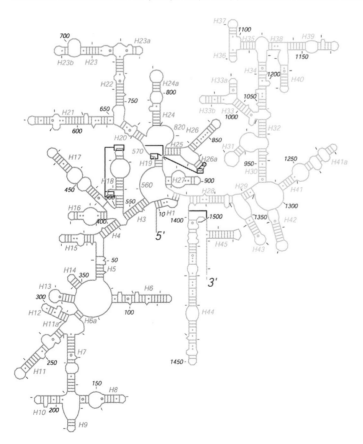

Figure 19.16 Secondary structure of 16S rRNA. This structure is based on optimal base-pairing and on x-ray crystallography of 30S ribosomal subunits from *Thermus thermophilus.* The numbering of bases in black corresponds to the *E. coli* 16S rRNA. The helices are numbered H1–H45 in color. Red, 5′-domain; green, central domain; yellow, 3′-major domain; cyan, 3′-minor domain. (*Source:* Wimberly, B. et al. Structure of the 30S ribosomal subunit. *Nature* 407 (21 Sep 2000) f. 2a, p. 329. © MacMillan Magazines Ltd.)

group led by François Franceschi determined the same structure to 3.3 Å resolution. The structure of Ramakrishnan and colleagues contained all of the ordered regions of the 16S rRNA (over 99% of the RNA molecule) and of 20 ribosomal proteins (95% of the protein). The parts of the proteins missing from the structure were only at their disordered ends.

Figure 19.17a is a stereo diagram of the 16S rRNA alone, and the RNA clearly outlines all of the important parts of the ribosome, including the familiar head, platform, and base (called the **body** here). In addition, we can see a **neck** joining the head to the body, a **beak** (sometimes called a **nose**) protruding to the left from the head, and a **spur** at the lower right of the body. Figure 19.17b shows front and back views of the 30S subunit with proteins added to the RNA. The proteins do not cause major changes in the overall shape of the subunit. In other words, the proteins do not contribute exclusively to any of the major parts of the subunit. These statements do not mean that the 16S rRNA would take the shape shown here in the absence of proteins, just that the rRNA is such a major part of the 30S subunit that its shape in the intact subunit resembles a skeleton of the subunit itself. The locations of most of the proteins agree well with the locations determined by neutron diffraction. However, S13, S14, S16, and S19 are shifted somewhat, and S20 is in a completely different location than that determined by neutron diffraction. S20 actually lies at the bottom of the body, opposite the spur.

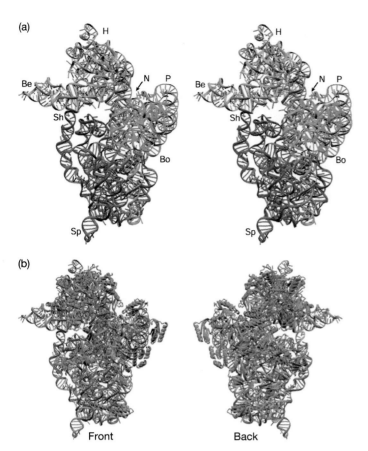

(a)

(b)

Front   Back

Figure 19.17   Crystal structure of the 30S ribosomal subunit.
(a) Stereo diagram of the 16S rRNA portion of the 30S subunit from
Thermus thermophilus. The major features are identified as follows:
H, head; Be, beak; Sh, shoulder; N, neck; P, platform; Bo, body; and
Sp, spur. **(b)** Front and back views of the 30S subunit with the proteins
(purple) added to the RNA (gray). The front is conventionally
recognized as the side of the 30S subunit that interacts with the 50S
subunit. Note that these are two different views of the ribosome, not a
stereo diagram. (*Source:* Wimberly, B. et al. Structure of the 30S ribosomal
subunit. *Nature* 407 (21 Sep 2000) f. 2b, p. 329. © MacMillan Magazines Ltd.)

SUMMARY Immunoelectron microscopy and neutron diffraction, among other methods, provided a
first look at the arrangement of proteins in the
E. coli 30S subunit (and in the 50S subunit, as
well). Sequence studies of 16S rRNA led to a proposal for the secondary structure (intramolecular
base pairing) of this molecule. X-ray crystallography studies have confirmed and extended most of
the conclusions of these studies. They show a 30S
subunit with an extensively base-paired 16S rRNA
whose shape essentially outlines that of the whole
particle. The x-ray crystallography studies have
also confirmed the locations of most of the 30S ribosomal proteins. However, S20 is located near
the bottom of the body of the subunit, far from the
spot assigned to it by neutron diffraction analysis.

Interaction of the 30S Subunit with Antibiotics Ramakrishnan and colleagues also obtained the crystal structures of the 30S subunit bound to three different antibiotics: spectinomycin, which inhibits translocation;
streptomycin, which causes errors in translation; and
paromomycin, which increases the error rate by another
mechanism. These data, together with the structure of the
30S subunit by itself, gave further insights about the
mechanism of translation.

First, Ramakrishnan and coworkers superimposed on
their 30S subunit structure the positions of the three
aminoacyl-tRNAs from the structure of the whole 70S ribosome (recall Figure 19.8). Figure 19.18a and b show
two different views of the positions of the anticodon stem
loops of the aminoacyl-tRNAs, and codons of a hypothetical mRNA, bound to the A, P, and E sites on the 30S
subunit. It is striking that the codons and anticodons in
the A and P sites lie in a region near the neck of the 30S
subunit that is almost devoid of protein. Thus, codon–
anticodon recognition occurs in an environment that is
surrounded by segments of the 16S rRNA, and very little
protein. Figure 19.18c shows which parts of the 16S
rRNA are involved at each of the three sites.

The positions of the three antibiotics on the 30S subunit yield further insight into the two activities of the 30S
subunit: translocation and **decoding** (codon–anticodon
interactions). The geometry of the 30s subunit suggests
that translocation must involve movement of the head
relative to the body. **Spectinomycin** is a rigid three-ring
molecule that inhibits translocation. Its binding site on
the 30S subunit lies near the point around which the
head presumably pivots during translocation. Thus, it is
in position to block the turning of the head that is necessary for translocation.

Streptomycin increases the error rate of translation by
interfering with initial codon–anticodon recognition and
with proofreading. The position of streptomycin on the
30S subunit (Figure 19.19) provides some clues about
how this antibiotic works. Streptomycin lies very close to
the A site, where decoding occurs. In particular, it makes
a close contact with A913 in helix H27 of the 16S rRNA.

This placement of streptomycin is significant because the
H27 helix is thought to have two alternative base-pairing
patterns during translation, and these patterns affect accuracy. The first is called the **ram** state (from **r**ibosome
ambiguity). As its name implies, this base-pairing scheme
for H27 stabilizes interactions between codons and anticodons, even noncognate anticodons, so accuracy is low
in the *ram* state. (The crystal structures obtained by Ramakrishnan and colleagues contain the H27 helix in the
ram state.). The alternative base-pairing pattern is **restrictive,** and it demands accurate pairing between codon and
anticodon. If the ribosome is locked into the *ram* state
it accepts noncognate aminoacyl-tRNAs too readily

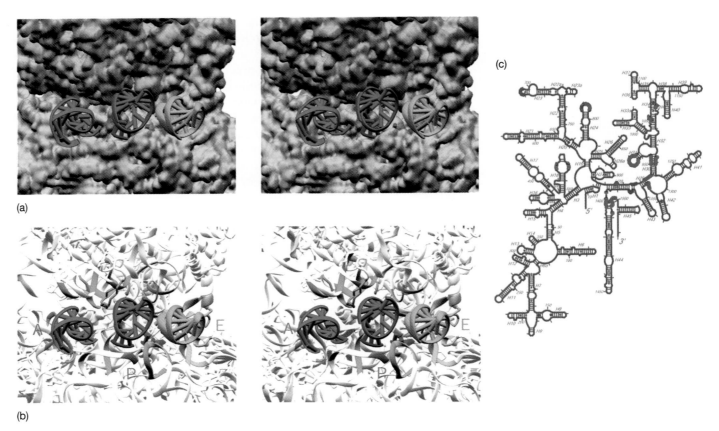

Figure 19.18   Locations of the A, P, and E sites on the 30S ribosomal subunit. (a) and **(b)** Two different stereo views of the inferred placement of the anticodon stem loops and mRNA codons on the 30S ribosomal subunit. The anticodon stem loops are colored magenta (A site), red (P site), and yellow (E site). The mRNA codons are colored green (A site) blue, (P site), and dotted magenta (E site). **(c)** Secondary structure of the 16S rRNA showing the regions involved in each of the three sites, color-coded the same as the anticodon stem loops in parts **(a)** and **(b)**: magenta, A site; red, P site; and yellow, E site. (*Source:* Carter, A. et al. Functional insights from the structure of the 30S ribosomal subunit and its interactions with antibiotics. *Nature* 407 (21 Sep 2000) f. 1, p. 341. © MacMillan Magazines Ltd.)

and cannot switch to the restrictive state required for proofreading. As a result, translation is inaccurate. On the other hand, if the ribosome is locked into the restrictive state, it is hyperaccurate–it rarely makes mistakes, but aminoacyl-tRNAs have a difficult time binding to the A site, so translation is inefficient.

The interactions between streptomycin and the 30S subunit indicate the antibiotic stabilizes the *ram* state. This would reduce accuracy in two ways. First, it would favor the *ram* state during decoding and thereby encourage pairing between a codon and noncognate aminoacyl-tRNAs. Second, it would inhibit the switching to the restrictive state that is necessary for proofreading.

Mutations in the ribosomal protein S12 can confer streptomycin resistance or even streptomycin dependence. Almost all of these S12 mutations are in regions of the protein that stabilize the 908–915 part of H27 and the 524–527 part of H18. These are also parts of the 16S rRNA that stabilize the *ram* state. These considerations led Ramakrishnan and colleagues to propose the following two-part hypothesis: First, S12 mutations that cause streptomycin resistance destabilize the *ram* state enough to counteract the *ram* state stabilization produced by the antibiotic. The result is a ribosome that works properly even in the presence of streptomycin. Second, S12 mutations that cause streptomycin dependence destabilize the *ram* state so much that the mutant ribosomes need the antibiotic to confer normal stability to the *ram* state. The result is a ribosome that cannot carry out normal translation without streptomycin.

In other words, translation that is both accurate and efficient depends on a balance between the *ram* state and the restrictive state of the ribosome. Streptomycin can tip the balance toward inaccuracy and efficiency by favoring the *ram* state, and mutations in S12 can tip the balance toward accuracy and inefficiency by favoring the restrictive state. To test this hypothesis, investigators will need the structure of the 30S subunit in the restrictive state, so they can compare it to the structure in the *ram* state that Ramakrishnan and colleagues have provided.

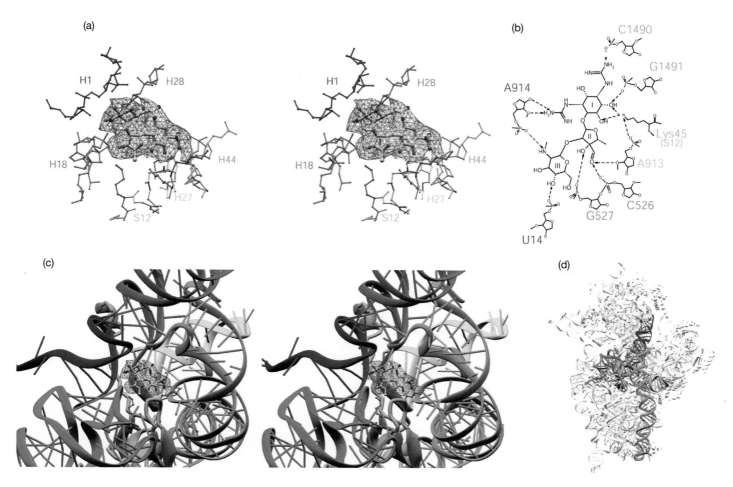

Figure 19.19   Interaction of streptomycin with the 30S ribosomal subunit. (a) Stereo diagram of streptomycin and its nearest neighbors in the 30S subunit. The streptomycin molecule is shown as a ball-and-stick model within a cage of electron density (actually the difference in density between 30S subunits with and without the antibiotic). The nearby helices of the 16S rRNA are shown. Notice especially the H27 helix (yellow), which is crucial for the activity of this antibiotic. Notice also the position of the only protein near the A site—S12 (salmon and red), which is also important in streptomycin activity. Amino acids of S12 that are altered in streptomycin-resistant cells are shown in red. **(b)** Interactions of specific groups of streptomycin (containing rings numbered I, II, and III) with neighboring atoms on the 30S subunit. Notice the interactions with A913 of H27 and Lys45 of S12. **(c)** Another stereo view of streptomycin and its nearest neighbors. Color coding is the same as in panel **(a)**. Notice again H27 (yellow) and S12 (salmon). **(d)** Location of the streptomycin-binding site on the whole 30S subunit. Streptomycin is shown as a small, red space-filling model at the point where all the colored 16S rRNA helices converge. (*Source:* Carter, A. et al. Functional insights from the structure of the 30S ribosomal subunit and its interactions with antibiotics. *Nature* 407 (21 Sep 2000) f. 5, p. 345. © MacMillan Magazines Ltd.)

Paromomycin also decreases accuracy of translation by binding to the A site. Figure 19.20 shows that this antibiotic binds in the major groove of the H44 helix and "flips out" bases **A1492** and **A1493.** That is, it forces these bases out of the major groove and puts them in position to interact with the minor groove between the codon and anticodon in the A site. Bases A1492 and A1493 are universally conserved and are absolutely required for translation activity. Mutations in either of these two bases are lethal.

These considerations led Ramakrishnan and coworkers to propose the following hypothesis: During normal decoding, bases A1492 and A1493 flip out and form H bonds with the 2′-OH groups of the sugars in the minor groove formed by the codon–anticodon base pairs in the A site. This helps to stabilize the interaction between codon and anticodon, which is important because the three base pairs would otherwise provide little stability. Flipping these two bases out ordinarily requires energy but paromomycin eliminates this energy requirement by

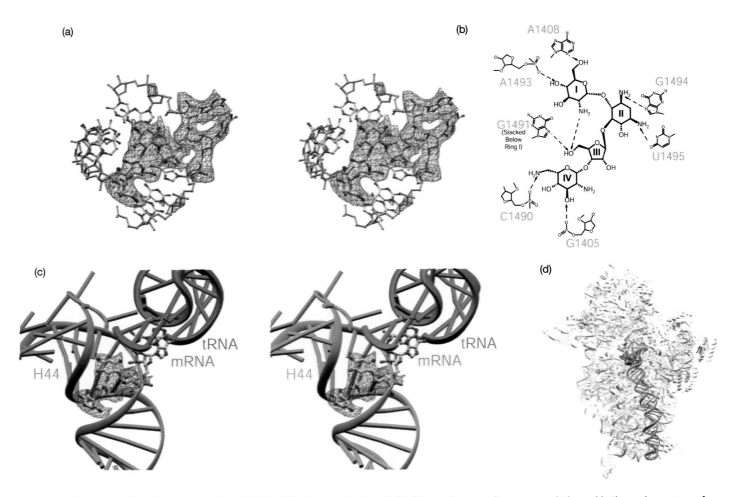

Figure 19.20 Interaction of paromomycin with the 30S ribosomal subunit. (a) Stereo diagram of paromomycin bound in the major groove of the H44 helix of the 16S rRNA. The beige electron density cage represents the difference map between 30S subunits with and without paromomycin. Thus, we can see not only the cage corresponding to the antibiotic, but also a cage at upper right surrounding the flipped out bases A1492 and A1493. These bases are forced out of the major groove by the antibiotic. (b) Model of paromomycin showing its interactions with neighboring groups in 16S rRNA. (c) Another stereo view of paromomycin bound to H44, with hypothetical interactions between the flipped out bases A1492 and A1493 and mRNA and tRNA in the A site. (d) Location of the paromomycin-binding site on the whole 30S subunit. Paromomycin is shown as a small, red space-filling model bound near the top of the H44 helix. (*Source:* Carter, A. et al. Functional insights from the structure of the 30S ribosomal subunit and its interactions with antibiotics. *Nature* 407 (21 Sep 2000) f. 6, p. 346. © MacMillan Magazines Ltd.)

forcing the bases to flip out. In this way, paromomycin stabilizes binding of aminoacyl-tRNAs, including noncognate aminoacyl-tRNAs, to the A site and thereby increases the error rate.

No codon or anticodon were preset in the crystal structure of the 30S subunit with paromomycin, so there is no direct evidence for the proposed interactions between bases A1492 and A1493 on the one hand, and the minor groove of the codon–anticodon duplex on the other. However, there are instances of these same kinds of interactions within the 16S rRNA itself. Figure 19.21 presents two examples of two adjacent adenines inserting into the minor groove of another part of the RNA, forming hydrogen bonds across the groove, and stabilizing the base pairing. To determine whether this same mechanism applies to codon–anticodon pairing in the A site, investigators will need a structure of the restrictive form of the 30S subunit bound to a tRNA and an mRNA.

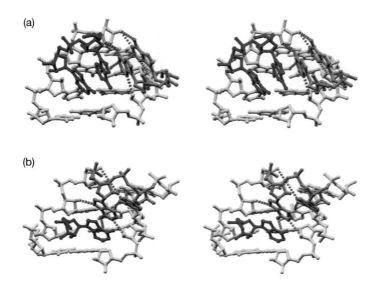

Figure 19.21 Two models for stabilization of an RNA double helix by two adjacent adenines inserting into the RNA minor groove. Both stereo diagrams (**a**) and (**b**) show the RNA double helix in yellow, the inserting adenines in red, the hydrogen bonds between adenines and sugars in the minor groove in gray, and neighboring stabilizing RNA residues in blue. (*Source:* Carter, A. et al. Functional insights from the structure of the 30S ribosomal subunit and its interactions with antibiotics. *Nature* 407 (21 Sep 2000) f. 7, p. 347. © MacMillan Magazines Ltd.)

SUMMARY The 30S ribosomal subunit plays two roles. It facilitates proper decoding between codons and aminoacyl-tRNA anticodons, including proofreading. It also participates in translocation. Crystal structures of the 30S subunit with three antibiotics that interfere with these two roles shed light on translocation and decoding. Spectinomycin binds to the 30S subunit near the neck, where it can interfere with the movement of the head that is required for translocation. Streptomycin binds near the A site of the 30S subunit and stabilizes the *ram* state of the ribosome. This reduces fidelity of translation by allowing noncognate aminoacyl-tRNAs to bind relatively easily to the A site and by preventing the shift to the restrictive state that is necessary for proofreading. Paromomycin binds in the minor groove of the 16S rRNA H44 helix near the A site. This flips out A1492 and A1493, so they can presumably stabilize base pairing between codon and anticodon. This flipping-out process normally requires energy, but paromomycin forces it to occur and keeps the stabilizing bases in place. This state of the 30S subunit interacts more readily with anticodons, including anticodons on noncognate aminoacyl-t-RNAs, so fidelity declines.

Fine Structure of the 50S Subunit

In 2000, Peter Moore and Thomas Steitz and their colleagues achieved a milestone in the study of ribosomal structure, and in the field of x-ray crystallography, by determining the crystal structure of a 50S ribosomal subunit at 2.4 Å resolution. They performed these studies on 50S subunits from the archaebacterium *Haloarcula marismortui,* because crystals of 50S subunits suitable for x-ray diffraction could be prepared from this organism. The structure, shown in Figure 19.22, includes 2833 of 3045 nucleotides in the rRNAs of the subunit (all 122 of the 5S rRNA nucleotides), and 27 of the 31 proteins. The other proteins were not well ordered and could not be located accurately.

One clear difference between the two subunits lies in the tertiary structures of their rRNAs. Whereas the 16S rRNA in the 30S subunit assumed a three-domain structure, the 23S rRNA of the 50S subunit is a monolithic structure with no clear boundaries between domains. Moore, Steitz, and colleagues speculate that the reason for this difference is that the structural domains of the 30S subunit have to move relative to one another, whereas those in the 50S subunit do not.

The smaller structures in Figure 19.22 show the locations of the proteins in the 50S subunit. As in the case of the 30S subunit, the proteins in the 50S subunit are generally missing from the interface between the two subunits, particularly in the center, where the peptidyl transferase active site is thought to lie. This was a provocative finding because some controversy (Chapter 18) surrounded the question whether the peptidyl transferase activity lies in the RNA or protein of the 50S subunit.

To determine whether proteins are present at the peptidyl transferase active site, one needs to identify the active site in a crystal structure. To accomplish this goal, Moore, Steitz, and coworkers soaked crystals of 50S subunits with two different peptidyl transferase substrate analogs, then performed x-ray crystallography and calculated electron difference maps. This located the electron densities corresponding to the substrate *analogs,* and therefore to the active site. One analog (*CCdAp-puromycin*) was designed by Michael Yarus to resemble the transition state, or intermediate, during the peptidyl transferase reaction, as illustrated in Figure 19.23. Thus, it is called the "Yarus analog." The other analog (see Figure 19.23) resembles an aminoacyl-tRNA and therefore binds to the ribosomal A site. It contains a stem loop, like the tRNA acceptor stem, ending in CC coupled to puromycin.

Figure 19.24 shows that the Yarus analog lies in the cleft in the face of the 50S subunit, right where the active site was predicted to be. And no proteins are around, only RNA. The same behavior was observed for the other

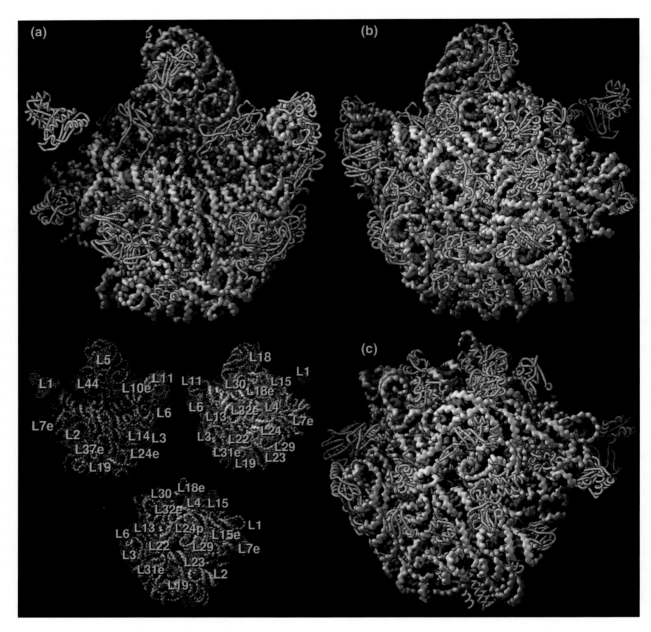

Figure 19.22   Crystal structure of the 50S ribosomal subunit. The three large structures show the subunit in three different orientations: (a) front, or "crown" view (so named because of the resemblance to a three-pointed crown); (b) back view (crown view rotated 180 degrees); (c) bottom view, showing the end of the polypeptide exit tunnel at center. The RNA is gray and the proteins are gold. The three small structures at lower left are the same three orientations, with the proteins identified. The letter "e" after some numbers designates archaebacterial proteins that have only eukaryotic (not eubacterial) homologs. (*Source:* Ban, N. et al. The complete atomic structure of the large ribosomal subunit at 2.4 Å resolution. *Science* 289 (11 Aug 2000) f. 7, p. 917. © AAAS.)

analog. Figure 19.25 is a model of the active site with all RNA removed, so we can see just how far the proteins are from the phosphate of the Yarus analog, which corresponds to the tetrahedral carbon atom at the very center of the transition state in the active site. The nearest protein is L3, which is more than 18 Å away from this active center–much too far to play any direct role in catalysis.

If RNA is the only molecule at the active site besides the substrate, it must have the enzymatic activity. The crystal structure reveals that adenine 2486 (A2486), which corresponds to A2451 in *E. coli,* is closest to the tetrahedral carbon at the active center. This base is conserved in ribosomes from every species examined from all three kingdoms of life, which indicates it plays a crucial role. Furthermore, chloramphenicol and carbomycin, which inhibit peptidyl transferase, bind at or near A2451 in *E. coli*. And *E. coli* cells with mutations in A2451 are chloramphenicol-resistant, further implicating this base in the reaction.

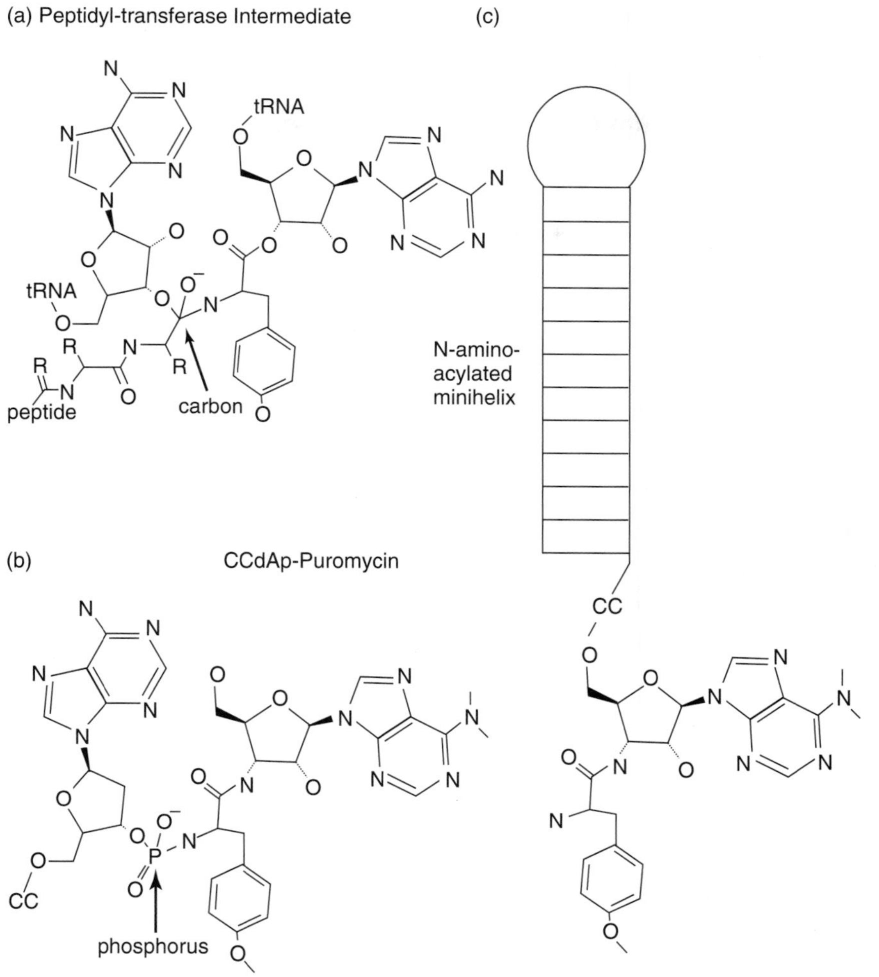

(a) Peptidyl-transferase Intermediate   (c)

(b)   CCdAp-Puromycin

Figure 19.23   Structure of peptidyl transferase intermediate and two analogs. (a) The peptidyl transferase intermediate (transition state), showing the tetrahedral carbon (arrow). (b) The Yarus analog (CCdAp-Puromycin). (c) The *N*-aminoacylated minihelix, which resembles an amino acid bound to the acceptor stem of a tRNA. (*Source:* Reprinted with permission from Nissen et al., *Science* 289:921, 2000. Copyright © 2000 American Association for the Advancement of Science.)

How could this base participate in the peptidyl transferase reaction? Figure 19.26 shows Moore, Steitz, and coworkers' hypothesis: A ring nitrogen (N3) of A2486 acts as a base by accepting a proton from the aminoacyl-tRNA. This makes the amino group of the aminoacyl-tRNA a better nucleophile, so it can attack the ester bond linking the peptide to the tRNA in the P site. During this reaction, the protonated N3 of A2486 hydrogen bonds to the oxyanion of the tetrahedral intermediate and stabilizes this intermediate. Finally, as the bond between the peptide and the tRNA in the P site breaks, the tRNA 3'-OH accepts the proton from N3 of A2486, returning the catalyst to its original condition.

For this scheme to work, the pK_a of N3 of A2486 must be approximately neutral so it can accept a proton readily. But free adenines have a pK_a of about 3.5, which is much too low. On the other hand, it is known that cer-tain bases in intact RNAs can have much higher pK_as, due to their environment within the RNA. Thus, it is very interesting that Scott Strobel and colleagues have demonstrated that A2486 has a pK_a of 7.6, which is well within the appropriate range for it to participate as an acid–base catalyst as proposed. The evidence therefore supports Moore, Steitz, and colleagues' hypothesis.

As the polypeptide product grows, it is thought to exit the ribosome through a tunnel in the 50S subunit. Moore, Steitz, and coworkers' studies also shed considerable light on this issue. Figure 19.27 shows a model of the 50S subunit cleaved in half through the central protuberance to reveal the exit tunnel. The peptidyl transferase center has been marked, and a polypeptide modeled in the tunnel. The tunnel has an average diameter of 15 Å and narrows in two places to as little as 10Å, just wide enough to accommodate a protein α-helix, so any further folding of

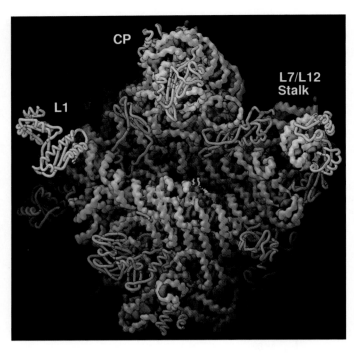

Figure 19.24   Location of the peptidyl transferase active site. This is a crown view of the 50S subunit as in Figure 19.19, with the location of the Yarus analog, which should be at the peptidyl transferase active site, in green. Notice the absence of proteins (gold) close to the active site. (*Source:* Ban, N. et al. The complete atomic structure of the large ribosomal subunit at 2.4 Å resolution. *Science* 289 (11 Aug 2000) f. 2, p. 907. © AAAS.)

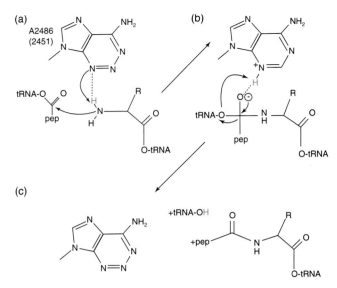

Figure 19.26   A model for the role of the 23S rRNA in the peptidyl transferase reaction. (**a**) N3 of A2486 accepts a proton from the amino group of the aminoacyl-tRNA in the A site. (**b**) The loss of the proton enhances the nucleophilic character (negative charge) of the amino group, enabling it to attack the bond between the peptide and the tRNA in the P site. Moreover, the proton on N3 of A2486 can hydrogen bond to the oxyanion in the tetrahedral intermediate, stabilizing it and thereby accelerating the reaction. (**c**) The 3′-hydroxyl group of the tRNA in the P site accepts the proton from N3 of A2486, returning the catalyst to its original condition. (*Source:* Reprinted with permission from Nissen et al., *Science* 289:925, 2000. Copyright © 2000 American Association for the Advancement of Science.)

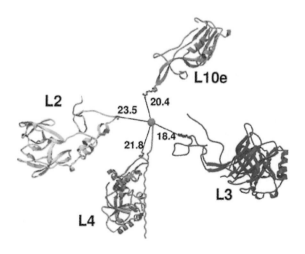

Figure 19.25   Peptidyl transferase active site with all RNA removed. The phosphate of the Yarus analog, at the center of the active site, is rendered in magenta (dark pink), with a long magenta tail representing a growing polypeptide. The four proteins closest to the active site are pictured, along with measurements of the closest approach of each protein to the active site. (*Source:* Nissen, P. et al. The structural basis of ribosome activity in peptide bond synthesis. *Science* 289 (11 Aug 2000) f. 6b, p. 924. © AAAS.)

the nascent polypeptide is unlikely. Much of the tunnel wall is made of hydrophilic RNA, so the exposed hydrophobic residues in a nascent polypeptide are not likely to find much in the tunnel wall to which to bind and retard the exit process.

SUMMARY The crystal structure of the 50S ribosomal subunit has been determined to 2.4 Å resolution. This structure reveals relatively few proteins at the interface between ribosomal subunits, and no protein within 18 Å of the peptidyl transferase active center tagged with a transition state analog. A specific base, A2486, may serve as an acid–base catalyst in the peptidyl transferase reaction. The exit tunnel through the 50S subunit is just wide enough to allow a protein α-helix to pass through. Its walls are made of RNA, whose hydrophilicity is likely to allow exposed hydrophobic side chains of the nascent polypeptide to slide through easily.

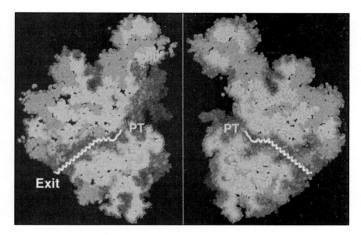

Figure 19.27 The polypeptide exit tunnel. The 50S subunit is pictured as if it were a fruit cut through the middle and opened up. This view reveals the exit channel leading away from the peptidyl transferase site (PT). A white α-helix is placed in the channel to represent an exiting polypeptide. (*Source:* Nissen, P. et al. The structural basis of ribosome activity in peptide bond synthesis. *Science* 289 (11 Aug 2000) f. 11a, p. 927. © AAAS.)

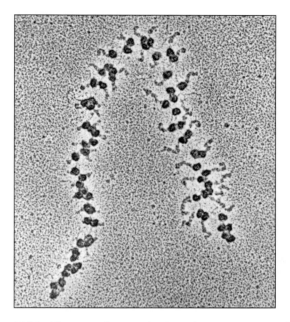

Figure 19.28 Electron micrograph of a polysome from the midge *Chironomus*. The 5′-end on the mRNA is at lower left, and the mRNA bends up and then down to the 3′-end at lower right. The dark blobs attached to the mRNA are ribosomes. The fact that many (74) of them are present is the reason for the name *polysome*. Nascent polypeptides extend away from each ribosome and grow longer as the ribosomes approach the end of the mRNA. The faint blobs on the nascent polypeptides are not individual amino acids but domains containing groups of amino acids. (*Source:* From Franke et al., Electron microscopic visualization of a discreet class of giant translation units in salivary glands of *Chironomus tetans. EMBO Journal* 1 (1-6) 1982, pp. 59–62. European Molecular Biology Organization.)

Polysomes

We have seen in previous chapters that more than one RNA polymerase can transcribe a gene at a time. The same is true of ribosomes and mRNA. In fact, it is common for many ribosomes to be traversing the same mRNA in tandem at any given time. The result is a polyribosome, or **polysome,** such as the one pictured in Figure 19.28. In this polysome we can count 74 ribosomes translating the mRNA simultaneously. We can also tell which end of the polysome is which by looking at the nascent polypeptide chains. These grow longer as we go from the 5′-end (where translation begins) to the 3′-end (where translation ends). Therefore, the 5′-end is at lower left, and the 3′-end is at lower right.

The polysome in Figure 19.28 is from a eukaryote (the midge). Because transcription and translation occur in different compartments in eukaryotes, polysomes will always occur in the cytoplasm, independent of the genes. Prokaryotes also have polysomes, but the picture in these organisms is complicated by the fact that transcription and translation of a given gene and its mRNA occur simultaneously and in the same location. Thus, we can see nascent mRNAs being synthesized and being translated by ribosomes at the same time. Figure 19.29 shows just such a situation in *E. coli*. We can see two segments of the bacterial chromosome running parallel from left to right. Only the segment on top is being transcribed. We can tell that transcription is occurring from left to right in this picture because the polysomes are getting longer as they move in that direction; as they get longer, they have room for more and more ribosomes. Do not be misled by the difference in scale between Figures 19.28 and 19.29; the ribosomes appear smaller, and the nascent protein chains are not visible in the latter picture. Remember also that the strands running across Figure 19.29 are DNA, whereas that in Figure 19.28 is mRNA. The mRNAs are more or less vertical in Figure 19.29.

SUMMARY Most mRNAs are translated by more than one ribosome at a time; the result, a structure in which many ribosomes translate an mRNA in tandem, is called a polysome. In eukaryotes, polysomes are found in the cytoplasm. In prokaryotes, transcription of a gene and translation of the resulting mRNA occur simultaneously. Therefore, many polysomes are found associated with an active gene.

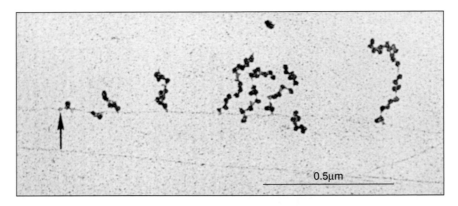

Figure 19.29 Simultaneous transcription and translation in E. coli. Two DNA segments stretch across the picture from left to right. The top segment is being transcribed from left to right. As the mRNAs grow, more and more ribosomes attach and carry out translation. This gives rise to polysomes, which are arrayed perpendicular to the DNA. The nascent polypeptides are not visible in this picture. The arrow at left points to a faint spot, which may be an RNA polymerase just starting to transcribe the gene. Other such spots denoting RNA polymerase appear at the bases of some of the other polysomes, where the mRNAs join the DNA. (*Source:* O. L. Miller, Visualization of bacterial genes in action. *Science* 169 (July 1970) p. 394. © AAAS.)

19.2 Transfer RNA

In 1958, Francis Crick postulated the existence of an adaptor molecule, presumably RNA, that could serve as a mediator between the string of nucleotides in DNA (actually in mRNA) and the string of amino acids in the corresponding protein. Crick favored the idea that the adaptor contained two or three nucleotides that could pair with nucleotides in codons, although no one knew the nature of codons, or even of the existence of mRNA, at that time. Transfer RNA had already been discovered by Paul Zamecnik and coworkers a year earlier, although they did not realize that it played an adaptor role.

The Discovery of tRNA

By 1957, Zamecnik and colleagues had worked out a cell-free protein synthesis system from the rat. One of the components of the system was a so-called pH 5 enzyme fraction that contained the soluble factors that worked with ribosomes to direct translation of added mRNAs. Most of the components in the pH 5 enzyme fraction were proteins, but Zamecnik's group discovered that this mixture also included a small RNA. Of even more interest was their finding that this RNA could be coupled to amino acids. To demonstrate this, they mixed the RNA with the pH 5 enzymes, ATP, and [14C]leucine. Figure 19.30a shows that the more labeled leucine these workers added to the mixture, the more was attached to the RNA. Furthermore, when they left out ATP, no reaction occurred. We now know that this reaction was the charging of tRNA with an amino acid.

Not only did Zamecnik and his coworkers show that the small RNA could be charged with an amino acid, they also demonstrated that it could pass its amino acid to a

growing protein. They performed this experiment by mixing the [14C]leucine-charged pH 5 RNA with microsomes—small sections of endoplasmic reticulum containing ribosomes. Figure 19.30b shows a near-perfect correspondence between the loss of radioactive leucine from the pH 5 RNA and gain of the leucine by the protein in the microsomes. This represented the incorporation of leucine from leucyl-tRNA into nascent protein on ribosomes.

SUMMARY Transfer RNA was discovered as a small RNA species independent of ribosomes that could be charged with an amino acid and could then pass the amino acid to a growing polypeptide.

tRNA Structure

To understand how a tRNA carries out its functions, we need to know the structure of the molecule, and tRNAs have a surprisingly complex structure considering their small size. Just as a protein has primary, secondary, and tertiary structure, so does a tRNA. The primary structure is the linear sequence of bases in the RNA; the secondary structure is the way different regions of the tRNA base-pair with each other to form stem loops; and the tertiary structure is the overall three-dimensional shape of the molecule. In this section, we will survey tRNA structure and its relationship to tRNA function.

In 1965, Robert Holley and his colleagues completed the first determination ever of the base sequence of a natural nucleic acid, an alanine tRNA from yeast. This primary sequence suggested at least three attractive secondary structures, including one that had a cloverleaf shape. By 1969, 14 tRNA sequences had been determined, and it became clear that, despite considerable differences in

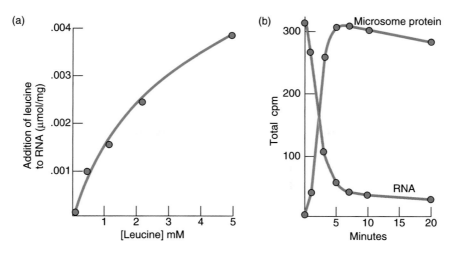

Figure 19.30   Discovery of tRNA. (a) The tRNA can be charged with leucine. Zamecnik and colleagues added labeled leucine to the tRNA-containing fraction and plotted the binding of leucine to the RNA as a function of labeled leucine added. **(b)** The charged tRNA can donate its amino acid to nascent protein. Zamecnik and colleagues followed the radioactivity (cpm) lost from the RNA (blue) and gained by the nascent proteins (red) in the microsomes, which contained the ribosomes. The reciprocal relationship between these curves suggested that the RNA was donating its amino acid to the growing protein. (*Source:* From Held et al., *Journal of Biological Chemistry* 231:244 & 252, 1958. Copyright © 1958 The American Society for Biochemistry & Molecular Biology, Bethesda, MD. Reprinted by permission.)

primary structure, all could assume essentially the same "cloverleaf" secondary structure, as illustrated in Figure 19.31a. As we study this structure we should bear in mind that the real three-dimensional structure of a tRNA is not cloverleaf-shaped at all; the cloverleaf merely describes the base-pairing pattern in the molecule.

The cloverleaf has four base-paired stems that define the four major regions of the molecule (Figure 19.31b). The first, seen at the top of the diagram, is the **acceptor stem,** which includes the two ends of the tRNA, which are base-paired to each other. The 3′-end, bearing the invariant sequence CCA, protrudes beyond the 5′-end. On the left is the **dihydrouracil loop (D loop),** named for the modified uracil bases this region always contains. At the bottom is the **anticodon loop,** named for the all-important anticodon at its apex. As we learned in Chapter 3, the anticodon base-pairs with an mRNA codon and therefore allows decoding of the mRNA. At right is the T loop, which takes its name from a nearly invariant sequence of three bases: TψC. The ψ stands for a modified nucleoside in tRNA, **pseudo-uridine.** It is the same as normal uridine, except that the base is linked to the ribose through the 5-carbon of the base instead of the 1-nitrogen. The region between the anticodon loop and the T loop in Figure 19.31 is called the **variable loop** because it varies in length from 4 to 13 nt; some of the longer variable loops contain base-paired stems.

Transfer RNAs contain many modified nucleosides in addition to dihydrouridine and pseudo-uridine. Some of the modifications are simple methylations. Others are more elaborate, such as the conversion of guanosine to a nucleoside called **wyosine,** which contains a complex three-ring base called the Y base (Figure 19.32). Some tRNA modifications are general. For example, virtually all tRNAs have a pseudo-uridine in the same position in the T loop, and most tRNAs have a hypermodified nucleoside such as wyosine next to the anticodon. Other modifications are specific for certain tRNAs. Figure 19.32 illustrates some of the common modified nucleosides in tRNAs.

The modification of tRNA nucleosides raises the question: Are tRNAs made with modified bases, or are the bases modified after transcription is complete? The answer is that tRNAs are made in the same way that other RNAs are made, with the four standard bases. Then, once transcription is complete, multiple enzyme systems modify the bases. What effects, if any, do these modifications have on tRNA function? At least two tRNAs have been made in vitro with the four normal, unmodified bases, and they were unable to bind amino acids. Thus, at least in these cases, totally unmodified tRNAs were nonfunctional. Although these studies suggested that the sum of all the modifications is critical, each individual base modification probably has more subtle effects on the efficiency of charging and tRNA usage.

In the 1970s, Alexander Rich and his colleagues used x-ray diffraction techniques to reveal the tertiary structure of tRNAs. Because all tRNAs have essentially the same secondary structure, represented by the cloverleaf model, it is perhaps not too surprising that they all have essentially the same tertiary structure as well. Figure 19.33 illustrates this L-shaped structure for yeast tRNA^{Phe}. Perhaps the most important aspect of this structure is that it maximizes the lengths of its base-paired stems by stacking

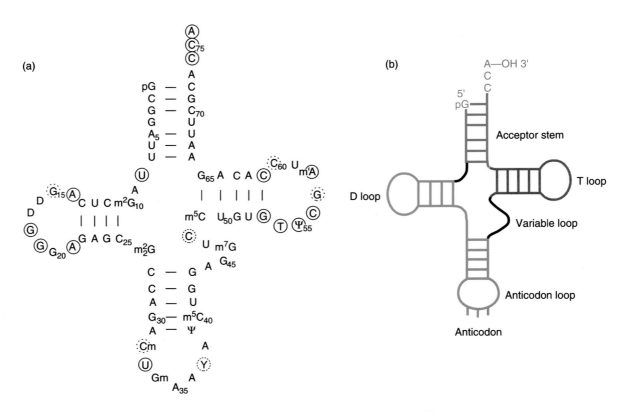

Figure 19.31 Two views of the cloverleaf structure of tRNA. (a) Base sequence of yeast tRNA[Phe], shown in cloverleaf form. Invariant nucleotides are circled. Bases that are always purines or always pyrimidines are in dashed circles. **(b)** Cloverleaf structure of yeast tRNA[Phe]. At top is the acceptor stem (magenta), where the amino acid binds to the 3′-terminal adenosine. At left is the dihydro U loop (D loop, blue), which contains at least one dihydrouracil base. At bottom is the anticodon loop (green), containing the anticodon. The T loop (right, gray) contains the virtually invariant sequence TψC. Each loop is defined by a base-paired stem of the same color. (*Source: (a)* From Kim et al., *Science* 185:435, 1974. Copyright © 1974 American Association for the Advancement of Science, Washington, DC. Reprinted by permission.)

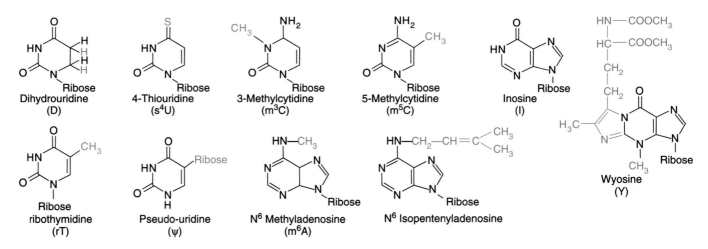

Figure 19.32 Some modified nucleosides in tRNA. Magenta indicates the variation from one of the four normal RNA nucleosides. Inosine is a special case; it is a normal precursor to both adenosine and guanosine.

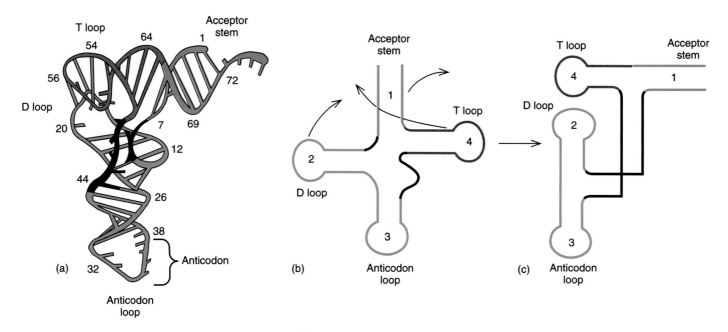

Figure 19.33 Three-dimensional structure of tRNA. (a) A planar projection of the three-dimensional structure of yeast tRNA^{Phe}. The various parts of the molecule are color-coded to correspond to (b) and (c). (b) Familiar cloverleaf structure of tRNA with same color scheme as part (a). Arrows indicate the contortions this cloverleaf would have to go through to achieve the approximate shape of a real tRNA, shown in part (c).

them in sets of two to form relatively long extended base-paired regions. One of these regions lies horizontally at the top of the molecule and encompasses the acceptor stem and the T stem; the other forms the vertical axis of the molecule and includes the D stem and the anticodon stem. Even though the two parts of each stem are not aligned perfectly and the stems therefore bend slightly, the alignment allows the base pairs to stack on each other, and therefore confers stability. The base-paired stems of the molecule are RNA–RNA double helices. As we learned in Chapter 2, such RNA helices should assume an A-helix form with about 11 bp per helical turn, and the x-ray diffraction studies verified this prediction.

Figure 19.34 is a stereo diagram of the yeast tRNA^{Phe} molecule. The base-paired regions are particularly easy to see in three dimensions, but you can even visualize them in two dimensions in the T stem-acceptor region because they are depicted almost perpendicular to the plane of the page, so they appear as almost-parallel lines.

As we have seen, a tRNA is stabilized primarily by the secondary interactions that form the base-paired regions, but it is also stabilized by dozens of tertiary interactions between regions. These include base–base, base–backbone, and backbone–backbone interactions. Figure 19.35 summarizes the base–base tertiary interactions that involve hydrogen bonds. Most of these occur between invariant or semi-invariant bases (the semi-invariant bases are always purines or always pyrimidines). Because these interactions allow the tRNA to fold into the proper shape, it makes sense that the bases involved tend not to vary; any variance would hinder the proper folding and hence

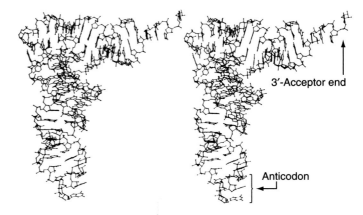

Figure 19.34 Stereo view of tRNA. To see the molecule in three dimensions, use a stereo viewer, or force the two images to merge either by relaxing your eyes as if focusing on something in the distance (the "magic eye" technique) or by crossing your eyes slightly. It may take a little time for the three-dimensional effect to develop.
(*Source:* Quigley & Rich, Structural domains of transfer RNA molecules. *Science* 194 (19 Nov 1976) f. 2, p. 798. © AAAS.)

the proper functioning of the tRNA. Only one of the base–base interactions is a normal Watson–Crick base pair (G19–C56). All the others are extraordinary. Consider, for example, the G15–C48 pair, which joins the D loop to the variable loop. This cannot be a Watson–Crick base pair because the two strands are parallel here, rather than antiparallel. We call this a *trans*-pair. Several examples also occur of one base interacting with two other bases. Figure 19.35 shows one of these, involving U8, A14, and A21. Now that the tertiary interactions

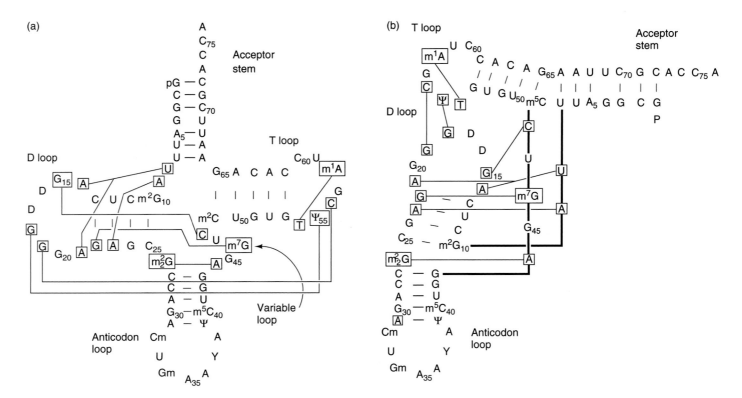

Figure 19.35   Tertiary interactions in a tRNA. Based on their x-ray crystallography data, Rich and colleagues proposed the base–base interactions indicated by the thin lines in panel (**a**) the cloverleaf model, and (**b**) the more realistic L-shaped model. Thick lines in panel (**b**) represent phosphodiester bonds linking nucleotides in the region of the tRNA that bridges between the horizontal and vertical stems of the molecule. Note the slight departure from linearity of the vertical and horizontal stems in panel (**b**). (*Source:* From Kim et al., *Science* 185:437, 1974. Copyright © 1974 American Association for the Advancement of Science, Washington, DC. Reprinted by permission.)

have been highlighted, you can look again at Figure 19.33a and see them in a more realistic form. Note for example the interactions between bases 18 and 55, and between bases 19 and 56. At first glance, these look like base pairs within the T loop; on closer inspection we can now see that they link the T loop and the D loop.

One other striking aspect of tRNA tertiary structure is the structure of the anticodon. Figure 19.34 demonstrates that the anticodon bases are stacked, but this stacking occurs with the bases projecting out to the right, away from the backbone of the tRNA. This places them in position to interact with the bases of the codon in an mRNA. In fact, the anticodon backbone is already twisted into a partial helix shape, which presumably facilitates base pairing with the corresponding codon. If this shape really does encourage codon–anticodon binding, then we would predict that two complementary anticodons would readily bind to each other. Josef Eisinger tested this prediction by mixing two tRNAs with complementary anticodons (Figure 19.36) and showing that they bound to each other 2–3 orders of magnitude more tightly than two complementary trimers with similar sequences. The two tRNAs in question were tRNA^Glu, with the anticodon thioUUC, and tRNA^Phe, with the anticodon GAA. Remember that we need to invert one of these anticodons to make the two antiparallel; then we can see that they are perfectly

complementary and should base-pair with each other. If they do, this would place the anticodon of the tRNA^Glu close to the wyosine (Y) adjacent to the 3′-A in the anticodon of the tRNA^Phe. The close approach of the two anticodons allows electronic interactions between the two tRNAs that quench almost all of the wyosine's fluorescence. This allowed the experimenters to detect RNA–RNA interaction by measuring the quenching of the fluorescence of the wyosine, as illustrated in Figure 19.36.

SUMMARY All tRNAs share a common secondary structure represented by a cloverleaf. They have four base-paired stems defining three stem loops (the D loop, anticodon loop, and T loop) and the acceptor stem, to which amino acids are added in the charging step. The tRNAs also share a common three-dimensional shape, which resembles an inverted L. This shape maximizes stability by lining up the base pairs in the D stem with those in the anticodon stem, and the base pairs in the T stem with those in the acceptor stem. The anticodon of the tRNA protrudes from the side of the anticodon loop and is twisted into a shape that readily base-pairs with the corresponding codon in mRNA.

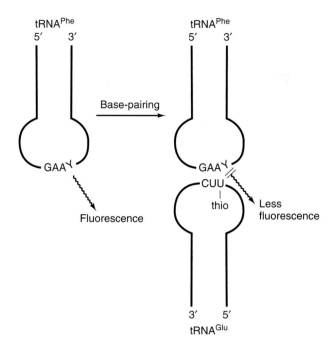

Figure 19.36   A fluorescence-quenching assay for tRNA–tRNA interaction. The wyosine (Y) in tRNA^{Phe} is fluorescent. When the anticodon in tRNA^{Phe} base-pairs with the anticodon in tRNA^{Glu}, however, the thioU (s^{4}U) in the latter tRNA quenches some of the fluorescence of the wyosine in the former. The strength of the binding is related to the amount of fluorescence quenching, which provides the basis for the binding assay.

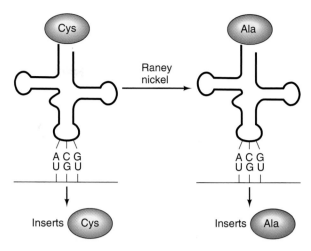

Figure 19.37   The ribosome responds to the tRNA, not the amino acid of an aminoacyl-tRNA. Lipmann, Ehrenstein, Benzer, and colleagues started with a cysteinyl-tRNA^{Cys}, which inserted cysteine (Cys, blue) into a protein chain, as shown at left. They treated this aminoacyl-tRNA with Raney nickel, which reduced the cysteine to alanine (Ala, red), but had no effect on the tRNA. This alanyl-tRNA^{Cys} inserted alanine into a protein chain at a position normally occupied by cysteine, as depicted at right. Thus, the nature of the amino acid attached to the tRNA does not matter; it is the nature of the tRNA that matters, because its anticodon has to match the mRNA codon.

Recognition of tRNAs by Aminoacyl-tRNA Synthetase: The Second Genetic Code

In 1962, Fritz Lipmann, Seymour Benzer, Günter von Ehrenstein, and colleagues demonstrated that the ribosome recognizes the tRNA, not the amino acid, in an aminoacyl-tRNA. They did this by forming cysteinyl-tRNA^{Cys}, then reducing the cysteine with Raney nickel to yield alanyl-tRNA^{Cys}, as illustrated in Figure 19.37. (Notice the nomenclature here. In cysteinyl-tRNA^{Cys} [Cys-tRNA^{Cys}] the first Cys tells what amino acid is actually attached to the tRNA. The second Cys [in the superscript] tells what amino acid *should be* attached to this tRNA. Thus, alanyl-tRNA^{Cys} is a tRNA that *should* bind cysteine, but in this case is bound to alanine.) Then they added this altered aminoacyl-tRNA to an in vitro translation system, along with a synthetic mRNA that was a random polymer of U and G, in a 5:1 ratio. This mRNA had many UGU codons, which encode cysteine, so it normally caused incorporation of cysteine. It should not cause incorporation of alanine because the codons for alanine are GCN, where N is any base, and the UG polymer contained no C's. However, in this case alanine was incorporated because it was attached to a tRNA^{Cys}. This showed that ribosomes do not discriminate among amino acids at-

tached to tRNAs; they recognize only the tRNA part of an aminoacyl-tRNA.

This experiment pointed to the importance of fidelity in the aminoacyl-tRNA synthetase step. The fact that ribosomes recognize only the tRNA part of an aminoacyl-tRNA means that if the synthetases make mistakes and put the wrong amino acids on tRNAs, then these amino acids will be inserted into proteins in the wrong places. That could be very damaging because a protein with the wrong amino acid sequence is likely not to function properly. Thus, it is not surprising that aminoacyl-tRNA synthetases are very specific for the tRNAs and amino acids they bring together. This raises a major question related to the structure of tRNAs: Given that the secondary and tertiary structures of all tRNAs are essentially the same, what base sequences in tRNAs do the synthetases recognize when they are selecting one tRNA out of a pool of over 20? This set of sequences has even been dubbed the "second genetic code" to highlight its importance. This question is complicated by the fact that some **isoaccepting species** of tRNA can be charged with the same amino acid by the same synthetase, yet they have different sequences, and even different anticodons.

If we were to guess about the locations of the tRNA elements that an aminoacyl-tRNA synthetase recognizes, two sites would probably occur to us. First, the acceptor stem seems a logical choice, because that is the locus on the tRNA that accepts the amino acid and is therefore likely to lie at or near the enzyme's active site as it is being charged. Because the enzyme presumably makes such intimate contact with the acceptor stem, it should be able to discriminate among tRNAs with different base sequences in the acceptor stem. Of course, the last three bases are irrelevant for this purpose because they are the same, CCA, in all tRNAs. Second, the anticodon is a reasonable selection, because it is different in each tRNA, and it has a direct relationship to the amino acid with which the tRNA should be charged. We will see that both these predictions are correct in most cases, and some other areas of certain tRNAs also play a role in recognition by aminoacyl-tRNA synthetases.

The Acceptor Stem   In 1972, Dieter Söll and his colleagues noticed a pattern in the nature of the fourth base from the 3'-end, position 73 in most tRNAs. That is, this base tended to be the same in tRNAs specific for a certain class of amino acids. For example, virtually all the hydrophobic amino acids are coupled to tRNAs with A in position 73, regardless of the species in which we find the tRNA. However, this obviously cannot be the whole story because one base does not provide enough variation to account for specific charging of 20 different classes of tRNAs. At best, it fills the role of a rough discriminator.

B. Roe and B. Dudock used another approach. They examined the base sequence of all the tRNAs from several species that could be charged by a single synthetase. This included some tRNAs that were charged with the wrong amino acid, in a process called *heterologous mischarging*. This term refers to the ability of a synthetase from one species to charge an incorrect tRNA from another species, although this mischarging is always slower and requires a higher enzyme concentration than normal. For example, yeast phenylalanyl-tRNA synthetase (PheRS) can charge tRNA^{Phe} from *E. coli*, yeast, and wheat germ correctly, but it can also charge *E. coli* tRNA^{Val} with phenylalanine.

Because all these tRNAs can be charged by the same synthetase, they should all have the elements that the synthetase uses to tell it which tRNAs to charge. So Roe and Dudock compared the sequences of all these tRNAs, looking for things they have in common, but are not common to all tRNAs. Two features stood out: base 73, and nine nucleotides in the D stem.

In 1973, J. D. Smith and J. E. Celis studied a mutant suppressor tRNA that inserted Gln instead of Tyr. In other words, the wild-type suppressor tRNA was charged by the GlnRS, but some change in its sequence caused it to be charged by the TyrRS instead. The only difference between the mutant and wild-type tRNAs was a change in base 73 from G to A.

In 1988, Hou and Schimmel used genetic means to demonstrate the importance of a single base pair in the acceptor stem to charging specificity. They started with a tRNA^{Ala} that had its anticodon mutated to 5'-CUA-3' so it became an amber suppressor capable of inserting ala-

nine in response to the amber codon UAG. Then they looked for mutations in the tRNA that changed its charging specificity. Their assay was a convenient one they could run in vivo. They built a *trpA* gene with an amber mutation in codon 10. This mutation could be suppressed only by a tRNA that could insert an alanine (or glycine) in response to the amber codon. Any other amino acid in position 10 yielded an inactive protein. Finally, they challenged their mutants by growing them in the absence of tryptophan. If the mutant could suppress the amber mutation in the *trpA* gene, it probably had a suppressor tRNA that could be charged with alanine. If not, the suppressor tRNA was probably charged with another amino acid. They found that the cells that grew in the absence of tryptophan had a G in position 3 of the suppressor tRNA and a U in position 70, so a G3-U70 wobble base pair could form in the acceptor stem three bases from the end of the stem.

This experiment suggested that the G3-U70 base pair is a key determinant of charging by AlaRS. If so, these workers reasoned, they might be able to take another suppressor tRNA that inserted another amino acid, change its bases at positions 3 and 70 to G and U, respectively, and convert the charging specificity of the suppressor tRNA to alanine. They did this with two different suppressor tRNAs: tRNA$^{Cys/CUA}$ and tRNA$^{Phe/CUA}$, where the CUA designation refers to the anticodon, which recognizes the UAG amber codon. Both of the tRNAs had a C3-G70 base pair in their acceptor stems. However, when Hou and Schimmel changed this one base pair to G3–U70, they converted the tRNAs to tRNA$^{Ala/CUA}$, as indicated by their ability to suppress the amber mutation in codon 10 of the *trpA* gene.

Did these altered amber suppressor tRNAs really insert alanine into the TrpA protein? Amino acid sequencing revealed that they did. Furthermore, these altered tRNAs could be charged with alanine in vitro. Thus, even though these two tRNAs differed from natural tRNA$^{Ala/CUA}$ in 38 and 31 bases, respectively, changing just one base pair from C–G to G–U changed the charging specificity from Cys or Phe to Ala.

SUMMARY Biochemical and genetic experiments have demonstrated the importance of the acceptor stem in recognition of a tRNA by its cognate aminoacyl-tRNA synthetase. In certain cases, changing one base pair in the acceptor stem can change the charging specificity.

The Anticodon   In 1973, LaDonne Schulman pioneered a technique in which she treated tRNA$_f^{Met}$ with bisulfite, which converts cytosines to uracils. She and her colleagues found that many of these base alterations had no

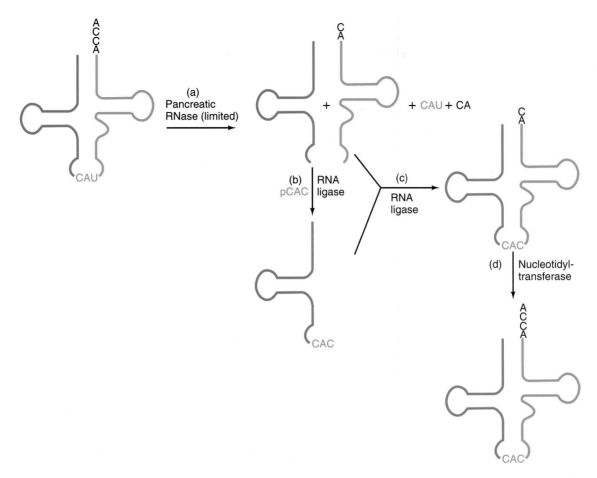

Figure 19.38 Method for generating tRNA$_f^{Met}$ molecules with mutations in their anticodons. (a) Schulman and Pelka subjected tRNA$_f^{Met}$ to mild digestion with pancreatic RNase, generating two half-molecules, the 5′-half (green), and the 3′-half (blue), plus the anticodon (CAU, red) and the 3′-terminal CA. (b) They used RNA ligase to attach a new anticodon (CAC, red) to the 5′-fragment. (c) They used RNA ligase again to join the 3′-fragment to the 5′-fragment with its new anticodon. (d) They used tRNA nucleotidyltransferase to reconstitute the terminal CCA. The result is a natural tRNA$_f^{Met}$ having an altered anticodon. This procedure can be used with different oligonucleotides in step (b) to make any changes desired in the anticodon, even adding or subtracting bases.

effect, but some destroyed the ability of the tRNA to be charged with methionine. One such change was a C→U change in base 73; another was a C→U change in the anticodon. Since then, Schulman and her colleagues have amassed a large body of evidence that shows the importance of the anticodon in charging specificity.

In 1983, Schulman and Heike Pelka developed a method to change specifically one or more bases at a time in the anticodon of the initiator tRNA, tRNA$_f^{Met}$, as described in Figure 19.38. First, they cut the wild-type tRNA in two with a limited digestion with pancreatic RNase. This removed the anticodon from the tRNA 5′-fragment, and also cut off the last two nucleotides of the CCA terminus of the 3′-fragment. Then they used T4 RNA ligase to attach a small oligonucleotide to the 5′-fragment that would replace the lost anticodon, with one or more bases altered, then they ligated the two halves of the molecule back together and then added back the lost

terminal CA with tRNA nucleotidyltransferase. Finally, they tested the tRNAs with altered anticodons in charging reactions in vitro. Table 19.1 shows that changing one base in the anticodon of tRNA$_f^{Met}$ was sufficient to lower the rate of charging with Met by at least a factor of 10^5. The first base in the anticodon (the "wobble" position) was the most sensitive; changing this one base always had a drastic effect on charging. Thus, the anticodon seems to be required for charging of this tRNA in vitro.

In 1991, Schulman and Leo Pallanck followed up the earlier in vitro studies with an in vivo study of the effects of altering the anticodon. Again, they changed the anticodon of the tRNA$_f^{Met}$, but this time they tested the ability of the altered tRNA to be mischarged with the amino acid corresponding to the new anticodon. They tested mischarging with a reporter gene encoding dihydrofolate reductase (DHFR), which is easy to isolate in highly purified form. Here is an example of how the assay worked:

Table 19.1 Initial Rates of Aminoacylation of tRNA$_f^{Met}$ Derivatives

tRNA*	Mol Met-tRNA/mol Met-tRNA synthetase per min	Relative rate, CAU/other
tRNA$_f^{Met}$	28.45	0.8
tRNA$_f^{Met}$ (gel)†	22.80	1
CAU	22.15	1
CAUA	1.59	14
CCU	4.0×10^{-1}	55
CUU	2.6×10^{-2}	850
CUA	2.0×10^{-2}	1,100
CAG	1.7×10^{-2}	1,300
CAC	1.2×10^{-3}	18,500
CA	0.5×10^{-3}	44,000
C	$<10^{-4}$	$>10^5$
ACU	$<10^{-4}$	$>10^5$
UAU	$<10^{-4}$	$>10^5$
AAU	$<10^{-4}$	$>10^5$
GAU	$<10^{-4}$	$>10^5$

*The oligonucleotide inserted in the anticodon loop of synthesized tRNA$_f^{Met}$ derivatives is indicated.
†Control sample isolated from a denaturing polyacrylamide gel in parallel with the synthesized tRNA$_f^{Met}$ derivatives.

Source: From L. H. Schulman and H. Pelka, "Anticodon loop size and sequence requirements for recognition of formylmethionine tRNA by methionyl-tRNA synthetase," *Proceedings of the National Academy of Sciences* 80:6755–6759, November 1993.

They altered the gene for tRNA$_f^{Met}$ so its anticodon was changed from CAU to GAU, which is an isoleucine (Ile) anticodon. Then they placed this mutant gene into *E. coli* cells, along with a mutant DHFR gene bearing an AUC initiation codon.

Ordinarily, AUC would not work well as an initiation codon, but in the presence of a tRNA$_f^{Met}$ with a complementary anticodon, it did, as shown in Table 19.2. Sequencing of the resulting DHFR protein demonstrated that the amino acid in the first position was primarily Ile. Some Met occurred in the first position, showing that the endogenous wild-type tRNA$_f^{Met}$ could recognize the AUC initiation codon to some extent.

Pallanck and Schulman used the same procedure to change the tRNA$_f^{Met}$ anticodon to GUC (valine, Val) or UUC (phenylalanine, Phe). In each case, they made a corresponding change in the DHFR initiation codon so it was complementary to the anticodon in the altered tRNA$_f^{Met}$. Table 19.2 demonstrates that in both cases, the gene functioned significantly better in the presence than in the absence of the complementary tRNA$_f^{Met}$. More importantly, this table also shows that the nature of the initiating amino acid can change with the alteration in the tRNA anticodon. In fact, with the tRNA$_f^{Met}$ bearing the valine anticodon, valine was the only amino acid found at the amino terminus of the DHFR protein. This means that a change of the tRNA$_f^{Met}$ anticodon from CAU to GAC altered the charging specificity of this tRNA from methionine to valine. Thus, in this case, the anticodon seems to be the crucial factor in determining the charging specificity of the tRNA.

On the other hand, notice that changing the anticodon of the tRNA$_f^{Met}$ always reduced its efficiency. In fact, the alterations listed at the bottom of Table 19.2 yielded tRNA$_f^{Met}$ molecules whose efficiency was too low to analyze further, even in the presence of complementary initiation codons. Thus, some aminoacyl-tRNA synthetases could charge a noncognate tRNA with an altered anticodon, but others could not. These latter enzymes apparently required more cues than just the anticodon.

SUMMARY Biochemical and genetic experiments have shown that the anticodon, like the acceptor stem, is an important element in charging specificity. Sometimes the anticodon can be the absolute determinant of specificity.

Structures of Synthetase–tRNA Complexes X-ray crystallography studies of complexes between tRNAs and their cognate aminoacyl-tRNA synthetases have shown

Table 19.2   Synthesis of DHFR from Methionine and Nonmethionine Initiation Codons in the Presence and Absence of Complementary tRNA$_f^{Met}$ Anticodon Mutants

| Initiation Codon | tRNA$_f^{Met}$ Anticodon | DHFR, units | | Efficiency Relative to AUG (%) | NH$_2$-terminal Amino Acids (%) |
		+ complementary tRNA$_f^{Met}$	–complementary tRNA$_f^{Met}$		
5′-AUG-3′	5′-CAU-3′	1085 ± 94	1109 ± 133	100	Met (100)
AUC	GAU	298 ± 26	104 ± 10	23	Ile (84), Met (16)
GUC	GAC	201 ± 3	2.7 ± 0.1	18	Val (100)
UUC	GAA	61 ± 11	2.0 ± 0.5	4.3	Phe (76), Met (21), Ile (3)
UGC	GCA	5.0 ± 0.3	0.5 ± 0.08	0.4	—
AAC	GUU	4.1 ± 0.1	0.5 ± 0.04	0.3	—
ACC	GGU	3.8 ± 0.4	0.5 ± 0.05	0.3	—
UUA	UAA	3.4 ± 0.3	1.2 ± 0.2	0.2	—
GGC	GCC	2.8 ± 0.2	0.5 ± 0.04	0.2	—
GAC	GUC	0.7 ± 0.3	0.4 ± 0.03	0.03	—

Source: From L. Pallanck and L. H. Schulman, "Anticodon-dependent aminoacylation of a noncognate tRNA with isoleucine, valine, and phenylalanine *in vivo,*" *Proceedings of the National Academy of Sciences* 88:3872–3876, May 1991.

that both the acceptor stem and the anticodon have docking sites on the synthetases. Thus, these findings underline the importance of the acceptor stem and anticodon in synthetase recognition. In 1989, Dieter Söll and Thomas Steitz and their colleagues used x-ray crystallography to determine the first three-dimensional structure of an aminoacyl-tRNA synthetase (*E. coli* GlnRS) bound to its cognate tRNA. Figure 19.39 presents this structure. Near the top, we see a deep cleft in the enzyme that enfolds the acceptor stem, including base 73 and the 3–70 base pair. At lower left, we observe a smaller cleft in the enzyme into which the anticodon of the tRNA protrudes. This would allow for specific recognition of the anticodon by the synthetase. In addition, we see that most of the left side of the enzyme is in intimate contact with the inside of the L of the tRNA, which includes the D loop side and the minor groove of the acceptor stem.

About half the synthetases, including GlnRS, are in a group called **class I.** These are all structurally similar and initially aminoacylate the 2′-hydroxyl group of the terminal adenosine of the tRNA. The other half of the synthetases are in **class II;** they are structurally similar to other members of their group, but quite different from the members of class I, and they initially aminoacylate the 3′-hydroxyl groups of their cognate tRNAs. In 1991, D. Moras and colleagues obtained the x-ray crystal structure of a member of this group, yeast AspRS, together with tRNA^{Asp}. Figure 19.40 contrasts the structures of the class I and class II synthetase–tRNA complexes. Several differences stand out. First, although the synthetase still contacts the inside of the L, it does so on the tRNA's op-

posite face, including the variable loop and the major groove of the acceptor stem. Also, the acceptor stem, including the terminal CCA, is in a regular helical conformation. This contrasts with the class I structure, in which the first base pair is broken and the 3′-end of the molecule makes a hairpin turn. Thus, x-ray crystallography has corroborated the major conclusions of biochemical and genetic studies on synthetase–tRNA interactions: Both the anticodon and acceptor stem are in intimate contact with the enzyme and are therefore in a position to determine specificity of enzyme-tRNA interactions.

SUMMARY X-ray crystallography has shown that synthetase–tRNA interactions differ between the two classes of aminoacyl-tRNA synthetases. Class I synthetases have pockets for the acceptor stem and anticodon of their cognate tRNAs and approach the tRNAs from the D loop and acceptor stem minor groove side. Class II synthetases also have pockets for the acceptor stem and anticodon, but approach their tRNAs from the opposite side, which includes the variable arm and major groove of the acceptor stem.

Proofreading and Editing by Aminoacyl-tRNA Synthetases

As good as aminoacyl-tRNA synthetases are at recognizing the correct (cognate) tRNAs, they have a more difficult job recognizing the cognate amino acids. The reason

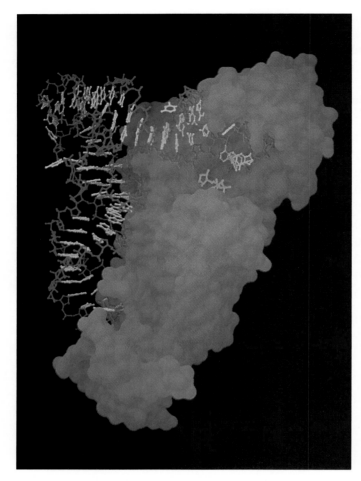

Figure 19.39 Three-dimensional structure of glutaminyl-tRNA synthetase complexed with tRNA and ATP. The synthetase is shown in blue, the tRNA in red and yellow, and the ATP in green. Note the three areas of contact between enzyme and tRNA: (1) the deep cleft at top that holds the acceptor stem of the tRNA, and the ATP; (2) the smaller pocket at lower left into which the tRNA's anticodon protrudes; and (3) the area in between these two clefts, which contacts much of the inside of the L of the tRNA. (*Source:* Courtesy T. A. Steitz; from Rould, Perona, Vogt, and Steitz, *Science* 246 (1 Dec 1989) cover. © AAAS.)

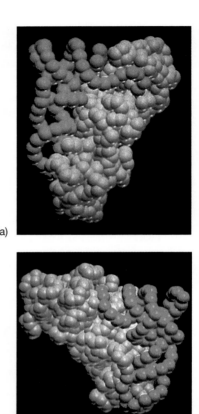

Figure 19.40 Models of (a) a class I complex: GlnRS-tRNA^{Gln}, and (b) a class II complex: yeast AspRS-tRNA^{Asp}. For simplicity, only the phosphate backbones of the tRNAs (red) and the α-carbon backbones of the synthetases (blue) are shown. Notice the approach of the two synthetases to the opposite sides of their cognate tRNAs. (*Source:* Ruff et al, Class II aminoacyl transfer RNA synthetases: Crystal structure of yeast aspartyl-tRNA synthetase complexed with tRNA^{Asp}. *Science* 252 (21 June 1991) f. 3, p. 1686. © AAAS.)

is clear: tRNAs are large, complex molecules that vary from one another in nucleotide sequence and in nucleoside modifications, but amino acids are simple molecules that resemble one another fairly closely—sometimes very closely. Consider isoleucine and valine, for example. The two amino acids are identical except for an extra methylene (CH_2) group in isoleucine. In 1958, Linus Pauling used thermodynamic considerations to calculate that isoleucyl-tRNA synthetase (IleRS) should make about one-fifth as much incorrect Val-tRNA^{Ile} as correct Ile-tRNA^{Ile}. In fact, only one in 150 amino acids activated by IleRS is valine, and only one in 3000 aminoacyl-tRNAs produced by this enzyme is Val-tRNA^{Ile}. How does isoleucyl-tRNA synthetase prevent formation of Val-tRNA^{Ile}?

As first proposed by Alan Fersht in 1977, the enzyme uses a **double-sieve** mechanism to avoid producing tRNAs with the wrong amino acid attached. Figure 19.41 illustrates this concept. The first sieve is accomplished by the **activation site** of the enzyme, which rejects substrates that are too large. However, substrates such as valine that are too small can fit into the activation site and so get activated to the aminoacyl adenylate form and sometimes make it all the way to the aminoacyl-tRNA form. That is where the second sieve comes into play. Activated amino acids or, less commonly, aminoacyl-tRNAs that are too small are hydrolyzed by another site on the enzyme: the **editing site**.

For example, IleRS uses the first sieve to exclude amino acids that are too large, or the wrong shape. Thus, the enzyme excludes phenylalanine because it is too large and leucine because it is the wrong shape. (One of the

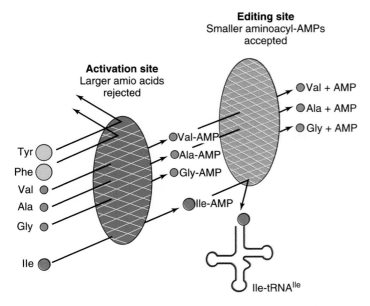

Editing site
Smaller aminoacyl-AMPs
accepted

Activation site
Larger amio acids
rejected

Figure 19.41 The double sieve of isoleucine-tRNA synthetase.
The activation site is the coarse sieve in which large amino acids, such as Tyr and Phe, are excluded because they don't fit. The editing (hydrolytic) site is the fine sieve, which accepts activated amino acids smaller than Ile-AMP, such as Val-AMP, Ala-AMP, and Gly-AMP, but rejects Ile-AMP because it is too large. As a result, the smaller activated amino acids are hydrolyzed to AMP and amino acids, whereas Ile-AMP is converted to Ile-tRNA^{Ile}. (*Source:* Reprinted with permission from Fersht, *Science* 280:541, 1998. Copyright © 1998 American Association for the Advancement of Science.)

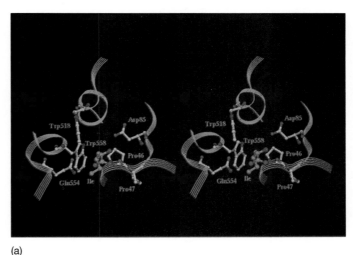

(a)

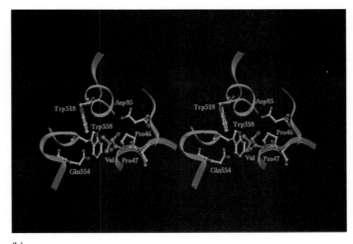

(b)

Figure 19.42 Stereo views of isoleucine and valine in the activation site of IleRS. The backbone of the enzyme is represented by turquoise ribbons, with the carbons of amino acid side chains in yellow. The carbons of the substrates (isoleucine, panel **a**; valine, panel **b**) are rendered in green. Oxygens of all amino acids are in red and nitrogens are in blue. Note that both isoleucine and valine fit into the activation site. (*Source:* Nureki, O. et al. Enzyme structure with two catalytic sites for double-sieve selection of substrate. *Science* 280 (24 Apr 1998) f. 2, p. 579. © AAAS.)

terminal methyl groups of leucine cannot fit into the activation site.) But what about smaller amino acids such as valine? In fact, they do fit into the activation site of IleRS, and so they become activated. But then they are transported to the editing site, where they are recognized as incorrect and degraded. This second sieve is called either **proofreading** or **editing**.

Shigeyuki Yokoyama and colleagues have obtained the crystal structure of the *Thermus thermophilus* IleRS alone, coupled to its cognate amino acid, isoleucine, and to the noncognate amino acid valine. These structures have amply verified Fersht's elegant hypothesis. Figure 19.42 shows the structure of the activation site, with either (a) isoleucine, or (b) valine bound. We can see that both amino acids fit well into this site, although valine makes slightly weaker contact with two of the hydrophobic amino acid side chains (Pro46 and Trp558) that surround the site. On the other hand, it is clear that this site is too small to admit large amino acids such as phenylalanine, and even leucine would be sterically hindered from binding by one of its two terminal methyl groups. This picture is fully consistent with the coarse sieve part of the double sieve hypothesis.

The enzyme has a second deep cleft comparable in size to the cleft of the activation site, but 34 Å away. This sec-

ond cleft is thought to be the editing site, based in part on the fact that a fragment of the enzyme containing this cleft still retains editing activity. The crystal structure confirms this hypothesis: When Yokoyama and colleagues prepared crystals of the IleRS with valine, they found a molecule of valine at the bottom of the deep cleft. However, when they prepared crystals with isoleucine, no amino acid was found in the cleft. Thus, because the cleft seems to be specific for valine, it appears to be the editing site. Furthermore, inspection of the pocket in which valine is found, shows that the space in between the side chains of Trp232 and Tyr386 is just big enough to accommodate valine (Figure 19.43), but too small to admit isoleucine.

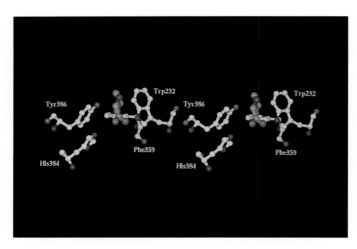

Figure 19.43   Stereo view of valine bound to the editing site of IleRS. Colors are as in Figure 19.42. (*Source:* Nureki, O. et al. Enzyme structure with two catalytic sites for double-sieve selection of substrate. *Science* 280 (24 Apr 1998) f. 2, p. 579. © AAAS.)

If this really is the editing site, we would expect that its removal would abolish editing. Indeed, when Yokoyama and colleagues removed 47 amino acids from this region, including Trp232, they abolished editing activity while retaining full activation activity. Thus, the second cleft really does appear to be the editing site. Several amino acid side chains are particularly close to the valine in the cleft, and Thr230 and Asn237 are well-positioned to take part in the hydrolysis reaction that is the essence of editing. To test this hypothesis, Yokoyama and coworkers changed the amino acids in the *E. coli* IleRS (Thr243 and Asn250) that correspond to Thr230 and Asn237 in the *T. thermophilus* enzyme. Sure enough, when they changed these two amino acids to alanine, the enzyme lost its editing activity, but retained its activation activity. All these data are consistent with the hypothesis that the second cleft is the editing site, and that hydrolysis of noncognate aminoacyl-AMPs such as Val-AMP occurs there.

SUMMARY The amino acid selectivity of at least some aminoacyl-tRNA synthetases is controlled by a double-sieve mechanism. The first sieve is a coarse one that excludes amino acids that are too big. The enzyme accomplishes this task with an active site for activation of amino acids that is just big enough to accommodate the cognate amino acid, but not larger amino acids. The second sieve is a fine one that degrades aminoacyl-AMPs that are too small. The enzyme accomplishes this task with a second active site (the editing site) that admits small aminoacyl-AMPs and hydrolyzes them. The cognate aminoacyl-AMP is too big to fit into the editing site, so it escapes being hydrolyzed. Instead, the enzyme transfers the activated amino acid to its cognate tRNA.

SUMMARY

The *E. coli* 70S ribosome is composed of two parts, a 50S and a 30S subunit. Electron microscopy has revealed the overall shapes and several structural features of these two subunits (the head, platform, and cleft of the 30S particle, and the ridge, valley, central protuberance, and stalk of the 50S particle). EM work has also revealed how the two subunits fit together to make the whole ribosome. Cryoelectron microscopy has elucidated more details of the ribosomal structure, especially a canyon at the interface between the two ribosomal particles. X-ray crystallography studies have shown that the tRNAs occupy the canyon between the two subunits. They interact with the 30S subunit through their anticodon ends, and with the 50S subunit through their acceptor stems. Eukaryotic cytoplasmic ribosomes are larger and more complex than their prokaryotic counterparts, but eukaryotic organellar ribosomes are smaller than prokaryotic ones.

The *E. coli* 30S subunit contains a 16S rRNA and 21 proteins (S1–S21). The 50S subunit contains a 5S rRNA, a 23S rRNA, and 34 proteins (L1–L34). Eukaryotic cytoplasmic ribosomes are larger and contain more RNAs and proteins than their prokaryotic counterparts. Assembly of the 30S ribosomal subunit in vitro begins with 16S rRNA. Proteins then bind sequentially and cooperatively, with proteins joining early in the process helping later proteins to bind to the growing particle.

Immunoelectron microscopy and neutron diffraction, among other methods, have given us a model of the arrangement of proteins in the *E. coli* 30S ribosomal subunit. Sequence studies of 16S rRNA led to a proposal for the secondary structure (intramolecular base pairing) of this molecule. X-ray crystallography studies have confirmed and extended most of the conclusions of these studies. They show a 30S subunit with an extensively base-paired 16S rRNA whose shape essentially outlines that of the whole particle. The x-ray crystallography studies have also confirmed the locations of most of the 30S ribosomal proteins. However, S20 is located near the bottom of the body of the subunit, far from the spot assigned to it by neutron diffraction analysis.

The 30S ribosomal subunit plays two roles. It facilitates proper decoding between codons and aminoacyl-tRNA anticodons, including proofreading. It also participates in translocation. Crystal structures of the 30S subunit with three antibiotics that interfere with these two roles shed light on translocation and decoding. Spectinomycin binds to the 30S subunit near the neck, where it can interfere with the movement of the head that is required for translocation. Streptomycin binds near the A site of the 30S subunit and stabilizes the *ram* state of the ribosome. This reduces fidelity of translation by allowing noncognate aminoacyl-tRNAs to bind relatively easily to the A site and by preventing the shift to the restrictive state that is necessary for proofreading. Paromomycin binds in the minor groove of the 16S rRNA H44 helix near the A site. This flips out A1492 and A1493, so they can presumably stabilize base-pairing between codon and anticodon, including anticodons on noncognate aminoacyl-tRNAs, so fidelity declines.

The crystal structure of the 50S ribosomal subunit has been determined to 2.4 Å resolution. This structure reveals relatively few proteins at the interface between ribosomal subunits, and no protein within 18 Å of the peptidyl transferase active center tagged with a transition state analog. A specific base, A2486, may serve as an acid–base catalyst in the peptidyl transferase reaction. The exit tunnel through the 50S subunit is just wide enough to allow a protein α-helix to pass through. Its walls are made of RNA, whose hydrophilicity is likely to allow exposed hydrophobic side chains of the nascent polypeptide to slide through easily.

Most mRNAs are translated by more than one ribosome at a time; the result, a structure in which many ribosomes translate an mRNA in tandem, is called a polysome. In eukaryotes, polysomes are found in the cytoplasm. In prokaryotes, transcription of a gene and translation of the resulting mRNA occur simultaneously. Therefore, many polysomes are found associated with an active gene.

Transfer RNA was discovered as a small RNA species independent of ribosomes that could be charged with an amino acid and could then pass the amino acid to a growing polypeptide. All tRNAs share a common secondary structure represented by a cloverleaf. They have four base-paired stems defining three stem loops (the D loop, anticodon loop, and T loop) and the acceptor stem, to which amino acids are added in the charging step. The tRNAs also share a common three-dimensional shape that resembles an inverted L. This shape maximizes stability by lining up the base pairs in the D stem with those in the anticodon stem, and the base pairs in the T stem with those in the acceptor stem. The anticodon of the tRNA protrudes from the side of the anticodon loop and is twisted into a shape that readily base-pairs with the corresponding codon in mRNA.

The acceptor stem and anticodon are important cues in recognition of a tRNA by its cognate aminoacyl-tRNA synthetase. In certain cases, each of these elements can be the absolute determinant of charging specificity. X-ray crystallography has shown that synthetase–tRNA interactions differ between the two classes of aminoacyl-tRNA synthetases. Class I synthetases have pockets for the acceptor stem and anticodon of their cognate tRNAs and approach the tRNAs from the D loop and acceptor stem minor groove side. Class II synthetases also have pockets for the acceptor stem and anticodon, but approach their tRNAs from the opposite side, which includes the variable arm and major groove of the acceptor stem.

The amino acid selectivity of at least some aminoacyl-tRNA synthetases is controlled by a double-sieve mechanism. The first sieve is a coarse one that excludes amino acids that are too big. The enzyme accomplishes this task with an active site for activation of amino acids that is just big enough to accommodate the cognate amino acid, but not larger amino acids. The second sieve is a fine one that degrades aminoacyl-AMPs that are too small. The enzyme accomplishes this task with a second active site (the editing site) that admits small aminoacyl-AMPs and hydrolyzes them. The cognate aminoacyl-AMP is too big to fit into the editing site, so it escapes being hydrolyzed. Instead, the enzyme transfers the activated amino acid to its cognate tRNA.

REVIEW QUESTIONS

1. Draw rough sketches of the *E. coli* 30S and 50S ribosomal particles and show how they fit together to form a 70S ribosome. Point out the major structural features of the ribosome.

2. Describe the process of two-dimensional gel electrophoresis described in this chapter. In what way is two-dimensional superior to one-dimensional electrophoresis?

3. Describe the technique of immunoelectron microscopy and show how it provides information about the locations of specific ribosomal proteins.

4. Outline the technique of neutron diffraction and describe how it can be used to complement immunoelectron microscopy to check the relative positions of ribosomal proteins.

5. How can x-ray diffraction data rule out ribosomal proteins as the active site in peptidyl transferase?

6. Diagram a mechanism by which an adenine in 23S rRNA can participate in the peptidyl transferase reaction. Outline the evidence that implicates this base in the reaction.

7. Present plausible hypotheses to explain how the following antibiotics interfere with translation. Present evidence for each hypothesis.
   a. Spectinomycin
   b. Streptomycin
   c. Paromomycin

8. Draw a diagram of a polysome in which nascent protein chains are visible. Identify the 5′- and 3′-ends of the mRNA and use an arrow to indicate the direction the ribosomes are moving along the mRNA.

9. Describe the experiments that led to the discovery of tRNA.

10. How was the "cloverleaf" secondary structure of tRNA discovered?

11. Draw the cloverleaf tRNA structure and point out the important structural elements.

12. Describe and give the results of an experiment that shows that two tRNAs with complementary anticodons bind more

strongly together than two complementary trinucleotides do. What is the probable basis for this extra attraction?

13. Describe and give the results of an experiment that shows that the ribosome responds to the tRNA part, not the amino acid part, of an aminoacyl-tRNA.

14. Describe and give the results of an experiment that shows that the G3–U70 base pair in a tRNA acceptor stem is a key determinant in the charging of the tRNA with alanine.

15. Present at least one line of evidence for the importance of the anticodon in the recognition of a tRNA by an aminoacyl-tRNA synthetase.

16. Based on x-ray crystallographic studies, what parts of a tRNA are in contact with the cognate aminoacyl-tRNA synthetase?

17. Diagram a double-sieve mechanism that ensures amino acid selectivity in aminoacyl-tRNA synthetases.

18. Outline the evidence for the double sieve in the isoleucine–tRNA synthetase that excludes larger and smaller amino acids.

SUGGESTED READINGS

General References and Reviews

Cech, T.R. 2000. The ribosome is a ribozyme. *Science* 289:878–79.

Fersht, A.R. 1998. Sieves in sequence. *Science* 280:541.

Hill, W.E., P.B. Moore, A. Dahlberg, D. Schlessinger, R.A. Garrett, and J.R. Warner. eds. 1990. *The Ribosome: Structure, Function and Evolution.* Washington, D.C.: American Society for Microbiology.

Noller, H.F. 1990. Structure of rRNA and its functional interactions in translation. In Hill, W.E., et al., eds. *The Ribosome: Structure, Function and Evolution.* Washington, D.C.: American Society for Microbiology, chapter 3, pp. 73–92.

Saks, M.E., J.R. Sampson, and J.N. Abelson. 1994. The transfer RNA identity problem: A search for rules. *Science* 263:191–97.

Waldrop, M.M. 1990. The structure of the "second genetic code." *Science* 246:1122.

Zimmermann, R.A. 1995. Ins and outs of the ribosome. *Nature* 376:391–92.

Research Articles

Ban, N., P. Nissen. J. Hansen, P.B. Moore, and T.A. Steitz. 2000. The complete atomic structure of the large ribosomal subunit at 2.4 Å resolution. *Science* 289:905–20.

Brimacombe, R., B. Greuer, P. Mitchell, M. Osswald, J. Rinke-Appel, D. Schüler, and K. Stade. 1990. Three-dimensional structure and function of *Escherichia coli* 16S and 23S rRNA as studied by cross-linking techniques. In Hill, W.E. et al., eds. *The Ribosome: Structure, Function and Evolution.* Washington, D.C.: American Society for Microbiology, chapter 4, pp. 93–106.

Capel, M.S., H. Kjeldgaard, D.M. Engelman, and P.B. Moore. 1988. Positions of S2, S13, S16, S17, S19, and S21 in the 30S ribosomal subunit of *Escherichia coli. Journal of Molecular Biology* 200:65–87.

Carter, A.P., W.M. Clemons, D.E. Brodersen, R.J. Morgan-Warren, B.T. Wimberly, and V. Ramakrishnan. 2000. Functional insights from the structure of the 30S ribosomal subunit and its interactions with antibiotics. *Nature* 407:340–48.

Cate, J.H., M.M. Yusupov, G.Zh. Yusupova, T.N. Earnest, and H.F. Noller. 1999. X-ray crystal structures of 70S ribosome functional complexes. *Science* 285:2095–2104.

Frank, J., J. Zhu, P. Penczek, Y. Li, S. Srivastava, A. Verschoor, U. Rodermacher, R. Grassucci, R.K. Lata, and R.K. Agrawal. 1995. A model of protein synthesis based on cryo-electron microscopy of the *E. coli* ribosome. *Nature* 376:441–44.

Gutell, R.R. 1985. Comparative anatomy of 16S-like ribosomal RNA. *Progress in Nucleic Acid Research and Molecular Biology* 32:155–216.

Held, W.A., B. Ballou, S. Mizushima, and M. Nomura. 1974. Assembly mapping of 30S ribosomal proteins from *Escherichia coli:* Further studies. *Journal of Biological Chemistry* 249:3103–11.

Hoagland, M.B., M.L. Stephenson, J.F. Scott, L.I. Hecht, and P.C. Zamecnik. 1958. A soluble ribonucleic acid intermediate in protein synthesis. *Journal of Biological Chemistry* 231:241–57.

Holley, R.W., J. Apgar, G.A. Everett, J.T. Madison, M. Marquisee, S.H. Merrill, J.R. Penswick, and A. Zamir. 1965. Structure of a ribonucleic acid. *Science* 147:1462–65.

Kaltschmidt, E. and H.G. Wittmann. 1970. Ribosomal proteins XII: Number of proteins in small and large ribosomal subunits of *Escherichia coli* as determined by two-dimensional gel electrophoresis. *Proceedings of the National Academy of Sciences USA* 67:1276–82.

Kim, S.H., F.L. Suddath, G.J. Quigley, A. McPherson, J.L. Sussman, A.H.J. Wang, N.C. Seeman, and A. Rich. 1974. Three-dimensional tertiary structure of yeast phenylalanine transfer RNA. *Science* 185:435–40.

Lake, J.A. 1976. Ribosome structure determined by electron microscopy of *Escherichia coli* small subunits, large subunits and monomeric ribosomes. *Journal of Molecular Biology* 105:131–59.

Lake, J.A. and L. Kahan. 1975. Ribosomal proteins S5, S11, S13, and S19 localized by electron microscopy of antibody-labeled subunits. *Journal of Molecular Biology* 99:631–44.

Miller, O., B.A. Hamkalo, and C.A. Thomas, Jr. 1970. Visualization of bacterial genes in action. *Science* 169:392–95.

Mizushima, S. and M. Nomura. 1970. Assembly mapping of 30S ribosomal proteins from *E. coli. Nature* 226:1214–18.

Muth, G.W., L. Ortoleva-Donnelly, and S.A. Strobel. 2000. A single adenosine with a neutral pK_a in the ribosomal peptidyl transferase center. *Science* 289:947–50.

Nissen, P., J. Hansen, N. Ban, P.B. Moore, and T.A. Steitz. The structural basis of ribosome activity in peptide bond synthesis. 2000. *Science* 289:920–30.

Nureki, O., D.G. Vassylyev, M. Tateno, A. Shimada, T. Nakama, S. Fukai, M. Konno, T.L. Henrickson,

P. Schimmel, and S. Yokoyama. 1998. Enzyme structure with two catalytic sites for double-sieve selection of substrates. *Science* 280:578–82.

Pallanck, L. and L.H. Schulman. 1991. Anticodon-dependent aminoacylation on a noncognate tRNA with isoleucine, valine, and phenylalanine in vivo. *Proceedings of the National Academy of Sciences USA* 88:3872–76.

Quigley, G.J. and A. Rich. 1976. Structural domains of transfer RNA molecules. *Science* 194:796–806.

Rould, M.A., J.J. Perona, D. Söll, and T.A. Steitz. 1989. Structure of *E. coli* glutaminyl-tRNA synthetase complexed with tRNA^{Gln} and ATP at 2.8 Å resolution. *Science* 246:1135–42.

Ruff, M., S. Krishnaswamy, M. Boeglin, A. Poterszman, A. Mitschler, A. Podjarny, B. Rees, J.C. Thierry, and D. Moras. 1991. Class II aminoacyl transfer RNA synthetases: Crystal structure of yeast aspartyl-tRNA synthetase complexed with tRNA^{Asp}. *Science* 252:1682–89.

Schluenzen, F., A. Tocilj, R. Zarivach, J. Harms, M. Gluehmann, D. Janell, A. Bashan, H. Bartels, I. Agmon, F. Franceschi, and A. Yonath. 2000. Structure of functionally activated small ribosomal subunit at 3.3 Å resolution. *Cell* 102:615–23.

Schulman, L.H. and H. Pelka. 1983. Anticodon loop size and sequence requirements for recognition of formylmethionine tRNA by methionyl-tRNA synthetase. *Proceedings of the National Academy of Sciences USA* 80:6755–59.

Stern, S., B. Weiser, and H.F. Noller. 1988. Model for the three-dimensional folding of 16S ribosomal RNA. *Journal of Molecular Biology* 204:447–81.

Stöffler-Meilicke, M., and G. Stöffler. 1990. Topography of the ribosomal proteins from *Escherichia coli* within the intact subunits as determined by immunoelectron microscopy. In Hill, W.E., et al., eds. *The Ribosome: Structure, Function and Evolution.* Washington, D.C.: American Society for Microbiology, chapter 7, pp. 123–33.

Wimberly, B.T., D.E. Brodersen, W.M. Clemons Jr., R.J. Morgan-Warren, A.P. Carter, C. Vonrhein, T. Hartsch, and V. Ramakrishnan. 2000. Structure of the 30S ribosomal subunit. *Nature.* 407:327–39.

CHAPTER 20

DNA Replication I: Basic Mechanism and Enzymology

In Chapter 3 we learned that genes have three main activities. One is to carry information, and we have spent most of the intervening chapters examining how cells decode this information through transcription and translation. Another activity of genes is to participate in replication. The next two chapters will examine this process in detail.

Computer generated model of an Okazaki fragment, a short segment of single-stranded DNA synthesized during DNA replication. © Ken Eward/SS/Photo Researchers, Inc.

20.1 General Features of DNA Replication

Let us first consider the general mechanism of DNA replication. The double-helical model for DNA includes the concept that the two strands are complementary. Thus, each strand can in principle serve as the template for making its own partner. As we will see, this semiconservative model for DNA replication is the correct one. In addition, molecular biologists have uncovered the following interesting general features of DNA replication: It is half discontinuous (made in short pieces that are later stitched together); it requires RNA primers; and it is usually bidirectional. Let us look at each of these features in turn.

Semiconservative Replication

The Watson–Crick model for DNA replication (introduced in Chapter 2) assumed that as new strands of DNA are made, they follow the usual base-pairing rules of A with T and G with C. The model also proposed that the two parental strands separate and that each then serves as a template for a new progeny strand. This is called **semiconservative** replication because each daughter duplex has one parental strand and one new strand (Figure 20.1a). In other words, one of the parental strands is "conserved" in each daughter duplex. However, this is not the only possibility. Another potential mechanism (Figure 20.1b) is **conservative** replication, in which the two parental strands stay together and somehow produce another daughter helix with two completely new strands. Yet another possibility is random **dispersive** replication, in which the DNA becomes fragmented so that new and old DNAs coexist in the same strand after replication (Figure 20.1 c).

In 1958, Matthew Meselson and Franklin Stahl performed a classic experiment to distinguish among these three possibilities. They labeled *E. coli* DNA with heavy nitrogen (^{15}N) by growing cells in a medium enriched in this nitrogen isotope. This made the DNA denser than normal. Then they switched the cells to an ordinary medium containing primarily ^{14}N, for various lengths of time. Finally, they subjected the DNA to CsCl gradient ultracentrifugation to determine the density of the DNA. Figure 20.2 depicts the results of a control experiment that shows that ^{15}N- and ^{14}N-DNAs are clearly separated by this method.

What outcomes would we expect after one round of replication according to the three different mechanisms? If replication is conservative, the two dense parental strands will stay together, and another, newly made DNA duplex will appear. Because this second duplex will be made in the presence of light nitrogen, both its strands will be less dense. The dense/dense (D/D) parental duplex and less

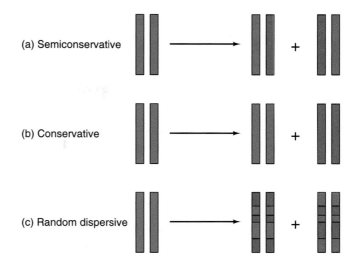

Figure 20.1   Three hypotheses for DNA replication.
(a) Semiconservative replication gives two daughter duplex DNAs, each of which contains one old strand (blue) and one new strand (red). **(b)** Conservative replication yields two daughter duplexes, one of which has two old strands (blue) and one of which has two new strands (red). **(c)** Dispersive replication gives two daughter duplexes, each of which contains strands that are a mixture of old and new.

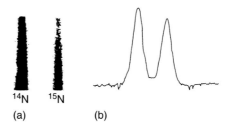

Figure 20.2   Separation of DNAs by cesium chloride density gradient centrifugation. DNA containing the normal isotope of nitrogen (^{14}N) was mixed with DNA labeled with a heavy isotope of nitrogen (^{15}N) and subjected to cesium chloride density gradient centrifugation. The two bands had different densities, so they separated cleanly. **(a)** A photograph of the spinning tube under ultraviolet illumination. The two dark bands correspond to the two different DNAs that absorb ultraviolet light. **(b)** A graph of the darkness of each band, which gives an idea of the relative amounts of the two kinds of DNA. (*Source:* Meselson and Stahl, The replication of DNA in *Escherichia coli. PNAS* 44 (1958) p. 673, f. 2.)

dense/less dense (L/L) progeny duplex will separate readily in the CsCl gradient (Figure 20.3a). On the other hand, if replication is semiconservative, the two dense parental strands will separate and each will be supplied with a new, less dense partner. These D/L hybrid duplexes will have a density halfway between the D/D parental duplexes and L/L ordinary DNA (Figure 20.3b). Figure 20.4 shows that this is exactly what happened; after the first DNA doubling, a single band appeared midway between the labeled D/D DNA and a normal L/L DNA. This ruled out conservative replication, but was still consistent with either semiconservative or random dispersive replication.

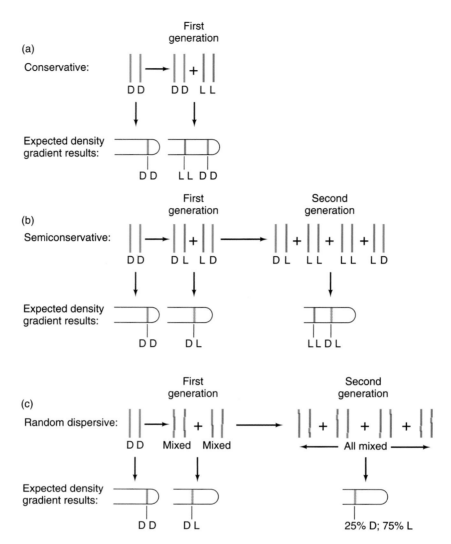

Figure 20.3   Three replication hypotheses. The (**a**) conservative model predicts that after one generation equal amounts of two different DNAs (dense/dense [D/D] and less dense/less dense [L/L]) will occur. Both the (**b**) semiconservative and (**c**) random dispersive models predict a single band of DNA with a density halfway between the D/D and L/L densities. Meselson and Stahl's results confirmed the latter prediction, so the conservative mechanism was ruled out. The random dispersive model predicts that the DNA after the second generation will have a single density, corresponding to molecules that are 25% D and 75% L. This should give one band of DNA halfway between the L/L and the D/L band. The semiconservative model predicts that equal amounts of two different DNAs (L/L and D/L) will be present after the second generation. Again, the latter prediction matched the experimental results, supporting the semiconservative model.

The results of one more round of DNA replication ruled out the random dispersive hypothesis. Random dispersive replication would give a product with one-fourth ^{15}N and three-fourths ^{14}N after two rounds of replication in a ^{14}N medium. Semiconservative replication would yield half of the products as D/L and half as L/L (see Figure 20.3b). In other words, the hybrid D/L products of the first round of replication would each split and be supplied with new, less dense partners, giving the 1:1 ratio of D/L to L/L DNAs. Again, this is precisely what occurred (see Figure 20.4). To make sure that the intermediate-density peak was really a 1:1 mixture of the dense and less dense DNA, Meselson and Stahl mixed pure ^{15}N-labeled DNA with the DNA after 1.9 generations in ^{14}N medium,

then measured the distances among the peaks. The middle peak was centered almost perfectly between the other two (50% ± 2% of the distance between them). Therefore, the data strongly supported the semiconservative mechanism.

Figure 20.3c showed a special case of dispersive replication in which the distribution of old and new DNA in the daughter duplexes is random. It is still possible to imagine a nonrandom dispersive distribution scheme that would yield the same gradient centrifugation result (one-half L/L and one-half D/L, after two generations) as would semiconservative replication. To eliminate the possibility of such dispersive replication, Meselson and Stahl took DNA after one round of replication and heated it to separate its strands. Then they subjected the separated

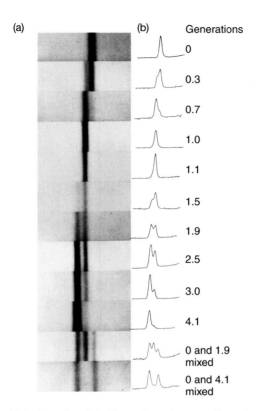

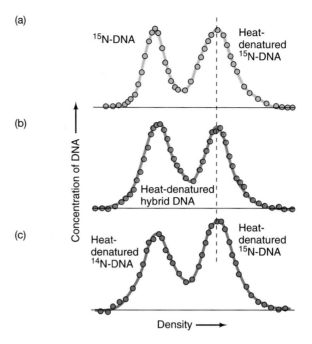

Figure 20.4   Results of CsCl gradient ultracentrifugation experiment to demonstrate semiconservative DNA replication. Meselson and Stahl shifted 15N-labeled *E. coli* cells to a 14N medium for the number of generations given at right, then subjected the bacterial DNA to CsCl gradient ultracentrifugation. (a) Photographs of the spinning centrifuge tubes under ultraviolet illumination. The dark bands correspond to dense DNA (right) and less dense DNA (left). A band of intermediate density was also observed between these two and is virtually the only band observed at 1.0 and 1.1 generations. This band corresponds to duplex DNAs in which one strand is labeled with 15N, and the other with 14N, as predicted by the semiconservative replication model. After 1.9 generations, Meselson and Stahl observed approximately equal quantities of the intermediate band (D/L) and the L/L band. Again, this is what the semiconservative model predicts. After three and four generations, they saw a progressive depletion of the D/L band, and a corresponding increase in the L/L band, again as we expect if replication is semiconservative. (b) Densitometer tracings of the bands in panel (a), which can be used to quantify the amount of DNA in each band. (*Source:* From M. Meselson and F. W. Stahl, "The Replication of DNA in *Escherichia coli,*" *Proceedings of the National Academy of Sciences* 44:675, 1958.)

Figure 20.5   Behaviors of heat-denatured DNAs in ultracentrifugation. Meselson and Stahl subjected the following DNAs to CsCl gradient ultracentrifugation and then plotted the DNA concentration versus density, as in Figure 20.4b. (a) Native and denatured 15N-DNA. (b) Denatured hybrid DNA after growing 15N-labeled *E. coli* cells for one generation in a 14N medium. (c) A mixture of denatured 14N-DNA and denatured 15N-DNA. The two peaks in panel (b) exactly coincided with the two peaks in (c), demonstrating that the hybrid DNA in panel (b) really did consist of one strand each of 14N-DNA and 15N-DNA. This is what the semiconservative model requires, but is incompatible with any dispersive model. (*Source:* From M. Meselson and F. W. Stahl, "The Replication of DNA in *Escherichia coli,*" *Proceedings of the National Academy of Sciences* 44:680, 1958.)

strands to gradient centrifugation. The semiconservative model predicts that half the strands should be as dense as 15N-DNA, and half should be as dense as 14N-DNA. These should form separate bands on centrifugation, coincident with standards prepared from pure denatured 15N- and 14N-DNAs. On the other hand, any dispersive model predicts that at least some strands should be mixed and therefore should not form bands coincident with pure 15N- or 14N-DNA strands. Figure 20.5 shows that the two strands were indeed readily separable and behaved as completely 15N-DNA and completely 14N-DNA strands,

just as the semiconservative replication model demands. The lack of mixed (14N/15N) strands allowed Meselson and Stahl to rule out the dispersive hypothesis.

SUMMARY DNA replicates in a semiconservative manner. When the parental strands separate, each serves as the template for making a new, complementary strand.

Semidiscontinuous Replication

If we were charged with the task of designing a DNA-replicating machine, we might come up with a system such as the one pictured in Figure 20.6a. DNA would unwind to create a fork, and two new DNA strands would be synthesized continuously in the same direction as the moving fork. However, this scheme has a fatal flaw. It demands that the replicating machine be able to make DNA in both the 5′→3′ and 3′→5′ directions. That is because of the antiparallel nature of the two strands of DNA; if one

(a) **Continuous:**

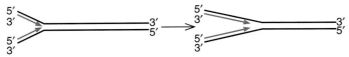

(b) **Semidiscontinuous:**

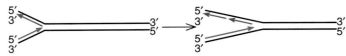

(c) **Discontinuous:**

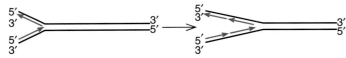

Figure 20.6 Continuous, semidiscontinuous, and discontinuous models of DNA replication. (a) Continuous model. As the replicating fork moves to the right, both strands are replicated continuously in the same direction, left to right (blue arrows). The top strand grows in the 3′→5′ direction, the bottom strand in the 5′→3′ direction. **(b)** Semidiscontinuous model. Synthesis of one of the new strands (the leading strand, bottom) is continuous (blue arrow), as in the model in panel **(a)**; synthesis of the other (the lagging strand, top) is discontinuous (red arrows), with the DNA being made in short pieces. Both strands grow in the 5′→3′ direction. **(c)** Discontinuous model. Both leading and lagging strands are made in short pieces (i.e., discontinuously; red arrows). Both strands grow in the 5′→3′ direction. In real life, DNA is replicated semidiscontinuously, as in model **(b)**.

runs 5′→3′ left to right, the other must run 3′→5′ left to right. But the DNA synthesizing part (**DNA polymerase**) of all natural replicating machines can make DNA in only one direction: 5′→3′. That is, it inserts the 5′-most nucleotide first and extends the chain toward the 3′-end by adding nucleotides to the 3′-end of the growing chain.

Following this line of reasoning, Reiji Okazaki concluded that both strands could not replicate continuously. DNA polymerase could theoretically make one strand (the **leading strand**) **continuously** in the 5′→3′ direction, but the other strand (the **lagging strand**) would have to be made **discontinuously** as shown in Figure 20.6b and c. The discontinuity of synthesis of the lagging strand comes about because its direction of synthesis is opposite to the direction in which the replicating fork is moving. Therefore, as the fork opens up and exposes a new region of DNA to replicate, the lagging strand is growing in the "wrong" direction, away from the fork. The only way to replicate this newly exposed region is to restart DNA synthesis at the fork, behind the piece of DNA that has already been made. This starting and restarting of DNA synthesis occurs over and over again. The short pieces of

DNA thus created would of course have to be joined together somehow to produce the continuous strand that is the final product of DNA replication.

Okazaki's model of semidiscontinuous replication made two predictions that his research team tested experimentally: (1) Because at least half of the newly synthesized DNA appears first as short pieces, one ought to be able to label and catch these before they are stitched together by allowing only very short periods (pulses) of labeling with a radioactive DNA precursor. (2) If one eliminates the enzyme (**DNA ligase**) responsible for stitching together the short pieces of DNA, these short pieces ought to be detectable even with relatively long pulses of DNA precursor.

For his model system, Okazaki chose replication of phage T4 DNA. This had the advantage of simplicity, as well as the availability of T4 ligase mutants. To test the first prediction, Okazaki and colleagues gave shorter and shorter pulses of ³H-labeled thymidine to *E. coli* cells that were replicating T4 DNA. To be sure of catching short pieces of DNA before they could be joined together, they even administered pulses as short as 2 sec. Finally, they measured the approximate sizes of the newly synthesized DNAs by ultracentrifugation.

Figure 20.7a shows the results. Already at 2 sec, some labeled DNA was visible in the gradient; within the limits of detection, it appeared that all of the label was in very small DNA pieces, 1000–2000 nt long, which remained near the top of the centrifuge tube. With increasing pulse time, another peak of labeled DNA appeared much nearer the bottom of the tube. This was the result of attaching the small, newly formed pieces of labeled DNA to much larger, preformed pieces of DNA that were made before labeling began. These large pieces, because they were unlabeled before the experiment began, did not show up until enough time had elapsed for DNA ligase to join the smaller, labeled pieces to them; this took only a few seconds. It is appropriate that the small pieces of DNA that are the initial products of replication have come to be known as **Okazaki fragments.**

The discovery of Okazaki fragments provided evidence for at least partially discontinuous replication of T4 DNA. This hypothesis was supported by the demonstration that these small DNA fragments accumulated to very high levels when the stitching enzyme, DNA ligase, did not operate. Okazaki's group performed this experiment with the T4 mutant containing a defective DNA ligase gene. Figure 20.7b shows that the peak of Okazaki fragments predominated in this mutant. Even after a full minute of labeling, this was still the major species of labeled DNA, suggesting that Okazaki fragments are not just an artifact of very short labeling times.

The predominant accumulation of small pieces of labeled DNA could be interpreted to mean that replication proceeded discontinuously on *both strands,* as pictured in

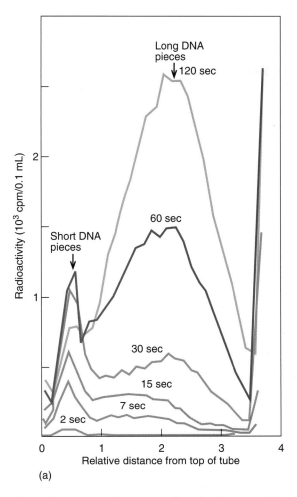

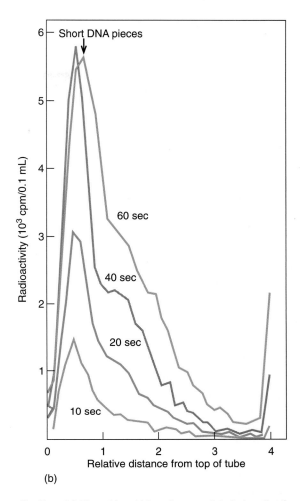

Figure 20.7   Experimental demonstration of at least semidiscontinuous DNA replication. (**a**) Okazaki and his colleagues labeled replicating phage T4 DNA with very short pulses of radioactive DNA precursor and separated the product DNAs according to size by ultracentrifugation. At the shortest times, the label went primarily into short DNA pieces (found near the top of the tube), as the discontinuous model predicted. (**b**) When these workers used a mutant phage with a defective DNA ligase gene, short DNA pieces accumulated even after relatively long labeling times (1 min in the results shown here).

Figure 20.6c. How else can we explain the lack of labeling of large DNA that should occur during continuous replication of the leading strand? The real explanation seems to be that the small DNA pieces are created by a DNA repair system that removes dUMP residues incorporated into DNA. UTP is an essential precursor of RNA, but the cell also makes dUTP, which can be accidentally incorporated into DNA (as dUMP) in place of dTMP. Two enzymes help to minimize this problem. One of these, **dUTPase**—the product of the *dut* gene, degrades dUTP. The other, **uracil N-glycosylase**—the product of the *ung* gene, removes uracil bases from DNA, creating "abasic sites" that are subject to breakage as part of the repair process. Thus, this process ensures that replicating DNA will contain a certain component of short DNA pieces regardless of whether it is made continuously or discontinuously. The question is, what proportion of the Okazaki

fragments observed in experiments such as those depicted in Figure 20.7 are due to discontinuous replication and what proportion to the repair of misincorporated dUMP residues?

To answer this question, Okazaki and colleagues performed their experiments with *ung⁻ E. coli* mutants defective in uracil N-glycosylase. These cells should experience very little DNA breakage due to removal of dUMP residues. However, about 50% of newly labeled DNA in these cells was still in short pieces. This is what we would expect of semidiscontinuous replication. Furthermore, some cell-free replication systems are largely free of uracil *N*-glycosylase, and they show the same behavior: about 50% of newly labeled DNA as Okazaki fragments. Thus, replication really appears to be **semidiscontinuous:** One strand replicates continuously, the other discontinuously.

SUMMARY DNA replication in *E. coli* (and in other organisms) is semidiscontinuous. One strand is replicated continuously in the direction of the movement of the replicating fork; the other is replicated discontinuously as 1–2 kb Okazaki fragments in the opposite direction. This allows both strands to be replicated in the 5′-3′-direction.

Priming of DNA Synthesis

We have seen in previous chapters that RNA polymerase initiates transcription simply by starting a new RNA chain; it puts the first nucleotide in place and then joins the next to it. But DNA polymerases cannot perform the same trick with initiation of DNA synthesis. If we supply a DNA polymerase with all the nucleotides and other small molecules it needs to make DNA, then add either single-stranded or double-stranded DNA with no strand breaks, the polymerase will make no new DNA. What is missing?

We now know that the missing component here is a **primer,** a piece of nucleic acid that the polymerase can "grab onto" and extend by adding nucleotides to its 3′-end. This primer is not DNA, but a short piece of RNA. Figure 20.8 shows a simplified version of this process. First, a replicating fork opens up; next, short RNA primers are made; next, DNA polymerase adds deoxyribonucleotides to these primers, forming DNA, as indicated by the arrows.

The first line of evidence supporting RNA priming was the finding that replication of M13 phage DNA by an *E. coli* extract is inhibited by the antibiotic rifampicin. This was a surprise because rifampicin inhibits *E. coli* RNA polymerase, not DNA polymerase. The explanation is that M13 uses the *E. coli* RNA polymerase to make RNA primers for its DNA synthesis. However, this is not a general phenomenon. Even *E. coli* does not use its own RNA polymerase for priming; it has a special enzyme system for that purpose.

Perhaps the best evidence for RNA priming was the discovery that DNase cannot completely destroy Okazaki fragments. It leaves little pieces of RNA 10–12 bases long. Most of this work was carried out by Tuneko Okazaki, Reiji Okazaki's wife and scientific colleague. She and her coworkers' first estimate of the primer size was too low— only 1–3 nt. Two problems contributed to this underestimation: (1) Nucleases had already reduced the size of the primers by the time they could be purified, and (2) the investigators had no way of distinguishing degraded from intact primers. In a second set of experiments, completed in 1985, Okazaki's group solved both of these problems and found that intact primers are really about 10–12 nt long.

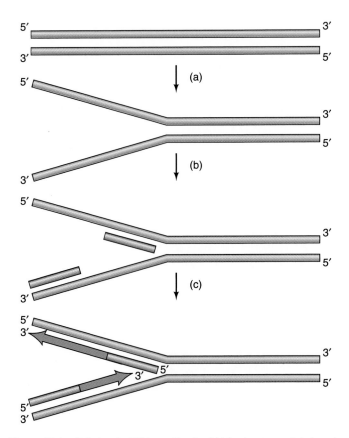

Figure 20.8 **Priming in DNA synthesis.** (a) The two parental strands (blue) separate. (b) Short RNA primers (red) are made. (c) DNA polymerase uses the primers as starting points to synthesize progeny DNA strands (green arrows).

To reduce nuclease activity, these workers used mutant bacteria that lacked ribonuclease H or the nuclease activity of DNA polymerase I, or both. This greatly enhanced the yield of the intact primer. To label only intact primer, they used the capping enzyme, guanylyl transferase, and [α-^{32}P]GTP, to label the 5′-ends of these RNAs. Recall from Chapter 15 that guanylyl transferase adds GMP to RNAs with 5′-terminal phosphates (ideally, a terminal diphosphate). If the primer were degraded at its 5′-end, it would no longer have these phosphates and would therefore not become labeled.

After radiolabeling the primers in this way, these investigators removed the DNA parts of the Okazaki fragments with DNase, then subjected the surviving labeled primers to gel electrophoresis. Figure 20.9 depicts the result. The primers from all the mutant bacteria produced clearly visible bands that corresponded to an RNA with a length of 11 ±1 nt. The wild-type bacteria did not yield a detectable band; nucleases had apparently degraded most or all of their intact primers. Further experiments actually resolved the broad band in Figure 20.9 into three discrete bands with lengths of 10, 11, and 12 nt.

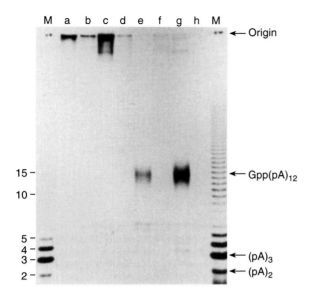

Figure 20.9 Finding and measuring RNA primers. Tuneko Okazaki isolated Okazaki fragments from wild-type and mutant *E. coli* cells lacking one or both of the nucleases that degrade RNA primers. Next, they labeled the intact primers on the Okazaki fragments with [^{32}P]GTP and a capping enzyme. They destroyed the DNA in the fragments with DNase, leaving only the labeled primers. They subjected these primers to electrophoresis and detected their positions by autoradiography. Lanes M are markers. Lanes a–d, before DNase digestion; lanes e–h, after digestion. Lanes a and e, cells were defective in RNase H; lanes b and f, cells were defective in the nuclease activity of DNA polymerase I; lanes c and g, cells were defective in both RNase H and the nuclease activity of DNA polymerase I; lanes d and h, cells were wild-type. The best yield of primers occurred when both nucleases were defective (lane g), and the primers in all cases were 12 ± 1 nt long. The position of the 13-mer Gpp(pA)$_{12}$ marker is indicated at right. (*Source:* Kitani et al., Evidence that discontinuous DNA replication in *Escherichia coli* is primed by approximately 10 to 12 residues of RNA starting with a purine. *J. Mol. Biol.* 184 (1985) p. 49, f. 2, by permission of Academic Press.)

SUMMARY Okazaki fragments in *E. coli* are initiated with RNA primers 10–12 nt long. Intact primers are difficult to detect in wild-type cells because of enzymes that attack RNAs.

Bidirectional Replication

In the early 1960s, John Cairns labeled replicating *E. coli* DNA with a radioactive DNA precursor, then subjected the labeled DNA to autoradiography. Figure 20.10a shows the results, along with Cairns's interpretation, which is explained further in Figure 20.10b. Figures 20.10a and b also attempt to explain the difference in labeling of the three loops (A, B, and C) of the DNA as follows: The DNA underwent one round of replication and part of a second, which means that all the DNA should have at least one labeled strand. During the second round of replica-

tion, the unlabeled parental strand will serve as the template to make a labeled partner; this duplex will therefore become singly labeled. Meanwhile, the labeled parental strand will obtain a labeled partner, making this duplex doubly labeled. This doubly labeled part (loop B) should expose the film more and therefore appear darker than the rest of the DNA (loops A and C).

The structure represented in Figure 20.10a is a so-called theta structure because of its resemblance to the Greek letter θ. Because it may not be immediately obvious that the DNA in Figure 20.10a looks like a theta, Figure 20.10c provides a schematic diagram of the events in the second round of replication that led to the autoradiograph. This drawing shows that DNA replication begins with the creation of a "bubble"—a small region where the parental strands have separated and progeny DNA has been synthesized. As the bubble expands, the replicating DNA begins to take on the theta shape. We can now recognize the autoradiograph as representing a structure shown in the middle of Figure 20.10c, where the crossbar of the theta has grown long enough to extend above the circular part.

The theta structure contains two **forks,** marked X and Y in Figure 20.10. This raises an important question: Does one of these forks, or do both, represent sites of active DNA replication? In other words, is DNA replication **unidirectional,** with one fork moving away from the other, which remains fixed at the origin of replication? Or is it **bidirectional,** with two replicating forks moving in opposite directions away from the origin? Cairns's autoradiographs were not designed to answer this question, but a subsequent study on *Bacillus subtilis* replication performed by Elizabeth Gyurasits and R. B. Wake showed clearly that DNA replication in that bacterium is bidirectional.

These investigators' strategy was to allow *B. subtilis* cells to grow for a short time in the presence of a weakly radioactive DNA precursor, then for a short time with a more strongly radioactive precursor. The labeled precursor was the same in both cases: [^{3}H]thymidine. Tritium (^{3}H) is especially useful for this type of autoradiography because its radioactive emissions are so weak that they do not travel far from their point of origin before they stop in the photographic emulsion and create silver grains. This means that the pattern of silver grains in the autoradiograph will bear a close relationship to the shape of the radioactive DNA. It is important to note that unlabeled DNA does not show up in the autoradiograph. The pulses of label in this experiment were short enough that only the replicating bubbles are visible (Figure 20.11a). You should not mistake these for whole bacterial chromosomes such as in Figure 20.10.

If you look carefully at Figure 20.11a, you will notice that the pattern of silver grains is not uniform. They are concentrated near both forks in the bubble. This extra labeling identifies the regions of DNA that were replicating

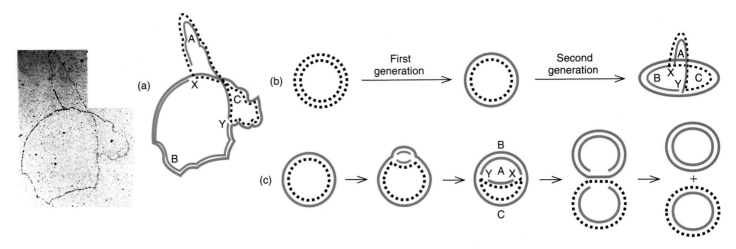

Figure 20.10   The theta mode of DNA replication in *Escherichia coli*. (a) An autoradiograph of replicating *E. coli* DNA with an interpretive diagram. The DNA was allowed to replicate for one whole generation and part of a second in the presence of radioactive nucleotides. The part of the DNA that has replicated only once will have one labeled strand (blue) and one "cold" (unlabeled) strand (dashed, black). The DNA that has replicated twice will have one part that is doubly labeled (loop B, blue and red) and one part with only one labeled strand (loop A, red and dashed black). Note that loop B contains two solid lines, indicating that it is the part of the DNA that has become doubly labeled. The fact that loop B is darker in the autoradiograph than loops A and C fits with this interpretation. (b) Interpretation of Cairns's autoradiograph, using the same dotted and solid line and color conventions as in panel (a). (c) More detailed description of the theta mode of DNA replication. The dotted and solid lines and colors have the same meaning as in panels (a) and (b). (*Source:* (a) Cairns, The chromosome of *Escherichia coli. CSH Symposia on Quantitative Biology* 28 (1963) p. 44.)

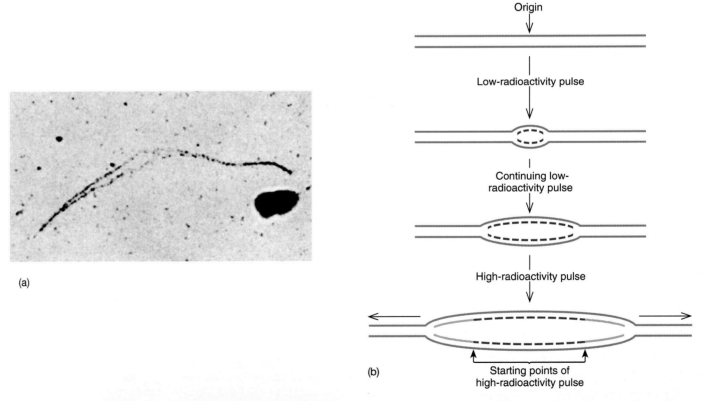

Figure 20.11   Experimental demonstration of bidirectional DNA replication. (a) Autoradiograph of replicating *Bacillus subtilis* DNA. Dormant bacterial spores were germinated in the presence of low-radioactivity DNA precursor, so the newly formed replicating bubbles immediately became slightly labeled. After the bubbles had grown somewhat, a more radioactive DNA precursor was added to label the DNA for a short period. (b) Interpretation of the autoradiograph. The purple color represents the slightly labeled DNA strands produced during the low-radioactivity pulse. The orange color represents the more highly labeled DNA strands produced during the later, high-radioactivity pulse. Because both forks picked up the high-radioactivity label, both must have been functioning during the high-radioactivity pulse. DNA replication in *B. subtilis* is therefore bidirectional. (*Source:* (a) Gyurasits and Lake, *J. Molec. Biol.* 73 (1973) p. 58, by permission of Academic Press.)

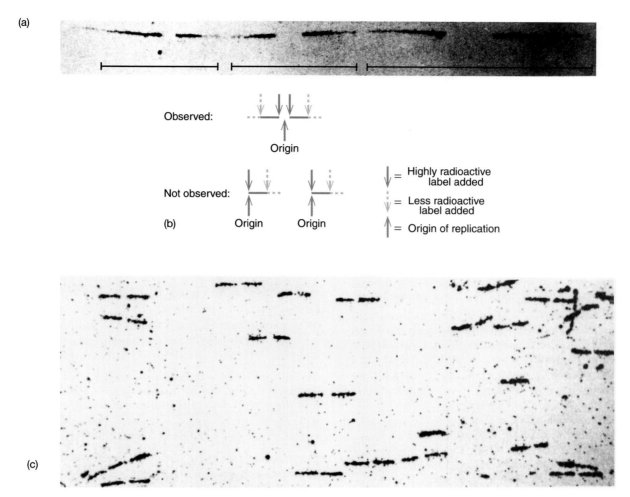

Figure 20.12   Bidirectional DNA replication in eukaryotes. (a) Autoradiograph of replicating *Drosophila melanogaster* DNA, pulse-labeled first with high-radioactivity DNA precursor, then with low. Note the pairs of streaks (denoted by brackets) tapering away from the middle. This reflects the pattern of labeling of a replicon with a central origin and two replicating forks. (b) Idealized diagram showing the patterns observed with high-, then low-radioactivity labeling, and the pattern expected if the pairs of streaks represent two independent unidirectional replicons whose replicating forks move in the same direction. The latter pattern was not observed. (c) Autoradiograph of replicating embryonic *Triturus vulgaris* DNA. Note the constant size and shape of the pairs of streaks, suggesting that all the corresponding replicons began replicating at the same time.
(*Source:* (a) Huberman and Tsai, *J. Molec. Biol.* 75 (1973) p. 8, by permission of Academic Press. (c) Callan, DNA replication in the chromosomes of eukaryotes. *Cold Spring Harbor Symposia on Quantitative Biology* 38 (1973) f. 4c, p. 195.)

during the "hot," or high-radioactivity, pulse period. Both forks incorporated extra label, showing that they were both active during the hot pulse. Therefore, DNA replication in *B. subtilis* is bidirectional; two forks arise at a fixed starting point—the **origin of replication**—and move in opposite directions around the circle until they meet on the other side. Later experiments employing this and other techniques have shown that the *E. coli* chromosome also replicates bidirectionally.

J. Huberman and A. Tsai have performed the same kind of autoradiography experiments in a eukaryote, the fruit fly *Drosophila melanogaster*. Here, the experimenters gave a pulse of strongly radioactive (high specific activity) DNA precursor, followed by a pulse of weakly radioactive (low specific activity) precursor. Alternatively, they reversed the procedure and gave the low specific ac-

tivity label first, followed by the high. Then they autoradiographed the labeled insect DNA. The spreading of DNA in these experiments did not allow the replicating bubbles to remain open; instead, they collapsed and appear on the autoradiographs as simple streaks of silver grains.

One end of a streak marks where labeling began; the other shows where it ended. But the point of this experiment is that the streaks always appear in pairs (Figure 20.12a). The pairs of streaks represent the two replicating forks that have moved apart from a common starting point. Why doesn't the labeling start in the middle, at the origin of replication, the way it did in the experiment with *B. subtilis* DNA? In the *B. subtilis* experiment, the investigators were able to synchronize their cells by allowing them to germinate from spores, all starting at the same time. That way they could get label into the cells before any of them had started making DNA (i.e., before germination). Such synchronization was not tried in the *Drosophila* experiments, where it would have been much more difficult. As a result, replication usually began before the label was added, so a blank area arises in the middle where replication was occurring but no label could be incorporated.

Notice the shape of the pairs of streaks in Figure 20.12a. They taper to a point, moving outward, rather like an old-fashioned waxed mustache. That means the DNA incorporated highly radioactive label first, then more weakly radioactive label, leading to a tapering off of radioactivity as we move outward in both directions from the origin of replication. The opposite experiment—"cooler" label first, followed by "hotter" label—would give a reverse mustache, with points on the inside. It is possible, of course, that closely spaced, independent origins of replication gave rise to these pairs of streaks. But we would not expect that such origins would always give replication in opposite directions. Surely some would lead to replication in the same direction, producing asymmetric autoradiographs such as the hypothetical one in Figure 20.12b. But these were not seen. Thus, these autoradiography experiments confirm that each pair of streaks we see really represents one origin of replication, rather than two that are close together. It therefore appears that replication of *Drosophila* DNA is bidirectional.

These experiments were done with *Drosophila* cells originally derived from mature fruit flies and then cultured in vitro. H. G. Callan and his colleagues performed the same type of experiment using highly radioactive label and embryonic amphibian cells. These experiments (with embryonic cells of the newt) gave the striking results shown in Figure 20.12c. In contrast to the pattern in adult insect cells, the pairs of streaks here are all the same. They are all approximately the same length and they all have the same size space in the middle. This tells us that replication at all these origins began simultaneously. This must be so, because the addition of label caught all the forks at the same point—the same distance away from their respective origins of replication. This phenomenon probably helps explain how embryonic newt cells complete their DNA replication so rapidly (in as little as an hour, compared to 40 h in adult cells). Replication at all origins begins simultaneously, rather than in a staggered fashion.

This discussion of origins of replication helps us define an important term: **replicon**. The DNA under the control of one origin of replication is called a replicon, even though it may require two replicating forks. The *E. coli* chromosome is a single replicon because it replicates from a single origin. Obviously, eukaryotic chromosomes have many replicons; otherwise, it would take far too long to replicate a whole chromosome. This would be true even at the breakneck speed of DNA synthesis in prokaryotes—up to a thousand nucleotides per second. In fact, eukaryotic DNA replicates much more slowly than this, which makes the need for multiple replicons even more obvious.

Unidirectional Replication

Do all genetic systems replicate bidirectionally? Apparently not. For example, consider the replication of a plasmid called **colE1**. A plasmid is a circular piece of DNA that replicates independently of the cell's chromosome (Chapter 4); some strains of *E. coli* harbor colE1, which explains how this plasmid got the "col" part of its name. Michael Lovett used an electron microscope to examine replicating molecules of colE1 and found that only one fork moves.

Of course, if the experimenters had simply looked at circular replicating molecules, they would have seen a replication bubble with two forks, but there would have been no way of knowing whether replication was occurring at both forks. They needed some kind of marker in the DNA to locate the two forks at different times during replication. Then, if both forks moved relative to the marker, they would know that both forks are active and DNA replication is bidirectional. On the other hand, if only one fork moves, replication must be unidirectional.

The marker Lovett and colleagues chose is the site at which colE1 DNA is cut by the restriction endonuclease *Eco*RI. As explained in Chapter 4, *Eco*RI recognizes the sequence 5′-GAATTC-3′. *Eco*RI has only one such recognition site in the whole colE1 molecule, so cutting colE1 DNA with *Eco*RI creates linear molecules whose ends are always at the same site: the *Eco*RI restriction site. Figure 20.13

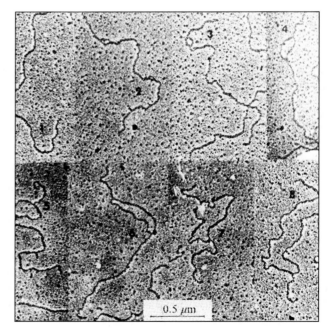

0.5 μm

Figure 20.13   Unidirectional colicin E1 DNA replication.
Molecule 1 has not begun to replicate, so no replicating bubble is seen. Molecules 2–8 show replicating bubbles of progressively increasing size. It is important to notice that the unreplicated stretch of DNA above the bubble in each picture remains the same, whereas that below the bubble grows shorter as the bubble grows larger. Clearly, only the lower fork is replicating; the other is stationary.
(*Source*: Lovett et al., Unidirectional replication of plasmid ColE1 DNA. *Nature* 251 (27 Sept 1974) p. 338. Macmillan Magazines Ltd.)

shows the results of the experiment. Clearly, only one fork moves relative to the ends of the linearized DNA molecule. Therefore, replication of colE1 DNA is unidirectional.

SUMMARY Most eukaryotic and prokaryotic DNAs replicate bidirectionally. ColE1 is an example of a DNA that replicates unidirectionally.

Rolling Circle Replication

Certain circular DNAs replicate, not by the θ mode we have already discussed, but by a mechanism called **rolling circle** replication. The *E. coli* phages with single-stranded circular DNA genomes, such as φX174, use a relatively simple form of rolling circle replication in which a double-stranded **replicative form** (RFI) gives rise to many copies of a single-stranded progeny DNA, as illustrated in simplified form in Figure 20.14. The intermediates (steps b and c in Figure 20.14) give this mechanism the rolling

circle name because the double-stranded part of the replicating DNA can be considered to be rolling counterclockwise and trailing out the progeny single-stranded DNA, rather like a roll of toilet paper unrolling as it speeds across the floor. This intermediate also somewhat resembles the lowercase Greek letter σ (sigma), so this mechanism is sometimes called the σ mode, to distinguish it from the θ mode.

The rolling circle mechanism is not confined to production of single-stranded DNA. Some phages (e.g., λ) use this mechanism to replicate double-stranded DNA. During the early phase of θ DNA replication, the phage follows the θ mode of replication to produce several copies of circular DNA. These circular DNAs are not packaged into phage particles; they serve as templates for rolling circle synthesis of linear λ DNA molecules that *are* packaged. Figure 20.15 shows how this rolling circle operates. Here, the replicating fork looks much more like that in *E. coli* DNA replication, with continuous synthesis on the leading strand (the one going around the circle) and discontinuous synthesis on the lagging strand. In λ, the progeny DNA reaches lengths that are several genomes long before it is packaged. The multiple-length DNAs are called **concatemers**. The packaging mechanism is designed to accept only one genome's worth of linear DNA into each phage head (the "head-full model"), so the concatemer must be cut enzymatically during packaging.

SUMMARY Circular DNAs can replicate by a rolling circle mechanism. One strand of a double-stranded DNA is nicked and the 3′-end is extended, using the intact DNA strand as template. This displaces the 5′-end. In phage φX174 replication, when one round of replication is complete, a full-length, single-stranded circle of DNA is released. In phage λ, the displaced strand serves as the template for discontinuous, lagging strand synthesis.

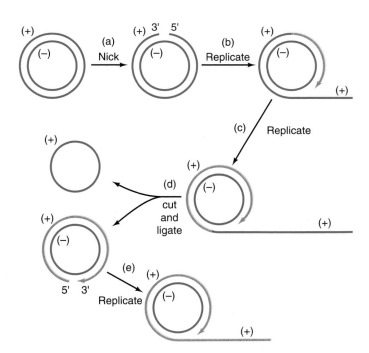

Figure 20.14 Schematic representation of rolling circle replication that produces single-stranded circular progeny DNAs. (**a**) An endonuclease creates a nick in the positive strand of the double-stranded replicative form. (**b**) The free 3′-end created by the nick serves as the primer for positive strand elongation, as the other end of the positive strand is displaced. The negative strand is the template. (**c**) Further replication occurs, as the positive strand approaches double length. The circle can be considered to be rolling counterclockwise. (**d**) The unit length of positive-strand DNA that has been displaced is cleaved off by an endonuclease. (**e**) Replication continues, producing another new positive strand, using the negative strand as template. This process repeats over and over to yield many copies of the circular positive strand.

Figure 20.15 Rolling circle model for phage λ DNA replication. As the circle rolls to the right, the leading strand (red) elongates continuously. The lagging strand (blue) elongates discontinuously, using the unrolled leading strand as a template. The progeny double-stranded DNA thus produced grows to many genomes in length (a concatemer) before one genome's worth is clipped off and packaged into a phage head.

20.2 Enzymology of DNA Replication

Over 30 different polypeptides cooperate in replicating the *E. coli* DNA. Let us begin by examining the activities of some of these proteins and their homologs in other organisms.

Strand Separation

In our discussion of the general features of DNA replication, we have been assuming that the two DNA strands at the fork somehow unwind. This does not happen automatically as DNA polymerase does its job; the two parental strands hold tightly to each other, and it takes energy and enzyme action to separate them.

Helicase   The enzyme that harnesses the chemical energy of ATP to separate the two parental DNA strands at the replicating fork is called a **helicase**. We have already seen an example of helicase action in Chapter 17, in our discussion of initiation of translation in eukaryotes. But that was an RNA helicase, which melted RNA hairpins in the leader regions of mRNAs, not a DNA helicase, which is needed to separate the DNA strands in advance of a replicating fork.

At least four DNA helicases have been identified in *E. coli* cells. The problem is finding which of these is involved in DNA replication. The first three to be investigated—the *rep* helicase, and DNA helicases II and III—could be mutated without inhibiting cellular multiplication. This made it unlikely that any of these three enzymes could participate in something as vital to cell survival as DNA replication; we would anticipate that defects in the helicase that participates in DNA replication would be lethal.

One way to generate mutants with defects in essential genes is to make the mutations conditional, usually temperature-sensitive. That way, one can grow the mutant cells at a low temperature at which the mutation is not expressed, then shift the temperature up to observe the mutant phenotype. As early as 1968, François Jacob and his colleagues discovered two classes of temperature-sensitive mutants in *E. coli* DNA replication (Figure 20.16). Type 1 mutants showed an immediate shut-off of DNA synthesis on raising the temperature from 30° C to 40° C, whereas type 2 mutants showed only a gradual decrease in the rate of DNA synthesis at elevated temperature.

One of the type 1 mutants was the *dnaB* mutant; DNA synthesis in *E. coli* cells carrying temperature-sensitive mutations in the *dnaB* gene stopped short as soon as the temperature rose to the nonpermissive level. This is what we would expect if *dnaB* encodes the DNA helicase required for replication. Without a functional helicase, the fork cannot move, and DNA synthesis must halt immedi-

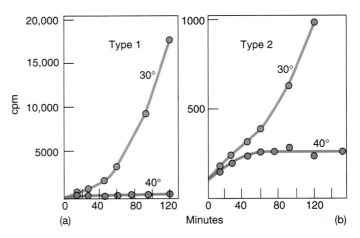

Figure 20.16   Two classes of *E. coli* DNA replication mutants. Jacob and colleagues tested mutants that were temperature-sensitive for DNA replication as follows: They shifted half of a culture to the nonpermissive temperature (40° C), and left the other half at the permissive temperature (30° C), then added [¹⁴C]thymidine to label newly synthesized DNA. They distinguished two classes of mutants; **(a)** type 1, in which DNA replication stopped abruptly upon shift to high temperature; and **(b)** type 2, in which DNA replication leveled off more gradually. (*Source:* From Y. Hirota, A. Ryter, and F. Jacob, "Thermosensitive Mutants of *E. coli* Affected in the Processes of DNA Synthesis and Cellular Division," *Cold Spring Harbor Symp. Quant. Biology* 33:678, 1968. Copyright © 1968 Cold Spring Harbor Laboratory Press, Cold Spring Harbor, NY. Reprinted by permission.)

ately. Furthermore, the *dnaB* product (DnaB) was known to be an ATPase, which we also expect of a DNA helicase, and the dnaB protein was found associated with the primase, which makes primers for DNA replication.

All of these findings suggested that DnaB is the DNA helicase that unwinds the DNA double helix during *E. coli* DNA replication. All that remained was to show that DnaB has DNA helicase activity. Jonathan LeBowitz and Roger McMacken did this in 1986. They used the helicase substrate shown in Figure 20.17a, which is a circular M13 phage DNA, annealed to a shorter piece of linear DNA, which was labeled at its 5′-end. Figure 20.17b shows how the helicase assay worked. LeBowitz and McMacken incubated the labeled substrate with DnaB, or other proteins, and then electrophoresed the products. If the protein had helicase activity, it would unwind the double-helical DNA and separate the two strands. Then the short, labeled DNA would migrate independently of the larger, unlabeled DNA, and would have a much higher electrophoretic mobility.

Figure 20.18 shows the results of the assay. DnaB alone had helicase activity, and this was stimulated by DnaG (which we will see in Chapter 21 is a primase), and by SSB, a single-stranded DNA-binding protein that we will introduce next. Neither DnaG nor SSB, by themselves or together, had any DNA helicase activity. Thus, DnaB is the helicase that unwinds the DNA at the replicating fork.

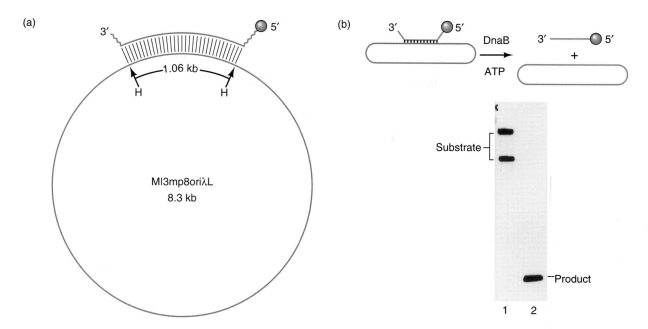

Figure 20.17 DNA helicase assay. (a) Substrate. LeBowitz and McMacken made a helicase substrate by ^{32}P-labeling a single-stranded 1.06-kb *Hinc*II DNA fragment (red) at its 5′-end and annealing the fragment to an unlabeled single-stranded recombinant M13 DNA bearing a complementary 1.06-kb region, bounded by two *Hinc*II sites (H). **(b)** Assay. Top: The dnaB protein, or any DNA helicase, can unwind the double-stranded region of the substrate and liberate the labeled short piece of DNA (red) from its longer, circular partner. Bottom: Electrophoresis of the substrate (lane 1) yields two bands, which probably correspond to linear and circular versions of the long DNA annealed to the labeled, short DNA. Electrophoresis of the short DNA by itself (lane 2) shows that it has a much higher mobility than the substrate (see band labeled "product"). (*Source:* LeBowitz & McMacken, The *Escherichia coli* dnaB replication protein is a DNA helicase. *J. Biol. Chem.* 261 (5 April 1986) figs. 2, 3, pp. 4740–41. American Society for Biochemistry and Molecular Biology.)

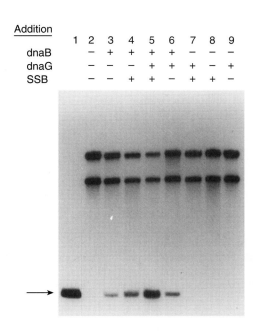

Figure 20.18 Helicase assay results. LeBowitz and McMacken performed the assay outlined in Figure 20.17 with the additions (DnaB, DnaG, and SSB) indicated at top. The electrophoresis results are given at bottom. Lane 1 is a control with the unannealed, labeled short DNA to show its electrophoretic behavior (arrow). Lane 3 shows that DnaB has helicase activity on its own, but lanes 4 and 5 demonstrate that the other proteins stimulate this activity. On the other hand, lanes 7–9 show that the other two proteins have no helicase activity without DnaB. (*Source:* LeBowitz & McMacken, The *Escherichia coli* dnaB replication protein is a DNA helicase. *J. Biol. Chem.* 261 (5 April 1986) figs. 2, 3, pp. 4740–41. American Society for Biochemistry and Molecular Biology.)

SUMMARY The helicase that unwinds double-stranded DNA at the replicating fork is encoded by the *E. coli dnaB* gene.

Single-Strand DNA-Binding Proteins

As we have just seen, another class of proteins, called **single-strand DNA-binding proteins (SSBs)**, participate in DNA strand separation during replication. These proteins do not catalyze strand separation, as helicases do. Instead, they bind selectively to single-stranded DNA as soon as it forms and coat it so it cannot anneal to re-form a double helix. The single-stranded DNA can form by natural "breathing" (transient local separation of strands, especially in A–T rich regions) or as a result of helicase action, then SSB catches it and keeps it in single-stranded form.

The best studied SSBs are prokaryotic. The *E. coli* protein is called SSB and is the product of the *ssb* gene. The T4 phage protein is **gp32,** which stands for "gene product *32*" (the product of gene *32* of phage T4). (see Table 20.1) The M13 phage protein is **gp5** (the product of the phage gene *5*). All of these proteins act cooperatively: The binding of one protein facilitates the binding of the next. For example, the binding of the first molecule of gp32 to single-stranded DNA raises the affinity for the next molecule a thousandfold. Thus, once the first molecule of gp32 binds, the second binds easily, and so does

Table 20.1   Properties of Three Prokaryotic SSBs

Property	T4 gp32	M13 gp5	E. coli SSB
Binding			
DNA > RNA	Yes	Yes	Yes
ss ≫ double strand	Yes	Yes	Yes
DNA strands in protein complex	One	Two	One
Nucleotides bound per monomer	10	4	8–16
Weight ratio, protein/DNA	12	8	4–8
Biological functions	Stimulates replication and recombination	Inhibits complementary strand synthesis	Stimulates replication and repair
Copies per cell	10,000	75,000	800
Copies per replication fork	170		270
Mass, native (kDa)	33.5	19.4	75.6
Native form	Monomer	Dimer	Tetramer

(*Source:* From *DNA Replication,* 2nd ed. by Kornberg and Baker © 1992 by W.H. Freeman and Company. Used with permission.)

the third, and so forth. This results in a chain of gp32 molecules coating a single-stranded DNA region. The chain will even extend into a double-stranded hairpin, melting it, as long as the free energy released in cooperative gp32 binding through the hairpin exceeds the free energy released by forming the hairpin. In practice, this means that relatively small, or poorly base-paired hairpins will be melted, but long, or well base-paired, ones will remain intact.

The gp32 protein binds to DNA as a chain of monomers, whereas gp5 binds as a string of dimers, and *E. coli* SSB binds as a chain of tetramers. Table 20.1 lists this and several other properties of these three well-studied prokaryotic SSBs. Notice the variation in molecular mass of these three proteins, from a monomer mass of 9.7 kD for gp5 to 33.5 kD for gp32, and the big difference in number of molecules per cell. The large number of molecules of the phage proteins per cell is probably necessitated by the large number of replicating phage DNA molecules, compared with the single replicating *E. coli* DNA molecule.

By now we have had some hints that the name "single-strand DNA-binding protein" is a little misleading. These proteins do indeed bind to single-stranded DNA, but so do many other proteins we have studied in previous chapters, including RNA polymerase. But the SSBs do much more. We have already seen that they trap DNA in single-stranded form; now we will see that they also stimulate their homologous DNA polymerases. Figure 20.19 depicts an experiment in which Kornberg and Bruce Alberts and their colleagues demonstrated that gp32 stimulates the T4 DNA polymerase. Moreover, this stimulation is specific; gp32 did not stimulate phage T7 polymerase or *E. coli* DNA polymerase I.

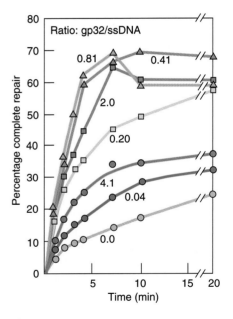

Figure 20.19   Stimulation of DNA synthesis (repair) by the T4 phage SSB, gp32. Kornberg, Alberts, and colleagues prepared a substrate for DNA synthesis during repair by partially (24%) degrading λ DNA with exonuclease III. They mixed this substrate with all four dNTPs, of which dTTP was α-^{32}P-labeled, and with T4 DNA polymerase. They measured incorporation of labeled dTTP into DNA and calculated the percentage complete repair based on the amount of DNA synthesized and the amount of DNA synthesis needed to fill all the gaps created by exonuclease III. They calculated the ratio of gp32 to ss (single-stranded) DNA assuming that one monomer of gp32 binds to 10 nt of ss DNA. The gp32 gave considerable stimulation of DNA synthesis, which peaked at a gp32/ss DNA ratio between 0.41 and 0.81 (purple and orange, respectively). (*Source:* Reprinted from *Journal of Molecular Biology,* vol. 62, J. A. Huberman and A. Kornberg, "Stimulation of T4 Bacteriophage DNA Polymerase by the Protein Product of T4 Gene *32*," 39–52, 1971, by permission of Academic Press Limited, London.)

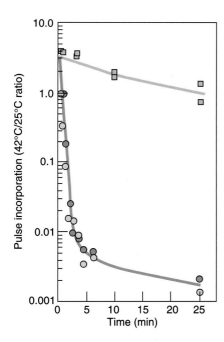

Figure 20.20   Temperature-sensitivity of DNA synthesis in cells infected by T4 phage with a temperature-sensitive mutation in the SSB (gp32) gene. Curtis and Alberts measured the relative incorporation of [³H]thymidine after 1 min pulses at 42° and 25° C in cells infected with the following T4 phage mutants: *am*B17 (gene *23*), blue; *ts*P7(gene *32*)-*am*B17 (gene *23*), red; *ts*P7 (gene *32*) -*am*E727 (gene *49*), yellow. The amber mutations in genes *23* and *49* have no effect on DNA synthesis, but prevent packaging of the T4 DNA into phage heads. Thus, the observed drop in DNA synthesis is due to the *ts* mutation in gene *32*. (*Source:* Reprinted from *Journal of Molecular Biology,* vol. 102, M. J. Curtis and B. Alberts, "Studies on the Structure of Intracellular Bacteriophage T4 DNA," 793–816, 1976, by permission of Academic Press Limited, London.)

Are the activities of the SSBs important? In fact, they are essential. Temperature-sensitive mutations in the *ssb* gene of *E. coli* render the cell inviable at the nonpermissive temperature. In cells infected by the tsP7 mutant of phage T4, with a temperature-sensitive gp32, phage DNA replication stops within 2 min after shifting to the nonpermissive temperature (Figure 20.20). Furthermore, the phage DNA begins to be degraded (Figure 20.21). This behavior suggests that one function of gp32 is to protect from degradation the single-stranded DNA created during phage DNA replication.

Based on the importance of the SSBs in prokaryotes, it is surprising that SSBs with similar importance have not yet been found in eukaryotes. However, a host SSB has been found to be essential for replication of SV40 DNA in human cells. This protein, called **RF-A,** or **human SSB,** binds selectively to single-stranded DNA and stimulates the DNA helicase activity of the viral large T antigen. Because this is a host protein, we assume that it plays a role in the uninfected human cell as well, but we do not know yet what that role is. We also know that virus-encoded

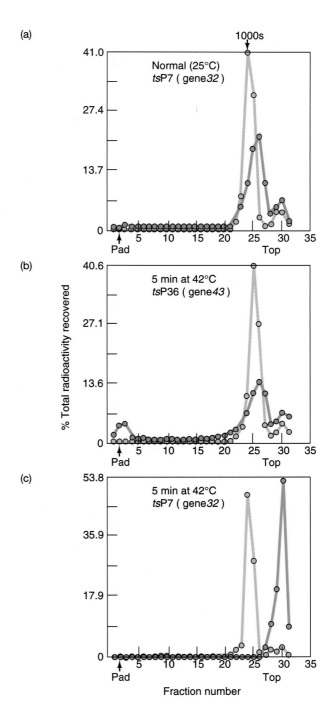

Figure 20.21   Degradation of T4 DNA in cells infected with T4 phage with a temperature-sensitive mutation in gene 32. Curtis and Alberts infected *E. coli* cells with the mutant phages at the temperatures listed in each panel. They also incubated the infected cells with [³H]thymidine to label the newly made T4 DNA. Then they subjected this labeled T4 DNA (red) to ultracentrifugation along with [³²P]labeled intact T4 DNA (blue) as a marker. (**a**) Cells infected with T4 phage with a temperature-sensitive gene *32* at 25° C. (**b**) Cells infected with T4 phage with a temperature-sensitive gene *43* at 42° C. (**c**) Cells infected with T4 phage with a temperature-sensitive gene *32* at 42° C. In panel (**c**) there is clear evidence of degradation of T4 DNA. (*Source:* Reprinted from *Journal of Molecular Biology,* vol. 102, M. J. Curtis and B. Alberts, "Studies on the Structure of Intracellular Bacteriophage T4 DNA," 793–816, 1976, by permission of Academic Press Limited, London.)

SSBs play a major role in replication of certain eukaryotic viral DNAs, including adenovirus and herpesvirus DNAs.

SUMMARY The prokaryotic single-strand DNA-binding proteins bind much more strongly to single-stranded than to double-stranded DNA. They aid helicase action by binding tightly and co-operatively to newly formed single-stranded DNA and keeping it from annealing with its partner. By coating the single-stranded DNA, SSBs also protect it from degradation. They also stimulate their homologous DNA polymerases. These activities make SSBs essential for prokaryotic DNA replication.

Topoisomerases

Sometimes we refer to the separation of DNA strands as "unzipping." We should not forget, when using this term, that DNA is not like a zipper with straight, parallel sides. It is a helix. Therefore, when the two strands of DNA separate, they must rotate around each other. Helicase could handle this task alone if the DNA were linear and unnaturally short, but closed circular DNAs, such as the *E. coli* chromosome, present a special problem. As the DNA unwinds at the replicating fork, a compensating winding up of DNA will occur elsewhere in the circle. This tightening of the helix will create intolerable strain unless it is relieved somehow. Cairns recognized this problem in 1963 when he first observed circular DNA molecules in *E. coli*, and he proposed a "swivel" in the DNA duplex that would allow the DNA strands on either side to rotate to relieve the strain (Figure 20.22). We now know that an enzyme known as DNA gyrase serves the swivel function. **DNA gyrase** belongs to a class of enzymes called **topoisomerases** that introduce transient single- or double-stranded breaks into DNA and thereby allow it to change its form, or topology.

To understand how the topoisomerases work, we need to look more closely at the phenomenon of supercoiled, or superhelical, DNA mentioned in Chapter 6. All naturally occurring, closed circular, double-stranded DNAs studied so far exist as supercoils. Closed circular DNAs are those with no single-strand breaks, or nicks. When a cell makes such a DNA, it causes some unwinding of the double helix; the DNA is then said to be "underwound." As long as both strands are intact, no free rotation can occur around the bonds in either strand's backbone, so the DNA cannot relieve the strain of underwinding except by supercoiling. The supercoils introduced by underwinding are called "negative," by convention. This is the kind of supercoiling found in most organisms; however, positive supercoils do exist in extreme thermophiles, which have a reverse DNA gyrase that introduces positive super-

Figure 20.22 Cairns's swivel concept. As the closed circular DNA replicates, the two strands must separate at the fork (F). The strain of this unwinding would be released by a swivel mechanism. Cairns actually envisioned the swivel as a machine that rotated actively and thus drove the unwinding of DNA at the fork.

coils, thus stabilizing the DNA against the boiling temperatures in which these organisms live.

You can visualize the supercoiling process as follows: (Figure 20.23). Take a medium-sized rubber band, hold it at the top with one hand, and cut through one of the sides about halfway down. With your other hand, twist one free end of the rubber band one full turn and hold it next to the other free end. You should notice that the rubber band resists the turning as strain is introduced, then relieves the strain by forming a supercoil. The more you twist, the more supercoiling you will observe: one superhelical turn for every full twist you introduce. Reverse the twist and you will see supercoiling of the opposite handedness or sign. DNA works the same way. Of course, the rubber band is not a perfect model. For one thing, you have to hold the severed ends together; in a circular DNA, chemical bonds join the two ends. For another, the rubber band is not a double helix.

If you release your grip on the free ends of the rubber band in Figure 20.23, of course the superhelix will relax. In DNA, it is only necessary to cut one strand to relax a supercoil because the other strand can rotate freely, as demonstrated in Figure 20.24.

Unwinding DNA at the replicating fork would form positive rather than negative supercoils if no other way for relaxing the strain existed. That is because replication

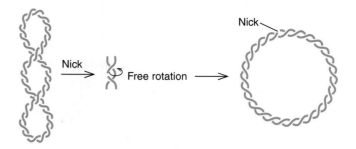

Figure 20.24   Nicking one strand relaxes supercoiled DNA. A nick in one strand of a supercoiled DNA (left) allows free rotation around a phosphodiester bond in the opposite strand (middle). This releases the strain that caused the supercoiling in the first place and so allows the DNA to relax to the open circular form (right).

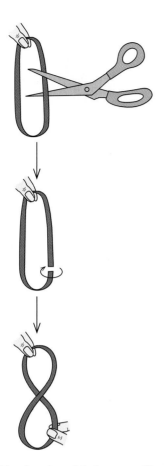

Figure 20.23   Rubber band model of supercoiling in DNA. If you cut the rubber band and twist one free end through one complete turn while preventing the other end from rotating, the rubber band will relieve the strain by forming one superhelical turn.

permanently unwinds one region of the DNA without nicking it, forcing the rest of the DNA to become over-wound, and therefore positively supercoiled, to compensate. To visualize this, look at the circular arrow ahead of the replicating fork in Figure 20.22. Notice how twisting the DNA in the direction of the arrow causes unwinding behind the arrow but overwinding ahead of it. Imagine inserting your finger into the DNA just behind the fork and moving it in the direction of the moving fork to force the DNA strands apart. You can imagine how this would force the DNA to rotate in the direction of the circular arrow, which overwinds the DNA helix. This overwinding strain would resist your finger more and more as it moved around the circle. Therefore, unwinding the DNA at the replicating fork introduces positive superhelical strain that must be constantly relaxed so that replication will not be retarded. You can appreciate this when you think of how the rubber band increasingly resisted your twisting as it became more tightly wound. In principle, any enzyme that is able to relax this strain could serve as a swivel. In fact, of all the topoisomerases in an *E. coli* cell, only one, DNA gyrase, appears to perform this function.

Topoisomerases are classified according to whether they operate by causing single- or double-stranded breaks in DNA. Those in the first class (**type I topoisomerases,** e.g., topoisomerase I of *E. coli*) introduce temporary single-stranded breaks. Enzymes in the second class (**type II topoisomerases,** e.g., DNA gyrase of *E. coli*) break and reseal both DNA strands. Why is *E. coli* topoisomerase I incapable of providing the swivel function needed in DNA replication? Because it can relax only negative supercoils, not the positive ones that form in replicating DNA ahead of the fork. Obviously, the nicks created by these enzymes do not allow free rotation in either direction. But DNA gyrase pumps negative supercoils into closed circular DNA and therefore counteracts the tendency to form positive ones. Hence, it can operate as a swivel.

Not all forms of topoisomerase I are incapable of relaxing positive supercoils. Topoisomerases I from eukaryotes and archaea (the so-called eukaryotic-like topoisomerases I) use a different mechanism from the prokaryotic-like topoisomerases I, and can relax both positive and negative supercoils.

There is direct evidence that DNA gyrase is crucial to the DNA replication process. First of all, mutations in the genes for the two polypeptides of DNA gyrase are lethal and they block DNA replication. Second, antibiotics such as novobiocin, coumermycin, and nalidixic acid inhibit DNA gyrase and thereby prevent replication.

The Mechanism of Type II Topoisomerases   Martin Gellert and colleagues first purified DNA gyrase in 1976. To detect the enzyme during purification, they used an assay that measured its ability to introduce superhelical turns into a relaxed circular DNA (the colE1 plasmid we discussed earlier in this chapter). Then they added varying amounts of DNA gyrase, along with ATP. After an hour, they electrophoresed the DNA and stained it with ethidium bromide so it would fluoresce under ultraviolet light.

Figure 20.25 depicts the results of one such assay. In the absence of gyrase (lane 2) or in the absence of ATP

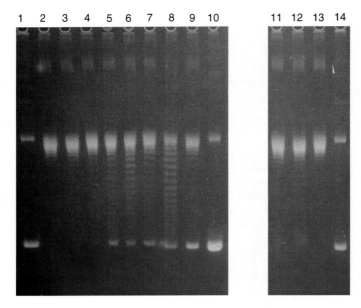

Figure 20.25 Assay for a DNA topoisomerase. Gellert and colleagues incubated relaxed circular ColE1 DNA with varying amounts of *E. coli* DNA gyrase, plus ATP, spermidine, and MgCl$_2$, except where indicated. Lane 1, supercoiled ColE1 DNA as isolated from cells; lane 2, no DNA gyrase; lanes 3–10, DNA gyrase increasing as follows: 24 ng, 48 ng, 72 ng, 96 ng, 120 ng, 120 ng, 240 ng, and 360 ng. Lane 11, ATP omitted; lane 12, spermidine omitted; lane 13, MgCl$_2$ omitted; lane 14, supercoiled ColE1 DNA incubated with 240 ng of gyrase in the absence of ATP. (*Source:* Gellert et al. DNA gyrase: An enzyme that introduces superhelical turns into DNA. *PNAS* 73 (976) fig 1, p. 3873.)

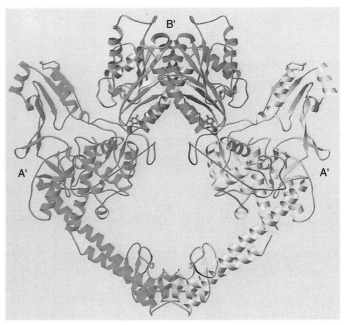

Primary dimer interface

Figure 20.26 Crystal structure of yeast topoisomerase II. The monomer on the left is represented in green and orange, and the monomer on the right is in yellow and blue. The domains of each monomer corresponding to prokaryotic A subunits are in green and yellow (and labeled A′), and the domains corresponding to prokaryotic B subunits are in orange and blue (and labeled B′). The B′ domains, with ATPase activity, form an upper "jaw" of the enzyme, and A′ domains form a lower jaw. The jaws are closed in this representation. The active-site tyrosines that become linked to DNA during the reaction are represented by purple hexagons near the interfaces between the A′ and B′ domains. The primary contact between the dimers is indicated at bottom. (*Source:* James C. Wang, Moving one DNA double helix through another by a type II DNA topoisomerase: the story of a simple molecular machine. *Quarterly Reviews of Biophysics* 31, 2(1998), f. 8, pp. 107–144. ©1998 Cambridge University Press.)

(lane 11) we see essentially only the low-mobility relaxed circular form of the plasmid. On the other hand, as the experimenters added more and more DNA gyrase (lanes 3–10), they observed more and more of the high-mobility form of the plasmid with many superhelical turns. At intermediate levels of gyrase, intermediate forms of the plasmid appeared as distinct bands, with each band representing a plasmid with a different, integral number of superhelical turns.

This experiment demonstrates the dependence of DNA gyrase on ATP, but the enzyme does not use as much ATP as you might predict based on all the breaking and reforming of phosphodiester bonds. The reason for this modest energy requirement is that the gyrase itself (not a water molecule) is the agent that breaks the DNA bonds, so it forms a covalent enzyme–DNA intermediate. This intermediate conserves the energy in the DNA phosphodiester bond so it can be reused when the DNA ends are rejoined and the enzyme is released in its original form.

What is the evidence for the enzyme–DNA bond? James Wang and colleagues trapped DNA–gyrase complexes by denaturing the enzyme midway through the breaking–rejoining cycle and found DNA with nicks in both strands, staggered by four bases, with the gyrase covalently linked to each protruding DNA end. In 1980,

Wang and colleagues went on to show that the covalent bond between enzyme and DNA is through a tyrosine on the enzyme. They incubated [32P]DNA with DNA gyrase, trapped the DNA-gyrase complex as before by denaturing the enzyme, then isolated the complex. They digested the DNA in the complex exhaustively with nuclease, and finally isolated [32P]enzyme, with the label in the A subunits. (DNA gyrase, like all prokaryotic DNA topoisomerase IIs, is a tetramer of two different subunits: A$_2$B$_2$). The fact that the enzyme's A subunits became labeled with 32P strongly suggested that these subunits had been linked through one of their amino acids to the 32P[DNA]. Which amino acid in the enzyme was linked to the DNA? Wang and colleagues digested the labeled A subunit in boiling HCl to break it down into its component amino acids. Then they purified the labeled amino acid, which copurified with phosphotyrosine. Thus, the enzyme is linked covalently through a tyrosine residue in each A subunit to the DNA.

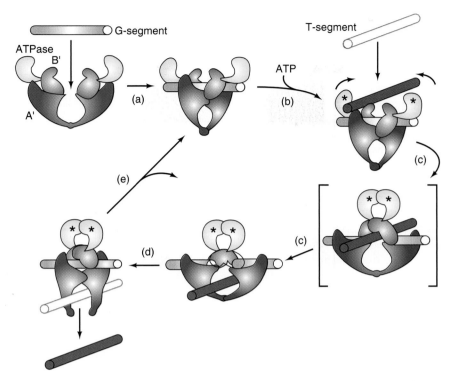

Figure 20.27 Model of the segment-passing step in the topoisomerase II reaction. Based on the crystal structure of the enzyme, and other evidence, Wang and colleagues proposed the following model: (**a**) The upper jaws of the enzyme open to bind the DNA G-segment, which is the one that will break to form a gate that will allow the other DNA segment to pass through. This binding of DNA induces a conformational change in the enzyme that brings the active-site tyrosines on the B′ domain into position to attack the DNA. (**b**) The ATPase domain of each upper jaw binds ATP (represented by an asterisk), and the upper jaw also binds the DNA T-segment, which will be passed through the G-segment. (**c**) In a series of conformational changes, including a hypothetical intermediate (in brackets), the active site breaks the DNA G-segment, and allows the T-segment to pass through into the lower jaws. The front B′ domain at the end of step (**c**) is uncolored so the DNA behind it can be seen. (**d**) The lower jaws open to release the T-segment. (**e**) The enzyme hydrolyzes the bound ATP, returning the enzyme to a state in which it can accept another T-segment and repeat the segment-passing process. (*Source:* Reprinted by permission from *Nature* 379:231, 1996. Copyright © 1996 Macmillan Magazines Limited.)

How do DNA gyrase and the other DNA topoisomerase IIs perform their task of introducing negative superhelical turns into DNA? The simplest explanation is that they allow one part of the double helix to pass through another part. Figure 20.26 shows a representation of the structure of yeast topoisomerase II, based on x-ray crystallography. Like all eukaryotic topoisomerase IIs, it is a dimer of identical subunits, and each monomer contains domains corresponding to the A and B subunits of the prokaryotic topoisomerase IIs. Yeast topoisomerase II is a heart-shaped protein made out of two crescent-shaped monomers. The protein can be considered as a double-jawed structure, with one jaw at the top and the other at the bottom.

Figure 20.27 presents a model for how these two jaws could cooperate in the DNA segment-passing process. The upper jaw binds one DNA segment, called the **G-segment** because it will contain the gate through which the other segment will pass. Then, after activation by ATP, the upper jaws bind the other DNA segment, called the **T-segment** because it will be transported through the G-segment. The two segments are perpendicular to each

other. The enzyme breaks the G-segment to form a gate, and the T-segment passes through into the lower gate, from which it is ejected.

SUMMARY One or more enzymes called helicases use ATP energy to separate the two parental DNA strands at the replicating fork. As helicase unwinds the two parental strands of a closed circular DNA, it introduces a compensating positive supercoiling force into the DNA. The stress of this force must be overcome or it will resist progression of the replicating fork. The name given to this stress-release mechanism is the swivel. DNA gyrase is the leading candidate for this role in *E. coli*.

Three DNA Polymerases in *E. coli*

Arthur Kornberg discovered the first *E. coli* DNA polymerase in 1958. Because we now know that it is only one of three DNA polymerases, we call it **DNA polymerase I**

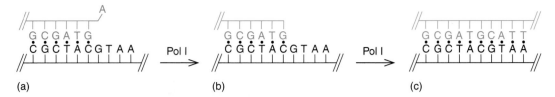

Figure 20.28   Proofreading in DNA synthesis. (a) An adenine nucleotide (red) has been mistakenly incorporated across from a guanine. This destroys the perfect base pairing required at the 3′-end of the primer, so the replicating machinery stalls. **(b)** This pause then allows Pol I to use its 3′ → 5′ exonuclease function to remove the mispaired nucleotide. **(c)** With the appropriate base paring restored, Pol I is free to continue DNA synthesis.

(pol I). In the absence of evidence for other cellular DNA polymerases, many molecular biologists assumed that pol I was the polymerase responsible for replicating the bacterial genome. As we will see, this assumption was incorrect. Nevertheless, we begin our discussion of DNA polymerases with pol I because it is relatively simple and well understood, yet it exhibits the essential characteristics of a DNA synthesizing enzyme.

Pol I   Although pol I is a single 102-kD polypeptide chain, it is remarkably versatile. It catalyzes three quite distinct reactions. It has a DNA polymerase activity, of course, but it also has two different exonuclease activities: a 3′→5′, and a 5′→3′ exonuclease activity. Why does a DNA polymerase also need two exonuclease activities? The 3′→5′ activity is important in **proofreading** newly synthesized DNA (Figure 20.28). If pol I has just added the wrong nucleotide to a growing DNA chain, this nucleotide will not base-pair properly with its partner in the parental strand and should be removed. Accordingly, pol I pauses and the 3′→5′ exonuclease removes the mispaired nucleotide, allowing replication to continue. This greatly increases the fidelity, or accuracy, of DNA synthesis. The 5′→3′ exonuclease activity allows pol I to degrade a strand ahead of the advancing polymerase, so it can remove and replace a strand all in one pass of the polymerase, at least in vitro. This DNA degradation function is useful because pol I seems to be involved primarily in DNA repair (including removal and replacement of RNA primers), for which destruction of damaged or mispaired DNA (or RNA primers) and its replacement by good DNA is required. Figure 20.29 illustrates this process for primer removal and replacement.

Another important feature of pol I is that it can be cleaved by mild proteolytic treatment into two polypeptides: a large fragment (the **Klenow fragment**), which has the polymerase and proofreading (3′→5′ exonuclease) activities and a small fragment with the 5′→3′ exonuclease activity. The Klenow fragment is frequently used in molecular biology when DNA synthesis is required and destruction of one of the parental DNA strands, or the primer, is undesirable. For example, the Klenow fragment is often used to perform primer extension to measure the prevalence or start site of a transcript, or to sequence a DNA. On the other hand, the whole pol I is used to per-

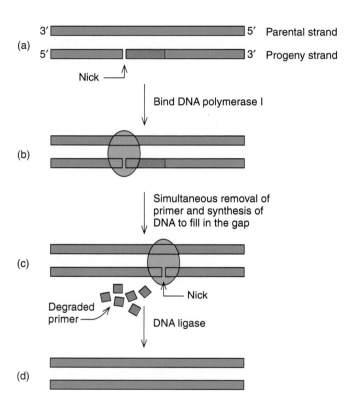

Figure 20.29   Removing primers and joining nascent DNA fragments. (a) We start with two adjacent progeny DNA fragments, the right-hand one containing an RNA primer (red) at its 5′-end. The two fragments are separated by a single-stranded break called a nick. **(b)** DNA polymerase I binds to the double-stranded DNA at the nick. **(c)** DNA polymerase I simultaneously removes the primer and fills in the resulting gap by extending the left-hand DNA fragment rightward. It leaves degraded primer in its wake. **(d)** DNA ligase seals the remaining nick by forming a phosphodiester bond between the left-hand and right-hand progeny DNA fragments.

form nick translation to label a probe in vitro, because nick translation depends on 5′→3′ degradation of DNA ahead of the moving fork.

Thomas Steitz and colleagues determined the crystal structure of the Klenow fragment in 1987, giving us our first look at the fine structure of a DNA-synthesizing machine. Figure 20.30 illustrates their findings. The most obvious feature of the structure is a great cleft between the helices labeled I and O. This is the presumed binding site

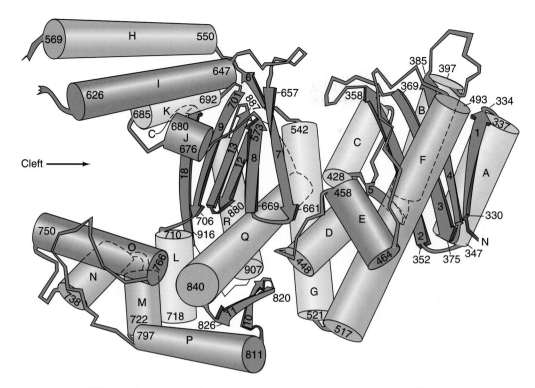

Figure 20.30   Crystal structure of Klenow fragment. Cylinders represent α-helices, with the darker helices closer to the observer. Beta sheets are represented by blue ribbon-arrows. The cleft between helices I and O is the likely site of interaction with DNA. (*Source:* Reprinted from *Trends in Biochemical Sciences*, vol. 12, C. M. Joyce and T. Steitz, "DNA polymerase I: From crystal structure to function via genetics," 290. Copyright © 1987 with permission of Elsevier Science.)

for the DNA that is being replicated. In fact, all of the known polymerase structures, including that of T7 RNA polymerase, are very similar, and have been likened to a hand. In the Klenow fragment, the O helix is part of the "fingers" domain, the I helix is part of the "thumb" domain, and the β-pleated sheet between them (blue arrows in Figure 20.30) is part of the "palm" domain. The palm domain contains three conserved aspartate residues that are essential for catalysis. They are thought to coordinate magnesium ions that catalyze the polymerase reaction.

To show the way DNA fits into this structure, Steitz and colleagues made a cocrystal of *Taq* polymerase and a model double-stranded DNA template containing 8 bp and a blunt end at the 3′-end of the nontemplate (primer) strand. *Taq* polymerase is the polymerase from the thermophilic bacterium *Thermus aquaticus* that is widely used in PCR (Chapter 4). Its polymerase domain is very similar to that of the Klenow fragment—so much so that it is called the "KF portion," for "Klenow fragment" portion, of the enzyme. Figure 20.31 shows the results of x-ray crystallography studies on the *Taq* polymerase–DNA complex. The primer strand (red) has its 3′-end close to the three essential aspartate residues in the palm domain, but not quite close enough for magnesium ions to bridge between the carboxyl groups of the aspartates and the 3′-hydroxyl group of the primer strand. Thus, this structure is not exactly like a catalytically productive one, perhaps in part because the magnesium ions are missing.

In 1969, Paula DeLucia and John Cairns isolated a mutant with a defect in the *polA* gene, which encodes pol I. This mutant (*polA*1) lacked pol I activity, yet it was viable, strongly suggesting that pol I was not really the DNA-replicating enzyme. Instead, pol I seems to play a dominant role in repair of DNA damage. It fills in the gaps left when damaged DNA is removed. The finding that pol I is not essential spurred a renewed search for the real DNA replicase, and in 1971, Thomas Kornberg and Malcolm Gefter discovered two new polymerase activities: **DNA polymerases II and III (pol II and pol III)**. We will see that pol III is the actual replicating enzyme.

SUMMARY Pol I is a versatile enzyme with three distinct activities: DNA polymerase; 3′→5′ exonuclease; and 5′→3′ exonuclease. The first two activities are found on a large domain of the enzyme, and the last is on a separate, small domain. The large domain (the Klenow fragment) can be separated from the small by mild protease treatment, yielding two protein fragments with all three activities intact. The structure of the Klenow fragment shows a wide cleft for binding to DNA. This polymerase active site is remote from the 3′→5′ exonuclease active site on the Klenow fragment.

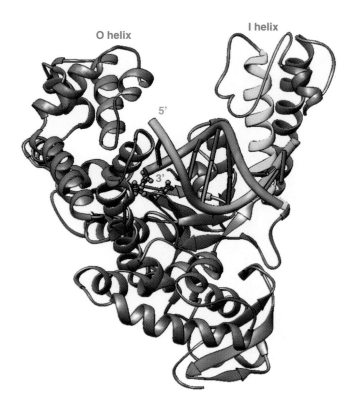

Figure 20.31 Cocrystal structure of *Taq* DNA polymerase with a double-stranded model DNA template. The O helix and I helix of the "fingers" and "thumb" of the polymerase "hand" are in green and yellow, respectively. The template and primer strands of the model DNA are in orange and red, respectively. The three essential aspartate side chains in the "palm" are represented by small red balls near the 3′-end of the primer strand. (*Source:* Eom, S. H., Wang, J., and Steitz, T. A. Structure of *Taq* polymerase with DNA at the polymerase active site. *Nature* 382 (18 July 1996) f. 2a, p. 280. © Macmillan Magazines, Ltd.)

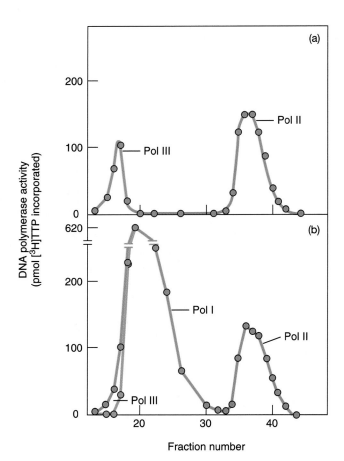

Figure 20.32 Phosphocellulose chromatography of DNA polymerases from *polA*1 and wild-type cells. Gefter and colleagues separated DNA polymerases from (**a**) *polA*1 and (**b**) wild-type *E. coli* cells by phosphocellulose chromatography. Pol II is well separated from pol III (**a**) and from pol I (**b**), but pol III is hidden by pol I in wild-type cells (**b**). The existence of pol III can be detected in the presence of pol I by assaying the fractions in panel (**b**) in the presence of *N*-etnylmaleimide (blue line). This reagent inhibits pol III, but not pol I. Thus, the pol III activity is the difference between the red and blue lines in panel (**b**). (*Source:* From M. L. Gefter et al., "Analysis of DNA Polymerases II and III in Mutants of *Escherichia coli* Thermosensitive for DNA Synthesis," *Proceedings of the National Academy of Sciences* 68(12):3150–3153, December 1971.)

Pol II and Pol III Pol II could be readily separated from pol I by phosphocellulose chromatography, but pol III had been masked in wild-type cells by the preponderance of pol I, as illustrated in Figure 20.32. Next, Gefter and Thomas Kornberg and colleagues used genetic means to search for the polymerase that is required for DNA replication. They tested the pol II and III activities in 15 different *E. coli* strains that were temperature-sensitive for DNA replication. Most of these strains were *polA*1⁻, which made it easier to measure pol III activity after phosphocellulose chromatography because there was no competing pol I activity. In those few cases where pol I was active, Gefter and colleagues used N-ethylmaleimide to knock out pol III so its activity could be measured as the difference between the activities in the presence and absence of the inhibitor.

Table 20.2 presents the results. The most striking finding was that there were five strains with mutations in the *dnaE* gene. In four of these, the pol III activity was very temperature-sensitive, and in the fifth it was slightly temperature-sensitive. On the other hand, none of the mu-

tants affected pol II at all. These results led to three conclusions: First, the *dnaE* gene encodes pol III. Second, the *dnaE* gene does not encode pol II, and pol II and pol III are therefore separate activities. Third, because defects in the gene encoding pol III interfere with DNA replication, pol III is indispensable for DNA replication. It would have been nice to conclude that pol II is *not* required for DNA replication, but that was not possible because no mutants in the gene encoding pol II were tested. However, in separate work, these investigators isolated mutants with inactive pol II, and these mutants were still viable, showing that pol II is not necessary for DNA replication. Thus, pol III is the enzyme that replicates the *E. coli* DNA.

SUMMARY Of the three DNA polymerases in *E. coli* cells, pol I, pol II, and pol III, only pol III is required for DNA replication. Thus, this polymerase is the enzyme that replicates the bacterial DNA.

The Pol III Holoenzyme   The enzyme that carries out the elongation of primers to make both the leading and lag-

Table 20.2   DNA Reaction Rate 45°C/30°C

Strain	DNA Polymerase II	DNA Polymerase III
1 (*dnaA*)	1.85	1.45
2 (*dnaB*)	1.73	1.74
3 (*dnaB*)	1.95	1.55
4 (*dnaC*)	1.86	1.20
5 (*dnaD*)	1.99	1.47
6 (*dnaE*)	1.87	1.0
7 (*dnaE*)	1.70	—*
8 (*dnaE*)	2.00	—*
9 (*dnaE*)	1.95	<0.1
10 (*dnaE*)	1.70	<0.1
11 (*dnaF*)	1.55	1.33
12 (*dnaF*)	1.77	1.43
13 (*dnaG*)	1.75	1.57
14 (*dna+*)	1.92	1.70
15 (*dna+*)	1.90	1.55

* A dash indicates that the enzyme activity was not detectable at 30°C.

(*Source:* From M. L. Gefter et al., "Analysis of DNA Polymerases II and III in Mutants of *Escherichia coli* Thermosensitive for DNA Synthesis," *Proceedings of the National Academy of Sciences* 68(12):3150–3153, December 1971.)

ging strands of DNA is called **DNA polymerase III holoenzyme (pol III holoenzyme).** The "holoenzyme" designation indicates that this is a multisubunit enzyme, and indeed it is: As Table 20.3 illustrates, the holoenzyme contains 10 different polypeptides. On dilution, this holoenzyme dissociates into several different subassemblies, also as indicated in Table 20.3. Each pol III subassembly is capable of DNA polymerization, but only very slowly. This suggested that something important is missing from the subassemblies because DNA replication in vivo is extremely rapid. The replicating fork in *E. coli* moves at the amazing rate of 1000 nt/sec. (Imagine the sheer mechanics involved in correctly pairing 1000 nt with partners in the parental DNA strands and forming 1000 phosphodiester bonds every second!) In vitro, the holoenzyme goes almost that fast: about 700 nt/sec, suggesting that this is the entity that replicates DNA in vivo. The other two DNA polymerases in the cell, pol I and pol II, are not ordinarily found in holoenzyme forms, and they replicate DNA much more slowly than the pol III holoenzyme does.

Charles McHenry and Weldon Crow purified DNA polymerase III to near-homogeneity and found that three polypeptides compose the core of pol III: the α, ε, and θ-subunits. These have molecular masses of 130, 27.5, and 10 kD, respectively. The rest of the subunits of the holoenzyme dissociated during purification, but the core subunits were bound tightly together. In this section, we will examine the pol III core more thoroughly, but we will save our discussion of the other polypeptides in the pol III holoenzyme for Chapter 21 because they play important roles in initiation and elongation of DNA synthesis.

The α-subunit of the pol III core has the DNA polymerase activity, but this was not easy to determine because the α-subunit is so difficult to separate from the

Table 20.3   Subunit Composition of E. coli DNA Polymerase III Holoenzyme

Subunit	Molecular Mass (kD)	Function	Subassemblies			
α	129.9	DNA polymerase	Core	Pol III′	Pol III*	Pol III holoenzyme
ε	27.5	3′→5′ exonuclease				
θ	8.6	Stimulates ε exonuclease				
τ	71.1	Dimerizes core / Binds γ complex				
γ	47.5	Binds ATP	γ complex (DNA-dependent ATPase)			
δ	38.7	Binds to β				
δ′	36.9	Binds to γ and δ				
χ	16.6	Binds to SSB				
ψ	15.2	Binds to χ and γ				
β	40.6	Sliding clamp				

(*Source:* Reprinted from Herendeen and Kelly, *Cell* 84:6, 1996. Copyright 1996, with permission from Elsevier Science.)

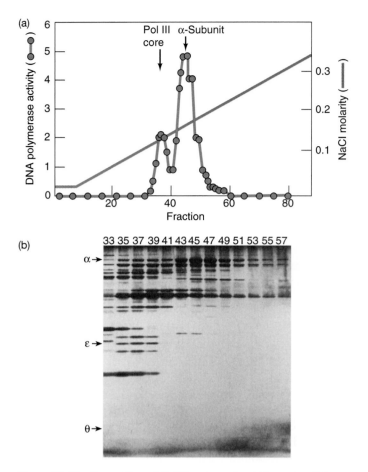

Figure 20.33 Separation of Pol III core and pol III α-subunit from a cell that over-expresses the α-subunit. Maki and Kornberg partially purified DNA polymerase III from cells that overexpressed the α-subunit, and then subjected this preparation to purification by DEAE-Sephacel chromatography. **(a)** Profile of DNA polymerase activity. **(b)** SDS-PAGE to determine the polypeptide composition of selected fractions across the peaks of polymerase activity. The second peak, labeled α-subunit, is rich in α-subunit but contains no detectable amount of ε- or θ-subunit. (*Source:* Maki & Kornberg, The polymerase subunit of DNA polymerase III of *Escherichia coli. J. Biol. Chem.* 260 no. 24 (25 Oct 1985) f. 1B, p. 12989. American Society for Biochemistry and Molecular Biology.)

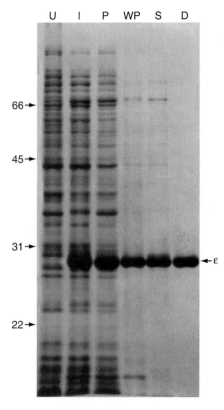

Figure 20.34 Purification of the Pol III ε-subunit. Scheuerman and Echols started with bacteria bearing a plasmid with the *dnaO* gene, which encodes the ε-subunit, under the control of a heat-inducible λ phage P_L promoter. (The promoter is heat-inducible because the λ repressor encoded by this phage is temperature-sensitive.) These investigators made extracts of uninduced and induced cells, lanes U and I, respectively, and subjected them to SDS-PAGE; then stained the proteins in the gel with Coomaasie Blue. They purified the ε-subunit through successive steps and electrophoresed the protein after each step. Each lane is labeled to designate the last purification step before SDS-PAGE, as follows: P, pellet from low-speed centrifugation; WP, low-speed pellet after washing with high-ionic strength buffer; S, soluble fraction after denaturation–renaturation; D, fraction from DEAE-Sephacel chromatography. (*Source:* Scheuermann & Echols, A separate editing exonuclease for DNA replication in the a subunit of *Escherichia coli* DNA polymerase III holoenzyme. *PNAS* 81 (1984) f. 1, p. 7749.)

other core subunits. When Hisaji Maki and Arthur Kornberg cloned and overexpressed the gene for the α-subunit, they finally paved the way for purifying the polymerase activity because the overproduced α-subunit was in great excess over the other two subunits. Figure 20.33 shows a DEAE-Sephacel chromatographic separation of the pol III core and the α-subunit, and the polypeptides present in each fraction. The peak α-subunit fractions appear to be devoid of ε- and θ-subunits. Hisaji Maki and Arthur Kornberg pooled these α-subunit fractions and purified the α-subunit through two more steps, by which time it was almost homogeneous. When they tested this purified α-subunit for DNA polymerase activity, they found that it had activity similar to the same amount of core. Thus, the α-subunit contributes the DNA polymerase activity to the core.

The pol III core has a 3′→5′ exonuclease activity that removes mispaired bases as soon as they are incorporated, allowing the polymerase to proofread its work. This is similar to the 3′→5′ exonuclease activity of the pol I Klenow fragment. Scheuermann and Echols used the overexpression strategy to demonstrate that the core ε-subunit has this exonuclease activity. They overexpressed the ε-subunit (the product of the *dnaQ* gene) and purified it through various steps, as detailed in Figure 20.34. After the last step, DEAE-Sephacel chromatography, the ε-subunit was essentially pure (lane D). Next, Scheuermann and Echols tested this purified ε-subunit, as well as core pol III, for

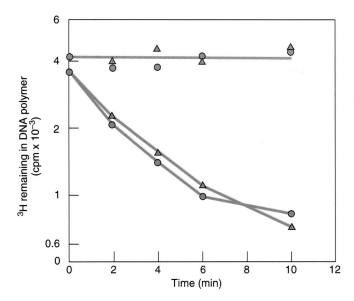

Figure 20.35 Exonuclease activity of ε-subunit and pol III core with substrates that are perfectly base-paired or that have mistmatches. Scheuerman and Echols incubated the purified ε-subunit with ^{3}H-labeled synthetic DNAs and measured the amount of radioactivity remaining in the DNAs after increasing lengths of time Symbols: triangles, pol III core; circles ε-subunit; blue, perfectly paired DNA substrate; red, DNA substrate with mismatches. (*Source:* From R. H. Scheuermann and H. Echols, "A separate editing exonuclease for DNA replication," *Proceedings of the National Academy of Sciences* 81:7747–7751, December 1984.)

exonuclease activity. Figure 20.35 shows that the core and the ε-subunit both have exonuclease activity, and they are both specific for mispaired DNA substrates, having no measurable activity on perfectly paired DNAs. This is what we expect for the proofreading activity. This activity also explains why *dnaQ* mutants are subject to excess mutations (10^3–10^5 more than in wild-type cells). Without adequate proofreading, many more mismatched bases fail to be removed and persist as mutations. Thus, we call *dnaQ* mutants **mutator** mutants, and the gene has even been referred to as the *mutD* gene because of this mutator phenotype.

Relatively little work has been performed on the θ-subunit of the core. Its function is unknown. However, it is clear that the α- and ε-subunits cooperate to boost each other's activity in the core polymerase. The DNA polymerase activity of the α-subunit increases by about two-fold in the core, compared with the free subunit, and the activity of the ε-subunit increases by about 10–80-fold when it joins the core.

> **SUMMARY** The pol III core is composed of three subunits, α, ε, θ. The α-subunit has the DNA polymerase activity. The ε-subunit has the $3' \rightarrow 5'$ exonuclease activity that carries out proofreading. The role of the θ-subunit is not yet clear.

Fidelity of Replication

The proofreading mechanism of pol III (and pol I) greatly increases the fidelity of DNA replication. The pol III core makes about one pairing mistake in one hundred thousand in vitro—not a very good record, considering that even the *E. coli* genome contains over four million base pairs. At this rate, replication would introduce errors into a significant percentage of genes every generation. Fortunately, proofreading allows the polymerase another mechanism by which to get the base pairing right. The error rate of this second pass is presumably the same as that of the first pass, or about 10^{-5}. This predicts that the actual error rate with proofreading would be $10^{-5} \times 10^{-5} = 10^{-10}$, and that is close to the actual error rate of the pol III holoenzyme in vivo, which is 10^{-10}–10^{-11}. (The added fidelity comes at least in part from mismatch repair, which we will discuss later in this chapter.) This is a tolerable level of fidelity. In fact, it is better than perfect fidelity because it allows for mutations, some of which may equip the organism to adapt to a changing environment.

Consider the implications of the proofreading mechanism, which removes a mispaired nucleotide at the $3'$-end of a DNA progeny strand (recall Figure 20.28). DNA polymerase cannot operate without a base-paired nucleotide to add to, which means that it cannot start a new DNA chain unless a primer is already there. That explains the need for primers, but why primers made of RNA? The reason seems to be the following: Primers are made with more errors, because their synthesis is not subject to proofreading. Making primers out of RNA guarantees that they will be recognized, removed, and replaced with DNA by extending the neighboring Okazaki fragment. The latter process is, of course, relatively error-free, because it is catalyzed by pol I, which has a proofreading function.

> **SUMMARY** Faithful DNA replication is essential to life. To help provide this fidelity, the *E. coli* DNA replication machinery has a built-in proofreading system that requires priming. Only a base-paired nucleotide can serve as a primer for the pol III holenzyme. Therefore, if the wrong nucleotide is incorporated by accident, replication stalls until the $3' \rightarrow 5'$ exonuclease of the pol III holoenzyme removes it. The fact that the primers are made of RNA may help mark them for degradation.

Multiple Eukaryotic DNA Polymerases

Much less is known about the proteins involved in eukaryotic DNA replication, but we do know that multiple DNA polymerases take part in the process, and we also have a good idea of the roles these enzymes play.

Table 20.4 lists the known mammalian DNA polymerases and their probable roles.

It had been thought that polymerase α synthesized the lagging strand because of the low processivity of this enzyme. **Processivity** is the tendency of a polymerase to stick with the replicating job once it starts. The *E. coli* polymerase III holoenzyme is highly processive. Once it starts on a DNA chain, it remains bound to the template, making DNA for a long time. Because it does not fall off the template very often, which would require a pause as a new polymerase bound and took over, the overall speed of *E. coli* DNA replication is very rapid. Polymerase δ is much more processive than polymerase α. Thus, it was proposed that the less processive DNA polymerase α synthesized the lagging strand, which is made in short pieces. However, it now appears that polymerase α, the only eukaryotic DNA polymerase with primase activity, make the primers for both strands. Then the highly processive DNA polymerase α elongates both strands.

Actually, much of the processivity of polymerase δ comes, not from the polymerase itself, but from an associated protein called **proliferating cell nuclear antigen,** or **PCNA.** This protein, which is enriched in proliferating cells that are actively replicating their DNA, enhances the processivity of polymerase δ by a factor of 40. That is, PCNA causes polymerase δ to travel 40 times farther elongating a DNA chain before falling off the template. PCNA works by physically clamping the polymerase δ onto the template. We will examine this clamping phenomenon more fully when we consider the detailed mechanism of DNA replication in *E. coli* in Chapter 21.

In marked contrast, polymerase β is not processive at all. It usually adds only one nucleotide to a growing DNA chain and then falls off, requiring a new polymerase to bind and add the next nucleotide. This fits with its postulated role as a repair enzyme that needs to make only short stretches of DNA to fill in gaps created when primers or mismatched bases are excised. In addition, the level of polymerase β in a cell is not affected by the rate of division of the cell, which suggests that this enzyme is not involved in DNA replication. If it were, we would expect

it to be more prevalent in rapidly dividing cells, as polymerases δ and α are.

Polymerase γ is found in mitochondria, not in the nucleus. Therefore, we conclude that this enzyme is responsible for replicating mitochondrial DNA.

> **SUMMARY** Mammalian cells contain five different DNA polymerases. Polymerases δ and α appear to participate in replicating both DNA strands: α by priming DNA synthesis and δ by elongating both strands. Polymerase β and ε seem to function in DNA repair. Polymerase γ probably replicates mitochondrial DNA.

20.3 DNA Damage and Repair

DNA can be damaged in many different ways, and this damage, if left unrepaired, can lead to mutations: changes in the base sequence of a DNA. This distinction is worth emphasizing at the outset: *DNA damage is not the same as mutation.* DNA damage is simply a chemical alteration to DNA. A mutation is a change in a base pair. For example, the change from a G–C pair to a methyl-G–C pair is DNA damage; the change from a G–C pair to any other natural base pair (A–T or T–A or C–G) is a mutation. Let us look at two common examples of DNA damage: base modifications caused by alkylating agents and pyrimidine dimers caused by ultraviolet radiation. Then we will examine the mechanisms that prokaryotic and eukaryotic cells use to deal with such damage. Most of these mechanisms involve DNA replication.

Damage Caused by Alkylation of Bases

Some substances in our environment, both natural and synthetic, are **electrophilic,** meaning electron - (or negative charge-) loving. Thus, **electrophiles** seek centers of negative charge in other molecules and bind to them. Many other environmental substances are metabolized in the body to electrophilic compounds. One of the most obvious centers of negative charge in biology is the DNA molecule. Every nucleotide contains one full negative charge on the phosphate and partial negative charges on the bases. When electrophiles encounter these negative centers, they attack them, usually adding carbon-containing groups called **alkyl groups.** Thus, we refer to this process as **alkylation.**

Figure 20.36 shows the centers of negative charge in DNA. Aside from the phosphodiester bonds, the favorite sites of attack by alkylating agents are the N7 of guanine and the N3 of adenine, but many other targets are available, and different alkylating agents have different preferences for these targets.

Table 20.4	Probable Roles of Eukaryotic DNA Polymerases
Enzyme	**Probable role**
DNA polymerase α	Priming of replication of both strands
DNA polymerase δ	Elongation of both strands
DNA polymerase β	DNA repair
DNA polymerase ε	DNA repair
DNA polymerase γ	Replication of mitochondrial DNA

What are the consequences of alkylations at these DNA sites? Consider the two predominant sites of alkylation, the N7 of guanine and the N3 of adenine. N7 alkylation of guanine does not change the base-pairing properties of the target base and is generally harmless. Alkylation of the N3 of adenine is more serious because it creates a base (3-methyl adenine [3mA]) that cannot base-pair properly with any other base—a so-called **noncoding base.** Because a DNA polymerase does not recognize any base-pair involving 3mA as correct, it stops at the 3mA damage, stalling DNA replication. Such blockage of DNA replication can kill a cell, so we say it is **cytotoxic.** On the other hand, as we will see later in this chapter, such stalled replication can be resumed without repairing the damage, but the mechanism of such resumption is error-prone and therefore leads to mutations.

Moreover, all of the nitrogen and oxygen atoms involved in base-pairing (see Figure 20.36) are also subject to alkylation, which can directly disrupt base pairing and

lead to mutation. The alkylation target that leads to most mutations is the O6 of guanine. Even though this atom is relatively rarely attacked by alkylating agents, such alkylations are very mutagenic because they allow the product to base-pair with thymine rather than cytosine. For example, consider the alkylation of the O6 of guanine by the common laboratory mutagen ethylmethane sulfonate (EMS), which transfers ethyl (CH_3CH_2) groups to DNA (Figure 20.37). The alkylation of the guanine O6 changes the tautomeric form (the pattern of double bonds) of the guanine so it base-pairs naturally with thymine. This leads to the replacement of a G-C base pair by an A-T base pair.

Many environmental **carcinogens,** or cancer-causing agents, are electrophiles that act by attacking DNA and alkylating it. As we have just seen, this can lead to mutations. If the mutations occur in genes that control or otherwise influence cell division, they can cause a cell to lose control over its replication and therefore change into a cancer cell.

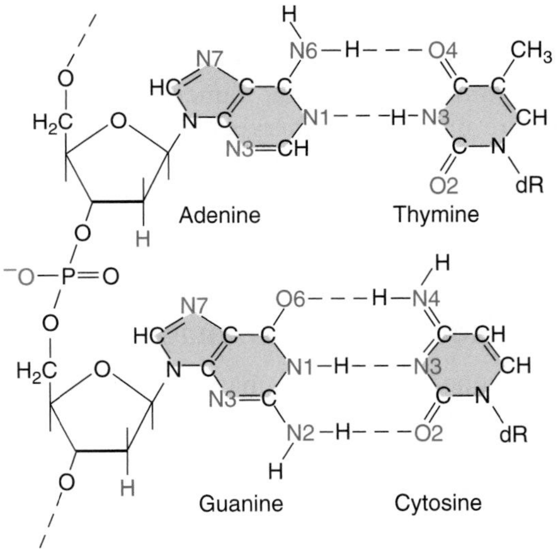

> **SUMMARY** Alkylating agents like ethylmethane sulfonate add alkyl groups to bases. Some of these alkylations do not change base-pairing, so they are innocuous. Others cause DNA replication to stall, so they are cytotoxic, and can lead to mutations if the cell attempts to replicate its DNA without repairing the damage. Other alkylations change the base-pairing properties of a base, so they are mutagenic.

Damage Caused by Ultraviolet Radiation

Ultraviolet (UV) **radiation** cross-links adjacent pyrimidines on the same DNA strand, forming **pyrimidine dimers.** Figure 20.38 shows the structure of a thymine dimer and illustrates how it interrupts base-pairing between the two DNA strands. These dimers block DNA replication because the replication machinery cannot tell which bases to insert opposite the dimer. As we will see, replication sometimes proceeds anyway, and bases are inserted without benefit of the base-pairing that normally

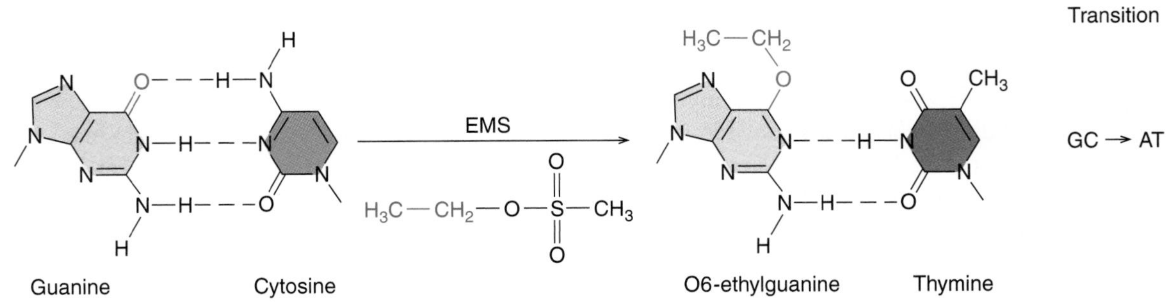

Figure 20.36 Electron-rich centers in DNA. The targets most commonly attacked, by electrophiles are the phosphate groups, N7 of guanine and N3 of adenine (red); other targets are in blue.

Figure 20.37 Alkylation of guanine by EMS. At the left is a normal guanine–cytosine base pair. Note the free O6, oxygen (red) on the guanine. Ethylmethane sulfonate (EMS) donates an ethyl group (blue) to the O6 oxygen, creating O6-ethylguanine (right), which base-pairs with thymine instead of cystosine. After one more round of replication, an A-T base pair will have replaced a G-C pair.

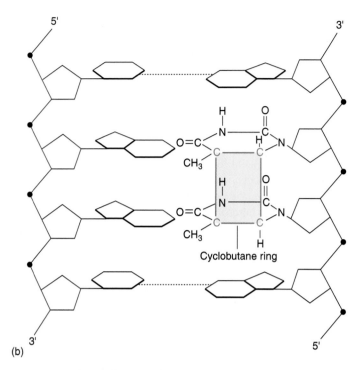

Figure 20.38   Pyrimidine dimers. (a) Ultraviolet light cross-links two pyrimidine bases (thymines in this case) on the top strand. This distorts the DNA so that these two bases no longer pair with their adenine partners. **(b)** The two bonds joining the two pyrimidines form a four-member cyclobutane ring (red).

provides accuracy. If these are the wrong bases, a mutation results.

Ultraviolet radiation has great biological significance; it is present in sunlight, so most forms of life are exposed to it to some extent. The mutagenicity of UV radiation explains why sunlight can cause skin cancer: Its UV component damages the DNA in skin cells, which leads to mutations that sometimes cause those cells to lose control over their division.

Given the dangers of UV radiation, we are fortunate to have a shield—the ozone layer—in the earth's upper atmosphere to absorb the bulk of such radiation. However, scientists have noticed alarming holes in this protective shield—the most prominent one located over Antarctica. The causes of this ozone depletion are still somewhat controversial, but they probably include compounds traditionally used in air conditioners and in plastics. Unless we can arrest the destruction of the ozone layer, we are destined to suffer more of the effects of UV radiation, including skin cancer.

Damage Caused by Gamma and X-Rays

The much more energetic **gamma rays** and **x-rays,** like ultraviolet rays, can interact directly with the DNA molecule. However, they cause most of their damage by ionizing the molecules, especially water, surrounding the DNA. This forms **free radicals,** chemical substances with an unpaired electron. These free radicals, especially those containing oxygen, are extremely reactive, and they immediately attack neighboring molecules. When such a free radical attacks a DNA molecule, it can change a base, but it frequently causes a single- or double-stranded break. Single-stranded breaks are ordinarily not serious because they are easily repaired, just by rejoining the ends of the severed strand, but double-stranded breaks are very difficult to repair properly, so they frequently cause a lasting mutation. Because ionizing radiation can break chromosomes, it is referred to not only as a mutagen, or mutation-causing substance, but also as a **clastogen,** which means "breaker."

SUMMARY Different kinds of radiation cause different kinds of damage. Ultraviolet rays have comparatively low energy, and they cause a moderate type of damage: pyrimidine dimers. Gamma and x-rays are much more energetic. They ionize the molecules around DNA and form highly reactive free radicals that can attack DNA, altering bases or breaking strands.

Directly Undoing DNA Damage

One way to cope with DNA damage is to repair it, or restore it to its original, undamaged state. There are two basic ways to do this: (1) Directly undo the damage, or (2) remove the damaged section of DNA and fill it in with new, undamaged DNA. Let us begin by looking at two methods *E. coli* cells use to directly undo DNA damage.

In the late 1940s, Albert Kelner was trying to measure the effect of temperature on repair of ultraviolet damage to DNA in the bacterium *Streptomyces.* However, he noticed that damage was repaired much faster in some bacterial spores than in others kept at the same temperature. Obviously, some factor other than temperature was operating. Finally, Kelner noticed that the spores whose damage was repaired fastest were the ones kept most directly exposed to light from a laboratory window. When he performed control experiments with spores kept in the dark, he could detect no repair at all. Renato Dulbecco soon observed the same effect in bacteria infected with UV radiation-damaged phages. It now appears that most forms of life share this important mechanism of repair, which is termed **photoreactivation,** or **light repair.** However, placental mammals, including humans, do not have a photoreactivation pathway.

It was discovered in the late 1950s that photoreactivation is catalyzed by an enzyme called **photoreactivating enzyme** or **DNA photolyase**. This enzyme operates by the mechanism sketched in Figure 20.39. First, the enzyme detects and binds to the damaged DNA site (a pyrimidine dimer). Then the enzyme absorbs light in the UV-A to blue region of the spectrum, which activates it so it can break the bonds holding the pyrimidine dimer together. This restores the pyrimidines to their original independent state. Finally, the enzyme dissociates from the DNA; the damage is repaired.

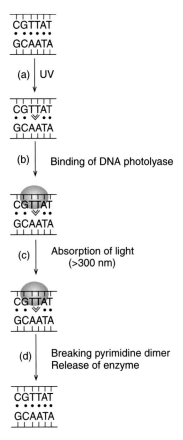

Figure 20.39   Model for photoreactivation. (a) Ultraviolet radiation causes a pyrimidine dimer to form. **(b)** The DNA photolyase enzyme (red) binds to this region of the DNA **(c)** The enzyme absorbs near UV to visible light. **(d)** The enzyme breaks the dimer and finally dissociates from the repaired DNA.

Organisms ranging from *E. coli* to human beings can directly reverse another kind of damage, alkylation of the O6 of guanine. After DNA is methylated or ethylated, an enzyme called **O6-methylguanine methyl transferase** comes on the scene to repair the damage. It does this by accepting the methyl or ethyl group itself, as outlined in Figure 20.40.

The acceptor site on the enzyme for the alkyl group is the sulfur atom of a cysteine residue. Strictly speaking, this means that the methyl transferase does not fulfill one part of the definition of an enzyme—that it be regenerated unchanged after the reaction. Instead, this protein seems to be irreversibly inactivated, so we call it a "suicide enzyme" to denote the fact that it "dies" in performing its function. The repair process is therefore expensive; each repair event costs one protein molecule.

One more property of the O6-methylguanine methyl transferase is worth noting. The enzyme, at least in *E. coli,* is induced by DNA alkylation. This means bacterial cells that have already been exposed to alkylating agents are more resistant to DNA damage than cells that have just been exposed to such mutagens for the first time.

> **SUMMARY** Ultraviolet radiation damage to DNA (pyrimidine dimers) can be directly repaired by a DNA photolyase that uses energy from near-UV to blue light to break the bonds holding the two pyrimidines together. O6 alkylations on guanine residues can be directly reversed by the suicide enzyme O6-methylguanine methyl transferase, which accepts the alkyl group onto one of its amino acids.

Excision Repair in Prokaryotes

The percentage of DNA damage products that can be handled by direct reversal is necessarily small. Most such damage products involve neither pyrimidine dimers nor O6-alkylguanine, so they must be handled by a different mechanism. Most are removed by a process called **excision repair.** The damaged DNA is first removed, then replaced with fresh DNA, by one of two mechanisms: base excision repair or nucleotide excision repair. Base excision repair is more prevalent and usually works on common,

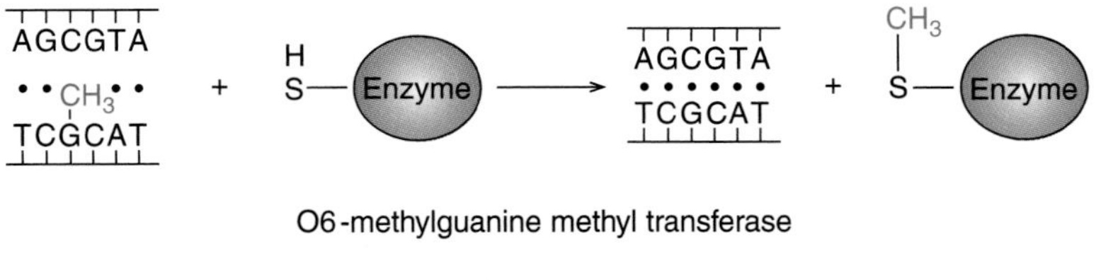

Figure 20.40   Mechanism of O6-methylguanine methyl transferase. A sulfhydryl group of the enzyme accepts the methyl group (blue) from a guanine on the DNA, thus inactivating the enzyme.

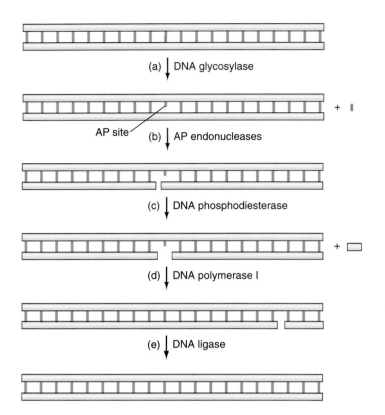

Figure 20.41   Base excision repair in *E. coli.* (a) DNA glycosylase removes the damaged base (red), leaving an apurinic or apyrimidinic site on the bottom DNA strand. (b) An AP endonuclease cuts the DNA on the 5′-side of the AP site. (c) DNA phosphodiesterase removes the AP-deoxyribose-phosphate (yellow block at right) that was left by the DNA glycosylase, (d) DNA polymerase I fills in the gap and continues repair synthesis for a few nucleotides downstream, degrading DNA and simultaneously replacing it. (e) DNA ligase seals the nick left by the DNA polymerase.

relatively subtle changes to DNA bases. Nucleotide excision repair generally deals with more drastic changes to bases, many of which distort the DNA double helix. These changes tend to be caused by mutagenic agents from outside of the cell. A good example of such damage is a pyrimidine dimer caused by ultraviolet light.

Base Excision Repair   In **base excision repair** (**BER**), a damaged base is recognized by an enzyme called **DNA glycosylase,** which breaks the **glycosidic bond** between the damaged base and its sugar (Figure 20.41). This leaves an **apurinic** or **apyrimidinic site** (**AP site**), which is a sugar without its purine or pyrimidine base. Once the AP site is created, it is recognized by an **AP endonuclease** that cuts, or **nicks,** the DNA strand on the 5′-side of the AP site. (The "endo" in endonuclease means the enzyme cuts inside a DNA strand, not at a free end; (Greek *endo,* meaning within.) In *E. coli,* DNA phosphodiesterase removes the AP sugar phosphate, then DNA polymerase I performs repair synthesis by degrading DNA in the 5′ → 3′ direction, while filling in with new DNA. Finally, DNA

ligase seals the remaining nick to complete the job. Many different DNA glycosylases have evolved to recognize different kinds of damaged bases. Humans have at least eight of these enzymes.

Nucleotide Excision Repair   Bulky base damage, including thymine dimers, can also be removed directly, without help from a DNA glycosylase. In this **nucleotide excision repair** (**NER**) pathway (Figure 20.42), the incising enzyme system makes cuts on either side of the damage, removing an oligonucleotide with the damage. The key enzyme *E. coli* cells use in this process is called the uvrABC endonuclease because it contains three polypeptides, the products of the *uvrA, uvrB,* and *uvrC* genes. This enzyme generates an oligonucleotide that is 12–13 bases long, depending on whether the damage affects one nucleotide (alkylations) or two (thymine dimers). A more general term for the enzyme system that catalyzes nucleotide excision repair is excision nuclease, or **excinuclease.** As we will see later in this chapter, the excinuclease in eukaryotic cells removes an oligonucleotide about 24–32 nt long, rather than a 12- to 13-mer. In any case, DNA polymerase fills in the gap left by the excised oligonucleotide and DNA ligase seals the final nick.

SUMMARY   Most DNA damage in *E. coli* is corrected by excision repair, in which the damaged DNA is removed and replaced with normal DNA. This can occur by two different mechanisms. (1) The damaged base can be clipped out by a DNA glycosylase, leaving an apurinic or apyrimidinic site that attracts the DNA repair enzymes that remove and replace the damaged region. (2) The damaged DNA can be clipped out directly by cutting on both sides with an endonuclease to remove the damaged DNA as part of an oligonucleotide. DNA polymerase I fills in the gap and DNA ligase seals the final nick.

Excision Repair in Eukaryotes

Much of our information about repair mechanisms in humans has come from the study of congenital defects in DNA repair. These repair disorders cause a group of human diseases, including Cockayne's syndrome and **xeroderma pigmentosum** (**XP**). Most XP patients are thousands of times more likely to develop skin cancer than normal people if they are exposed to the sun. In fact, their skin can become literally freckled with skin cancers. However, if XP patients are kept out of sunlight, they suffer only normal incidence of skin cancer. Even if XP patients are exposed to sunlight, the parts of their skin that are shielded from light have essentially

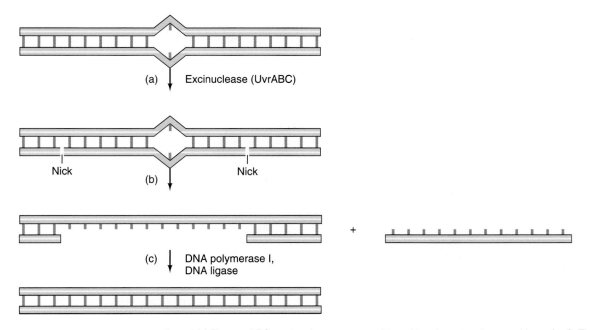

Figure 20.42 Nucleotide excision repair in *E. coli* (a) The UvrABC excinuclease cuts on either side of a bulky damaged base (red). This causes removal **(b)** of an oligonucleotide 12 nucleotides long. If the damage were a pyrimidine dimer, then the oligonucleotide would be a 13-mer instead of a 12-mer. **(c)** DNA polymerase I fills in the missing nucleotides, using the top strand as template, and then DNA ligase seals the nick to complete the task, as in base excision repair.

no cancers. These findings underscore the potency of sunlight as a mutating agent.

Why are XP patients so extraordinarily sensitive to sunlight? XP cells are defective in NER and therefore cannot repair helix-distorting DNA damage, including pyrimidine dimers, effectively. Thus, the damage persists and ultimately leads to mutations, which ultimately lead to cancer. Because NER is also responsible for repairing chemically induced DNA damage that is helix-distorting, we would expect XP patients to have a somewhat higher than average incidence of internal cancers caused by chemical mutagens, and they do. However, the incidence of such cancers in XP patients is only marginally higher than that in normal people. This suggests that most internal DNA damage in humans is not helix-distorting and we have an alternative pathway for correcting that milder kind of damage: the BER pathway. But we have no alternative pathway for correcting UV damage because we do not have a photoreactivation system.

Nucleotide excision repair takes two forms in eukaryotes. It can involve all lesions throughout the genome (**global genome NER, or GG-NER**), or it can be confined to the transcribed strands in genetically active regions of the genome (**transcription-coupled NER, or TC-NER**). The mechanisms of these two forms of NER share many aspects in common, but the method of recognition of the damage differs, as we will see. Let us examine both processes as they occur in humans.

Global Genome NER   What repair steps are defective in XP cells? There are at least eight answers to this question. The problem has been investigated by fusing cells from different patients to see if the fused cells still show the defect. Frequently they do not; instead, the genes from two different patients complement each other. This probably means that a different gene was defective in each patient. So far, seven different complementation groups affecting excision repair have been identified this way. In addition, some patients have a variant form of XP (**XP-V**) in which excision repair is normal, and the patients' cells are only slightly more sensitive to ultraviolet light than normal cells are. We will discuss the gene responsible for XP-V later in this chapter. Taken together, these studies suggest that the defect can lie in any of at least eight different genes. Seven of these genes are responsible for excision repair, and they are named **XPA–XPG**. Most often, the first step in excision repair, incision, or cutting the affected DNA strand, seems to be defective.

The first step in human global genome NER (Figure 20.43) is the recognition of a distortion in the double helix caused by DNA damage. This is where the first XP protein (**XPC**) gets involved. XPC, together with another protein called hHR23B, recognizes a lesion in the DNA, binds to it, and causes melting of a small DNA region around the damage. This role in melting DNA is supported by in vitro studies performed in 1997 with templates that contain lesions surrounded by or adjacent to a small "bubble" of melted DNA. These templates do not

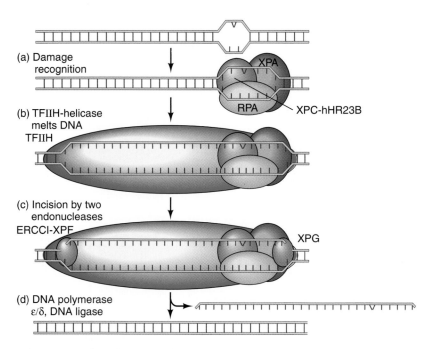

Figure 20.43 Human global genome NER. (a) In the damage recognition step, the XPC–hHR23B complex recognizes the damage (a pyrimidine dimer in this case), binds to it, and causes localized DNA melting. XPA also aids this process. RPA binds to the undamaged DNA strand across from the damage. (b) The DNA helicase activity of TFIIH causes increased DNA melting. (c) RPA helps position two endonucleases (the ERCC1–XPF complex and XPG) on either side of the damage, and these endonucleases clip the DNA. (d) With the damaged DNA removed on a fragment 24–32 nt long, DNA polymerase fills in the gap with good DNA and DNA ligase seals the final nick.

require XPC, suggesting that this protein's job had already been performed when the DNA was melted. Also, Jan Hoeijmakers and colleagues used DNase footprinting in 1998 to show that XPC binds directly to a site of helix distortion in DNA and causes a change in the DNA's conformation (presumably a strand separation).

XPA, which has an affinity for damaged DNA, is also involved in an early stage of damage recognition. Because both XPC and XPA can bind to damaged DNA, why do we believe that XPC is the first factor on the scene? Competition studies performed by Hoeijmakers and colleagues, with different sized templates, support this hypothesis. These workers incubated XPC with one damaged template, and all the other factors *except* XPC with the other damaged template. Then they mixed the two together. Repair began first on the template that was originally incubated with XPC alone, suggesting that XPC binds first to the damaged DNA. Then what is the role of XPA? It can bind to many of the other factors involved in NER, so it may verify the presence of a DNA lesion in DNA that is already denatured (by XPC or by other means), and help to recruit the other NER factors.

At first, it may seem surprising to learn that two of the other XP genes—**XPB** and **XPD**—code for two subunits of the general transcription factor TFIIH, implicating this general transcription factor in NER. However, we now know that these two polypeptides have the DNA helicase

activity inherent in TFIIH (Chapter 11). So one role of TFIIH is to enlarge the region of melted DNA around the damage. But TFIIH is required for NER in vitro even with damaged DNAs that have large melted regions, so this protein must have a function beyond providing DNA helicases. The fact that TFIIH interacts with a number of other NER factors suggests that it serves as an organizer of the NER complex.

The melting of the DNA by TFIIH attracts nucleases that nick one strand on either side of the damage, excising a 24–32-nt oligonucleotide that contains the damage. Two excinucleases make the cuts on either side of the damaged DNA. One is the **XPG** product, which cuts on the 3′-side of the damage. The other is a complex composed of a protein called **ERCC1** plus the **XPF** product, which cuts on the 5′-side. These nucleases are ideally suited for their task: They specifically cut DNA at the junction between double-stranded DNA and the single-stranded DNA created by the TFIIH around the damage. Another protein known as **RPA** helps position the two excinucleases for proper cleavage. RPA is a single-strand-binding protein that binds preferentially to the undamaged strand across from the lesion. The side of RPA facing toward the 3′-end of this DNA strand binds the ERCC1–XPF complex, and the other side of RPA binds XPG. This automatically puts the two excinucleases on the correct sides of the lesion.

Once the defective DNA is removed, DNA polymerase ε or δ fills in the gap, and DNA ligase seals the remaining nick. The role of **XPE** is not clear yet. Although it is important in vivo and gives rise to XP symptoms when it is defective, it is not essential in vitro. This makes it more difficult to pin down its function.

Transcription-Coupled NER Transcription-coupled NER uses all of the same factors as does global genome NER, except for XPC. Because XPC appears to be responsible for initial damage recognition and limited DNA melting in GG-NER, what plays these roles in TC-NER? The answer is RNA polymerase. When RNA polymerase encounters a distortion of the double helix caused by DNA damage, it stalls. This places the bubble of melted DNA, which is created by the polymerase, at the site of the lesion. At that point, XPA could recognize the lesion in the denatured DNA and recruit the other factors. From that point on, these factors would behave much as they do in GG-NER, enlarging the melted region, clipping the DNA in two places, and removing the piece of DNA containing the lesion.

Consider the usefulness of RNA polymerase as a DNA damage detector. It is constantly scanning the genome as it transcribes, and lesions block its passage, demanding attention. Lesions in parts of the DNA that are not transcribed (or even on the nontranscribed strand in a transcribed region) would not be detected this way, but they can wait longer to be repaired because they are not blocking gene expression. Thus, the fact that noncoding lesions such as pyrimidine dimers and 3mA block transcription as well as DNA replication is useful to the cell in that these lesions stall the transcribing polymerase, which recruits the repair machinery.

SUMMARY Eukaryotic NER follows two pathways. In GG-NER, a complex composed of XPC and hHR24B initiates repair by binding to a lesion anywhere in the genome and causing a limited amount of DNA melting. This protein apparently recruits XPA and RPA. TFIIH then joins the complex, and two of its subunits (XPB and XPD) use their DNA helicase activities to expand the melted region. RPA binds two excinucleases (XPF and XPG) and positions them for cleavage of the DNA strand on either side of the lesion. This releases the damage on a fragment between 24 and 32 nt long. TC-NER is very similar to GG-NER, except that RNA polymerase plays the role of XPC in damage sensing and initial DNA melting. In either kind of NER, DNA polymerase ε or δ fills in the gap left by the removal of the damaged fragment, and DNA ligase seals the DNA.

Mismatch Repair

So far, we have been discussing repair of bulky DNA damage caused by mutagenic agents. What about DNA that simply has a mismatch due to incorporation of the wrong base and failure of the proofreading system? At first, it would seem tricky to repair such a mistake because of the apparent difficulty in determining which strand is the newly synthesized one that has the mistake and which is the parental one that should be left alone. At least in *E. coli* this is not a problem because the parental strand has identification tags that distinguish it from the progeny strand. These tags are methylated adenines, created by a methylating enzyme that recognizes the sequence GATC and places a methyl group on the A. Because this 4-base sequence occurs approximately every 250 bp, one is usually not far from a newly created mismatch.

Moreover, GATC is a palindrome, so the opposite strand also reads GATC in the 5′→3′ direction. This means that a newly synthesized strand across from a methylated GATC is also destined to become methylated, but a little time elapses before that can happen. The **mismatch repair** system (Figure 20.44) takes advantage of this delay; it uses the methylation on the parental strand as a signal to leave that strand alone and correct the nearby mismatch in the unmethylated progeny strand. This process must occur fairly soon after the mismatch is created, or both strands will be methylated and no distinction between them will be possible. Eukaryotic mismatch repair is not as well understood as that in *E. coli*. The genes encoding the mismatch recognition and excision enzymes (MutS and MutL) are very well conserved, so the mechanisms that depend on these enzymes are likely to be similar in eukaryotes and prokaryotes. However, the gene encoding the strand recognition protein (MutH) is not found in eukaryotes, so eukaryotes appear not to use the methylation recognition trick. It is not clear yet how eukaryotic cells distinguish the progeny strand from the parental strand at a mismatch.

SUMMARY The *E. coli* mismatch repair system recognizes the parental strand by its methylated adenines in GATC sequences. Then it corrects the mismatch in the complementary (progeny) strand. Eukaryotes use part of this repair system, but they rely on a different, uncharacterized method for distinguishing the strands at a mismatch.

Failure of Mismatch Repair in Humans

Failure of human mismatch repair has serious consequences, including cancer. One of the most common forms of hereditary cancer is **hereditary nonpolyposis colon cancer (HNPCC)**. Approximately 1 American in 200 is

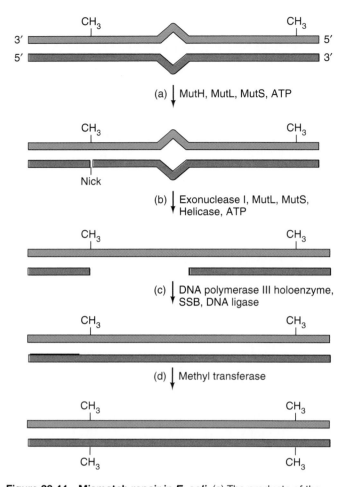

Figure 20.44   Mismatch repair in *E. coli.* (a) The products of the *mutH, L,* and *S* genes along with ATP, recognize a base mismatch (center), identify the newly synthesized strand by the absence of methyl groups on GATC sequences, and introduce a nick into that new strand, across from a methylated GATC and upstream of the incorrect nucleotide. **(b)** Exonuclease I, along with MutL, MutS, DNA helicase, and ATP, removes DNA downstream of the nick, including the incorrect nucleotide. **(c)** DNA polymerase III holoenzyme, with help from single-stranded binding protein (SSB), fills in the gap left by the exonuclease, and DNA ligase seals the remaining nick **(d)** A methyl transferase methylates GATC sequences in the progeny strand across from methylated GATC sequences in the parental strand. Once this happens, mismatch repair nearby cannot occur because the progeny and parental strands are indistinguishable.

affected by this disease, and it accounts for about 15% of all colon cancers—20,000 cases per year in the United States. One of the characteristics of HNPCC patients is microsatellite instability, which means that DNA **microsatellites,** tandem repeats of 1–4-bp sequences, change in size (number of repeats) during the patient's lifetime. This is unusual; the number of repeats in a given microsatellite may differ from one normal individual to another, but it should be the same in all tissues and remain constant throughout the individual's lifetime. The relationship between microsatellite instability and mismatch repair is that the mismatch repair system is responsible for recog-

nizing and repairing the "bubble" created by the inaccurate insertion of too many or too few copies of a short repeat because of "slippage" during DNA replication. When this system breaks down, such slippage goes unrepaired, leading to mutations in many genes whenever DNA replicates in preparation for cell division. This kind of genetic instability presumably leads to cancer, by mechanisms involving mutated genes (oncogenes and tumor suppressor genes) that are responsible for control of cell division.

SUMMARY The failure of human mismatch repair leads to microsatellite instability, and ultimately to cancer.

Coping with DNA Damage without Repairing It

The direct reversal and excision repair mechanisms described so far are all true repair processes. They eliminate the defective DNA entirely. However, cells have other means of coping with damage that do not remove it but simply skirt around it. These are sometimes called repair mechanisms, even though they really are not. A better term might be **damage tolerance mechanism.** These mechanisms come into play when a cell has not performed true repair of a lesion, but has either replicated its DNA or both replicated its DNA and divided before repairing the lesion. At each of these steps (DNA replication and cell division), the cell loses attractive options for dealing with DNA damage and is increasingly faced with more dangerous options.

Recombination Repair   Recombination repair is the most important of these mechanisms. It is also sometimes called **postreplication repair** because replication past a pyrimidine dimer can leave a problem: a gap opposite the dimer that must be repaired. Excision repair will not work any longer because there is no undamaged DNA opposite the dimer—only a gap—so recombination repair is one of the few alternatives left. Figure 20.45 shows how recombination repair works. First, the DNA is replicated. This creates a problem for DNA with pyrimidine dimers because the dimers stop the replication machinery. Nevertheless, after a pause, replication continues, leaving a gap (a **daughter strand gap**) across from the dimer. (A new primer is presumably required to restart DNA synthesis.) Next, recombination occurs between the gapped strand and its homologue on the other daughter DNA duplex. This recombination depends on the *recA* gene product, which exchanges the homologous DNA strands (Chapter 22). The net effect of this recombination is to fill in the gap across from the pyrimidine dimer and to create

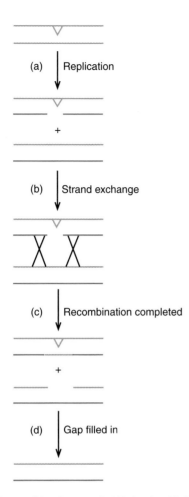

Figure 20.45   Recombination repair. We begin with DNA with a pyrimidine dimer, represented by a V shape. **(a)** During replication, the replication machinery skips over the region with the dimer, leaving a gap; the complementary strand is replicated normally. The two newly synthesized strands are shown in red. **(b)** Strand exchange between homologous strands occurs. **(c)** Recombination is completed, filling in the gap opposite the pyrimidine dimer, but leaving a gap in the other daughter duplex. **(d)** This last gap is easily filled, using the normal complementary strand as the template.

a new gap in the other DNA duplex. However, because the other duplex has no dimer, the gap can easily be filled in by DNA polymerase and ligase. Note that the DNA damage still exists, but the cell has at least managed to replicate its DNA. Sooner or later, true DNA repair could presumably occur.

Error-Prone Bypass   So-called **error-prone bypass** is another way of dealing with damage without really repairing it. In *E. coli*, this pathway is induced by DNA damage, including ultraviolet damage, and depends on the product of the *recA* gene. We have encountered *recA* before—in the preceding paragraph and in our discussion of the induction of a lambda prophage during the SOS response (Chapter 8)—and we will encounter it again in our consideration of recombination in Chapter 22. Indeed,

error-prone repair is also part of the SOS response. The chain of events seems to be as follows (Figure 20.46): Ultraviolet light or another mutagenic treatment somehow activates the RecA coprotease activity. This coprotease has several targets. One we have studied already is the λ repressor, but its main target is the product of the *lexA* gene. This product, **LexA**, is a repressor for many genes, including repair genes; when it is stimulated by RecA coprotease to cleave itself, all these genes are induced.

Two of the newly induced genes are *umuC* and *umuD,* which make up a single operon (*umuDC*). The product of the *umuD* gene (UmuD) is clipped by a protease to form **UmuD′**, which associates with the *umuC* product, **UmuC,** to form a complex **UmuD′$_2$C**. This complex has DNA polymerase activity, so it is also referred to as **DNA pol V.** (DNA pol IV is the product of another SOS response gene, *dinB*.) Pol V can cause error-prone bypass of pyrimidine dimers in vitro on its own, but pol III holoenzyme and RecA protein stimulate this process considerably. Such bypass involves replication of DNA across from the pyrimidine dimer even though correct "reading" of the defective strand is impossible. This avoids leaving a gap, but it frequently puts the wrong bases into the new DNA strand (hence the name "error-prone"). When the DNA replicates again, these errors will be perpetuated.

If the *umu* genes are really responsible for error-prone bypass, we might expect mutations in one of these genes to make *E. coli* cells *less* susceptible to mutation. These mutant cells would be just as prone to DNA damage, but the damage would not be as readily converted into mutations. In 1981, Graham Walker and colleagues verified this expectation by creating a **null allele** of the *umuC* gene (a version of the gene with no activity), and showing that bacteria harboring this gene were essentially unmutable. In fact, "umu" stands for "unmutable."

These workers established an *E. coli* strain carrying the *umuC* mutant, and an ochre *his⁻* mutation that is ordinarily revertable by UV radiation. Then they challenged this bacterial strain with UV radiation and counted the *his⁺* revertants. The more revertants, the more mutation was allowed because a reversion is just a back-mutation. Figure 20.47 shows the results. A reasonable number of revertants occurred in wild-type cells (about 200 at the highest UV dose). By stark contrast, in *umuC⁻* cells almost no revertants occur. Furthermore, addition of a plasmid bearing the *muc* gene, which can suppress the unmutable phenotype of *umuC⁻* cells, caused a dramatic increase in the number of revertants (about 500, even at a relatively low UV dose).

The null allele in this experiment was created by insertion of the *lac* structural genes, without the *lac* promoter, into the *umuC* gene, then screening for *lac⁺* cells. The cells were originally *lac⁻*, so the appearance of *lac⁺* cells indicated that the *lac* genes had inserted downstream of a promoter—the *umuDC* promoter, in this case. The fact

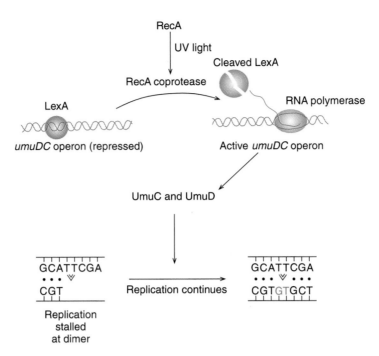

Figure 20.46 Error-prone (SOS) bypass. Ultraviolet light activates the RecA coprotease, which stimulates the LexA protein (purple) to cleave itself, releasing it form the *umuDC* operon. This results in synthesis of UmuC and UmuD proteins, which allow DNA synthesis across from a pyrimidine dimer, even though mistakes (blue) will frequently be made.

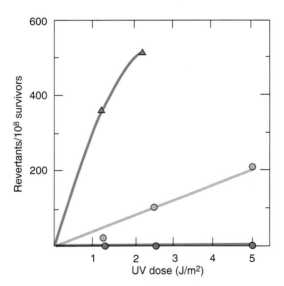

Figure 20.47 An *umuC* strain of *E. coli* is unmutable. Walker and colleagues tested three *his⁻* strains of bacteria for the ability to generate *his⁺* revertants after UV irradiation. The strains were: wild-type with respect to *umuC* (blue), a *umuC⁻* strain (red), and a *umuC⁻* strain supplemented with a plasmid containing the *muc* gene (green). (*Source:* From Bagg et al., *Proceedings of the National Academy of Sciences* 78:5750, 1981.)

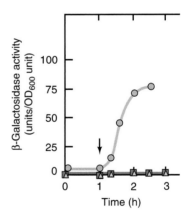

Figure 20.48 The *umuDC* promoter is UV-inducible. Walker and colleagues irradiated cells with the *lac* genes under control of the *umuDC* promoter with a UV dose of 10 J/m². They performed the irradiation at 1 h, as indicated by the arrow. Then they measured the accumulation of β-galactosidase activity (blue curve) per OD_{600} unit (an index of turbidity and therefore of cell density). They also performed the same experiment in *lexA* mutant (yellow triangles) and *recA⁻* cells (red squares). The *lexA* mutant was an "uninducible" one encoding a LexA protein that cannot be cleaved and therefore cannot be removed from the *umuDC* operator. (*Source:* From Bagg et al., *Proceedings of the National Academy of Sciences* 78:5751, 1981.)

that the *lac* genes fell under control of the *umuDC* promoter allowed Walker and colleagues to test the inducibility of this promoter by UV radiation, simply by measuring β-galactosidase activity. Figure 20.48 shows that the promoter was indeed inducible by UV radiation at a dose of 10 J/m² (blue curve). But the promoter was not inducible in *lexA* mutant or *recA⁻* cells (red curve). The *lexA* mutant cells used in this experiment encoded a LexA protein that was not cleavable and therefore could not be removed from the *umuDC* operator.

Wild-type *E. coli* cells can tolerate as many as 50 pyrimidine dimers in their genome without ill effect because of their active repair mechanisms. Bacteria lacking one of the *uvr* genes cannot carry out excision repair, so their susceptibility to UV damage is greater. However, they are still somewhat resistant to DNA damage. On the other hand, double mutants in *uvr* and *recA* can perform neither excision repair nor recombination repair, and they are very sensitive to UV damage, perhaps because they have to rely on error-prone repair. Under these conditions, only one to two pyrimidine dimers per genome is a lethal dose.

Obviously, if bacterial cells had evolved without the error-prone bypass, they would be subject to many fewer mutations. If that is the case, then why have they retained this mutation-causing mechanism? It is likely that the error-prone bypass system does more good than harm by allowing an organism to replicate its damaged genome even at the risk of mutation. This is especially obvious if the price for failure to replicate is death, as would be the case after a cell replicates its damaged DNA and then divides without repairing the damage. This chain of events would produce one daughter cell with a DNA gap across from a lesion. By this time, excision repair and even recombination repair are no longer possible. So the last resort is error-prone bypass to stave off cell death.

It is also true that a certain level of mutation is good for a species because it allows the genomes of a group of organisms to diverge so they do not all have equal susceptibility to disease and other insults. That way, when a new challenge arises, some of the members of a population have resistance and can survive to perpetuate the species.

Error-Prone and Error-Free Bypass in Humans Human cells also have an error-prone system to bypass pyrimidine dimers. It involves a distinct DNA polymerase known as **DNA polymerase ζ** (zeta) that inserts bases at random to get past the pyrimidine dimer. But human cells also have a relatively error-free bypass system that automatically inserts two dAMPs into the DNA strand across from a pyrimidine dimer. Thus, even though the bases in the dimer cannot base-pair, this system is able to make the correct choice if both bases in the dimer are thymines—which is often the case. The DNA polymerase involved in this relatively error-free bypass is **DNA polymerase η** (eta).

In 1999, Fumio Hanaoka and colleagues discovered that the defective gene in patients with the variant form of XP (XP-V) is the gene that codes for DNA polymerase η. Thus, these patients cannot carry out the comparatively error-free bypass of pyrimidine dimers catalyzed by DNA polymerase η and must therefore rely on the error-prone bypass catalyzed by DNA polymerase ζ. This error-prone system introduces mutations during replication of pyrimidine dimers not removed by the excision repair system. However, because these patients have normal excision repair, few dimers are left for the error-prone system to deal with. This argument accounts for the relatively low sensitivity of XP-V cells to ultraviolet radiation.

Although the term "error-free" for DNA polymerase η is justified in terms of its ability to deal with thymine dimers, it is really a misnomer. When Hanaoka, Thomas Kunkel, and colleagues tested the fidelity of this enzyme in vitro, using a double-stranded DNA with a gap in it, they found that DNA polymerase η had a lower fidelity than any other template-dependent DNA polymerase ever studied: one mistake per 18–380 nt incorporated. Thus, it is a good thing that cells normally have the NER system. Without it, DNA polymerase η would be a very poor backstop for dealing with anything but thymine dimers—as typical XP patients can attest.

SUMMARY Cells can employ nonrepair methods to circumvent DNA damage. One of these is recombination repair, in which the gapped DNA strand across from a damaged strand recombines with a normal strand in the other daughter DNA duplex after replication. This solves the gap problem but leaves the original damage unrepaired. Another mechanism to deal with DNA damage, at least in *E. coli,* is to induce the SOS response, which causes the DNA to replicate even though the damaged region cannot be read correctly. This results in errors in the newly made DNA, so the process is called error-prone bypass.

SUMMARY Humans have a relatively error-free bypass system that inserts dAMPs across from a pyrimidine dimer, thus replicating thymine dimers (but not dimers involving cytosines) correctly. This system uses DNA polymerase η. When the gene for this enzyme is defective, DNA polymerase ζ takes over. But this polymerase inserts random nucleotides across from a pyrimidine dimer, so it is error-prone. These errors in correcting UV damage lead to a variant form of XP known as XP-V. Even DNA polymerase η is not really error-free. With a gapped template it is the least accurate template-dependent polymerase known.

SUMMARY

Several principles apply to all (or most) DNA replication: (1) Double-stranded DNA replicates in a semiconservative manner. When the parental strands separate, each serves as the template for making a new, complementary strand. (2) DNA replication in *E. coli* (and in other organisms) is semidiscontinuous. One strand is replicated continuously in the direction of the movement of the replicating fork; the other is replicated discontinuously, forming 1–2 kb Okazaki fragments in the opposite direction. This allows both strands to be replicated in the 5′→3′ direction. (3) Initiation of DNA replication requires a primer. Okazaki fragments in *E. coli* are initiated with RNA primers 10–12 nt long. (4) Most eukaryotic and prokaryotic DNAs replicate bidirectionally. ColE1 is an example of a DNA that replicates unidirectionally.

Circular DNAs can replicate by a rolling circle mechanism. One strand of a double-stranded DNA is nicked and the 3′-end is extended, using the intact DNA strand as template. This displaces the 5′-end. In phage λ, the displaced strand serves as the template for discontinuous, lagging strand synthesis.

The helicase that unwinds double-stranded DNA at the replicating fork is encoded by the *E. coli dnaB* gene. The prokaryotic single-strand DNA-binding proteins bind much more strongly to single-stranded than to double-stranded DNA. They aid helicase action by binding tightly and cooperatively to newly formed single-stranded DNA and keeping it from annealing with its partner. By coating the single-stranded DNA, SSBs also protect it from degradation. They also stimulate their homologous DNA polymerases. These activities make SSBs essential for prokaryotic DNA replication.

As a helicase unwinds the two parental strands of a closed circular DNA, it introduces a compensating positive supercoiling force into the DNA. The stress of this force must be overcome or it will resist progression of the replicating fork. A name given to this stress-release mechanism is the swivel. DNA gyrase, a bacterial topoisomerase, is the leading candidate for this role in *E. coli*.

Pol I is a versatile enzyme with three distinct activities: DNA polymerase; 3′→5′ exonuclease; and 5′→3′ exonuclease. The first two activities are found on a large domain of the enzyme, and the last is on a separate, small domain. The large domain (the Klenow fragment) can be separated from the small by mild protease treatment, yielding two protein fragments with all three activities intact. The structure of the Klenow fragment shows a wide cleft for binding to DNA. This polymerase active site is remote from the 3′→5′ exonuclease active site on the Klenow fragment.

Of the three DNA polymerases in *E. coli* cells, pol I, pol II, and pol III, only pol III is required for DNA replication. Thus, this polymerase is the enzyme that replicates the bacterial DNA. The pol III core is composed of three subunits, α, ε, and θ. The α-subunit has the DNA polymerase activity. The ε-subunit has the 3′→5′ activity that carries out proofreading.

Faithful DNA replication is essential to life. To help provide this fidelity, the *E. coli* DNA replication machinery has a built-in proofreading system that requires priming. Only a base-paired nucleotide can serve as a primer for the pol III holenzyme. Therefore, if the wrong nucleotide is incorporated by accident, replication stalls until the 3′→5′ exonuclease of the pol III holoenzyme removes it. The fact that the primers are made of RNA may help mark them for degradation.

Mammalian cells contain five different DNA polymerases. Polymerases δ and α appear to participate in replicating both DNA strands Polymerase α makes the primers for both strands, and polymerase δ catalyzes elongation. Polymerases β and ε seem to function in DNA repair. Polymerase γ probably replicates mitochondrial DNA.

Alkylating agents like ethylmethane sulfonate add bulky alkyl groups to bases, either disrupting base-pairing directly or causing loss of bases, either of which can lead to faulty DNA replication or repair.

Different kinds of radiation cause different kinds of damage. Ultraviolet rays have comparatively low energy, and they cause a moderate type of damage: pyrimidine dimers. Gamma and x-rays are much more energetic. They ionize the molecules around DNA and form highly reactive free radicals that can attack DNA, altering bases or breaking strands.

Ultraviolet radiation damage to DNA (pyrimidine dimers) can be directly repaired by a DNA photolyase that uses energy from visible light to break the bonds holding the two pyrimidines together. O6 alkylations on guanine residues can be directly reversed by the suicide enzyme O6-methylguanine methyl transferase, which accepts the alkyl group onto one of its amino acids.

Most DNA damage in *E. coli* is corrected by excision repair, in which the damaged DNA is removed and replaced with normal DNA. This can occur by two different mechanisms. (1) In base excision repair (BER), the damaged base can be clipped out by a DNA glycosylase, leaving an apurinic or apyrimidinic site that attracts the DNA repair enzymes that remove and replace the damaged region. (2) In nucleotide excision repair, the damaged DNA can be clipped out directly by cutting on both sides with an endonuclease to remove the damaged DNA as part of an oligonucleotide. DNA polymerase I fills in the gap and DNA ligase seals the final nick.

Eukaryotic NER follows two pathways. In global genome NER (GG-NER), a complex composed of XPC and hHR24B initiates repair by binding to a lesion anywhere in the genome and causing a limited amount of DNA melting. This protein apparently recruits XPA and RPA. TFIIH then joins the complex and two of its subunits (XPB and XPD) use their DNA helicase activities to expand the melted region. RPA binds two excinucleases (XPF and XPG) and positions them for cleavage of the DNA strand on either side of the lesion. This releases the damage on a fragment between 24 and 32 nt long. Transcription-coupled NER (TC-NER) is very similar to global genome NER, except that RNA polymerase plays the role of XPC in damage sensing and initial DNA melting. In either kind of NER, DNA polymerase ε or δ fills in the gap left by the removal of the damaged fragment, and DNA ligase seals the DNA.

Errors in DNA replication leave mismatches that can be detected and repaired. The *E. coli* mismatch repair system recognizes the parental strand by its methylated adenines in GATC sequences. Then it corrects the mismatch in the complementary (progeny) strand. The failure of human mismatch repair leads to microsatellite instability, and ultimately to cancer.

Cells can employ nonrepair methods to circumvent DNA damage. One of these is recombination repair, in which the gapped DNA strand across from a damaged strand recombines with a normal strand in the other daughter DNA duplex after replication. This solves the gap problem but leaves the original damage unrepaired. Another mechanism to deal with DNA damage, at least in *E. coli,* is to induce the SOS response, which causes the DNA to replicate even though the damaged region cannot be read correctly. This results in errors in the newly made DNA, so the process is called error-prone bypass.

Humans have a relatively error-free bypass system that inserts AMPs across from a pyrimidine dimer, thus replicating thymine dimers (but not dimers involving cytosines) correctly. This system uses DNA polymerase η. When the gene for this enzyme is defective, DNA polymerase ζ takes over. But this polymerase inserts random nucleotides across from a pyrimidine dimer, so it is error-prone. These errors in correcting UV damage lead to a variant form of XP known as XP-V. Even DNA polymerase η is not really error-free. With a gapped template it is the least accurate template-dependent polymerase known.

REVIEW QUESTIONS

1. Compare and contrast the conservative, semiconservative, and dispersive mechanisms of DNA replication.

2. Describe and give the results of an experiment that shows that DNA replication is semiconservative. Be sure to rule out all kinds of dispersive mechanisms.

3. Compare and contrast the continuous, discontinuous, and semidiscontinuous modes of DNA replication.

4. Describe and give the results of an experiment that shows that DNA replication is semidiscontinuous.

5. Explain the fact that short pulses of labeled nucleotides appear only in short DNA fragments in wild-type *E. coli* cells. In other words, why does DNA replication appear to be fully discontinuous instead of semidiscontinuous in these cells?

6. Why is it improbable that we will ever observe fully continuous DNA replication in nature?

7. Describe and give the results of an experiment that measures the size of Okazaki fragments.

8. Present electron microscopic evidence that DNA replication of the *B. subtilis* chromosome is bidirectional, whereas replication of the colE1 plasmid is unidirectional.

9. Diagram the rolling circle replication mechanism used by the λ phage.

10. Compare and contrast the activity of a helicase with that of a topoisomerase.

11. Diagram a substrate used in a DNA helicase assay, explain the assay, and show sample results.

12. List the roles that SSBs play in DNA replication.

13. Present evidence that an SSB is essential for DNA synthesis.

14. Present evidence that an SSB protects replicating DNA from degradation.

15. Explain why nicking one strand of a supercoiled DNA removes the supercoiling.

16. Describe an assay for a DNA topoisomerase. Show sample results.

17. How do we know that DNA gyrase forms a covalent bond between an enzyme tyrosine and DNA? What is the advantage of forming this bond?

18. Present a model, based on the structure of yeast DNA topoisomerase II, for the DNA segment-passing step.

19. Diagram the proofreading process used by *E. coli* DNA polymerases.

20. What activities are contained in *E. coli* DNA polymerase I? What is the role of each in DNA replication?

21. How does the Klenow fragment differ from the intact *E. coli* DNA polymerase I? Which enzyme would you use in nick translation? Primer extension? Why?

22. Of the three DNA polymerases in *E. coli,* which is essential for DNA replication? Present evidence.

23. Which pol III core subunit has the DNA polymerase activity? How do we know?

24. Which pol III core subunit has the proofreading activity? How do we know?

25. Explain how the necessity for proofreading rationalizes the existence of priming in DNA replication.

26. List the eukaryotic DNA polymerases and their roles. Outline evidence for these roles.

27. Compare and contrast the concepts of DNA damage and mutation.

28. Compare and contrast the DNA damage done by UV rays and x-rays or gamma rays.

29. What two enzymes catalyze direct reversal of DNA damage? Diagram the mechanisms they use.

30. Compare and contrast base excision repair and nucleotide excision repair. Diagram both processes. For what types of damage is each primarily responsible?

31. How does transcription-coupled NER differ from global genome NER?

32. What DNA repair system is missing in most cases of xeroderma pigmentosum? Why does that make XP patients so sensitive to UV light? What is the primary backup system for these patients?

33. What DNA repair system is missing in XP-V patients? Why is the incidence of skin cancer lower in these people than in typical XP patients? What is the backup system for lesions missed by the NER system is XP-V patients?

34. Diagram the mismatch repair mechanism in *E. coli*.

35. Diagram the recombination repair mechanism in *E. coli*.

36. Diagram the error-prone bypass system in *E. coli*.

37. Explain why recombination repair and error-prone bypass are not real repair systems.

SUGGESTED READINGS

General References and Reviews

Citterio, E., W. Vermeulen, and J.H.J. Hoeijmakers. 2000. Transcriptional healing. *Cell* 101:447–50.

de Latt, W.L., N.G.J. Jaspers, and J.H.J. Hoeijmakers. 1999. Molecular mechanism of nucleotide excision repair. *Genes and Development* 13:768–785.

Friedberg, E.C. and V.L. Gerlach. 1999. Novel DNA polymerases offer clues to the molecular basis of mutagenesis. *Cell* 98:413–16.

Herendeen, D.R. and T.J. Kelly. 1996. DNA polymerase III: Running rings around the fork. *Cell* 84:5–8.

Joyce, C.M. and T.A. Steitz. 1987. DNA polymerase I: From crystal structure to function via genetics. *Trends in Biochemical Sciences* 12:288–92.

Kornberg, A. and T. Baker. 1992. *DNA Replication.* New York: W.H. Freeman and Company.

Lindahl, T. and R.D. Wood. 1999. Quality control by DNA repair. *Science* 286:1897–1905.

Maxwell, A. 1996. Protein gates in DNA topoisomerase II. *Nature Structural Biology.* 3:109–12.

Radman, M. and R. Wagner. 1988. The high fidelity of DNA replication. *Scientific American* (August):40–46.

Sharma, A., and A. Mondragón. 1995. DNA topoisomerases. *Current Opinion in Structural Biology* 5:39–47.

Wood, R.D. 1997. Nucleotide excision repair in mammalian cells. *Journal of Biological Chemistry* 272:23465–68.

Wood, R.D. 1999. Variants on a theme. *Nature* 399:639–70.

Research Articles

Bagg, A., C.J. Kenyon, and G.C. Walker. 1981. Inducibility of a gene product required for UV and chemical mutagenesis in *Escherichia. coli. Proceedings of the National Academy of Sciences, USA* 78:5749–53.

Berger, J.M., S.J. Gamblin, S.C. Harrison, and J.C. Wang. 1996. Structure and mechanism of DNA topoisomerase II. *Nature* 379:225–32.

Cairns, J. 1963. The chromosome of *Escherichia coli. Cold Spring Harbor Symposium on Quantitative Biology* 28:43–46.

Curtis, M.J. and B. Alberts. 1976. Studies on the structure of intracellular bacteriophage T4 DNA. *Journal of Molecular Biology* 102:793–816.

Drapkin, R., J.T. Reardon, A. Ansari, J.-C. Huang, L. Zawel, K. Ahn, A. Sancar, and D. Reinberg. 1994. Dual role of TFIIH in DNA excision repair and in transcription by RNA polymerase II. *Nature* 368:769–72.

Eom, S.H., T. Wang, and T.A. Steitz. 1996. Structure of *Taq* polymerase with DNA at the active site. *Nature* 382:278–281.

Gefter, M.L., Y.Hirota, T.Kornberg, J.A. Wechster, and C. Barnoux. 1971. Analysis of DNA polymerases II and III in mutants of *Escherichia coli* thermosensitive for DNA synthesis. *Proceedings of the National Academy of Sciences USA* 68:3150–53.

Gellert, M., K. Mizuuchi, M.H. O'Dea, and H.A. Nash. 1976. DNA gyrase: An enzyme that introduces superhelical turns into DNA. *Proceedings of the National Academy of Sciences USA* 73:3872–76.

Gyurasits, E.B. and R.J. Wake. 1973. Bidirectional chromosome replication in *Bacillus subtilis. Journal of Molecular Biology* 73:55–63.

Hirota, G.H., A. Ryter, and F. Jacob. 1968. Thermosensitive mutants in *E. coli* affected in the processes of DNA synthesis and cellular division. *Cold Spring Harbor Symposia on Quantitative Biology* 33:677–93.

Huberman, J.A., A. Kornberg, and B.M. Alberts. 1971. Stimulation of T4 bacteriophage DNA polymerase by the protein product of T4 gene 32. *Journal of Molecular Biology* 62:39–52.

Kitani, T., K.-y. Yoda, T. Ogawa, and T. Okazaki. 1985. Evidence that discontinuous DNA replication in *Escherichia coil* is primed by approximately 10 to 12 residues of RNA starting with a purine. *Journal of Molecular Biology* 184:45–52.

LeBowitz, J.H., and R. McMacken. 1986. The *Escherichia coli* dnaB replication protein is a DNA helicase. *Journal of Biological Chemistry* 261:4738–48.

Maki, H., and A. Kornberg. 1985. The polymerase subunit of DNA polymerase III of *Escherichia coli. Journal of Biological Chemistry* 260:12987–92.

Masutani, C., R. Kusumoto, A. Yamada, N. Dohmae, M. Yokoi, M. Yuasa, M. Araki, S. Iwai, K. Takio, and

F. Hanaoka. 1999. The *XPV* (xeroderma pigmentosum variant) gene encodes human DNA polymerase η. *Nature* 399: 700–04.

Matsuda, T., K. Bebenek, C. Masutani, F. Hanaoka, and T.A. Kunkel. 2000. Low fidelity DNA synthesis by human DNA polymerase-η. *Nature* 404:1011–13.

Meselson, M. and F. Stahl. 1958. The replication of DNA in *Escherichia coli. Proceedings of the National Academy of Sciences USA* 44:671–82.

Okazaki, R., T. Okazaki, K.Sakabe, K.Sugimoto, R. Kainuma, A. Sugino, and N. Iwatsuki. 1968. In vivo mechanism of DNA chain growth. *Cold Spring Harbor Symposia on Quantitative Biology* 3:129–43.

Scheuermann, R.H. and H. Echols. 1984. A separate editing exonuclease for DNA replication: The ε subunit of *Escherichia coli* DNA polymerase III holoenzyme. *Proceedings of the National Academy of Sciences USA* 81:7747–57.

Sugasawa, K., J.M.Y. Ng, C. Masutani, S. Iwai, P.J. van der Spek, A.P.M. Eker, F. Hanaoka, D. Bootsma, and J.H.J. Hoeijmakers. 1998. Xeroderma pigmentosum group C protein complex is the initiator of global genome nucleotide excision repair. *Molecular Cell* 2:223–32.

Tse, Y.-C., K. Kirkegaard, and J.C. Wang. 1980. Covalent bonds between protein and DNA. *Journal of Biological Chemistry* 255:5560–65.

Wakasugi, M. and A. Sancar. 1999. Order of assembly of human DNA repair excision nuclease. *Journal of Biological Chemistry* 274:18759–68.

CHAPTER 21

DNA Replication II: Detailed Mechanism

Schematic computer graphic representation of the replication of DNA. © Clive Freeman, The Royal Institution/SPL/Photo Researchers, Inc.

We learned in Chapter 20 that DNA replication is semidiscontinuous and requires the synthesis of primers before DNA synthesis can begin. We have also learned about some of the major proteins involved in DNA replication in *E. coli*. Thus, we know that DNA replication is complex and involves more than just a DNA polymerase. This chapter presents a close look at the mechanism of this process in *E. coli* and in eukaryotes. We will begin with a short discussion of the speed of replication in *E. coli*, then look at the three stages of replication—initiation, elongation, and termination—in a variety of systems.

21.1 Speed of Replication

Minsen Mok and Kenneth Marians performed one of the studies that measured the rate of fork movement in vitro with the pol III holoenzyme. They created a synthetic circular template for rolling circle replication, as shown in Figure 21.1. This template contained a ^{32}P-labeled, tailed, full-length strand with a free 3'-hydroxyl group for priming. Mok and Marians incubated this template with either holoenzyme plus preprimosomal proteins and SSB, or plus DnaB helicase alone. At 10-sec intervals, they removed the labeled product DNAs and measured their lengths by electrophoresis. Panels (**a**) and (**b**) in Figure 21.2 depict the results with the two reactions, and 21.2c shows plots of the rates of fork movement with the two reactions. Both plots yielded rates of 730 nt/sec, close to the in vivo rate.

Furthermore, the elongation in these reactions with holoenzyme was highly processive. As we have mentioned, processivity is the ability of the enzyme to stick to

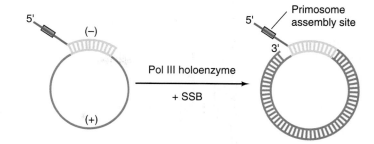

Figure 21.1   Synthesis of template used to measure fork velocity in vitro. Mok and Marians started with the 6702-nt positive strand (red) from the f1 phage and annealed it to a primer (green) that hybridized over a 282-nt region (yellow). This primer contained a primosome assembly site (orange). They elongated the primer with pol III holoenzyme and single-stranded binding protein (SSB) to create the negative strand (blue). The product was a double-stranded template for multiple rounds of rolling circle replication, in which the free 3'-end could serve as the primer. (*Source:* From Mok and Marians, *Journal of Biological Chemistry* 262:16645, 1987. Copyright © 1987 The American Society for Biochemistry & Molecular Biology, Bethesda, MD. Reprinted by permission.)

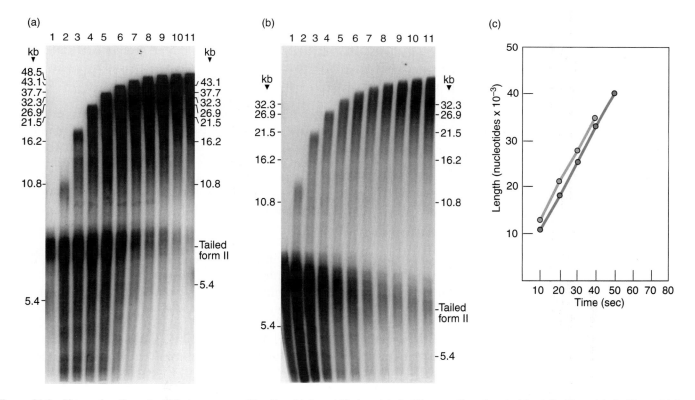

Figure 21.2   Measuring the rate of fork movement in vitro. Mok and Marians labeled the negative strand of the tailed template in Figure 21.1 and used it in in vitro reactions with pol III holoenzyme plus: (**a**) the preprimosomal proteins (the primosomal proteins minus DnaG); or (**b**) DnaB alone. They took samples from the reactions at 10-sec intervals, beginning with lanes 1 at zero time and lanes 2 at 10 sec, electrophoresed them, and then autoradiographed the gel. Panel (**c**) shows a plot of the results from the first five and four time points from panels (**a**) (red) and (**b**) (blue), respectively. (*Source:* Mok & Marians, The *Escherichia coli* preprimosome and DNA B helicase can form replication forks that move at the same rate. *J. Biol. Chem.* 262 no. 34 (5 Dec 1987) f. 6a-b, p. 16650. American Society for Biochemistry and Molecular Biology.)

its job a long time without falling off and having to reinitiate. This is essential because reinitiation is a time-consuming process, and little time can be wasted in DNA replication. To measure processivity, Mok and Marians performed the same elongation assay as described in Figure 21.2, but included either of two substances that would prevent reinitiation if the holoenzyme dissociated from the template. These substances were a competing DNA, poly(dA), and an antibody directed against the β-subunit of the holoenzyme. In the presence of either of these competitors, the elongation rate was just as fast as in their absence, indicating that the holoenzyme did not dissociate from the template throughout the process of elongation of the primer by at least 30 kb. Thus, the holoenzyme is highly processive in vitro, just as it is in vivo.

> **SUMMARY** The pol III holoenzyme synthesizes DNA at the rate of about 730 nt/sec in vitro, just a little slower than the rate of almost 1000 nt/sec observed in vivo. This enzyme is also highly processive, both in vitro and in vivo.

21.2 Initiation

As we have seen, initiation of DNA replication means primer synthesis. Different organisms use different mechanisms to make primers; even different phages that infect E. coli (coliphages) use quite different primer synthesis strategies. The coliphages were convenient tools to probe E. coli DNA replication because they are so simple that they have to rely primarily on host proteins to replicate their DNAs.

Priming in E. coli

As mentioned in Chapter 20, the first example of coliphage primer synthesis was found by accident in M13 phage, when this phage was discovered to use the host RNA polymerase as its **primase** (primer-synthesizing enzyme). But E. coli and its other phages do not use the host RNA polymerase as a primase. Instead, they employ a primase called **DnaG**, which is the product of the E. coli dnaG gene. Arthur Kornberg noted that E. coli and most of its phages need at least one more protein (**DnaB**, the product of the E. coli dnaB gene) to form primers, at least on the lagging strand.

Kornberg and colleagues discovered the importance of DnaB with an assay in which single-stranded φX174 phage DNA (without SSB) is converted to double-stranded form. Synthesis of the second strand of phage DNA required primer synthesis, then DNA replication. The DNA replication part used pol III holoenzyme, so the other required proteins should be the ones needed for

primer synthesis. Kornberg and colleagues found that three proteins: DnaG (the primase), DnaB, and pol III holoenzyme were required in this assay. Thus, DnaG and DnaB were apparently needed for primer synthesis. Kornberg coined the term **primosome** to refer to the collection of proteins needed to make primers for a given replicating DNA. Usually this is just two proteins, DnaG and DnaB, although other proteins may be needed to assemble the primosome.

The φX174 primosome is mobile and can repeatedly synthesize primers as it moves around the uncoated circular phage DNA. As such, it is also well suited for the repetitious task of priming Okazaki fragments on the lagging strand of E. coli DNA. This contrasts with the activity of RNA polymerase or primase alone, which prime DNA synthesis at only one spot—the origin of replication.

Two different general approaches were used to identify the important components of the E. coli DNA replication system, with DNA from phages φX174 and G4 as model substrates. The first approach was a combination genetic–biochemical one, the strategy of which was to isolate mutants with defects in their ability to replicate phage DNA, then to complement extracts from these mutants with proteins from wild-type cells. The mutant extracts were incapable of replicating the phage DNA in vitro unless the right wild-type protein was added. Using this system as an assay, the protein can be highly purified and then characterized. The second approach was the classical biochemical one: Purify all of the components needed and then add them all back together to reconstitute the replication system in vitro. Kornberg, the chief practitioner of the reconstitution strategy summed up his support for it this way: "I remain faithful to the conviction that anything a cell can do, a biochemist should be able to do. He (or she) should do it even better. . . . Put another way, one can be creative more easily with a reconstituted system."

The Origin of Replication in E. coli   Before we discuss priming further, let us consider the unique site at which DNA replication begins in E. coli: *oriC*. An origin of replication is a DNA site at which DNA replication begins and which is essential for proper replication to occur. We can locate the place where replication begins by several means, but how do we know how much of the DNA around the initiation site is essential for replication to begin? One way is to clone a DNA fragment, including the initiation site, into a plasmid that lacks its own origin of replication but has an antibiotic resistance gene. Then we can select for autonomously replicating plasmids with the antibiotic. Any cell that replicates in the presence of the antibiotic must have a plasmid with a functional origin (Figure 21.3). Once we have such an *oriC* plasmid, we can begin trimming and mutating the DNA fragment containing *oriC* to find the minimal effective DNA sequence.

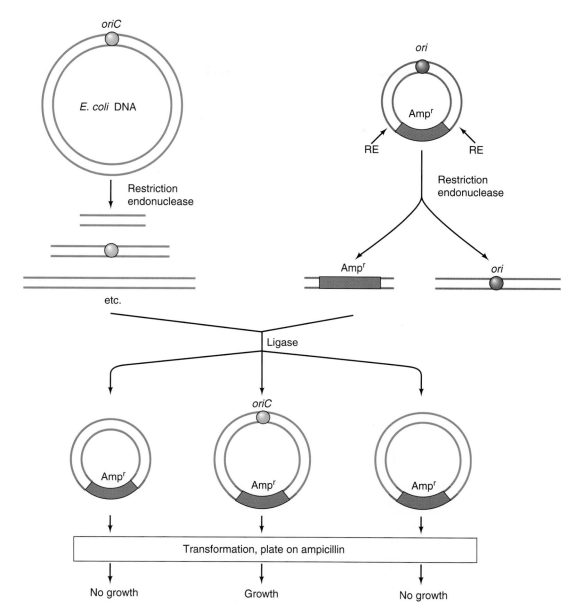

Figure 21.3   Construction of an *oriC* plasmid. Begin by cleaving the *E. coli* chromosomal DNA (red) with a restriction enzyme. The size of the *E. coli* DNA is minimized, and the positions of the restriction sites are not given because they are too numerous. At the same time, cleave a plasmid (blue) with the same restriction endonuclease. This places the plasmid origin of replication (green) and the ampicillin resistance gene (Amp^r, purple) on separate restriction fragments. Next, ligate the fragment bearing Amp^r to the many fragments from the *E. coli* chromosomal DNA, generating recombinants as pictured at bottom. Some of these will have irrelevant *E. coli* DNA, but at least some should receive *oriC*. Finally, transform the plasmids at random into *E. coli* cells and plate them on ampicillin plates to select for those that can replicate in the presence of this antibiotic. Only cells that received a plasmid with *oriC* should grow, because plasmids without an origin of replication cannot replicate and therefore cannot confer ampicillin resistance.

Figure 21.4 shows this sequence from *E. coli* and several other enteric bacteria. The minimal origin in *E. coli* is 245 bp long. We can also see that some features of the origins are conserved in all these bacteria, and the spacing between them is also conserved.

Figure 21.5 illustrates the steps in initiation at *oriC*. The origin includes four nine-mers with the consensus sequence TTATCCACA. Two of these are in one orientation, and two are in the opposite orientation. DNase footprinting shows that these nine-mers are binding sites for the ***dnaA* product (DnaA)**. These nine-mers are therefore sometimes called ***dnaA* boxes.** DnaA appears to facilitate the binding of DnaB to the origin.

DnaA helps DnaB bind at the origin by stimulating the melting of three 13-mer repeats at the left end of *oriC* to form an **open complex.** This is analogous to the open promoter complex we discussed in Chapter 6. DnaB can then bind to the melted DNA region. As in the assembly

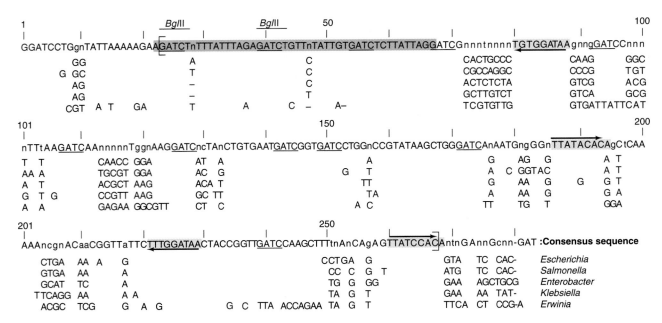

Figure 21.4 **Sequence of *oriC,* and comparison with origin sequences in other enteric bacteria.** The consensus sequence is on top, followed by the sequences from all five species where they differ from one another. Capital letters denote bases conserved in all five species of bacteria; lowercase letters denote bases conserved in only three species. The n denotes any base, or a deletion. The blue box contains highly conserved 13-mers. Pink boxes are the DnaA -binding sequences. The 245-bp minimal origin of replication is enclosed in brackets. (*Source:* From J. W. Zyskind et al., "Chromosomal replication origin from the marine bacterium *Vibrio harveyi* functions in *Escherichia coli,*" *Proceedings of the National Academy of Sciences* 80:1164–1168, March 1983.)

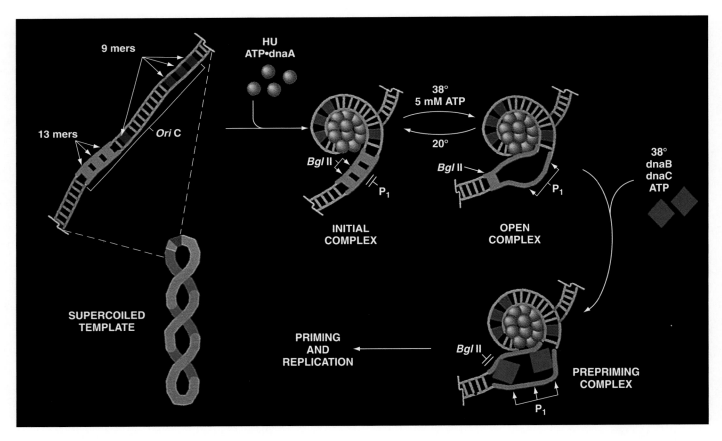

Figure 21.5 **Priming at *oriC.* First, DnaA binds ATP and forms a multimer.** Along with the HU protein, the DnaA/ATP complex binds to the DNA, encompassing the four nine-mers. In all, this complex covers about 200 bp. HU protein probably induces the bend in the DNA pictured here. The binding of DnaA, along with the bending induced by HU protein, apparently destabilizes the adjacent 13-mer repeats and causes local DNA melting there. This allows the binding of DnaB protein to the melted region. DnaC binds to the DnaB protein and helps deliver it to the DNA. Finally, primase binds to the prepriming complex and converts it to the primosome, which can make primers to initiate DNA replication. (*Source:* Kornberg & Baker, *DNA Replication,* plate 15. W. H. Freeman & Co., 1990.)

of the φX174 primosome, DnaC binds to DnaB and helps deliver it to the origin.

The evidence also strongly suggests that DnaA directly assists the binding of DnaB. Here is one line of evidence that points in this direction. A *dnaA* box resides in the stem of a hairpin stem loop in a plasmid called R6K. We can bind DnaA to this DNA, whereupon DnaB (with the help of DnaC) can also bind. Here, no DNA melting appears to occur, so we infer that DnaA directly affects binding between DNA and DnaB.

At least two other factors participate in open complex formation at *oriC*. The first of these is RNA polymerase. This enzyme does not serve as a primase, as it does in M13 phage replication, but it still serves an essential function. We know the RNA polymerase product does not serve as a primer because it can be terminated with 3′-dATP, which lacks the 3′-hydroxyl group essential to a primer, and priming still occurs. But we know RNA polymerase action is required, because rifampicin blocks primosome assembly. The role of RNA polymerase seems to be to synthesize a short piece of RNA that creates an R loop. The R loop can be adjacent to *oriC*, rather than within it. The second factor is **HU protein**. This is a small basic DNA-binding protein that can induce bending in double-stranded DNA. This bending, together with the R loop, presumably destabilizes the DNA double helix and facilitates melting of the DNA to form the open complex.

Finally, DnaB stimulates the binding of the primase (DnaG), completing the primosome. Priming can now occur, so DNA replication can get started. The primosome remains with the replisome as it carries out elongation, and serves at least two functions. First, it must operate repeatedly in priming Okazaki fragment synthesis to build the lagging strand. Second, DnaB serves as the helicase that unwinds DNA to provide templates for both the leading and lagging strands. To accomplish this task, DnaB moves in the 5′→3′ direction on the lagging strand template—the same direction in which the replicating fork is moving. This anchors the primosome to the lagging strand template, where it is needed for priming Okazaki fragment synthesis.

SUMMARY Primer synthesis in *E. coli* requires a primosome composed of DnaB and the primase, DnaG Primosome assembly at the origin of replication, *oriC*, occurs as follows: DnaA binds to *oriC* at sites called *dnaA* boxes and cooperates with RNA polymerase and HU protein in melting a DNA region adjacent to the leftmost *dnaA* box. DnaB then binds to the open complex and facilitates binding of the primase to complete the primosome. The primosome remains with the replisome, repeatedly priming Okazaki fragment synthesis on the lagging strand. DnaB also has a helicase activity that unwinds the DNA as the replisome progresses.

Priming in Eukaryotes

Eukaryotic replication is considerably more complex than the prokaryotic replication we have just studied. One complicating factor is the much bigger size of eukaryotic genomes. This, coupled with the slower movement of eukaryotic replicating forks, means that each chromosome must have multiple origins. Otherwise, replication would not finish within the time allotted—the S phase of the cell cycle—which can be as short as a few minutes. Because of this multiplicity and other factors, identification of eukaryotic origins of replication has lagged considerably behind similar work in prokaryotes. However, when molecular biologists face a complex problem, they frequently resort to simpler systems such as viruses to give them clues about the viruses' more complex hosts. Scientists followed this strategy to identify the origin of replication in the simple monkey virus SV40 as early as 1972. Let us begin our study of eukaryotic origins of replication there, then move on to origins in yeast.

The Origin of Replication in SV40 Two research groups, one headed by Norman Salzman, the other by Daniel Nathans, identified the SV40 origin of replication in 1972 and showed that DNA replication proceeded bidirectionally from this origin. Salzman's strategy was to use *Eco*RI to cleave replicating SV40 DNA molecules at a unique site. (Although this enzyme had only recently been discovered and characterized, Salzman knew that SV40 DNA contained only a single *Eco*RI site.) After cutting the replicating SV40 DNA with *Eco*RI, Salzman and colleagues visualized the molecules by electron microscopy (Figure 21.6). They observed only a single replicating bubble, which indicated a single origin of replication. Furthermore, as they followed the growth of this bubble, they found that it grew at both ends, showing that both replicating forks were moving away from the single origin. This analysis revealed that the origin lies 33% of the genome length from the *Eco*RI site. But which direction from the *Eco*RI site? Because the SV40 DNA is circular, and these pictures contain no other markers besides the single *Eco*RI site, we cannot tell. But Nathans used another restriction enzyme (*Hin*dII), and his results, combined with these, placed the origin at a site overlapping the SV40 control region, adjacent to the GC boxes and the 72-bp repeat enhancer we discussed in Chapters 10 and 12 (Figure 21.7).

The minimal *ori* sequence (the *ori* core) (Figure 21.8) is 64 bp long and includes several essential elements (1) four pentamers (5′-GAGGC-3′), which are the binding site for **large T antigen**, the major product of the viral early region; (2) a 15-bp palindrome, which is the earliest region melted during DNA replication; and (3) a 17-bp region consisting only of A–T pairs, which probably facilitates melting of the nearby palindrome region.

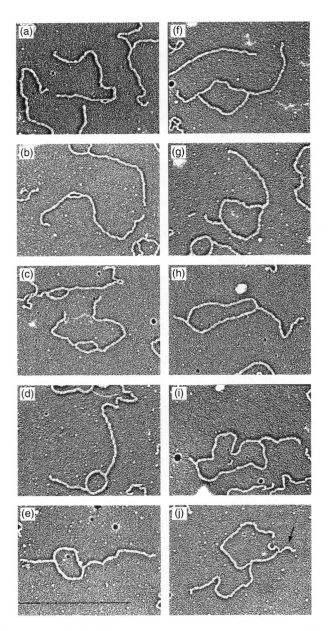

Figure 21.6   SV40 DNA molecules at various stages of replication. The DNAs are arranged from (a) to (j) in order of increasing degree of replication. In all panels except (j), the short arm of the DNA is oriented to the left. In panel (j), replication has passed beyond the *Eco*RI site, so the short arm no longer exists. The arrow in panel (j) denotes a short, extraneous piece of DNA. The bar (panel e) represents 1 μm. (*Source:* Fareed et al., Origin and direction of Simian virus 40 DNA replication. *J. Virology* 10 (1972) f. 4, p. 489, by permission of American Society of Microbiology.)

Other elements surrounding the *ori* core also participate in initiation. These include two additional large T antigen-binding sites, as indicated in Figure 21.8 and the GC boxes to the left of the *ori* core. The GC boxes provide about a 10-fold stimulation of initiation of replication. If the number of GC boxes is reduced, or if they are moved only 180 bp away from *ori*, this stimulation is reduced or eliminated. This effect is somewhat akin to the participa-

Figure 21.7   Location of the SV40 *ori* in the transcription control region. The core *ori* sequence (green) encompasses part of the early region TATA box and the cluster of early transcription initiation sites. Pink arrows denote bidirectional replication away from the replication initiation site. Black arrows denote transcription initiation sites.

tion of RNA polymerase in initiation at *ori*C in *E. coli.* One difference: At the SV40 *ori,* no transcription need occur; binding of the transcription factor Sp1 to the GC boxes is sufficient to stimulate initiation of replication.

Once large T antigen binds at the SV40 *ori,* its DNA helicase activity unwinds the DNA and prepares the way for primer synthesis. The primase in eukaryotic cells is associated with DNA polymerase α, and this also serves as the primase for SV40 replication.

> **SUMMARY** The SV40 origin of replication is adjacent to the viral transcription control region. Initiation of replication depends on the viral large T antigen, which binds to a region within the 64-bp *ori* core, and at two adjacent sites, and exercises a helicase activity, which opens up a replication bubble within the *ori* core. Priming is carried out by a primase associated with the host DNA polymerase α.

The Origin of Replication in Yeast So far, yeast has provided most of our information about eukaryotic origins of replication. This is not surprising, because yeasts are among the simplest eukaryotes, and they lend themselves well to genetic analysis. As a result, yeast genetics are well understood. As early as 1979, Hsiao and Carbon discovered a yeast DNA sequence that could replicate independently of the yeast chromosomes, suggesting that it contains an origin of replication. This DNA fragment contained the yeast *ARG4+* gene. Cloned into a plasmid, it transformed *arg4−*, yeast cells to *ARG4+*, as demonstrated by their growth on medium lacking arginine. Any yeast cells that grew must have incorporated the *ARG4+* gene of the plasmid and, furthermore, must be propagating that gene somehow. One way to propagate the gene would be by incorporating it into the host chromosomes by recombination, but that was known to occur with a low frequency—about 10^{-6}–10^{-7}. Hsiao and Carbon obtained *ARG4+* cells at a much higher frequency—about 10^{-4}. Furthermore, shuttling the plasmid back and forth between yeast and *E. coli* caused no change in the plasmid structure, whereas recombination with the yeast genome would have changed it noticeably. Thus, these investigators

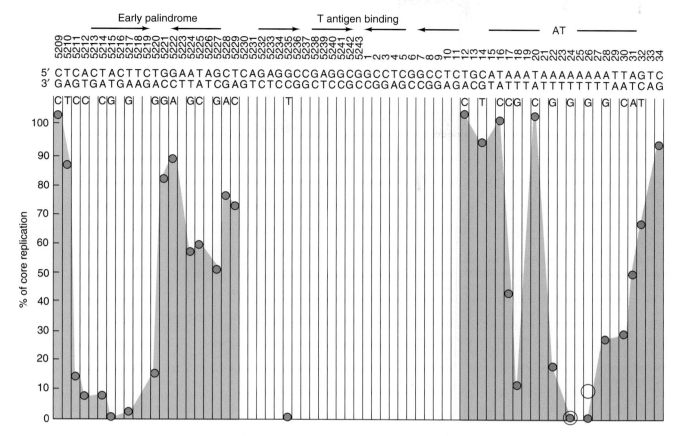

Figure 21.8   Structure of the SV40 *ori* core. The sequence of the *ori* core is given just below the numbering of the base pairs in the SV40 DNA. The core extends from base pair 5211 to base pair 31. The positions of the palindrome, the T antigen-binding sites, and the A-T region are given at top. The orientations of the four pentamers in the T antigen-binding site are indicated by arrows. The effects of point mutations in many of the positions are given by the bar graph. The red circles indicate percentage of wild-type replication observed when the indicated base change was made. The open circles at positions 24 and 26 denote effects of A → T changes. (*Source:* From Deb et al., *Molecular and Cellular Biology* 6:4582, 1986. Copyright © 1986 American Society for Microbiology, Washington, DC. Reprinted by permission.)

concluded that the yeast DNA fragment they had cloned in the plasmid probably contained an origin of replication. Also in 1979, R.W. Davis and colleagues performed a similar study with a plasmid containing a yeast DNA fragment that converted *trp⁻* yeast cells to *TRP⁺*. They named the 850-bp yeast fragment **autonomously replicating sequence 1,** or **ARS1.**

Although these early studies were suggestive, they failed to establish that DNA replication actually begins in the ARS sequences. To demonstrate that ARS1 really does have this key characteristic of an origin of replication, Bonita Brewer and Walton Fangman used two-dimensional electrophoresis to detect the site of replication initiation in a plasmid bearing ARS1. This technique depends on the fact that circular and branched DNAs migrate more slowly than linear DNAs during gel electrophoresis, especially at high voltage or high agarose concentration. Brewer and Fangman prepared a yeast plasmid bearing ARS1 as the only origin of replication. They allowed this plasmid to replicate in synchronized yeast cells and then isolated replication intermediates (RIs). They linearized these RIs with a

restriction endonuclease, then electrophoresed them in the first dimension under conditions (low voltage and low agarose concentration) that separate DNA molecules roughly according to their sizes. Then they electrophoresed the DNAs in the second dimension using higher voltage and agarose concentration that cause retardation of branched and circular molecules. Finally, they Southern blotted the DNAs in the gel to a membrane and probed it with a labeled plasmid-specific DNA.

Figure 21.9 shows an idealized version of the behavior of various branched and circular RIs of a hypothetical 1-kb fragment. Simple Y's (panel a) begin as essentially linear 1-kb fragments with a tiny Y at their right ends; these would behave almost like linear 1-kb fragments. As the fork moves from right to left, the Y grows larger and the mobility of the fragment in the second (vertical) dimension slows. Then, as the Y grows even larger, the fragment begins to look more and more like a linear 2-kb fragment, with just a short stem on the Y. This is represented by the horizontal linear forms with short vertical stems in panel (a). Because these forms resemble linear

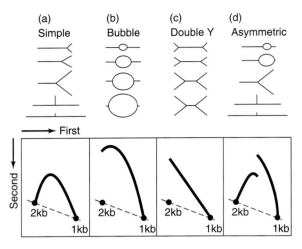

Figure 21.9   Theoretical behaviors of various types of replication intermediates on two-dimensional gel electrophoresis. The top parts of panels **a–d** are cartoons showing the shapes of growing simple Y's, bubbles, double Y's, and asymmetric bubbles that convert to simple Y's as replication progresses. The bottom parts of each panel are cartoons that depict the expected deviation of the changing mobilities of each type of growing RI from the mobilities of linear forms growing progressively from 1 to 2 kb (dashed lines). (*Source:* Reprinted from Brewer and Fangman, *Cell* 51:464, 1987. Copyright 1987, with permission from Elsevier Science.)

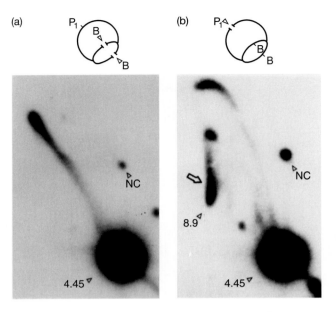

Figure 21.10   Locating the origin of replication in ARS-1. (**a**) Results of cleaving 2-μm plasmid with *Bgl*II. Top: cartoon showing the shape expected when an RI is cut with *Bgl*II, assuming the origin lies adjacent to the *Bgl*II site within ARS1. The bubble contains DNA that has already replicated, so there are two copies of the *Bgl*II site (arrowheads labeled B), both of which are cut to yield the double-Y intermediate depicted. Bottom: experimental results showing the straight curve expected of double-Y intermediates. (**b**) Results of cleaving the plasmid with *Pvu*I. Top: cartoon showing the shape expected when an RI is cut with *Pvu*I, assuming the origin lies almost across the circle from the *Pvu*I site within ARS1. Bottom: experimental results showing the rising arc, with a discontinuity near the end. This is what we expect for a bubble-shaped RI that converts to a nearly linear Y as one of the replication forks passes a *Pvu*I site. Both of these results confirm the expectations for an origin of replication within ARS1. (*Source:* Brewer & Fangman, The localization of replication origins on ARS plasmids in *S. cerevisiae. Cell* 51 (6 Nov 1987) f. 8, p. 469.)

shapes more and more as the fork moves, their mobility increases correspondingly, until the fork has nearly reached the end of the fragment. At this point, they have a shape and mobility that is almost like a true linear 2-kb fragment. This behavior gives rise to an arc-shaped pattern, where the apex of the arc corresponds to a Y that is half-replicated, at which point it is least like a linear molecule.

Figure 21.9b shows what to expect for a bubble-shaped fragment. Again, we begin with a 1-kb linear fragment, but this time with a tiny bubble right in the middle. As the bubble grows larger, the mobility of the fragment slows more and more, yielding the arc shown at the bottom of the panel. Panel (c) shows the behavior of a double Y, where the RI becomes progressively more branched as the two forks approach the center of the fragment. Accordingly, the mobility of the RI decreases almost linearly. Finally, panel (d) shows what happens to a bubble that is asymmetrically placed in the fragment. It begins as a bubble, but then, when one fork passes the restriction site at the right end of the fragment, it converts to a Y. The mobilities of the RIs reflect this discontinuity: The curve begins like that of a bubble, then abruptly changes to that of a Y, with an obvious discontinuity showing exactly when the fork passed the restriction site and converted the bubble to a Y.

This kind of behavior is especially valuable in mapping the origin of replication. In panel (d), for example, we can see that the discontinuity occurs in the middle of the curve, when the mobility in the first dimension was

that of a 1.5-kb fragment. This tells us that the arms of the Y are each 500 bp long. Assuming that the two forks are moving at an equal rate, we can conclude that the origin of replication was 250 bp from the right end of the fragment.

Now let us see how this works in practice. Brewer and Fangman chose restriction enzymes that would cleave the plasmid with its ARS1 just once, but in locations that would be especially informative if the origin of replication really lies within ARS1. Figure 21.10 shows the locations of the two restriction sites, at top, and the experimental results, at bottom. The first thing to notice about the autoradiographs is that they are simple and correspond to the patterns we have seen in Figure 21.9. This means that there is a single origin of replication; otherwise, there would have been a mixture of different kinds of RIs, and the results would have been more complex.

The predicted origin within ARS1 lies adjacent to a *Bgl*II site (B, in panel a). Thus, if the RI is cleaved with this enzyme, it should yield double-Y RIs. Indeed, as we see in the lower part of panel (a), the autoradiograph is nearly linear—just as we expect for a double-Y RI. Panel (b) shows that a *Pvu*I site (P) lies almost halfway around the plasmid from the predicted origin. Therefore, cleaving with *Pvu*I should yield the bubble-shaped RI shown at the top of panel (b). The autoradiograph at the bottom of panel (b) shows that Brewer and Fangman observed the discontinuity expected for a bubble-shaped RI that converts at the very end to a very large single Y, as one fork reaches the *Pvu*I site, then perhaps to a very asymmetric double Y as the fork passes that site. Both of these results place the origin of replication adjacent to the *Bgl*II site, just where we expect it if ARS1 contains the origin.

York Marahrens and Bruce Stillman performed linker scanning experiments to define the important regions within ARS1. They constructed a plasmid very similar to the one used by Brewer and Fangman, containing (1) ARS1 in a 185-bp DNA sequence; (2) a yeast centromere; and (3) a selectable marker—*URA3*—which confers on *ura3-52* yeast cells the ability to grow in uracil-free medium. Then they systematically substituted an 8-bp *Xho*I linker for the normal DNA at sites spanning the ARS1 region. They transformed yeast cells with each of the linker scanning mutants and selected for transformed cells with uracil-free medium. Some of the transformants containing mutant ARS1 sequences grew more slowly than those containing wild-type ARS1 sequences. Because the centromere in each plasmid ensured proper segregation of the plasmid, the most likely explanation for poor growth was poor replication due to mutation of ARS1.

To check this hypothesis, Marahrens and Stillman grew all the transformants in a non-selective medium containing uracil for 14 generations, then challenged them again with a uracil-free medium to see which ones had not maintained the plasmid well. The mutations in these unstable plasmids presumably interfered with ARS1 function. Figure 21.11 shows the results. Four regions of ARS1 appear to be important. These were named A, B1, B2, and B3 in order of decreasing effect on plasmid stability. Element A is 15 bp long, and contains an 11-bp ARS consensus sequence:

$$5'-{}_A^T TTTA_{CG}^{TA} TTT_A^T-3'$$

When it was mutated, *all* ARS1 activity was lost. The other regions had a less-drastic effect, especially in selective medium. However, mutations in B3 had an apparent effect on the bending of the plasmid, as assayed by gel electrophoresis. The stained gel below the bar graph shows increased electrophoretic mobility of the mutants in the B3 region. Marahrens and Stillman interpreted this as altered bending of the ARS1.

The existence of four important regions within ARS1 raises the question whether these are also *sufficient* for ARS function. To find out, Marahrens and Stillman constructed a synthetic ARS1 with wild-type versions of all

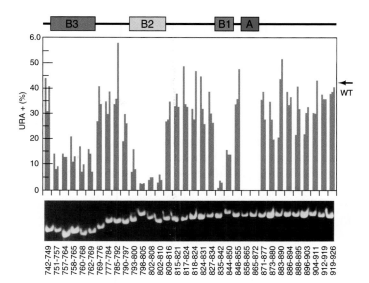

Figure 21.11   Linker scanning analysis of ARS1. Marahrens and Stillman substituted linkers throughout an ARS1 sequence within a plasmid bearing a yeast centromere and the *URA3* selectable marker. To test for replication efficiency of the mutants, they grew them for 14 generations in nonselective medium, then tested them for growth on selective (uracil-free) medium. The vertical bars show the results of three independent determinations for each mutant plasmid. Results are presented as a percentage of the yeast cells that retained the plasmid (as assayed by their ability to grow). Note that even the wild-type plasmid was retained with only 43% efficiency in nonselective medium (arrow at right). Four important regions (A, B1, B2, and B3) were identified. The regions that were mutated are identified by base number at bottom. The stained gel at bottom shows the electrophoretic mobility of each mutant plasmid. Note the altered mobility of the B3 mutant plasmids, which suggests altered bending. (*Source:* Marahrens & Stillman, A yeast chromosomal origin of DNA replication defined by multiple functional elements. *Science* 255 (14 Feb 1992) f. 2, p. 819. © AAAS.)

four regions, spaced just as in the wild-type ARS1, but with random sequences in between. A plasmid bearing this synthetic ARS1 was almost as stable under nonselective conditions as one bearing a wild-type ARS1. Thus, the four DNA elements defined by linker scanning are sufficient for ARS1 activity. Finally, these workers replaced the wild-type 15-bp region A with the 11-bp ARS consensus sequence. This reduced plasmid stability dramatically, suggesting that the other 4 bp in region A are also important for ARS activity.

SUMMARY The yeast origins of replication are contained within autonomously replicating sequences (ARSs) that are composed of four important regions (A, B1, B2, and B3). Region A is 15 bp long and contains an 11-bp consensus sequence that is highly conserved in ARSs. Region B3 may allow for an important DNA bend within ARS1.

21.3 Elongation

Once a primer is in place, real DNA synthesis (elongation) can begin. We have already identified the pol III holoenzyme as the enzyme that carries out elongation in *E. coli*, and DNA polymerase δ as the enzyme that elongates both the lagging and leading strands in eukaryotes. The *E. coli* system is especially well characterized, and the data point to an elegant method of coordinating the synthesis of lagging and leading strands in a way that keeps the pol III holoenzyme engaged with the template so replication can be highly processive, and therefore very rapid. Let us focus on this *E. coli* elongation mechanism.

The Pol III Holoenzyme and Processivity of Replication

The core by itself is a very poor polymerase. It puts together about 10 nt and then falls off the template. Then it has to spend about a minute reassociating with the template and the nascent DNA strand. This contrasts sharply with the situation in the cell, where the replicating fork moves at the rate of almost 1000 nt/sec. Obviously, something important is missing from the core.

That "something" is an agent that confers processivity on the holoenzyme, allowing it to remain engaged with the template while polymerizing at least 50,000 nt before stopping—quite a contrast to the 10 nt polymerized by the core before it stops. Why such a drastic difference? The holoenzyme owes its processivity to a "**sliding clamp**" that holds the enzyme on the template for a long time. The **β-subunit** of the holoenzyme performs this sliding clamp function, but it cannot associate by itself with the preinitiation complex (core plus DNA template). It needs a **clamp loader** to help it join the complex, and a group of subunits called the **γ complex** provides this help. The γ complex includes the γ, δ, δ′, χ, and ψ subunits. In this section, we will examine the activities of the β clamp and the clamp loader.

The β clamp One way we can imagine the β-subunit conferring processivity on the pol III core is by binding both the core complex and DNA. That way, it would tie the core to the DNA and keep it there—hence the term β **clamp.** In the course of probing this possibility, Mike O'Donnell and colleagues demonstrated direct interaction between the β and α subunits. They mixed various combinations of subunits, then separated subunit complexes from individual subunits by gel filtration. They detected subunits by gel electrophoresis, and activity by adding the missing subunits and measuring DNA synthesis. Figure 21.12 depicts the results. It is clear that α and ε bind to each other, as we would expect, because they are both part of the core. Furthermore, α, ε, and β form a com-

plex, but which subunit does β bind to, α or ε? Panels (d) and (e) show the answer: β binds to α alone, but not to ε alone. Thus, α is the core subunit to which β binds.

This scheme demands that β be able to slide along the DNA as α and ε together replicate it. This in turn suggests that the β clamp would remain bound to a circular DNA, but could slide right off the ends of a linear DNA. To test this possibility, O'Donnell and colleagues performed the experiment reported in Figure 21.13. The general strategy of this experiment was to load 3H-labeled β dimers onto circular, double-stranded phage DNA with the help of the γ complex, then to treat the DNA in various ways to see if the β dimers could dissociate from the DNA. The assay for β-binding to DNA was gel filtration.

In panels (a) and (b), the DNA was treated with *Sma*I or *Bam*HI, respectively, to linearize the DNA, then examined to see whether the β clamp had slid off. In both cases, the β clamp remained bound to circular DNA, but had dissociated from both linearized DNAs, apparently by sliding off the ends. Panel (c) demonstrates that the nick in the circular DNA is not what caused retention of the β dimer, because the nick can be removed with DNA ligase, and the β dimer remains bound to the DNA. The inset shows electrophoretic evidence that the ligase really did remove the nick because the nicked form disappeared and the closed circular form was enhanced. Panel (d) shows that adding more β-subunit to the loading reaction increased the number of β dimers bound to the circular DNA. In fact, more than 20 molecules of β-subunit could be bound per molecule of circular DNA. This is what we would expect if many holoenzymes can replicate the DNA in tandem.

If the β dimers are lost from linear DNA by sliding off the ends, one ought to be able to prevent their loss by binding other proteins to the ends of the DNA. O'Donnell's group did this in two ways. Panel (e) shows what happened when they attached binding sites for the eukaryotic protein EBNA1 to a region flanking an *Eco*RV restriction site. Then they added EBNA1 and linearized the DNA with *Eco*RV. The *Eco*RV reaction was complete, as shown by the electrophoretic results in the insert. In this case, the β dimer remained bound to the linear DNA—the EBNA1 protein apparently prevented the β dimer from sliding off the DNA ends. Panel (f) shows the results of a similar experiment in which the ends of the DNA were made single-stranded and bound to SSB. Again, the β dimer remained bound to this linear DNA. Even without SSB, most of the β dimers remained bound to the linear DNA with long single-stranded tails, suggesting that the β clamp does not slide over long stretches of even naked single-stranded DNA. However, when the DNA–protein complex was cut in two with *Hin*dIII, it provided the β clamp with the opportunity to slide off the linear DNA, but only if it slid in the direction opposite the one it would take during DNA synthesis. The blue line in

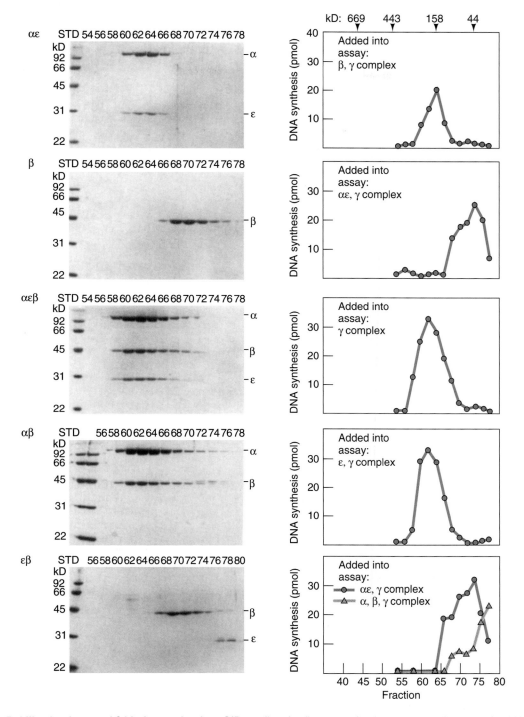

Figure 21.12 The Pol III subunits α and β bind to each other. O'Donnell and colleagues mixed various combinations of pol III subunits, subjected the mixtures to gel filtration to separate complexes from free subunits, then electrophoresed fractions from the gel filtration column to detect complexes. The electrophoresis results are presented on the left. At right, the same gel filtration fractions were supplemented with the missing pol III holoenzyme subunits and assayed for DNA synthesis to verify the presence of complexes. For example, in the top left panel, α and ε were in the complex eluted from the gel filtration column, so the other essential factors, β and γ, were added to the polymerase assay. Notice that α, ε, and β formed a complex, as did α and β, but ε and β did not. (*Source:* Stukenberg et al., Mechanism of the sliding b-clamp of DNA polymerase III holoenzyme. *J. Biol Chem.* 266 no. 17(15 June 1991) figs. 2a-e, 3, pp. 11330-31. American Society for Biochemistry and Molecular Biology.)

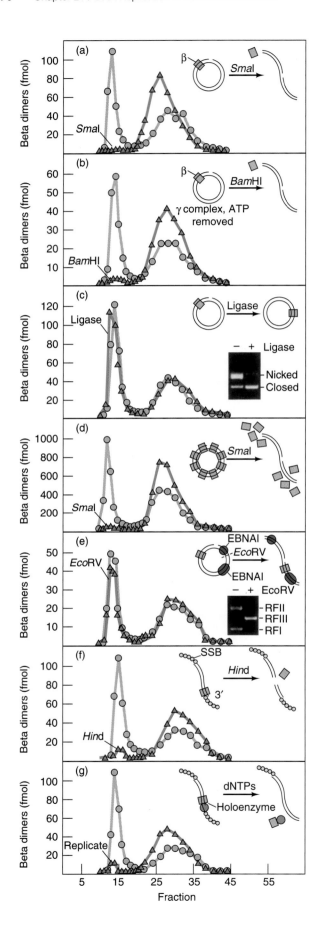

Figure 21.13   The β clamp can slide off the ends of a linear DNA.
O'Donnell and colleagues loaded ^{3}H-labeled β-dimers onto various
DNAs, with the help of the γ-complex, then treated the complexes in
various ways as described. Finally, they subjected the mixtures to gel
filtration to separate protein–DNA complexes (which were large and
eluted quickly from the column, around fraction 15), from free protein
(which was relatively small and eluted later, around fraction 28).
(**a**) Effect of linearizing the DNA with *Sma*I. DNA was cut once with
*Sma*I and then assayed (red). Uncut DNA was also assayed (blue).
(**b**) Effect of linearizing the DNA with *Bam*HI. To make sure the effect
in panel (**a**) was not due to some special property of the *Sma*I enzyme,
such as the fact that it leaves blunt ends, DNA was linearized with
*Bam*HI, which leaves 4-nt 5′-overhangs, then assayed (red). Again,
uncut DNA was also assayed (blue). (**c**) Effect of removing a nick in the
template. The nick in the template was removed with DNA ligase
before assay (red), or left alone (blue). The inset shows the results of
electrophoresis of DNAs before and after the ligase reaction. (**d**) Many
β dimers can be loaded onto the DNA and then lost when it is
linearized. The ratio of β-dimers loaded onto DNA templates was
increased by raising the concentration of β-subunits and lowering the
concentration of DNA templates. Then the DNA was either cut with
*Sma*I before assay (red) or not cut (blue). (**e**) Proteins on the ends of
linear DNA can block the exit of the β clamp. Binding sites (brown) for
the eukaryotic protein EBNA1 were added to the DNA, flanking an
*Eco*RV site, then EBNA1 proteins (green) were bound and the DNA
was cut with *Eco*RV. This time, the β-subunits stayed on the DNA. The
inset shows the results of electrophoresis of uncut DNA and DNA cut
by *Eco*RV. The *Eco*RV reaction appeared to be complete, as RFIII
corresponds to linear DNA. (**f**) Effect of SSB bound to the ends of a
linear DNA. Long single-stranded regions at both ends of the DNA
were created with exonuclease III, which nibbles away at 3′-ends of
double-stranded DNAs. Then SSB was added to these single-
stranded ends. The complexes were assayed directly (blue) or after
cleavage with *Hin*dIII (red), which cuts the DNA in two and allows the β
clamp to slide off if it moves in the direction opposite that taken by a
replicating fork. (**g**) Effect of forming a holoenzyme on loss of the β
clamp from a linear template whose ends are blocked by SSB. The
same template as in panel (**f**) was created and loaded with
holoenzyme (pol III* plus β), instead of just with β. The holoenzyme
was allowed to replicate the template in the presence of all four
dNTPs (red), or was prevented from carrying out replication by
supplying it with only two dNTPs (blue). As long as replication was
possible, the β clamp was lost from the template, even though its
ends were blocked by SSB. (*Source:* Stukenberg et al., Mechanism of the
sliding β-clamp of DNA polymerase III holoenzyme. *J. Biol Chem.* 266 no. 17(15
June 1991) figs. 2a-e, 3, pp. 11330-31. American Society for Biochemistry and
Molecular Biology.)

panel (f) shows that it did indeed slide off. Finally, panel (g)
shows that β can slide off of SSB-coated single-stranded
DNA if it is part of a holoenzyme replicating that DNA.
Here, O'Donnell and colleagues added pol III* (holoenzyme without the β-subunit) to the same β–DNA complex as in panel (f), and the labeled β clamp did indeed
slide off the end (red), but only if all four dNTPs were supplied so that replication could occur. The blue line shows
that the β clamp remained on the DNA (with the holoenzyme) when only two dNTPs were present.

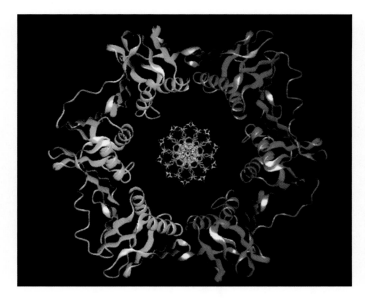

Figure 21.14 Model of the β dimer/DNA complex. The β dimer is depicted by a ribbon diagram in which the α-helices are coils and the β-sheets are flat ribbons. One β monomer is yellow and the other is red. A DNA model, seen in cross section, is placed in a hypothetical position in the middle of the ring formed by the β dimer. (*Source:* Kong et al., Three-dimensional structure of the beta subunit of *E. coli* DNA polymerase III holoenzyme: A sliding DNA clamp. *Cell* 69 (1 May 1992) f. 1, p. 426. Reprinted by permission of Elsevier Science.)

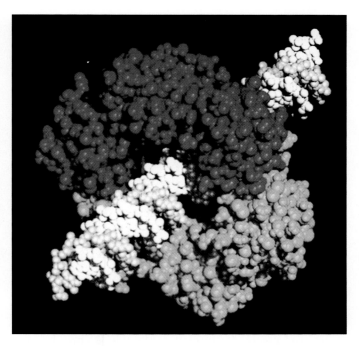

Figure 21.15 Space-filling model of the β dimer/DNA complex. The β monomers are shown in yellow and red, with the DNA in green and white. Again, the position of the DNA is hypothetical, but highly plausible. (*Source:* Kong et al., *Cell* 69 (1 May 1992) f. 3, p. 427. Reprinted by permission of Elsevier Science.)

O'Donnell and John Kuriyan used x-ray crystallography to study the structure of the β clamp. The pictures they produced provide a perfect rationale for the ability of the β clamp to remain bound to a circular DNA but not to a linear one: The β dimer forms a ring that can fit around the DNA. Thus, like a ring on a string, it can readily fall off if the string is linear, but not if the string is circular. Figure 21.14 is one of the models O'Donnell and Kuriyan constructed; it shows the ring structure of the β dimer, with a scale model of B-form DNA placed in the middle. Figure 21.15 is a space-filling model of the hypothetical β dimer/DNA complex, again showing how neatly the β dimer could fit around a double-stranded DNA.

As mentioned in Chapter 20, eukaryotes also have a processivity factor called PCNA, which performs the same function as the bacterial β clamp. The primary structure of PCNA bears no apparent similarity to that of the β clamp, and the eukaryotic protein is only 2/3 the size of its prokaryotic counterpart. Nevertheless, x-ray crystallography performed by John Kuriyan and his colleagues demonstrates that yeast PCNA forms a trimer with a structure arrestingly similar to that of the β clamp dimer: a ring that can encircle a DNA molecule, as shown in Figure 21.16.

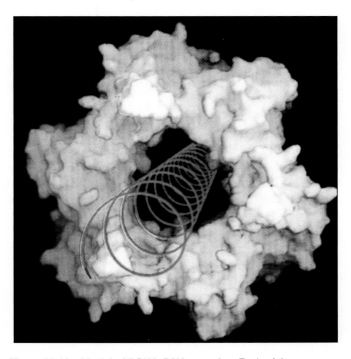

Figure 21.16 Model of PCNA–DNA complex. Each of the monomers of the PCNA trimer is represented by a different pastel color. The shape of the trimer is based on x-ray crystallography analysis. The red helix represents the probable location of the sugar–phosphate backbone of a DNA associated with the PCNA trimer. (*Source:* Krishna et al., Crystal structure of the eukaryotic DNA polymerase processivity factor PCNA. *Cell* 79 (30 Dec 1994) f. 3b, p 1236. Reprinted by permission of Elsevier Science.)

SUMMARY The Pol III core (αε or αεθ) does not function processively by itself, so it can replicate only a short stretch of DNA before falling off the template. By contrast, the core plus the β-subunit can replicate DNA processively at a rate approaching 1000 nt/sec. The β-subunit forms a dimer that is ring-shaped. This ring fits around a DNA template and interacts with the α-subunit of the core to tether the whole polymerase and template together. This is why the holoenzyme stays on its template so long and is therefore so processive. The eukaryotic processivity factor PCNA forms a trimer with a similar ring shape that can encircle DNA and hold DNA polymerase on the template.

The Clamp Loader O'Donnell and his colleagues demonstrated the function of the clamp loader in an experiment presented in Figure 21.17. These workers used the α- and ε-subunits instead of the whole core, because the θ-subunit was not essential in their in vitro experiments. As template, they used a single-stranded M13 phage DNA annealed to a primer. They knew that highly processive holoenzyme could replicate this DNA in about 15 sec but that the αε core could not give a detectable amount of replication in that time. Thus, they reasoned that a 20-sec pulse of replication would allow all processive polymerase molecules the chance to complete one cycle of replication, and therefore the number of DNA circles replicated would equal the number of processive polymerases. Figure 21.17a shows that each femtomole (fmol, or 10^{-15} mol) of γ complex resulted in about 10 fmol of circles replicated in the presence of αε core and β-subunit. Thus, the γ complex acts catalytically: One molecule of γ complex can sponsor the creation of many molecules of processive polymerase. The inset in this figure shows the results of gel electrophoresis of the replication products. As expected of processive replication, they are all full-length circles.

This experiment suggested that the γ complex itself is not the agent that provides processivity. Instead, the γ complex adds something else to the core polymerase that makes it processive. Because β was the only other polymerase subunit in this experiment, it is the likely processivity-determining factor. To confirm this, O'Donnell and colleagues mixed the DNA template with ³H-labeled β-subunit and unlabeled γ complex to form preinitiation complexes, then subjected these complexes to gel filtration to separate the complexes from free proteins. They detected the preinitiation complexes by adding αε to each fraction and assaying for labeled double-stranded circles formed (RFII, green). Figure 21.17b demonstrates that only a trace of γ complex (blue) remained associated with the DNA, but a significant fraction of the labeled β-subunit

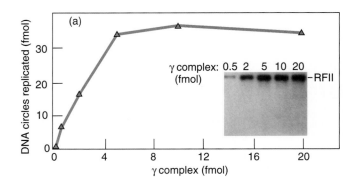

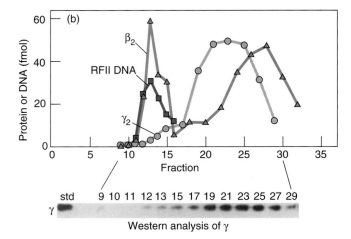

Figure 21.17 Involvement of β and γ complex in processivity. (a) The γ complex acts catalytically in forming a processive polymerase. O'Donnell and coworkers added increasing amounts of γ complex (indicated on the x axis) to a primed M13 phage DNA template coated with SSB, along with αε core, and the β-subunit of pol III holoenzyme. Then they allowed a 20-sec pulse of DNA synthesis in the presence of [α-³²P]ATP to label the DNA product. They determined the radioactivity of part of each reaction and converted this to fmol of DNA circles replicated. To check for full circle replication, they subjected another part of each reaction to gel electrophoresis. The insert shows the result: The great majority of each product is full-circle size (RFII). (b) The β subunit, but not the γ complex associates with DNA in the preinitiation complex. O'Donnell and colleagues added ³H-labeled β-subunit and unlabeled γ complex to primed DNA coated with SSB, along with ATP to form a preinitiation complex. Then they subjected the mixture to gel filtration to separate preinitiation complexes from free proteins. They detected the β-subunit in each fraction by radioactivity, and the γ complex by Western blotting, with an anti-γ antibody as probe (bottom). The plot shows that the β-subunit (as dimers) bound to the DNA in the preinitiation complex, but the γ complex did not. (*Source:* Stukenberg et al., Mechanism of the sliding [beta]-clamp of DNA polymerase III holoenzyme. *J. Biol. Chem.* 266 (15 June 1991) f. 1a&c, p. 11329. American Society for Biochemistry and Molecular Biology.)

(red) remained with the DNA. (The unlabeled γ complex was detected with a Western blot using an anti-γ antibody, as shown at the bottom of the figure.)

This experiment also allowed O'Donnell and colleagues to estimate the stoichiometry of the β-subunit in

Table 21.1   Activation of Holoenzyme Activity on Poly(dA) · oligo(dT₁₂)

Experiment	Nucleotide Addition	K_M (μM)	K_1(μM)	SSB	$[NaCl]_{eq}$ mM	DNA Synthesis (pmol/min)
1	None			−	28	1.0
	ATP or dATP	10		−	28	10
	None			+	34	9.4
	ATP or dATP	10		+	34	33
2	None			−	110	<0.1
	ATP or dATP	8		−	110	8.2
	None			+	110	<0.1
	ATP or dATP	8		+	110	37
	ATPγS		8	+	110	<0.1
	AMPPNP, dAMPPNP		>1000	+	110	<0.1
	GTP, CTP, UTP, dGTP, dCTP, dTTP		>1000	+	110	<0.1
	ADP		60	+	110	<0.1

(*Source:* From Burgers and Kornberg, *Journal of Biological Chemistry* 257:11469, 1982. Copyright © 1982 The American Society for Biochemistry & Molecular Biology, Bethesda, MD. Reprinted by permission.)

the preinitiation complex. They compared the fmol of β with the fmol of complex, as measured by the fmol of double-stranded circles produced. This analysis yielded a value of about 2.8 β-subunits/complex, which would be close to one β-*dimer*/complex, in accord with other studies that suggested that β acts as a dimer.

Implicit in the discussion so far is the fact that ATP is required to load the β clamp onto the template. Peter Burgers and Kornberg demonstrated the necessity for ATP (or dATP) with an assay that did not require dATP for replication. The template in this case was poly(dA) primed with oligo(dT). Table 21.1 shows that ATP or dATP is required for high-activity elongation of the oligo(dT) primer, especially in the presence of a relatively high salt concentration.

> **SUMMARY** The β-subunit needs help from the γ complex (γ, δ, δ′, χ, and ψ) to load onto the DNA template. The γ complex acts catalytically in forming this processive αεβ-complex, so it does not remain associated with the complex during processive replication. Clamp loading is an ATP-dependent process.

Lagging Strand Synthesis Structural studies on pol III* (holoenzyme minus the β clamp) have shown that the enzyme consists of two core polymerases, linked through a

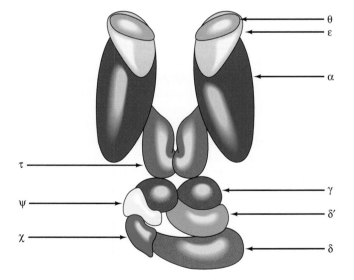

Figure 21.18   Model of the pol III* subassembly. Note that two cores and two τ-subunits are present, but only one γ complex (γ₂, δ, δ′, χ,and ψ). (*Source:* Reprinted from Herendeen and Kelly, *Cell* 84:6, 1996. Copyright 1996, with permission from Elsevier Science.)

dimer of the τ-subunit to a clamp loader, as illustrated in Figure 21.18. The following reasoning suggests that the τ-subunit serves as a dimerizing agent for the core enzyme: The α-subunit is a monomer in its native state, but τ is a dimer. Furthermore, τ binds directly to α, so α is automatically dimerized by binding to the two τ-subunits. In turn, ε is dimerized by binding to the two α-subunits, and θ is dimerized by binding to the two ε-subunits. The fact that the holoenzyme contains two core polymerases fits very nicely with the fact that two DNA strands need to be replicated. This leads directly to the suggestion that each of the core polymerases replicates one of the strands as the holoenzyme follows the moving fork. Figure 21.19 presents one model for such simultaneous synthesis of the two progeny strands.

Because discontinuous synthesis of the lagging strand must involve repeated dissociation and reassociation of the core polymerase from the template, this model raises two important questions: First, how can discontinuous synthesis of the lagging strand possibly keep up with continuous synthesis of the leading strand? If the pol III core really dissociated from the template after making each Okazaki fragment of the lagging strand, it would take a long time to reassociate and would fall hopelessly behind the leading strand. A second, related question is this: How is repeated dissociation and reassociation of the pol III core from the template compatible with the highly processive nature of DNA replication? After all, the β clamp is essential for processive replication, but once it clamps onto the DNA, how can the core polymerase dissociate every 1–2 kb as it finishes one Okazaki fragment and jumps forward to begin elongating the next?

The answer to the first question seems to be that the pol III core making the lagging strand does not really dissociate completely from the template. It remains tethered to it by its association with the core that is making the leading strand. Thus, it can release its grip on its template strand without straying far from the DNA. This enables it to find the next primer and reassociate with its template within a fraction of a second, instead of the many seconds that would be required if it completely left the DNA.

The second question requires us to look more carefully at the way the β clamp interacts with the clamp loader and with the core polymerase. We will see that these two proteins compete for the same binding site on the β clamp, and that the relative affinities of the clamp for one or the other of them shifts back and forth to allow dissociation and reassociation of the core from the DNA. We will also see that the clamp loader can act as a clamp unloader to facilitate this cycling process.

Theory predicts that the pol III* synthesizing the lagging strand must dissociate from one β clamp as it finishes one Okazaki fragment and reassociate with another β clamp to begin making the next Okazaki fragment, as depicted in Figure 21.20. But does dissociation of pol III* from its β clamp actually occur? To find out, O'Donnell and his colleagues prepared a primed M13 phage template (M13mp18) and loaded a β clamp and pol III* onto it. Then they added two more primed phage DNA templates, one (M13Gori) pre-loaded with a β clamp and the other (φX174) lacking a β clamp. Then they incubated the

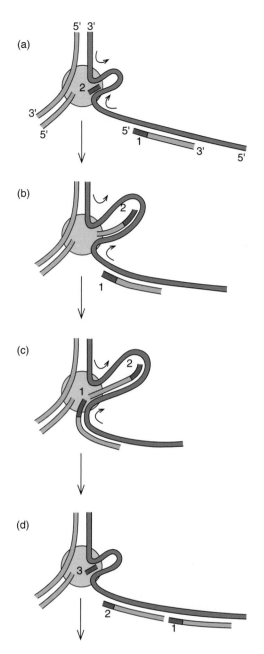

Figure 21.19 A model for simultaneous synthesis of both DNA strands. (a) The lagging template strand (blue) has formed a loop through the replisome (gold), and a new primer, labeled 2 (red), has been formed by the primase. A previously synthesized Okazaki fragment (green, with red primer labeled 1) is also visible. The leading strand template and its progeny strand are shown at left (gray), but the growth of the leading strand is not considered here. **(b)** The lagging strand template has formed a bigger loop by feeding through the replisome from the top and bottom, as shown by the arrows. The motion of the lower part of the loop (lower arrow) allows the second Okazaki fragment to be elongated. **(c)** Further elongation of the second Okazaki fragment brings its end to a position adjacent to the primer of the first Okazaki fragment. **(d)** The replisome releases the loop, which permits the primase to form a new primer (number 3). The process can now begin anew.

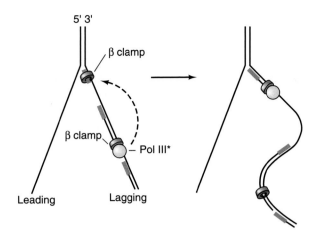

Figure 21.20 A model for cycling of the pol III* complex between β clamps on the lagging strand. When the pol III* complex (yellow) finishes one Okazaki fragment, it encounters a nick before the primer for the next completed fragment. This is its cue to dissociate from the β clamp (blue) and move (dashed arrow) to the β clamp assembled at the primer (red) for the Okazaki fragment to be made next. (*Source:* Reprinted from Stukenberg et al., *Cell* 78:878, 1994. Copyright 1994, with permission from Elsevier Science.)

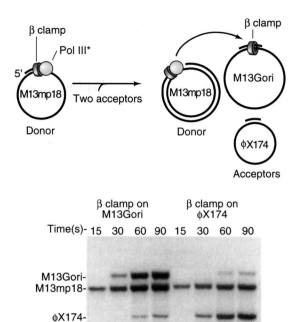

Figure 21.21 Test of the cycling model. If one assembles a pol III* complex with a β clamp on one primed template (M13mp18, top left) and presents it with two acceptor primed templates, one with a β clamp (M13Gori) and one without (ϕX174), the pol III* complex should choose the template with the clamp (M13Gori, in this case) to replicate when it has finished replicating the original template. O'Donnell and colleagues carried out this experiment, allowing enough time to replicate both the donor and acceptor templates. They also included labeled nucleotides so the replicated DNA would be labeled. Then they electrophoresed the DNAs and detected the labeled DNA products by gel electrophoresis. The electrophoresis of the replicated DNA products (bottom) show that the acceptor template with the β clamp was the one that was replicated. When the β clamp was on the M13Gori acceptor template, replication of this template predominated. On the other hand, when the β clamp was on the ϕX174 template, this was the one that was favored for replication. The positions of the replicated templates are indicated at left. (*Source:* Stukenberg et al., An explanation for lagging strand replication: Polymerase hopping among DNA sliding clamps. *Cell* 78 (9 Sept 1994) f. 2, p. 878. Reprinted by permission of Elsevier Science.)

templates together under replication conditions long enough for the original template and secondary template to be replicated. We know we will see replicated M13mp18 DNA, but the interesting question is this: Which secondary template will be replicated, the one with, or the one without, the β clamp. Figure 21.21 (lanes 1–4) demonstrates that replication occurred preferentially on the M13Gori template—the one with the β clamp. What if we put the β clamp on the other template instead? Lanes 5–8 show that in that case, the other template (ϕX174) was preferentially replicated. If the pol III* kept its original β clamp, it could have begun replicating either secondary template, regardless of which was pre-loaded with a β clamp. Thus, the results of this experiment imply that dissociation of pol III* from the template, and its β clamp, really does happen, and the enzyme can bind to another template (or another part of the same template), if another β clamp is present.

To check this conclusion, these workers labeled the β clamp with ^{32}P by phosphorylating it with [γ-^{32}P]ATP, then labeled pol III* with ^{3}H in either the θ- or τ-subunits, or in the γ complex. Then they allowed these labeled complexes to either idle on a gapped template in the presence of only dGTP and dCTP or to fill in the whole gap with all four dNTPs and thus terminate. Finally, they subjected the reaction mixtures to gel filtration and determined whether the two labels had separated. When the polymerase merely idled, the labeled β clamp and pol III* stayed together on the DNA template. By contrast, when termination occurred, the pol III* separated from its β clamp, leaving it behind on the DNA. O'Donnell and coworkers observed the same behavior regardless of which subunit of pol III* was labeled, so this whole entity, not just the core enzyme, must separate from the β clamp and DNA template upon termination of replication.

The *E. coli* genome is 4.6 Mb long, and its lagging strand is replicated in Okazaki fragments only 1–2 kb long. This means that over 2000 priming events are required on each template, so at least 2000 β clamps are needed. Because an *E. coli* cell holds only about 300 β dimers, the supply of β clamps would be rapidly exhausted if they could not recycle somehow. This would require that they dissociate from the DNA template. Does this happen? To find out, O'Donnell and colleagues assembled several β clamps onto a gapped template, then removed all other protein by gel filtration. Then they added pol III* and reran the gel filtration step. Figure 21.22a shows that, sure enough, the β clamps dissociated in the presence of pol III*, but not without the enzyme. Figure 21.22b demonstrates that these liberated β clamps were also competent to be loaded onto an acceptor template.

It is clear from what we have learned so far that the β clamp can interact with both the core polymerase and the γ complex (the clamp loader). It must associate with the core during synthesis of DNA to keep the polymerase on the template. Then it must dissociate from the template so it can move to a new site on the DNA where it can interact with another core to make a new Okazaki fragment. This movement to a new DNA site, of course, requires the β clamp to interact with a clamp loader again. One crucial question remains: How does the cell orchestrate the shifting back and forth of the β clamp's association with core and with clamp loader?

To begin to answer this question, it would help to show how and when the core and the clamp loader interact with the β clamp. O'Donnell and associates first answered the "how" question, demonstrating that the α-subunit of the core contacts β, and the δ-subunit of the clamp loader also contacts β. One assay these workers used to reveal these interactions was **protein footprinting**. This method works on the same principle as DNase footprinting, except the starting material is a labeled protein instead of a DNA, and protein-cleaving reagents are used instead of DNase. In this case, O'Donnell and colleagues introduced a six-amino acid protein kinase recognition sequence into the C-terminus of the β-subunit by manipulating its gene. They named the altered product $β^{PK}$. Then they phosphorylated this protein in vitro using protein kinase and labeled ATP (an ATP derivative with an oxygen in the γ-phosphate replaced by ^{35}S); this labeled the protein at its C-terminus. (Note that this is similar to labeling a DNA at one of its ends for DNase footprinting.) First they showed that the δ-subunit of the clamp loader and the α-subunit of the core could each protect $β^{PK}$ from phosphorylation, suggesting that both of these proteins contact $β^{PK}$.

Protein footprinting reinforced these conclusions. O'Donnell and colleagues mixed labeled $β^{PK}$ with various proteins, then cleaved the protein mixture with two proteolytic enzymes: pronase E and V8 protease. Figure 21.23 depicts the results. The first four lanes at the bottom of each panel are markers formed by cleaving the labeled β-subunit with four different reagents that cleave at known positions. Lane 5 in both panels shows the end-labeled peptides created by cleaving β in the absence of another protein. We observe a typical ladder of end-labeled products. Lane 6 in panel (a) shows what happens in the presence of δ. We see the same ladder as in lane 5, with the exception of the smallest fragment (arrow),

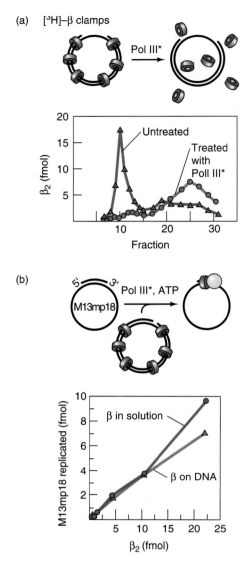

Figure 21.22   Pol III* has clamp unloading activity. (a) Clamp unloading. O'Donnell and colleagues used the γ complex to load β clamps (blue, top) onto a gapped circular template, then removed the γ complex by gel filtration. Then they added pol III* and performed gel filtration again. The graph of the results (bottom) shows β clamps that were treated with pol III* (red line) were released from the template, whereas those that were not treated with pol III* (blue line) remained associated with the template. **(b)** Recycling of β clamps. The β clamps from a β clamp–template complex treated with pol III* (red line) were just as good at rebinding to an acceptor template as were β clamps that were free in solution (blue line). (*Source:* Reprinted from Stukenberg et al., *Cell* 78:883, 1994. Copyright 1994, with permission from Elsevier Science.)

which is either missing or greatly reduced in abundance. This suggests that the δ-subunit binds to β near its C-terminus and blocks a protease from cleaving there. If this δ–β interaction is specific, one should be able to restore cleavage of the labeled $β^{PK}$ by adding an abundance of unlabeled β to bind to δ and prevent its binding to the labeled $β^{PK}$. Lane 7 shows that this is what happened. Lanes 8 and 9 in panel (a) are similar to 6 and 7, except that O'Donnell and coworkers used whole γ complex

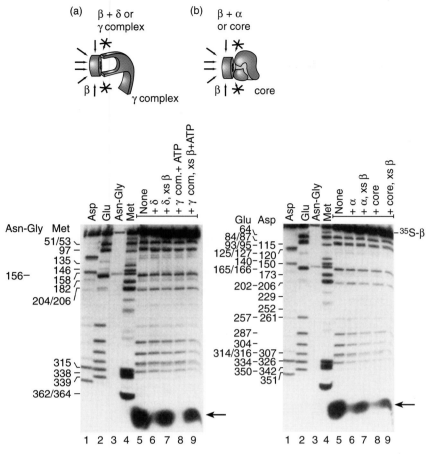

Figure 21.23   Protein footprinting of β with the γ complex and core polymerase. O'Donnell and colleagues labeled β^{PK} at its C-terminus by phosphorylation with protein kinase and [³⁵S]ATP. Then they mixed this end-labeled β with either δ or the whole γ complex (panel **a**) or with either α or the whole core (panel **b**). Then they subjected the protein complexes to mild cleavage with a mixture of pronase E and V8 protease to generate a series of end-labeled digestion products. Finally, they electrophoresed these products and autoradiographed the gel to detect them. The first four lanes in each panel are digestion products that serve as markers. The amino acid specificity of each treatment is given at top. Thus, in lane 1, the protein was treated with a protease that cleaves after aspartate (Asp) residues. Lane 5 in both panels represents β^{PK} cleaved in the absence of other proteins. Lanes 6–9 in both panels represent β^{PK} cleaved in the presence of the proteins listed at the top of each lane. The δ- and α-subunits and the γ and core complexes all protect the same site from digestion. Thus, they reduce the yield of the fragment indicated by the arrow at the bottom of the gel. The drawings at top illustrate the binding between the β clamp and either the γ complex (**a**) or the core (**b**), emphasizing that both contact the β clamp at the same places near the C-terminus of each β monomer and prevent cleavage there (arrows with Xs). (*Source:* Naktinis et al., A molecular switch in a replicating machine defined by an internal competition for protein rings. *Cell* 84 (12 June 1996) f. 3ab bottoms, p. 138. Reprinted by permission of Elsevier Science.)

instead of purified δ. Again, the γ complex protected a site near the C-terminus of β^{PK} from cleavage, and unlabeled β prevented this protection.

Panel (b) of Figure 21.23 is just like panel (a), except that the investigators used the α-subunit and whole core instead of the δ-subunit and whole γ-complex to footprint labeled β^{PK}. They observed exactly the same results: α and whole core protected the same site from cleavage as did δ and whole γ-complex. This suggests that the core and the clamp loader both contact β at the same site, and that the α- and δ-subunits, respectively, mediate these contacts. In a further experiment, these workers used whole pol III* to footprint β^{PK}. Because pol III* contains both the core and the clamp loader, one might have expected it to yield a larger footprint than either subassembly separately. But it did not. This is consistent with the hypothesis that pol III* contacts β through either the core or the clamp loader, but not both at the same time.

If the β clamp can bind to the core or the clamp loader, but not both simultaneously, which does it prefer? O'Donnell and colleagues used gel filtration to show that when the proteins are free in solution, β prefers to bind to the clamp loader, rather than the core polymerase. This is satisfying because free β needs to be loaded onto DNA by the γ-complex before it can interact with the core polymerase.

However, that situation should change once the β clamp is loaded onto a primed DNA template; once that happens, β needs to associate with the core polymerase and begin making DNA. To test this prediction, O'Donnell and colleagues loaded ³⁵S-labeled β clamps onto primed M13 phage DNA and then added either ³H-labeled clamp loader (γ complex) and unlabeled core (Figure 21.24a), or ³H-labeled core and unlabeled γ-complex (Figure 21.24b). Then they subjected these mixtures to gel filtration to separate DNA–protein complexes from free proteins. Under these conditions, it is clear that the β clamp on the DNA prefers to associate with the core polymerase. Almost no γ complex bound to the β clamp–DNA complex.

Once the holoenzyme has completed an Okazaki fragment, it must dissociate from the β clamp and move to a new one. Then the original β clamp must be removed from the template so it can participate in the synthesis of another Okazaki fragment. We have already seen the pol III* has clamp-unloading activity, but we have not seen what part of pol III* has this activity. O'Donnell and associates performed gel filtration assays that showed that the γ complex has clamp-unloading activity. Figure 21.25 illustrates this experiment. The investigators loaded β clamps onto a nicked DNA template, then removed all other proteins. Then they incubated these DNA–protein

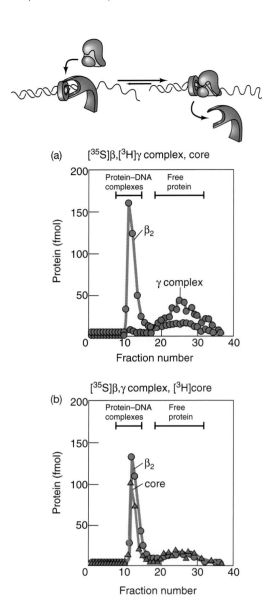

Figure 21.24   The β clamp binds core in preference to the γ complex in the presence of a primed template. O'Donnell and associates loaded a β^{PK} clamp (labeled with ^{35}S) onto a primed M13 phage template, as shown at top left. Then they added ^{3}H-labeled γ complex (red) and unlabeled core (green) and subjected the mixture to gel filtration to see whether the labeled γ complex remained with the β clamp–DNA complex. Panel (**a**) shows the results. The labeled γ complex (red) migrated independently of the β clamp–DNA complex (blue). Alternatively, these workers added ^{3}H-labeled core and unlabeled γ complex and repeated the gel filtration. Panel (**b**) shows the results. The labeled core (green) comigrated with the β clamp–DNA complex (blue), showing that core binds to the β clamp when it is associated with a primed DNA template. (*Source:* Reprinted from Naktinis et al., *Cell* 84:140, 1996. Copyright 1996, with permission from Elsevier Science.)

complexes in the presence and absence of the γ complex. We can see that the β clamps are unloaded from the nicked DNA much faster in the presence of the γ complex and ATP than in their absence.

Thus the γ complex is both a clamp loader and a clamp unloader. But what determines when it will load

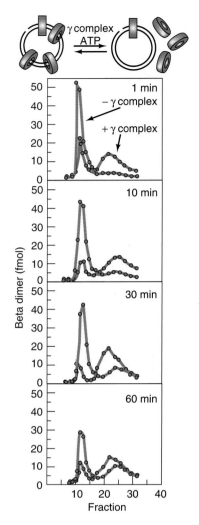

Figure 21.25   Clamp unloading activity of the γ complex. O'Donnell and coworkers loaded β clamps onto a nicked circular DNA template, as shown at top, then incubated these complexes in the presence (red) or absence (blue) of the γ complex and ATP for the times indicated. Finally, they subjected the mixtures to gel filtration to determine how much β clamp remained associated with the DNA and how much had dissociated. The cartoon at top interprets the results: The γ complex and ATP served to accelerate the unloading of β clamps from the nicked DNA. (*Source:* Reprinted from Naktinis et al., *Cell* 84:141, 1996. Copyright 1996, with permission from Elsevier Science.)

clamps and when it will unload them? The state of the DNA seems to throw this switch, as illustrated in Figure 21.26. Thus, when β clamps are free in solution and there is a primed template available, the clamps associate preferentially with the γ complex, which serves as a clamp loader to bind the β clamp to the DNA. Once associated with the DNA, the clamp binds preferentially to the core polymerase and sponsors processive synthesis of an Okazaki fragment. When the fragment has been synthesized, and only a nick remains, the core loses its affinity for the β clamp. The clamp reassociates with the γ complex, which now acts as a clamp unloader, removing the

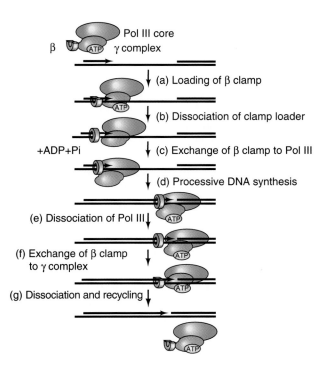

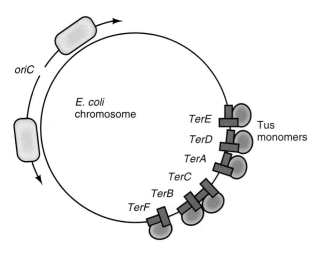

Figure 21.27   The termination region of the *E. coli* genome. Two replicating forks with their accompanying replisomes (yellow) are pictured moving away from *oriC* toward the terminator region on the opposite side of the circular *E. coli* chromosome. Three terminator sites operate for each fork: *TerE, TerD,* and *TerA* stop the counterclockwise fork; and *TerF, TerB,* and *TerC* stop the clockwise fork. The Tus protein binds to the terminator sites and helps arrest the moving forks. (*Source:* Reprinted from Baker, *Cell* 80:521, 1995. Copyright 1995, with permission from Elsevier Science.)

Figure 21.26   Summary of lagging strand replication. We begin with a β clamp associated with the γ complex part (red) of a pol III*. (**a**) The γ complex loads the β clamp (blue) onto a primed DNA template. (**b**) The γ complex, or clamp loader, dissociates from the β clamp. (**c**) The core (green) associates with the clamp. (**d**) The core and clamp cooperate to processively synthesize an Okazaki fragment, leaving just a nick between two Okazaki fragments. (**e**) The polymerase core dissociates from the clamp. (**f**) The γ complex reassociates with the β clamp. (**g**) The γ complex acts as a clamp unloader, removing the β clamp from the template. Now it is free to repeat the process, recycling to another primer on the template. (*Source:* Reprinted from Herendeen and Kelly, *Cell* 84:7, 1996. Copyright 1996, with permission from Elsevier Science.)

clamp from the template so it can recycle to the next primer and begin the cycle anew.

SUMMARY The pol III holoenzyme is double-headed, with two core polymerases attached through two τ-subunits to a γ complex. One core is responsible for continuous synthesis of the leading strand, the other performs discontinuous synthesis of the lagging strand. The γ complex serves as a clamp loader to load the β clamp onto a primed DNA template. Once loaded, the β clamp loses affinity for the γ complex and associates with the core polymerase to help with processive synthesis of an Okazaki fragment. Once the fragment is completed, the β clamp loses affinity for the core polymerase and associates with the γ complex, which acts as a clamp unloader, removing the clamp from the DNA. Now it can recycle to the next primer and repeat the process.

21.4   Termination

Termination of replication is relatively straightforward for λ and other phages that produce a long, linear concatemer. The concatemer simply continues to grow as genome-sized parts of it are snipped off and packaged into phage heads. But for bacteria and eukaryotes, where replication has a definite end as well as a beginning, the mechanisms of termination are more complex and more interesting. In bacterial DNA replication, the two replication forks approach each other in the terminus region, which contains 22-bp terminator sites that bind specific proteins. In *E. coli,* the terminator sites are *TerA–TerF,* and they are arranged as pictured in Figure 21.27. The *Ter* sites bind proteins called **Tus** (for terminus utilization substance). Replicating forks enter the terminus region and pause before quite completing the replication process. This leaves the two daughter duplexes entangled. They must become disentangled before cell division occurs, or they cannot separate to the two daughter cells. Instead, they would remain caught in the middle of the cell, cell division would fail, and the cell would probably die. These considerations raise the question: How do the daughter duplexes become disentangled? For eukaryotes, we would like to know how cells fill in the gaps left by removing primers at the 5′-ends of the linear chromosomes. Let us examine each of these problems.

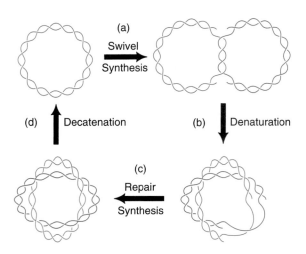

Figure 21.28   A two-stage model for decatenation. (a) When a circular bacterial DNA replicates by the θ mode, using the swivel activity of DNA gyrase, it reaches a stage where replication stops, leaving several double-helical turns of the parental DNA strands still entwined. (b) Denaturation of the remaining parental double helix occurs, leaving stretches of single-stranded DNA in the daughter DNAs, which remain entwined. (c) Repair synthesis fills in the single-stranded gaps, but the two daughter duplexes are still linked together in a catenane. The number of nodes, or times one duplex crosses the other in the catenane, is the same as the number of crossings in the unreplicated parental duplex after step (a)—four in this case. (d) Decatenation separates the two daughter duplexes. (*Source:* Reprinted from Adams et al., *Cell* 71:278, 1992. Copyright 1992, with permission from Elsevier Science.)

Decatenation: Disentangling Daughter DNAs

Figure 21.28 illustrates the problem faced by bacteria near the end of DNA replication. Because of their circular nature, the two daughter duplexes remain entwined as two interlocking rings, a type of **catenane.** For these interlocked DNAs to move to the two daughter cells, they must be unlinked, or **decatenated.** If decatenation occurs before repair synthesis, a single nick will suffice to disentangle the DNAs, and a type I topoisomerase can perform the decatenation. However, if repair synthesis occurs first, a type II topoisomerase, which passes a DNA duplex through a double-stranded break, is required. *Salmonella typhimurium* and *E. coli* cells contain four topoisomerases: topoisomerases I–IV (topo I–IV). Topo I and III are type I enzymes, and topo II and IV are type II. The question is: Which topoisomerase is involved in decatenation?

Because DNA gyrase (topo II) acts as the swivel during DNA replication, many molecular biologists assumed that it also decatenates the daughter duplexes. But Nicholas Cozzarelli and his colleages demonstrated that **topo IV** is really the decatenating enzyme. They tested various temperature-sensitive mutant strains of *Salmonella typhimurium,* a close relative of *E. coli,* for ability to decatenate dimers of the plasmid pBR322 in vivo at the permis-

sive and nonpermissive temperatures. The mutations lay in *gyrA,* the gene that encodes the A subunit of topo II, in *parC* and *parE,* the genes that encode the two subunits of topo IV, and in *parF,* a gene linked to *parC* and *parE.* They also tested the effect of norfloxacin, a potent inhibitor of DNA gyrase, and therefore of DNA replication.

To assay for decatenation, these workers grew cells at 30° or 44°C, in the presence or absence of norfloxacin, then isolated DNA from the cells. They nicked the DNA with DNase I to relax any supercoils, then electrophoresed and Southern blotted it. Finally, they hybridized the Southern blot to a pBR322 probe. Catenanes appear in the autoradiograph of the blot as a series of bands between the nicked dimer and nicked monomer markers. Each band corresponds to a catenane with a different number of **nodes,** or crossings. (For example, the structure at the lower left of Figure 21.28 contains four nodes where the two plasmids cross each other.) Notice that the higher the number of nodes, the higher the electrophoretic mobility of the catenane. The autoradiograph in Figure 21.29 shows that the mutants with lesions in the *parC* and *parE* genes failed to decatenate the plasmid at the nonpermissive temperature (44°C) in the absence of norfloxacin. This suggests that topo IV is important in decatenation. Norfloxacin, by blocking DNA gyrase, halted DNA replication and presumably allowed subsequent decatenation by the small amount of residual topo IV, or by another topoisomerase. By contrast, the strain with the mutant DNA gyrase did not accumulate catenanes at the nonpermissive temperature, either in the presence or absence of norfloxacin, suggesting that this enzyme does not participate in decatenation. When they tested temperature-sensitive *parC* and *parE* mutants of *E. coli,* Cozzarelli and colleagues observed similar behavior, indicating that topo IV also participates in decatenation in *E. coli.*

The catenane at lower left in Figure 21.28 has four nodes because the two daughter duplexes at upper right contained four remaining interlocking turns at the end of DNA replication. The two-stage mechanism pictured in Figure 21.28 makes several predictions. First, the catenane will be what we call a **torus,** in which the two daughter plasmids are helically wound around each other. (A torus, by the way, is a doughnut-shaped structure.) Second, because the DNA is a right-handed helix, the helical turns by which the two daughter plasmids coil around each other must be right-handed. Third, we should observe only even numbers of nodes. Finally, because the two daughter plasmids have the same orientation, the catenane is said to be parallel. Only the two-stage model will produce right-handed, parallel torus catenanes.

To check the characteristics of the catenanes left in the mutant cells at elevated temperature, Cozzarelli and colleagues coated them with RecA protein to make them easy to see and subjected the coated catenanes to electron microscopy. Figure 21.30 shows a variety of torus

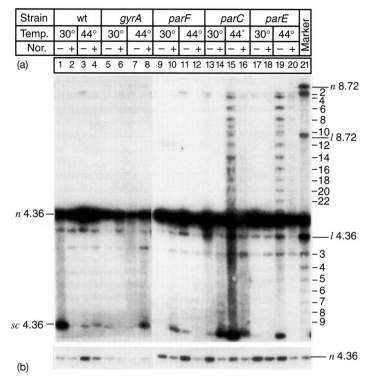

Strain	wt		gyrA		parF		parC		parE		Marker
Temp.	30°	44°	30°	44°	30°	44°	30°	44°	30°	44°	
Nor.	− +	− +	− +	− +	− +	− +	− +	− +	− +	− +	
(a)	1 2	3 4	5 6	7 8	9 10	11 12	13 14	15 16	17 18	19 20	21

Figure 21.29 Catenated plasmid accumulation in *Salmonella* mutants. (a) Autoradiograph of blot. Cozzarelli and colleagues grew wild-type, and four different temperature-sensitive mutant strains of *S. typhimurium* at the permissive temperature (30° C) or the nonpermissive temperature (44° C), and in the presence and absence of norfloxacin, as indicated at top. All strains harbored the pBR322 plasmid. They isolated plasmid DNA from the bacterial cells, nicked the DNAs to relax supercoils, and electrophoresed them on an agarose gel, which distinguished catenanes having different numbers of nodes. They Southern blotted the DNAs and probed the blot with a labeled pBR322 DNA, The markers labeled *n* and *l* are nicked and linear plasmids having the size (in kilobases) indicated at right and left. *SC* marks the position of the supercoiled plasmid. The *parC* and *parE* mutants failed to decatenate the daughter plasmids at the nonpermissive temperature, but all the other strains, including the *gyrA* strain that has a temperature-sensitive DNA gyrase, resolved the catenanes at the nonpermissive temperature. **(b)** Underexposure of blot of nicked pBR322 (designated *n* 4.36) to show the relative amounts loaded in each lane. (*Source:* (a-b) Adams et al., The role of topoisomerase IV in partitioning bacterial replicons and the structure of catenated intermediates in DNA replication. *Cell* 71 (16 Oct 1992) f. 2, p. 279. Reprinted by permission of Elsevier Science.)

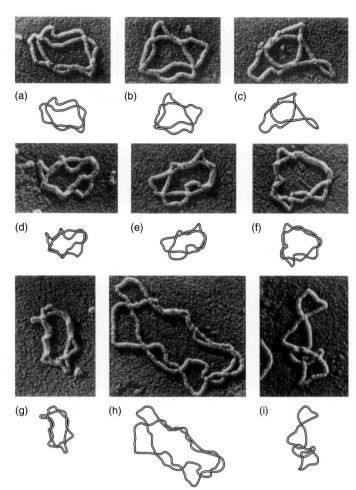

Figure 21.30 Electron micrographs of torus catenanes resulting from replication of pBR322 plasmid in *S. typhimurium.* Cozzarelli and colleagues isolated catenanes from *parC* **(b)**, **(c)**, **(d)**, **(e)**, and **(g)**, and *parE* **(a)**, **(f)** and **(h)** mutants cultured for 20 min at the nonpermissive temperature. They nicked the DNAs to relax supercoils, coated them with RecA protein to make them easier to see, spread them on EM grids, and photographed them. The micrographs are each accompanied by an interpretive diagram. All but **(i)** are right-handed torus catenanes with the following numbers of nodes: **(a)–(c)**, 4; **(d)** and **(e)**, 6; **(f)** and **(g)**, 8. Panel **(i)** shows a singly linked catenane in which the lower plasmid is knotted. (*Source:* Adams et al., *Cell* 71 (16 Oct 1992) f. 7, p. 284. Reprinted by permission of Elsevier Science.)

catenanes with even numbers of nodes. (Figure 21.29 also showed only even numbers of nodes in the catenanes.) Moreover, careful examination of the micrographs shows that the torus catenanes are right-handed, as demanded by the two-stage model. It is impossible to tell from these pictures whether the two daughter plasmids are parallel, but further, more complex analysis demonstrated that they are indeed parallel. Thus, all of the conditions of the two-stage model are met, and the data strongly favor that model, with topo IV playing the decatenation role with plasmids. But does topo IV play the same role in decatenating bacterial chromosomes? It appears that is does, be-

cause topo IV mutations are lethal, and the dead bacteria exhibit incompletely separated chromosomes.

Eukaryotic chromosomes are not circular, but they have multiple replicons, so replication forks from neighboring replicons approach one another just as the two replication forks of a bacterial chromosome approach each other near the termination point opposite the origin of replication. Apparently, this inhibits completion of DNA replication, so eukaryotic chromosomes also form catenanes that must be disentangled. Eukaryotic topo II resembles bacterial topo IV more than it does DNA gyrase, and it is a strong candidate for the decatenating enzyme.

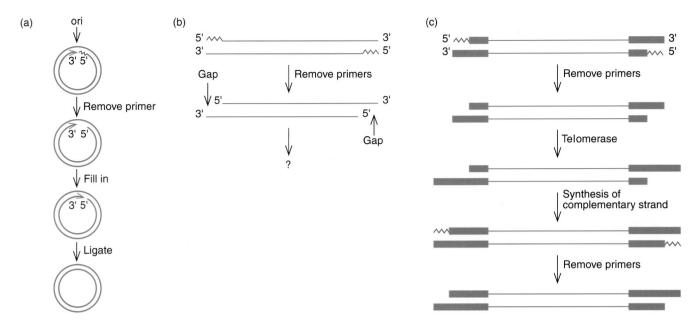

Figure 21.31   Coping with the gaps left by primer removal. (a) In prokaryotes, the 3'- end of a circular DNA strand can prime the synthesis of DNA to fill in the gap left by the first primer (red). For simplicity, only one replicating strand is shown. **(b)** Hypothetical model to show what would happen if primers were simply removed from the 5'-end of linear DNA strands with no telomerase action. The gaps at the ends of chromosomes would grow longer each time the DNA replicated. **(c)** How telomerase can solve the problem. In the first step, the primers (red) are removed from the 5'-ends of the daughter strands, leaving gaps. In the second step, telomerase adds extra telomeric DNA (green boxes) to the 3'-ends of the other daughter strands. In the third step, DNA synthesis occurs, using the newly made telomeric DNA as a template. In the fourth step, the primers used in step three are removed. This leaves gaps, but the telomerase action has ensured that no net loss of DNA has occurred. The telomeres represented here are not drawn to scale with the primers. In reality, human telomeres are thousands of nucleotides long. (*Source:* (c) Greider & Blackburn, Identification of a specific telomere terminal transferase activity in tetramere extracts. *Cell* 43 (Dec Pt1 1985) f. 1A, p. 406. © Cell Press.)

> **SUMMARY** At the end of replication, circular bacterial chromosomes form catenanes that are decatenated in a two-step process. First, the remaining, unreplicated double-helical turns linking the two strands are denatured. Then repair synthesis fills in the gaps. This leaves a right-handed parallel torus catenane with an even number of nodes that is decatenated by topoisomerase IV. Linear eukaryotic chromosomes also require decatenation during DNA replication.

Termination in Eukaryotes

Eukaryotes face a difficulty at the end of DNA replication that prokaryotes do not: filling in the gaps left when RNA primers are removed. With circular DNAs, such as those in bacteria, there is no problem filling all the gaps because another DNA 3'-end is always upstream to serve as primer (Figure 21.31a). But consider the problem faced by eukaryotes, with their linear chromosomes. Once the first primer on each strand is removed (Figure 21.31b), there is no way to fill in the gaps because DNA cannot be extended in the 3' → 5' direction, and no 3'-end is upstream, as there would be in a circle. If this were actually the situation, the DNA strands would get shorter every time they replicated. This is a termination problem in that it deals with the formation of the ends of the DNA strands, but how do cells solve this problem?

Telomere Maintenance   Elizabeth Blackburn and her colleagues provided the answer, which is summarized in Figure 21.31c. The **telomeres,** or ends of eukaryotic chromosomes, are composed of repeats of short, GC-rich sequences. The G-rich strand of a telomere is added at the very 3'-ends of DNA strands, not by semiconservative replication, but by an enzyme called **telomerase.** The exact sequence of the repeat in a telomere is species-specific. In *Tetrahymena,* it is TTGGGG/AACCCC; in vertebrates, including humans, it is TTAGGG/AATCCC. Blackburn showed that this specificity resides in the telomerase itself and is due to a small RNA in the enzyme that serves as the template for telomere synthesis. This solves the problem: The telomerase adds many repeated copies of its characteristic sequence to the 3'-ends of chromosomes. Priming can then occur within these telomeres to make the C-rich strand. There is no problem when terminal primers are removed and not replaced, because only telomere sequences are lost, and these can always be replaced by telomerase and another round of telomere synthesis.

Blackburn made a clever choice of organism in which to search for telomerase activity: *Tetrahymena*, a ciliated protozoan. *Tetrahymena* has two kinds of nuclei: (1) micronuclei, which contain the whole genome in five pairs of chromosomes that serve to pass genes from one generation to the next; and (2) macronuclei, in which the five pairs of chromosomes are broken into more than 200 smaller fragments used for gene expression. Because each of these minichromosomes has telomeres at its ends, *Tetrahymena* cells have many more telomeres than human cells, for example, and they are loaded with telomerase, especially during the phase of life when macronuclei are developing and the new minichromosomes must be supplied with telomeres. This made isolation of the telomerase enzyme from *Tetrahymena* relatively easy.

In 1985, Carol Greider and Blackburn succeeded in identifying a telomerase activity in extracts from synchronized *Tetrahymena* cells that were undergoing macronuclear development. They assayed for telomerase activity in vitro using a synthetic primer with four repeats of the TTGGGG telomere sequence and included a radioactive nucleotide to label the extended telomere-like DNA. Figure 21.32 shows the results. Lanes 1–4 each contained a different labeled nucleotide (dATP, dCTP, dGTP, and dTTP, respectively), plus all three of the other, unlabeled nucleotides. Lane 1, with labeled dATP showed only a smear, and lanes 2 and 4 showed no extension of the synthetic telomere. But lane 3, with labeled dGTP, exhibited an obvious periodic extension of the telomere. Each of the clusters of bands represents an addition of one more TTGGGG sequence (with some variation in the degree of completion, which accounts for the fact that we see *clusters* of bands, rather than *single* bands. Of course, we should observe telomere extension with dTTP, as well as with dGTP. Further investigation showed that the concentration of dTTP was too low in this experiment, and that dTTP could be incorporated into telomeres at higher concentration. Lanes 5–8 show the results of an experiment with one labeled, and only one unlabeled nucleotide. This experiment verifed that dGTP could be incorporated into the telomere, but only if unlabeled dTTP was also present. This is what we expect because this strand of the telomere contains only G and T. Controls in lanes 9–12 showed that an ordinary DNA polymerase, Klenow fragment, cannot extend the telomere. Further controls in lanes 13–16 demonstrated that telomerase activity depends on the telomere-like primer.

How does telomerase add the correct sequence of bases to the ends of telomeres without a complementary DNA strand to read? It uses its own RNA constituent as a template. Greider and Blackburn demonstrated in 1987 that telomerase is a ribonucleoprotein with essential RNA and protein subunits. Then in 1989 they cloned and sequenced the gene that encodes the 159-nt RNA subunit of

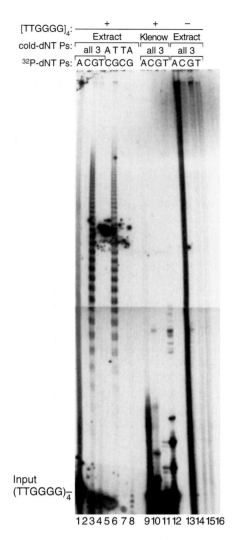

Figure 21.32 Identification of telomerase activity. Greider and Blackburn synchronized mating of *Tetrahymena* cells and let the offspring develop to the macronucleus development stage. They prepared cell-free extracts and incubated them for 90 min with a synthetic oligomer having four repeats of the TTGGGG telomere repeat sequence, plus the labeled and unlabeled nucleotides indicated at top. After incubation, they electrophoresed the products and detected them by autoradiography. Lanes 9–12 contained the Klenow fragment of *E. coli* DNA polymerase I instead of *Tetrahymena* extract. Lanes 13–16 contained extract, but no primer. Telomerase activity is apparent only when both dGTP and dTTP are present.
(*Source:* Greider & Blackburn, Identification of a specific telomere terminal transferase activity in tetramere extracts. *Cell* 43 (Dec Pt1 1985) f. 1A, p. 406. Reprinted by permission of Elsevier Science.)

the *Tetrahymena* telomerase and found that it contains the sequence CAACCCCAA. In principle, this sequence can serve as template for repeated additions of TTGGGG sequences to the ends of *Tetrahymena* telomeres as illustrated in Figure 21.33.

Blackburn and her colleagues used a genetic approach to prove that the telomerase RNA really does serve as the template for telomere synthesis. They showed that mutant telomerase RNAs gave rise to telomeres with corresponding alterations in their sequence. In particular, they

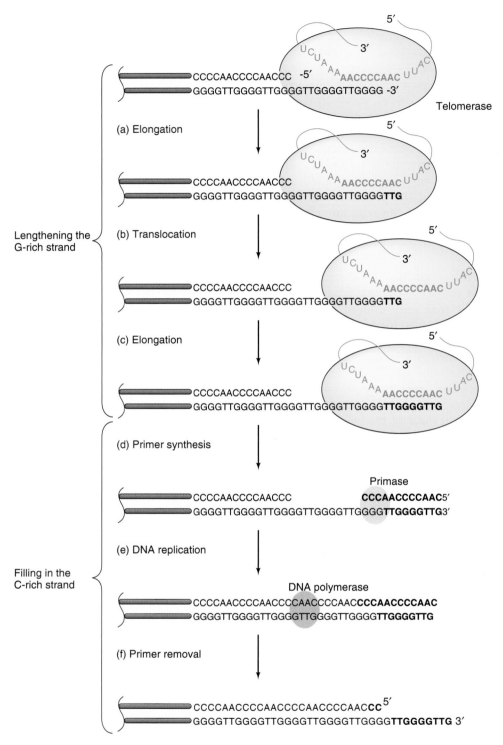

Figure 21.33 Forming telomeres in *Tetrahymena*. (a) Telomerase (yellow) promotes hybridization between the 3′-end of the G-rich telomere strand and the template RNA (red) of the telomerase. The telomerase uses three bases (AAC) of its RNA as a template for the addition of three bases (TTG, boldface) to the 3′-end of the telomere. (b) The telomerase translocates to the new 3′-end of the telomere, pairing the left-hand AAC sequence of its template RNA with the newly incorporated TTG in the telomere. (c) The telomerase uses the template RNA to add six more nucleotides (GGGTTG, boldface) to the 3′-end of the telomere. Steps (a) through (c) can repeat indefinitely to lengthen the G-rich strand of the telomere. (d) When the G-rich strand is sufficiently long (probably longer than shown here), primase (orange) can make an RNA primer (boldface), complementary to the 3′-end of the telomere's G-rich strand. (e) DNA polymerase (green) uses the newly made primer to prime synthesis of DNA to fill in the remaining gap on the C-rich telomere strand. (f) The primer is removed, leaving a 12–16-nt overhang on the G-rich strand.

changed the sequence 5'-CAACCCCAA-3' of a cloned gene encoding the *Tetrahymena* telomerase RNA as follows:

wt: 5'-CAACCCCAA-3'
1: 5'-CAA<u>C</u>CCCCAA-3'
2: 5'-CAACC<u>T</u>CAA-3'
3: 5'-C<u>G</u>ACCCCAA-3'

The underlined bases in each of the three mutants (1, 2, and 3) denote the base changed (or added, in 1). They introduced the wild-type or mutated gene into *Tetrahymena* cells in a plasmid that ensured the gene would be overexpressed. Even though the endogenous wild-type gene remained in each case, the overexpression of the transplanted gene swamped out the effect of the endogenous gene. Southern blotting of telomeric DNA from cells transformed with each construct showed that a probe for the telomere sequence expected to result from mutants 1 (TTGGGGG) and 3 (GGGGTC) actually did hybridize to telomeric DNA from cells transformed with these mutant genes. On the other hand, this did not work for mutant 2; no telomeric DNA that hybridized to a probe for GAGGTT was observed.

These results suggested that mutant telomerase RNAs 1 and 3, but not 2, served as templates for telomere elongation. To confirm this suggestion, Blackburn and colleagues sequenced a telomere fragment from cells transformed with mutant telomerase RNA 3. They found the following sequence:

5'-CTTTTACTCAATGTCAAAGAAATTATTAAATT(GGGGTT)$_{30}$
(GGGGTC)$_2$GGGGTT(GGGGT<u>C</u>)$_8$GGGGTTGGGGT<u>C</u>(GGGGTT)$_N$-3'

where the underlined bases must have been encoded by the mutant telomerase RNA. This nonuniform sequence differs stikingly from the normal, very uniform telomeric sequence in this species. The first 30 repeats appear to have been encoded by the wild-type telomerase RNA before transformation. These are followed by 11 mutant repeats interspersed with 2 wild-type repeats, then by all wild-type repeats. The terminal wild-type sequences may have resulted from recombination with a wild-type telomere, or from telomere synthesis after loss of the mutant telomerase RNA gene from the cell. Nevertheless, the fact remains that a significant number of repeats have exactly the sequence we would expect if they were encoded by the mutant telomerase RNA. Thus, we can conclude that the telomerase RNA does serve as the template for telomere synthesis, as Figure 21.33 suggests.

Interestingly, the somatic cells of higher eukaryotes, including humans, lack telomerase activity, whereas germ cells retain this activity. Furthermore, cancer cells also have telomerase activity. These findings have profound implications for the characteristics of cancer cells, and perhaps even for their control (see Box 21.1).

SUMMARY Eukaryotic chromosomes have special structures known as telomeres at their ends. One strand of these telomeres is composed of many tandem repeats of short, G-rich regions whose sequence varies from one species to another. The G-rich telomere strand is made by an enzyme called telomerase, which contains a short RNA that serves as the template for telomere synthesis. The C-rich telomere strand is synthesized by ordinary RNA-primed DNA synthesis, like the lagging strand in conventional DNA replication. This mechanism ensures that chromosome ends can be rebuilt and therefore do not suffer shortening with each round of replication.

Telomere Structure Besides protecting the ends of chromosomes from degradation, telomeres play another critical role: They prevent the DNA repair machinery from recognizing the ends of chromosomes as chromosome breaks and sticking chromosomes together. This inappropriate joining of chromosomes would be potentially lethal to the cell. If telomeres really looked the way they are pictured in Figures 21.31 and 21.33, little would distinguish them from real chromosome breaks. How do telomeres allow the cell to recognize the difference?

For years, molecular biologists pondered this question and, as telomere-binding proteins were discovered, they theorized that these proteins bind to the ends of chromosomes and in that way identify the ends. One problem with this hypothesis is that the mammalian telomere-binding proteins, such as the **TTAGGG repeat-binding factors TRF1** and **TRF2** bound duplex DNA in the telomeres specifically. No one could find a mammalian protein that was specific for the very end of the telomere (the single-stranded G-rich overhang illustrated in Figure 21.33). Then, in 1999, Jack Griffith and Titia de Lange and their colleagues discovered that telomeres are not linear, as had been assumed, but form a DNA loop they called a **t loop** (for telomere loop). These loops are unique in the chromosome and therefore quite readily set the ends of chromosomes apart from breaks that occur in the middle and would yield linear ends to the chromosome fragments.

What is the evidence for t loops? Griffith, de Lange and colleagues started by making a model mammalian telomeric DNA with about 2 kb of repeating TTAGGG sequences, and a 150–200-nt single stranded 3'-overhang at the end. They added one of the telomere-binding proteins, TRF2, then subjected the complex to electron microscopy. Figure 21.34a shows that a loop really did form, with a ball of TRF2 protein right at the loop–tail junction. Such structures appeared about 20% of the

BOX 21.1

Telomeres, the Hayflick Limit, and Cancer

Everyone knows that organisms, including humans, are mortal. But biologists used to assume that cells cultured from humans were immortal. Each individual cell would ultimately die, of course, but the cell *line* would go on dividing indefinitely. Then in the 1960s Leonard Hayflick discovered that ordinary human cells are not immortal. They can be grown in culture for a finite period—about 50 generations. Then they enter a period of senescence, and finally they all die. This ceiling on the lifetime of normal cells is known as the Hayflick limit. But cancer cells do not obey any such limit. They *do* go on dividing generation after generation, indefinitely.

Investigators have discovered a significant difference between normal cells and cancer cells that may explain why cancer cells are immortal and normal cells are not: Human cancer cells contain telomerase, whereas normal somatic cells lack this enzyme. (Germ cells must retain telomerase, of course, to safeguard the ends of the chromosomes handed down to the next generation.) Thus, we see that cancer cells can repair their telomeres after every cell replication, but most normal cells cannot. Therefore, cancer cells can go on dividing without degrading their chromosomes, whereas normal cells' chromosomes grow shorter with each cell division. Sooner or later this loss of genetic material stops the replication of normal cells, so they die. But this does not happen to cancer cells; telomerase saves them from that fate.

One of the essential changes that must occur in a cell to make it cancerous is the reactivation of the telomerase gene. This is required for the immortality that is the hallmark of cancer cells. This discussion also suggests a potential treatment for cancer: Turn off the telomerase gene in cancer cells or, more simply, administer a drug that in-

hibits telomerase. Such a drug should not harm most normal cells because they have no telomerase to begin with. Cancer researchers are hard at work on this strategy.

Some signs indicate that simply inhibiting the telomerase of cancer cells may not cause the cells to die. For one thing, knockout mice totally lacking telomerase activity survive and reproduce for at least six generations, though eventually the loss of telomeres leads to sterility. However, cells from these telomerase knockout mice can be immortalized, they can be transformed by tumor viruses, and these transformed cells can give rise to tumors when transplanted to immunodeficient mice. Thus, the presence of telomerase is not an absolute requirement for the development of a cancer cell. It may be that mouse cells have a way of preserving their telomeres without telomerase. We will have to see whether human cells behave differently.

Finally, immortalizing human cells in culture leads to the idea of immortalizing human beings themselves. Could it be that reactivating telomerase activity in human somatic cells would lengthen human lifetimes? Or would it just make us more susceptible to cancer? To begin answering this question, Serge Lichtsteiner, Woodring Wright, and their colleagues transplanted a human telomere reverse transcriptase gene into human somatic cells in culture, so these cells were forced to express telomerase activity. The results were striking: The telomeres in these cells grew longer and the cells went on dividing far past their normal lifetimes. They remained youthful in appearance and in their chromosome content. Furthermore, they did not show any signs of becoming cancerous. These findings were certainly encouraging, but they do not herald a fountain of youth. For now, that remains in the realm of science fiction.

time. By contrast, when these workers cut off the single-stranded 3'-overhang, or left out TRF2, they found a drastic reduction in loop formation.

One way for a telomere to form such a loop would be for the single-stranded 3'-overhang to invade the double-stranded telomeric DNA upstream, as depicted in Figure 21.35. If this hypothesis is correct, one should be able to stabilize the loop with psoralen and UV radiation, which cross-link thymines on opposite strands of a double-stranded DNA. Because the invading strand base-pairs with one of the strands in the invaded DNA this creates double-stranded DNA that is subject to cross-linking and

therefore stabilization. Figure 21.34b shows the results of an experiment in which Griffith, de Lange, and coworkers cross-linked the model DNA with psoralen and UV, then deproteinized the complex, then subjected it to electron microscopy. The loop is still clearly visible, even in the absence of TRF2, showing that the DNA itself has been crosslinked, stabilizing the t loop.

Next, these workers purified natural telomeres from several human cell lines and from mouse cells and subjected them to psoralen–UV treatment and electron microscopy. They obtained the same result as in Figure 21.34b, showing that t loops appear to form in vivo.

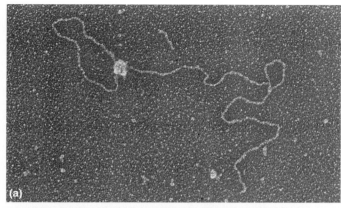

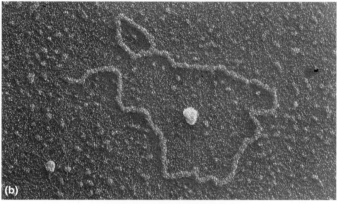

Figure 21.34 Formation of t loops in vitro. (a) Direct detection of loops. Griffith and colleagues mixed a model DNA with a telomere-like structure with TRF2, then spread the mixture on an EM grid, shadowed the DNA and protein with tungsten, and observed the shadowed molecules with an electron microscope. An obvious loop appeared, with a blob of TRF2 at the junction between the loop and the tail. **(b)** Stabilization of the loop by cross-linking. Griffith and coworkers formed the t loop as in panel (a), then crosslinked double-stranded DNA with psoralen and UV radiation, then removed the protein, spread the cross-linked DNA on an EM grid, shadowed with platinum and paladium, and visualized the shadowed DNA with an electron microscope. Again, an obvious loop appeared. The bar represents 1 kb. (*Source:* Griffith, J. et al. Mammalian Telomeres end in a large duplex loop. *Cell* 97 (14 May 1999) f. 1, p. 504. Reprinted by permission of Elsevier Science.)

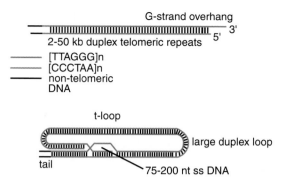

Figure 21.35 A model of a mammalian t loop. The single stranded 3′-end of the G-rich strand (red) invades the double-stranded telomeric DNA upstream, forming a long t loop and a 75–200-nt displacement loop at the junction between the loop and the tail. A short subtelomeric region (black) is pictured adjoining the telomere (blue and red). (*Source:* Reprinted from Griffith et al., *Cell*, 97:511, 1999. Copyright 1999, with permission from Elsevier Science.)

Chapter 20) if the displaced DNA is long enough. Figure 21.36b demonstrates that SSB is indeed visible, right at the tail–loop junction. That is just where the hypothesis predicts we should find the displaced DNA.

What roles do TRF1 and TRF2 play in t loop formation? Figure 21.34b showed that TRF2 is found at the loop–tail junction where strand invasion occurs, and we know that no loops form without TRF2, so Griffiths, de Lange, and coworkers speculated that this protein plays a role in strand invasion and remains bound at the site of strand invasion. Figure 21.36a showed that TRF1 binds to the double-stranded telomeric DNA in the t loop. Furthermore, TRF1 is known to induce double-stranded DNA to bend, and also to pair telomeric duplex DNAs. Thus, Griffiths, de Lange and colleagues speculated that this protein plays a role in bending the telomeric DNA and holding it in the looped conformation, thus facilitating strand invasion. Figure 21.37 summarizes these speculations. The net result of these gyrations is that the sensitive end of the chromosome is tucked away in a displacement loop where it is protected from DNA repair enzymes that might recognize it as a broken end of a chromosome and try to attach it to another chromosome end—with disastrous consequences.

Furthermore, the sizes of these putative t loops correlated well with the known lengths of the telomeres in the human or mouse cells, reinforcing the hypothesis that these loops really do represent telomeres.

To test further the notion that the loops they observed contain telomeric DNA, Griffith, de Lange and colleagues added TRF1, which is known to bind very specifically to double-stranded telomeric DNA, to their looped DNA. They observed loops coated with TRF1, as shown in Figure 21.36a.

If the strand invasion hypothesis in Figure 21.35 is valid, the single-stranded DNA displaced by the invading DNA (the displacement loop, or D loop) should be able to bind *E. coli* single-strand-binding protein (SSB, recall

SUMMARY TRF1 and TRF2 appear to help telomeric DNA in mammalian cells form a t loop, in which the single-stranded 3′-end of the telomere invades the double-stranded telomeric DNA upstream. TRF1 may help bend the DNA into shape for strand invasion, and TRF2 binds at the point of strand invasion and may stabilize the displacement loop. The t loop structure protects the end of the chromosome from inappropriate repair in which it would be attached to another chromosome end.

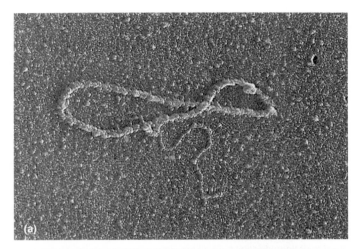

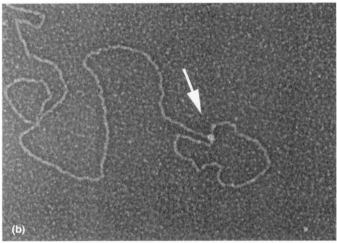

Figure 21.36   Binding of TRF1 and SSB to t loops. (a) TRF1. Griffith, de Lange, and colleagues purified natural HeLa cell t loops, cross-linked them with psoralen and UV radiation, and added TRF1, which binds specifically to double-stranded telomeric DNA. Then they shadowed the loop with platinum and paladium and performed electron microscopy. The t loop, but not the tail, is coated uniformly with TRF1. **(b)** SSB. These workers followed the same procedure as in panel **(a)**, but substituted *E. coli* SSB for TRF1. SSB should bind to single-stranded DNA, and it was observed at the loop–tail junction (arrow), where the the single-stranded displacement loop was predicted to be. The bar represents 1 kb. (*Source:* Griffith, J. et al. Mammalian Telomeres end in a large duplex loop. *Cell* 97 (14 May 1999) f. 5, p. 510. Reprinted by permission of Elsevier Science.)

SUMMARY

The pol III holoenzyme synthesizes DNA at the rate of about 730 nt/sec in vitro, just a little slower than the rate of almost 1000 nt/sec observed in vivo. This enzyme is also highly processive, both in vitro and in vivo.

Primer synthesis in *E. coli* requires a primosome composed of DnaB and the primase, DnaG. Primosome assembly at the origin of replication, *oriC*, occurs as

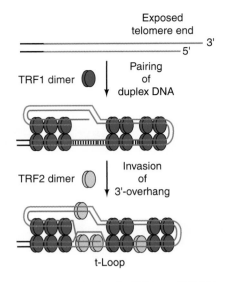

Figure 21.37   Hypothetical roles for TRF1 and TRF2 in mammalian t loop formation. TRF1 (green), by binding to the double-stranded part of the telomere, bends the DNA and holds it in an appropriate conformation for strand invasion. TRF2 (yellow) may also bind to the double-stranded region, but its main role appears to be to bind at the site of strand invasion and stabilize the displacement loop, thereby stabilizing the t loop. (*Source:* Reprinted from Griffith et al., *Cell*, 97:511, 1999. Copyright 1999, with permission from Elsevier Science.)

follows: DnaA binds to *oriC* at sites called *dnaA* boxes and cooperates with RNA polymerase and HU protein in melting a DNA region adjacent to the leftmost *dnaA* box. DnaB then binds to the open complex and facilitates binding of the primase to complete the primososme. The primosome remains with the replisome, repeatedly priming Okazaki fragment synthesis on the lagging strand. DnaB also has a helicase activity that unwinds the DNA as the replisome progresses.

The SV40 origin of replication is adjacent to the viral transcription control region. Initiation of replication depends on the viral large T antigen, which binds to a region within the 64-bp minimal *ori,* and at two adjacent sites, and exercises a helicase activity, which opens up a replication bubble within the minimal *ori*. Priming is carried out by a primase associated with the host DNA polymerase α.

The yeast origins of replication are contained within autonomously replicating sequences (ARSs) that are composed of four important regions (A, B1, B2, and B3). Region A is 15 bp long and contains an 11-bp consensus sequence that is highly conserved in ARSs. Region B3 may allow for an important DNA bend within ARS1.

The pol III core (αε or αεθ) does not function processively by itself, so it can replicate only a short stretch of DNA before falling off the template. By contrast, the core plus the β-subunit can replicate DNA processively at a rate approaching 1000 nt/sec. The β-subunit forms a dimer that is ring-shaped. This ring fits around a DNA template and interacts with the α-subunit of the core to tether the whole polymerase and template together. This is why the holoenzyme stays on its template so long and is therefore so processive. The eukaryotic processivity factor PCNA forms a trimer with a similar ring shape that can encircle DNA and hold DNA polymerase on the template.

The β-subunit needs help from the γ complex (γ, δ, δ′, χ, and ψ) to load onto the complex. The γ complex acts catalytically in forming this processive αε β complex, so it does not remain associated with the complex during processive replication. Clamp loading is an ATP-dependent process.

The pol III holoenzyme is double-headed, with two core polymerases attached through two τ-subunits to a γ complex. One core is responsible for continuous synthesis of the leading strand, the other performs discontinuous synthesis of the lagging strand. The γ complex serves as a clamp loader to load the β clamp onto a primed DNA template. Once loaded, the β clamp loses affinity for the γ complex and associates with the core polymerase to help with processive synthesis of an Okazaki fragment. Once the fragment is completed, the β clamp loses affinity for the core polymerase and associates with the γ complex, which acts as a clamp unloader, removing the clamp from the DNA. Then it can recycle to the next primer and repeat the process.

At the end of replication, circular bacterial chromosomes form catenanes that are decatenated in a two-step process. First, denaturation of the remaining, unreplicated double-helical turns linking the two strands are denatured. Then repair synthesis fills in the gaps. This leaves a right-handed parallel torus catenane with an even number of nodes that is decatenated by topoisomerase IV. Linear eukaryotic chromosomes also require decatenation during DNA replication.

Eukaryotic chromosomes have special structures known as telomeres at their ends. One strand of these telomeres is composed of many tandem repeats of short, G-rich regions whose sequence varies from one species to another. The G-rich telomere strand is made by an enzyme called telomerase, which contains a short RNA that serves as the template for telomere synthesis. The C-rich telomere strand is synthesized by ordinary RNA-primed DNA synthesis, like the lagging strand in conventional DNA replication. This mechanism ensures that chromosome ends can be rebuilt and therefore do not suffer shortening with each round of replication.

TRF1 and TRF2 appear to help telomeric DNA in mammalian cells form a t loop, in which the single-stranded 3′-end of the telomere invades the double-stranded telomeric DNA upstream. TRF1 may help bend the DNA into shape for strand invasion, and TRF2 binds at the point of strand invasion and may stabilize the displacement loop. The t loop structure protects the end of the chromosome from inappropriate repair in which it would be attached to another chromosome end.

REVIEW QUESTIONS

1. Describe and give the results of an experiment that shows the rate of elongation of a DNA strand in vitro.

2. Describe a procedure to check the processivity of DNA synthesis in vitro.

3. Describe an assay to locate and determine the minimal length of an origin of replication.

4. List the components of the E. coli primosome and their roles in primer synthesis.

5. Outline a strategy for locating the SV40 origin of replication.

6. Outline a strategy for identifying an autonomously replicating sequence (ARS1) in yeast.

7. Outline a strategy to show that DNA replication begins in ARS1 in yeast.

8. Which subunit of the pol III holoenzyme provides processivity? What proteins load this subunit (the clamp) onto the DNA? To which core subunit does this clamp bind?

9. Describe and give the results of an experiment that shows the different behavior of the β clamp on circular and linear DNA. What does this behavior suggest about the mode of interaction between the clamp and the DNA?

10. What mode of interaction between the β clamp and DNA do x-ray crystallography studies suggest?

11. What mode of interaction between PCNA and DNA do x-ray crystallography studies suggest?

12. Describe and give the results of an experiment that shows that the clamp loader acts catalytically. What is the composition of the clamp loader?

13. How can discontinuous synthesis of the lagging strand keep up with continuous synthesis of the leading strand?

14. Describe and give the results of an experiment that shows that pol III* can dissociate from its β clamp.

15. Describe a protein footprinting procedure. Show how such a procedure can be used to demonstrate that the pol III core and the clamp loader both interact with the same site on the β clamp.

16. Describe and give the results of an experiment that shows that the γ complex has clamp-unloading activity.

17. Describe how the β clamp cycles between binding to the core pol III and to the clamp unloader during discontinuous DNA replication.

18. Why is decatenation required after replication of circular DNAs?

19. Describe and give the results of an experiment that shows that topoisomerase IV is required for decatenation of plasmids in *Salmonella typhimurium* and *E. coli*.

20. Why do eukaryotes need telomeres, but prokaryotes do not?

21. Diagram the process of telomere synthesis.

22. Why was *Tetrahymena* a good choice of organism in which to study telomerase?

23. Describe an assay for telomerase activity and show sample results.

24. Describe and give the results of an experiment that shows that the telomerase RNA serves as the template for telomere synthesis.

25. Diagram the t loop model of telomere structure.

26. What evidence supports the existence of t loops?

27. What evidence supports the strand-invasion hypothesis of t loop formation?

SUGGESTED READINGS

General References and Reviews

Baker, T.A. 1995. Replication arrest. *Cell* 80:521–24.

Blackburn, E.H. 1990. Telomeres: Structure and synthesis. *Journal of Biological Chemistry* 265:5919–21.

Blackburn, E.H. 1994. Telomeres: No end in sight. *Cell* 77:621–23.

Greider, C.W. 1999. Telomeres do D-loop-T-loop. Cell 97:419–22.

Herendeen, D.R. and T.J. Kelly. (1996). DNA polymerase III: Running rings around the fork. *Cell* 84:5–8.

Kornberg, A. and T.A. Baker. 1992. *DNA Replication*, 2nd ed. New York: W.H. Freeman.

Marx, J. 1994. DNA repair comes into its own. *Science* 266:728–30.

Marx, J. 1995. How DNA replication originates. *Science* 270:1585–86.

Newlon, C.S. 1993. Two jobs for the origin replication complex. *Science* 262:1830–31.

Stillman, B. 1994. Smart machines at the DNA replication fork. *Cell* 78:725–28.

Wang, J.C. 1991. DNA topoisomerases: Why so many? *Journal of Biological Chemistry* 266:6659–62.

West, S.C. (1996). DNA helicases: New breeds of translocating motors and molecular pumps. *Cell* 86:177–80.

Zakian, V.A. 1995. Telomeres: Beginning to understand the end. *Science* 270:1601–6.

Research Articles

Adams, D.E., E.M. Shekhtman, E.L. Zechiedrich, M.B. Schmid, and N.R. Cozzarelli. 1992. The role of topoisomerase IV in partitioning bacterial replicons and the structure of catenated intermediates in DNA replication. *Cell* 71:277–88.

Arai, K. and A. Kornberg. 1979. A general priming system employing only *dnaB* protein and primase for DNA replication. *Proceedings of the National Academy of Sciences USA* 76:4309–13.

Arai, K., R. Low, J. Kobori, J. Shlomai, and A. Kornberg. 1981. Mechanism of *dnaB* protein action V. Association of *dnaB* protein, protein n', and other prepriming proteins in the primosome of DNA replication. *Journal of Biological Chemistry* 256:5273–80.

Blackburn, E.H., 1990. Functional evidence for an RNA template in telomerase. *Science* 247:546–52.

Bouché, J.-P., L. Rowen, and A. Kornberg. (1978). The RNA primer synthesized by primase to initiate phage G4 DNA replication. *Journal of Biological Chemistry* 253:765–69.

Brewer, B.J. and W.L. Fangman. 1987. The localization of replication origins on ARS plasmids in *S. cerevisiae*. *Cell* 51:463–71.

Burgers, P.M.J. and A. Kornberg. (1982). ATP activation of DNA polymerase III holoenzyme of *Escherichia coli*. *Journal of Biological Chemistry* 257:11468–73.

Deb, S., A.L. DeLucia, A. Koff, S. Tsui, and P. Tegtmeyer. (1986). The adenine-thymine domain of the simian virus 40 core antigen directs DNA bending and coordinately regulates DNA replication. *Molecular and Cellular Biology* 6:4578–84.

Fareed, G.C., C.F. Garon, and N.P. Solzman. (1972). Origin and direction of SV40 deoxyribonucleic acid replication. *Journal of Virology* 10:484–91.

Greider, C.W., and E.H. Blackburn. (1985). Identification of a specific telomere terminal transferase activity in *Tetrahymena* extracts. *Cell* 43:405–13.

Greider, C.W. and E.H. Blackburn. (1989). A telomeric sequence in the RNA of *Tetrahymena* telomerase required for telomere repeat synthesis. *Nature* 337:331–37.

Griffith, J.D., L. Comeau, S. Rosenfield, R.M Stansel, A. Bianchi, H. Moss, and T. de Lange. 1999. Mammalian telomeres end in a large duplex loop.

Kong, X.-P., R. Onrust, M. O'Donnell, and J. Kuriyan. 1992. Three-dimensional structure of the β subunit of *E. coli* DNA polymerase III holoenzyme: A sliding DNA clamp. *Cell* 69:425–37.

Krishna, T.S.R., X.-P. Kong, S. Gary, P.M. Burgers, and J. Kuriyan. 1994. Crystal structure of the eukaryotic DNA polymerase processivity factor PCNA. *Cell* 79:1233–43.

Marahrens, Y. and B. Stillman. 1992. A yeast chromosomal origin of DNA replication defined by multiple functional elements. *Science* 255:817–23.

Mok, M. and K.J. Marians. (1987). The *Escherichia coli* preprimosome and DNA B helicase can form replication forks that move at the same rate. *Journal of Biological Chemistry* 262:16644–54.

Naktinis, V., J. Turner, and M. O'Donnell. (1996). A molecular switch in a replication machine defined by an internal competition for protein rings. *Cell* 84:137–45.

Stukenberg, P.T., P.S. Studwell-Vaughan, and M.O'Donnell. 1991. Mechanism of the sliding β-clamp of DNA polymerase III holoenzyme. *Journal of Biological Chemistry* 266:11328–34.

Stukenberg, P.T., J. Turner and M. O'Donnell. 1994. An explanation for lagging strand replication: Polymerase hopping among DNA sliding clamps. *Cell* 78:877–87.

Zyskind, J.W., J.M. Cleary, W.S.A. Brusilow, W.E. Harding, and D.W. Smith. (1983). Chromosomal replication origin from the marine bacterium *Vibrio harveyi* functions in *Escherichia coli*: *oriC* consensus sequence. *Proceedings of the National Academy of Sciences USA* 80:1164–68.

Homologous Recombination

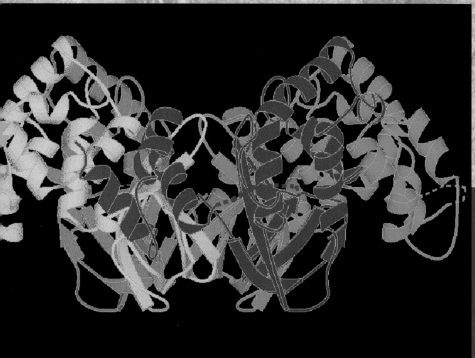

Crystal structure of the RuvA tetramer, with each monomer represented by a different color. RuvA binds to Holliday junctions and facilitates branch migration during recombination in *E. coli*.
Rafferty et al., Crystal structure of DNA recombination protein RuvA and a model for its binding to the Holliday junction. Science *274 (18 Oct 1996) f. 2e, p. 417.* © AAAS.

Geneticists have known for a long time that sexual reproduction gives offspring a different genetic makeup from their parents. Some of this variation comes from independent assortment of parental chromosomes. Most of the rest results from homologous recombination, which occurs between homologous chromosomes during meiosis. This process scrambles the genes of maternal and paternal chromosomes, so nonparental combinations occur in the offspring. This scrambling is valuable because the new combinations sometimes allow the progeny organisms a better chance of survival than their parents. Furthermore, meiotic recombination forms physical links between homologous chromosomes that allow the chromosomes to align properly during meiotic prophase so they separate properly during meiotic metaphase. Also, as we saw in Chapter 20, homologous recombination plays an important role in allowing cells to deal with DNA damage by so-called recombination repair.

Figure 22.1 illustrates several variations on the theme of homologous recombination. Each variation is characterized by a crossover event that joins DNA segments that were previously separated. This does not mean the two segments must start out on separate DNA molecules. Recombination can be intramolecular, in which case crossover between two sites on the same chromosome either removes or inverts the DNA segment in between. On the other hand, bimolecular recombination involves crossover between two independent DNA molecules. Ordinarily, recombination is reciprocal—a two-way street in which the two participants trade DNA segments. DNA molecules can undergo one crossover event, or two, or more, and the number of events strongly influences the nature of the final products.

Our current models for the mechanism of homologous recombination share the following essential steps, although not necessarily in this order: (1) breaking two of the homologous DNA strands; (2) pairing of two homologous DNAs (DNAs having identical or very similar sequences); (3) re-forming phosphodiester bonds to join the two homologous strands (or regions of the same strand in intramolecular recombination); (4) breaking the other two strands and joining them. ■

22.1 Models for Homologous Recombination

Several models have been proposed to explain how these events occur. Let us consider the most popular of these hypotheses.

The Holliday Model

In 1964, Robin Holliday proposed the model presented in Figure 22.2. This scheme calls for pairing of two homologous DNA duplexes and creation of nicks in corresponding positions in two homologous DNA strands. Once both homologous strands are broken, the free ends can exchange and join with ends in the other duplex, forming intermolecular bonds, rather than rejoining to form the old intramolecular bonds.

When the two strands have exchanged and DNA ligase has sealed the new intermolecular phosphodiester bonds, we have a **Holliday junction**—also known as a half-chiasma or chi structure—a cross-shaped structure pictured in several ways in Figure 22.2. The branch in the Holliday junction can migrate in either direction simply

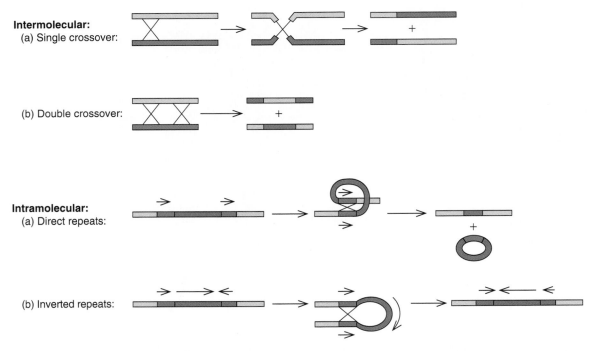

Figure 22.1 Examples of recombination. The X's represent crossover events between the two chromosomes or parts of the same chromosome. To visualize how these work, look at the intermediate form of the reciprocal recombination on the top line. Imagine the DNAs breaking and forming new, interstrand bonds as indicated by the arms of the X. This same principle applies to all the examples shown.

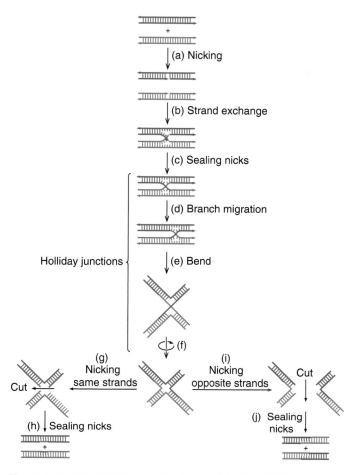

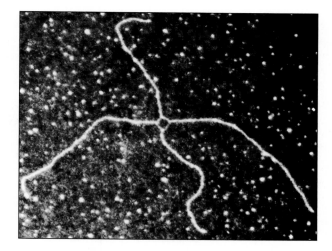

Figure 22.3 Holliday junction generated during recombination. This corresponds to the intermediate generated in Figure 22.2f, after some base pairs have opened up around the central "diamond" to produce short stretches of single-stranded DNA. (*Source:* Courtesy Dr. Huntington Potter and Dr. David Dressler.)

Figure 22.2 The Holliday model of recombination. (a) Nicks occur in the same places in homologous chromosomes (blue and red). **(b)** Strands of the two chromosomes exchange. **(c)** Nicks are sealed, permanently joining the two chromosomes through two of their four strands and yielding a Holliday junction. **(d)** Branch migration occurs by breaking some base pairs and re-forming others. **(e)** and **(f)** These are simply bends and twists introduced into this illustration of the Holliday junction to make the subsequent events easier to understand. **(g)** One kind of resolution of the Holliday junction. The same strands are nicked as were nicked in panel (a). **(h)** Sealing these nicks leaves two DNA duplexes with short stretches of "heteroduplex," containing one strand from each of the recombining partners. This is called non-crossover recombination. **(i)** The alternative resolution of the Holliday junction. The strands opposite those nicked originally are nicked. **(j)** When the nicks are sealed, two crossover recombinant DNA duplexes emerge.

by breaking old base pairs and forming new ones in a process called **branch migration.** The Holliday junction is first shown with two of its strands exchanged, but we can also imagine the Holliday junction after a 180-degree rotation of either the top or bottom half, which uncrosses these strands to give an intermediate with a diamond

shape at the center where all the strands meet. If a few base pairs come apart at the junction point, the diamond, which is composed of single-stranded DNA, will expand. Figure 22.3 is an electron micrograph of recombining plasmid DNAs, which demonstrates beautifully that such structures really do form.

Of course, the two duplexes participating in the Holliday junction cannot remain linked indefinitely. The other two DNA strands must also break and rejoin to resolve the Holliday junction into two independent DNA duplexes. This can happen in two different ways. If the same strands that broke the first time break again a **non-crossover recombination** results, yielding two DNA duplexes with patches of **heteroduplex.** These patches contain one strand from each of the DNAs involved in the Holliday junction. On the other hand, if the strands that did *not* break the first time are involved in the second break, **crossover recombination** occurs. The products have exchanged chromosome arms flanking the Holliday junction.

The Meselson–Radding Model

One problem with Holliday's model is that it assumes two breaks in exactly corresponding positions in homologous DNA strands. Single-strand breaks occur frequently, so imagining one break appearing eventually in any given position in a DNA molecule is easy. Nevertheless, it seems unlikely that two breaks, conveniently located right across

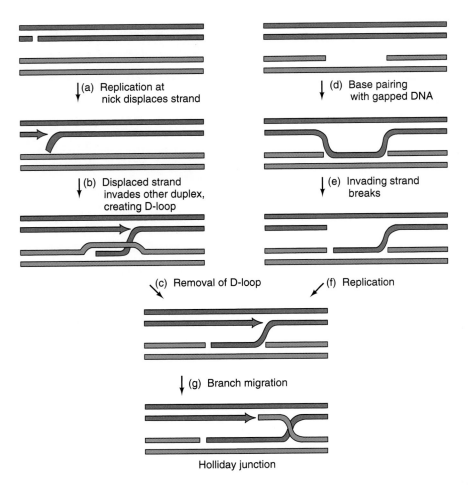

Figure 22.4 Models to explain the creation of a Holliday junction. The original Meselson–Radding model called for replication to displace a single strand of DNA, which invaded a homologous duplex DNA. The resulting structure in the center after steps (c) and (f) could then be converted to a Holliday junction by branch migration to the right in (**g**). The two remaining nicks in the Holliday junction can easily be sealed by DNA ligase. Two alternative routes exist to reach the structure in the middle. In the first route (left), steps (**a**) and (**b**) show the displacement of the end of a single-stranded DNA by replication from a nick, and invasion of a duplex DNA by this free end, forming a D-loop. Step (**c**) shows the removal of the D-loop. In the second route (right), steps (**d**) and (**e**) show the invasion of a gapped duplex by one of the strands of another DNA duplex, followed by breakage of the invading strand. In step (**f**), replication from one of the ends formed during the break yields the same structure as does step (**c**).

from each other in two paired molecules, would occur spontaneously. Other investigators have therefore proposed alternative models to explain the origin of the second break. One of these is the Meselson–Radding model (Figure 22.4), first proposed by Matthew Meselson and Charles Radding in 1975. This hypothesis suggests that (a) the first free end is created by DNA replication in one of the duplexes, with displacement of one of the strands. (Alternatively, it could be created by nicking and breathing of the duplex.) (b) This free end invades the other duplex, base-pairing with one of the strands and displacing the other. The displaced strand forms a loop called a **D loop,** where the D stands for "displacement." According to this scheme, (c) the act of forming the D loop allows degradation of the looping strand. Later, Radding modified the original model to allow for (d) pairing between a gapped duplex and one strand of another duplex, with (e) breakage of the invading strand. (f) Replication from the resulting free end created by the break would yield the same intermediate seen after step (c). Now branch migra-

tion, as in step (g), can give rise to a Holliday junction. The nicks in this Holliday junction can easily be sealed with DNA ligase.

The RecBCD Pathway

A recombination pathway (the **RecBCD pathway,** Figure 22.5) actually used by *E. coli* is a variation on the Meselson–Radding model. This recombination process begins when the **RecBCD protein,** the product of the *recB, –C,* and *–D* genes, creates a nick near the 3′-end of a so-called **chi site, χ,** which has the sequence 5′-GCTGGTGG-3′. Chi sites are found on average every 5000 bp in the *E. coli* genome. RecBCD also has helicase activity, so it can unwind the DNA adjacent to the nick near a χ site, producing a single-stranded tail, which can then be coated by **RecA protein** (the product of the *recA* gene) and SSB. The SSB helps keep the tail from re-forming a DNA duplex, and RecA allows the tail to invade a double-stranded DNA duplex and search for a region of homology. Once the tail

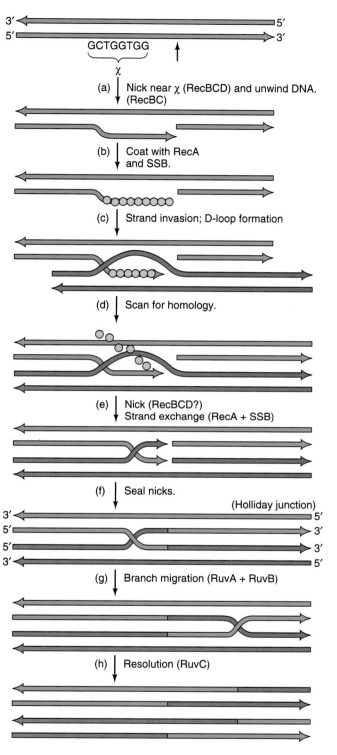

finds a homologous region, a nick occurs in the D-looped DNA, possibly with the aid of RecBCD. This nick allows RecA and SSB to create a new tail that can pair with the gap in the other DNA. DNA ligase seals both nicks to generate a Holliday junction. (The RecBCD protein was originally thought to be encoded by just the *recB* and *recC* genes, so it was called RecBC. We now know that it also contains the *recD* product, RecD.)

Branch migration does not occur at a useful rate spontaneously. Just as in DNA replication, DNA unwinding is required, and this in turn requires helicase activity and energy from ATP. Two proteins, **RuvA** and **RuvB**, collaborate in this function. Both have DNA helicase activity, and RuvB is an ATPase, so it can harvest energy from ATP for the branch migration process. Finally, two DNA strands must be nicked to resolve the Holliday junction into hetero-duplexes or recombinant products. The **RuvC** protein carries out this function.

SUMMARY Homologous recombination in *E. coli* may begin with invasion of a duplex DNA by a single-stranded DNA from another duplex. A free end can be generated by nicking and DNA helicase action or by DNA replication. The invading strand either pairs with an available gap in a recipient DNA, or forms a D loop. Subsequent degradation of the D-loop strand, if a D loop has formed, leads to the formation of a branched intermediate. Branch migration in this intermediate yields a Holliday junction with two strands exchanging between homologous chromosomes. Finally, the Holliday junction can be resolved by nicking two of its strands. This can yield two noncrossover recombinant DNAs with patches of heteroduplex, or two crossover recombinant DNAs that have traded flanking DNA regions. In the RecBCD pathway, the RecBCD DNase activity could create the first nick, and possibly the second. The invading strand is coated with RecA protein and SSB, and the RecA protein facilitates base pairing between the invading strand and a complementary strand in the recipient duplex. RuvA and RuvB have helicase activity that makes branch migration possible. RuvC nicks the Holliday junction to resolve it into recombinant DNAs.

Figure 22.5   The RecBCD pathway of homologous recombination. (a) The RecBCD protein nicks one strand to the 3′-side of the χ sequence. The DNA helicase activity of RecBC then unwinds the DNA through the χ-site forming a single-stranded tail. **(b)** RecA and SSB (both represented as yellow spheres) coat the tail, preventing its reassociation with the complementary strand. **(c)** RecA promotes invasion of another DNA duplex, forming a D loop. **(d)** RecA helps the invading strand scan for a region of homology in the recipient DNA duplex. Here, the invading strand has base-paired with a homologous region, releasing SSB and RecA. **(e)** Once a homologous region is found, a nick in the looped-out DNA appears, perhaps caused by RecBCD. This permits the tail of the newly nicked DNA to base-pair with the single-stranded region in the other DNA, probably aided by RecA and SSB. **(f)** The remaining nicks are sealed by DNA ligase, yielding a four-stranded complex with a Holliday junction. **(g)** Branch migration occurs, sponsored by RuvA and RuvB. Notice that the branch has migrated to the right. **(h)** Nicking by RuvC resolves the structure into two molecules, crossover recombinants in this example.

22.2 Experimental Support for the RecBCD Pathway

Now that we have seen a brief overview of the RecBCD pathway, let us look at the experimental evidence that supports this important prokaryotic recombination mechanism.

RecA

We have encountered RecA before, in our discussion of induction of the λ phage (Chapter 8). Indeed, this is a protein of many functions, but it was first discovered in the context of recombination, and that is how it got its name. In 1965, Alvin Clark and Ann Dee Margulies isolated two *E. coli* mutants that could accept F plasmids, but could not integrate their DNA permanently by recombination. These mutants were also highly sensitive to ultraviolet light, presumably because they were defective in recombination repair of UV damage. Characterizing these mutants led ultimately to the discovery of two key proteins in the RecBCD pathway: RecA and RecBCD.

The *recA* gene has been cloned and overexpressed, so abundant RecA protein is available for study. It is a 38-kD protein that can promote a variety of strand exchange reactions in vitro. One of these, pictured in Figure 22.6, is the pairing of a single-stranded circular phage DNA (+) strand with the (−) strand in a linear phage duplex DNA to form a double-stranded circle with a nick in the (−) strand. This leaves a linear single-stranded (+) strand as a by-product. Using this convenient assay, Radding discerned three stages of participation of RecA in strand exchange:

1. **Presynapsis,** in which RecA, (and SSB) coat the single-stranded DNA.

2. **Synapsis,** or alignment of complementary sequences in the single-stranded and double-stranded DNAs that will participate in strand exchange.

3. **Postsynapsis,** or **strand exchange,** in which the single-stranded DNA replaces the (+) strand in the double-stranded DNA to form a new double helix. An intermediate in this process is a **joint molecule** in which strand exchange has begun and the two DNAs are intertwined with each other.

Presynapsis The best evidence for the association between RecA and single-stranded DNA is visual. Radding and colleagues constructed a linear, double-stranded phage DNA with single-stranded tails, incubated this DNA with RecA, spread the complex on an electron microscope grid and photographed it. Figure 22.7a shows that RecA bound preferentially to the single-stranded ends, leaving the double-stranded DNA in the middle un-

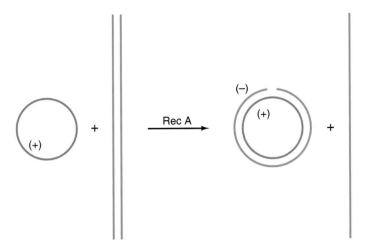

Figure 22.6   Assay for RecA involvement in strand exchange. We begin with a single-stranded DNA viral (+) strand (blue) and a double-stranded linear form of the same viral DNA (red). RecA promotes pairing between the circular single-stranded DNA and its complementary strand in the double-stranded linear form. This yields a nicked double-stranded circular DNA and a linear single-stranded DNA.

coated. These workers also incubated single-stranded circular M13 phage DNA with RecA and subjected these complexes to the same procedure. Figure 22.7b shows extended DNA circles coated uniformly with RecA. The magnification in panels (a) and (b) is the same, so the thickness of the circular fiber in panel (b), compared with the naked DNA and RecA–DNA complex in panel (a), demonstrates clearly that these circular DNAs are indeed coated with RecA.

We have also indicated that SSB is important in forming the coated DNA fiber in the presynapsis process. Figure 22.8 shows the DNA–protein complexes formed by mixing single-stranded M13 phage DNA with SSB alone (panel [a]), or SSB plus RecA (panel [b]). The appearances of the two complexes are clearly different, and the DNA–protein complexes with both SSB and RecA strongly resemble those with RecA alone in Figure 22.7. Furthermore, SSB accelerates the formation of the coated DNA. Figure 22.9 depicts the kinetics of formation of extended circular filaments, such as those in Figure 22.8b. In the presence of SSB plus RecA, formation of these filaments was complete after only 10 min (panel [b]). By contrast, the process had barely begun after 10 min when SSB was absent (panel [a]).

Because RecA can coat a single-stranded DNA by itself, what is the role of SSB? It seems to be required to melt secondary structure (hairpins) in the single-stranded DNA. Evidence for this notion comes from several sources. Radding and colleagues assayed for strand exchange when RecA was incubated with single-stranded DNA at low and high concentrations of $MgCl_2$. Low $MgCl_2$ concentrations destabilize DNA secondary

(a)

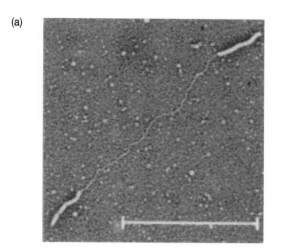

(b)

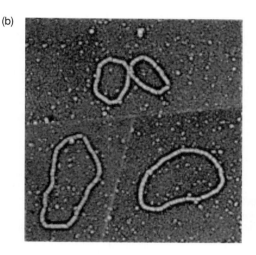

Figure 22.7   Binding of RecA to single-stranded DNA. Radding and colleagues prepared (**a**) a linear double-stranded DNA with single-stranded ends and (**b**) a circular single-stranded phage DNA. Then they added RecA, allowed time for a complex to form, spread the complexes on coated electron microscope grids, and photographed them. The bar in panel (**a**) represents 500 nm for both panels. (*Source:* Radding et al., *Cold Spring Harbor Symposia of Quantitative Biology* 47 (1982) f. 3 f&j, p. 823.)

(a)

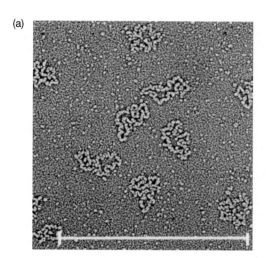

(b)

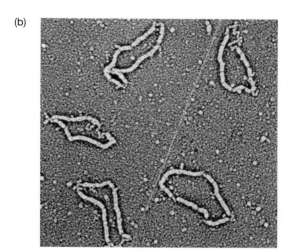

Figure 22.8   Binding of SSB and SSB plus RecA to single-stranded DNA. Radding and colleagues bound single-stranded M13 DNA to (**a**) SSB alone, or (**b**) to SSB plus RecA, then spread and photographed the complexes. The bar represents 500 nm for both panels. (*Source:* Radding et al., *CSH Symp. of Quant. Biol.* 47 (82) f. 4, p. 824.)

structure, but high $MgCl_2$ concentrations stabilize it. In this experiment, SSB was required under high, but not low, $MgCl_2$ concentration conditions, which is the result we expect if SSB is needed to relax DNA secondary structure.

Later in this section, we will see that ATP hydrolysis is normally required for strand exchange, but I.R. Lehman and coworkers showed that ATPγS, an unhydrolyzable analog of ATP, will support a limited amount of strand exchange if SSB is present. Because ATPγS causes RecA to bind essentially irreversibly to both single-and double-stranded DNA, Radding and colleagues posed the hypothesis that ATPγS causes RecA to trap DNA in secondary structure that is unfavorable for strand exchange. If this is true, then SSB should be able to override this difficulty by removing secondary structure if it is added to DNA before RecA. As expected, SSB did indeed accelerate strand exchange if it was added before RecA. This provided more evidence that the role of SSB is to unwind secondary structure in a single-stranded DNA participating in recombination.

SUMMARY In the presynapsis step of recombination, RecA coats a single-stranded DNA that is participating in recombination. SSB accelerates the recombination process, apparently by melting secondary structure and preventing RecA from trapping any secondary structure that would inhibit strand exchange later in the recombination process.

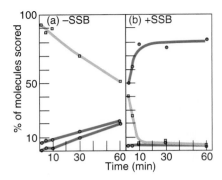

Figure 22.9 Effect of SSB on formation of extended circular DNA–protein filaments. Radding and coworkers followed the formation of extended circular DNA–protein filaments, such as those pictured in Figure 22.8b (red); collapsed filaments, with little protein attached (blue); and aggregates or linear filaments (green). They performed their experiments (**a**) with single-stranded M13 phage DNA and RecA alone, or (**b**) with DNA, RecA, and SSB. SSB caused a great acceleration in the rate of forming extended DNA–protein filaments. (*Source:* From C. M. Radding et al., "Three Phases in Homologous Pairing," *Cold Spring Harbor Symp. Quant. Biology* 47:824, 1982. Copyright © 1982 Cold Spring Harbor Laboratory Press, Cold Spring Harbor, NY. Reprinted by permission.)

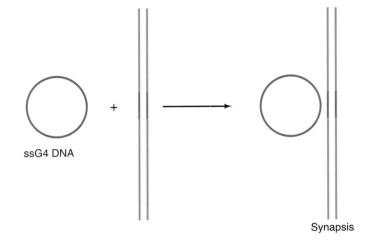

Figure 22.10 Synapsis. Synapsis is shown between circular single-stranded G4 phage DNA (red) and linear double-stranded M13(G4) DNA (M13 phage DNA [blue] with a 274-bp insert of G4 phage DNA [red]). The synapsis does not involve any intertwining of the linear and circular DNAs.

Synapsis: Alignment of Complementary Sequences We will see later in this section that RecA stimulates strand exchange, which involves invasion of a duplex DNA by a single strand from another DNA. In this process, the invading strand forms a new double helix with one of the strands of the other duplex. But the step that precedes strand exchange entails a simple alignment of complementary sequences, without the formation of an intertwined double helix. This process yields a less stable product and is therefore more difficult to detect than strand exchange. Nevertheless, Radding and colleagues presented good evidence for synapsis as early as 1980.

As in the presynapsis experiments, electron microscopy was a key technique in the first demonstration of synapsis. Radding and coworkers used a favorite pair of substrates for their synapsis experiments: a single-stranded circular phage DNA and a double-stranded linear DNA. However, in this case, the single-stranded circular DNA was G4 phage DNA, and the double-stranded linear DNA was M13 DNA with a 274-bp G4 DNA insert near the middle. Because this target for the single-stranded G4 DNA lay thousands of base pairs from either end, and nicks were very rare in this DNA, it was unlikely that true strand exchange could happen. Instead, simple synapsis of complementary sequences could occur, as illustrated in Figure 22.10.

To measure synapsis between the two DNAs, Radding and colleagues mixed the two DNAs in the presence and absence of RecA and subjected the mixture to electron microscopy. They found a significant proportion of the DNA molecules undergoing synapsis, as depicted in Figure 22.11. In most of the aligned molecules, the length of the aligned regions was appropriate, and the position of the aligned region within the linear DNA was correct. Furthermore, synapsis was reduced by 20–40-fold with two DNAs that did not share a homologous region. Synapsis also failed in the absence of RecA.

Although the percentage of nicked double-stranded DNAs was very low in this experiment, it was conceivable that nicks could create free ends in the linear DNA that could allow formation of a true, intertwined **plectonemic double helix**. If this had happened, the linkage between the two DNAs would have been stable to temperatures approaching the melting point of DNA. However, the alignments shown in Figure 22.11 were destroyed by heating at 20° C below the melting point for 5 min. Thus, the synapsis observed here does not involve the formation of a base-paired Watson–Crick double helix. Instead, it probably involves a **paranemic helix** in which the two aligned DNA strands are side by side, but not intertwined. Further support for the notion that nicks are not required for synapsis comes from the finding that supercoiled DNA (unnicked by definition) works just as well as linear double-stranded DNA in these experiments.

How much homology is necessary for synapsis to occur? David Gonda and Radding provided an estimate by showing that a 151-bp homologous region gave just as efficient synapsis as did 274 bp of homology. But DNAs with just 30 bp of homology gave only background levels of aligned DNA molecules. Thus, the minimum degree of homology for efficient synapsis is somewhere between 30 and 151 bp.

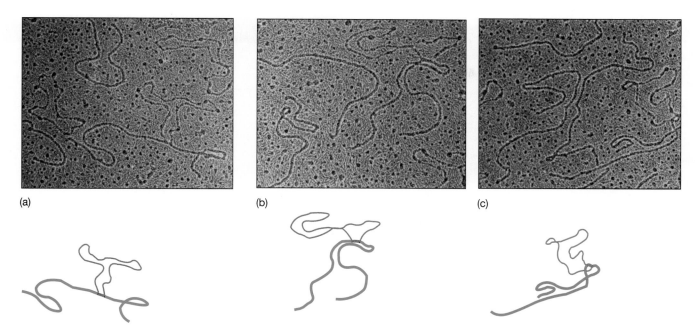

Figure 22.11   Demonstration of RecA-dependent synapsis in vitro. Radding and colleagues mixed the circular single-stranded and linear double-stranded DNAs described in Figure 22.10 with RecA and then examined the products by electron microscopy. Panels (**a–c**) show three different examples of aligned DNA molecules. Below each electron micrograph is an interpretive diagram, showing the linear double-stranded DNA in blue and the circular single-stranded DNA in red. Thick red lines denote the zones of synapsis between the two DNAs in each case. (*Source:* DasGupta et al., The topology of homologous pairing promoted by *rec*A protein. *Cell* 22 (Nov 1980 Pt2) f. 9 d-f, p. 443. Reprinted by permission of Elsevier Science.)

SUMMARY Synapsis occurs when a single-stranded DNA finds a homologous region in a double-stranded DNA and aligns with it. No intertwining of the two DNAs occurs at this point.

Postsynapsis: Strand Exchange   We have learned that RecA is required for the first two steps in strand exchange: presynapsis and synapsis. Now we will see that it is also required for the last: postsynapsis, or strand exchange itself. Lehman and colleagues measured strand exchange between double-stranded and single-stranded phage DNA by a filter-binding assay for D-loop formation as follows: They incubated ^{3}H-labeled duplex P22 phage DNA with unlabeled single-stranded P22 DNA in the presence or absence of RecA. They used high-salt and low-temperature conditions that restricted branch migration, which would have completely assimilated the single-stranded DNA and eliminated the D loops. Then they removed protein from the DNA with a detergent (either Sarkosyl or sodium dodecyl sulfate). Finally, they filtered the mixture through a nitrocellulose filter. If D loops formed in the duplex DNA, the single-stranded D loop would cause the complex to bind to the filter, so labeled DNA would be retained. If no D loop formed, the unlabeled single-stranded DNA would stick to the filter, but

the labeled duplex DNA would flow through. The detergent prevented DNA from sticking to the filter simply because of its association with RecA. Lehman and colleagues also performed this assay with supercoiled duplex M13 phage DNA and linear M13 DNA. In both cases, about 50% of the DNA duplexes formed D loops, but only in the presence of RecA. Without RecA, retention of D-looped DNA was less than 1%. Also, with a nonhomologous single-stranded DNA, retention of D looped DNA was only 2%.

To verify that D loops had actually formed, these workers treated the complexes with S1 nuclease to remove single-stranded DNA, then filtered the product. This greatly reduced the retention of the labeled DNA on the filter, suggesting that a D loop was really involved. To be sure, they directly visualized the D loops by electron microscopy. Figure 22.12 shows the results: D loops are clearly visible in both kinds of DNA—linear and supercoiled. This experiment thus had the added benefit of demonstrating that supercoiled DNA is not required for strand exchange.

Table 22.1 shows the effects of nucleotides on D loop formation, as measured by retention by nitrocellulose filtration. We see that ATP is required for D loop formation, and its role cannot be performed by GTP, UTP, or ATPγS. The fact that ATPγS cannot substitute for ATP in D loop formation indicates that ATP hydrolysis is required. In fact, it appears that ATP hydrolysis permits

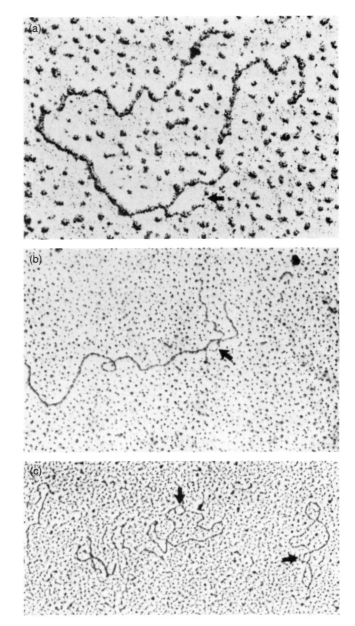

Figure 22.12   RecA-dependent D loop formation between duplex and single-stranded phage DNAs. Lehman and colleagues mixed duplex and single-stranded phage DNAs with RecA, salts, and ATP under conditions that limited branch migration. Then they removed protein from the DNA with a detergent and visualized the resulting D loops by electron microscopy. Panels (**a**) and (**b**) show D loops in linear P22 phage DNA at two different magnifications. Panel (**c**) shows D loops in supercoiled M13 phage DNA. D loops are denoted by arrows. (*Source:* McEntee et al., Initiation of general recombination catalyzed *in vitro* by the *recA* protein of *Escherichia coli. PNAS* 76 (1979) f. 1, p. 2616.)

RecA to dissociate from DNA, which permits the new base pairing that must occur in strand exchange.

Thus, this experiment demonstrated that ATP hydrolysis is essential for D loop formation. RecA, an incredibly versatile protein, has ATPase activity, which cleaves ATP as RecA falls off the DNAs to allow the D loops to form.

Table 22.1 Requirements for D Loop Formation		
Duplex DNA	**Reaction components**	**D loops formed (%)**
P22	Complete	100
	−RecA	<1
	−ATP	<1
	−ATP + GTP	<1
	−ATP + UTP	<1
	−ATP + ATPγS	<1
M13	Complete	100
	−RecA	1
	−ATP	1
	−ATP + GTP	2

Radding and colleagues also used electron microscopy to produce striking visual evidence of strand exchange between two DNAs. They used a single-stranded circular G4 phage DNA coated with RecA, and a linear double-stranded restriction fragment of M13*Gori*1 DNA containing 1200 bp of G4 sequence, which should base-pair with the circular DNA, and 800 bp of M13 sequence, which should not base-pair. Figure 22.13a is an electron micrograph of the joint molecule formed by strand exchange, and Figure 22.13b is an interpretive diagram. Where a new double-stranded region formed between the linear and circular DNAs, RecA was removed and the DNA appears naked. Surprisingly, the displaced single strand from the linear DNA did not complex well with RecA, but it can be seen with some RecA attached.

SUMMARY RecA and ATP collaborate to promote strand exchange between a single-stranded and double-stranded DNA. ATP is necessary to clear RecA off the synapsing DNAs to make way for formation of double-stranded DNA involving the single strand and one of the strands of the DNA duplex.

RecBCD

In our discussion of RecA, we have been considering model reactions involving a single-stranded and a duplex DNA. The reason, of course, is that RecA requires a single-stranded DNA to initiate strand exchange. But naturally recombining DNAs are usually both double-stranded, so how does RecA get the single strand it needs? The answer is probably that RecBCD provides it.

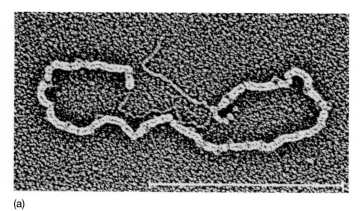

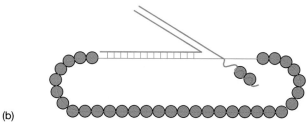

(a)

(b)

Figure 22.13   Demonstration of strand exchange between single-stranded circular and double-stranded linear DNAs. Radding and colleagues mixed a single-stranded circular G4 phage DNA, coated with RecA, with a 2.0-kb linear double-stranded DNA (M13*Gori*1) having 800 bp of G4 sequence. Then they subjected the complex to electron microscopy. **(a)** Experimental results. Bar represents 500 nm. **(b)** Interpretive diagram, showing strand exchange between the two DNAs. (*Source:* Radding et al., Three phases in homologous pairing: Polymerization of *recA* protein on single-stranded DNA, synapsis, and polar strand exchange. *CSH Symp. of Quant. Biol.* 47 (82) f. 6, p. 824.)

RecBCD, like RecA, is a surprisingly versatile enzyme. It contains:

1. A DNA-dependent ATPase activity that is stimulated by both single-stranded and double-stranded DNAs.

2. A DNA nuclease that acts on both single-stranded and double-stranded DNAs and cleaves specifically at χ sites.

3. A double-stranded DNA helicase activity.

Returning to Figure 22.5, we can see two steps at which these activities would be needed. First, in step (a), which creates the free single-stranded DNA, the nuclease could nick the DNA downstream of a χ site, and the helicase, with energy provided by the ATPase, could unwind the DNA upstream of the nick, producing the free single-stranded DNA end that can invade the other DNA duplex. Second, in step (e), the nuclease could come into play again to nick the single-stranded D loop, allowing a reciprocal strand exchange.

Let us focus on step (a). Chi sites were discovered in genetic experiments with bacteriophage λ. Lambda *red*

gam phages lacked χ sites, but their efficient replication depended on recombination by the RecBCD pathway. Because, as we will see, the RecBCD pathway depends on χ sites, these mutants made small plaques. Franklin Stahl and colleagues showed that certain λ *red gam* mutants made large plaques, suggesting that these mutants had more active RecBCD recombination. Stahl and colleagues then discovered that recombination was enhanced near the point of the mutation and named these mutated sites Chi, for "crossover hotspot instigator." The fact that the mutations promoted recombination nearby suggested that these mutations did not behave like ordinary gene mutations that occur in coding regions and change the structures of gene products. Instead, it appeared that these mutations created new χ sites that stimulated recombination nearby.

We know that χ sites stimulate the RecBCD pathway, but not the λ Red (homologous) pathway, the λ Int (site-specific) pathway, or the *E. coli* RecE and RecF pathways (both homologous). This points strongly to the participation of the RecBCD protein at χ sites, because this is the only component of the RecBCD pathway not found in any of the others. In fact, because RecBCD has endonuclease activity, one attractive hypothesis is that it nicks DNA near χ sites to initiate recombination.

Gerald Smith and his colleagues found evidence for this hypothesis. They generated a 3′-end-labeled, double-stranded fragment of plasmid pBR322 with a χ site near its end. The labeled 3′-end lay only about 80 bp from the χ site, as shown in Figure 22.14a. Then they added purified RecBCD protein. After heat-denaturing the DNAs in some of the reactions, they electrophoresed the DNA products and looked for the 80-nt fragment that would be generated by nicking at the χ site. (They had to run the reaction briefly to avoid general degradation of the DNA by the nonspecific RecBCD nuclease activity.) Figure 22.14b shows that the 80-nt product was indeed observed and that its appearance depended on the presence of the RecBCD protein. Heat-denaturation of the DNA was not necessary to yield the 80-nt single-stranded DNA, which suggested that the RecBCD protein was not only nicking, but also unwinding the DNA beyond the nick. Smith and colleagues also mapped the exact cleavage sites by running the labeled ≈80-nt fragment alongside sequencing lanes generated by chemical cleavage of the same labeled substrate (Maxam-Gilbert sequencing). They observed two bands, one nucleotide apart, that showed that RecBCD cut this substrate in two places, as indicated by the asterisks in this sequence:

5′-GCTGGTGGGTT*G*CCT-3′

Thus, RecBCD cut this substrate 4 and 5 nt to the 3′-side of the GCTGGTGG χ site (underlined). Another substrate could be cut in three places, 4, 5, and 6 nt to the 3′-side of the χ site. Thus, the exact cleavage sites depend on the substrate.

It was conceivable that a contaminant in the RecBCD preparation, not RecBCD itself, was causing the nicking in these experiments. To rule this out, Smith and coworkers performed the same assay with RecBCD purified from a *recC*⁻ mutant that was genetically identical to the strain used in the previous experiment in every way but in the *recC* gene. This enzyme did not generate the short single-stranded fragment. Because the only change in this experiment was in RecBCD, it is very likely that RecBCD was really responsible for the nicking seen with wild-type preparations.

The nick in the experiment in Figure 22.14 occurred in the strand bearing the χ sequence 5′-GCTGGTGG-3′. To see whether the opposite strand could also be nicked, Smith and colleagues prepared another substrate with label at the end of a DNA bearing the complementary sequence: 3′-CGACCACC-5′. No short fragment was produced under these conditions, indicating that nicking is specific for the strand having the 5′-GCTGGTGG-3′ sequence.

These findings supported the idea that RecBCD nicks DNA near a χ site and also suggested that RecBCD can unwind the DNA, starting at the nick. Further support for the role of RecBCD in unwinding DNA came from work by Stephen Kowalczykowski and colleagues. Figure 22.15 presents the results of one of their experiments, which demonstrated that: (1) RecA alone, or even RecA plus SSB, could not cause pairing between two homologous double-stranded DNAs. (2) However, with RecBCD in addition to RecA and SSB, strand exchange, which depends on DNA unwinding, occurred rapidly, as long as the two DNAs were homologous. (3) RecBCD was dispensible if one of the DNAs was heat-denatured. This last finding implied that one function of RecBCD is to unwind one of the DNAs to provide a free DNA end; RecA and SSB can coat and then use this free DNA end to initiate strand invasion.

Stuart Linn and colleagues illustrated the DNA helicase activity of RecBCD using electron microscopy to detect the unwound DNA products. Figure 22.16 shows the results of incubating T7 phage DNA for increasing lengths of time with purified RecBCD in the presence and absence of SSB. Panel (a) shows the result when the experimenters used RecBCD for 30 min in the absence of SSB. The DNA appears to be duplex throughout, which fits the following hypothesis: RecBCD unwound the DNA, but, with no SSB to trap the single-stranded DNA produced, the two strands wound back up again to form the duplex DNA we observe. If this is true, then adding RecBCD for a long time, followed by a short incubation with RecBCD and SSB, should yield a short bubble of melted DNA—

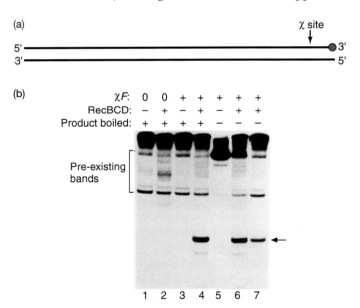

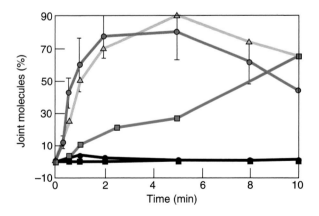

Figure 22.14   Chi-specific nicking of DNA by RecBCD.
(a) Substrate for nicking assay. Smith and colleagues prepared a 1.58-kb *Eco*RI-*Dde*I restriction fragment with a χ site about 80 bp from the *Dde*I end. They 3′-end-labeled the *Dde*I end by end-filling with [³²P] nucleotide (red). (b) Nicking assay. Smith and colleagues incubated the end-labeled DNA fragment in panel (a), designated "+" in the top line, or a similar fragment lacking the χ site, designated "0" in the top line, with or without RecBCD (designated "+" or "–" in the middle line) for 30 sec. Then they terminated the reactions and electrophoresed the products. Some of the reaction products were boiled for 3 min as indicated at top. The arrow at right denotes the 80-nt labeled fragment released by nicking at the χ site. The appearance of this product depended on RecBCD and a χ site, but not on boiling the product. (*Source:* (b) Ponticelli et al., Chi-dependent DNA strand cleavage by *recBC* enzyme. *Cell* 41 (May 1985) f. 2, p. 146. Reprinted by permission of Elsevier Science.)

Figure 22.15   RecBCD-dependence of strand exchange between two duplex DNAs. Kowalczykowski and coworkers incubated two duplex DNAs with RecA, RecBCD, and SSB (red) and assayed for joint molecules (strand exchange) by filter binding or by gel electrophoresis. (The joint molecules have a lower electrophoretic mobility than the nonrecombining DNAs.) They also assayed for joint molecules without RecA or without RecBCD (black lines). The blue line shows the results when RecBCD was omitted, but one of the DNAs was heat-denatured. The green line shows the results when all components except RecA were preincubated together, then RecA was added to start the reaction. RecBCD was added last in all other reactions. (*Source:* From L. J. Roman et al., "RecBCD-dependent joint molecule formation promoted by the *Escherichia coli* RecA and SSB proteins," *Proceedings of the National Academy of Sciences* 88:3367–3371, April 1991.)

trapped by SSB—within the duplex. Panel (b) shows such a bubble created under just such circumstances. On the other hand, if the experimenters added SSB and RecBCD together, they observed forked DNAs with duplex DNA adjoining two single strands. This implied that RecBCD began unwinding at the end of the duplex, and SSB trapped the two single-stranded DNAs that were generated. As expected, the forks grew longer with time (panels c–e). Presumably, RecBCD could begin unwinding DNA at an internal nick, perhaps even a nick of its own creation.

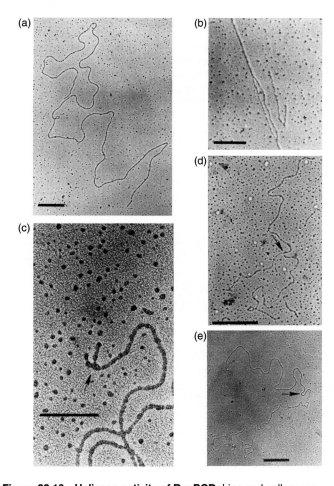

Figure 22.16 Helicase activity of RecBCD. Linn and colleagues incubated double-stranded T7 DNA with purified RecBCD under the following conditions: (**a**) 30 min, no SSB added; no unwinding is detectable. The bar represents 500 nm. (**b**) 2 min without SSB, then SSB added for 20 sec; a 1550-bp bubble is visible. (**c**) 10 sec with SSB added together with RecBCD at the beginning; 420-nt single-stranded tails are seen (branch at arrow). The bars for panels (**b**) and (**c**) represent 250 nm. (**d**) and (**e**) 15 min with SSB and RecBCD added together at the beginning; (**d**) shows single-stranded tails of 6300 and 9000 nt (branch at arrow); (**e**) shows a single-stranded tail of 20,000 nt. The arrow denotes the junction between single- and double-stranded DNA. The bars for panels (**d**) and (**e**) represent 1 μm.
(*Source:* Rosamond et al., Modulation of the action of the *recBC* enzyme of *Escherichia coli* K-12 by Ca²⁺. *J. Biol. Chem.* 254 no. 17 (10 Sept 1979) f. 3, p. 8649. American Society for Biochemistry and Molecular Biology.)

> **SUMMARY** RecBCD has a DNA nuclease activity that can introduce nicks into double-stranded DNA, especially near χ sites, and an ATPase-driven DNA helicase activity that can unwind double-stranded DNAs from their ends. It can thus provide the single-stranded DNA ends that RecA needs to initiate strand exchange.

RuvA and RuvB

RuvA and RuvB form a DNA helicase that catalyzes the branch migration of a Holliday junction. We have seen that Holliday junctions can be created in vitro; in fact they are a by-product of experiments that measure the effect of RecA on strand exchange. Early work on RuvA and RuvB used such RecA products as the Holliday junctions that could interact with RuvA and RuvB. Later, Stephen West and his colleagues devised a method for using four synthetic oligonucleotides whose sequences required that they base-pair in such a way as to form a Holliday junction, as illustrated in Figure 22.17.

Carol Parsons and West end-labeled such a synthetic Holliday junction and used a gel mobility shift assay to measure binding of RuvA and RuvB to the Holliday junction. Because branch migration was known to require ATP, they used the unhydrolyzable ATP analog, ATPγS. In principle, this should allow assembly of RuvA and RuvB on the DNA, but should prevent the branch migration that would dissociate the Holliday junction. They were successful in demonstrating a complex between

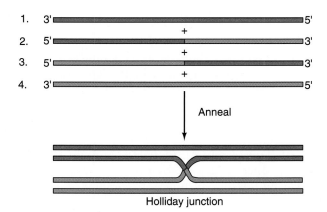

Figure 22.17 Forming a synthetic Holliday junction. Oligonucleotides 1–4 are mixed under annealing conditions so the complementary parts of each can base-pair. The 5′-end of oligo 2 is complementary to the 3′-end of oligo 1, so those two half-molecules can base-pair, but the 3′-end of oligo 2 is complementary to the 5′-end of oligo 4, so those two half-molecules form base pairs. Similarly, the two ends of oligo 3 are complementary to the other ends of oligos 1 and 4, so oligo 3 crosses over in its base pairing, in a manner complementary to that of oligo 2. The result is a synthetic Holliday junction.

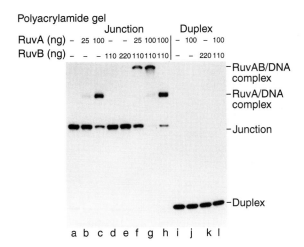

Figure 22.18 **Detecting a RuvA–RuvB–Holliday junction complex.** Parsons and West constructed a labeled synthetic Holliday junction and mixed it with varying amounts of RuvA and RuvB, as indicated at top. All mixtures contained ATPγS except the one in lane h. Parsons and West then treated the mixtures with glutaraldehyde to cross-link proteins in the same complex and prevent their dissociation. Finally, they subjected the complexes to polyacrylamide gel electrophoresis and autoradiography to detect the labeled complexes. (*Source:* Parsons & West, Formation of a RuvAB-Holliday junction complex *in vitro. J. Mol. Biol.* 232 (1993) f. 2, p. 400, by permission of Academic Press.)

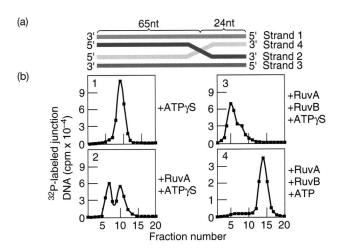

Figure 22.19 **Detecting a RuvA–RuvB–Holliday junction complex by glycerol gradient ultracentrifugation.** (a) The synthetic Holliday junction. This construct is asymmetric, with one long end (left) and one short end (right). This should cause asymmetric binding of RuvA and RuvB. (b) Glycerol gradient ultracentrifugation. Hiom and West end-labeled the synthetic Holliday junction from panel (a) and mixed it with ATPγS (1), RuvA and ATPγS (2), RuvA, RuvB, and ATPγS (3), or RuvA, RuvB, and ATP (4). Then they subjected the complexes to glycerol gradient ultracentrifugation to measure their sizes. The bottom of the gradient is at left in each graph, so larger complexes are found closer to the left. (*Source:* Reprinted from Hiom and West, *Cell* 80:788, 1995. Copyright 1995, with permission from Elsevier Science.)

RuvA and the Holliday junction, but did not see a supershift with RuvB, which would have indicated a RuvA–RuvB–Holliday junction complex. This suggested that this ternary complex was too unstable under these experimental conditions. To stabilize the putative complex, they added glutaraldehyde, which should cross-link the proteins in the complex and prevent them from dissociating during gel electrophoresis.

Figure 22.18 demonstrates cooperative binding between RuvA and RuvB. At low RuvA concentration (lane b), little, if any, binding to the Holliday junction occurred. On the other hand, at high concentration, abundant binding occurred. Furthermore, RuvB by itself, even at high concentration, could not bind to the Holliday junction (lane e), but both proteins together could bind, even at a concentration of RuvA that could not bind well by itself (lanes f and g). Only RuvA could bind to the Holliday junction in the absence of ATPγS; the ternary RuvA–RuvB–Holliday junction complex did not form (lane h). Finally, neither RuvA, nor RuvB, nor both together could bind to an ordinary duplex DNA of the same length as the Holliday junction (lanes j–l), so these proteins bind specifically to Holliday junctions.

These studies showed that RuvA can bind by itself to a Holliday junction, and it allows RuvB to bind. Assuming that RuvA binds to the middle of the Holliday junction implies some kind of polarity, with RuvB on one side or the other of RuvA. But the synthetic Holliday junctions used in the earlier studies were symmetric, so any such

polarity would be difficult to detect. To get around this problem, Kevin Hiom and West constructed the asymmetric Holliday junction pictured in Figure 22.19a. This allowed them to examine the target sites for both RuvA and RuvB on the DNA, and thus determine the polarity of the complex.

First, they used glycerol gradient ultracentrifugation to detect protein–DNA binding. Because this is a milder procedure than electrophoresis, they did not have to cross-link the complexes with glutaraldehyde. Figure 22.19b shows that RuvA can bind by itself, and that RuvA and RuvB both bind to the Holliday junction to form an even larger complex. Finally, this experiment showed that ATP (instead of ATPγS) allowed branch migration, which dissociated the Holliday junction into two DNA duplexes that sedimented much more slowly than the Holliday junction.

Next, Hiom and West used DNase footprinting to find the binding sites for both RuvA and RuvB in the complex. This allowed them to show that the complex really was asymmetrical, as predicted. Figure 22.20 shows the results of DNase footprinting with each of the four strands of the Holliday junction labeled in turn. In each case, they footprinted the RuvA complex, and the RuvA–RuvB complex, separately. They found that RuvA always protected an area centered on the crossover point, indicated by the star in each footprint. Furthermore, the RuvA-RuvB-Holliday junction complex always overlapped the RuvA–Holliday

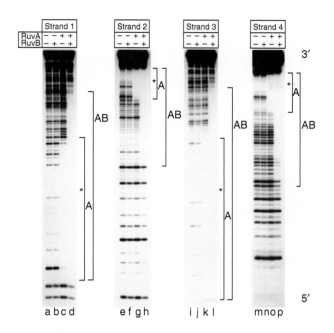

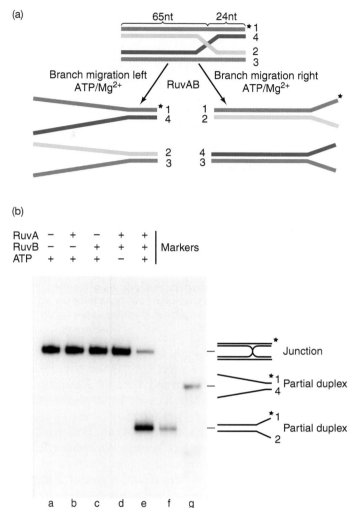

Figure 22.20 Footprinting RuvA and RuvA + RuvB on a synthetic Holliday junction. Hiom and West 5′-end-labeled each strand of the synthetic Holliday junction in turn, as indicated at top. Then they added RuvA, or RuvB, or both, also as indicated at top, and performed DNase footprinting. RuvA binds to a region centered on the point where the two arms of the Holliday junction cross (indicated by stars). The RuvA and RuvA + RuvB complexes always ended at the same point on the right, but the RuvA + RuvB complex always extended farther to the left (see Figure 22.21). (*Source:* Hiom & West, Branch migration during homologous recombination assembly of a RuvAB-Holliday junction complex *in vitro. Cell* 80 (10 Mar 1995) f. 2, p. 788. Reprinted by permission of Elsevier Science.)

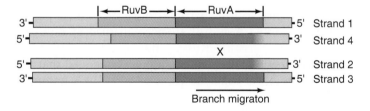

Figure 22.21 Interpretation of the results of DNase footprinting the synthetic Holliday junction with RuvA and RuvB. The regions protected by RuvA and RuvB are indicated by yellow and blue boxes, respectively. The indistinct end of the RuvA footprint on the right in strands 2 and 4 denotes the lack of precision in these footprints. The crossover point is represented by an X. The direction of migration determined in Figure 22.22 is indicated at bottom. (*Source:* Reprinted from Hiom and West, *Cell* 80:789, 1995. Copyright 1995, with permission from Elsevier Science.)

Figure 22.22 Determining the polarity of branch migration. (**a**) Strategy of the experiment. Hiom and West labeled strand 1 of the synthetic Holliday junction at its 5′-end, as indicated by the star. Then they added RuvA, RuvB, and ATP to allow branch migration. The diagrams on the left and right show the DNA duplexes that would result if the Holliday junction dissociated by branch migration to the left or right, respectively. (**b**) Results. Hiom and West added RuvA, RuvB, and ATP, as indicated at top, then incubated the mixtures, deproteinized them, and electrophoresed them along with Y-shaped partial duplex markers as indicated. Autoradiography showed the position of the partial duplex product matched that of the strand 1/strand 2 partial duplex marker. (*Source:* Hiom & West, *Cell* 80 (10 Mar 1995) f. 5b, p. 790. Reprinted by permission of Elsevier Science.)

junction complex perfectly at one end, but extended farther at the other end, just as we would expect if RuvB always bound to one side of RuvA in the complex. In this case, it was to the left of the crossover point, as indicated in Figure 22.21.

Next, Hiom and West asked which direction the complex moved as it catalyzed branch migration. Figure 22.22a shows what would happen if the Holliday junction dissociated into two DNA duplexes by branch migration either to the left or the right. Given that strand 1 was labeled at its 5′-end, the experimenters reasoned as follows: Because strand 1 is complementary to strand 4 only to the right of the crossover point, branch migration to the left would yield a labeled strand 1/strand 4 duplex held together by only 24 bp on the right. On the other hand, because strand 1 is complementary to strand 2 only to the left of the crossover point, branch migration to the

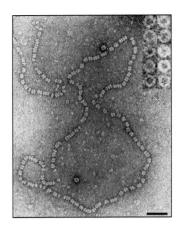

Figure 22.23 Electron microscopy of RuvB–duplex DNA complex.
West, Egelman, and colleagues mixed RuvB with relaxed double-stranded φX174 DNA, then stained the complex with uranyl acetate and subjected it to electron microscopy. RuvB can be seen binding as double rings with the DNA going through the center of the rings. Inset: Top views of 10 RuvB rings, showing the hole through the center. The bar represents 40 nm. (*Source:* Stasiak et al., *PNAS* 91 (1994) f. 1, p. 7619. © National Academy of Sciences, USA.)

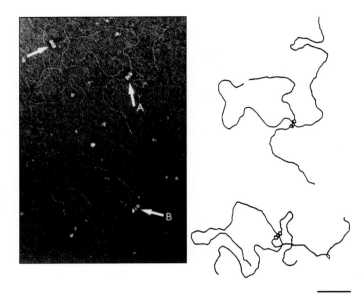

Figure 22.24 Electron microscopy of RuvA–RuvB–Holliday junction complexes. West and coworkers mixed RuvA, RuvB, and Holliday junctions to form complexes, then stained the complexes with uranyl acetate and visualized them by electron microscopy. Arrows point to tripartite complexes presumably composed of one tetramer of RuvA in the center, flanked by two hexameric circles of RuvB. Complexes A and B are accompanied by interpretive diagrams at right. Bar represents 100 nm. (*Source:* Parsons et al., Structure of a multi-subunit complex that promotes DNA branch migration. *Nature* 374 (23 Mar 1995) f. 2, p. 376. © Macmillan Magazines Ltd.)

right would yield a labeled strand 1/strand 2 duplex held together by 65 bp on the left. These two labeled duplexes are easily distinguished by gel electrophoresis: Because the strand 1/strand 4 duplex is a Y-shaped complex with much longer arms, it has lower electrophoretic mobility than the strand 1/strand 2 duplex. In fact, as Figure 22.22b (lane e) shows, the strand 1/strand 2 duplex was observed, demonstrating branch migration to the right, away from the RuvB protein.

To get more insight into the geometry of binding between RuvB and DNA, Edward Egelman and West and their colleagues used electron microscopy to examine RuvB–DNA complexes. Figure 22.23 shows an electron micrograph of RuvB bound to ordinary duplex DNA (not a Holliday junction). We can see RuvB rings bound in pairs to the DNA. We can tell that the rings are formed around the DNA because they look like beads on a string: all aligned with the DNA double-helical axis. If they were merely attached to the outside of the DNA, their orientation would not be so consistent. It would also be possible to see this kind of picture if the DNA were wound around RuvB, in a manner analogous to nucleosomes. However, the contour length of this DNA is exactly what we expect for nonwrapped DNA. Because it would be considerably shortened if it engaged in any wrapping, we can conclude that no such wrapping occurs. Furthermore, the inset in Figure 22.23 shows top views of several RuvB double rings. We can see holes through the middle of them that could accommodate a DNA duplex. With a little imagination, we can also see the hexameric nature of these rings.

West and colleagues have also performed an electron microscopic examination of the geometry of the

RuvA–RuvB–Holliday junction complex. Figure 22.24 is an electron micrograph depicting three such complexes. All three appear to be tripartite (composed of three parts), and two (marked A and B) are accompanied by interpretive diagrams. Because RuvA binds to the center of the Holliday junction, West and colleagues suggested that the center blob in the tripartite complex is a RuvA tetramer, and the two flanking blobs are RuvB hexamer rings. Because these two RuvB rings are diametrically opposed, rather than lying at an angle, West and colleagues proposed that the Holliday junction, when complexed with RuvA and RuvB, is in a square planar conformation, as illustrated in Figure 22.25. This model explains how RuvB could cause branch migration by catalyzing DNA duplex unwinding and rewinding as indicated by the arrows.

RuvB can drive branch migration by itself if it is present in high enough concentration, so it has the DNA helicase and attendant ATPase activity. What, then, is the role of RuvA? It binds to the center of a Holliday junction and facilitates binding of RuvB, so branch migration can take place at a much lower RuvB concentration. Furthermore, as we will soon see, it appears to hold the Holliday junction in a square planar conformation that is favorable for rapid branch migration.

What is the nature of the binding between RuvA and the Holliday junction? David Rice and colleagues have

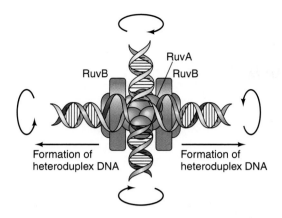

Figure 22.25 Model of the RuvA–RuvB–Holliday junction complex. RuvA binds to the middle of the Holliday junction, which is in a square planar conformation, with each branch of the junction at right angles to the next. The RuvB rings are asymmetrical, with their convex sides pointing outward. Horizontal arrows indicate that branch migration in either direction will create heteroduplex DNA. Circular arrows indicate the directions the DNAs must wind or unwind to permit branch migration. (*Source:* Reprinted with permission from *Nature* 374:377, 1995. Copyright © 1995 Macmillan Magazines Limited.)

bolstered the hypothesis of a square planar conformation for the RuvA–Holliday junction complex by performing x-ray crystallography on RuvA tetramers and showing that they have a square planar shape. This is illustrated in Figure 22.26a, in which we see that each monomer is roughly L-shaped, containing a leg and a foot connected by a flexible loop of indeterminate shape, represented by a dashed colored line. The foot of the L of one monomer interacts with the leg of the next to form a lobe; one of the four lobes is surrounded by a white dashed line in the figure. These lobes are arranged in a fourfold symmetrical pattern, with a natural groove between each pair of lobes; white dashed lines lie in two of these grooves. Figure 22.26b shows a side view of the tetramer, which reveals a concave surface on top and a convex surface on the bottom.

Molecular modeling showed that this RuvA tetramer could mate naturally with a Holliday junction in a corresponding square planar conformation, as shown in Figure 22.27. Note the neat fit between the DNA and the concave surface of the protein. The four branches of the Holliday junction could lie in the four grooves on the surface of the protein. Four β-turns, one on each monomer, form a hollow-looking pin that protrudes through the center of the Holliday junction. This square planar shape would allow rapid branch migration, as Figure 22.25 suggested. Any deviation from this shape would slow branch migration, which emphasizes the importance of the square planar shape of RuvA.

West and Egelman, along with Xiong Yu, performed further electron microscopy of the RuvAB-Holliday junction complex. They made 100 micrographs of the complex, scanned them, and combined them to create an aver-

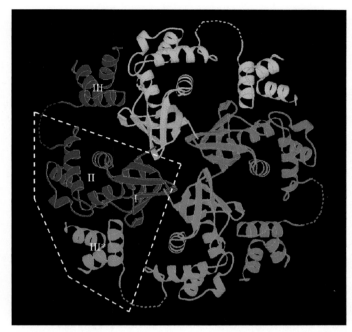

(a)

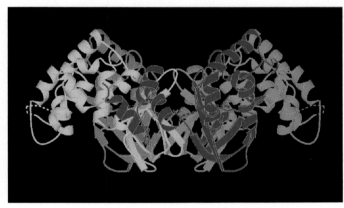

(b)

Figure 22.26 Structure of RuvA tetramer as revealed by x-ray crystallography. (a) Top view. The four monomers are represented by different colored ribbons, and one of the four lobes in the square planar structure is outlined with a dashed white line. The three domains of the blue monomer are numbered, as is the third domain (the "foot" of the L) of the green monomer. **(b)** Side view. The same structure, represented by the same colored ribbons, is shown from the side. The concave and convex surfaces at top and bottom, respectively, are evident. (*Source:* Rafferty et al., Crystal structure of DNA recombination protein RuvA and a model for its binding to the Holliday junction. *Science* 274 (18 Oct 1996) f. 2 d-e, p. 417. © AAAS.)

age image. Figure 22.28a presents a model based on this image, with color-coded DNA strands added. Just as in the hypothetical model in Figure 22.25, the RuvA tetramer is in the center at the junction, with two RuvB hexameric rings flanking it. Panel (b) shows what happens on bending two of the arms of the complex (with the RuvA tetramer removed for clarity), and panel (c) shows

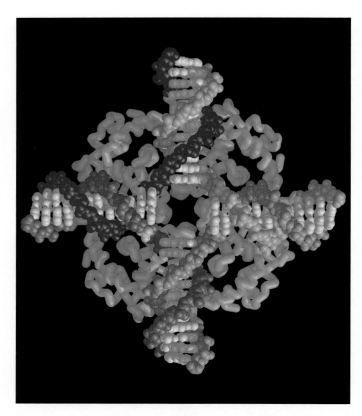

Figure 22.27 Model for the interaction between RuvA and a Holliday junction. The RuvA monomers are represented by green tubes that trace the α-carbon backbones of the polypeptides. The DNAs in the Holliday junction are represented by space-filling models containing dark and light red and blue backbones and silver base pairs. The yellow balls denote the phosphate groups of one of the two pairs of sites that can be cut by RuvC to resolve the Holliday junction. (*Source:* Rafferty et al., *Science* 274 (18 Oct 1996) f. 3d, p. 418. © AAAS.)

the result of rotating the bottom DNA duplex through 180 degrees out of the plane of the paper (again the RuvA tetramer is removed for clarity). The RuvB rings on this familiar Holliday junction are poised to catalyze branch migration by moving in the direction of the arrows.

> **SUMMARY** RuvA and RuvB form a DNA helicase that can drive branch migration. A RuvA tetramer with square planar symmetry recognizes the center of a Holliday junction and binds to it. This presumably induces the Holliday junction itself to adopt a square planar conformation, and promotes binding of hexamer rings of RuvB to two diametrically opposed branches of the Holliday junction. Then RuvB uses its ATPase to drive DNA unwinding and rewinding that is necessary for branch migration.

RuvC

After branch migration, we have a Holliday junction such as the one pictured in Figure 22.29. As we saw in Figure 22.2, there are two ways to resolve this junction. Figure 22.29 presents these two possibilities again but in a different format. Here we see that cutting the two strands involved in the original crossover (sites 1 and 2) resolve the Holliday junction into two non-crossover recombinant duplex DNAs with patches of heteroduplex DNA. On the other hand, cutting the other two strands at sites 3 and 4 resolves the structure into two spliced (crossover) recombinant DNAs.

What nuclease is responsible for making the cuts that resolve Holliday junctions? West and colleagues showed

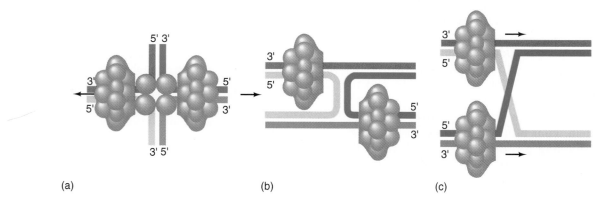

(a)   (b)   (c)

Figure 22.28 Model for RuvAB–Holliday junction complex based on combining EM images of the complex. (a) Complex with DNA branches perpendicular to each other. (b) The blue-yellow and red-green branches from panel (a) have been rotated 90 degrees in the plane of the paper, and the RuvA tetramer has been removed so we can see the center of the junction. (c) The blue-green and blue-yellow branches from panel (b) have been interchanged by rotating the lower limb of the junction 180 degrees out of the plane of the paper. This produces a familiar Holliday junction with RuvB hexamer rings in position to catalyze branch migration by moving in the direction of the arrows. (*Source:* Reprinted from *Journal of Molecular Biology,* vol. 266, X. Yu, S. C. West, and E. H. Egelman, "Structure and Subunit Composition of the RuvAB-Holliday Junction Complex," 217–222, 1997, by permission of Academic Press Limited, London.)

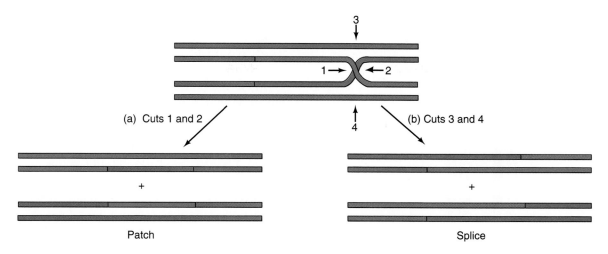

Figure 22.29 Resolution of a Holliday junction. The Holliday junction pictured at top can be resolved in two different ways, as indicated by the numbered arrows. **(a)** Cuts 1 and 2 yield two duplex DNAs with patches of heteroduplex whose length corresponds to the distance covered by branch migration before resolution. **(b)** Cuts 3 and 4 yield crossover recombinant molecules with the two parts joined by a staggered splice.

in 1991 that it is RuvC. They built the ^{32}P-labeled synthetic Holliday junction pictured in Figure 22.30a, with a short (12 bp) homologous region (J) at the joint, but the rest of the structure composed of nonhomologous regions. Next, they used a gel mobility shift assay to test the ability of RuvC to bind to the Holliday junction and to a linear duplex DNA. Figure 22.30b shows the results: As they added more and more RuvC to the Holliday junction, West and colleagues observed more and more DNA-protein complex, indicating RuvC–Holliday junction binding. But the same assay showed no binding between RuvC and a linear duplex DNA made from strand 1 and its complement.

Thus, RuvC binds specifically to a Holliday junction, but can it resolve the junction? Figure 22.30c shows that it can. West and coworkers added increasing amounts of RuvC to the labeled Holliday junction, or to the duplex DNA. They found that RuvC caused resolution of the Holliday junction to a labeled species the same length as the duplex DNA, which is what we expect for resolution. More complex experiments that can distinguish between patch or splice resolution showed that the splice products predominate, at least in vitro.

Thanks to x-ray crystallography studies performed by Kosuke Morikawa and colleagues, we now know the three-dimensional structure of RuvC. It is a dimer, with its two active sites 30 Å apart. That puts them right in position to cleave the square planar Holliday junction at two sites, as shown in Figure 22.31a. Figure 22.31b presents a more detailed representation of a RuvC-Holliday junction complex. Does RuvC act alone, as this model implies, or does it act on a Holliday junction already bound to RuvB or RuvA plus RuvB?

The evidence strongly suggests the latter: that RuvC acts on a Holliday junction already bound to RuvA and RuvB. West and colleagues have reconstituted a system

that carries out the intermediate to late stages of recombination in vitro and have shown that monoclonal antibodies against RuvA, RuvB, and RuvC each block resolution of Holliday junctions. One way to explain this result is that RuvA, RuvB, and RuvC work together. If that is true, then the proteins probably naturally associate with one another, and one should be able to cross-link them. So West and colleagues prepared various mixtures of the two proteins, added glutaraldehyde to cross-link them, and electrophoresed them to detect cross-links. They found that RuvA and RuvB could be cross-linked, as expected, and RuvB and RuvC could also be cross-linked, but RuvA and RuvC could not. Thus, RuvB can bind to both RuvA and RuvC, suggesting that all three proteins can bind together to a Holliday junction.

This hypothesis of concerted action by RuvA, RuvB, and RuvC is consistent with the notion that branch migration is necessary during resolution to help RuvC find its preferred sites of cleavage. It is also consistent with x-ray crystallography data showing that RuvA can associate with a Holliday junction as a tetramer, as we have already seen, or as an octamer, with tetramers on either side of the DNA. West hypothesized that the complex involving the RuvA octamer is specific for efficient branch migration (Figure 22.32a). Later, RuvC could replace one of the RuvA tetramers to form the RuvABC–junction complex, or "resolvasome" (Figure 22.32b) that is specific for resolution of the Holliday junction.

Mutations in *ruvA, ruvB,* and *ruvC* all produce the same phenotype: heightened sensitivity to UV light, ionizing radiation, and the antibiotic mitomycin C because of defective recombination repair. But RuvA and B promote branch migration, whereas RuvC catalyzes resolution of Holliday junctions. Why should defects in all three of these proteins have the same end result? One way to

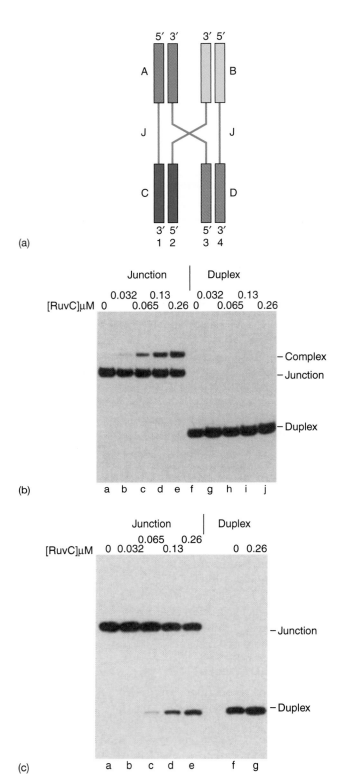

(a)

Junction | Duplex

[RuvC]μM 0 0.032 0.13 0 0.032 0.13
              0.065 0.26   0.065 0.26

— Complex
— Junction

— Duplex

(b)   a b c d e f g h i j

Junction | Duplex

[RuvC]μM 0 0.032 0.065 0.13 0.26 0 0.26

— Junction

— Duplex

(c)   a b c d e f g

answer this question would be to show that resolution depends on branch migration. Then defective RuvA or RuvB would block resolution indirectly by blocking branch migration.

West and colleagues did not exactly do that, but they did show that hotspots for RuvC resolution occur, which implies that branch migration is needed to reach those hotspots. They started with the ^{32}P-end-labeled Holliday junction created with the help of RecA, as shown in Figure 22.33a. As usual, this Holliday junction can be resolved in two ways: Cuts at sites a and c yield a labeled crossover recombinant dimer, and cuts at b and d yield labeled non-crossover recombinant "patch" products—nicked circles and gapped linear DNA duplexes. To obtain resolution, West and colleagues treated the Holliday junction with RuvC, or with the phage T4-encoded resolvase, **endonuclease VII.** They added these resolvases either directly or after spontaneous branch migration

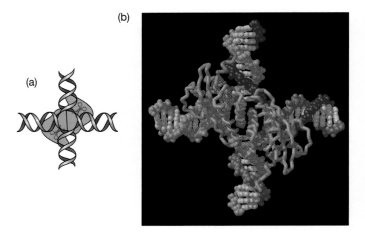

Figure 22.31   Model for the interaction between RuvC and a Holliday junction. (a) Schematic model showing the RuvC dimer (gray) bound to the square planar Holliday junction. Scissors symbols (green) denote the active sites on the two RuvC monomers. Note how the location of these active sites fits with the positioning of the DNA strands to be cleaved in resolving the complex. (b) Detailed model. Gray tubes represent the carbon backbone of the RuvC dimer. The Holliday junction is represented by the same blue and red backbone and silver base pairs as in Figure 22.27. (*Source:* Rafferty et al., Crystal structure of DNA recombination protein RuvA and a model for its binding to the Holliday junction. *Science* 274 (18 Oct 1996) f. 3e, p. 418. © AAAS.)

Figure 22.30   RuvC can resolve a synthetic Holliday junction. (a) Structure of the synthetic Holliday junction. Only the 12-bp central region (J, red) is formed from homologous DNA. The other parts of the Holliday junction (A, B, C, and D) are nonhomologous, as indicated by the different colors. (b) Binding of RuvC to the synthetic Holliday junction. West and colleagues end-labeled the Holliday junction (and a linear duplex DNA) and bound them to increasing amounts of pure RuvC under noncleavage conditions (low temperature and absence of MgCl$_2$). Then they electrophoresed the products. RuvC binds to the Holliday junction, but not to ordinary duplex DNA. (c) Resolution of the Holliday junction by RuvC. West and coworkers mixed the labeled Holliday junction (or linear duplex DNA) with increasing concentrations of RuvC under cleavage conditions (37° C and 5 mM MgCl$_2$). Then they electrophoresed the products. RuvC resolved some of the Holliday junction to a linear duplex form. (*Source:* Dunderdale et al., Formation resolution of recombination intermediates by *E. coli recA* and RuvC proteins. *Nature* 354 (19-26 Dec 1991) f. 5b-c. p. 509. © Macmillan Magazines Ltd.)

(a) Alternate RuvAB–junction complex

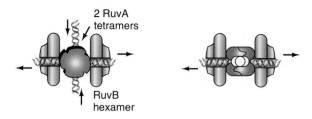

(b) Putative RuvABC–junction complex

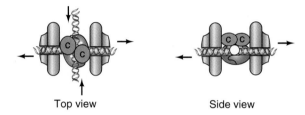

Top view   Side view

Figure 22.32   Models of Ruv protein–junction complexes.
(a) A RuvAB–junction complex discovered by Pearl and colleagues. In contrast to the complex with a RuvA tetramer described by other workers, this one contains a RuvA octamer at the Holliday junction. This could be the form the complex takes during active migration.
(b) West's model of the RuvABC–junction complex. This could be the form the complex takes during resolution. (*Source:* Reprinted from West, *Cell* 94:700, 1998. Copyright 1998, with permission from Elsevier Science.)

induced by raising the temperature to 55° C for 1 h. The extent of branch migration was limited by a 643-bp stretch of heterologous DNA, represented by the orange block in Figure 22.33a. Figure 22.33b shows that RuvC yielded both products in about equal amounts. So did endonuclease VII.

To detect any heterogeneity in the cutting sites, West and colleagues subjected the same reaction products to denaturing gel electrophoresis on a polyacrylamide gel. This separated the DNA strands, so the lengths of the short labeled fragments generated by cuts at sites a and b could be determined. Figure 22.33c shows the results. In the absence of branch migration, both RuvC and endonuclease VII gave a restricted set of fragments. On the other hand, after branch migration, RuvC yielded a discrete set of about eight major fragments, and some minor ones, whereas endonuclease VII gave a smear of many fragments. This suggested that RuvC prefers to cut at particular sites (hotspots), and endonuclease VII has little sequence specificity.

To determine the sequences at the RuvC cutting sites, West and coworkers performed primer extension analysis on the RuvC products, using the same primers for DNA sequencing of the same DNAs. In all, they identified 19 cutting sites. The sequences of 11 of these (sites 9–19) are presented in Figure 22.34. We see a clear consensus sequence: 5′-(A/T)TT ↓ (G/C)-3′. RuvA and B are presum-

ably needed to catalyze branch migration in vivo to reach such consensus sites. This hypothesis also implies that resolution to patch or splice products depends on the frequencies of the RuvC resolution sequence in the two DNA strands. Overall, this should be a 50/50 mix.

> **SUMMARY** Resolution of Holliday junctions is catalyzed by the RuvC resolvase. This protein acts as a dimer to clip two DNA strands to yield either patch or splice recombinant products. This clipping occurs preferentially at the consensus sequence 5′-(A/T)TT↓(G/C)-3′. Branch migration is essential for efficient resolution of Holliday junctions, presumably because it is essential to reach the prefered cutting sites. Accordingly, RuvA, B, and C appear to work together in a complex to locate and cut those sites.

22.3 Meiotic Recombination

As mentioned early in this chapter, meiosis in most eukaryotes is accompanied by recombination. This process shares many characteristics in common with homologous recombination in bacteria. In this section, we will examine the mechanism of meiotic recombination in yeast, with special emphasis on the initiating event, which is quite different from the models of initiation of bacterial homologous recombination presented earlier in this chapter.

The Mechanism of Meiotic Recombination: Overview

Figure 22.35 presents a hypothesis for meiotic recombination in yeast, where it has been most thoroughly studied. The process starts with a chromosomal lesion: a double-stranded break. Next, an exonuclease recognizes the break and digests the 5′-ends of two of the strands, creating 3′ single-stranded overhangs. One of these single-stranded ends can then invade the other DNA duplex, forming a D loop as we observed in bacterial homologous recombination. Next, DNA repair synthesis fills in the gaps in the top duplex, expanding the D loop in the process. Next, branch migration can occur in both directions, leading to two Holliday junctions. Finally, the Holliday junctions can be resolved to yield either a noncrossover recombinant with two sections of heteroduplex, or a crossover recombinant that has exchanged flanking DNA regions.

There is good evidence for most of these steps, but a few features of the hypothesis are contradicted by experiment. In particular, the model predicts that hybrid DNA will be produced on both sides of the double-stranded

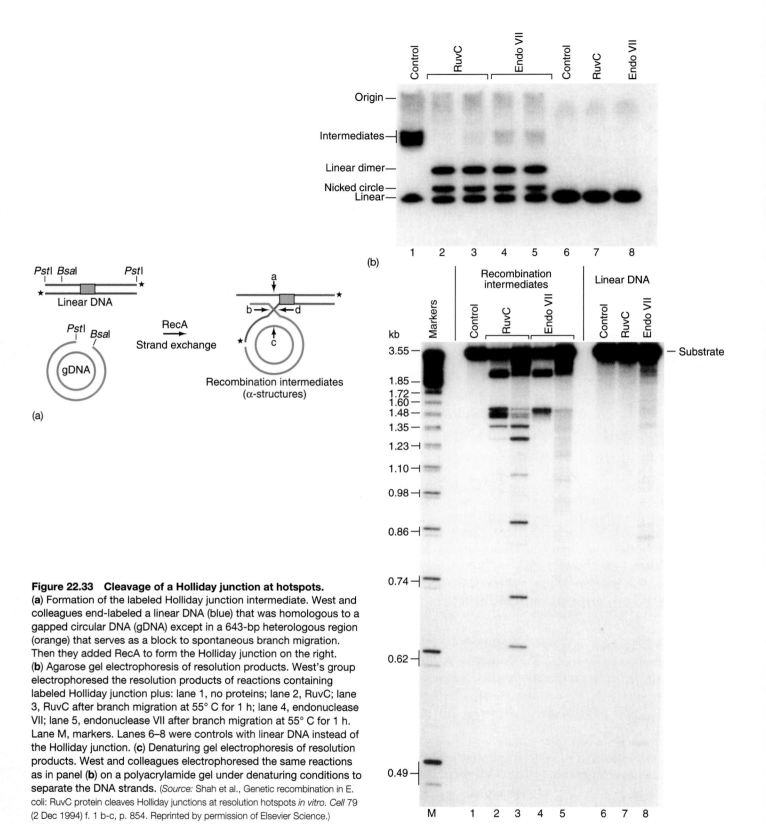

(b)

(a)

(c)

Figure 22.33 Cleavage of a Holliday junction at hotspots.
(a) Formation of the labeled Holliday junction intermediate. West and colleagues end-labeled a linear DNA (blue) that was homologous to a gapped circular DNA (gDNA) except in a 643-bp heterologous region (orange) that serves as a block to spontaneous branch migration. Then they added RecA to form the Holliday junction on the right.
(b) Agarose gel electrophoresis of resolution products. West's group electrophoresed the resolution products of reactions containing labeled Holliday junction plus: lane 1, no proteins; lane 2, RuvC; lane 3, RuvC after branch migration at 55° C for 1 h; lane 4, endonuclease VII; lane 5, endonuclease VII after branch migration at 55° C for 1 h. Lane M, markers. Lanes 6–8 were controls with linear DNA instead of the Holliday junction. (c) Denaturing gel electrophoresis of resolution products. West and colleagues electrophoresed the same reactions as in panel (b) on a polyacrylamide gel under denaturing conditions to separate the DNA strands. (*Source:* Shah et al., Genetic recombination in E. coli: RuvC protein cleaves Holliday junctions at resolution hotspots *in vitro. Cell* 79 (2 Dec 1994) f. 1 b-c, p. 854. Reprinted by permission of Elsevier Science.)

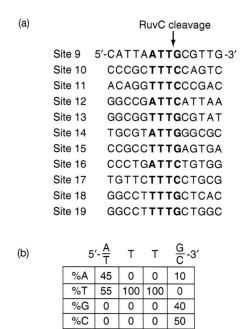

(a)

RuvC cleavage
↓

Site 9   5'-CATTA**ATT**GCGTTG-3'
Site 10   CCCGC**TTT**CCAGTC
Site 11   ACAGG**TTT**CCCGAC
Site 12   GGCCG**ATT**CATTAA
Site 13   GGCGG**TTT**GCGTAT
Site 14   TGCGT**ATT**GGGCGC
Site 15   CCGCC**TTT**GAGTGA
Site 16   CCCTG**ATT**CTGTGG
Site 17   TGTTC**TTT**CCTGCG
Site 18   GGCCT**TTT**GCTCAC
Site 19   GGCCT**TTT**GCTGGC

(b)

$$5'\text{-}\frac{A}{T}\ T\ T\ \frac{G}{C}\text{-}3'$$

%A	45	0	0	10
%T	55	100	100	0
%G	0	0	0	40
%C	0	0	0	50

Figure 22.34   Consensus sequence of RuvC cleavage sites. (a) The sequences surrounding 11 of the cleavage sites determined by primer extension are presented, with boldface denoting the consensus region surrounding the cleavage site (arrow). (b) The percentage of occurrence of each base in the four positions of the consensus sequence in 23 different RuvC cleavage sites. (*Source:* Reprinted from Shah et al., *Cell* 79:857, 1994. Copyright 1994, with permission from Elsevier Science.)

break. However, when this prediction was tested genetically, hybrid DNA was usually found only on one side of the break. In the few cases where hybrid DNA was found on both sides of the break, it was in the same chromatid, not in both chromatids as the model predicts. Thus, more data are needed to resolve this discrepancy and perhaps amend the hypothesis.

The Double-Stranded DNA Break

How do we know that recombination in yeast initiates with a double-stranded DNA break (**DSB**)? Jack Szostak and colleagues laid the groundwork for answering this question in 1989 by mapping a recombination initiation site in the *ARG4* gene of the yeast *Saccharomyces cerevisiae*. They did not look at recombination per se, but at meiotic gene conversion, which depends on meiotic recombination in yeast. Because both gene conversion and recombination initiate at the same site, these workers could use gene conversion as a surrogate for recombination. We will examine the mechanism of gene conversion later in this chapter.

Szostak and coworkers verified earlier work that had shown that meiotic gene conversion in the *ARG4* locus was polar: It was common (about 9% of total meioses) near the 5'-end of the gene, and relatively rare (about 0.4% of total meioses) at the 3'-end. This behavior suggested that the initiation site of recombination lies near the 5'-end of the gene. So Szostak and colleagues made deletions in this region to try to remove the initiation site and therefore to block gene conversion. They found that deletions having their 3'-ends in the −316 to +1 region all greatly decreased gene conversion rates, suggesting that the initiation site for recombination lies in the promoter region of the *ARG4* gene.

This information allowed these workers to look for a DNA break, either single- or double-stranded, in a very restricted region of the yeast genome. Accordingly, they cloned a 15-kb fragment of DNA, including the *ARG4* gene, into a yeast plasmid, and introduced it into a strain of yeast that carries out synchronous meiosis immediately on transfer to sporulation medium. They extracted plasmid DNA at various times after induction of sporulation and subjected it to electrophoresis. Figure 22.36 depicts the results of the electrophoresis. At time zero, we can see mostly supercoiled monomers, with some supercoiled dimers, relaxed circular monomers, and a band of lower mobility that is probably some form of dimer. These same bands appeared throughout the time course after induction of sporulation. The one novelty during sporulation was a relatively faint band of linear monomer that first appeared at 3 h, peaked at 4 h, and decreased after that time. This linear DNA must have been created by a double-stranded break in the plasmid. The timing of this DSB coincided with the timing of commitment to meiotic recombination in these cells (2.5–5 h), and with the appearance of recombination products (4 h). These findings are all consistent with the hypothesis that the first step in meiotic recombination is the formation of a DSB.

Szostak and colleagues used restriction mapping to demonstrate that DSBs occurred in three different locations in the plasmid, indicated by arrows in Figure 22.36a. One of these breaks (site 2) lies within a 216-bp restriction fragment in the control region just 5' of the *ARG4* gene. Szostak and colleagues' previous work had shown that a 142-bp deletion in this same region depressed the level of meiotic gene conversion in the *ARG4* gene, so they tested the same deletion for effect on the DSB. They found that this deletion did indeed eliminate the DSB at site 2, but had no effect on the DSBs at sites 1 and 3. Thus, the ability to form the DSB at site 2 is correlated with the efficiency of meiotic gene conversion just downstream in the *ARG4* gene.

If a DSB is the initiating event in meiotic recombination, then it should occur in a yeast chromosome, not just in a plasmid. Therefore, Szostak and colleagues used restriction mapping in cells lacking the plasmid to search for these same DSBs. They found that a DSB at site 2 (and at site 1) also occurred in yeast chromosomal DNA, and that the timing of appearance of these DSBs was the same as in the plasmid.

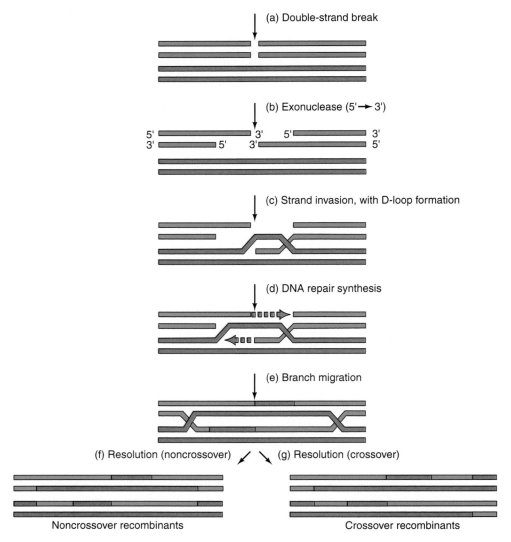

Figure 22.35 Model for meiotic recombination in yeast. (a) A double-stranded break occurs in one DNA duplex (blue), which is paired with another DNA duplex (red). (b) An exonuclease digests the DNA 5′-ends at the newly created break. (c) A single-stranded 3′-end of the top duplex invades the bottom duplex, creating a D loop. (d) DNA repair synthesis extends the free 3′-ends, with enlargement of the D loop. (e) Branch migration occurs both leftward and rightward to yield two Holliday junctions. (f) The Holliday junctions are resolved by cleaving the inside strands at both Holliday junctions, yielding noncrossover recombinant DNAs with patches of heteroduplex, but no exchange of DNA arms beyond the Holliday junctions. (g) The Holliday junctions are resolved by cleaving the inside strands at the left Holliday junction and the outside strands at the right Holliday junction. This yields crossover recombinant DNAs with exchange of DNA arms to the right of the right Holliday junction.

Nancy Kleckner and colleagues demonstrated similar double-stranded breaks in a yeast chromosome into which they had inserted a *LEU2* gene next to a *HIS4* gene to create a hotspot for meiotic recombination. Actually, two DSBs occurred close together at this hotspot. They also discovered a nonnull mutation in the *RAD50* gene (*rad50S*) that caused a buildup of the fragments caused by the DSBs. This mutation apparently blocked a step downstream of DSB formation, so it allowed the DSBs to accumulate.

In 1995, Scott Keeney and Kleckner discovered that the 5′-ends created by a DSB are covalently bound to a protein in *rad50S* mutants. One attractive hypothesis to explain this behavior is that the catalytic protein that created the DSB would normally dissociate from the DNA

immediately, but in this mutant the protein remained bound to the DNA ends it had created. If that is the case, then identifying the protein bound to the DSB ends would identify a prime suspect for the endonuclease that created the DSB.

Accordingly, Kleckner and colleagues set out to identify the protein or proteins covalently bound to the DSB. They began by isolating nuclei from meiotic *rad50S* cells. Because of their accumulation of the protein–DSB complex, these cells should provide a rich source of the DSB-bound protein. To purify the bound protein, Kleckner and coworkers used a two-stage screening procedure: First, they extracted the nuclei and denatured the protein with guanidine and detergent, then purified DNA and DNA–protein complexes by $CsCl_2$ gradient ultracentrifugation. Any

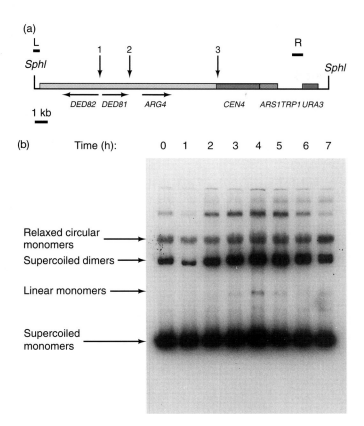

(a)

DED82 DED81 ARG4   CEN4   ARS1 TRP1 URA3

1 kb

(b) Time (h):   0 1 2 3 4 5 6 7

Relaxed circular monomers

Supercoiled dimers

Linear monomers

Supercoiled monomers

Figure 22.36   Detecting a double-stranded DNA break in a plasmid bearing a recombination initiation site. (a) Map of the plasmid used to detect the DSB. The yellow bar represents the 15-kb insert of yeast DNA containing the recombination initiation site. The other colored bars represent loci within the vector, including the centromere (*CEN4*). The locations of genes are indicated below the bars. L and R are the locations of probes used to hybridize to blots. The arrows marked 1, 2, and 3, are the locations of DSBs mapped in this experiment. (b) Electrophoresis results. Szostak and colleagues transformed yeast cells with the plasmid depicted in panel (a), induced sporulation, then electrophoresed samples of plasmid collected at the indicated times after induction. Finally, they Southern blotted the DNAs and probed them with a ^{32}P-labeled probe that hybridized to the vector. The identities of the bands are indicated at left. (*Source:* Sun, H. et al., Double-strand breaks at an initiation site for meiotic gene conversion. *Nature* 338 (2 Mar 1989) f. 1, p. 88. © Macmillan Magazines Ltd.)

tions, and their appearance depended on nuclease treatment of DNA–protein complexes.

Next, Kleckner and coworkers excised the two candidate bands (M_r = 34 kD and 45 kD) from a preparative gel, subjected them to trypic digestion, then sequenced some of the trypic peptides. The short protein sequences obtained from these peptides yielded the corresponding DNA sequences. Then, because the sequence of the entire yeast genome was already known, these short DNA sequences allowed for easy identification of the corresponding genes. The 45-kD protein corresponded to (coincidentally) the *spo11* gene product, **Spo11**, and the 34-kD protein corresponded to a mixture of five different proteins, including two ribosomal proteins.

Because Spo11 was already known to be required for meiosis, it was an attractive candidate for the DSB-bound protein. To reinforce this hypothesis, Kleckner and colleagues demonstrated that Spo11 is bound specifically to DSBs, and not to bulk DNA. To do this, they used an epitope tagging approach. They created a *Spo11* gene fused to the coding region for an epitope of the protein hemagglutinin. The protein product of this gene (Spo11-HA), and any DNA attached, could therefore be immunoprecipitated with an antihemagglutinin antibody.

In a preliminary experiment, these workers isolated DNA (with any covalently attached proteins) from meiotic *rad50S* cells containing the same *HIS4LEU2* recombination hotspot as in their previous studies. They cut the DNA with *Pst*I, electrophoresed it, blotted the fragments, and probed the blot with a DNA probe for the hotspot region. Figure 22.37a depicts a map of the hotspot region, showing the two sites where DSBs occur during meiotic recombination, two *Pst*I sites flanking the DSB sites, and the location to which the probe hybridizes downstream of *HIS4LEU2*. Thus, if no DSBs occur, only the parent *Pst*I fragment should be observed. On the other hand, if the DSBs do occur, two additional smaller fragments, corresponding to DSB sites I and II, should appear. Figure 22.37b demonstrates that both these smaller fragments do indeed appear, both in wild-type cells (SPO11+) and in SPO11-HD cells.

Next, Kleckner and colleagues checked to see whether Spo11-HD was specifically bound to these fragments created by DSBs. They repeated the experiment we have just discussed, but this time they immunoprecipitated Spo11-HA–DNA complexes after cutting the DNA with *Pst*I. Figure 22.37c presents the results. It is clear that the two DNA fragments created by DSBs (but very little of the parental fragment) were immunoprecipitated along with Spo11-HA. However, they were not immunoprecipitated in the absence of the anti-HA antibody, nor were they precipitated from a yeast strain with the wild-type *SPO11* gene with no HA tag attached. Further analysis showed that these fragments were not immunoprecipitated from a wild-type *RAD50* strain that did not accumulate DSBs, nor from a mutant strain that did not form DSBs at all.

proteins bound to DNA under these denaturing conditions should be covalently attached. Then, they passed this mixture through a glass fiber filter that bound the DNA–protein complexes but allowed pure DNA to flow through. The material bound to the filter should be highly enriched in covalent DNA–protein complexes, so Kleckner and colleagues digested the DNA in these complexes with a nuclease and subjected the liberated proteins to SDS-PAGE. They observed several bands, two of which appeared in *rad50S* cells, but not in a *spo11Δ* mutant that was blocked in DSB formation. These same two bands appeared in preparative-scale, as well as pilot-scale prepara-

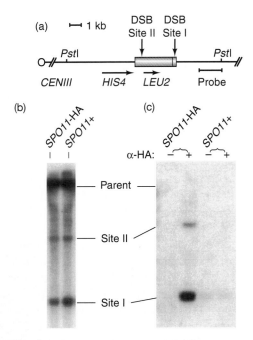

Figure 22.37   Association of Spo11 with DSB fragments. (a) Map of the hotspot region. A fragment (red and blue) with the *LEU2* gene (red) has been inserted adjacent to the *HIS4* gene in yeast chromosome III. The centromere (*CENIII*), the locations of the DSB sites I and II, and the site to which the Southern blot probe hybridizes are shown, along with the locations of two *Pst*I sites flanking the DSBs. **(b)** Southern blot of total DNA, cut with *Pst*I, electrophoresed, blotted, and hybridized to the probe. The parent fragment, as well as the subfragments generated by DSBs, are present. **(c)** Southern blot of DNA, cut with *Pst*I, then immunoprecipitated with an anti-HA antibody, then blotted and probed as in panel (b). The subfragments generated by DSBs are greatly enriched relative to the parent fragment. (*Source:* Keeney, S., Giroux, C., and Kleckner, N. Meiosis-specific DNA double-strand breaks are catalyzed by Spo11, a member of a widely conserved protein family. *Cell* 88 (Feb 1997) f. 3, p. 378. Reprinted by permission of Elsevier Science.)

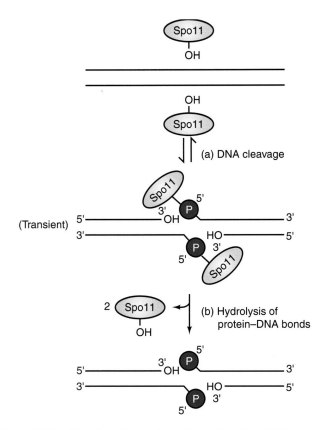

Figure 22.38   Model for the participation of Spo11 in DSB formation. (a) DNA cleavage. Two molecules of Spo11, with active site tyrosines represented by their OH groups, attack the two DNA strands at slightly offset positions. This transesterification reaction breaks phosphodiester bonds within the DNA strands and creates new phosphodiester bonds between the new DNA 5′-ends and the Spo11 tyrosines. **(b)** Hydrolysis of protein–DNA bonds. The phosphodiester bonds linking the enzyme molecules to the DNA ends are hydrolyzed, releasing the enzyme, and leaving a DSB.

If Spo11-HD merely bound nonspecifically to DNA, it should have been attached to the parental DNA fragment as well as the two subfragments created by DSBs, but the subfragments were enriched over 600-fold relative to the parental DNA in the immunoprecipitates. Thus, Spo11 appears to bind specifically to DSBs and is likely to be the catalytic part of the enzyme that created the DSBs. Furthermore, Spo11 is homologous to proteins in other organisms, including an archaebacterium, a fission yeast, and a roundworm. All four proteins have only one conserved tyrosine, which is likely to be the catalytic amino acid that becomes covalently attached to the DSB. In this way, it would resemble the active site tyrosine of a topoisomerase (see Chapter 23). The conserved tyrosine in Spo11 is Tyr-135 and, as expected, it is essential for activity. Accordingly, one can propose a model such as the one in Figure 22.38, which calls for the participation of two molecules of Spo11, one to attack each strand of the DNA at slightly offset positions. This process creates the DSB, and leaves a transient intermediate with a molecule of SPO11 covalently attached through its active site tyrosine to the newly created 5′-phosphate on each strand. Thus, the creation of DSBs appears not to occur by a simple hydrolysis, but by a **transesterification,** in which the attacking group is a tyrosine residue of the enzyme, rather than a water molecule.

The covalent association between Spo11 and the DSB ends is only transient, so the two molecules of Spo11 must be removed somehow. This process probably occurs by direct hydrolysis of the protein–DNA bonds, as illustrated in Figure 22.38. In principle, it could also occur by the action of an endonuclease that would remove the proteins along with a short stretch of DNA from each end. However, this latter mechanism seems much less likely, as we will see in the next section.

SUMMARY The DSB model of meiotic recombination in yeast begins with a double-strand break. DSBs can be directly observed in a plasmid DNA or in chromosomal DNA. DSBs accumulate in *rad50S* mutants, where Spo11 can be found covalently bound to the DSB ends. Two molecules of Spo11 collaborate to create DSBs by cleaving both strands at closely spaced sites. The cleavage operates through transesterification reactions involving active site tyrosines on the two molecules of Spo11.

Creation of Single-Stranded Ends at DSBs

Once Spo11 has created a DSB, the new 5′-ends are digested to yield free 3′-ends that can invade another DNA duplex. Szostak and coworkers first discovered the single-stranded ends in 1989 when they examined the structure of the DNA termini created by DSBs. They digested the DNAs with S1 nuclease, which specifically degrades single-stranded DNA, but leaves double-stranded DNA intact. The S1 nuclease had no effect on the intensities of the bands, but it reduced the lengths of all three DNAs. This result is exactly what the model in Figure 22.35 predicts: After the DSB occurs, an exonuclease digests the two 5′-ends at the break, creating single-stranded DNA that would then be susceptible to S1 nuclease digestion. This S1 nuclease digestion would reduce the length of the DNA fragment.

Kleckner and colleagues also found evidence for digestion, or resection, of one strand at DSBs. They discovered that the fragments that accumulated in wild-type cells produced diffuse bands on gel electrophoresis, as if the ends had been nibbled to varying extents. By contrast, the bands were discrete in *rad50S* mutants that blocked resection of the DSB ends.

Mutations in *RAD50, MRE11,* and *COM1/SAE2* obstruct DSB resection. Actually, null alleles of *RAD50* and *MRE11* block DSB formation altogether, so only certain nonnull alleles of these genes permit DSB formation but impede DSB resection. Thus, the products of both of these genes are required for both DSB formation and resection. Evidence for the role of these two gene products comes from a comparison with *E. coli*. The *E. coli* proteins SbcC and SbcD are homologous to yeast Rad50 and Mre11, respectively. Furthermore, the two bacterial proteins work as an SbcC/SbcD dimer, which has exonuclease activity on double-stranded DNA. This finding suggests that Rad50 and Mre11 also work together to resect the 5′-ends of DSBs in yeast.

SUMMARY Formation of the DSB in meiotic recombination is followed by 5′ → 3′ exonuclease digestion of the 5′-ends at the break, yielding overhanging 3′-ends that can invade another DNA duplex. Rad50 and Mre11 probably collaborate to carry out this resection.

22.4 Gene Conversion

When fungi such as pink bread mold (*Neurospora crassa*) sporulate, two haploid nuclei fuse, producing a diploid nucleus that undergoes meiosis to give four haploid nuclei. These nuclei then experience mitosis to produce eight haploid nuclei, each appearing in a separate spore. In principle, if one of the original nuclei contained one allele (*A*) at a given locus, and the other contained another allele (*a*) at the same locus, then the mixture of alleles in the spores should be equal: four *A*'s and four *a*'s. It is difficult to imagine any other outcome (five *A*'s and three *a*'s, for example), because that would require conversion of one *a* to an *A*. In fact, aberrant ratios *are* observed about 0.1% of the time, depending on the fungal species. This phenomenon is called **gene conversion.** We discuss this topic under the heading of recombination because the two processes are related.

The mechanism of meiotic recombination discussed in this chapter, and illustrated in Figure 22.35, suggests a mechanism for gene conversion during meiosis. Figure 22.39 depicts this hypothesis in the case of *Neurospora crassa*. We start with a nucleus in which DNA duplication has already occurred, so it contains four chromatids. In principle, two chromosomes should bear the *A* allele and two the *a* allele. But in this case, strand exchange and branch migration have occurred, followed by resolution yielding two chromosomes with patches of heteroduplex, as illustrated in Figure 22.39a. These heteroduplex regions just happen to be in the region where alleles *A* and *a* differ at one base, so each chromosome has one strand with one allele and the other strand with the other allele. If DNA replication occurred immediately, this situation would resolve itself simply, yielding two *A* duplexes and two *a* duplexes. However, before replication, one or both of the heteroduplexes may attract the enzymes that repair base mismatches. In the example shown here, only the top heteroduplex is repaired, with the *a* being converted to *A*. This leaves three strands with the *A* allele, and only one with the *a* allele. Now DNA replication will produce three *A* duplexes and only one *a* duplex. When we add the two *A* and two *a* duplexes resulting from the chromosomes that did not undergo heteroduplex formation, a final ratio of five *A*'s and only three *a*'s results.

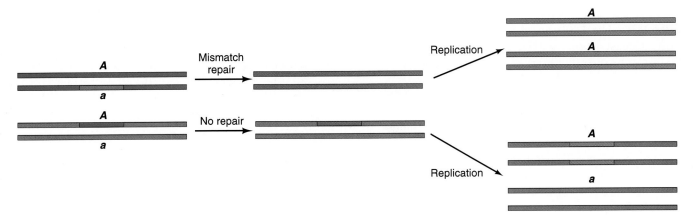

Figure 22.39   A model for gene conversion in sporulating *Neurospora*. A strand exchange event with branch migration during sporulation has resolved to yield two duplex DNAs with patches of heteroduplex in a region where a 1-base difference occurs between allele *A* (blue) and allele *a* (red). The other two daughter chromosomes are homoduplexes, one pure *A* and one pure *a* (not shown). The top heteroduplex undergoes mismatch repair to convert the *a* strand to *A*; the bottom heteroduplex is not repaired. Replication of the repaired DNA yields two *A* duplexes; replication of the unrepaired DNA yields one *A* and one *a*. Thus, the sum of the daughter duplexes pictured at right is three *A*'s and one *a*. Replication of the two DNAs not pictured yields two *A*'s and two *a*'s. The sum of all daughter duplexes is therefore five *A*'s and three *a*'s, instead of the normal four of each.

Figure 22.40 presents another pathway by which gene conversion can occur, this time without mismatch repair. We start with the situation after step (c) in Figure 22.35, just after strand invasion and D-loop formation. The region in which allele *A* differs from allele *a* is indicated at the top. Blue indicates *A* and red indicates *a*. This scheme differs from Figure 22.35 in that the invading strand is subject to partial resection, with shrinkage of the D loop, before repair synthesis begins. That resection allows a longer stretch of repair synthesis to occur, which converts more of the invading strand from *A* to *a*. And the conversion occurs in the exact region where alleles *A* and *a* differ. After branch migration and resolution (either crossover or noncrossover resolution), we have all four DNA strands representing the *a* allele in the significant region, whereas we started with two of each. Gene conversion has occurred.

Gene conversion is not confined to meiotic events. It is also the mechanism that switches the mating type of baker's yeast (*Saccharomyces cerevisiae*). This gene conversion event involves the transient interaction of two versions of the *MAT* locus, followed by conversion of one gene sequence to that of the other.

SUMMARY When two similar but not identical DNA sequences interact, the possibility exists for gene conversion—the conversion of one DNA sequence to that of the other. The sequences participating in gene conversion can be alleles, as in meiosis, on nonallelic genes, such as the *MAT* genes that determine mating type in yeast.

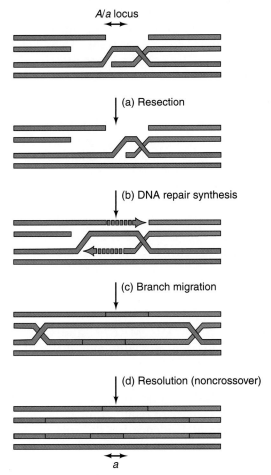

Figure 22.40   A model for gene conversion without mismatch repair. We begin in the middle of the DSB recombination scheme illustrated in Figure 22.35, just after strand invasion. **(a)** This time, the invading strand is partially resected, which causes partial collapse of the D loop. **(b)** DNA repair synthesis is more extensive because of the resection, which produces a region (indicated at top and bottom) in which all four DNA strands are allele *a* (red). **(c)** and **(d)** Branch migration and resolution do not change the nature of the four DNA strands in the region in which alleles *A* and *a* differ: All are allele *a*. Thus, this process has converted a DNA duplex that was allele *A* to allele *a*.

SUMMARY

Homologous recombination is essential to life. In eukaryotic meiosis, it locks homologous chromosomes together so they separate properly. It also scrambles parental genes in offspring. In all forms of life, homologous recombination helps to cope with DNA damage.

Homologous recombination by the RecBCD pathway in *E. coli* begins with invasion of a duplex DNA by a single-stranded DNA from another duplex. A free end can be generated by the nicking and helicase activities of RecBCD, which prefers to nick DNA at special sequences called χ sites. The invading strand becomes coated with RecA and SSB. RecA helps this invading strand pair with its complementary strand in a homologous DNA, forming a D loop. SSB accelerates the recombination process, apparently by melting secondary structure and preventing RecA from trapping any secondary structure that would inhibit strand exchange later in the recombination process. Subsequent nicking of the D-loop strand, probably by RecBCD, leads to the formation of a branched intermediate called a Holliday junction. Branch migration, catalyzed by the RuvA–RuvB helicase, brings the crossover of the Holliday junction to a site that is favorable for resolution. Finally, the Holliday junction can be resolved by RuvC, which nicks two of its strands. This can yield two DNAs with patches of heteroduplex (noncrossover recombinants), or two crossover recombinant DNAs.

Meiotic recombination in yeast begins with a double-strand break (DSB). Two molecules of Spo11 collaborate to create DSBs by cleaving both strands at closely spaced sites. The cleavage operates through transesterification reactions involving active site tyrosines on the two molecules of Spo11. Formation of the DSB in meiotic recombination is followed by $5' \rightarrow 3'$ exonuclease digestion of the 5'-ends at the break. Rad50 and Mre11 probably collaborate to carry out this resection. Next, the newly generated 3'-overhang invades the other DNA duplex, creating a D loop. DNA repair synthesis and branch migration yield two Holliday junctions that can be resolved to produce either noncrossover or crossover recombinants.

When two similar but not identical DNA sequences interact, the possibility exists for gene conversion—the conversion of one DNA sequence to that of the other. The sequences participating in gene conversion can be alleles, as in meiosis, or nonallelic genes, such as the *MAT* genes that determine mating type in yeast.

REVIEW QUESTIONS

1. Draw a diagram of a Holliday junction. Starting with that diagram, illustrate:
   a. Branch migration
   b. Resolution to yield a short heteroduplex
   c. Resolution to yield crossover recombinant DNAs

2. List the three steps in homologous recombination in which RecA participates, with a short explanation of each.

3. What evidence indicates that RecA coats single-stranded DNA? What role does SSB play in the interaction between RecA and single-stranded DNA?

4. Describe and give the results of an experiment that shows that RecA is required for synapsis at the beginning of recombination.

5. How would you show that the apparent synapsis you observe by electron microscopy is really synapsis, rather than true base pairing.

6. Describe and give the results of an experiment that shows that RecA is required for strand exchange in vitro.

7. Describe and give the results of an experiment that shows that RecBCD nicks DNA near a χ site. How could you demonstrate that it is RecBCD, and not a contaminant, that causes the nicking?

8. Describe and give the results of an experiment that shows that RecBCD has DNA helicase activity.

9. Show how you could use a gel mobility shift assay to demonstrate that RuvA can bind to a Holliday junction by itself at high concentration, that RuvB cannot bind by itself at all, and that RuvA and RuvB can bind cooperatively at relatively low concentrations. What is the function of glutaraldehyde in this experiment?

10. Show how you could use DNase footprinting to demonstrate that RuvA binds to the center of the Holliday junction, and that RuvB binds to the upstream side relative to the direction of branch migration.

11. Describe and give the results of an experiment that shows that branch migration occurs in the direction away from the RuvB protein in a RuvAB–Holliday junction complex.

12. Draw a diagram of the RuvAB–Holliday junction complex, with the Holliday junction in its familiar cross-shaped form, ready for branch migration. Include the RuvB rings, but not the RuvA tetramer.

13. Describe and give the results of an experiment that shows that RuvC can resolve a Holliday junction.

14. What evidence suggests that RuvA, B, and C are all together in a complex with a Holliday junction.

15. Present a model for meiotic recombination in yeast.

16. Describe and give the results of an experiment that shows that a DSB forms during meiotic recombination in yeast.

17. Describe and give the results of an experiment that shows that Spo11 is covalently attached to a DSB during meiotic recombination in yeast.

18. Present a model for meiotic gene conversion.

SUGGESTED READINGS

General References and Reviews

Fincham, J.R.S. and P. Oliver. 1989. Initiation of recombination. *Nature* 338:14–15.

McEntee, K. 1992. RecA: From locus to lattice. *Nature* 355:302–3.

Meselson, M. and C.M. Radding. 1975. A general model for genetic recombination. *Proceedings of the National Academy of Sciences USA* 72:358–61.

Roeder, G.S. 1997. Meiotic chromosomes: It takes two to tango. *Genes and Development* 11:2600–21.

Smith, G.R. 1991. Conjugational recombination in *E. coli*: Myths and mechanisms. *Cell* 64:19–27.

West, S.C. 1998. RuvA gets x-rayed on Holliday. *Cell* 94:699–701.

Research Articles

Cao, L., E. Alani, and N. Kleckner. 1990. A pathway for generation and processing of double-strand breaks during meiotic recombination in *S. cerevisiae*. *Cell* 61:1089–1101.

Dasgupta, C., T. Shibata, R.P. Cunningham, and C.M. Radding. 1980. The topology of homologous pairing promoted by RecA protein. *Cell* 22:437–46.

Dunderdale, H.J., F.E. Benson, C.A. Parsons, G.J. Sharples, R.G. Lloyd, and S.C. West. 1991. Formation and resolution of recombination intermediates by *E. coli* RecA and RuvC proteins. *Nature* 354:506–10.

Eggleston, A.K., A.H. Mitchell, and S.C. West. 1997. In vitro reconstitution of the late steps of genetic recombination in *E. coli*. *Cell* 89:607–17.

Hiom, K. and S.C. West. 1995. Branch migration during homologous recombination: Assembly of a RuvAB–Holliday junction complex in vitro. *Cell* 80:787–93.

Honigberg, S.M., D.K. Gonda, J. Flory, and C.M. Radding. 1985. The pairing activity of stable nucleoprotein filaments made from RecA protein, single-stranded DNA, and adenosine 5′-(γ-thio)triphosphate. *Journal of Biological Chemistry* 260:11845–51.

Keeney, S, C.N. Giroux, and N. Kleckner. 1997. Meiosis-specific DNA double-strand breaks are catalyzed by Spo11, a member of a widely conserved protein family. *Cell* 88:375–84.

McEntee, K., G.M. Weinstock, and I.R. Lehman. 1979. Initiation of general recombination catalyzed in vitro by the recA protein of *Escherichia coli*. *Proceedings of the National Academy of Sciences USA* 76:2615–19.

Parsons, C.A. and S.C. West. 1993. Formation of a RuvAB–Holliday junction complex in vitro. *Journal of Molecular Biology* 232:397–405.

Parsons, C.A., A. Stasiak, R.J. Bennett, and S.C. West. 1995. Structure of a multisubunit complex that promotes DNA branch migration. *Nature* 374:375–78.

Ponticelli, A.S., D.W. Schultz, A.F. Taylor, and G.R. Smith. 1985. Chi-dependent DNA strand cleavage by RecBC enzyme. *Cell* 41:145–51.

Radding, C.M. 1991. Helical interactions in homologous pairing and strand exchange driven by RecA protein. *Journal of Biological Chemistry* 266:5355–58.

Radding, C.M., J. Flory, A. Wu, R. Kahn, C. DasGupta, D. Gonda, M. Bianchi, and S.S. Tsang. 1982. Three phases in homologous pairing: Polymerization of *recA* protein on single-stranded DNA, synapsis, and polar strand exchange. *Cold Spring Harbor Symposia of Quantitative Biology* 47:821–28.

Rafferty, J.B., S.E. Sedelnikova, D. Hargreaves, P.J. Artymiuk, P.J. Baker, G.J. Sharples, A.A. Mahdi, R.G. Lloyd, and D.W. Rice. 1996. Crystal structure of DNA recombination protein RuvA and a model for its binding to the Holliday junction. *Science* 274:415–21.

Roman, L.J., D.A. Dixon, and S.C. Kowalczykowski. 1991. RecBCD-dependent joint molecule formation promoted by the *Escherichia coli* RecA and SSB proteins. *Proceedings of the National Academy of Sciences USA* 88:3367–71.

Rosamond, J., K.M. Telander, and S. Linn. 1979. Modulation of the action of the *recBC* enzyme of *Escherichia coli* K-12 by Ca²⁺. *Journal of Biological Chemistry* 254:8646–52.

Shah, R., R.J. Bennett, and S.C. West. 1994. Genetic recombination in *E. coli*: RuvC protein cleaves Holliday junctions at resolution hotspots in vitro. *Cell* 79:853–64.

Stasiak, A., I.R. Tsaneva, S.C. West, C.J.B. Benson, X. Yu, and E.H. Egelman. 1994. The *Escherichia coli* branch migration protein forms double hexameric rings around DNA. *Proceedings of the National Academy of Sciences USA* 91:7618–22.

Sun, H., D. Treco, N.P. Schultes, and J.W. Szostak. 1989. Double-strand breaks at an initiation site for meiotic gene conversion. *Nature* 338:87–90.

Yu, X., S.C. West, and E.H. Egelman. 1997. Structure and subunit composition of the RuvAB–Holliday junction complex. *Journal of Molecular Biology* 266:217–22.

CHAPTER 23

Site-Specific Recombination and Transposition

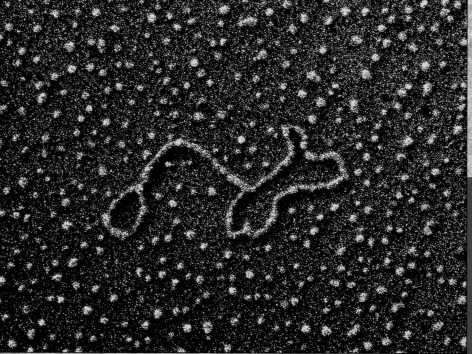

False color micrograph of a genetically engineered transposon DNA (×140,000)

In this chapter, we will consider two more types of recombination. Site-specific recombination requires much less sequence homology between recombining DNAs than does homologous recombination. It also almost always involves recombination between defined DNA sequences, which explains why we call it site-specific. A prominent example of this kind of recombination is the insertion of λ phage DNA into the *E. coli* host DNA, or excision of the λ DNA back out again. Transposition has little if any requirement for homology between recombining DNAs and is therefore not site-specific. Let us begin our discussion with site-specific recombination, and then move on to transposition.

23.1 Site-Specific Recombination

In contrast to homologous recombination, site-specific recombination requires only limited homology between the recombining species, and these species are usually targeted to specific sites. We will consider two examples: (1) integration of λ phage DNA into the *E. coli* host DNA, and the excision of the phage DNA out again during prophage induction; and (2) the inversion of a DNA element that switches the "phase," or type of protein in the flagella of *Salmonella typhimurium*.

Lambda Phage Integration and Excision

As we learned in Chapter 8, lysogeny of *E. coli* cells by phage λ requires the *int* gene, which encodes an integrase that can incorporate the λ genome into the host genome. This fusion of the two genomes is a site-specific recombination event that is outlined in Figure 23.1a. It requires two proteins: **Int**, the product of the λ *int* gene, and a host protein called **IHF** that is encoded by two host genes (*himA* and *hip*). The reverse reaction—excision of the λ genome (Figure 23.1b) occurs during induction of the prophage and requires IHF plus two λ proteins: the *int* product and the product (**Xis**) of another phage gene, *xis*.

The earliest evidence for the participation of the λ phage proteins was genetic. Ethan Signer, Harrison Echols, and others isolated λ mutants that were unable to form stable lysogens because they could not integrate their DNA into host bacterial DNA. They mapped these mutations to *int*. Then they found other mutants that could form lysogens, but these lysogens could not be induced because the phage DNA could not be excised from the host DNA. These mutations mapped to *xis*. On the other hand, IHF was discovered in biochemical experiments that showed that a cell-free extract of uninfected *E. coli* cells could stimulate integration of λ DNA into host DNA. Later, genetic experiments revealed the two genes that encode IHF.

The DNA Sites   Site-specific recombination, by definition, takes place at certain specific DNA sites. Genetic experiments first localized these **attachment sites** (***att sites***) as regions in phage and host DNA in which mutations could block proper insertion. The phage DNA site is called ***att*P** (P stands for "phage") and the host DNA site is called *att*B (B stands for "bacterial"). More detailed analysis, made possible by DNA sequencing, has shown that the sequence homology between *att*P and *att*B is actually confined to a small area in the middle of each sequence, called **O**. This required a more elaborate scheme for integration and excision, as presented in Figure 23.2. Here we can see the pairing between the O sequences in the two DNAs (initiated by protein binding) and how the O regions divide *att*P and *att*B into three regions: P, O, and P'; and B, O,

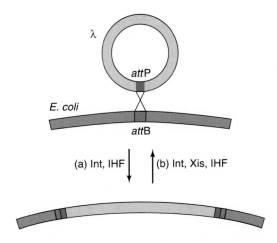

Figure 23.1   Integration and excision of λ DNA. (a) Integration. Int and IHF promote integration of the λ DNA (yellow) into the host DNA (blue). After the λ DNA circularizes by annealing of its cohesive ends, homologous regions of *att*P in the phage DNA and *att*B in the host DNA line up. Strand exchange and resolution then allow insertion of the phage DNA into the host DNA. **(b)** Excision. A reversal of the steps in panel **(a)** leads to excision of the λ DNA in circular form. This requires Int and IHF, as well as another phage product, Xis.

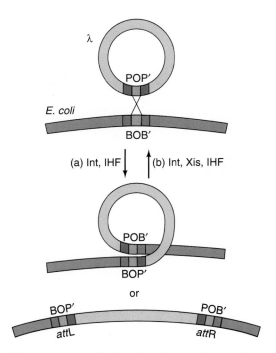

Figure 23.2   A more detailed model of integration and excision of λ DNA. This model is similar to that in Figure 23.1 except that it expands *att*P and *att*B to include three sections, P, O, and P′, and B, O, and B′, respectively. Recombination depends on sequence homology between the O regions (orange). The integrated form is drawn two ways: linearly (bottom), as in Figure 23.1, and coiled (middle) to emphasize the joining between the circular phage DNA and the segment of the much larger host DNA.

and B′, respectively. We also see that, after insertion, the recombinant *att* regions are called *att*L and *att*R, for the regions flanking the inserted prophage on the left and right, as written here. Deletion analysis in the *att* sites has shown that the sequence between positions −152 and +82 is required for full *att*P function. By contrast, *att*B function requires only the 15-bp common core plus about 5 bp on either side.

Footprinting analysis has shown where each of the three proteins, Int, Xis, and IHF, bind on *att*P and *att*B. (The latter attachment site is quite short, covering less than 40 bp, but the former encompasses much more DNA—almost 240 bp. A 15-bp core region is common between the two DNAs.

Mechanism   Site-specific recombination in λ phage-infected *E. coli* cells begins with binding of proteins, including Int, to *att*P. This complex then binds to *att*B on the bacterial DNA through interaction with the *att*P-bound Int. This process brings *att*P and *att*B together so they can per-

form strand exchange, or crossing over the first two strands to form a Holliday junction. Finally, resolution, or crossing over the other two DNA strands occurs, yielding the recombinant molecule(s). Of these, the strand exchange steps deserve special attention because they use a mechanism quite different from that of RecBCD recombination.

Unlike RecBCD recombination, site-specific recombination occurs in vitro in the absence of an energy source such as ATP. This means that the energy released by breaking phosphodiester bonds in DNA strands must be trapped somehow and used when new phosphodiester bonds are formed. There is precedent for such a mechanism: Topoisomerases (and Spo11) can trap the energy of phosphodiester bonds by forming a covalent protein–DNA bond with a terminal phosphate group created by transiently breaking a DNA strand. Then they contribute the energy of this protein–DNA bond to the formation of a new DNA–DNA phosphodiester bond, as illustrated in Figure 23.3a. This

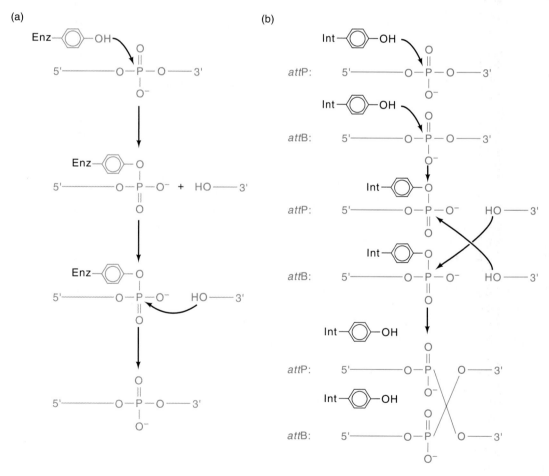

Figure 23.3   Covalent enzyme–DNA bonds in DNA cleavage reactions. (a) Topoisomerase. A tyrosine side chain on a topoisomerase attacks a phosphodiester bond, cleaving the bond, and forming a new phosphoester bond between the DNA's 3′-phosphate and the enzyme's tyrosine. The nick created by the topoisomerase is only transient, so the 5′-hydroxyl group of the right-hand DNA fragment can attack the bond between the enzyme and the left-hand fragment, restoring the original DNA phosphodiester bond. **(b)** λ integrase. Int is a topoisomerase, and it has a tyrosine residue that can break phosphodiester bonds. Here, two tyrosine side chains are breaking two homologous strands simultaneously. As the DNA phosphodiester bonds are broken, new bonds form between DNA and enzyme, conserving the bond energy. Now the right-hand fragments can cross over and attack the enzyme–DNA bonds on homologous fragments, creating a Holliday junction, and releasing the enzyme.

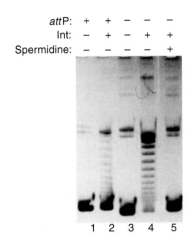

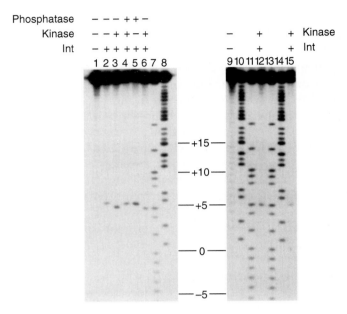

Figure 23.4   Assay for topoisomerase activity of Int. Kikuchi and Nash incubated Int with highly supercoiled plasmids, then electrophoresed the DNAs to detect relaxation of the supercoils. The reactions in lanes 1–5 contained a plasmid with an *att*P sequence (+), or no *att*P sequence (–) (pBR322 DNA), as indicated at top. The presence or absence of Int and spermidine is also indicated at top. (*Source:* Kikuchi and Nash, *PNAS* 76 (1979) f. 2a, p. 3761.)

Figure 23.5   Locating the site of Int cleavage in *att*B. Craig and Nash 3′-end-labeled a restriction fragment containing *att*B, then cleaved it with Int, as indicated at top. The fragments produced by Int were then treated with protein kinase or alkaline phosphatase (or both), as indicated at top, and subjected to gel electrophoresis. Lanes 7, 8, 10, 11, 13, and 14 are sequencing lanes generated by Maxam–Gilbert sequencing of the original labeled restriction fragment. The nucleotide sequence is numbered (–5 to +15) where nucleotide 0 lies in the middle of *att*B. (*Source:* Craig & Nash, The mechanism of phage lambda site-specific recombination: site-specific breakage of DNA by Int topoisomerase. *Cell* 35 (Dec 1983 Pt2) f. 4, p. 798. Reprinted by permission of Elsevier Science.)

has great relevance for site-specific recombination because the λ integrase is a type I topoisomerase. Thus, in principle, it could cleave two homologous DNA strands, form covalent bonds with two of the ends, then exit when new, intermolecular bonds are formed, as depicted in Figure 23.3b.

How do we know Int acts as a topoisomerase? Yoshiko Kikuchi and Howard Nash demonstrated in 1979 that it can relax superhelical DNA by nicking and resealing it—the characteristic reaction catalyzed by a topoisomerase. They mixed Int with plasmid DNAs containing negative supercoils, then electrophoresed them to detect relaxation. Highly supercoiled DNA has a higher electrophoretic mobility than relaxed DNA. Moreover, the loss of each superhelical turn is detectable as a decrease in electrophoretic mobility, so we see a ladder, in which each "rung" represents a DNA with a different degree of supercoiling. Adjacent rungs in the ladder differ by just one superhelical turn. Figure 23.4 shows what happened to superhelical DNA upon incubation with Int: It yielded a ladder including highly supercoiled, partially relaxed, and fully relaxed circles. Such a ladder is a signature of a topoisomerase. A simple nuclease would not give partially relaxed DNA molecules. The DNAs would either remain highly supercoiled, or, if nicked, would relax completely. This experiment also showed that the topoisomerase activity of Int is nonspecific. It relaxed a plasmid lacking an *att* sequence just as successfully as one containing an *att* sequence.

Although the topoisomerase activity of Int is not sequence-specific, its recombination activity is very specific. Nancy Craig and Nash showed that it cuts *att*B and *att*P in vitro within the homologous overlap region, precisely at those sites that break and rejoin during recombi-

nation in vivo. They performed this analysis by 3′-end labeling one of the strands of a DNA fragment bearing an *att*B sequence, then cleaving this labeled DNA with Int. After cleavage, they treated the DNA with proteinase K to remove any covalently bound Int and electrophoresed the product with markers generated by sequencing the same DNA fragment (Figure 23.5). They were not sure whether their labeled fragment contained a 5′-phosphate, so they treated it with polynucleotide kinase or alkaline phosphatase (or both) before electrophoresis. They found that treatment with polynucleotide kinase (and ATP) caused an increase in electrophoretic mobility, as we would expect from the addition of negative charge. Further treatment with alkaline phosphatase reduced the electrophoretic mobility to that of the initial DNA fragment. This behavior indicated that the fragment was originally produced with a free 5′-hydroxyl group, which can be phosphorylated by polynucleotide kinase. The phosphorylated product can be compared directly with the DNA fragments in the sequencing lanes, because they also contain 5′-phosphates. In this case, we can see that the cleavage in *att*B occurred between positions +4 and +5. This and similar experiments with the other strand in *att*B and both strands in *att*P allowed Craig and Nash to show that the sites of cleavage were the same as those observed in vivo.

Craig and Nash also showed that 5′-end-labeled fragments had to be treated with proteinase K before electrophoresis to yield a DNA product that would enter the gel. This suggested that a covalent bond existed between the DNA fragment and Int. Thus, the protein had to be degraded, or it would severely retard the electrophoretic mobility of the DNA. Presumably, the proteinase K left an amino acid or small peptide still attached to the DNA, but it did not interfere dramatically with the mobility of the DNA. But some uncertainty remained because the yield of labeled fragment was so small. The reason for this low yield is presumably that the fragment very quickly engages in recombination, so its lifetime, and therefore its concentration, is small.

Arthur Landy and his colleagues designed an experiment to circumvent this problem. They used a ^{32}P-labeled "suicide substrate" with a nick in the middle of the *att* region. Why is this called a suicide substrate? By cleaving it, Int introduces a second nick close to the first. This causes the substrate to fall apart and prevents recombination from progressing any further. Thus, the Int–DNA intermediate accumulates to a high enough concentration to permit careful analysis. Landy and colleagues verified Craig and Nash's finding that Int is bound covalently to its DNA substrate, and they went further to show that the linkage is a phosphotyrosine.

To demonstrate the phosphotyrosine linkage, these workers first hydrolyzed the labeled Int-DNA complex with P1 nuclease to remove DNA. This yielded ^{32}P-labeled Int, so the terminal labeled phosphate group from the DNA was still bound to Int. Next, they treated the ^{32}P-labeled Int with 6 N HCl for 2 h at 110° C. This procedure breaks peptide bonds and phosphonucleotide bonds, but, for the most part, leaves phosphotyrosine (or phosphoserine and phosphothreonine) bonds intact. Finally, they performed two-dimensional electrophoresis on the labeled product of this reaction, along with unlabeled phosphotyrosine, phosphoserine, and phosphothreonine markers. Figure 23.6 shows that the labeled product comigrated with the phosphotyrosine marker. Thus, as Int cleaves its DNA substrate, it binds to the newly formed 3′-DNA phosphate group through a tyrosine residue.

To find out which tyrosine, Landy and colleagues broke the ^{32}P-labeled Int into fragments with formic acid and showed that the C-terminal peptide had most of the label. Because this peptide contains two tyrosines, it was necessary to show which of these is the active one that binds to the substrate. Accordingly, Landy and colleagues mutagenized the *int* gene to convert each tyrosine separately to phenylalanine. When they tested the altered gene products, they found that changing Tyr-318 to Phe had no effect on the enzyme's activity, but changing Tyr-342 to Phe completely destroyed both topoisomerase and recombinase activity. Thus, Tyr-342 is very likely the active site tyrosine that attacks and breaks the phosphodiester bond in *att*, resulting in the transient covalent bond between enzyme and substrate.

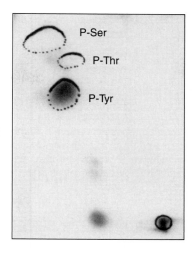

Figure 23.6 Demonstrating a phosphotyrosine linkage between Int and DNA. Landy and colleagues incubated Int with a ^{32}P-labeled suicide substrate, forming a covalent bond between enzyme and cleaved DNA substrate. They treated the enzyme–DNA adduct with P1 nuclease to degrade the DNA, then treated the ^{32}P-labeled enzyme with 6 N HCl to degrade the protein. This procedure yielded a phosphorylated amino acid, which represents the linkage between enzyme and DNA. To determine which amino acid is involved in this linkage, they performed two-dimensional electrophoresis on the labeled phosphoamino acid, along with unlabeled markers, and followed this with autoradiography to visualize the labeled species. The partially dotted outlines represent the positions of the indicated phosphoamino acids. The bulk of the label (dark spot) comigrated with phosphotyrosine. The circled spot at lower right marks the origin. Minor spots probably represent incomplete digestion products. (*Source:* Pargellis et al., Suicide recombination substrates yield covalent lambda integrase-DNA complexes and lead to identification of the active site tyrosine. *J Biol. Chem.* 263 no. 15 (5 Jun 1988) f. 3, p. 7680. American Society for Biochemistry and Molecular Biology.)

SUMMARY The mechanism of strand exchange catalyzed by Int is different from that of RecBCD homologous recombination. Int breaks specific phosphodiester bonds within *att*P and *att*B and forms a covalent bond between the new 3′-phosphate group on the DNA and tyrosine-342 in the enzyme. This conserves the energy of the broken phosphodiester bond, so it can be used to form a new phosphodiester bond with the homologous DNA during strand exchange.

Bacterial Use of Site-Specific Recombination

In 1922, F. W. Andrewes discovered that the intestinal bacterium *Salmonella typhimurium* could spontaneously change the protein in its **flagella,** the whip-like tails with which it propels itself. He could grow a colony of bacteria, all derived from one cell and all showing the same kind of flagellar protein, **flagellin.** Once this colony grew large enough, Andrewes isolated cells that grew up into new colonies. Most of these had the original type of

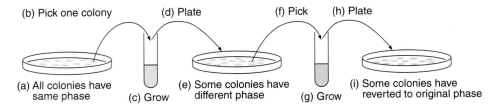

Figure 23.7   Detection of phase variation in *Salmonella*. (a) Start with a group of bacterial colonies, all showing the same phase. (b) Pick one colony, all of whose members came from a single cell and should be genetically identical; then (c) allow the cells to multiply in a test tube. (d) Plate the cells by spreading them on a new plate. (e) Some of the resulting colonies have a different phase (blue). (f) Pick one of the colonies showing the new phase, then (g) grow these cells in a test tube. (h) Plate these cells as before. (i) Some colonies have reverted to the original phase (yellow).

flagellin, but some had a new variety. Clearly, some of the bacteria had replaced one kind of protein in their flagella with another. This is referred to as a **phase variation** (Figure 23.7). Furthermore, a colony with the new type of flagellin, or **phase**, could give rise to subcolonies with the original phase. A shift in phase occurs about once every thousand cell divisions.

Joshua Lederberg and P. R. Edwards showed in 1953 that the two phases in *Salmonella* are caused by two widely separated genes, *H1* and *H2,* that code for two different flagellins. Later work showed that when *H2* is active, *H1* is repressed and the bacteria exhibit the phase 2 phenotype. When *H2* is inactive, *H1* is derepressed and the bacteria exhibit the phase 1 phenotype. Transcription of *H2* is important in this switch because *H2* is linked to another gene, *rh1,* whose product represses the *H1* gene. Thus *H2* and *rh1* are transcribed on the same mRNA; when one is expressed, the other is too. Furthermore, when DNA containing active *H2* and *rh1* genes was transferred to a cell with an active *H1* gene, the *H2* and *rh1* genes remained on, repressing *H1*. This could be done with pure DNA, which means that something about the active *H2* DNA itself, not any protein factors, dictates that the gene should be turned on.

Subsequent work has shown that the switch from active to inactive *H2* involves an inversion of a 995-bp segment of DNA, called the **H segment,** just upstream of the *H2* coding region (Figure 23.8). The H segment includes the *H2* promoter and the *hin* gene, which codes for the recombinase that catalyzes H segment inversion. When this DNA is in one orientation, the promoter is adjacent to the *H2* coding region, and transcription of *H2* and *rh1* occurs. However, when this DNA flips over, the promoter is moved hundreds of base pairs upstream and inverted so that it is aimed the wrong way. Transcription of *H2* and *rh1* becomes impossible. Now, because no repressor is made, transcription of *H1* occurs and the phase changes. Thus this switch in gene activity involves a repressor, but the primary switch depends on an inversion.

The inversion of the DNA in the *H2* flanking region can take place in cells that lack active genes for homologous recombination. This suggests that another kind of

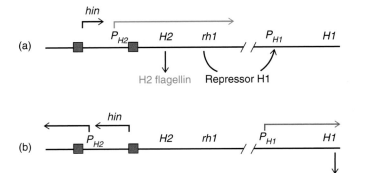

Figure 23.8   Molecular basis of phase variation. (a) With *hin* and the *H2* promoter (P_{H2}) in the rightward orientation, *H2* and *rh1* are expressed (blue arrow). The *H2* product is H2 flagellin (blue). The *rh1* product is a repressor that binds near the *H1* promoter (P_{H1}) and prevents expression of the *H1* gene. (b) With the *hin-H2* promoter segment (the H segment) inverted, *H2* and *rh1* cannot be expressed; thus, no H2 flagellin or H1 repressor is made. With no repressor, the *H1* gene is active (red arrow), so H1 flagellin is made (red). The red boxes denote inverted repeats at the borders of the invertible segment.

recombination operates in the phase variation. Furthermore, the H segment is flanked by two inverted repeats that contain the breakage sites used in the recombination event. Moreover, the H segment codes for a recombinase called **Hin** (for "H-inversion") that participates in the recombination, and inversion also depends on at least one host function. This is clear from in vitro experiments with purified DNA and recombinase in which inversion is greatly stimulated by crude extracts from cells that lack the recombinase. These extracts must provide the needed host function(s). The inversion of the H segment is independent of the homologous recombination system, but requires specific sequences that flank the segment. Thus, this is another example of site-specific recombination.

Controlling genes by inverting a DNA segment by site-specific recombination is not unique to the phase transition in *Salmonella*. Phage Mu has a G segment that contains two sets of phage tail fiber genes (Figure 23.9). When these segments are in one orientation, one of the

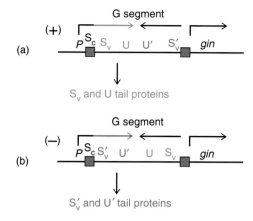

Figure 23.9 A model for G segment inversion in phage Mu. Transcription of the G segment begins at the promoter (*P*) to the left. (**a**) With the G segment in the G(+) orientation, genes *S* and *U* (blue) are next to the promoter and oriented left to right, so they are expressed (blue arrow). This gives S_V and U tail proteins (blue), which allow the phage to infect *E. coli* strain K12 but not C. (**b**) With the G segment in the G(−) orientation, genes S'_V and *U'* (red) are next to the promoter and oriented left to right, so they are expressed (red arrow). This gives S'_V and U' tail proteins (red), which allow the phage to infect *E. coli* strain C, but not K12. S_C refers to a constant region transcribed the same way regardless of the orientation of the invertible segment. The red boxes are inverted repeats at the borders of the invertible segment.

sets of tail fiber genes lies next to a promoter just outside the invertible segment, so this set is expressed. When the segment inverts, the other set of tail fiber genes is placed next to the promoter, so that set is expressed. Because the tail fibers are important in determining what strain of bacteria can be infected, this inversion changes the host range of the phage. Phages P1 and P7 have a similar system. All of the inversion mechanisms just described are very similar. The recombinases can be substituted for one another, and the inverted repeats have similar sequences.

SUMMARY The bacterium *Salmonella typhimurium* exists in two phases, depending on the kind of protein in its flagella. The two proteins are products of two genes: *H1* and *H2*. The activity of *H2* is regulated by a site-specific recombination event that causes inversion of a 995-bp piece of DNA (the H segment) just upstream of the gene. When this DNA is in one orientation, the gene is active. Furthermore, a companion gene, *rh1*, which codes for a repressor for the *H1* gene, is also active, so *H1* is repressed. When the H segment is inverted, *H2* and *rh1* are turned off, so *H1* can be expressed. Phages Mu, P1, and P7 take advantage of a related inversion scheme to switch from one kind of tail fiber protein to another.

23.2 Bacterial Transposons

The third major type of recombination is transposition, in which a **transposable element,** or **transposon,** moves from one DNA address to another. Barbara McClintock discovered transposons in the 1940s in her studies on the genetics of maize. Since then, transposons have been found in all kinds of organisms, from bacteria to humans. We will begin with a discussion of the bacterial transposons.

Discovery of Bacterial Transposons

James Shapiro and others provided the experimental background for the discovery of bacterial transposons with their discovery in the late 1960s of mutations that did not behave normally. For example, they did not revert readily the way point mutations do, and the mutant genes contained long stretches of extra DNA. Shapiro demonstrated this by taking advantage of the fact that a λ phage will sometimes pick up a piece of host DNA during lytic infection of *E. coli* cells, incorporating the "passenger" DNA into its own genome. He allowed λ phages to pick up either a wild-type *E. coli* galactose utilization gene (*gal*⁺) or its mutant counterpart (*gal*⁻), then measured the sizes of the recombinant DNAs, which contained λ DNA plus host DNA (Figure 23.10). He measured the DNA sizes by measuring the densities of the two types of phage using cesium chloride gradient centrifugation (Chapter 20). Because the phage coat is made of protein and always has the same volume, and because DNA is much denser than protein, the more DNA the phage contains the denser it will be. It turned out that the phages harboring the *gal*⁻ gene were denser than the phages with the wild-type gene and therefore held more DNA. The simplest explanation is that foreign DNA had inserted into the *gal* gene and thereby mutated it. Indeed, later experiments revealed 800–1400-bp inserts in the mutant *gal* gene, which were not found in the wild-type gene. In the rare cases when such mutants did revert, they lost the extra DNA. These extra DNAs that could inactivate a gene by inserting into it were the first transposons discovered in bacteria. They are called **insertion sequences (ISs).**

Insertion Sequences: The Simplest Transposons

Bacterial insertion sequences contain only the elements necessary for transposition. The first of these elements is a set of special sequences at a transposon's ends, one of which is the inverted repeat of the other. The second element is the set of genes that code for the enzymes that catalyze transposition.

Because the ends of an insertion sequence are inverted repeats, if one end of an insertion sequence is 5′-ACCGTAG,

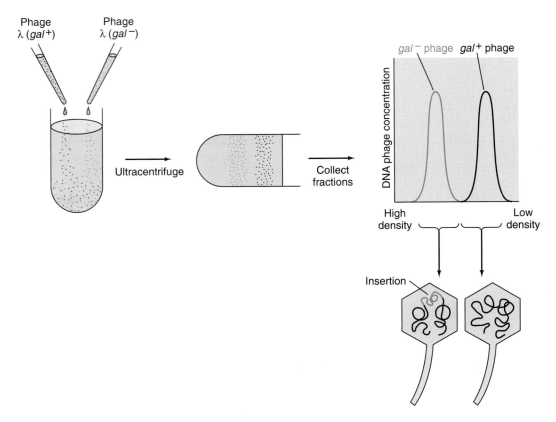

Figure 23.10   Demonstration of mutation by insertion. Lambda phage bearing a wild-type (*gal*⁺) or mutant (*gal*⁻) gene from *E. coli* was subjected to CsCl gradient centrifugation. Two bands of phage separated clearly, the denser (red) bearing the mutant phage. Because the denser phage must contain more DNA, the mutation in the *gal* gene must have been caused by an insertion of extra DNA.

the other end of that strand will be the reverse complement: CTACGGT-3′. The inverted repeats given here are hypothetical and are presented to illustrate the point. Typical insertion sequences have somewhat longer inverted repeats, from 15 to 25 bp long. IS1, for example, has inverted repeats 23 bp long. Larger transposons can have inverted repeats hundreds of base pairs long.

Stanley Cohen provided one graphic demonstration of inverted repeats at the ends of a transposon with the experiment illustrated in Figure 23.11. He started with a plasmid containing a transposon with the structure shown at the upper left in Figure 23.11a. The original plasmid was linked to the ends of the transposon, which were inverted repeats. Cohen reasoned that if the transposon really had inverted repeats at its ends, he could separate the two strands of the recombinant plasmid, and get the inverted repeats on one strand to base-pair with each other, forming a stem loop structure as shown on the right in Figure 22.11a. The stems would be double-stranded DNA composed of the two inverted repeats: the loops would be the rest of the DNA in single-stranded form. The electron micrograph in Figure 23.11b shows the expected stem loop structure.

The main body of an insertion sequence codes for at least two proteins that catalyze transposition. These proteins are collectively known as **transposase;** we will discuss their mechanism of action later in this chapter. We know that these proteins are necessary for transposition because mutations in the body of an insertion sequence can render that transposon immobile.

One other feature of an insertion sequence, shared with more complex transposons, is found just outside the transposon itself. This is a pair of short direct repeats in the DNA immediately surrounding the transposon. These repeats did not exist before the transposon inserted; they result from the insertion process itself and tell us that the transposase cuts the target DNA in a staggered fashion rather than with two cuts right across from each other. Figure 23.12 shows how staggered cuts in the two strands of the target DNA at the site of insertion lead automatically to direct repeats. The length of these direct repeats depends on the distance between the two cuts in the target DNA strands. This distance depends in turn on the nature of the insertion sequence. The transposase of IS1 makes cuts 9 bp apart and therefore generates direct repeats that are 9 bp long.

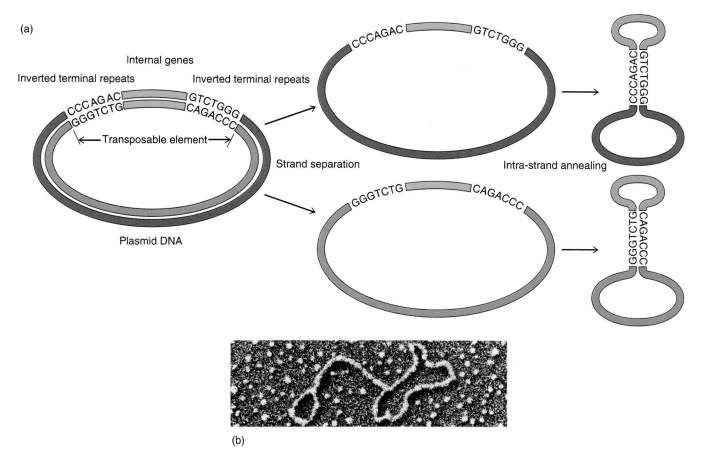

Figure 23.11   Transposons contain inverted terminal repeats. (a) Schematic diagram of experiment. The two strands of a transposon-bearing plasmid were separated and allowed to anneal with themselves separately. The inverted terminal repeats will form a base-paired stem between two single-stranded loops corresponding to the internal genes of the transposon (small loop, green) and host plasmid (large loop, purple and red). **(b)** Experimental results. The DNA was shadowed with heavy metal and subjected to electron microscopy. The loop-stem-loop structure is obvious. The stem is hundreds of base pairs long, demonstrating that the inverted terminal repeats in this transposon are much longer than the 7 bp shown for convenience in part **(a)**. (*Source:* (b) Courtesy Stanley N. Cohen, Stanford University.)

SUMMARY Insertion sequences are the simplest of the transposons. They contain only the elements necessary for their own transposition; short inverted repeats at their ends and at least two genes coding for an enzyme called transposase that carries out transposition. Transposition involves duplication of a short sequence in the target DNA; one copy of this short sequence flanks the insertion sequence on each side after transposition.

More Complex Transposons

Insertion sequences and other transposons are sometimes called "selfish DNA," implying that they replicate at the expense of their hosts and apparently provide nothing useful in return. However, some transposons do carry genes that are valuable to their hosts, the most familiar being genes for antibiotic resistance. Not only is this a clear benefit to the bacterial host, it is also valu-able to molecular biologists, because it makes the transposon much easier to track.

For example, consider the situation in Figure 23.13, in which we start with a donor plasmid containing a gene for kanamycin resistance (Kan^r) and harboring a transposon (**Tn3**) with a gene for ampicillin resistance (Amp^r); in addition, we have a target plasmid with a gene for tetracycline resistance (Tet^r). After transposition, Tn3 has replicated and a copy has moved to the target plasmid. Now the target plasmid confers both tetracycline and ampicillin resistance, properties that we can easily monitor by transforming antibiotic-sensitive bacteria with the target plasmid and growing these host bacteria in medium containing both antibiotics. If the bacteria survive, they must have taken up both antibiotic resistance genes; therefore, Tn3 must have transposed to the target plasmid.

Mechanisms of Transposition

Because of their ability to move from one place to another, transposons are sometimes called "jumping genes."

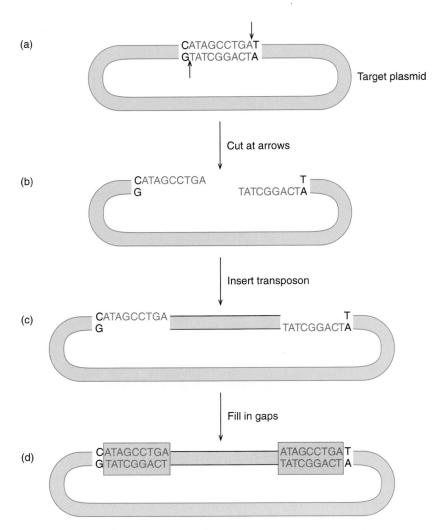

Figure 23.12 Generation of direct repeats in host DNA flanking a transposon. (a) The arrows indicate where the two strands of host DNA will be cut in a staggered fashion, 9 bp apart. (b) After cutting. (c) The transposon (yellow) has been ligated to one strand of host DNA at each end, leaving two 9-bp base gaps. (d) After the gaps are filled in, 9-bp repeats of host DNA (pink boxes) are apparent at each end of the transposon.

However, the term is a little misleading because it implies that the DNA always leaves one place and jumps to the other. This mode of transposition does occur and is called **nonreplicative transposition** because both strands of the original DNA move together from one place to the other without replicating. However, transposition frequently involves DNA replication, so one copy of the transposon remains at its original site as another copy inserts at the new site. This is called **replicative transposition** because a transposon moving by this route also replicates itself. Let us discuss how both kinds of transposition take place.

Replicative Transposition of Tn3 Tn3, whose structure is shown in Figure 23.14, illustrates one well-studied mechanism of transposition. In addition to the *bla* gene, which encodes ampicillin-inactivating β-**lactamase,** Tn3 contains two genes that are instrumental in transposition. Tn3 transposes by a two-step process, each step of which

requires one of the Tn3 gene products. Figure 23.15 shows a simplified version of the sequence of events. We begin with two plasmids; the donor, which harbors Tn3, and the target. In the first step, the two plasmids fuse, with Tn3 replication, to form a **cointegrate** in which they are coupled through a pair of Tn3 copies. This step requires recombination between the two plasmids, which is catalyzed by the product of the Tn3 transposase gene *tnpA.* Figure 23.16 shows a detailed picture of how all four DNA strands involved in transposition might interact to form the cointegrate. Figures 23.15 and 23.16 illustrate transposition between two plasmids, but the donor and target DNAs can be other kinds of DNA, including phage DNAs or the bacterial chromosome itself.

The second step in Tn3 transposition is a **resolution** of the cointegrate, in which the cointegrate breaks down into two independent plasmids, each bearing one copy of Tn3. This step, catalyzed by the product of the **resolvase** gene

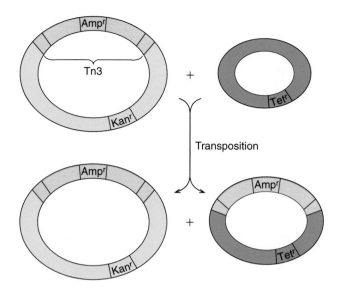

Figure 23.13   Tracking transposition with antibiotic resistance genes. We begin with two plasmids: The larger (blue) encodes kanamycin resistance (Kan^r) and bears the transposon Tn3 (yellow), which codes for ampicillin resistance (Amp^r); the smaller (green) encodes tetracycline resistance (Tet^r). After transposition, the smaller plasmid bears both the Tet^r and Amp^r genes.

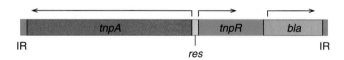

Figure 23.14   Structure of Tn3. The *tnpA* and *tnpR* genes are necessary for transposition; *res* is the site of the recombination that occurs during the resolution step in transposition; the *bla* gene encodes β-lactamase, which protects bacteria against the antibiotic ampicillin. This gene is also called Amp^r. Inverted repeats (IR) are found on each end. The arrows indicate the direction of transcription of each gene.

tnpR, is a recombination between homologous sites on Tn3 itself, called *res* sites. Several lines of evidence show that Tn3 transposition is a two-step process. First, mutants in the *tnpR* gene cannot resolve cointegrates, so they cause formation of cointegrates as the final product of transposition. This demonstrates that the cointegrate is normally an intermediate in the reaction. Second, even if the *tnpR* gene is defective, cointegrates can be resolved if a functional *tnpR* gene is provided by another DNA molecule—the host chromosome or another plasmid, for example.

Nonreplicative Transposition   Figures 22.15 and 23.16 illustrate the replicative transposition mechanism, but transposition does not always work this way. Some transposons (e.g., Tn10) move without replicating, leaving the donor DNA and appearing in the target DNA. How does this occur? It may be that nonreplicative transposition starts out in the same way as replicative transposition, by nicking

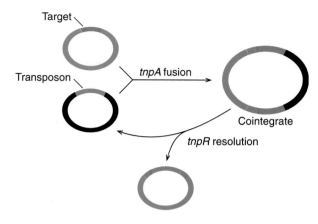

Figure 23.15   Simplified scheme of the two-step Tn3 transposition. In the first step, catalyzed by the *tnpA* gene product, the plasmid (black) bearing the transposon (blue) fuses with the target plasmid (green, target in red) to form a cointegrate. During cointegrate formation, the transposon replicates. In the second step, catalyzed by the *tnpR* gene product, the cointegrate resolves into the target plasmid, with the transposon inserted, plus the original transposon-bearing plasmid.

and joining strands of the donor and target DNAs, but then something different happens (Figure 23.17). Instead of replication occurring through the transposon, new nicks appear in the donor DNA on either side of the transposon. This releases the gapped donor DNA but leaves the transposon still bound to the target DNA. The remaining nicks in the target DNA can be sealed, yielding a recombinant DNA with the transposon integrated into the target DNA. The donor DNA has a double-stranded gap, so it may be lost or, as shown here, the gap may be repaired.

SUMMARY Many transposons contain genes aside from the ones necessary for transposition. These are commonly antibiotic resistance genes. For example, Tn3 contains a gene that confers ampicillin resistance. Tn3 and its relatives transpose by a two-step process (replicative transposition). First the transposon replicates and the donor DNA fuses to the target DNA, forming a cointegrate. In the second step, the cointegrate is resolved into two DNA circles, each of which bears a copy of the transposon. An alternative pathway, not used by Tn3, is conservative transposition, in which no replication of the transposon occurs.

23.3 Eukaryotic Transposons

It would be surprising if prokaryotes were the only organisms to harbor transposable elements, especially because these elements have powerful selective forces on their side.

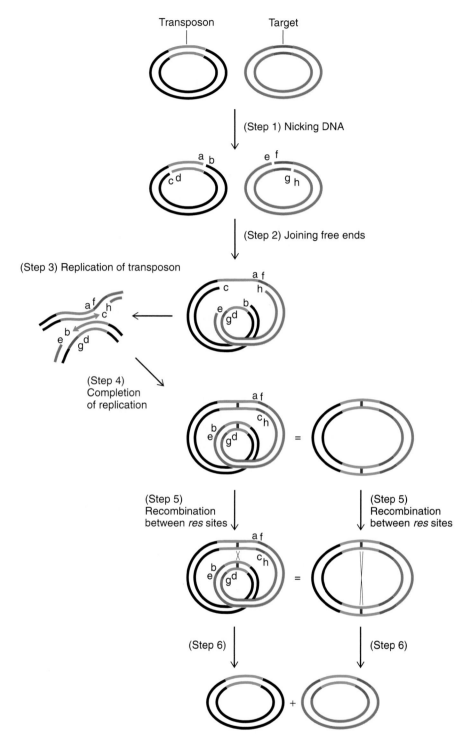

Figure 23.16   Detailed scheme of Tn3 transposition. Step 1: The two plasmids are nicked to form the free ends labeled a–h. Step 2: Ends a and f are joined, as are g and d. This leaves b, c, e, and h free. Step 3: Two of these remaining free ends (b and c) serve as primers for DNA replication, which is shown in a blowup of the replicating region. Step 4: Replication continues until end b reaches e and end c reaches h. These ends are ligated to complete the cointegrate. Notice that the whole transposon (blue) has been replicated. The paired *res* sites (purple) are shown for the first time here, even though one *res* site existed in the previous steps. The cointegrate is drawn with a loop in it, so its derivation from the previous drawing is clearer; however, if the loop were opened up, the cointegrate would look just like the one in Figure 23.15 (shown here at right). Steps 5 and 6: A crossover occurs between the two *res* sites in the two copies of the transposon, leaving two independent plasmids, each bearing a copy of the transposon. This process is shown for the two equivalent forms of the cointegrate, at right and at left.

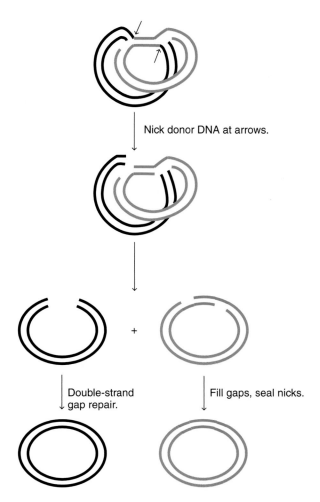

Figure 23.17 Nonreplicative transposition. The first two steps are just like those in replicative transposition, and the structure at the top is the same as that between steps 2 and 3 in Figure 23.16. Next, however, new nicks occur at the positions indicated by the arrows. This liberates the donor plasmid minus the transposon, which remains attached to the target DNA. Filling gaps and sealing nicks completes the target plasmid with its new transposon. The free ends of the donor plasmid may or may not join. In any event, this plasmid has lost its transposon.

First, many transposons carry genes that are an advantage to their hosts. Therefore, their host can multiply at the expense of competing organisms and can multiply the transposons along with the rest of their DNA. Second, even if transposons are not advantageous to their hosts, they can replicate themselves within their hosts in a "selfish" way. Indeed, transposable elements *are* also present in eukaryotes. In fact, they were first identified in eukaryotes.

The First Examples of Transposable Elements: *Ds* and *Ac* of Maize

Barbara McClintock discovered the first transposable elements in a study of maize (corn) in the late 1940s. It had been known for some time that the variegation in color

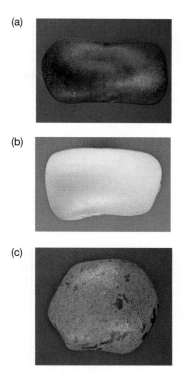

Figure 23.18 Effects of mutations and reversions on maize kernel color. (a) Wild-type kernel has an active *C* locus that causes synthesis of purple pigment. (b) The *C* locus has mutated, preventing pigment synthesis, so the kernel is colorless. (c) The spots correspond to patches of cells in which the mutation in *C* has reverted, again allowing pigment synthesis. (*Source:* F. W. Goro, from Fedoroff, N, *Scientific American* 86 (June 1984).)

observed in the kernels of so-called Indian corn was caused by an unstable mutation. In Figure 23.18a, for example, we see a kernel that is colored. This color is due to a factor encoded by the maize *C* locus. Figure 23.18b shows what happens when the *C* gene is mutated; no purple pigment is made, and the kernel appears almost white. The spotted kernel in Figure 23.18c shows the results of reversion in some of the kernel's cells. Wherever the mutation has reverted, the revertant cell and its progeny will be able to make pigment, giving rise to a dark spot on the kernel. It is striking that so many spots occur in this kernel. That means the mutation is very unstable: It reverts at a rate much higher than we would expect of an ordinary mutation.

In this case, McClintock discovered that the original mutation resulted from an insertion of a transposable element, called **Ds** for "dissociation," into the *C* gene (Figure 23.19a and b). Another transposable element, **Ac** for "activator," could induce *Ds* to transpose out of *C*, causing reversion (Figure 23.19c). In other words, *Ds* can transpose, but only with the help of *Ac*. *Ac* on the other hand, is an autonomous transposon. It can transpose itself and therefore inactivate other genes without help from other elements.

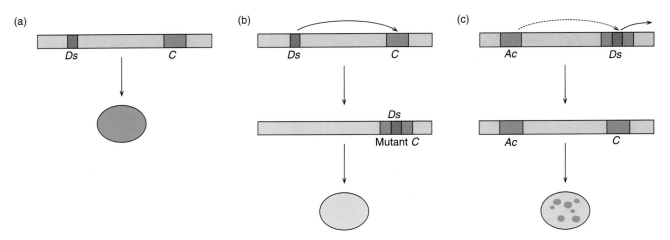

Figure 23.19   Transposable elements cause mutations and reversions in maize. (a) A wild-type maize kernel has an uninterrupted, active *C* locus (blue) that causes synthesis of purple pigment. (b) A *Ds* element (red) inserts into *C*, inactivating it and preventing pigment synthesis. The kernel is therefore colorless. (c) *Ac* (green) is present, as well as *Ds*. This allows *Ds* to transpose out of *C* in many cells, giving rise to groups of cells that make pigment. Such groups of pigmented cells account for the purple spots on the kernel. Of course, *Ds* must have transposed into *C* before it (*Ds*) became defective, or else it had help from an *Ac* element.

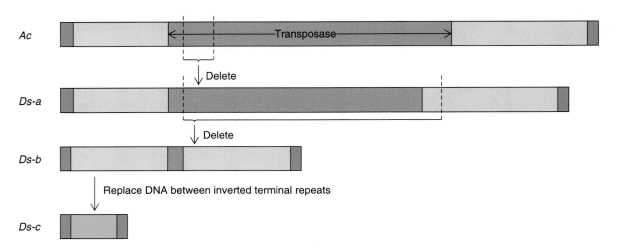

Figure 23.20   Structures of *Ac* and *Ds*. *Ac* contains the transposase gene (purple) and two imperfect inverted terminal repeats (blue). *Ds-a* is missing a 194-bp region from the transposase gene (dashed lines); otherwise, it is almost identical to *Ac*. *Ds-b* is missing a much larger segment of *Ac*. *Ds-c* has no similarity to *Ac* except for the inverted terminal repeats.

Now that molecular biological tools are available, we can isolate and characterize these genetic elements decades after McClintock found them. Nina Fedoroff and her collaborators obtained the structures of *Ac* and three different forms of *Ds*. *Ac* resembles the bacterial transposons we have already studied (Figure 23.20). It is about 4500 bp long, is bounded by short, imperfect inverted repeats, and contains a transposase gene. The various forms of *Ds* are derived from *Ac* by deletion. *Ds-a* is very similar to *Ac*, except that a piece of the transposase gene has been deleted. This explains why *Ds* is unable to transpose itself. *Ds-b* is more severely shortened, retaining only a small frag-

ment of the transposase gene, and *Ds-c* retains only the inverted repeats in common with *Ac*. These inverted repeats are all that *Ds-c* needs to be a target for transposition directed by *Ac*.

It is interesting that the first pea gene described by Mendel himself (*R* or *r*), which governs round versus wrinkled seeds, seems to involve a transposable element. We now know that the *R* locus encodes an enzyme (starch branching enzyme) that participates in starch metabolism. The wrinkled phenotype results from a malfunction of this gene; this mutation is in turn caused by an insertion of an 800-bp piece of DNA that seems to be a member of the *Ac/Ds* family.

SUMMARY The variegation in the color of maize kernels is caused by multiple reversions of an unstable mutation in the *C* locus, which is responsible for the kernel's coat color. The mutation and its reversion result from a *Ds* (dissociator) element, which transposes into the *C* gene, mutating it, and then transposes out again, causing it to revert to wild-type. *Ds* cannot transpose on its own; it must have help from an autonomous transposon called *Ac* (for activator), which supplies the transposase. *Ds* is an *Ac* element with more or less of its middle removed. All *Ds* needs in order to be transposed is a pair of inverted terminal repeats that the *Ac* transposase can recognize.

P Elements

The phenomenon called **hybrid dysgenesis** illustrates another obvious kind of mutation enhancement caused by a eukaryotic transposon. In hybrid dysgenesis, one strain of *Drosophila* mates with another to produce hybrid offspring that suffer so much chromosomal damage that they are dysgenic, or sterile. Hybrid dysgenesis requires a contribution from both parents; for example, in the **P-M system** the father must be from strain **P** (paternal contributing) and the mother must be from strain **M** (maternal contributing). The reverse cross, with an M father and a P mother, produces normal offspring, as do crosses within a strain (P × P or M × M).

What makes us suspect that a transposon is involved in this phenomenon? First of all, any P male chromosome can cause dysgenesis in a cross with an M female. Moreover, recombinant male chromosomes derived in part from P males and in part from M males usually can cause dysgenesis, showing that the P trait is carried on multiple sites on the chromosomes.

One possible explanation for this behavior is that the P trait is governed by a transposable element, and that is why we find it in so many different sites. In fact, this is the correct explanation, and the transposon responsible for the P trait is called the **P element.** Margaret Kidwell and her colleagues investigated the P elements that inserted into the *white* locus of dysgenic flies. They found that these elements had great similarities in base sequence but differed considerably in size (from about 500 to about 2900 bp). Furthermore, the P elements had direct terminal repeats and were flanked by short direct repeats of host DNA, both signatures of transposons. Finally, the *white* mutations reverted at a high rate by losing the entire P element—again a property of a transposon.

If P elements act like transposons, why do they transpose and cause dysgenesis only in hybrids? The answer is that the P element also encodes a repressor of transposition, which accumulates in the cytoplasm of the developing germ cells. Thus, in a cross of a P or M male with a P female, the female cytoplasm contains the repressor, which binds to any P elements and prevents their transposition. But in a cross of a P male with an M female, the early embryo contains no repressor, and none is made at first, because the P element becomes active only in developing germ cells. When the P element is finally activated, the transposase and repressor are both made, but the transposase alone goes to the nucleus, where it freely stimulates transposition.

Hybrid dysgenesis may have important consequences for speciation—the formation of new species that cannot interbreed. Two strains of the same species (such as P and M) that frequently produce sterile offspring will tend to become genetically isolated—their genes no longer mix as often—and eventually will be so different that they will not be able to interbreed at all. When this happens, they have become separate species.

P elements are now commonly used as mutagenic agents in genetic experiments with *Drosophila*. One advantage of this approach is that the mutations are easy to locate; we just look for the P element and it leads us to the interrupted gene. We saw an example of this approach in Chapter 10.

SUMMARY The P-M system of hybrid dysgenesis in *Drosophila* is caused by the conjunction of two factors: (1) a transposable element (P) contributed by the male, and (2) M cytoplasm contributed by the female, which allows transposition of the P element. Hybrid offspring of P males and M females therefore suffer multiple transpositions of the P element. This causes damaging chromosomal mutations that render the hybrids sterile. P elements have practical value as mutagenic agents in genetic experiments with *Drosophila*.

Rearrangement of Immunoglobulin Genes

Rearrangement of the mammalian genes that produce **antibodies,** or **immunoglobulins,** uses a process that closely resembles transposition. Even the recombinases involved in the two processes have similar structures.

As mentioned in Chapter 3, an antibody is composed of four polypeptides: two heavy chains and two light chains. Figure 23.21 illustrates an antibody schematically and shows the sites that combine with an invading antigen. These sites, called **variable regions,** vary from one antibody to the next and give these proteins their specificities; the rest of the molecule (the **constant region**) does not vary from one antibody to another within an antibody class, though some variation occurs between the few

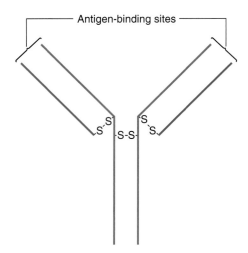

Figure 23.21   Structure of an antibody. The antibody is composed of two light chains (blue) bound through disulfide bridges to two heavy chains (red), which are themselves held together by a disulfide bridge. The antigen-binding sites are at the amino termini of the protein chains, where the variable regions lie.

classes of antibodies. Any given immune cell can make antibody with only one kind of specificity. Remarkably enough, humans have immune cells capable of producing antibodies to react with virtually any foreign substance we would ever encounter. That means we can make many millions of different antibodies.

Does this imply that we have millions of different antibody genes? That is an untenable hypothesis; it would place an impossible burden on our genomes to carry all the necessary genes. So how do we solve the antibody diversity problem? As unlikely as it may seem, a maturing B cell, a cell that is destined to make an antibody, rearranges its genome to bring together separate parts of its antibody genes. The machinery that puts together the gene selects these parts at random from heterogeneous groups of parts, rather like ordering from a Chinese menu ("Choose one from column A and one from column B"). This arrangement greatly increases the variability of the genes. For instance, if 40 possibilities are present in "column A" and 5 in "column B," the total number of combinations of A + B is 40×5 or 200. Thus, from 45 gene fragments, we can assemble 200 genes. And this is just for one of the antibody polypeptides. If a similar situation exists for the other, the total number of antibodies will be the product of the numbers of the two polypeptides. This description, though correct in principle, is actually an oversimplification of the situation in the antibody genes; as we will see, they have somewhat more complex mechanisms for introducing diversity, which lead to an even greater number of possible antibody products.

Studies on mammalian antibodies have revealed two families of antibody light chains called kappa (κ) and lambda (λ). Figure 23.22 illustrates the arrangement of the gene parts for a human κ light chain. "Column A" of this Chinese menu contains 40 variable region parts (V); "Column B" contains 5 **joining region** parts (J). The J segments actually encode the last 12 amino acids of the variable region, but they are located far away from the rest of the V region and close to a single constant region part. This is the situation in the germ cells, before the antibody-producing cells differentiate and before rearrangement brings the two unlinked regions together. The rearrangement and expression events are depicted in Figure 23.22.

First, a recombination event brings one of the V regions together with one of the J regions. In this case, V_3 and J_2 fuse together, but it could just as easily have been V_1 and J_4; the selection is random. After the two parts of the gene assemble, transcription occurs, starting at the beginning of V_3 and continuing until the end of C. Next, the splicing machinery joins the J_2 region of the transcript to C, removing the extra J regions and the intervening sequence between the J regions and C. It is important to remember that the rearrangement step takes place at the DNA level, but this splicing step occurs at the RNA level by mechanisms we studied in Chapter 14. The messenger RNA thus assembled moves into the cytoplasm to be translated into an antibody light chain with a variable region (encoded in both V and J) and a constant region (encoded in C).

Why does transcription begin at the beginning of V_3 and not farther upstream? The answer seems to be that an enhancer in the intron between the J regions and the C region activates the promoter closest to it: the V_3 promoter in this case. This also provides a convenient way of activating the gene after it rearranges; only then is the enhancer close enough to turn on the promoter.

The rearrangement of the heavy chain gene is even more complex, because there is an extra set of gene parts in between the V's and J's. These gene fragments are called D, for "diversity," and they represent a third column on our Chinese menu. Figure 23.23 shows that the heavy chain is assembled from 65 V regions, 27 D regions, and 6 J regions. On this basis alone, we could put together $65 \times 27 \times 6$, or 10,530 different heavy-chain genes. Furthermore, 10,530 heavy chains combined with 200 light chains yield more than 2 million different antibodies.

But there are even more sources of diversity. The first derives from the fact that the mechanism joining V, D, and J segments, which we call **V(D)J joining,** is not precise. It can add or delete bases on either side of the joining

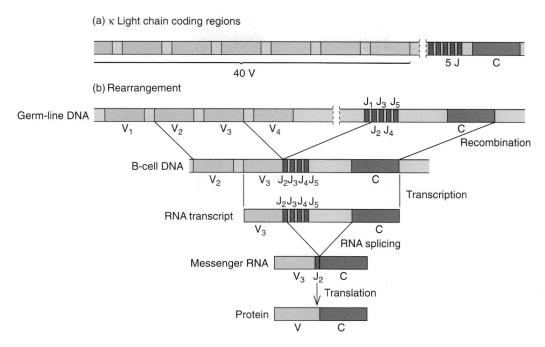

Figure 23.22   Rearrangement of an antibody light chain gene. (a) The human κ-antibody light chain is encoded in 40 variable gene segments (V; light green), five joining segments (J; red), and one constant segment (C; blue). (b) During maturation of an antibody-producing cell, a DNA segment is deleted, bringing a V segment (V_3, in this case) together with a J segment (J_2 in this case). The gene can now be transcribed to produce the mRNA precursor shown here, with extra J segments and intervening sequences. The material between J_2 and C is then spliced out, yielding the mature mRNA, which is translated to the antibody protein shown at the bottom. The J segment of the mRNA is translated into part of the variable region of the antibody.

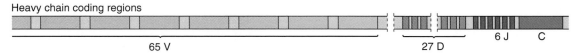

Figure 23.23   Structure of antibody heavy chain coding regions. The human heavy chain is encoded in 65 variable segments (V; light green), 27 diversity segments (D; purple), 6 joining segments (J; red), and 1 constant segment (C; blue).

site. This leads to extra differences in antibodies' amino acid sequences.

Another source of antibody diversity is somatic mutation, or mutation in an organism's somatic (nonsex) cells. In this case, the mutations occur in antibody genes, probably at the time that a clone of antibody-producing cells proliferates to meet the challenge of an invader. This offers an even wider variety of antibodies and, therefore, a better response to the challenge. Assuming that the extra diversity introduced by imprecise joining of antibody gene segments and somatic mutation is a factor of about 100, this brings the total diversity to at least hundreds of millions, surely enough different antibodies to match any attacker.

SUMMARY The immune systems of vertebrates can produce many millions of different antibodies to react with virtually any foreign substance. These immune systems generate such enormous diversity by three basic mechanisms: (1) assembling genes for antibody light chains and heavy chains from two or three component parts, respectively, each part selected from heterogeneous pools of parts; (2) joining the gene parts by an imprecise mechanism that can delete bases or even add extra bases, thus changing the gene; and (3) causing a high rate of somatic mutations, probably during proliferation of a clone of immune cells, thus creating slightly different genes.

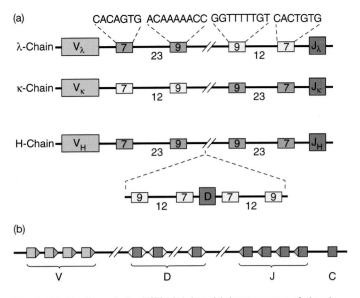

Figure 23.24 Signals for V(D)J joining. (a) Arrangement of signals around coding regions for immunoglobulin κ and λ light chain genes and heavy chain gene. Boxes labeled "7" or "9" are conserved heptamers or nonamers, respectively. Their consensus sequences are given at top. The 12-mer and 23-mer spacers are also labeled. Notice the arrangement of the 12 signals and 23 signals such that joining one kind to the other naturally allows assembly of a complete gene. **(b)** Schematic illustration of the arrangement of the 12 and 23 signals in an immunoglobulin heavy chain gene. The yellow triangles represent 12 signals, and the orange triangles represent 23 signals. Notice again how the 12/23 rule guarantees inclusion of one of each coding region (V, D, and J) in the rearranged gene. (*Source:* (a) Reprinted with permission from *Nature* 302:577, 1983. Copyright © 1983 Macmillan Magazines Limited.)

Recombination Signals How does the recombination machinery determine where to cut and paste to bring together the disparate parts of an immunoglobulin gene? Susumu Tonegawa examined the sequences of many mouse immunoglobulin genes (encoding κ and λ light chains, and heavy chains) and noticed a consistent pattern (Figure 23.24a): Adjacent to each coding region lies a conserved palindromic heptamer (7-mer), with the consensus sequence 5'-CACAGTG-3'. This heptamer is accompanied by a conserved nonamer (9-mer) whose consensus sequence is 5'-ACAAAAACC-3'. The heptamer and nonamer are separated by a nonconserved spacer containing either 12 bp (a **12 signal**) or 23 (±1) bp (a **23 signal**). The arrangement of these **recombination signal sequences (RSSs,** Figure 23.24b) is such that recombination always joins a 12 signal to a 23 signal. This **12/23 rule** stipulates that 12 signals are never joined to each other, nor are 23 signals joined to each other, and thus ensures that one, and only one, of each coding region is incorporated into the mature immunoglobulin gene.

Aside from the existence of consensus RSSs, what is the evidence for their importance? Martin Gellert and colleagues have systematically mutated the heptamer and nonamer by substituting bases, and the spacer regions by

adding or subtracting bases, and observed the effects of these alterations on recombination. They measured recombination efficiency in the following way: They built a recombinant plasmid with the construct shown in Figure 23.25. The first element in this construct is a *lac* promoter. This is followed by a 12 signal, then a prokaryotic transcription terminator, then a 23 signal, and finally a CAT reporter gene. They made mutations throughout these RSSs, then introduced the altered plasmids into a pre-B cell line. Finally, they purified the plasmids from the pre-B cells and introduced them into chloramphenicol-sensitive *E. coli* cells and tested them for chloramphenicol resistance. If no recombination took place, the transcription terminator prevented CAT expression, and therefore chloramphenicol resistance was almost nonexistent. On the other hand, if recombination between the 12 signal and the 23 signal occurred, the terminator was either inverted or deleted, and therefore inactivated. In that case, CAT expression occurred under control of the *lac* promoter, and many chloramphenicol-resistant colonies formed. This experiment showed that many alterations in bases in the heptamer or nonamer reduced recombination efficiency to background level. The same was true of insertions and deletions of bases in the spacer regions. Thus, all these elements of the RSSs are important in V(D)J recombination.

SUMMARY The recombination signal sequences (RSSs) in V(D)J recombination consist of a heptamer and a nonamer separated by either 12 bp or 23 bp spacers. Recombination occurs only between a 12 signal and a 23 signal, which guarantees that only one of each coding region is incorporated into the rearranged gene.

The Recombinase David Baltimore and his colleagues searched for the gene(s) encoding the V(D)J recombinase using a recombination reporter plasmid similar to the one we just discussed, but designed to operate in eukaryotic cells by conferring resistance to the drug mycophenolic acid. They introduced this plasmid, along with fragments of mouse genomic DNA, into NIH 3T3 cells, which lack V(D)J recombination activity, and tested for recombination by assaying for drug-resistant 3T3 cells. This led to the identification of a **recombination-activating gene (RAG-1)** that stimulated V(D)J joining activity in vivo.

However, the degree of stimulation by a genomic clone containing most of *RAG-1* was modest—no more than that obtained with whole genomic DNA. Furthermore, cDNA clones containing the whole *RAG-1* sequence did no better, so something seemed to be missing. Baltimore's group sequenced the whole genomic fragment containing most of *RAG-1* and found another whole gene

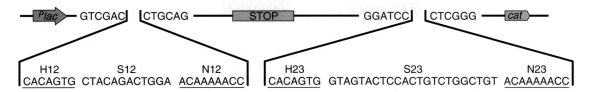

Figure 23.25   Structure of reporter construct used to measure effects of mutations in RSSs on recombination efficiency. Gellert and coworkers made a recombination reporter plasmid containing a *lac* promoter and CAT gene separated by an insert containing a transcription terminator flanked by a 12 signal and a 23 signal. Recombination between the two RSSs either inverts or deletes the terminator, allowing expression of CAT. Transformation of bacterial cells with the rearranged plasmid yields many CAT-producing colonies that are chloramphenicol-resistant. On the other hand, transformation of bacteria with the unrearranged plasmid yields almost no chloramphenicol-resistant colonies. (*Source:* From J. Hesse et al., "V(D)J recombination: a functional definition of the joining signals," *Genes & Development* 3:1053–1061, 1989. Copyright © 1989 Cold Spring Harbor Laboratory Press, Cold Spring Harbor, NY. Reprinted by permission.)

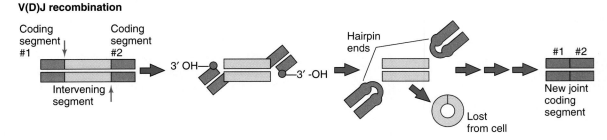

Figure 23.26   Mechanism of cleavage at RSSs. Nicking of opposite strands (vertical arrows) occurs at RSSs at the junctions between coding regions (red) and the intervening segment (yellow). The new 3′-hydroxyl groups (blue) attack and break the opposite strands, forming hairpins and releasing the intervening segment, which is lost. Finally, the hairpins open, and the two coding regions are joined by an imprecise mechanism. (*Source:* From Craig, *Science* 271:1512, 1996. Copyright © 1996 American Association for the Advancement of Science, Washington, DC. Reprinted by permission.)

tightly linked to it. They wondered whether this other gene might also have something to do with V(D)J joining, so they tested this genomic fragment plus a *RAG-1* cDNA in the same transfection experiment. When they introduced the two DNAs together into the same cell, they found many more drug-resistant cells. In this way, they discovered that two genes are responsible for V(D)J recombination, and they named the second *RAG-2*

RAG-1 and *RAG-2* are expressed only in pre-B and pre-T cells, where V(D)J joining of immunoglobulin and **T-cell receptor** gene segments, respectively, are occurring. The T-cell receptors are membrane-bound antigen-binding proteins with an architecture similar to that of the immunoglobulins. The genes encoding the T-cell receptors rearrange according to the same rules that apply to the immunoglobulin genes, complete with RSSs containing 12 signals and 23 signals. Thus, *RAG-1* and *RAG-2* are apparently involved in both immunoglobulin and T-cell receptor V(D)J joining.

Mechanism of V(D)J Recombination   V(D)J joining is imprecise, which contributes to the diversity of products from the process. Both loss of bases and addition of extra bases at the joints are frequently observed. This is good for immunoglobulin and T-cell receptor production, be-

cause it adds to the variety of proteins that can be made from a limited repertoire of gene segments.

How do we explain this imprecision? Figure 23.26 illustrates the mechanism of cleavage at the RSSs flanking an intervening segment between two coding segments. We see that the products of the *RAG-1* and *RAG-2* genes, **Rag-1** and **Rag-2,** respectively, first nick the DNAs at the joints. Then the new 3′-hydroxyl groups attack phosphodiester bonds on the complementary strands, liberating the intervening segment and forming hairpins at the ends of the coding segments. These hairpins are the key to the imprecision of joining; they can open up on either side of the apex of the hairpin, and bases can then be added or subtracted to make the DNA ends blunt for joining. The Rag-1 and Rag-2 proteins hold both hairpins together in a complex so they can join covalently with each other.

How do we know hairpins form? They were first found in vivo, but in very low concentration. Gellert and his colleagues later developed an in vitro system in which they could be readily observed. Figure 23.27a illustrates one of the labeled substrates these workers used. It was a 50-mer labeled at one 5′-end with ^{32}P. It contained a 12 signal, represented by a yellow triangle, flanked by a 16-bp segment on the left; the right-hand end of the fragment was, therefore, a 34-bp segment, which included the

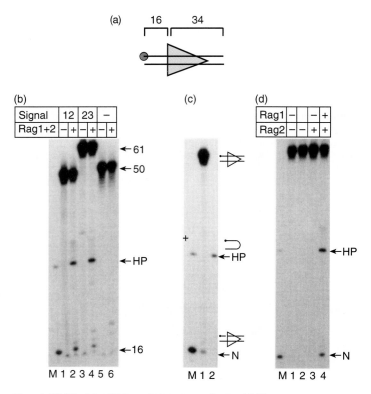

Figure 23.27   Identifying cleavage products. (a) Cleavage substrate. Gellert and colleagues constructed this labeled 50-mer, which included 16 bp of DNA on the left, then a 12 signal (yellow), included in a 34-bp segment on the right. The single 5′-end label is indicated by the red dot. These workers also made an analogous 61-mer substrate with a 23 signal. **(b)** Identifying the hairpin product. Gellert and coworkers incubated Rag-1 and Rag-2 proteins, as indicated at top, with either the labeled 12-signal or 23-signal substrate, also as indicated at top. After the incubation, they subjected the products to nondenaturing gel electrophoresis and autoradiographed the gel to detect the labeled products. The positions of the 61-mer and 50-mer substrates, the hairpin (HP), and the 16-mer are indicated at right. **(c)** Identifying the products from a nondenaturing gel. Gellert and colleagues recovered the labeled products (apparently uncleaved 50-mer substrate and 16-mer fragment) from the bands of a nondenaturing gel. They then electrophoresed these DNAs again in lanes 1 and 2, respectively, of a denaturing gel, along with markers (identified with diagrams at right) corresponding to the uncleaved substrate, the 16-bp hairpin (HP), and the single-stranded 16-mer released by denaturing the nicked substrate. **(d)** Requirement for Rag-1 and Rag-2. This experiment was very similar to the one in panel (b) except that the presence of Rag-1 and Rag-2 proteins (indicated at top) were the only variables. "N" denotes the position of the 16-mer released from the nicked species. (*Source:* McBlane et al., Cleavage at a V(D)J recombination signal requires only RAG1 and RAG2 proteins and occurs in two steps. *Cell* 83 (3 Nov 1985) f. 4 a-c, p. 390. Reprinted by permission of Elsevier Science.)

12 signal. A similar substrate contained the same flanking segments, but had a 23 signal instead of a 12 signal. Thus, it was 61 bp long.

Gellert and colleagues incubated these substrates with Rag-1 and Rag-2 proteins, then electrophoresed the products under nondenaturing conditions to see if any

DNA cleavages had occurred (Figure 23.27b). They found a 16-mer, demonstrating that a double-stranded cleavage had occurred. However, nondenaturing gel electrophoresis could not distinguish between a true double-stranded 16-mer and a 16-mer with a hairpin end, so these workers subjected the same products to denaturing polyacrylamide gel electrophoresis in the presence of urea and at an elevated temperature (Figure 23.27c). Under these conditions, a double-stranded 16-mer would give rise to two single-stranded 16-mers. On the other hand, a 16-mer with a hairpin at the end would give rise to a single-stranded 32-mer. This is what Gellert and coworkers observed whenever the DNA contained either a 12 signal or a 23 signal and both Rag-1 and Rag-2 proteins were present. A DNA with no 12 or 23 signal gave no product, hairpin or otherwise, and reactions lacking either Rag-1 or Rag-2 protein gave no product (Figure 23.27d). Thus, Rag-1 and Rag-2 recognize both the 12 signal and the 23 signal and cleave the DNA adjacent to the signal, forming a hairpin at the end of the coding segment.

Moreover, the 16-mer product from the nondenaturing gel yielded only hairpin product on the denaturing gel, demonstrating that no simple double-stranded 16-mer formed. But labeled DNA migrating with the substrate in the nondenaturing gel yielded a small amount of 16-mer in the denaturing gel. This cannot have come from a double-stranded break, or it would not have remained with the substrate in the nondenaturing gel. Thus, it must have come from a nick in the labeled strand. The 16-mer created by the nick would have remained base-paired to its partner during nondenaturing electrophoresis, but would have migrated independently as a 16-mer during denaturing electrophoresis. Thus, single-stranded nicking is apparently also part of the action of Rag-1 and Rag-2 proteins.

To investigate further the relationship between nicking and hairpin formation, Gellert and colleagues ran a time-course study in which they incubated the substrate for increasing lengths of time with Rag-1 and Rag-2 proteins and then subjected the products to denaturing gel electrophoresis. They found that the nicked species appeared first, followed by the hairpin species. This suggested that the nicked species is a precursor of the hairpin species. To test this hypothesis, they created nicked intermediates and incubated them with Rag-1 and Rag-2. Sure enough, the Rag-1 and Rag-2 converted the nicked DNAs to hairpins. Subsequent work by Gellert's group has shown the sequence of events seems to be: Rag-1 and Rag-2 nick one DNA strand adjacent to a 12 signal or a 23 signal; then the newly formed hydroxyl group attacks the other strand in a transesterification reaction, forming the hairpin, as was illustrated in Figure 23.26.

SUMMARY Rag-1 and Rag-2 introduce single-strand nicks into DNA adjacent to either a 12 signal or a 23 signal. This leads to a transesterification in which the newly created 3'-hydroxyl group attacks the opposite strand, breaking it, and forming a hairpin at the end of the coding segment. The hairpins then break in an imprecise way, allowing joining of coding regions with loss of bases or gain of extra bases.

Retrotransposons

McClintock's maize transposons are examples of so-called cut-and-paste or copy-and-paste transposons, similar to the bacterial transposons we discussed earlier in this chapter. If DNA replication is involved, it is direct replication. Humans also carry transposons in this class, which constitute about 1.6% of the human genome. The most prevalent example is called **mariner,** but all of the mariner elements studied so far have been defective in transposition. Eukaryotes also carry many more transposons of another kind: **retrotransposons,** which replicate through an RNA intermediate. In this respect, the retrotransposons resemble **retroviruses,** some of which cause tumors in vertebrates, and some of which (the human immunodeficiency viruses, or HIVs) cause AIDS. As an introduction to the replication scheme of the retrotransposons, let us first examine the replication of the retroviruses.

Retroviruses The most salient feature of a retrovirus, indeed the feature that gives this class of viruses its name, is its ability to make a DNA copy of its RNA genome. This reaction, RNA → DNA, is the reverse of the transcription reaction, so it is commonly called **reverse transcription.** In 1970, Howard Temin and, simultaneously, David Baltimore convinced a skeptical scientific community that this reaction takes place. They did so by finding that the virus particles contain an enzyme that catalyzes the reverse transcription reaction. Inevitably, this enzyme has been dubbed **reverse transcriptase.** A more proper name is **RNA-dependent DNA polymerase.**

Figure 23.28 illustrates the retrovirus replication cycle. We start with a virus infecting a cell. The virus contains two copies of its RNA genome, linked together by base pairing at their 5'-ends (for simplicity, only one copy is shown). When the virus enters a cell, its reverse transcriptase makes a double-stranded DNA copy of the viral RNA, with **long terminal repeats (LTRs)** at each end. This DNA recombines with the host genome to yield an integrated form of the viral genome called the **provirus.** The host RNA polymerase II transcribes the provirus, yielding viral mRNAs, which are then translated to viral proteins.

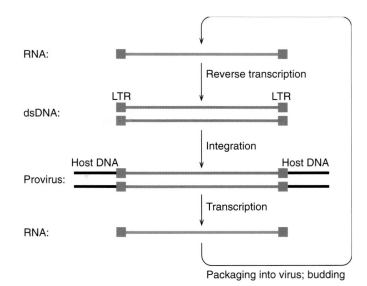

Figure 23.28 Retrovirus replication cycle. The viral genome is an RNA, with long terminal repeats (LTRs, green) at each end. Reverse transcriptase makes a linear, double-stranded DNA copy of the RNA, which then integrates into the host DNA (black), creating the provirus form. The host RNA polymerase II transcribes the provirus, forming genomic RNA. The viral RNA is packaged into a virus particle, which buds out of the cell and infects another cell, starting the cycle over again.

To complete the replication cycle, polymerase II also makes RNA copies of the provirus, which are new viral genomes. These genomic RNAs are packaged into virus particles (Figures 23.29 and 23.30) that bud out of the infected cell and go on to infect other cells.

Evidence for Reverse Transcriptase The skepticism about the reverse transcription reaction arose from the fact that no one had ever observed it, and the notion that it violated the "central dogma of molecular biology" promulgated by Watson and Crick, which said that the flow of genetic information is from DNA to RNA to protein, not the reverse. Crick has since stated that the DNA→RNA arrow was intended to be double-headed, but that was clearly not the popular perception at the time. What evidence did Baltimore and Temin bring to bear to dispel this skepticism?

Figure 23.31 shows the result of one of Baltimore's experiments. He incubated purified retrovirus particles (Raucher mouse leukemia virus, or R-MLV) with all four dNTPs, including [3H]dTTP, then measured the incorporation of the labeled TTP into a polymer (DNA) that could be precipitated with acid. He observed a clear incorporation (red curve) that could be inhibited by including RNase in the reaction (blue curve), and inhibited even more by preincubating with RNase (green curve). This sensitivity to RNase was compatible with the hypothesis that RNA is the template in the reverse transcription reaction.

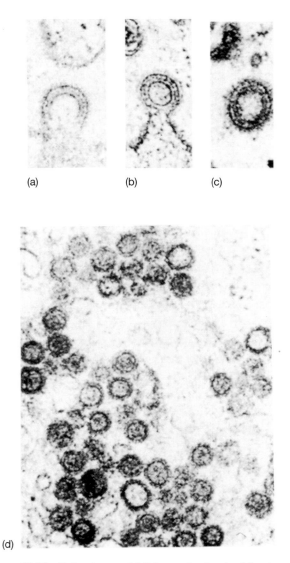

(a)   (b)   (c)

(d)

Figure 23.29   Retroviruses. (a) B-type retrovirus budding from a mouse mammary tumor cell. **(b)** Later in the budding process. **(c)** Free, but still immature B-type retrovirus. **(d)** Intracellular, A-type particles in a mouse mammary tumor. (*Source:* T. H. Maugh, RNA virus: The age of innocence ends. *Science* 183 (22 Mar 1974) p. 181–1185. © AAAS.)

Baltimore also examined the product of the reaction and showed that it was insensitive to RNase and base hydrolysis, but sensitive to DNase. Furthermore, the virions could support the incorporation of dNTPs only. Ribonucleotides, including ATP, could not be incorporated. Thus, the product behaved like DNA, and the enzyme behaved like an RNA-dependent DNA polymerase—a reverse transcriptase. Baltimore and Temin both performed similar experiments on Rous sarcoma virus particles, with very similar results. Thus, it appeared that all RNA tumor viruses probably contained reverse transcriptase and behaved according to the provirus hypothesis. This has proven to be true.

Evidence for a tRNA Primer   As molecular biologists began to investigate the molecular biology of reverse tran-

scription, they discovered that the viral reverse transcriptase is like every other DNA polymerase known: It requires a primer. In 1971, Baltimore and colleagues found RNA primers attached to the 5′-ends of nascent reverse transcripts using the following strategy: They labeled the nascent reverse transcripts in avian myeloblastosis virus (AMV) by the same method Baltimore and Temin had used—incubating virus particles with labeled dNTPs. Then they subjected the products to Cs_2SO_4 gradient ultracentrifugation to separate RNA from DNA based on their densities (RNA being denser than DNA).

In the first experiment, Baltimore and colleagues isolated the nucleic acids from the virus particles and subjected them immediately to ultracentrifugation. Figure 23.32a shows the results: a peak of labeled DNA that appeared to have the density of RNA. This finding is consistent with the hypothesis that the nascent DNA is still base-paired to the much bigger RNA template, so the whole complex behaves like RNA. If this hypothesis is true, then heating the RNA–DNA hybrid should denature it and release the DNA product as an independent molecule. When Baltimore and colleagues performed that experiment, they observed the behavior in Figure 23.32b: Now the nascent DNA product had a density much closer to that of DNA, but still a little too dense, as if there were still some RNA attached.

That behavior could be explained if the nascent DNA still had an RNA primer covalently attached to it. To check this possibility, Baltimore and coworkers treated the nascent DNA wtih RNase and again subjected it to ultracentrifugation. This time, the density of the product behaved exactly as expected for pure DNA. Thus, the nascent reverse transcript appears to be primed by RNA. But what RNA?

In the process of making an inventory of all the molecules within the retrovirus particle, molecular biologists had discovered some tRNAs, one of which, host tRNA[Trp], appeared to be partially base-paired to the viral RNA. Could this be the primer? If so, it should bind to the reverse transcriptase. To see if it does, Baltimore, James Dahlberg, and colleagues labeled host tRNA[Trp], or the tRNA[Trp] from virus particles, with ^{32}P and mixed these labeled tRNAs with AMV reverse transcriptase. Then they subjected these mixtures to gel filtration on Sephadex G-100 (Chapter 5). By itself, tRNA[Trp] was included in the gel and eluted in a peak centered at about fraction #25. However, both host and virion tRNAs, when mixed with reverse transcriptase eluted with the enzyme in a peak centered at about fraction #20. Thus, this reverse transcriptase binds tRNA[Trp]. Together with the data we have already discussed, the binding data strongly suggest that tRNA[Trp] serves as the primer for this enzyme. The virus does not encode a tRNA, so the primer must be picked up from the host cell.

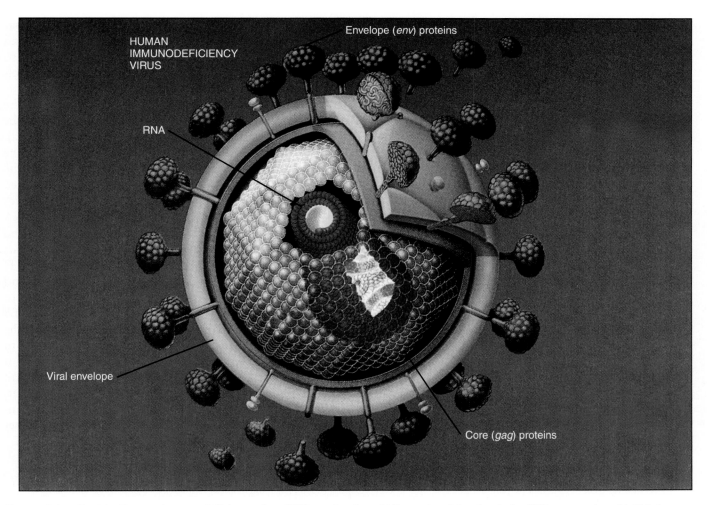

Figure 23.30 Model of human immunodeficiency virus (HIV), a retrovirus. At the center of the virus is the RNA genome (purple). This is surrounded by a coat made of core proteins (gray), which are products of the *gag* gene. The viral core is enclosed in an envelope (yellow) derived from the cellular plasma membrane. This envelope is decorated with envelope proteins (green), which are products of the viral *env* gene. Homologous genes are found in RNA tumor viruses, and the RNA tumor virus particles can be represented by this same model. (*Source:* Courtesy Coulter Corporation.)

The Mechanism of Retrovirus Replication The initial product of reverse transcription in vitro is a short piece of DNA called **strong-stop DNA.** The reason for the strong-stop is obvious when we consider the site on the viral RNA to which the tRNA primer hybridizes (the **primer-binding site,** or **PBS**). It is only about 150 nt (depending on the retrovirus) from the 5′-end of the viral RNA. This means that the reverse transcriptase will synthesize DNA for just 150 nt or so before reaching the end of the RNA template and stopping. This raises the interesting question: What happens next?

That question is related to another paradox of retro-virus replication, illustrated in Figure 23.33. The provirus is longer than the viral RNA, yet the viral RNA serves as the template for making the provirus. In particular, the LTRs in the viral RNA are incomplete. The left LTR contains a redundant region (**R**) plus a 5′-untranslated region (**U5**), whereas the right LTR contains an R region plus a

3′-untranslated region (**U3**). How can the provirus have complete LTRs on each end while its template is missing a U3 region at its left end and a U5 region at its right end? Harold Varmus proposed an answer based on the important fact that reverse transcriptase has another distinct activity: an RNase activity. The RNase inherent in reverse transcriptase is **RNase H,** which specifically degrades the RNA part of an RNA–DNA hybrid.

Varmus's hypothesis is illustrated in Figure 23.34 First, (a) the reverse transcriptase uses the tRNA to prime synthesis of strong-stop DNA. This appears at first to be the end of the line, but then (b) RNase H recognizes a stretch of RNA hybrid between the strong-stop DNA and the RNA template, and degrades the R and U5 parts of the RNA. The removal of this RNA leaves a tail of DNA (blue) that can hybridize through its R region with the RNA at the other end of the RNA template, or with another RNA template (c). This hybridization to another

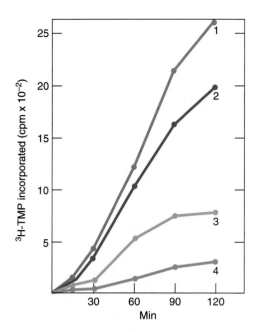

Figure 23.31   Effect of RNase on reverse transcriptase activity. Baltimore incubated R-MLV particles with the four dNTPs, including [^{3}H]dTTP, under various conditions, then acid-precipitated the product and measured the radioactivity of the product by liquid scintillation counting. Treatments: red, no extra treatment; purple, preincubation for 20 min with water; blue, RNase included in the reaction; green, preincubated with RNase. (*Source:* Reprinted with permission from *Nature New Biology* 223:133, 1971. Copyright © 1971 Macmillan Magazines Limited.)

R region is called the "first jump." Here we assume that the DNA jumps to the other end of the same RNA, and this could be facilitated by looping the RNA around so the strong-stop DNA does not even need to leave the left end of the RNA to pair with the right end.

After the first jump, the strong-stop DNA is at the right end of the template and can serve as a primer for the reverse transcriptase to copy the rest of the viral RNA (d). Notice that the first jump has allowed the right LTR to be completed. The U5 and R regions were copied from the left LTR of the viral RNA and the U3 region was copied from the right LTR. In step (e), the RNase H removes most of the viral RNA, but it leaves a small piece of RNA adjacent to the right LTR to serve as a primer for second strand synthesis (f). After the reverse transcriptase extends this primer to the end, including the PBS region, RNase H removes the remaining RNA (g)—the second strand primer and the tRNA—both of which were paired to DNA. This sets up the second jump (h), in which the PBS region on the right pairs with the one on the left. Like the first jump, the second jump can be visualized as a jump to another molecule, or the same molecule as shown here. If the same molecule is involved in the jump, the DNA can loop around to allow the two PBS regions to base-pair. After the second jump, the stage is set for reverse transcriptase, which can use DNA as a template, or another

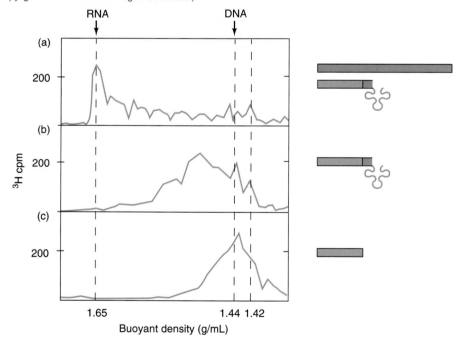

Figure 23.32   Reverse transcripts contain an RNA primer. Baltimore and colleagues labeled reverse transcripts in AMV particles with [^{3}H]dTTP, then subjected them to Cs_2SO_4 gradient ultracentrifugation after the following treatments: (**a**) no treatment; (**b**) heating to denature double-stranded polynucleotides; and (**c**) heating and RNase to remove any primers attached to the reverse transcripts. Interpretive drawings at right provide an explanation for the results: (**a**) The untreated material has a high density like RNA because the reverse transcript is short and is base-paired to a much longer viral RNA template. (**b**) The heated material has a density closer to that of DNA because the RNA template has been removed, but it is still denser than pure DNA because of an RNA primer that is covalently attached. (**c**) The heated and RNase-treated material has the density of a pure DNA because the RNase has removed the RNA primer. The approximate densities of pure RNA and DNA are indicated at top. (*Source:* Reprinted with permission from *Nature New Biology* 223:133, 1971. Copyright © 1971 Macmillan Magazines Limited.)

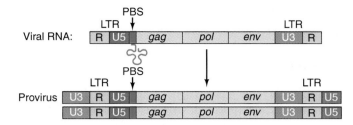

Figure 23.33 Structures of retroviral RNA and provirus DNA. This is a nondefective retroviral RNA that contains all the genes necessary for replication: a coat protein gene (*gag*), a reverse transcriptase gene (*pol*), and an envelope protein gene (*env*). In addition, it contains long terminal repeats (LTRs) at both ends, but these repeats are not identical. The left LTR contains an R and a U5 region, but the right LTR contains a U3 and an R region. On the other hand, the proviral DNA, made using the viral RNA as a template, contains full LTRs (U3, R, and U5) at each end.

DNA polymerase to complete both strands (i), using the long single-stranded overhangs at each end as templates.

Once the provirus is synthesized, it can be inserted into the host genome by an **integrase**. This enzyme is originally part of a **polyprotein** derived from the *pol* gene, which we have seen also encodes reverse transcriptase and RNase H. The integrase is cut from the polyprotein by a **protease,** which also starts out as part of the same polyprotein. The protease also cuts itself out of the polyprotein. (It is worth noting that some of the most promising drugs for combatting AIDS are protease inhibitors that target the HIV version of this enzyme.) Once the provirus is integrated into the host genome, it is transcribed by host RNA polymerase II to yield viral RNAs.

SUMMARY Retroviruses replicate through an RNA intermediate. When a retrovirus infects a cell, it makes a DNA copy of itself, using a virus-encoded reverse transcriptase to carry out the RNA→DNA reaction, and an RNase H to degrade the RNA parts of RNA–DNA hybrids created during the replication process. A host tRNA serves as the primer for the reverse transcriptase. The finished double-stranded DNA copy of the viral RNA is then inserted into the host genome, where it can be transcribed by host polymerase II.

Retrotransposons All eukaryotic organisms appear to harbor transposons that replicate through an RNA intermediate and therefore depend on reverse transcriptase. These **retrotransposons** fall into two groups with different modes of replication. The first group includes the retrotransposons with LTRs, which replicate in a manner very similar to retroviruses, except that they do not pass from cell to cell in virus particles. Not surprisingly, these are called **LTR-containing**

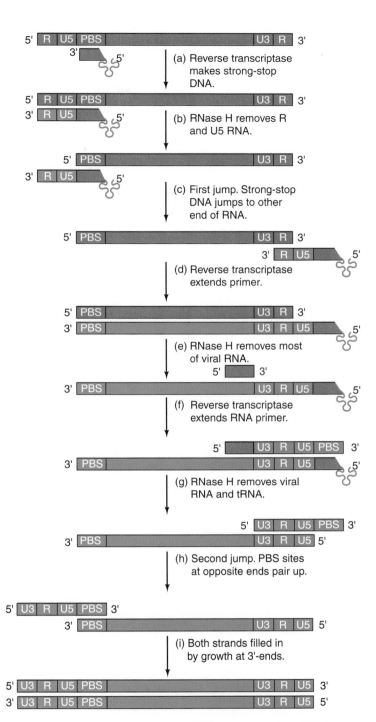

Figure 23.34 A model for the synthesis of the provirus DNA from a retroviral RNA template. RNA is in red and DNA is in blue, throughout. The tRNA primer is represented by a cloverleaf with a 3′-tag that hybridizes to the primer-binding site (PBS) in the viral RNA. The steps are described more fully in the text.

retrotransposons. The second group includes the retrotransposons that lack LTRs (the **non-LTR retrotransposons**). We will consider these two kinds of retrotransposons in turn.

LTR-Containing Retrotransposons The first examples of retrotransposons were discovered in the fruit fly

(*Drosophila melanogaster*) and yeast (*Saccharomyces cerevisiae*). The prototype *Drosophila* transposon is called *copia* because it is present in the genome in copious quantity. In fact, *copia* and related transposons called *copia*-like elements account for about 1% of the total fruit fly genome. Similar transposable elements in yeast are called **Ty,** for "transposon yeast." These transposons have LTRs that are very similar to the LTRs in retroviruses, which suggests that their transposition resembles the replication of a retrovirus. Indeed, several lines of evidence indicate that this is true. Here is a summary of the evidence that the Ty*1* elements replicate through an RNA intermediate, just as retroviruses do:

1. Ty*1* encodes a reverse transcriptase. The *tyb* gene in Ty codes for a protein with an amino acid sequence closely resembling that of the reverse transcriptases encoded in the *pol* genes of retroviruses. If the Ty1 element really codes for a reverse transcriptase, then this enzyme should appear when Ty*1* is induced to transpose; moreover, mutations in *tyb* should block the appearance of reverse transcriptase. Gerald Fink and his colleagues have performed experiments that bear out both of these predictions.

2. Full-length Ty*1* RNA and reverse transcriptase activity are both associated with particles that closely resemble retrovirus particles (Figure 23.35). These particles appear only in yeast cells that are induced for Ty*1* transposition.

3. In a clever experiment, Fink and colleagues inserted an intron into a Ty*1* element and then analyzed the element again after transposition. The intron was

gone! This finding is incompatible with the kind of transposition prokaryotes employ, in which the transposed DNA looks just like its parent. But it is consistent with the following mechanism (Figure 23.36): The Ty element is first transcribed, intron and all; then the RNA is spliced to remove the intron; and finally, the spliced RNA is reverse transcribed within a virus-like particle and the resulting DNA is inserted back into the yeast genome at a new location.

4. Jef Boeke and colleagues demonstrated that the host tRNA$_i^{Met}$ serves as the primer for Ty*1* reverse transcription. First, they mutated 5 of 10 nucleotides in the Ty*1* element's PBS that are complementary to the host tRNA$_i^{Met}$. These changes abolished transposition, presumably because they made it impossible for the tRNA primer to bind to its PBS. Then Boeke and coworkers made five compensating mutations in a copy of the host tRNA$_i^{Met}$ gene that restored binding to the mutated PBS. These mutations restored transposition activity to the mutant Ty*1* element. As we have seen many times throughout this book, this kind of mutation suppression is powerful evidence for the importance of interaction between

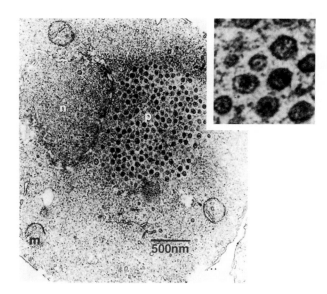

Figure 23.35   Retrovirus-like particles in a yeast cell induced for Ty transposition. The particles in this electron micrograph are labeled p. The inset shows the particles at higher magnification. The nucleus and a mitochondrion are labeled n and m, respectively. (*Source:* (*inset*) Courtesy Stanley N. Cohen, Stanford University.)

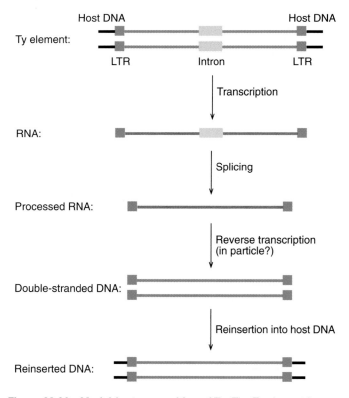

Figure 23.36   Model for transposition of Ty. The Ty element has been experimentally supplied with an intron (yellow). The Ty element is transcribed to yield an RNA copy containing the intron. This transcript is spliced, and then the processed RNA is reverse-transcribed, possibly in a virus-like particle. The resulting double-stranded DNA then reinserts into the yeast genome. *Abbreviation:* LTR = long terminal repeat.

two molecules: in this case, interaction between the tRNA$_i^{Met}$ primer and its binding site in the Ty*1* element.

Copia and its relatives share many of the characteristics we have described for Ty, and it is clear that they also transpose in the same way as Ty. Humans also have LTR-containing retrotransposons, but they lack a functional *env* gene. The most prominent examples are the **human endogenous retroviruses (HERVs)**, which make up 1–2% of the genome. So far, no transposition-competent HERVs are known, so the HERVs may be relics of previous retrotransposition.

> **SUMMARY** Several eukaryotic transposons, including Ty of yeast and *copia* of *Drosophila*, apparently transpose in a similar way. They start with DNA in the host genome, make an RNA copy, then reverse transcribe it—probably within a virus-like particle—to DNA that can insert in a new location. HERVs probably transposed in the same way until most or all of them lost the ability to transpose.

Non-LTR Retrotransposons Retrotransposons that lack LTRs are much more abundant than those with LTRs, at least in mammals. The most abundant of all are the **long interspersed elements (LINEs)**, one of which (L1) is present in at least 100,000 copies and makes up about 15% of the human genome, although about 97% of the copies of L1 are missing more or less of their 5′-ends and the great majority have mutations that prevent their transposition. The prevalence of L1 elements means that this retrotransposon, which we would probably classify as "junk DNA," occupies about five times as much of the genome as all the human exons do. Figure 23.37 is a map of an intact L1 element, showing its two ORFs. ORF 1 encodes an RNA-binding protein (p40), and ORF2 encodes a protein with two activities: an endonuclease and a reverse transcriptase. L1, like all retrotransposons in this class, is polyadenylated.

We have just seen that the LTR is crucial for replication of most retrotransposons with LTRs, so how do non-LTR retrotransposons replicate? In particular, what do

they use for a primer? The answer is that their endonuclease creates a single-stranded break in the target DNA and their reverse transcriptase uses the newly formed DNA 3′-end as a primer. Our best information on this mechanism comes from Thomas Eickbush and colleagues' studies on **R2Bm,** a LINE-like element from the silkworm *Bombyx mori.* This element resembles the mammalian LINEs in that it encodes a reverse transcriptase, but no RNase H, protease, or integrase, and it lacks LTRs. But it differs from the LINEs in that it has a specific target site—in the 28S rRNA gene of the host. This latter property made the insertion mechanism easier to investigate.

Eichbush and colleagues first showed that the single ORF of R2Bm encodes an endonuclease that specifically cleaves the 28S rDNA target site. Next, they purified the endonuclease (and an RNA cofactor that was required for activity) and added it to a supercoiled plasmid containing the target site. If a single strand of the plasmid is cut, the supercoiled plasmid will be converted to a relaxed circle. If both strands are cut, a linear DNA should appear. Figure 23.38a shows the rapid appearance of relaxed circles, followed by the slower appearance of linear DNA. Thus, the R2Bm endonuclease rapidly cleaves one of the DNA strands at the target site, then much more slowly cleaves the other strand. This cleavage is specific: The nuclease cannot cut even one strand of a plasmid that lacks the target site.

Next, these workers removed the RNA cofactor and showed that the protein by itself still caused rapid single-stranded nicking of the target site, but barely detectable cutting of the other strand (Figure 23.38b). They also showed that the linear DNA could be recircularized by T4 DNA ligase, which requires a 5′-phosphate group. Thus, cleavage by the R2Bm endonuclease leaves a 5′-phosphate and a 3′-hydroxyl group. Next, they used the endonuclease to create single-stranded nicks and showed by primer extension analysis that the transcribed strand is the one that is nicked. (The nick in the transcribed strand stopped the DNA polymerase in the primer extension experiment, but primer extension on the other strand proceeded unimpeded by nicks.) With more precise primer extension experiments on DNA cut in both strands, they located the cut sites exactly and found the two strands are cut 2 bp apart.

To see if the nicked target DNA strand really does serve as the primer, Eickbush and colleagues performed an in vitro reaction with a short piece of pre-nicked target DNA as primer, R2Bm RNA as template, R2Bm reverse transcriptase, and all four dNTPs, including [^{32}P]dATP. They electrophoresed and autoradiographed the products to see if they were the right size. Figure 23.39a shows what should happen at the molecular level, and panel b shows the results. When a nonspecific RNA was added as template, no product was made (lane 1), but when the R2Bm RNA was added, a strong band at 1.9 kb appeared. Is this what we expect? It is hard to know because

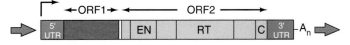

Figure 23.37 Map of the L1 element. The subregions within ORF2 (yellow) are designated EN (endonuclease), RT (reverse transcriptase), and C (cysteine-rich). The purple arrows at each end indicate direct repeats of host DNA, and the A$_n$ on the right indicates the poly(A).

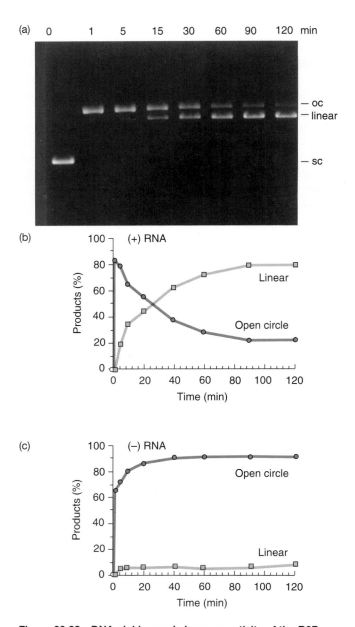

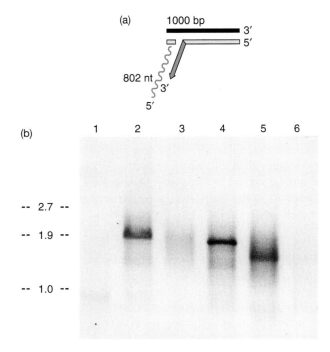

Figure 23.39   Evidence for target priming of reverse transcription of R2Bm. (**a**) Model of the product we expect if the R2Bm endonuclease makes a nick near the left end of a 1-kb target DNA and uses the new 3′-end to prime reverse transcription of an 802-nt transposon RNA. The reverse transcript (blue) is covalently attached to the primer (yellow). The rest of the lower DNA strand is also rendered in yellow at left. The opposite DNA strand is black. (**b**) Experimental results. Eickbush and colleagues started with a 1-kb target DNA with the target site close to the left end. They added R2Bm RNA and the ORF2 product and dNTPs, including [^{32}P]dATP to allow labeled reverse transcripts to be formed. Then they electrophoresed the products and autoradiographed them. Lane 1, a nonspecific RNA was used instead of R2Bm RNA; lanes 2–6, R2Bm RNA was used; lane 3, dideoxy CTP was included in the reverse transcription reaction; lane 4, the product was treated with RNase A before electrophoresis; lane 5, the product was treated with RNase H before electrophoresis; lane 6, a nonspecific target DNA was used. (*Source:* Luan, D. et al. Reverse transcription of R2Bm RNA is primed by a nick at the chromosomal target site: a mechanism for non-LTR retrotransposition *Cell* 72 (Feb 1993) f. 4, p. 599. Reprinted by permission of Elsevier Science.)

Figure 23.38   DNA nicking and cleavage activity of the R2Bm endonuclease. Eickbush and colleagues mixed a supercoiled plasmid bearing the target site for the R2Bm retrotransposon with the purified R2Bm endonuclease, with or without its RNA cofactor, then electrophoresed the plasmid to see if it had been nicked (relaxed to an open circular form) or cut in both strands to yield a linear DNA. (**a**) Electrophoretic gel stained with ethidium bromide. The positions of the supercoiled plasmid (sc), the open circular plasmid (oc), and the linear plasmid (linear) are indicated at right. (**b**) Graphical representation of the results from panel (**a**). (**c**) Results from a similar experiment in which the RNA cofactor was omitted. (*Source:* Luan, D. et al. Reverse transcription of R2Bm RNA is primed by a nick at the chromosomal target sige: a mechanism for non-LTR retrotrasposition. *Cell* 72 (Feb 1993) f. 2, p. 597. Reprinted by permission of Elsevier Science.)

we do not know exactly how far the reverse transcriptase traveled and we are dealing with a slightly branched polynucleotide, but it is close because the primer is 1 kb long and the template is 802 nt long. To investigate the nature of the product further, Eickbush and colleagues in-

cluded dideoxy-CTP in the reaction (lane 3). As expected, it caused premature termination of reverse transcription at a number of sites, leading to a fuzzy band. In another reaction, they treated the product with RNase A to remove any part of the template not base-paired to the reverse transcription product before electrophoresis. Lane 4 shows that this sharpened the product to a 1.8-kb band, suggesting that about 100 nt had been removed from the 5′-end of the RNA template, so the reverse transcriptase had apparently not completed its task in the majority of cases. These workers also treated the product with RNase H prior to electrophoresis and obtained a diffuse band of about 1.5 kb. This procedure should remove the RNA template because it is in a hybrid with the product. The fact that the band is still longer than 1 kb indicates

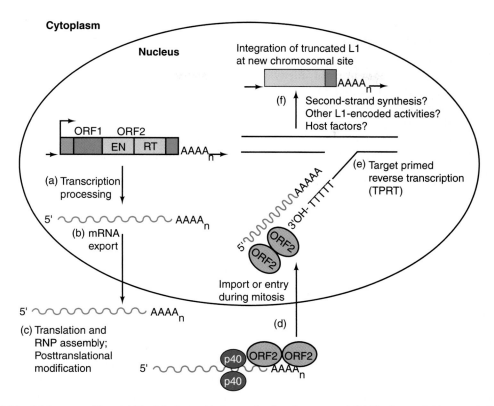

Figure 23.40 A model for L1 transposition. (a) The L1 element is transcribed and processed. **(b)** The transcript is exported from the nucleus. **(c)** The mRNA is translated to yield the ORF1 product (p40) and the ORF2 product, with endonuclease and reverse transcriptase activities. These proteins associate with the mRNA. **(d)** The ribonucleoprotein reenters the nucleus. **(e)** The endonuclease nicks the target DNA (anywhere in the genome), and the reverse transcriptase uses the new DNA 3′-end to prime synthesis of the reverse transcript. **(f)** In a series of unspecified steps, the second L1 strand is made and the whole element is ligated into the target DNA. (*Source:* Reprinted from Kazazian and Moran, "The impact of L1 retrotransposons on the human genome," *Nature Genetics* 19:19–24, 1998. Copyright 1998, with permission from Nature Publishing Group.)

that a strand of DNA has been extended. Lane 6 is another negative control in which a nonspecific DNA was used instead of the target DNA.

Similar experiments with a target DNA that extended farther to the left (with the target site in the middle) showed a predominance of large, Y-shaped products (recall Figure 23.39), suggesting that reverse transcription occurred before second-strand cleavage. If second-strand cleavage had occurred first, the products would have been linear and smaller. To confirm that the target DNA was serving as the primer, Eickbush and coworkers performed PCR with primers that hybridized to the target DNA and to the reverse transcript, and obtained PCR products of the expected size and sequence.

Based on these and other data, H. H. Kazazian and John Moran proposed the model of L1 transposition presented in Figure 23.40. First, the transposon is transcribed and the transcript is processed. The processed mRNA leaves the nucleus to be translated in the cytoplasm. It associates with its two products, p40 and the ORF2 product, and reenters the nucleus. There, the endonuclease activity of the ORF2 product nicks the target DNA. For L1, the target can be any region of the DNA. Then the reverse

transcriptase activity of the ORF2 product uses the target DNA 3′-end created by the endonuclease as a primer to copy the L1 RNA. Finally, in steps that are still poorly understood, the second strand of L1 is made, the second strand of the target is cleaved, and the L1 element is ligated into its new home.

You would suspect that anything as prevalent as L1 is in the human genome must have some negative consequences, and indeed a number of L1-mediated mutations have been discovered that have led to human disease. In particular, copies of L1 have been found: in the blood clotting factor VIII gene, causing hemophilia; in the *DMD* gene, causing Duchenne muscular dystrophy; and in the *APC* gene, helping to cause adenomatous polyposis coli, a kind of colon cancer. In this last case, the patient's cancer cells had the L1 element in their *APC* gene, but the normal cells did not. Thus, this transposition had occurred during the patient's lifetime as a somatic mutation.

What is more surprising is that the L1 may actually have beneficial consequences as well. For example, significant homology occurs between the reverse transcriptase of L1 and human telomerase, suggesting that L1 may have been the origin of the enzyme that maintains the ends of our chromosomes (although the reverse may also have been true). But the most plausible beneficial aspect of L1 is that it may facilitate exon shuffling, the exchange of exons among genes. This happens because the polyadenylation signal of L1 is weak, so the polyadenylation machinery frequently bypasses it in favor of a polyadenylation site downstream in the host part of the transcript. RNAs polyadenylated in that way will include a piece of human RNA attached to the L1 RNA, and this human RNA will be incorporated as a reverse transcript wherever the L1 element goes next. This is bound to have deleterious consequences sometimes, but it also creates new genes out of parts of old genes, and that can give rise to proteins with new and useful characteristics.

Why are the polyadenylation signals of L1 elements weak? Moran offers the following explanation: If the polyadenylation signals were strong, insertion of these elements into the introns of human genes would cause premature polyadenylation of transcripts, so all the exons downstream would be lost. That would probably inactivate the gene and might well lead to the death of the host. And, unlike retroviruses, which can move from one individual to another, the L1 elements live and die with their hosts. On the other hand, weak polyadenylation signals allow these elements to insert into introns of human genes without disrupting a very high percentage of the transcripts of these genes. Thus, because the amount of DNA devoted to introns is much higher than that devoted to exons, the L1 elements have a large area of the human genome to colonize relatively safely.

SUMMARY LINEs and LINE-like elements are retrotransposons that lack LTRs. These elements encode an endonuclease that nicks the target DNA. Then the element takes advantage of the new DNA 3′-end to prime reverse transcription of element RNA. After second-strand synthesis, the element has become replicated at its target site. A new round of transposition begins when the LINE is transcribed. Because the LINE polyadenylation signal is weak, transcription of a LINE frequently includes one or more downstream exons of host DNA.

Nonautonomous Retrotransposons Members of another class of non-LTR retrotransposons (**nonautonomous retrotransposons**) encode no proteins, so they are not autonomous like the transposition-competent LINEs. Instead, they depend on other elements, probably the LINEs due to their prevalence, to supply the proteins, including the reverse transcriptase they need to transpose. The best studied of these nonautonomous retrotransposons are the **Alu elements**, so-called because they contain the sequence AGCT that is recognized by the restriction enzyme *Alu*I.

These are about 300 bp long and are present in up to a million copies in the human genome. Thus, they have been even more successful than the LINEs. One reason for this success may be that the transcripts of the Alu elements contain a domain that resembles the 7SL RNA that is normally part of the signal recognition particle that helps attach certain ribosomes to the endoplasmic reticulum. Two signal recognition particle proteins bind tightly to Alu element RNA and may carry it to the ribosomes, where the L1 RNA is being translated. This may put the Alu element RNA in a position to help itself to the proteins it needs to be reverse transcribed and inserted at a new site.

The LINEs have probably also played a role in shaping the human genome by facilitating the creation of **processed pseudogenes**. Ordinary **pseudogenes** are DNA sequences that resemble normal genes, but for some reason cannot function. Sometimes they have internal translation stop signals; sometimes they have inactive or missing splicing signals; sometimes they have inactive promoters; usually a combination of problems prevents their expression. They apparently arise by gene duplication and subsequently accumulate mutations. This process has no deleterious effect on the host because the original gene remains functional.

Processed pseudogenes also arise by gene duplication, but apparently by way of reverse transcription. We strongly suspect that RNA is an intermediate in the formation of processed pseudogenes because: (1) these pseudogenes frequently have short poly(dA) tails that seem to have derived from poly(A) tails on mRNAs; and (2) processed pseudogenes lack the introns that their progenitor genes usually have. As in the case of the Alu elements, which are not derived from mRNAs, the LINEs could provide the molecular machinery that allows mRNAs to be reverse transcribed and inserted into the host genome.

SUMMARY Nonautonomous retrotransposons include the very abundant Alu elements in humans and similar elements in other vertebrates. They cannot transpose by themselves because they do not encode any proteins. Instead they take advantage of the retrotransposition machinery of other elements, such as LINEs. Processed pseudogenes probably arose in the same way: mRNAs were reverse-transcribed by LINE machinery and then inserted into the genome.

Group II introns In Chapter 14 we learned that group II introns, which inhabit bacterial, mitochondrial, and chloroplast genomes, are self-splicing introns that form a lariat intermediate. In 1998, Mariene Belfort and

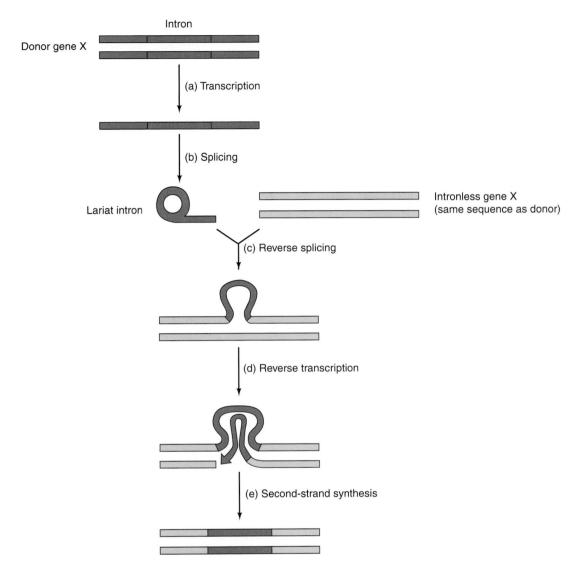

Figure 23.41 Retrohoming. (a) The donor gene X (blue) bearing a group II intron (red) is transcribed. **(b)** The transcript is spliced, yielding a lariat-shaped intron. **(c)** The intron reverse-splices itself into another copy of gene X (yellow) that has the same or similar sequence as the first except that it lacks the intron. **(d)** The intron-encoded reverse transcriptase makes a DNA copy of the intron, using a nick in the bottom DNA strand as primer. **(e)** The second strand (DNA version) of the intron is made, replacing the RNA intron in the top strand. This completes the retrohoming process.

colleagues discovered that a group II intron in a particular gene could insert into an intronless version of the same gene somewhere else in the genome. This process, called **retrohoming,** appears to occur by the mechanism outlined in Figure 23.41 The gene bearing the intron is first transcribed, then the intron is spliced out as a lariat. This intron can then recognize an intronless version of the same gene and invade it by reverse-splicing. Reverse transcription creates a cDNA of the intron, and second-strand synthesis replaces the RNA intron with a second strand of DNA.

In 1991, Phillip Sharp proposed that group II introns could be the ancestors of modern spliceosomal introns, in part because of their very similar mechanisms of splicing. In 2000, Belfort and colleagues showed how this could have happened. They detected true retrotransposition, not just retrohoming, of a bacterial group II intron. Thus, the intron moved to a variety of new sites, not just to an intronless copy of the intron's home gene.

To detect such retrotransposition, Belfort and colleagues engineered a plasmid containing a modified version of the *Lactococcus lactis* L1.LtrB intron, as

(a)

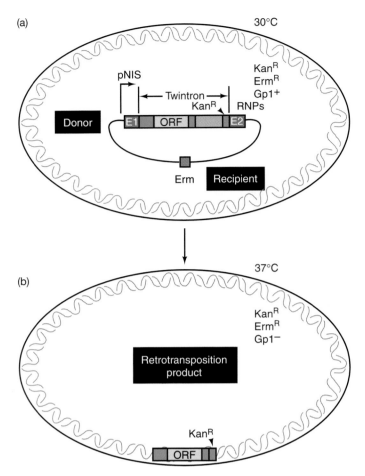

Figure 23.42   Detecting retrotransposition of a group II intron.
(a) Before transposition. A bacterial cell contains a temperature-sensitive donor plasmid with the following elements. 1. an erythromycin-resistance gene (Em^R, purple); 2. a gene driven by a nisin-inducible promoter, containing two exons (E$_1$ and E$_2$, blue) and a twintron; 3. a twintron containing a group II intron (red, with yellow ORF), a group I intron (orange); and a kanamycin resistance gene (Kan^R). The cell also contains a recipient chromosome (double helix around the perimeter). This cell is cultured at 30° C to maintain the temperature-sensitive plasmid. The code at upper right indicates that this cell is kanamycin-resistant (Kan^R), erythromycin-resistant (Erm^R), and contains the group I intron (GpI$^+$). (b) After transposition. The cell has been cured of the temperature-sensitive plasmid by growing it at 37° C, so it is erythromycin-sensitive (Erm^S), but the group II intron has retrotransposed to the bacterial chromosome, so the cell remains kanamycin-resistant (Kan^R). Because the retrotransposition went through an RNA intermediate, the group I intron spliced itself out and was lost (GpI$^-$). (*Source:* Reprinted by permission from Cousineau et al., "Retrotransposition of a bacterial group II intron," *Nature* 404:1019, 2000. Copyright © 2000 Macmillan Magazines Ltd.)

pictured in Figure 23.42. The plasmid contained the following features:

1. It was temperature-sensitive, so it could replicate at 30° C, but not at 37° C.

2. It had an erythromycin resistance gene to allow screening for cells that had lost the plasmid.

3. It had an inducible promoter driving expression of a gene containing the group II intron, and other elements. This allowed the induction of transposition by stimulating the promoter with the inducer, nisin.

4. The group II intron was itself interrupted by a self-splicing group I intron. This allowed verification of an RNA intermediate because if transposition went through an RNA intermediate the group I intron would splice itself out and be lost. The whole dual intron structure was dubbed a "twintron."

5. The group II intron also contained a kanamycin resistance gene to allow selection for cells that had experienced retrotransposition and then had lost the plasmid.

Belfort and coworkers transformed cells with this plasmid, induced transposition with nisin, then increased the temperature to 37° C to cure cells of the plasmid—that is, to prevent the plasmid from replicating so it would be lost from the replicating cells. They grew the cured cells in the presence of kanamycin to select for cells in which the group II intron had transposed to the cell's chromosome. Such a transposition would carry the kanamycin resistance gene to the chromsome so it would remain even after the plasmid had been lost. They also screened these cells for sensitivity to erythromycin to verify that the plasmid really had been lost. Finally, they tested clones of kanamycin-resistant, erythromycin-sensitive cells for DNA containing the group II intron, but lacking the group I intron by colony hybridization with [^{32}P]DNA probes specific for each intron. The clones that had all these characteristics should have experienced a retrotransposition of the group II intron from the plasmid to the bacterial chromosome.

Next, Belfort and colleagues isolated intron-containing genomic DNA fragments from eight such clones and sequenced the DNA flanking the intron. They found eight different target sites, each of which had some sequence in common with the intron, especially in a 13-nt region adjacent to the insertion site. This sequence presumably attracted the intron to its target. Also, each insertion was into a transcribed region of a gene, and each was into the nontranscribed DNA strand (i.e., the strand with the same sense and orientation as the gene transcript).

Unlike retrohoming, the retrotransposition observed here did not require the intron's endonuclease activity, but it did require the host's recombination machinery. This finding indicates that retrotransposition and retrohoming use different mechanisms, and Belfort and colleagues proposed the mechanism in Figure 23.43 for retrotransposition. First, the intron is transcribed and splices itself out of the transcript. Next, it is translated to yield its protein product, which binds to the intron to form a ribonucleoprotein (RNP). Then the RNP catalyzes the reverse splicing of the intron into the RNA product of

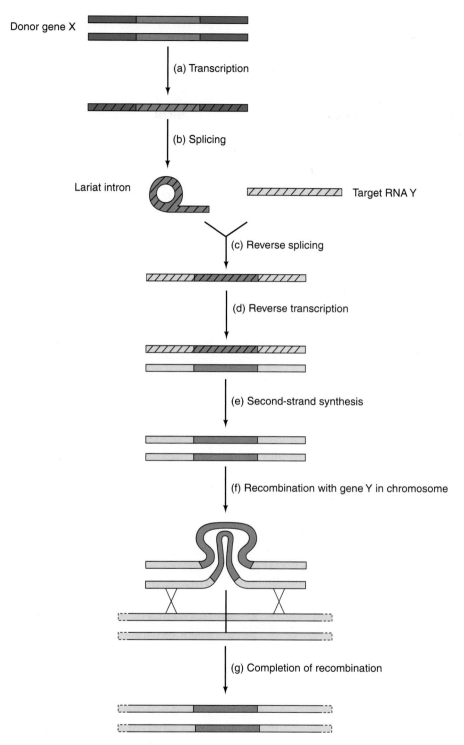

Figure 23.43 A model for retrotransposition of a group II intron. (a) The donor gene X (blue) bearing a group II intron (red) is transcribed. (RNAs are shaded, throughout, but DNAs are not.) **(b)** The transcript is spliced, yielding a lariat-shaped intron. **(c)** The intron reverse-splices itself into an unrelated transcript Y (yellow). **(d)** The intron-encoded reverse transcriptase makes a DNA copy of the target RNA with the intron in it. **(e)** Second-strand DNA is synthesized. **(f)** Recombination begins with a crossover between the newly synthesized double-stranded DNA and an intronless copy of the target gene. **(g)** Recombination is completed, and the intron has inserted into a new gene.

the target gene. This is the major difference from retro-homing, where reverse splicing takes the intron into the gene itself rather than into the gene's RNA product. Next, reverse transcription creates a DNA copy of the RNA with its new intron, and second-strand synthesis completes the double-stranded DNA copy of the RNA. Finally, the host's recombination system allows recombination with an unrelated gene, inserting the intron into a new location.

Several lines of evidence suggest that RNA, not DNA is the target for retrotransposition of the group II introns. First, all the insertions were in the "sense" orientation. This would have to be the case if insertion occurred at the RNA level, but would not have to be the case for an insertion into DNA. Second, two of the eight transpositions resulted in insertion of the intron into a 23S rRNA gene. This high percentage would be expected if RNA were the target, because 23S rRNA is very abundant in the cell. Third, we know that reverse splicing into RNA occurs in vitro and in vivo. Fourth, the lack of dependence on the intron-encoded endonuclease is consistent with the lack of target specificity in retrotransposition.

Once a group II intron has retrotransposed, it retains its ability to splice itself out, so the target gene should usually continue to function. Thus, spread of group II introns may have occurred readily and with relative safety in the precursors to modern eukaryotes. Ultimately, eukaryotes appear to have developed spliceosomes to make the splicing process more efficient.

SUMMARY Group II introns can retrohome to intronless copies of the same gene by insertion of an RNA intron into the gene, followed by reverse transcription and second-strand synthesis. Group II introns can also undergo retrotransposition by insertion of an RNA intron into an unrelated transcript, followed by reverse transcription and recombination with the target gene. This kind of retrotransposition of group II introns may have provided the ancestors of modern-day eukaryotic spliceosomal introns and may account for their widespread appearance in higher eukaryotes.

SUMMARY

The mechanism of strand exchange catalyzed by the λ phage protein Int is different from that of homologous recombination. Int breaks specific phosphodiester bonds within attP and attB and forms a covalent bond between the new 3′-phosphate group on the DNA and tyrosine 342 in the enzyme. This conserves the energy of the broken phosphodiester bond, so it can be used to form a new phosphodiester bond with the homologous DNA during strand exchange.

Another kind of site-specific recombination controls the type of protein in the flagella of the bacterium *Salmonella typhimurium*. This protein varies, depending on whether it is produced by the *H1* or the *H2* gene. The control of these two genes is linked to the inversion of a 995-bp DNA element (the H segment) between them. This inversion is a site-specific recombination event. Phages Mu, P1, and P7 use a similar inversion scheme to switch from one kind of tail fiber to another, and therefore to switch their host range.

Transposable elements, or transposons, are pieces of DNA that can move from one site to another. Some transposable elements replicate, leaving one copy at the orignal location and placing one copy at a new site; others transpose without replication, leaving the original location altogether. Bacterial transposons include the following types: (1) insertion sequences such as IS1 that contain only the genes necessary for transposition, flanked by inverted terminal repeats; (2) transposons such as Tn3 that are like insertion sequences but contain at least one extra gene, usually a gene that confers antibiotic resistance.

Eukaryotic transposons use a wide variety of replication strategies. The DNA transposons, such as *Ds* and *Ac* of maize and the P elements of *Drosophila* behave like the DNA transposons, such as Tn3, of bacteria.

The immunoglobulin genes of mammals even rearrange using a mechanism that resembles transposition. Vertebrate immune systems create enormous diversity in the kinds of immunoglobulins they can make. The primary source of this diversity is the assembly of genes from two or three component parts, each selected from a heterogeneous pool of parts. This assembly of gene segments is known as V(D)J recombination. The recombination signal sequences (RSSs) in V(D)J recombination consist of a heptamer and a nonamer separated by either 12-bp or 23-bp spacers. Recombination occurs only between a 12 signal and a 23 signal, which ensures that only one of each coding region is incorporated into the rearranged gene. Rag-1 and Rag-2 are the principal players in V(D)J recombination. They introduce single-strand nicks into DNA adjacent to either a 12 signal or a 23 signal. This leads to a transesterification in which the newly created 3′-hydroxyl group attacks the opposite strand, breaking it, and forming a hairpin at the end of the coding segment. The hairpins then break and join with each other in an imprecise way, allowing joining of coding regions with loss of bases or gain of extra bases.

The retrotransposons come in two different types. The LTR-containing retrotransposons replicate like retroviruses, as follows: Retroviruses replicate through an RNA intermediate. When a retrovirus infects a cell, it makes a DNA copy of itself, using a virus-encoded reverse transcriptase to carry out the RNA→DNA reaction, and an RNase H to degrade the RNA parts of RNA–DNA hybrids created during the replication process. A host tRNA serves as the primer for the reverse transcriptase. The finished double-stranded DNA copy of the viral RNA is then inserted into the host genome, where it can be transcribed by host polymerase II. The retrotransposons Ty of yeast and *copia* of *Drosophila* replicate in much the same way. They start with DNA in the host genome, make an RNA copy, then reverse-transcribe it—probably within a virus-like particle—to DNA that can insert in a new location.

The other class of eukaryotic retrotransposons are the non-LTR retrotransposons, and they use different methods of priming reverse transcription. For example, LINEs and LINE-like elements encode an endonuclease that nicks the target DNA. Then the element takes advantage of the new DNA 3′-end to prime reverse transcription of element RNA. After second-strand synthesis, the element has become replicated at its target site. A new round of transposition begins when the LINE is transcribed. Because the LINE polyadenylation signal is weak, transcription of a LINE frequently includes one or more downstream exons of host DNA and this can transport host exons to new locations in the genome.

Nonautonomous, non-LTR retrotransposons include the very abundant Alu elements in humans and similar elements in other vertebrates. They cannot transpose by themselves because they do not encode any proteins. Instead, they take advantage of the retrotransposition machinery of other elements, such as LINEs. Processed pseudogenes probably arose in the same way: mRNAs were probably reverse-transcribed by LINE machinery and then inserted into the genome.

Group II introns represent another class of non-LTR retrotransposons found in both bacteria and eukaryotes. They can retrohome to intronless copies of the same gene by insertion of an RNA intron into the gene, followed by reverse transcription and second-strand synthesis. Group II introns can also undergo retrotransposition by insertion of an RNA intron into an unrelated transcript, followed by reverse transcription and recombination with the target gene. This kind of retrotransposition of group II introns may have provided the ancestors of modern-day eukaryotic spliceosomal introns and may account for their widespread appearance in higher eukaryotes.

REVIEW QUESTIONS

1. Draw a simple diagram of an overview of the processes by which λ DNA integrates into the host DNA and then is excised out again.

2. Present models for DNA strand cleavage and rejoining by the active tyrosine of (a) an ordinary topoisomerase, and (b) the topoisomerase of Int.

3. Describe and give the results of an experiment that shows that Int has topoisomerase activity.

4. Describe and give the results of an experiment that shows where Int cleaves in *att*B.

5. Describe and give the results of an experiment that shows that Int binds to DNA through a phosphotyrosine linkage. How do we know which tyrosine is involved?

6. Using diagrams, explain how inversion of a DNA segment can turn off one *Salmonella* flagellin gene and turn on another.

7. Using diagrams, explain how inversion of a phage Mu gene segment can turn off one tail protein gene and turn on another.

8. Describe and give the results of an experiment that shows that bacterial transposons contain inverted terminal repeats.

9. Use a diagram to explain how transposition can create direct repeats of host DNA flanking an inserted transposon.

10. Compare and contrast the genetic maps of the bacterial transposons IS1 and Tn3, and the eukaryotic transposon *Ac*.

11. Describe a method for tracking the transposition of Tn3 from one plasmid to another.

12. Diagram the mechanism of Tn3 transposition, first in simplified form, then in detail.

13. Diagram a mechanism for conservative transposition.

14. Explain how transposition can give rise to speckled maize kernels.

15. Draw a sketch of an antibody protein, showing the light and heavy chains.

16. Explain how thousands of immunoglobulin genes can give rise to many millions of antibody proteins.

17. Diagram the rearrangement of immunoglobulin light- and heavy-chain genes that occurs during B-lymphocyte maturation.

18. Explain how the signals for V(D)J joining ensure that each one of the parts of an immunoglobulin gene will be included in the mature, rearranged gene.

19. Diagram a reporter plasmid designed to test the importance of the heptamer, nonamer, and spacer in a recombination signal sequence. Explain how this plasmid detects recombination.

20. Present a model for cleavage and rejoining of DNA strands at immunoglobulin gene recombination signal sequences. How does this mechanism contribute to antibody diversity?

21. Describe and give the results of an experiment that shows that cleavage at an immunoglobulin recombination signal sequence leads to formation of a hairpin in vitro.

22. Present evidence for a reverse transcriptase activity in retrovirus particles and the effects of RNase on this activity.

23. Describe and show the results of an experiment that demonstrates that strong-stop reverse transcripts in retroviruses are base-paired to the RNA genome and covalently attached to an RNA primer.

24. Illustrate the difference between the structures of the LTRs in genomic retroviral RNAs and retroviral proviruses.

25. Diagram the conversion of a retrovirus RNA to a provirus. Show how this explains the difference in question 24.

26. Compare and contrast the mechanisms of retrovirus replication and retrotransposon transposition.

27. Summarize the evidence that retrotransposons transpose via an RNA intermediate.

28. Describe and show the results of an experiment that demonstrates that the endonuclease of a LINE-like element can specifically nick one strand of the element's target DNA.

29. Describe and show the results of an experiment that demonstrates that a LINE-like element can use a nicked strand of its target DNA as a primer for reverse transcription of the element.

30. Present a model for retrotransposition of a LINE-like element.

31. Compare and contrast with diagrams the proposed retrohoming and retrotransposition mechanisms used by group II introns.

32. Describe a method to detect retrotransposition of a group II intron. How can the experimenter tell that the retrotransposition is not retrohoming?

SUGGESTED READINGS

General References and Reviews

Baltimore, D. 1985. Retroviruses and retrotransposons: The role of reverse transcription in shaping the eukaryotic genome. *Cell* 40:481–82.

Cohen, S.N. and J.A. Shapiro. 1980. Transposable genetic elements. *Scientific American* 242 (February):40–49.

Craig, N.L. 1985. Site-specific inversion: Enhancers, recombination proteins, and mechanism. *Cell* 41:649–50.

Craig, N.L. 1988. The mechanism of conservative site-specific recombination. *Annual Reviews of Genetics* 22:77–105.

Craig, N.L. 1996 V(D)J recombination and transposition: Closer than expected. *Science* 271:1512.

Doerling, H.-P. and P. Starlinger. 1984. Barbara McClintock's controlling elements: Now at theDNA level. *Cell* 39:253–59.

Eickbush, T.H. 2000. Introns gain ground. *Nature* 404:940–41.

Engels, W.R. 1983. The P family of transposable elements in *Drosophila*. *Annual Review of Genetics* 17:315–44.

Federoff, N.V. 1984. Transposable genetic elements in maize. *Scientific American* 250(June):84–99.

Grindley, N.G.F. and A.E. Leschziner. 1995. DNA transposition: From a black box to a color monitor. *Cell* 83:1063–66.

Kazazian, H.H., Jr. and J.V. Moran. 1998. The impact of L1 retrotransposons on the human genome. *Nature Genetics* 19:19–24.

Levin, K.L. 1997. It's prime time for reverse transcriptase. *Cell* 88:5–8.

Lewis, S.M. 1994. The mechanism of V(D)J joining: Lessons from molecular, immunological, and comparative analyses. *Advances in Immunology* 56:27–50.

Mitzuuchi, K. and R. Craigie, 1986. Mechanism of bacteriophage Mu transposition. *Annual Review of Genetics* 20:385–429.

Tonegawa, S. 1983. Somatic generation of antibody diversity. *Nature* 302:575–81.

Voytas, D.F. 1996. Retroelements in genome organization. *Science* 274:737–38.

Weisberg, R.A. and A. Landy. 1983. Site-specific recombination in phage lambda. In R.W. Hendrix, J.W. Roberts, F.W. Stahl, and R.A. Weisberg, eds., *Lambda II*. Cold Spring Harbor, NY: Cold Spring Harbor Laboratory Press, pp. 211–50.

Research Articles

Baltimore, D. 1970. Viral RNA-dependent DNA polymerase. *Nature* 226:1209–11.

Boland, S. and N. Kleckner, 1996. The three chemical steps of Tn10/IS10 transposition involve repeated utilization of a single active site. *Cell* 84:223–33.

Burgin, A.B., Jr. and H.A. Nash. 1992. Symmetry in the mechanism of bacteriophage λ integrative recombination. *Proceedings of the National Academy of Sciences USA* 89:9642–46.

Chapman, K.B., A.S. Byström, and J.D. Boeke. 1992. Initiator methionine tRNA is essential for Ty1 transcription. *Proceedings of the National Academy of Sciences USA* 89:3236–40.

Cousineau, B., S. Lawrence, D. Smith, and M. Belfort. 2000. Retrotransposition of a bacterial group II intron. *Nature* 404:1018–21.

Craig. N.L. and H.A. Nash. 1983. The mechanism of phage λ site-specific recombination: Site-specific breakage of DNA by Int topoisomerase. *Cell* 35:795–803.

Davies, D.R., I.Y. Goryshin, W.S. Reznikoff, and I. Rayment. 2000. Three-dimensional structure of the Tn5 synaptic complex transposition intermediate. *Science* 289:77–85.

Difilippantonio, M.J., C.J. McMahan, Q.M. Eastman, E. Spanopoulou, and D.G. Schatz. 1996. RAG1 mediates signal sequence recognition and recruitment of RAG2 in V(D)J recombination. *Cell* 87:253–62.

Garfinkel, D.J., J.F. Boeke, and G.R. Fink. 1985. Ty element transposition: Reverse transcription and virus-like particles. *Cell* 42:507–17.

Hesse, J.E., M.R. Lieber, K. Mizuuchi, and M. Gellert. 1989. V(D)J recombination: A functional definition of the joining signals. *Genes and Development* 3:1053–61.

Kikuchi, Y. and H.A. Nash. 1979. Nicking-closing activity associated with bacteriophage λ *int* gene product. *Proceedings of the National Academy of Sciences USA* 76:3760–64.

Luan, D.D., M.H. Korman, J.L. Jakubczak, and T.H. Eickbush. 1993. Reverse transcription of R2Bm RNA is primed by a nick at the chromosomal target site: A mechanism for non-LTR retrotransposition. *Cell* 72:595–605.

Oettinger, M.A., D.G. Schatz, C. Gorka, and D. Baltimore. 1990. RAG-1 and RAG-2, adjacent genes that synergistically activate V(D)J recombination. *Science* 248:1517–22.

Panet, A., W.A. Haseltine, D. Baltimore, G. Peters, F. Harada, and J.E. Dahlberg. 1975. Specific binding of tryptophan transfer RNA to avian myeloblastosis virus RNA-dependent DNA polymerase (reverse transcriptase). *Proceedings of the National Academy of Sciences USA* 72:2535–39.

Pargellis, C.A., S.E. Nunes-Düby, L. Moitoso de Vargas, and A. Laudy. 1988. Suicide recombination substrates yield covalent λ integrase–DNA complexes and lead to identification of the active site tyrosine. *Journal of Biological Chemistry* 263:7678–85.

Temin, H.M. and Mizutani, S. 1970. RNA-dependent DNA polymerase in virions of Rous sarcoma virus. *Nature* 226:1211–13.

Wessler, S.R. 1988. Phenotypic diversity mediated by the maize transposable elements *Ac* and *Spm*. *Science* 242:399–405.

CHAPTER 24

Genomics

A DNA microarray used to measure expression of thousands of genes at a time. *Inset:* A technician analyzes expression of particular genes. © *IncyteGenomics.*

Throughout most of this book, we have been concerned with the activities of genes—taken one gene at a time. But, with the advent of rapid and relatively inexpensive sequencing methods, molecular biologists have been able to obtain the base sequences of entire genomes, and a new subdiscipline has been born: genomics, the study of the structure and function of whole genomes. In this chapter we will examine the techniques scientists use to sequence DNA on a massive scale, and the ways in which these sequences can be used.

24.1 The First Sequenced Genomes

The first genome to be sequenced, as you might expect, was a very simple one: The small DNA genome of an *E. coli* phage called φX174. Frederick Sanger, the inventor of the dideoxy chain termination method of DNA sequencing, obtained the sequence of this 5375-nt genome in 1977.

What kind of information can we glean from this sequence? First, we can locate exactly the coding regions for all the genes. This tells us the spatial relationships among genes and the distances between them to the exact nucleotide. How do we recognize a coding region? It contains an **open reading frame (ORF)**, a sequence of bases that, if translated in one frame, contains no stop codons for a relatively long distance—long enough to code for one of the phage proteins. Furthermore, the open reading frame must start with an ATG (or occasionally a GTG) triplet, corresponding to an AUG (or GUG) translation initiation codon, and end with a stop codon (UAG, UAA, or UGA). In other words, an open reading frame is the same as a gene's coding region.

The base sequence of the phage DNA also tells us the amino acid sequence of all the phage proteins. All we have to do is use the genetic code to translate the DNA base sequence of each open reading frame into the corresponding amino acid sequence. This may sound like a laborious process, but a personal computer can do it in a split second.

Sanger's analysis of the open reading frames of the φX174 DNA revealed something unexpected and fascinating: Some of the phage genes overlap. Figure 24.1a shows that the coding region for gene B lies within gene A and the coding region for gene E lies within gene D. Furthermore, genes D and J overlap by 1bp. How can two genes occupy the same space and code for different proteins? The answer is that the two genes are translated in different reading frames (Figure 24.1b). Because entirely different sets of codons will be encountered in these two frames, the two protein products will also be quite different.

This was certainly an interesting finding, and it raised the question of how common this phenomenon would be. So far, major overlaps seem to be confined almost exclusively to viruses, which is not surprising because these simple genetic systems have small genomes in which the premium is on efficient use of the genetic material. Moreover, viruses have prodigious power to replicate, so enormous numbers of generations have passed during which evolution has honed the viral genomes.

With the advent of automated sequencing, geneticists have added much larger genomes to the list of total known sequences. In 1995, Craig Venter and Hamilton Smith and colleagues determined the entire base sequences of the genomes of two bacteria: *Haemophilus influenzae* and *Mycoplasma genitalium*. The *H. influenzae* (strain Rd) genome contains 1,830,137 bp and it was the first genome from a free-living organism to be completely sequenced. The *M. genitalium* genome is the smallest of any known free-living organism and contains only about 470 genes. In April 1996, the leaders of an international consortium of laboratories announced another milestone: The 12-million-bp genome of baker's yeast (*Saccharomyces cerevisiae*) had been sequenced. This was the first eukaryotic genome to be entirely sequenced. Later in 1996, the first genome of an organism (*Methanococcus jannaschii*) from the third domain of life, the archaea, was sequenced. Then, in 1997, the long-awaited sequence of the 4.6 million-bp *E. coli* genome was reported. This is

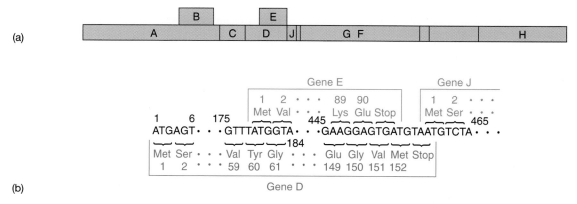

Figure 24.1 The genetic map of phage φX174. (a) Each letter stands for a phage gene. **(b)** Overlapping reading frames of φX174. Gene D (red) begins with the base numbered 1 in this diagram and continues through base number 459. This corresponds to amino acids 1–152 plus the stop codon TAA. Dots represent bases or amino acids not shown. Only the nontemplate strand is shown. Gene E (blue) begins at base number 179 and continues through base number 454, corresponding to amino acids 1–90 plus the stop codon TGA. This gene uses the reading frame one base to the right, relative to the reading frame of gene D. Gene J (gray) begins at the base number 459 and uses the reading frame one base to the left, relative to gene D.

Table 24.1	Milestones in Genomic Sequencing		
Genome (Importance)		**Size**	**Year**
Phage φX174 (first genome)		5,375	1977
Hemophilus infiuenzae (bacterium, first organism)		1,830,000	1995
Mycoplasma genitalium (bacterium, smallest genome)		580,000	1995
Saccharomyces cerevisiae (yeast, first eukaryote)		12,068,000	1996
Methanococcus jannaschii (first archaeon)		1,660,000	1996
Escherichia coli (best studied bacterium)		4,639,221	1997
Borrelia burgdorferi (the spirochete that causes Lyme disease)		910,725	1997
Caenorhabditis elegans (first animal, roundworm)		97,000,000	1998
Arabidopsis thaliana (first plant, mustard family)		120,000,000	2000
Human chromosome 22 (first human chromosome)		53,000,000	1999
Drosophila melanogaster (a favorite genetic model)		180,000,000	2000
Human (working draft of the "holy grail" of genomics)		3,200,000,000	2001

only about one-third the size of the yeast genome, but the importance of *E. coli* as a genetic tool made this a milestone as well. In 1998, the sequence of the first animal genome, from the roundworm *Caenorhabditis elegans,* was reported. The first plant genome (from the mustard family member *Arabidopsis thaliana*) was completed in 2000. Also in 2000, the eagerly-awaited rough draft of the human genome sequence was announced. By 2001, this "working draft" of the human genome was published and dozens of bacterial genomes had been sequenced. Table 24.1 presents a time line of the most important achievements in genome sequencing. In the following sections we will discuss the lessons we have learned from these sequences.

SUMMARY The base sequences of viruses and organisms ranging from simple phages to bacteria to simple animals and plants have been obtained. A rough draft of the human genome has also been obtained.

The Human Genome Project

In 1990, American geneticists embarked on an ambitious quest: to map and ultimately sequence the entire human genome. This effort, which quickly became an international program, was somewhat controversial at first, partly because of the enormous effort and cost of carrying it through to its ultimate goal: knowing the entire base sequence of every one of the human chromosomes. The reason for the high cost, of course, is that the human genome is huge—more than 3 billion bp. To get an idea of the magnitude of this task, consider that if all 3 billion bp were written down, it would take about 500,000 pages of the journal *Nature* to contain all the information. If you

could stand the boredom, it would take you about 60 years, working 8 h/day, every day, at 5 bases a second, to read it all. Assuming a 1990 cost of about a dollar a base, the project would consume more than $3 billion, vastly more than we are used to devoting to a single biological project. But now, more efficient sequencing methods are allowing the project to be completed much sooner and at much lower cost than originally estimated.

The original plan for the Human Genome Project was systematic and conservative: First, geneticists would prepare genetic and physical maps of the genome. These would contain the markers, or signposts, that would allow DNA sequences to be pieced together in the proper order. The bulk of the sequencing would be done only after the mapping was complete and clones representing all points on the map were in hand—systematically stored in freezers around the world. The original target date for completion of the sequence was 2005.

Then, in May of 1998, Craig Venter, who had established a private company, Celera, to sequence the human genome (and other genomes), shocked the genomics community by announcing that Celera would complete a rough draft of the human genome by the end of 2000. That timetable was astonishing enough, but the method by which he proposed to do the sequencing was even more arresting. Instead of relying on a map, with the ordered clones used to build it, Venter proposed a **shotgun sequencing** approach in which the whole human genome would be chopped up and cloned, then the clones would be sequenced at random, and finally the sequences would be pieced together using powerful computer programs. It was not long before Francis Collins, director of the publicly financed Human Genome Project, rose to Venter's challenge and promised that he and his colleagues would also produce a rough draft by the end of 2000, and a polished final draft by 2003, using the map-then-sequence strategy.

The upshot of this race was a tie of sorts. Venter and Collins appeared with President Clinton and other dignitaries at a ceremony in the East Room of the White House on June 26, 2000, to announce the completion of a rough draft of the human genome. In truth, the Celera group's rough draft was more complete than the public group's: 99% vs. 85%. Still, a rough draft was indeed complete. We will examine the two approaches to sequencing large genomes: mapping, then sequencing (clone by clone); and shotgun sequencing. But first, let us examine the cloning vectors that have been developed for massive projects like the Human Genome Project.

Vectors for Large-Scale Genome Projects

No matter which sequencing strategy is used, one must first clone fragments of the genome in appropriate vectors, and large fragments are particularly valuable. We will describe two of the most popular here: yeast artificial chromosomes and bacterial artificial chromosomes. The early mapping work relied on yeast artificial chromosomes, so we will begin with those.

Yeast Artificial Chromosomes   The main problem with the cloning tools described in Chapter 4 is that they do not hold enough DNA for large-scale physical mapping of the human genome. Even the cosmids accommodate DNA inserts up to only about 50 kb, which is too small for efficient mapping of regions spanning more than a million bases.

Vectors called **yeasts artificial chromosomes,** or **YACs,** were very useful in mapping the human genome because they could accommodate hundreds of thousands of kilobases each. YACs containing a megabase or more are known as "megaYACs." A YAC contains a left and right yeast chromosomal telomere (Chapter 21), which are both necessary for proper chromosomal replication, and a yeast centromere, which is necessary for segregation of sister chromatids to opposite poles of the dividing yeast cell. The centromere is placed adjacent to the left telomere, and a huge piece of human (or any other) DNA can be placed in between the centromere and the right telomere, as shown in Figure 24.2. The large DNA inserts are prepared by slightly digesting long pieces of human DNA with a restriction enzyme. The YACs, with their huge DNA inserts, can then be introduced into yeast cells, where they will replicate just as if they were normal yeast chromosomes.

Using YACs, geneticists made great strides in the mapping phase of the Human Genome Project. They produced a genetic map of the whole genome that provided an average resolution of 0.7 centimorgan. A **centimorgan (cM)** is the distance that yields a 1% recombination frequency between two markers and corresponds to an average of about 1 Mb in humans. These researchers also produced

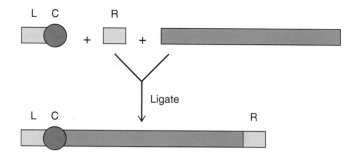

Figure 24.2 Cloning in yeast artificial chromosomes. We begin with two tiny pieces of DNA from the two ends of a yeast chromosome. One of these, the left arm, contains the left telomere (yellow, labeled L) plus the centromere (red, labeled C). The right arm contains the right telomere (yellow, labeled R). These two arms are ligated to a large piece of foreign DNA (blue)—several hundred kilobases of human DNA, for example—to form the YAC, which can replicate in yeast cells along with the real chromosomes.

relatively high-resolution physical maps of two of the smallest chromosomes, 21 and Y. These maps were especially useful in that they represented long stretches of overlapping DNA segments cloned in YACs. Thus, in the days before the human genome was sequenced, if you were interested in a disease gene that mapped to one of these chromosomes, you had a much simplified task. You needed only to discover two markers flanking the gene of interest, look on the map to find which YAC or YACs contained these markers, obtain the YACs, and begin your final search for the gene.

Bacterial Artificial Chromosomes   Despite all the success they made possible in human genome mapping, YACs suffer from several serious drawbacks: They are inefficient (not many clones are obtained per microgram of DNA); they are hard to isolate from yeast cells; they are unstable; and they tend to contain scrambled inserts that are really composites of DNA fragments from more than one site. **Bacterial artificial chromosomes (BACs)** solve all of these problems and have therefore been the vector of choice for the sequencing phase of the Human Genome Project.

BACs are based on a well-known natural plasmid that inhabits *E. coli* cells: the **F plasmid.** This plasmid allows conjugation between bacterial cells. In some conjugation events, the F-plasmid itself is transferred from a donor **F⁺ cell** to a recipient **F⁻ cell,** converting the latter to an F⁺ cell. In other events, a small piece of host DNA is transferred as an insert in the F plasmid (which is called an **F′ plasmid** if it has an insert of foreign DNA). And in still other events, the F plasmid inserts into the host chromosome and mobilizes the whole chromosome to pass from the donor cell to the recipient cell. Thus, because the *E. coli* chromosome contains over 4 million bp, the F plasmid can obviously accommodate a large insert of DNA. In practice, BACs usually have inserts less than

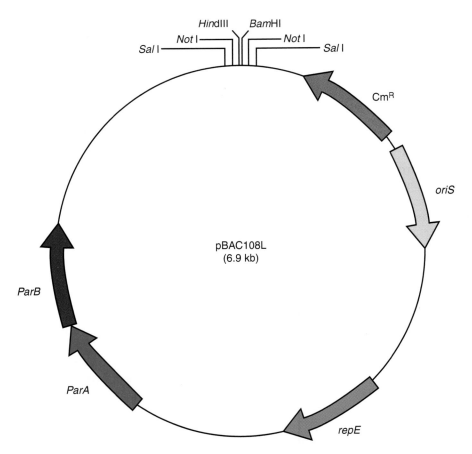

Figure 24.3   Map of the BAC vector, pBAC108L. Key features include the cloning sites *Hin*dIII and *Bam*HI, at top; the chloramphenicol resistance gene (Cm^R), used as a selection tool; the origin of replication (*oriS*); and the genes governing partition of plasmids to daughter cells (*ParA* and *ParB*).

300,000 bp (average about 150,000 bp), and these plasmids are stable in vivo and in vitro. Unlike the linear YACs, which tend to break under shearing forces, the circular, supercoiled BACs resist breakage.

Figure 24.3 shows the map of one of the first BACs, which was developed by Melvin Simon and colleagues in 1992. It has an origin of replication, a cloning site with two restriction sites (for *Hin*dIII and *Bam*HI) into which large DNA fragments may be inserted. It also has genes (the Par genes) that govern plasmid partition to the daughter cells that keep the plasmid copy number at about two per cell. This contributes to the stability of the plasmid.

SUMMARY Two high-capacity vectors have been used extensively in the Human Genome Project. Much of the mapping work was done with yeast artificial chromosomes (YACs), which can accept inserts of a million or more base pairs. Most of the sequencing work is being performed with bacterial artificial chromosomes (BACs) which can accept up to about 300,000 bp. The BACs are more stable and easier to work with than the YACs.

The Clone-by-Clone Strategy

This strategy has inherent appeal because it is so systematic. First, the whole genome is mapped by finding markers regularly spaced along each chromosome. A byproduct of the mapping is a collection of clones corresponding to the markers. Because we already know the order of these clones, we can sequence each one and put that sequence in its proper place in the whole genome. Thus, this method is commonly called the **clone-by-clone sequencing** strategy. Aside from their usefulness in cloning, genetic and physical maps have another important benefit: They give us signposts to use when searching for the genes responsible for diseases. We will return to this issue when we discuss positional cloning later in this chapter. In the next section, we will consider some of the most powerful methods used in mapping large genomes.

Restriction Fragment Length Polymorphisms   In the late twentieth century, we knew the locations of relatively few human genes, so the likelihood of finding one of these close to a new gene we were trying to map was small. Another approach, which does not depend on finding linkage with a known gene, is to establish linkage with an

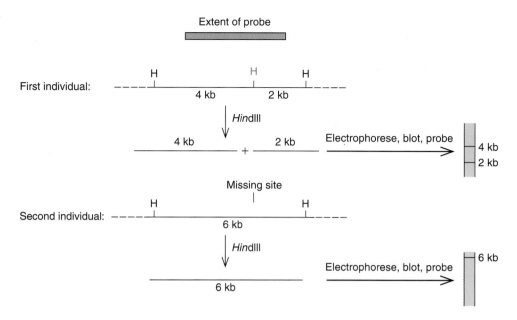

Figure 24.4   Detecting a RFLP. Two individuals are polymorphic with respect to a *Hind*III restriction site (red).The first individual contains the site, so cutting the DNA with *Hind*III yields two fragments, 2 and 4 kb long, that can hybridize with the probe, whose extent is shown at top. The second individual lacks this site, so cutting that DNA with *Hind*III yields only one fragment, 6 kb long, which can hybridize with the probe. The results from electrophoresis of these fragments, followed by blotting, hybridization to the radioactive probe, and autoradiography, are shown at right. The fragments at either end, represented by dashed lines, do not show up because they cannot hybridize to the probe.

"anonymous" stretch of DNA that may not even contain any genes. We can recognize such a piece of DNA by its pattern of cleavage by restriction enzymes.

Because each person differs genetically from every other, the sequences of their DNAs will differ somewhat, as will the pattern of cutting by restriction enzymes. Consider the restriction enzyme *Hind*III, which recognizes the sequence AAGCTT. One individual may have three such sites separated by 4 and 2 kb, respectively, in a given region of a chromosome (Figure 24.4). Another individual may lack the middle site but have the other two, which are 6 kb apart. This means that if we cut the first person's DNA with *Hind*III, we will produce two fragments, 2 kb and 4 kb long, respectively. The second person's DNA will yield a 6-kb fragment instead. In other words, we are dealing with a **restriction fragment length polymorphism (RFLP).** Polymorphism means that a genetic locus has different forms, or alleles (Chapter 1), so this clumsy term simply means that cutting the DNA from any two individuals with a restriction enzyme may yield fragments of different lengths. The abbreviated term, RFLP, is usually pronounced "rifflip."

How do we go about looking for a RFLP? Clearly, we cannot analyze the whole human genome at once. It contains hundreds of thousands of cleavage sites for a typical restriction enzyme, so each time we cut the whole genome with such an enzyme, we release hundreds of thousands of fragments. No one would relish sorting through that morass for subtle differences between individuals.

Fortunately, there is an easier way. With a Southern blot (Chapter 5) one can highlight small portions of the total genome with various probes, so any differences are easy to see. However, there is a catch. Because each labeled probe hybridizes only to a small fraction of the total human DNA, the chances are very poor that any given one will reveal a RFLP linked to the gene of interest. We may have to screen thousands of probes before we find the right one. As laborious as it is, this procedure at least provides a starting point, and it has been a key to finding the genes responsible for several genetic diseases.

Variable Number of Tandem Repeats   The greater the degree of polymorphism of a RFLP, the more useful it will be. If only 1 person in a hundred has one form of the RFLP (the 6-kb fragment, for example), and the other 99 have the other form (the 4-kb and 2-kb fragments), we will have to screen many individuals before we find the one rare variant. This makes mapping very tedious. However, some RFLPs, called **variable number tandem repeats,** or **VNTRs,** are more useful. These derive from **minisatellites,** stretches of DNA that contain a short core sequence repeated over and over in tandem (head to tail). Because the number of repeats of the core sequence in a VNTR is likely to be different from one individual to another, VNTRs are highly polymorphic, and therefore relatively easy to map. However, VNTRs have a disadvantage as genetic markers: They tend to bunch together at the ends of chromosomes, leaving the interiors of the chromosomes relatively devoid of markers.

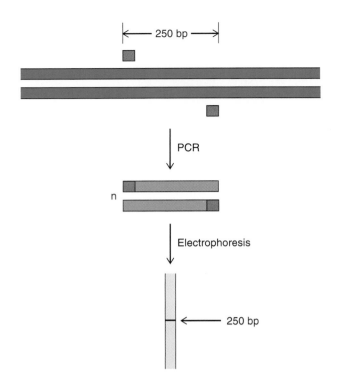

Figure 24.5 Sequence-tagged sites. We start with a large cloned piece of DNA, extending indefinitely in either direction. The sequences of small areas of this DNA are known, so one can design primers that will hybridize to these regions and allow PCR to produce double-stranded fragments of predictable lengths. In this example, two PCR primers (red) spaced 250 bp apart have been used. Several cycles of PCR generate many double-stranded PCR products that are precisely 250 bp long. Electrophoresis of this product allows one to measure its size exactly and confirm that it is the correct one.

Sequence Tagged Sites Another kind of anonymous genetic marker, which is very useful to genome mappers, is the **sequence-tagged site (STS).** STSs are short sequences, about 60–1000 bp long, that can be detected by PCR. Figure 24.5 illustrates how to use PCR to detect an STS. One must first know enough about the DNA sequence in the region being mapped to design short primers that will hybridize a few hundred base pairs apart and cause amplification of a predictable length of DNA in between. One can then apply PCR with these two primers to any unknown DNA; if the proper size amplified DNA fragment appears, then the unknown DNA has the STS of interest. Notice that hybridization of the primers to the unknown DNA is not enough; they must hybridize a specific number of base pairs apart to give the right size PCR fragment. This provides a check on the specificity of hybridization. One great advantage of STSs as a mapping tool is that no DNA must be cloned and examined and kept in someone's freezer. Instead, the sequences of the primers used to generate an STS are published and then anyone in the world can order those same primers and find the same STS in an experiment that takes just a few hours.

Microsatellites STSs are very useful in physical mapping or locating specific sequences in the genome. But they are worthless as markers in traditional genetic mapping unless they are polymorphic. Only then can we use them to determine genetic linkage. Fortunately, geneticists have discovered a class of STSs called **microsatellites** that are highly polymorphic. Microsatellites are similar to minisatellites in that they consist of a core sequence repeated over and over many times in a row. However, whereas the core sequence in typical minisatellites is a dozen or more base pairs long, the core in microsatellites is much smaller—usually only 2–4 bp long. In 1992, Jean Weissenbach and his colleagues produced a linkage map of the entire human genome based on 814 microsatellites containing a C–A dinucleotide repeat. They isolated cloned DNAs containing these microsatellites and used their sequences to design PCR primers that flank the repeats at each locus. A given pair of primers yielded a PCR product whose size depended on the number of C–A repeats in a given individual's microsatellite at that locus. Happily, the number of repeats varied quite a bit from one individual to another. Besides the fact that microsatellites are highly polymorphic, they are also widespread and relatively uniformly distributed in the human genome. Thus, they are ideal as markers for both linkage and physical mapping.

Genetic (linkage) mapping with microsatellites is done by the same technique outlined in Chapter 1 for traditional genetic markers in fruit flies. Instead of determining the recombination frequency between, say, wing shape and eye color, geneticists would determine the recombination frequency between two microsatellites. For example, consider an example in which a man's DNA yields a microsatellite at one locus that is 78 bp long and a microsatellite at a nearby locus that is 42 bp long. His wife has a microsatellite at the first locus that is 103 bp long and a microsatellite at the second locus that is 35 bp long. Within limits, the more their children show nonparental combinations of these two markers (e.g., a child with microsatellites that are 78 and 35 bp long, respectively), the more recombination has occurred between the markers, and the farther apart the markers are on the chromosome.

Geneticists interested in physically mapping or sequencing a given region of a genome aim to assemble a set of clones called a **contig,** which contains contiguous (actually overlapping) DNAs spanning long distances. This is rather like putting together a jigsaw puzzle; the bigger the pieces, the easier the puzzle. Thus, it is essential to have vectors like BACs and YACs that hold big chunks of DNA. Assuming we have a BAC library of the human genome, we need some way to identify the clones that contain the region we want to map. This can be done in several ways. We could hybridize BAC DNA to a labeled DNA probe corresponding to the region of interest, but this is subject to some uncertainty due to possible nonspecific hybridization. A more reliable method is to look for

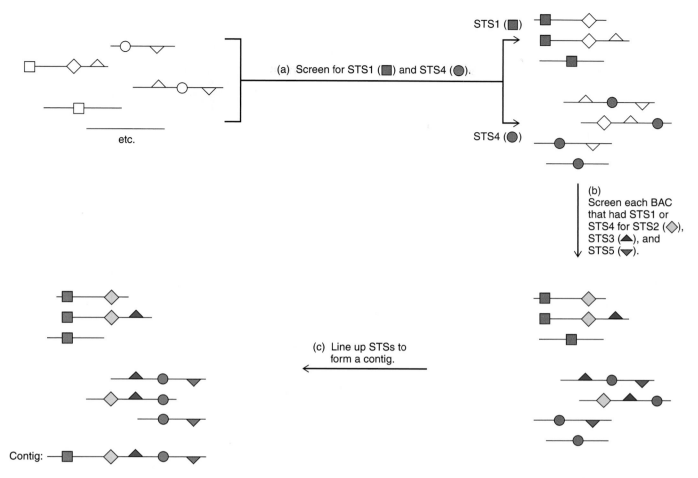

Figure 24.6 Mapping with STSs. At top left, several representative BACs are shown, with different symbols representing different STSs placed at specific intervals. In step **(a)** of the mapping procedure, screen for two or more widely spaced STSs. In this case screen for STS1 and STS4. All those BACs with either STS1 or 4 are shown at top right. The identified STSs are shown in color. In step **(b)**, each of these positive BACs is further screened for the presence of STS2, STS3, and STS5. The colored symbols on the BACs at bottom right denote the STSs detected in each BAC. In step **(c)**, align the STSs in each BAC to form the contig. Measuring the lengths of the BACs by pulsed-field gel electrophoresis helps to pin down the spacing between pairs of BACs.

STSs in the BACs. It is best to screen the BAC library for at least two STSs, spaced hundreds of kilobases apart, so BACs spanning a long distance are selected.

After we have found a number of positive BACs, we begin mapping by screening them for several additional STSs, so we can line them up in an overlapping fashion as shown in Figure 24.6. This set of overlapping BACs is our new contig. We can now begin finer mapping, and even sequencing, of the contig.

Radiation Hybrid Mapping This procedure of mapping with BACs sounds straightforward, but it presents difficulties. One of the most important is that BACs are so small relative to a whole human chromosome that creating a BAC contig of a whole chromosome would be unbearably laborious. So we need a method to find linkage between STSs that are even farther apart than those that could fit into a single BAC. **Radiation hybrid mapping**

provides a way. We begin by irradiating human cells with lethal doses of ionizing radiation, such as x-rays or gamma rays, which break the human chromosomes into pieces. Next, we fuse these doomed human cells with hamster cells to form hybrid cells that contain only some of the human chromosome fragments. Then, we examine clones of hybrid cells to see which STSs tend to be found together in the hybrid cells. The more often they are together, the closer together they are likely to be on a human chromosome.

In 1996, an international consortium of geneticists, including G. D. Schuler, published a human map based on STSs mapped by this technique. It contained more than 16,000 STS markers, plus about a thousand genetic markers mapped by classical linkage methods (family studies), which provided an overall framework for the map. The STS markers used in this study were a special class called **expressed sequence tags** (**ESTs**). These are STSs that are generated by starting with mRNAs and using the enzyme reverse transcriptase to make short corresponding cDNAs. These short cDNAs are then amplified by PCR and cloned. Thus, ESTs represent genes that are expressed in the cell from which the mRNAs were isolated. Because the STS (or EST) method amplifies only a small part of a gene, a given gene may be represented by many different ESTs in an EST database. To minimize such duplications, the mapping consortium confined their mapping to ESTs that represented the 3′-untranslated regions of genes. By 1998, the international consortium (P. Deloukas et al.) had refined and extended the map to include over 30,000 genes.

> **SUMMARY** Mapping the human genome requires a set of landmarks to which we can relate the positions of genes. Some of these markers are genes, but many more are anonymous stretches of DNA, such as RFLPs, VNTRs, STSs (including ESTs), and microsatellites. The latter two are regions of DNA that can be identified by formation of a predictable length of amplified DNA by PCR with pairs of primers. By 1998, geneticists had constructed STS-based maps that included over 30,000 genes.

Shotgun Sequencing

The shotgun-sequencing strategy (Figure 24.7), first proposed by Craig Venter, Hamilton Smith, and Leroy Hood in 1996, bypasses the mapping stage and goes right to the sequencing stage. The sequencing starts with a set of BAC clones containing very large DNA inserts, averaging about 150 kb. The insert in each BAC is sequenced on both ends using an automated sequencer that can usually read about 500 bases at a time, so 500 bases at each end of the clone will be determined. Assuming that 300,000 clones of human DNA are sequenced this way, that would generate 300 million bases of sequence, or about 10% of the total human genome, and the 500-base sequenced regions would therefore occur on average every 5 kb in the genome. These 500-base sequences serve as an identity tag, called a **sequence-tagged connector** (STC), for each BAC clone. On average, assuming an average clone size of 150 kb, and an STC every 5 kb, 30 clones (150 kb/5 kb = 30) should share a given STC somewhere within their span. This is the origin of the term *connector*—each clone should be "connected" via its STCs to about 30 other clones.

The next step is to **fingerprint** each clone by digesting it with a restriction enzyme. This serves two important purposes. First, it tells the insert size (the sum of the sizes of all the fragments generated by the restriction enzyme). Second, it allows one to eliminate aberrant clones whose fragmentation patterns do not fit the consensus of the overlapping clones.

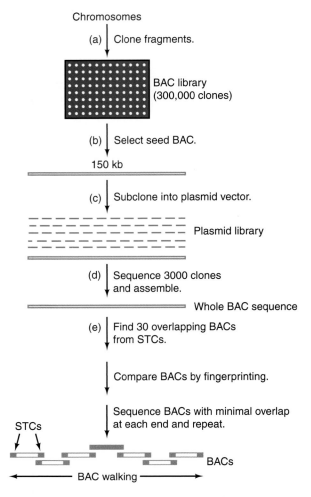

Figure 24.7 Shotgun-sequencing method. (a) Chromosomes are cloned into a BAC vector, yielding a collection of 300,000 BAC clones. A 96-well microtiter plate is shown with 96 of the clones. **(b)** A seed BAC is selected for sequencing. **(c)** The seed BAC is subcloned into a plasmid vector, yielding a plasmid library. **(d)** Three thousand of the plasmid clones are sequenced, and the sequences are ordered by their overlaps, producing the sequence of the whole 150-kb BAC. **(e)** Find the BACs (about 30) with overlapping STCs, then compare them by fingerprinting to find those with minimal overlaps, and sequence them. This process, known as BAC walking, can in principle create a contig covering the whole chromosome. (*Source:* Reprinted with permission from Venter et al., "A new strategy for genome sequencing," *Nature* 381:365, 1996. Copyright © 1996 Macmillan Magazines Limited.)

The next step is to obtain the entire sequence of a BAC that looks interesting (a **seed BAC**). This is done by subdividing the BAC into smaller clones—frequently in a pUC vector with inserts averaging only about 2 kb. This whole BAC sequence allows the identification of the 30 or so other BACs that overlap with the seed: They are the ones with STCs that occur somewhere in the seed BAC.

Next, one selects other BACs with minimal overlap with the original one and proceeds to sequence them. Then this process is repeated with other BACs with minimal overlap with the second set, and so forth. This strategy, called **BAC walking,** would in principle allow one laboratory to sequence the whole human genome—given enough time.

But we do not have that much time, so Venter and colleagues modified the procedure by sequencing BACs at random until they had about 35 billion nt of sequence. In principle that should cover the human genome ten times over, giving a high degree of coverage and accuracy. Then they fed all the sequence into a computer with a powerful program that found areas of overlap between clones and fit their sequences together, building the sequence of the whole genome.

As mentioned a little earlier, the bulk of the sequencing is done with pUC clones with relatively small inserts—only about 2 kb each. But these small inserts would not provide enough overlaps to piece together the whole genome. This drawback is especially apparent in regions of repeated DNA. A 2-kb cloned sequence from a 10-kb region of tandem DNA repeats would give no clues about where the cloned sequence fit within the larger repeat region—one part looks the same as another. That is one way the BAC clones come in handy: They are large enough to cover almost any repeated region. They also provide overlaps spanning large DNA regions, so they can help to organize the smaller cloned fragments. This job was also facilitated by the physical maps, especially the STS maps, that were already available. So the shotgun strategy for sequencing the human genome was in practice a hybrid of a pure shotgun and a map-then-sequence strategy.

Any strategy to sequence over 3 billion bp depends on a high-volume, low-cost sequencing method. We now have sequencing devices that perform electrophoresis of DNA fragments in capillary tubes instead of the traditional thin gel slabs. These instruments are fully automated and each can handle about 1000 samples per day with only 15 min of human attention. Another of Venter's companies, The Institute for Genome Research (TIGR), has 230 such instruments; together, they can produce about 100 Mb of DNA sequence every day, with a relatively low labor cost.

> **SUMMARY** Massive sequencing projects can take two forms: (1) In the map-then-sequence strategy, one produces a physical map of the genome including STSs, then sequences the clones (mostly BACs) used in the mapping. This places the sequences in order so they can be pieced together. (2) In the shotgun approach, one assembles libraries of clones with different size inserts, then sequences the inserts at random. This method relies on a computer program to find areas of overlap among the sequences and piece them together. In practice, a combination of these methods is being used to sequence the human genome.

Sequencing Standards

What do we mean by "rough draft," or "working draft," and "final draft" of a genome? That depends on whom you ask. Most investigators agree that a working draft may be only 90% complete and may have an error rate of up to 1%. Although there is less agreement about what qualifies as a final draft, there is consensus that it should have an error rate of less than 1/10,000 (0.01%) and should have as few gaps as possible. Some molecular biologists insist that a genome is not completely sequenced until every last gap is filled, but it will be very difficult to eliminate all gaps in the human genome. As we will see in the next section, some regions of DNA, for mostly unexplained reasons, resist cloning. The cost of overcoming the obstacles to cloning these regions will likely be prohibitively high, so the task of filling in the last few million bases of the human genome may never be done. As detailed in the next section, the consortium that sequenced human chromosome 22 decided that their sequence was "functionally complete" when they had obtained all the sequence possible with the cloning and sequencing tools currently available, even though significant gaps remained.

Progress in Sequencing the Human Genome

At the end of 1999, we tasted the first fruit of the Human Genome Project: The final draft of human chromosome 22. Then, in mid-2000 an international consortium published the final draft of human chromosome 21. These are the two smallest human autosomes, but their sequences have been very valuable. In February 2001, the Venter group and the public consortium each published their versions of a working draft of the whole human genome. In this section we will look at the lessons we have learned from the finished sequences of the two small chromosomes and the working draft of the whole genome. Before we begin, one lesson worth noting is that the finished sequences of both chromosomes came from the more orderly clone-by-clone approach. This strategy yields the final draft sequences of whole chromosomes as soon as the groups sequencing each chromosome complete their work. On the other hand, the raw sequence in the shotgun sequencing approach is not pieced together until the very end, when the computer finds the overlaps necessary to build contigs. Thus, this strategy may not yield the final draft sequence of any chromosome until the whole genome is finished.

Chromosome 22 In reality, only the long arm (22q) of the chromosome was sequenced; the short arm (22p) is composed of pure heterochromatin and is thought to be devoid of genes. Also, 11 gaps remained in the sequence. Ten of these were gaps between contigs that could not be filled with clones—presumably due to "unclonable"

DNA. The other corresponded to a 1.5-kb region of cloned DNA that resisted sequencing. The reasons that some DNAs, sometimes called "poison regions," are unclonable are not completely clear, but it is known that DNAs with unusual secondary structure or repetitive sequences are frequently lost from bacterial cells.

What have we learned from the sequence of chromosome 22? Several findings are interesting. First, we are going to have to learn to live with gaps in our sequence of the human genome, although perhaps not as many as first appeared in the sequence of this chromosome. Already by the summer of 2000, one of the gaps had been filled, and in time more probably will be. Still, it is very likely that the same problems encountered in spanning the gaps in chromosome 22 will bedevil investigators sequencing the other chromosomes. Table 24.2 lists the sequenced contigs in chromosome 22 and the gaps in between them. The contigs account for 33,464 kb, or about 97% of the chromosome, and they have been sequenced with very high accuracy—estimated at less than one error per 50,000 bases. It is interesting that all of the gaps occur in the regions of the chromosome close to the centromere and telomere. In between gaps 4 and 5 is an enormous contig composed of 23,006 kb that covers more than two-thirds of chromosome 22q.

The second major finding is that chromosome 22 contains 679 **annotated genes** (genes or gene-like sequences that are at least partially identified). These can be categorized as follows: **known genes,** whose sequences are identical to known human genes or to the sequences deduced from known human proteins; **related genes,** whose sequences are homologous to known genes of human or other species or which have regions of similarity to known genes; **predicted genes,** which contain sequences homologous to ESTs (so we are fairly sure they are expressed); and **pseudogenes,** whose sequences are homologous to known genes, but they contain defects that preclude proper expression. There are 247 known genes, 150 related genes, 148 predicted genes, and 134 pseudogenes in chromosome 22q. Computer analysis of the sequence has predicted another 325 genes, but such analyses are still very inaccurate because the algorithms depend on finding exons, and the many long introns in human genes make exons hard to spot. The best guess is that about 100 of the 325 additional genes predicted by computation are actually genes, so we can estimate that human chromosome 22 contains about 800 genes.

The third major finding is that the coding regions of genes account for only a tiny fraction of the length of the chromosome. Even counting introns, the annotated genes account for only 39% of the total length of 22q, and the exons account for only 3%. By contrast, fully 41% of 22q is devoted to repeat sequences, especially Alu sequences and LINES (Chapter 23). Table 24.3 lists the interspersed repeat elements found in chromosome 22 and their prevalences.

Table 24.2 Sequence Contigs on Chromosome 22

Contig*	Size (kb)	
AP000522–AP000529	234	
gap		**1.9**
AP000530–AP000542	406	
gap		**~150**
AP000543–AC006285	1,394	
gap		**~150**
AC008101–AC007663	1,790	
gap		**~100**
AC007731–AL049708	23,006	
gap		**~50**
AL118498–AL022339	767	
gap		**~50–100**
Z85994–AL049811	1,528	
gap		**~150**
AL049853–AL096853	2,485	
gap		**~50**
AL096843–AL078607	190	
gap		**~100**
AL078613–AL0117328	993	
gap		**~100**
AL080240–AL022328	291	
gap		**~100**
AL096767–AC002055	380	
Total sequence length	33,464	
Total length of 22q	34,491	

*Contigs are indicated by the first and last sequence in the orientation centromere to talomere and are named by their Genbank/EMBL/DDBJ accession numbers.
Source: Reprinted with permission from *Nature* 402:491, 1999. Copyright © 1999 Macmillan Magazines Limited.

A fourth major finding is that the rate of recombination varies across the chromosome, with long regions in which recombination is relatively low interspersed with short regions of relatively high rates of recombination (Figure 24.8). As we have seen earlier in this chapter, geneticists had already made a genetic map of the human genome, including chromosome 22, based on microsatellites. This map was based on recombination frequencies between microsatellites and was therefore calibrated in centimorgans. The chromosome 22 sequencing team was able to find these microsatellites in the sequence and measure the real physical distance between them. Figure 24.8 shows that a plot of the genetic distance between markers versus the physical distance between the same markers is not linear. The shaded areas are regions of high rates of recombination, and therefore high apparent genetic

Table 24.3   The Interspersed Repeat Content of Human Chromosome 22

Repeat Type	Total Number	Coverage (bp)	Coverage (%)
Alu	20,188	5,621,998	16.80
HERV	255	160,697	0.48
Line 1	8,043	3,256,913	9.73
Line 2	6,381	1,273,571	3.81
LTR	848	256,412	0.77
MER	3,757	763,390	2.28
MIR	8,426	1,063,419	3.18
MLT	2,483	605,813	1.81
THE	304	93,159	0.28
Other	2,313	625,562	1.87
Dinucleotide	1,775	133,765	0.40
Trinucleotide	166	18,410	0.06
Tetranucleotide	404	47,691	0.14
Pentanucleotide	16	1,612	0.0048
Other tandem	305	102,245	0.31
Total	55,664	14,024,657	41.91

Source: Reprinted with permission from *Nature* 402:491, 1999. Copyright © 1999 Macmillan Magazines Limited.

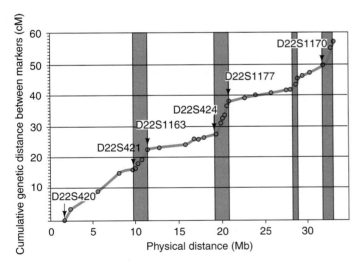

Figure 24.8   Genetic distance plotted against physical distance in chromosome 22q. The cumulative genetic distance between markers (in cM) is graphed versus the physical distance between the same markers (in Mb). Some of the genetic markers are labeled. The shaded bars delineate areas of relatively high rates of recombination (as reflected in the steeply rising curves). (*Source:* Reprinted with permission from *Nature* 402:492, 1999. Copyright © 1999 Macmillan Magazines Limited.)

distance, separated by longer regions of relatively low rates of recombination. The average ratio of genetic distance to physical distance in this chromosome is 1.87 cM/Mb. Of course, we should remember that the *y* axis represents *cumulative* genetic distance, and that the actual genetic distance between widely separated markers is not the same as the sum of the distances between intervening markers. That is because *multiple* recombination events are more probable between distant markers, which makes them appear closer together than they really are.

A fifth major finding is that chromosome 22q has several local and long-range duplications. The most obvious involves the immunoglobulin λ locus. Clustered together at this locus are 36 gene segments that are at least potentially able to encode λ variable regions (V-λ gene segments), as well as 56 V-λ pseudogenes and 27 partial V-λ pseudogenes known as "relics." Other duplications are separated by long distances. In one striking example, a 60-kb region is duplicated with greater than 90% fidelity almost 12 Mb away. Compared with the interspersed repeats, such as Alu sequences and LINES, these duplications are found in few copies, so they are known as **low-copy repeats** or **LCRs.** Seven of the eight previously described LCR22s in the centromeric end of 22q were sequenced; the eighth (LCR22-1) probably lies in the sequence gap closest to the centromere.

The sixth major finding is that large chunks of human chromosome 22q are conserved in several different mouse chromosomes. The sequencing team found 113 human genes whose mouse orthologs had been mapped to mouse chromosomes. (**Orthologs** are homologous genes in different species that have evolved from a common ancestral gene. **Paralogs,** by contrast, are homologous genes that have evolved by gene duplication within a species. **Homologs** are any kind of homologous genes—orthologs or paralogs.) These mouse orthologs clustered into eight regions on seven different mouse chromosomes, as shown in Figure 24.9. The mouse chromosomes represented in human 22q are chromosomes 5, 6, 8, 10, 11, 15, and 16. Mouse chromosome 10 is represented in two regions of human 22q. As the two species have diverged, their chromosomes have rearranged, but linkage among many markers has been preserved. Clearly, our knowledge of the sequence of the human genome will speed the process of determining the sequence of the mouse genome.

SUMMARY Human chromosome 22q has been sequenced to high accuracy, but the sequence still has 11 gaps that cannot be filled with available methods. There are 679 annotated genes, but the great bulk of the chromosome is made up of noncoding DNA, over 40% of it in interspersed repeats such as Alu sequences and LINES. The rate of recombination varies across the chromosome, with long regions of low rates of recombination punctuated by short regions with relatively high rates. The chromosome contains several examples of local and long-range duplications. The human chromosome contains large regions where linkage among genes has been conserved with that in seven different mouse chromosomes.

Chromosome 21   Despite its number (next to last), chromosome 21 is actually a little smaller than 22, but it is of special interest to geneticists because of its involvement in Down syndrome: People who inherit three copies of chromosome 21 (trisomy 21), instead of the usual two copies, are afflicted with the mental and physical abnormalities of Down syndrome. What have we learned from the sequence of chromosome 21, and how does it compare with chromosome 22?

Again, the long arm (21q) was the main target of the sequencing effort because the short arm is thought to harbor few genes and it varies considerably in length from one person to another. However, in this case, the consortium did sequence 281 kb from the short arm and found at least one gene amid many highly repetitive sequences. Also, as with chromosome 22, gaps occur in the sequence, but this time there are only three gaps, totaling only 100 kb. This means that the total amount of chromosome 21 sequenced (33.55 Mb) is a little more than the total amount of chromosome 22 sequenced (33.46 Mb) and amounts to 99.7% of chromosome 21q.

The most obvious difference in the two chromosomes revealed by sequencing is the relatively low gene density in chromosome 21. Whereas the chromosome 22 consortium identified 545 genes, the chromosome 21 consortium found only 224. One hundred twenty-seven of these are

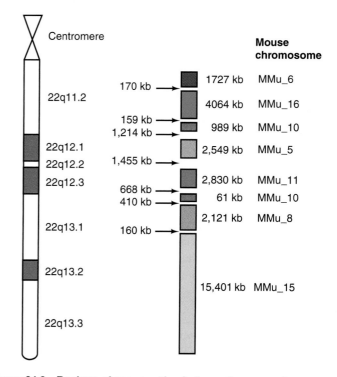

Human chromosome 22

Centromere

Mouse chromosome

170 kb →	1727 kb MMu_6
	4064 kb MMu_16
159 kb →	
1,214 kb →	989 kb MMu_10
	2,549 kb MMu_5
1,455 kb →	
	2,830 kb MMu_11
668 kb →	
410 kb →	61 kb MMu_10
	2,121 kb MMu_8
160 kb →	
	15,401 kb MMu_15

22q11.2
22q12.1
22q12.2
22q12.3
22q13.1
22q13.2
22q13.3

Figure 24.9   Regions of conservation between human and mouse chromosomes. Human chromosome 22 is depicted on the left, with the centromere near the top, and prominent bands in white and brown. The regions of seven different mouse chromosomes containing orthologs in conserved order are shown on the right. Colors correspond to conserved blocks of orthologs in the mouse chromosomes listed at far right. Gaps between blocks are labeled with the distance in kilobases between the last ortholog in one block and the first in the next. (*Source:* Reprinted with permission from *Nature* 402:494, 1999. Copyright © 1999 Macmillan Magazines Limited.)

known genes and 98 are predicted genes; 59 pseudogenes were also identified. Part of this large difference may be explained by the different techniques the two groups used to locate genes. The chromosome 22 group relied in part on matches to ESTs to identify genes. But the chromosome 21 team demanded a more stringent criterion: The ESTs whose sequences matched those of regions of the chromosome must also show signs of having been spliced from a larger precursor, as most real human transcripts are. So this more conservative approach probably underestimated the number of genes in chromosome 21.

On the other hand, the chromosome 21 consortium took a more liberal approach to estimating genes by computer algorithm (in silico predictions). Whereas the chromosome 22 group included a correction to subtract false-positives from their in silico predictions, the chromosome 21 group used no such correction. This liberal method probably overestimated the number of genes in chromosome 21. Thus, the two methods used by the chromosome

21 team, one conservative, the other liberal, probably cancelled each other, so we can feel fairly safe in concluding that chromosome 21q has many fewer genes per megabase of DNA than 22q. This may help explain why an extra copy of chromosome 21 can be tolerated, though it does cause Down syndrome. The low number of genes reduces the chances that having an extra copy of one would be lethal.

Can we use the estimated number of genes in the two chromosomes to predict the total number of human genes? That would be a risky extrapolation, because it is based on just 2% of the total human genome. Nevertheless, the chromosome 21 consortium did make such a prediction: They concluded that the human genome contains as few as 40,000 genes—much fewer than the 100,000 or so that geneticists had been expecting to find. This is close to the estimate of about 30,000 genes based on the rough draft of the whole human genome.

Another interesting insight from the chromosome 21 sequence is that all 24 genes in mouse chromosome 10 that are known to be homologous to genes in human chromosome 21 occur in the same order in both chromosomes. This is another indication of the high degree of evolutionary conservation of genetic information between the two species.

Finally, the sequence of chromosome 21 is sure to accelerate the pace of discovery of the genes involved in genetic disorders, not only Down syndrome, but maladies that are governed by defects in single genes in chromosome 21. These include a type of Alzheimer's disease and amyotrophic lateral sclerosis (Lou Gehrig's disease).

SUMMARY Human chromosome 21q, and a little of 21p, have been sequenced, leaving relatively few and short gaps. The sequence reveals a relative poverty of genes—only about 225, plus 59 pseudogenes. All 24 genes known to be shared between mouse chromosome 10 and human chromosome 21 are in the same order in both chromosomes.

Working draft of the human genome In February 2001, the Venter group and the public consortium published separately their own versions of the working draft of the whole human genome. The drafts of the human genome presented by the two groups were by no means complete. They had many gaps and inaccuracies, but they also contained a wealth of information that will keep scientists busy for years analyzing and extending it. Furthermore, the public draft, at least, will continue to improve as groups working on its separate parts complete the laborious finishing phase that eliminates gaps and corrects errors. Within two or three years, the accuracy and completeness of the whole genome should resemble that of chromosomes 21 and 22. Here are some of the most important findings so far.

The most striking discovery from both groups was the low number of genes in the genome. The Venter group found 26,588 genes for which there were at least two independent lines of evidence, and about 12,000 more potential genes. These potential genes were identified computationally, but there were no other supporting data. Venter and colleagues assumed that most of these latter sequences were false positives. The public consortium estimated that the human genome contains 30,000 to 40,000 genes.

Thus, contrary to earlier estimates, the number of human genes seems to be only about twice as large as the number of genes in a lowly roundworm or a fruit fly. Clearly, the complexity of an organism is not directly proportional to the number of genes it contains. How then can we explain human complexity? One emerging explanation is that the *expression* of genes in humans is more complex than it is in lower organisms. For example, it is estimated that at least 40% of human transcripts experience alternative splicing (Chapter 14). Thus, a relatively small number of gene regions encoding domains and motifs of proteins can be shuffled in different ways to give a rich variety of proteins with different functions. Moreover, posttranslational modification of proteins in humans seems more complex than that in lower organisms, and this also gives rise to a greater variety of protein functions.

Another important finding is that about half of the human genome appears to have come from transposable elements duplicating themselves and carrying human DNA from place to place within the genome. However, even though transposons have contributed so greatly to the genome, the vast majority of them are now inactive. In fact, all of the DNA transposons are inactive, and all of the LTR-containing retrotransposons seem to be.

Hundreds of human genes appear to have come via horizontal transmission from bacteria, and dozens more from new transposons entering human cells. Thus, the human genome has been shaped not entirely by internal mutations and rearrangements, but also by importation of genes from the outside world.

The total size of the human genome appears to be close to the 3 billion bp (3 Gb) predicted for many years. The Venter group has sequenced about 2.9 Gb, and the public consortium predicts that the total size of the genome is about 3.2 Gb.

SUMMARY The working draft of the human genome reported by two separate groups shows that the genome probably contains fewer genes than anticipated: only about 25,000 to 40,000. About half of the genome has derived from the action of transposons, and transposons themselves have contributed dozens of genes to the genome. In addition, bacteria appear to have donated hundreds of genes.

24.2 Applications of Genomics

So far in this chapter we have been dealing with the process of finding out the sequences of genomes. We can call that process **structural genomics** because of its focus on the structure of genomes. Now we need to ask this question: Of what use is the sequence of a genome? There are many applications, but two of the most important are: (1) probing the pattern of gene expression in a given cell type at a given time (**functional genomics**); and (2) finding genes involved in genetic traits, especially genetic diseases (**positional cloning**). It is quite possible to do both functional genomics and positional cloning studies without knowing the complete sequence of an organism's genome, but the sequence greatly facilitates both processes. In this section we will examine the techniques of functional genomics, and in the next, positional cloning. Then we will consider some other applications.

Functional Genomics Techniques

To discover the pattern of expression of a gene in a given tissue over time, we can perform a dot blot analysis as described in Chapter 5. That is, we make spots containing a single-stranded DNA from the gene in question on filters and then hybridize these dot blots to labeled RNAs made in the tissue in question at different times. But suppose we want to know the pattern of expression of all the genes in that tissue over time. In principle, we could make a large dot blot with at least 100 thousand single-stranded DNAs corresponding to all the potential mRNAs in a cell and hybridize labeled cellular RNAs to that monster dot blot. But the sheer size of that blot would present a serious problem. Fortunately, molecular biologists have devised some methods to miniaturize such blots, and some novel methods to analyze the expression of whole genomes. We will look first at DNA arrays and gene microchips, and then at a more exotic method.

DNA Microarrays and Microchips To circumvent the problem of size, molecular biologists have devised ways of spotting tiny volumes of DNA on a chip, so the dots of the dot blot are very small. This allows many different DNAs to be spotted on one chip, called a **DNA microarray**. One system, developed by Vivian Cheung and colleagues, uses a robot with 12 parallel pens, each of which can squirt out a tiny volume of DNA solution: 0.25–1.0 nL (billionth of a liter). The spots are exquisitely small, only 100–150 μm in diameter, and the centers of the spots are only 200–250 μm apart. The result looks like the schematic diagram in Figure 24.10, which represents a DNA microarray with 5808 DNA spots on a common microscope slide. After spotting, the DNAs are air dried, and covalently attached by ultraviolet radiation to a thin silane layer on top of the glass.

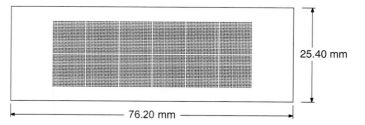

Figure 24.10 Schematic diagram of a DNA microarray. This drawing represents a standard, 1″ × 3″ glass microscope slide with an array of 5808 tiny spots of DNA. Each dot is 200 μm in diameter, and the distance between the dot centers is 400 μm. This is by no means the highest density of spots presently attainable. It is actually possible to place more than 10,000 spots on a slide of this size. (*Source:* Cheung, V. et al. Making and reading microarrays. *Nature Genetics Supplement* Vol. 21 (1999) f. 2, p. 17. © Macmillan Magazines Ltd.)

Another strategy for reducing the size of a blot has been to synthesize many oligonucleotides simultaneously, right on the surface of a chip. Steven Fodor and his colleagues pioneered this method in 1991, using the same kind of photolithographic techniques employed in computer chip manufacture, to build short DNAs (oligonucleotides) on tiny, closely spaced spots on a small glass microchip. In a 1999 version of this technique (Figure 24.11), these workers started with a small glass slide coated with a synthetic linker that was blocked with a photoreactive group that can be removed by light. They masked some of the areas of the slide and illuminated it, so the blocking agent was removed only from the unmasked areas. Then they added a nucleotide (also blocked with a photoreactive group) and chemically coupled it to all the areas of the slide that had been unblocked in the previous step. The result: A nucleotide was attached to a subset of the tiny spots on the chip. Next, they masked a different subset of spots, illuminated the others to remove the blocking groups, and attached another nucleotide. On the spots that were unmasked in both steps, dinucleotides were formed. By repeating this process, they could build up different oligonucleotides on each spot.

The resulting chip is known as a **DNA microchip** or **oligonucleotide array**. The technology is so miniaturized that about 300,000 oligonucleotides can be attached to a chip only 1.28 × 1.28 cm (about 1/2″ square). And the process is so efficient that a set of 4^n different oligonucleotides can be built in only $4 \times n$ cycles. So if our goal is to generate all the possible 9-mers (4^9, or about 250,000 different oligonucleotides), we can do it in only $4 \times 9 = 36$ cycles. How long must an oligonucleotide be to uniquely identify one human gene product in a mixture of all the others? Knowing the sequence of the human genome will help us answer this question with great accuracy. However, we can already do a calculation to give us a minimum estimate. A given sequence of n bases will occur in a DNA about every 4^n bases. In other words, a DNA sequence

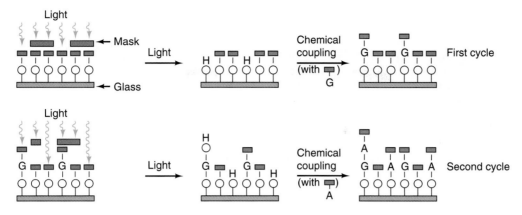

Figure 24.11   Growing oligonucleotides on a glass substrate. The glass is coated with a reactive group that is blocked with a photosensitive agent (red). This blocking agent can be removed with light, but parts of the plate are masked (blue) so the light cannot get through. In the first cycle, four of the six spots pictured are masked, so the light reaches only two unmasked spots and removes the blocking agent. Then a blocked guanosine nucleotide is chemically coupled to the unblocked spots. In the second cycle, three spots are masked, and the other three are therefore exposed to the light. This removes the blocking agent from three spots, including the first one, which already has a G attached. Thus, after a blocked adenosine nucleotide is chemically coupled to the three unblocked spots, the first spot has a G–A dinucleotide, the third and sixth spots have an A mononucleotide, the fourth has a G mononucleotide, and the second and fifth spots, which were masked in both cycles, have no nucleotides attached yet. As the cycle is repeated over and over with different masking patterns and different nucleotides, unique oligonucleotides are built up in each spot.

needs to be n bases long to occur about once in a DNA 4^n bases long. Thus, we need to solve the following equation for n to find the minimum size of an oligonucleotide we would expect to find only once in the whole human genome, which may be as much as 3.5×10^9 bases long:

$$4^n = 3.5 \times 10^9$$

The answer is that if $n = 16$, $4^n > 3.5 \times 10^9$. So our oligonucleotides need to be at least 16 bases long. Again, however, this is a minimum estimate, so it would be a good idea to start with longer oligonucleotides to be reasonably sure that they occur only once in the human genome and therefore uniquely identify human genes.

Even before the publication of the sequence of the first human chromosome, scientists at Affymetrix, Inc. were already producing microchips containing 25-mers designed to recognize single genes. They based their design on the sequence that was available, including the many ESTs already in the database. To enhance the reliability of their chips, they included multiple oligonucleotides designed to hybridize to single transcripts, so the results obtained with each of these oligonucleotides could be checked against one another.

The oligonucleotides on a microchip or the cDNAs on a microarray can be hybridized to labeled RNA isolated from cells (or to corresponding cDNAs) to see which genes in the cell were being transcribed. For example, consider a study by Patrick Brown and colleagues in which they used the DNA microarray technique to examine the effect of serum on the RNAs made by a human

cell. They isolated RNA from cells grown in the presence and absence of serum, then reverse transcribed the two RNA samples in the presence of nucleotides tagged with fluorescent dyes, so the cDNA products would be labeled with the fluorescent tags. They used a green-fluorescing nucleotide to label the cDNA from serum-deprived cells, and a red-fluorescing nucleotide to label the cDNA from serum-stimulated human cells. Then they mixed the cDNAs, hybridized them to DNA microarrays containing unlabeled cDNAs corresponding to 8613 different human genes, and detected the resulting fluorescence. Figure 24.12 shows the same region of the microarray from triplicate hybridizations. The red spots correspond to genes that are turned on by serum, and the green spots represent genes that are active in serum-deprived cells. The yellow spots result from hybridization of both probes to the same spot (the green and red fluorescence together produce a yellow color). Thus, the yellow spots correspond to genes that are active in both the presence and absence of serum.

Once we know the sequence of all the genes in a human cell, we will be able to design a comprehensive DNA chip to assay transcription from every gene, with no guesswork about the oligonucleotides we need to put on the chip. In fact, our knowledge of the complete yeast genome sequence has already allowed molecular biologists to use DNA chips to analyze the expression of every yeast gene at once, under a variety of conditions. Thus, the power of this kind of functional genomics increases with our knowledge of the sequence of the organism's genome.

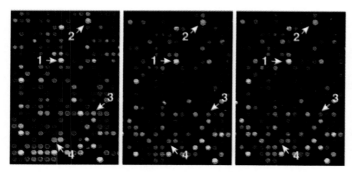

Figure 24.12   Using a DNA chip. Brown and colleagues made cDNAs from RNAs from serum-starved and serum-stimulated human cells. They labeled the cDNAs corresponding to RNAs from serum-starved cells with a green fluorescent nucleotide; they labeled the cDNAs corresponding to RNAs from serum-stimulated cells with a red fluorescent nucleotide. Then they hybridized these fluorescent cDNAs together to DNA chips containing cDNAs corresponding to over 8600 human genes. The figure shows the same part of the DNA chip from three different hybridizations. The red spots (e.g., spots 2 and 4) correspond to genes that are more active in the presence of serum. The green spots (e.g., spot 3) correspond to genes that are more active in the absence of serum. The yellow spots (e.g., spot 1) correspond to genes that are roughly equally active in the presence or absence of serum. (*Source:* Iyer, V. et al. The transcriptional program in the response of human fibroblasts to serum. *Science* 283 (1 Jan 1999) f. 1, p. 83. © AAAS.)

SUMMARY One approach to functional genomics is to create DNA microarrays or DNA microchips, holding thousands of cDNAs or oligonucleotides, respectively, then to hybridize labeled RNAs (or corresponding cDNAs) from cells to these arrays or chips. The intensity of hybridization to each spot reveals the extent of expression of the corresponding gene.

Serial Analysis of Gene Expression   In 1995, Victor Velculescu, working with Kenneth Kinzler and colleagues, developed a novel method of analyzing the range of genes expressed in a given cell. They called this method **serial analysis of gene expression (SAGE)**. The underlying strategy of SAGE is to synthesize short cDNAs, or **tags**, from all the mRNAs in a cell, and then link these tags together in clones that can be sequenced to learn the nature of the tags, and therefore the nature of the genes expressed in the cell, and the extent of expression of each gene.

Figure 24.13 shows how Velculescu and colleagues carried out this strategy. First, they used a biotinylated oligo(dT) primer to prime reverse transcription of the mRNAs present in human pancreatic tissue, yielding double-stranded cDNAs. The goal was to reduce the size of the cDNAs to short tags that could be ligated together

and sequenced readily. Because of the shortness of the tags (9 bp in the example given here), it is important to confine them to a small region of the cDNAs to increase the chance that they will uniquely identify one cDNA. To begin the shortening process, Velculescu and colleagues cleaved the biotinylated cDNAs with an **anchoring enzyme (AE)** to chop off a short 3'-terminal fragment. They chose as their anchoring enzyme *Nla*III, which recognizes 4-base restriction sites and therefore yields fragments averaging 250 bp long. They bound these biotinylated 3'-fragments to streptavidin beads, which bind biotin.

Next, they divided the bead-bound cDNA fragments into two pools and ligated one pool to a linker (A) and the other pool to a second linker (B). Both linkers contained the recognition site for a type IIS restriction endonuclease (the **tagging enzyme [TE]**) that cuts 20 bp downstream of this recognition site. The result of cleavage of the cDNA fragments with the tagging enzyme *Fok*I was a set of short fragments, each containing the linker (A or B) followed by the 4-bp anchoring enzyme site, followed by 9 bp from the cDNA. That 9-bp piece of cDNA is the tag. If the tagging enzyme leaves overhangs, these can be filled in to yield blunt ends.

Velculescu and colleagues' next task was to ligate the tags together, along with defined DNA so they could tell where one tag left off and another began. To do this, they blunt-end-ligated the tagged fragments together to form fragments with two tags abutting each other in the middle (forming a **ditag**) and linkers on each end. The linkers contain sites that are complementary to a pair of primers that can be used to amplify the whole fragment by PCR. After the PCR amplification, Velculescu and colleagues cleaved the products with the anchoring enzyme, ligated these restriction fragments together, and cloned the products. Now the ditags can be easily identified because each one is flanked by the 4-bp anchoring enzyme recognition sites. And, of course, half of each ditag belongs to one tag, and half to the other. Clones with at least 10 tags (some had more than 50) can be identified by PCR analysis and sequenced. If enough clones are sequenced, we can get an idea of the range of genes expressed, and tags that show up repeatedly indicate genes that are very actively expressed.

Velculescu and colleagues' SAGE analysis of expression in the human pancreas had predictable, and therefore encouraging, results. The most common tags (GAGCACACC and TTCTGTGTG) corresponded to the genes for procarboxypeptidase A1 and pancreatic trypsinogen 2, respectively. These are two abundantly expressed pancreatic proenzymes, which, after cleavage to the mature enzyme form, digest proteins in the small intestine. Many other familiar pancreatic genes were identified among the plentiful tags, but many of the tags did not match any gene sequences in the database, so their identities were unknown. As the database expands

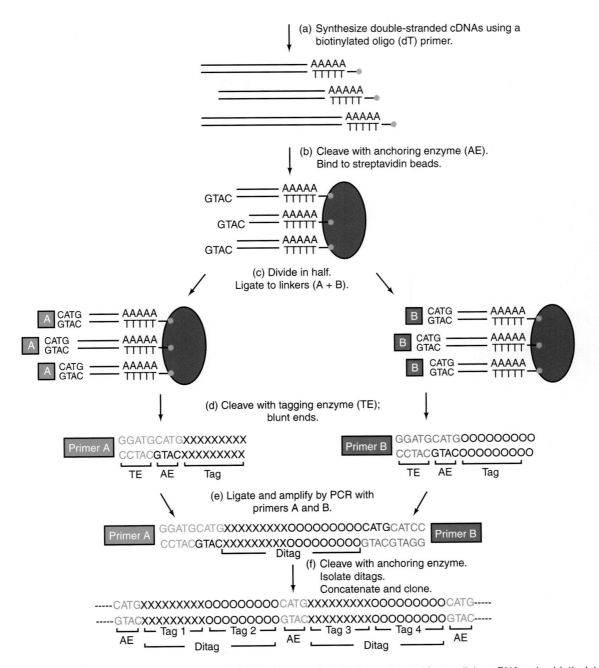

Figure 24.13   Serial analysis of gene expression (SAGE). (a) Double-stranded cDNAs are formed from cellular mRNA, using biotinylated oligo(dT) to prime first-strand cDNA synthesis. Orange balls represent biotin. **(b)** Biotinylated cDNAs are cleaved with an anchoring enzyme (AE, *Nla*III in this case), and the biotinylated 3′-end fragments are bound to streptavidin beads (purple). **(c)** The bead-bound fragments are divided into two pools; the fragments in one pool are ligated to linker A (blue) and the fragments in the other pool are ligated to linker B (red). **(d)** The fragments are cleaved with the tagging enzyme (TE), and ends are filled in if necessary to create blunt ends. In this case, the tagging enzyme is *Fok*I, which leaves 9-bp tags attached to the linkers. The tag attached to linker A is represented by X's, and the tag attached to linker B is represented by O's. **(e)** Tag-containing fragments are blunt-end-ligated together and amplified by PCR with primers that hybridize to primer A and primer B regions in each linker. Only fragments ligated with tags joined tail to tail (ditags) will be amplified by PCR. **(f)** The amplified ditag-containing fragments are cleaved with the anchoring enzyme to yield ditags with sticky ends. The ditags are ligated together to form concatemers, which are cloned. Part of a concatemer of ditags is shown, with the 4-base recognition sites for the anchoring enzyme shown in green. Note that these 4-base sites set off each ditag so it can be recognized easily. The clones are then sequenced to discover which tags are represented, and in what quantity. This tells which genes are expressed, and how actively. (*Source:* From Veculescu et al., *Science* 270:484, 1995. Copyright © 1995 American Association for the Advancement of Science, Washington, DC. Reprinted by permission.)

to include all human genes, all tags should at least be correlated to genes, even if the functions of some of those genes remain obscure.

SUMMARY SAGE allows us to determine which genes are expressed in a given tissue and the extent of that expression. Short tags, characteristic of particular genes, are generated from cDNAs and ligated together between linkers. The ligated tags are then sequenced to determine which genes are expressed and how abundantly.

Positional Cloning

Before the genomics era, geneticists seeking the genes responsible for human genetic disorders frequently faced a problem: They did not know the identity of the defective protein, so they were looking for a gene without knowing its function. Thus, they had to identify the gene by finding its position on the human genetic map, and this process therefore came to be called **positional cloning.**

The strategy of positional cloning begins with the study of a family or families afflicted with the disorder, with the goal of finding one or more markers that are tightly linked to the gene causing the disease. Because the position of the marker is known, the disease gene can be pinned down to a relatively small region of the genome. However, that "relatively small" region usually contains about a million base pairs, so the job is not over. The next step is to search through the million or so base pairs to find a gene that is the likely culprit. Some tools have traditionally been used in the search, and we will describe two here. These are: (1) finding exons with exon traps; and (2) locating the CpG islands that tend to be associated with genes. We will see how these tools have been used as we discuss two classical positional cloning experiments—finding the genes responsible for Huntington's disease and cystic fibrosis—in the next section of this chapter.

But these tools are rapidly being superceded by the Human Genome Project. Now, once a scientific team has done the initial linkage study, and they have located the gene to a small region of the genome, they can simply look up that genome region in the database and find the genes it contains, along with their sequences. That should usually provide all the clues needed to find the gene of interest.

Exon Traps   Once we have a contig stretching over hundreds of kilobases, how do we sort out the genes from the other DNA? If that DNA region has not yet been sequenced, we can sequence it and look for ORFs, but that is very laborious. Several more efficient methods are available, including a procedure invented by Alan Buckler called exon amplification or **exon trapping.** Figure 24.14

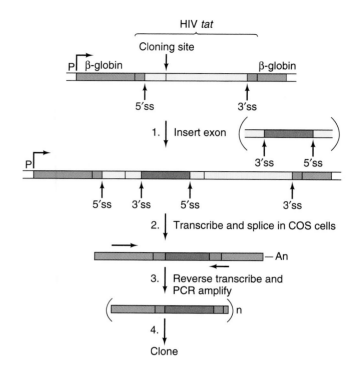

Figure 24.14   Exon trapping. Begin with a cloning vector, such as pSPLI, shown here in slightly simplified form. This vector has an SV40 promoter (P), which drives expression of a hybrid gene containing the rabbit β-globin gene (orange), interrupted by part of the HIV *tat* gene, which includes two exon fragments (blue) surrounding an intron (yellow). The exon–intron borders contain 5'- and 3'-splice sites (ss). The *tat* intron contains a cloning site, into which random DNA fragments can be inserted. In step I, an exon (red) has been inserted, flanked by parts of its own introns, and its own 5'- and 3'-splice sites. In step 2, insert this construct into COS cells, where it can be transcribed and then the transcript can be spliced. Note that the foreign exon (red) has been retained in the spliced transcript, because it had its own splice sites. Finally (steps 3 and 4), subject the transcripts to reverse transcription and PCR amplification, with primers indicated by the arrows. This gives many copies of a DNA fragment containing the foreign exon, which can now be cloned and examined. Note that a nonexon will not have splice sites and will therefore be spliced out of the transcript along with the intron. It will not survive to be amplified in step 3, so one does not waste time studying it.

shows how an exon trap works. We begin with a plasmid vector such as pSPL1, which Buckler designed for this purpose. This vector contains a chimeric gene under the control of the SV40 early promoter. The gene was derived from the rabbit β-globin gene by removing its second intron and substituting a foreign intron from the human immunodeficiency virus (HIV), with its own 5'- and 3'-splice sites. We splice human genomic DNA fragments into a restriction site within the intron of this plasmid, then insert the recombinant vector into monkey cells (COS-7 cells) that can transcribe the gene from the SV40 promoter. Now if any of the genomic DNA fragments we placed into the intron are complete exons, with their own 5'- and 3'-splice sites, this exon will become part of the processed transcript in the COS cells. We purify the RNA made by the COS cells, reverse transcribe it to make cDNA, then subject this cDNA to amplification by PCR, using primers designed to amplify any new exon. Finally, we clone the PCR products, which should represent only exons. Any other piece of DNA inserted into the intron will not have splicing signals; thus, after being transcribed, it will be spliced out along with the surrounding intron and will be lost.

CpG Islands   Another gene-finding technique takes advantage of the fact that active human genes tend to be associated with unmethylated CpG sequences, whereas the CpGs in inactive regions are almost always methylated. Furthermore, the restriction enzyme *Hpa*II cuts at the sequence CCGG, but only if the second C is unmethylated. In other words, it will cut active genes that have unmethylated CpGs within CCGG sites, but it will leave inactive genes (with methylated CCGGs) alone. Thus, geneticists can scan large regions of DNA for "islands" of sites that could be cut with *Hpa*II in a "sea" of other DNA sequences that could not be cut. Such a site is called a **CpG island,** or an **HTF island** because it yields *Hpa*II tiny fragments.

> **SUMMARY** Several methods are available for identifying the genes in a large contig. One of these is the exon trap, which uses a special vector to help clone exons only. Another is to use methylation-sensitive restriction enzymes to search for CpG islands—DNA regions containing unmethylated CpG sequences.

Applications of Functional Genomics

Let us conclude with two classic examples of the use of functional genomics: Pinpointing the genes for Huntington's disease and cystic fibrosis.

Huntington's Disease   Huntington's disease (HD) is a progressive nerve disorder. It begins almost imperceptibly with small tics and clumsiness. Over a period of years, these symptoms intensify and are accompanied by emotional disturbances. Nancy Wexler, an HD researcher, describes the advanced disease as follows: "The entire body is encompassed by adventitious movements. The trunk is writhing and the face is twisting. The full-fledged Huntington patient is very dramatic to look at." Finally, after 10–20 years, the patient dies.

Huntington's disease is controlled by a single dominant gene. Therefore, a child of an HD patient has a 50:50 chance of being affected. People who have the disease could avoid passing it on by not having children, except that the first symptoms usually do not appear until after the childbearing years.

Because they did not know the nature of the product of the HD gene (*HD*), geneticists could not look for the gene directly. The next best approach was to look for a gene or other marker that is tightly linked to *HD*. Michael Conneally and his colleagues spent more than a decade trying to find such a linked gene, but with no success.

In their attempt to find a genetic marker linked to *HD,* Wexler, Conneally, and James Gusella turned next to RFLPs. They were fortunate to have a very large family to study. Living around Lake Maracaibo in Venezuela is a family whose members have suffered from HD since the early nineteenth century. The first member of the family to be so afflicted was a woman whose father, presumably a European, carried the defective gene. So the pedigree of this family can be traced through seven generations, and the number of individuals is unusually large, a typical nuclear family having 15–18 children.

Gusella and colleagues knew they might have to test hundreds of probes to detect a RFLP linked to *HD*, but they were amazingly lucky. Among the first dozen probes they tried, they found one (called G8) that detected a RFLP that is very tightly linked to *HD* in the Venezuelan family. Figure 24.15 shows the locations of *Hin*dIII sites in the stretch of DNA that hybridizes to the probe. We can see seven sites in all, but only five of these are found in all family members. The other two, marked with asterisks and numbered 1 and 2, may or may not be present. These latter two sites are therefore polymorphic, or variable.

Let us see how the presence or absence of these two restriction sites gives rise to a RFLP. If site 1 is absent, a single fragment 17.5 kb long will be produced. However, if site 1 is present, the 17.5-kb fragment will be cut into two pieces having lengths of 15 kb and 2.5 kb, respectively. Only the 15-kb band will show up on the autoradiograph because the 2.5-kb fragment lies outside the region that hybridizes to the G8 probe. If site 2 is absent, a 4.9-kb fragment will be produced. On the other hand, if site 2 is present, the 4.9-kb fragment will be subdivided into a 3.7-kb fragment and a 1.2-kb fragment.

There are four possible **haplotypes** (clusters of alleles on a single chromosome) with respect to these two polymorphic *Hin*dIII sites, and they have been labeled *A–D*:

Haplotype	Site 1	Site 2	Fragments Observed
A	Absent	Present	17.5; 3.7; 1.2
B	Absent	Absent	17.5; 4.9
C	Present	Present	15.0; 3.7; 1.2
D	Present	Absent	15.0; 4.9

The term *haplotype* is a contraction of *haploid genotype,* which emphasizes that each member of the family will inherit two haplotypes, one from each parent. For example, an individual might inherit the *A* haplotype from one parent and the *D* haplotype from the other. This person

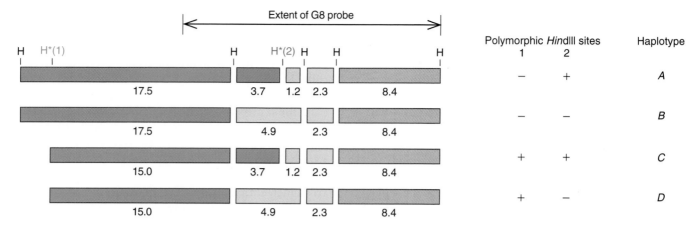

Figure 24.15   The RFLP associated with the Huntington's disease gene. The *Hind*III sites in the region covered by the G8 probe are shown. The families studied show polymorphisms in two of these sites, marked with an asterisk and numbered 1 (blue) and 2 (red). Presence of site 1 results in a 15-kb fragment plus a 2.5-kb fragment that is not detected because it lies outside the region that hybridizes to the G8 probe. Absence of this site results in a 17.5-kb fragment. Presence of site 2 results in two fragments of 3.7 and 1.2 kb. Absence of this site results in a 4.9-kb fragment. Four haplotypes (*A–D*) result from the four combinations of presence or absence of these two sites. These are listed at right, beside a list of polymorphic *Hind*III sites and a diagram of the *Hind*III restriction fragments detected by the G8 probe for each haplotype. For example, haplotype *A* lacks site 1 but has site 2. As a result, *Hind*III fragments of 17.5, 3.7, and 1.2 are produced. The 2.3- and 8.4-kb fragments are also detected by the probe, but we ignore them because they are common to all four haplotypes.

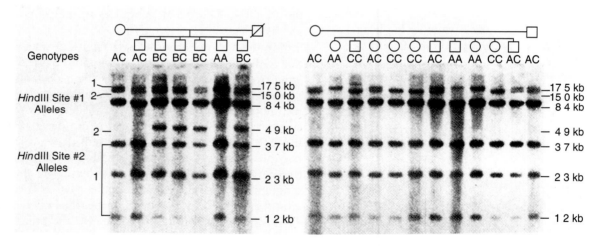

Figure 24.16   Southern blots of *Hind*III fragments from members of two families, hybridized to the G8 probe. The bands in the autoradiographs represent DNA fragments whose sizes are listed at the right. The genotypes of all the children and three of the parents are shown at the top. The fourth parent was deceased, so his genotype could not be determined. From James F. Gusella, et al., *Nature* 306:236. © 1983 Macmillan Magazines Limited.

would have the *AD* genotype. Sometimes different genotypes (pairs of haplotypes) can be indistinguishable. For example, a person with the *AD* genotype will have the same RFLP pattern as one with the *BC* genotype because all five fragments will be present in both cases. However, the true genotype can be deduced by examining the parents' genotypes. Figure 24.16 shows autoradiographs of Southern blots of two families, using the radioactive G8 probe. The 17.5- and 15-kb fragments migrate very close together, so they are difficult to distinguish when both are present, as in the *AC* genotype; nevertheless, the *AA* geno-

type with only the 17.5-kb fragment is relatively easy to distinguish from the *CC* genotype with only the 15-kb fragment. The *B* haplotype in the first family is obvious because of the presence of the 4.9-kb fragment.

Which haplotype is associated with the disease in the Venezuelan family? Figure 24.17 demonstrates that it is *C*. Nearly all individuals with this haplotype have the disease. Those who do not will almost certainly develop it later. Equally telling is the fact that no individual lacking the *C* haplotype has the disease. Thus, this is a very accurate way of predicting whether a member of this family is

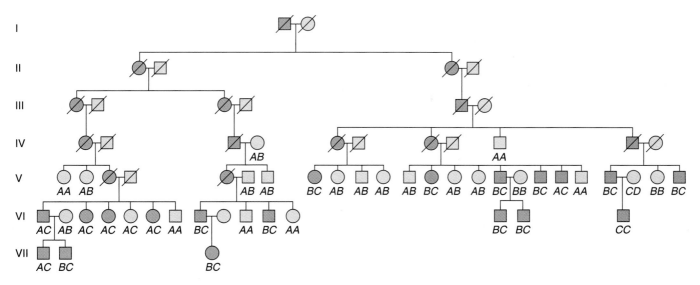

Figure 24.17   Pedigree of the large Venezuelan family with Huntington's disease. Family members with confirmed disease are represented by purple symbols. Notice that most of the individuals with the *C* haplotype already have the disease, and that no sufferers of the disease lack the *C* haplotype. Thus, the *C* haplotype is strongly associated with the disease, and the corresponding RFLP is tightly linked to the Huntington's disease gene.

carrying the Huntington's disease gene. A similar study of an American family showed that, in this family, the *A* haplotype was linked with the disease. Therefore, each family varies in the haplotype associated with the disease, but within a family, the linkage between the RFLP site and *HD* is so close that recombination between these sites is very rare. Thus we see that a RFLP can be used as a genetic marker for mapping, just as if it were a gene.

Aside from its use as a genetic counseling tool, what does this analysis reveal? For one thing, it allowed Gusella and colleagues to locate *HD* on chromosome 4. They did this by making mouse–human hybrid cell lines, each containing only a few human chromosomes. They then prepared DNA from each of these lines and hybridized it to the radioactive G8 probe. Only the cell lines having chromosome 4 hybridized; the presence or absence of all other chromosomes did not matter. Therefore, human chromosome 4 carries *HD*.

At this point, the *HD* mapping team's luck ran out. One long detour arose from a mapping study that indicated the gene lay far out at the end of chromosome 4. This made the search much more difficult because the tip of the chromosome is a genetic wasteland, full of repetitive sequences, and apparently devoid of genes. Finally, after wandering for years in what he calls a genetic "junkyard," Gusella and his group turned their attention to a more promising region. Some mapping work suggested that *HD* resided, not at the tip of the chromosome, but in a 2.2-Mb region several megabases removed from the tip. Over 2 Mb is a tremendous amount of DNA to sift through to find a gene, so Gusella decided to focus on a 500-kb region that was highly conserved among about

one-third of *HD* patients, who seem to have a common ancestor.

To find genes in the 500-kb region, Gusella and colleagues used an exon trapping strategy and identified a handful of exon clones. They then used these exons to probe a cDNA library to identify the DNA copies of mRNAs transcribed from the target region. One of the clones, called IT15, for "interesting transcript number 15," hybridized to cDNAs that identified a large (10,366 nt) transcript that codes for a large (3144 amino acid) protein. The presumed protein product did not resemble any known proteins, so that did not provide any evidence that this is indeed *HD*. However, the gene had an intriguing repeat of 23 copies of the triplet CAG (one copy is actually CAA), encoding a stretch of 23 glutamines.

Is this really *HD*? Gusella's team's comparison of the gene in affected and unaffected individuals in 75 HD families demonstrated that it is. In all unaffected individuals, the number of CAG repeats ranged from 11 to 34, and 98% of these unaffected people had 24 or fewer CAG repeats. In all affected individuals, the number of CAG repeats had expanded to at least 42, up to a high of about 100. Thus, we can predict whether an individual will be affected by the disease by looking at the number of CAG repeats in this gene. Furthermore, the severity, or age of onset of the disease correlates at least roughly with the number of CAG repeats. People with a number of repeats at the low end of the affected range (now known to be 38–40) generally survive well into adulthood before symptoms appear, whereas people with a number of repeats at the high end of the range tend to show symptoms in childhood. In one extreme example, an individual with

BOX 24.1

Problems in Genetic Screening

Now that we have identified the genes for Huntington's disease, cystic fibrosis, and other serious maladies, how will we use this information? We are in a position to identify carriers of genetic diseases or people who are at risk for developing these debilitating conditions. This sounds like a good thing; in general, it undoubtedly will be, but there are some ethical problems.

Consider, for example, the actual case of an adoption agency that was trying to place a baby girl whose mother suffered from HD. Understandably, potential adoptive parents wanted to know if the baby had inherited the gene. However, Michael Conneally, the geneticist consulted in the case, decided not to test the child for the following reason: If the results were positive and were later shared with the girl, she would know that she was doomed to get the disease. Many people at risk of developing this terrible disease feel that they cannot cope with knowing for sure that they have the gene, so they elect not to take the test. In this case, the child was too young to choose for herself.

This choice about whether or not to know if one is carrying the *HD* gene has recently become more complex. Now, by counting the number of CAG repeats in the gene, we can predict the age at which the disease will begin. Before, one could cling to the hope that, even if one carried the gene, the disease might not appear until very late in life. Now a genetic test could remove even that hope.

Another dilemma would occur in the following situation: A woman is at risk of developing HD because her father has it. She has decided she cannot bear to know that she has inherited the gene, so she will not submit to a test.

Then she gets pregnant. She and her husband want the child, but not if it carries the defective gene. They could check the fetus's genotype by amniocentesis. A negative result would end their worries about the child and allow the mother to cling to hope for her own future, but a positive result would confirm her worst fears, not only for the baby but for herself.

In the case of the autosomal–recessive condition cystic fibrosis (CF), a different set of problems arises. Cystic fibrosis is the most common lethal genetic disease afflicting American children. One in 20 whites is a *CF* carrier, and therefore more than 1 in 2000 children will be homozygous for *CF* and suffer the disease. Now that the *CF* gene has been identified, one would predict that millions of people would want to be tested to help them with their family planning. Nevertheless, in 1989 the American Society for Human Genetics called for a moratorium on widespread screening, and though small-scale screening has been tried, caution is still the watchword.

There are two main reasons for this reluctance. First, in spite of the expectation that CF would be caused by only a small number of different mutations, the actual number had already risen to 170 by 1992. About 20 of these mutations are common, so we would probably need to test for all 20, and even then we would only discover 85%–90% of all carriers. Even assuming a 90% detection rate, we would still detect only 81% of all couples at risk ($0.9 \times 0.9 = 0.81$). That seems better than nothing, but missing almost 20% of couples likely to have an affected child is certainly far from perfect, and testing services may fear giving such imperfect advice.

---

the highest number of repeats detected (about 100) started showing disease symptoms at the extraordinarily early age of 2. Finally, two people were affected, even though their parents were not. In both cases, the affected individuals had expanded CAG repeats, whereas their parents did not. New mutations (expanded CAG repeats), although a rare occurrence in HD, apparently caused both these cases of disease.

Another way of demonstrating that this gene is really *HD* would be to deliberately mutate it and show that the mutation has neurological effects. Obviously, one cannot perform such an experiment in humans, but it would be feasible in mice, if the gene corresponding to *HD* is known. Fortunately, *HD* is conserved in many species, including the mouse, where the gene is known as *Hdh*. In 1995, a team of geneticists led by Michael Hayden cre-

ated knockout mice with a targeted disruption in exon 5 of *Hdh*. Mice that are homozygous for this mutation die in utero. Heterozygotes are viable, but they show loss of neurons with corresponding lowering of intelligence. This reinforces the notion that *Hdh*, and therefore *HD*, plays an important role in the brain—exactly what we would expect of the gene that causes *HD*.

How can we put this new knowledge to work? One obvious way is to perform accurate genetic screening to detect people who will be affected by the disease. In fact, by counting the CAG repeats, we may even be able to predict the age of onset of the disease. However, that kind of information is a mixed blessing, as it can be psychologically devastating (see Box 24.1). What we really need, of course, is a cure, but that may be a long way off.

Even if the test were perfect, problems would remain. The major difficulty arises from the sheer volume of the task. Because CF is so common, most prospective parents might want to be tested. But in the United States alone, about 150 million people are in the reproductive age group. Even assuming we could test that many people, someone has to explain to them what the results mean, and not nearly enough professional genetic counselors are available to do that.

Our experience with screening African-Americans for sickle cell disease teaches us what can happen to an inadequate program. In the early 1970s, many states passed laws requiring testing of African-Americans for the sickle cell gene, but did not provide enough counseling to make the program successful. This had several unfortunate outcomes. For one, many carriers of the sickle cell gene came to the mistaken and distressing conclusion that they actually had the disease. Also, the suggestion to African-American couples who were both carriers that they forego having children, coming as it frequently did from whites, provoked charges of racism. Moreover, in some cases confidentiality was not respected. This led to the denial of health insurance to some carriers. The result of all this was a loss of confidence in the screening process among those it was designed to help. Thus, few people took part, and even fewer used the information they received to help in their family planning.

These cases exemplify the tough decisions that people must make about their own genetic testing and the psychological impact such tests may have on them. Quite a different kind of question arises concerning the mandatory testing of others. Would it be ethical, for example, for insurance companies to demand genetic tests? Those who test positive for such conditions as HD or Alzheimer's disease would either be denied health insurance or forced to pay exorbitant premiums. We have already seen that this happened to carriers of the sickle cell gene. If it is not fair to burden people because of their genes, over which they have no control, would it be fair to require them to submit to tests before obtaining life insurance? Keep in mind that life insurance companies have traditionally required proof of good health before issuing insurance and that genetic screening could simply be interpreted as an extension of this practice.

Let us take the problem one step further and assume that a life insurance company has paid for genetic testing of an individual and found him likely to develop a debilitating disease such as Alzheimer's. Would it be ethical for the insurance company to share such information with others—the person's employer or prospective employer, for instance? This information would certainly give an indication of the person's long-term fitness for employment, but it would just as certainly place an enormous burden on that person and maybe even render him unemployable. The problem becomes more difficult if the insurer is also the employer, as is the case with some large corporations.

These thorny questions have no easy answers, but the development of more and more tests for genetic diseases will force us to deal with them. Ultimately, if our genetic studies really do lead to effective treatments, these questions should recede in importance.

To find a cure for HD, we need to answer at least two questions. First, what is the normal function of the *HD* product (now called huntingtin)? To get at that question, geneticists are constructing transgenic mice carrying the mutant form of *HD*. This should shed light on the function of the gene in the brain and in other tissues, where it is also active. Second, we need to know how the expansion of the CAG repeat causes disease. Gusella speculated that the extra glutamines in the protein give it a new function that somehow disrupts brain activity. We now know that abnormal huntingtin binds to another protein with a string of glutamines: CREB-binding protein (CBP, Chapter 12). By binding and inactivating this mediator, abnormal huntingtin may disturb nerve function.

SUMMARY Using RFLPs, geneticists mapped the Huntington's disease gene (*HD*) to a region near the end of chromosome 4. Then they used an exon trap to identify the gene itself. The mutation that causes the disease is an expansion of a CAG repeat from the normal range of 11–34 copies, to the abnormal range of at least 38 copies.

Cystic Fibrosis Cystic fibrosis (CF) is the most common lethal genetic disease afflicting whites. It is caused by an autosomal-recessive mutation carried by 1 in 20 people of European descent (over 12 million people in the United

States alone). This means that 1 in 400 white couples will both be carriers, and 1 in 4 of their children will have the disease. In the United States, over one thousand children are born with cystic fibrosis every year.

The disease affects tissues called secretory epithelia, which are responsible for transporting water and salt at the interface between the bloodstream and the external environment (e.g., in the lungs, intestine, and sweat glands). The abnormal secretory epithelia in CF patients fail to carry out this transport properly, which causes buildup of thick mucus in the affected organs. This in turn causes the clinical symptoms of the disease: failure of the pancreas to secrete digestive enzymes into the intestine; bacteria that find a fertile environment for growth in the thick mucus in the lungs; and high levels of salt in the sweat. The last symptom is not dangerous, but it is a convenient marker that physicians use to diagnose the disease. The lung infections, on the other hand, are very serious. It is primarily because of these that many CF patients do not survive past their twenties.

Because this disease is so prevalent and so devastating, it has been the subject of intense research. Geneticists knew that if they could identify the defective gene and its product, this knowledge might suggest new avenues to a treatment for the disease. At least it would provide a test for the mutation so couples would know in advance if they were at risk of having affected children. The test could also be used prenatally to determine whether a fetus was homozygous for the mutation.

As in the case of HD, geneticists trying to find the CF gene did not know the identity of its protein product. They knew that something about salt transport was disrupted, but nothing about the nature of the defective protein. In spite of the enormous difficulties of this kind of searching for a "needle in a haystack," geneticists led by Lap-Chee Tsui, Francis Collins, and John Riordan used a powerful combination of classical and molecular approaches and in 1989 finally located the cystic fibrosis gene (CF).

The first step in the search for the gene was to establish its linkage to known markers, especially those whose chromosomal location was known. One of the first markers to be found linked to CF was a variability in serum activity of an enzyme called paraoxonase; unfortunately, this marker had not been mapped, and it was not even known what chromosome it resided on. However, using RFLP mapping techniques similar to those employed in the search for the HD gene, Tsui and colleagues found linkage between the CF gene and a more promising marker—a RFLP called DOCRI-917.

Next, these workers showed that this RFLP is located on chromosome 7. To do this mapping, Tsui and colleagues employed somatic cell hybridization techniques. They obtained a panel of hamster–human hybrid cells, with known human chromosome content, from another laboratory. Then they screened DNA from these cells for the ability to hybridize to a radioactive probe for DOCRI-917. Whenever the cell contained human chromosome 7, its DNA hybridized to the probe. Whenever the cell did not contain any part of human chromosome 7, its DNA would not hybridize to the probe. Thus, the RFLP is located on chromosome 7, and because CF is linked to the RFLP, we know that it is on chromosome 7 as well.

At about the same time, Ray White and coworkers discovered tight linkage between the CF gene and a RFLP in a gene called met. This was valuable information because met had been mapped to the middle third of the long arm of chromosome 7. This placed CF in the same vicinity.

Robert Williamson, Jorg Schmidtke, and colleagues strengthened this conclusion by showing linkage between CF and two RFLPs. The first is an anonymous RFLP called pJ3.11; the second is a RFLP in a gene called TCRb. Schmidtke and his coworkers used somatic cell hybrid panels to map pJ3.11 to the top (band q22) of the long arm of chromosome 7, and other workers mapped the other RFLP to the bottom (band q3) of the long arm.

Linkage analysis on a number of different markers culminated in the assignment of CF to band q31 of chromosome 7, between two closely linked markers: met and D7S8 (defined by the pJ3.11 RFLP). The recombination frequency between met and CF in males was 0.013, whereas the recombination frequency between CF and D7S8 was only 0.009. This placed met and D7S8 only about 1–2 million bp apart, thus confining the search to a relatively narrow region of the human genome. Still, even a million base pairs was a lot of DNA to sift through looking for a single gene, so creative approaches were needed.

The first step in searching for a gene in a megabase pair region of DNA is to clone at least part of the DNA in question. This task has been simplified by techniques such as YAC and BAC cloning. In this case, the investigators used another technique called chromosome walking, which generates a collection of clones with overlapping DNAs that cover the whole region of interest (Figure 24.18). To perform a chromosome walk from left to right in the example illustrated here, we start with a genomic library and a clone at the left end of the sequence we wish to examine. We cut a piece from the right end of this starting clone and label it; then we use this radioactive fragment to probe the library by plaque hybridization. Any clone that hybridizes will overlap the right end of the starting clone and is very likely to contain additional DNA to the right. This process is repeated with a probe from the right end of the second clone to find an overlapping clone farther to the right. We continue this process until we have a contig containing overlapping clones representing the whole region. Finally, we can sequence all our clones to obtain the base sequence of the whole region.

(a) Chromosome walking

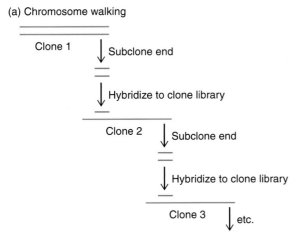

(b) Physical mapping (restriction sites and STSs)

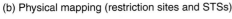
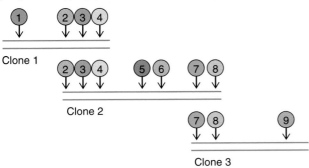

Figure 24.18 Mapping by chromosome walking. (a) Chromosome walking. To start the walk, choose a cloned piece of DNA (clone 1) and subclone one end of it. Then use this small end piece (red) as a probe to identify an overlapping clone (clone 2) in a library. Repeating the process, subclone the far end of clone 2 to generate a probe to identify yet another overlapping clone (clone 3). Repeat this cycle as many times as needed to build a set of overlapping clones spanning large stretches of DNA. **(b)** Physical mapping of restriction sites or STSs in each clone allows one to align the overlapping DNAs and build a map of the whole contig.

Chromosome walking works very well for examining relatively short stretches of a chromosome (100 kb or so), but runs into trouble when we need to canvass hundreds of kilobases of DNA, as in the present case. The biggest problem is that certain regions of DNA are unclonable, as we saw in our discussion of the Human Genome Project. Still other regions can be cloned but are unstable and are lost.

Collins worked out a clever way around these problems: a modification of chromosome walking called chromosome jumping. This technique (Figure 24.19) allows the investigator to jump over unclonable regions and start the chromosome walk anew. The key to the procedure is to prepare a library of clones containing about 100 kb of DNA, then to form circles of these DNAs by ligating in a short piece of DNA containing a selectable marker—for example, the *supF* gene, which suppresses amber muta-

tions. This brings together two DNA regions that had been separated by about 100 kb. Next we cut the circular DNA with a restriction enzyme and subclone the fragments into a λ phage with an amber mutation in a critical gene. This ensures that only clones with the *supF* gene will survive, and this gene will mark the place where circle formation took place. We select a clone by plaque hybridization to a probe for a given site in our search (say the *met* locus). Because this cloned DNA contains *met*, and a circle-joining site marked by *supF*, the chances are high that the DNA lying on the other side of the *supF* gene originally lay about 100 kb away from *met*. This can then be used as a new start site for chromosome walking. We have just performed a chromosome jump of 100 kb.

In the end, Riordon, Tsui, Collins, and their colleagues cloned almost 300 kb of DNA, contained within a region of about 500 kb. They suspected that this region contained the *CF* gene, but where was it?

One important clue in this sort of mystery is to find a stretch of DNA that is conserved in several different species. If the DNA is conserved, it probably codes for something and is therefore a gene. To find conserved DNA quickly, the research team blotted fragmented genomic DNA from human, cow, mouse, and chicken and hybridized this "zoo blot" to radioactive probe DNAs from different parts of the cloned human DNA region. Four radioactive probes cross-hybridized to DNA from other species besides human. These probes identified four candidate regions in which to focus the search.

The first candidate region mapped very close to *met*, and far from D7S8, so it could not contain *CF*. The second did not contain any ORFs, so it was also eliminated from consideration. The third region was used as a probe to look for mRNAs with the same sequence, but none was found. Thus, it appeared that this region is not transcribed and therefore could not contain the *CF* gene.

Fortunately, the fourth candidate region did contain *CF*. When the research team determined the sequence of this region they found an encouraging sign: the presence of a CpG island. It took exhaustive screening of many cDNA libraries with a probe for this region, but finally the researchers were rewarded with a cDNA (clone 10-1) containing a 920-bp piece of DNA encoding part of an mRNA from the sweat gland of a normal (non-CF) individual. This meant that the CG-rich region was transcribed in normal sweat glands, as *CF* ought to be. Clone 10-1 also detected a 6.5-kb transcript in a Northern blot of human RNAs, again demonstrating the expression of this region in human cells. Finally, using cDNAs to hybridize to genomic clones, this group was able to show that *CF* spans approximately 250 kb of DNA and includes at least 24 exons.

How do we know this gene is really *CF*? Several lines of evidence supported this conclusion. First, Northern blotting experiments showed that the gene is transcribed

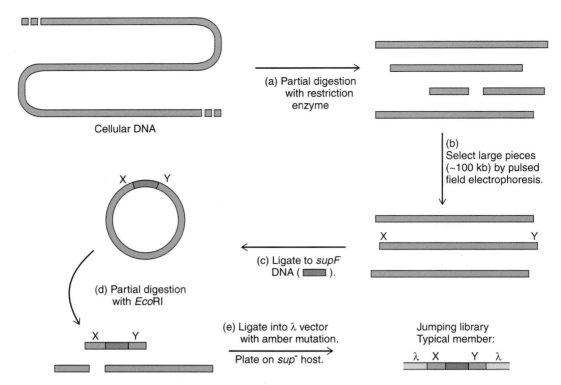

Figure 24.19   Chromosome jumping. (a) Start with high-molecular-weight cellular DNA and partially digest it with a restriction enzyme to produce large DNA fragments. **(b)** Subject the fragments to pulsed-field gel electrophoresis and select those that are about 100 kb long. The ends of one of these fragments are labeled X and Y, so they can be tracked in subsequent steps. **(c)** Ligate these long DNA fragments to a cloned *supF* gene (red) that allows suppression of amber mutations. The *supF* gene inserts between the former free ends (X and Y) of the long DNA fragments. **(d)** Partially digest the circularized DNA with *Eco*RI. Some of the resulting fragments will contain *supF* flanked by X and Y. **(e)** Ligate the *Eco*RI fragments into a vector that has a lethal amber mutation, package the DNA, and plate the resulting phage particles on a *supF* host. Only the phages with the *supF* gene will replicate, and these constitute the jumping library. Each cloned DNA will have a *supF* gene, flanked by two regions (X and Y) that originally lay about 100 kb apart in the cellular DNA. By adjusting the size fragments selected in step **(b)**, one can alter the size of the jump in the jumping library. Assuming that X is a known region encompassed by a previous chromosome walk, one can use Y to begin a new chromosome walk 100 kb away.

in all tissues affected by CF. Second, the base sequence of the gene showed that it encodes a protein with a so-called membrane-spanning domain and so is very likely to be a membrane protein, as the *CF* product is predicted to be because it is involved in membrane transport. Further sequence analysis suggested at first that the product of the *CF* gene was probably not itself a channel for chloride ions, but somehow regulated the transport of these ions across the membrane. This led to renaming the protein product of *CF* as the cystic fibrosis transmembrane conductance regulator (**CFTR**). However, we now know that CFTR is in fact a chloride channel because it can create such channels when it is the only protein added to an artificial membrane.

A third finding linking this gene with *CF* is that most CF patients have a 3-bp deletion in this gene, which would result in a mutant protein with one amino acid (a phenylalanine) missing. Work by Tsui and his colleagues showed that about 70% of the first group of CF patients tested had this mutation. The rest presumably had mutations elsewhere in *CF*. Finally, and most tellingly, two teams of workers, including Michael Welsh, as well as

Tsui and Collins, have shown that expression of a transplanted wild-type *CF* gene in CF cells cures the defect in chloride ion transport. Thus, the gene encoding CFTR really is *CF*, now usually called the *CFTR* gene.

Now that we know the identity of the *CFTR* gene, how can we use this knowledge to find treatments for the disease? One way is to develop an animal model so we can examine in detail the effects of the disease on lungs and other tissues as the disease progresses. The supply of lung tissue from CF patients who die in infancy is small; by the time most patients die, infections have damaged their lungs so badly that their condition in the early stages of the disease is impossible to determine. Two research groups have bred *CFTR* knockout mice that should help circumvent these problems. Now we can see precisely what is happening in diseased tissues from birth (or before) until death. And we can try out therapies on these mice before risking them on human patients. One problem, however, is that the knockout mice do not show exactly the same phenotype as human CF patients, so we will have to be careful in interpreting experiments.

Another way knowledge of the *CFTR* gene can help is in designing therapies for the disease. So far, the genetic findings have suggested several new therapies. The most obvious idea is to supply a wild-type gene to CF patients. We could use any of several viruses or liposomes to carry the *CFTR* gene to the place where it would do the most good—the lungs—perhaps as simply as by using an aerosol inhaler.

Instead of providing a new gene, we may be able to ameliorate the disease symptoms simply by giving regular doses of the *CFTR* protein by inhaler. One danger would be that the patients could develop an allergy to the protein, but it appears that they have nonfunctional protein in their lungs already, and the difference between mutant and wild-type proteins (only one amino acid) is so slight that the body may not recognize the wild-type protein as foreign, and so might tolerate it. Another strategy is to stimulate a latent activity of the mutant protein with drugs, and CF specialists are already considering some drugs that may be able to do this.

SUMMARY Using a combination of RFLP mapping and somatic cell genetic techniques, geneticists located *CF* to a 1-2 Mb region of band q31 of chromosome 7. Then they pinpointed *CF* by first building a clone library encompassing the DNA region where the gene was presumed to be located. Contiguous clones in this library were identified by chromosome walking and jumping. DNA–DNA hybridization identified four candidate genes that were well conserved among mammalian species. Of these, one had promising characteristics, including a CpG island. This was identified as *CF* based on (1) its pattern of expression in organs, such as sweat glands, that are involved in CF; (2) its sequence; (3) the fact that CF patients have mutations in this gene; and (4) the fact that CF cells can be cured of their chloride transport deficiency by introducing the wild-type version of this gene.

Other Applications

Now that we have a rough draft of the human genome sequence, we can begin to look for differences among individuals. So far, most of these are differences in single nucleotides, so we call them **single-nucleotide polymorphisms,** or **SNPs** (pronounced "snips"). If we can link these SNPs to human diseases, we could then screen individuals for the tendency to develop those diseases. We might also be able to find sets of SNPs that associate with polygenic traits, such as intelligence, and thus pin down the genes responsible for these traits.

We may also be able to identify SNPs that correlate with good or poor response to certain drugs. Using this information, physicians should be able to screen a patient for key SNPs, then custom design a drug treatment program for that patient based on his or her predicted responses to a range of drugs. This field of study is called **pharmacogenomics.**

However, these tasks will not be easy. Already, geneticists are discovering that the vast majority of SNPs are not in genes at all, but in intergenic regions of DNA. Even when they are found within genes, they tend to be silent mutations that do not alter the structure of the protein product, and thus do not cause any malfunction that could lead to a disease. The reason for this situation is clear: Polymorphisms caused by mutations that change the products of genes are generally deleterious, and are therefore selected against. That is, the individuals with these damaging mutations generally die before they can reproduce and thus the mutations are lost.

We can detect SNPs correlating with disease or other traits in any given individual by hybridizing that person's DNA to DNA microarrays containing oligonucleotides with the wild-type and mutated sequences. Such knowledge can be useful in helping to prevent or treat disease. However, as Box 24.1 points out, it can also be abused.

How do SNPs differ from RFLPs? RFLPs are identical to SNPs if the single-nucleotide difference between two individuals lies in a restriction site, as we observed in the RFLPs involving *Hin*dIII sites in Huntington's disease patients. In such a case, a single-nucleotide difference makes a difference in the pattern of restriction fragments. However, RFLPs can also result from insertion of a chunk of DNA between two restriction sites in one individual, but not another—VNTRs, for example. That would not be a SNP because it involves more than just a single-nucleotide difference.

The complete sequences of genomes of lower organisms can also be important in understanding and treating human diseases. For example, as soon as the complete yeast genome had been sequenced, molecular biologists began systematically mutating every one of the 6000 yeast genes to see what effects those mutations would have. They also began systematically screening all 18 million possible protein–protein interactions using a yeast two-hybrid screen (Chapter 14). The results of such experiments can tell us much about the activities of gene products that are still uncharacterized. And knowing the activities of all the proteins in an organism, and the other proteins with which they interact, should lead to greater understanding of biochemical pathways, such as the ones that metabolize drugs. This understanding, in turn, should give us important clues about how these pathways work in humans. Moreover, yeast cells can be used as human surrogates to test the effects of knocking out the yeast ortholog of a known human disease gene.

SUMMARY Single-nucleotide polymorphisms can probably account for many genetic conditions caused by single genes, and even multiple genes. They might also be able to predict a person's response to drugs. But sorting out the important SNPs from those with no effect will be very difficult.

Bioinformatics

As our databases swell with billions of bases of sequence from the human and other genomes, one crucial problem will be to access and manipulate all those data. Accordingly, a new specialty has arisen, known as **bioinformatics.** Practitioners of bioinformatics must understand both biology and computerized data processing, so they can manage the data collecting during genome sequencing and then provide for useful access to the data.

Two types of databases are already established. First, we have generalized databases that include DNA and protein sequences from all organisms. Two generalized databases for DNA sequences are GenBank and EMBL. Swissprot is a generalized protein sequence database. Second, we have specialized databases that deal with a particular organism. For example, FlyBase is a database of the genome of the fruitfly *Drosophila melanogaster.* You can access it online at http://flybase. bio.indiana.edu:82, and search it for genetic maps, genes, DNA sequences, and other information. The Genome Database (GDB) is a website (*http://gdbwww. gdb.org*) that allows you to search the human genome for any gene you are interested in. You must select a database to search, and two of the most popular are EMBL and GenBank. The Institute for Genomics Research (TIGR), has a website (*http://www.tigr.org/ tdb/*) that links to the TIGR databases, including the complete genomes of many microbes.

The problem, as William Gelbart has pointed out, is that we are functional illiterates in understanding the genomic sequence. He uses a language analogy: We know a few of the "nouns," or polypeptide coding regions of the genome, but we don't know the "verbs," "adjectives," and "adverbs" that tell when and how much of each gene to express. And we don't know the "grammar" that tells how polypeptides assemble into complexes to do their jobs, such as catalyzing biochemical pathways. Bioinformatics will supply the databases and annotation that will be needed to understand genomes fully.

SUMMARY Bioinformatics is the building and manipulation of biological databases. The most complex of these databases contain the DNA sequences of genomes. Bioinformatics is essential to understanding the expression patterns of complex genomes.

Proteomics

One important application of genomics is the discovery of the levels of expression of all the genes in an organism at any one time, using DNA chips or similar tools that measure levels of mRNAs. This kind of functional genomics strategy usually focuses on the gene transcripts in a cell at a given time. If we were to consider all the transcripts an organism can make in its lifetime, we might call that the organism's **transcriptome,** by analogy with the term *genome,* which refers to all the genes in an organism. And functional genomics approaches that measure the levels of RNAs produced from many genes at a time would be called **transcriptomics.** In fact, these two terms are gaining in popularity.

A similar, but much more complex, quest is to learn about an organism's **proteome,** that is, the properties and activities of all the proteins that organism makes in its lifetime. This field is therefore called **proteomics.** Whereas the task of analyzing an organism's genome, or even its transcriptome, is relatively straightforward, the task of analyzing an organism's proteome is anything but simple, in large part because of the complexity of proteins relative to nucleic acids. Indeed, with current techniques, proteomics is impossible for any but the simplest organisms.

Given this difficulty, why are scientists even interested in proteomics, when they already have transcriptomics, in which they can probe the expression of vast numbers of genes simultaneously by looking at the levels of their transcripts? The answer is that the level of transcription of a gene gives only a rough idea of the real level of expression of that gene. For one thing, an mRNA may be produced in abundance, but degraded rapidly, or translated inefficiently, so the amount of protein produced is minimal. For another, many proteins experience posttranslational modifications that have a profound effect on their activities. For example, some proteins are not active until they are phosphorylated. Thus, if the cell is not phosphorylating such a protein at a given time, production of a large amount of mRNA for that protein would give a misleading picture of the true level of expression of the corresponding gene. Finally, many transcripts give rise to more than one protein—through alternative splicing, or

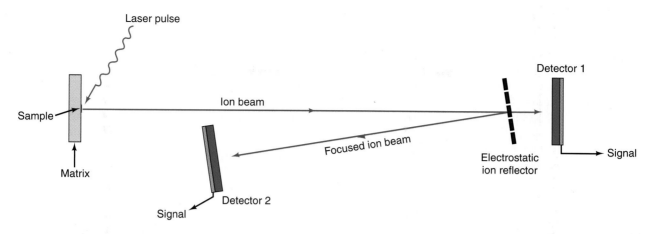

Figure 24.20   Principle behind the MALDI-TOF mass spectrometry. Place a sample (a peptide in this case) on the matrix at left and ionize it with a laser pulse. A potential difference between the matrix and the sample then accelerates the ionized sample toward detector 1. The time it takes the ions to reach detector 1 depends on their masses, so one can learn much about their masses by analyzing the time of flight to detector 1. Alternatively, one can turn on an electrostatic ion reflector in front of detector 1 to focus the ions and reflect them toward detector 2. This detector gives even more precise data about the masses of the ions, according to their times of flight.

alternative posttranslational modification. So measuring the level of a gene transcript doesn't necessarily tell what protein products will be made.

Therefore, if we want to measure real gene expression, we must look at the protein level. To analyze all the proteins in an organism, we need to do two things: First, we need to separate all those proteins from one another. Second, we have to analyze each protein by identifying it and measuring its activity. How do we do these things?

Protein Separations   The best separation tool available is two-dimensional gel electrophoresis, which was invented in the 1970s (Chapter 5). As powerful as that technique is, it is not up to the job of resolving all the tens of thousands of proteins in the human proteome. An average 2-D gel can resolve only about 2000 proteins, and even the best gel in the best hands can resolve only about 11,000 proteins. This problem is compounded by the fact that the performance of 2-D electrophoresis is unpredictable, and it frequently seems to be more art than science. Another problem is that many very interesting membrane proteins are too hydrophobic to dissolve in the buffers used in 2-D electrophoresis, so they cannot be seen at all. Finally, many proteins are present in such tiny quantities in the cell that a 2-D gel cannot detect them. Most of these problems are presently intractable, but scientists have dealt with 2-D gel resolution problem by analyzing different cellular compartments separately. For example, they can start with just the nucleus, or even a subcompartment like the nucleolus or a protein assembly like the nuclear pore complex. With many fewer proteins to separate, resolu-

tion is not such a serious problem. However, these limited studies fall far short of true proteomics.

Protein Analysis   Once the proteins are separated and quantified, how are they analyzed? First, they have to be identified, and the best method now available works like this: Individual spots are cut out of the gel and cleaved into peptides with proteolytic enzymes. These peptides can then be identified by **mass spectrometry**. Figure 24.20 illustrates a popular technique known by the cumbersome title matrix-assisted laser desorption-ionization time-of-flight (MALDI-TOF) mass spectrometry system. In this procedure, a peptide is placed on a matrix, which causes the peptide to form crystals. Then the peptide on the matrix is ionized with a laser beam (the matrix helps the peptide ionize), and an increase in voltage at the matrix is used to shoot the ions toward a detector. Assuming all the ions have just one charge (and almost all do), the time it takes an ion to reach the detector depends on its mass. The higher the mass, the longer the time of flight of the ion. In a MALDI-TOF mass spectrometer, the ions can also be deflected with an electrostatic reflector that also focuses the ion beam. Thus, we can determine the masses of the ions reaching the second detector with high precision, and these masses can reveal the exact chemical compositions of the peptides, and therefore their identities. If the sequence of the whole genome is known, we know what proteins to expect, so a computer can use the information from the mass spectrometer to match each spot on the 2-D gel with one of the genes in the genome, and therefore predict the sequence of the whole protein. However, knowing the sequence of a protein does not necessarily tell us that protein's activity, so further research will be necessary to determine the activities of many proteins.

You may be thinking that it would be nice to make a microchip that could identify thousands of proteins at once, as DNA chips identify thousands of RNAs at once in functional genomics studies. That would remove the need to separate the proteins because a mixture of many proteins could simply be incubated with the chip to see what binds. One such strategy would be to produce antibodies that can recognize proteins specifically and quantitatively and place them on microchips. But many obstacles stand in the way of realizing that dream. To begin with, antibodies are much more expensive and time-consuming to produce than oligonucleotides. In fact, the task of generating antibodies for every human protein is unthinkably vast at present. Moreover, the task of detecting low-abundance proteins, already impossible for many proteins using 2-D gels, would only be exacerbated by the miniaturization of microchip technology. On the other hand, the technology to complete the human genome in a reasonable period of time was not available when that project was first proposed in the mid-1980s, but the project stimulated the development of the technology. Perhaps we will experience a similar phenomenon if a full-scale human proteome project is initiated.

SUMMARY The sum of all RNAs produced by an organism is its transcriptome, and the study of these transcripts is transcriptomics. Similarly, the sum of all proteins produced by an organism is its proteome, and the study of these proteins is proteomics. Current research in proteomics requires first that proteins be resolved on a massive scale, for which the best available tool is 2-D gel electrophoresis. Then the proteins must be identified, and the best method for doing that involves digestion of the proteins one by one with proteases, and identifying the resulting peptides by mass spectrometry. Someday microchips with antibodies attached may allow analysis of proteins in complex mixtures without separation.

Solved Problems

Problem 1

Here is the physical map of a region of DNA you are mapping by RFLP analysis.

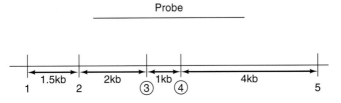

The vertical lines with numbers at the bottom represent restriction sites for the enzyme *Pst*I. Sites 3 and 4 are polymorphic; the others are not. The distances between *Pst*I sites are given. The extent of a probe you are using to detect the RFLP is indicated by a horizontal line at top. You cut DNA from different individuals with *Pst*I, electrophorese the fragments, blot them to a membrane, and hybridize the blot to the labeled probe.

a. Draw a picture of what the blot will look like when the DNA comes from individuals homozygous for the following haplotypes:

Haplotype	Site 3	Site 4
A	Present	Present
B	Present	Absent
C	Absent	Present
D	Absent	Absent

b. What effect would presence or absence of site 1 have on the results?

Solution

a. Haplotype *A:* If sites 3 and 4 are both present, all possible fragments will be produced. In order from left to right these are: 1.5 kb, 2 kb, 1 kb, and 4 kb. Of these, the 2-kb, 1-kb, and 4-kb fragments will hybridize to the probe and therefore will be visible. Haplotype *B:* Site 3 is present, so the 2-kb fragment will still be produced; site 4 is absent, so the 1-kb fragment will be fused to the 4-kb fragment, yielding a 5-kb fragment. Haplotype *C:* Site 3 is absent, so the 2-kb fragment will be fused to the 1-kb fragment, yielding a 3-kb fragment; site 4 is present, so the 4-kb fragment will be produced. Haplotype *D:* Sites 3 and 4 are both absent, so the 2-kb, 1-kb, and 4-kb fragments will all be fused together, yielding a 7-kb fragment. The following figure shows a summary of the results you will obtain:

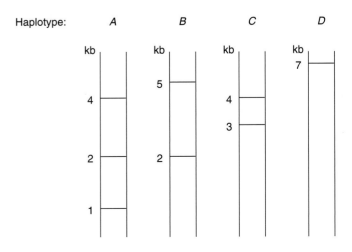

b. The presence or absence of site 1 will make no difference in the results because the probe does not extend to the left of site 2, which is always present. Thus, even though the absence of site 1 will yield a longer fragment to the left of site 2, you will not be able to detect this fragment with your probe.

Problem 2

You are mapping the gene responsible for a human genetic disease. You find that the gene is linked to a RFLP detected with a probe called B-101. You hybridize labeled B-101 DNA to DNAs from a panel of mouse–human hybrid cells. The following table shows the human chromosomes present in each hybrid cell line, and whether the probe hybridized to DNA from each. Which human chromosome carries the disease gene?

Cell Line	Human Chromosome Content	Hybridization to Probe
A	1, 2, 17	−
B	3, 5, 9, 12	+
C	5, 18, 22	−
D	1, 3, 9	+
E	2, 9, 18	−
F	2, 3, 7	+

Solution

Look at the human chromosome content in each of the positive cell lines (B, D, and F). These are:

B: 3, 5, 9, 12;

D: 1, 3, 9; and

F: 2, 3, 7.

Now find the one chromosome common to all three positive lines. It is chromosome 3. Therefore, the RFLP maps to chromosome 3, and so must the disease gene that is linked to the RFLP.

SUMMARY

Rapid, automated DNA sequencing methods have allowed molecular biologists to obtain the base sequences of viruses and organisms ranging from simple phages to bacteria to yeast and simple animals and plants. Much of the mapping work in the Human Genome Project was done with yeast artificial chromosomes (YACs), vectors that contain a yeast origin of replication, a centromere, and two telomeres. Foreign DNA up to 1 million bp long can be inserted between the centromere and one of the telomeres. It will then replicate along with the YAC. On the other hand, because of their superior stability and ease of use, most of the sequencing work in the Human Genome Project has been done with bacterial artificial chromosomes (BACs). BACs are vectors based on the F-plasmid of *E. coli*. They can accept inserts up to about 300 kb, but their inserts average about 150 kb.

Mapping the human genome, or any large genome, requires a set of landmarks (markers) to which we can relate the positions of genes. Genes can be used as markers in mapping, but markers are usually anonymous stretches of DNA such as RFLPs, VNTRs, STSs (including ESTs), and microsatellites. RFLPs (restriction fragment length polymorphisms) are differences in the lengths of restriction fragments generated by cutting the DNA of two or more different individuals with a restriction endonuclease. RFLPs can be caused by the presence or absence of a restriction site in a particular place, but they can also be caused by a variable number of tandem (head-to-tail) repeats (VNTRs) between two restriction sites. STSs (sequence-tagged sites) are regions of DNA that can be identified by formation of a predictable length of amplified DNA by PCR with pairs of primers. ESTs (expressed sequence tags) are a subset of STSs generated from cDNAs, so they represent expressed genes. Microsatellites are a subset of STSs generated by PCR with pairs of primers flanking tandem repeats of just a few nucleotides (usually 2–4 nt).

Radiation hybrid mapping allows mapping of STSs and other markers that are too far apart to fit on one BAC. In radiation hybrid mapping, human cells are irradiated to break chromosomes, then these dying cells are fused with hamster cells. Each hybrid cell has a different subset of human chromosome fragments. The closer together two markers are, the more likely they are to be found in the same hybrid cell.

Massive sequencing projects can take two forms: The map-then-sequence (clone-by-clone) approach or the shotgun approach. Actually, a combination of these methods is being used to sequence the human genome. The clone-by-clone strategy calls for production of a physical map of the genome including STSs, then sequencing the overlapping clones (mostly BACs) used in the mapping. This places the sequences in order so they can be pieced together. The shotgun strategy calls for the assembly of libraries of clones with different size inserts, then sequencing the inserts at random. This method relies on a computer program to find areas of overlap among the sequences and piece them together.

Sequencing of human chromosome 22q has revealed:

1. 11 gaps that cannot be filled with available methods.
2. 679 annotated genes.
3. The great bulk (about 97%) of the chromosome is made up of noncoding DNA.

4. Over 40% of the chromosome is in interspersed repeats such as Alu sequences and LINES.

5. The rate of recombination varies across the chromosome, with long regions of low rates of recombination punctuated by short regions with relatively high rates.

6. Several examples of local and long-range duplications.

7. Large regions where linkage among genes has been conserved with that in seven different mouse chromosomes.

Chromosome 21q, and a little of 21 p, have been sequenced, leaving relatively few and short gaps. The sequence reveals a relative poverty of genes—only about 225, plus 59 pseudogenes. All 24 genes known to be shared between mouse chromosome 10 and human chromosome 21 are in the same order in both chromosomes.

Functional genomics is the study of expression of all the genes in an organism. One approach to functional genomics is to create DNA microarrays or DNA microchips, holding thousands of cDNAs or oligonucleotides, respectively, then to hybridize labeled RNAs (or corresponding cDNAs) from cells to these arrays or chips. The intensity of hybridization to each spot reveals the extent of expression of the corresponding gene.

SAGE (serial analysis of gene expression) allows us to determine which genes are expressed in a given tissue and the extent of that expression. Short tags, characteristic of particular genes, are generated from cDNAs and ligated together between linkers. The ligated tags are then sequenced to determine which genes are expressed and how abundantly.

Several methods are available for identifying the genes in a large region of DNA. One of these is the exon trap, which uses a special vector to help clone exons only. Another is to use methylation-sensitive restriction enzymes to search for CpG islands, DNA regions containing unmethylated CpG sequences. But the best, and most modern, method is to look at the sequence of the DNA and find the open reading frames.

To implicate a gene as a causative agent in a disease, we must first show that it lies close to a marker that is linked to the disease state. Then we should show that it is expressed in the tissue(s) affected by the disease, it is mutated in individuals with the disease, and a wild-type version of the gene can cure the defect in diseased cells.

The sum of all RNAs produced by an organism is its transcriptome, and the study of these transcripts is transcriptomics. Similarly, the sum of all proteins produced by an organism is its proteome, and the study of these proteins is proteomics. Current research in proteomics requires first that proteins be resolved on a massive scale, the best available tool for which is 2-D gel electrophoresis. Then the proteins must be identified, and the best method for doing that involves digestion of the proteins one by one with proteases and identifying the resulting peptides by mass spectrometry. Someday microchips with antibodies attached may allow analysis of proteins in complex mixtures without separation.

REVIEW QUESTIONS

1. What is an open reading frame(ORF)? Write a DNA sequence containing a short ORF.

2. What are the essential elements of a YAC vector?

3. On what plasmid are the BAC vectors based? What essential elements do they contain?

4. Compare and contrast the clone-by-clone sequencing strategy and the shotgun sequencing strategy for large genomes.

5. Describe the procedure for finding an STS in a genome.

6. Describe microsatellites and minisatellites. Why are microsatellites better tools for linkage mapping than minisatellites?

7. Show how to use STSs in a set of BAC clones to form a contig. Illustrate with a diagram different from the one given in the text.

8. Describe the use of radiation hybrid mapping to map STSs.

9. How does an expressed sequence tag (EST) differ from an ordinary STS?

10. What major conclusions can we draw from the sequence of human chromosome 22?

11. What major conclusions can we draw from the sequence of human chromosome 21?

12. Compare and contrast "structural genomics" and "functional genomics."

13. Describe the process of making a DNA microchip (oligonucleotide array).

14. List in order the steps you would perform to make a DNA microchip in which two of the spots contain the dinucleotides AC and AT. You may ignore all of the other spots.

15. Describe a hypothetical experiment using a DNA microarray or a DNA microchip to measure the transcription from viral genes at two stages of infection of cells by the virus. Present sample results.

16. Describe a SAGE experiment to measure transcription in cancer cells of a certain type. Show how the production of a ditag works, with actual sequences of your own invention.

17. Will the following DNA fragments be detected by an exon trap? Why or why not?
   a. An intron
   b. Part of an exon
   c. A whole exon with parts of introns on both sides
   d. A whole exon with part of an intron on one side

18. List three ways of detecting a gene in a megaYAC clone.

19. a. What kind of mutation gave rise to Huntington's disease?

   b. What is the evidence that the gene identified as HD is really the gene that causes HD?

20. The following is a physical map of a region you are mapping by RFLP analysis.

Extent of probe

The numbered vertical lines represent restriction sites recognized by *Sma*I. The circled sites (2 and 3) are polymorphic, the others are not. You cut the DNA with *Sma*I, electrophorese the fragments, blot them to a membrane, and probe with a DNA whose extent is shown at top. Give the sizes of bands you will detect in individuals homozygous for the following haplotypes with respect to sites 2 and 3.

Haplotype	Site 2	Site 3
A	Present	Present
B	Present	Absent
C	Absent	Present
D	Absent	Absent

21. You are mapping the gene responsible for a human genetic disease. You find that the gene is linked to a RFLP detected with a probe called X-21. You hybridize labeled X-21 DNA to DNAs from a panel of mouse–human hybrid cells. The following shows the human chromosomes present in each hybrid cell line, and whether the probe hybridized to DNA from each. Which human chromosome carries the disease gene?

Cell Line	Human Chromosome Content	Hybridization to X-21
A	1, 5, 21	+
B	6,7	–
C	1, 22, Y	–
D	4, 5, 18, 21	+
E	8, 21, Y	–
F	2, 5, 6	+

22. a. What kind of *CF* mutation is most prevalent in cystic fibrosis?

   b. What is the evidence that the gene that encodes CFTR is really CF?

23. What are SNPs? Why are most of them unimportant? How can some of them be useful? How can they be abused?

24. Compare and contrast the techniques used in, and information obtained in, transcriptomics and proteomics.

SUGGESTED READINGS

General References and Reviews

Abbott, A. 1999. A post-genomic challenge: Learning to read patterns of protein synthesis. *Nature* 402:715–20.

Cheung, V.G., M. Morley, F. Aguilar, A. Massimi, R. Kucherlapati, and G. Childs. 1999. Making and reading microarrays. *Nature Genetics Supplement* 21:15–19.

Collins, F.S., M.S. Guyer, and A. Chakravarti. 1997. Variations on a theme: Cataloging human DNA sequence variation. *Science* 278:1580–81.

Goffeau, A. 1995. Life with 482 genes. *Science* 270:445–46.

Hieter, P. and Boguski, M. 1997. Functional genomics: It's all how you read it. *Science* 278:601–02.

Lipshutz, R.J., S.P.A. Fodor, T.R. Gingeras, and D.J. Lockhart. 1999. High density synthetic oligonucleotide arrays. *Nature Genetics Supplement* 21:20–24.

Morell, V. 1996. Life's last domain. *Science* 273:1043–45.

Murray, T.H. 1991. Ethical issues in human genome research. *FASEB Journal* 5:55–60.

Reeves, R.H. 2000. Recounting a genetic story. *Nature* 405:283–4.

Service, R.F. 1998. Microchip arrays put DNA on the spot. *Science* 282:396–99

Venter, J.C., H.O. Smith, and L. Hood. 1996. A new strategy for genome sequencing. *Nature* 381:364–66.

Young, R.A. 2000. Biomedical discovery with DNA arrays. *Cell* 102:9–15.

Research Articles

Blattner, F.R., et al. 1997. The complete genomic sequence of *Escherichia coli* K12. *Science* 277:1453–62.

Bult, C.J., et al. 1996. Complete genome sequence of the methanogenic archaeon, *Methanococcus jannaschii*. *Science* 273:1058–73.

C. elegans Sequencing Consortium, The. 1998. Genome sequence of the nematode *C. elegans*: A platform for investigating biology. *Science* 282:2013–18.

Deloukas, P., et al. 1998. A physical map of 30,000 human genes. *Science* 282:744–46.

Dunham, I., et al. (The Chromosome 22 Sequencing Consortium) 1999. The DNA sequence of human chromosome 22. *Nature* 402:489–95.

Fraser, C.M., et al. 1997. Genomic sequence of a Lyme disease spirochete, *Borrelia burgdorferi*. *Nature* 390: 580–86.

Goffeau, A., et al. 1996. Life with 6000 genes. *Science* 274:546–67.

Gusella, J.F., N.S. Wexler, P.M. Conneally, S.L. Naylor, M.A. Anderson, R.E. Tauzi, P.C. Watkins, K. Ottina, M.R. Wallace, A.Y. Sakaguchi, A.B. Young, I. Shoulson, E. Bonilla, and J.B. Martin. 1983. A polymorphic DNA marker genetically linked to Huntington's disease. *Nature* 306:234–38.

Hattori, M., et al. (The Chromosome 21 Sequencing Consortium) 2000. The DNA sequence of human chromosome 21. Nature 405:311–19.

Hudson, T.J., et al. 1995. An STS-based map of the human genome. *Science* 270:1945–54.

International Human Gene Sequencing Consortium. 2001. Initial sequencing and analysis of the human genome. *Nature* 409:860–921.

Iyer, V.R., et. al. 1999. The transcriptional program in the response to human fibroblasts to serum. *Science* 283:83–87.

Schuler, G.D., et al. 1996. A gene map of the human genome. *Science* 274:540–46.

Shizuya, H., B. Birren, U.-J. Kim, V. Mancino, T. Slepak, Y. Tachiiri, and M. Simon. 1992. Cloning and stable maintenance of 300-kilobase-pair fragments of human DNA in *Escherichia coli* using an F-factor-based vector. *Proceeding of the National Academy of Sciences USA* 89:8794–97.

Velculescu, V.E., L. Zhang, B. Vogelstein, and K.W. Kinsler. 1995. Serial analysis of gene expression. *Science* 270:484–87.

Venter, J.C., et al. 2001. The sequence of the human genome. *Science* 291:1304–51.

Wang, D.G., et al. 1998. Large-scale identification, mapping, and genotyping of single-nucleotide polymorphisms in the human genome. *Science* 280:1077–82.

A site   The ribosomal site to which new aminoacyl-tRNAs (except the first one) bind.

abortive transcripts   Very short transcripts (about 6 nt long) synthesized at prokaryotic promoters before promoter clearance occurs.

acceptor stem   The part of a tRNA molecule formed by base-pairing between the 5′- and 3′-ends of the molecule. The 3′-end can "accept" an amino acid by charging, hence the name.

acidic domain   A transcription activation domain rich in acidic amino acids.

aconitase   An enzyme whose apoprotein form (lacking iron) binds to iron response elements in mRNAs and controls their translation or degradation.

activation region I (ARI)   The region of CAP that is thought to bind to the carboxyl-terminal domain (CTD) of the α subunit of *E. coli* RNA polymerase.

activation region II (ARII)   The region of CAP that is thought to bind to the amino-terminal domain (NTD) of the α subunit of *E. coli* RNA polymerase.

activation site   The site on an aminoacyl-tRNA synthetase that activates the amino acid, forming an aminoacyl adenylate.

activator   A protein that binds to an enhancer (or activator-binding region) and activates transcription from a nearby promoter. In eukaryotes, stimulates formation of a preinitiation complex.

activator interference   *See* squelching.

activator site   The DNA site to which a prokaryotic transcription activator binds (e.g., the CAP–cAMP-binding site in a catabolite repressible operon).

A-DNA   A form of DNA found at low relative humidity, with 11 bp per helical turn. The form assumed in solution by an RNA–DNA hybrid.

affinity chromatography   Chromatography that relies on affinity between a substance one wishes to purify, and another substance immobilized on a resin. For example, oligohistidine fusion proteins can be purified by affinity chromatography on a nickel resin because of the affinity between nickel ions and oligohistidine.

affinity labeling   Labeling one substance by covalently attaching a reactive compound with specific affinity for the first substance (e.g., labeling the active site of an enzyme by linking a substrate analog covalently to the enzyme).

α-amanitin   A toxin produced by poisonous species of mushrooms in the *Amanita* genus. Inhibits RNA polymerase II at very low concentration, and RNA polymerase III at higher concentration. Usually does not inhibit RNA polymerase I at all.

alkylation   The addition of carbon-containing groups to other molecules. Alkylation of DNA bases constitutes DNA damage that can lead to mutations.

alternative splicing   Splicing the same pre-mRNA in two or more ways to yield two or more different mRNAs that produce two or more different protein products.

Alu element   A human nonautonomous retrotransposon that contains the AGCT sequence recognized by the restriction enzyme *Alu*I. Present in about 1 million copies in the human genome.

amber codon   UAG, coding for termination.

amber mutation   *See* nonsense mutation.

amber suppressor   A tRNA bearing an anticodon that can recognize the amber codon (UAG) and thereby suppress amber mutations.

amino acid   The building block of proteins.

aminoacyl-tRNA synthetase   The enzyme that links a tRNA to its cognate amino acid.

   a. **class I** An aminoacyl-tRNA synthetase that charges the 2′-OH of the tRNA.

   b. **class II** An aminoacyl-tRNA synthetase that charges the 3′-OH of the tRNA.

amino tautomer   The normal tautomer of adenine or cytosine found in nucleic acids.

amino terminus   The end of a polypeptide with a free amino group. The end at which protein synthesis begins.

amplification   Selective replication of a gene to produce more than the normal single copy in a haploid genome.

anabolic metabolism   The process of building up substances from relatively simple precursors. The *trp* operon encodes anabolic enzymes that build the amino acid tryptophan.

anaphase   A short stage of nuclear division in which the chromosomes move to the poles.

aneuploid   Not having the normal diploid number of chromosomes.

annealing of DNA   The process of bringing back together the two separate strands of denatured DNA to re-form a double helix.

annotated genes   Genes or gene-like sequences from a genomic sequencing project that are at least partially identified.

antibody   A protein with the ability to recognize and bind to a substance, usually another protein, with great specificity. Helps the body's immune system recognize and trigger an attack on invading agents.

anticodon   A 3-base sequence in a tRNA that base-pairs with a specific codon.

anticodon loop   The loop, conventionally drawn at the bottom of tRNA molecule, that contains the anticodon.

antigen   A substance recognized and bound by an antibody.

antiparallel   The relative polarities of the two strands in a DNA double helix; if one strand goes 5′→3′, top to bottom, the other goes 3′→5′. The same antiparallel relationship applies to any double-stranded polynucleotide or oligonucleotide, including the RNAs in a codon–anticodon pair.

antirepression   Prevention of repression by histones or other transcription-inhibiting factors. Antirepression is part of a typical activator's function.

antisense RNA   An RNA complementary to an mRNA.

antiserum   Serum containing an antibody or antibodies directed against a particular substance.

antiterminator   A protein, such as the λ N gene product, that overrides a terminator and allows transcription to continue.

AP1   A transcription activator composed of one molecule each of Fos and Jun. (Jun–Jun homodimers also have AP1 activity.) Mediates response to the mitogenic phorbol esters.

AP endonuclease   An enzyme that cuts a strand of DNA on the 5′-side of an AP site.

AP site   An apurinic or apyrimidinic site in a DNA strand.

aporepressor   A repressor in an inactive form, without its corepressor.

apurinic site (AP site)   A deoxyribose in a DNA strand that has lost its purine base.

apyrimidinic site (AP site)   A deoxyribose in a DNA strand that has lost its pyrimidine base.

AraC   The negative regulator of the *ara* operon.

archaea   The kingdom of prokaryotic organisms whose biochemistry and molecular biology resemble those of eukaryotes as well as those of bacteria. The archaea typically live in extreme environments that are very hot or very salty. Some of the archaea are strict anaerobes that generate methane.

architectural transcription factor   A protein that does not activate transcription by itself, but helps DNA bend so other activators can stimulate transcription.

assembly factor   A transcription factor that binds to DNA early in the formation of a preinitiation complex and helps the other transcription factors assemble the complex.

assembly map   A scheme showing the order of addition of ribosomal proteins during self-assembly of a ribosomal particle in vitro.

asymmetric transcription   Transcription of only one strand of a given region of a double-stranded polynucleotide.

ATPase   An enzyme that cleaves ATP, releasing energy for other cellular activities.

attachment sites   *See att* sites.

***att*B**   The *att* site on the *E. coli* genome.

attenuation   A mechanism of transcription control that involves premature transcription termination.

attenuator   A region of DNA upstream from one or more structural genes, where premature transcription termination (attenuation) can occur.

***att*P**   The *att* site on the λ phage genome.

***att* sites**   Sites on phage and host DNA where recombination occurs, allowing integration of the phage DNA into the host genome as a prophage.

A-type particles   Intracellular, noninfectious particles that resemble retroviruses. May be vehicles for transposition of retrotransposons.

autonomously replicating sequence 1 (ARS1)   A yeast origin of replication.

autoradiography   A technique in which a radioactive sample is allowed to expose a photographic emulsion, thus "taking a picture of itself."

autoregulation   The control of a gene by its own product.

autosome   Any chromosome except the sex chromosomes.

auxotroph   An organism that requires one or more substances in addition to minimal medium.

BAC   *See* bacterial artificial chromosome.

BAC walking   Sequencing a BAC with minimal overlap with a seed BAC, then sequencing another BAC with minimal overlap to the second, and so forth until all BACs in a contig are sequenced.

back mutation   *See* reversion.

bacterial artificial chromosome (BAC)   A vector based on the *E. coli* F plasmid, capable of holding inserts up to 300,000 bp (average insert size, about 150,000 bp).

bacteriophage   *See* phage.

Barr body   Structure formed when the inactivated X chromosome is heterochromatinized.

basal level transcription   A very low level of class II gene transcription achieved with general transcription factors and polymerase II alone.

base   A cyclic, nitrogen-containing compound linked to deoxyribose in DNA and to ribose in RNA.

base excision repair   An excision repair pathway that begins by removing a damaged base by a DNA glycosylase and continues by cleaving the 5′ side of the resulting AP site by an AP endonuclease, then removing the AP sugar phosphate and concludes with removing downstream bases and filling in the gap by DNA polymerase and DNA ligase.

base pair   A pair of bases (A–T or G–C), one in each strand, that occur opposite each other in a double-stranded DNA.

β clamp   A dimer of β subunits of the DNA pol III holoenzyme that clamps around DNA, tethering the holoenzyme to the DNA and thereby conferring processivity.

B-DNA   The standard Watson–Crick model of DNA favored at high relative humidity and in solution.

BER   *See* base excision repair.

β-galactosidase   An enzyme that breaks the bond between the two constituent sugars of lactose.

bHLH domain   An HLH motif coupled to a basic motif. When two bHLH proteins dimerize through their HLH motifs, the basic motifs are in position to interact with a specific region of DNA. The bHLH protein grasps the major groove of a DNA, like a pair of tongs.

bHLH–ZIP domain   A dimerization and DNA-binding domain. A basic region is coupled with both an HLH and a leucine zipper (ZIP) domain.

bidirectional DNA replication   Replication that occurs in both directions at the same time from a common starting point, or origin of replication. Requires two active replicating forks.

bioinformatics   The building and manipulation of biological databases. In the context of genomics, this means managing massive amounts of sequencing data and providing useful access to and interpretation of the data.

β-lactamase   An enzyme that breaks down ampicillin and related antibiotics and renders a bacterium resistant to the antibiotic.

branch migration   Lateral motion of the branch of a Holliday junction during recombination.

branchpoint-bridging protein (BBP)   A protein essential for splicing that binds to U1 snRNP at the 5′-end of the intron, and to Mud2p at the 3′-end of the intron.

bromodeoxyuridine (BrdU)   An analog of thymidine that can be incorporated into DNA in place of thymidine and can then cause mutations by forming an alternative tautomer that base-pairs with guanine instead of adenine.

bZIP domain   A leucine zipper motif coupled to a basic motif. When two bZIP proteins dimerize through their leucine zippers, the basic motifs are in position to interact with a specific region of DNA. The bZIP protein grasps the major groove of a DNA, like a pair of tongs.

C value   The amount of DNA in picograms (trillionths of a gram) in a haploid genome of a given species.

C value paradox   The fact that the C value of a given species is not necessarily related to the genetic complexity of that species.

cAMP response element (CRE)   The enhancer that responds to cAMP.

cap   A methylated guanosine bound through a 5′-5′ triphosphate linkage to the 5′-end of an mRNA, an hnRNA, or an snRNA.

CAP (catabolite activator protein)   A protein which, together with cAMP, activates operons that are subject to catabolite repression. Also known as CRP.

cap-binding protein (CBP)   A protein that associates with the cap on a eukaryotic mRNA and allows the mRNA to bind to a ribosome. Also known as eIF4F.

carboxyl-terminal domain (CTD)   The carboxyl-terminal region of the largest subunit of RNA polymerase II. Consists of dozens of repeats of a heptamer rich in serines and threonines.

carboxyl terminus   The end of a polypeptide with a free carboxyl group.

carcinogen   A cancer-causing agent (usually radiation or chemical).

catabolic metabolism   The process of breaking substances down into simpler components. The *lac* operon encodes catabolic enzymes that break down lactose into its component parts, galactose and glucose.

catabolite activator protein   *See* CAP.

catabolite repression   The repression of a gene or operon by glucose or, more likely, by a catabolite, or breakdown product of glucose.

catenane   A structure composed of two or more circles linked in a chain.

CBP   *See* CREB-binding protein.

CCAAT-binding transcription factor (CTF)   A transcription activator that binds to the CCAAT box.

CCAAT box   An upstream motif, having the sequence CCAAT, found in many eukaryotic promoters recognized by RNA polymerase II.

cDNA   A DNA copy of an RNA, made by reverse transcription.

centimorgan (cM)   The genetic distance that yields a 1% recombination frequency between two markers.

centromere   Constricted region on the chromosome where spindle fibers are attached during cell division.

CF I and CF II   *See* cleavage factors I and II.

CFTR   *See* cystic fibrosis transmembrane conductance regulator.

charging   Coupling a tRNA with its cognate amino acid.

Charon phages   A set of cloning vectors based on λ phage.

(χ)   *See* chi site.

chi site (χ)   An *E. coli* DNA site with the sequence 5′-GCTGGTGG-3′. RecBCD cuts at the 3′-end of a chi site to initiate homologous recombination.

chi structure   *See* Holliday junction.

CHIP   *See* chromatin immunoprecipitation.

chloramphenicol   An antibiotic that kills bacteria by inhibiting the peptidyl transferase reaction catalyzed by 50S ribosomes.

chloramphenicol acetyl transferase (CAT)   An enzyme whose bacterial gene is frequently used as a reporter gene in eukaryotic transcription and translation experiments. This enzyme adds acetyl groups to the antibiotic chloramphenicol.

chloroplast   The photosynthetic organelle of green plants and other photosynthetic eukaryotes.

chromatids   Copies of a chromosome produced in cell division.

chromatin   The material of chromosomes, composed of DNA and chromosomal proteins.

chromatin immunoprecipitation (CHIP)   A method for purifying chromatin containing a protein of interest by immunoprecipitating the chromatin with an antibody directed against that protein or against an epitope tag attached to the protein.

chromatography   A group of techniques for separating molecules based on their relative affinities for a mobile and a stationary phase. In ion-exchange chromatography, the charged resin is the stationary phase, and the buffer of increasing ionic strength is the mobile phase.

chromogenic substrate   A substrate that produces a colored product when acted on by an enzyme.

chromosome   The physical structure, composed largely of DNA and protein, that contains the genes of an organism.

chromosome theory of inheritance   The theory that genes are contained in chromosomes.

***cis*-acting**   A term that describes a genetic element, such as an enhancer, a promoter, or an operator, that must be on the same chromosome in order to influence a gene's activity.

***cis*-dominant**   Dominant, but only with respect to genes on the same piece of DNA. For example, an operator constitutive mutation in one copy of the *lac* operon of a merodiploid *E. coli* cell is dominant with respect to that copy of the *lac* operon, but not to the other. That is because the operator controls the operon that is directly attached to it, but it cannot control an unattached operon.

cistron   A genetic unit defined by the *cis–trans* test. For all practical purposes, it is synonymous with the word gene.

clamp loader   The γ complex portion of the DNA pol III holoenzyme that facilitates binding of the β clamp to the DNA.

class II promoter   A promoter recognized by RNA polymerase II.

clastogen   An agent that causes DNA strand breaks.

cleavage factors I and II (CF I and CF II)   RNA-binding proteins that are important in cleavage of a pre-mRNA at the polyadenylation site.

cleavage poly(A) specificity factor (CPSF)   A protein that recognizes the AAUAAA part of the polyadenylation signal in a pre-mRNA and stimulates cleavage and polyadenylation.

cleavage stimulation factor (CstF)   A protein that recognizes the GU-rich part of the polyadenylation signal in a pre-mRNA and stimulates cleavage.

clone-by-clone sequencing   A systematic method of sequencing large genomes. First the whole genome is mapped, then clones corresponding to known regions of the genome are sequenced.

clones   Individuals formed by an asexual process so that they are genetically identical to the original individual. Also, colonies of cells or groups of viruses that are genetically identical.

closed promoter complex   The complex formed by relatively loose binding between RNA polymerase and a prokaryotic promoter. It is "closed" in the sense that the DNA duplex remains intact, with no "opening up," or melting of base pairs.

coactivators   Factors that have no transcription–activation ability of their own, but help other proteins stimulate transcription.

codon   A 3-base sequence in mRNA that causes the insertion of a specific amino acid into protein or causes termination of translation.

coiled coil   A protein motif in which two α-helices (coils) wind around each other. When the two a-helices are on separate proteins, the formation of a coiled coil causes dimerization.

cointegrate   An intermediate in transposition of a transposon such as Tn3 from one replicon to another. The transposon replicates, and the cointegrate contains the two replicons joined through the two transposon copies.

colE1   A plasmid found in certain strains of *E. coli,* which codes for a bacteriocidal toxin known as a colicin. The colE1 DNA replicates unidirectionally.

colony hybridization   A procedure for selecting a bacterial clone containing a gene of interest. DNAs from a large number of clones are simultaneously tested with a labeled probe that hybridizes to the gene of interest.

commitment complex (CC)   A complex containing at least nuclear pre-mRNA and U1 snRNP that is committed to splicing out the intron to which the U1 snRNP has bound.

complementary polynucleotide strands   Two strands of DNA or RNA that have complementary sequences; that is, wherever one has an adenine the other has a thymine, and wherever one has a guanine the other has a cytosine.

composite transposon   A bacterial transposon composed of two types of parts: two arms containing IS or IS-like elements, and a central region comprised of the genes for transposition and one or more antibiotic resistance genes.

concatemers   DNAs of multiple genome length.

conditional lethal   A mutation that is lethal under certain circumstances, but not under others (e.g., a temperature-sensitive mutation).

conjugation (of bacteria)   The movement of genetic material from an F⁺ to an F⁻ bacterium.

consensus sequence   The average of several similar sequences. For example, the consensus sequence of the –10 box of an *E. coli* promoter is TATAAT. This means that if you examine a number of such sequences, T is most likely to be found in the first position, A in the second, and so forth.

conservative replication   DNA (or RNA) replication in which both parental strands remain together, producing a progeny duplex both of whose strands are new.

conservative transposition   Transposition in which both strands of the transposon DNA are conserved as they leave their original location and move to the new site.

constant region   The region of an antibody that is more or less the same from one antibody to the next.

constitutive   Always turned on.

constitutive mutant   An organism containing a constitutive mutation.

constitutive mutation   A mutation that causes a gene to be expressed at all times, regardless of normal controls.

contig   A group of cloned DNAs containing contiguous or overlapping sequences.

copia   A transposable element found in *Drosophila* cells.

core element   An element of the eukaryotic promoter recognized by RNA polymerase I. Includes the bases surrounding the transcription start site.

core histones   All the nucleosomal histones except H1. The histones inside the DNA coils of a nucleosome.

core particle   The remnant of a nucleosome left after nuclease digestion. Contains 146 bp of DNA wrapped around an octamer of core histones.

core polymerase   *See* RNA polymerase core.

corepressor   A substance that associates with an aporepressor to form an active repressor (e.g., tryptophan is the corepressor of the *trp* operon). A protein that works in conjunction with other proteins to repress gene transcription. Histone deacetylases can act as corepressors.

core promoter (class II)   Whatever promoter elements are near the transcription initiation site in a particular promoter. Could include up to three elements: a TATA box, an initiator, and a downstream element.

core promoter elements   The minimal elements of a promoter (e.g., the –10 and –35 boxes in bacterial promoters).

cos   The cohesive ends of the linear λ phage DNA.

cosmid   A vector designed for cloning large DNA fragments. A cosmid contains the cos sites of λ phage, so it can be packaged into λ heads, and a plasmid origin of replication, so it can replicate as a plasmid.

C_0t   The product of DNA concentration and time of annealing in a DNA renaturation experiment.

counts per minute (cpm)   The average number of scintillations detected per minute by a liquid scintillation counter. Generally, this is dpm times the efficiency of the counter.

CpG island   A region of DNA containing many unmethylated CpG sequences. Usually associated with active genes.

CPSF   *See* cleavage poly(A) specificity factor.

CRE   *See* cAMP response element.

CREB   *See* CRE-binding protein.

CREB-binding protein (CBP)   A coactivator (or mediator) that binds to phosphorylated CREB at the CRE and then contacts one or more general transcription factors to stimulate assembly of a preinitiation complex.

CRE-binding protein (CREB)   The activator that is phosphorylated and thus activated by cAMP-stimulated protein kinase A. Binds to CRE and, together with CBP, stimulates transcription of associated genes.

Cro   The product of the λ *cro* gene. A repressor that binds preferentially to O_R3 and turns off the λ repressor gene (*cI*).

cross-linking   A technique for probing the interaction between two species (e.g., a protein and a DNA). The two species are chemically cross-linked as they form a complex, then the nature of the cross-linked species is examined.

crossing over   Physical exchange between DNAs that occurs during recombination.

CRP   Cyclic-AMP receptor protein. *See* CAP.

CRSP   A coactivator that collaborates with Sp1 to activate transcription.

cryptogene   A gene coding for the unedited version of an RNA that requires editing.

CstF   *See* cleavage stimulation factor.

CTD   *See* carboxyl-terminal domain.

α-CTD The carboxyl-terminal domain of the α-subunit of bacterial RNA polymerase.

cyanobacteria (blue-green algae) Photosynthetic bacteria. The ancestors of modern cyanobacteria are thought to have invaded eukaryotic cells and evolved into chloroplasts.

cyclic-AMP (cAMP) An adenine nucleotide with a cyclic phosphodiester linkage between the 3′ and 5′ carbons. Implicated in a variety of control mechanisms in prokaryotes and eukaryotes.

cystic fibrosis transmembrane conductance regulator gene (*CFTR*) The gene that is defective in cystic fibrosis. Encodes a chloride channel.

cytidine A nucleoside containing the base cytosine.

cytoplasmic polyadenylation element (CPE) A sequence in the 3′-UTR of an mRNA (consensus, UUUUUAU) that is important in cytoplasmic polyadenylation.

cytoplasmic ribosomes The ribosomes that translate the mRNAs encoded in the nucleus.

cytosine (C) The pyrimidine base that pairs with guanosine in DNA.

cytotoxic Having the ability to kill cells.

DAI *See* double-stranded RNA-activated inhibitor of protein synthesis.

***dam* methylase** The deoxyadenosine methylase that adds methyl groups to the A in the sequence GATC in the DNA of *E. coli* cells. The mismatch repair system inspects the methylation of GATC sequences to determine which strand is newly synthesized, and therefore unmethylated.

daughter strand gap The gap left by the DNA replication machinery after it skips over a noncoding base or a pyrimidine dimer.

deadenylation The removal of AMP residues from poly(A) in the cytoplasm.

deamination of DNA The removal of an amino group (NH₂) from a cytosine or adenine in DNA, in which the amino group is replaced by a carbonyl group (C–O). This converts cytosine to uracil and adenine to hypoxanthine.

decatenation The process of unlinking the circles in a catenane.

decoding Interactions between codons and anticodons on the ribosome that lead to binding of the correct aminoacyl-tRNA.

defective virus A virus that is unable to replicate without a helper virus.

degenerate code A genetic code, such as the one employed by all life on Earth, in which more than one codon can stand for a single amino acid.

delayed early genes Phage genes whose expression begins after that of the immediate early genes. Their transcription depends on at least one phage protein. Also, eukaryotic viral genes transcribed before viral DNA replication begins, but after immediate early transcription begins. Delayed early transcription depends on at least one immediate early viral product.

deletion A mutation involving a loss of one or more base pairs; the case in which a chromosomal segment or gene is missing. Also called *deficiency*.

denaturation (DNA) Separation of the two strands of DNA.

denaturation (protein) Disruption of the three-dimensional structure of a protein without breaking any covalent bonds.

densitometer An instrument that measures the darkness of a spot on a transparent film (e.g., an autoradiograph).

deoxyribose The sugar in DNA.

dideoxyribonucleotide A nucleotide, deoxy at both 2′- and 3′-positions, used to stop DNA chain elongation in DNA sequencing.

dihydrouracil loop *See* D loop.

dimer (protein) A complex of two polypeptides. These can be the same (in a homodimer), or different (in a heterodimer).

dimerization domain The part of a protein that interacts with another protein to form a dimer (or higher multimer).

dimethyl sulfate (DMS) An agent used for methylating DNA. After methylation, the DNA can be chemically cleaved at the methylated sites.

diploid The chromosomal number of the zygote and other cells (except the gametes). Symbolized as *2n*.

directional cloning Insertion of foreign DNA into two different restriction sites of a vector, such that the orientation of the insert can be predetermined.

disintegrations per minute (dpm) The average number of radioactive emissions produced each minute by a sample.

dispersive replication A hypothetical mechanism in which the DNA becomes fragmented so that new and old DNA coexist in the same strand after replication.

distributive Opposite of processive. Unable to continue a task without repeatedly dissociating and reassociating with the substrate or template.

D loop A loop formed when a free DNA or RNA end "invades" a double helix, base-pairing with one of the strands and forcing the other to "loop out."

DMS footprinting A technique similar to DNase footprinting that uses DMS methylation and chemical cleavage rather than DNase cleavage of DNA.

DNA (deoxyribonucleic acid) A polymer composed of deoxyribonucleotides linked together by phosphodiester bonds. The material of which most genes are made.

DnaA The first protein to bind to *oriC* in forming an *E. coli* primosome.

***dnaA* box** A 9-mer within *oriC* to which DnaA binds in forming an *E. coli* primosome.

DnaB A key component of the *E. coli* primosome. Helps assemble the primosome by facilitating primase binding. Also has DNA helicase activity to unwind the parental DNA strands prior to primer synthesis.

DNA-binding domain The part of a DNA-binding protein that makes specific contacts with a target site on the DNA.

DNA fingerprints The use of highly variable regions of DNA to identify particular individuals.

DnaG The *E. coli* primase.

DNA glycosylase An enzyme that breaks the glycosidic bond between a damaged base and its sugar.

DNA gyrase A topoisomerase that pumps negative superhelical turns into DNA. Relaxes the positive superhelical strain created by unwinding the *E. coli* DNA during replication.

DNA ligase An enzyme that joins two double-stranded DNAs end to end.

DNA melting *See* denaturation (DNA).

DNA microarray A chip containing many tiny spots of DNA. Used as a dot blot to measure expression of many genes at once.

DNA microchip A chip with many oligonucleotides built in separate tiny spaces on the chip. Used as a dot blot to measure expression of many genes at once.

DNA photolyase The enzyme that catalyzes photoreactivation by breaking pyrimidine dimers.

DNA polymerase η   A relatively error-free polymerase (compared with pol ζ) that inserts two dAMPs across from a pyrimidine dimer.

DNA polymerase I   One of three different DNA-synthesizing enzymes in *E. coli;* used primarily in DNA repair.

DNA polymerase III holoenzyme   The enzyme within the replisome, which actually makes DNA during replication.

DNA polymerase V   *See* UmuD′₂C.

DNase   Deoxyribonuclease, an enzyme that degrades DNA.

DNase footprinting   A method of detecting the binding site for a protein on DNA by observing the DNA region this protein protects from degradation by DNase.

DNase hypersensitive sites   Regions of chromatin that are about a hundred times more susceptible to attack by DNase I than bulk chromatin. These usually lie in the 5′-flanking regions of active or potentially active genes.

DNA typing   The use of molecular techniques, especially Southern blotting, to identify a particular individual.

dominant   An allele or trait that expresses its phenotype when heterozygous with a recessive allele; for example, *A* is dominant over *a* because the phenotypes of *AA* and *Aa* are the same.

domain (protein)   An independently folded part of a protein.

double helix   The shape two complementary DNA strands assume in a chromosome.

double-stranded RNA-activated inhibitor of protein synthesis (DAI)   A protein kinase that responds to interferon and double-stranded RNA by phosphorylating eIF-2a, strengthening its binding to eIF-2B and thereby blocking translation initiation. This prevents viral protein synthesis in an infected cell.

down mutation   A mutation, usually in a promoter, that results in less expression of a gene.

downstream element   A DNA region downstream of the transcription start site that is important in the efficiency of transcription from a few class II promoters.

Drosophila melanogaster   A species of fruit fly used widely by geneticists.

Ds   A defective transposable element found in corn, which relies on an *Ac* element for transposition.

DSB   Double-strand break in DNA. Required for initiation of meiotic recombination.

dsRNA   Double-stranded RNA.

dUTPase   An enzyme that degrades dUTP and thereby prevents its incorporation into DNA.

editing site   The site on an aminoacyl-tRNA synthetase that examines aminoacyl adenylates and, sometimes, aminoacyl-tRNAs and hydrolyzes those whose amino acids are too small.

EF-2   The eukaryotic homologue of EF-G.

EF-G   The prokaryotic translation elongation factor that, along with GTP, fosters translocation.

EF-Ts   An exchange factor that exchanges GDP on EF-Tu for GTP.

EF-Tu   The prokaryotic translation elongation factor that, along with GTP, carries aminoacyl-tRNAs (except fMet-tRNA$_f^{Met}$ to the ribosomal A site.

EGF   *See* epidermal growth factor.

eIF1   A eukaryotic initiation factor that stimulates scanning to locate the proper initiation codon.

eIF2   A eukaryotic initiation factor that is responsible for binding Met-tRNA$_i^{Met}$ to the 40S ribosomal subunit.

eIF2B   A eukaryotic exchange factor that exchanges GTP for GDP on eIF2.

eIF3   A eukaryotic initiation factor that binds to 40S ribosomal subunits and prevents their reuniting prematurely with 60S subunits.

eIF4A   One of the subunits of eIF4F; a DEAD family RNA-binding protein with RNA helicase activity. In conjunction with eIF4B, eIF4A can bind to the leader region of an mRNA and remove hairpins in advance of the scanning ribosomal subunit.

eIF4E   The cap-binding component of eIF4F.

eIF4F   A cap-binding eukaryotic initiation factor composed of three parts: eIF4A, eIF4E, and eIF4G.

eIF4G   One of the subunits of eIF4F. Serves as an adapter by binding to two different proteins: Binds to eIF4E, which is bound to the cap; also binds to eIF3, which is bound to the 40S ribosomal particle. In this way, it brings the 40S particle together with the 5′-end of an mRNA, where it can begin scanning.

eIF5   A eukaryotic initiation factor that promotes association between a 40S initiation complex and a 60S ribosomal subunit.

eIF6   A eukaryotic initiation factor with activity similar to that of eIF3.

electron density map   A three-dimensional representation of the electron density in a molecule or complex of molecules. Usually determined by x-ray crystallography.

electrophile   A molecule that seeks centers of negative charge in other molecules and attacks them there.

electrophoresis   A procedure in which voltage is applied to charged molecules, inducing them to migrate. This technique can be used to separate DNA fragments, RNAs, or proteins.

electroporation   The use of a strong electric current to introduce DNA into cells.

elongation factor   A protein that is necessary for either the aminoacyl-tRNA binding or the translocation step in the elongation phase of translation.

encode   To contain the information for making an RNA or polypeptide. A gene can encode an RNA or a polypeptide.

end-filling   Filling in the recessed 3′-end of a double-stranded DNA using deoxynucleoside triphosphates and a DNA polymerase. This technique can be used to label the 3′-end of a DNA strand.

endonuclease   An enzyme that makes cuts within a polynucleotide strand.

endonuclease VII   The resolvase of phage T4.

endoplasmic reticulum (ER)   Literally, "cellular network"; a network of membranes in the cell on which proteins destined for export from the cell are synthesized.

endospores   Dormant spores formed within a cell, as in *Bacillus subtilis.*

enhanceosome   The complex formed by enhancers coupled to their activators.

enhancer   A DNA element that strongly stimulates transcription of a gene or genes. Enhancers are usually found upstream of the genes they influence, but they can also function if inverted or moved hundreds or even thousands of base pairs away.

enhancer-binding protein   *See* activator.

enhanson   Minimal unit within an enhancer that is capable of binding to an activator.

enol tautomer   An abnormal tautomer of uracil, thymine, or guanine, found only rarely in nucleic acids. Base-pairs abnormally and can therefore cause mutations.

enzyme   A molecule—usually a protein, but sometimes an RNA—that catalyzes, or accelerates and directs, a biochemical reaction.

epidermal growth factor   A protein that binds to a transmembrane receptor, signaling the cell to divide.

epigenetic change   A change in the expression of a gene, not in the structure of the gene itself.

epitope tagging   Using genetic means to attach a small group of amino acids (an epitope tag) to a protein. This enables the protein to be purified readily by immunoprecipitation with the antibody that recognizes the epitope tag.

ERCC1   Together with XPF, cuts on the 5′-side of DNA damage during human NER.

eRF1   The eukaryotic release factor that recognizes all three stop codons and releases the finished polypeptide from the ribosome.

eRF3   The eukaryotic release factor with ribosome-dependent GTPase activity that collaborates with eRF1 in releasing finished polypeptides from the ribosome.

error-prone bypass   A mechanism cells use to replicate DNA with pyrimidine dimers or noncoding bases. An error-prone DNA polymerase is recruited to insert nucleotides at random across from the lesion.

Escherichia coli (E. coli)   An intestinal bacterium; the favorite subject for bacterial molecular biology.

E site   The exit site to which deacylated tRNAs bind on their way out of the ribosome.

EST   *See* expressed sequence tag.

euchromatin   Chromatin that is extended, accessible to RNA polymerase, and at least potentially active. These regions stain either lightly or normally and are thought to contain most of the genes.

eukaryote   An organism whose cells have nuclei.

excinuclease   An endonuclease that participates in cutting out the oligonucleotide containing the damage in human NER.

excision repair   Repair of damaged DNA that involves removing the damage and replacing it with normal DNA.

exon   A region of a gene that is ultimately represented in that gene's mature transcript. The word refers to both the DNA and its RNA product.

exon trapping   A method for cloning exons by placing random DNA fragments into a vector that will express them only if they are complete exons.

exonuclease   An enzyme that degrades a polynucleotide from the end inward.

expressed sequence tag (EST)   An STS generated by amplifying cellular mRNA by RT-PCR.

expression site   A locus on a chromosome where a gene can be moved to be expressed efficiently. For example, the expression site in trypanosomes is at the end (telomere) of a chromosome.

expression vector   A cloning vector that allows expression of a cloned gene.

F⁻ cell   A cell lacking an F plasmid; acts as the recipient in bacterial conjugation.

F⁺ cell   A cell bearing an F plasmid; acts as the donor in bacterial conjugation.

F plasmid   An *E. coli* plasmid that allows conjugation between bacterial cells.

F′ plasmid   An F plasmid that has picked up a piece of host DNA.

F_1   The progeny of a cross between two parental types that differ at one or more genes; the first filial generation.

F_2   The progeny of a cross between two F_1 individuals or the progeny of a self-fertilized F_1; the second filial generation.

ferritin   An intracellular iron storage protein.

fingerprint (protein)   The specific pattern of peptide spots formed when a protein is cut into pieces (peptides) with an enzyme (e.g., trypsin), and then the peptides are separated by chromatography.

FISH   Fluorescence in situ hybridization. A means of hybridizing a fluorescent probe to whole chromosomes to determine the location of a gene or other DNA sequence within a chromosome.

Fis sites   Bacterial enhancers that bind the activator Fis.

5′-end   The end of a polynucleotide with a free (or phosphorylated or capped) 5′-hydroxyl group.

flagellin   The protein of which a flagellum is composed.

flagellum   A whip-like structure on the surface of a cell, used for propulsion (Latin: *flagellum*, little whip; plural, flagella).

fluor   A substance that emits photons when excited by a radioactive emission.

fMet   *See* N-formyl methionine.

Fos   One of the two subunits of the activator AP-1 (the other is Jun).

fragment reaction   A substitute for the peptidyl transferase reaction using simpler substrates. A 6-nt fragment of $tRNA_f^{Met}$ linked to fMet substitutes for a peptidyl-tRNA, and puromycin substitutes for an aminoacyl tRNA. The product is fMet-puromycin released from the ribosome.

frameshift mutation   An insertion or deletion of one or two bases in the coding region of a gene, which changes the reading frame of the corresponding mRNA.

free radicals   Very reactive chemical substances with an unpaired electron. Can attack and damage DNA.

functional genomics   The study of the pattern of genome-wide gene expression at various times or under various conditions.

fusidic acid   An antibiotic that blocks the release of EF-G from the ribosome after GTP hydrolysis and thereby blocks translation after the translocation step.

fusion protein   A protein resulting from the expression of a recombinant DNA containing two open reading frames (ORFs) fused together. One or both of the ORFs can be incomplete.

G protein   A protein that is activated by binding to GTP and inactivated by hydrolysis of the bound GTP to GDP by an inherent GTPase activity.

G segment   The segment of DNA that breaks to form a gate through which the T segment passes during topoisomerase II activity.

galactoside permease   An enzyme encoded in the *E. coli lac* operon that transports lactose into the cell.

GAGA box   Element of certain *Drosophila* insulators.

GAL4   A transcription factor that activates the galactose utilization (GAL) genes of yeast by binding to an upstream control element (UAS_G).

gamete   A haploid sex cell.

γ complex　The complex of the γ, δ, δ′, χ, and ψ subunits of the pol III holoenzyme. Has clamp loader activity.

gamma rays　Very high energy radiation that ionizes cellular components. The ions then can cause chromosome breaks.

GAP　*See* GTPase activator protein.

GC box　A hexamer having the sequence GGGCGG on one strand, which occurs in a number of mammalian structural gene promoters. The binding site for the transcription factor Sp1.

GDPCP　An unhydrolyzable analog of GTP with a methylene linkage between the β- and γ-phosphate groups.

gel filtration　A column chromatographic method for separating substances according to their sizes. Small molecules enter the beads of the gel and so take longer to move through the column than larger molecules, which cannot enter the beads.

gel mobility shift assay　An assay for DNA–protein binding. A short labeled DNA is mixed with a protein and electrophoresed. If the DNA binds to the protein, its electrophoretic mobility is greatly decreased.

gene　The basic unit of heredity. Contains the information for making one RNA and, in most cases, one polypeptide.

gene cloning　Generating many copies of a gene by inserting it into an organism, such as a bacterium, where it can replicate along with the host.

gene cluster　A group of related genes grouped together on a eukaryotic chromosome.

gene conversion　The conversion of one gene's sequence to that of another.

gene expression　The process by which gene products are made.

gene mutation　A mutation confined to a single gene.

general transcription factors　Eukaryotic factors that participate, along with one of the RNA polymerases, in forming a preinitiation complex.

genetic code　The set of 64 codons and the amino acids (or terminations) they stand for.

genetic linkage　The physical association of genes on the same chromosome.

genetic mapping　Determining the linear order of genes and the distances between them.

genetic marker　A mutant gene or other peculiarity in a genome that can be used to "mark" a spot in a genome for mapping purposes.

genome　One complete set of genetic information from a genetic system; e.g., the single, circular chromosome of a bacterium is its genome.

genomic library　A set of clones containing DNA fragments derived directly from a genome, rather than from mRNA.

genomics　The study of the structure and function of whole genomes.

genotype　The allelic constitution of a given individual. The genotypes at locus *A* in a diploid individual may be *AA, Aa,* or *aa.*

GG-NER　*See* global genome NER.

G-less cassette　A double-stranded piece of DNA lacking a G in the nontemplate strand. A G-less cassette can be placed under the control of a promoter and used to test transcription in vitro in the absence of GTP. Transcripts of the G-less cassette will be produced because GTP is not needed, but nonspecific transcripts originating elsewhere will not grow beyond very small size due to the lack of GTP.

global genome NER (GG-NER)　NER that can remove lesions throughout the genome.

glucose　A simple, six-carbon sugar used by many forms of life as an energy source.

glutamine-rich domain　A transcription-activating domain rich in glutamines.

glycosidic bond (in a nucleoside)　The bond linking the base to the sugar (ribose or deoxyribose) in RNA or DNA.

Golgi apparatus　A membranous organelle that packages newly synthesized proteins for export from the cell.

gp5　The product of the phage M13 gene 5. The phage single-strand DNA-binding protein.

gp28　The product of phage SP01 gene 28. A phage middle gene-specific σ-factor.

gp32　The product of the phage T4 gene 32. The phage single-strand DNA-binding protein.

gp33 and gp34　The products of the phage SP01 genes 33 and 34. Together they constitute a phage late gene-specific σ factor.

gpA　The product of the phage φX174 *A* gene. Plays a central role in phage DNA replication as a nuclease to nick one strand of the RF and as a helicase to unwind the double-stranded parental DNA.

GRB2　An adapter protein with an SH2 domain that recognizes phosphotyrosines on signal transduction proteins, and an SH3 domain that binds proline-rich helices in other signal transduction proteins, thus passing on the signal.

g-RNAs　*See* guide RNAs (editing).

group I introns　Self-splicing introns in which splicing is initiated by a free guanosine or guanosine nucleotide.

group II introns　Self-splicing introns in which splicing is initiated by formation of a lariat-shaped intermediate.

GTPase activator protein (GAP)　A protein that activates the inherent GTPase of a G protein and thereby inactivates the G protein.

GTPase-associated site　A site (or sites) on the ribosome that interacts with the G protein initiation, elongation, and termination factors and stimulates their GTPase activities.

guanine (G)　The purine base that pairs with cytosine in DNA.

guanine nucleotide exchange protein　A protein that replaces GDP with GTP on a G protein, thereby activating the G protein.

guanosine　A nucleoside containing the base guanine.

guide RNAs (editing)　Small RNAs that bind to regions of an mRNA precursor and serve as templates for editing a region upstream.

guide sequences (splicing)　Regions of an RNA that bind to other RNA regions to help position them for splicing.

hairpin　A structure resembling a hairpin (bobby pin), formed by intramolecular base pairing in an inverted repeat of a single-stranded DNA or RNA.

half-C_0t　The C_0t at which half of the DNA molecules in an annealing solution will be renatured.

half-life　The time it takes for half of a population of molecules to disappear.

haploid　The chromosomal number in the gamete (*n*).

haplotype　A cluster of alleles on a single chromosome.

HAT　*See* histone acetyl transferase.

HAT-A　A HAT that acetylates core histones and plays a role in gene regulation.

HAT-B　A HAT that acetylates histones H3 and H4 prior to their assembly into nucleosomes.

HCR   *See* heme-controlled repressor.

heat shock genes   Genes that are switched on in response to environmental insults, including heat.

helicase   An enzyme that unwinds a polynucleotide double helix.

helix-loop-helix domain (HLH domain)   A protein domain that can cause dimerization by forming a coiled coil with another HLH domain.

helix-turn-helix   A structural motif in certain DNA-binding proteins, especially those from prokaryotes, that fits into the DNA major groove and gives the protein its binding capacity and specificity.

helper virus (or phage)   A virus that supplies functions lacking in a defective virus, allowing the latter to replicate.

heme-controlled repressor (HCR)   A protein kinase that phosphorylates eIF-2α, strengthening its binding to eIF-2B and thereby blocking translation initiation.

hemoglobin   The red, oxygen-carrying protein in the red blood cells.

hereditary nonpolyposis colon cancer (HNPCC)   A common form of human hereditary colon cancer, caused by failure of mismatch repair.

heterochromatin   Chromatin that is condensed and inactive.

heterocyst   A nitrogen-fixing cyanobacterial cell that develops during conditions of nitrogen starvation.

heteroduplex   A double-stranded polynucleotide whose two strands are not completely complementary.

heterogeneous nuclear RNA (hnRNA)   A class of large, heterogeneous-sized RNAs found in the nucleus, including unspliced mRNA precursors.

heterozygote   A diploid genotype in which the two alleles for a given gene are different; for example, A_1A_2.

histone acetyl transferase (HAT)   An enzyme that transfers acetyl groups from acetyl CoA to histones.

histone fold   A structural motif found in histones, consisting of three helices linked by two loops.

histones   A class of five small, basic proteins intimately associated with DNA in most eukaryotic chromosomes.

HLH domain   *See* helix-loop-helix domain.

HMG domain   A domain resembling a domain common to the HMG proteins and found in some architectural transcription factors.

HMG protein   A nuclear protein with a high electrophoretic mobility (high-mobility group). Some of the HMG proteins have been implicated in control of transcription.

HNPCC   *See* hereditary nonpolyposis colon cancer.

hnRNA   *See* heterogeneous nuclear RNA.

Holliday junction   The branched DNA structure formed by the first strand exchange during recombination.

homeobox   A sequence of about 180 bp found in homeotic genes and other development-controlling genes in eukaryotes. Encodes a homeodomain.

homeodomain   A 60-amino acid domain of a DNA-binding protein that allows the protein to bind tightly to a specific DNA region. Resembles a helix-turn-helix domain in structure and mode of interaction with DNA.

homeotic gene   A gene in which a mutation causes the transformation of one body part into another.

homologous chromosomes   Chromosomes that are identical in size, shape, and, except for allelic differences, genetic composition.

homologous recombination   Recombination that requires extensive sequence similarity between the recombining DNAs.

homologs   Genes that have evolved from a common ancestral gene. Includes orthologs and paralogs.

homozygote   A diploid genotype in which both alleles for a given gene are identical; for example, A_1A_1 or *aa*.

hormone response elements   Enhancers that respond to nuclear receptors bound to their ligands.

housekeeping genes   Genes that code for proteins needed for basic processes in all kinds of cells.

HTF island   *See* CpG island.

human immunodeficiency virus (HIV)   The retrovirus that causes acquired immune deficiency syndrome (AIDS).

HU protein   A small, DNA-binding protein that induces bending of *oriC,* thereby encouraging formation of the open complex.

hybrid dysgenesis   A phenomenon observed in *Drosophila* in which the hybrid offspring of two certain parental strains suffer so much chromosomal damage that they are sterile, or dysgenic.

hybridization competition   A method for determining the identity of an unknown RNA by determining which known RNA competes with it for hybridization to a DNA.

hybridization (of polynucleotides)   Forming a double-stranded structure from two polynucleotide strands (either DNA or RNA) from different sources.

hybrid polynucleotide   The product of polynucleotide hybridization.

hydroxyl radicals   OH units with an unpaired electron. They are highly reactive and can attack DNA and break it. They are therefore useful reagents for footprining.

hyperchromic shift   The increase in a DNA solution's absorbance of 260-nm light on denaturation.

ICR   *See* internal control region.

identity elements   Bases or other DNA elements recognized by a DNA-binding domain.

IF-1   The prokaryotic initiation factor that promotes dissociation of the ribosome after a round of translation. Also augments the activities of the other two initiation factors.

IF-2   The prokaryotic initiation factor responsible for binding fMet-tRNA$_f^{Met}$ to the ribosome.

IF-3   The prokaryotic initiation factor responsible for binding mRNA to the ribosome and for keeping ribosomal subunits apart once they have separated after a round of translation.

imino tautomer   An abnormal tautomer of adenine or cytosine, found only rarely in nucleic acids. Base-pairs abnormally and so can cause mutations.

immediate early genes   Phage genes that are expressed immediately on infection. They are transcribed even if phage protein synthesis is blocked, so their transcription depends only on host proteins. Also, eukaryotic viral genes that are expressed immediately on infection of host cells. Transcription of these genes does not depend on prior synthesis of any viral products.

immunity region   The control region of a λ or λ-like phage, containing the gene for the repressor as well as the operators recognized by this repressor.

immunoblotting   *See* Western blotting.

immunoelectron microscopy   A technique for locating a particular molecule in a large complex of molecules. Especially useful in locating a ribosomal protein in an intact ribosomal particle. The experimenter reacts an antibody with a specific ribosomal protein, then uses electron microscopy to locate the antibody on the surface of the ribosome. This indicates the position of the protein with which the antibody interacts.

immunoglobulin (antibody)   A protein that binds very specifically to an invading substance and alerts the body's immune defenses to destroy the invader.

immunoprecipitation   A technique in which labeled proteins are reacted with a specific antibody or antiserum, then precipitated by centrifugation. The precipitated proteins are usually detected by electrophoresis and autoradiography.

in cis   A condition in which two genes are located on the same chromosome.

in trans   A condition in which two genes are located on separate chromosomes.

incision   Nicking a DNA strand with an endonuclease.

inclusion bodies   Insoluble aggregates of protein frequently formed on high-level expression of a foreign gene in *E. coli*. The protein in these inclusion bodies is usually inactive, but can frequently be reactivated by controlled denaturation and renaturation.

independent assortment   A principle discovered by Mendel, which states that genes on different chromosomes behave independently.

inducer   A substance that releases negative control of an operon.

initiation factor   A protein that helps catalyze the initiation of translation.

initiator   A site surrounding the transcription start site that is important in the efficiency of transcription from some class II promoters, especially those lacking TATA boxes.

inosine (I)   A nucleoside containing the base hypoxanthine, which base-pairs with cytosine.

insertion sequence (IS)   A simple type of transposon found in bacteria, containing only inverted terminal repeats and the genes needed for transposition.

insulator   A DNA element that shields a gene from the positive effect of an enhancer or the negative effect of a silencer.

Int   The product of the λ *int* gene. The topoisomerase that binds and cleaves *att*B and *att*P sites during integration of λ DNA into the host genome and excision of λ DNA during induction of the prophage.

integrase   An enzyme that integrates one nucleic acid into another; for example, the provirus of a retrovirus into the host genome.

intensifying screen   A screen that intensifies the autoradiographic signal produced by a radioactive substance. The screen contains a fluor that emits photons when excited by radioactive emissions.

intercalate   To insert between two base pairs in DNA.

interferon   A double-stranded RNA-activated antiviral protein with various effects on the cell.

intergenic suppression   Suppression of a mutation in one gene by a mutation in another.

intermediate   A substrate–product in a biochemical pathway.

internal control region (ICR)   The part of a eukaryotic promoter recognized by RNA polymerase III that lies inside the gene's coding region.

internal ribosome entry sequence (IRES)   A sequence to which a ribosome can bind and begin translating in the middle of a transcript, without having to scan from the 5′-end.

interphase   The stage of the cell cycle during which DNA is synthesized but the chromosomes are not visible.

intervening sequence (IVS)   *See* intron.

intracistronic complementation   Complementation of two mutations in the same gene. Can occur by cooperation among different defective monomers to form an active oligomeric protein.

intrinsic terminator   A bacterial terminator that does not require help from rho.

intron   A region that interrupts the transcribed part of a gene. An intron is transcribed, but is removed by splicing during maturation of the transcript. The word refers to the intervening sequence in both the DNA and its RNA product.

inverted repeat   A symmetrical sequence of DNA, reading the same forward on one strand and backward on the opposite strand. For example:

GGATCC

CCTAGG

IRE   *See* iron response element.

iron response element (IRE)   A stem-loop structure in an untranslated region of an mRNA that binds a protein that in turn influences the lifetime or translatability of the mRNA.

isoaccepting species (of tRNA)   Two or more species of tRNA that can be charged with the same amino acid.

isoelectric focusing   Electrophoresing a mixture of proteins through a pH gradient until each protein stops at the pH that matches its isoelectric point. Because the proteins have no net charge at their isoelectric points, they can no longer move toward the anode or cathode.

isoelectric point   The pH at which a protein has no net charge.

isoschizomers   Two or more restriction endonucleases that recognize and cut the same restriction site.

Iswi   A family of coactivators that help remodel chromatin by moving nucleosomes.

joining region (J)   The segment of an immunoglobulin gene encoding the last 13 amino acids of the variable region. One of several joining regions is joined by a chromosomal rearrangement to the rest of the variable region, introducing extra variability into the gene.

joint molecule   An intermediate in the postsynapsis phase of homologous recombination in *E. coli* in which strand exchange has begun and the two DNAs are intertwined with each other.

Jun   One of the two subunits of the activator AP-1 (the other is Fos).

keto tautomer   The normal tautomer of uracil, thymine, or guanine found in nucleic acids.

kilobase pair (kb)   One thousand base pairs.

kinetic experiment   An experiment that measures the kinetics (speed) of a reaction. Because chemical reactions take place on a very short time scale, such experiments require rapid measurements.

kinetoplasts   The mitochondria of trypanosomes. Genome consists of many minicircles and maxicircles.

kink   A sharp bend in a double-stranded DNA made possible by base unstacking.

Klenow fragment   A fragment of DNA polymerase I, created by cleaving with a protease, that lacks the 5′→3′ exonuclease activity of the parent enzyme.

known genes   Genes from a genomic sequencing project whose sequences are identical to previously characterized genes.

L1   An abundant human LINE, present in at least 100,000 copies, which occupies about 15% of the human genome.

lacA   The *E. coli* gene that encodes galactoside transacetylase.

lacI   The *E. coli* gene that encodes the *lac* repressor.

lac **operon**   The operon-encoding enzymes that permit a cell to metabolize the milk sugar lactose.

lac **repressor**   A protein, the product of the *E. coli lacI* gene, that forms a tetramer that binds to the *lac* operator and thereby represses the *lac* operon.

lactose   A two-part sugar, or disaccharide, composed of two simple sugars, galactose and glucose.

lacY   The *E. coli* gene that encodes galactoside permease.

lacZ   The *E. coli* gene that encodes β-galactosidase.

lagging strand   The strand that is made discontinuously in semidiscontinuous DNA replication.

λ gt11   An insertion cloning vector that accepts a foreign DNA into the *lacZ* gene engineered into a λ phage.

λ phage   A temperate phage of *E. coli*. Can replicate lytically or lysogenically.

λ repressor   The protein that forms a dimer and binds to the lambda operators O_R and O_L, thus repressing all other phage genes except the repressor gene itself.

large T antigen   The major product of the SV40 viral early region. A DNA helicase that binds to the viral *ori* and unwinds DNA in preparation for primer synthesis. Also causes malignant transformation of mammalian cells.

lariat   The name given the lasso-shaped intermediate in certain splicing reactions.

late genes   The last phage genes to be expressed. By definition, their transcription begins after the onset of phage DNA replication. Phage structural proteins are generally encoded by late genes. Also, eukaryotic viral genes transcribed after viral DNA replication begins.

leader   A sequence of untranslated bases at the 5′-end of an mRNA (the 5′-UTR).

leading strand   The strand that is made continuously in semidiscontinuous DNA replication.

LEF-1   Lymphoid enhancer-binding factor. An architectural transcription factor.

lethal   An allele that causes mortality. A recessive lethal causes mortality only when homozygous.

leucine zipper   A domain in a DNA-binding protein that includes several leucines spaced at regular intervals. Appears to permit formation of a dimer with another leucine zipper protein. The dimer is then empowered to bind to DNA.

LexA   The product of the *E. coli lexA* gene. A repressor that represses, among other things, the *umuDC* operon.

light repair   *See* photoreactivation.

limited proteolysis   Mild treatment with a protease that can break a protein down into its component domains.

LINEs   *See* long interspersed elements.

liposome   A lipid-bounded vessicle. Can be used to introduce DNA into cells.

liquid-scintillation counting   A technique for measuring the degree of radioactivity in a substance by surrounding it with scintillation fluid, a liquid containing a fluor that emits photons when excited by radioactive emissions.

locus (loci, pl.)   The position of a gene on a chromosome, used synonymously with the term gene in many instances.

long interspersed elements (LINEs)   The most abundant non-LTR retrotransposons in mammals.

long terminal repeats (LTRs)   Regions of several hundred base pairs of DNA found at both ends of the provirus of a retrovirus.

low-copy repeats   Large segments of DNA repeated a few times throughout a genome.

LTRs   *See* long terminal repeats.

LTR-containing retrotransposon   A retrotransposon with LTRs at both ends. Replicates in a manner identical to that of retroviruses except that no transmissible virus is involved.

luciferase   An enzyme that converts luciferin to a chemiluminescent product that emits light and is therefore easily assayed. The firefly luciferase gene is used as a reporter gene in eukaryotic transcription and translation experiments.

luxury genes   Genes that code for specialized cell products.

Lyonized chromosome   An X chromosome in a female mammal that is composed entirely of heterochromatin and is genetically inactive.

lysis   Rupturing the membrane of a cell, as by a virulent phage.

lysogen   A bacterium harboring a prophage.

Mad-Max   A mammalian repressor.

mal **regulon**   The group of genes coding for proteins that enable bacterial cells to metabolize maltose.

MalT   A protein that cooperates with CAP in regulating the *mal* regulon.

MAP kinase kinase   An enzyme that phosphorylates MAPK in response to a signal transduction pathway initiated by a mitogen.

MAPK   *See* mitogen-activated protein kinase.

MAPKK   *See* MAP kinase kinase.

marker   A gene or mutation that serves as a signpost at a known location in the genome.

mass spectrometry   A high-resolution analytical technique that ionizes molecules and shoots them toward a target. Assuming they are all singly charged, their times of flight to the target are related to their masses. The masses give important information about the nature of the molecules.

maternal message   An mRNA produced in an oocyte before fertilization. Many maternal messages are kept in an untranslated state until after fertilization.

maternal mRNA   *See* maternal message.

mediator   A coactivator that binds to an activator and helps it stimulate assembly of a preinitiation complex.

megabase pair (Mb)   One million base pairs.

meiosis   Cell division that produces gametes (or spores) having half the number of chromosomes of the parental cell.

Mendelian genetics   *See* transmission genetics.

merodiploid   A bacterium that is only partially diploid—that is, diploid with respect to only some of its genes.

message   *See* mRNA.

messenger RNA   *See* mRNA.

metaphase   Intermediate stage of nuclear division in which the chromosomes move to the equatorial plane.

methylation interference assay   A means of detecting the sites on a DNA that are important for interacting with a particular protein. These are the sites whose methylation interferes with binding to the protein.

micrococcal nuclease   A nuclease that degrades the DNA between nucleosomes, leaving the nucleosomal DNA alone.

microsatellite   A short DNA sequence (usually 2–4 bp) repeated many times in tandem. A given microsatellite is found in varying lengths, scattered around a eukaryotic genome.

minisatellite   A short sequence of (usually) 12 or more bp repeated over and over in tandem.

minus ten box (–10 box)   An *E. coli* promoter element centered about 10 bp upstream of the start of transcription.

minus thirty-five box (–35 box)   An *E. coli* promoter element centered about 35 bp upstream of the start of transcription.

mismatch repair   The correction of a mismatched base incorporated by accident—in spite of the editing system—into a newly synthesized DNA.

missense mutation   A change in a codon that results in an amino acid change in the corresponding protein.

mitogen   A substance, such as a hormone or growth factor, that stimulates cell division.

mitogen-activated protein kinase (MAPK)   A protein kinase that is activated by phosphorylation as a result of a signal transduction pathway initiated by a mitogen such as a growth factor.

mitosis   Cell division that produces two daughter cells having nuclei identical to the parental cell.

MO15/CDK7   The protein kinase that associates with TFIIH.

M1 RNA   The catalytic RNA subunit of an RNase P.

mRNA (messenger RNA)   A transcript that bears the information for making one or more proteins.

multiple cloning site (MCS)   A region in certain cloning vectors, such as the pUC plasmids and M13 phage DNAs, that contains several restriction sites in tandem. Any of these can be used for inserting foreign DNA.

mutagen   A mutation-causing agent.

mutant   An organism (or genetic system) that has suffered at least one mutation.

mutation   The original source of genetic variation caused, for example, by a change in a DNA base or a chromosome. Spontaneous mutations are those that appear without explanation, while induced mutations are those attributed to a particular mutagenic agent.

mutator mutants   Mutants that accumulate mutations more rapidly than wild-type cells.

N utilization site (*nut* site)   A site in the immediate early genes of phage λ that allows N to serve as an antiterminator. Transcription of the *nut* sites gives rise to a site in the corresponding transcripts that binds N. N can then interact with several proteins bound to RNA polymerase, converting the polymerase to a juggernaut that ignores the terminators at the ends of the immediate early genes.

N-CoR/SMRT   Mammalian corepressors that work in conjunction with nuclear receptors.

nearest neighbor analysis   A technique for measuring how often each nucleotide is the nearest neighbor on the 5′-side of each of the four nucleotides.

negative control   A control system in which gene expression is turned off unless a controlling element (e.g., a repressor) is removed.

NER   *See* nucleotide excision repair.

Neurospora crassa   A common bread mold, developed by Beadle and Tatum as a subject for genetic investigation.

neutron-diffraction analysis   A technique in which an experimenter passes a beam of neutrons through a sample and measures the diffraction of neutrons by the sample. Diffraction of neutrons by proteins and nucleic acids (e.g., in a ribosome) is affected by the substitution of deuterium for hydrogen in the sample or in the solvent. Measuring the effects of such substitutions gives structural information about the sample.

N-formyl methionine (fMet)   The initiating amino acid in prokaryotic translation.

nick   A single-stranded break in DNA.

nick translation   The process by which a DNA polymerase simultaneously degrades the DNA ahead of a nick and elongates the DNA behind the nick. The result is the movement (translation) of the nick in the 3′-direction in a DNA strand.

nif genes   Bacterial genes that code for nitrogen fixation.

nitrocellulose   A type of paper that has been changed chemically so it binds single-stranded DNA and proteins. Used for blotting DNA prior to hybridizing with a labeled probe. Also used for blotting proteins prior to probing with antibodies.

nodes   Points where two circles cross each other in a catenane.

nonautonomous retrotransposon   A non-LTR retrotransposon that encodes no proteins, so it depends on other retrotransposons for transposition activity.

noncoding base   A DNA base (e.g., 3-methyl adenine) that cannot base-pair properly with any natural base.

non-LTR retrotransposon   A retrotransposon that lacks LTRs and replicates by a mechanism different from that used by the LTR-containing retrotransposons.

nonpermissive conditions   Those conditions under which a conditional mutant gene cannot function.

nonsense codons UAG, UAA, and UGA   These codons tell the ribosome to stop protein synthesis.

nonsense mutation   A mutation that creates a premature stop codon within a gene's coding region. Includes amber mutations (UAG), ochre mutations (UAA), and opal mutations (UGA).

nontemplate DNA strand   The strand complementary to the template strand. Sometimes called the coding strand or sense strand.

nontranscribed spacer (NTS)   A DNA region lying between two rRNA precursor genes in a cluster of such genes.

Northern blotting   Transferring RNA fragments to a support medium (*see* Southern blotting).

nt   *See* nucleotide.

nTAFs (neural TAFs)   TAFs associated with TRF1.

α-NTD   The amino-terminal domain of the α-subunit of bacterial RNA polymerase.

nuclear receptor   A protein that interacts with hormones, such as the sex hormones, glucocorticoids, or thyroid hormone, or other substances, such as vitamin D or retinoic acid, and binds to an enhancer to stimulate transcription. Some nuclear receptors remain in the nucleus, but some meet their ligands in the cytoplasm, form a complex, and then move into the nucleus to stimulate transcription.

nucleic acid   A chain-like molecule (DNA or RNA) composed of nucleotide links.

nucleocapsid   A structure containing a viral genome (DNA or RNA) with a coat of protein.

nucleolus   A cell organelle found in the nucleus that disappears during part of cell division. Contains the rRNA genes.

nucleoside   A base bound to a sugar—either ribose or deoxyribose.

nucleosome   A repeating structural element in eukaryotic chromosomes, composed of a core of eight histone molecules with about 200 bp of DNA wrapped around the outside and one molecule of histone H1, also bound outside the core histone octamer.

nucleosome core particle   The part of the nucleosome remaining after nuclease has digested all but about 145 bp of nucleosomal DNA. Contains the core histone octamer, but lacks histone H1.

nucleosome positioning   The establishment of specific positions of nucleosomes with respect to a gene's promoter.

nucleotide (nt)   The subunit, or chain-link, in DNA and RNA, composed of a sugar, a base, and at least one phosphate group.

nucleotide excision repair   An excision repair pathway in which enzymes cut the DNA strand on either side of a damaged base, removing an oligonucleotide that contains the damage. The gap is then filled in with DNA polymerase and DNA ligase.

null allele   An allele that lacks activity.

nutritional mutation   A mutation that makes an organism dependent on a substance, such as an amino acid or vitamin, that it previously could make for itself.

O6-methylguanine methyl transferase   A suicide enzyme that accepts methyl or ethyl groups from alkylated DNA bases and thereby reverses the DNA damage.

ochre codon   UAA, coding for termination.

ochre mutation   *See* nonsense mutation.

ochre suppressor   A tRNA bearing an anticodon that can recognize the ochre codon (UAA) and thereby suppress ochre mutations.

Okazaki fragments   Small DNA fragments, 1000–2000 bases long, created by discontinuous synthesis of the lagging strand.

oligo(dT)-cellulose affinity chromatography   A method for purifying poly(A)⁺ RNA by binding it to oligo(dT) cellulose in buffer at relatively high ionic strength, and eluting it with water.

oligomeric protein   A protein that contains more than one polypeptide subunit.

oligonucleotide   A short piece of RNA or DNA.

oligonucleotide array   *See* DNA microchip.

oncogene   A gene whose product can contribute to the transformation of cells to a malignant phenotype.

one gene-one polypeptide hypothesis   The hypothesis, now generally regarded as valid, that one gene codes for one polypeptide.

oocyte 5S rRNA genes   The 5S rRNA genes (haploid number about 19,500 in *Xenopus laevis*) that are expressed only in oocytes.

opal codon   UGA, coding for termination.

opal mutation   *See* nonsense mutation.

opal suppressor   A tRNA bearing an anticodon that can recognize the opal codon (UGA) and thereby suppress opal mutations.

open complex   A complex of *dnaA* protein and *oriC* in which three 13-mers within *oriC* are melted.

open promoter complex   The complex formed by tight binding between RNA polymerase and a prokaryotic promoter. It is "open" in the sense that approximately 10 bp of the DNA duplex open up, or separate.

open reading frame (ORF)   A reading frame that is uninterrupted by translation stop codons.

operator   A DNA element found in prokaryotes that binds tightly to a specific repressor and thereby regulates the expression of adjoining genes.

operon   A group of genes coordinately controlled by an operator.

O region   The small region of homology between *att*P and *att*B.

ORF   *See* open reading frame.

oriC   The *E. coli* origin of replication.

origin of replication   The unique spot in a replicon where replication begins.

orthologs   Homologous genes in different species that have evolved from a common ancestral gene.

P_{RE} The λ promoter from which transcription of the repressor gene occurs during the establishment of lysogeny.

P_{RM} The lambda promoter from which transcription of the repressor gene occurs during maintenance of the lysogenic state.

P site   The ribosomal site to which a peptidyl tRNA is bound at the time a new aminoacyl-tRNA enters the ribosome.

PAB I   *See* poly(A)-binding protein I.

PAB II   *See* poly(A)-binding protein II.

palindrome   *See* inverted repeat.

panediting   Extensive editing of a pre-mRNA.

paralogs   Homologous genes that have evolved by gene duplication within a species.

paromomycin   An antibiotic that binds to the A site of the ribosome and decreases accuracy of translation.

pas **site**   *See* primosome assembly site.

pathway (biochemical)   A series of biochemical reactions in which the product of one reaction (an intermediate) becomes the substrate for the following reaction.

pause sites   DNA sites where an RNA polymerase pauses before continuing elongation.

pBR322   One of the original plasmid vectors for gene cloning.

PBS   *See* primer-binding site.

PCNA   Proliferating cell nuclear antigen. A eukaryotic protein that confers processivity on DNA polymerase δ during leading strand synthesis.

PCR   *See* polymerase chain reaction.

P element   A transposable element of *Drosophila*, responsible for hybrid dysgenesis. Can be used to mutagenize *Drosophila* deliberately.

peptide bond   The bond linking amino acids in a protein.

peptidyl transferase   An enzyme that is an integral part of the large ribosomal subunit and catalyzes the formation of peptide bonds during protein synthesis.

permissive conditions   Those conditions under which a conditional mutant gene product can function.

phage   A bacterial virus.

phage 434   A lambdoid (λ-like) phage with its own distinct immunity region.

phagemid   A plasmid cloning vector with the origin of replication of a single-stranded phage, which gives it the ability to produce single-stranded cloned DNA on phage infection.

phage Mu   A temperate phage of *E. coli* that replicates lytically by transposition.

phage P1   A lytic phage of *E. coli* used in cloning large pieces of DNA.

phage P22 A lambdoid (λ-like) phage with its own distinct immunity region.

phage T7 A relatively simple DNA phage of *E. coli*, in the same group as phage T3. These phages encode their own single-subunit RNA polymerases.

pharmacogenomics Using a patient's SNPs to predict his or her reaction to various drugs so that therapy can be custom designed.

phase variation The replacement of one type of protein in a bacterium's flagella with another type.

phenotype The morphological, biochemical, behavioral, or other properties of an organism. Often only a particular trait of interest, such as weight, is considered.

phorbol ester A compound capable of stimulating cell division by a cascade of events, including the activation of AP1.

phosphodiester bond The sugar-phosphate bond that links the nucleotides in a nucleic acid.

phosphorimager An instrument that performs phosphorimaging.

phosphorimaging A technique for measuring the degree of radioactivity of a substance (e.g., on a blot) electronically, without using film.

photoreactivating enzyme *See* DNA photolyase.

photoreactivation Direct repair of a pyrimidine dimer by DNA photolyase.

physical map A genetic map based on physical characteristics of the DNA, such as restriction sites, rather than on locations of genes.

plaque A hole that a virus makes on a layer of host cells by infecting and either killing the cells or slowing their growth.

plaque assay An assay for virus (or phage) concentration in which the number of plaques produced by a given dilution of virus is determined.

plaque-forming unit (pfu) A virus capable of forming a plaque in a plaque assay.

plaque hybridization A procedure for selecting a phage clone that contains a gene of interest. DNAs from a large number of phage plaques are simultaneously tested with a labeled probe that hybridizes to the gene of interest.

plasmid A circular DNA that replicates independently of the cell's chromosome.

point mutation An alteration of one, or a very small number, of contiguous bases.

poly(A) Polyadenylic acid. The string of about two hundred A's added to the end of a typical eukaryotic mRNA.

poly(A)-binding protein I (PAB I) A protein that binds the poly(A) tails on mRNAs and apparently helps confer translatability on the mRNAs.

poly(A)-binding protein II (PAB II) A protein that binds to nascent poly(A) at the end of a pre-mRNA and stimulates lengthening of the poly(A).

polyadenylation Addition of poly(A) to the 3′-end of an RNA.

polyadenylation signal The set of RNA sequences that govern the cleavage and polyadenylation of a transcript. An AAUAAA sequence followed 20–30 nt later by a GU-rich region is the canonical cleavage signal. After cleavage, the AAUAAA sequence is the polyadenylation signal.

poly(A) polymerase (PAP) The enzyme that adds poly(A) to an mRNA or to its precursor.

polycistronic message An mRNA bearing information from more than one gene.

polymerase chain reaction (PCR) Amplification of a region of DNA using primers that flank the region and repeated cycles of DNA polymerase action.

polynucleotide A polymer composed of nucleotide subunits; DNA or RNA.

polypeptide A single protein chain.

polyprotein A long polypeptide that is processed to yield two or more smaller, functional polypeptides; for example, the *pol* polyprotein of a retrovirus.

polyribosome *See* polysome.

polysome A messenger RNA attached to (and presumably being translated by) several ribosomes.

positional cloning Locating genes involved in particular genetic traits.

positive control A control system in which gene expression depends on the presence of a positive effector such as CAP (and cAMP).

positive strand The strand of a viral genome with the same sense as the viral mRNAs.

positive strand phage (or virus) An RNA phage (or virus) whose genome also serves as an mRNA.

postsynapsis The phase of homologous recombination in *E. coli* in which the single-stranded DNA replaces a strand in the double-stranded DNA to form a new double helix.

postreplication repair *See* recombination repair.

posttranscriptional control Control of gene expression that occurs during the posttranscriptional phase when transcripts are processed by splicing, clipping, and modification.

posttranscriptional gene silencing *See* RNA interference.

posttranslational modification The set of changes that occur in a protein after it is synthesized.

predicted genes Genes from a genomic sequencing project that contain sequences homologous to ESTs.

preinitiation complex The combination of RNA polymerase and general transcription factors assembled at a promoter just before transcription begins.

preprimosome An incompletely assembled primosome.

presynapsis The phase of homologous recombination in *E. coli* in which RecA (and SSB) coat the single-stranded DNA prior to invasion of the DNA duplex.

Pribnow box *See* minus ten box (–10 box).

primary structure The sequence of amino acids in a polypeptide, or of nucleotides in a DNA or RNA.

primary transcript The initial, unprocessed RNA product of a gene.

primase The enzyme within the primosome that actually makes the primer.

primer A small piece of RNA that provides the free end needed for DNA replication to begin.

primer-binding site (PBS) The site on a retroviral RNA to which the tRNA primer binds to start reverse transcription.

primer extension A method for quantifying the amount of a transcript in a sample, and also for locating the 5′-end of the transcript. A labeled DNA primer is hybridized to a particular mRNA in a mixture, extended to the 5′-end of the transcript with reverse transcriptase, and the DNA product electrophoresed to determine its size and abundance.

primosome   A complex of about 20 polypeptides, which makes primers for *E. coli* DNA replication.

primosome assembly site   A site on φX174 DNA where primosome assembly occurs.

probe (nucleic acid)   A piece of nucleic acid, labeled with a tracer (traditionally radioactive) that allows an experimenter to track the hybridization of the probe to an unknown DNA. For example, a radioactive probe can be used to identify an unknown DNA band after electrophoresis.

processed pseudogene   A pseudogene that has apparently arisen by retrotransposon-like activity: transcription of a normal gene, processing of the transcript, reverse transcription, and reinsertion into the genome.

processing (of RNA)   The group of cuts that occur in RNA precursors during maturation, including splicing, 5′- or 3′-end clipping, or cutting rRNAs out of a large precursor.

processivity   The tendency of an enzyme to remain bound to one or more of its substrates during repetitions of the catalytic process. Thus, the longer a DNA or RNA polymerase continues making its product without dissociating from its template, the more processive it is.

prokaryotes   Microorganisms that lack nuclei. Comprising bacteria, including cyanobacteria (blue-green algae), and archaea.

proliferating cell nuclear antigen (PCNA)   A protein that associates with eukaryotic DNA polymerase δ and enhances its processivity.

proline-rich domain   A transcription activation domain rich in prolines.

promoter   A DNA sequence to which RNA polymerase binds prior to initiation of transcription—usually found just upstream of the transcription start site of a gene.

promoter clearance   The process by which an RNA polymerase moves away from a promoter after initiation of transcription.

proofreading (aminoacyl-tRNA synthetase)   The process by which aminoacyl adenylates and, less commonly, aminoacyl-tRNAs are hydrolyzed if their amino acids are too small for the sythetase.

proofreading (DNA)   The process a cell uses to check the accuracy of DNA replication as it occurs and to replace a mispaired base with the right one.

proofreading (protein synthesis)   The process by which aminoacyl-tRNAs are double-checked on the ribosome for correctness before the amino acids are incorporated into the growing protein chain.

prophage   A phage genome integrated into the host's chromosome.

prophase   Early stage of nuclear division in which chromosomes coil, condense, and become visible.

protease   An enzyme that cleaves proteins; for example, the protease that cleaves a retroviral polyprotein into its functional parts.

protein   A polymer, or polypeptide, composed of amino acid subunits. Sometimes the term *protein* denotes a functional collection of more than one polypeptide (e.g., the hemoglobin protein consists of four polypeptide chains).

protein footprinting   A method, analogous to DNase footprinting, for determining the site at which one protein contacts another. One protein is end-labeled, bound to the other protein, then mildly digested with a protease. If the bound protein protects part of the labeled protein from digestion, a band will be missing when the proteolytic fragments are electrophoresed.

protein kinase A   A serine–threonine-specific protein kinase whose activity is stimulated by cAMP.

protein sequencing   Determining the sequence of amino acids in a protein.

proteolytic processing   Cleavage of a protein into pieces.

proteome   The structures and activities of all the proteins an organism can make in its lifetime.

proteomics   The study of proteomes.

provirus   A double-stranded DNA copy of a retroviral RNA, which inserts into the host genome.

Prp28   A protein component of the U5 snRNP; required for exchange of U6 for U1 snRNP at the 5′-splice site.

pseudogene   A nonallelic copy of a normal gene, which is mutated so that it cannot function.

pseudouridine   A nucleoside, found in tRNA, in which the ribose is joined to the 5-carbon instead of the 1-nitrogen of the uracil base.

pulse-chase   The process of giving a short period, or "pulse," of radioactive precursor so that a substance such as RNA becomes radioactive, then adding an excess of unlabeled precursor to "chase" the radioactivity out of the substance.

pulsed-field gel electrophoresis (PFGE)   An electrophoresis technique in which the electric field is repeatedly reversed. Allows separation of very large pieces of DNA, up to several Mb in size.

pulse labeling   Providing a radioactive precursor for only a short time. For example, DNA can be pulse labeled by incubating cells for a short time in radioactive thymidine.

purine   The parent base of guanine and adenine.

puromycin   An antibiotic that resembles an aminoacyl-tRNA and kills bacteria by forming a peptide bond with a growing polypeptide and then releasing the incomplete polypeptide from the ribosome.

pyrimidine   The parent base of cytosine, thymine, and uracil.

pyrimidine dimers   Two adjacent pyrimidines in one DNA strand linked covalently, interrupting their base pairing with purines in the opposite strand. The main DNA damage caused by UV light.

Q utilization site (*qut* site)   A Q-binding site overlapping the λ late promoter. When Q binds to *qut* it allows RNA polymerase to ignore the nearby terminator and extend transcription into the late genes.

quaternary structure   The way two or more polypeptides interact in a complex protein.

quenching   Quickly chilling heat-denatured DNA to keep it denatured.

R   A "redundant" region in the LTR of a retrovirus that lies between the U3 and U5 regions.

R2Bm   A LINE-like element from the silkworm *Bombyx mori*.

RACE   *See* rapid amplification of cDNA ends.

RAD25   The subunit of yeast TFIIH that has DNA helicase activity.

radiation hybrid mapping   A mapping technique in which human cells are treated with ionizing radiation to fragment their chromosomes, then these cells are fused with hamster cells to form hybrids with varying contents of human chromosome fragments. Genetic markers that are close together on a chromosome tend to be found in the same hybrid cells.

Rag-1 Product of the human *RAG-1* gene. Cooperates with Rag-2 in cleaving immature immunoglobulin and T cell receptor genes at RSSs so the various gene segments can recombine with each other.

Rag-2 Product of the human *RAG-2* gene. Cooperates with Rag-1 in cleaving immature immunoglobulin and T cell receptor genes at RSSs so the various gene segments can recombine with each other.

ram **state** Ribosome ambiguity state of the H27 helix of the 16S rRNA of *E. coli* ribosomes. In this state, the base pairing in the H27 helix stabilizes pairing between codons and anticodons (even noncognate anticodons), so decoding accuracy is decreased.

RAP1 A yeast telomere-binding protein that binds to a specific telomeric DNA sequence and recruits other telomeric proteins, including the SIR proteins.

rapid amplification of cDNA ends (RACE) A method for extending a partial cDNA to its 5′- or 3′- end.

rapid turnover determinant The set of structures in the 3′-UTR of the TfR mRNA that ensure the mRNA will have a short lifetime unless iron starvation conditions cause stabilization of the mRNA.

rare cutter A restriction endonuclease that cuts only rarely, because its recognition site is uncommon.

Ras Product of the *ras* oncogene. In its GTP-bound, activated form, it activates Raf, which passes on the signal to turn on genes that stimulate cell division.

Ras exchanger A protein that replaces GDP with GTP on Ras, thereby activating Ras.

reading frame One of three possible ways the triplet codons in an mRNA can be translated. For example, the message CAGUGCUCGAC has three possible reading frames, depending on where translation begins: (1)CAG UGC UCG; (2)AGU GCU CGA; (3)GUG CUC GAC. A natural mRNA generally has only one correct reading frame.

RecA The product of the *E. coli recA* gene. Along with SSB, coats a single-stranded DNA tail and allows it to invade a DNA duplex to search for a region of homology in homologous recombination. Also functions as a coprotease during the SOS response.

recA The *E. coli* gene that encodes the RecA protein.

RecBCD pathway The major homologous recombination pathway in *E. coli,* initiated by the RecBCD protein.

RecBCD protein The protein that nicks one of the strands at a chi site to initiate homologous recombination in *E. coli.*

recessive An allele or trait that does not express its phenotype when heterozygous with a dominant allele; for example, *a* is recessive to *A* because the phenotype for *Aa* is like *AA* and not like *aa.*

recognition helix The α-helix in a DNA-binding motif of a DNA-binding protein that fits into the major groove of its DNA target and makes sequence-specific contacts that define the specificity of the protein. In effect, the recognition helix recognizes the specific sequence of its DNA target.

recombinant DNA The product of recombination between two (or more) fragments of DNA. Can occur naturally in a cell, or be fashioned by molecular biologists in vitro.

recombination Reassortment of genes or alleles in new combinations. Occurs by crossing over between or within DNAs.

recombination repair A mechanism that cells use to replicate DNA containing pyrimidine dimers. First, the two strands are replicated, leaving a gap across from the dimer. Next, recombination between the progeny duplexes places the gap across from normal DNA so it can be filled in.

recombination signal sequence (RSS) A specific sequence at a recombination junction, recognized by the recombination apparatus during immunoglobulin and T-cell receptor gene maturation.

recruitment Encouraging the binding of a substance to a complex. Usually refers to enhancing the binding of RNA polymerase or transcription factors to a promoter.

regulon A group of noncontiguous genes that are coordinately controlled.

related genes Genes from a genomic sequencing project whose sequences are homologous to known genes, or parts of genes, of the same or other species.

release factor A protein that causes termination of translation at stop codons.

renaturation of DNA *See* annealing of DNA.

repetitive DNA DNA sequences that are repeated many times in a haploid genome.

rep helicase The product of the *E. coli rep* gene.

replacement vector A cloning vector derived from λ phage, in which a significant part of the phage DNA is removed and must be replaced by a segment of foreign DNA of similar size.

replicating fork The point where the two parental DNA strands separate to allow replication.

replicative form The double-stranded version of a single-stranded RNA or DNA phage or virus that exists during genome replication.

replicative transposition Transposition in which the transposon DNA replicates, so one copy remains in the original location as another copy moves to the new site.

replicon All the DNA replicated from one origin of replication.

replisome The large complex of polypeptides, including the primosome, which replicates DNA in *E. coli.*

reporter gene A gene attached to a promoter or translation start site, and used to measure the activity of the resulting transcription or translation. The reporter gene serves as an easily assayed surrogate for the gene it replaces.

repressed Turned off. When an operon is repressed, it is turned off, or inactive.

resolution The second step in transposition through a cointegrate intermediate; it involves separation of the cointegrate into its two component replicons, each with its own copy of the transposon. Also, the final step in recombinatin, in which the second pair of strands is broken.

resolvase The enzyme that catalyzes resolution of a cointegrate; an endonuclease that nicks two DNA strands to resolve a Holliday junction after branch migration.

res **sites** Sites on the two copies of a transposon in a cointegrate, between which crossing over occurs to accomplish resolution.

restriction endonuclease An enzyme that recognizes specific base sequences in DNA and cuts at or near those sites.

restriction fragment A piece of DNA cut from a larger DNA by a restriction endonuclease.

restriction fragment length polymorphism (RFLP) A variation from one individual to the next in the number of cutting sites for a given restriction endonuclease in a given genetic locus.

restriction map A map that shows the locations of restriction sites in a region of DNA.

restriction–modification system (R-M system) The combination of a restriction endonuclease and the DNA methylase that recognizes the same DNA site.

restriction site A sequence of bases recognized and cut by a restriction endonuclease.

restrictive conditions *See* nonpermissive conditions.

restrictive state Alternative to the *ram* state of the H27 helix of the 16S rRNA of *E. coli* ribosomes. In this state, required for proofreading, the base pairing in the H27 helix demands accurate pairing between codons and anticodons, so decoding accuracy is increased.

retrohoming The process by which a group II intron in one gene can transpose into an intronless version of the same gene somewhere else in the genome.

retrotransposon A transposable element such as *copia* or Ty that transposes via a retrovirus-like mechanism.

retrovirus An RNA virus whose replication depends on formation of a provirus by reverse transcription.

reverse transcriptase RNA-dependent DNA polymerase; the enzyme, commonly found in retroviruses, that catalyzes reverse transcription.

reverse transcriptase PCR (RT-PCR) A PCR procedure that begins with the synthesis of cDNA from an mRNA template, using reverse transcriptase. The cDNA then serves as the template for conventional PCR.

reverse transcription Synthesis of a DNA using an RNA template.

reversion A mutation that cancels the effects of an earlier mutation in the same gene.

RF (replicative form) The circular double-stranded form of the genome of a single-stranded DNA phage such as φX174. The DNA assumes this form in preparation for rolling circle replication.

RF-1 The prokaryotic release factor that recognizes UAA and UAG stop codons.

RF-2 The prokaryotic release factor that recognizes UAA and UGA stop codons.

RF-3 The prokaryotic release factor with ribosome-dependent GTPase activity. Along with GTP, helps RF-1 and RF-2 bind to the ribosome.

RF-A A human single-strand DNA-binding protein that is essential for SV40 virus DNA replication.

RFLP *See* restriction fragment length polymorphism.

rho (ρ:) A protein that is needed for transcription termination at certain terminators in *E. coli* and its phages.

rho-dependent terminator A terminator that requires rho for activity.

rho-independent terminator *See* intrinsic terminator.

rho loading site A site on a growing mRNA to which rho can bind and begin pursuing the RNA polymerase.

ribonuclease H (RNase H) An enzyme that degrades the RNA part of an RNA–DNA hybrid.

ribonucleoside triphosphates The building blocks of RNA: ATP, CTP, GTP, and UTP.

ribose The sugar in RNA.

ribosomal RNA *See* rRNA.

ribosome An RNA–protein particle that translates mRNAs to produce proteins.

ribozyme A catalytic RNA (RNA enzyme).

rifampicin An antibiotic that blocks transcription initiation by *E. coli* RNA polymerase.

R-looping A technique for visualizing hybrids between DNA and RNA by electron microscopy. A classic R loop is formed when an RNA hybridizes to one strand of a DNA and displaces the other strand as a loop. R-looping can also be performed with single-stranded DNA and RNA; in this procedure, classic R loops do not form, but loops can still be observed if the DNA contains information not found in the RNA.

RNA (ribonucleic acid) A polymer composed of ribonucleotides linked together by phosphodiester bonds.

RNA-dependent DNA polymerase *See* reverse transcriptase.

RNA groove A groove extending away from the DNA channel of RNA polymerase II that is wide enough to hold the RNA product.

RNA interference (RNAi) Control of gene expression by specific mRNA degradation caused by insertion of a transgene into a cell.

RNA ligase An enzyme that can join two pieces of RNA, such as the two pieces of a pre-tRNA created by cutting out an intron.

RNA polymerase The enzyme that directs transcription, or synthesis of RNA.

RNA polymerase core The collection of subunits of a prokaryotic RNA polymerase having basic RNA chain-elongation capacity, but no specificity of initiation; all the RNA polymerase subunits except the σ-factor.

RNA polymerase IIA A form of polymerase II with the CTD in an unphosphorylated or underphosphorylated condition.

RNA polymerase II holoenzyme The combination of polymerase II, transcription factors, and other proteins that can be purified as a unit using mild techniques.

RNA polymerase IIO A form of polymerase II with the CTD in a highly phosphorylated condition.

RNase E The enzyme that removes the *E. coli* 5S rRNA from its precursor RNA.

RNase H An RNase that is specific for the RNA part of an RNA–DNA hybrid. One of the activities of a retroviral reverse transcriptase.

RNase mapping A variation on S1 mapping in which the probe is RNA; RNase, instead of S1 nuclease, is used to digest the single-stranded RNA species.

RNase P The enzyme that cleaves the extra nucleotides from the 5′-end of a tRNA precursor. Most forms of RNase P have a catalytic RNA subunit.

RNase protection assay *See* RNase mapping.

RNase III The enzyme that performs at least the first cuts in processing of *E. coli* rRNA precursors.

RNA splicing The process of removing introns from a primary transcript and attaching the exons to one another.

rolling circle replication A mechanism of replication in which one strand of a double-stranded circular DNA remains intact and serves as the template for elongation of the other strand at a nick.

rRNA (ribosomal RNA) The RNA molecules contained in ribosomes.

***rrn* genes** Bacterial genes encoding rRNAs.

RSS   *See* recombination signal sequence.

run-off transcription assay   A method for quantifying the extent of transcription of a particular gene in vitro. A double-stranded DNA containing a gene's control region and the 5′-region of the gene is transcribed in vitro with labeled ribonucleoside triphosphates to label the product. The RNA polymerase "runs off" the end of the truncated gene, giving a short RNA product of predictable length. The abundance of this run-off product is a measure of the extent of transcription of the gene in vitro.

run-on transcription assay   A method for measuring the amount of transcription of a particular gene in vivo. Nuclei are isolated, with RNA polymerases caught in the act of elongating various RNA chains. These chains are elongated in vitro in the presence of labeled nucleotides to label the RNAs. Then the labeled RNAs are hybridized to Southern blots or dot blots containing unlabeled samples of DNAs representing the genes to be assayed. The extent of hybridization to each band or spot on the blot is a measure of the number of elongated RNA chains made from the corresponding genes, and therefore of the extent of transcription of these genes.

RuvA   Together with RuvB, forms a DNA helicase that promotes branch migration during homologous recombination in *E. coli.*

RuvB   Together with RuvA, forms a DNA helicase that promotes branch migration during homologous recombination in *E. coli.* Contains the ATPase that provides energy to the helicase.

RuvC   The resolvase of the RecBCD pathway of homologous recombination.

Saccharomyces cerevisiae   Baker's yeast.

SAGA   A transcription adapter complex with histone acetyltransferase activity. Mediates the effects of certain activators.

SAGE   *See* serial analysis of gene expression.

scanning   A model of translation initiation in eukaryotes that invokes a 40S ribosomal subunit binding to the 5′-end of the mRNA and scanning, or sliding along, the mRNA until it finds the first start codon in a good context for initiation.

scintillation   A burst of light created by a radioactive emission striking a fluor in a liquid-scintillation counter.

scintillation fluid   *See* liquid-scintillation counting.

screen   A genetic sorting procedure that allows one to distinguish desired organisms from unwanted ones, but does not automatically remove the latter.

secondary structure   The local folding of a polypeptide or RNA. In the latter case, the secondary structure is defined by intramolecular base pairing.

second-site reversion   A mutation in a site different from that of an original mutation, which reverses the effect of the original mutation.

sedimentation coefficient   A measure of the rate at which a molecule or particle travels toward the bottom of a centrifuge tube under the influence of a centrifugal force.

seed BAC   A BAC chosen for complete sequencing during shotgun sequencing of a large genome.

selection   A genetic sorting procedure that eliminates unwanted organisms, usually by preventing their growth or by killing them.

semiconservative replication   DNA replication in which the two strands of a parental duplex separate completely and pair with new progeny strands. One parental strand is therefore conserved in each progeny duplex.

semidiscontinuous replication   A mechanism of DNA replication in which one strand is made continuously and the other is made discontinuously.

sequence-tagged connector (STC)   A sequence of about 500 bp obtained from the end of a large clone, such as a BAC, during large-scale genomic sequencing.

sequence-tagged site (STS)   A short stretch of DNA that can be identified by amplifying it using PCR with defined primers.

sequencing   Determining the amino acid sequence of a protein, or the base sequence of a DNA or RNA.

serial analysis of gene expression (SAGE)   A method for determining the levels of expression of many genes at once. Uses short cDNAs, or tags, from many mRNAs that are linked together, cloned, and sequenced. Tags that are most frequently found are expressed most actively.

70S initiation complex   The complex of 70S ribosome, mRNA, and fMet-tRNA$_f^{Met}$ that is poised to initiate translation.

SH2 domain   A phosphotyrosine-binding domain found in many signal transduction proteins.

SH3 domain   A proline-rich-helix-binding domain that causes protein–protein interactions.

Shine–Dalgarno (SD) sequence   A G-rich sequence (consensus = AGGAGGU) that is complementary to a sequence at the 3′-end of *E. coli* 16S rRNA. Base pairing between these two sequences helps the ribosome bind an mRNA.

shotgun sequencing   Genomic sequencing performed by chopping the genome up into small pieces that are cloned and sequenced at random. Later these sequences are pieced together to give the sequence of the whole genome.

shuttle vector   A cloning vector that can replicate in two or more different hosts, allowing the recombinant DNA to shuttle back and forth between hosts.

sickle cell disease   A genetic disease in which abnormal β-globin is produced. Because of a single amino acid change, this blood protein tends to aggregate under low-oxygen conditions, distorting red blood cells into a sickle shape.

sigma (σ)   The prokaryotic RNA polymerase subunit that confers specificity of transcription—that is, ability to recognize specific promoters.

σ^{43}   The principal σ-factor of *B. subtilis.*

σ^{70}   The principal σ-factor of *E. coli.*

signal peptide   A stretch of about 20 amino acids, usually at the amino terminus of a polypeptide, that helps to anchor the nascent polypeptide and its ribosome in the endoplasmic reticulum. Polypeptides with a signal peptide are destined for packaging in the Golgi apparatus and are usually exported from the cell.

SII   A protein that stimulates transcription elongation by RNA polymerase II by limiting pausing at pause sites.

silencer   A DNA element that can act at a distance to decrease transcription from a eukaryotic gene.

silencing   Repression of eukaryotic gene activity. Can occur by forming heterochromatin in the region of the gene, or by more localized mechanisms, including tightening the binding between particular nucleosomes and DNA.

silent mutations   Mutations that cause no detectable change in an organism, even in a haploid organism or in homozygous condition.

SIN3   A yeast corepressor.

SIN3A and SIN3B   Mammalian counterparts of the yeast corepressor SIN3.

single-copy DNA   DNA sequences that are present once, or only a few times, in a haploid genome.

single-nucleotide polymorphism (SNP)   A single-nucleotide difference between two or more individuals at a particular genetic locus.

single-strand DNA-binding protein   *See* SSB.

SIR2, SIR3, and SIR4   Proteins associated with, and required for, the formation of yeast heterochromatin, including telomeric heterochromatin.

site-directed mutagenesis   A method for introducing specific, predetermined alterations into a cloned gene.

site-specific recombination   Recombination that always occurs in the same place and depends on limited sequence similarity between the recombining DNAs.

SL1   A class I transcription factor that contains TBP and three TAF$_I$s. Acts synergistically with UBF to stimulate polymerase I binding to DNA and transcription.

sliding clamp   The clamp-like structure of RNA polymerase II between the tip of the arm and the underlying shelf that keeps the enzyme from dissociating from the DNA template, thus enhancing processivity.

Slu7   A splicing factor required for selection of the proper AG at the 3′-splice site.

small nuclear RNAs (snRNAs)   A class of nuclear RNAs a few hundred nucleotides in length. These RNAs, together with tightly associated proteins, make up **snRNPs (small nuclear ribonucleoproteins)** that participate in splicing, polyadenylation, and 3′-end maturation of transcripts.

SNP   *See* single-nucleotide polymorphism.

solenoid   The 25-nm diameter hollow tube model of a coiled string of nucleosomes. Represents the second order of chromatin folding in eukaryotes (Greek: *solenoeides,* pipe-shaped).

somatic cells   Nonsex cells.

somatic 5S rRNA genes   The 5S rRNA genes (haploid number about 400 in *Xenopus laevis*) that are expressed in both somatic cells and oocytes.

somatic mutation   A mutation that affects only somatic cells, so it cannot be passed on to progeny.

S1 mapping   A method for quantifying the amount of a transcript in a sample, and also for locating the ends of the transcript. The transcript is hybridized to a labeled single-stranded DNA probe, the single-stranded DNA and RNA are cleaved with S1 nuclease, and the protected fragment of the probe remaining after S1 nuclease treatment is electrophoresed to determine its size and abundance.

S1 nuclease   A nuclease that specifically degrades single-stranded RNA and DNA. Used in S1 mapping.

Sos   A Ras exchanger.

SOS response   The activation of a group of genes, including recA, that helps *E. coli* cells respond to environmental insults such as chemical mutagens or radiation.

Southern blotting   Transferring DNA fragments separated by gel electrophoresis to a suitable support medium such as nitrocellulose, in preparation for hybridization to a labeled probe.

spacer DNA   DNA sequences found between, and sometimes within, repeated genes such as rRNA genes.

spectinomycin   An antibiotic that inhibits the translocation step in translation.

spliced leader (SL)   The independently synthesized 35-nt leader that is *trans*-spliced to surface antigen mRNA coding regions in trypanosomes.

spliceosome   The large RNA–protein body on which splicing of nuclear mRNA precursors occurs.

spliceosome cycle   The process of forming the spliceosome, splicing, then dissociation of the spliceosome.

splicing   The process of linking together two RNA exons while removing the intron that lies between them.

splicing factors   Proteins besides snRNP proteins that are essential for splicing nuclear pre-mRNAs.

Spo11   The endonuclease that creates the DSBs that initiate meiotic recombination in yeast.

spore   (1) A specialized haploid cell formed sexually by plants or fungi, or asexually by fungi. The latter can either serve as a gamete, or germinate to produce a new haploid cell. (2) A specialized cell formed asexually by certain bacteria in response to adverse conditions. Such a spore is relatively inert and resistant to environmental stress.

sporulation   Formation of spores.

squelching   Inhibition of one activator by increasing the concentration of a second one. Presumably caused by competition for a scarce common factor.

SR proteins   A group of RNA-binding proteins having an abundance of serine (S) and arginine (R).

SSB   Single-strand DNA-binding protein, used during DNA replication. Binds to single-stranded DNA and keeps it from base-pairing with a complementary strand.

STC   *See* sequence-tagged connector.

sticky ends   Single-stranded ends of double-stranded DNAs that are complementary and can therefore base-pair and stick together.

stop codon   One of three codons (UAG, UAA, and UGA) that code for termination of translation.

stopped-flow apparatus   An apparatus for performing kinetic experiments, in which reagents are forced together very rapidly, enabling their reaction to be measured.

strand exchange   *See* postsynapsis.

streptavidin   A protein made by *Streptomyces* bacteria that binds avidly to biotin.

streptomycin   An antibiotic that kills bacteria by causing their ribosomes to misread mRNAs.

stringency (of hybridization)   The combination of factors (temperature, salt, and organic solvent concentration) that influence the ability of two polynucleotide strands to hybridize. At high stringency, only perfectly complementary strands will hybridize. At reduced stringency, some mismatches can be tolerated.

strong-stop DNA   The initial product of reverse transcription of a retroviral RNA. It initiates at the primer-binding site and terminates about 150 nt later, at the 5′-end of the viral RNA.

structural genomics   The study of the sequences of genomes.

STS   *See* sequence-tagged site.

superbead   A model for the 30-nm chromatin fiber, with relatively disorganized collections of nucleosomes making up beads with an approximate 30-nm diameter.

supercoil   *See* superhelix.

superhelix   A form of circular double-stranded DNA in which the double helix coils around itself like a twisted rubber band.

supershift   The extra gel mobility shift observed when a new protein joins a protein–DNA complex.

suppression   Compensation by one mutation for the effects of another.

suppressor mutation   A mutation that reverses the effects of a mutation in the same or another gene.

SV40   Simian virus 40; a DNA tumor virus with a small circular genome, capable of causing tumors in certain rodents.

Swi/Snf   A family of coactivators that help remodel chromatin by disrupting nucleosome cores.

synapsis   Alignment of complementary sequences in the single-stranded and double-stranded DNAs that will participate in strand exchange during homologous recombination in *E. coli*.

synthetic lethal screen   A screen for interacting genes that uses cells with a nonlethal mutation (e.g., a conditional lethal mutation) in one gene to search for other genes in which ordinarily nonlethal mutations are lethal. Presumably, this means that the products of these two genes interact in some way; a defect in one or the other is nonlethal, but a defect in both is lethal.

TAF   *See* TBP-associated factor.

T antigen   The major product of the early region of the DNA tumor virus SV40. A DNA-binding protein with DNA helicase activity; has the ability to transform cells and thereby cause tumors.

Taq polymerase   A heat-resistant DNA polymerase obtained from the thermophilic bacterium *Thermus aquaticus*.

TATA box   An element with the consensus sequence TATAAAA that begins about 30 bp upstream of the start of transcription in most eukaryotic promoters recognized by RNA polymerase II.

TATA-box-binding protein (TBP)   A subunit of SL1, TFIID, and TFIIIB in class I, II, and III preinitiation complexes, respectively. Binds to the TATA box in class II promoters that have a TATA box.

tautomeric shift   The reversible change of one isomer of a DNA base to another by shifting the locations of hydrogen atoms and double bonds.

tautomerization   The shift of a base from one tautomeric form to the other. Also called tautomeric shift.

tautomers   Two forms of the same molecule that differ only in the location of a hydrogen atom and a double bond. For example, two tautomers of adenine are possible, an amine form and an imine form.

t loop   A loop formed in the telomere at the end of a eukaryotic chromosome.

T loop   The loop in a tRNA molecule, conventionally drawn on the right, that contains the nearly invariant sequence TψC, where ψ is pseudouridine.

TBP   *See* TATA box-binding protein.

TBP-associated factor   A protein associated with TBP in SL1, TFIID, or TFIIIB.

TBP-free TAF$_{II}$-containing complex (TFTC)   An alternative TFIID that lacks TBP.

TBP-related factor 1 (TRF1)   An alternative TBP found in *Drosophila* and active during neural development.

TC-NER   *See* transcription-coupled NER.

T-DNA   The tumor-inducing part of the Ti plasmid.

telomerase   An enzyme that can extend the ends of telomeres after DNA replication.

telomere   A structure at the end of a eukaryotic chromosome, containing tandem repeats of a short DNA sequence.

telomere position effect (TPE)   Silencing of genes near a telomere.

telophase   The last stage of nuclear division, in which the nuclear membrane forms and encloses the chromosomes in the daughter cells.

temperate phage   A phage that can enter a lysogenic phase in which a prophage is formed.

temperature-sensitive mutation   A mutation that causes a product to be made that is defective at high temperature (the nonpermissive temperature) but functional at low temperature (the permissive temperature).

template   A polynucleotide (RNA or DNA) that serves as the guide for making a complementary polynucleotide. For example, a DNA strand serves as the template for ordinary transcription.

template DNA strand   The DNA strand of a gene that is complementary to the RNA product of that gene; that is, the strand that served as the template for making the RNA. Sometimes called the anticoding strand or antisense strand.

teratogen   A substance that causes abnormal development of an organism.

terminal transferase   An enzyme that adds deoxyribonucleotides, one at a time, to the 3′-end of a DNA.

terminal uridylyl transferase (TUTase)   An enzyme that adds UMP residues to pre-mRNAs during RNA editing.

terminator   *See* transcription terminator.

***Ter* sites**   DNA sties within the *E. coli* DNA replication termination region. There are six sites, called *TerA, TerB, TerC, TerD, TerE,* and *TerF.*

tertiary structure   The overall three-dimensional shape of a polypeptide or RNA.

tetramer (protein)   A complex of four polypeptides.

TFIIA   A class II general transcription factor that stabilizes binding of TFIID to the TATA box.

TFIIB   A class II general transcription factor that binds to a promoter after TFIID in vitro. Helps TFIIF plus RNA polymerase II bind to the promoter.

TFIID   A class II general transcription factor that binds first to TATA-box-containing promoters in vitro and serves as a nucleation site around which the preinitiation complex assembles. Contains TATA-box-binding protein (TBP) and TBP-associated factors (TAFIIs).

TFIIE   A class II general transcription factor that binds to the preinitiation complex after TFIIF and RNA polymerase and before TFIIH in vitro.

TFIIF   A class II general transcription factor that binds to the preinitiation complex cooperatively with RNA polymerase II in vitro after TFIIB has bound.

TFIIH   The last class II general transcription factor to bind to the preinitiation complex in vitro. Associated with protein kinase and DNA helicase activities.

TFIIS   *See* SII.

TFIIIA   A general transcription factor that works with TFIIIC to help activate eukaryotic 5S rRNA genes by facilitating binding of TFIIIB.

TFIIIB   A general transcription factor that activates genes transcribed by RNA polymerase III by binding to a region just upstream of the gene.

TFIIIC   A general transcription factor that stimulates binding of TFIIIB to classical class III genes.

TfR   *See* transferrin receptor.

TFTC   *See* TBP-free TAF$_{II}$-containing complex.

3′-end   The end of a polynucleotide with a free (or phosphorylated) 3′-hydroxyl group.

thermal cycler   An instrument that performs PCR reactions automatically by repeatedly cycling among the three temperatures required for primer annealing, DNA elongation, and DNA denaturation.

30S initiation complex   The complex composed of a 30S ribosomal particle, an mRNA, an fMet-tRNA$_f^{Met}$, initiation factors, and GTP. This complex is ready to join with a 50S ribosomal subunit.

thymidine   A nucleoside containing the base thymine.

thymine (T)   The pyrimidine base that pairs with adenine in DNA.

thymine dimer   Two adjacent thymines in one DNA strand linked covalently, whose base pairing with adenines in the opposite strand is interrupted.

thyroid hormone receptor   A nuclear receptor. In the absence of thyroid hormone, it acts as a repressor; in the presence of thyroid hormone, it acts as an activator.

thyroid hormone response element (TRE)   An enhancer that responds to the thyroid hormone receptor plus thyroid hormone.

Ti plasmid   The tumor-inducing plasmid from *Agrobacterium tumefaciens*. Used as a vector to carry foreign genes into plant cells.

toeprint assay   A primer-extension assay that locates the edge of a protein bound to a DNA or RNA.

topo IV   The topoisomerase that decatenates the daughter duplexes at the end of *E. coli* DNA replication.

topoisomerase   An enzyme that changes a DNA's superhelical form, or topology.

   1. type I topoisomerase A topoisomerase that introduces transient single-strand breaks into substrate DNAs.

   2. type II topoisomerase A topoisomerase that introduces transient double-strand breaks into substrate DNAs.

torus   A donut-shaped structure.

TPA   12-*O*-tetradecanoylphorbol-13-acetate. One of the phorbol esters.

TPE   *See* telomer position effect.

trailer   The untranslated region of bases at the 3′-end of an mRNA between the termination codon and the poly(A). Also called the 3′-UTR.

trans-acting   A term that describes a genetic element, such as a repressor gene or transcription factor gene, that can be on a separate chromosome and still influence another gene. These *trans*-acting genes function by producing a diffusible substance that can act at a distance.

transcribed spacer   A region encoding a part of an rRNA precursor that is removed during processing to produce the mature rRNAs.

transcript   An RNA copy of a gene.

transcription   The process by which an RNA copy of a gene is made.

transcription-activating domain   The part of a transcription activator that stimulates transcription.

transcription bubble   The region of locally melted DNA that follows the RNA polymerase as it synthesizes the RNA product.

transcription-coupled NER (TC-NER)   NER that can remove lesions only in transcribed strands because the RNA polymerase detects the lesions and attracts the NER apparatus.

transcription factor   A protein that stimulates transcription of a eukaryotic gene by binding to a promoter or enhancer element.

transcription terminator   A specific DNA sequence that signals transcription to terminate.

transcription unit   A region of DNA bounded by a promoter and a terminator that is transcribed as a single unit. May contain multiple coding regions, as in the major late transcription unit of adenovirus.

transcriptome   The sum of all the different transcripts an organism can make in its lifetime.

transcriptomics   The global study of an organism's transcripts.

transducing virus (or phage)   A virus that can carry host genetic information from one cell to another.

transductants   Cells that have received genetic information from a transducing virus.

transduction   The use of a phage (or virus) to carry host genes from one cell to another cell of different genotype.

transesterification   A reaction that simultaneously breaks one ester bond and creates another. For example, the formation of the lariat intermediate in nuclear pre-mRNA splicing is a transesterification reaction.

transfection   Transformation of eukaryotic cells by incorporating foreign DNA into the cells.

transferrin   An iron-carrier protein that imports iron into cells via the transferrin receptor.

transferrin receptor (TfR)   A membrane protein that binds transferrin and allows it to enter the cell with its payload of iron.

transfer RNA   *See* tRNA.

transformation (genetic)   An alteration in a cell's genetic makeup caused by introducing exogenous DNA.

transgene   A foreign gene transplanted into an organism, making the recipient a transgenic organism.

transgenic organism   An organism into which a new gene or set of genes has been transferred.

transition   A mutation in which a pyrimidine replaces a pyrimidine, or a purine replaces a purine.

translation   The process by which ribosomes use the information in mRNAs to synthesize proteins.

translocation   The translation elongation step, following the peptidyl transferase reaction, that involves moving an mRNA one codon's length through the ribosome and bringing a new codon into the ribosome's A site.

transmission genetics   The study of the transmission of genes from one generation to the next.

transposable element   A DNA element that can move from one genomic location to another.

transposase   The name for the collection of proteins, encoded by a transposon, that catalyze transposition.

transposition   The movement of a DNA element (transposon) from one DNA location to another.

transposon   *See* transposable element.

trans-splicing   Splicing together two RNA fragments transcribed from separate transcription units.

transversion   A mutation in which a pyrimidine replaces a purine, or vice versa.

TRE   *See* thyroid hormone response element.

TRF1 and TRF2 (TTAGGG repeat-binding factors)   Telomere-binding proteins that bind specifically to double-stranded DNA within telomeres.

TRF1   *See* TBP-related factor 1.

Trl   A GAGA-box-binding protein.

tRNA (transfer RNA)   A relatively small RNA molecule that binds an amino acid at one end and "reads" an mRNA codon at the other, thus serving as an "adapter" that translates the mRNA code into a sequence of amino acids.

tRNA$_i$ *See* tRNA$_i^{Met}$.

tRNA$_f^{Met}$ The tRNA responsible for initiating protein synthesis in bacteria.

tRNA$_i^{Met}$ The tRNA responsible for initiating protein synthesis in eukaryotes, analogous to tRNA$_f^{Met}$.

tRNA$_m^{Met}$ The tRNA that inserts methionines into the interiors of proteins.

tRNA endonuclease   The enzyme that cuts an intron out of a tRNA precursor.

***trp* operon**   The operon that encodes the enzymes needed to make the amino acid tryptophan.

***trp* RNA-binding attenuation protein (TRAP)**   A *Bacillus subtilis* protein that is activated by binding to tryptophan, and then binds to the *B. subtilis trp* attenuator and causes it to form a terminator.

trypanosomes   Protozoa that parasitize both mammals and tsetse flies; the latter spread the disease by biting mammals.

T segment   The segment of DNA that passes through the G segment gate during topoisomerase II activity.

Tus   *Ter* utilization substance. An *E. coli* protein that binds to *Ter* sites and participates in replication termination.

TUTase   *See* terminal uridylyl transferase.

12 signal   An RSS composed of a conserved heptamer and nonamer separated by a nonconserved 12-bp sequence.

12/23 rule   The recombination scheme used in immunoglobulin and T-cell receptor gene maturation, in which a 12 signal is always joined to a 23 signal, but like signals are never joined to each other.

23 signal   An RSS composed of a conserved heptamer and nonamer separated by a nonconserved 23-bp sequence.

two-dimensional gel electrophoresis   A high-resolution method for separating proteins. First the proteins are separated in the first dimension by isoelectric focusing. Then they are separated in the second dimension by SDS-PAGE.

2 μm plasmid   A yeast plasmid that serves as the basis for yeast cloning vectors.

Ty   A yeast transposon that transposes via a retrovirus-like mechanism.

U1 snRNP   The first snRNP that recognizes the 5-splice site in a nuclear pre-mRNA.

U2AF35 and U2AF65   The two subunits of U2AF.

U2-associated factor (U2AF)   A splicing factor that helps recognize the correct AG at the 3′-splice site by binding to both the polypyrimidine tract in the 3′-splice signal and the AG.

U2 snRNP   The snRNP that recognizes the branchpoint in a nuclear pre-mRNA.

U3   The 3′-untranslated region in the LTR of a retrovirus.

U4 snRNP   The snRNP that base-pairs with U6 snRNP until U6 snRNP is needed for splicing a nuclear pre-mRNA.

U5   The 5′-untranslated region in the LTR of a retrovirus.

U5 snRNP   The snRNP that associates with both 5′ and 3′ exon–intron junctions, thus helping to bring the two exons together for splicing.

U6 snRNP   The snRNP that base-pairs with both the 5′-splice site and with U2 snRNP in the spliceosome.

UAS$_G$ *See* upstream activating sequence.

UBF   *See* upstream binding factor.

UCE   *See* upstream control element.

ultraviolet (UV) radiation   Relatively low energy radiation found in sunlight. Causes pyrimidine dimers in DNA.

UmuC   One of the components of the UmuD′$_2$C complex.

umuC   One of the genes of the *umuDC* operon, which is induced by the SOS response to DNA damage.

UmuD   Clipped by a protease to form UmuD′, one of the components of the UmuD′$_2$C complex.

umuD   One of the genes of the *umuDC* operon, which is induced by the SOS response to DNA damage.

UmuD′$_2$C   Also known as DNA polymerase V, which can cause error-prone bypass of pyrimidine dimers.

undermethylated region   A region of a gene or its flank that is relatively poor in, or devoid of, methyl groups.

unidirectional DNA replication   Replication that occurs in one direction, with only one active replicating fork.

untranslated region (UTR)   A region at the 5′- or 3′-end of an mRNA that lies outside the coding region, so is not translated.

UP element   An extra promoter element found upstream of the −35 box in certain strong bacterial promoters. Allows extra strong interaction between polymerase and promoter.

up mutation   A mutation, usually in a promoter, that results in more expression of a gene.

upstream activating sequence (UAS$_G$)   An enhancer for yeast galactose-utilization genes that binds the activator GAL4.

upstream binding factor (UBF)   A class I transcription factor that binds to the upstream control element (UCE). Acts synergistically with SL1 to stimulate polymerase I binding and transcription.

upstream control element (UCE)   An element of eukaryotic promoters recognized by RNA polymerase I. Includes bases between positions approximately −100 to −150.

uracil (U)   The pyrimidine base that replaces thymine in RNA.

uracil N-glycosylase   An enzyme that removes uracil from a DNA strand, leaving an abasic site (a sugar without a base).

uridine   A nucleoside containing the base uracil.

UTR   An untranslated region on an mRNA—either at the 5′- or 3′-end.

3′-UTR   3′-untranslated region. *See* trailer.

variable loop   The loop, or stem, in a tRNA molecule, which lies between the anticodon and T loops.

variable number tandem repeats (VNTR)   A type of RFLP that includes tandem repeats of a minisatellite between the restriction sites.

variable region   The region of an antibody that binds specifically to a foreign substance, or antigen. As its name implies, it varies considerably from one kind of antibody to another.

variable surface glycoprotein (VSG)   The antigen that coats a trypanosome. The cell can vary the nature of this coat by switching on new versions of the VSG gene and switching off the old versions.

vector   A DNA (a plasmid or a phage DNA) that serves as a carrier in gene cloning experiments.

vegetative cell   A cell that is reproducing by division, rather than sporulating or reproducing sexually.

viroid   An infectious agent (of plants) containing only a small circular RNA.

virulent phage   A phage that lyses its host.

VNTR   *See* variable-number tandem repeats.

void volume   The fraction in a gel filtration experiment that contains the large molecules that cannot enter the pores in the gel at all.

VP16   A herpesvirus transcription factor with an acidic transcription-activating domain, but no DNA-binding domain.

Western blotting   Electrophoresing proteins, then blotting them to a membrane and reacting them with a specific antibody or antiserum. The antibody is detected with a labeled secondary antibody or protein A.

wobble   The ability of the third base of a codon to shift slightly to form a non-Watson–Crick base pair with the first base of an anticodon, thus allowing a tRNA to translate more than one codon.

wobble base pair   A base pair formed by wobble (a G–U or A–I pair).

wobble hypothesis   Francis Crick's hypothesis that invoked wobble to explain how one anticodon could decode more than one codon.

wobble position   The third base of a codon, where wobble base pairing is permitted.

wyosine   A highly modified guanine nucleoside found in tRNA.

xeroderma pigmentosum   A disease characterized by extreme sensitivity to sunlight. Even mild exposure leads to many skin cancers. Caused by a defect in nucleotide excision repair.

Xis   The product of the λ *xis* gene. Responsible for excision of λ DNA from host DNA.

XPA   Verifies damaged DNA that is already bound to XPC and helps assemble the other components of the human NER complex.

XPA–XPG   The human genes involved in NER. Mutations in any of these genes can cause XP.

XPB   One of the two subunits of the human TFIIH DNA helicase. Necessary for DNA melting during NER.

XPC   Together with another protein, can recognize DNA lesions and initiate human GG-NER.

XPD   One of the two subunits of the human TFIIH DNA helicase. Necessary for DNA melting during NER.

XPF   Together with ERCC1, cuts on the 5′-side of DNA damage during human NER.

XPG   An endonuclease that cuts on the 3′-side of DNA damage during human NER.

XP-V   A variant form of xeroderma pigmentosum caused by mutations in the DNA polymerase gene.

x-ray crystallography   *See* x-ray diffraction analysis.

x-ray diffraction analysis   A method for determining the three-dimensional structure of molecules by measuring the diffraction of x-rays by crystals of the molecule or molecules.

x-rays   High-energy radiation that is diffracted by crystals. The pattern of x-ray diffraction can then be used to determine the shape of the molecule(s) in the crystal. X-rays also ionize cellular components and can therefore cause chromosome breaks.

YAC   *See* yeast artificial chromosome.

yeast artificial chromosome   A high-capacity cloning vector consisting of yeast left and right telomeres and a centromere. DNA placed between the centromere and one telomere becomes part of the YAC and will replicate in yeast cells.

yeast two-hybrid assay   An assay for interaction between two proteins. One protein (the bait) is produced as a fusion protein with a DNA-binding domain from another protein. The other protein (the target, or prey) is produced as a fusion protein with a transcription-activation domain. If the two fusion proteins interact in the cell, they form an activator that can activate one or more reporter genes.

Z-DNA   A left-handed helical form of double-stranded DNA whose backbone has a zigzag appearance. This form is stabilized by stretches of alternating purines and pyrimidines.

zinc finger   A DNA-binding motif that contains a zinc ion complexed to four amino acid side chains, usually the side chains of two cysteines and two histidines. The motif is roughly finger-shaped and inserts into the DNA major groove, where it makes specific protein–DNA contacts.